The Environmental Challenges for Japan and Germany

WORK – TECHNOLOGY – ORGANIZATION – SOCIETY

Edited by Wiking Ehlert, György Széll and Heinz Sünker

ARBEIT – TECHNIK – ORGANISATION – SOZIALES

Herausgegeben von Wiking Ehlert, György Széll und Heinz Sünker

Band 27

PETER LANG

Frankfurt am Main · Berlin · Bern · Bruxelles · New York · Oxford · Wien

György Széll
Ken'ichi Tominaga
(eds.)

The Environmental Challenges for Japan and Germany

Intercultural and Interdisciplinary Perspectives

PETER LANG
Europäischer Verlag der Wissenschaften

Bibliographic Information published by Die Deutsche Bibliothek
Die Deutsche Bibliothek lists this publication in the Deutsche Nationalbibliografie; detailed bibliographic data is available in the internet at <http://dnb.ddb.de>.

Gedruckt mit Unterstützung von/published with the support by
V.A.U.S.T., Osnabrück.

ISSN 0946-8811
ISBN 3-631-51238-4
US-ISBN 0-8204-6472-4

www.peterlang.de

Contents

Foreword

GYÖRGY SZÉLL
KEN'ICHI TOMINAGA

This book, entitled "The Environmental Challenges for Germany and Japan", is a collection of papers presented at the Congress "Environment in Natural and Socio-Cultural Context" of the German-Japanese Society for Social Sciences, held at the Faculty of Environmental and Information Studies (Yokohama Campus), Musashi Institute of Technology, on October 5 and 6, 2002. Two other related papers have been added.

Of the two editors, György Széll is an internationally oriented active German sociologist, specialised in sociological theory, labour relations and environmental problems, visiting Japan frequently for lectures, speeches and international conferences. The other editor Ken'ichi Tominaga is similarly an internationally oriented Japanese sociologist, specialised in theoretical and economic sociology as well as modernisation theory, also visiting Germany for lectures and international conferences. They have been, one from the German side and the other from the Japanese side, taking the role of promoting the organisation of the symposium and of editing this book.

The German-Japanese Society for Social Sciences (the German name is Deutsch-Japanische Gesellschaft für Sozalwissenschaften e.V.) is an international association established in 1990 aiming for German-Japanese and sociological/psychological exchange. It is, thus in combining the two axes of "Germany and Japan" and "sociology and psychology", mainly composed of German sociologists and psychologists who are interested to Japan on the one hand, and by Japanese sociologists and psychologists who are interested to Germany on the other, some members are coming from political science and pedagogy.

The German-Japanese Society held a Congress every other year alternatively in Japan and in Germany. Each Congress has been organised by volunteer organisers of different universities of the two countries. The Congresses 1 to 7 up to now have been held: the first in 1990 at Keio University, Tokyo, Japan; the second in 1992 at the University of Saarbruecken, Germany; the third in 1994 at Kansai University, Osaka, Japan; the fourth in

1996 at the University of Konstanz, Germany; the fifth in 1998 at Waseda University, Tokyo, Japan; the sixth in 2000 at the University of Kassel, Germany; and the seventh, this time, in 2002 at Musashi Institute of Technology, Yokohama, Japan.

At the Yokohama congress a new contrivance was introduced, that is, to organise a symposium by selecting a particular subject to which both sociologists and psychologists can be interested. This time the subject "Environment" was selected, because, firstly, the Congress was held at the Yokohama Campus of Musashi Institute of Technology in which the entire Campus is occupied by the Faculty of Environmental and Information Studies, and secondly, Germany and Japan are amongst one of the most advanced countries in the world in regard to environmental policies.

In the viewpoint of analysing the problem of the environment, the 18 articles included in this book are so diverse that it is significant to classify them into several categories.

The first category are the theoretical ones. In the opening introductory paper of Ken'ichi Tominaga, one of the editors of this book, it is suggested that the common basis of sociology and psychology is the concept of action, and that a possible common theoretical viewpoint for these disciplines is the paradigm of the theory of action and interaction. The human being is an "acting" animal, and for the concept of action the process of interaction with the environment that includes many elements such as the physical environment, plant environment, animal environment and human environment are indispensable.

The system of actions from the sociological viewpoint is the social system surrounded by the environment. The paper of Friedrich Fürstenberg points out that there is a "symbiotic" relation between the social system and its environment. However, the social system thinking so far takes the viewpoint of a "domination model", putting the closed boundary between the social system and the environment, and regarding the environment as resources to be utilised efficiently. What we now need, according to Fürstenberg, is to relinquish such a false viewpoint and instead to shift to "cooperation model" between the social system and the environment.

From the related theoretical point of view, the paper of Harutoshi Funabashi presents the key concept of the "environmental control system". In the pre-industrial stage, according to Funabashi, such a control system was not necessary, because the relation between the social system and the nature was in harmony. But with entering into the industrialisation stage the economic growth brought the worsening of the environment so that the need of the control system became necessary. Funabashi develops a scheme of four stages from A to D in the intervention of environmental control, and of seven paths enabling the transition from stage C to stage D.

The paper of Akira Tokuyasu surveys the process in which the highlight of consciousness of ecology and risk, connected with a system-theoretical analysis, brought the formation of the concept SRI (Socially Responsible Investing) in the United States. While the concept of SRI introduces the notion of "investment" from the economic theory, what the investment brings in this case is not the monetary return, but the realisation of social values such as the environmental improvement and the evasion of risk.

The second category includes the empirical studies. These are contrasted to the theoretical studies of the first category.

The late Erwin Scheuch's paper is a typical representative of the second category. While Scheuch on the one hand criticises the American idea of selling and buying of the so-called emission rights between advanced and developing countries of the world instituted in the U.N. Kyoto Protocol, he surveys and criticises on the other hand the 40 German publications of environmental studies written by younger generation sociologists, mostly doctoral dissertations, as not scientific researches but "aesthetics of nature" or "theology of environment" on the other. Scheuch' s point of emphasis is to assert that what is really necessary is, in the former, to make clear that the idea of "externalising" environmental pollution as the problem of production costs is to hinder the real solution of environmental pollution, and, in the latter, to produce the data that clarify the questions about empirical fact: "Does global warming really exist?" and "To what extent is the quality of environment really worsened?".

The paper of György Széll, the other editor of this book, presents the findings of a comparative research project in Germany and Japan. Based on the empirical social research of five small and medium sized German and Japanese companies each, Széll finds out that, while in both countries employees, recognising the importance of the environmental problem, commonly wrestle with the task of environmental "kaizen" (improvement), the German companies take the policy for the representatives of works council to participate in the supervisory board, whereas the Japanese companies follow a policy for all the employees on the working group level to participate in the project for promoting the "team spirit" based on the Japanese cultural tradition of "ie" (house) structure.

The contribution by György Széll and Wiking Ehlert treats an innovative subject, i.e. Environmental Business Associations in Germany. In the

1980s more and more companies and managers started to be aware of problems of the environment and in the global debate around sustainable development several associations were created. The article presents the results of an empirical study, which was conducted between 1997 and 2000 together with the three main of these associations as partners. The outcome of this study may be summarised as follows: Although strong support has been given to these associations, and even some the main global players, like Volkswagen, DaimlerChrysler, Siemens etc. are members of these associations they have not really grown since, and even face a certain crisis. The several hundreds of members these associations have are a small minority within the hundreds of thousands of companies in Germany. And the associations were not yet able to fundamentally change the politics and policies of the mainstream. Environmental concern has even decreased over the last couple of years not only within the general public, but also within the business world – although the problems have increased substantially. And there has been nearly no spreading to other countries, not to speak of the other big economies like the USA and Japan.

Shuji Yamada is concerned with the Japanese business world, and insofar complements nicely the other two contributions in this sector. Environmental concern started in Japan also for the companies with the so-called Kogai diseases, which are the "Minamata" or "Itaiitai" disease. The article discusses the environmental measures and environmental management in the case of Japanese enterprises, which developed out of environmental concern and legislation. Both strategies play a great role for social change to a sustainable society. In conclusion, it is argued that enterprises should take a central position in Japanese society to educate people in the sense of environmental protection.

The third category is also dedicated to empirical studies but on the specific aspects of the environmental problem such as urban planning, new-town design, food industry, and forest industry.

Takashi Namba's paper picks up the problem of Germany-Japan comparison of the urban renewal. It is at a glance clear by viewing from the air plane before landing that German cities are in good order while Japanese cities are disordered. Namba compares the urban renewal of the Prenzlauer Berg in ex-East Berlin and Toyonaka Shonai area near Osaka, and finds out the three big points of difference: first, the length of renewal is longer in Germany than in Japan (renewal of the Prenzlauer Berg took ten years), second, there is "gentrification" (displacement of the residents) in Germany while this is not the case in Japan, and third, German people have the consciousness of urban landscape to admit public control while this is not the case in Japan.

Kazuo Iwamura's paper, written from the viewpoint of an architect, reports the new-town residents' "unconscious consciousness" from the result of group interviews in Kohoku New-Town in Yokohama. Iwamura finds out that the residents have almost no interest in regard to the aesthetic quality of the urban landscape and of the public space, which interests the city planners most.

Yoshiki Otsuka's paper deals with the problem of the safety of food products such as BSE, dioxin, O-157 as the matter of many people's concern. Otsuka suggests that the orientation on the "scientific and market rationality" so far emphasised is, in the case of food industry, not enough and instead the criteria of "public and social rationality" are necessary.

Seiichi Nishikawa' s paper reports the present state of the forest environment in Japan. According to Nishikawa, the lumber produced in Japan is so expensive that it cannot be competitive with the imported lumber. The owners of the forest, therefore, give up the management of forestry industry, so that the rate of domestic supply in lumber is very low. Nishikawa analyses the causes and the effects of such a failure of the forestry industry in Japan.

The fourth category includes three articles that deal with the environmental problem as policy.

The paper of Gesine Foljanty-Jost analyses the results of a comparative questionnaire survey of the national organisations (governmental units, independent research units, NGOs, business organisations, and mass media) in Germany and Japan. According to her the big differences between Germany and Japan are three: while in Germany there is a close co-operation and information exchange among the policy organisations, in Japan they are scarce; while in Germany NGOs are influential, in Japan they are not; while in Japan business organisations and mass media are influential, in Germany they are not.

Koichi Hasegawa' paper is a trend report of the development of environmental sociology in Japan, in which Hasegawa observes that while the Japanese environmental sociology has been so far mostly antagonistic to the environmental policy of the Government, recently it comes to stand on the viewpoint of "active environmentalism" to connect the regional and cultural conditions, family life, residents' social organisation, and the social value with the policy problem.

"Transnational environmental policies: selected cases of Japan and Germany" is the title of the article by Rüdiger Kühr, insofar it is at the very core of this book. He tackles first indicators for responsibility and discerns the following: imports of forest and agricultural products, of resources, treats the case of aluminium and gas emissions. Then he differentiates between multilateral approaches on the one side (examples the Whaling Convention, and the Kyoto Protocol) and bilateral approaches. He comes to the conclusion that both nations are in regard to their transnational environmental policies bound by the production systems, and have therefore only limited interest in implementing strong environmental standards, whereas Japan is even more restrictive, as can be seen by the two examples mentioned. Both are also far away from a second generation kind of policy, i.e. sustainable development. The good example to follow, however, is the World Summit on Sustainable Development, which may be regarded as a landmark for further actions to be taken.

The fifth category consists of the two articles that see the environment as culture and one more article that sees the environment as a "social construct" from the psychological viewpoint.

The article by Helmut Loiskandl is the only one that deals with religion as an environment. He maintains that a challenge to the environment is connected with the problem of value, and that the religion is the provider of the value. According to Loiskandl, Christianity had no interest in "nature", because it occurred in cities. However, it is important that since the reformation Calvinism came to be familiar with modern industrialism. We know that Max Weber saw Benjamin Franklin's asceticism in saving time and money as the "spirit of capitalism". Loiskandl suggests that as the environmental problem has been caused by industrialism, the protestant ethics need to be extended to the environmental ethics.

We have two articles coming from psychology. Among them the paper of Makoto Kobayashi seizes the internal aspect of the environment as "welfare", and expresses it by the concept of Quality of Life (QOL). QOL means the state, in which the desirableness of human life is realised, so that the degree of satisfaction is the indicator of QOL. Kobayashi refers to survey data of the World Youth Survey and points out that both in Germany and Japan among 11 countries, including advanced and developing ones, the level of satisfaction is very low. But it is not clarified in Kobayashi's article how he interprets the big gap between the subjective degree of satisfaction and the objective living standard.

The other psychological article, Volker Linneweber's, regarding the concept of environment as a "social construct" in the psychological sense, presents a data analysis of what he refers to as "thematic career" in the last

40 years using data of a sample survey by Kuckartz and Grunenberg (2003). Linneweber interprets, based on Festinger's theory of cognitive dissonance, that people evaluate the change of environment during these 40 years as more stable in the case of the closer environment to one's own city than the global environment. But here again it is not clarified in Linneweber's article how he interprets the distance between the evaluations of the close and the global environment.

In conclusion, we may summarise, sociological research into the environment came with several decades of delay. Ecology has been developed more than one hundred years ago by biologists. The sociologists only woke up after the ecological movements had established themselves. So it is no hazard that these movements had become one of the major research areas of environmental sociology. For Japan the Minamata case has been especially important. We have different approaches in sociology in general and in environmental questions especially. A Marxian approach is criticised by Weberians or postmodernists. In so far it is not correct to speak of "the" sociological perspective in regard to the environment as the title of the first session is formulated. We should therefore speak of different sociological perspectives, as actually was the case. That leads us to some fundamental epistemological questions and to history. There was a discussion until recently if climate change is due to human interference into natural cycles. However, at the moment it is no question anymore that today's problems in regard to climate and pollution are man-made. The discussion only goes about the degree, i.e. 60 or 80 %.

The role of social sciences is a critical assessment of human action, i.e. in regard to society and its structure. The question of social inequality stands in its foreground. But we should never forget that this approach has take account of the really existing capitalism, in which we live today. Or you can name it market-economy. Definitely these relations of production and the interests underlying them are determining our perception and the action following out of it. Even the Brundtland-Report "Our Common Future" of the United Commission for Environment and Development from 1987 is a compromise of divergent interests between the North and the South, the companies and ecological movements. And the famous report to the Club of Rome by Dennis Meadows and others in 1972 was titled "The Limits to Growth". A clear indication of the approach dominated by economic interests. Erwin Scheuch organised a meeting of the Research Council of the International Sociological Association with the title "Zero-Growth" near Cologne in 1984. Except the initiative on Zero-Emission by the United Nations University in

Tokyo this approach seems to have completely disappeared from the public and scientific debate. Even the ecological movements and the Green Parties – especially when they are in power – do not mention it anymore.

Nature as we see and treat it is a social construction (Maurer and Höll 2003). That does not mean that it does not exist without human beings, but only that the way we perceive it is determined by our *Weltanschauung* and our interests behind it. "La nature de la nature" (the nature of the nature) as the French sociologist Edgar Morin defines it is biased by us. So for e.g. some Japanese architects say that "concrete is beautiful", which may be contested by many lovers of flowers and gardens. Actually what we are confronted with in general is a "Secondary Nature" – the last primary forests are in the state of vanishing forever.

As value criteria go into our assessment of nature a central issue is the quality of life. What is the quality of life in mega-cities like Tokyo, Mexico, Peking or San Paulo? It seems that our ancestors had often a more balanced relationship to nature, i.e. a sustainable one: not using more nature resources than nature can replace by itself. The Aborigines in Australia were presented as an example of such a culture. But it seems that we cannot sustain this assumption anymore. The desertification of Australia stems largely to their intervention. And the wonderful hot baths of the Romans – besides ship-building – contributed largely to the deforestation of Italy, more than 2,000 years ago.

The role of social scientists is to develop warnings before the crisis happens. When the crisis is there it mostly very expensive to overcome or even too late to intervene effectively. By the way the biggest environmental polluter is the industry, which is closest to nature, i.e. the agriculture. There rural sociology has its role, however, badly managed. For sure there are different societal models in the USA, in Europe and Japan, which have also an effect on their relationship to nature. The vast areas of the USA, Canada or Brazil lead entrepreneurs and politicians to the attitude that there is nature of plenty, and has not to be taken care of.

That leads us to a final remark. The Danish sociologist Bent Flyvbjerg has published in 2001 a book with the title "Making Social Science Matter. Why social inquiry fails and how it can succeed again". There he comes back to the old Aristotlean approach of "Phronesis", i.e. the search for the "Good society" – contrasted with *episteme* and *techne*, the today dominant and seemingly only scientific approaches. We are convinced, sustainability can only be realised, if we follow this approach.

Osnabrück and Tokyo, February 2004

References

Flyvbjerg, B. (2001), *Making Social Science Matter. Why social inquiry fails and how it can succeed again*, Cambridge University Press, Cambridge.

Maurer, M. and Höll, Ö. (eds) (2003), *Natur als Politikum*, Rosa Luxemburg Institute, Vienna.

Meadows, D. et al. (1974), *The Limits to Growth*, Universe, New York, 2nd pr. 1974.

Morin, E. (1981), *La Méthode – La nature de la nature*, Seuil, Paris.

Széll, G. (1994), 'Technology, Production, Consumption and the Environment', *International Social Science Journal*, pp. 215–225.

Part I:
Introduction

1 Environment from the Point of View of Action Theory

KEN'ICHI TOMINAGA

The articles collected in this book are for the most part those presented at the Symposium "Environment in Natural and Socio-Cultural Context" at the Yokohama Congress of the German-Japanese Society for Social Sciences held at the Faculty of Environmental and Information Studies (Yokohama Campus), Musashi Institute of Technology on October 5–6, 2002.

The German-Japanese Society for Social Sciences is the Society, which organised the co-operation of German and Japanese sociologists and psychologists since 1990. The naming of "Social Sciences" of this society is actually a little bit too big, as it is centring only around two disciplines of sociology and psychology.

In the United States the term "behavioral sciences" have been used for combining sociology and psychology, but it is not fit for our case, because this term tends to emphasise the objective viewpoint associated with the tradition of "behaviourism" on the one hand, and to be interested to the common aspect of human and animal behaviour on the other. In addition the word behavioural sciences spread neither in Germany nor in Japan. Thus in this introductory paper it was decided to conceptualise the common basis of sociology and psychology by the word "action" theory rather than "behavioural" theory.

The theme of the Symposium "Environment" was selected because the Yokohama Congress was organised by the Faculty of Environmental and Information Studies, Musashi Institute of Technology, of which I was the Dean (now retired). This Faculty has started only recently in 1997 as a multidisciplinary organisation of connecting many disciplines from natural to social sciences for studying current problems in an age of environmental and information problems. But so as to fit the theme to the German-Japanese Society we specified the subject as "Environment in Natural and Socio-Cultural Context".

The Action Theory

The action theory starts from seeing the human being as an "acting" animal. Psychologically it can be said that action is the process beginning from the needs and ending in satisfaction of these needs. However, to use the term "needs and satisfaction" is not enough, as these terms belong to the physiological level of the human action. A more important point of view is to take it into account that the human being is a "thinking" animal.

Human action is, in contrast to animal behaviour, characterised by the big part of intervening function of the thinking between the recognition as input from the outside and the performance as output from the inside. Thinking is the internal working of the human brain, so that the action essentially requires the internal process. Therefore, as Max Weber put it (Weber 1921), action is closely related to the world of "subjective meaning" of the actor. By "subjective" here we mean the "internal" side of the actor.

But to say simply that human action is concerned with the world of actor's subjective meaning is still not enough. Action is never the "solipsistic" process of the actor-subject. It is the process of expressing one's internal subjective world to external others. The important point is that the subjective world of the human actor is not closed but open to the other actor. The same is of course true for the actor-object, so that human action is possible, as was said by Talcott Parsons, only as "*inter*action" between the two subjective worlds of the actor-subject and the actor-object (Parsons 1951; Parsons and Shils 1951).

The world of subjective meaning in Weber's sense corresponds to what G.H. Mead conceptualised as "mind" (Mead 1934). Mind is not inborn, but is gradually being formed with one's growth after birth by "taking the attitude of the other" in Mead's term. In other words, it is formed through the interactions with many other actors by recognising and learning their internal subjective world. This is also the case for the side of other actors. It is important, therefore, that the formation of the mind develops through the process of interactions in the following two aspects.

The one is the objective, that is, the social-structural side. As formulated by Parsons, the continued process of interaction between *ego* and *alter* comes to the role relations between them. Whenever they have a common will to co-operate with each other in attaining the joint goal, these role relations develop to the institutionalisation of an organisation possibly including many others who share such a joint goal. Under the institution of organisation the role relations come to the social structure among these actors.

The other is the subjective, that is, ego's internal side. Mead made it clear that the formation of mind is never possible in the internal process of a single actor alone. It is the common process of ego and alter intermediated through the "significant symbol" to each other. By "significant" is meant that actors give the common meaning to them, and what bears that common meaning is the "symbol".

While thinking refers to the subjective world of a particular actor, it can be objectified in the process of being commonly shared among plural actors. It is essential that the ego and the alter can "understand" each other through the mediation of significant symbols. Among the significant symbols language is the most important. No animal other than the human being has a language. Language is essential both in *intra*personal and in *inter*personal processes. In intrapersonal terms the human being cannot even think without language. In interpersonal terms ego and alter cannot understand each other's internal subjective world without it.

The symbol is the basic category that constitutes the culture. There is, in addition to the language symbol, the category of non-verbal symbol such as sound, drawing, and gesture. Culture that consists of language is, first, cognitive culture such as science, philosophy, ideology, and religion, second, expressive culture such as novel, poem, and drama, third, evaluative culture such as law and moral. Non-verbal culture is such as music, picture, sculpture, dancing, architecture, dressing, and food. There is also the type of mixed culture of verbal and non-verbal such as drama and cinema (Tominaga 1997).

Action and the Environment

The fact that one's world of subjective meaning is not what the human being has at the birth is illustrated by the well-known case of the wild boy "Victor" reported in detail by Jean Itard (Itard 1894), who was found at the forest of Aveyron, France in 1799 at about 12 years of age lonely in the forest. He could not learn the language in his environment, as there was no other human actor with whom he could have interaction. As the result that he had no language, he could not have the internal world of subjective meaning to be expressed to, and to be understood by other actors.

There are also well-known similar cases such as two wild girls of "Amara" and "Kamara", estimated age 1 and 8, found in 1920 in East India, reported by the clergy Singh (Singh and Zingg 1942), and a girl "Anna" aged six in USA who was left alone in a shed of a farmhouse in Pennsylvania, reported and analysed by the sociologist Kingsley Davis (Davis 1940, 1947). In Japan there is the case of a sister and a brother, aged six and five respectively in Yamanashi Prefecture, reported and analysed by the psychologist Tamotsu Fujinaga (Fujinaga et al. 1988). The common feature of all these cases is that these wild children have no language and no socialised self.

Talcott Parsons in his general theory of action constructed the conceptual scheme named "action frame of reference" in terms of the three elements: actors, situation of action, and the orientation of actors to that situation. In this frame of reference Parsons used the word "situation" instead of the word "environment", which was divided into three categories of social objects, cultural objects, and natural objects. It seems that Parsons brought the word "situation" from W.I. Thomas. It indicates, as Thomas' well-known phrase "definition of the situation" suggests (Thomas and Znaniecki 1918), the external things that the actor "internalises" into one's subjective world through learning.

The word environment is, unlike situation, the external world as it exists objectively outside of actors. Though Parsons did not introduce this word in his action frame of reference in 1951, he began to use the word environment in his later formulation of the conceptual scheme of "human condition" (Parsons 1978). It might have something to do with the influence from the environmental sociology that started in the 1970s. It is important that the concept of environment came to be located by Parsons in the context of his theory of action in this way.

While the inorganic physical thing like stone does not interact with the environment, the living thing does interact. A well-known example is that the plant takes in carbon dioxide with solar energy and water, and takes out oxygen. Human being takes in oxygen, water, and food, while it takes out carbon dioxide, feces and urine. In the case of the human being it is important to interact, in addition to the material things, with other humans as actor-object through mediation of the significant symbol in the formation of the subjective world of mind.

The human being uses stones, for example, as building material, but this relation is only in one-way, as a stone does not return interaction to humans. The plant, in a sense like a stone, does not act, so that interaction with human beings is not possible; but unlike stones, there is a coexistent relation between human beings and the plant as an element of the ecosystem, such that they exchange oxygen and carbon dioxide, or that the human being breeds the plant for appreciating it or eating it.

The animal, at least the higher animal, unlike stones and plants, acts like a human being, so that interaction is possible between a human being and an animal, for example a pet dog, in which an animal can understand human sentiments as an element of a quasi-social-system. While an animal, as it has, unlike a human being, no language, therefore cannot respond to a human being with language, a human being breeds the domestic animal such as horse and cattle, for making use of them for riding or pulling a cart on the one hand, and for getting milk or eating on the other. This is also a coexistent relation between a human being and the animal again as an element of the ecosystem, though the ecosystem is not the social system (Tominaga 2002).

We can conclude from the above that human action is not possible by a single individual. It requires other actors (persons), physical objects (things), plants and animals (living organisms), and symbols (culture) as constituent elements as environment outside of the human actor. The environment is thus an indispensable category in the context of analysis of action and interaction: interaction requires other actors; human action requires many kinds of physical objects for satisfying needs as resources; the human being as a living organism coexists with many other non-human living organisms of the plants and the animals; interaction is not possible without cultural symbols. The environment consists of all these elements. In the following we divide them into three categories: the social environment, the cultural environment, and the natural environment.

The Social Environment

Insofar as the human action is usually not a single action but an interaction with another actor or actors, the first important element of the environment for the human being is the collectivity of other actors. When an infant is born, the environment that comes into the eyes of the baby is the mother as the first "significant other" who gives the baby milk, who talks to the baby, and who takes care of the baby. As the baby grows older, other family members such as the father and the siblings, and then non-family members like neighbours, friends, and teachers come to join the group of significant others. The "social" environment is the environment made of the co-operation with these other actors.

The human being as a single person is so powerless that one cannot live without close co-operation with others on the various levels of "society", like family, school, firm, community, and national society. Human behaviour that can be attained without dependence on the social environment is limited to, as a matter of facts, such physiological need satisfaction as breathing, sleeping, and evacuation. Even in human evacuation the toilet training by the mother is essential. Eating is in itself a physiological need satisfaction, but the human appetite is closely connected with sociable action in a family, in a school, in a firm, in a community, and with friends.

School offers the social environment appropriate for studying. The firm provides the social environment for working, which is indispensable especially in the highly developed system of division of labour in the production and in the marketing. The city as modern community is the social environment necessary for the consumer who buys everything in the market as produced by others in the industrial society. The national society is the social environment made of those who share the language, culture, and the consciousness of identity (Shimizu 1940).

Until the stage of feudal society in middle-age Europe as well as in Tokugawa Japan, the social environment was enclosed in the village community. Villagers were born and died in the same village, and rarely went out of it. With the start of modernisation in the 18–19th centuries in Europe as well as in the Meiji Restoration of 1868 in Japan, when the system of feudal lords was disorganised and their territories came to be integrated into the nation state, the range of the social environment was extended to the entire national society. This was the formation of the nation state and the national society in the age of modernity.

Following the political process of the formation of the nation state did the technological-economic innovation of the industrialisation start as another aspect of modernisation. While the industrial revolution was on the one hand the technological progress from human and animal muscle to mechanical power, and was on the other the economic progress from self-sufficiency to market economy, the big effect brought by these progresses on the social environment was the continuous enlargement of the scale of the human society from the village to the city, and from the city to the national society, and even to the global society.

The core of modernity was in this sense the realisation of integration of the national society as the largest social environment. This integrative function is what Talcott Parson referred to by the concept of the "societal community" (Parsons 1971). As the modern social system is characterised by the differentiated "role pluralism", the regulation of the national society by the loyalty to the societal community is important. The ideology of the loyalty to the national society has been emphasised as nationalism. In the United States especially as the "First New Nation" (Lipset 1963) which is characterised by the plurality of ethnicity, religion and culture, the idea of the societal community is above all essential.

Now that the age of "globalisation" is extending from the national society to the global society beyond the age of simple modernity, one can say no more that the national society is the largest social environment, in the sense that many people, many things and much information come in everyday from outside of the nation state. Of course it is apparent that the boundary of the nation state still exists, as shown in the daily experiences such that no one can depart from, nor enter into one's own nation state without being controlled by passport, and that no one can escape from the obligation of tax payment to one's own nation state. But one can say that the boundary is somewhat relativised, to the extent that the scope of the social environment is enlarged at a stroke into a global scale, through the networks of the not-so-expensive jet plane, of the satellite broadcasting in home TV, and of electronic mail in the personal computer.

It is true that globalisation started, as was pointed out by Saskia Sassen (Sassen 1996), in the 1960s from the trade (GATT and WTO) and the diplomacy (UN and NGO). It soon extended, however, from the economic-political level to the social-cultural level. As Roland Robertson suggests, the concern with culture is amplified and the value that was once on the local and ethnic level now spreads into the generalised culture (Robertson 1992; Tomlinson 1999). This means that the consideration on the enlargement of the social environment is connected with that of the cultural environment.

The Cultural Environment

The second important element of the environment for the human being is the culture. While the social environment is directly the collectivity of other actors, the cultural environment is indirectly the accumulation of what the human being has interpretatively produced through actions. There are two kinds of such product of human action. One is the symbol in abstract, and the other is the material expression of the symbol on the concrete thing.

The symbol means what is significantly created and commonly shared by human being as an instrument of thinking and interpersonal communication. The symbol is carried by the medium such as sound in the spoken language, letter in the written language, figure in the graphic. Symbols and media make it possible to communicate actions between actors. One can share the culture commonly with others by communication. It is by interaction using these symbols and media that human beings can constitute the social system.

The symbol is divided into two categories: the language symbol and the non-language symbol. The language symbol is either the spoken language in which the medium is the pronunciation as sound or the written language in which the medium is the letter. There are many non-language symbols such as gesture, dance, music, pictures, figures, sculptures, architecture, clothing, and cooking.

The language culture makes (1) the cognitive culture such as science, philosophy, ideology, and religion, (2) the expressive culture such as novel, poem, and drama, and (3) the evaluative culture such as law and moral. The non-language culture creates ceremony, pantomime, dance, music, pictures, sculptures, architecture, clothing, and cooking. The mixture of language and non-language culture makes the drama, opera, movies, and animation.

While the symbol makes the culture in abstract, culture can take the concrete form of "thing". For example, science and novels are published as a book; music is sold as record and CD; a picture is written on paper or canvas; a drama, opera, movie, and animation circulates as Video and DVD; a sculpture is carved on stone, wood, or metal; a building is made of wood, stone, iron, or concrete, and so on. These can be regarded as cultural resources embodied as material (Tominaga 1997).

A human being has many different kinds of language, and many different styles of music and architecture. Philosophy, ideology, and religion are in many cases peculiar to a nation and ethnicity. In this sense culture is localised and particularised. However, on the other hand, science is universal over the world, and the same religion and art is commonly accepted by different nations and ethnicities.

Because of this duality the globalisation of culture has tended to produce two opposite directions, that is, the conflicting tendency among different cultural spheres on the one hand, and the generalising tendency among them on the other. During the cold war period until 1990 the element of ideological conflict had been so crucially in front between the two poles of the United States and Soviet Russia, that the element of religious conflict had looked to be in behind. Of this latter point two remarkable concepts in the Christianity serve as good illustrations. The one is the "ecumenism" advocated by Pope John XXIII in 1963, as the movement of dissolution of conflict among Catholicism, Protestantism, and Judaism (Parsons 1978). The other is the movement of the secularised non-church form of Christianity named by Robert Bellah as the "civil religion" (Bellah 1970).

After the end of the cold war, however, a new religious conflict between the Christian and the Islamic World is coming to the front. There was an explosion of this conflict in the 9/11, 2001 which brought the recognition that the movement of religious peace came to an end and that the world is again divided into two. Although the suicide bombers' attack did not in itself take a form of religious attack, it cannot be denied that it was a link of the religious confrontation in the post-cold-war-regime.

The Natural Environment

The third important element of the environment for human being is the nature. Food for human being is obtained from plants and animals *plus* the working of micro-organism in the soil. Clothes, residence and other necessities of life that a human being requires are made from plants, animals and physical things. All these organic and inorganic materials which human beings use as resources come from the natural environment.

The science studying the interdependent relation among many kinds of plants and animals is a branch of biology called the ecology. Plants such as a tree, flower, and grass are called by ecologists "producers", in the sense that they synthesise organic substances from inorganic materials in the circulation of the natural world. Animals such as beast, bird, and fish, similarly, are called by ecologists "consumers", in the sense that they take in nutrition by eating plant and other animals in this material circulation. Micro-organism in the soil such as microbes and bacteria are called by ecologists "mediators" between the producer and the consumer, in the sense that they decompose excrements and dead bodies of animals into the nutrition of plants (Odum 1983).

The ecologist does not deal with human beings. From the enlarged point of view of the ecology, however, human beings, as it is one species of animals, can be said to be the biggest "consumer" in the above-mentioned material circulation, taking plants and animals as food on the one hand, and breathing oxygen in and carbon dioxide out on the other. Thus it is clear that human beings cannot live without dependence on all these elements of the natural environment. At the same time, however, we must add immediately that in the material circulation of the ecosystem every living organism is interdependent, so that not only for human beings but also for every organism on the earth all other organisms are elements of the natural environment to each other.

Also, the physical-chemical objects such as water, soil, iron, coal, stone and many other mineral resources are given to human beings on the earth. Human beings, by utilising them, have produced many kinds of industrial equipment and instruments by these resources. As a result, the modern city today is physically composed of many industrial products such as roads, bridges, buildings, automobiles, subways, television, computers and all forms of "artificial" environment. Formerly the environment of today's city was composed of woods, hills, rivers, and coasts, but now such natural environment is replaced for the most part by the above-enumerated artificial environment by technology invented by human beings. But originally the natural environment such as woods, rivers, hills and coasts were enjoyed by animals and plants, and even by the human beings in the pre-modern age.

Today the natural environment has almost disappeared from the modern city, and is contaminated by industrialisation: air, water and soil are polluted; woods and hills are destroyed; rivers and coasts are buried; the ecosystem of plants and animals is thus destructed; electric generation by nuclear power threatens inhabitants; the temperature of the earth is rising; the ozone layer is disrupted. Ulrich Beck called such a situation by the name of the "risk society", and emphasised the need of a "reflexive" modernisation (Beck 1986). How is it possible to overcome the situation of these risks and to realise the reflexive modernisation is the contemporary task with which the human being is now confronted?

A social action theorist, however, is not the researcher of plants and animals, air, water, and soil, the temperature of the earth, nor the ozone layer. What is then the activity that the social action theorist can contribute to the study for the prevention of contamination of the natural environment?

Human beings produce physical-chemical objects in industry. Industry deals with material things taken from the natural environment. But at the same time we can say that the industrial activity is a social action of organising people. Environmental problems have occurred as the result of these human social actions in co-operating with other organisation members. Reflexive modernisation in Beck's sense means the reflection toward the ways, in which the human social actions in modern industrialism is organised. This is the context in which the action theory can include the elements of the natural environment into consideration. Then, what is the viewpoint of the action theory that makes it possible to draw the problem of natural environment into that context? This is the central point to be argued in the remaining two sections in this introductory article.

The Viewpoint of Including the Elements of Natural Environment into the Context of the Action Theory

Among the natural environment plants and animals are the living organisms similar to human beings. The science studying the interdependent relations among plants, animals, and physical-chemical materials is the ecology, we mentioned above. Our problem here is to add the human being to these interdependent relations.

According to Amos Hawley, an urban sociologist, the basic elements for ecology are the ecosystem, population, and the environment. The ecosystem is the system that consists of the interdependent relation among living things and their non-living environment. Population is the gathering of living things as "society". Environment is the non-living elements outside of living things (Hawley 1986). While Hawley's "society" seems to designate the interdependent relations of plants and animals, we would possibly add the society of human beings to Hawley's concept of society.

The relation among plants, animals, physical-chemical materials and human beings has so far not appeared in action theory, as it has been thought that they have no or limited interaction between human beings. Non-living things do not act, have no consciousness, and develop no living activity. Plants do not act at least overtly and have no consciousness, but they develop purposive activity for living such as spreading a branch with leaves for taking solar energy and carbon dioxide in, and extending roots underground for absorbing nutrition and water. An animal does not interact with human beings at least on the language level, but at least the higher animals have consciousness, and interact with human beings, though without language.

A possible viewpoint that connects the natural environment to the concept of action is in the notion of "resources". As human action is the purposive process in which the rational means is chosen for attaining the end, the actor takes useful resources from outside in the process of need satisfaction. Anything outside of the actor is environment, so that the element in the environment that is useful as means for attaining the end can be the resources.

This view, while it is a way of formulating the connection between the natural environment and the human action, was criticised by Catton and Dunlap, environmental sociologists, pointing out that it brings the "human-centered paradigm" of industrialism in which human beings, in utilising the natural environment as resources one-sidedly, dominate the nature. The point of this problem is that the relation between the natural environment and the human being is at a glance unsymmetrical so that it looks as if human beings can exploit nature without sanction (Catton and Dunlap 1978, 1980; Humphrey and Buttel 1982).

A substitute formulation presented from the point of view of the environmental sociology was the "New Ecological Paradigm". The point in this formulation was that the system of human actions, the present writer interprets it, must be seen as a part of the ecosystem constituted by nature, and that human beings and nature must be seen on an equal footing, as interdependent to each other. This can be interpreted as substantially equal to Ulrich Beck's formulation (Beck 1986, 1997) that the modernisation must be seen as divided into two stages in such a way that the modernisation of the traditional society (simple modernisation) must be followed by the modernisation of the industrial society (reflexive modernisation).

Both of these formulations can be reinterpreted in the term of our human action theory, such that the natural environment and the system of human social actions are both the elements of the ecosystem. For example, if human action breaks forests, the social system as system of human social action also suffers from the sanction by nature, in the sense that the recovery of the air from contamination is impeded. It thus assumes that the social system could be connected to the concept of the ecological system, so as to adapt the social system to plant-animal relations.

Up to the stage of agricultural society, economic production was in a farm as natural field, so that the viewpoint of the social system as the system of human social actions was not separated from the ecological system as plant-animal relations. The separation of the social system and the ecological system was the product of modern industrialism. Industrialisation brought the thought that modern technology "conquered" the nature. Based on such a thought the modern city has hitherto destroyed the nature and built the artificial environment made of iron and concrete afterwards. The result was the acceleration of destruction of nature, accompanied by increasing environmental problems on the entire earth scale.

A Sociological Example: The Need of the Community

In Japan, rural villages before the starting of the high rate economic growth in 1955 were in the situation called "natural village" by Eitaro Suzuki, which meant the continuation of the traditional structure of rural community, such as collaborative work of the paddy irrigation, mutual help of the transplantation of the rice seedlings (*yui*), and the village ownership of the forest since the Tokugawa age (Suzuki 1940). While the Japanese family structure was in the urban area already for the most part the nuclear family (Toda 1937), the component of the pre-war rural community was the traditional patriarchal family, which has been in Japan characterised by the name of the "house institution" (*ie seido*). In the post-war civil law reform (1947–50) the house institution was deprived of the legal foundation, so that the patriarchal family basis of the rural community was decisively disorganised.

The important point in interpreting the post-war Japanese social change is that, it was just after the completion of the post-war reform that the high rate of economic growth began. The economic growth brought the acceleration of the outflow of the younger rural population into cities. On the one hand, there were plenty of employment opportunities in cities, which drew the rural younger people by the rapidly rising income level. On the other hand, as the traditional house institution had already legally disappeared, it was impossible for the farmhouse to secure the next generation under the economically declining condition. Thus the result was the disorganisation of the traditional rural community (Hasumi 1990; Tominaga 1990).

Now that the period of high economic growth was replaced by that of long-term depression since 1990, there is recently a movement of re-evaluating the importance of the community. The community here means, of course, something other than the simple revival of the traditional rural community before the high economic growth, which is absolutely impossible. While there are many kinds of definition of the concept of community, the point is here at least that it is the region in which there is a strong mutually dependent connection by the consciousness of solidarity among human beings including the relation among plants, animals, and physical-chemical materials.

Recently there is a tendency of family crisis in the sense that the functions performed by the family are reduced by such a situation as a high rate of the aged, small numbers of children, few full-time housewives, and a rapidly increasing ratio of single households. As the result of all these trends, the family becomes more and more dependent on the community. Especially the role of community is coming to be reconfirmed in many relations such as in care of the aged, in mutual help at the time of disaster, in disposal of waste, and in citizens' movements for the protection of the ecosystem of plants and animals. Recent tendencies toward a revival of Toennies' word *Gemeinschaft* is related to this trend. Gemeinschaft is distinguished from Gesellschaft in that the media in human social interaction in Gemeinschaft is not money in the market exchange, but symbols in the *social* exchange. The ecosystem that consists of the plants and animals has no media, as it is not the social system. But the mutually interdependent relation is common to both the social system and the ecosystem.

What follows is the record of the Yokohama Symposium entitled "Environment in the natural and socio-cultural context" analysed from the wide interest as mentioned above by sociologists and psychologists of the German-Japanese Society for Social Sciences. Among these papers various points of view such as environment protection, ecology, theory of social system, environment policy, urban environment, cultural environment are crossed to each other. While the Congress of this Society has been so far held by a double program of the symposium and the free session, the papers included in this book are those presented to the symposium "environment" and the related ones. But the present writer as an organiser of the Congress, being satisfied with the agreement of all the participants to accept the theme of environment, wishes to appreciate the kindness of presenting so many papers and to eulogise the efforts of all the paper writers.

References

Beck, U. (1986), *Risikogesellschaft*, Suhrkamp, Frankfurt/M.

Beck, U. (1997), *Was ist Globalisierung?*, Suhrkamp, Frankfurt/M.

Bellah, R. (1970), *Beyond Belief*, Harper & Row, New York.

Catton, W.R. and Dunlap, R.E. (1976), 'Environmental Sociology: A New Paradigm', *American Sociologist*, vol. 13, pp. 41–49.

Catton, W.R. and Dunlap, R.E. (1980), 'A New Ecological Paradigm for Post-exuberant Sociology', *American Behavioral Scientist*, vol. 24, pp. 15–47.

Davis, K. (1940), 'Extreme Isolation of a Child', *American Journal of Sociology*, vol. 45, pp. 554–565.

Davis, K. (1947), 'Final Note on a Case of Extreme Social Isolation', *American Journal of Sociology*, vol. 52, pp. 432–437.

Fujinaga, T. et al. (1988), *Ningen Hattatsu to Shoki Kankyo* (Human Development and the Early Environment), Yuhikaku, Tokyo.

Hasumi, O. (1990), *Kuno-suru Noson* (Rural Villages in Distress), Yushin-do, Tokyo.

Humphrey, C.R. and Buttel, F.R. (1982), *Environment, Energy, and Society*, Wadsworth, Belmont.

Hawley, A.H. (1986), *Human Ecology: A Theoretical Essay*, University of. Chicago Press, Chicago.

Itard, J.M.G. (1894), *Rapports et memoires sur le sauvage de l'Aveyron*, F. Alcan, Paris.

Lipset, S.M. (1963), The First New Nation, Basic Books, New York.

Mead, G.H. (1934), *Mind, Self and Society*, University of Chicago Press, Chicago.

Odum, E. (1983), *Basic Ecology*, CBS College Publishing, New York.

Parsons, T. (1951), *The Social System*, Free Press, Glencoe.

Parsons, T. (1971), *The System of Modern Society*, Prentice Hall, Englewood Cliffs/New York.

Parsons, T. (1978), *Action Theory and the Human Condition*, Free Press, New York.

Parsons, T. and Shils, E.A. (eds) (1951), *Toward a General Theory of Action*, Harvard UP, Cambridge/Mass.

Robertson, R. (1992), *Globalization*, Sage Publications, London.

Sassen, S. (1996), *Losing Control? Sovereignty in an Age of Globalization*, Columbia UP, New York.

Shimizu, I. (1940), *Shakaiteki Ningen Ron* (On the Social Nature of Humanity), Kawade-shobo, Tokyo.
Singh, J.A.L. and Zingg, R.M. (1942), *Wolf-Children and Feral Man*, Harper & Bros, New York.
Suzuki, E. (1940), *Nihon Noson Shakaigaku Genri* (Principles of Rural Society of Japan), Jicho-sha, Tokyo.
Thomas, W.I. and Znaniecki, F.W. (1918–20), *The Polish Peasant in Europe and America*, 5 vols., University of Chicago Press, Chicago.
Toda, T. (1937), *Kazoku Kosei* (Family Construction), Kobundo, Tokyo.
Tomlinson, J. (1999), *Globalization and Culture*, Polity Press, Cambridge.
Tominaga, K. (1990), *Nihon no Kindaika to Shakaihendo* (Modernization and Social Change of Japan), Kodan-sha, Tokyo.
Tominaga, K. (1997), *Kankyo to Joho no Shakaigaku* (Sociology of Environment and Information), Nikkagiren Shuppan, Tokyo.
Tominaga, K. (2002), 'Shakai-System ron kara mita Kankyo-Joho-gaku' (Environmental and Information Study from a Viewpoint of the Social System Theory), *Kankyo-Joho-gakubu Kiyo*, no.3, pp. 7–26.
Weber, M. (1972), *Wirtschaft und Gesellschaft*, J.C.B.Mohr, Tübingen.

Part II: Epistemological and General Reflections

2 Nature as an Issue in Environmental Protection

ERWIN K. SCHEUCH

Introduction

Environmental protection is a major international topic, and agreements between governments on such documents as the Kyoto Protocol have become the subject of intense controversies. But for all this attention in practical politics the major concepts relevant in the controversies remain empty phrases. What is ecological farming? What constitutes improper conditions in keeping animals? Above all what is natural about the nature in which we live and which we should preserve?

As nature is both a consequence of human action, and as our bond to it is a social relation, one would expect a considerable body of social science literature, and also of empirical research on this topic. As concern for the quality of the environment is especially pronounced in Germany, we expected a sizeable volume of social science literature and research specifically in the German speaking language area (Federal Republic, Austria, Switzerland). Indeed the body of writing is vast but predominantly it deals with popular concerns, and these largely reflect the changes in the focus of attention in the media. It is quite instructive how the foci have changed since the explosion in Tschernobyl on April 26, 1986 (Peters et al. 1987: 764–782). Currently it is dangers to health from foodstuff that tops the list of concerns. There are always new dangers revealed, often because instruments of research are becoming ever more finely tooled. The level of concerns, and the volume of media attention was fairly constant with a decline in the second half of the 90's only, however, with large changes in the subject of worries.

In preparation of the Yokohama conference I systematically searched in the scholarly repositories for German language material.[1] This is what I came up with:

---

[1] In the Federal Republic, these scholarly sources are part of a social science network, GESIS.

1.	Search in a survey data archive	1 study
2.	Search in a social science data bank for as yet unpublished projects	4 projects
3.	Search in a data bank specialising in studies on the environment	4 projects
4.	Search in the online-abstract register for social science literature	18 titles
5.	Search in the online-abstract register for environmental literature	22 titles

The one major survey with a scientific focus was part of the ISSP (International Social Science Program), and towards the end of this article I shall refer to the German part of it. The other projects were quite specialised, focus on rather marginal aspects of the field. Thus, we are informed on the basis of participant observations about the way in which city dwellers experience nature in an urban context.[2] In another theoretical project we are promised as a new speciality "Social Ecology".[3] A third project deals with contemporary life philosophy.[4] Only two of the altogether eight projects promised interesting information: a collection of advertisement in Switzerland with the theme "Nature and Man in Advertising since 1920",[5] and a study of TV advertising in the United Kingdom. More about these later.

A review of the authors of 40 publications reveals that the more substantial publications are based on doctoral dissertations. It is a common practice in German academia that the heavy-weights, the full professors, let candidates for a diploma or a doctorate work on aspects of a not yet fully established field before involving themselves in a major way. A next step are

2 This is an as yet unpublished project of the University of Halle, combining observational data and interview material. It is application oriented with the expectation to provide a better basis for action in nature protection. The character of the immediate neighbourhood and the age of persons are presumed to determine the perception and use of nature within an urban space.

3 A Vienna-based research institute wants to contribute to the criteria in applying the label "sustainable development". It hopes to achieve this by determining the conditions for co-operation between a natural science approach and the understanding of the interaction between man and nature in the social sciences. Based on an analysis of the literature the goal is the establishment as a new field in the sciences, "social ecology".

4 In a project at the University of Chemnitz the project leader seeks to clarify the relation between nature and culture. This is approached from contemporary perspectives of "Lebensphilosophie" (life philosophy).

5 The portrayal of man in natural settings as it is communicated via advertisements since 1920 is the basis for this project at the University of Berne. The observations are stored in machine readable format, and the resulting disc is to be used also for training purposes.

calling conferences, and the more substantial publications of a second rank of importance are anthologies based on conferences. If we view the collective of publications from a sociology of science perspective we can conclude: a new field of social ecology is in the making (Diekmann and Preisendörfer 2001; Diekmann and Jäger 1996; Huber 2001 and 2002).

An example of dissertations is the monograph by Guenther Sandner: "Nature and its Reverse". It investigates the discussion in two associations affiliated with the social democratic movement in Germany and Austria; the "Naturfreunde" and the "Arbeiter Abstinentenbund". The author approaches nature in a dialectical way by understanding as nature both the environment of man, and man's nature itself (Sandner 1999). With his orientation the author wants to show the influence of power in a society, and for this theoretical orientation he borrows from Antonio Gramsci, Jürgen Habermas and Michel Foucault.

Anette Braun in her "Perception of Forests and Nature" (Braun 2000) identifies two contrary approaches to nature in the last two hundred years. On the one hand it is man who forms nature to serve his purposes, and in the other approach nature is valued as unfolding according to its own determinants. On the basis of qualitative interviews, the author identifies five different clusters of attitudes towards nature, and specifically of the perspective towards forests – which are highly emotional objects in German Culture (Lehmann and Schriewer 2000) – but the dominant axis remains (1) nature as a resource or (2) as a value in its own right.

Other examples of books based on dissertations are "Nature in a Distance" (Solies 1998). Based on Georg Simmel's concept of distance, the author demonstrates how city dwellers relate to the unfamiliar nature by viewing it in an aesthetic mode as "landscape". "The Controversy about an Ecological Future" by Andreas Brenner is an analysis of the writings of three contemporary philosophers Hans Jonas, Robert Spaemann, and K.M. Meyer-Abich (Brenner 1994). They all reject a tenet central to the Enlightenment namely that culture means an overcoming of nature. All three attempt a "new ethic" which would render culture to be part of and subservient to nature.

The ethical question as part of a ecological movement is an important topic in the literature: Guenther Altner argues in "Steps Toward a Reconciliation of Nature and Culture" that man needs to become part of nature again (Altner 1994). Man has to limit himself in order to meet the responsibilities for future generations in dealing with nature. Gernot Boehme argues in "Natural Nature" that it is unclear whether there is something like nature in the naïve sense of the term (Boehme 1992). The "constructionists" maintain that reality, were it to exist outside our perceptions of it, is or would be socially irrelevant.[6] Relevant is merely what we communicate as presumed reality (van den Deale 1992). The late Niklas Luhmann may be considered as constructivist as he maintains that even damage to our health due to environmental problems is socially irrelevant as long as we do not communicate about it (Renn 1996: 40).

Unless one is philosophically a solipsist this post-modern stance is not worth discussing. Beyond the competence of a social scientist is also the assertion that nature has an existence something like that of a person. This calls for a new perspective on nature, and as a first step we are supposed to develop an aesthetics of nature. Klaus Michael Meyer-Abich as an early proponent of the movement for environmental protection has chosen ecological philosophy as the centre of his activities (Meyer-Abich 1987). The Leitmotiv is "make peace with nature" which implies that nature is of equal status with man; otherwise this is of course a nonsensical metaphor (Meyer-Abich 1984). Currently Meyer-Abich is developing a formal constitution codifying the permitted behaviour of man towards nature.

Even a psychologist as Walter Molt who maintains that the live-space of man is a culturally changed nature and not a natural nature, wants to limit economic activities (Molt 1984). He asserts that economic growth is destructive to nature which leads him to the problem: how can we lead a life worth living without economic growth?

From a left position Jochen Reiche argues against a prevailing trend in the literature (Reiche 1984). A turning-away from culture and civilisation in favour of a presumed "natural nature" would constitute a regression. When currently ecology is advocated as the central discipline then it is with such representatives as Carl Amery a camouflaged theology. The so-called natural order is not natural at all but the result of man's interference. Consequently, the destructive potential of industrial civilisation is neutralised by further interventions in nature based on our technical competence (Puntscher Riekmann 1996).

We cited Jochen Reiche last because he is a counter-point to what is otherwise the tenor of the social science literature on nature. Part of this literature is indeed theological in character, even though this is not the theology of churches but some kind of return to animism (Brenner 1994). References to empirical observations are few, the prevailing tone is philosophical, and the treatment of topics is "Geisteswissenschaft" (Goy 1992). What is most

---

6 This "post modern" environmentalism is of little importance in German language literature, but had a major impact on American sociology between the mid-seventies and mid-nineties.

striking is the bypassing of developments in the real world, to which we shall turn now.

Recent Developments

One of the most significant developments during the last about two decades was the transforming of ecological controversies into technical issues. The one form of this development was monitoring, the other the agreement on emission rights. Both should presuppose a knowledge of the mechanisms for hazards to the environment, but these are in many areas largely unknown. Up to this time our knowledge of processes endangering the safety of ground water is woefully incomplete, as is our knowledge of the contribution of winds to air pollution. The political way of liberating ourselves from inactivity is to agree on standards as mere conventions. These can always be changed if our knowledge improves.

The idea central to emission rights is simple: If emissions are a harmful by-product of industry, then the perpetrators should pay for it. Emissions as such are considered an unavoidable by-product of modern economies, and for the acceptance of a modern economy it is necessary to agree on overall levels of in principle harmful emissions that are considered as still tolerable. Shares are then sold which entitle companies to produce a certain volume of emissions. In this way emissions are turned into part of the cost of production. If these entitlements prove to be too costly for a given company, then it is out of business.

For a given country this convention would control the current harmful practice of externalising the true cost of production. However, this trade in emission rights is ineffective internationally, if there is no global convention. As the USA so far have refused to even sign the Kyoto Protocol it is hard to imagine that such a convention would come about. However, it would be helpful already if it were to become effective even just for a region such the European Union.

In this case, however, another problem would become obvious: The less developed economies of Europe such as Greece or Portugal would find it harder to meet emission standards. Contrary to the imagery of the environmentalists: The lower the level of development the higher the level of pollution for a given production-line. An example is China where low-grade coal is the chief fuel for heavy industry resulting in a level of Pollution that was last tolerated in Europe more than 100 years earlier. In addition, the mix of various industries in a nation would for countries such as China be disadvantageous, as in Europe the "smokeless industries" are an ever increasing part of production. Thus, emission rights as a policy would lead to international political conflicts – which is not necessarily a general argument against such a tool.

When the condition of the environment was finally recognised as a problem, and especially after this problem was understood calling for action the need for data increased exponentially. Does "global warming" really exist? How bad is air pollution? Is the condition of rivers improving? Environmental politics obviously requires data in order to set priorities. Data collection by public administration and by non-governmental organisations, on the national and the international level, enables us to carry out controversies in a less emotional manner then at the time, when the environment first became a concern. By now this need for data has led to the development of monitoring systems, although large areas are still not covered.

As already mentioned such monitoring was extended to cover bio-topes in urban spaces. Already towards the end of the seventies the cities of Munich and Berlin were the objects of such data collection (Tessin 1991). There are some differences in the methods of spatial representation of data, but the major issues are political (Lachmund 2002). The emphasis in collecting information was on plant life. The condition of animals was largely bypassed, even though it is reported in an aside that there are surprisingly numerous animal populations in cities. The major conflicts arose with the classifications that included evaluative categories. A prime example is the characterisation of open spaces and unused areas in urban communities, which were often coded with a view of their potential as habitat for plants, or alternatively for housing.

In general there are frequently conflicts in assigning levels of damage to such objects as rivers or forests. In Germany, it was especially the grading of damage to forests that became the topic of heated controversies. Should the reporting of damages include deviations from a norm that are visible only to specialists, or merely obvious damages? After all, no person is completely healthy if standards of perfection are applied to humans. The emotional label "Waldsterben" (dying forests) would, however, only been justified if an extensive damage could by documented. The controversies were further intensified when government programs of compensation to owners of damaged wooden areas were under discussion. Thus, the mere existence of fact-finding programs in the field of environmental protection could intensify conflicts over the state of nature. However, by a large monitoring tended to lead to a dampening of passions.

All devices for conflict reduction in disagreements over the state of the environmental fail, if there is dissent whether the presumed danger exist at all. An example for Germany is the claim that something like "electro-smog" exists – invisible rays from overland high tension electric wires. By now yet another invisible culprit is added to the dangers that envelop us presumably, rays from the antennas for mobile phones. Some of these invisible and elusive dangers are culture-specific. Thus, it is mainly in England that people proclaim they are adversely affected by rays from their microwave-ovens. One may take these claims of environmental hazards as indicators for the level of generalised anxiety in a country. And here the famous Thomas-theorem is applicable: if people believe something to be real, it is real in its consequences.

German Nature

There is considerable agitation about the threat of natural catastrophes. The flood in the Eastern part of Germany in August/September 2002 not only led to hysteric overreactions on a large scale but also predictions that worse is to come soon. And of course these were to be catastrophes caused by man, e.g. as a result of this elusive global warming. As a result of this, the Netherlands are presumed to become a part of the North Sea within this century.

We can treat such assumptions as indicators for states of knowledge and states of affairs that are otherwise not readily discernible. Germany may be a country with a sometime deplorable climate (although it is neither really hot or really cold), her soil is not the most fertile, and by now the earlier mineral resources are nearly all depleted. It may also be true that an active volcanic underground exists in the Eifel region, and in the valley of the Rhine between Freiburg and Karlsruhe. However, in historic times no volcano erupted, and no major earthquakes occurred. We ourselves experienced an earthquake of the strength 5.8 on the Richter Scale in April 1992 in Cologne that resulted merely in cosmetic damages to buildings.[7] In a survey by the "Deutsches Komitee für die internationale Dekade für Katastrophenvorbeugung" (German Committee for the International Decade for Catastrophe Prevention), listing all major natural catastrophes since 1980, Germany does not appear even once. There is hardly ever a storm exceeding 100 km an hour – at most around 120 km –, and the ample rain is merely an inconven-

7 In a developing country earth tremors of this strength would have a devastating effect in built-up areas.

ience. It appears to us that precisely because of the absence of a hostile nature, nature is venerated (Mainusch 1996). As we mentioned earlier, some authors bestow it with sacred character. A damage to nature is in German called "Naturfrevel" – insufficiently translated as "outrage" or "transgression". Nature is good to man! And indeed, there are no more predatory animals dangerous to man, just one or two kinds of very rare somewhat poisonous snakes. One can stroll through forests without fear of poison ivy as this makes American forests uncomfortable, nor does one need to be aware of deadly water moccasins when bathing, as one has to be in the US. There is no wilderness area in all of the country!

This Germany has a nature that is mostly friendly to man, because man changed it for his own benefit (Groß 2001). There is no wood in this fairly wooded country that is primeval growth. The famous Black Forest is only around two hundred years old, and at that time pine trees were planted that do not go well with the local kind of soil (Braun 2000). There is not a single larger river in the country that flows in its original bed, not a single hill that has its original shape (Krieger and Jäggi 1997). The Alps alone have by and large retained their original natural appearance. There is nothing natural in the literal meaning of the word about the landscape that Germans live in (Großklaus 1983).

By nature, Germany was a far less habitable country than it is today. Its areas were mostly covered by woods that were fairly impenetrable. A large percentage of the land was swampy. The upper part of the Rhine valley was such a marshland, where until about two hundred years ago malaria mosquito's felt at home (Krieger and Jäggi 1997).

Farming was and is a major factor in producing the landscape that we are used to look at. Farming is part of living in Central Europe since about 7,000 to 8,000 years. West of the Elbe this meant a mix of plant cultivation and animal husbandry. Between around 1,000 and 1,300 B.C. there was a population explosion, and at that time the tops of hills were cleared of woods. In a second thrust of opening more land to farming the swampy river valleys were settled, leaving for woods the slopes of hills. This is readily discernible today by looking down through an air-plane window.

The type of farming is since the end of the medieval times the family farm that was largely self-sufficient. Marketable products were until recently of only marginal importance, and there was no road system allowing for transport on a larger scale. Such a farm was a delicately balanced enterprise: a mix of pigs, cows, horses for wealthier peasants, geese, and chicken. The relation of animals to each other and to the time required for caring was also determined by the need for manure for fertilising plants. The familiar mix between meadows with grazing cows, various fields, and farmyards with animals is by no means "nature" in an emphatic sense. Viewed as an economic enterprise the continuation of this kind of farming is no longer profitable (Elsen 1999). Consequently, some spokesmen of farmers in Germany demand subsidies as "landscape gardeners". The argument: If you want the pleasant landscape that follows from our kind of farming, you the public should pay for it.

The woods that are an object of special importance for those self-appointed protectors of nature were some 250 years ago endangered. Farmers exploited woods as a mere resource, and cut down the trees as this was their major kind of fuel. As England was economically ahead of Germany, deforestation proceeded far more quickly than on the Continent. Coal was abundant, but it would not be mined beyond a depth of approximately 45 feet because there was no way to effectively control ground water. The steam engine was developed and with it as a means of drainage fuel became available in unlimited quantities. By that time already half of the wooded areas in Germany had been deforested. Subsequently, systematic reforestisation resulted in the current land use: Not quite one third of Germany is forest – none of it primeval.

Graph 1: Distribution of land uses (in percent)

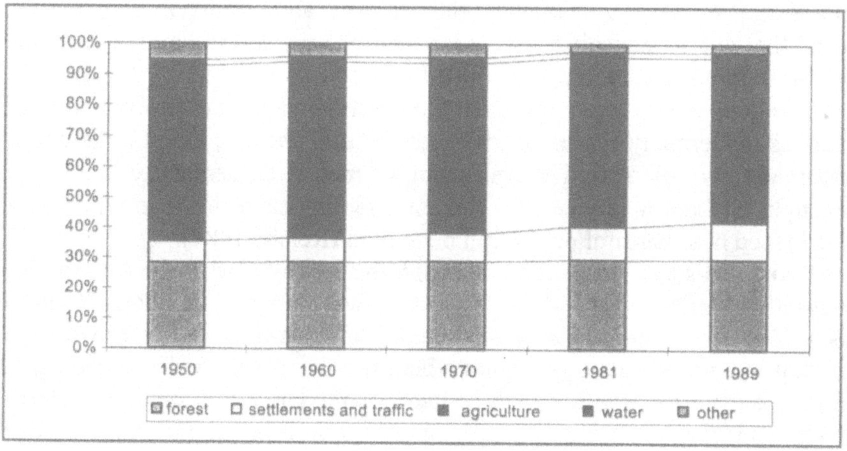

In asking for a better protection of nature, climate is assumed to be constant (Stehr 1999). This is gravely an error! Between 1000 and 1300 Europe including Scandinavia was considerably warmer than today. This was one of the factors that led to the raids by the Vikings as South Sweden (Värmland) became overpopulated. During the first half of the 18th century, temperatures fell and one spoke of a "small ice age". These variations in average yearly temperature exceeded considerably the variations that are currently such an object of concern. These observations are not meant to discourage attempts to reduce air pollution, but to suggest more sobriety in setting goals. Thus, the target cannot primarily be the climate but the health and comfort of our compatriots.

With the use of chemical agents, heavy machinery, and specialisation, conflicts between agriculture and environmentalists increase (Ribbe 2001). As an alternative to a rationalised mode of producing foods, the "Greens" advocate "bio-farming". At this point I want to bypass the controversy whether bio-agriculture generally results in healthier food, and concentrate on the confusion that exists in the debate of environmentalism.

What is Natural?

When we criticise the breeding of chicken or pigs in containers, what is the benchmark in evaluation? We can readily agree that animals should feel comfortable as the criterion – but what is it that makes them comfortable? In Northrhine-Westfalia a "Green" minister was able to have a regulation passed that not only specifies the space that each pig has to have in its stable but also requires that each pig is entitled to have a toy.

The goal central to environmental protection cannot be the protection of nature as in Germany we do not live in a "natural" nature, but in an environment made over by man. The well-being of man is consequently the proper yardstick, and not a service to a romantic feeling-state of those who never experienced how harmful nature can be to man (Reiche 1984).

Some years ago we travelled through the East of Canada to Alaska. For thousands of miles to the left and right there was nothing but woods. Then in this wilderness we drove past vast stretches of dried-up trees. Later we were told that is a normal state of affairs where the dying of wooded locations is the prerequisite to strong growth in later years. There is a natural cycle of growth and of decay which is completely ignored in the German concern over "Waldsterben". Somewhat later we drove past a clear stream that looked inviting to us thirsty travellers. Alas, as we walked closer we were warned: do not drink – it's dangerous. Later we learned that foxes in this wilderness had digestive problems, and their feces were polluting the well-waters. It is not only man who causes nature to be unpalatable! (Glaeser 1992)

When the Eastern part of Germany was hit in 2002 by an unusual flood this immediately triggered the search for human culprits. Historians recalled that some 150 years earlier there was a flood of even somewhat greater dimensions, and at that time man had not yet done much to change river courses, straighten the river beds, and build dams cutting off the waters from surrounding meadows that could serve as retention areas to maintain the overflow. No, it had to be man's intervention into a natural state of affairs that caused damage to people – fortunately limited – and to landscape plus housing estimated at 22 billion Euros. Specialists calculated that at most 10 to 20% of the flooding could have been accommodated if the rivers of the tributary system of the Elbe had been left in their natural condition. A human culprit was required to restore the perception that nature if left alone is good to man. When man tries to improve nature to serve his interest, he is presumably damaging something perfect – something nearly sacred.

Some time ago a wilderness area in Hamburg was chosen as a site for family homes. Vociferous opposition arose almost immediately as it was maintained that this area was home to a very rare kind of bird. When proponents of the housing project argued that no one had ever seen this particular bird the response was, that this was proof how rare this precious bird was. In arguing about the need to fell some trees for a new access road to a motorway, we suggested that an equal number of trees should be planted elsewhere to make up for the loss. The already mentioned Meyer-Abich objected: One should not treat a piece of land as a commodity; each had a dignity in its own right. These are not the prevailing sentiments but they do express a feeling among a minority that any change to an existing state of nature has to be considered as a possible violation of some quasi-sacred order (Mainzer 1990).

While in paintings up to the Baroque style landscapes were often the mere backdrop to something else, with Romanticism landscape became a motif in its own right. In the British version of Romanticism, such as with Gainsborow and later Constable, landscapes are "lovely"; in Germany with Böcklin they are mysterious. With the French impressionists landscapes then become the locations for leisure. In the 20th century photography and advertising took over much of the earlier functions of painting. It is especially advertising that is the provider of much of our visual imagery: Nature is attractive (Dinnebier 1996).

While land until the beginning of the 19th century was a resource and commodity, with the overcoming of hunger and with life in urban areas, land turned into a resource for recreation. The same mechanism produced a sentimental relation of modern man in the West to those animals that became his companions, and by now this includes most higher animals. No wonder that feelings of protectionism for something vulnerable developed. Our appreciation of nature comes close to viewing it as an object of luxury. In environmental protection in its more sober forms we intervene in order to retain this nature as a comfort to man (Bartmann and John 2000).

As with many fields of behaviour these reactions, too are contingent. By now it would be possible for a firm such as Nestlé to produce a great variety of foodstuffs all based on soybeans. By adding the proper chemicals food engineers could make soybeans look and taste like a steak. But why should we want to eat such a synthetic food? There is a sufficient supply of true steaks! The attitude of persons living on the verge of starvation, however, is likely to be quite different (Boehme and Riple 1995).

There is a whole body of empirically based literature on the readiness in the population to live with risks. The studies agree that a very rare danger with grave consequences is less acceptable than many small risks even though the total damage of the latter may exceed the consequences of one large damage. This is quite the reverse of risk taking in lotteries, where the very unlikely jackpot is more attractive than many more small winnings (Lüdemann 1999).

An evident expression of using man-made nature as luxury objects are gardens for pleasure only (Sperlich 1979). Gardens are developed only in highly advanced cultures primarily for the more advantaged elements of a population. As such gardens can be looked at as a projective test of a culture. In Europe the two extreme cases are the highly stylised French gardens, and the seemingly "natural" English style gardens. In parts of gardening in China nature is artificially created by overemphasising features of nature such as grotesque rock formations. There is even a specialised gardener there called "rock builder". It could be one way to ascertain the imagery that members of different cultures have of nature by confronting them with pictures of gardens from Roman times, French gardens, Florentine gardens, English gardens, Chinese gardens, Persian gardens and Japanese gardens. "Nature" appears to be in no small way also culture specific.

Genetic engineering deeply disturbs the naive appreciation of nature. There is strong opposition to genetically changed plants, even though by now it is quite difficult to avoid the results of genetic engineering. The object of such engineering is hardly ever a total plant. Most of the genetic changes are directed at components of plants such as enzymes, or at yeast, to increase the resistance to pests. There are, however, widespread fears that unpredictable harm may result from genetically changing plants. Scientists may argue that genetic engineering is nothing else than an extremely rapid alternative to breeding. This does little to alleviate fears of unpredictable consequences.

Public Perception

"Save nature" is part of a movement of gaining control over one's life (Ott and Gorke 2000). People want to know what they are eating, how they can control their state of health, how they can learn about their own limits of competence. Studies of risk acceptance demonstrate, that the prime cause for concern is not the actual risk from a particular danger but the impossibility of influencing this risk through own actions.

Thus, there is fear of air-plane accidents but much less fear of damage in a car accident, fear of chemicals in food and little fear of the consequences of smoking. High risk sports as hang gliding or bungee, or wilderness canoeing, or climbing under extreme conditions, all have big growth rates. Riding heavy motor cycles is considered a proof of male-bravado among the young, while in the same age group there is fear of damage from contaminated drinking water.

Since 1972 there was an increase in the concern with the quality of the environment. The focus of attention has been shifting during that period, with the concern about the survival of the forests (Waldsterben) drawing internationally most attention. Currently, fear of food contamination through an agriculture based on fertilisers and with breeders using hormones is rising, and peaked when there was fear of the BSE contamination. In between concern with clean air dominated. As a general goal all political parties now embrace an "ecological market economy" that allows living with an "undamaged nature".

There is a widespread tendency also in the social sciences to attribute the perception of reality specifically in the highly developed countries to the media, and here especially to advertising. This does not hold for images of nature. A content analysis of TV advertising between March 6 and April 23, 2000 in Britain resulted in the following distribution of images (Hansen 2002: 505):

Table 1: Primary images/uses of nature in television advertisements in the UK

Categories	Frequency	Percent
Nature as intrinsically good (e.g. healthy, fresh)	43	32.6
Human mastery/power over nature	28	21.2
Nature as a nice place to be	25	18.9
Recreational function of nature	11	8.3
Nature as distance/space traversed/ the in-between/obstacle	9	6.8
Nature as a threat	5	3.8
Nature as metaphor for life's journey	4	3.0
Nature as symbol of freedom	2	1.5
Nature as spectacle, packaged (TV) spectacle	2	1.5
Nature as global, big, awesome, impressive	2	1.5
Nature as challenge/sport/manhood/endurance	1	.8
Total	132	100.0

The authors emphasise "the virtual complete absence of images of nature as vulnerable, fragile, or in need of protection" (oedem loco). They assert that this is in contrast to the public debate in the media. "It thus seems that the environment is quite simply not present as a social problem in British prime time television advertising." (Hansen 2002: 506)

This is the perception of the German population in East and West-Germany in complaining about their environments:

Table 2: Environmental concerns in Germany

Year	high noise levels		air pollution		lack of parks	
	West	East	West	East	West	East
1994	10 %	17 %	9 %	18 %	3 %	10 %
1999	7 %	10 %	6 %	8 %	3 %	4 %

Source: Federal Office of Statistics. Data Report 2002, p. 514–515

At the time of reunification, the environment was in a sorry state in the East, and according to all counts it has improved considerably. There is an increasing sensitivity to noise as a nuisance, while annoyance about dirt has declined dramatically. As factors endangering the quality of the environment, air pollution through industry tops the list, followed by fear of global warming. There is little that can be done about the latter complaint as it is the kind of concern that "Waldsterben" was earlier. Altogether the percentage of those reporting grave concerns about the environment has declined from 61% in 1990 to 23% in 2000 (Socioeconomic panel 1990–2000).

Graph 2: Environmental protection as one of the most important problems

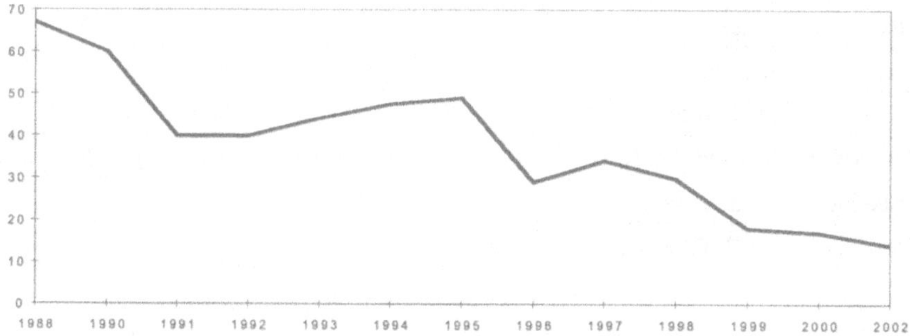

Source: 1988–1999 EMNID, 2000 and 2002 UBA – in %

The rank order of problems, and even the percentages for worries remain quite stable over the last 24 months. In the following table there are two exceptions: the concern over the flooding of part of East Germany, and the worry about war. The latter increased in response to President Bush's speeches, and the criticism of it during the closing phase of the election campaign. (The elections for a new parliament took place on September 22, 2002) With the exception of unemployment – the rate hovers around 9.5 to 10 % – there needs to be an acute event in order to temporarily focus concerns for the about 15 top concerns during the last two years.

Table 3: Major worries in the general public in Germany

	August 2002		September 2002	
	1st half	2nd half	1st half	2nd half
Unemployment	81 %	72 %	79 %	85 %
Flood in East Germany	-	38	16	4
Economic condition of Germany	13	12	13	14
Anger at politics/corruption	8	7	5	5
Terrorism/danger of war	-	-	5	16
Environmental protection	8	7	11	6
Taxes/tax increases	5	5	7	6
Quality of education	6	4	6	8
Immigration/asylum seekers	4	5	6	5
State of families and youth	3	4	5	5
Price increases/the Euro	6	3	-	-
Forschergruppe Wahlen: Politbarometer				

Graph 3: Evaluating the quality of the environment I (Germany)

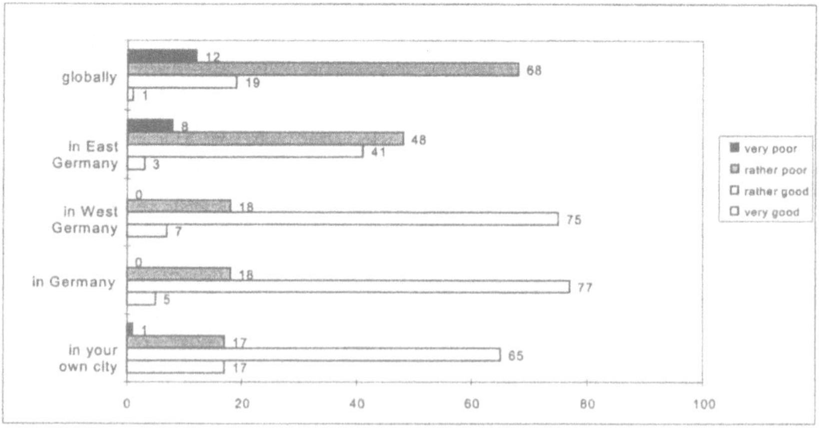

Source: Kuckartz: Jan/Feb 2002 survey, n=2.361 (answers in %)

Graph 4: Evaluating the quality of the environment II (Germany)

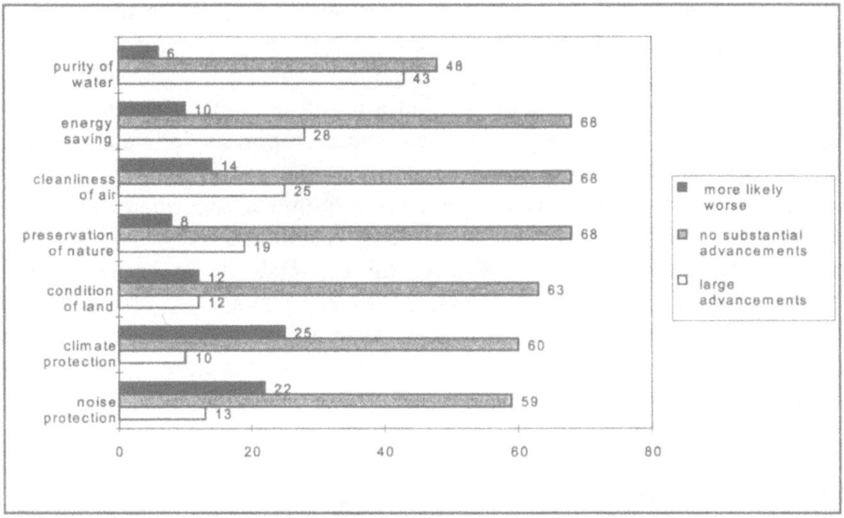

Source: Kuckartz: Jan/Feb 2002 survey, n=2.361 (answers in %)

Religion and Science

While in general surveys concern about environmental problems generally do not dominate other issues, and in the absence of an acute problem rank behind bread-and-butter concerns, given the speed of BSE as a result of beef consumption, environmental worries can erupt any time. During such eruptions these concerns are not easily amenable to sober argumentation.

The ALLBUS study of 2000, concentrating on environmental concerns, revealed that the population in Germany is deeply divided about the intensity of a presumed environmental problem, and on its general priority among other issues. Asked whether economic growth was harmful to the environment 29.7 % agreed, 22.5 % were undecided, and 37.8 % disagreed (ALLBUS 2000). Another question with the opposite meaning confirmed this divisiveness. "To protect the environment, Germany needs economic growth": Yes 57.3 %, undecided 16.6 %, no growth needed 26.1 %.

Unemployment is the no. 1 concern of the German public since many years. Therefore the following question was a very strong test for those responding on other questions with priority given to the quality of the environment: "We worry too much about the future of the environment, and too little about prices and jobs now.": It is right to worry first about the environment: 35.8 %, undecided 15.6 %, not enough worry about jobs 48.4 %. In a further question 37.3 % declared themselves willing to tolerate some reduction in the standard of living if this were to help the environment, but 34.4 % opposed this. Even if one takes into consideration that such declarations in an interview express primarily a mood, and are not necessarily to be taken as predicting a respondents later reactions, these responses indicate that the feelings expressed here have deep roots.

In earlier surveys we had concluded that religiously Germans belong to three camps: Christians with varying intensity, non-believers, and "pantheists". German cemeteries are indicative of this tendency to bestow nature with a transcendental meaning: we bury the dead among tree groves. Thus, we suspected that the environment as an issue activates this quasi-religious attitude to nature, bestows the "defence" of nature with a transcendental significance.

Table 4: Religion and Nature

The ALLBUS 2000 posed the following question: "Which of these statement comes closest to your own views?"

Nature is holy as a creation by God (= God's creation)	15.5 %
Nature is holy by itself ("heilig an sich")	22.4 %
Nature is important but not holy	62.1 %
Total	100.0 %

While a majority has a rather sober appreciation of nature, it is a sizeable minority that holds a more exalted view.

We wanted to clarify how this correlated with religious sentiments, and for this purpose we used three indicators of religiosity:

1. Frequency of church attendance as a criterion for institutionalised religiosity;
2. Belief in God as showing the agreement with the teachings of organised religion;
3. Self-rating of the respondent as of his degree of religiousness.

Of course, all three indicators inter-correlate – but they are relating to the dependent variable "view of nature" in somewhat different ways. This allows us a first tentative interpretation of the sentiments of those who hold a pantheistic view of nature.

Table 5: Views of nature by frequency of church attendance in Germany

	Views of Nature			
Church Attendance	**God's creation**	**Holy by itself**	**Not Holy**	**Marginals**
At least weekly	28.1 %	7.1 %	6.7 %	10.1 %
Monthly and more	19.0 %	10.0 %	5.7 %	8.7 %
At least once a year	38.0 %	55.4 %	49.8 %	49.2 %
Never	14.9 %	27.6 %	37.8 %	32.0 %
Sum	100	100	100	100

Attachment to the churches is strongest among those seeing in nature the creation of God. That was to be expected, although somewhat surprisingly we witness that 14.9 % of those never attend church services. It confirmed our expectations that those seeing nature as being merely natural have the loosest relation to organised religion. What interested us most were the values for those who view nature as holy by itself. Dominant is here a loose relationship: Religious feelings but with a distance to organised religion.

Table 6: Views of nature by belief in God in Germany

	Views of Nature			
Belief in God	**God's creation**	**Holy by itself**	**Not Holy**	**Marginals**
Non believers	8.3 %	21.7%	37.9 %	29.4 %
There is some "Higher Power"	11.4 %	26.1%	22.2 %	21.4 %
Belief with doubts	17.1 %	37.3 %	26.0 %	27.2 %
Unconditional belief	63.2 %	14.9 %	14.1 %	22.1 %
Sum	100	100	100	100

The pattern visible already in the previous table is identifiable here again – only somewhat clearer. Those venerating nature as holy by itself show in around two thirds of all cases an unstructured religious sentiment. In an admittedly bold interpretation we diagnose that nature is replacing institutionalised religion.

Table 7: Views of nature by self-rating as religious in Germany

	Views of Nature			
Self-rating as religious (scale 1–10)	**God's creation**	**Holy by itself**	**Not Holy**	**Marginals**
not at all religious (1,2,3)	8.5 %	32.3 %	45.1 %	36.6 %
Intermediate/so-so (4,5,6,7)	33.5 %	40.4 %	33.8 %	35.2 %
Quite religious (8,9,10)	58.0 %	27.3 %	21.1 %	28.2 %
Sum	100	100	100	100

Much more distinctly than with the previous breakdowns the view of nature as God's creation is tied to religious sentiment in conformity with ecclesiastical teaching. Those that interpret nature as holy by itself are in terms of their religious self-identification quite heterogeneous. To us the most instructive finding here are the 1/3 professing not to be religious at all, and at the same time imputing to nature a holy character. "Nature as holy" is not just another manifestation of familiar religiousness, but in a large part an alternative to religion institutionalised as churches.

Recently the sociologist Nico Stehr argued that the history of human civilisation can be understood as a growing emancipation from hidden and uncontrollable forces of nature (Neue Zürcher Zeitung, 22.09.02, p. 56). For hundreds of years technology seemed to make man increasingly the master of nature. But now a minority understands man to be the servant of nature.

Table 8: Science and the environment

In the ALLBUS 2000 we asked: "Modern science will solve our environmental problems, requiring only small modifications in our life style":

strongly agree	5.29 %
agree somewhat	28.67 %
undecided	22.71 %
rather disagree	33.06 %
disagree strongly	10.28 %
Total	100.00 %

The result should not be read as mirroring a rejection of modern science. It is rather a scepticism against the promise of an omnipotence of science. Thus, when we asked "All in all modern science does more harm than good" we observed the following distribution.

Table 9: Modern science

"All in all modern science does more harm than good":

strongly agree	2.40 %
agree somewhat	12.28 %
undecided	16.65 %
rather disagree	41.57 %
disagree strongly	27.10 %
Total	100.00 %

The gap between the personnel debating the more philosophical aspects of environmental protection, and the place of this topic among the concerns of the general public, is increasing. The reflections about the kind and the place of the need for environmental protection are getting ever more ideological. The arguments tend to fuse with the older and well known culture criticism of the consumer society and culminate in calls for a different economic order, or at least in a change of priorities.

A minority, quite vocal in the media of high culture, has a way of perceiving nature which is at odds with the large majority. In these milieus nature is viewed in an anthropomorphic way, and from this perspective nature as part of modern civilisation is turned into "our sick fellow-being". This is a consequence of a lack of confidence in the individual and especially collective ability to control events.

The thrust of our own argumentation was of course not to criticise the concern for the environment as such, but recommend a change in the kind of concern. We strongly believe in environmental protection as long as it serves the long term interest of man. Environmental protection as an attempt to comfort nature as the sick fellow-being is to miss the proper goal of environmentalism.

References

Bartmann, H. and John, K.D. (eds) (2000), *Natur und Umwelt*, Shaker, Aachen.

Baumol, W.J. and Oates, W.E. (1975), *The Theory of Environmental Policy. Externalities, Public Outlays, and the Quality of Life*, Prentice Hall, Englewood Cliffs.

Biese, A. (1926), *Das Naturgefühl im Wandel der Zeiten*, Quelle und Meyer, Leipzig.

Birnbacher, D. (ed.) (1986), *Ökologie und Ethik*, Reclam, Stuttgart.

Bloch, J.R. and Maier, W. (eds) (1984), *Wachstum der Grenzen: Selbstorganisation in der Natur und die Zukunft der Gesellschaft*, Sendler, Frankfurt.

Boehme, G. and Riple, W. (1995), 'Ohne Mensch wertlos? Wie läßt sich Natur dauerhaft schützen?', *Politische Ökologie*, vol. 43, pp. 25–27.

Bohrer, K.H. (1987), 'Nach der Natur – Ansichten einer Moderne jenseits der Utopie', *Merkur*, vol. 41, pp. 631–645.

Bookchin, M. (1982), *Natur und Bewußtsein*, Winddruck, Siegen.

Braun, A. (2000), *Wahrnehmung von Wald und Natur*, Leske + Budrich, Opladen.

Brenner, A. (1994), *Streit um die ökologische Zukunft: neue Ethik und Kulturalisierungskritik*, Köningshausen & Neumann, Würzburg.

Buttel, F. (1987), 'New Directions in Environmental Sociology', *Annual Review of Sociology*, vol. 13, pp. 465–488.

Diekmann, A. and Jaeger, C.C. (eds) (1996), 'Umweltsoziologie', *Kölner Zeitschrift für Soziologie und Sozialpsychologie*. Special Issue, no. 36.

Diekmann, A. and Preisendörfer, P. (2001), *Umweltsoziologie*, Rowohlt, Reinbek.

Dinnebier, A. (1996), *Die Innenwelt der Außenwelt – Die schöne Landschaft als Problem*, TU Berlin, Berlin.

Dischner, G. (1972), *Ursprünge der Rheinromantik in England. Zur Geschichte der romantischen Ästhetik*, Klostermann, Frankfurt.

Dunlap, R. et al. (1994), 'Understanding Environmental Problems – A Sociological Perspective', in Beat Burgenmeier (ed.), *Economy, Environment, and Technology*, M.E. Sharpe, Armonk (USA), pp. 27–49.

Dunlap, R. and Michelson, W. (eds) (1998), *Handbook of Environmental Sociology*, Westport (Ct.).

Elsen, T. van (1999), 'Kann Landwirtschaft die Natur "ernähren"? Bewußtes Wahrnehmen als Ausgangspunkt individueller Landschaftsentwicklung', *Lebendige Erde*, vol. 50, pp. 334–337.

Fischer, H. (1991), 'Draußen vom Walde. Die Einstellung zum Wald im Wandel der Geschichte', in *Naturwälder in Niedersachsen*, o.V., Schneverdingen, pp. 92–100.

Forschungsgruppe Wahlen (1997–2002), *Politbarometer*, Mannheim.

Funabashi, H. (1992a), 'Social Mechanisms of Environmental Destruction: Social Dilemmas and the Separate-Dependent Ecosystem', in H. Krupp (ed.), *Energy Politics and Schumpeter Dynamics*, Springer, Verlag, Tokio, pp. 265–275.

Funabashi, H. (1992b), 'Environmental Problems in Postwar Japanese Society', *International Journal of Japanese Sociology*, no. 1, pp. 3–18.

Glaeser, B. (1992), *Natur in der Krise? Ein kulturelles Mißverständnis*, Wissenschaftszentrum Berlin, Berlin.

Goy, K. (1992), 'Organologische Technik oder technische Natürlichkeit – das Programm einer Einheit von Technik und Natur', *Deutsche Zeitschrift für Philosophie*, vol. 40, pp. 490–502.

Groß, M. (2001), *Die Natur der Gesellschaft. Eine Geschichte der Umweltsoziologie*, Juventa, Weinheim.

Großklaus, G. (1983), 'Der Naturtraum des Kulturbürgers', in G. Großklaus and E. Oldemeyer (eds), *Natur als Gegenwelt*, von Loeper Verlag, Karlsruhe, pp. 169–196.

Hackl, F. (2000), 'Der Wahnsinn der Ökonomie oder der Unsinn der Ökologie: eine epistemologische Analyse', *Zeitschrift für Umweltpolitik & Umweltrecht*, vol. 23, pp. 165–185.

Halfmann, J. (1986), 'Autopoiesis und Naturbeherrschung', in H.-J. Unverferth (ed.), *System und Selbstproduktion*, Lang, Frankfurt/M., pp. 192–229.

Hansen, A. (2002), 'Discourses of Nature in Advertising', *European Journal of Communication Research*, vol. 27, pp. 499–511.

Huber, J. (2001), *Allgemeine Umweltsoziologie*, Westdeutscher Verlag, Opladen.

Huber, J. (2002), 'Environmental Sociology in Search of a Profile', *Soziologie*, no. 3, pp. 23–36.

Kliment, T. (1993), *Kernkraftprotest und Medienreaktionen*, Dissertation in Sozialwissenschaften, Ruhr-Universität Bochum.

Kliment, T. and Renn, O. (1993), *Die Chancen und Risiken der Gentechnologie aus der Sicht der Bevölkerung*, Akademie für Technikfolgenabschätzung in Baden-Württemberg, Stuttgart.

Kreul, W. (1997), *Die Ernte der Agronomen*, Ed. Interfrom, Osnabrück.

Krieger, D.J. and Jäggi, C.J. (1997), *Natur als Kulturprodukt: Kulturökologie und Umweltethik*, Birkhäuser Verlag, Basel.

Kruse-Graumann, L. (1996), *Umweltschutz aus psychologischer Perspektive: Bewußtsein und Verhalten, Beiträge zur Ökosystemforschung und Umwelterziehung III*, German MAB National Comittee, Bonn, pp. 171–178.

Kuckartz, U. (2000), *Umweltbewußtsein in Deutschland*, Bundesumweltministerium, Berlin.

Kükelhaus, H. (1993), *Organismus und Technik*, Fischer, Frankfurt.

Lachmund, J. (2002), 'Die registrierte Natur – zur Konstruktion von Expertenwissen bei ökologischen Bestandsaufnahmen', *Soziale Welt*, vol. 53, pp. 209–236.

Lehmann, A. and Schriewer, K. (eds) (2000), *Der Wald – ein deutscher Mythos*, Reimer, Berlin.

Linneweber, V. (2002), *Environment: thematic career of a social construct*, Unpublished paper, University of Magdeburg.
Lüdemann, C. (1999), 'Verhalten, Schwellenwerte und deren Determinanten: Die empirische Überprüfung einiger Hypothesen im Bereich Umweltverhalten', *Kölner Zeitschrift für Soziologie und Sozialpsychologie*, vol. 51, pp. 532–549.
Mainusch, H. (1996), 'Natur als Ideologie', *Mut*, no. 343, pp. 74–87.
Mainzer, K. (2002), 'Natur, Technik und Kultur: Zur Philosophie moderner Lebenswelt', *Konstanzer Blätter für Hochschulfragen*, vol. 26, pp. 36–45.
Meulemann, H., (2001), *Naturalistische Weltbilder und die Frage nach der Umdeutung sozialer Prozesse – Ergebnisse aus der Umfrageforschung*, Unpublished paper, University of Cologne.
Meyer-Abich, K.M. (1984), 'Frieden mit der Natur', *Frankfurter Hefte*, vol. 39, special issue, pp. 46–53.
Meyer-Abich, K. M. (1987), 'Menschliche Wahrnehmung der natürlichen Mitwelt', in J. Calliess and R.E. Lob (eds), *Handbuch Praxis der Umwelt- und Friedenserziehung*, vol. 1, Schwann, Düsseldorf, pp. 309-317.
Molt, W. (1984), 'Psychologie, Gesellschaft, Umweltkrise', in Helmut Moser and Siegfried Preiser (eds), *Umweltprobleme und Arbeitslosigkeit*, Beltz, Weinheim, pp. 61–66.
Ott, K. and Gorke, M. (eds) (2000), *Spektrum der Umweltethik*, Metropolis, Marburg.
Peters, H. et al. (1987), 'Die Reaktion der Bevölkerung auf die Ereignisse in Tschernobyl', *Kölner Zeitschrift für Soziologie und Sozialpsychologie*, vol. 39, pp. 764–782.
Poferl, A. et al. (1997), *Umweltbewußtsein und Alltagshandeln*, Leske + Budrich, Opladen.
Puntscher Riekmann, S. (1996), 'Von der Beherrschung zur Überwindung der Natur – Versuch einer Theorie der Macht über die Natur', *Österreichische Zeitschrift für Politikwissenschaft*, vol. 25, pp. 121–135.
Redclift, M. and Woodgate, G. (eds) (1997), *The International Handbook of Environmental Sociology*, Elgar Publ., London.
Reiche, J. (1984), *Ökologie und Zivilisation – der Mythos von den natürlichen Kreisläufen, Die Linke neu Denken,* Wagenbach, Berlin.
Renn, O. (1996), 'Rolle und Stellenwert der Soziologie in der Umweltforschung', in A. Diekmann and C.C. Jaeger (eds), Umweltsoziologie, *Kölner Zeitschrift für Soziologie und Sozialpsychologie*, special issue, no. 36., pp. 28–58.
Ribbe, L. (2001), 'Die Wende in der Landwirtschaft', *Aus Politik und Zeitgeschichte*, vol. 24, pp. 30–38.
Rink, D. (2002), *Naturbilder und Naturvorstellungen sozialer Gruppen*, unpublished, pp. 23–39.
Sandner, G. (1999), *Die Natur und ihr Gegenteil: politische Diskurse der sozialdemokratischen Kulturbewegung*, Peter Lang Verlag, Frankfurt/M.
Scheuch, E.K. (1990), 'Vom schmerzlichen Werden einer modernen Gesellschaft', *Hamburger Jahrbuch für Wirtschafts- und Gesellschaftspolitik,* Hamburg, pp. 27–51.
Scheuch, E.K. (1996), 'The Puzzle of "Quality of Life"', in W. D'Antonio et al. (eds), *Ecology, Society & the Quality of Social Life*, Transaction Publishing, New Brunswick (USA), pp. 81–111.
Schramm, E. (1985), 'Die Zukunft der Ökologie im Lichte der Wissenschaftsforschung', in G. Boehme and E. Schramm, *Soziale Naturwissenschaft – Wege zu einer Erweiterung der Ökologie*, Fischer, Frankfurt, pp. 108–122.
Siegrist, M. (1996), 'Fragebogen zur Erfassung der ökozentrischen und anthropozentrischen Umwelteinstellung', *Zeitschrift für Sozialpsychologie*, vol. 27, pp. 290–294.

Solies, D. (1998), *Natur in der Distanz: Zur Bedeutung von Georg Simmels Kulturphilosophie für die Landschaftsästhetik*, Dissertation, Mainz.

Sperlich, M. (1979), 'Das neue Arkadien. Der Garten als utopische Landschaft', *Neue Heimat Monatshefte*, vol. 26, pp. 10–23.

Stehr, N. (1999), *Klima – Wetter – Mensch*, C.H. Beck, München.

Tessin, W. (1991), 'Bedingungen großstädtischen Naturerlebens', *Gartenamt*, vol. 40, pp. 15–20.

van den Daele, W. (1992), 'Concepts of Nature in Modern Societies', in M. Dierkes and B. Bieverts (eds), *European Social Science in Transition*, Westview Press/Campus, Boulder/Frankfurt/M., pp. 526–560.

Zentralarchiv für empirische Sozialforschung (2001), *ALLBUS 2000*, Köln, pp. 449–470.

3 From Boundary to Symbiosis: Limits of the Systems Approach to Environmental Problems

FRIEDRICH FÜRSTENBERG

There is wide consensus, that any perception of reality implies the construction of concepts provided by a given language, including the use of mathematical symbols. Thereby our experience is structured in a way which permits meaningful connections between perception, knowledge and action. Fundamentally important is the relative status thereby assigned to the perceiving actor and the perceived situation: Is it marked by total independence, by mutual dependence or by domination? In other words: How is the relationship between man and environment structurally shaped by the relevant conceptual framework?

Harmonious and Dualistic Perceptions of the World

Problems connected with the use of the construct "environment" become more concrete when comparing the *attitude towards nature*, predominant in Western culture, with traditional Asian approaches. Joseph Needham, the eminent researcher on science and society in East and West, has tried to explain the fact of an early and stable predominance of Chinese science, being later replaced by the dynamics of Western innovations. He stressed the traditional Chinese world view, characterised by the harmonious co-operation of all beings and the Taoist principle of non-interference (*wu wei*). And he concluded: "It seems probable that a society like this would be favourable to reflection upon the world of Nature. Man should try to penetrate as far as possible into the mechanisms of the natural world and to utilise the sources of power which it contained while intervening directly as little as possible, and utilizing 'action at a distance'" (1979: 210). "If the Chinese were ... the greatest observers among all ancient peoples, was it not perhaps precisely because of the encouragement of non-interventionist principles?" (1979:

211). Sharply contrasting with this is the Western interventionism which marks the scientific revolution since the Age of Renaissance.

The famous Japanese physicist Yukawa Hideki, who was awarded the Nobel prize in 1949, has given a widely accepted characteristic and explanation of this relationship which shall be quoted from his contribution to an East-West Philosophers' Conference, held in Honolulu 1959:

> "Under the severe circumstances of the Western zone, adventurous effort to secure survival leads to the establishment of artificial environments for human beings and consequently gives rise to a dynamic idea of a non-directional progress. In order to compromise with Nature, man stands in need, not only of autonomous action to modify external conditions, but also of effort to modify himself. This implies that some existing conceptions are renounced, and then one is to conform to a new concept selected from an ensemble of alternative possibilities. This is the very origin of the rational mode of thinking, and, furthermore, of the scientific spirit to discover physical laws in natural phenomena. Science is the outcome of that rational contact of man with external Nature." (1959: 55f.)

Daisetz Suzuki proceeds along this line of argument by characterising Western thinking as revolving around a dualistic axis (1967: 137). Instead of trying to attain a complete union of man and Nature, the West lives in a *dualistic reality*, separated into two terms: subject and object (1967: 63f.).

Of course, the complex development of Western rational thinking cannot be grasped in such a simplifying manner. However, there is little doubt that the notion of "environment" as a domain of human conquest can only emerge in cultures experiencing mentally the fundamental dividing line between actors and the objects of their actions, which means that there is a basic alienation between the actor and his environment, which eventually enables him to make unlimited use of it. Self-determination within cultural boundaries then corresponds with uninhibited availability of external resources. It may be mentioned that such divides also exist in non-Western cultures, but are expressed rather on the basis of socio-cultural entities than on individual awareness and experience. The idea that pre-modern China was located in the middle of the world, surrounded by barbarians, and Japan's often claimed insular mentality, linked with feelings of superiority, reflect such perceptions of an external world where own socio-cultural standards cannot and need not to be applied.

Systems Thinking and the Concept of Environment

In modern Western societies the meanwhile globalised model of thinking along borderlines between the acting subject and a principally disposable environment has been further developed in a constant process of "rationalisation". Due to technical and economic advances, the actors are no longer perceived only as actors in terms of individuals or groups and their integration into organisations, institutions and nations but instead also as depersonalised systems, operating within a specified environment. They are the predominant tools to utilise and even dominate the intelligible reality. Thus we may forward the thesis, that the present state of advanced societies is characterised by the steady procurement of system technologies, linked by communication systems and operated within a financially globalised system of capitalism, whose representatives contrast sharply with the traditional Weberian entrepreneur, motivated by the Protestant Ethic.

Systems are conceived as persisting sets of interactions, linked by specific functions on the basis of a system-specific rationality. They owe their existence to characteristic boundaries, separating them from the external world. While within systems the provision and maintenance of functional requirements grants stability, the environment is conceived as an infinite zone of uncertainty, if not chaos. Systems as islands of rationality within a sea of irrationality mark the gradual conquest of reality in almost all domains of human thought. Thus we speak of technical, economic and social systems, the latter ones being highly subdivided, in e.g. political, legal, educational, health, communication, industrial relations systems etc.

Compared with the traditional pattern of Western dual thinking, which often included a dialectic relationship between subject and object, marked by mutual dependence and interference (e.g. exposed in the philosophical system of Georg Wilhelm Friedrich Hegel), this relatively new systems thinking usually is built upon Darwinist evolutionary principles: Within their realm there is no alternative to strictly functional operations within fixed boundaries. Niklas Luhmann in his systems theory time and again has clearly exposed the inherent logic of such systems which leads to an almost total neglect of mutual intercourse with environmental structures and processes. Furthermore, the systems model marks a trend towards an increased fragmentation of the environment. As each system has its own peculiar environment, a universal concept of Nature is obsolete. Likewise the specific functionality of each social system makes the notion of Society obsolete.

A sociologist dealing with environmental problems in a given social context has to consider this relatively new pattern of thought and corresponding action. We can no longer reflect a relationship between such entities as "Man" and "Environment", "Culture" and "Nature" or "Subject" and "Object". They mark a state of perception of reality which no longer corresponds with actors' strategies directed towards system building. These aim at separating domains for controllable action by submitting them to depersonalised rules of functionality. From the point of systems theory, the external world is relevant only to the extent that system functionality is at stake.

But exactly this is increasingly happening. Systems are by no means self-sufficient and "autopoietic" in their development. They depend upon exchange, which never can be limited to an "interpenetration" between systems. Intercourse with non-systematised reality is unavoidable. All this calls for a revision of the systems approach which may be considered as representative of an advanced perspective on the rationalisation of human action fields. Originally they were considered to develop rationally by modelling reality according to discoveries of causal and functional connections. Gradually they became perceived as autonomous artefacts.

Revision of System Boundaries

In order to counterbalance Darwinist exploitation and exhaustion of "external" resources, including human beings, "*sustainability*" has been advocated. As guiding principle especially for technical and economic claims to a given environment it advocates self-restriction of resource utilisation in order to preserve chances for future development. Such a model of environmental protection, however, still is based upon the assumption that there is a distinct dividing line between the centre of actions, usually conceived as a system, and a basically exploitable external area. Instead, another approach shall be discussed in greater detail, being founded upon a revision of the *boundary concept*, which is constitutive for systems thinking.

Life may be characterised by continuous exchange operations both material and communicative and even imaginative, if consciousness is involved. Taking this into account, the mental division between closed systems and their environment needs to be qualified especially when considering relations created and started by human beings. This includes of course each interaction between man and his technical artefacts, an interaction which may be interpreted as well as an exchange procedure causing structural change. We may think along this line of any artefact's "history of utilisation". There we encounter quite different utilisation contexts for the same product, if the users or the terms of application change, e.g. by widening the access or by transfer of property rights. Therefore, the so-called environment for this specific artefact, which may be a banana plantation, a traffic system or even the worldwide web, is in a constant process of structural change, having a thorough impact upon the system and its components.

These exchange relations take quite different shape depending upon the consideration of environment merely as "space of perception" (Merkwelt) or as "space of action" (Wirkwelt), using a distinction introduced by Jakob Johann von Uexkuell (1864–1944) founder of environmental research and the first related institute (1926). Especially the active utilisation of an artefact, e.g. by starting its operations, never is neutral to an environment, in the sense that there is no impact. In fact, the intentions of inventors, producers and users show effects far beyond system boundaries. To a large extent system building also means not only influencing but also shaping the environment.

If we now consider particularly the social components of systems, we encounter the widely discussed problem of not intended and unforeseen effects. The control of artefacts and their integration into systems only extends to the validity and observance of the instructions for their use. Such control results in a "factual restraint" (Sachzwang), modified in such situations, where utilisation, as intended and anticipated by the innovator, competes with quite different needs, interests and values. This accounts also largely for the external risks inherent in the operation of systems. Within the modern world, environment is not only the remaining untouched nature but to a growing extent a man-made area of external systems, competing or cooperating, and thereby augmenting, diminishing or altering system functions and effects. In other words: social systems encounter not only reactions but also actors in their environment. Human actions and artefacts are embedded in structures surpassing system boundaries and forming specific ties with their environment. This poses the question how to deal with these external (if not externalised) relationships.

"Environmental Relations"

We may distinguish different models of relations between systems and their environment. Most common is the *domination model*, which is marked by a prevailing strategy to shape and utilise environmental structures and processes according to systems' goals. Jürgen Habermas, obviously influenced by Wilhelm Diltheys harmonising notion of a "contented existence", has characterised such an asymmetric relationship as "colonisation of living conditions" ("Kolonisierung der Lebenswelt").

From the point of view of further domination, an improvement towards sustainability is mainly conceived as the product of further rationalisation, aiming at "environmental compatibility". This is usually interpreted as recognition of respective utilisation requirements and conditions to be ascertained by scientific investigation. This not only means respecting the laws of nature but also socio-cultural standards. The term "waste" e.g. shows quite different connotations according to the relevant level and context of civilisation. Each system produces its specific waste, even the internet. But within the domination model, "environment" cannot claim a voice of its own. System restraints are voluntary and due to enlightened strategies of responsible stewardship of external resources.

The *exchange model* is based upon the understanding, that utilisation risks open up only in the process of interaction between actors, comprising potential actors concerned. The system planners never are able to fully anticipate possible effects upon an environment, which is never totally controlled. This means that any system design, of course including social systems too, cannot be limited to (socio-)technical construction according to a functionalistic ideals concept. As any human action, those actions related to environment should be perceived as chains of events, enabling reciprocal experience and understanding. Where human beings are involved, each shaping of environmental relations is basically dialogic. In this sense it cannot be evaluated only according to the technical criterion of efficient practicability and the economic test of profitable utilisation. It needs to be guided as well by the socio-psychological criterion of conflict-free acceptance. Thus, environmental relations become an open-end exchange process, which ultimately alters the perceptions of "system" and "environment". The concept of "networks", characterised by actors connected by structured, but rather fragile relationships, has emerged from such observations.

The *co-operation model* relies upon strategies, taking environment into account as co-existent reality, as "co-habitat" (Mitwelt) in its own claim (for the philosophical foundation of such a perspective see Meyer-Abich 1990). This orientation helps to overcome system-blindness, being an attempt to perceive system procedures no longer as centres of events, prescribed by experts, but rather as servers within a larger context of often joint performance, which then may be perceived as a more or less stable set of networks.

A by far too narrow systems concept denounces such orientation as depreciation of the system's self-performance. But if we gain a more open view upon perceived system boundaries and their possible dysfunction for an efficient resources management, we cannot deny the importance replacing model-construction through isolating abstraction by context orientation. This would open the chance for emerging alternative concepts. They are marked by evaluating the system performance in terms of linking different processes, both internal and external to the relevant system, to a *symbiosis*: System and environment are linked by mutually stabilised co-existence. In the case of human beings involved this amounts to a basic and lasting partnership. Within the sphere of market products and services we find already manifold examples for such a successful transformation of exploitation into exchange and of exchange into symbiosis. And when regarding human history at large, we even realise that all lasting major socio-cultural achievements document a rather harmonious solution of a system's demarcation problem. "Chinese Walls" as well as suppressive colonialisation of the outside world time and again have proven inefficient in the course of time.

Dissolution of System Boundaries into Symbiotic Structures

We may perceive socio-technical systems and even transnational networks, with the tendency towards global extension, as "open" systems with the need for co-operative exchange within environments perceived as action fields. These are not dominated by systems but by rather diffuse sets of more or less stable relations. Then we face the problem of relevant structures for co-operative action in and between such action fields, including – from the systems' perspective – the establishment of symbiotic environmental relations.

Such relations may be shaped by non-directive evolution, depending upon erratic chances, by socio-cultural traditions, depending upon the internalisation of normative patterns, or by strategic planning, depending upon the rational use of resources. As this latter pattern coincides with the emergence of social systems and is presently considered as a possible way to cope with environmental problems, it shall now be discussed in greater detail.

Strategic planning of symbiotic environmental relations is mostly conceived by analogy to law-making in terms of norm-setting and their enforcement. But such authoritative procedure rarely leads to voluntary motivation. Another starting point is the renunciation of central guidance and regulation in favour of decentralised planning in competing action units according to the market model. As unequal chances for such planning prevail among the different units, there will be always the danger of asymmetrical relations based upon the dominance of the most powerful. Finally remains the shaping of symbiotic environmental relations through co-operative problem-solving, which, however, needs to be founded in adequate procedures, embedded in structures enabling co-operation. A relevant model, widely applied in social systems, e.g. industrial relations, is marked by the definition of planning and development as a multi-stage decision-making with gradated chances for participation of those concerned. There is the immanent danger, that such procedure deteriorates into a permanent discourse with considerable delay of decisions. This may be avoided by preparing alternatives, which then are presented for discussion and harmonising. But such applied social technology is no safeguard against a reduced perception of "environment" as seen by the representatives of particular interests. In order to comprehend the symbiotic aspect of environmental relations, mediating and counterbalancing evaluation is needed. This may be achieved pragmatically by combining authoritative norm-orientation with market-oriented bargaining and with participative action, always with the chance of a public, system-overstepping discourse.

All this reflects the experience, that co-operative action related to symbiotic environmental relations always takes place within a social tension-field with multiple orientations. This is the toll to be paid for avoiding the dominance of unilateral system-planning or the regression of open exchange into fragmented system operations. As a result, problem solutions reflect the wide variety of prerequisites for the present and further existence of environmental components. From the point of view of scientific research procuring applicable knowledge, a fundamental prerequisite is the continuous dialogue between Man- and Nature-oriented sciences, aiming at the accurate assessment of the context of human action.

Criteria for Symbiotic Effects

Symbiosis may be conceived as a pattern of advanced co-existence, founded upon mutual benefits through exchange and resulting in mutual dependence. This exchange reflects the basic modes of social control: It may be the result of market operations, of socio-cultural conventions or of autocratic orders. Basically, any persisting social phenomenon is symbiotic. This holds especially true for socio-cultural patterns. Their historical decline has been proven to be due to a large extent to a disruption of their symbiotic ties. In the long run the same may be said about the relation between any system and its environment.

Therefore, the thesis may be maintained that any human-made entity, also when perceived as a social system, owes its stability to a *symbiotic transformation*, to the founding of its functions as contributions within a larger context.

This symbiotic transformation is marked by shaping environmental relations,

- which meet not only functional requirements defined within systems,
- but also stabilise exchange processes with a given natural environment
- as well as demands from a larger socio-cultural context, defining particular action fields,
- thereby granting social acceptance.

Special, but important historical examples are the cultivation of plants, the domestication of animals, the foundation of government by transforming dominating power into authoritative rule within an emerging legal system, and, in modern times, the establishment of telecommunication networks. Results meeting these symbiotic requirements may gain symbolic quality as successful cultural patterns within a universe of value orientations and then become part of meaningful traditions. Within tradition-bound societies, this has opened the way for continuous imitation. In the course of modernisation, however, socio-cultural traits are gradually replaced by short-lived life styles, according to a growing influence of the market model for social action and social control. Thus, any attempt to symbiosis faces the rapidly changing demands upon an increasingly disposable environment. This leads to varying system strategies which correspond with manipulated and individualised reactions. A development is likely which furthermore externalises adaptation risks of system management, thereby enlarging the margins for system planning at the cost of possible partners in symbiotic processes. They become ideologically dominated parts of an amorphous so-called "environment" which resembles more and more a battlefield for competing interests, euphemistically called "system requirements".

We have reached a phase where "risk management" of systems seems to function as a mere strategy of cost-externalisation, meaning the lack of self-restraint, as far as environmental relations are concerned. This of course results in the generation of problem areas (environmental disruption, food contamination, destruction of social ties, alienation and ultimately terrorism). They call for successful legal and administrative regulation. But such activities do not change the underlying mechanism of problem generation and increasing regulation demands can be observed. Thus we have already reached the stage of globalisation of environmental problems, and this not only in technical and economic terms. World civic policies are demanded with the aim of global environmental governance (Wehling 2001: 740). But contrasting to such ambitious projects, with relation to whole societies we get used to conceive the world increasingly divided by cultural boundaries. The regression into a war-like state between different value systems appears to become a quite realistic scenario, as is indicated e.g. in the partly militant objection against World Bank policies and the outcome of G-8 Summits. This again seems to call for enforced interference in order to make the world compatible with dominant political and social systems. But at least at global level the chances for consensual, symbiotic regulation appear to be small as is demonstrated by the outcome of the 1997 Kyoto Conference and the attempts to establish an International Penal Court. All these experiences call for a change of perspectives. System-induced thinking in boundaries needs a revision, which opens the understanding and respect of natural and cultural complexity beyond an increasingly fragile exploitation of both natural and human resources. Ultimately we need culture-generated restraints for power potentials due to system-building. This system construction and system defence happens increasingly at the cost of an externalised "environment", which not only comprises non-human nature, but also powerless and even deprived human beings. Therefore, scientists of all faculties should avoid such kind of boundary thinking, which ultimately would lead to the maintenance of closed systems, operated by people with "closed minds" (Rokeach 1960). This awareness implies to accept a universal principle of obligation and responsibility, whose foundations have been elaborated by the German philosopher Hans Jonas (1979). Such an orientation may well include greater attention to symbiotic processes and the chances for their promotion.

References

Brand, K.-W. (ed.) (1997) *Nachhaltige Entwicklung. Eine Herausforderung an die Soziologie,* Leske + Budrich, Opladen.

Brunkhorst, H. (2002), *Solidarität. Von der Bürgerfreundschaft zur globalen Rechtsgenossenschaft*, Suhrkamp, Frankfurt/M.

Fürstenberg, F. (1995), *Soziale Handlungsfelder*, Leske + Budrich, Opladen.

Jonas, H. (1979), *Das Prinzip Verantwortung. Versuch einer Ethik für die technologische Zivilisation*, Suhrkamp, Frankfurt/M.

Meyer-Abich, K.M. (1990), *Aufstand für die Natur. Von der Umwelt zur Mitwelt*, Hanser, München/Wien.

Moore, C.A. (ed.) (1967), *The Japanese Mind*, University of Hawaii Press, Honolulu.

Needham, J. (1979), *The Grand Titration. Science and Society in East and West*, Allen Unwin, London.

Piechocki, R. (2001), ''Ein Regenwurm kennt nur Regenwurmdinge'. Über die Karriere der "Umwelt" und ihrer Bedeutungen', *Frankfurter Allgemeine Zeitung*, no. 232, 6.10.2001.
Rokeach, M. (1960), *The Open and Closed Mind*, Basic Books, New York.
Schneider, V. (1992), 'Kooperative Akteure und vernetzte Artefakte', in G. von Bechmann and W. Rammert (eds), *Technik und Gesellschaft*, Jahrbuch 6, Campus, Frankfurt/New York, pp. 113–139.
Suzuki, D. (1967a), 'Reason and Intuition in Buddhist Philosophy', in C. A. Moore (ed.), *The Japanese Mind*, University of Hawaii Press, Honolulu, pp. 66–109.
Suzuki, D. (1967b), 'An Interpretation of Zen Experience', in C. A. Moore (ed.), *The Japanese Mind*, University of Hawaii Press, Honolulu, pp. 122–142.
v. Uexkuell, J. J. (1909), *Umwelt und Innenwelt der Tiere*, 2nd ed. 1921, Springer, Berlin.
Wehling, P. (2001), 'Die Konstruktion ökologischer Globalität: Globale Umweltprobleme und transnationale Umweltpolitik', in J. Allmendinger (ed.), *Gute Gesellschaft? Verhandlungen des 30. Kongresses der Deutschen Gesellschaft für Soziologie in Köln 2000*, Leske + Budrich, Opladen, pp. 723–747.
Yukawa, Hideki (1967), 'Modern Trend of Western Civilization and Cultural Peculiarities in Japan', in C.A. Moore (ed.), *The Japanese Mind*, University of Hawaii Press, Honolulu, pp. 52–65.

4 Environmental Sociology in Japan: Problems, Topics and Major Characteristics*

KOICHI HASEGAWA

The Turning Point for the Second Stage

Many years have passed since the establishment of the predecessor of the Japanese Association for Environmental Sociology with 53 participants in May, 1990 and since then research in this field has been institutionalised and actively pursued. Many young researchers as well have been attracted to this speciality. Now the association has over 600 members, 12 times higher than the number at the starting point. [Fig. 1]

The accomplishments till date are well documented in the following publications: the Journal of Environmental Sociology, an official journal of the association, which published its number 8 issue in October 2002; the five volumes of Environmental Sociology in Japan, which were completed and published from Yuhikaku in September 2001; and the six volumes of the Environmental Sociology in Japan series published by Shinyousha. We held two international symposia on "Asian Society and Environmental Issues" (1993) and the Kyoto International Environmental Sociology Conference in October 2001. Japanese Environmental Sociology has had a very productive founding stage during the first twelve years since its establishment. Our association proved to be a good example for Korean scholars in the field. They founded the Korean Association for Environmental Sociology in 1995 and published their journal ECO in 2000.

Among sociological research fields internationally and also in Japan, environmental sociology is one of the most active research fields along with gender studies, ethnic studies, and historical sociological studies. Environmental sociology in Japan is now standing at a turning point after its first decade, it is now about to enter its second stage, to develop during the next decade. I think we need to work on four strategic focus points: (1) policy

---

* Translation by Rick Davis

orientation, (2) interdisciplinary orientation, (3) international orientation, (4) sociological theory orientation. [Fig. 2 and 3]

In this article, I would like to stress the focus on policy study (see for a more detailed argument Hasegawa [2001]). This is because during the first decade environmental sociologists in Japan gave relatively little attention to investigations of environmental policies and the policy-making process.

Study on Environmental Policies

The Weakness of the Policy-oriented Focus in Japanese Environmental Sociology

In studying environmental policy, I believe there is an angle specific to environmental sociology, that is, the interaction between policy and environmental movements and the influence of environmental movements on governmental policies.

Considering the two keywords "environmental movement" and "environmental policy", a great deal has been accomplished in Europe, the US and in Japan on the study of environmental movements. This is because environmental sociology takes a unique look at the many dimensions of possible achievements by environmental movements. These dimensions are hardly noticed by fields like environmental economics and environmental law. It can be said that the environmental movements of the 1960s and '70s served as the mother of environmental sociology in the US and in Japan, and perhaps also in Germany.

At a symposium on environmental study a Japanese environmental movement leader, Mr. Suma Harumi, criticised the current environmental sociology: "I don't see how environmental sociology is assisting in solving actually existing problems. I've began to think that environmental sociology might be only a study of taxonomy and ex post facto explanation." (2001: 74) The Japanese Association for Environmental Sociology calls for "contributions to environmental problem solving" in section 2 of its by-laws, but is still being criticised for having produced no visible contribution. We must take this criticism seriously.

Of course at the root of many environmental problems lie certain policies for development, environmental, industrial, transportation, energy, waste, city planning, and agricultural projects. Thus in most cases – whether it is the study of pollution, development problems, organic farming, waste problems, or the preservation of historical towns – sociological research on environmental problems intrinsically includes the assessment and often criticism of governmental policies.

However, until now, debates over policies in the field of environmental sociology have mainly focused on criticising environmental policies as the source and structural factor behind environmental damages and risks. Study of the intrinsic problems of environmental policies themselves of the policy making process, and research aiming to provide alternative policy suggestions have been relatively rare.

The joint study by Harutoshi Funabashi, myself and others on pollution from bullet trains and rail line construction process (Funabashi et al. [1985, 1988]), the study on governmental river management by Shigeru Tanaka (1997), the study on nuclear energy problems by Koichi Hasegawa (1996, 1999), and the joint study by Funabashi, myself and Iijima on a failure of large-scale industrial development projects and construction sites for nuclear waste dumps (Funabashi, Hasegawa and Iijima eds. [1998]), are examples of the few studies, which have been clearly focused on policy-oriented research.

In contrast, we can observe that the Japanese academic society mainly comprised of environmental economists calls itself the "Society for Environmental Economics and Policy Studies", and the society of environmental law specialists calls itself the "Japan Association for Environmental Law and Policy". Research on environmental policy is a main component of study from the perspectives of environmental economics and environmental law. Both societies are highly focused on policy-oriented research.

Japan's environmental sociology has been aiming for a focus on practicality. However, it is unclear how much environmental sociology has contributed to solving environmental problems. Moreover, the possibility of a contribution is not methodologically clear either, and no thorough discussion has been held within or outside the environmental sociology association on this point. Contributions from environmental sociology can be hoped for on themes like environmental assessment, local agendas for environmental improvement stimulated by Agenda 21 of the 1992 Earth Summit, and far-reaching discussions on local environmental programmes. But to date there has been little offered by sociologists (Seki 2001).

Why has environmental sociology been weak on a policy-oriented focus? This weakness is in fact a common tendency of sociological studies in Japan, not the first particular characteristic of environmental sociology. Not only within environmental sociology but also in general, sociology in Japan has been remarkably lacking in policy-oriented studies.

I assume that, as in the case of Japan's environmental sociology, a similar weakness in policy-oriented studies exists in the US and Europe. Even in the well-known text book for sociology written by Humphery and Buttel, which is often read in Japan, environmental policies are merely discussed in an introductory manner in the last chapter (Humphery and Buttel 1982).

Weakness of Policy-oriented Studies in Japanese Sociology in General

The lack of policy-oriented studies in Japanese sociology and the reasons behind such lack can be analysed from various historical and structural aspects:

(1) Sociology in Japan has theoretical and holistic characteristics. The philosopher of science Karl Popper criticised Karl Mannheim and asserted that sociology should be based on "piecemeal social engineering". There was, however, a strong Marxist influence on sociology in Japan, and thus it did not develop in the course laid out by Popper.
(2) Theoretical studies dominate, and usually are limited to introducing the most recent theoretical trends in Europe and America. Japanese sociology has been short on efforts to tackle recent social problems and challenges regarding Japanese society and its policies, and weak on efforts to theorise such analyses.
(3) Considerable effort has been made on collecting qualitative and quantitative data for empirical studies, but little focus has been given to forming policy suggestions based on such data.
(4) Compared to detailed knowledge held by legal specialists and public administration specialists, the interest of empirical sociologists lies mostly in matters related to the actual life scene, and interest in systems and policies has been relatively low.
(5) As termed sometimes as the "spell of Weber" and the "spell of Marx", which binds Japanese sociologists, the strong influence of Max Weber and Karl Marx over sociology in Japan can be assumed as one of the primary factors (Kosaka 2000). It can be said that Weber's ascetic scholastic outlook and Marxist revolution-orientation and holistic characteristic has had a restraining effect on possible collaboration between environmental sociologists and policy authorities flourishing under the one-party dominance of the Liberal Democratic Party.
(6) There were few connections among policy authorities and a considerable distance between researchers and authorities. Because the department of sociology is usually situated under the faculty of arts and letters in Japan, graduates from the faculty of law and the faculty of economics are not very familiar with sociology and sociologists. Also because the department of sociology was put under the division of philosophy for a

long period of time, economists and law experts hold a strong image that sociology is an abstract study with a highly speculative flavour, or a critics' study based mainly on opinions/statements.

(7) Interviews and studies on policy authorities are indispensable for a sociological policy study. But in Japan authorities have long resisted the disclosure of official information. Until now the central government, local municipalities, and companies have only sporadically responded to demands for information disclosure and requests for personal interviews by researchers. In general, most officials have been defensive and cautious towards critical researchers who belong to the faculty of arts and letters. Such tendency was especially strong with issues that were being fought in court.

Significance of Policy Study for Environmental Sociology

The most important challenge for environmental sociology now is to strengthen policy-oriented studies by building upon accomplishments made so far and the characteristics of sociological analysis.

First of all, there are expectations for research experts to strengthen their interdisciplinary problem-solving skills regarding real environmental issues. Environmental economists stress the necessity for internalising social costs, but they are criticised that "there are few studies that offer tangible suggestions, including realistic means, to internalise social costs" (2001: 74). This criticism is directed at environmental economics, but it applies to environmental sociology as well. Such criticism shows the strong desire of activists for environmental sociologists to provide research, which can be used to reform environmental laws and systems. Here lies the challenge to achieve collaboration between environmental economics, environmental law and environmental sociology.

Second, there is a lack of scholars and research in the field of environmental politics and environmental administration studies in Japan. One of the challenges, which must be taken up by environmental sociology is to study the environmental policy-making process and the enforcement process exhibited by the environmental administration of central and local governments.

Third, in the new 21st century, expectations for reforms are heightening, and fresh thoughts and views are desired. People are reflecting upon "the lost decade" of the '90s and through such reflection, the problems with Japan's bureaucrat-led and closed policy-making process have become clear.

Fourth, attention should be given to the significant shift in Japan's environmental movement towards a more policy-oriented flavour during the last decade. For a long time, there were few contacts between environmental activists and environmental policy-makers and the environmentalists were hostile towards policy authorities. In recent years, however, environmentalists have become increasingly influential over policy authorities. The time is ripe for environmental sociology, which has a credible ability for the analysis of environmental activism, to conduct studies, which shed light on the collaboration process of environmental activism and environmental policies, and the possibilities of environmental activism helping bring about improvements in environmental policies.

Characteristics of Environmental Sociology and Policy Study

What kind of merits does environmental sociology have over environmental economics and environmental law when conducting policy studies? I will now present the sociology on environmental problems.

There are incredibly varied understandings as to how to conceptualise sociological methods. However, I suggest the following characteristics:

(1) Under a sociological approach, social phenomena are understood as a collection of actions. Thus, particular attention is given to people's decisions and choices due to the understanding that social phenomena are changeable and always evolving.
(2) A sociological approach gives special importance to field surveys. This is a significant characteristic. That is because economists often base their proof on statistics provided by governments, and law specialists often base their proof on court proceedings, rulings, and laws. They have a strong tendency to be content with using secondary resources. In comparison, sociologists place significant importance on first-hand data which they themselves collect from the "scene". The expectations on sociological analysis are directed at this ability for conducting field investigations.
(3) Great emphasis is placed on the viewpoints of citizens who live and work in the area. Most sociological methods and sociological researches value such citizens' understanding of the issues, and focus on local communal groups that are the scene of everyday life.
(4) Problems are analysed in a holistic manner, and importance is placed on understanding the relationships between various aspects of the issue.

Analysing market structures is the premise of environmental economics. For example, how to quantitatively evaluate environmental value, or how to internalise the costs of an outside non-economic factor like environmental pollution. This is how the issue is set by environmental economists. On the other hand, the centre of the interests of environmental law specialists lie with whether there has been an intrusion of environmental rights based on law models and laws in practice, who or where the legal responsibility lies, how to recover rights or attain legal relief. Issues are limited and are solved technically.

Compared to such limited approaches by environmental economics and environmental law, environmental sociology is based on all-encompassing viewpoints, and oriented towards grasping a holistic understanding of the relationships between the various aspects of an issue. Only environmental sociology looks into local sociological conditions and cultural conditions, individual and family life, residential organisations, citizens groups, and social activities as factors forming values, models and the structure of the environmental dispute.

Thus the sociological understanding of human activity patterns and characteristics, groups and organisations, and local societies form the base of sociological policy suggestions. In other words, the tools and characteristics of sociology lie in the ability to form and carry out social investigations, which place importance on the experiences of the victims, sufferers, residents, and general citizens, and a holistic grasping of various aspects of the environmental dispute. Also sociology can use the same methods to investigate the social and cultural organisation of authorities of decision-making process.

For example, the interest of law specialists in an issue like pollution from the bullet train and railway construction would be limited to the debates in court, the possibility of proving the illegality of actions, whether there was infringement of rights, or the legality of asserting environmental rights. The interest of modern economists would be limited to internalising of social costs, evaluating economical effects from development and assessing costs vs. benefits when environmental damage is suffered at the site and its vicinity. Sociologists' research interests would lie in the following four aspects: (1) Damage (emphasis is put not only on direct damage like illness, but on various damages being experienced by family and the local community like disconnected human relationships and re-organisation of life after moving); (2) The mechanism causing the damage (including the mechanism of inaction that avoids forming active response and persistence on previous examples); (3) Social activism (including residents' activism and the significance and challenges of contesting at court); (4) Policies (including comparison of sociological meanings of damage response taken at the site and the vicinity, and critical evaluation of social benefits, effects of development and the benefits for the public from the project). (Funabashi et al. 1985, 1988)

Policy-making processes, values and judgements made by authorities can be investigated from the viewpoint of sociological institutionalism and sociological constructionism. Sociology can also be used to analyse the process of building up public opinion as a part of the public sphere.

The Three Dimensions of Environmental Policy Analysis

Decision Process, Measures and Policy Value

In light of social planning theory, which is a field of sociology concerned with public policy, environmental policy studies may be thought to comprise three dimensions of analysis: (1) environmental policy decision process, (2) environmental policy and measures, and (3) environmental policy value. Our concerns here are: "Who (actor) implements environmental policy, what is the purpose (goal), at what is it directed (object), and how is it accomplished (means)? Further, how do we evaluate the results of policy?" Environmental policy too should be subjected to scrutiny in the five criteria of actor, goal, object, means, and evaluation. These five criteria can be categorised into the (1) environmental policy-decision process, (2) environmental policy and measures, and (3) environmental policy value.

In discussing environmental policy decision processes, everything from setting goals and developing plans and policies, to implement and evaluate those plans and policies, is examined along the lines of their associated circumstances: connections with actors, social relationships, and the procedures and pathways involved from planning to implementation. Specific areas of concern include the types of environmental policy actors and agents, procedures for social decision-making on environmental policy, consensus-building for environmental policy, environmental impact assessments, the process by which environmental problems are discovered and become issues, power and participation, environmental movements, the role of environmental NGOs/NPOs, democracy, the public sphere, and responsibility for planning and policy. One might think of this as environmental policy studies as seen in the context of political sociology.

Analysis of environmental policy and measures focuses mainly on the rationality of means for planning and policy meant for a given environmental goal, and the consideration of spill-over effects and social impacts. Items examined include the technical methods of environmental planning and policy, methods of evaluating the suitability and effectiveness of goals and policy means, and the resources that can be used for attaining goals. Topics include the study of existing environment-related legislation and institutions, environmental taxes and emissions trading, policy mix methods, how to internalise social costs, the contingent valuation method (CVM), willingness to pay (WTP), ability to pay and payment by beneficiaries, green purchasing, environmental auditing and environmental accounting, life cycle assessment, and feasibility studies. Environmental policy studies from the perspective of environmental economics and environmental law places its main emphasis in these areas.

The study of environmental policy value could be called "meta-policy studies", and its major topics are the study of policy goals and policy postulates, as well as their premises: social values, environmental ethics, desirability, people's attitudes on the environment and nature, and environment-related rights. It entails taking another look at the very values that underpin environmental policy, and especially important is an examination of criteria including environmental and social justice, environmental rights, the rights of nature, publicness, transparency, accountability, amenity, the polluter pays principle, expanded producer responsibility, and the prevention principle. This involves frequent normative considerations of what kind of environmental values and environmental policies we should have. Discussions of environmental ethics take place mainly on this level. Many contributions have also been made by Awaji Takehisa (1980) and other environmental law scholars. Particularly in environmental economics, Miyamoto Ken'ichi (1989) has emphasised such perspectives.

In the work of Japanese environmental sociologists to date, Iijima Nobuko's *Theory of the Structure of Environmental Damage* (Iijima 1984/1993) as well as *A Social History of Environmental Problems* (Iijima 2000), and the work of Funabashi Harutoshi et al. (1985, 1988) are mainly in the realm of environmental policy decision processes, while the issues of environmental justice raised by Toda Kiyoshi (1994) have contributed to environmental policy value analysis. The criticism by Ikeda Kanji (1995) and others of the model presented by Garrett Hardin in *Tragedy of the Commons* raises questions in terms of policy value. The life environmentalism of Torigoe Hiroyuki, Kada Yukiko, and others (Torigoe and Kada 1984; Torigoe 1989) could be considered discussion on the level of policy value analysis if one goes so far as to find correspondences to the three dimensions. A study by Nakano Yasute et al. on waste volume reduction (Nakano et al. 1996) limits its focus to factors that encourage and impede waste reduction efforts, and therefore belongs to the realm of policy and measures. Until now environmental sociology has made significant contributions on the level of policy value, but one can hope that it will contribute in the future to all three dimensions.

The first thing to be done when examining the debate on environmental policy is to look closely to see, which of the three dimensions is emphasised in terms of the debate's agenda-setting and its analysis viewpoint. Environmental sociologists, environmental economists, environmental law scholars, environmental ethicists, and others should plan and carry out joint research on specific environmental policies to work for a better environment.

The Principle of Expanded Producer Responsibility

In short, discussion on a given environmental policy should include: (1) Why that policy was developed, against the backdrop of what environmental issues and policy challenges, through what decision-making process, and in connection with what social interests? (Identifying issues pertaining to policy decision process analysis.) (2) How effective are the policy means in achieving policy goals, and what spill-over effects and social impacts will result? (Identifying issues pertaining to policy and measures analysis) And (3) What policy goals are cited, what values underlie those goals and policy means, and what stakeholder rights do the goals presume? (Identifying issues pertaining to policy value analysis) Only when these three perspectives are all included can environmental policy be discussed in a more integrated and comprehensive manner.

For simplification the foregoing discussion has cited the main contemporary topics in environmental policy while matching them up with the three dimensions, but let us take a look at, for example, extended producer responsibility (EPR) by integrating the three dimensions of analysis. EPR is a principle that seeks to place the responsibility for reducing waste generation primarily on the producer, instead of on distributors, consumers, waste management companies, or the national and local governments. The idea is to motivate producers to reduce wastes, set up avenues to recycle and sell recyclable products, and take other actions. This principle attempts to internalise the social costs of wastes at the producer/manufacturer end, and its approach is that to reduce the generation of wastes it is most effective to impose restrictions on production, at the far upstream end of the sequence comprising production, distribution, consumption, and disposal. It means that the producer, who stands to profit the most, is made to assume the cost, so this could also be called a beneficiary pays principle. Japan's Container and Packaging

Waste Recycle Law, which in April 2000 took complete effect for all containers and packaging, goes only so far as to recognise that manufacturers are responsible for recycling, but such producer responsibility is already being widely institutionalised in EU member states and other countries, as in Germany's Cyclical Economy and Waste Management Act, and the OECD is pushing it.

As this shows, it is consideration based on policy value analysis that inquires into EPR's basic principles, philosophy, and underlying values. There are questions relating to system design, and to policy and measures, which arise over the matter of specifically what sort of institutions will allow one to implement the EPR principle. Also, a decision-making process analysis is needed to determine why such institutionalisation-based principles are making progress in Germany and other EU member states, but not in Japan. For example, out of Germany's population of 82,070,000, there are 550,000 Greenpeace members (0.7% of the population) and 360,000 supporting members of the World Wide Fund for Nature (WWF, 0.4%).[1] Additionally, the nation-wide environmental organisation 'Bund für Umwelt und Naturschutz Deutschland' (BUND) has 250,000 members (0.3%). The Netherlands has a population of 15,600,000, of which 600,000 are Greenpeace members (3.8% of the population) and 830,000 are WWF members (5.3%). As a proportion of the population, Holland's figures are 5 to 12 times those of Germany. Japan's memberships in those organisations are a mere 5,000 and 40,000, respectively (all data are 1997 figures). There are enormous differences between the social influence of environmental NGOs in EU states and Japan, and the government-NGO divide is far wider in Japan than in EU states.

The Role of Environmental Movements in the Environmental Policymaking Process

To discuss "the dynamism of environmental movements and environmental policy" is to see importance in: (1) the role of environmental movements as influential actors in the policy decision process, (2) the role of environmental movements as evaluators of policy means, and (3) the role of environmental movements as submitters of policy value.

An example would be a case study by the author and others, and detailed in Funabashi Harutoshi et al. (1985), in which a community movement and lawsuit opposing Shinkansen train noise pollution in Nagoya made a social issue out of this pollution, and induced the institutionalisation of pre-

---

1 The figures for Greenpeace and WWF members by country are according to Matsumoto Yasuko (2001: 187).

viously non-existent measures to mitigate that pollution (a role in the decision-making process). But the former Japan National Railways (JNR) offered only a limited selection of remedial measures, while it was reduced train speed in residential areas that the community most sought (evaluation of policy means). With the support of the former JNR Labour Union, this demand was met for a nine-year period beginning in 1974 by about two-thirds of the trains running through the track-side residential areas involved in the dispute. But the demand was subsequently turned down by two court decisions, and trains have not slowed down since. The movement against Shinkansen pollution stood in opposition to the high-speed, high-volume transport advocated by JNR, and was a movement that advocated the countervailing value of "quiet" (to live without the disturbance of traffic noise) to be enjoyed by track-side residents (submission of policy value).

Essentially, environmental study is interdisciplinary research. It is a field where the validity of established disciplines and academism have been strongly questioned. Building on its academic characteristics, environmental sociology can contribute to themes such as issues regarding waste, energy and global environment – and especially global warming – in which up until now the contribution of Japanese environmental sociologists have been limited. The contribution of environmental sociology in the area of policy-making processes will further increase the role of environmental social activism and environmental NGO/NPOs.

Figure 1: Membership of Japanese Association of Environmental Sociology (1990–2002)

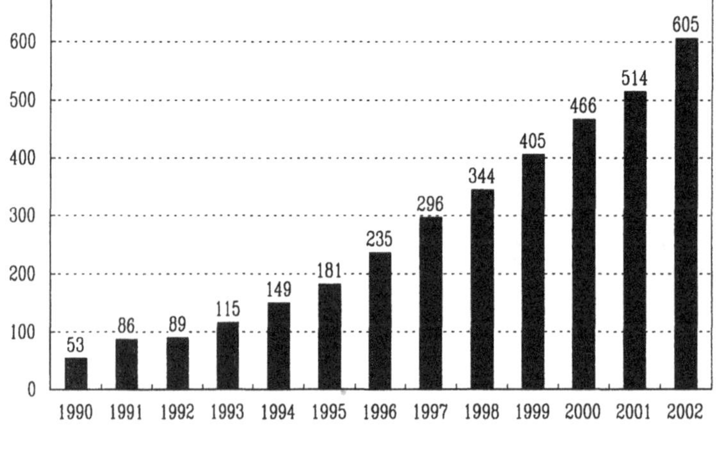

Figure 2: Focus of the first stage of Japanese environmental sociology

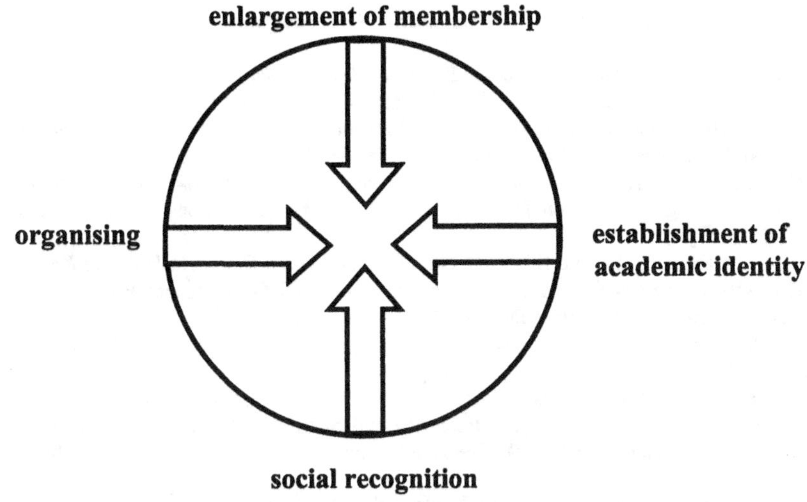

Figure 3: Focus of the second stage of Japanese environmental sociology

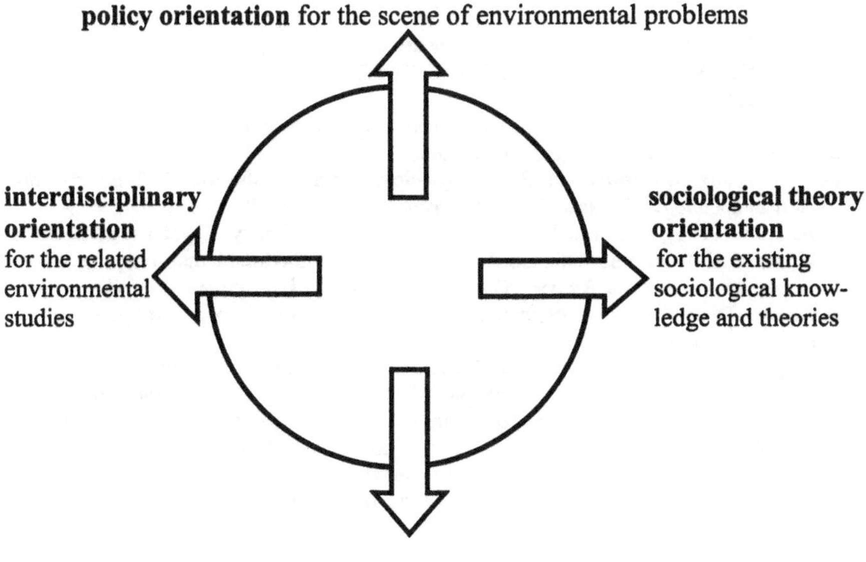

References

Awaji, T. (1980), *Kankyo Ken no Hori to Saiban* (Environmental Rights and Lawsuit), Yuhikaku, Tokyo.

Awaji, T., K. Ueta and K. Hasegawa (eds) (2001), *Kankyo Seisaku Kenkyu no Furontia* (The Frontier of Environmental Policy Studies), Keizai Shinpo-sha, Tokyo.

Funabashi, H., Hasegawa, K., Hatanaka, M., and Katsuta, H. (1985), *Shinkansen Kogai* (Sociological Study on Noise Pollution from Bullet Trains), Yuhikaku, Tokyo.

Funabashi, H., Hasegawa, K., Hatanaka, M. and Kajita, T. (1988), *Kosoku Bunmei no Tiiki Mondai* (Sociological Study on The New Bullet Trains Rail Line Construction Process), Yuhikaku, Tokyo.

Funabashi, H., Hasegawa, K., and Iijima, N. (eds) (1998), *Kyodai Chiiki Kaihatsu no Koso to Kikeksu* (Vision versus Results in a Large-scale Industrial Development Project in the Mutsu-Ogawara District), Tokyo Daigaku Syuppankai, Tokyo.

Hasegawa, K. (1996), *Datsu Genshiryoku Shakai no Sentaku* (A Choice for Post-nuclear Society), Shinyo-sha, Tokyo.

Hasegawa, K. (1999), 'Genshiryoku Hatsuden wo meguru Nihon no Seiji/Keizai/Shakai (Nuclear Energy Issues and Politics, Economy and Society in Japan)', in Y. Sakamoto (ed.), *Kaku to Taiketsu suru 20 Seiki*(The 20th Century against Nuclear), Iwanami Shoten, Tokyo, pp. 281–337.

Hasegawa, K. (2001), 'Kankyo Undo to Kankyo Seisaku (Environmental Movement and Environmental Policy Studies)', in K. Hasegawa (ed.), *Koza Kankyo Shakaigaku vol. 4 Kankyo Seisaku to Undo no Dainamizumu* (Dynamism of Environmental Movement and Policy (Environmental Sociology in Japan vol. 4)), Yuhikaku, Tokyo, pp. 1–34.

Humphrey, C.R. and F. Buttel (1982), *Environment, Energy and Society*, Wadsworth, Belmont.

Iijima, N. (1984/1993), *Kankyo Mondai to Higaisya Undo revised ed* (Environmental Problems, the Victims' Voices and Their Movement), Gakubun-sha, Tokyo.

Iijima, N. (2000), *Kankyo Mondai no Shakai-shi* (The Social History of Environmental Problems), Yuhikaku, Tokyo.

Ikeda, K. (1995), 'Kankyo Shakaigaku no Syoyuronteki Perspective (Property as a Perspective of Environmental Sociology)', pp. 21–38.

Kosaka, K. (2000), 'Midoruman no Susume (Sociologists as Middle Man)', *Bulletin of the Faculty of Sociology, Kwansei Gakuin University*, vol. 87, pp. 197–206.

Matsumoto, Y. (2001), 'Kokusai Kankyo NGO and Kokusai Kankyo Kyotei (International Environmental NGO and International Agreement on Environmental Protection)', in K. Hasegawa (ed.), *Koza Kankyo Shakaigaku Vol. 4 Kankyo Seisaku to Undo no Dainamizumu* (Dynamism of Environmental Movement and Policy (Environmental Sociology in Japan vol. 4)), pp. 179–210.

Miyamoto K. (1989), *Kankyo Keizai Gaku* (Environmental Economics), Iwanami Shoten.

Nakano, Y., et al. (1996), 'Shakaiteki Jirenma toshiteno Gomi Mondai (Environmental Problems as Social Dilemmas)', *Journal of Environmental Sociology*, vol. 2, pp. 123–139.

Seki, R. (2001), 'Kankyoken no Shiso to Kodo (Thought and Activities on Environmental Rights)', in K. Hasegawa (ed.), *Koza Kankyo Shakaigaku Vol. 4 Kankyo Seisaku to Undo no Dainamizumu* (Dynamism of Environmental Movement and Policy (Environmental Sociology in Japan, vol. 4)), pp. 211–236.

Tanaka, S. (1997), 'Kasen Kankyo Jigyo toshiteno "Tashizengata Kawa Zukuri" ("Renaturalization of the River" as a River Environment Enterprize)', *Kankyo Shakaigaku Kenkyu* (Journal of Environmental Sociology), vol. 3, pp. 58–71.

Toda, K. (1994), *Kankyoteki Kousei wo Motomete* (Searching for Environmental Justice), Shinyo-sha, Tokyo.

Torigoe, T. (ed.) (1989), *Kankyo Mondai no Shakai Riron* (A Social Theory on Environmental Issues), Ochanomizu Shobo, Tokyo.

Torigoe, T. and Kada, Y. (eds) (1984), *Mizu to Hito tono Kankyoshi* (Environmental History of Water and Residents), Ochanomizu Shobo, Tokyo.

5 Environment: Thematic Career of a Social Construct

VOLKER LINNEWEBER

Some Results of Recent Surveys

In a recent survey (N = 2.361 subjects) we will refer to repeatedly in the next section, Kuckartz and Grunenberg (2003b) have shown that in German public, the evaluation of the environmental quality is related to the dimension under consideration: the closer to the individual the better the quality is evaluated (Fig. 1).

Figure 1: Evaluating the quality of the environment I

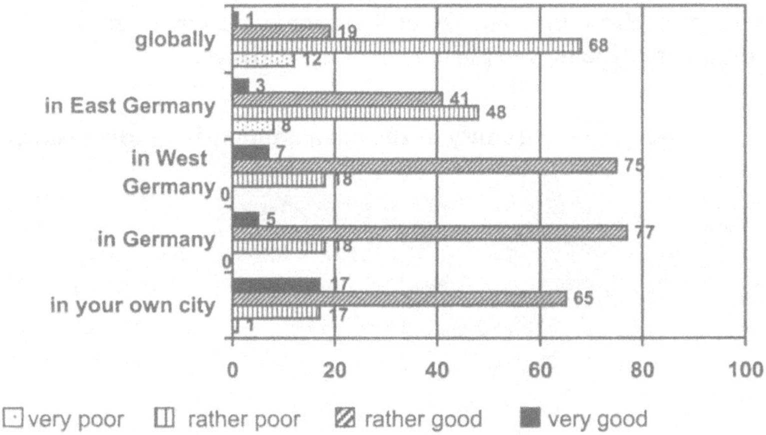

Source: Kuckartz: Jan/Feb 2002 survey, N = 2.361 (answers in %)

Psychologically, this is highly plausible since it reduces cognitive dissonance (Festinger 1957): accepting that the proximate surrounding is less attractive than the wider or even global, would be highly uncomfortable. This would include a social comparison people intend to avoid. One has to admit however, that an empirical basis may also be true: compared with at least certain regions known to be in a very bad condition, Germany's environmental situation is not too bad and there has been a significant improvement during the last decades concerning e.g. freshwater in rivers and lakes. A certain awareness regarding global developments is indicated by answers, too. There is obviously some kind of concern concerning global environmental change if addressed directly. Maybe, discrepancies between the evaluation of the proximate vs. the global environment explains why environmental engagement still may be improved: "Global" is more close to what others have to, while proximate is associated with the own concern. Evaluating that the "own" environment is in a better shape than that of others justifies own inactivity or only moderate activity conserving the environment. Additionally, evaluating the own environment as being in a better condition as the average (global), is psychologically attractive since it contributes in a good conscience.

The fact that conditions in East Germany are evaluated as worse that those in West Germany may reflect stereotypes resulting from conditions in the GDR. Significant improvements are better known to those living in East Germany or travelling there. Since the majority of the sample lives in West Germany, the prevailing stereotype may explain most of the variance.

Figure 2: Evaluating the quality of the environment II: improvement

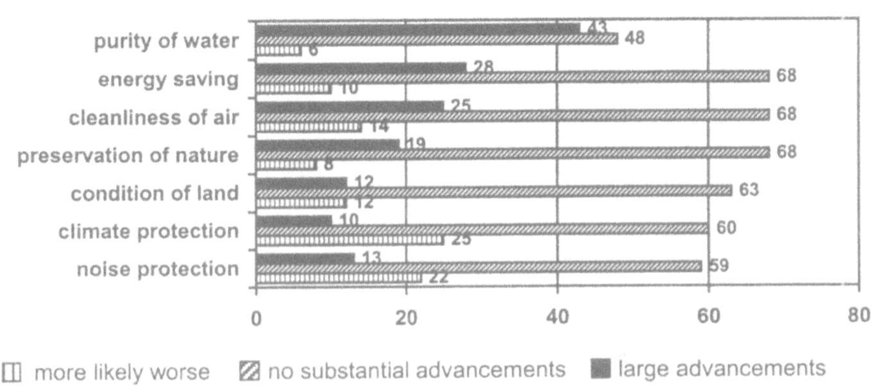

Source: Kuckartz: Jan/Feb 2002 survey, N = 2.361 (answers in %)

A closer look to perceived improvements underscores the interpretation (Fig. 2). While improvements are seen concerning particularly water and other more proximate variables, climate protection is more likely to be considered as a problem. However, also noise protection appears to be evaluated as not being improved substantially or even becoming worse. This may indeed reflect actual developments caused by still increasing traffic particularly by transportation of goods on roads.

Figure 3: Evaluating the quality of the environment III: impact on health

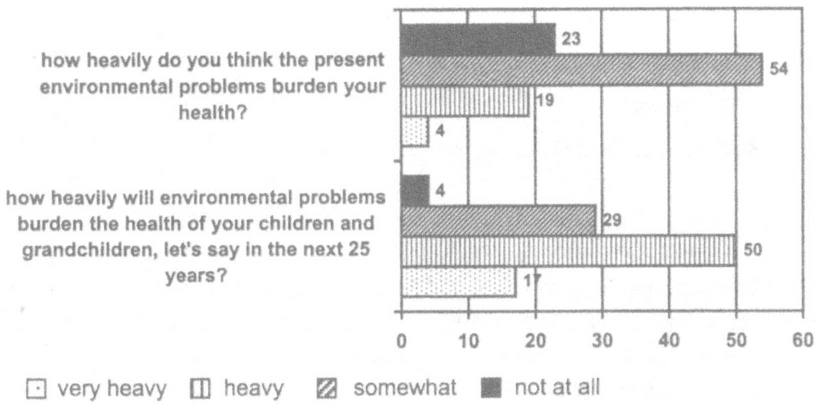

Source: Kuckartz: Jan/Feb 2002 survey, N = 2.361 (answers in %)

The tendency to cognitively or psychologically protect the own person is also indicated by comparing answers to questions concerning the impact of environment on health (Fig. 3). Present impacts of environment on health indicating the own affectedness are evaluated as only moderate, but future generations may be affected more severely.

Figure 4: Expected outcomes of future scenarios

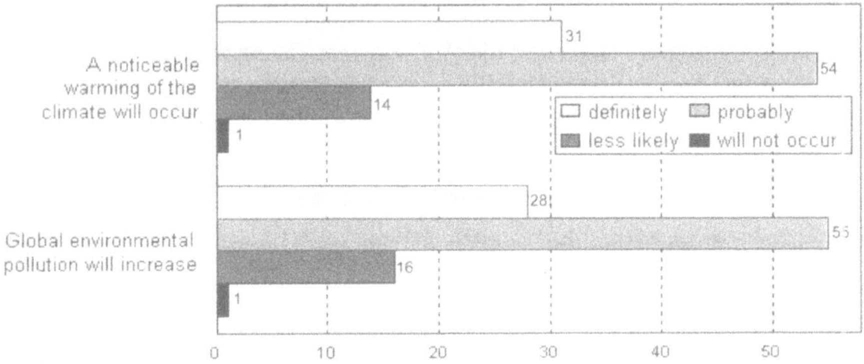

Question: At the beginning of a new millenium one can venture a look into the future. How likely do you view the following events to occur within a time span of the next 20 to 50 years?

Source: Kuckartz: Jan/Feb 2002 survey, N = 2.361 (answers in %)

Generally, the "naive" version of what is called "earth system analysis" in science (Schellnhuber and Wenzel 1998) indicates that only a minority refuses global warming and increasing environmental pollution (Fig. 4). Although there is still some uncertainty in science being recognised in public, too, meanwhile there seems to be a consensus in Germany that environmental protection is important particularly with respect to future generations.

Figure 5 shows that the core concept of the debate on protecting the environment, sustainable development still is a "terra incognita". However, a comparison between the 2000 and 2002 results indicates some change.

Figure 5: Sustainable development: still terra incognita?

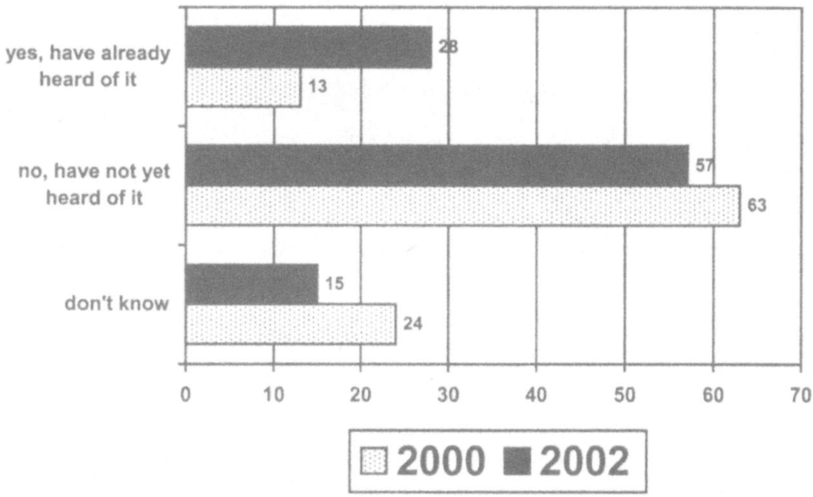

Source: Kuckartz: 2000 vs. 2002 survey, N = 2.018 vs. 2.361 (answers in %). Question: as a leading principle for environmental protection occasionally the term "sustainable development" turns up. Have you already heard of the term or have you not yet heard of it?

In a recent study concerning a particularly fragile area, the North Sea coastline, analyses with experts on coastal zone management, regional development and other aspects related to the environment, confirm these results (Hartmuth, Deising, Fritsche, and Linneweber 2002; Linneweber, Fritsche, and Hartmuth, in press; Linneweber, Hartmuth, Deising, and Fritsche 2001; Linneweber, Hartmuth, Deising, and Fritsche 2002; Linneweber, Hartmuth, and Fritsche 2001). Although the sample is not representative by intention, expert interviews concerning the future of a highly fragile island (Sylt) being artificially stabilised in its present position and shape with immense expenses (up to 10 Mio € per year), are predictive. In social systems, experts are in the position of "opinion leaders", and they explain the social system's functioning in interactions with the environment. This general position justifies in depth analyses with small selected samples (Dörner, Hofinger, and Tisdale 1999; Hofinger 2001).

Figure 6: Case-study "Sylt" (Linneweber et al. 2000): subjective likelihood of tidal waves resulting from greenhouse effect (N = 50 Ss.)

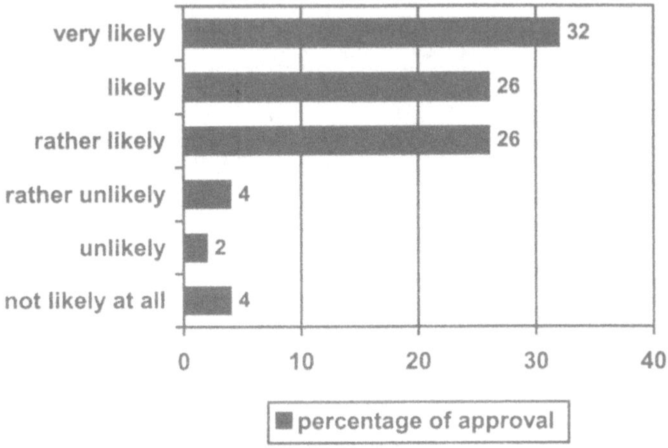

Figure 7: Case-study "Sylt" (Linneweber et al. 2000): certainty for likelihood of tidal waves resulting from greenhouse effect (N = 50 Ss.)

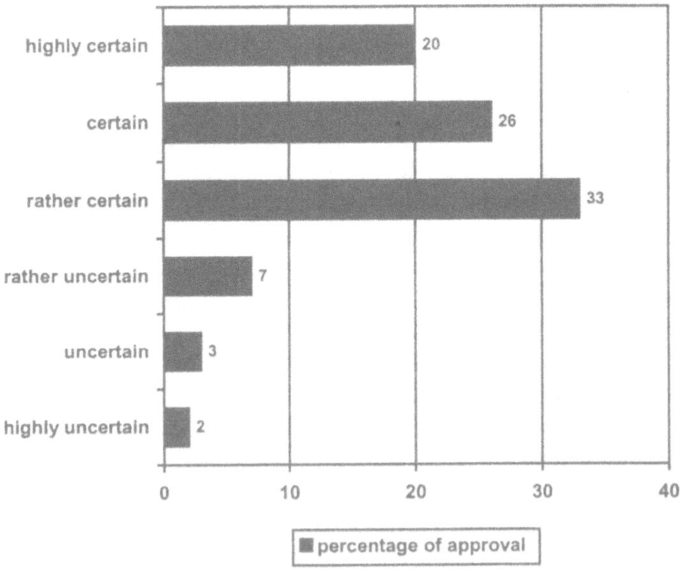

The majority expresses a high subjective probability of tidal waves resulting from greenhouse effect (Fig. 6), and they are quite certain about this (Fig. 7).

Figure 8: Case-study "Sylt" (Linneweber et al., 2000): evaluation of risk by increasing tidal waves (N = 50 Ss.)

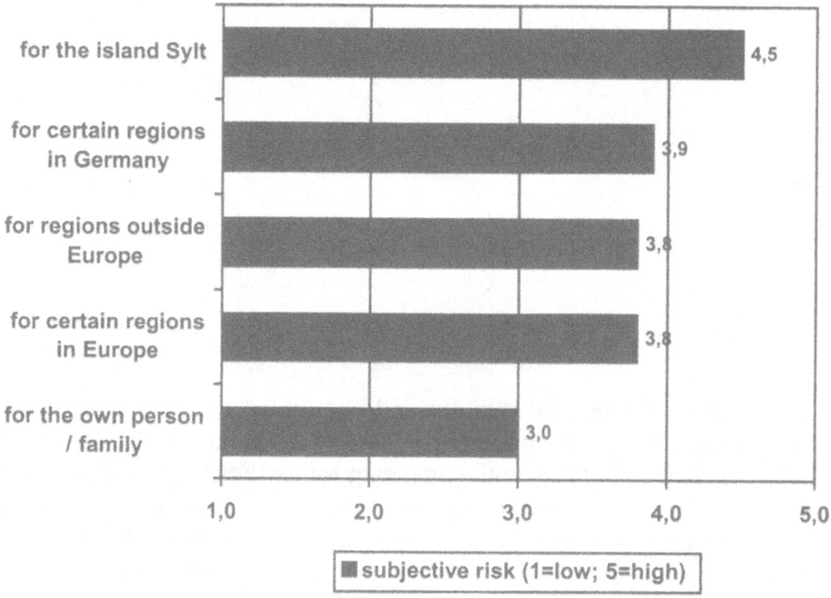

Interestingly, we again find what may be called a "positive illusion" (Taylor 1989; Taylor and Brown 1988, 1994) concerning the own potential affectedness: Although the island is considered as being severely affected, even more than other regions nation-wide as well as world-wide, the own person or family is considered as being less susceptible (Fig. 8).

Figure 9: Environmental protection vs. Crime protection as most important problem (Brochure of the Federal Ministry of Environment)

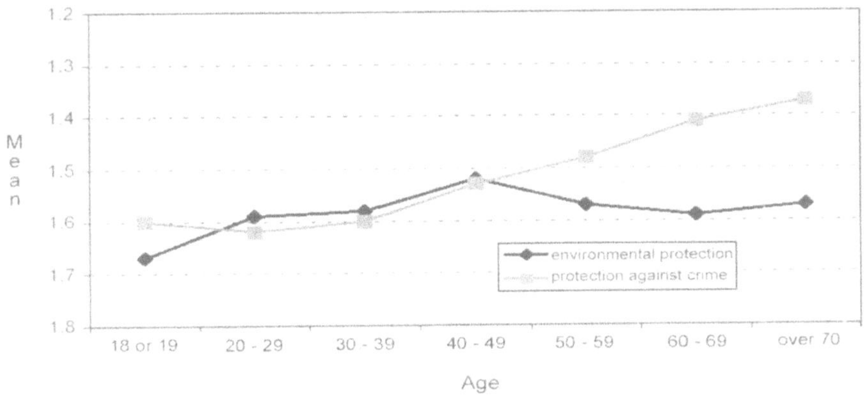

A comprehensive view on attitudes and cognitions or "social representations" concerning the environment (Linneweber et al. 2002) necessarily should include comparisons with competing topics. For the majority in Germany, protecting the environment is an accepted orientation (which must not be confused with corresponding behavioural consequences) and hence socially desirable. Figure 9 shows a comparison between two important topics, crime protection and environmental protection broken down by age groups. The graph indicates that for the elderly, the former topic (crime protection increasingly gains importance). Due to the design of the study, we may not attribute this to age-related change of attitudes since it may also be a cohort effect. Nevertheless, we have a first comparison to further topics of public interest.

Figure 10: Voting preferences of environmentally engaged citizens (Kuckartz 2002)

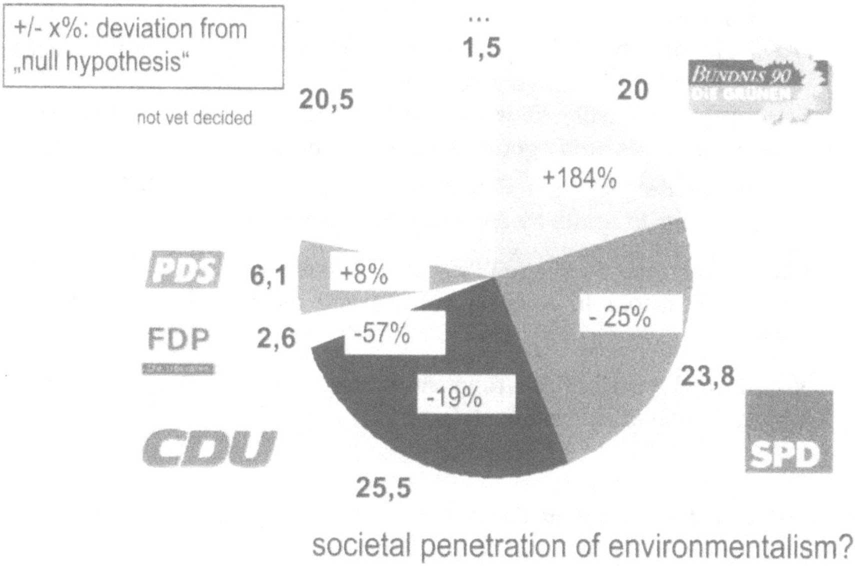

This contextualising environmental protection by competitive topics is continued by looking at voting preferences of environmentally engaged citizens. We find voting preferences for all political parties (Fig. 10, numbers outside the pie). The percent values inside the pie indicate deviations from the "null hypothesis" (based on the September 2002 election). A value of 0% would indicate no pronounced preference. We see that particularly the percentage of 184% indicates a specific preference for voting "green". Interestingly, particularly for CDU and SPD only 19% or 25% less votes that expected under the null hypothesis appear. We may interpret this as a societal penetration of environmentalism.

Trying to summarise the snapshot we intended in the first part, or first interim summary says:

- Presently (2002), the evaluation of the environment shows a differentiated picture:

- While the proximate environment (e.g. own town) is considered as pretty good, the quality of the environment world-wide is considered less positive and future problems are anticipated with high certainty.
- Two reasons may explain this stable pattern:
  - The quality of the immediate environment in Germany in fact is evaluated as relatively good.
  - Psychologically, it is plausible to consider the proximate environment as pretty good, since this reduces cognitive dissonance.
- The latter explanation also serves to explain the expectation of serious threats to health by environmental problems in the future.

- Interestingly, the orientation "sustainable development" – although frequently discussed in science, politics and media – is not very well known.
- Voting preferences of environmentally engaged citizens indicate societal penetration of environmentalism.

Looking Back: The Last 40 Years

Following this snapshot, we will briefly outline developments during the last four decades. To support an orientation, in Table 1 various aspects of relevant topics are indicated in the columns, various levels refer to these examples in the rows of Table 1. In the fields, examples for the discourse are inserted.

This two-dimensional order may help to locate relevant aspects trying to understand the thematic career of our topic. Although there are comprehensive analyses on the genesis of the term environment in general, our intent analysing the thematic career (or "issue attention cycle") allows to concentrate on the public debate in Germany.

Table 1: Levels and topics of the societal discourse on the environment

Level of aggregation	Examples of Discourse			
	Activity	**Impact**	**Adaptation**	**Mitigation**
World Community	Consumption of fossil fuels, water, resources	Global warming, depletion of ozone layer	Land use, development of adaptive structures	Various Restrictions
Generation	Consumption of fossil fuels, water, resources	Disadvantaging future generations	Developing alternatives	Negotiation, Conventions, Commitment
Nation/ Culture	Economy	Emission, consumption	Appropriate ways	Taxes
Region	Water consumption	Shortage, health	Appropriate ways	Regional sustainability
Community	Land use	Sealing of soil	Banks, dikes	Local Agenda 21
Economic actor	Production, Service	Emission, consumption	Product lines: mobile ACs	Zero-emission vehicles
Type of individuals	Life style	Emission, consumption	Life style	Life style
Single individual	Eco-balance	Emission, consumption	Migration	Investments

Figure 11 intends to present a first approximation. Basically, we intend to indicate that an inverse U-shape results when locating topics on the vertical dimension proximal (individual) to distal (global). Starting with relatively proximal topics of attention (polluted lakes and rivers) in the 60s, "environment was associated with acid rain in the 70s. Global environmental change became a topic in the 80s and has – at least in Germany – lost part of its attention or is accompanied by another process of interest with the unification. Since the 90s, a series of significant conferences on climate and the global environment repeatedly focused public attention, the public continued debating on this topic. However, recently, we observe some kind of concentration on the proximate environment caused by extreme events (storms, floods) in the last couple of years. Causal relationships between local events and global states or developments are increasingly met in media reports on the occasion of extreme events.

Figure 11: "Environment": contextualised thematic carreer

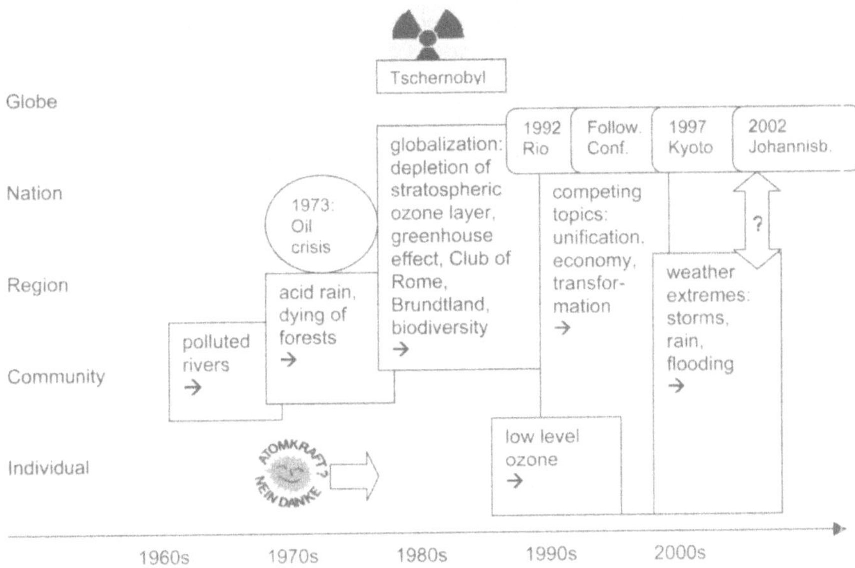

With a more pronounced interpretation Poferl, Schilling, and Brand (1997) differentiate four phases since 1969 and characterise the contextualised thematic career like follows:

- Phase I (1969–1974):
  - marginal public debate
  - "top town", politics-based debate by government (SPD/FDP coalition)
  - basics of environmental laws
  - consensus on necessity of means to protect the environment

- Phase II (1975–1982):
  - Discrepancy between
    - dominating tendency towards economic innovations and
    - dramatic scenarios concerning "limits of growth"
    - Polarisation between
    - bottom-up "movement" and
    - established political-industrial "system"

- Phase III (1982–1990):
  - Green party moving into the parliament
  - institutionalisation of the environmental movement and concerns of environmental policy
  - "Greenisation" of public life and industry
  - efforts for consensus and dialogue between environmentalists, state, political and economic actors

- Phase IV (since 1991/92):
  - discourse on economic crisis
  - competing social and economic problems by German re-unification
  - new conflicts between interests concerning economic growth, social justice and environment
  - at the same time: maintenance of awareness for environmental problems and state of the debate, diffusion of the concept "sustainable development"

Poferl, Schilling, and Brand (1997) also outline relations between societal discourse and everyday-life (Table 2). As Tendencies of the societal discourse, they identify science-orientation, moralisation, and "politisation". This interacts with processes (social definition of environmental problems/ debate on definition), which interact with offers for symbolic meaning. Poferl, Schilling, and Brand's (1997) argumentation corresponds with the one presented here: "Environment" is conceptualised as a social construct. In order to understand the societal interaction with various environmental components, we must try to understand how "environment" is socially constructed as well as represented. With respect to the last 40 years, we have seen considerable developments.

Table 2: Poferl et al. (1997):

Relation between societal discourse and everyday-life

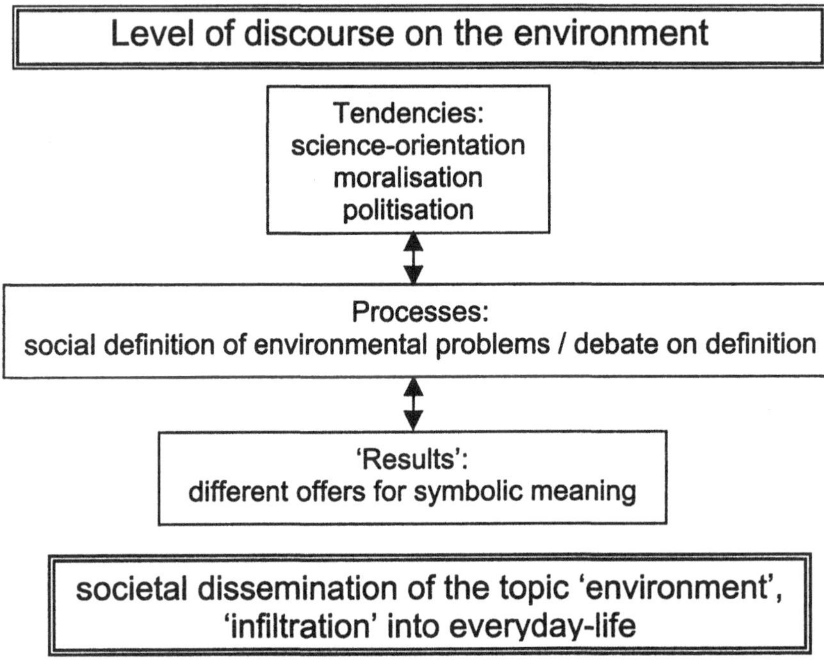

Concerning the last 40 years, we come to a second interim summary:

- In the last 40 years, the term environment has changed its meaning:
  - While aspects immediately perceivable (e.g. polluted rivers) dominated in the 60s,
  - Air pollution and acid rain indicated trans-regional impacts in the 70s,
  - Global environmental change has been discussed in the 80s,
  - The attention to environmental problems apparently declined in the 90s.
- The change of meaning was accompanied by political and societal developments as well as significant events (e.g. Tschernobyl)

- Also, international developments in science and policy (e.g. Report of the Club of Rome) had impacts on the thematic career.
- The topic "environment" is increasingly contextualised: debate on "economy vs. ecology"

Looking Back: The Last Decade

Due to the "new" topic global environmental change and with respect to future developments, the last decade is of particular interest, so we will now focus on this period.

Earth system analyses confront us with alarming signs: CO_2-emissions (Fig. 12) and consequently the global temperature (fig. 13) are increasing in an unprecedented extent.

Figure 12: Earth system analysis I: CO2-emissions

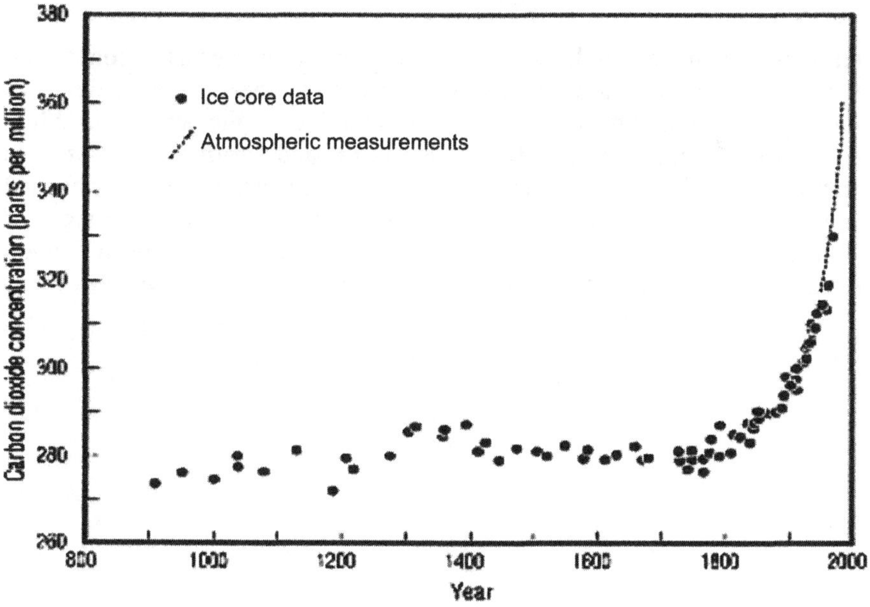

Source: http://www.linmpi.mpg.de/~natalie/climate/glchange.html

Figure 13: Earth system analysis II: temperature

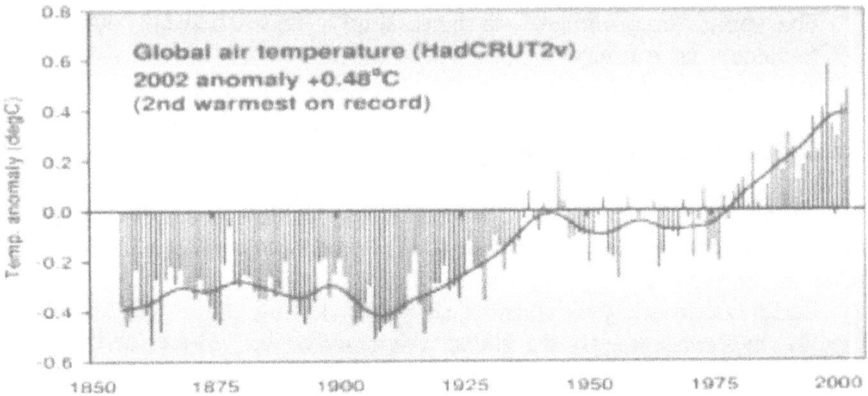

Source: http://www.cru.uea.ac.uk/

Irrespective of interdisciplinary debates and differing scenarios concerning the extent of global warming, sea-level rise and other alarming developments, graphs like this have caught considerable public attention. While some years ago, local events like floods, storms and weather extremes have primarily been discussed concentrating on the event itself, this has changed since a couple of years. Increasingly, these events are discussed as signs of the greenhouse effect. Environmental protection and particularly mitigation strategies like reducing CO_2-emissions are elements of public awareness.

Figure 14: Environmental protection as one of the most important problems

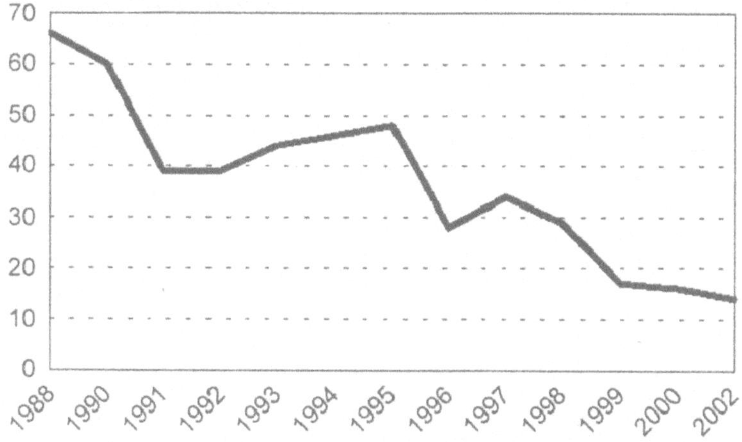

Source: 1988 to 1999 EMNID; 2000 and 2002 UBA, indications in %

Figure 15: Contextualising environmental protection by competitive topics II:

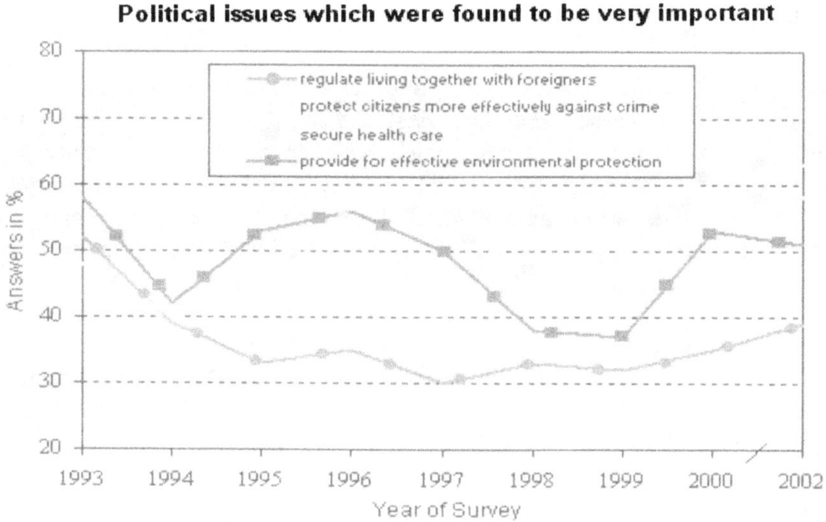

Figure 16: Contextualising environmental protection by competitive topics III:

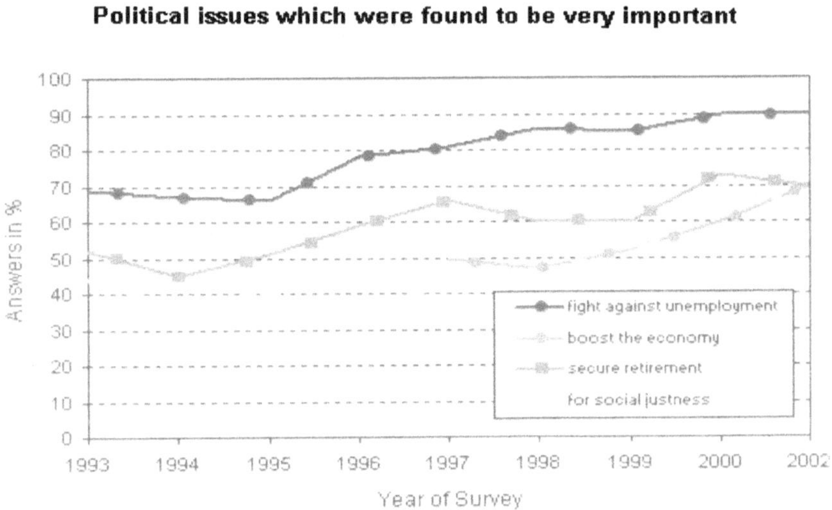

Source: Kuckartz, U. and Grunenberg, H.: Umweltbewusstsein in Deutschland 2002

In the light of these "surrounding" analyses, only the exclusive status as "one of the most important topics" faces some decline (Fig. 14).

From a psychological point of view, another contextualisation is surprising. In Germany, comparisons between East and West are still conducted frequently. This concerns the demographic development, the economy, and also environmental conditions. In a recent brochure of the Federal Ministry of Environment, data are presented, which allow analyses based on social psychological studies concerning social identity and inter-group relations (van Knippenberg and Ellemers 1990).

Figure 17 shows the percentage of persons evaluating the environment as good and very good. The figure indicates that generally, the proportion of persons evaluating the environment as good and very good is increasing. Persons from West Germany clearly show this tendency. They moreover evaluate West Germany's environmental quality ("their own environment") prevailingly "better" than the East German environment. Both lines representing evaluations of persons from West Germany are surprisingly parallel.

Figure 17: Evaluating the quality of the environment
% of persons evaluating the environment as good and very good.
Evaluating (east vs. west) the own (circle) vs. the other area (square)

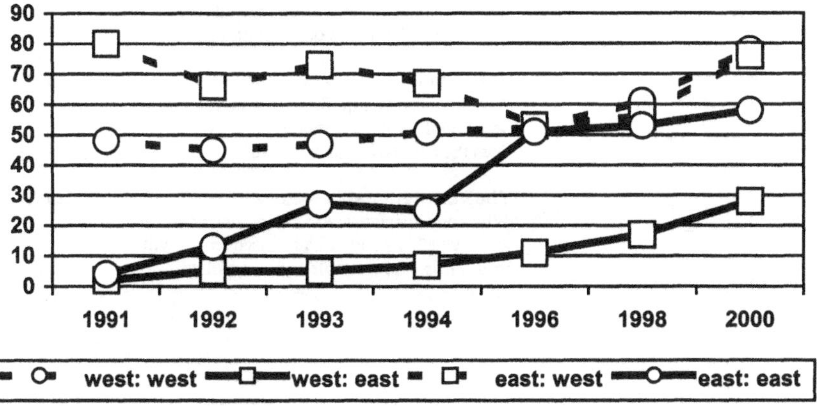

Source: Data from Fed. Min. Env.

In contrast to this, evaluations of persons from East Germany show an approximation from both sides: From 1991 to 1996 West Germany's environment is evaluated by and by less positive while East Germany's (the own proximate environment's) evaluation continuously improves. From 1996 to 2000 the increase of those evaluating the own environment in East Germany is less pronounced but positive evaluations of West Germany rise.

In terms of research on social comparisons in inter-group relations, West Germans clearly show an "ingroup-favoritism". Judgements from East Germany show some kind of disillusion concerning the West but at the same time an acknowledgement of improvements of the own territory. The fact that people from West Germany do not recognise the fact that conditions in East Germany improve and that the contrast by and by disappears, may also be attributed to a lack of immediate experience.

Figure 18: "Worldview": myths of nature and their acceptance (Kuckartz 2000)

			supporters (survey 2000)
1		**The "Sturdy Nature"** Fundamentally nature is so organized that it always returns to equilibrium. No matter what one does, the ball always returns to its original position.	4 %
2		**The "Sensitive Nature"** Nature is very sensitive to every form of interference. Even small amounts interference cause the ball to move out of control.	23 %
3		**The "Limited Tolerance Nature"** At certain levels, interference in nature can occur. Only when a certain level is exceeded does the ball move out of control.	53 %
4		**The "Unpredictable Nature"** When one interferes with nature, one does not know whether good or bad consequences will result. It is unpredictable how the ball will move.	20 %

With respect to "worldviews" concerning myths of nature, the 2000 survey of the series mentioned above (Kuckartz and Grunenberg 2003a) shows that in Germany, 53% recognise nature (and hence environment) as having a limited tolerance. While there is nearly no support to the "sturdy nature" model, the "sensitive nature" (23%) and the unpredictable nature (20%) find some acceptance. This result may be interpreted as a general view that nature is fragile and nature as well as environment have limited carrying capacities concerning anthropogenic impacts.

Our third interim summary concerning the last decade says:

- A close look to the last decade further indicates a "contextualisation" of environment:
  - In surveys, environmental protection as "one of the most important problems" has dropped from 65% in 1988 to 15% in 2002.

- Closer analyses show that "environment" as important societal topic found prominent company by attention on:
  - integrating immigrants,
  - health care,
  - crime protection,
  - unemployment,
  - securing pension,
  - instigating the economy,
  - social justice.
- At the same time,
  - The quality of the immediate environment is considered as increasingly positive
  - Global environmental change is considered as increasingly problematic and the worldview of the "limited tolerance nature" is preferred.

Figure 19: Societal discourse on the environment

Features:
•media-based, variable.
Predictors for variance in visibility:
– competing topics
– significant events

Integrated Model

In the following section, we intend to present a first model integrating our considerations (Fig. 19). The societal discourse on the environment develops in various societal sections: particularly in economy, science, politics and

NGOs. Only part of this discourse is visible, and the "window of visibility" changes in size, position and transparency. This window is of course media-based (Bell 1994) and hence subject to various impacts from outside particularly by competing topics. Another source of variance of the window of visibility are significant events.

As predictors of variance in visibility we hence may identify:

- competing topics
  - related to other central values and derived "orientors" (Bossel 1990)
    - existence: prosperity, health
    - freedom of action: individual freedom, freedom of partial systems, societal independence
    - security: related to supplies, nation, threats
    - efficacy: in steering, transformation
    - flexibility: diversity, ability for innovations
    - responsibility: (international) sociability, sustainability
- significant events
  - (natural) disasters
    - particularly in proximity like weather extremes,
    - significant human impacts (e.g. pollution, oil spills),
    - and – increasingly – maladaptive structures.
  - actor-related "news"
    - economy: new technology (e.g. CFC substitutes, zero/low emission vehicles)
    - science: new scientific knowledge
    - policy: changes in environmental policy
    - NGOs: sensational activities

The Future

Some years ago, environmental protection was associated with renunciation of consumption. A contrast between economy and ecology has been discussed. This position has been revised by and by. Several examples show that environmental protection is highly compatible with economic development and the number of those sticking to this stereotype is declining. It may be anticipated that in future, the compatibility of environmental protection and economic progress will be even more visible. Linneweber (2000) consequently proposed a combination of two dimensions environment (orientation "sustainability") and society (orientation "social identity") (Fig. 20).

Figure 20: Utilisation of environmental resources vs. environmental protection

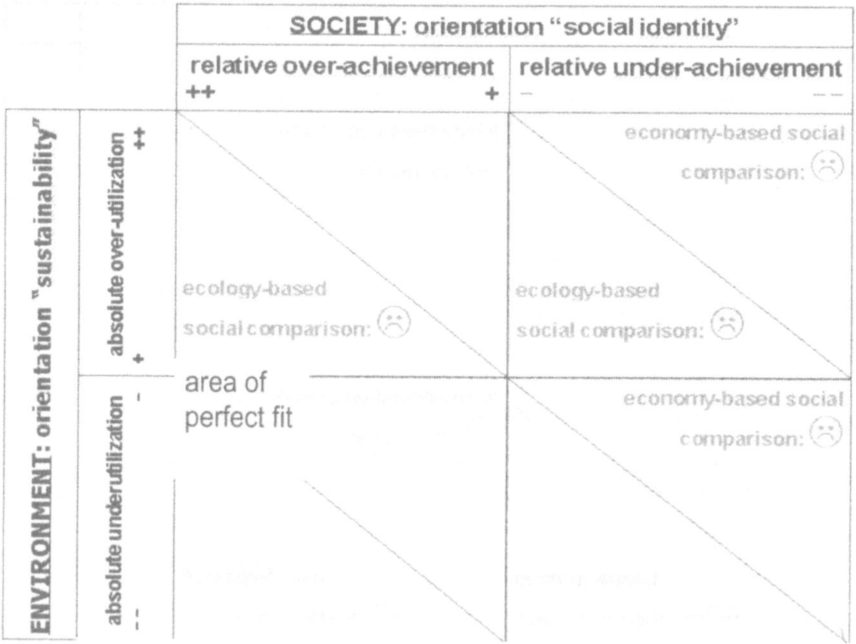

The latter is meant to include economy but also orientations like quality of life, security and others mentioned by Bossel (1990). To simplify, we differentiate four fields (2 x 2), although not categories but dimensions should be imagined. In line with our argumentation, we conceptualise social comparison processes on both dimensions. With respect to the environment, an optimal utilisation related to others is desirable. It should not exceed the carrying capacity of ecosystems (sustainability); of course the definition is a matter of social construction, too (Linneweber 1998). With respect to other dimensions of social comparison, particularly those resulting from economic variables, a maximum of relative over-achievement should be attained. Hence, an "area of perfect fit" results: environmental resources are perfectly utilised and compared to other actors (nations, societies, groups), an optimum is reached economically and on further dimensions.

Interrelated with environment, new topics will shift into public attention: international security and environment (water, land use, and migration, environmental refugees). Although already mentioned (Bossel 1990) and under debate in science, presently the latter are not yet object of public debate.

Final Summary

- After about 40 years of establishing the term environment in public, the present societal discourse on the environment is primarily driven by forces located in the
  - economy
  - policy
  - sciences
  - NGOs.
- The resulting "window of visibility" is
  - unstable in size, position and transparency,
  - media-based and variable.
- Predictors for variance are
  - competing topics, related to other societal "orientors" and
  - significant events like natural disasters and actor-related "news".
- The future debate will further contextualise environmental protection, particularly by considerations on economy (e.g. reducing unemployment by environmentally friendly technologies; optimal utilisation of environmental resources).
- The relation between global environmental change and local events will shift into the focus of attention by decreasing scientific uncertainty
- "New" topics will be: international security and environment (water, land use, and migration, environmental refugees).
- Presently the latter are not yet object of public debate.

References

Bell, A. (1994), 'Climate of opinion: public and media discourse on the global environment', *Discourse & Society*, vol. 5, pp. 33–64.

Bossel, H. (1990), *Umweltwissen: Daten, Fakten, Zusammenhänge*, Springer, Berlin [et al.].

Dörner, D., Hofinger, T., and Tisdale, T. (1999), *Forschungsvorhaben "Umweltbewußtsein, Umwelthandeln, Werte und Wertewandel". Endbericht*. Bamberg: Otto-Friedrich-Universität Bamberg, Institut für Theoretische Psychologie.

Festinger, L. (1957), *A theory of cognitive dissonance*, Stanford University Press, Stanford/California.

Hartmuth, G., Deising, S., Fritsche, I., and Linneweber, V. (2002), 'Globaler Wandel im lokalen Kontext: Sylter Perspektiven auf Klimaänderungen', in A. Daschkeit and P. Schottes (eds), *Sylt. Klimafolgen für Mensch und Küste*, Springer, Berlin, pp. 219–249.

Hofinger, G. (2001), *Denken über Umwelt und Natur*, Beltz, Weinheim.

Kuckartz, U. and Grunenberg, H. (2003a), *Environmental awareness in Germany 2000*, retrieved 03-03-10, from http://staff-www.uni-marburg.de/%7Ecmperzwi/englisch/hauptframe.htm

Kuckartz, U. and Grunenberg, H. (2003b), *Environmental awareness in Germany 2002*, retrieved 03-03-10, from http://www.umweltbewusstsein.de/

Linneweber, V. (1998), 'Nachhaltige Entwicklung' als unscharfes Prädikat', *Umweltpsychologie*, vol 2, no.1, pp. 66–76.

Linneweber, V. (2000), 'Sozialpsychologische Aspekte umweltbezogener Emotionen', in E. Kals, N. Platz and R. Wimmer (eds), *Emotionen in der Umweltdiskussion*, Deutscher Universitäts-Verlag, Wiesbaden, pp. 115–125.

Linneweber, V., Fritsche, I., and Hartmuth, G. (in press), 'Coastal zone protection and climate change: towards a contextualized analysis of environmental identity', in S. Clayton and S. Opotow (eds), *Identity and the Natural Environment*, MIT-Press, Cambridge.

Linneweber, V., Hartmuth, G., Deising, S., and Fritsche, I. (2001), 'Akteure in der Umweltbewertung und -nutzung: Eine feldorientierte Typisierung', in G. de Haan, E.-D. Lantermann, V. Linneweber, and F. Reusswig (eds), *Typenbildung in der sozialwissenschaftlichen Umweltforschung*, Leske + Budrich, Opladen, pp. 209–225.

Linneweber, V., Hartmuth, G., Deising, S., and Fritsche, I. (2002), 'Soziale Repräsentationen der Gefährdung Sylts angesichts möglicher Klimaänderungen', in K.S. Radke (ed.), *Zweites Forum Katastrophenvorsorge: "Extreme Naturereignisse – Folgen, Vorsorge, Werkeuge"*, Deutsches Komitee für Katastrophenvorsorge (DKKV), Bonn/Leipzig, p. 406.

Linneweber, V., Hartmuth, G., and Fritsche, I. (2001), *Coastal zone protection, climate change, and socioeconomic dynamics: what humans value*. Paper presented at the 2001 Open Meeting of the Human Dimensions of Global Environmental Change Research Community, October 6–8, Rio de Janeiro/Brazil.

Poferl, A., Schilling, K., and Brand, K.-W. (1997), *Umweltbewußtsein und Alltagshandeln*, Leske + Budrich, Opladen.

Schellnhuber, H.-J., and Wenzel, V. (eds) (1998), *Earth System Analysis: Integrating Science for Sustainability*, Springer, New York.

Taylor, S.E. (1989), *Positive illusions: creative self-deception and the healthy mind*, Basic Books, New York.

Taylor, S.E. and Brown, J.D. (1988), 'Illusion and well-being: A social psychological perspective on mental health', *Psychological Bulletin*, vol. 103, pp. 193–210.

Taylor, S.E. and Brown, J.D. (1994), 'Illusion" of mental health does not explain positive illusions', *American Psychologist*, vol. 49, pp. 972–973.

van Knippenberg, A. and Ellemers, N. (1990), 'Social identity and intergroup differentiation processes', in W. Stroebe and M. Hewstone (eds), *European review of social psychology*, vol. 1, Wiley, Chichester, pp. 137–169.

Part III: Politics

6 Environmental NGOs in Germany – The Question of Power and Influence*

GESINE FOLJANTY-JOST

Introduction

In the past decade Germany has become a well-studied case of proactive environmental policies. The introduction of the eco-tax, the phasing out of nuclear power and the revision of the energy supply law characterise the kind of policies, which have made Germany one of the environmental pioneers in global climate change policy.

In this article the main assumption is that this success is at least partly due to the high degree of integrating ecological interest groups into the policy-making process. The assumption derives from the debate on the preconditions for nation states' capacity for proactive environmental policy (Jänicke 1996a, Jänicke/Weidner 1995). A wide range of literature indicates that the influence of environmental NGOs in the eco-network is crucial to the output of the political decision-making process. The integration of NGOs into the policy network may improve the legitimisation of political decisions, it may enlarge political choices, improve political solutions and their acceptance in society, and minimise social conflict (Bichsel 1996, Brunnengräber, Walk 1997, Princen/Finger 1994). NGOs may serve as supporters, critics, experts, and advocators (Rucht 1996: 37). The comparison of the environmental policy performance of various countries supports the assumption that co-operation between state actors and centralised, professional environmental organisations improves policy success (Jänicke 1996b: 17, 1996a: 75–76).

There are, however, preconditions necessary to provide civil environmental organisations with the means of becoming an integral part of the policy-making process. We may distinguish at least the following factors:

---

* Englishing by Teresa Gehrs

- institutional framework: this defines participation rights and conditions of access to the policy-making process,
- availability of resources: this relates to the financial, informational, and human resources necessary to make use of participation opportunities,
- capacity of participation in terms of the will and skill of the actors involved in environmental policy-making to integrate and co-operate with NGOs as well as the capacity of NGOs to co-operate with other actors.

Using these considerations as a background, this article will discuss the role the big national environmental organisations play in German environmental policy. By using quantitative data from interviews with leading representatives of NGOs in Germany in 2000, the article will analyse the position NGOs hold in the policy-making process.[1]

The article will start with an overview concerning the position of NGOs in the ecological network in Germany and the role they are considered to have played until now. In part 2 the preconditions of the participation of NGOs will be analysed in terms of resources available, access to information, and co-operation patterns. In a final step the data will be discussed comprehensively with regard to the question of power and influence by focussing on the terms of integration of NGOs into the German environmental network.

NGOs in the Environmental Policy Network

The rise of civil movements in environmental politics dates back to the 1970s, when former West Germany experienced strong civil movements against the construction of nuclear power plants. These movements were locally based, but received support from all over the country. Mostly directly mobilising the general public against the government and the electricity pro-

---

1 This research was part of a joint research project with Yutaka Tsujinaka of Tsukuba University. The Ministry of Education in Japan sponsored the empirical research in Germany. In Germany we used the Japanese questionnaire and adapted it to German circumstances. Klaus Jacob of the Free University of Berlin prepared the German sample. It consists of 92 national organisations. 53 granted a complete interview, i.e. the return rate was 57.6%. The questionnaire consists of three categories of variables:

- variables concerning the history and resources of the interviewed organisation,
- variables concerning information exchange, co-operation and opposition with regard to 132 national and international actors in the field of global environment policy,
- variables concerning the influence, interest and activities of the interviewed organisation.

ducers, these movements stood explicitly outside the established system of policy-making and mainly relied on direct action. From the late 1970s, however, these movements became well-accepted actors at the local level in the course of turning from protest to co-operation (Sachverständigenrat für Umweltfragen 1996: 226).

During the 1980s environmental interests met favourable conditions for gaining influence. Society was supportive of civil movements, which were engaged not only in environmental issues but also in social, feminist, or town-planning issues. Mainly due to the appearance of the Green Party in parliaments of some *Länder*, the environment issue became part of competition among the leading political parties. Since the Green Party of those days did not strive for participation in government but served at least in part as the voice of civil movements in society, the party was predestined to serve as a driving force for NGO interests (Raschke 1993). Moreover, in the 1980s new environmental research institutes were founded with some, such as the Öko-Institut Freiburg, explicitly in support of the environmental movements. This politically and socially favourable environment supported their progress towards institutionalisation. By improving their expertise and becoming stable institutions, environmental interest groups became part of the national environmental policy-making network (Sachverständigenrat für Umweltfragen 1996: 230).

After the reunification of Germany in 1989 new environmental interest groups mainly appeared in former East Germany due to the new experience of democracy and democratic opportunities of participation. However, after some years of consolidation these movements are partly dissolute, and have partly joined national environmental interest groups. Based in former West Germany, these groups have integrated environmentalists from the East by building up local branches in the former GDR. However, compared with former West Germany, membership numbers remain low.

According to our interviews, nine environmental interest groups are currently considered to be part of the German environmental policy-making process. This number is small compared with the number of actors from business and labour, foundations, and corporations, which also represent the non-governmental sector, and of whom 27 are considered to be part of the policy-making process. However, the nine NGOs, which were named for the sample are highly professionalised organisations with a long tradition in environmental policy, like Greenpeace Germany, BUND, WWF, NABU and

BBU (Bund Bürgerinitiativen Umwelt).[2] Combined their membership amounts to more than 1.3 million people.

Goals and Actions

With regard to the climate change issue the data indicate a strong engagement of NGOs. This is not particularly surprising since among the actors in the environmental policy-making network only NGOs may be considered as single-purpose organisations representing general environmental concerns but not particular interests. In the case of the introduction of the eco-tax, all other groups of actors in the eco-network were divided into for and against. While the conservative parties, the BDI, and the trade unions in the sector of chemistry and mining (IG Chemie/IG Bergbau) joined forces in their opposition, the trade unions of metal workers union and of public services (IG Metall/ÖTV) supported the NGOs in their total dedication to the introduction of the eco-tax as a means of influencing energy consumption as well as to the implementation of the Kyoto protocol (Krebs/Reiche 1998: 127). In contrast, desertification and soil pollution attract less attention.

Table 1: Interest in and network organisation's position towards different environmental issues

	Resolution of Kyoto		**Ecological Tax Reform**	
	Interest (a)	support	Interest (a)	support
Governmental Units	2.6	72.73%	2.7	45.45%
Research Units	2.4	61.54%	2.2	46.15%
NGOs	3.0	100.00%	3.4	100.00%
Other interest groups	1.3	50.00%	2.9	43.75%

(a) average based on a scale between "0" (no interest) and "5" (very strong)

2 Out of nine NGOs eight granted an interview. The interviews lasted between 32 and 159 minutes.

When asked what kind of activities they had employed to reach their goals, contacting governmental institutions and the government were the most frequently mentioned activities.

Table 2: What are the most effective activities for influencing global environmental policies (% of group members)

	Governmental Units	Research Units	NGOs	Other interest groups
Contacting the ruling parties	16.67	18.75	37.50	37.50
Contacting the opposition parties	8.33	6.25	0.00	0.00
Contacting governmental institutions	58.33	75.00	62.50	68.75
Appealing through influential people	8.33	12.50	12.50	6.25
Helping to draft bills	66.67	31.25	25.00	75.00
Presenting research results	25.00	62.50	0.00	18.75
Sending representatives to councils/advisory bodies	8.33	37.50	0.00	12.50
Letters/phone calls through members	0.00	0.00	12.50	12.50
Protest campaigns	0.00	6.25	37.50	6.25
Organising mass meetings	0.00	0.00	12.50	12.50
Informing the media	58.33	31.25	50.00	37.50
Running advertisements in the media	0.00	0.00	0.00	6.25
Press conferences	16.67	6.25	37.50	25.00
Contributions to political campaigns	8.33	6.25	25.00	12.50
Other	0.00	0.00	0.00	6.25

The pattern of activities to reach political goals indicates the high institutionalisation of the NGOs as well as their roots as social movements. In fact, addressing the Ministry of Environment and the Federal Environmental Agency is mentioned most frequently and demonstrates that NGOs do not differ in terms of action profile from other interest groups. What differs is that unlike the other actors, NGOs combine targeting the federal environmental administration with a broad range of other activities. Most frequently, various means of mobilising the general public are named. Demonstrations and political campaigning as well as contacting the mass media, holding press conferences, etc. are considered as important means for reaching political goals. By employing a combination of typical lobbying activities and "soft" means of mobilising and informing the general public, NGOs gain media coverage that is not lower than that of other interest groups.

Table 3: Coverage in the mass media over the past 3 years (%) (a)

	up to 100 times	up to 1000 times	over 1000 times
Governmental Units	60.00	40.00	0.00
Research Units	72.73	18.18	9.09
NGOs	20.00	80.00	0.00
Business Associations	55.56	33.33	11.11

(a) related to global environmental policies; based on valid per cent

Combining the data concerning the activities of NGOs with media coverage support, the argument of Rucht (1996: 38) according to which the mass media are crucial to the high public support of NGOs but that sympathetic reporting is also due to the great dedication of NGOs in dealing with the media, is confirmed.

With regard to the results of their dedication, the data demonstrate that NGOs in Germany consider themselves to be an important part of the environmental policy-making network and are also considered influential by the other actors. When asking the actors involved in the process about the influence other actors have in the decision-making process, NGOs are mentioned most frequently besides institutions of the federal environmental administration. Among those 10 organisations with the highest reputation in terms of influence are Greenpeace and the Federal Environment Agency. On a scale from 1 (no influence) to 5 (very influential) Greenpeace received a mean value of 3.37, while the mean value of the Federal Environment Agency was 2.96.

The good reputation of the NGOs is supported by case studies on the introduction of the eco-tax. According to Krebs/Reiche (1998a), NGOs played a leading role in influencing the public by forming coalitions, and finally succeeded in gaining support from research institutes, parts of political parties and interest groups in promoting the introduction of the eco-tax.

Preconditions of Influence: Supply with Resources

It has been argued that besides legal provisions, financial and human resources and access to information are decisive for gaining influence. The most important human resource is members. In their capacity as voters they may exert political pressure as long as they succeed in politicising environmental issues. Moreover, the number of members determines the income of NGOs in terms of membership fees and influences the chances of attracting donors. The availability of financial resources determines the number of technical and professional staff and may influence the forms of activities as well as the scope of campaigns. This is why it has become commonplace to argue that the crucial point for stable and equal participation in environmental policy is the availability of sufficient resources.

In order to explore resources we asked NGOs and other interest groups from the eco-network about their budget, their membership numbers, and their staff. Table 4 corresponds to the expected results: NGOs are weaker in terms of staff and budget, even though the data for the annual environmental budget are rarely comparable with other interest groups from labour or business.

Table 4: Supply with resources

	employees (a)	membership (a)		organisation's budget (million DM) (b)
		individual	organisational	
NGOs	172.38	180,095	43	2.79
Other interest groups	14,497.00	5,302,525.63	572.27	11.77

(a) Mean value of number of people (b) 1999

We found no NGOs with more than 50 staff; the majority relies on less than 10. The same is true in the case of employees who are in charge with the collection and analysis of technical data. All NGOs with technical staff rely on less than 50 people, while for instance 6.3% of the other interest groups have more than 101 employees. Not included are volunteers and staff on the unemployment program of the government (ABM) or young men who have been accepted to join the staff for a year or so instead of serving the army. Rucht et al. (1997) call such additional staff a kind of public subsidy, which definitely serves to compensate weak resources, since those taking part as volunteer staff in NGOs are typically well motivated and highly dedicated regarding environmental issues.

Budgets are also low compared with other interest groups, since NGOs mainly depend on membership fees and donations. Even though the legal conditions for NGOs, like tax exemption for donors and membership fees, are favourable, at least compared with countries like Japan, data indicate that these sources of income do not balance out the unequal financial situation between NGOs and other interest groups, especially from business.

In total, the data demonstrate that those who have argued that the most crucial problem facing NGOs is the restricted availability of resources are right. At least in the well-known case of the introduction of the eco-tax, the influence of NGOs was restricted due to the imbalance of power between NGOs and other interest groups from labour and business in terms of financial and human resources. Reich and Krebs (1998: 291) therefore describe NGO activities in support of the introduction of the eco-tax as "low-budget" activities.

The organisations themselves consider the quality of their scientific and technical staff as the most important aspect to gain success. Especially technical staff became important in recent years since electronic communication has become an important tool for campaigning under the conditions of restricted financial and human resources. In contrast, capable policy staff play a lesser role, while resources like membership, quality, and quantity of the general staff, legal staff, and the budget are considered to be less important than expertise and closeness to the government.

We may therefore argue that even though weak resources are crucial to political influence, it would be too easy to reduce the question of influence and access to the political process to resources alone. Even if resources play a considerable role, it should be taken into account that mobilising the public, forming strategic coalitions or gaining support from epistemic communities might compensate weak resources. Moreover, new means of electronic communication allow visibility under restricted working conditions (Rucht

1996: 37). With regard to activities we have already stated that addressing the general public is a well-employed means of exerting influence. In order to explore coalitions and information exchange as a means of compensating weak resources, we will now take a look at information exchange and co-operation patterns.

Preconditions of Influence: Information Exchange

A limited number of professional staff implies restrictions in developing one's own expertise and increases dependence on external support. Limited budgets may restrict the chances of employing professional research institutes for exploration. Access to information in terms of participating in informal information exchange can therefore be considered to be of considerable importance, since the availability of a wide variety of information sources is crucial to the professionalisation and quality of political decisions. Moreover, information exchange promotes knowledge diffusion and is an important step towards co-operation.

We asked the environmental NGOs about their involvement in information exchange. We distinguish between the number of information contacts NGOs mentioned, the frequency with which NGOs were named as information sources and the number of organisations with whom they exchange information, i.e. the number of information contacts with organisations that also named NGOs as an information source. This is to explore not only information sources for NGOs, but also the role they play in the information network.

NGOs and the governmental units named the highest number of organisations from which they obtain information. Among the ten organisations with the highest number of information sources are the Ministry of the Environment, the Foreign Ministry, the Federal Environment Agency, the Green Party, Greenpeace, BUND, Germanwatch, the Wuppertal Institute for Climate, Environment and Energy, but only the Association of Electricity Suppliers (VDEW) from the business sector.

Table 5: Information exchange

	Mean number of organisations mentioned as an information source	Mean frequency of being mentioned as an information source	Mean number of information exchange relations
Governmental Units n = 11	32.27	32.18	24.90
Research Units n = 13	21.53	24.92	14.15
NGOs n = 8	31.87	29.00	23.37
Other interest groups n = 17	26.11	24.94	17.35

Most of these organisations are among those who were also most frequently named by other members of the network as an information source. Among the top ten are again the Ministry of Environment, the Federal Environment Agency and the Green Party, and the country's three most important NGOs, leaving the interest groups from business and labour behind them. In both perspectives NGOs and governmental units are mentioned more frequently than research units and interest groups. At least in the case of the research institutes this is surprising, since these institutes represent a highly professionalised epistemic community with strong environmental expertise. The lower frequency with which they are mentioned as an information source may indicate that they represent professional information sources that carry out research funded by public institutions without directly participating in the policy-making process. But even though their direct participation in information exchange is lower than that of other groups their mere research provides expert knowledge to society by generating publications on environmental issues. Moreover, from a comparative point of view it is unique to the German research institutes that they hold independent ecological positions.

When looking at inter-group information contacts, we find the highest relative frequency of contacts between the government units and the NGOs. The frequency reflects the increase in institutionalisation of regular contacts between government and ecological interest groups since the formation of the red-green coalition in 1998. NGOs do not depend highly on information from the research institutes in order to compensate weak resources, but have even more contacts with other interest groups, while the other interest groups also contact governmental units and NGOs more than research institutes.

Table 6: Relative frequency of information contacts

	Governmental Units	Research Units	NGOs	Other interest groups
Governmental Units	0.669	0.636	0.739	0.631
Research Units	0.615	0.497	0.375	0.312
NGOs	0.761	0.500	0.797	0.625
Other interest groups	0.631	0.439	0.566	0.526

We can therefore conclude that NGOs are equally well integrated into the information network as the other groups, or even better. They not only compensate weak resources with a high number of information sources, but are also accepted as a valuable information source. We may therefore conclude that the network is highly inclusive with no significant differences between the groups in terms of contacting each other.

Preconditions of Influence: Co-operation Pattern

Beyond mere information exchange we also asked organisations about their co-operative relations[3] with other members of the network. This question was mainly aimed at enlightening the potential for coalition-building in the national policy-making process. As expected, the correlation between information exchange and co-operation is high, i.e. those who have many information exchange contacts have many supportive relations, too. Again, we differentiate between the number of organisations, which each organisation mentions as being supportive, the frequency, with which the same organisation is mentioned as being supportive by other groups, and the co-operation relations, i.e. the number of organisations, which were named by organisations they themselves consider as being supportive. Again, the governmental

3 Since the expression "co-operation" implies an equal relation between at least two partners, we prefer to use the term "support" to indicate that an organisation may provide support without gaining any vice versa.

units responsible for environmental policy are in the centre of the network, followed by the NGOs.

Table 7: Co-operation pattern

	Mean number of organisations mentioned as being supportive	Mean frequency of being mentioned as being supportive	Mean number of co-operation relations
Governmental Units n = 11	15.36	16.27	4.72
Research Units n = 13	10.69	12.00	4.69
NGOs n = 8	10.62	15.37	5.62
Other interest groups n = 17	14.41	11.00	5.52

Looking at the co-operation pattern between the groups demonstrates that NGOs rely on the highest inter-group co-operation frequency among the actors. Since co-operative relations constitute the capacity of increasing expertise and joining forces in political action, the number indicates the existence of powerful means to compensate relatively weak numbers of staff. However, if we qualify the co-operative relations it becomes obvious that they most frequently co-operate with other NGOs. Co-operation among NGOs has a long tradition deriving from joint actions against nuclear power and environmental protection. Joining forces behind the same goals increases political pressure by mobilising the public, while co-operation with governmental units and research institutes play a lesser role.

Table 8: Relative frequency of co-operation between the groups

	Governmental Units	Research Units	NGOs	Other interest groups
Governmental Units	0.421	0.294	0.330	0.262
Research Units	0.280	0.361	0.125	0.113
NGOs	0.170	0.144	0.453	0.191
Other interest groups	0.415	0.183	0.406	0.320

The low frequency of co-operative relations with the research institutes may again only reflect a lack of informal co-operation, since there are well-known cases, in which research institutes have prepared surveys in the name of NGOs. In these cases, research institutes serve more as business partners with professional but well-paid expertise than as partners in policy-making co-operation. In contrast, co-operation between NGOs and other interest groups is better established, even though especially interest groups from business more often than not take a controversial position in issues like phasing out nuclear energy or the introduction of the eco-tax. But since the interest groups in the eco-network do not represent a homogeneous body, parts of them serve as a potential for building coalitions with NGOs like interest groups from labour, humanitarian organisations like Misereor and other civil organisations.

Preconditions of Influence: Political Orientation

Comparing Germany with other countries would invite an analysis of the role party politics play in the integration of the eco-network and the role NGOs might play. The research of Jänicke (1996a:76) and others suggests that the role of the dominant party does not matter for environmental success. However, pioneer decisions in German environmental politics, like phasing out nuclear power, the commercial and industrial waste management legislation, or the introduction of the eco-tax came into existence under a red-green coalition. Moreover, due to the similar political roots of the Green Party and environmental NGO activists, we can assume that at least co-operation between the Green Party in government and the NGOs has improved and that NGOs have gained a stronger position than ever in the political process. According to our data, 20% of the representatives of NGOs we asked called their membership progressive, while 80% considered their membership to be equally divided into conservatives and progressives. Since 'n' is too small the data do not allow any further interpretation. However, when recalling the origins of the environmental movement in Germany, it is not unlikely that especially in the field of nature conservation we can find a high percentage of members with conservative orientations. In terms of forms of action, information exchange and co-operation data indicate that in any case the relation between the government and NGOs is relatively close. Regular meetings between NGOs and the Environmental Ministry have been institutionalised under the red-green coalition, improving NGO access to the policy-making process. However, as Raschke gleefully described, the relation is not without problems. According to him, the Green Party is under pressure to remain supportive of environmental interests without selling them out in favour of staying in power (Raschke 2001). However, by gaining power in government, we may assume that due to the Green Party the system has become more open for environmental professionals from research institutes and NGOs to enter the environmental administration as well as political organisations. These green experts, who cross borders from the civil to the public sector and vice versa, can be considered as forming an informal but influential small network of experts linking the various parts of the eco-network together by personal ties.

Conclusions: Explaining Political Output by Policy Networks

The German contribution to global climate change politics has been highly appreciated internationally. German environmental policies have become a well-studied model by environmentalists abroad. Our data demonstrate that levels of conflict are relatively low and co-operation and information exchange is not fragmented. The network is open to co-operation coalitions. NGOs are a well-accepted part of the network and enjoy widespread integration into information and co-operation networks. This might explain why the political output in environmental politics has only been criticised by environmentalists as a compromise from time to time. Especially after the Green Party succeeded in becoming a coalition partner in government in 1998, many felt disappointed because of the unexpected watering down of green positions (Rucht 2001: 157, 166–167). German critics of German environmental policy continue to complain about the power of interest groups in the environmental policy-making process. They argue that, irrespective of the party affiliation of the ruling party, every government has to co-operate with interest groups from business in order to survive. In the case of the introduction of the eco-tax it was a success for environmental interests that the tax was introduced and that its introduction was decoupled from the activities of the EU. However, from the radical environmentalist position it has failed to become a means of ecological modernisation, since the tax income will not be spent for environmental protection purposes (Krebs/Reiche 1998: 146).

When referring to the situation in other countries, the critics might be placated: Germany is currently regarded as one of the world leaders in environmental politics. Both the data and the case studies tell us that the cleavages are not as deep as expected. The German network is open and pluralistic. All groups participate in the policy-making process. The data support qualitative research, according to which NGOs are well integrated and an accepted part of the network. Even though, as in other countries too, NGOs are "low-budget" organisations, that which makes a difference are the means to compensate weak resources. NGOs have been able to become an influential and well-integrated part of the environmental policy-making process due to

- a favourable environment in society, since environmental protection is still high on the agenda,[4]
- a supportive mass media,
- coalitions with research institutes and labour unions,
- access to the environmental administration.

The level of co-operation is as high as information exchange, providing the NGOs with favourable preconditions for professionalisation and joint actions. Even though the implementation of policies, such as the introduction of the eco-tax, has been controversial, the NGOs have proved to be a critical, flexible, and well-informed partner in the policy-making process, open to new coalitions and co-operation patterns. A study by the famous German Institute for Economic Research (DIW), which was conducted on behalf of Greenpeace in 1994, is said to have been extremely influential in the decision-making process. The labour unions joined forces with NGOs in favour of its introduction. Greenpeace, for instance, co-operated with the big unions of public services (ÖTV). Last but not least, influential members of environmental NGOs and research institutes have become environmental professionals in parliamentary and governmental institutions. Besides the openness of the system and a well-developed capacity for policy co-operation, the diffusion of environmental experts into all parts of the ecological decision-making process might be considered as an indirect impact of NGOs, which should attract further attention.

References

Bichsel, A. (1996), 'NGOs as agents of public accountability and democratization in intergovernmental forums', in Lafferty, William M., Meadowcroft, James (eds), *Democracy and the Environment, Problems and Prospects*, Elgar, Cheltenham, pp. 234–255.

---

4 In 2000, 53% of Germans considered environmental protection to be "very important" (Rucht 2001: 147)

Brunnengräber, A., Walk, H. (1997), 'Die Erweiterung der Netzwerktheorien. Nicht-Regierungs-Organisationen verquickt mit Markt und Staat', in E. Altvater et al. (eds), *Vernetzt und verstrickt. Nicht-Regierungs-Organisationen als gesellschaftliche Produktivkraft*, Westfälisches Dampfboot, Münster, pp. 65–84.

Jänicke, M. (1996a), 'Democracy as a condition for environmental policy success: the importance of non-institutional factors', in W.M. Lafferty, J. Meadowcroft (eds), *Democracy and the Environment. Problems and Prospects*, Elgar, Cheltenham, pp. 71–85.

Jänicke, M. (1996b), 'Erfolgsbedingungen von Umweltpolitik', in M. Jänicke (ed.), *Umweltpolitik der Industrieländer*, ed sigma, Berlin, pp. 9–28.

Jänicke, M. and Weidner, H. (1995), *Successful Environmental Policy, A Critical Evaluation of 24 Cases*, ed sigma, Berlin.

Krebs, C. and Reiche, D.T. (1998), *Der mühsame Wege zu einer "Ökologischen Steuerreform", Ein Beitrag zur Systematisierung der Debatte*, Lang, Berlin.

Krebs, C. and Reiche, D.T. (1998a), 'Der Agenda-Setting Prozeß der Ökologischen Steuerreform', *Zeitschrift für Umweltpolitik und Umweltrecht*, vol. 2, pp. 177–212

Priewe, J. (1998), *Die Ökosteuer-Diskussion. Positionen und Kontroversen. Eine Bilanz*, ed sigma, Berlin.

Princen, T. and Finger, M. (1994), *Environmental NGOs in World Politics: Linking the Local and the Global*, Routledge, London/New York.

Raschke, J. (1993), *Die Grünen – Wie sie wurden, was sie sind*, Bund-Verlag, Köln.

Raschke, J. (2001), *Die Zukunft der Grünen. "So kann man nicht regieren"*, Campus, Frankfurt/New York.

Reiche, D.T. and Krebs, C. (1998), *Der Einstieg in die ökologische Steuerreform*, Lang, Berlin/Frankfurt,Main.

Rucht, D., Blattert, B. and Rink, D. (1997), *Soziale Bewegungen auf dem Weg zur Institutionalisierung*, Campus, Frankfurt/New York.

Rucht, D. (1996), 'Multinationale Bewegungsorganisationen. Bedeutung, Bedingungen, Perspektiven', *Forschungsjournal, Neue soziale Bewegungen*, vol. 9, no. 2., pp. 30–41.

Sachverständigenrat für Umweltfragen (1996), *Zur Umsetzung einer dauerhaften – umweltgerechten Entwicklung – Umweltgutachten 1996*, Deutscher Bundestag Drucksache 13/4108 (Langfassung).

7 Intervention of the Environmental Control System in the Economic System and the Environmental Cluster*

HARUTOSHI FUNABASHI

Introduction

I try to present a theoretical framework, which provides a comprehensive perspective concerning historical change of environmental problems and development of environmental policy. At first, I show the harmonious relationship between human society and nature in pre-industrial society, namely in the original stage. However this relationship was destroyed by industrialisation. In order to reconstruct this harmonious relationship in industrial society, we need to establish the environmental control system and its intervention in the economic system. I present four logical stages necessary for this intervention. In order to resolve contemporary environmental problems, we must push our society as a whole into the fourth stage of intervention where environmental concern is incorporated into the economic system as a primary administrative task. I will explain seven channels that promote this transition into the fourth stage. At last, I present the notion of environmental cluster and four factors that contribute to reinforce its social influence.

Harmonious Relationship between Human Society and Nature in Pre-industrial Society

In order to understand the actual situation of environmental problems, I begin to examine a harmonious relationship between human society and nature

---

* Translation by Graham Page and Dylan Lee

in pre-industrial society. I name it the original stage. Figure O represents the original stage.

Of course, there was not always a harmonious relationship between pre-industrial societies and nature. According to recent studies (Perlin 1989; Ponting 1991), many civilisations in pre-industrial societies declined and perished because they destroyed their environment. However, we know a lot of good examples of harmonious relationships in pre-industrial societies, and we can present the model of original stage by extracting their particular characteristics. What are characteristics of this original stage?

First, pre-industrial societies were based on hunting and gathering, or agricultural economies, which are both intimately related to the ecosystem. In so far as a pre-industrial society succeeded in maintaining ecological conditions for hunting, gathering and agriculture, it could be sustainable. Production and consumption based on the market mechanism were not prevalent. A local community mainly used only its own local resources.

Second, in this stage, the circulation of materials between human society and nature was linked with the circulation of materials within the ecosystem. All environmental elements were related in the chains of circulation. Human society was also embedded in these chains. In other words, production and consumption of a society were limited within the carrying capacity of its environment. The amount of waste did not exceed the purificative capacity of nature. People's use of natural resources remained within its regenerative capacity.

Third, effective social norms, ecologically sound production skills and adequate lifestyles are necessary conditions for the original stage. Effective social norms must control production, distribution and consumption in a local community. Production skill must utilise natural resources by maintaining material circulation in the ecosystem. And the lifestyle must be characterised by temperance and sensibility for the value of nature.

However, this harmonious relationship between society and nature changed when industrialisation began. Market mechanisms and industrial firms came to play an important role in the production of all kinds of goods. But, at the initial epoch of industrialisation, there was no effective regulation that could control firms for the protection of the environment. Consequently, firms disposed of various pollutants into the environment while simultaneously exhausting its resources. Such action causes serious damage to the environment and destroys the harmonious relationship between society and nature. Chains of circulation that link environmental elements are now broken everywhere. For example, in Japan, the mining industry caused a lot of environmental destruction due to rapid industrialisation in the Meiji era.

Factories in large cities also polluted air and water by various pollutants in the first half of the 20th century (Iijima 1979).

In order to analyse this change that caused many environmental problems, and the social process concerning their solution, we must introduce the notion of "the environmental control system". Using this concept and analysing different stages of its intervention in the economic system, we can sociologically approach historical development of environmental problems, environmental movements, and environmental policy.

Four Logical Stages of Intervention of the Environmental Control System into the Economic System in Industrial Society

The environmental control system consists of government offices charged with environmental policies and various environmental movements. The economic system consists of the market system and the governmental control system on economic activities. Main actors in the economic system are firms, consumers and government offices charged with economic policy. The fundamental idea of my perspective is that the macro trend of contemporary social change can be defined as the formation of a more powerful environmental control system and its deepening intervention in the economic system. The progress of environmental policy can be interpreted as a deepening process of this intervention. Figures A, B, C and D show the images of different degrees of this intervention.

In this context, what is the function of the environmental control system? We can interpret an environmental problem as a structural strain in social structure. People experience such structural strain as degradation of the environment and as damage to their quality of life. In order to resolve an environmental problem, it is necessary to change social structures. The basic function of the environmental control system is to convert structural strains, which appear as environmental problems, into the pressure required to force a solution and to create effective efforts that bring these solutions.

We can present four logical stages of intervention of the environmental control system into the economic system. The history of environmental problems and environmental policy can be divided into these four stages, namely, lack of constraints, imposition of constraints, incorporation of environmental concern as a secondary administrative task and finally, incorporation of environmental concern as a primary administrative task. Each transition to the next stage means the progress of environmental policies. In every stage, there are factors that push the transition to the next stage and other factors that resist this transition.

Let us examine characteristics of each stage briefly by comparing four figures. Symbols used in these figures can be classified in four groups. The first group of symbols in Figure A indicates actors and their areas of concern. The small white circle represents interests or administrative tasks for protecting the environment, whereas the small black circle indicates interests or administrative tasks concerning economic goals. The second group of symbols shows the characteristics of the interrelation of these actors. The third group of symbols represents different types of political arenas between the actors. The fourth group indicates material flow concerning the ecosystem.

Lack of Constraints on the Economic System

When the harmonious relationship between society and nature in the original stage began to be destroyed by industrialisation, the transition of society into stage A occurred. In this stage, the chains of material circulation within the eco-system as well as between society and nature were cut off everywhere. Economic activities caused unregulated disposal of pollutants and exploitative exhaustion of resources. This stage (stage A) is characterised by a lack of constraints. Only those who were directly affected took action, namely only victims' movement tried to intervene into economic activity. However, governmental regulations for protecting the environment is lacking completely. In other words, the environmental control system is in germ.

In Japan, this stage corresponds to the historical situation before the mid 1960s. Acute environmental destruction such as the Minamata disease, the Yokkaichi asthma, and the Morinaga milk poisoning illustrate this stage. Both national and local governments took no effective measures to stop pollution. Often these organisations even protected unscrupulous firms. In those days, mass media reports were very poor, and public opinion did not recognise the importance of environmental destruction. As Figure A indicates, the political arena is characterised by the phase of oppressive exclusion. Perpetrators refused to adequately compensate victims or even enter into a dialogue with them. A profound difference concerning the definition of the situation and the definition of legitimacy exists between two camps. Japanese society as a whole remained in this stage before the mid 1960s.

Two fundamental social mechanisms that degrade the environment in contemporary society have appeared in this stage. Namely, the mechanism of social dilemma and the external imposition of the environmental burden

(Funabashi 1992, 1998). In order to overcome these mechanisms, more powerful intervention in the economic system is necessary.

Imposition of Constraints on the Economic System

Figure B shows the full-scale appearance of the environmental control system and the beginning of its systematic intervention in the economic system. In this stage, the Environment Agency and environmental movements try to impose effective constraints on firms. Confrontation between the two camps becomes possible in the political arena. Based on the 14 laws enacted in December 1970, the Japanese Government established environmental quality standards concerning air pollution, water pollution, soil contamination, noise, etc. These environmental quality standards imply constraints that prohibit, to a certain degree, actors in the economic system from damaging the environment and paying damages to the victims.

The Polluter Pays Principle formulated by the OECD is a representative policy principle in stage B. In the 1950s and 1960s, many firms refused to take adequate measures to avoid pollution under the pretext of hard market competition and consequently destroyed the environment. However, introducing new constraints based on this principle can define the new condition of fair competition for all firms in the market.

What contributing factors promoted the transition from stage A to stage B? I would like to highlight the important role of various anti-pollution movements at a grass-root level, and the initiatives of local governments. In 1964, a residents' movement of Numazu, Mishima and Shimizu in Shizuoka prefecture succeeded in stopping construction of a big petro-chemical complex for the first time in order to avoid pollution. Victims of environmental destruction successively filed suits against perpetrative firms. These suits enabled groups of victims to get an effective arena of confrontation against perpetrators. Four grand suits concerning the Niigata Minamata disease, the Yokkaichi asthma, the Itai-itai disease and the Kumamoto Minamata disease that had been filed from 1967 to 1969 resulted in victory for victims from 1971 to 1973. Reformist governors and mayors, especially in large cities, frequently introduced stricter measures than the national government in the late 1960s and in the beginning of the 1970s. In addition, the mass media began to campaign actively against pollution. Thus, the prevention of pollution became a widely shared and legitimate value at the beginning of the 1970s.

Incorporation of the Environmental Concern as a Secondary Administrative Task

Stage B changes into the third stage (stage C) when actors in the economic system begin to accept the reduction of pollution and other kinds of environmental burden not only as a constraint imposed on them but also as a goal that they must achieve. Small white circles in actors of the economic system symbolise this change. Firms began to mobilise their resources to achieve this goal. Negotiation between two camps now becomes possible in the political arena.

One example of a stage C incident is the reduction of pollutants in automobile exhausts in the late 1970s. Under severe regulation by the government, Japanese automobile companies incorporated the reduction of the environmental burden as a secondary administrative task in their organisation. These companies did their best to develop new technologies for reducing pollutants in automobile exhausts. After several years' effort, they succeeded, one after another, in producing new low polluting cars that could eliminate approximately 90 % of pollutants such as CO, HC and NOx. This case shows that in stage C the business world and government offices in the economic system incorporated the environmental concern in the economic activity as an administrative task. Firms began to pay attention to environmental problems and they tried to improve their technology and production methods in order to avoid environmental destruction. They sharply increased the investment for pollution-preventing equipment in the 1970s. The sum of such investment was 20 billion yen in 1970, but increased to 70 billion by 1975.

However, in stage C administrative tasks for the protection of the environment rest as a secondary task. If there is a conflict between economic goals and environmental ones, the priority tended to be given to economic goals.

Incorporation of the Environmental Concern as a Primary Administrative Task

With the global environmental concerns that arose in the latter half of the 1980s, mankind entered into a new historical epoch that is defined by the universalisation of environmental problems. All processes of production and consumption came to be considered as responsible for increasing the environmental burden. Although the environmental burden produced by a person or an organisation may be slight, its accumulation will surely cause various kinds of environmental damage. In order to cope with this situation, the transition from stage C to stage D becomes necessary. Figure D shows stage D. Mankind became aware of the danger of the environmental catastrophe, which will inevitably occur in the future if we do not establish adequate environmental policies (Gruhl 1975). In order to avoid the catastrophe, most countries in the world came to support the idea of sustainable development. This idea requires reducing the environmental burden in all areas of social and economic activity. Then, the reduction of the environmental burden must be incorporated by all firms and households as a primary administrative task in the 21st century. Inspired by the notion of sustainability, the Basic Environment Law, enacted in 1993, and the Basic Environment Plan on national as well as local levels show comprehensively a new Japanese administrative framework. With more strict constraints imposed, firms tend to incorporate more deeply environmental goals.

As Figure D shows, the basic feature of this stage is an interpenetration of the environmental control system and the economic system. If environmental consideration is incorporated in a firm as a primary administrative task, prosperity of the firm and protection of the environment coincide with each other. Several innovative concepts appeared that showed this coincidence. The notion of eco-business shows that some environmental problems can be resolved, or at least ameliorated, within the framework of private enterprise and market mechanism. The construction of wind power plants illustrates an eco-business. The idea of zero-emission (Capra and Pauli 1995) shows another example of the practice in stage D. It implies an incorporation of the reduction of the environmental burden as a primary administrative task. The principle of Extended Producer Responsibility (EPR) formulated by the OECD in the 1990s (OECD Environment policy committee 1998) also shows another radical idea that is necessary to resolve today's environmental problems, especially the problem of waste disposal. And the environment management system is an effective tool to promote the deeper incorporation of environmental concern into various organisations.

Differently from stage C, firms in stage D situate environmental goals as a primary task in their organisation and give priority to them. The leaders of such firms can share environment oriented values with leaders of the environment movement. So, in this stage, dialogue and partnership among different groups become possible.

This stage shows a certain similarity with pre-industrial society. In this stage, as with pre-industrial society, sustainable circulation of materials and a harmonious relationship exist between society and nature. The economic system must restrain its use of resources and waste's quality as well as its amount within purificative and regenerative capacity of ecosystem. Differently from pre-industrial society, stage D must achieve this restriction under the condition of market mechanism, industrialisation and a certain quality of life.

In order to push our society as a whole into stage D, the role of technology is very important. However, it is noteworthy that, in order to reach stage D, we must abandon certain kinds of technology characterised by inevitable risk and extreme amounts of environmental burden. For example, we cannot continue to use nuclear power plants because they inevitably produce a lot of radioactive waste. This causes dilemmas concerning safekeeping. Also because the risk of grave accident exists. Instead, we must develop natural energy such as wind power plants and solar batteries that are suitable for stage D (Scheer 1999).

Contrastive Characteristics of Different Stages

Now I would like to present comprehensive remarks about the different stages.

First, social climate profoundly changes according to different stages of intervention. When a transition from one stage to the next occurs, we experience a different personal reality in Alvin Gouldner's sense (Gouldner 1970). In order to realise the new situation that a different stage presents us, we need new view-points, new key-words and a new frame of reference. Previous common sense may lose its validity, and we must revaluate things that were previously self-evident.

Second, related with the change of social climate, different theoretical viewpoints appear that correspond to different stages of intervention. It is noteworthy that adequate sociological theory inevitably varies according to different stages. For example, the theory of life-environmentalism that was formulated by a group of Japanese researchers (Torigoe 1989; Kada 1991) corresponds to the transition from the original stage in pre-industrial society to the next stage (stage A) caused by industrialisation. The theory of social structure of pollution victims created by Iijima (Iijima 1992) is especially sensible to the situation of stages A and B. The theory of environmental justice (Bullard 1994) created in the USA also tries to analyse problems in stage A and to seek the path to stage B. The theory of the benefits and victimised zones was invented by another group of Japanese sociologists (Funabashi et al. 1985) through their case studies on social conflicts situated in stages A, B and C. By contrast, the theory of social dilemma (Hardin 1968; Funabashi 1992) is useful to analyse the social mechanism of environmental destruction in the period of universalisation of environmental problems. It seems just as valid to environmental problems situated in stages C and D, and it is useful to clarify the necessary conditions for the transition to stage D. Similarly, in order to promote the transition into stage D, the notion of ecological modernisation is very suggestive (Spaagaren 2000; Mol 2000).

Third, according to different stages, different policy principles and environmental movements that are suitable to each stage become necessary. For example, the Polluter Pays Principle emerged in the first half of the 1970s in order to push the transition from stage A to B, and the principle of Extended Producer Responsibility became necessary in the 1990s to reach stage D. Similarly, adequate style and valid strategy of environmental movements change profoundly according to different stages. In stage A, some movements did not hesitate to take direct action such as sit-ins or making picket lines to prevent some destructive development projects. And filing suit is an effective measure to impose constraints on economic activity in stage B. By contrast, environmental movements oriented toward to dialogue and co-operation with business world can be effective in stages C and D.

These four stages are logical stages. And various parts of a society do not change simultaneously. So we find usually plural stages in a society of a historical period. But each historical period can be characterised by its dominant stage. In order to solve contemporary environmental problems, we must push our society as a whole into stage D.

The Seven Paths which Promote Intervention

How can we promote this transition? I find seven channels that can push this intervention more deeply. Figure D presents these seven paths that enable deeper intervention of the environmental control system in the economic system.

Pressure for Environmental Concern by Environmental Movements

The first channel is the pressure toward environmental concern produced by the environmental movement. Their perpetual pressure have forced the firms to adopt more effective measures for the protection of the environment. In fact, environmental movements have always been the most powerful sources of pressure toward a solution to various environmental problems. There are various kinds of environmental movements such as pollution victims' movement, residents' movement, ecologist movement, co-operative movement and green consumer movement. All movements are based on a particular social context and have a suitable strategy and style to their respective contexts.

Pressure for Environmental Concern by Governmental Offices

The second channel is the pressure produced by governmental offices. They can persuade firms and consumers to have deeper concern for the environment and to reduce the environmental burden. In Japan, the Basic Environment Law enacted in 1993 and the Basic Environment Plan formulated by the national government in 1994 are typical of this pressure. Similarly, local governments have successively introduced the basic environmental ordinance and the basic environmental plan on the local level.

Economic Incentives such as Environmental Taxes

When the governmental pressure escalates to the creation of economic incentives such as environmental taxes, the third channel of transition can appear. The third channel is defined as the imposition of economic incentives. Its representative method is environmental taxation. For example, the government can impose a higher tax burden on high CO_2 emission cars than on those with low emission. Economic incentives can change the meaning of rational action in the market for the firms as well as for the consumers. Once environmental taxes are introduced, production and consumption with more environmental concern becomes economically competitive. Under such conditions, economic interests motivate firms and consumers to choose more environmentally sound actions.

Coincidence between Environmental Concern and Economic Goals

The fourth channel is the coincidence between the environmental concern and economic goals in particular industrial domains. For example, wind power plants, solar power plants, organic agriculture and some recycling industries represent this coincidence. In these industrial domains, environmental concern can harmonise with economic success, and economic achievements can directly contribute to the reduction of the environmental burden. In such domains, the notion "eco-business" becomes a key word, and firms can overcome the contradiction between environmental concern and economic profits. Their economic prosperity can also imply environmental amelioration.

Incorporation of Environmental Concern Through Individuals with Value Rationality

The fifth channel is the incorporation of environmental concern into the economic system through the internalisation of environmental values in certain individuals. Value rationality as described by Max Weber can play an important role in this channel. Certain industrial leaders give priority to environmental concerns. They take initiatives in order to incorporate environmental values in the economic system in spite of constraints and economic handicaps, which are often raised at least in the short term. Another example of this path is the foundation of firms by activists within environmental movements. Some leaders of environmental movements, who were opposed to firms before, now approve of a positive role for firms to improve the environment. Sometimes, they even try to found an environment oriented firms by themselves. According to them, it is necessary to create new types of firms in order to incorporate environmental concerns in the economic system. In these cases, internalisation of environmental values in a key person can produce a firm, which tries to achieve environmental goals as a primary administrative task.

Pressure for Environmental Concern Produced by Other Firms

The sixth possibility is the pressure for environmental concern produced by commercial transactions with other firms. In this case, a firm that has already incorporated environmental concern as a primary administrative task tries to change the attitude of other firms by imposing environment oriented conditions in transactions. Some electronics companies and automobile companies in Japan, for example, demand that other firms, which transact with them to create an environment management system and obtain certification of ISO 14001. They publish their basic policy about environmental concern, and refuse to buy articles from firms that cannot satisfy these conditions. Those other firms are obliged to incorporate environmental concern themselves in order to avoid loss of market share.

Pressure Produced by the Green Consumer Movement

The green consumer movement represents the seventh path. Green consumers are defined as consumers who select environment oriented articles consciously. They are a new type of consumer with strong environmental concerns. In Figure D, a small white circle in the symbol for green consumers represents the internalisation of environmental concern. The accumulation of their choices can change the demand structure in the market. Firms must always adapt themselves to consumers' attitudes. Through daily purchases, consumers can influence the policy of enterprises. If consumers prefer the articles, which necessitate less environmental burden, firms are motivated to make and sell such articles. The green consumer guide can be an effective measure in this context. In Europe, the USA as well as in Japan, many groups have published the green consumer guide (Elkington and Hailes 1989). Firms neglect this situation at their peril.

As Figure D shows, through these seven paths, environmental concern can be incorporated into the economic system as a primary administrative task. In other words, the transition into stage D can be promoted through these seven channels. However, the transition into stage D is not easy, because resistance to this transition always stubbornly exists. In order to protect direct economic interests, firms and consumers tend to resist the deeper intervention of the environmental control system. How can actors of the environmental control system overcome these resistances and push the transition into stage D? The notion of the environmental cluster seems to be very important in this situation.

Environmental Cluster and its Empowerment

Environmental Cluster and its Elements

What conditions are crucial for the promotion of this transition from stage C to stage D through the seven channels? My basic hypothesis is that the environment oriented cluster of various actors will play a very important role. As the concept of "industrial cluster" demonstrates, the notion of cluster implies an accumulation of elements which are interrelated so intimately that they can produce emergent excellent quality. As Figure E indicates, the environmental cluster consists of the environment oriented public and six groups of organisations, that is, governmental offices, political parties, social movements, organisations for education and research, organisations of mass media and industrial organisations.

Governmental offices contain the Ministry of the Environment and environmental departments of municipalities. In Japan, the first creation of these organisations began in the 1960s, and they were totally established in the early 1970s.

Political parties can also play an important role in the environmental cluster. Unlike several European countries, Japan has no powerful Green party. Instead, groups of environment oriented assembly-persons with other political parties have contributed to improve environmental problems. Sometimes even, only one member of parliament can play a decisive role to resolve an environmental problem.

There are various kinds of social movements for the environment, such as, the anti-pollution movement, the victims' movement, the residents' movement, the ecologist movements, the green consumer movement, the co-operative movement and the environment oriented labour movement. These movements have different emphasis, but as a whole, they contribute enormously to improve environmental problems.

Recently, more and more organisations for education and research have appeared to be exhibiting strong environmental concern. All levels of educational establishments, such as the elementary school, junior high school, high school and university have prepared various teaching programs for environmental education. Their mission is very important for the diffusion of correct knowledge, and for teaching the value of a good environment. The number of research organisations with environmental concern such as think tank, consultant, research centres in universities and associations for environmental science, continues to grow. These organisations create knowledge and theories to understand environmental problems, and can contribute to the presentation of effective ideas for their solution.

Other important actors belong to the mass media sector. They include publishing companies, newspapers, journals and broadcasting stations of radio and television. They have a powerful influence on the social climate and social conscience. Every journalist and every editor can contribute to make the public opinion more environmentally concerned.

Producers with serious environmental concern in various industrial domains are now increasing. Farmers who produce organic agricultural products, refuse to use chemical pesticides. Some companies eagerly try to develop technologies for natural resource energy, such as solar energy and wind power. Other firms adopt the concept of zero-emission in order to reduce all kinds of wastes. Ultimately, it is their action that enables the incorporation of environmental concern into the economic system as a primary administrative task.

The public and these groups interact with each other. Figure E also indicates interaction between them. For the growth of these groups and organisations, intimate interrelations with the public are crucial. The public plays different roles according to the social context, as tax payers they interact with governmental offices, as voters with political parties, as participants and supporters with social movements, as clients with schools and research organisations, as an audience with mass media., and as workers and consumers with industrial companies. In every social context, the public can contribute to reinforce the environment oriented action of these organisations. For example, when they buy merchandise in the market, they can be green consumers, and when they vote, they can support politicians with environmental concern. Collaboration between these groups and organisations can make their activities more effective. These actors reinforce and support each other through networks between them.

Four Factors for the Empowerment of Environmental Clusters

What factors contribute to the further empower the environmental cluster? Four changes can promote the growth of environmentally oriented clusters. Figure F shows this possibility. Transition from the left to the right in Figure F indicates the effects of four factors.

The first condition is the increase of an environmentally concerned public. The environmentally oriented public is always the necessary and basic component of the environmental cluster. All groups and organisations in six domains of the environmental cluster need a positive relationship with the public. Without such relations, no groups nor organisations can obtain sufficient resources. They cannot take effective action without concern, support and co-operation from the public.

The second factor is the increase and enlargement of environment-oriented groups and organisations in every domain. Their number and scale directly influence their political and economic power. For example, groups of resident movements increased sharply in the latter half of the 1960s in Japan and this change was one necessary condition in order to push Japanese society from stage A to B. In order to cope with various environmental problems, sufficient research must be done in natural as well as social sciences. More powerful research activity depends upon the increase and enlargement of research groups.

The third condition is the enrichment of the interrelationship between all actors with environmental concern, namely all individuals, groups and organisations belonging to the environmental cluster. It is well known that co-operation between plural actors can create emergent effects. This is more effective than the simple sum of an individual's effect produced separately. When individuals establish an association and take action as its members, they can obtain more power than when they act separately. Similarly, an environmental oriented group can exert more influence by associating with other groups and individuals. For example, a group of environmental movements can obtain richer knowledge through co-operative relationship with organisations for research and acquire more powerful influence by making a link with some assembly-persons as well as other groups of the social movement.

The fourth element is the improvement of qualities and abilities of individuals belonging to the environmental cluster. The power of all groups and organisations is ultimately based on the qualities and abilities of their members as individuals. Consequently, it is necessary to continuously ameliorate various qualities and abilities of individual people. Perpetual efforts must be done to increase the level of knowledge, skills, sensibility and sense of justice etc.

I order to empower the environmental cluster, all its members should be aware of these four factors. When a group or an organisation with environmental concern decides its policy and strategy, it is desirable to choose adequate ways from this viewpoint.

Growth of the environmental cluster seems to be imperative to avoid environmental catastrophe in the future. Inspired by the notion of the public sphere presented by Jürgen Habermas (1990), this growth also can be interpreted as the formation of one type of public sphere. When interaction between actors belonging to environmental cluster becomes rich, and they succeed in organising a continuous debate in a public arena, it can be said that they construct an environmental public sphere. It will be an indispensable component of the public sphere in future society.

Conclusion

I have presented five logical stages of environmental problems, that is the original stage in pre-industrial society and the four stages of intervention of the environmental control system in the industrialised economic system. I have highlighted the importance of stage D, where the economic system incorporates environmental concern as a primary administrative task, and the seven paths that enable the transition from stage C to D as well as the notion of the environmental cluster, which promotes this transition. Sustainability will be possible only when intervention of the environmental control system reaches stage D through empowerment of environmental cluster. In order to sufficiently understand contemporary environmental problems and find adequate solutions, these theoretical perspectives would seem to be necessary.

References

Bullard, D.R. (1994), *Dumping in Dixie: Race, Class and Environmental Quality*, Westview Press, Boulder.
Capra, F. and G. Pauli (eds) (1995), *Steering Business Toward Sustainability,* United Nations University Press, Tokyo.
Elkington, J. and Hailes, J. (1989), *The Green Consumer Guide: From Shampoo to Champagne,* Victor Gollancz Ltd., London.
Funabshi, H. (1992), 'Social Mechanisms of Environmental Destruction: Social Dilemmas and the Separate-Dependent Ecosystem', in H. Krupp (ed.), *Energy Politics and Schumpeter Dynamics,* Springer-Verlag, Tokyo/Berlin, pp. 265–275.
Funabashi, H. and Hasegawa, K., Hatanaka, M. and Katsuta, H. (1985), *Shinkansen-kogai (Environmental Destruction by Bullet Express Train)*, Yuhikaku, Tokyo.
Funabashi, H. (1998), 'Kankyo-mondai no Mirai to Shakai-hendo (Future of Environmental Problems and Social Change) ', in H. Funabashi and N. Iijima (eds), *Environmental Sociology (Sociology in Japan 12)*, University of Tokyo Press, Tokyo.
Gouldner, A. (1970), *The Coming Crisis of Western Sociology*, Basic Books, New York.
Gruhl, H. (1975), *Ein Planet wird geplündert,* S. Fischer Verlag, Frankfurt.
Habermas, J.(1990), *Strukturwandel der Öffentlichkeit: Untersuchungen zu einer Kategorie der bürgerlichen Gesellschaft*, Suhrkamp Verlag, Frankfurt.
Iijima, N. (1979), *Pollution Japan: Historical Chronology*, Asahi Evening News, Tokyo.
Iijima, N. (1992), 'Social Structures of Pollution Victims', in Jun Ui (ed.), *Industrial Pollution in Japan,* United Nations University Press, Tokyo, pp. 154–172.
Kada, Y. (1991), 'Understanding the State of Lake Environment in the Context of Socio-Cultural Perspective', *Guidelines of Lake Management*, vol.2, pp. 7–18.
Mol, A.P.J. (2000), 'Globalization and Environment: between apocalypse-blindness and ecological modernization', in G. Spaargaren, A.P.J. Mol and F.H. Buttel (eds), *Environment and Global Modernity*, SAGE Publications, London.
OECD Environment policy committee (1998), *Extended and Shared Producer Responsibility Phase2*, OECD, Paris.
Perlin, J. (1988), *A Forest Journey: The Role of Wood in the Development of Civilization,* W.W. Norton & Company, New York/London.
Ponting, C. (1992), *A Green History of the World,* St. Martin's Press, New York.
Scheer, H. (1999), *Solare Weltwirtschaft*, Verlag Antje Kunstmann, München.
Spaargaren, G. (2000), 'Ecological Modernization Theory and the Changing Discourse on Environment and Modernity', in G. Spaargaren, A.P.J. Mol and F.H. Buttel (eds), *Environment and Global Modernity,* SAGE Publications, London.
Torigoe, H. (1989), 'Environmental Problems and Everyday Life', *Kwansei Gakuin University Annual Studies*, vol. 38.

Figure O Original stage of the harmonious relationship between human society and nature

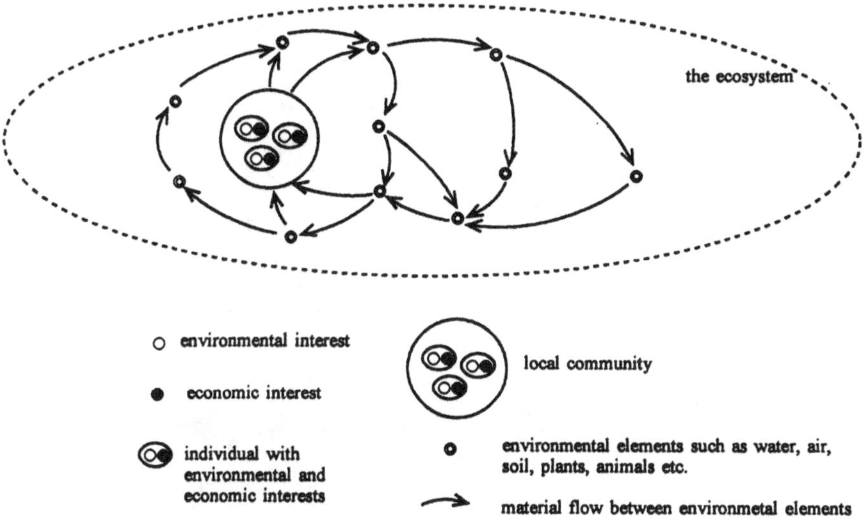

Figure A Lack of constraints on the economic system

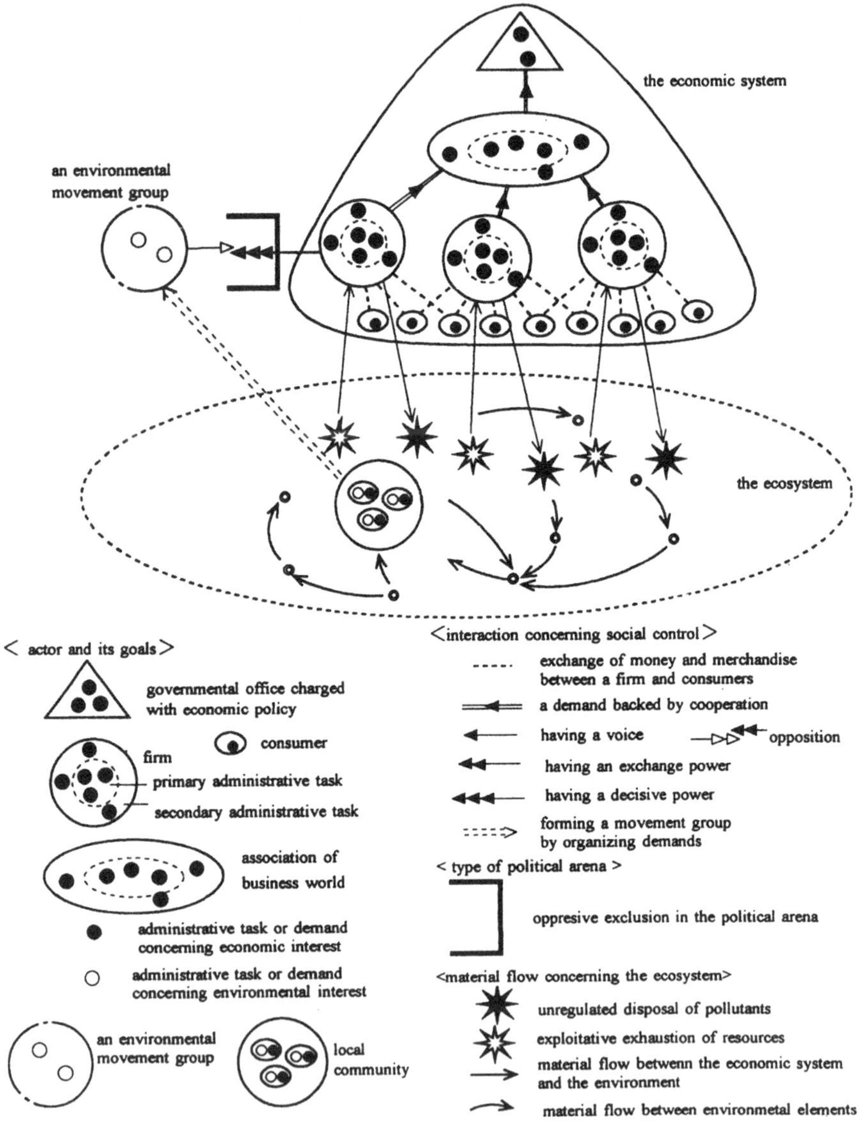

Figure B Imposition of constraints on the economic system

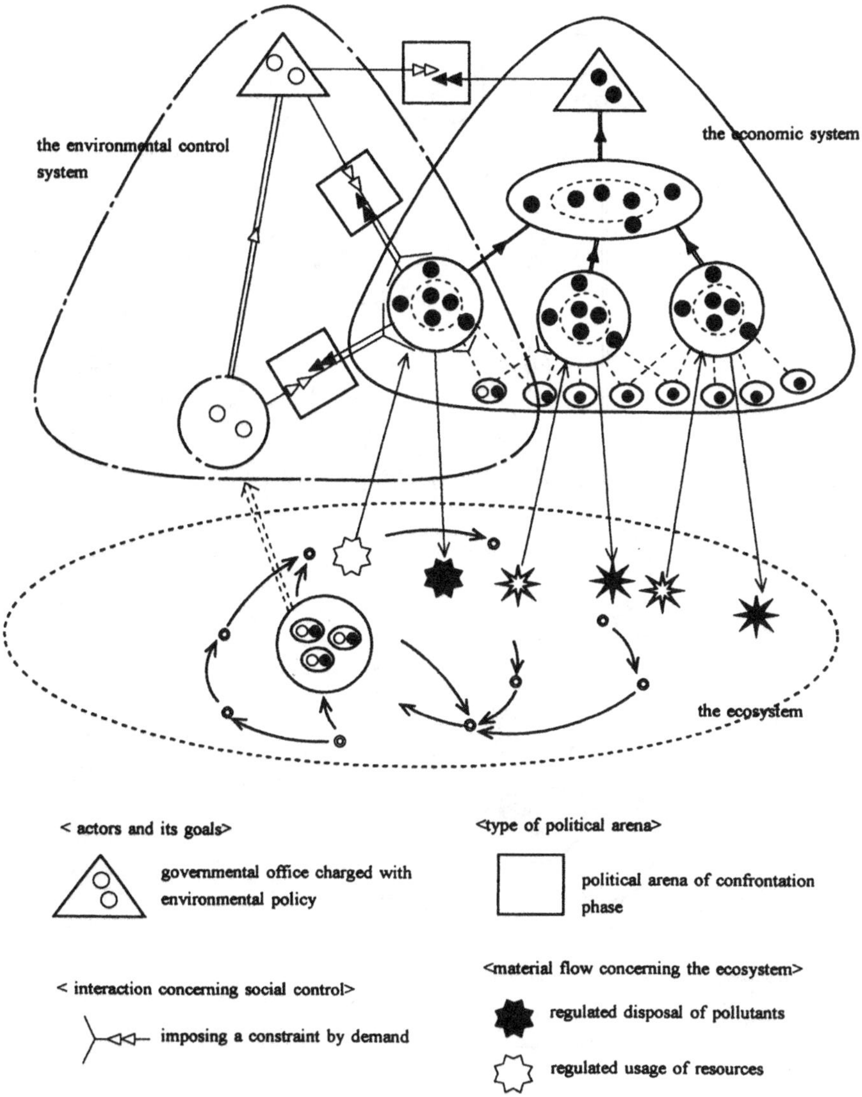

Figure C Incorporation of the environmental concern as a secondary administrative task

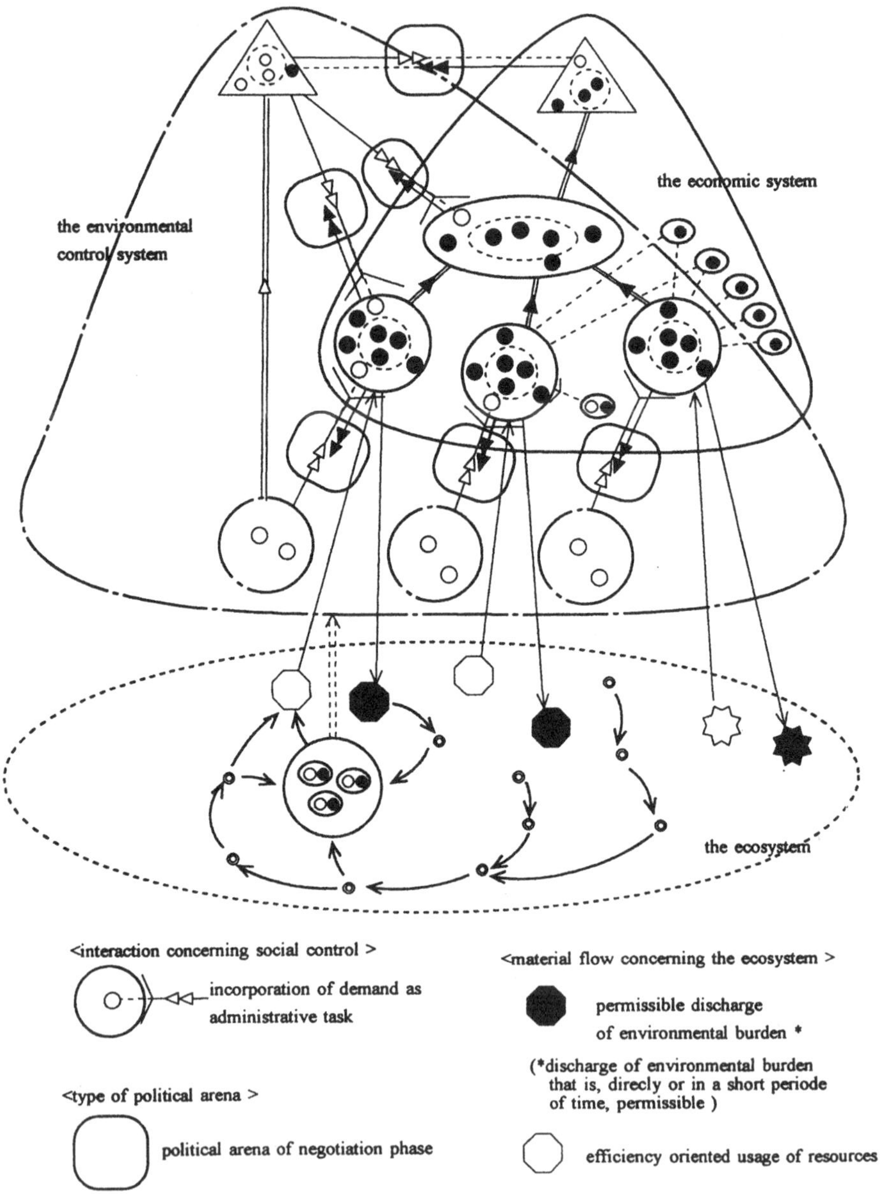

Figure D Incorporation of the environmental concern as a primary administrative task

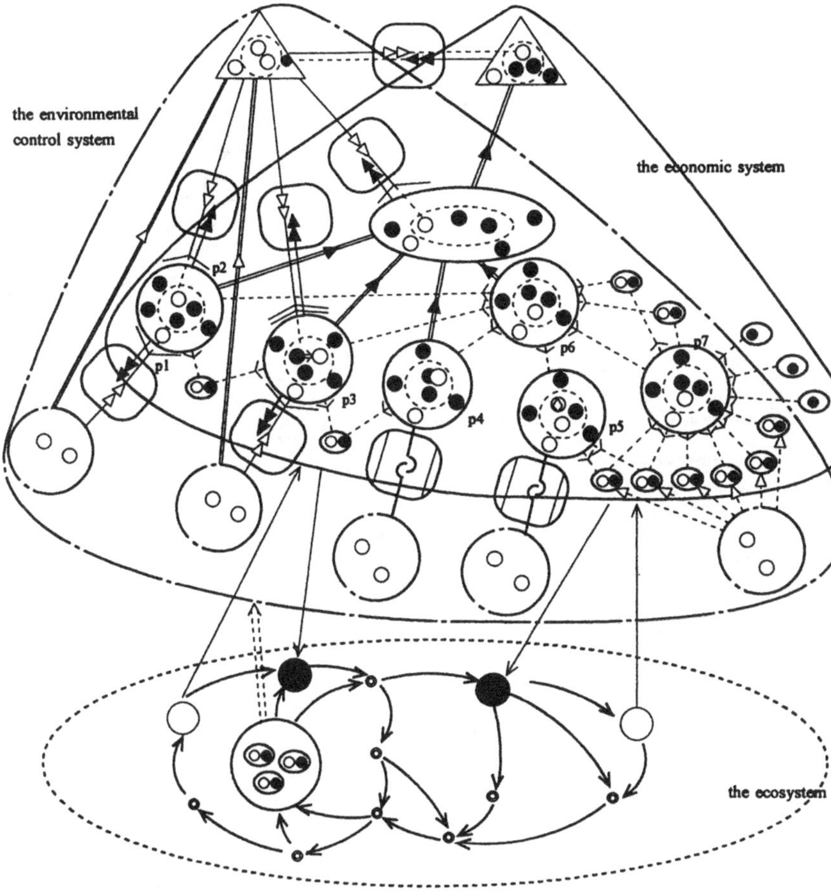

<interaction concerning social control >

p1:pressure for environmental concern by environmental movements
p2:pressure for environmental concern by governmental offices
p3:economic incentives such as environmental taxes ( )
p4:coincidence between environmentalconcern and economic goals ( )
p5:incorporation of environmental concern through internalization of environmental values in individuals ( )
p6:pressure for environmental concern produced by other firms
p7:pressure produced by the green consumer movement

— imposition of constraints by economic incentive

<type of political arena >

exchanges of opinions in consensus phase of political arena

<material flow concerning the ecosystem >

environmentally harmonious discharge of environmental burden

environmentally harmonious usage of resources

Figure E Environmental cluster

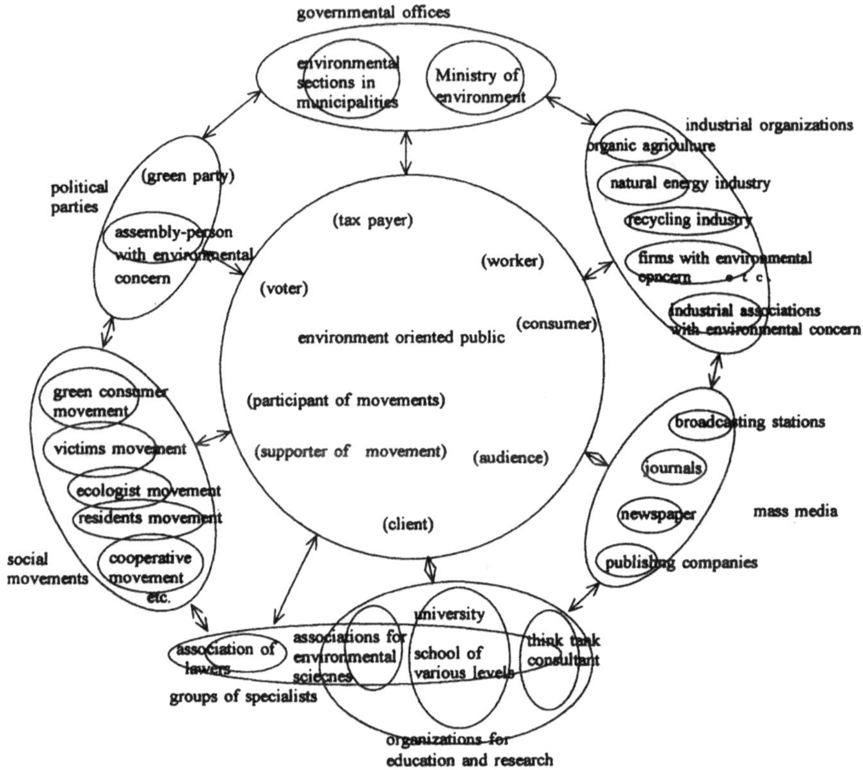

Figure F Growth of the environmental cluster

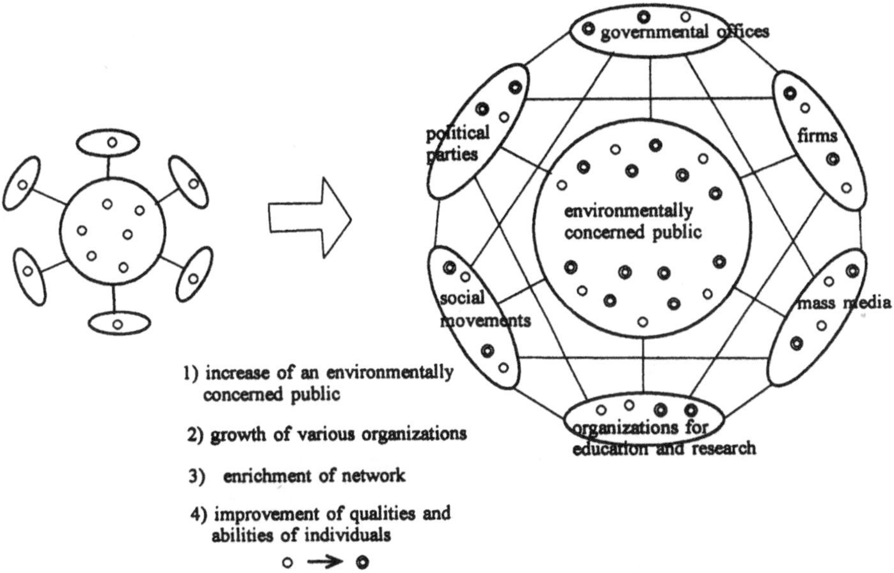

8 Transnational Environmental Policies: Selected Cases of Japan and Germany

RÜDIGER KÜHR

Introduction – Transnationalisation and Nation States

Transnational environmental problems are a relatively new area of public discussion and politics. Just a few decades ago it was unimaginable that local problems can cause global effects not stopping at national borderlines. But growing numbers of analyses verify the existence of a network of complex ecological interdependences hardly to grasp. Simultaneously the knowledge and understanding about an environmental sound society has grown and models have been developed describing different paths toward this aim (Robèrt 2002). But the traditional problem solving attempts of states have proved weak in the face of the threats of an industrial society as environmental problems are increasingly seen to be threatening the present and future of current and coming generations around the globe. There is an obvious lack of governmental striking force and political self-assertion. This phenomenon is mainly a specific problem of national transnational environmental policies under the paradigm of sustainability and matters more than in other policy fields.

In essence one can fix nine problem areas of concern in transnational environmental relations, all of them with relatively clearly identifiable transnational and global effects: (i) climate change,[1] (ii) the loss of stratospheric

---

1 In June 1999 two islands of the Pacific nation of Kiribati have been swallowed by the ocean. Tebua Tarawa and Abanuea were both uninhabited. The former was used by fishermen; the latter is known as "the beach which is long lasting". Not in the age of global warming!

The South Pacific Regional Environment Program says other islands of Kiribati and Tuvalu are in danger. Coasts are eroding in the Maldives, and in the Marshall Islands where, according to the BBC, it would cost $100 million to build a seawall for just one ozone, (iii) the loss of biodiversity, (iv) the loss of forests, (v) the degradation of international waters, (vi) the pollution of air,[2] (vii) the exploitation and export of limited natural resources, (viii) the increase and export of harmful wastes, and (ix) the increasing desertification.[3] There is unanimous agreement that especially the problem of climate change is already with global effect. At the sight of increasing consequences, whose extent and effects will be demonstrated with increasing and improved analytics in the future, for most analysts it appears appropriate to describe all these trends with "transnationalisation[4] of environmental problems" of two categories: a) transnational environmental problems of global character and b) transnational environmental problems of local and/or regional character.

A growing understanding of the origins, facets, developments and problem solving approaches regarding environmental problems does not only demonstrate the necessity for transnational action, but also provides the perspective for an increased efficiency through mutual sustained measures. Although transnationalisation brings certainly also along many transnational and global problems, it offers also new and long-time unexpected problem solving at-

---

atoll. The BBC quotes Maldives President Maumoon Abdul Gayoom: "Sea-level rise is not a fashionable scientific hypothesis. It is a fact."

2 Air pollution must be considered as a clear transnational environment problem. For example the burning of lignite in the East of China and the North of the Czech Republic cause acid rain at Japan's West Coast or in the Thuringian (Thüringer) Forest in the South-East of Germany. But as the World Bank has developed six problem areas of concern in global environmental relations, which are supplemented through "waste" by Simonis (1996) an analysis of the transnational environmental problems must also include the general topic "air pollution", as appropriate cleansing technology through e.g. filter systems are not yet exhaustively installed. Finally also the export of limited natural resources has to be added to this list. The waste of these resources often leads to enormous environmental impacts in the exporting countries, resulting *also* from the demand of other countries.

3 The World Bank does only identify six problem areas (World Bank 1997: 5f). Despite the implementation of several cleaner technologies waste avoidance has not yet become a real social duty neither in most developed nor in developing countries (Simonis 1996: 20).

4 It does not appear appropriate to describe these developments as "globalisation of environmental problems". Following the German Advisory Council on Global Change (WBGU) global environmental problems could only be distinguished from those of local or regional character through the necessity of international measures against them (1993). But not all environmental problems are of global dimension or even proofing a globalising tendency. Moreover bilateral agreements between two states or even multilateral settlements of more than two states are not necessarily of global dimension, although they are called international. This means, that international environmental agreements are not necessarily only dealing with global environmental changes. Global environmental problems are a sub-category of transnational environmental problems.

tempts through e.g. the simple just-in-time information exchange via internet and policies of rapprochement. These allow new perspectives for sustainable solutions and also a well-founded advice of experts thousands of kilometres away from the local/regional problem.

For some transnational environmental problems are the empirical evidence of state decline as it demonstrates the breakdown in the distinction between foreign and domestic affairs in a modernised and interdependent international system (Morse 1970: 172). Arguments for the diminishing of state's loss of effective control and autonomy, if it ever existed, and sovereignty are exchanged,[5] but focus predominately on the declining of national-international distinctions through globalisation (e.g. Albrow 1998, Jones 1995, Hirst and Thompson 1996, Robertson 1992, Rosenau 1996, Waters 1995). Increased co-operation efforts among states at the international level, as well as the rise of non-state actors in international environmental and development politics among other indicators, have been put forward by analysts for the erosion of state's sovereignty (e.g. Edelmann 1976: 189, Lindblom 1980: 124, Mayntz 1987: 106, Richter 1992; Altvater 1998; Messner 1998). In their eyes the states are no longer perceived to be the only legitimate actor in these fields (eroded by environmental developments) or may even be enhanced through collective actions.

But as long as modern industrial nations are organised as democracies with a parliamentary system, the state and its machinery should play the key-role to solve life-threatening problems for society, even if under the complex conditions of modern industrial societies modified functions are coming up. On the other hand, there are rather radical discourses which move beyond the international to global politics, but assume that transnationalisation diminishes the state element of "governance". But obviously they are basing their assessment on the classic definition of the modern state, which appears inappropriate under the changing framework. While this literature sees that governance now involves more than the nation-state, it implies also that this should lead to replace a state-perspective with a perspective of governance. Max Weber specified that

> "A compulsory political organisation with continuous operations will be called a 'state' insofar as its administrative staff successfully upholds the claim to the monopoly of the legitimate use of physical force in the enforcement of its order." (Mann 1986: 55)

---

5 These are the two central features of the Westphalian "Temple" (Zacher 1992: 60f)

Following Weber, Anthony Giddens (1985) defines the modern nation-state as "bordered power container". But many states of today are small, weak, with problematic national coherence, and above all only limited autonomy in any sense. Latest with the demise of the imperial European state of the last centuries, the nation-state has become a more or less universal political form, spreading first to the rest of Europe, then what became known as the "Third World", and finally to the remains of the Soviet Union. But with becoming more universal, nation-states have also been shorn of the key characteristics of autonomous state power. Consequently most nation-states are no longer autonomous classic nation-states as characterised in Weber's and Gidden's definition. And with the end of the Cold War especially the western states, including Japan, are becoming transnational centres, and have been able to utilise legitimate transnational institutions, notably the United Nations, to underwrite its own transnational projection of power. Certainly, western leaders including the Japanese and Germans have developed a rhetoric of global responsibility, but there is still a great reluctance to commit real resources or thought to the development of global institutions or to form a global social change, which might offer greater stability.

The responsibilities, claims and realities of Japan's and Germany's transnational environmental policy are comparatively examined in this chapter. The responsibility deduces out of the economic super-power of both nations with huge financial resources and well-developed technologies, what is correlated with the concentration of pollution and big pressure on the environment and thus also the available natural resources. The claim to being an environmental flagship refers to the increasing environmental policy engagements of both nations in the early 1990s, what encourages prominent Japanese and Germans to express their will for transnational environmental leadership during international events. But is this claim of two nations for leadership in the transnational environmental area justified taking their recent policies into consideration? Or do they declare misjudge the realities, in which the main driving forces towards progress are with transnational organisations, Non Governmental Organisations or even companies, so that have already taken over as "head goose" of flying geese?

Indicators for Responsibility

Germany's (82 million) and Japan's (126.4 million) national share of the world population is relatively small. The 208.4 million inhabitants of both countries made up only 3.3 per cent of almost 6 billion mortals in 1998. Also the landmass of both nations – Germany: 357 tkm^2 + Japan 378 tkm^2 = 735 tkm^2 – is only 0.54 per cent of the world's 135,641 tkm^2. But despite their small stake in the world's population and area, Germany's US$ 1,873 billion and Japan's US$ 4,245 billion make almost one fifth of the world's US$ 31,283 billion Gross Domestic Product (GDP), the sum of the economic production within a country (World Bank 2002). These figures include expenditures for curative measures against environmental pollution and damages through natural catastrophes, which are finally also identified as economic yields.

Table 1: Gross Domestic Product (2002)

Country	GDP in 2002 (in billion US$)	% of total world-wide GDP	Real GDP* in 2002
Japan	4,245	13.61	-0.7
USA	10,171	32.61	2.3
Germany	1,873	6.00	0.4
Total OECD	n.a.	n.a.	1.5
WORLD	31,283	100.00	n.a.

*** percentage change from previous period**

Sources: World Bank 2002 and OECD 2002

Being international industrial and economic world powers, Japan's and Germany's firms have found their way into most countries around the world. Accordingly, both nations exert also a sizeable influence on the state of the local, regional and transnational environment. In 2001 the world's merchandised exports reached US$ 6,155 billion, out of which Germany's share is 9.3% (US$ 570.7 billion) and Japan's 6.6% (US$ 403.5 billion) (WTO 2003). Furthermore environmental statistics demonstrate the enormous impact of both states as resource poor importers of raw materials and oil, coal and lignite, metals, timber, ore, and especially in the case of Japan foodstuffs etc. in order to produce goods for the export and the domestic market.

Imports of Forest and Agriculture Products

Although rich of forest resources, table 2 shows that Japan (US$ 17.184 billion) is in total numbers the second largest importer of forest industry products behind the USA (US$ 24.120 billion) but before Germany (US$ 12.127 billion) as third in this ranking (OECD 1999: 130). Timber is largely used as a raw material in forest industries that process sawn-wood and sleepers, wood-based panels, wood pulp, and paper and paperboard, playing an important role in both Japan and Germany. But in sharp contrast to Germany and the USA, which have been campaigning against imports of tropical timber, Japan is in total amounts number one importer of cork and wood from tropical countries, but on per capita basis with US$ 9.1/cap. far behind Portugal (US$ 19.9/cap.), Belgium (US$ 16.5/cap.) and Korea (US$ 11.1/cap.). The harvesting of wood from tropical forests, and the export of a part of this harvest is a significant cause of tropical deforestation. Tropical wood harvesting has also an impact on the world's genetic resources and the increasing CO_2-concentration in the atmosphere. Japan's total amount of imports of cork and wood from tropical countries has, however, decreased from US$ 2.697 billion in 1995 to US$ 1.145 billion in 1998 (OECD 1999: 134), which is equal to a 58 per cent decrease within three years, but nevertheless remains at the summit of list. One important reason for this reduction, as Japanese officials explain, is the fact that one-way chopsticks are now predominately produced with domestic timber.

Table 2: Timber imports (1998)

Country	Import of industrial timber products (in billion US$)	Import of tropical timber (in billion US$)	total per capita (US$/cap.)
Japan	17.184	1.145	9.1
USA	24.120	0.426	1.6
Germany	12.127	0.157	1.9
Total OECD	113.001	5.586	5.1
WORLD	145.477	n.a.	n.a.

Sources: OECD 1999

To satisfy the needs of agricultural products that Japan and Germany consume throughout one year, both nations have rather high import quota. Already in 1992 the Japan National Institute for Environmental Studies (NIES) undertook a test calculation to demonstrate Japan's reliance upon the soil of other countries. This study came to result that about 10.3 million hectares of land was taken up all over the world for the purpose of producing wheat, barley, sorghum, maize, soybeans, coffee beans, cotton and natural rubber for Japan. This was an area equivalent to 2.3 times the total cultivated land area in Japan and about 27 per cent of the nation's total land area (Environmental Agency 1995: 155). Especially declining land utilisation rates and abandoned cultivation land is resulting in inadequate soil management in Japan. Consequently Japan imported in 1999 11.4% of the total amount of world-wide imports (US$ 308 billion) of agriculture products (FAO 2000) – all EU countries 17.8%.

Although it is an agricultural products exporting country, Germany imported 3.4 times more of those products than it exported in 2001 (Federal Statistical Office of Germany 2003). About 36 per cent of Germany's total area is arable acreage, 16 per cent pasture. But in contrast to Japan Germany is in the position to produce a large part of its foodstuff essentials – up to 80 per cent.[6] Furthermore Germany does not have to import neither meat nor dairy products, but through transnational trade under international trade agreements and thus especially through intra-EU trades, those products are increasingly getting onto the German market.

This became apparent by October 1996, when BSE[7] had been first reported from 10 countries and areas outside the United Kingdom, including Germany, and the reappearance of the Foot-and-Mouth-Disease (FMD) in February 2001 also in the UK. The first BSE case in Japan, with 28% the world-wide largest importer of meat, was discovered in Chiba Prefecture in September 2001 with another 5 cases confirmed since then.

---

6 See for example: http://www.straubing.baynet.de/~k.czauderna/BRD.htm

7 Bovine Spongiform Encephalopathy (BSE) first came to the attention of the scientific community in November 1986 with the appearance in cattle of a newly-recognised form of neurological disease in the United Kingdom (UK). Between November 1986 and 31 May 1996 approximately 160,000 cases of this newly-recognised cattle disease were confirmed in the UK. Epidemiological studies conducted in the United Kingdom suggested that the source of the disease was cattle feed prepared from carcasses of dead ruminants, and that changes in the process of preparing cattle feed introduced in 1981–1982 may have been a contributing risk factor. A newly recognised variant of the Creutzfeldt-Jakob Disease affecting human brains is expected to be linked to BSE. (WHO 1996)

As can be seen, excess dependence for both exporting and importing countries is causing problems with regional effects, but crossing national borders and thus must also be considered as transnational environmental problems.

Imports of Resources

The environmental impact of industrial activities may vary considerably with structural changes, also Japan and Germany experienced in recent years. The decline of the so-called traditional industries, such as iron and steal, in most OECD countries has typically reduced quantitative pressures on the environment in this environmentally significant industrial sector. But this turndown was accompanied with an increase of the total production in the electrical machinery industry (9%), petroleum refineries (4%) and motor vehicles (3%). And surprisingly Germany's iron and steal industry kept its production level from 1995 also in 1998 due to governmentally backed up subsidiary support. But considerable increase can be recognised in the chemical product industry (20%), electrical machinery (12%) and motor vehicle industry (27%) (OECD 1999: 255f). The growth in these so-called new industries can create not only qualitative pressures, such as soil and groundwater contamination by trace toxic chemicals, but also requires immense imports of necessary resources.

Modern information and communication technology (ICT) plays a crucial role in today's economies and societies not only of Japan and Germany. World-wide sales in the IT sector in 2002 totalled US$ 1,104 billion, well on the way to crossing the US$ 1,400 billion line in 2003 (EITO 2002: 391). Of the total, 41.7 per cent was generated in the US, 13.1 per cent in Japan, and 6.7 per cent in Germany and thus contribute a lot to the real economic growth in the both nations under comparison here. But on a per-unit basis, a study by Atlantic Consulting (1998) for the European Union estimated that the final production of a personal computer (including material production, manufacture, and distribution) would lead to the release of 0.19 metric tons of greenhouse gases, 36 kg of overall waste, and would require 3.6 GJ of energy. Consequently, ICT use also has implications for resource inputs other than energy. Those of the most significance from an environmental perspective are the metal and materials sectors. While there is a continued migration towards plastic, rather than metal parts, the critical functions of this type of equipment are achieved by advances in uses of metals and materials. Metals still used in ICTs are aluminium, lead, cadmium, and nickel.

Table 3: Imports of selected raw materials (1997/8)

	Japan		Germany	
	Quantity (in 1,000t)	**Value (in mill. US$)***	**Quantity (in 1,000t)**	**Value (in mill. US$)****
Coal	131,764	8,010	24,565$^{(1)}$	84.7$^{(1)}$
Salt	7,914	290	1,705	3.4
Phosphate Rock	976	110	223	11.3
Oil$^{(a)}$	280,300	--	137,400	--
Iron Ore	120,782	3,980	47,597	1,231
Natural Rubber	669	740	345	239

* calculation on the basis of JP¥ 100 = US$ 1
** calculation on the basis of DM 2 = US$ 1
(a) Net oil imports in 1997, source: OECD 1999
(1) lignite + hard coal

Composed out of data of OECD (1999); Japan Statistical Yearbook (2000) and Federal Statistical Office Germany (2000)

The Case of Aluminium

The imports of aluminium and aluminium scraps amounted in 2000 2.473 million tons in Japan (Japan Statistical Yearbook 2003) and 1.090 million tones in Germany (Drasdo 2001: 10), most of those from countries with a relatively lax environmental legislation, such as Russia or Indonesia, but also Australia. Aluminium ore, most commonly bauxite, occurs mainly in tropical and sub-tropical areas. Bauxite is refined into aluminium oxide trihydrate (alumina) and then electrolytically reduced into metallic aluminium. Bauxite is generally extracted by open cast mining from strata, typically some 4–6 metres thick under a shallow covering of topsoil and vegetation. In most cases the topsoil is removed and stored having a enormous impact on the landscape.

The production of primary aluminium uses a great deal of electricity. The environmental profile for this part of the aluminium's life cycle is therefore highly dependent on how the electricity used is generated – and Germany's aluminium production on the basis of lignite and coal results already to 15 kg CO_2 and 45 kwh primary energy per kilogram of aluminium. In environmental terms, the most important emissions from the aluminium production process are then the emission of CO_2 and CF gases, which produce the greenhouse effect, and SO_2, which results in acidification, as well as the emission of fluorides and polyaromatic hydrocarbons (PAH compounds), which are toxic to humans and the environment. And also in response to these environmental impacts Japan's former Ministry of Trade and Industry (MITI) has long-time proposed to build aluminium smelter overseas. And German governmental officials are not getting tired to emphasise the importance of coal and lignite – making up 25% of Germany's energy production – in order to secure Germany's energy demands.

Gas Emissions

Since the beginning of the 1980s a large percentage of Japanese industry has been manufacturing and storing its products in other Asian countries. This has been aimed at benefiting from the strength of the yen, the low storage costs and, importantly, the less rigid environmental controls in other countries. And with the completely enacting of Germany's law on pollutants[8] in 1986, which e.g. resulted in investments of EUR 10 billion of Germany's energy producers in order to install effective filter-systems in their power-plants using fossil resources, also many German companies shifted parts of their production to other countries. And Germany's metal-producers are until today only using high concentrated pre-processed ores.

Further examples for Germany's and Japan's impact on the transnational environment and thus also for their transnational environmental responsibility result from statistics on emissions of gases.

Table 4 illustrates the man-made emissions of Nitrogen Oxides (NO_x) and Sulphur Oxides (SO_x). This table refers to the major categories of emission sources of these pollutants, which include motor vehicles, power stations, fuel combustion, pollutants emitted in manufacturing, waste incineration, and agricultural burning, etc. Nevertheless Japan holds only the second lowest place in a direct comparison of per capita emissions of NO_x from OECD countries. Japan is also located behind Norway and the Netherlands in regard to its SO_x emissions. And Germany is also more than 50 per cent under the OECD average. But one should note, that these emissions of rather traditional pollutants result not only from strict domestic environmental leg-

8 In German: Bundes-Immissionsschutzgesetz (BImSchG)

islation, but also from the shifting of certain high polluting industries to outside the country.

Table 4: Emissions of Greenhouse Gases, NO_X and SO_X (in late 1990s)

In 1,000 tonnes

	Greenhouse Gases						
Country	CH_4	N_2O	CO_2			NO_X (in 1995)	SO_X (in 1995)
			From energy use	From industrial processes	Total		
Japan	1,389	66	1,172,641	59,501	1,232,141	1,476	876
USA	31,357	1,290	5,470,494	65,155	5,535,648	19,758	16,619
Germany	3,564	216	883,992	25,000	908,992	2,210	2,995
Total OECD	71,000	3,000	12,235,100	409,500	12,644,600	--	--
WORLD	--	--	22,561,600	828,400	23,390,000	--	--

Composed out of data of OECD (1999) and OECD (1997)

Behind the USA and the Russian Federation Japan and Germany are taking up place three and four in the hierarchy of top-greenhouse gases emitants (OECD 1999: 53). Man-made greenhouse gases to the atmosphere disturb the balance of the earth's radiative energy budget. This leads to an increase in the earth's surface temperature and to related effects on e.g. the climate, a rising of the sea-level, desertification etc. CO_2 is the gas that contributes the largest share, most resulting from energy use. Japan's CO_2 emissions from industrial processes are negligible smaller than that of the USA, but Germany's are less than half of it.

The above given examples are some of many indicators proofing Japan's and Germany's impact on the transnational environment as economic and industrial world powers. This results not only from industrial production processes, but also from consumption manners in both nations, which are unsustainable in many ways and thus also having an impact through e.g. the exploitation of natural resources, the production of metals, the extensive burning of fossil fuels for energy production and certainly also the shortening life-spans of many products. And all this has influences extensively not only the global environment, but also local/regional environments far away from Japan and Germany.

These above given examples for Japan's and Germany's responsibility in the transnational environmental sphere demonstrate that some interstate interactions may involve governments, but it may not involve only governments. Industrial actors, financial actors, international actors and non-governmental actors can also play a significant role in here. Consequently interstate interactions, however, involving all other actors groups – individuals or organisations – are considered as "transnational" (Menderhausen 1969). And based on the growing importance of non-governmental actors in the field of international environmental policies and development policies,[9] and especially also in the case of environmental technology transfer through private public partnerships (PPP), it appears justified to speak also from an ongoing "transnationalisation". Transnationalisation is characterised through the growing role individuals and organisations play in world politics vis-à-vis foreign government or foreign societies and thus bypass their own governments (Singer 1969: 24). It became clear, that the globalisation and transnationalisation processes perforate border lines drawn by nation states, which finally leads to a polycentric distribution of power addressed by George Rosenau in 1990. This means a turning away from the sovereignty of Hobbes's Leviathan and Herz's hard shell, which understand the statesperson as the only source of governance represented through a government (Hobbes 1996: 144; Herz 1976: 100). Additionally it also reflects one key error in these days' globalisation debates: the identification of the modern state with the nation state. But are their any indicators in Japan's and Germany's transnational environmental policies that this transnationalisation necessarily leads to a constant loss of control and governance ability of states, making their claim for "being a flagship" ridiculous?

Transnational Environmental Policies

Today multilateral and bilateral environmental treaties are dealing with nearly every kind of transnational environmental problems. Not just only the existence of such contracts makes clear the changed consciousness of justice and norms, but also the perception of environmental problems and political strategies for their solution. Remarkable is the predominance of marine, lakes

9 See e.g. Weiß, Jens (2000): Umweltpolitik als Akteurshandeln, Marburg.

and river related agreements. In the truest sense of the word border lines are fluid in water and thus permeable for every kind of pollution. Moreover seas, lakes and rivers always represent a sensitive, indispensable, but limited source for food of a big part of the world population. Thus especially the fishing in international waters required transnational conventions finally leading to common efforts for the protection of the water and its flora and fauna.

The geographical position of Japan and Germany is not without consequences for its transnational environmental policies. In Germany regional transnational conflicts with neighbouring countries are regularly on the agenda. Examples are the contamination of the Rhine at its upper run so that this influences the drinking water processing in both Germany and the Netherlands. Another example are effects of the necessary ground-water lowering for the brown coal open mining Garzweiler II on the conservation areas close to the borderline of the Netherlands, or the acid rain – released by brown coal firing in the north of the former Czechoslovakia, – which lead in the 80's to a substantial forest dying in the German Erzgebirge. Of special socially and political explosiveness are transports for the reprocessing of spent fuel rods of nuclear power plants to e.g. the French reprocessing plant La Hague.[10]

Although Japan has a large-spatial geographical position, regionally limited transnational and similarly bilateral environmental problems are not out of question, but still relatively rare. Primarily they concern the marine pollution in the Sea of Japan and thus neighbouring Russia, China and the Korea Peninsula. In recent years the acid rain problem has been widely observed in Japan and led to a substantial forest dying at Japan's west coast. Since this region sparsely populated with few smaller industrial centres the measured acid deposits, which are concentrated along the west coast, must be resulting from the burning of fossil fuels in neighbouring countries adjacent to the Sea of Japan.

10 These transports were suspended in May 1998 due to yearlong conciliation from limit value exceeding through radioactive surface contamination of transport containers and other means of transport. After removal of these obvious deficits the German Federal Office for Radiation Protection approved on 22.09.2000 for the first time again eight transports. The disposal of radioactive wastes from the operation of nuclear power plants shall be restricted to direct final storage from 1 July 2005.

Multilateral Approaches

The "Agreement for the Protection of Birds useful for Agriculture"[11] from March 1902 can be considered as the first transnational environmental agreement between sixteen European states including Germany. The purpose of this agreement was to extensively support the well-being of birds through e.g. the dam-up of the mass murder of migrating birds, the creation of nesting places and the feeding in winter. Impetus came among others from the German Society for Bird Protection (Bund für Vogelschutz [BfV]) with Lina Hähnle in the chair. Ms. Hähnle, wife of the entrepreneur and liberal member of the German Reichstag, Hans Hähnle, aimed at a broad natural conservation under the flag of bird protection (NABU 1999: 2).[12]

Germany was also a party of the "Police Regulation of Fishing in the North Sea out of the Costal Waters" (1882) and the "Contract concerning the Regulation of Salomon Fishing in the Rhine River" (1886). Although these agreements deal with limited natural resources, the economic interests clearly prevail ecological consideration here. The securing of certain national fishing quota triggered these transnational contracts, instead of joint efforts against over-fishing and thus a massive decline of the fish stocks, which is still a major challenge in transnational environmental co-operation these days, as the fish stocks are dramatically shrinking and the share of young fishes in the nets, which were unable to reproduce, is extensively increasing.

In November 1933 it came to the first classical environmental convention – the "Convention Relative to the Preservation of Fauna & Flora into their Natural State". This convention was primarily decided to introduce a special regime for the preservation of flora and fauna, particularly of Africa, through e.g. the constitution of national parks. But neither Germany, whose colonies were distributed among the victorious nations of the First World War, nor Japan, which was never in command of any territories in Africa, were among the signing governments. And also the "Convention on Nature Protection and Wild Life Preservation in the Western Hemisphere" (1940) is outside Japan's and Germany's interests, as it is an agreement of American republics.

---

11 Translated by the author. The German title of this agreement is "Übereinkunft vom 19.3.1902 zum Schutz der für die Landwirtschaft nützlichen Vögel".

12 The foundation of the German Society for Bird Protection (Bund für Vogelschutz [BfV]) was basically a reaction to the industrialisation with its enormous impacts on the environment and the social structure. Consequently the BfV was a Wilhelmian reform movement, which sets a backward oriented worldview, idealising the agrarian society against the modernity.

To maintain and increase international co-operation in controlling pests and diseases of plants and plant products, and in preventing their introduction and spread across national boundaries Japan and Germany ratified the "International Plant Protection Convention" (1952). Among others Japan and Germany became also parties of the "Convention on the High Seas" (1958), "The Antarctic Treaty" (1959), "Convention on Wetlands of International Importance Especially as Waterfowl Habitat (Ramsar Convention)" (1971), and the "Convention for the Protection of the World Cultural and Natural Heritage" (1972).

But only with the "Convention on Long-Range Transboundary Air Pollution" in 1979 it came to the first multilateral environmental treaty, which devotes transnational environmental problems of air-pollution in Europe. Especially the growing acid deposition in Europe and North America with far-reaching impacts on drinking water, fishes and other aquatic life forms, with threatening impact on forests, and which accelerate and aggravate the decline and decay of monuments and buildings, led to this treaty. Thus, this convention expressed a common commitment on the part of the signing parties to reduce acid precursor emissions and jointly monitor emissions and precipitation chemistry. But it is amazing that with a history of more than 30 years of experience in developing regimes against transnational environmental problems and with cases in industrial countries of partly catastrophic air-pollution not necessarily stopping at national border-lines,[13] clearly demonstrating a connection between extremely high levels of pollution and day-to-day excessive mortality (EPA 1996), it came only to the first multilateral environmental agreement on air-pollution at the end of the 70ties.[14] Until then the fundamental conviction of the majority of decision makers from politics and industry was in line with that of the people, that national initiatives are only necessary to prevent extreme situations through air pollution. But smoking chimneys were still regarded in both Germany

---

13 In winter 1952 about four thousand persons died because of smog resulting from the burning of low-cost but high of sulphur coal in nearly each household of Great Britain's capital London.

Additionally to the water-pollution through e.g. ammoniac, chlorine, saltpetre, which was since the beginning of the century recognised as a major environmental load and resulted in many diseases including a typhus-epidemic in 1901 in Gelsenkrichen. In winter 1962 the air pollution is worsening the situation in Germany's Ruhr Valley with a sight of less than ten metres during some nights resulting in an increasing mortality.

14 In 1285, London's air was so polluted by coal burning, that Edward I established the world's first air pollution commission. And 22 years later the same king made it illegal to burn lignite, so that one Londoner was even executed. But the ban as such was not successful.

and Japan as symbols of growth and prosperity, so that drastic countermeasures put these at risk. Furthermore national initiatives were long-time regarded as sufficient and easier to channel and control. But this assumption turned out to be a fallacy in the mid 70s, when the dying of trees and respectively also the reasons for this moved into the focus, what had one of its climax with a story of three parts in the German weekly magazine *Der Spiegel* (11/1981) under the title: "Acid Rain: there is something in the air."[15] Conspicuously the acid rain through air pollution in some affected regions must have originated from across the border lines, because of the predominant wind direction. And latest since the mid 90s acid rain, resulting from e.g. high sulphur lignite burning in China, shows its effects along Japan's west coast.

Further international agreements signed by Japan and Germany are e.g. the "International Tropical Timber Agreement" (1983), the "Protocol on Substances that Deplete the Ozone Layer" (1987), and the "United Nations Framework Convention on Climate Change" (1992), and the various agreements forged as part of the United Nations Conference on Environment and Development (UNCED) in 1992.

The Case of the "Whaling Convention"

With the establishment of numerous UN organisations after the Second World War and a changing assignment of mandates many agreements and conventions were made dealing with the conservation and the protection of the environment.[16] The "International Convention for the Regulation of Whaling" (1946) aiming to protect all species of whales from over-fishing and safeguard for future generations and to establish a proper system of conservation and development of whale stocks is the first multilateral environmental agreement, which entered into force in Japan in 1951. Germany together with many more basically non-whaling nations agreed with this convention only 31 years later in 1982, when the International Whaling Commission (IWC) without the agreement of its Scientific Committee agreed on a moratorium for the commercial fishing of all kinds of whales from 1985.[17, 18] Japan, Norway, Iceland and few more nations have rose objections

15 German original title: "Säureregen: Da liegt was in der Luft"

16 For a detailed chronological list on conventions please check http://fletcher.tufts.edu/multi/chrono.html

17 The Scientific Committee asserted the moratorium as unnecessary (Nagasaki 1995).

18 The moratorium was decided on condition that latest in 1990 a comprehensive check should analyse the effects. On the basis of this a modification of the moratorium or new fishing quota should be discussed. But the IWC decided against these discussions and

against this moratorium. But the U.S. placed especially pressure on Japan using the Packwood-Magnuson Amendment. This Amendment to the US Fishery Conservation and Management Act of 1976, allows the United States to reduce or suspend fishing privileges in US waters for nations acting contrary to IWC guidelines. And as Japan was concerned about its own fishing industry and its enormous trade surplus toward the US at that time (JWA 1988), Japan withdrew the objections from the IWC Convention and terminated commercial whaling. In spite of the US's promise to refrain from imposing sanctions on Japan, the US executed the Packwood-Magnuson Amendment on Japan in 1988. This political decision reflects a cultural bias that whales are not resources to be exploited against. But although Japan officially accepted the ban of commercial whaling, it is accused to continuously violate the momentum.

The IWC Convention contains a provision allowing member nations to self-issue permits to kill whales for scientific research. When the commercial whaling moratorium took effect, Japan, Iceland, and in 1993 Norway self-issued research permits. Thus Japan conducts a research program on mink whales in the Antarctic, killing about 300 whales annually. In addition, Japan has initiated a program in the western north Pacific, annually killing 100 mink whales. But critics say Japan's research program is merely an approach for supplying expensive restaurants with whale meat, which is a delicacy in Japan (Greenpeace 2002). And in the US and Europe a moral and philosophical outlook developed, condemning the consumption of whale meat. Popular movies like "Flipper" and "Free Willy" also contributed to the presentation of whales as the most specialised mammals, being sentient, being intelligent, having their own community and that they can suffer (Skare 1994).

The Japanese always have taken whales for food due to the Buddhist belief, people must not eat meat of the "four-legged animals" until the middle of the 19th century. Additionally the Japanese culture developed through the decades in close relation with the sea and its resources. However, since whales were regarded as fish by the Japanese, they took also whales and ate whale meat. Consequently whaling is considered as an integral part of Japan's history and culture. Unlike western countries where whales have been caught and killed for products like oil, Japanese whaling has been predominately for human consumption. Consequently the Japanese have different views about whale conservation, viewing exploitation as little different from

---

confronted whaling nations with a new ban of commercial whaling. Thus the Norwegian government, active member of the IWC since 1946, resumed whaling in 1993 again.

cattle consumed by US citizens or Europeans. Thus Japanese officials are increasingly campaigning to resume whaling, especially referring to Japan's and Norway's estimates of mink whale population abundance (760,000 in the Antarctic and 118,000 in the Northeast Atlantic, respectively). This indicated in the eyes of the officials that these populations could sustain a limited commercial harvest (Buck 1997). But the Japanese estimates are in question. Singly the IWC assumes the mink whale populations fifty per cent smaller. Simultaneously NGOs including Greenpeace accuse Japanese officials to bribe especially developing countries through promises of Official Development Assistance (ODA) and thus to support Japanese efforts to abolish the moratorium (Süddeutsche Zeitung 2002; Tageszeitung 2002; n-tv 2002: 21.05.2002).

Despite all ideological discussions, the threat of some species of whales is without quest and accepted by all whaling and anti-whaling nations. It is also without quest that the limited harvests of Japan and Norway these days are not seriously endangering the whale populations, but the growing pollution of the oceans. But the slow recovery of the whale population under harder environmental conditions would come to an abrupt end with an official legalisation of commercial harvests. A great demand would also lead to a growing shooting of poachers ignoring all quotas and are hard to catch in violating international agreements.

But additionally to the transnational ecological and economic dimension of whaling, these problems illustrate also the growing pressure from actors outside the national territory influencing or at least trying to influence national policies into certain directions. This way of transnationalisation results in a countermovement, highlighting the endangerment of cultural and national characteristics, becoming a plaything in a game dominated by a hegemon and thus consequently to apparently lose the ability to act for and efficiently control the well-being of a nation. Obviously Japanese politicians are confronted with immense pressure from outside their country. But on the other hand this pressure has led to domestic and international discussions and certain sensitivity as far as the loss of national and cultural characteristics, so that it is not necessarily impossible, that the Japanese policy is re-gaining power in other political fields and spheres again.

The Case of the "Kyoto Protocol"

As an approach to stabilise atmospheric concentrations of greenhouse gases at safe levels the United Nations Framework Convention on Climate Change (UNFCC) came into force in March 1994. Parties of the UNFCC meet regularly at the annual conference of the Parties (COPs) to review the implementation of the convention and continue talks on how to best tackle climate change. Decisions made during the COPs make up a comprehensive "rule book" for the implementation of the convention. And key-decisions are sometimes given a more high-profile title – such as the "Berlin Mandate" and the "Kyoto Protocol". [19]

The so called "Kyoto Protocol" was completed in the famous Japanese town of Kyoto in 1997 and signed by 160 states. It aims to limit or reduce the emissions of the six greenhouse gases carbon dioxide (CO_2), nitrous oxide (N_2O), methane (CH_4), hydrofluorocarbons (HFCs), perflurocarbons (PFCs), and sulphur hexafluoride (SF_6) to individual, legally-binding targets adding up to a total cut of at least 5 per cent from 1990 levels in the commitment period 2008-2012.[20] Some specified activities in the land-use, land-use change and forestry sector, namely afforestation, deforestation and reforestation, which emit or remove and thus absorb CO_2 from the atmosphere, are also covered. Additionally the protocol establishes three innovative mechanisms, known as (i) joint implementation, (ii) emissions trading and the (iii) clean development mechanism (CDM), which are designed to help some parties to reduce the costs of meeting their emission targets by achieving or acquiring reductions more cheaply in other countries than at home. The clean

19 Parties agreed that specific commitments in the convention for annex I parties were not adequate. They therefore launched a new round of talks to decide stronger and more detailed commitments for these countries. And after two and a half years of intense negotiations, the Kyoto Protocol was adopted at COP 3 in December 1997.

20

Country	Target
EU-15, Bulgaria, Czech Republic, Estonia, Latvia, Liechtenstein, Lithuania, Monaco, Romania, Slovakia, Slovenia, Switzerland	-8 %
USA	-7 %
Canada, Hungary, Japan, Poland	-6 %
Croatia	-5 %
New Zealand, Russian Federation, Ukraine	0
Norway	+1 %
Australia	+8 %
Iceland	+10 %

Source: Kyoto Protocol of the UNFCC, Annex B.

development mechanism also aims to assist developing countries in achieving sustainable development by promoting environmentally-sound investment in their economies from industrialised country governments and businesses.

Major decisions on finalising the operational rules of the Kyoto Protocol were expected at the sixth COP in The Hague. It was aimed to come to final decisions on how much of a nation's commitment could be met through emissions trading systems, consequences for non-compliances, and to which extent one could take credit for carbon absorbed in the sinks. But finally the COP-6 negotiations collapsed, when the United States and the European Union reportedly led by Denmark and Germany failed to reach agreement on these key-issues, with particular controversy centring on how much credit nations could be allowed for carbon uptake – sequestration – by forests. In late March 2001 the Administration under US President George W. Bush considered the Kyoto protocol "dead" in terms of US policy. This initiated a high-level effort by European nations with Germany at lead and also by Japan playing a major role to re-engage the USA in the Kyoto process, which is emitting a quarter of all CO_2 emissions originating world-wide, or 40 percent of all industrialised countries.

Although Japan and the EU principally agreed about the importance of the Kyoto Protocol, Japan took an active part in softening the EU's stance on certain points, also in order to maximise the chances of bringing the US back to the Kyoto fold. But these efforts were without success and in mid-June Bush outlined his preferred approach and confirmed the rejections of the US. Nevertheless to the surprise of many, negotiators at the resumed "COP-6 bis" meeting in Bonn in July 2001 and during COP-7 in Marrakech in November 2001 reached agreement on most of the outstanding political issues and thus provided a sufficient basis for the countries to ratify.

Both, the Japanese and the German government represent their individual interests with stress and considerably influenced the negotiations and the final decisions on key-aspects of the Kyoto Protocol. Although both nations played a key-role in trying to re-engage the USA in the Kyoto Protocol negotiations in order to ensure the effectiveness of measures against climate change, and although Japan and Germany agreed on not intending to delay the progress of the negotiations, Japan took an active part in softening the EU's stance on certain points. And this request was obviously not only guided by the attempt to maximise the chances of bringing the USA back to the Kyoto fold, but also to lighten the load through commitments out of the Kyoto Protocol for Japan. Therefore Japan supported the United States with the position that there should be no limits to the amount of emission reductions that are allowed toward a country's obligations through emissions trading or joint implementation. Japan's Minister of Environment justified this standpoint with the argument that her country's CO_2 emission per Gross-Domestic Product (GDP) is already among the lowest of the industrialised world (19.07.2001). Respectively Japan's total consumption of energy per GDP is less than 70 per cent of the USA and less than 80 per cent that of Germany (OECD 1999: 221). This demonstrates in the eyes of the government that Japan has achieved already the highest level of energy efficiency. But this cannot obscure the fact that Japan's efforts to reduce greenhouse gas emissions have not gone well. Japan's greenhouse gas emissions fell in 1997 and 1998 due to the economic crisis. And according to ministry officials, they exploded in 1999, with private households increasing emissions by 5.3 percent, from industry by 4.2 percent, from office buildings by 3.3 percent and by 23 percent from the transportation sector (Asahi News Service, 11 July 2001).

As shown above CO_2 emissions from energy use are by far the biggest source of greenhouse gases. Attempts to domestically intensify the measures, although it is officially intended by the government, will be assessed by the rating of Japan's initiatives in international comparison, the expected benefit and certainly the costs.

At the face of this position of the Japanese government it can be expected that Japan will extensively make use of special decisions made under the Kyoto Protocol such as the mechanisms of CDM and emissions trading, but against which the EU members and others offered long-time resistance in order to force nations to take more extensive domestic action to reduce emissions. There are three different indications for this: First, the Japanese Minister of Environment did never become tired of emphasising the so called "Kyoto Initiative" as the COPs approached.[21] Under this initiative Japan has committed itself to providing on average annually about US$ 2.4 billion in highly concessional loans for projects related to climate change in developing countries, and furthermore financial and technical assistance amounting to US$ 7.4 billion since 1998 (Kawaguchi 2002[ab]). Thus Japan assigns its Official Development Assistance (ODA) through the provision of loans, the transfer of environmental technology and through environmental capability building a major role in responding to the Kyoto process.

Second, throughout the negotiations Japan and Russia, assisted by Canada and Australia insisted on lax criteria. And Russia, by far the largest credit

---

21 For a more detailed description on the background and philosophy behind the Kyoto Initiative please refer to MOFA (2000).

supplier on the emissions market, is essential for filling Japan's gap of the Kyoto target through buying credits.

Third, Japan has the obligation to reduce greenhouse gas emissions of 6 per cent below 1990 levels for the first commitment period of 2008–2012. But independent observers consider this as a rather low reduction target, which results furthermore mainly because Japan's Ministry of Internal Trade and Industry (MITI) insisted actively that it is impossible to reduce carbon dioxide emissions below this, even if realistic policies and measures are implemented (Ayukawa and Mizutani 2002). And within the Japanese Government's New Climate Policy the first period heavily depends on industry's current Voluntary Action Plan, and excluding possibilities of any mandatory CO_2 emission reduction measure to be introduced before 2004. But as Japan's CO_2 equivalent emissions have already risen by 6.8 per cent by the end of 1999, this measure does not appear effective, so that this policy initiative is primarily a compliance with the industry. Additionally the new climate policy plans to expand nuclear power generation by 30 per cent by 2010, what is associated with unresolved serious environmental problems and hampers the development of renewable alternatives.

In contrast to Japan Germany's role regarding the detailed operational rules for the Kyoto Protocol is less controversial. This results from Germany's stubborn position against the soaking of especially compliance measures and sinks rule, which gave rise to criticism from e.g. the USA and Japan referring to the German unification and the restructuring of the British coal-industry as reasons for the success in reducing of greenhouse gas emissions in the European Union. But a recent study of several German and British institutes for the German Federal Environment Agency (UBA) comes to the conclusion, that the greenhouse gas reductions in Germany and the United Kingdom are at 50 per cent a coincidence and 50 per cent policy induced. But at the same time Germany and the United Kingdom are said to be on the right path to meet the Kyoto target as a part of Europe's climate change programme advocating a twin-track approach. This includes a multi-stakeholder consultative process for policy-formulation and an internal EU greenhouse gas emission trading. Furthermore Germany has found its way in actively supporting renewable energy alternatives and additionally its way out of nuclear power generation within the next decades.

Certainly, every country acts to protect their national interests and has to respond to specific national background conditions – so does Japan, so does Germany and so are other EU members also no exception. But there is one fundamental reservation in comparing Japan and Germany: While the EU states have been pursuing for their economic interests they are trimming their economy more sustainable in order to converge their economic security with the environmental one. Japan so far keeps obviously economic fundamentals set to counter the environment. This became clear through Japan's present policy initiatives and statements and role during the Kyoto process, what simultaneously convey, that it will extensively make use of the flexible mechanisms through the transfer of huge amounts of money for loans and the transfer of environmental technology to developing countries, what has also the positive effect of securing its position on a growing market. Impetus on domestic efforts for reducing greenhouse gas emissions is missing as it might put a risk at the recovery of Japan's economy.

The above given disputes about the "Whaling Convention" and the "Kyoto Protocol" show three different aspects relevant for his study:

1. The tendency through the growing transnationalisation to prescribe certain sights, resulting in major conflicts. These conflicts occur out of the lacking consideration and scrutinising of the different interests, so that they end up in an opposition against a majority or a supposed hegemon.
2. The massive possibilities to influence the decision of governments from abroad, especially when they are not taking a clear position against, because
3. their decision to ratify and respond to a certain Multilateral Environmental Agreement (MEA) seems to be prevailed in many cases by e.g. economic and trade interests rather than eco-political motives.

Bilateral Approaches

In 1959 it came to the first bilateral environmental treaty of the Federal Republic of Germany (FRG) with a neighbouring country in the broader sense. In the "Agreement to fight Animal Epidemics in the border area" Germany and the Netherlands agreed on joint efforts (Bundesanzeiger (Banz) 244/59) to jointly fight disease, spreadening through the extensive transnational trade of cattle. And in 1961 the FRG and France bilaterally agreed on the "Protocol concerning the Constitution of an International Commission for the Protection of the Saar-River against Pollution" (Bundestag-Drucksachen 12/5446, 6617), to secure the water supply in the necessary quality for the people and industry along the course of the Saar River. In the following years until today Germany has signed numerous bilateral environmental agreements, many of them very concrete in nature, other rather generally targeting co-operation in the environmental protection (Auswärtiges Amt 1998). Additionally Germany also agreed, that environmental problems are more effective to solve them transnationally within the European Community

than by single initiatives of nation states. The Single European Act[22] (1987) formed a basis for environmental policies followed by the Maastricht Treaty of the European Union[23] (1992) and the Amsterdam Treaty (1997), naming sustainability a key-target of the European Union.
Just as the FRG also the German Democratic Republic (GDR) signed bilateral environmental agreements with its neighbours, predominately to the East. The treaty on the "Cooperation in Plant Protection" from 1950 and the protocol on the "Protection of the fishes in the Oder Bay" from 1954 between the GDR and the People's Republic of Poland can be considered as first such bilateral agreements (Bundesgesetzblatt (BGB) II 1993: 1080; Bundesarchiv DK 5 VA 3741).[24] Consequently they were signed several years before FRG's first initiatives in the transnational environmental policy sector. These treaties of the GDR are followed by the agreement on the "Peaceful Usage of Nuclear Energy" (BGB II 1991: 1077) with Czechoslovakia, and the "Basis-Relations Treaty" (Grundlagenvertrag) between the FRG and GDR in 1972, specifying environmental protection in article 7 (BGB II 1973: 421ff).[25] Finally, in the context of the unification of the FRG and the GDR and thus the unification contract (BGB I 1990: 885), all environmental agreements of the GDR expired, whereas those of FRG remained effective or were modified through necessary negotiations.[26]

The rapid expansion and large catches of the Japanese high-seas driftnet fisheries for salmon in the early 1950s helped lead to a treaty between Canada, Japan, and the United States establishing a provisional eastern boundary line in 1954 for the Japanese high seas salmon fisheries in the North Pacific Ocean at 175°W longitude and required the scientific determination of which line would best divide Asian and North American salmon in any areas of intermingling. But Japan's first bilateral environmental agreement came only

---

22 http://europa.eu.int/abc/obj/treaties/en/entr14a.htm#C_Single_European_Act

23 http://europa.eu.int/en/record/mt/top.html

24 Schreiber (1986) is describing the bilateral transnational environmental problems between the People's Republic of Poland and the German Democratic Republic in more depth.

25 The documentation of bilateral treaties, especially in the starting time of the GDR, is rather insufficient due to lax publication practice. Although this study incorporated the available documents of Germany's Federal Archive in Berlin, it cannot be ruled out, that further treaties were existing, but are not detectable.

26 This is the result of comprehensive studies of the relevant original documents of the GDR put into the archives of the German Federal Archive in Berlin in comparison to legislative changes published in the Bundesgesetzblatt. As a follow-up to this result the German Federal Ministry Environment conducted a detailed analysis of the bilateral environmental treaties of the GDR and confirmed it later-on. But the results have not yet been published.

into force in 1956. This was concluded with the Soviet Union and aimed at regulating Japanese salmon driftnet vessels fishing on Soviet stocks. Further agreements on fisheries came into force in 1965 with the Republic of Korea, with the People's Republic of China in 1975 and New Zealand in 1978. Furthermore bilateral environmental agreements with neighbouring countries are the Agreement between Japan and the USA on co-operation in the field of environmental protection (1975), the China-Japan Agreement for the Protection of Migratory Birds (1981), the Japan-USSR Agreement on Fishery Cooperation (1985), the Agreement between the Government of Japan and the Government of the Republic of Korea on Cooperation in the Field of Environmental Protection (1993), Japan-China Agreement on Cooperation in the Field of Environmental Protection (1994), and the Japan-German Agreement on the Cooperation in the Field of Environmental Protection (1997).[27] Consequently Japan concluded only few bilateral environmental treaties in the narrower sense, most of them dealing with the environmental medium sea and its flora. But obviously they aim especially at channelling the commercial fishery, so that the national interest in certain harvest quota is secured. But bilateral approaches to long-term protection of the flora and fauna of the seas are obviously of less importance or do not have the necessary lobby in Japan. Long-time bilateral environmental agreements have not been on the political agenda, simply because environmental problems were regarded predominately as home-made and thus domestically to solve. But e.g. extreme weather conditions, a growing forest-dying at Japan's west coast through acid rain, shrinking fish-stocks through over-fishing and the occurrence of BSE through international trade demonstrates that also island states are increasingly affected by transnational environmental problems calling for appropriate political consequences.

Conclusions

Japan's and Germany's impacts on the transnational environment are by large resulting from its production and consumption processes, which are unsustainable in many ways and driven through motives dominating over environmental considerations. Since the UNCED in Rio in 1992 global awareness on cleaner production and sustainable consumption has been created, many demonstration projects have been conducted, National Cleaner

---

27 with Korea: http://www.mofa.go.jp/policy/economy/apec/1995/member/info/9.html
with China: http://www.mofa.go.jp/policy/economy/apec/1995/member/info/5.html
with USSR: http://www.earthtrust.org/dnpaper/history.html

Production Centres have been established and an International Declaration on Cleaner Production launched. But neither prominent Japanese politicians nor prominent German politicians are yet passionate enough for cleaner production and sustainable consumption. Yet, it seems that politicians, as well as the industry and the mass media, have forgotten about the importance and threat of transnational environmental issues with the collapse of Japan's bubble economy, the present German economic crisis and at the face of military conflicts in the Middle East and on the Korean Peninsula. A boost, however, of the economy through an increased consumption has top priority, whereas environmental considerations in this regard are of less importance.

In Japan political decisions are made mainly by countless, and almost nameless, bureaucrats who are not willing to make their work more transparent. Japanese bureaucracy, with its origins in serving the Emperor, does not always serve the public effectively. Thus the present Japanese system does not contribute to governance with long-term concepts, as it is required to solve environmental problems in a sustainable way. Moreover, solution strategies that are easy to grasp and control are often favoured. This explains Japan's interest in technical solutions, while often forgetting about the social dimension of sustainability that would require a much more holistic approach. Likewise, an emphasis on technical solutions presents the possibility to secure shares in the growing market of environmental technologies. Furthermore, the interests and motivations involved in environmental policy are so disparate that consensus would often be at the cost of political leadership and with mediocre results. In addition to this Japanese transnational policies are predominately orientated along those of the USA. This is another reason, why it will take time until Japan can again play a leading role regarding environmental issues, as for this to occur, it will also be important that the USA pro-actively intensify its own levels of environmental engagement.

In contrast to Japan Germany has taken a leading role against softening the Kyoto Protocol. And it was pro-active at the World Summit on Sustainable Development (WSSD) to make regenerative energies a vital part of future energy concepts. But nevertheless the realities of a first social-democratic-green government are behind the high expectations in the environmental field. For example the substantial resistance of the German automobile industry against the "old car regulation" (which forces the industry to take back old cars free of charge), as well as the rebellion of another German industry against the deposit on beverage cans to guarantee high return quotas for recycling, clarifies substantially the problems with the implementation of single regulations against powerful actor groups. And this resistance is supported by the still relatively low prices of natural resources on the world market, favouring completely new production, without the partly complex recycling procedures also demanding virgin natural resources. Additionally, the investments and costs required for the take-back systems need to be recognised.

But neither Germany nor Japan have yet developed an environmental policy of the so called second generation taking the three dimensions of sustainable development – ecology, economy and social – mutual into account. Their policies are still dominated by curative countermeasures uncoupling development and social satisfaction instead of a holistic preventive strategy towards a determaterialisation of products and services.

One very positive outcome of the WSSD is the new partnership between government, civil society, industry and the United Nations in areas such as corporate responsibility and environmental standards, as stated by UNEP's Executive Director Klaus Töpfer (UNEP/DTIE 27.09.2002). But this means also, that from a national perspective governments, civil society and industry must become partners in developing an appropriate strategy towards sustainable development. But one can already easily confess inter-ministerial disagreement about the best strategy towards sustainable development in both Japan and Germany. The present Japanese and German systems do not contribute to governance with long-term concepts, as it is required to solve environmental problems in a sustainable way. Moreover, solution strategies that are easy to grasp and control are often favoured. This explains the interest of both nations in technical solutions, while often forgetting about the social dimension of sustainability that would require a much more holistic approach. Likewise, an emphasis on technical solutions presents the possibility to secure shares in the growing market of environmental technologies. Furthermore, the interests and motivations involved in environmental policy are so disparate that consensus would often be at the cost of political leadership and with mediocre results.

Despite many official announcements and some changes in its policies, the above give cases described that both Japan and Germany have a long way to go before they can claim to be a transnational environmental flagships. And the political will must be backed up by an efficient and constructive cooperation with all other major actors' groups, what is a novel challenge for politics in a transnationalising world. It will require a reorganisation of political decision making processes, but offers also the chances for more influence on a transnational level.

References

Albrow, M. (1997), *The global age: state and society beyond modernity*, Stanford University Press, Stanford/Calif.

Altvater, E. and B. Mahnkopf (1997), *Grenzen der Globalisierung. Ökonomie, Ökologie und Politik in der Weltgesellschaft*, Westfälisches Dampfboot, Münster, 3rd Edition.

Ayukawa, Y. and Y. Mizutani (2002), *Japan ... Could Do Better*, WWF-Japan, Tokyo.

Buck, E.H. (1997), 'Whale Conservation and Whaling', in *Oceans & Coastal Resources: A briefing Book, Congressional Research Service Report 97-588 ENR*, Washington D.C.

Danish Environmental Protection Agency (2002), *Aluminium seen in a life-cycle*, Danish Environmental Protection Agency, Copenhagen.

Drasdo, P. et al. (2001), *Belastungen der deutschen Industrie durch das Erneuerbare Energien Gesetz*, Universität Köln, Köln.

Edelman, M. (1976), *Politik als Ritual. Die symbolische Funktion staatlicher Institutionen und politischen Handelns*, Campus, Frankfurt/New York.

Environment Agency (1995), *Quality of the Environment in Japan*, Environmental Agency, Tokyo.

European Information Technology Observatory (EITO) (2002), EITO, Frankfurt am Main.

FAO (2000), *FAOSTAT agriculture database for statistics on international trade*, (http://www.fao.org/cgi.bin/nph-db.pl?subset=agriculture – last accessed 10 March 2003)

Federal Statistical Office of Germany (2003), *Imports and exports (special trade) by division of the national Product Classification for Production Statistics 2001*, (http://www.destatis.de/basis/e/aussh/ausshtab2.htm – last accessed on 23 February 2003)

German Advisory Council on Global Change – WBGU (1993), *Welt im Wandel – Grundstruktur globaler Mensch-Umwelt-Beziehungen, Jahresgutachten 1993*, Economica, Bonn.

Giddens, A. (1985), *The Nation-State and Violence*, The University of California Press, Cambridge.

Greenpeace (2002), *http://www.greenpeace.org*, last accessed on 28 February 2003.

Herz, J.H. (1976), *The Nation-State and the Crisis of World Politics*, Longman May, New York.

Hirst, P. and Thompson, G. (1996), *Globalisation in Question: The International Economy and the Possibilities of Governance*, Polity, Oxford.

Hobbes, T. (1976), *Leviathan or the matter, form, and power of a commonwealth, ecclesiastical and civil*, Penguin, New York.

Japan Statistical Yearbook (2003), *Supply and Demand of non-ferrous metal products (1980-2000) (1980–2000)* (http://www.stat.go.jp/english/data/nenkan/zuhyou/b0713000.xls – last accessed on 1 March 2003)

Jones, R.J.B. (1995), *Globalisation and Interdependence in the International Political Economy: Rhetoric and Reality*, Pinter Publishers, London.

Kawaguchi, Y. (2001a), *Statement of the Minister of Environment of Japan*, COP-6bis, Bonn, 23. July 2001.

Kawaguchi, Y. (2001b), *Statement of the Minister of Environment of Japan at the COP6 Resumed Session*, 19. July 2001.

Lindblom, C.F. (1980), *The Policy-Making Process*, Prentice Hall, New Jersey.

Mann, M. (1986), *The Source of Social Power*, Cambridge UP, Cambridge.

Mayntz, R. (1987), 'Politische Steuerung und gesellschaftliche Steuerungsproblem – Anmerkungen zu einem theoretischen Paradigma', in T. Ellwein et al. (eds), Jahrbuch zur Staats- und Verwaltungswissenschaft, vol. 1, Nomos, Baden Baden, pp. 89–110.

Messner, D. (1998), 'Die Transformation von Staat und Politik im Globalisierungsprozeß', in D. Messner (ed.), Die Zukunft des Staates und der Politik. Möglichkeiten und Grenzen politischer Steuerung in der Weltgesellschaft, Dietz, Bonn, pp. 14–43.

Morse, E.L. (1970), 'The Transformation of Foreign Policies: Modernization, Interdependence, and Externalization', *World Politics*, vol.22, no.3, pp. 371–392.

Nagasaki, F. (1995), *Research on whales*, Tokyo, (http://www.geocities.com/Athens/Styx/9189/whaligcon.html – last accessed on 02 February 2004).

Naturschutzbund Deutschland e.V. (NABU) (1999), *100 Jahre NABU – Historischer Abriss*, NABU, Bonn.

Organisation for Economic Co-Operation and Development (2002), *Economic Outlook No. 72, December 2002*, OECD, Paris.

Organisation for Economic Co-Operation and Development (1999), *Environmental Data Compendium 1999*, OECD, Paris.

Organisation for Economic Co-Operation and Development (1997), *Environmental Data Compendium 1997*, OECD, Paris.

Richter, E. (1992), *Der Zerfall der Welteinheit. Vernunft und Globalisierung der Moderne*, Campus, Frankfurt.

Robèrt, K.-H. et al. (2002), 'Strategic Sustainable Development. Selection, Design and Synergies of Applied Tools', *Journal of Cleaner Production*, vol. 10, no. 3, pp. 197–214.

Robertson, R. (1992), *Globalization: Social Theory and Global Culture*, Sage, London.

Rosenau, J.N. (1997), *Along the Domestic-Foreign Frontier: Exploring Governance in a Turbulent World*, Cambridge UP, Cambridge.

Simonis, U.E. (1996), *Globale Umweltpolitik: Ansätze und Perspektiven*, Meyer, Mannheim.

Skare, M. (1994), 'Whaling: a sustainable use of natural resources or a violation of animal rights?', *Environment*, vol. 36, pp. 12–26.

Waters, M. (1995), *Globalization*, Routledge, London.

World Bank (1997), *Environment Matters*, Washington.

World Bank (2003), *World Development Indicator Database, August 2002*, Washington.

World Trade Organisation (2003), *World merchandised exports by region and selected economy*, (http:// www.wto.org/english/res_e/statis_e/its2002_e/appendix_e/a04.xls – last accessed on 23 February 2003).

Zacher, M.W. (1992), 'The Decaying Pillars of the Westphalian Temple: Implications for International Order and Governance', in J.N. Rosenau and E.-O. Czempiel (eds), Governance without Government: Order and Change in World Politics, Cambridge UP, Cambridge.

Part IV:
Built Environment and Culture

9 Subconscious Awareness of Residents on Urban Design: An Analysis of the Group Interviews in Kohoku New Town, Yokohama

KAZUO IWAMURA

Cyclical Design Process for Sustainable Development

The author has developed the following design process for the sake of transparent and ever-cyclical improvement of the built environment for sustainable future.

Pre-Design Process

The proposed design process refers first to evaluation of the site from diverse view-points, in order to explore and discuss the concept to be shared by the stakeholders, determining and assessing the specific features of the project, according to *1) Natural Environment*, *2) Socio-cultural Environment*, and *3) Project Circumstances*, in order to elaborate *the concept* and *objectives* to be shared by all the stakeholders of the project.

Design Process

The pre-design process is followed by the implementation process of interpreting the prior knowledge into space and program, developing the best practice of the concept and objectives in both hardware and software at every life-stage of the project from *1) Ecological*, *2) Technological*, *3) Regenerative*, *4) Socio-cultural*, *5) Aesthetic*, and *6) Economic* view-points.

Post-Design Process

Then, the design should be implemented on site, to be followed by *1) Construction Management and Inspection*, and *2) Post-occupancy Investigations and Evaluations* during the operation stage, for every *Cyclical Design Improvement.*

This cyclical design process, as shown below, could be applied globally, and at the same time reveal many of the local aspects to be respected and considered.

Figure 1: Cyclical Design Process

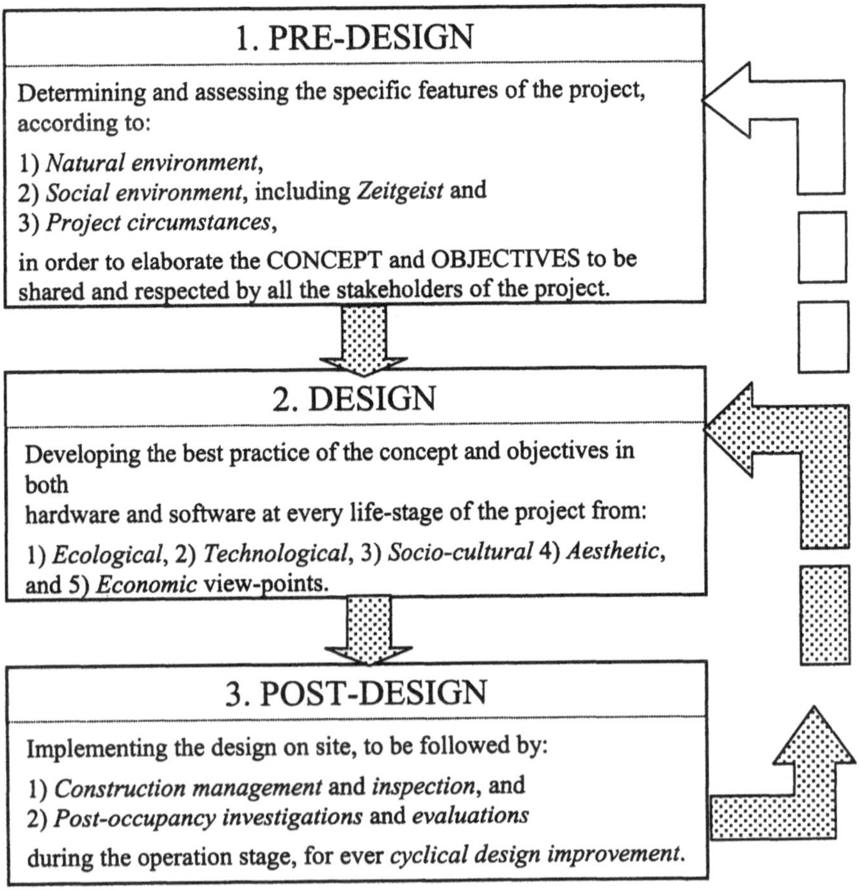

Pre-Design Phase

During the first Pre-design phase, designers are requested to make minute effort to investigate the project's circumstances including the natural and socio-cultural environment, in order to elaborate the concept to be shared by all the stakeholders of the project. This paper refers to a part of investigation to reveal the unconscious awareness of the residents in the Kohoku New Town (NT hereinafter), in the northern part of Yokohama, where drastic changes of the physical and social environment have been experienced during the past two decades of development in large scale as the regional policy drive of Yokohama City.

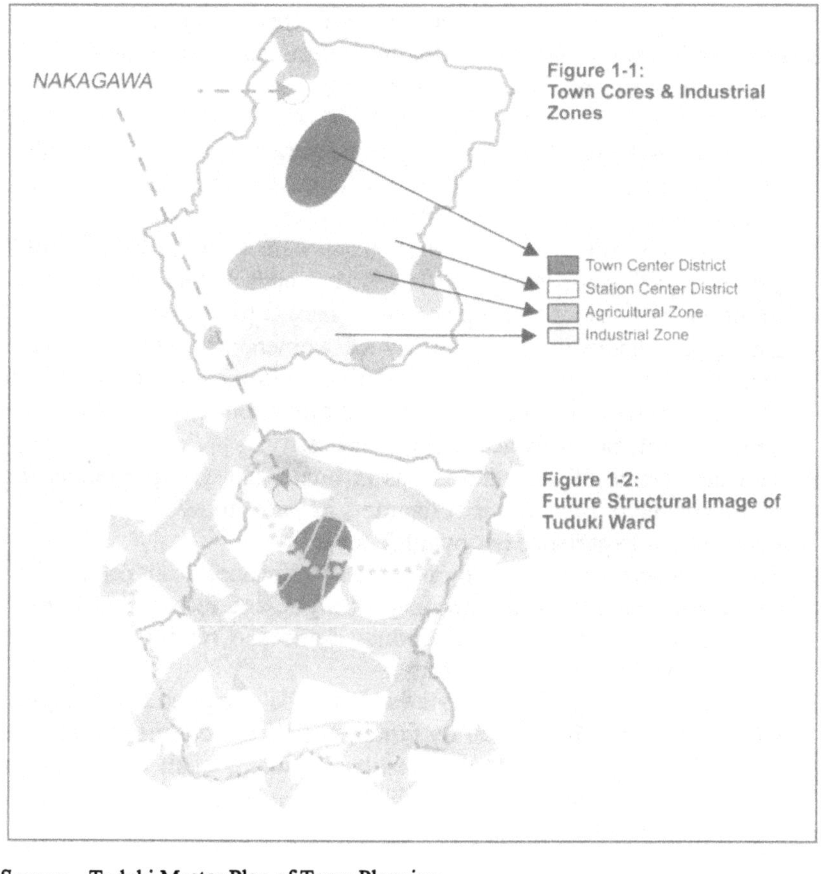

Source: Tuduki Master Plan of Town Planning

Main interests of the research have been focused on the unconscious awareness of the residents, whose average age is only 35, according to their property characteristics that had been categorised by prior researches in the designated area. This research was executed, as both pre-design and post-design of NT, in September and October 1998, with the thorough support of Tuduki Ward, Yokohama, in preparation for elaborating Ward's Town Planning Master Plan.

Group Interview Method

The author has been applying a qualitative research method, called "group interview method", to such above issues of pre-design and post-design phases of architectural and urban design projects since more than 20 years. The procedure of this method is as follows:

1) Analysis on the residents' property characteristics in and around NT.
2) Selection of a key or typical person representing a property, through local authority or other organisations.
3) Forming five to ten groups to be interviewed, with the help of the key person in selecting her (or his) closed friends belonging to the same property. The size of a group should be max. 5 to 6 persons.
4) The group members are preferably all women, who are well aware of everyday life issues, including shopping, children's education, administration services, social welfare, local information etc. Men's groups are added complementarily according to necessity.
5) The best venue of the interview is at home of a group member, or in members' favourite café or restaurant, in order to avoid formal atmosphere and to experience her own life-style.
6) The interview is not focused on some specific topics but on usual daily life of each member from morning till night, and recorded only by noting instead of using any electronic devices. Creating a casual and friendly atmosphere is top priority.
7) The results of the group interviews should be compiled and analysed as soon as possible, first group by group, and then according to issues discovered through crossing each result. Keywords shall be mapped and grouped on the theme map, which illustrates the structure of issues regarding the residents' lifestyle.

The Results: Themes, Emerged from the Group Interview

The relevant research has been executed to the following seven groups, representing the major specific types of residents and the results have been compiled as follows:

Properties of the Groups Interviewed

Group 1: Morning Market Group
House wives community created through the acquaintance at a Morning Market, sharing the problems of taking care of handicapped children. Mixture of old and new residents is characteristic.

- One of the motivations to move here was the planned wide road network, however, the present traffic jam is a cause of dissatisfaction.
- Priority for their residence is the traffic and shopping convenience. Little concern and awareness about the town-scaping.
- The preserved green in NT is highly evaluated, and enjoyed.
- They like the NT and their own life. Little knowledge about the drastic change of the old residents' life.
- They observe the society and community through their handicapped children, and created a mutual network to help themselves.
- They seek the relationship with the existing community, which, however, is often provided by the local administration.
- Personal contact with the old residents is requested, but in vain. They feel to be rejected by them.

Group 2: Welfare Commissioner Group
Typical aged house wives of old residents, living originally on farming, and now on the basis of new real estate business. They lead an active life, contributing to the local community as welfare commissioner.

- They quite enjoy their own life.
- Their life with big family is appreciated. They had been taking care of their parents-in-law as a wife, and are now living with the wife of their sons.
- They have their own indigenous community since long time, therefore, little requirement to NT.

- While they received financial advantages from the NT development, their traditional life-style was lost. They feel somehow to be violated by the newcomers, therefore, their relationship became more tight.
- They have their own extensive land and commodious house surrounded by rich green. Public parks are, therefore, no priority for them, and although they feel sad when green disappears, it is no serious problem for them.
- Their old inconvenient life-style is now their sweet memory, and they are even proud of it.
- They prefer rather to live with their family and to sustain it within the existing community around, than to challenge to begin with something new.

Group 3: Double Income Group
Very active working mothers of new double income residents living in privately-owned flats of the Governmental Housing Corporation since 7 to 8 years. With high academic background, they are very conscious of the rights of residents, sharing a house for kids by themselves. They are absolutely short of time.

- They happened to move in NT, just because they won the lottery for a condominium provided by the Housing Corporation. No special interest in NT at the beginning.
- They all have a high academic background, and share the common sense that double-income is just normal.
- Strong desire for further learning.
- They have scarce gender gap, and share their household and kids' education with their husband.
- Typical active life-style of "the more busy, the more active".
- Very little concern about the town-scaping, but very responsive to the information for kids.
- Their behaviour is oriented by clear policies like strong concern about the environmental problems including healthy food.
- Although sharing a house for kids by themselves costs a lot, they keep their job as profession.
- It is not the town as a whole, but their life in condominium that attracts them.

Group 4: House Wife Group
House wives of new residents living in privately-owned flats of the Governmental Housing Corporation. They have a quite different consciousness from Group 3, despite their similar residential circumstances.

- House wives, almost fully engaged in their kids rearing.
- They find their human networks through their kids' relations.
- Although they have some dissatisfactions, no active opinion is addressed to the local administration and community.
- The longest time to be with their kids is their priority for the time being.
- They long to move in a detached house, however, it is not affordable in NT.
- They find condominium life very convenient and safe, where they are spending most of their time.
- They have no contact with the old residents, and closed human network of their same age.
- Nuclear family is the centre.

Group 5: Farmers' Wife Group
House wives of old farmer residents with the age of 50's to 70's. They are the core members of the women division of Farmers' Cooperative Organisation, actively and comfortably living on their husbands' income from farming and non-farming business such as real estates operations.

- They try to get accustomed to their present life, keeping good memories of old days.
- They think that their men's life has been more drastically changed than their own. They tend to keep their great liking agriculture-centred life with earth, although the surrounding situation has been totally changed.
- Due to the NT development, their family structure has been distorted, causing fundamental changes.
- During the early stage of the development, there were harsh struggles between the pros and the cons within their men's society.
- They still live within their old community, and glance at NT from time to time as a foreign world.
- There are some who really enjoy the new life, recognising themselves as bad wives.

- Although they were obliged to change their life-style during quite long time in their temporary house due to NT construction, their agriculture-centred life has been quickly recovered after they came back home.
- Their children, grown up in the temporary houses, have no experience of the old farmer's houses, which is a pity for the parents.
- Living as farmer's wife, they understand their husband through co-operative works in the field.
- They experienced the early days as wife, expressing no opinion on their family and community. Even now, they tend to speak after men's words.
- Within the clear community they understand well each other and share a fellowship feeling beyond their generation gap.

Group 6: Volunteer Housewife Group

House wives of new residents with the age of 50's, who have already finished raising their kids and have therefore much time for community activities. These people are the core member of residents' initiatives of the Kohoku New Town.

- They have an ample situation both spiritually and financially.
- Provided with the benefits of the Bubble Economy, they could have moved to a better residence several times.
- Their human network is based upon the acquaintances at events for children.
- They are active core members of PTA, expressing vitally their own opinions.
- Their children are grown up and their husbands are at business. Therefore they seek their own world outside of the family, keeping the relationship with friends as a most precious resource for them.
- As the place of children's education is not NT any more, they have no serious need for the contact with the local community. However, they worry about the local information gap between old residents and new residents.
- They came from other regions and created a nuclear family here. Scarce contact with their relatives, except at yearly ritual events.
- They belong to the generation, preparing for the forthcoming life with their own parents who live alone in their home town.
- They have very firm concepts and opinions.

Group 7: U21 Girl Group

Girls under 21 years old, mostly teenagers, who constitute a considerable proportion of the New Town, the average residents' age is 33 years.

- For them NT is only the base camp of their life, and their activity fields are in Tokyo and Yokohama.
- Therefore, they have no clear need and favourable image for NT.
- Enjoying diverse urban acquaintances and stimulations, they give efforts to make up themselves.
- They are fond of NT. It's convenient location to both Tokyo and Yokohama is very much appreciated.
- However, they find the district's public bus transportation inconvenient.
- They feel scared when walking at night alone.
- They stay at home only at night and during the weekend. Their daily activities are based at their school or jobbing place.
- They are all grown up in NT. Therefore, they have many friends of primary school and junior high school to spend time together at the weekend.
- Their information source of fashion is magazine, and they do not watch TV so often.
- They are more or less interested in volunteer activities to take care of children or the handicapped. Some of them are practising such activities.

Figure 3: Group Mapping

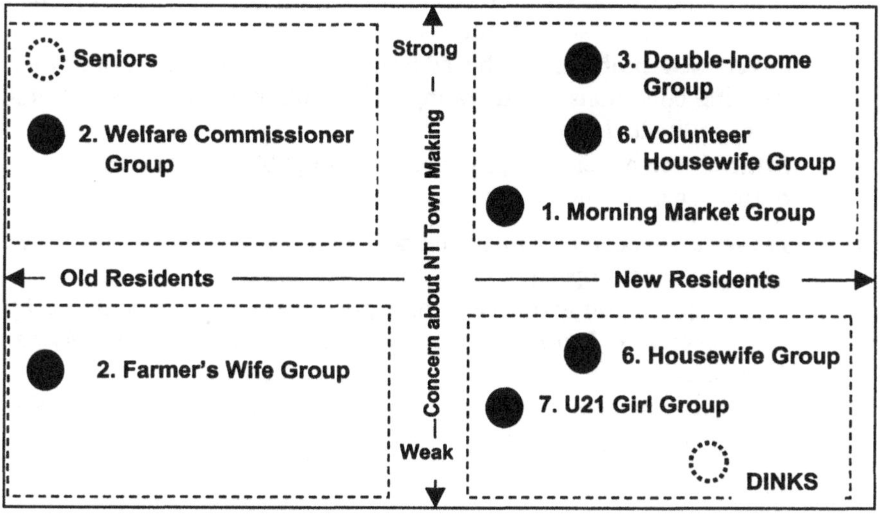

Results of the Group Interviews

The results of these series of group interviews have been analysed and compiled into the following three themes and twelve issues that revealed the unconscious awareness of residents of the designated area.

Human Environmental Issues:

1) Psychological barriers between old and new residents, as well as among each of them
   - Time lag in coping with their 'life' and 'change', and the difference of financial background
   - Desire for the mutual touch and exchange → *Creation of spaces and opportunities for daily communication.*
2) Old inhabitants as bearers of the indigenous culture and the former farm-village's life styles
   - Witness of the indigenous and living culture of the place → *Town making with historical elements to be inherited.*
3) Strong needs for daily town information as well as intellectual ones, in terms of contents, way of circulation and its speed
   - Gap between demand and supply of community information, during the maturing process of a new town → *Needs for sustainable communication site.*
4) Itemised communities and citizen's activities, crystallised from and around their own personal problems
   - Diversified problems to be addressed by the residents, including working conditions, child rearing, education, nursing etc, which initiated individual activities.
   - Little choice of services provided by both public and private sectors, during the maturing process
5) Desire for self-accomplishment and socialisation, beyond the difference of generation and property
   - Little information and access for participation, to be recognised by the residents → *Improvement of the community information services is required.*
   - Gap between demand and supply of volunteer activities → *Networking the potential residents' activities is needed.*

Figure 4: A chart example, representing the relationship between old and new residents

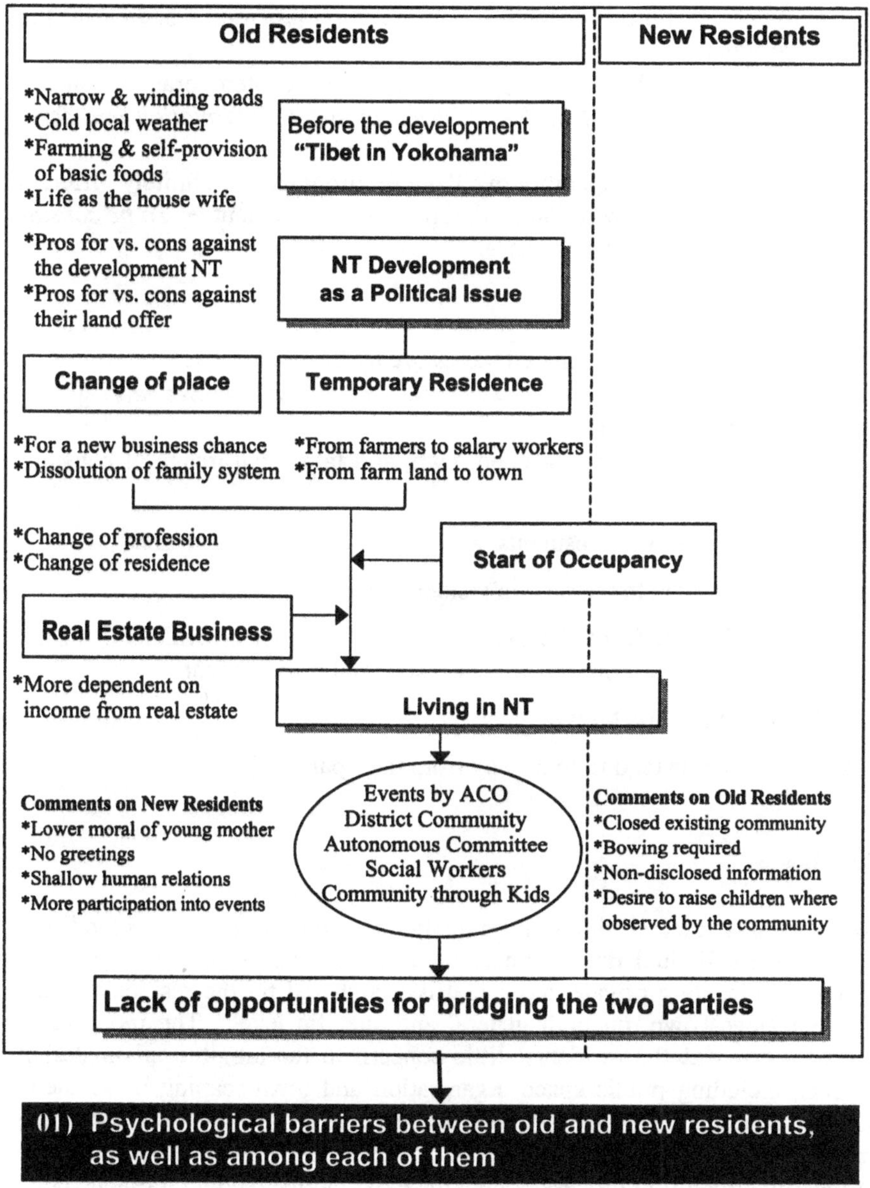

Time Environmental Issues:

6) Diversification and acceleration of the time environment
   - Available time is recognised as the most precious resource by the residents.
   - Awareness of time shortage dominates our contemporary urban life → *To be reflected in the provision of space, information and services.*
   - The quality and quantity of "Time Environment", definitely different according to generation and property of the resident → *To be considered as the basis of lifestyle.*

Space Environmental Issues:

7) Planning concept, not shared nor aware by the inhabitants
   - Bird's eye, human's eye and insect's eye level → *Eye level gap in observing the town.*
   - Stereotype image of a good town, and fabricated image of consumers.
8) Spatial gap between residents' area and the administrative area
9) Little interest in townscape in residents' mind
10) Satisfaction grade of the present home and town is evaluated on the basis of prior experiences of their residential environment
11) The present home forever?
12) New residents tend to live away from their parent(s)

Conclusion

Although this research through the qualitative group interview method was quantitatively limited due to time and finances, we could overview very vividly the existing private and social issues shared by the residents in the Kohoku New Town, on both general and specific levels. The most unexpected result was the residents' little concern to the tangible urban design aspects, including public space organisation and town-scaping of aesthetic quality, which are always major assignments of urban designers. After the growing stage of 20 years, however, the New Town will experience the forthcoming maturing process, where the existing diverse cultural and social activities of the residents seem to contribute to the collaboration towards a sustainable Old Town.

References

Iwamura, K.(1999), *Residents and Town-scaping*, Housing Industry Foundation, Yokohama.

Iwamura, K. et al. (2002), *Towards the Architecture for a Global Environment*, Shokokusha, Tokyo

Tuduki Master Plan of Town Planning (2002), Tuduki Ward, Yokohama.

10 A Comparative Study on the Problems of Gentrification in German and Japanese Cities

TAKASHI NAMBA

Introduction

When you enter an urban area after arriving at a Japanese airport from Europe, what you probably first notice is that the streets are not lined with attractive, neatly arranged houses and buildings. Recently I was asked by a German, "Is there any city planning in Japan?" In Japanese urban areas, harmony and coherence in the design, height, colour and appearance of buildings can rarely be seen. On the other hand, in European cities, especially those in Germany, the beautiful, orderly urban landscapes and the extensive, well maintained residential infrastructure including sidewalks, bicycle paths cannot be compared with what we find in Japan.

In 1980, the Japanese City Planning law was amended referring to the German Bebauungsplan (B-Plan). This law, which was aimed at the redevelopment of urban blocks established detailed planning regulations concerning land use, public facilities, and the maintenance of buildings. The law also required the participation of the residents in redevelopment planning for the first time. The 1992 revision of this law established a long-term vision of what cities will look like in the future. Moreover it introduced a requirement for city master-plans. More than ten years have passed since the revision of the law. Unfortunately, however, city master-plans have not been implemented smoothly in Japan.

Two cases of urban redevelopment are taken up in this article. I have attempted to make a bi-national comparison of the problems which accompany redevelopment in Japan and Germany. In addition, I have searched for a method of redevelopment in which gentrification – a process in which a residential area is radically transformed when its original residents are supplanted by people belonging to another socio-economic class – does not occur.

In order to investigate the German case, I researched the redevelopment in the Prenzlauer Berg area in Berlin over a five week period in 2002. My overview of the German redevelopment here is based on the material, which I obtained at that time. And, in my examination of the Japanese case, I have incorporated the results of research I have carried out since 1994 on the redevelopment of the Shonai area in Toyonaka City. Although the two cases are not representative of all urban renewal schemes in Germany and Japan, they allow us to discover many common points and differences in the redevelopment schemes in the two countries. In this article I have placed emphasis on the following five points:

(1) The comparison of the social mix before and after the redevelopment
(2) The expense burden in the redevelopment
(3) Complex ownership patterns
(4) The agreement formation process of the redevelopment
(5) The ideal of redevelopment / Overcoming gentrification problems

The Reality of Redevelopment in Germany – The Case of the Prenzlauer Berg Area in Berlin

First, I want to take up the Prenzlauer Berg area in Berlin as a case of urban redevelopment in Germany. The Prenzlauer Berg area, which was a part of the former East Berlin, is located in the centre (Mitte) of Berlin. It is conveniently located about 10 minutes by street car and bus from Alexanderplatz and Friedrichstraße. (Fig. 1)

Before and during the 1920s, numerous dwellings for workers were built in this area to cope with a large influx of workers to Berlin. The number of residences built at that time reached 70% of the present number. When it was part of East Germany, the area underwent a period of neglect and decay under unfavourable conditions that included its being the area nearest to the Berlin Wall. .

Redevelopment has progressed rapidly in the Prenzlauer Berg area since German reunification. Many new restaurants, boutiques, theatres, beer halls and other establishments testify to the area's speedy development. (Fig. 2)

The first characteristic of the redevelopment in Prenzlauer Berg is that it is not a scrap-and-built type or bulldozer type of redevelopment, but rather a remodelling/refurbishing type. In other words, the frames of the buildings, which were built during and before the 1920s are kept as they are, and the appearance of the buildings from the streets is unchanged. This technique, which is relatively effective in controlling the increase in rental costs and which is claimed to seldom induce gentrification, has been adopted in many cities in Germany.

Now let us take a look at the change in the population structure that has accompanied the redevelopment of the Prenzlauer Berg area. Table 1 shows the change in the mobility and the population of the Prenzlauer Berg area. From this table we can see that the redevelopment scheme has resulted in a gradually increasing influx of people accompanied by a decreasing outflow of people since 1998 when the redevelopment was completed.

Table 2 shows the change of the population structure by age in the Prenzlauer Berg area. We can see a decrease in the number of inhabitants aged 25 years and under along with those aged 45 years and over. At the same time we notice an increase in the 25 to 45 age category. This indicates that married couples with children have left the area, and have been replaced by singles and married couples without children.

Table 3 shows the change in the residents' income distribution. We can see a remarkable decrease in low income inhabitants, and an increase in residents in high income brackets.

Let us examine the conditions of the expenses burden in the redevelopment. Table 4 shows the change of the redevelopment promotion program expenses in Berlin from 1993 to 1999. The expense burden decreased for all items except for the item of residential policy measures. Actually, public subsidies for all redevelopment programs were completely cut off. The decrease of the public subsidies has slowed the pace of residential redevelopment while accelerating the displacement of the original residents. This supports the idea that gentrification is occurring in the Prenzlauer Berg area.

The Reality of Redevelopment in Japan – The Case of the Toyonaka Shonai Area

The Shonai area is adjacent to Osaka, and is considered a part of the Osaka Metropolitan Area because of its location near the core of the metropolitan area, although for administrative purposes it is part of Toyonaka City. It is conveniently located about 10 minutes by train to the central part of Osaka City. (Fig. 3)

Starting in the 1960s during Japan's period of rapid economic growth, this area was densely built up with wooden frame tenement row houses without adequate city infrastructure. These dwellings, which are called "Mietskaserne" (barracks-type tenement houses) in Germany, have become superannuated and plagued by numerous urban problems. About 200 tenement houses collapsed when the Hanshi-Awaji earthquake hit in 1994, and 940 of the residents were evacuated. And in 1996, one tenement house inside the area burned down in a fire that spread rapidly, resulting in the death of 15 people. These tragedies were due to the fragile condition of the dwellings.

The first characteristic of the redevelopment in this area is that all the original buildings were razed and replaced with completely new buildings in accordance with the area plan. In contrast with the remodelling/refurbishing type of redevelopment we see in Berlin, this type of urban renewal is the "scrap and build" or "bulldozer" type. The residence redevelopment of the Shonai area involved replacing fragile, dilapidated dwellings with residences that are resistant to earthquakes and fire. The adoption of this method of redevelopment has resulted in high rents, which tends to lead to displacement of the original residents. (Fig. 4)

Complex Ownership Patterns

The complexity of ownership patterns is the third point I wish to take up. In Germany, usually one person or one company has ownership of one building. In comparison with this situation in Germany, ownership patterns are extremely complicated in Japan.

To make it easier to understand ownership patterns I wish to classify them according to the following ABC system. In the "AAA", pattern A owns both the land and the residence, and lives in that residence himself. In the "ABC" pattern, the land is owned by A. B leased the land from A. B built the residence, and C lives in that residence. In the "AAC" pattern, A owns the land, A built the residence and rented it to C. In the "ABB" pattern, A owns the land, B leased the land from A. B built the residence and lives in it himself. In the "A—" pattern, A owns the vacant land.

In the Prenzlauer Berg area, most dwellings can be classified as "AAC" or "ABC" types. By contrast, in the Shonai area, even in the case of tenement house residences, "ABB" "AAA" type coexists with the "AAC" and "ABC" types. In Japan as in Germany, undivided, shared ownership of the land based on the ratio of the apartment area to the total land area is required in the case of recently developed condominiums. But, as buildings get older and are torn down, in more than a few cases the land is divided into small lots. This happens even in the case of older apartment houses. This is one factor, which has obstructed the progress of residential redevelopment in Japan.

The Process of Reaching Agreement in Redevelopment

By what kind of processes is redevelopment actually carried out? Let us examine them systematically. In both Germany and Japan, reaching agreement in redevelopment involves a two-step process. Reaching agreement with the landowner/landlord is the first step. The second step is achieving agreement between the resident and the landowner/landlord.

In order to eliminate complex patterns of ownership, the redevelopment, method which Shonai area adopted was one, in which the state purchased the land, where ownership rights were complicated, and the city built municipal apartment complexes. The residents who could not afford to pay high rents were given priority in moving into the municipal apartment houses. Another feature is that the residents of the "ABB" and "AAA" patterns combined their ownership rights and replaced old buildings with high-rise condominiums. An enormous official fund totalling about two billion yen, has been spent on this method over a 30 year period.

In the Shonai first municipal officials first carried out negotiations with landowners/landlords. They visited the homes of the landowners/landlords, and persuaded them by explaining the benefits they would receive from redevelopment including the subsidy system. It took a considerable amount of time to establish relations of trust between the municipal officials and the landowners/landlords. A council of landlords and landowners was established in 1972.

The second step involves negotiations between landlords and residents. In most cases these are independent negotiations between landlords and tenants without the active participation of municipal officials in consideration of the complex ownership patterns mentioned above. Measures taken include a resident council that was established to cope with "AAA" and "ABB" type residents as well as informal social gatherings between municipal officials and resident representatives. The Japanese model of redevelopment is shown in Figure 7.

Since urban redevelopment policy has already been decided on in the *Bebauungsplan*, redevelopment is usually implemented under a top down system. However, since the Prenzlauer Berg area was located in former East Berlin and immediate measures were necessary, the usual German redevelopment method was not applied. The old model of urban renewal by Häußermann and Holm (2002) is shown in Figure 5. Negotiations with the residents, negotiations with the owner, all are entrusted to the redeveloper (Sanierungsträger) under this model. My research disclosed that the Prenzlauer Berg area has a two-step process like the one in Japan.

First, negotiations/decision making is carried out between the national government (Staat) and the owners (Eigentümer). In this process a private redevelopment agent acts as a mediator between the national government and the owners, and holds explanation meetings and individual negotiations as a representative of the national government. This process has reportedly helped owners to establish co-operative relations with each other. This "new" urban redevelopment model of Häußermann and Holm is shown in Figure 3.

The Ideal of the Redevelopment / Overcoming Gentrification Problems

From the preceding overview, we notice that the negotiation processes carried out in the Prenzlauer Berg area and the Shonai area have a lot in common. Although the redevelopment methods were set up and implemented independently from each other, negations involving conflicting interests seem to be proceeding in the generally the same direction in both Japan and Germany. However three differences can be mentioned.

The first is the redevelopment period. In the Prenzlauer Berg area redevelopment has been nearly completed in ten years. This is influenced by various factors such as the 'repair type' or a remodeling type of redevelopment, the area of the redeveloped blocks, and the degree of decay. Redevelopment is not something that can be accomplished by only spending a lot of time. We have a lot to learn from Germany, which is able to implement programs quickly and effectively.

The second point is that while displacement of the original residents was limited as much as possible in the redevelopment of the Shonai, urban renewal in the Prenzlauer Berg area has been accompanied by gentrification. The comment of an official of Prenzlauer Berg ward left a strong impression who said: "No displacement of the original residents has occurred in this redevelopment. This case will be a good model for Japan in the future." From now on, we will have to search for a future model of redevelopment, which does not entail a displacement of residents, and can be carried out in a short time.

Finally, I wish to mention the difference in consciousness between German and Japanese owners as to the harmony and coherence of the urban landscapes. Sometimes it is said in Japan: "The best piece of land is one where you may build any kind of buildings." While the Prenzlauer Berg area redevelopment was carried out rapidly under the control of the municipal administration, in the Shonai area where there is no effective means of administrative control, the only way to implement redevelopment schemes has been by offering subsidies to "sweeten the deal".

In this report, we have briefly examined the redevelopments of Berlin-Prenzlauer Berg area and Toyonaka-Shonai area. In the case of the Prenzlauer Berg area, I have analysed research materials only since August 2002. In the future, I will carry out a detailed analysis of negotiation history between residents and the tenant information bureau, and the history of negotiations between the owners and the redevelopment agent. I also intend to clarify the good/bad points of redevelopment projects in both Japan and Germany in detail.

References

Grubitzsch, P. (1995), 'Prenzlauer Berg', in W. Ribbe (ed.), *Geschichte der Berliner Verwaltungsbezirke*, Stapp Verlag, Berlin, p. 12.

Häußermann, H., Holm, A., and Zunzer, D. (2002), *Stadterneuerung in der Berliner Republik.* Beispiel Prenzlauer Berg. Leske + Budrich, Opladen.

Namba, T. (2000), 'Gentrification', in Tomita, H., Moriya T. (eds), *Shakaigaku Forum/ Ochitsukanai <Watashi> to <Shakai>*, Fukumura Shuppan.

Namba, T. (2002), 'Gentrification of Prewar Inner City Housing in Japan', in Teichler, U., Trommsdorff, G. (eds), *Challenge of the 21st Century in Japan and Germany*, Pabst Science Publishers, Lengerich, p. 175.

Figure 1: Berlin Prenzlauer Berg Area

Figure 2: The redeveloped blocks

Figure 3: Toyonaka Shonai Area

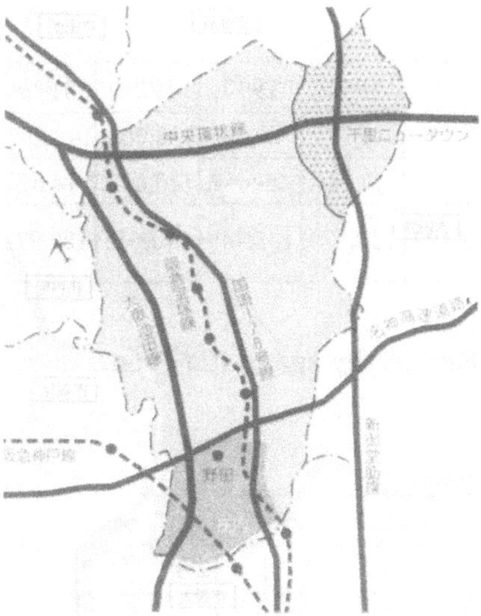

Figure 4: The redeveloped blocks

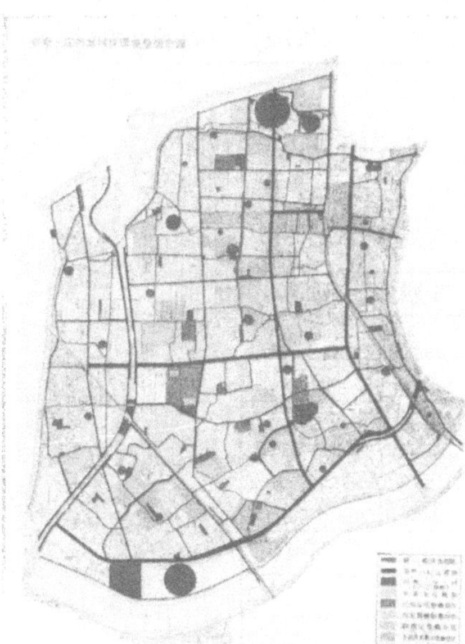

Table 1: Change in the mobility and the population of the Prenzlauer Berg area 1991–1999

	1991	1992	1993	1994	1995	1996	1997	1998	1999
inflow	10,375	11,412	13,123	12,965	13,558	15,315	16,445	18,068	18,202
outflow	8,184	8,431	11,029	13,633	15,572	17,717	20,617	20,622	18,639
popula-tion	143,277	145,082	145,900	144,567	142,197	138,878	133,449	130,664	129,971

Table 2: Change of the population structure by age in the Prenzlauer Berg area

%

	–5	18–24	25–29	30–34	35–39	40–44	45–64	65–	Total
1991	6.8	12.4	12.1	8.5	6.9	5.7	22.9	12.8	100.0
2000	4.2	10.5	13.7	15.3	10.7	6.4	19.4	11.1	100.0

Berlin total

	–5	18–24	25–29	30–34	35–39	40–44	45–64	65–	Total
1991	6.4	9.2	10.0	8.6	7.4	7.1	25.2	14.1	100.0
2000	5.0	8.3	6.9	9.0	9.5	7.6	27.3	14.5	100.0

Table 3: Change of the resident's income distribution in the Prenzlauer Berg area

Prenzlauer Berg %

	1991	1995	1999
Less than 1,000 DM	24.4	8.8	10.1
Less than 1,400 DM	63.3	16.9	21.2
Greater than 3,000 DM	2.9	19.4	22.2

Berlin total

	1991	1995	1999
Less than 1,000 DM	15.8	9.4	10.2
Less than 1,400 DM	34.7	17.0	18.2
Greater than 3,000 DM	17.0	25.8	29.9

Figure 5: The "old" model of urban renewal

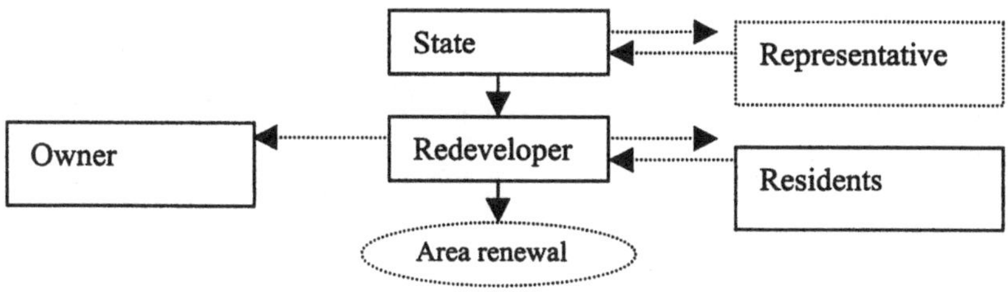

Figure 6: The model of the "new" urban renewal

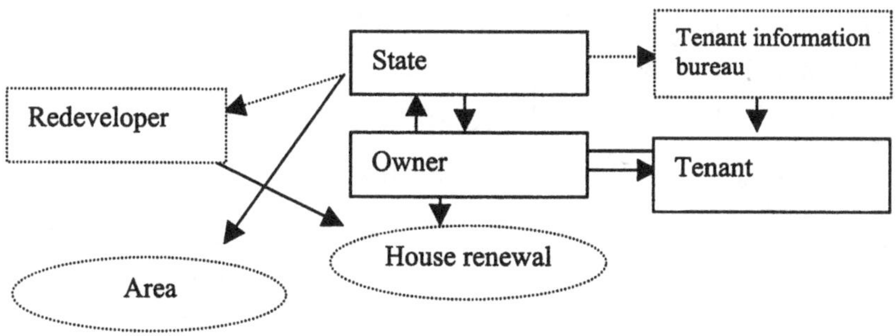

Figure 7: The model of urban renewal in Japan

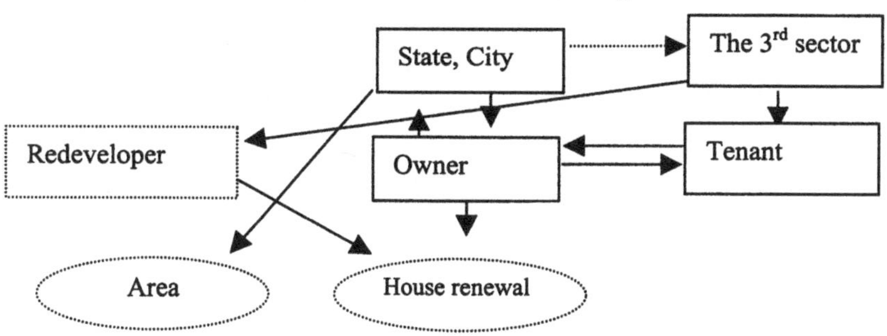

Table 4: Change of the re-development promotion program expenses

East Berlin	1993	1994	1995	1996	1997	1998	1999	Total
Pre-fabricated	119.6	287.6	335.4	284.8	149.6	104.3	55.1	1,336
Social city renewal	135.4	243.3	213.7	203.5	188.1	195.8	67.4	1,247
City-wide initiatives	149.2	155.8	108.3	140.6	54.9	94.1	23.6	727
Renter's modernisation	47.1	34.9	21.8	19.5	12.0	9.8	2.1	147
Housing policy initiatives	8.4	22.5	10.4	16.6	27.6	28.7	21.5	136
Other	259.1	137.5	134.7	52.3	58.8	66.8	38.1	747
East Berlin total	718.8	881.7	824.3	717.2	491.0	499.5	207.8	4,340
in % of governments funds	78.7	81.7	82.2	86.0	83.3	87.9	76.6	82.5
West Berlin total	194.3	196.8	178.6	116.4	98.6	69.0	63.4	917
in % of governments funds	21.3	18.3	17.8	14.0	16.7	12.1	23.4	17.5
Berlin total	913.1	1,078.5	1,002.9	833.7	589.7	568.5	271.2	5,258
in %	100	100	100	100	100	100	100	100

Figure 7: Complexity ownership patterns
Ownership patterns in the Shonai area

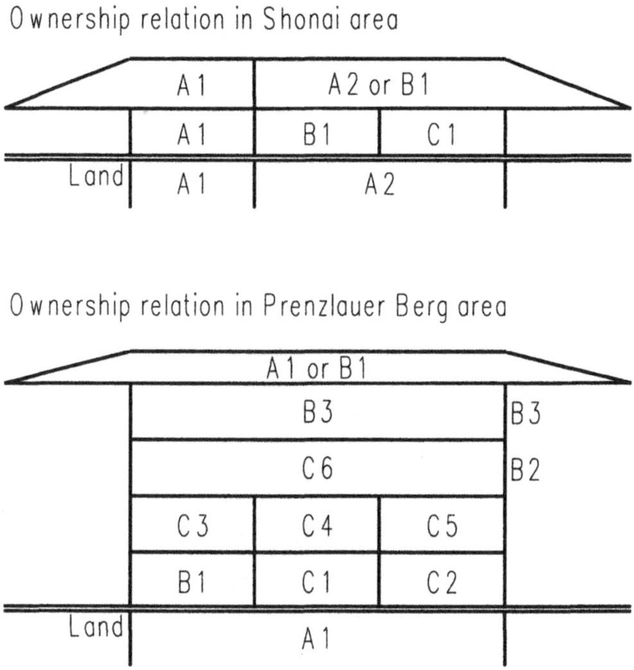

11 Christian Traditions and Ecological Challenge. Notes on Some Concepts in this Discussion

HELMUT LOISKANDL

Central concepts in the discussion to follow are obviously ecology, environment and ecological challenge. As they are used with different meanings at times, let me try to explain my own use of them. The word "ecology" seems a good point to start at. If we translate verbatim, ecology seems to be something like the science of housekeeping. "Oikos" is the Greek term for house, but can be extended to much of the humanly constructed environment. Environment is the world we live in. The astronomical environment is still safely outside of our grip, even if attempts at Moon and Mars have to be registered. Our "Oikos" is basically on our planet. And our planet provides us with the environment we live in: On it we have things given and things humanly transformed – and that is true for inanimate natural things as well as for things alive. Things humanly transformed are on the increase – no wonder, as all life forms share the tendency to transform what they touch. Due to rising numbers and increasing technology the human impact is growing stronger, with unanticipated consequences: The disappearance of non-renewable resources, the extinction of more and more blue-prints for life, and the transformation of an equilibrium we have taken for granted. No matter if we speak of limits to growth or about killing the goose with the golden eggs – it is clear that we face a problem. Our traditional approach to the world we live in does not work any longer. And while a low-density population with limited technology poses relatively manageable challenges for the environment (but even our stone-age ancestors apparently did their bit to get rid of megafauna like Mammoth or helped to bring about climate changes like in Australia over the last 50,000 years), the high-density population of today uses a more and more sophisticated technology in a clearly destructive way. In other words, our traditional value system for housekeeping does not work any longer. While one might prefer the culturally transformed environment of European parks, agricultural expanses or city-scapes to what is left of nature not touched by humans, the pressure of population and technology combined requires green planning even for the survival of the environment we have got used to. The extent of the human contribution to the disappearance of what once was the centre of Roman wheat production in North Africa, or of the city-scapes now under the sand of the Gobi-desert, might be under discussion, but there is no reasonable doubt that the desertification of vast areas in Africa and Asia is linked to human activities. Thus we have the ecological challenge: to find motivation for changing the rules in our encounter with the environment as we encounter definitive limits.

Can we find new motivational impetus in our religious traditions? Max Weber argued that the values leading to the capitalist approach to the environment can be clearly linked to one Christian tradition. Do we have to assume that Christianity thus is generally linked to an exploitative tradition, or could this religion (and religion generally) provide arguments for the new ecological consciousness as well? The following will look at this basic question. And while the following pages will mainly argue from the Weberian position that there exists a strong affinity between the capitalist spirit and the Protestant ethic, which again is just a reformulation of parsimony and time discipline typical of the monastic tradition, the point will be also made that the cosmic idea of peace presented in the biblical tradition together with changes in the appresentational framework of the medieval culture provide certainly a potential basis for a green ethic as well.

The Nature of the Issue in Question

Let us look at these traditions first, as in some recent discussions about ecological themes the environmental competence of Christian theology has been clearly accepted. As this ecological competence, however, is a fabric woven of different strands, an analysis of some of these strands seems in order.

It would have been of advantage if Christian traditions could have been compared to Buddhist traditions. A comparative approach to ecological values and their possible religious roots in Germany and in Japan would have provided an interesting project. However, as Buddhism has not really yet become a partner in the ecological dialogue, – a contribution by a Buddhist specialist envisaged at the Kyoto Environmental Sociology Conference (21/22 Oct. 2001) for instance did not make it into the final program –, it is hardly possible to try a full comparative presentation for various reasons. A glance at Germany's as well as Japan's tradition might lead us to the suspicion, though, that Christianity as well as Buddhism, having originated in pre-industrial city-cultures, did probably not try to answer ecological problems of a globalised industrial system – a suspicion which does not exclude the possibility that followers of those traditions tried to come to an arrangement with a changing situation. These arrangements, though, are of different nature. Christianity started out with a monopoly, Buddhism had engaged in a continued dialogue with a still potent nature-religion.

While Christianity in Europe was able to replace pre-urban religions to a large part, Buddhism in Japan had to come to a certain modus vivendi with Shinto, thus leaving itself open to continuing influences from pre-urban religiosity. A good example is Shugendo, an integration of Buddhist monasticism and traditional nature piety. On the other hand, though, Buddhism was successful in replacing the collection of plant life typical for traditional gardens with abstract non-life forms, with the sand- and stone-configurations of gardens in the Zen-tradition, stressing its distance to the generalised concept "life". This distance allows for expressions of compassion which seem rather strange to the western eye, the almost ritual consumption of fish or crayfish still alive for example. As the cruelty of natural designs becomes apparent, the potential of human designs to improve nature can be activated in different forms. The traditional Japanese garden which leaves almost nothing to nature is a good example. This short note should be enough to point out that European Christianity and Japanese traditional religions show at the one hand some general similarity in their approach to nature, while the details on the other hand are definitely quite different.

But the following presentation will deal with Christian traditions alone. The article will look especially at the conditions of early Christianity. Christianity started out as a city religion. Thus the goal of this movement was not to regain a Garden of Eden long lost but to wait for the new Heavenly City. The images used in the sacred scriptures of Christianity clearly indicate the city-orientation of this new religion. Consequently the early theologians also were not so much concerned with nature – which was assumed to be a "fallen nature" anyway – but with the "Civitas Dei", the "City of God". The fortunes of the new movement were clearly with city people; people in the countryside resisted for a long time – a reminder is the word "pagans". In other words, early Christianity showed the distrust for life outside the city typical for the city-dwellers of the Greco-Roman world. The monasteries in the desert, typical of this early period, were not seen as bucolic retreats but as encampments of the soldiers of God in an hostile environment, fighting the daemons and the evil forces of nature outside of the walls of the city. Because facets of nature had been seen as places of the epiphany of the sacred in some pre-Christian religions, the former hierophanies were reinterpreted as being evil; and for centuries a symbol of Christianity's approach to nature was the missionary Boniface and his cutting of giant trees venerated by the local populace.

The Graeco-Roman distrust of the non-human environment saw changes when, for instance, Petrarca discovered the beauty of venturing onto a mountain-top; and, at about the same time, Francis of Assisi changed the rules of engagement with non-human nature, proclaiming a brother- and sisterhood of creation including animate and inanimate nature.

But it seems evident that this new Christian attitude grew out of the encounter with a world in which the interest in a "natura naturans", a nature naturing, had become quite visible. A vivid feeling for nature in life and literature, from the troubadours to Petrarca, is reflected in the rejoicing of Francis in "the beauty of the country, the charm of the vineyards, all that was pleasing to the eye" (Joergensen 1955: 25). While Francis clearly takes a stand against the abuse of nature, his approach is definitely quite anthropocentric. One of the stories relating his relations to wild animals makes that obvious: In the case of the wolf of Gubbio his solution to the war between wild wolfs and the town-people is a peace based on imposing the morality of non-aggression on wild creatures in return for free feeding. But there are also glimpses of an almost modern ecological mentality when Francis orders his followers who were cutting wood in the forest to leave a part of the tree standing, so that there might be some hope of its putting forth branches again (Joergensen 1955: 256).

While Francis of Assisi on the one hand obviously is influenced by a new attitude to nature which significantly influenced the cultural landscape of his era, it cannot be denied that he also is an example for the resurgence of an older trend visible in the Bible as well as in the early history of Christianity: the cosmic idea of peace, which has its place not only in this tradition (Lutterbach 2001: 771). Early religions generally see the cosmos as a magic universe in which everything and everybody is part of an harmonious whole. In the biblical tradition the state of harmony between nature and man is limited to the period in the Garden of Eden, though. After that, and especially after the Great Flood, evil permeates the world – and God turns even into a partner of man when it comes to the eradication of wild animals. Only as promise for the future appears the vision of a world in which lamb, wolf, lion and humans live peacefully together again. In early Christianity this vision gains new potency, as the first Christians saw themselves as trying to live the life of the world to come already in the world present. The people who succeed as saints were understood as being able to realise the promises of the prophets as well – the wild animals peacefully follow the saint. Especially in Irish legends these stories of power over wild animals play a significant role. However, it seems fully justified to speak of a paradigm change in the case of Francis of Assisi (Lutterbach 2001: 778): His closeness to nature is not the consequence of power over nature gained through extreme ascetic action but the necessary outcome of a general compassion with everything existing. Everything created is brother and sister, everything counts equally. This compassion became a significant characteristic in the mystic theology to follow, but not of Christian theology in general. However, Christian groups at the periphery kept this compassion alive – visible expression of attitudes thus nourished was the intent of William Penn manifest in the creation of Pennsylvania, a state intended to be fair to nature, immigrants and natives (Lutterbach 2001: 784).

Thus the heritage of Christian tradition is ambivalent: On the one hand we have a continuation of the Graeco-Roman abhorrence of a nature not transformed by human input, on the other hand we have the Franciscan tradition proclaiming the brother- and sisterhood of creation, more looking for a restauration of the garden of Eden than for building the "New City of God". This article will just follow the Graeco-Roman tradition and its influence, though.

Nature in Christian Theology and the Traditions of Power

At least since Max Weber's "Protestant Ethic and the spirit of Capitalism" (Weber 1995) it has been more or less generally accepted that religions influenced the dominant value structures of society. An example is the ongoing discussion about the possible input of Japanese religious traditions on the development of Japanese capitalism (Bellah 1985, Lebra 1986, Loiskandl 1995, Otaka 1984, Yawata 1981).

Thus it is not surprising that European discussions of ecological competence look to Christian theologians as partners (Heid-Hoft-Rodax 2000: 195). Analysing their contributions, however, it becomes quite obvious that these theologians have the task to locate possible correlations between modern problems and a tradition which developed in a rather different social climate, and then to translate the general principles of a different time into the language of contemporary concerns. And while the basis of religious values might remain the same, their actual theological pronunciation is changing over time - an example could be easily produced in the changing

Christian attitude to money and related matters like banking. Different strands are thus woven into one more or less unified fabric of religious value orientation, providing a vast potential for opposing interpretation of specific problems.

But the beginning is clear: In the centre of nascent Christian writings looms the city. That is certainly true for the gospels, but here we will follow Patlagean's (1983: 107) attempt to uncover the hidden structure in the biographies of saints of this period. And to Patlagean the all-encompassing space in which the biographies are located is urban and civic; all events originate and lead to towns. The whole world appears as a collection of juxtaposed urban spaces with gaping voids between them. The urban environment is the proper place for human concerns; outside of the city walls are wild animals, daemons and savages, endangering the human power to pursue happiness.

Against this background we have three models imposed on the person to be a saint: a demonic, a scriptural and an ascetic one (Patlagean 1983: 105). The first one sees the typical journey of the hero as fight against daemon power which, while being all-pervasive, culminates in the desert. Thus the human to become a saint has to leave the relative security of the city, in which he is depicted as having partaken in all its civilised power. He can, as a rule, read and write; stems from a well-to-do family, and enjoys the advantages of socialised life. Having left the advantages of urban socialisation behind, he is now alone to do battle against nature and against the daemons lurking. Supreme assurance of victory comes from his ability to deny the need of sexual gratification and even of food. Having won his war the saint returns to the city to use his new powers in the service of people there; he becomes a miracle worker. While this model seems to predate Christianity – it could be found in the life of the Buddha as well –, it claims a special Christian dimension by explaining it as an imitation of Christ – easily done as the life of Christ is already depicted along the lines of the demonic model. And finally the aspect of ascetic training gains importance. The issue of obedience, so dear to all hagiographies, finds its historic climax in the introduction of time discipline in the monasteries of the desert.

It is obvious that this strand of tradition does not lead to ecological concerns. Actually it leads to what Max Weber claimed as the affinity of Protestant ethic and the spirit of capitalism: an attitude asking for domination over a world seen as being destined to be used, an attitude centring on discipline and the use of time. Let us follow Weber's argument, and from there have another look on early monasticism.

Protestant Ethic and Time – a Weberian Argument

It is interesting to note that already in Max Weber's analysis of the spirit of capitalism the attempt to turn time into a commodity is distinctly pointed out as a crucial fact, a development which in this analysis was closely related to the importance of time as a disciplinary instrument in the new ethical orientation based on religious arguments of the Reformation. However, in the ensuing heated discussion the time-dimension of the original argument was widely overlooked. Had it not been so, Max Weber probably would have come to a certain refinement of his argument which, it could be argued, is present in a hidden form anyway: As religion influences social developments, religion in turn is influenced by existing social constructions. So the boundary between religion, which obviously is a social construct itself, and other constructs is fluent, as the "epiphany of the sacred" (Mircea Eliade 1954) may be located in changing objects at different times, and this sacred quality of objects might be lost at other times, only shifting the explanation of its accepted importance, though.

Thus it could be argued that time has a central position in most religions, be it as celebration of the recurring seasons, of the changes of the night-sky or of the daily rising and disappearing of the sun. In the religions of all high cultures there are tendencies to transcend circadian rhythms (Aveni 1983: 339), though. It could be argued that especially in the Biblical tradition this trend to establish a theology not only of myth but increasingly one of history becomes more and more dominating. This theodicy of history could be seen as ethical rationalisation (Berger 1967: 120) leading eventually to the process of secularisation characteristic of modernisation, a process which is reflected in the affinity of Calvinism and early American capitalism. In speaking of the consequences of the Reformation and its influence on the development of a new ethic, it probably should not be overlooked that this movement led to an appropriation not only of Church land by a selected public but also to the appropriation of cultural properties, and one of these properties was the time-consciousness which had developed in Catholic monasteries. Weber is well aware that the monastic ethic was an antecedent of the Protestant ethic, and that even as monastic ethic it had influenced the economy (Winckelmann 1978: 153).

While it was understood in the theology of the Reformation that the special Christian calling to a life of perfection extended to every Christian, it had been assumed previously that most Christians could get away with a normal existence, and only a few were called to lead an exemplary life in a special invitation of Christ, following the strict discipline of the monastic lifestyle. This discipline did not only involve the imposition of the rules of chastity, poverty and obedience, but also the imposition of an effective time control. As the history of monasticism shows, however, this new life-style was not without experiences of frustration among its followers, and the ideal was not always executed in its pure form. No wonder that in its transformation to a general rule of behaviour time-consciousness and alienation developed in close interrelation.

Weber was fascinated – and frightened at same time – by the forces of alienation set free in the process of technological development and economic streamlining, forces which in his opinion were out of control. The "iron cage" he is talking about is not so much a consequence of material factors; it is the logical result of the decision to adopt these and to adapt to the necessities inherent in them; "economic rationality" is a human construct after all. And "economic rationality" is arguably at the basis of what we now experience as a veritable war against the ecology. Thus for Weber a crucial question is the question why and how this decision to embrace "economic rationality" even beyond the borders of sanity came about. Elements of capitalism had been around for a long time; from money economy to technical invention and production for profit, and not only in Europe but in India and China as well. But to join these elements and to make them work, it needed a new ingredient, a new set of mind, he argued, and he called it the "spirit of capitalism". A new set of mind, because it meant to do what humans up to this point had done only in exceptional cases. As Weber sees it, this spirit of capitalism as expressed by Franklin, appears from the point of view of the happiness of, or utility to, the single individual, entirely transcendental and absolutely irrational (Weber 1995: 50). Why should humans be dominated by the urge to save time and money and make this the ultimate purpose or their life? Why should they strictly avoid spontaneous enjoyment, why should they forego the little satisfactions and luxuries depending on the spending of these two commodities, time and money? The postponement of gratification as a generalised principle, as advocated by Franklin, who, according to Weber, was a "colourless deist" – this postponement seems to be closely related to some religious emotions remaining after the religious convictions had disappeared. In the religious context the rejection of quasi-instant gratification could be regarded as mere postponement; now in the secularised context it turns into indefinite deferment.

While limited, the postponement of gratification in one form or the other is present in all human cultures, postponement as a general principle indeed first became an option for a few in other-worldly asceticism. Thus Max Weber's suspicion was that the inner-worldly asceticism of capitalism as a deferment of gratification under the pretext of saving time and money, without proclaiming other worldly rewards, an asceticism, which reduced life to an existence in an iron cage (Weber 1995: 181), had to have its roots in attitudes growing from a rational and logical matrix. Or in other words: As inner-worldly asceticism in itself is not rational, this attitude may have to be explained as a remnant of an otherworldly asceticism which within its own frame of reference is rational. Benjamin Franklin's adhortation to save time and money, cited by Weber as an eminent example of the spirit of capitalism, draws its justification clearly not from capitalism as such, but from an obsession with stewardship alien to mature capitalism, though very dear to a specific religious tradition. For Max Weber, the unifying factor in the ethical attitudes of the leaders of emergent American capitalism is exactly this religious tradition: the tradition of Calvinism. And what he claims is not that Calvinism as such becomes directly instrumental in the development of capitalism, but that attitudes born out of religious rationality keep their potency even after the victory of secular rationality.

But, following his own declared intention, Weber did not choose to follow up on the emancipation of the spirit of capitalism from Protestant ethics; he was interested in the origin of this partnership or, more precisely, in the very origin of ascetic ethics in Protestantism, as mainstream early Protestantism could be claimed to have developed out of opposition to the ascetic tradition in Catholicism. And, doing so, Max Weber centred on the Calvinist paradigm. In the theory and praxis of Calvin he found clear indication that the concept of "calling", in Catholic theology reserved for a special, religiously authorised and organised lifestyle, for the lifestyle of the monk, had become the corner-stone for a reconstruction of asceticism, not quite otherworldly any longer, but not quite inner-worldly yet. The rejection of a monastic special calling went hand in hand with the expectation that every Christian was supposed to have this calling to perfection; every Christian had to live his life as strictly as a monk ever could. To avoid too close a link-up with the traditions rejected, this strictness was no longer advocated by the necessity to do good works but by a theology of predestination. Predestination, though, is not part of the main argument, but is only used to counter the consternation of the "sola fides" proponents – a connection Weber later clearly perceived, as is indicated by his decision to omit this concept from later discussions and to stress the motive of calling even more (Collins 1986: 33).

The theology of the calling and of predestination thus, Weber argued, were at the centre of Calvin's praxis as well; of a praxis which, while agreeing with Luther's "anti-Mammon" stance in theory, grew out of the conviction that "the Christian spirit could express itself and maintain its existence within the sphere of a society which was based essentially upon a money economy, upon trade and industry" (Troeltsch in Green 1959: 21). While Lutheran and Catholic theologians retained a sceptical position to the appearing paraphernalia of a society based on banking and industrial production, – the question of interest taking was just one issue among many – Calvin integrated features of this newly forming system into the structure of his intended ideal society in Geneva.

Thus Calvin's Geneva accepted industrial production based on money economy. Actually, Calvin himself became deeply involved in the process of industrial development in the city. In contrast to other reformators he did not see the agrarian patriarchal form of life as the ideal, but only as one option, together with professional work and industrial production within the money economy. While his followers in Geneva eventually embarked on the road to state capitalism – taxation, interest rates, state loans and the foundation of a state bank were approved by the ecclesiastics before being actually created by the politicians –, Calvin personally instigated the introduction of cloth and velvet manufacture into Geneva. Later on, when French competition grew too strong, he saw the watch industry as a promising replacement option for Geneva (Troeltsch in Green 1959: 22). So dear was this option to Calvin that this practised lawyer and ardent religious reformator even spent much time on drafting a plan for a watch (Chadwick 1964: 86).

This, it seems, is a crucial fact, even if Max Weber did not catch sight of it. It definitely strengthens his argument for the perceived affinity of Calvin and Franklin. It could be argued that time as commodity gains reality if it can be measured and metered out, thus the medium for such an interactional use of time is of basic importance. For the disciplinarian Calvin everything fell into place: The time discipline of the old monasteries could be surreptitiously transferred to the civil society, and time discipline would work, like in the old tradition, as training ground for generalised discipline. The portable watch as medium of time keeping would allow the transfer of centralised timekeeping to individual responsibility, while this individual responsibility would be under performance pressure, as the economic foundation of the communal life was to be more and more influenced by the criterion of timekeeping.

Calvin's option turned out to be a success. While in the home-town of the portable watch, in Nuremberg, the inventor of this new device died not exactly a rich man, Geneva and Switzerland prospered by adopting and propagating the new medium of interaction. And, just as a side-note, watches from Geneva are even today the big success-story in Tokyo, as keeping the time has become a world-wide obsession, not to mention that it is really the basis of interaction in an industrialised world-society; and the controlled use of time is more than just a symbol of the trend to use everything as commodity.

The discussion between Weber and his critics progressed never to the issue of time as such. It got caught up in determining what it really meant when contemporaries of the Reformation spoke about the obligation that every Christian now had to be a sort of a monk. Adversaries of the Weber thesis like Rachfahl (Winckelmann 1978: 91) insisted that this figure of speech did not really have much concrete meaning and definitely did not indicate any commensurability between Calvinist ethics and Catholic asceticism, while Weber pointed to the economic success of the old monasteries, which he explained as a consequence of an ascetic lifestyle, in which strict discipline streamlined activity in a rational way. Thus, he argued, Calvinist communities reduplicated both, discipline and economic success (Weber in Winckelmann 1978: 153).

Time Discipline and Monasticism: The Heritage of the Roman Army

The argument could be easily continued by shifting to the time dimension of discipline. It is well documented that the disciplinarian Calvin did not take lapses in time-discipline lightly, and a general acceptance of the portable watch – and that is only an unscientific suspicion – must have given him great pleasure. To establish this as a fact or to refute it as contrary to the historical evidence would require some in-depth historical studies which frankly have not been done in the preparation of this article. But even a more general reading provides some indication that for Calvin "the calling" included also a calling to punctuality, and monasteries belonged to the few places where this virtue was honoured before the Reformation. Thus it becomes clear again that attitudes instrumental in the development of industrial society and the exploitation of the environment can be easily related to some strands of Christian tradition. To understand the time-dimension of the postulate that everybody had the calling, that everybody had to lead a life as ascetic as that of a monk, we therefore have to add some comments on time-keeping in the early Christian church. Weber's thesis of the influence of religious ideas and values can draw strong support from the theology of the Middle Ages. Liturgy, including hourly activities like the "horae", was clearly understood as "memoriale Domini", as keeping time structures alive which were important in the life of Christ, by locating them firmly as time structures shaping the life of his followers. But this argument does not explain how the time structure in question came into existence. Actually, the gospels use timeframes existing at the time of their writing without theological explanation; existing time frames are used automatically to explain the larger historical context as well as the time of the day at crucial happenings.

It is important to keep in mind that Christianity developed in a cultural area shaped by the traditions of Mesopotamia and Egypt as well as those of Greece and Rome. Thus time-frames were already present in everyday life. Mesopotamia had developed a day consisting of twelve double-hours, Egypt a day of twelve hours and a night of twelve hours and the Roman army first, and then administrative bodies in the Roman empire as well, had established a three-hour long period as practical unit of activities (Roellig 1975, 1473–1475). While religious activities were connected to seasonal and circadian rhythms, short constructed time periods do not seem to have been related in a very visible way to the sphere of the sacred originally. Christianity, though marked a change. The gospels (except John) located the agony and the death of Christ in the twelve-hours time frame from the sixth to the ninth hour, thus laying the foundation for an integration of this time structure into the realm of the sacred. The first step to sacralisation was the daily commemoration of the death of Christ at the ninth hour in prayer (Martimort 1963, II: 329) leading eventually to a sacralisation of the three-hour-system as such.

To summarise again: Like other religions, Christianity adapted its ritual life to already established timeframes. This applied to yearly repeated celebrations, to weekly commemorations and to the developing order of daily prayers. While the justification for the timeframes applied was always theological, the actual origin was a mixture of religious and secular influences. The place of new developments was the daily structure of the prayers, which were expected to be mainly community prayers. The evening prayer seems to be a continuation of an established Jewish practice, while the prayers at first day light start out with a specific Christian content. As in Roman cities the third, sixth and ninth hour of the day were publicly announced, these markers received a theological interpretation already in the second century and were integrated into the official prayer cycle of the Christian day (Martimort 1963, II: 329–331). As this public praxis was strongly developed in what is now the Middle East, nascent Islam had a paradigm at hand for its own religious structuralisation of the day.

In western Christianity the paradigm of the ancient Christian East did not have as much public impact; here the daily discipline of time was left to the ascetic elite. The roots of their daily religious timekeeping, however, also point to the early Christian Orient. Here the monastic element had made additions to the prayer times of the day. Egypt, one of the cradles of Christian monasticism, seems to have added the night to the periods if prayer; not surprisingly though, as the monasteries had become attractive to retired soldiers. As "Miles Christi" they apparently retained the "vigiliae" they were used to in the Roman Army. Thus the prayer order of coptic monks knows a morning prayer, prayers at the third, sixth and ninth hour, the vesper prayer and two periods of prayers in the night, the first and the second completorium. The second completorium seems to be an adaptation to gain more uninterrupted sleeping time, as early sources know a "prayer at the first call of the cock" instead of the second completorium, a prayer which was located around three o'clock in the morning (Quecke 1970: 15). While researchers into early monasticism are split on this point – some prefer the possibility this early morning prayer might have been originally part of the midnight prayer, extended into the early hours at some stage, and then lost again –, on the whole this discussion presents only variations on the main thesis: The monastic order of prayer reflects the Roman constructed time of the day, which contains four periods, and of the night with its four "vigiliae". As the name indicates, the "vigiliae" were originally watch-periods in the Roman Army, each about three hours long, and they ended with a changing of the guard. It is tempting to assume that the retired soldiers turned monks just transferred a rhythm they had got used to into a religious ritual of change of duties.

The strict time discipline of Catholic monasteries, thus, can be traced back to the general appreciation of structured time in the Roman Empire. In Roman cities the main hours were publicly announced. And this appreciation is especially linked to time as a constitutive element of order in the Roman Army, as the beginning of monasticism is heavily located in the Egyptian Thebais, home of the famous Thebaic Legion, and as many of the early monks had been reputedly retired soldiers, including Pachomius, one of the early organisers of ascetic communities in nascent Christianity. Roman soldiers had been prominent among the first pagans joining the emerging Christian community, as already the New Testament indicates. It generally can be stated that religions promising immortality attracted substantial numbers in the Roman Army, be it the cult of Mithras or be it Christianity (Baus 1940: 166). By the year 200 this had become an issue of heated discussion among Christian theologians. Rigorists like Tertullian declared that Christians were not allowed to become soldiers, while he exempted from this rule those who already had been members of the army when they found the gift of faith (Esser 1915: 253). Mainstream theologians were more lenient, however. These arguments seem to indicate that soldiers constituted a sizeable percentage of Christianity around 200 and thus presented a problem for a rigoristic interpretation of Christianity. Generally, it seems, soldiers were willingly accepted into the fold; there is a surprisingly high number of reports relating the marter death of Christian soldiers. It is also significant that the last wave of persecution of Christians in the Roman Empire started in the army; and it was eventually the Christian soldiers in this army who gave victory in a civil war to Constantine, a victory which made Christianity the official religion of the Roman Empire (Baus 1940: 169). One of the units hardest hit in this last big persecution by Diocletian was the Thebaic Legion, according to traditional documents (Van Berchem 1965: 15), an army group recruited from the area around Thebes in Egypt, an area known for its high percentage of Christians. While there is some discussion about the historical reality of a Specific "Thebaic" legion, the fact is undisputed that this region of Egypt was the home of many Roman soldiers. From this area came Pachomius or Pachom as well. After some years of service in Constantine's army he had retired early and joined a colony of Christian ascetics in the desert (Mueller 1975: 288). These early colonies of ascetics were agglomerations of highly individualistic, saintly or net-yet-so-saintly dropouts, and it does not seem too strange that a former member of the army tried to reform this group by imposing rules stressing especially "discipline, uniformity and unity" (Gribomont 1962: 1331). Discipline in Pachom's understanding had to do with structured time; and the time structures given in his rules indeed reflect the time structure of the Roman army. The new connotation, though, is that constructed time is now experienced as a place of sacred experience.

Western monasticism accepted the prayer times thus established. Benedict of Nursia (1992: 47) integrated them into his rules which should become the backbone of monastic life in Europe. The names changed slightly: From Benedict onward the European monks and ecclesiastics know the official prayer times as Laudes; Prima, Tertia, Sexta, Nona Hora; Vespers and Completorium – the Vigiliae so important to coptic monks have lost their daily appearance and are only mentioned as preparation for special feasts. Being on time for these communal obligations is very important for Benedict; actually he sees it as basic for the discipline required of a monk. Thus his rules envisage a scale of punishments for members of the community lacking in time discipline, a scale surprisingly harsh, as he obviously wants to implant a structure quite alien to people of his era (Benedictus 1992: 148). But the continuous prayer thus structurally integrated turned into an obligation for a religious elite only (Schnitker 1996: 528–532). Religious interaction and general interaction in this elite were clearly structured along the lines of constructed time. The time-discipline developed over centuries by a monastic elite was then arguably transplanted in Geneva to a nascent capitalist society. But was it a case of religion influencing economic structures? Only in a limited way. Religion obviously played a role as internalising agent, but the ideas transferred had been transplanted into religious structures at a previous stage. It could be argued, however, that the beginning of time-keeping indeed had been correlated with religious ideas and religious institutions. But the structures of time in place 2000 years ago are rather secular. While historians point out that the attempt to use the constructed timeframe of Rome in a sacred way in early Christianity has its roots not in Judaism but in paganism (Schnitker 1996), this statement should consequently be interpreted by not defining paganism as religion but as the secular way of life of the pagan community. The appreciation of constructed time in Christianity thus passes from secular roots through monastic adaptations into the social life of a Calvinist community, to become eventually an ethical imperative of lapsed Calvinists in early capitalism.

In conclusion it could be argued that in any discussion about the origins of capitalism, a discussion virtually started by Max Weber, it becomes quite clear that changes in the constructed media of interaction have a decisive influence on their use. It can be argued that "Protestant Ethics" and the "Spirit of Capitalism" are actually such structures of interaction. The similarity of these two special structures in the West, though, culminates in a specific relation to time. Time is not understood any longer as majestic river flowing around us, but as something which has to be counted, to be kept, to be used, to be saved, in short, as a commodity. Weber's interest is in understanding the conditions leading to the adaptation of attitudes which really are counterproductive to the human quest for happiness, of attitudes leading to structures experienced at a later stage as an iron cage. While he is successful in showing the affinities of early capitalism and beginning Calvinism in their shared interest in the "calling" and time, he probably was only touching an issue dear to theologians, but not so important to ordinary people, when he brought in the issue of predestination. The theme of the "calling" really is enough to explain the adoption by Calvin of time consciousness as cultivated by an elite in monasteries, a time consciousness, though surviving only in the religious sphere, which originated clearly in a non-religious environment. Borders and boundaries, constructed in the process of building structures of domination, became integrated into a religious universe of meaning as places of hierophany; and as the emotive strength of this placing carries them through changes in structured interaction, even the collapse of the religious universe at hand might be survived by the structure thus empowered, turning them into seeds for new structures of domination.

We can argue, therefore, that some Christian traditions provided certainly for attitudes rather alien to the concerns of ecology today; but we could argue convincingly as well that religious tradition is also constantly appresented to changing frameworks of appresentation, thus reinventing itself again and again. Christian traditions thus are highly ambivalent. They developed within the framework of the pre-industrial city, and especially through the tool of time-discipline, one strand of this tradition turned out to be highly compatible with emerging industrialisation and capitalism. But on the other hand, we have the strand of compassion, leading from pre-city roots over visions of cosmic peace to a Christian mysticism shaped by Franciscan traditions and eventually to the concept of "stewardship of creation", typical of contemporary green discussions in Christian theology. Thus Christian theology justifiably could be accepted as partner in a discussion of present ecological challenges, as well as in efforts to delineate the structures of a business ethic embracing the impact of globalisation.

References

Aveni; A.F. (1989), *Empires of Times? Calendars, Clocks and Cultures*; Basic Books, New York.

Baus, K. (1940), *Der Kranz in Antike und Christentum*, P. Hanstein, Bonn.

Bellah, R. (1985), *Tokugawa Religion – The cultural roots of modern Japan*, The Free Press-Macmillan, New York (paperbook edition).

Benedictus a Nursia (1992), *Regul,* Beuroner Kunstverlag, Beuron.

Berger, P.L. (1967), *The Sacred Canopy – Elements of a Sociological Theory of Religion*, Doubleday, New York.

Chadwick, O. (1964), *The Reformation*, Penguin Books, Harmondsworth/New York.

Collins, R. (1986), *Weberian Sociological Theory*, Cambridge UP, Cambridge/New York.

Calpe, C. (1988), 'Zur Neubegründung einer Phänomenologie der Religionen und der Religion', in H. Zinser (ed.), Religionswissenschaft, Berlin.

Durkheim, E. (1915), *The Elementary Forms of the Religious Life*, George Allen & Unwin, London.

Eliade, M. (1954), *Die Religionen und das Heilige*, Mueller, Salzburg.

Esser, G. (ed.) (1915), *Tertullians Apologetische, Dogmatische und Montanistische Schriften*, Jos. Koesel, Kempten, München.

Freud, S. (1979), *Civilization and it's Discontents*, Hogarth Press, London.

Green, R.W. (ed.) (1959), *Protestantism and Capitalism – The Weber Thesis and its Critics*, Heath, Boston.

Gribomont, J. (1962), 'Pachomios', in Hoefer, J. and Rahner, K. (eds), *Lexikon für Theologie und Kirche*, Herder, Vol. 7 (ed. 2), Freiburg, pp. 1330–1331.

Heid, H., Hoft, E.-H., Rodax, K. (eds) (2000), *Jahrbuch Bildung und Arbeit – Ökologische Kompetenz*, Leske + Budrich, Opladen.

Joergensen, J. (1955), *St. Francis of Assisi*, Doubleday Imagebooks, New York.

Lebra, T.S. and Lebra, W.P. (1986), *Japanese Culture Behaviour – Selected Readings* (revised edition); University of Hawai Press, Honolulu.

Loiskandl, H. (1995), 'Religion, Ethics and Economic Interaction in Japan', in R. H. Roberts, *Religion and the Transformations of Capitalism*; Routledge, London/New York.

Lutterbach, H. (2001), 'Die Tiere folgen dem Heiligen – Eine kosmische Friedensidee im Christentum', in K. Garber, J. Held, F. Jürgensmeier, F. Krüger and U. Széll (eds), *Erfahrung und Deutung von Krieg und Frieden*, Wilhelm Fink Verlag, München, pp. 771–785

Martimort, A.-G. (ed.) (1963), *Handbuch der Liturgie-Wissenschaft*, St. Benno Verlag, Leipzig.

Mueller, D. (1975), 'Pachom', in J. Assfalg (ed.), *Kleines Wörterbuch des Christlichen Orients*; O. Harrassowitz, Wiesbaden.

Patlagean, E. (1983), 'Ancient Byzantine Hagiography and Social History', in S. Wilson (ed.), *Saints and their Cults*, Cambridge U.P., Cambridge.

Otaka, K. (1984), *Nipponteki Keiei (Japanese Management)*, Chuo-Koran-sha, Tokyo.

Quecke, H. (1970), *Untersuchungen zum Koptischen Stundengebet*, Institute Orientaliste, Louvain.

Roellig, W. (1975), 'Zeitrechnung', in K. Ziegler and W. Sontheimer (eds), Der Kleine Pauly – Lexikon der Antike, vol. 5, dtv, München, pp. 1473–1475.

Schnitker, T.A. (1996), 'Stundengebet', in Fahlbusch et al., *Evangelisches Kirchenlexikon tome 4*, Vandenhoek & Ruprecht, Göttingen.

Thompson, K. (1972) 'Religion and Economy', in The Open University (ed.), *Social Sciences, The Sociological Perspective, Unties 12–14: Beliefs and Religion*, The Open University Press, Berkshire.

Van Berchem, D. (1965), 'Thebaische Legion', in J. Hoefer and K. Rahner, *Lexikon für Theologie und Kirche*, vol. 10, ed. 2, Herder, Freiburg, p. 14.

Weber, M. (1995), *The Protestant Ethic and the Spirit of Capitalism*, Roxbury Publishing, Los Angeles.

Winckelmann, J. (ed.) (1978), *Max Weber – Kritiken und Antikritiken; Die Protestantische Ethik II*, Gütersloh, Gerd Mohn-Siebenstern.

Yawata, Y. (1981), 'Religiosity, communality and modern business organizations in Japan', *Listening*, vol. 16, no. 3, pp. 282–295.

12 Culture as an Environment: Possibility of the Developmental Study focused on the Contextuality

MAKOTO KOBAYASHI

Psychology of Welfare – A New trend of Psychological Study

As a new trend of psychological study, people pay increasing attention to the Psychology of Welfare or Welfare Psychology. This new discipline is concerned with psychological study on the conditions and processes which enable welfare of human beings through desirable adaptation to the natural, social and psychological environment. This is quite a new field of research, which came into being in the 1980s (Friedman and DiMatteo 1989). In Japan there is still very few literature about this domain of research interest. On the other hand, an increasing number of universities and colleges in Japan established or begin to establish a department of welfare studies in recent years. In such departments, Welfare Psychology constitutes an indispensable foundation of the discipline.

The German psychologist Kurt Lewin has already indicated in his famous formula B = f (P, E) that human behaviour is understood as the function of Person and Environment (Lewin 1936). So long as human behaviour shall be observed in relation to the environment, the concern about welfare and social well-being based on the optimal adaptation between human mind and the environment will get crucial position in the next generations. In this sense, Welfare Psychology is quite an actual topic, which corresponds with the new tendencies of human society today, such as globalisation, environmental problems including pollution, ageing society, a flood of information, and the structural change of education (cf. Siegal 1985).

There are two important factors, which lead to the formation of Welfare Psychology. One factor is the increase of the general interest in the welfare as a universal human right for everyone. Most studies of the social welfare policy have focused as far on the support for and care of the disabled, aged, sick or unemployed people who suffer from a marginal position in society. The guarantee of compensation for disability and the treatment of illness are of course very important components of social well-being, which require urgent improvement of the jurisdictional, institutional, medical and educational environment. At the same time, it has been gradually taken into account that welfare is not a special problem for the people who suffer from disadvantageous conditions, but basically a universal and casual right for all human beings, regardless of whether the person is confronted with such critical or depriving situations for the moment. In correspondence with the increasing rate of the aged people in the whole population, the society came to realise more clearly this universal aspect of human welfare, which shall be relevant for everyone. Health Psychology is another sub-discipline in the psychological studies, which is concerned with the psychological conditions for physical, mental and social health. Health Psychology represents this universal perspective on the human well-being (Taylor 1991). The emphasis of the health concept in this field implies the shift of focus from the treatment-oriented welfare to preventive welfare. It is why the quality of life (QL) has got such importance in actual discussions about the human welfare at all.

The other factor, which contributed to the development of this new discipline, is the increased awareness or interest in the inner aspects of welfare. Social welfare has basically two aspects:

1) institutional systems including law, organisation and the occupational qualification, and
2) psychological aspects, which have been considered in the concept "Quality of Life" (QL).

In the studies of social welfare, most attention has been paid to the institutional aspects as far. But after having achieved a certain level of living standards in the material and institutional sense, people began to pay more attention to the psychological aspect of welfare. Welfare is primarily a value concept, which indicates the desirable state of human life to be pursued and is based on the subjective experience in its essence (Sato and Yamane 1998). Institutional systems such as medical service, insurance, protection against unemployment etc. surely provide necessary conditions for the welfare, do not guarantee its realisation, however. To deal with the question of human welfare theoretically as well as politically, our inner, subjective world, particularly the value system, which provides us with life orientation, must be taken into consideration.

It implies also the enlargement of the concept "environment" itself. When it is spoken about the "welfare environment", it includes not only the political, economic and social environment, but also the psychological environment. As the human value containing moral judgement, happiness, or the ideal of life varies strongly after the cultural context (Trommsdorff 1993; Azuma 1997), the "culture" as an environment shall be also a critical factor, which takes substantial influence on the social welfare policy in the determination of its orientation. The state of mind shall be a central concern in the welfare concept, and it should be pointed out that the value system, which defines the question of happiness is a dependent variable of culture. It means that the integration of the institutional and the psychological perspectives in the welfare issue shall be intermediated through the cultural factor. In a concrete term, it can be formulated in a question as follows: Do Japanese people feel happy under the same conditions of life as German people feel happy?

Importance of the Concept "Quality of Life" (QL)

One of the most important key concepts in Welfare Psychology is the "Quality of Life" (QL). As already mentioned, welfare includes not only supplement or treatment against the existing problems disturbing human life, but also the preventive actions, which guarantee the basic conditions for the realisation of happiness. For this purpose, the enhancement in the quality of life is indispensable. It is also expressed in the concept "Culture of Welfare". Historically, the concept of welfare has been developed in accordance with the progress in the philosophy of human rights in the modern age. In the general understanding of human rights today, the right to enjoy a standard of living and to enjoy cultural life is widely acknowledged in various relevant documents as an essential component of human rights. As examples, two documents are illustrating in this concern: 1) Articles 25 and 27 of the Universal Declaration of Human Rights which has been proclaimed on 10 December 1948 by the General Assembly of the United Nations, and 2) Articles 13 and 25 of the Constitution of Japan, which was promulgated in 1947.

Universal Declaration of Human Rights (United Nations 1948)

Article 25.
(1) Everyone has the right to a standard of living adequate for the health and well-being of himself and of his family, including food, clothing, housing and medical care and necessary social services, and the right to security in

the event of unemployment, sickness, disability, widowhood, old age or other lack of livelihood in circumstances beyond his control.

Article 27.
(1) Everyone has the right freely to participate in the cultural life of the community, to enjoy the arts and to share in scientific advancement and its benefits.

The Constitution of Japan (1947)

Article 13.
All of the people shall be respected as individuals. Their right to life, liberty, and the pursuit of happiness shall, to the extent that it does not interfere with the public welfare, be the supreme consideration in legislation and in other governmental affairs.

Article 25.
(1) All people shall have the right to maintain the minimum standards of wholesome and cultured living.
(2) In all spheres of life, the State shall use its endeavours for the promotion and extension of social welfare and security, and of public health.

In these articles, social welfare combined with a standard of living and the cultural life is recognised obviously as a substantial part of human rights. The concept "Quality of Life" has also been originally derived from this proclaimed right to a standard of living including cultural quality.

According to the Centre for Health Promotion in the Department of Public Health Sciences, University of Toronto, the Quality of Life is defined as follows:

> "Quality of Life is an area of study that has attracted a great deal of interest over the past ten years, particularly in the areas of health and social services, but increasingly in medicine, education, and others. The study of Quality of Life is an examination of factors that contribute to the goodness and meaning of life, as well as people's happiness. It also explores the inter-relationships among these factors." (Centre for Health Promotion, University of Toronto)

The key points in this definition are "goodness" and "happiness", and these words are understood in relation to the "meaning of life". The Quality of Life is, therefore, a value concept, which indicates a certain orientation concerning the goal of life. It means the state of life conditions which guarantee human rights and the full development of the human personality in the sense of self realisation to all people in the society.

Welfare Psychology today points to 5 aspects of the Quality of Life (Masuko 2000):

1) Control of physical state.

This is concerned with the most fundamental level of welfare, which is closely connected with health. It includes the lack of pain or other unpleasant symptoms of the body as well as the lack of any physical or mental disability. However, as the WHO clearly states, health is not only the lack of disease or physical dysfunctions. It should also have some more positive meaning, namely a sufficient level of physical as well as mental activities.

WHO Definition of Health
"Health is a state of complete physical, mental and social well-being and not merely the absence of disease or infirmity." (Preamble to the Constitution of the World Health Organisation: adopted by the International Health Conference, New York, 19–22 June, 1946 and entered into force on 7 April 1948)

2) Guarantee of the minimum economic level

This is concerned with the economic aspects of the Quality of Life. It is often argued that the guarantee of the economic standards is a necessary, but not sufficient condition for the realisation of QL.

3) Fulfilment of mental activities

This demand deals with the mental satisfaction in life. Whether the person feels fulfilled in his or her mental activities contributes largely to the realisation of QOL.

4) Fulfilment of social life

It covers the quality and range of the social activities including occupational, regional and private life domains. The quality of human network including family, friends and colleagues has particular significance in this concern. It is well known that the support on account of such social networks is a significant contribution to the promotion of physical and mental health.

5) Spiritual fulfilment

Whether the person has a convincing explanation about the meaning of his or her own life has been dealt with as far in the term of "existential satisfaction". This existential or religious level of QL has been widely neglected in the Japanese social policy as far, requires however higher attention, particularly in connection with the hospice or the mental care for the dying patients. Fundamentally, this constitutes a substantial aspect of QL for everyone in everyday situations.

The question is whether the content of this goodness or happiness has some universal substance or is subject to the cultural and generational variability of value orientation.

Life Orientation Expressed in the World Youth Survey

The results of the World Youth Survey implemented by the Youth Affairs Administration of the Management and Coordination Agency every five years since 1972 provide us with some relevant information about the cultural differences of the value orientation of today's youth and its change (Youth Affairs Administration 1999). This survey covers various aspects of value and life style. With regard to the understanding of social welfare, two critical results shall be shown from this survey which seem to indicate the cultural variability of the welfare concept. As for the relationship between the country and the self, two questions were put:

1) Are you proud to be a citizen of your country?
2) Would you like to do something for your country?

As shown in Table 1, the rates of the Japanese and German subjects who answered "Yes" to these questions are quite low in comparison to the rate of other countries' youth. It reveals that the adolescence of these two countries keep greater psychological distance from their countries than the youth of other countries. If the society thinks about an appropriate measure to fulfil their need for social welfare, this great distance between the country and the personal self at German and Japanese youth should be taken into account. The content of the social well-being perceived by Japanese and German youth must concentrate more on the happiness experienced within their private life. This tendency is underlined by the results concerning the answer on the question:

How much interest do you take in politics in your country?

The rates of German and Japanese youth who were interested in the country's politics were not quite high, either (Table 2). On account of these results, it should be pointed out that the life orientation of the Japanese youth today cannot be simply described as "collectivist" due to the large distance of the individual from the country and politics. In order to clarify the cultural variety of the welfare concept, it is actually much more important to analyse, which criteria are relevant for the satisfaction with the life reality in each cultural context.

Interestingly, the level of satisfaction with the society at German and Japanese youth is rather low, either, as shown in the Table 3. Only the youth of Korea and of Russia shows still lower level of satisfaction. Besides the level of satisfaction, the reason why the person is discontent with the society also provides significant information about the culture-specific perception of welfare. Though the level of satisfaction with society at Japanese and German youth is quite similar, there is much difference between both groups with regard to the reason or dissatisfaction. The problems which Japanese youth pointed out most were: "Work/pay difference based on education" and "Political system is flawed", while the most serious problems for the German youth were: "Difficult to find work and many are unemployed" and "Gap between rich and poor too wide". The complaint about the insufficiency in the protection of the environment and about the inadequate social welfare was also characteristic for Japanese youth while these complaints were not expressed by German youth.

From this survey, we can already remark the cultural variability in the perception of welfare. To make clear the concrete feature of the QOL which is congruent with the respective social and cultural context, however, it is necessary to inquire about the more detailed contents which the youth of each country understand under the concept "welfare" and what they exactly pursue in their struggle with the existing reality.

Moral Dilemma and Culture

Another approach to find out the culture specific content of the welfare concept would be the analysis of the teaching materials used in the school context which particularly contain implications for moral value. It is connected with the question: What makes a moral dilemma in various cultural context? The teaching materials of moral education reveal the critical dimensions of moral value, which have relevance for human relations and the life orientation in the concerned society. If the main and typical pattern of conflict can be found out through the interpretation of such materials, it would lead to the insight into the essence of the welfare, which shall be protected by an effective solution of the conflict.

Here two examples of teaching materials will be shown, which are used in the moral education at Japanese elementary schools. The first one is a poem, and the second one is a short anecdote.

1) Big Catch (Akahori et al. 1995)

Big Catch	**Guter Fang**
The seashore in the morning glow There comes a cry of big catch A big catch of sardines We've made a good catch of them!	Der Strand im Morgenrot Man hat einen guten Fang gemacht Von den Sardinen Haben wir einen guten Zug getan!
The seashore is in festive mood With a joyful celebration While deep in the sea The funeral For thousands of sardines Proceeds in silent sorrow.	Am Strand herrscht die laute Freude Wie ein Fest Während unten tief im Meer Die Bestattung Für tausende von Sardinen In leiser Trauer vorgeht.

(Japanese Original: "Tairyô" by Misuzu KANEKO; English and German translation by the author)

2) The Bill of the Mother (Akahori et al. 1995)

On Sunday morning, a school boy Daisuke put a sheet of paper on the breakfast table. His mother was surprised as she got to know that it was a "bill" of Daisuke for his helping acts in household matters. At lunch Daisuke found 400 yen in front of his dish, the amount just required by him. Daisuke was very pleased that his demand on the "bill" was accepted, and thought he would like to write such bills more in the future. Under the money, however, Daisuke found a sheet of paper. He saw it was a bill of his mother for all the acts she had done for Daisuke; the amount of the helping acts done by the mother including cooking, nursing, education and all other household matters was much more than Daisuke's acts, of course. But Daisuke remarked that the amount of the mother's bill was 0 yen: the mother required nothing as reward for her helping acts. Daisuke read the bill of his mother again and again. As Daisuke came to realise the egoism in his behaviour and the selfless helping of his mother, his eyes were filled with tears. (Outlined translation by the author)

The poem "Big catch" written by Misuzu Kaneko is used concerning the educational objective "Respect for the dignity of life" (The Guideline to the Curriculum for Elementary School: Ministry of Education 1998). The moral message, which is emphasised in this poem is the importance of perspective taking and the pluralistic view. This is based on the insight that the affair, which offers pleasure to a person can be the cause of agony or suffer for another person. This is a special type of moral dilemma, which is less concerned with the conflict between moral norms or imperatives, but much more with the conflict between personal perspectives. It seems that the consideration of other persons' perspectives is particularly emphasised in the moral education at Japanese schools, but this hypothesis must be subjected to further empirical research through cross-cultural comparison of teaching materials for moral education.

The short story "The bill of the mother" is also based on the exchange of perspectives between the mother and her son Daisuke. Here it is implicitly expected that the children come to recognise the reciprocity of behaviour in human relations and the possibility of pure gratitude beyond the calculation of interest. Besides the cross-cultural comparison in the norms and objectives of the moral educational as itself, it is also very important to see the way how a certain moral norm is transmitted by the use of a certain material and what kind of view on human welfare and happiness is hidden behind such explicit moral norms. From these examples (To which extent they are representative for the moral education in Japan is another question.), we can suppose at least that the arrangement of the conditions, which guarantees flexible change and exchange of perspectives is a substantial psychological factor for the realisation of the QL in the context of the Japanese culture.

Conclusion

As far, the moral issue has not been treated systematically in the discussion on Welfare Psychology. But so long as the image of happy human life, the proper objective of social welfare, depends strongly on the value orientation of the recipients of welfare (Being happy is a form of goodness.), the analysis of the moral value must be integrated into the scientific research on Welfare Psychology. In this concern, it is recommendable to rediscover the actual sense of the religious belief from the perspectives of Welfare Psychology. It is connected to the fundamental question: Can a person really have a happy life without providing it with any ultimate meaning of life? The answer seems to be "no". Only with the deep belief in the ultimate meaning of human life, the person can reach the fulfilment of his or her personal, social and spiritual needs. And in this struggle, it is always taken into account, of which culture the concerned person is a resident, i.e. in which meaning system the person recognises and experiences moral and welfare values. It would lead to the enlargement of the welfare concept. Only with the cultural variability and morality of welfare being taken into consideration, the study on welfare policy will get actual validity, with which the welfare system can correspond with the universal, culture-specific and person-specific needs of each citizen in the present society.

References

Akahori, H. et al. (eds) (1995), *Dotoku-kyoiku no atarashii tenkai (New development of moral education)*, Shogakukan, Tokyo.

Azuma, H. (1997), 'Nihonjin no dotoku-ishiki (Moral consciousness of the Japanese)', in K. Kashiwagi, S. Kitayama and H. Azuma (eds), *Bunka shinrigaku (Cultural psychology)*, University of Tokyo Press, Tokyo, pp. 88-108.

Brown, I., Friefeld, S. and Schiller, C. (1993), *Quality of life for persons with developmental disabilities (An annotated bibliography)*, The Centre for Health Promotion, University of Toronto, Toronto.

Friedman, H.S. and DiMatteo, M.R. (1989), *Health psychology*, Prentice Hall, Englewood Cliffs, N.J.

Lewin, K. (1936), *Principles of topological psychology*, McGraw-Hill, New York.

Masuko, K. (2000), *Fukushi-bunka no kenkyu: Nyumon kara jissen e no shiten (A study of welfare culture: From introductory and practical perspectives)*, Hokuju-shuppan, Tokyo.

Ministry of Education (1998), *Shogakko Gakushu-shido-yoryo (The Guideline to the Curriculum for Elementary School)*, Ministry of Finance Press, Tokyo.

Otani, T. and Tokunaga, R. (1948), *The constitution of Japan*, Apolo-sha, Tokyo.

Sato, Y. and Yamane, R. (Eds.) (1998), *Fukushi shinrigaku (Welfare psychology)*, Gakugei-Tosho, Tokyo.

Siegal, M. (1985), *Children, parenthood, and social welfare in the context of developmental psychology*, Oxford University Press, Oxford.

Taylor, S.E. (1991), *Health Psychology*, McGraw-Hill, New York.

Trommsdorff, G. (1993), 'Entwicklung im Kulturvergleich', in A. Thomas (ed.), *Kulturvergleichende Psychologie: Eine Einführung*, Hogrefe, Göttingen, pp. 103–143.

United Nations (1948), *Universal Declaration of Human Rights*, United Nations Department of Public Information, New York.

Youth Affairs Administration (1999), *The Japanese Youth: A summary report of the sixth World Youth Survey, 1998*, Management and Coordination Agency, Tokyo.

Table 1: Attitude towards the country in cross-cultural comparison

Nationality[1)]	Proud of Country[2)]	Nationality[1)]	Service to Country[3)]
Thailand	99.8	Thailand	98.7
Philippines	99.4	Philippines	96.9
Brazil	91.9	Brazil	94.1
U.S.A.	91.7	Korea	82.9
Sweden	86.2	Sweden	70.8
U.K.	85.5	U.K.	66.5
Korea	84.2	Russia	66.1
France	77.4	France	61.8
Japan	77.1	U.S.A.	53.3
Russia	68.4	Japan	49.3
Germany	52.1	Germany	28.3

Source: The Japanese Youth: A Summary Report of the sixth World Youth Survey 1998. Youth Affairs Administration

Notes: 1) The nationalities are put on the ranking order respectively.
2) The percentage of respondents who said they are proud to be a citizen of their country.
3) The percentage of respondents who said they want do something for their country.

Table 2: Interest in politics of the country in cross-cultural comparison

Nationality[1)]	Interested in Politics[2)]
Thailand	77.0
Philippines	74.5
Korea	71.2
U.S.A.	59.5
Russia	45.7
France	41.1
Germany	40.5
Japan	37.2
U.K.	35.7
Brazil	32.0
Sweden	31.8

Source: The Japanese Youth: A Summary Report of the sixth World Youth Survey 1998. Youth Affairs Administration

Notes: 1) The nationalities are put on the ranking order.
2) The percentage of respondents who said they are interested in politics of their country (adding the answers "very interested" and "quite interested").

Table 3: Satisfaction with society in cross-cultural comparison

Nationality[1)]	Interested in Politics[2)]
Thailand	76.0
Philippines	72.7
Sweden	68.7
U.S.A.	67.8
U.K.	63.4
Brazil	41.5
France	40.4
Germany	40.0
Japan	35.2
Korea	25.7
Russia	24.9

Source: The Japanese Youth: A Summary Report of the sixth World Youth Survey 1998. Youth Affairs Administration)

Notes: 1) The nationalities are put on the ranking order.
2) The percentage of respondents who said they are satisfied with society in their country (adding the answers "satisfied" and "slightly satisfied").

Part V:
Economics and Business

13 Environmental *Kaizen*: a German-Japanese Comparison*

GYÖRGY SZÉLL

Introduction

The article is based on the findings of a comparative research project, which has been co-financed by the German Federal Environmental Foundation – the biggest of its kind in Europe. The research has been realised within ECOS Japan Consult, Osnabrück, Germany.

Within Germany as well as in Japan five small and medium sized companies in the industrial sector were investigated in regard to

- how far they are able to cope with the challenges of sustainable production,
- how far there are differences in this respect between German and Japanese companies,
- and how far similarities can be found.

The first hypothesis is: As well the German as the Japanese corporate governance have good preconditions due to their participative structures bring especially good preconditions for the development of sustainable production and distribution.

Second hypothesis: Historically speaking, the participative structures of the two countries have developed quite distinctively. On the one hand the continual improvement process "kaizen" and direct participation, as it was practised by Toyota and developed by its chief engineer Taiichi Ohno, which is

---

* I am especially grateful to Dr. Peter-Jörg Alexander and Diplom-Sozialwirt Wilhelm Meemken for their most valuable contributions to the success of the project. A follow-up conference took place in 2001 in Berlin (ECOS 2002).

based on the Ie-structure (Széll 2001), on the other hand codetermination with its representative structures of the works council and participation of employees in the supervisory board (Burschel 1995; Hildebrandt and Schmidt 2001).

The project was conceived as interactive, in which representatives of the company participated besides experts from institutions (like the RKW, Social Science Centre Berlin, the Centre for the Environment and Technology, Osnabrück, environmental business associations as well as the German Federal Environmental Foundation itself). This innovative method permitted to reach qualitatively new insights and to facilitate the implementation of the commonly won research results into the praxis.

After the ISO/EN 9000 norms to promote quality management had been introduced on a world-wide scale at the end of the eighties of last century, the ISO/EN 14000 norms for ecological management followed at the beginning of the nineties. At the moment work on the definition of the ISO/EN 21000 norms for Zero Emission is in progress, led by Japanese institutes, the United Nations University and Japanese companies.

The essential results of the project may be summarised as follows:

A precondition for successful environmental management is to involve from the very beginning all employees of the company into this process. The participative structures may be historically diverse, but it is decisive that they exist and function and are further developed within environmental management. The Japanese pump producer Ebara seems to be the most advanced with his project "Zero Emission City" in this regard. So, the results, which we obtained, are not only valid for the two countries we investigated, but are in principle valid for the whole world – with the necessary socio-cultural modifications.

The Context with the Overall Problems

The question of environmental management is central for the realisation of sustainable development and therefor also within socio-economic research. The topic of environmental kaizen is on the background of comparative studies between Germany and Japan of special interest, because it allows to demonstrate foremost the differences and similarities in regard to the company and social structure.

State of the Art in Environmental Sociology

Environmental sociology surprisingly has only established itself within the last decade as well on the international and the national levels in Germany and Japan. I myself founded during the 12th World Congress of Sociology in Madrid in 1990 first an ad-hoc group "Environment & Society" within the International Sociological Association (ISA), which then very quickly has been promoted in 1992 as a Research Committee 24, and which today is one of the most active within ISA.

A similar development took place at about the same time on the national level in Germany and Japan. In 2001 RC 24 of ISA organised for its tenth anniversary an international congress at Bukkyo-University in Kyoto, at which took part a large number of Japanese and German colleagues, of which quite a few are here today again.

However, it has to be stated that environmental sociology – in spite of all its global and fundamental importance – has still not yet been promoted as a part of the classical core areas of sociology, as also the topic and the structure of the XVth World of sociology in July 2002 in Brisbane proved.

To treat adequately the topic and the problems and because of the necessary interdisciplinarity of the approach besides sociology economics as well as psychology, anthropology, law and naturally also the natural sciences have to be included. This interdisciplinary approach demands also new methods. In so far there exist still a huge research and development need, to cope with the challenges for the survival of humankind.

> *Also a travel of one thousand miles*
> *begins with the first step*
> (Japanese wisdom)

Kaizen – a Survival Strategy

"With the help of Kaizen to success!" "Japan's key industries have the highest efficiency!" "Until 2010 Japan will become the leader in the environmental sector."

That are just three headlines, with which Japanese companies make publicity. Especially the Kaizen-concept has been mentioned recently more and more as the reason for their success. And even so, though big companies like Toyota started already shortly after World War II to introduce small, continuos, incremental improvements in the production process. So, it no surprise that just the Japanese car manufacturers are among the world-champions of productivity. It was often too late that American and European car manufacturers began to introduce Kaizen, to win back market shares from their Japanese competitors.[1] Not only the car manufacturers but also in the environmental technologies Japanese companies trust from the very beginning on Kaizen – with the target to overtake first all world leader Germany – to become 'ichiban' also in this growing sector of the world market.

What is Kaizen?

The word 'Kaizen' – also known as permanent improvement process – is formed by the words 'change' (kai) and 'good' (zen). In the core it is foremost only to be regarded as a collection of well known management instruments. Among them are: Quality circles, Zero-default principle, Zero-buffer process, Just-in-time deliveries, integration of research and development departments of the suppliers and the customers. Besides it is not a single or temporary management strategy, but a continuous, incremental rationalisation method, which asks for the involvement of *all* employees of the company. New is only the focus on co-operation and teamwork on the one side and customer orientation on the other side. And in regard to customer orientation not only Germany is a developing country. (Cf. figure 1)

In this paper the focus lies on the possibilities to gain advantage out of the combination of company environmental protection and Kaizen, to realise permanent cost reductions and process improvements – as some Japanese companies already do.

1 Certainly over the last couple of years due to financial mis-management as well on the national as well on the company level, the Japanese economy is in great difficulties and has lost a lot opf their competitive edge. This situation led that European and American car makers took control over the majority of Japanese car makers: Nissan by Renault, Mitsubishi by DaimlerChrysler, Isuzu by General Motors, Mazda by Ford. Only Toyota, Subaru and Honda remain so far independent.

Figure 1: KAIZEN – a comprehensive approach

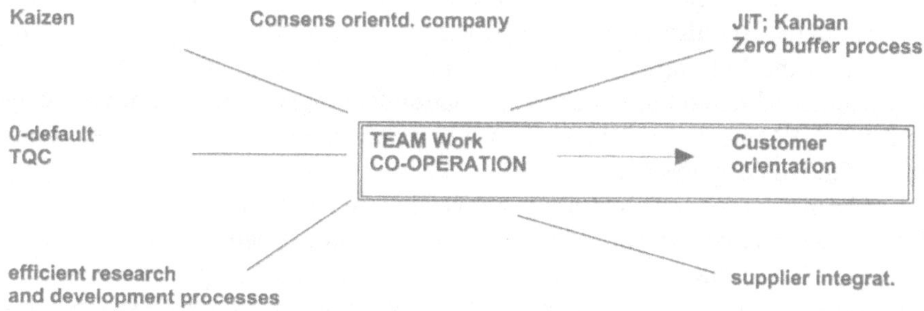

Company and Inter-company Fundamentals of Kaizen

Kaizen as a Social Process in the Company – the Team Spirit

The roots of Kaizen lie in the praxis of Total-Quality-Circle (TQC) activities, i.e. to work in a team. In the framework of Kaizen that does not merely mean statistical control to improve product quality, but in the extension of quality circles all over the company – starting with the top-management to the lowest levels [stairs have to be cleaned from the top to the ground!]: Japanese companies regard as their main priority in this context the improvement of the qualification of their employees. Because only when the employee is qualified, it is possible to think of improving high quality products and to satisfy the customers.

Naturally even Japanese employees are not motivated by themselves to take part in TQC-activities. Therefore the personnel department has not only to offer material incentives but also promotion within the company. This again has to be embedded into a long-term personnel development strategy, which leads to the result that Kaizen per se cannot be practised as a short- or medium-term strategy – as it is often practised in German companies – but as a long-term company perspective.

The same is particularly true in the same way for the Japanese environmental sector, which is the principle for Japanese companies in general: Supported from top-management – in Japan Kaizen is CEO-business! – besides Zero-default products the main concern lies in implementing environmental targets with the help of TQC-activities. A particular concern in this context is

'Zero emission', i.e. the target of 'Zero environmental pollution', however, in the meantime also the reduction of 'Just-in-time'-deliveries (JIT), which have led – due to the increasing traffic on Japanese roads – to heavy exhaust of pollutants. But also in this context the Japanese environmental sector tries – through the long-term practice of Kaizen – to enlarge their share on the environmental market against the international competition. (Japan, Germany and the USA are the world leaders with each about 20 % of the environmental technology market.)

German employees certainly are at least as well technologically qualified as their Japanese counterparts. On the contrary, still the German employee – due to the in some regards better vocational raining – can even better compete in this regard his Japanese competitors. However, German companies are often disadvantaged, what concerns team spirit in comparison to Japanese companies: It is self-speaking, when often it is asked for, to better train the core competence of 'team- and co-operation capability' of individually thinking German employees. In this case German companies should involve the members of the board as well as the concierge. Foremost a new approach seems to be necessary for the top-management; not only profit maximisation, but optimisation of the quality of performance and customer satisfaction should rank at the first place. That is how the leading Japanese companies see it – and the success is on their side.

How are Long-term TQM-Activities in German Companies Implemented?

TQC has first of all to be an affair of the top-management and be regarded as a permanent CEO-affair. The TQC-meetings are to be organised by a co-ordinator for environmental affairs, and Quality circles to be moderated and co-ordinated by employees – and not necessarily by the supervisors, only include few persons (in Japan four to six), last one to two hours per week, and at least temporarily take place during working-time. According to the Japanese approach they involve all employees. It is worthwhile to start only with volunteers. It can be important that topics are on the agenda, which have been proposed and chosen by the participants themselves (Figure 2).

Figure 2: Organisational model of TQC

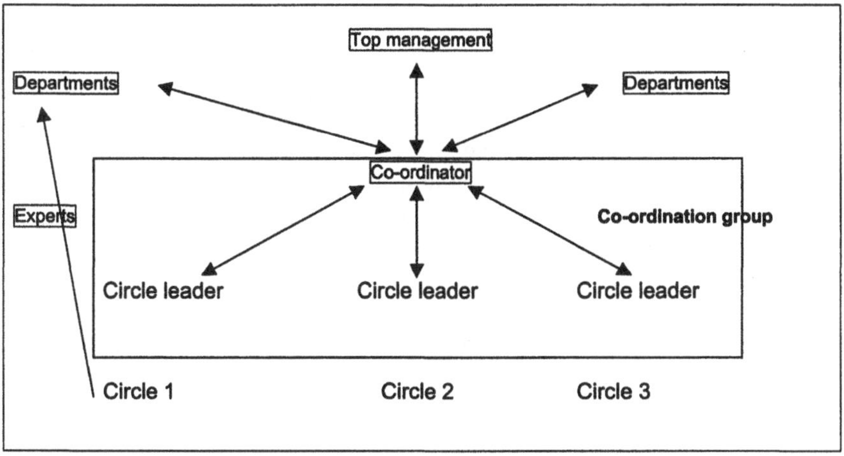

> „A basic knowledge of the participants over the interrelationship of their tasks with the environmental protection in the company is necessary. An effective participation in the quality circles is promoted if the technologies for problem solving, idea generation and creativity as well as the quality techniques to be applied are mastered. The investment in training before the first meeting is therefore always of benefit." *(Cf. Butterbrodt, p. 85)*

Kaizen as Link to the Market – Customer Orientation

A lack of customer orientation in regard to environmental problems has already often led to problems with the image of the company and to enormous, but evitable costs, as the case of Shell/Brent Spar has demonstrated in 1996. The increasing critique in regard to customer orientation in Germany has proven very clearly that only a few German companies apply the Kaizen-concept correctly. And even so that the also in Germany much read Guru of Kaizen, Masaaki Imai, has clearly expressed that Kaizen besides product quality includes also customer oriented delivery times and service. Kaizen means too, to fulfil customer wishes through communication, i.e. to respect them as well in the construction process and the environmental protection through integration of all suppliers. Above that this gives an explanation, why Japanese companies are bringing permanently product improvements on the market, and contribute to the effect that always shorter innovation cycles are enforced on the international competitors. Nevertheless German engineers still believe quite often – in face of the evident Japanese market success[2] – that Japanese companies are not able to realise big innovation jumps.

Closely linked to the Japanese Kaizen-concept are the Quality circles as well as customer orientation: all activities of TQC are customer orientated. So, e.g. Komatsu, a direct competitor of Caterpillar, has defined his supreme TQC-target as follows: "The satisfaction of all customers of Komatsu in the whole world has to be guaranteed by reasonable, cost orientated research and development as well as marketing efforts and service". No wonder that German business people, which have to do with Japanese companies, are without any exception charmed by the extreme friendliness and politeness of the Japanese.

Behind that phenomenon stand always practised rules for behaviour; good behaviour is still very important within Japanese companies. In the meantime this feature has been discovered by many German companies as well. Training courses for good behaviour are an indicator for this.

What Does Customer Orientation Mean?

Four factors are central in this context:

1. Customer orientation has to be an important management target, where all employees have to be involved.
2. Communication with the customer has to be a priority.
3. Customer satisfaction has to be evaluated regularly.
4. A permanent effort to improve customer satisfaction has to be a routine.

Back to the environmental problems: Customer satisfaction does not just signify image care – e.g. improvement of their packaging or waste policy. Customer satisfaction signifies too to be active within political, social, legal and other arenas and to try to develop environmental-friendly ways. In this context it should not be forgotten that what is today environmental-friendly, is not necessarily so tomorrow. Therefore environmental kaizen has to be flexible, has to adapt permanently to new developments and to evaluate them.

---

2 However, Japan is still in absolute numbers the second biggest export nation in the world.

Kaizen in German-Japanese Comparison

A comparison of the kaizen-praxis in Japan and Germany can only be roughly done here. However, there are some clear tendencies, where differences between Japanese and German companies can be demonstrated:

Company philosophy and time dimension The Japanese company is first of all determined by its company philosophy of continuity, the permanent improvement of its products, its client orientation as well as its long-term entrepreneurial perspective. In contrast to that in Germany during reorganisation often the old structures are completely overcome and the companies – in difference to the Japanese – in general plan only in a mid-term perspective. In regard to Kaizen this means foremost that in Japan improvements are introduced even when the market situation is satisfactory. Whereas in Germany improvements are only implemented when the market situation is not satisfactory – and then often the older positive elements of the existing structure are eliminated in a hectic atmosphere.

Employee orientation In Japanese companies the smallest unit is the working group, in German companies that is the individual. In so far the strength of the Japanese is their group-discipline, whereas for the Germans it is their self-discipline.

Project realisation In contrast to Germany during the project realisation much more time is spent in Japan for planning. The main reason for that lies in the participation and co-ordination in the preparation of decision-making and realisations with all those concerned (figure 3). This procedure has the advantage that in the realisation phase a few errors occur, and that there is much less resistance than in German companies. The results are evident: The realisation of projects can be accomplished quicker, smoother and above all with less costs.

Figure 3: Model of Japanese/German planning processes within the company in comparison

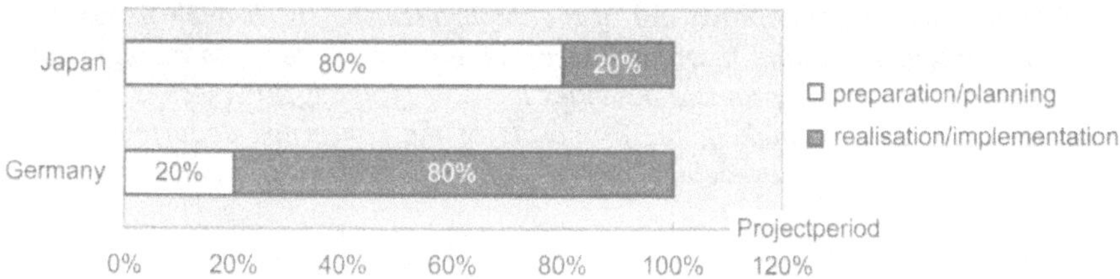

The example of EBARA Corporation[3]

Hiroyuki Fujimura, President of the EBARA Corporation, insisted already in his inaugural speech in 1989, that environmental management has to become more important within EBARA. By that time President Fujimura presented a medium- and long-term business plan, which enlarged the focus of the company on the areas water, air and the environment. With them the new company targets had been defined. Today EBARA Corporation is one of the most important companies within the environmental sector: 30 branches and a turn-over of more than 3.3 billion US $ made EBARA one of the world leaders in industrial plants. EBARA started in the year 1912 as a pump-producer. Today the production consists of system components, parts and services in the areas 'air', 'water' and 'environment'.

Environmental protection as company target: The implementation by EBARA

The continuation of the new business plan started intensive discussions within the different departments. The management received by it a feed-back on the situation and their new company targets. In the follow-up a department for environmental management was created for the implementation of the environmental concept. With its help the environmental targets and the results of the discussion with and between the employees were transformed into concrete action. The company could rely in this phase on the over the past years accumulated environmental protection know-how. In the meantime all embracing environmental concepts and activities – including so-called in-house audits and regular training of employees – are practised, and there exists for those employees, who bring forward improvement proposals for the environment, a bonus system. The result is to be seen: the EBARA Corp. has reached its first step in its environmental-kaizen strategy, and has been certified as one of the first Japanese companies according to EN ISO 14000 ff.

The Environment-management department

The Environment-management department is with its five collaborators the central institution for company environmental protection for the whole enterprise. Under its comprehensive direction are as well the central company environmental committee and the environmental committees of the different branches. Their members are composed out of the department heads, branch heads and factory directors. Beyond that the environmental committees create single special committees, which promote activities locally. Typically Japanese is the procedure: For decision-making the proposals of the concerned departments are discussed first of all in detail, before the management takes the decision.

[3] Headquarters are in Kawasaki.

Through this procedure a feed-back happens, which helps to secure the support of all employees.

The Zero-Emission-Project [4]

The core-element of the EBARA-environmental initiative is the Fujisawa-Project, where economic considerations on the one hand and the protection of the environment on the other hand are brought together in a reasonable way. The transformation of the future visions by the management with a total budget of 25 billion YEN (210 million US $) into a concrete concept was a task of working group, consisting of scientists and managers. Their target is the realisation of a city, in which there will be no more emissions, the 'Zero-Emission-City' In this Zero-Emission-City there will not only be integrated the employees of EBARA; but the idea is, to initiate a collective learning process for the co-existence of man and nature in daily life.

In the first phase of this project company own apartments have been transformed to gather first experiences. One result of the test-phase was: For EBARA even with today's technology the Zero-Emission-City is feasible But became clear too, that much commitment and big motivation of all those concerned is necessary, because even a well-thought ecological infrastructure is not a guarantee for environmental behaviour of the inhabitants.

For the second phase of the project more than six acres of the factory estate in Fujisawa have been provided. More than 450 apartments were planned as well as a training centre for the employees. Above that a series of infrastructure installations like those for biowaste and biogas with an integrated energy unit, waste incineration et al. as complements.

The Fujisawa-project makes it clear that environmental protection cannot be limited on the enterprise and the production process, but has to integrate the citizen as well. However, there are for EBARA Corporation also very concrete material benefits: Through the Fujisawa-project EBARA is creating its own important reference project with an astounding, lasting public media presence, i.e. public relations for free. That has led to the effect that EBARA in Japan today is regarded as well under technological and social aspects as an example for progressive environmental protection.

Kaizen is not to be transferred to Germany in a 1:1 relation; for that the cultures are too different. Besides there are strengths of German employees, and part of it is certainly their creativity, which is to be used. Here it is first of all the task of the co-ordinators for environmental kaizen that their team members are not only to be convinced of the importance of environmental kaizen, but also to use the creative potentials of the employees.

---

4 For more details cf. Kühr and Széll 1997

In consideration of the Japanese competition one thing has to be kept in mind: When Japanese companies in the environmental sector decide to become number one on the world market then they will do everything to reach this goal. Whereas in Germany there are often great visions, but a lack of implementation, because resistance within the company due to a lack of coordination or even resistance in the political, social, legal or cultural company environment occur. Kaizen with its customer orientation can begin here, to calculate resistance from the very beginning to overcome it.

Here follows the example of a successful German company:

From Quality management to Environmental management – the example of the international logistics company Hellmann –[5]

Parallel to the introduction and certification of the quality management system according to DIN EN ISO 9002 the working group environmental protection developed a system to improve the enterprise group environmental protection, for which the company already in 1992 received the environmental award of the DEKRA Environment GmbH.

Priorities were the respect of legal obligations, the development of environment-friendly transport systems, the controlled use of energy, environment-friendly purchase, separation of waste as well as environment-friendly shaping of the exterior of the company. For the sensibilisation resp. strengthening of the environment-consciousness of the employees, they were involved into this concept from the very beginning.

After the first draft of the "Handbook for the business environment protection" had been finished, the contents of the handbook was structured accordingly to that of the quality management, to use so the possibilities to comply with the demands of the EC-Eco-Audit ordinance or the prepare the foundations for the certification according to DIN EN ISO 14001.

So the preparations were made for a validation according to the EC Eco-Audit Ordinance and implemented, though the legal requirements (Eco-audit Law from 15 December 1995) did not allow yet that companies within the transport sector participated "voluntarily at a EC-system for environmental management and environmental certification".

Target of the top-management was therefore to apply for the certification according to DIN EN ISO 14001, because this certificate is internationally recognised.

So, as a complement to DIN EN ISO 9002 an environmental management system was established, where environmentally relevant elements of DIN EN ISO 14001 were integrated. The following figure shows the overlapping of the management systems.

5 Headquarters are in Osnabrück, Germany

Figure 4: Management systems

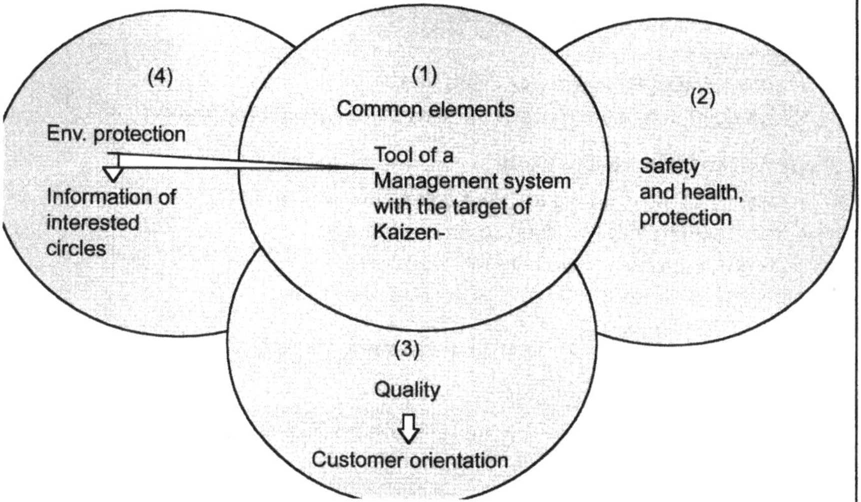

(1) Management systems within a dynamic-cyclical process
(2) Safety management according to ISO 17000 ff. with a specific target
(3) Quality management according to ISO 9000 ff. with a specific target
(4) Environmental management ISO 14001 with a specific target

The main idea of all management systems presented here – and especially DIN EN ISO 1400, starting with the basic elements of DIN EN ISO 9000 ff. – is to deliver a qualitatively high-valued contribution to business environmental protection.

The already existing quality-management system facilitates the introduction of the environmental management system, because 12 elements of the norm-structure, though with quality orientation and target, are by now included in DIN EN ISO 9000 ff.

Since 10 July 1996 the Gebr. Hellmann GmbH & Co. KG has been certified as the first company in the logistics sector in Germany resp. in Europe has been certified according to DIN EN ISO 14001.

With the establishment of quality management according to DIN EN ISO 9002 and the environmental management system according to DIN EN ISO 14001, which has been implemented after the main seat in Osnabrück to all other locations world-wide the company can offer its customers a larger, a broader, a more comprehensive future-oriented quality package, with which it takes over – according to the company slogan – "Responsibility for the future".

In an exemplary way a few environmental measures are enumerated:

Promotion of environmental communication

- motivation of the employees
- training
- integration of the employees' children
- introduction of a environment proposal system with rewards

Prevention and reduction of waste

- Change of the drinking automates from one-way to re-use bottles
- and from throw-away mugs to ceramics mugs
- Use of re-usable office materials

Waste

- Separate waste-collection in administration, stores and workshops

Optimisation of technology

- Introduction of material-saving spinning-machines, e.g. stretch-folios spinners

Development of new logistics concepts

- Introduction of the Playbox-Systems ⇒ less empty rides, cost-reduction
- Introduction of the Jumbo-change bridge ⇒ 20% more space

Urgency/Event planing

- Preventive measures, e.g. to prevent environmental accidents or hazards by water-polluting materials

Competitive Advantages and a Better Future through 'Environmental Kaizen'

For a better understanding of Japanese Kaizen three principles may be formulated. Kaizen is employee orientated, customer orientated, and it is a comprehensive management strategy. In regard to environmental kaizen this has the following significance:

1. *Employee orientated* – Environmental kaizen means that all employees have to be trained for environmental protection according to the principle 'people before products', starting from their workplace to work-satisfaction through qualification and career possibilities via participation at TQC-activities.
2. *Customer orientated* – Environmental kaizen includes not only the production of environmental-friendly, high quality products. The products and services have to be adapted to the wishes of the customers. Priority

of environmental kaizen is therefor not only the product-quality, but also customer satisfaction.
3. Environmental kaizen is defined as a *comprehensive management strategy* to reach the company's environmental targets – therefor it is top-management business – and should be extended to all company levels and functions.

Many products today nearly do not differ in regard to their technical features or price on the international markets; they are exchangeable. But through positive environmental aspects a competitive edge can be realised.

This is the challenge not only for German and Japanese companies, but for all companies around the world.

References

Alexander, P.-J. (1999), 'Umweltkaizen – Umweltschutz als Prozeß', *Japan Magazin*, number 9, pp. 9–10; number 10, pp. 8–9; number 11, pp. 7–9.

Burschel, C. (1995), *Umweltschutz als sozialer Prozeß,* Westdeutscher Verlag, Opladen/Wiesbaden.

Butterbrodt, D. and Tammler, U. (1996), *Techniken des Umweltmanagements,* Carl Hanser Verlag, München/Wien.

ECOS (2002), *Produktionsorientierter Umweltschutz,* Osnabrück.

Ehlert, W. (2002), *Unternehmen und betriebliche Umweltschutzverbände,* Peter M. Lang, Frankfurt/Bern/New York.

Greif, S. (1996), *Handbuch Selbstorganisiertes Lernen,* Verlag für Angewandte Psychologie, Göttingen.

Hallay, H. (ed.) (1990), *Die Ökobilanz – Ein betriebliches Informationssystem* (Schriftenreihe des IÖW 27/89), Institut für ökologische Wirtschaftsforschung, Berlin.

Hans-Böckler-Stiftung (2000), *Arbeit und Ökologie,* Düsseldorf.

Hans-Böckler-Stiftung und Projektgruppe Öko-Audit beim DGB-Bundesvorstand (eds) (1995), *Orientierungshilfe "Umwelt-Audit und Arbeitnehmerbeteiligung" – Leitfaden,* 2nd edition, Düsseldorf.

Hildebrandt, E. (1992), 'Ecology', in G. Széll (ed.), *Concise Encyclopaedia of Participation and Co-Management,* de Gruyter, Berlin/New York, pp. 242–257.

Hildebrandt, E. and E. Schmidt (2001), 'The Ecological Challenge', in G. Széll (ed.), *European Labour Relations,* vol. 1, pp. 116–136, Aldershot, Gower/Ashgate.

Hoffmann, J.; Matthies, H. and U. Mückenberger (eds) (1992), *Der Betrieb als Ort ökologischer Politik,* Verlag Westfälisches Dampfboot, Münster.

Imai, M. (1986), *Kaizen: The Key To Japan's Competitive Success.* Irwin/McGraw-Hill, New York [et. al.].

Ishikawa, A. et al. (eds) (2000), *Workers Firms and Unions,* Part 2, *The Development of Dual Commitment,* Peter Lang Verlag, Frankfurt [et al.].

Jänicke, M. and Weidner, H. (eds) (1995), *Successful Environmental Policy,* edition sigma rainer bohn verlag, Berlin.

Köhn, J. and Welfens, M. (1996), *Neue Ansätze in der Umweltökonomie,* Metropolis Verlag, Marburg.

Kühr, R. and Széll, G. (eds) (1997), *Zero Emissions Research Initiative Kongreß Osnabrück,* Dokumentation, Universität Osnabrück.

Leitschuh-Fecht, H. and Burmeister, K. (1994), *Die Zukunft der Unternehmen in einer ökologischen Wirtschaft. Verlag der Ökologischen Briefe,* Frankfurt am Main.

Martin, R. et al. (eds) (1998), *Workers Firms and Unions,* Part 1, *Industrial Relations in Transition,* Peter Lang Verlag, Frankfurt [et al.].

Öko-Institut Freiburg (ed.) (1996), *Öko-Audit – Leitfaden und Arbeitsmaterialien zur Zertifizierung,* Economica Verlag, Bonn.

Pauli, G.(1996), *Breakthroughs – what business can offer society,* Epsilon Press, Haslemere.

Rationalisierungs-Kuratorium der Deutschen Wirtschaft (RKW) e.V./Foth, M., Gerds, P., Nitschke, C., and Schmitz, A. (1994), *Umweltschutz ist wirtschaftlich,* Eschborn.

Schmidheiny, S.(1992), *Changing Course – A Global Perspective on Development and the Environment,* MIT, Harvard.

Simonis, U.E. (1990), *Beyond growth: elements of sustainable development,* Ed. Sigma Bohn, Berlin.

Simonis, U.E. (2000), 'Globalisation and ecological responsibility', in K. Busch et al. (eds), *Ways to social peace in Europe,* secolo, Osnabrück, pp. 115–123.

Simonis, U.E., Altner, G., Mettler-Maibom, B. and von Weizsäcker, E.U. (eds) (since 1992), *Jahrbuch Ökologie,* Beck, München.

Széll, G. (1992a) 'The environmental crisis at the turn of the millenium', *Revue Internationale de Sociologie,* vol. 1, pp. 173–199.

Széll, G. (1992b), 'Environment', in G. Széll (ed.), *Concise Encyclopaedia of Participation and Co-Management,* de Gruyter, Berlin/New York, pp. 318–325.

Széll, G. (1994a), 'High-Technology, Industrialization, and Problems of Development', in W.V. D'Antonio et al. (eds), *Ecology, Society & the Quality of Social Life,* Transaction Publishers, New Brunswick/London, pp. 149–168.

Széll, G. (1994b), 'Technology, Production, Consumption and the Environment', *International Social Science Journal,* # 140, pp. 215–225.

Széll, G. (2000), 'Arbeit, Bildung und Demokratie', *Jahrbuch Ökologie 2001,* Beck, München, pp. 174–182.

Széll, G.(2001), 'To learn from Japan?', in G. Széll (ed.), *European Labour Relations,* Gower/Ashgate, Aldershot, vol. 2, pp. 191–205.

Széll, G. (2002), 'Japanese Joint Ventures in China', in U. Teichler and G. Trommsdorff (eds), *Challenges of the 21st Century in Japan and Germany,* Pabst Science Publishers, Lengerich [et al.], pp. 215–228.

Tischler, K, (1994a), *Grundwissen Umwelt,* Ernst Klett Verlag, Stuttgart.

Tischler, K. (1994b), *Umweltökonomie,* R. Oldenbourg Verlag, München.

Tischler, K. (1996), *Ökologische Betriebswirtschaftslehre,* R. Oldenbourg Verlag, München.

Tischler, K. (1998), *Betriebliches Umweltmanagement als Lernprozeß. Theorie und Praxis organisationsorientierter Umweltbildung als Beitrag zur Steigerung der Fortschrittsfähigkeit der Unternehmung,* Peter Lang Verlag, Frankfurt.

Winkelmann, O. (2002), *Entwicklung der Mitbestimmung – Industrial Relations, Industrial Districts und Netzwerke. Das Beispiel der Betriebsräte-Netzwerke im Industrial District Ostwestfalen Lippe,* Peter Lang Verlag, Frankfurt.

Winter, G. (1993), *Das umweltbewußte Unternehmen,* Beck, München.

14 Industrial Food for Thought: Globalisation of Agro-Food-System and Environmental Concern

YOSHIKI OTSUKA

Introduction

During the last decade we have seen many public controversies concerning risks of food against human health and natural environment, which include BSE (bovine spongiform encephalopathy), dioxins and PCBs, endocrine disrupting hormones, *E.coli* O-157 bacteria and GM (genetically modified) food. A survey by the Japanese newspaper Asahi Shimbun (July 12, 2002) indicated that 77% of Japanese were feeling some uneasiness in the safety of food. The major concerns for the safety of food were about food additives (45%), producers and manufacturers (17%), organic or non-organic (17%) and GM or non-GM (13%). Thus these concerns seem to relate to industrialisation (i.e., food processing and additives, agrochemicals and GM technologies) and globalisation (import of foods, agricultural products and animal feeds). The trend also accompanies some alternative food movements and discourses, such as local food, farmers' markets, food miles, slow food, fair trade, etc., predominantly led by consumers' organisations, environmental NGOs, small farmers and retailers. These new actors in the agro-food system seem to reconstruct the previous discursive space with scientific and market rationalities that have been shaped by agricultural and food scientists, food manufacturers, agro-chemical companies and regulators in the governments. This presentation will discuss an analytical frame to theorise the reconstruction of discursive space, mainly based on my experiences in working with the "consensus conferences on GM crops in Japan" during the last two years and in interviewing with the participants of the consensus conferences, consumers' organisations and farmers.

Industrial Food: Its Invisibility and Boundaries of Discursive Space

One of the common characteristics of processed food, imported food and agricultural products with industrialised farming is their invisibility of production processes and of potential hazards from ordinary consumers (Adam 1998), i.e., most people cannot know by whom, where and how their daily food has been produced, processed, packed, labelled, distributed and contaminated. Industrialisation of food production entails the application of techno-scientific expertise and industrial organisations in the key processes of the agro-food chain. Globalisation of the food production enlarges geographical and temporal distances between producers and consumers. Thus, these spatial and knowledge gaps between the entire supply chain and ordinary consumers may have alienated (or "disentangled" in Michel Callon's terminology [1998]) the latter from participating in the shaping of the global agro-food system with techno-economic expertise.

The agro-food system is defined as social networks around a bundle of commodity chains from agricultural inputs to food products ready for consumption. The system has evolved with continuous and reflexive interactions between actors in the system. Many sociologists of agriculture and food have distinguished the transformation of the agro-food system from traditional model to Fordist or agro-industrial model (McMichael 1994; Allaire and Boyer 1995; Fonte 2002) in which industrialised farms produce substantially equivalent commodities for the global market and for transnational food corporations by means of numerous techno-economic intermediaries, such as agrochemicals, high yield varieties, contract farming, chilled transportation, marketing strategies, etc. Hence, the agro-industrial (or Fordist) model, though it has now been under restructuring, implies the dominance of techno-economic discourses in the agro-food chains. In other words, the agro-industrial model has demarcated the agro-food discursive space between the institutionalised discourse by techno-economic experts and utterance by the lay public.

Scientific and Market Rationality

The scientific and market discourses by experts in the agro-food system have been legitimating the rationality of practices of industrial production and consumption of food. On the one hand, in the "upstream" of agro-food chains, experts in agricultural research centres, universities, agricultural cooperatives as well as seed and agrochemical companies instructed farmers about scientifically sound and economically efficient farming oriented to mass-producing power. Protectionist regime of agricultural trade, which has been embodied in farm-support policies and institutions, may have helped rather uniform adoption of such techno-economic expertise. On the other hand, in the "down stream" of the chains, nutritional and food experts educated consumers what they should eat, while food industries (both manufacturers and retailers) contested for attracting consumers by varied values added to their products. Standardisation of food sanitary and nutrition may have functioned as a frame of references, in which consumers and food companies determine their behaviours, though the latter regularly participated in the shaping of such standards.

Scientific rationality is iteratively constructed through peer reviews of papers submitted to a journal that integrates a community of scientists of a certain field. In this regard, scientific rationality is a collective and reflexive construct by a "journal community" (Fujigaki 2002), which coincidently divides scientists and lay public who are allegedly scientifically irrational. Usually, however, such a collective process is not visible from the lay public. In contrast, market rationality seems more open to public. Market rationality can be defined as an iterative process of reflexive determinations of commodity values among various stakeholders. In the case of the agro-food system, the values include not only prices but also other varied qualities, such as tastes, colours, shelf life, brands, safety, nutritional values, producing locations and methods, etc. However, these determinations are structured with governmental and international regulations, standards and other institutions that bring techno-economic considerations into the process as well as market powers of large corporations.

Thought: Visualising Industrial Food and Transgressing Boundaries of Discursive Space

Public controversies concerning risks of food against human health and natural environment have occasionally visualised how the scientific rationales for regulations and standards of new technologies have been shaped, and how the economic rationales for the adoption of the technologies by actors within the agro-food system. A series of BSE scandals in Europe and also recently in Japan have brought uncertainties in the techno-scientific expertise and in entangled commodity chains. In the case of GM food, the environmental risk analysis of the EU has been reconstructed and scientific uncertainty has come in (Levidow 2001). These processes may involve significant debates on the manners of scientific experimentation, interpretation of data, publication of results and the relationship between risk assessment and risk management. Such debates and reconstructions inevitably destabilise the boundaries between scientific rationality and other social discourses or, if any, social rationality. The hypothetical social rationality may consist of a broader scope and more diverse participants than the scientific rationality. Similarly, emerging non-GM markets with identity-preserved handlings may reflect the transformation of market structures and involvement of new values and stakeholders.

Blurring Boundaries between Scientific Rationality and Other Social Discourses

Contested nature of the techno-scientific expertise on health and environmental risks of industrial production of food may be rooted, as some sociologists of science argue, in the epistemological ambiguity of the techno-scientific knowledge (Wynne 2002; van Zwanenberg and Millstone 2000). This ambiguity raises a dispute between epistemological realism and relativism or constructivism. The epistemological realists simply regard disagreements among experts as a lack of precise data. From this perspective a public controversy, even though it originates from disputes among scientists, indicates public misunderstanding of science. On the contrary, the epistemological relativist or constructivist perspective presumes the social embeddedness of the techno-scientific expertise and its interpretative flexibility. Although the perspective deconstructs the legitimacy of the risk policy based on scientific rationality, its strongest version may remain agnostic about what should count as robust or reliable knowledge.

The differences between these epistemological positions may appear as the bases of dissimilar policies on risk management and public participation in the decision-making of the development and regulation of new technologies with possible environmental concerns. For example, some versions of the precautionary principle in Europe and recent attempts of public participation in technology assessment world-wide may reflect the emergence of constructivist approach in the technology policy, which may blur the boundaries between scientific rationality and other social discourses. Besides, these epistemological positions sometimes cross over depending on the subjects (e.g., scrapie and GM food in the US and the EU; Busch 2002) or policy areas (e.g., risk assessment and labelling in proponents and critics of GM food; Klintman 2002). This indicates epistemologically fuzzy positions in the actual policy making, which may imply the constructive nature of both epistemological realism and strongest version of relativism. Scientific rationality alone does not seem to legitimate the decisions in the risk management any longer, though social determinism underlying the epistemological relativism cannot make any techno-scientific decisions either.

Transforming Market Rationality toward Identity Politics of Food

The agro-industrial model of the agro-food system has comprised of the global market for industrialised farm products with subsequently equivalent qualities in which local differences have been dispelled under rather indulgent standards. Given that market is iteratively constructed with definitions of values and identities of commodities, scientific assurance of an equivalent quality has played a vital role. The obscuration of the scientific rationality undermines the equivalence of agricultural products. Such scientific ambiguity may coerce food companies to incorporate more sophisticated techno-scientific expertise in order to make salient the differences of products rather than neglect them. After the BSE scandals and GM controversies, some pharmaceutical companies have already found their positions in the agro-food system as manufacturers of test kits for these food contaminants and as guarantors of the qualities.

These trends may help further differentiation of food qualities in the "downstream" of the agro-food system, for which food and retailing companies have contested by means of processing, packaging and marketing at the stage of satiety with food (Fonte 2002). Techno-scientific expertise that guarantees the "traceability" and hence the identity of food attempts to visualise at least some aspects of the food production. As a result, segmented "niche" markets with differentiated qualities of not only "downstream" food commodities but also "upstream" farm products have emerged within the global market. However, identity preservation of crops and livestock requires further techno-scientification of agriculture, which includes genetic and biochemical tests, computers, GPS (global positioning systems) and buyer-call-for-delivery system. An identity of food and of its consumer, be it "Green", "anti-GM" or "organic", should be guaranteed by the institutions utilising such techno-science, unless the consumer herself makes a trustworthy relationship with the producer, which becomes more and more difficult in the globalised agro-food system. Here comes a dilemma of consumers who wish to visualise the agro-food system. Introduction of new techno-science and new organisations to visualise and identify the food-producing process may make the process more invisible because of further dependence upon unfamiliar expertise and new industrial organisations.

Shaping of Social Rationality

Scientific rationality and market rationality do not seem sufficient to reconstruct socially acceptable technology in the agro-food system. Epistemological relativist and constructivist perspectives encouraged through food scandals and controversies have undermined the rationality of science. Market rationality may bring unintended expansion of the intellectual and organisational gaps between producer and consumer. It seems that the more both rationalities entangle techno-economic object, the more the agro-food system disentangled consumers and in some cases farmers who can not afford such technology. A sociological project Public Perceptions of Agricultural Biotechnologies in Europe (PABE) (Marris et al. 2001) has revealed that ordinary people were neither 'for' nor 'against' GM organisms (GMOs). In other words, public perceptions of GMOs were distinct from those anticipated by techno-scientific experts or by marketing managers of food industries that act as "progressive" in creating differentiated consumption. The underlying factors of their perceptions were not limited to the problems of misinformation and uneasiness about putative health risks but also about a broad range of ethical and socio-political considerations. Similarly, in the "consensus conference on GM crops" in Japan public participants expressed a wider range of concerns that include long-term effects on environment and especially biological diversity, effects on the self-sufficiency of food supply in Japan, relationships between research institutions and farmers, effects on the economic disparities between the developed and developing countries, issues on patenting living organisms, consistency of Japanese agricultural policies between introduction of GMOs and promotion of organic farming, etc., in contrast to the expert participants (mostly natural scientists) who attempted to narrow the scope of the discussion. The tendency suggests the need for a different discursive space with a wider scope of perspectives and participants. Social rationality may denote such discursive space, though the concept still remains hypothetical and needs further entanglement to shape and characterise it.

References

Adam, B. (1998), *Timescapes of modernity: the environment and invisible hazards,* Routledge, London.

Allaire, G. and Boyer, R. (eds) (1995), *La grande transformation de l'agriculture,* INRA, Economica, Paris.

Busch, L. (2002), 'The homiletics of risk', *Journal of Agricultural and Environmental Ethics,* vol. 15, pp. 17–29.

Callon, M. (ed.) (1998), *The laws of the markets,* Blackwell, Oxford.

Fonte, M. (2002), 'Food systems, consumption models and risk perception in late modernity', *International Journal of Sociology of Agriculture and Food,* vol. 10 (1), pp. 13–21.

Fujigki, Y. (2002), 'Science policy: science and public', in O. Kanamori and H. Nakajima (eds), *Science studies in front*, Keiso Shobo, Tokyo [in Japanese].

Klintman, M. (2002), 'The genetically modified (GM) food labelling controversy: Ideological and epistemic crossovers', *Social Studies of Science,* vol. 32 (1), pp. 71–91.

Levidow, L. (2001), 'Precautionary uncertainty: regulating GM crops in Europe', *Social Studies of Science,* vol. 31 (6), pp. 842–874.

Marris, C., Wynne, B., Simmons, P. and Weldon, S. (2001), 'Public Perceptions of Agricultural Biotechnologies in Europe', *Final Report of the PABE research project*, funded by the Commission of European Communities, Contract number: FAIR CT98-3844 (DG12 – SSMI), available at http://www.pabe.net.

McMichael, P. (ed.) (1994), *The global restructuring of agro-food systems,* Cornell University Press, Ithaca/NY.

Wynne, B. (2002), 'Risk and environment as legitimatory discourses of technology: reflexivity inside out?, *Current Sociology,* vol. 50 (3), pp. 459–477.

Zwanenberg, P. van and Millstone, E. (2000), 'Beyond sceptical relativism: evaluating the social constructions of expert risk assessments', *Science, Technology and Human Values,* vol. 25 (3), pp. 259–282.

15 Social Communications on Ecology and Risk: The Systems-Theoretic Analysis of Socially Responsible Investing

AKIRA TOKUYASU

Introduction

In the 1980s, the so-called Socially Responsible Investing appeared, and since then it has been growing rapidly in the North America, especially in the United States. Socially responsible investing is a kind of investing, that is an economic activity. It pursues, however, not only monetary returns, but also the realisation of certain societal values, such as environmental improvement, avoidance of the risks of alcohol and tobacco, reduction of military weapons, protection of human rights, good labour relations and employment, development of communities, and so on. Especially with respect to environmental improvement, we can consider socially responsible investing as a kind of social communication on ecology and risk.

Socially responsible investing has several aspects. It is an economic activity, because it pursues monetary profits. It is a kind of resource mobilisation, because it mobilises money as a special medium or resource to make business companies improve the environmental side effects of their economic activities. It is a kind of public discourse on ecology and risk, because it promotes conversations between shareholders and companies on the environmental aspects of the economic activities of these companies. And finally it is an ecological movement, because it tries to spread ecological values and to make enterprises change their attitudes toward the environmental problems and risks.

We can investigate this socially responsible investing from various perspectives. There are several conventional perspectives such as social movement, public discourse, and public policies. Theories of social movement are most popular, especially those of new social movements and those of resource mobilisation. Theories of public discourse are also popular, especially in relation to Habermas' theory of communicative action. They focus on the rational discourse and consensus on societal problems in the public sphere. Today emergence of new public spheres on the Internet is also argued. The arguments of public policies might be relatively different, for their perspective is, so to say, a top down perspective, that is, political and/or legal control directly over economy.

Instead of these perspectives we use here the systems-theoretic perspective as an alternative. One of the fundamental assumptions of the social systems theory is functional differentiation of the modern society. Talcott Parsons, the great founder of the social systems theory, emphasises societal integration in spite of differentiation. Niklas Luhmann, the great successor of the social systems theory, emphasises autonomy of differentiated subsystems and impossibility of societal integration. We combine these two points of view to interpret the societal function of socially responsible investing.

Fundamental Aspects of Socially Responsible Investing

What Is Socially Responsible Investing?

In the beginning it should be made clear what socially responsible investing is. The biannual report of the Social Investment Forum (2001) defines socially responsible investment as follows:

> Socially responsible investing (SRI) is an investment process that considers the social and environmental consequences of investments, both positive and negative, within the context of rigorous financial analysis. It is a process of identifying and investing in companies that meet certain baseline standards or criteria of Corporate Social Responsibility (CSR) and is increasingly practiced internationally.

Social investing, socially aware investing, ethical investing, mission based investing, and natural investing: all these terms describe the almost same concept as SRI. These terms are often used interchangeably to describe an approach to investing that integrates social and environmental concerns into investment decisions.

There are various standards or criteria for socially responsible investing. For example, the Social Investment Forum classifies twelve criteria: alcohol, tobacco, gambling, defence/weapons, animal testing, products/services, environment, human rights, labour relations, employment/equality, community investment, community relations. Usually several criteria are combined to evaluate company's activities. The environmental criterion is one of the most broadly used criteria. It is used in 50% or more screened portfolios.

Who Are Social Investors?

Social investors include individuals, businesses, universities, hospitals, foundations, pension funds, religious institutions, and other non-profit organisations. Social investors consciously put their money to work in ways designed to achieve specific financial goals while working to build a better, more just and sustainable economy. To put it another way, social investors put their money to work to build a better tomorrow while earning competitive returns today. Socially responsible investing requires investment managers to overlay a qualitative analysis of corporate policies, practices, and impacts onto the traditional qualitative analysis of profit potential.

The Three Strategies of SRI

Socially responsible investing incorporates three strategies that work together to promote socially and environmentally responsible business practices and, in turn, encourage improvements in the quality of life throughout society:

- *Screening*: it is the practice of including or excluding publicly traded securities from investment portfolios or mutual funds based on social and/or environmental criteria. Generally, social investors seek to own profitable companies that make positive contributions to society. Conversely, they avoid investing in companies whose products and business practices are harmful. The criteria of screening have been just mentioned before.
- *Shareholder Advocacy*: it describes the actions many socially aware investors take in their role as owners of corporate America. They include dialoguing with companies on issues of concern, as well as filing, co-filing, and voting proxy resolution. Proxy resolutions on social issues are generally aimed at influencing corporate behaviour toward a more responsible level of corporate citizenship, steering management toward action that enhances the well-being of all the company's stakeholders, and improving financial performance over time.
- *Community Investing*: it is financing that generates resources and opportunities for economically disadvantaged people in urban and rural

communities in the U.S. and abroad that are underserved by traditional financial institutions. This type of investment is rapidly growing in recent years, but still remains relatively small.

The History and Current Situations of Socially Responsible Investing

The Roots of SRI

The Social Investment Forum refers to some historical roots of socially responsible investing. In early biblical times, Jewish laws laid down many directives on how to invest ethically. In the mid-18th century, John Wesley, the founder of Methodism, noted the fact that the use of money was the second most important subject of New Testament teachings. As Quakers settled North America, they refused to invest in weapons and slavery.

The modern roots can be traced to the impassioned political climate of the 1960s. During that decade, a series of social and environmental movements, from civil rights and women's rights, to the anti-war and environmental movements, served to escalate awareness around issues of social responsibility. These concerns also broadened to include management and labour issue, and anti-nuclear sentiment. In the late 1970s, the concept of social investing began attracting a considerably larger group of American investors due to concerns about the racist system of Apartheid in South Africa.

The Contemporary Trends of SRI

Over the past 20 years, the Bhopal, Chernobyl, and Exxon Valdez incidents, along with vast amounts of new information about global warming, ozone depletion, and the concomitant risks to life on the planet, have brought the seriousness of environmental issues to the forefront of social investors' minds. Investors also began to look more deeply at the employment practices of companies in the U.S. Most recently, issues of human rights and healthy working conditions in factories around the world producing goods for U.S. consumption have become rallying points for investors.

In recent years, many socially responsible portfolios have moved beyond selecting companies that are working to halt their negative environmental and social impacts to choosing companies that are actively working to improve their social and environmental performance. In addition, there is a growing realisation among corporate leaders and academics that the adoption of principles of sustainability can co-exist with long-term corporate profitability.

The Current Situation of SRI

Here, we had better examine the economic impact of socially responsible investing. In 2001 the amount of socially responsible investing in the U.S. is $ 2.34 trillion. It accounts for nearly 12 percent of the total $ 19.9 trillion in investment assets under professional management in the U.S. Among the total amount, more than $ 2 trillion are invested through screening, and more than $ 900 billion through shareholder advocacy. Socially responsible investing is growing very rapidly in the U.S. It is even said that market slump is providing unexpected boost to socially responsible investing. The economic impact of socially responsible investing is considerably big (Social Investment Forum 1995, 1997, 1999, 2001).

Two Analytical Perspectives

Integrative Perspective

Now, we have to analyse socially responsible investing from the theoretical point of view. The first perspective is integrative. It assumes the hierarchical structure of the society. Talcott Parsons once provided the conception of cybernetic hierarchy of control, in which four functional subsystems are vertically ordered, that is, from L through I and G to A. In this hierarchical structure, cultural values are superior to economic interests. And in the broader scheme of the human condition, the telic system serves as a transcendental base of cultural values (Parsons/Smelser 1956; Parsons 1971).

In this perspective, socially responsible investing is considered as directly controlling the economic system through environmental ethics. It seems that environmental ethics require or even order the economic system to realise some cultural values, that is, substantial rationality. In the American society, so-called civil religion occupies the transcendental position in the scheme of the human condition. If it is true that socially responsible investing has deep historical roots in religious traditions, then particularistic values of these traditions have been generalised into the universalistic values of American civil religion. And it is these universalistic values that serve as a transcendental base of socially responsible investing.

Non-integrative Perspective

The second perspective is non-integrative. It assumes the heterarchical structure of society. Niklas Luhmann emphasised the autonomy of functionally differentiated subsystems. There is no vertical ordering of functional systems, nor hierarchical control mechanism, but only resonance between subsystems. It follows that cultural values cannot directly control the economic system, but only give some disturbance that causes self-referential operations within the economic system, the total consequences of which cannot be foreseen. Furthermore there is no transcendental foundation like Parsons' telic system (Luhmann 1982, 1989).

In this perspective, socially responsible investing is considered as the mere disturbance for the economic system that operates autonomously. Social investors cannot directly control the economic system through environmental ethics, but they should translate their ecological ideals into monetary terms. Only money can serve as an effective communication medium in the economic system. If socially responsible investing functions at all, it is not because of its ecological ideals, but because of the impact of invested money. It is not environmental ethics but money that has universalistic validity in the economic system. And at last there are no universalistic values that serve as a transcendental base of socially responsible investing.

Concluding Remarks

The argument of socially responsible investing in the U.S. strongly claims its effective influences on the improvement of companies' practices. It might seem that socially responsible investing is more and more realising Parsons' conception of integrative society. If the proportion of socially responsible investing might grow up to 20 or 30 percent in the near future, this conception would be realised more perfectly. Ecological thought might provide better ethics and realise a "good society".

But we should ask whether environmental values are superior to all other human values, or whether all the human values can be completely integrated. Socially responsible investing might be a better alternative to the conventional investing, but we cannot foresee all the side effects inside and outside of the economic system. It follows that we cannot foresee and manage future risks perfectly. In this sense we should not underestimate the significance of socially responsible investing, but at the same time we had better hope not so much that socially responsible investing will solve the environmental problems in the future.

References

Luhmann, N. (1982), *The Differentiation of Society*, Columbia University Press, New York.

Luhmann, N. (1989), *Ecological Communication*, Polity Press, Cambridge UK.

Parsons, T. (1971), *The System of Modern Societies*, Prentice-Hall, Englewood Cliffs, N.J.

Parsons, T. and Smelser, N.J. (1956), *Economy and Society: A Study in the Integration of Economic and Social Theory*, Free Press, New York.

Social Investment Forum (1995), *After South Africa: The State of Socially Responsible Investing in the United States*. Washington, DC.

Social Investment Forum (1997), *1997 Report on Socially Responsible Investing Trends in the United States*. Washington, DC.

Social Investment Forum (1999), *1999 Report on Socially Responsible Investing Trends in the United States*. Washington, DC.

Social Investment Forum (2001), *2001 Report on Socially Responsible Investing Trends in the United States*. Washington, DC.

16 Problems in Japanese Forestry: The Present Situation and Future Aspects

SEIICHI NISHIKAWA

Introduction

It is a well-known fact that forests do not only have an economic function, this being the production of timber, but also fulfil other communal functions such as water balance, soil protection, conservation of the countryside and so forth. Both kinds of functions are important topics in forest policy.

Still, regarding the present situation of Japanese planted forest stands, we have to admit that Japanese forest policy has been partially unsuccessful. In spite of the growth in area of planted forests as well as the growth of stands of planted forest inside Japan, the Japanese rate of self-supply with timber is very low. Since prices for timber went down, Japanese forestry cannot work profitably any more. Some forest-owners have given up their business.

Thinning is very important for a healthy young forest, therefore thinning is performed through governmental measures for some young Japanese forests in order to ensure the upkeep of these forests' communal functions. Such uniform measures are taken throughout Japan. Many of the thinnings are left behind in the forests; on the other hand a big part of our supply of timber is imported from other countries. We have to re-think the system of the utilisation of thinnings, and at an even more basic level, we have to think about the supply of native timber as well as forestry as a local culture within the present trend towards decentralisation.

The Present Situation of Japanese Forestry – Locating the Problem

The Japanese System of Forestry

The Japanese system of forestry is a centralised one. Its centre is the Agency for Forestry located within the Ministry of Agriculture, Forestry and Fishery.

Below the national level there are prefectural governments and municipalities, this structure being shown in Figure 1. Measures taken by the Agency for Forestry are passed on to the municipal and local administrations by the prefectural governments and subsequently they are carried out uniformly. This system is very rigid as well as reliable. For this reason more than 10 million hectares of planted forest stands have been realised in the post war period.

On the other hand the variety of Japanese forestry that was omnipresent in earlier days, has been lost through these uniform measures. The patronisation of prefectural, municipal and local administrations, all of them dependant on the central agency is another problem. For these authorities it is difficult to act autonomously.

The Political Failure of Afforestation in Japan

Figure 2 shows the areas of planted forest stands of all tree ages in Japan in the year 2000. Figure 3 shows the total volume of forests in Japan between 1966 and 1999. In Figure 2 there is a remarkable tip indicating that afforestation had been pursued more actively during the post-war years. Due to this post-war policy of afforestation the area of planted forest stands has increased to more than 10 million hectares and continues to rise.

Afforestation was promoted in order to enable mass production of native timber. Another less visible reason for afforestation to be promoted since about 1955 is the felling of hardwood forests for paper production. By the way, stands of planted forest are almost exclusively made up of Japanese cedar (Cryptomeria japonica) and Japanese cypress (Chamaecyparis obtusa).

Within 10 years the stands of planted forest timber will increase by about 400 or 500 million cubic metres (compare Figure 3; 40 to 50 million cubic metres per year). Figure 4 shows the supply with timber and the rate of self-supply in Japan between 1955 and 2000. The supply of timber had risen to about 100 million cubic metres around 1970. Comparing Figures 3 and 4 we can calculate that for about half of the timber supply we can use the increase in stands of planted forest. But in fact the rate of self-supply with timber has gradually decreased. In the year 2000 it was at 18.2 per cent. Here we have to ask why the Japanese rate of self-supply has gradually decreased in spite of the continuing increase in area of stands of planted forest to more than 10 million hectare.

Because prices for timber went down, Japanese forestry is no longer profitable. The price of foreign timber lowers the price of native timber. Here especially the price of the North-American hemlock controls the price of the

Japanese cedar. Felling costs are already higher than the market price of the timber (compare Figures 5 and 6). For this reason some forest owners who had been supported by the agency and consequently afforested their woods, have now given up their business.

Stands of timber of less than 50 years of age are too young for clearing. To ensure the growth of a healthy forest thinning is indispensable for forests of trees aged between 11 and 50 years. Stands of planted forest requiring thinning make up 85.5 per cent of all stands of planted forest. The health of the forest is crucial for the fulfilment of its different functions and thinning is indispensable for a healthy young forest. Therefore some younger Japanese forests are thinned through governmental measures to ensure the upkeep of their communal functions. But many thinnings are left behind in the forests; while at the same time we import a large part of our timber supply from foreign countries.

Japanese Forestry – Future Aspects. Can We be Confident about the Future of Japanese Forestry?

Opinion Poll – Some Results

In 1996 the Japanese government initiated an opinion poll about Japanese forest and forestry. The sample of the poll was 3,000 subjects, 76.1 per cent of the questionnaires were returned. Some of the results are shown in Figures 5, 6, 7 and 8.

Figure 5 shows opinions concerning the low Japanese rate of self-supply. 56 per cent think that the rate of self-supply with timber should be increased. Only 22.5 per cent advocate a more active import of foreign timber.

Figure 6 shows results concerning living and housing. 69 per cent prefer a living space made of native timber in a traditional architectural style. The most important reason for this is that such houses are well adapted to the Japanese climate.

Figure 7, based on the same opinion poll, represents the public opinion on the functions of the forest. An opinion stating that the production of timber is the main function of the forest is only supported by a relatively small group. According to Figure 8 we can state that general opinion regards economic forestry as a measure of overcoming stagnation in forestry negatively.

One might argue that many Japanese are egotistical concerning the forestry in their own country, or that they are not sufficiently informed about the present situation of Japanese forestry. If there were many Japanese with a profound knowledge of present-day Japanese forestry, then governmental measures dealing with the problems of thinning would be influenced to a greater extent by public opinion. Thinning of Japanese planted forests is indispensable to ensure the upkeep of their communal functions, but at the same time it is important to make use of the thinnings in order to increase the rate of Japanese self-supply with timber.

Tendencies toward Decentralisation and Incorporation.

Presently there are strong tendencies towards decentralisation and incorporation in Japan. The new Japanese laws on decentralisation came into effect in the year 2000. We hope that the degree of centralisation will be reduced and autonomy will be reinforced. Presently the Japanese government actively and uniformly supports incorporation throughout Japan.

But the situation of the now more than 3,200 municipalities differs greatly. 36.9 per cent of the municipalities are mountain villages with more than 75 per cent of forest (also see Figure 9).

Due to their lack of any big industry and their almost exclusive dependence on governmental support such mountain villages immediately encounter severe financial difficulties in case of cuttings in governmental financial support. For these villages faced with the present tendencies toward decentralisation, it is a topic of utmost importance of how to activate forestry in the long term.

One can state that the solution to the problems of Japanese forestry is the key to the success of Japanese decentralisation and incorporation. Proceedings concerning forestry in the municipalities must be relieved from uniform and centralist measures. One important idea in this context is the idea of forestry as local culture.

Final Remarks

Seen from an ecological point of view, the problem of thinning is as important as the economical use of stocks of timber and the restoration of the diversity of the forest. I also think it is important to realise that mountain villages will have to face some severe problems as the present situation drifts toward decentralisation and incorporation.

I sincerely hope that many ecological villages may be born in Japan, gaining autonomy and being surrounded by beautiful forests.

Figure 1: System of Japanese Forestry. A typical example of support

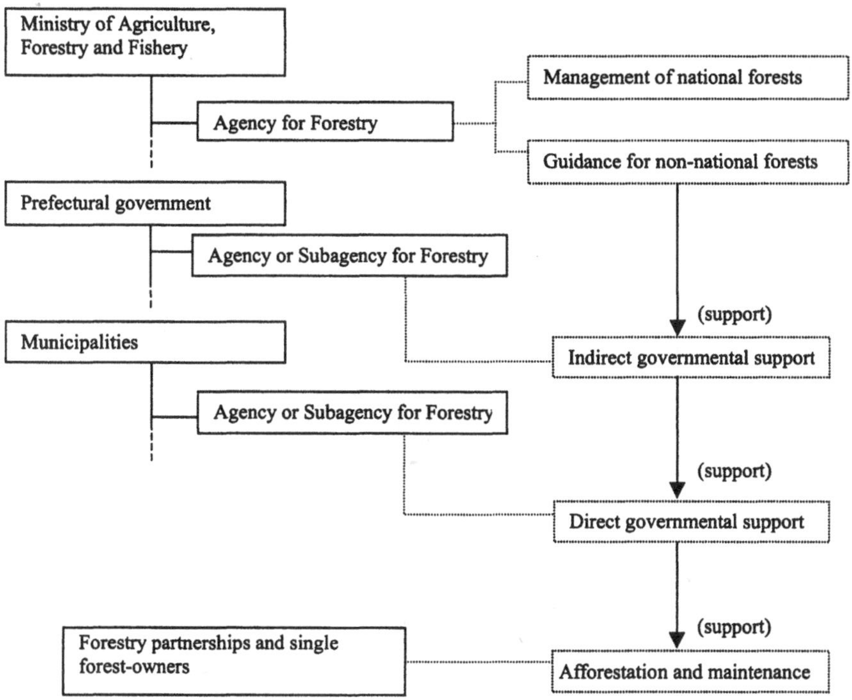

Figure 2: Total planted forest area of all tree ages in Japan, 2000

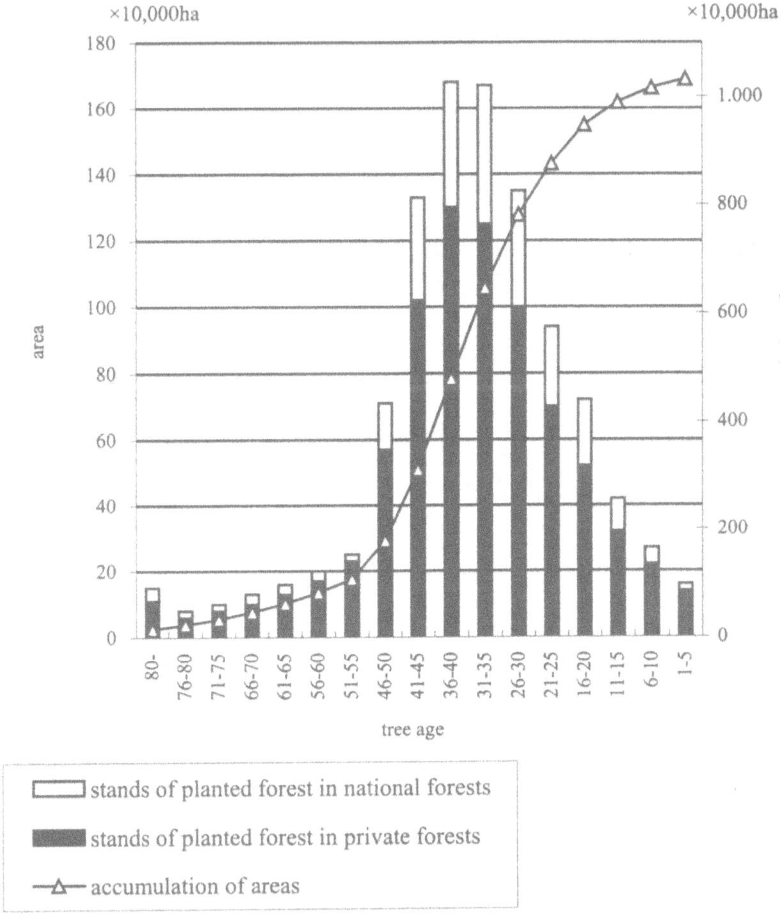

Figure 3: Total volume of forest in Japan, 1966–1999

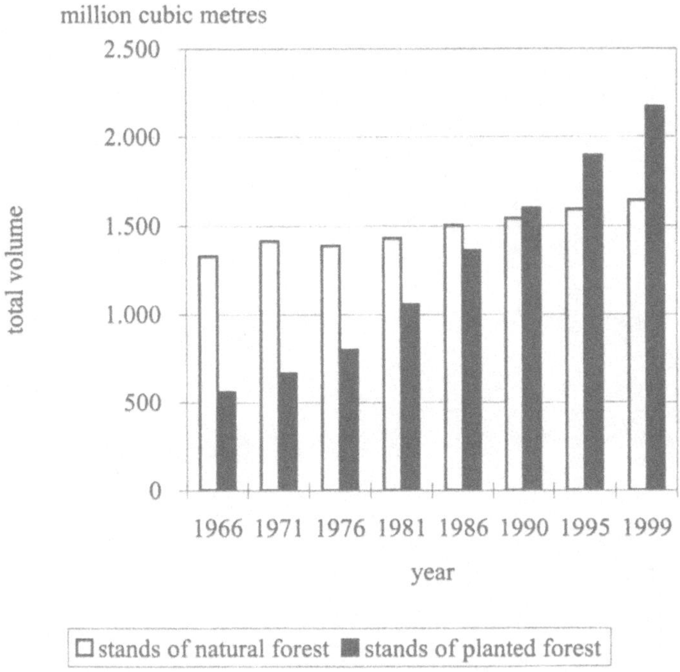

Figure 4: Timber supply and the rate of domestic production in Japan, 1955–2000

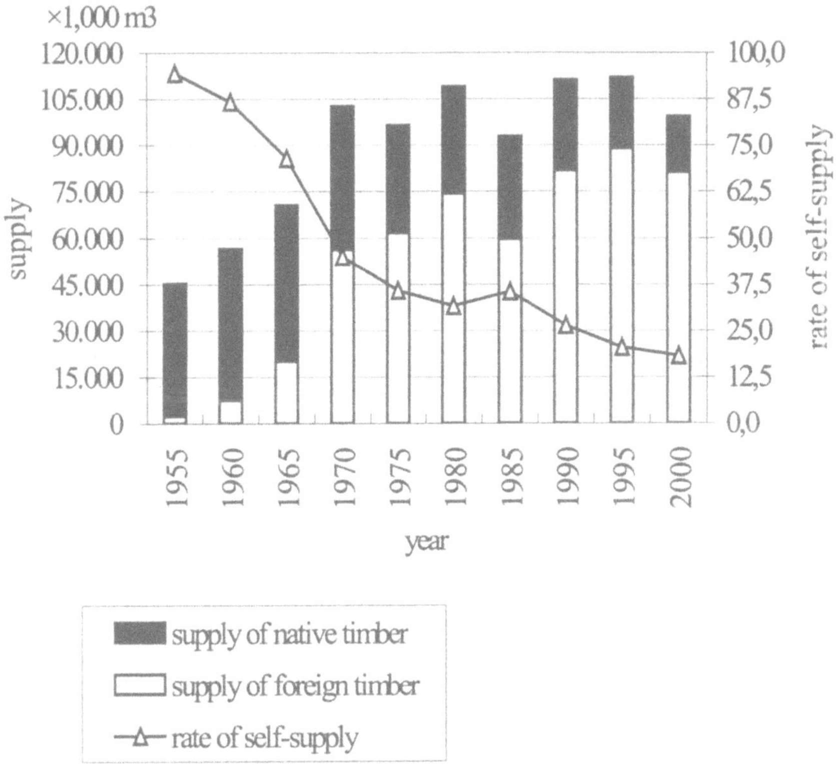

Figure 5: Opinions concerning the low rate of Japanese domestic timber production (from an opinion poll by the Japanese govt., 1996)

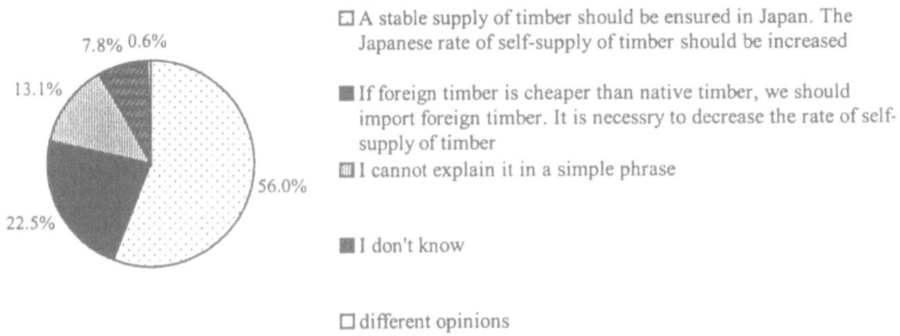

Figure 6: Opinions concerning living and housing (from an opinion poll by the Japanese govt., 1996)

When either purchasing or building a new place to live what kind of housing would you prefer, and why?

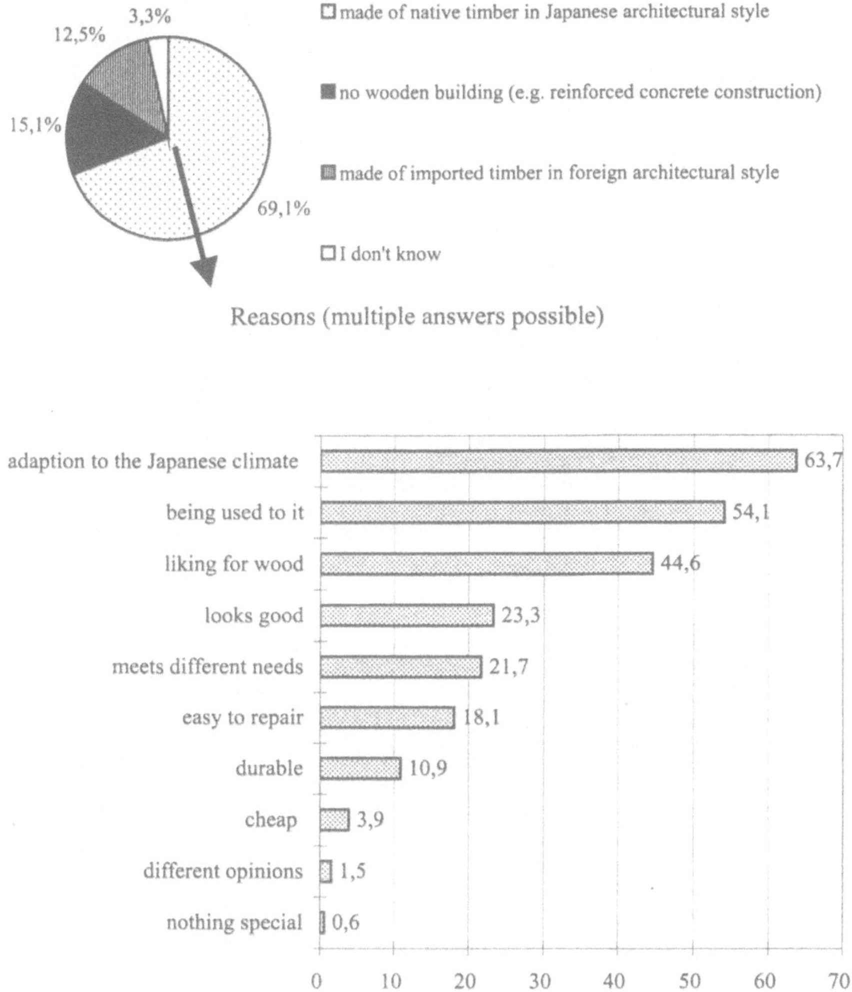

Figure 7: Opinions concerning the functions of the forest (from an opinion poll by the Japanese govt., 1996)

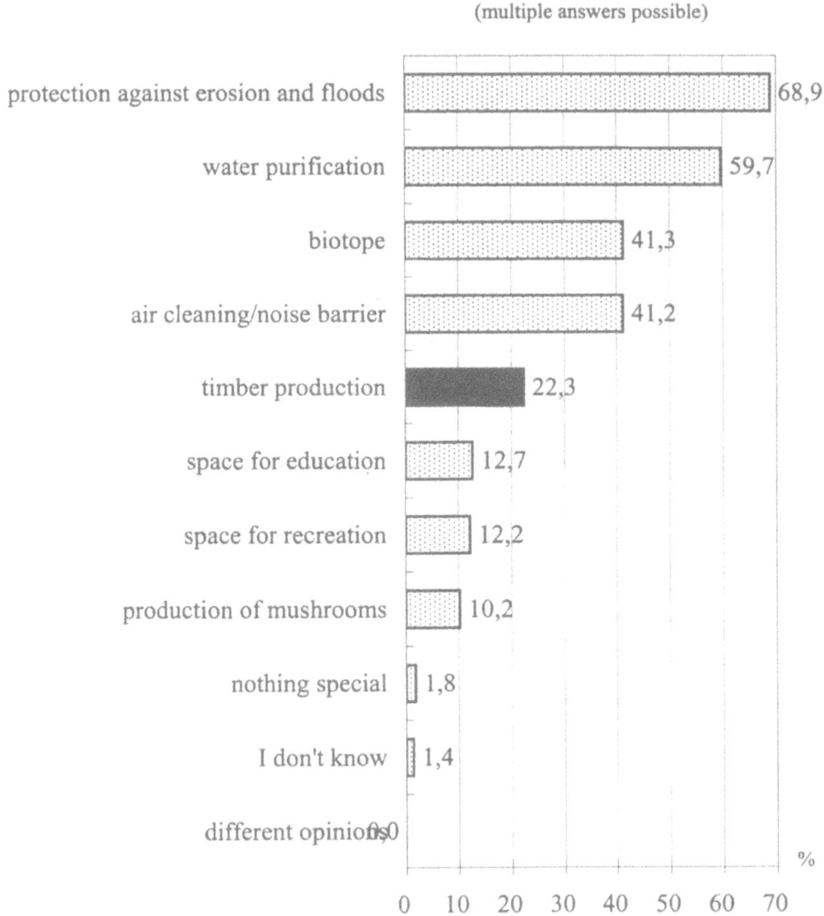

Figure 8: Opinions concerning the direction of forest policy (from an opinion poll by the Japanese govt., 1996)

What kind of measures should be taken in order to promote stagnant forestry? Which direction in forest policy do you consider more important, A or B?

A: The forest is a place for economic activity. Therefore priority should be given to economic forest policy.

B: Protection from erosion, protection from floods and so forth are important functions of the forest. Therefore forest policy should promote these functions, even if there is no financial gain.

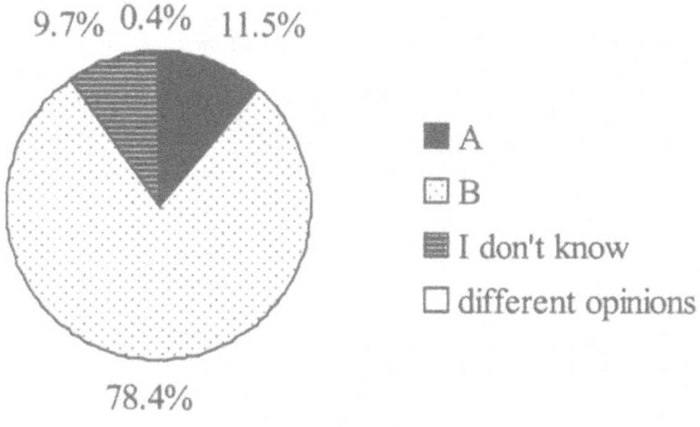

Figure 9: Mountain villages in Japan, 1999

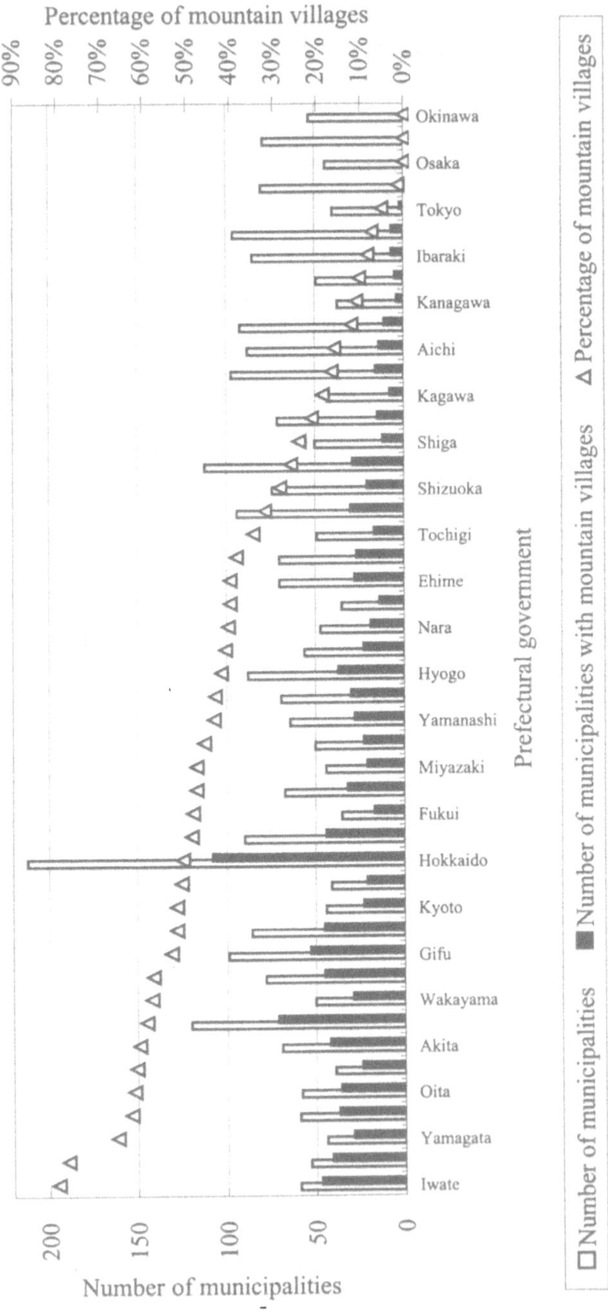

mountain village: more than 75% of forest; population less than 116 persons/km2 or 300

17 Environmental Measures in Japanese Enterprises – A Study from an Aspect of Socialisation for Employees

SHUJI YAMADA

Introduction

Since the 1990s, a new stage of the environmental movement had started in the whole world. It means that we have to not only think about both knowledge and action for global environmental protection on the citizen's level, but also the expansion of the Environmental Management in enterprises or firms.

After the Brundtland Report, "Sustainable Development" has become one of the most famous slogans in Japan too. Also in 1992, the UNCED Conference in Rio de Janeiro requested the actual environmental protection policies with the "Agenda 21". At the same time, many of the enterprises had started establishing their environmental measures and new management policies. Especially, Japanese enterprises are notorious in the experiences of "Kogai"[1] diseases, which are "Minamata"[2] or "Itaiitai"[3] diseases and so on, in the world. And many Japanese enterprises also made the same serious pollutions in Asian countries over a long time. In the name of the "Kogai Exports", they produced environmental disasters caused by factories' operations, and they were importing large amounts of raw materials from those econo-

1 This word means pollutions and its diseases in Japanese. Especially, Iijima in the Japanese Association for Environmental Sociology defined that "Kogai" (and Environmental Problems) is physical, chemical and natural environmental degradation, which is finally causing fundamental social and cultural change and crisis. In this situation, many people were disabled to make their daily life and all activities (Iijima 1995: 7–8).

2 Minamata disease, which is mercury toxicosis, was officially found in 1956 in Kumamoto prefecture and in 1965 in Niigata.

3 Itaiitai disease, which is cadmium pollution, was officially found in 1955 in Toyama prefecture. Including these diseases, we say "4 big pollutions" or "4 big pollution court actions", which are Minamata, Niigata-Minamata, Itaiitai, and Yokkaichi-air-pollution.

mies. In order to that, Japanese enterprises have to re-organise their operations and activities into integrated environmental-management-policy for maintaining, or not to lose their influence and presence in the global market.

This effort may just suit for another slogan: "Think globally, Act locally". However, we should examine them from a sociological point of view with the next questions: What kind of positive effects can we find from those environmental measures? And, how did most of the enterprises change?

This article will discuss about the environmental measures and environmental management in the case of Japanese Enterprises. And then, both of them have a great roll for social change to an environmentally suitable society. And in the end, it will be showed that enterprises should be a central element in our society to educate people in the sense of environmental protection.

Enterprise and Environment

An increasing awareness of environmental problems, which are especially connected with the "Kogai" experiences in Japan, made Japanese scientists themselves investigate the causes of the environmental disasters for the sake of its victims. As Iijima insisted on, Environmental Sociology has an important roll that is "research from the victims' point of view" (Iijima 1995: 215–216). In the same way, she had also explained as next; to know the mechanism of victimisation patterns is to know the mechanism of harming patterns. It seems to be true that all the problems had simple harming-victimisation social relations. Although, global environmental problems, like ozone depletion or global warming, have a much more complicated pattern. So, we need another way of thinking. The reason is why our society has too many social actors to get the real image of environmental degradations. Of course, the enterprise is one of those.

Unfortunately, we have only a few approaches to make clear the new roll of enterprise in Japan. For instance, those are the very peripheral themes: (1) what is the influence of recent environmental measures for society from economic view point? And (2) what is the influence of the work place from the managerial point?

As many economics and business ethics scholars said, we did not connect the measures for environmental protection with management. Also, in the beginning of the establishment of industrial sociology, like the "Hawthorne experiments" at Western Electric, they found the so-called "Human

Relations Approach" (Akimoto et al. 1990: 104–110). And Taylor insisted on the "Scientific Management", after that, "Fordism" was refined, and so on. These examples show us that industrial sociology was developed only in a sense of efficiency. Anyhow, it is no doubt that these were important findings.

It was also an important research how we could reform the organisation of the enterprise on the base of economic growth. Although this tendency was affected with the timing that means many of us wanted the economic development in the mean time, we lost the chance to think about the relationship between management and environmental protection. The "Kogai" disease in Japan was the very example to find it.

According to many economists, the environment was always analysed from the point of "external diseconomies", which is a disadvantage for good management in this process, the environment was only a hindrance factor, and it was not a main topic to consider management on the enterprise level.

As an environmental sociologist Hannigan said, the most impressive and successive environmental degradation image was to show the "ozone hole" pictures on the global level (Hannigan 1995: 45–48). After presenting them, the citizens who could see those photos started worrying about global environment. At the same time, some enterprises also started taking care for environmental protection.

In Japan, a manager in one famous electric company said that it was mid 1970s to start their environmental measures among large-scale enterprises. Without saying it, one big reason was the "Kogai" experience and its protection. And another one was that when some top managers in "Keidanren"[4] went to the USA, they saw that many US enterprises had environmental measures for the community. And many of them realised it is no doubt that many environmental measures were requested by community members in the near future (Yamada 1997: 73). In any case, the 70s were a big turning point to think about environment.

Appearance of Environmental Measures

On the enterprise level in Japan, we can find that environmental measures like anti-pollution were mostly established in the beginning of the 1970s. As we already argued, the enterprises, which were connected with the "Kogai"

---

4 "Keidanren" is "The Federation of Economic Organisations". Its proper name is "Nippon (= Japan) Keidanren" now.

disease, were requested to integrate anti-pollution measures by its victims mostly through local court actions.

Even if "Kogai" experiences were one of the most serious disasters, of course it was true, but it is significant until now. And sometimes thanks to these experiences, many Japanese enterprises started their environmental measures in the meaning of "Kogai" protection.[5] However we do not have to forget them forever.

Generally speaking, environmental measures mean that the basic idea is to combine enterprise's environmental protection and its integrated policy with organisational common sense and also with ethically correct social responsibility in management. In the beginning of the 1990s many rather large-scale enterprises established these measures. What was the reason?

Factors in Japan

Needless to say, the enterprise is an organisation to create values for society and it has an interest to make benefit by their activity. In the 1970s, there were some impacts to make enterprises change. Those were "Kogai" that means pollution and its disease, the "oil shock", risk of energy and raw material shortage in Japan. The image of environmental measures was connected with all of them. Also, the organisational structure was changed in the same logic. For instance, they established a "material section", a "section for protection of pollution" and an "energy reducing section" etc. in their enterprises (Kato 1991: 59).

Although, as an engineer said, this is the "end of pipe technology", it is not a fundamental solving but a temporary one (Széll 1996: 5). Actually, "end of pipe technology" only means no more than aiming at accomplishing small input and high output. However, it was not for environmental protection but sometimes for achieving low energy consumption or developing new technologies. Furthermore, it was good for increasing Japanese competitive-

---

5 Especially after the four big pollution court cases, the compensation costs were rapidly increasing. According to some estimates, the compensation was much higher than the environmental or Kogai protection cost. For example, annual costs for compensation of Yokkaichi-air-pollution was 21 trillion yen, but the annual cost of anti-pollution was 12.7 trillion yen. Similarly, the Minamata (Kumamoto) disease cost 12.6 trillion resp. 123 million yen; the Itaiitai disease cost 2.5 trillion resp. 602 million yen (Ministry of Environment 2002: 82). Even if we could say anti-pollution costs were not cheap, the compensation costs were much more expensive in all cases. Generally speaking, it is better for the enterprises to take anti-pollution measures, and it is better for community members too.

ness in a global market, especially by constructing anti-pollution technologies (Fujii 2002: 98–102).

At the same time, it was a pity that some Japanese enterprises caused the export of "Kogai" to Asian countries (Hiraoka 1993: 179–184).

We may conclude that the environmental measures of the "Kogai" protection type aimed at their benefit. In short, what they were thinking was to input much less materials and energy, to have as output much more high quality valuables. Without any doubt, it was more or less effective for environmental protection. The most important result was to increase Japanese enterprises' advantages and abilities.

To introduce environmental measures was to make them a success, however, advanced measures for global environmental risks did not arise in the 1970s. Moreover, many enterprises evaluate themselves in an optimistic way. Some environmental managers presented it like that "we have no problems because we do 'Kogai' protection with a safe process" (Yamada 2000b: 172).

Until the 1980s, not all enterprise activities were connected with global environmental protection. "Mass production" and "mass consumption" were agreed in a sense of development and Industrialisation. Also, most of Japanese people did not take care for it, and thought that it is symbol of development to use much paper and energy. It was only at the end of the 1980s, that global environmental measures took place in Japanese enterprises, because some environmental treaties were accepted on a world-wide level.

Jumping about a bit in time, a few arrangements and requests had a great role in establishing the environmental measures as a start line. Those were the "Global Environmental Charter (Chikyu Kankyo Kensho)" of the "Keidanren" in 1991, "Actions for Global warming – what we have to do now for future generations" of "Keizaidoyukai"[6] in 1992, and "Voluntary Action Plan for Environment" to enterprises from MITI[7] in 1992. With this as a turning point, enterprises introduced environmental measures as their strategies for achieving an advanced organisation.

Global Environmental Movements and ISO 14000 series

"Sustainable Development", a well-known slogan, had been launched in 1987 by the "Brundtland Report". After five years, the UNCED conference held in Rio de Janeiro in 1992 declared the "Agenda 21" to realise sustain-

---

6 "Keizaidoyukai" is "The Japan Association of Corporate Executives".

7 In those days, the MITI was the Ministry of International Trade and Industry. However, after the administrative reform in 2001, the name has changed into Ministry of Economy, Trade and Industry (METI).

able development. Before having had the UNCED, the "BCSD" established "SAGE", which discussed basic ideas for ISO 14000 series.

These slogans and discussions, of course, took effect in Japanese enterprises. Especially, it might be true that ISO 14000 is the most success certification in Japan.[8]

After ISO 14000 has come into effect in Japan, that was in September 1996, many rather large-scale enterprises moved to take ISO certification to organise their environmental management system (so-called "EMS"). At the beginning, companies of the electric machine industry performed as a leader.[9] Why was it? Because the electric machine industry had many relationships with European and American markets, which are susceptible to EMS and ISO14000, so many enterprises took the certification for keeping their business (Mitsuhashi 1997: 90–92).

Furthermore, there are some cases which are results of transactions between Europe and Japan or parent and subsidiary companies:

(1) Alps electric (in Japan) was demanded to take ISO 14000 certification by Volvo. Even if Alps electric delayed to take it, Volvo reminded them to show their environmental measures' plan for keeping their business (Adachi 1998: 43).
(2) Tohoku-Nihon-Denki received a letter of request to show the copy of ISO 14000 certification from Europe and American distributors (Adachi 1998: 43).
(3) Ricoh made a system of "Green partnership" which insisted on a guideline of Green Procurement to subsidiary companies. And Ricoh gives advice to them to take ISO 14000 certification (Ricoh Group annual Environmental Report: 1998).

As above-mentioned, we are able to say that the expanding global market accelerates the establishment of environmental management systems.

Needless to say, that ISO 14001 specifies the uninterrupted process, keeping "Plan-Do-Check-Action" cycle. And the certified enterprises are

---

8 According to the web page of "ISO World", total numbers of certified sites are 10,952 in December 2002 in Japan. So far as a number of ISO14001 certification, Japan is the leading country (ISO14000-BS7750 certificated sites.htm).

9 More than 50% (in fact, 56.3%) of certified sites were electric machine industries in December 1996. It was just after the establishment of ISO 14001 in Japan. In December 2001, this position had dramatically changed. The top three were the service (25.3%), electric machine (17.7) and construction industry (8.4) (ISO14000-BS7750 certificates. htm).

continuing to develop their environmental management. Each enterprise can freely decide their actual environmental measures according to their management or production process. The well-defined statement says that ISO 14001 means a preparation for refining each enterprise's EMS.

Moreover, it is said that ISO 14000 is not only for the large-scale enterprise but also for middle- or small-sized ones. Because small-sized enterprises are rather difficult to introduce a management system in general, it is better to improve their organisational system and to do their business in a big way introducing ISO 14000. For instance, they can expect high efficiency and increasing profits with cutting unreasonableness, unnecessary efforts and uneven work. According to a sociological investigation, there are in a reality some middle- or small-sized enterprises, which succeed to improve their management systems by introducing ISO 14000 (Ishii 1999: 9).

For most of Japanese enterprises, the reasons for certifying according to ISO 14000 are a better image to keep good relations with the parent company, and for global environmental protection (Yamada 2000a: 58–59).

Evaluation of Environmental Measures

It might be thought that environmental measures are divided into four categories:[10]

(A) cutting and blotting out the factors of environmental degradation from organisations, firms and production processes,
(B) environmental business,
(C) disclosing all environmental information,
(D) environmental education for employees and community people.

Both (A) and (B) are directly connected with production and profit. The enterprise is an organisation for profit seeking. If it gives it up, it loses its reason for existing (Haneda 1993: 49–50). However, environmental protection often requests a few conditions that are disadvantageous for profit seeking (Nogami 1997: 170). One clear solution was crystallised as the environmental business integrating profit into environmental management (Adachi 1997: 240–241, Sakoi 1997: 220–226).

---

10 For detail, see MITI-report (1994). In this report, every case is grouped in the same type of industry. However, there are similar cases in each measure, which are affected by the MITI's request for making a voluntary plan. The reason is that this request includes many advanced examples for easy establishing.

The latter (C) and (D) are not for production, but they connect with one confidence of each side and human resources. These two types are very important. These are expressions of keeping organisational ethics, which include elements of relief and confidence, from the enterprise to the consumer. This expression means social responsibility.

Every enterprise has its corporate policy or identity. Concerning environmental measures, it is also essential to define targets and goals in the management process. And after that, the results with those measures are evaluated from the points of growing concern and the aspect of environmental justice.

The Beginnings of the Environmental Market

Different from the indispensable change with EMS, the market mechanism is also a great factor to reorganise enterprises. Indeed, the beginnings of the environmental market were affected by many serious environmental risks. For instance, the reasons are increasing awareness of the environmental degradations as social problems, elevating consumer's interests with products and services, establishing regulations on the local, national and global level and so on.

Every time when an enterprise starts to reform anti-pollution measures, it always needs new machines, information and talented persons. It is thought that the growing environmental market is a good condition to promote environmental management.

Environmental Business in Japan

According to the report of the Tokyo metropolitan government (1998: 18), there are six categories for the environmental business in the Japanese market.[11]

1. Anti-pollution machine and business support
2. Waste management and recycling
3. Environmental repair and creation

---

11 On the other hand, according to Adachi (1997: 240–241), types of the environmental business were also argued. This discussion categorised them into four types: (1) anti-pollution type, (2) environmental preservation type, (3) creating and maintain the environment, (4) information type. In these categories, we can understand how enterprise thinks its strong point and how they think about environment in their business activity.

4. Renewable energies
5. New environmentally friendly products
6. New production processes

These categories show the multi-character of "environment" and many spheres in this business too. So, it might be that various industries can easily get into this market with various production and service strategies.

The sales volume of environmental business in the Japanese market was around 25 trillion yen in 1997. In the estimate of MoE,[12] it will be more than 40 trillion yen in 2010 (MoE 2002: 79). As these volumes show, the environmental market will rapidly grow in the next 6–7 years. The average growth rate is 3.7% per year. And it is expected that this tendency will open up a new labour market (Table 1):

Table 1: Scale of the environmental market in Japan

	Sales volume (million yen)	Employees
1997	24,742,600	695,145
2010	40,094,300	867,007

Source: MoE 2002: 79

Life Cycle of this Market

Typically in "Kogai" experiences, it is important to set regulations to promote an environmental market. For example, we can find its role in the beneath life cycle model (Figure 1).

The start line is facing with environmental problems as a social problem to solve. This stage marks a latent need to open up the environmental market. After "Kogai" court actions, many enterprises were requested by community members to introduce anti-pollution measures. With these demands anti-pollution technologies, which were desulfurisation equipment, energy-saving machines and water-purifying devices etc., had developed since the 1970s.

---

12 Ministry of Environment was reformed in 2001. The Environmental agency was founded in 1971.

Figure 1: Life Cycle model of the environmental market

Growth

1. Appearance of the environmental problems
2. Beginnings (of the market)
3. *Establishing regulations and institutions*
4. Formation (of the market)
5. Growth
6. Mature
7. Reduction (disappearance)

Source: Tokyo Metropolitan Government 1998: 17

On the contrary to the liberalist fashion, we can find not only the sequence of market mechanisms but also official regulations as an indispensable stage from this model. Although easing of official restrictions is asked for in Japan in these decades, it is very interesting to find that regulations are a key to develop the environmental market (Yamada 2003b: 67–68). And in the end, when environmental problems are solved, although it is too difficult, the vitality of the environmental market will decrease or disappear.

However, it is discussed that regulations are a key, this market is made by consumer's awareness of the environmental situation and an image of a desirable society. Because needs from users have an influence on the volume of distribution, the environmental market may have the same character as the ordinary market.

As mentioned above, the environmental market is characterised by the following four points (Tokyo Metropolitan Government 1998: 6; Yamada 2003b: 64).

1. The appearance of the environmental problems is a big business chance for almost all enterprises.
2. It establishes a new social and market system. It might promote the desirable society.
3. As we think to solve these environmental problems by the cost mechanism, the market limitations and its influence are increasing.
4. Regulations and institutions are often expanding this market.

Tasks of Management for the Environmental Business

Making enterprises for environmental market, it is requested to accomplish some tasks. Usually, to keep their competitiveness, it needs innovation and reorganisation. In the environmental market, five conditions are pointed out (Yamada 2003b: 65–66).

1. To know the strong points in each enterprise and to have a strategic management.
2. Cleverly using the existing management system and establishing a new system.
3. Constructing necessary networks inside and outside the enterprise.
4. Keeping a leading position in this business sphere.
5. Managing environmental risks and costs.

Although there are a couple of tasks, environmental business has a wide niche sphere. And it is also able to apply mature technology. From these reasons, it is said that middle- or small-sized enterprises can easily enter into this market (Tokyo Metropolitan Government 1998: 38). That is why Government and enterprises expect the environmental market to activate the whole Japanese market and to expand the labour market.

The Industrial Situation in Japan

Meaning of Work

In the sequence from Industrial Society to Post Industrial Society, as Bell argued, industrial structures also had changed (Bell 1995: 53). The main industry was transformed from the secondary (manufacturing) to the tertiary (service) sector. After the 1980s, as the figures below show (Table 2), the Japanese industrial structure is shifting much more to the tertiary industry. Especially the service sector is increasing much quicker than before.

Table 2: Japanese Industrial Structure **(%)**

	Secondary Industry		Tertiary Industry		
	Total	Manufacturing	Total	Retail/Trade	Service
1980	35	25	54	23	18
1995	34	23	60	22	23
2010	32	22	64	19	31

Source: EARC (Koyo Kaihatsu Center) 2001: 4

According to a White Paper by the Japanese Ministry of Health, Labour and Welfare (2002: 288), there were 67.66 million employees in Japan (Graph 1). In the last 30 years, the total number of the working population is still increasing.

This situation is showing that each employee is still working. At least during some hours, he/she spends time with doing his/her job or task. So far as the enterprise is the place for socialisation, around half of the total Japanese population develop a part of their personality at the work place. Moreover, according to the concept of "methodological individualism", employees are in many parts the social base. From these realities, we can argue that the function of socialisation at the work place can be said to be still active in contemporary society.

Graph 1: Work Force in Japan (1970–2010)

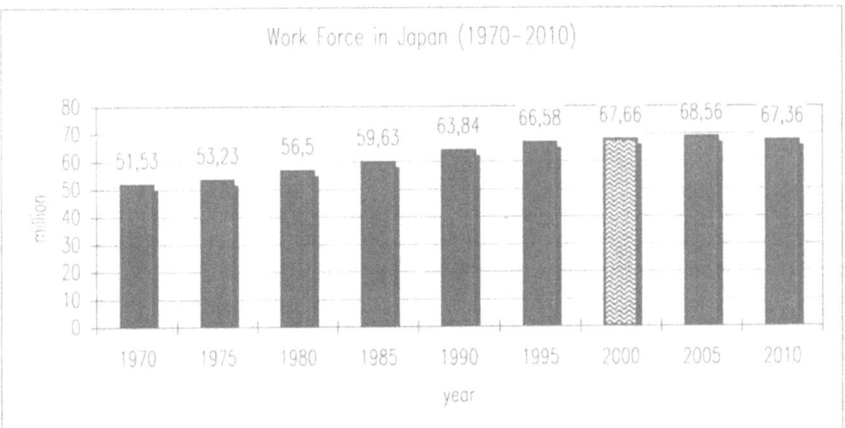

Source: Ministry of Health, Labour and Welfare 2002: 288

Problems

However, there are some big problems. First of all, the total number of the work force will be increasing until 2005, but after that, it will be decreasing. Second, it is pointed out that the younger generation, mostly less than 25 years old, will be decreasing in the same estimate. And the elder generation, mostly over 55 years old, will be increasing (Graph 2). These tendencies are called the imbalance of the work force.

It is also said that this imbalance of the work force is the result of the restructuring in each enterprise. So, it is meaningful to make sure that the organisational treatment for younger and elder generations is well functioning (Tajima 2001: 124–126; Ide 2001: 136–142). In short, even if developing environmental market makes enterprises optimistic, they have to integrate the management of human resources into the management of environmental risks under these economic conditions.

Graph 2: The Change of the Work Force in Japan

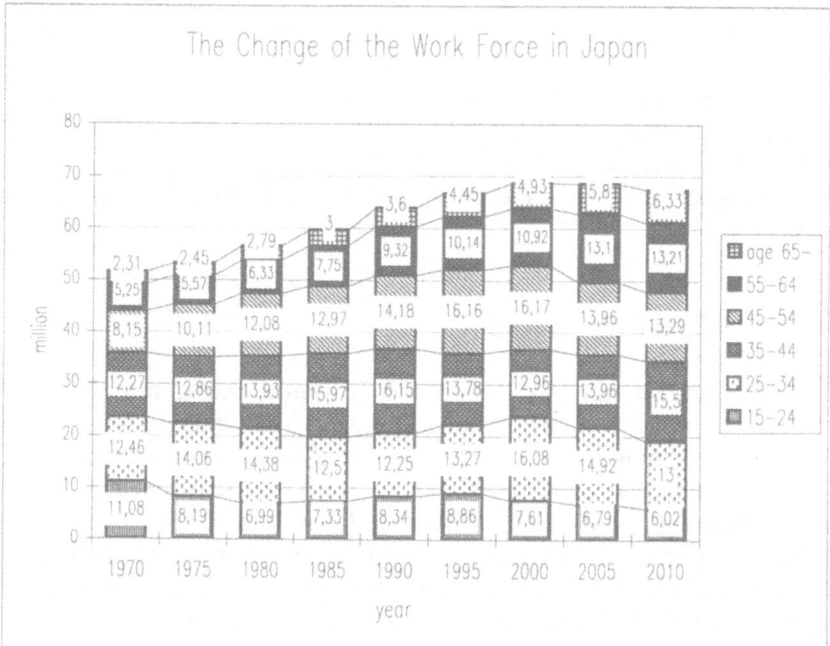

Source: Ministry of Health, Labour and Welfare, 2002: 288

One may be also worried about the high unemployment rate since the last decade until now. In detail, younger and elder generations are affected by it. As the character of younger generation is concerned, some are choosing unemployment by their own initiative. However, this is not the case for older persons (EARC 2001: 22–24).

With problems of unemployment, one of the most serious factors is connected with socialisation. In the Japanese management system, employees were accepting rules and abilities of each enterprise through engaging to the organisation, which is dominated by the seniority system. This problem is also related to the so-called post-Japanese management, introducing meritocracy.

According to the EARC research (2001: 41–43), although it is said that the use of "Chuto Saiyosha"[13] is increasing in general, considerably many enterprises take a strategy to keep the core staff in a system of lifelong employment. It shows that traditional long-time training with lifelong employment system is expected to keep each enterprise's competency.[14] This tendency is sometimes criticised by many enterprises, however, the system of lifelong employment is still used in some enterprises.

To consider these situations, we might say that the function of organisation, which obviously or latently educates people, is not rapidly disappearing. Even if we see the reality that is the growth of part-time work, employees who engage in an organisation are still being socialised by its training system.

Case of "Company A"

When an enterprise introduces and pushes forward its environmental measures, it has to introduce an environmental policy to the statutory form and fix it among the employees. So, in the process of introducing environmental policy, education and training for putting the policy into practice will be needed. Moreover, good relations between labour and management, which are fulfilling environmental policy with a feedback of the information from employees and ever-lasting improvement by management, are also essential. That is two-way, which means top-down and bottom-up, communication.

Introducing the case of "Company A", we will examine how they fix employees' environmental minds and activities.

---

13 "Chuto Saiyosha" means who has an experience to work somewhere else before hiring this company.

14 According to Széll (2002), he argued that Japanese management system and its crystallised style "Toyotism" are useful to establish environmental management because of long-term "Kaizen" activities. It is interesting to have a new understanding of traditional Japanese management system comparing with Western system.

Environmental Measures in Company A

This Company A[15] is a large-scaled machine industry, aiming at an Environmental Friendly enterprise. There are around 1,000 blue-collar and 4,000 white-collar workers, engineering is the largest volume of it. The main category of business was machinery in the beginning. But, after doing diversified business, chemical, electric and consulting businesses are assimilated.

As we glance over the history, it took a new turn of developing anti-pollution technologies in the 1960s and environmental preservation technologies in the 70s (Umeda 1997: 38). To sum up, in addition to the main machinery business, we can find that environmental business is a pillar of the company. In other words, Company A is always using a business chance that is "Kogai" and environmental degradation to grow up.

In the 1980s, introducing environmental policymaking, it was rather earlier than other enterprises. For real, it started environmental business in 1988. This reform included the strategy that was to make good use of its technologies, which were already well designed before. And the company president[16] proposed environmental principles in 1989. That is crystallised like this: "As a main business with water, air and environment, we contribute to healthy and comfortable life for all people in the world by offering our best technologies and services." The base of this idea was "in addition to our business related with water and air, we have to be connected in environment totally and aggressively" (from pamphlet of Company A). In this way, the first point of its success was that top managers had decided to establish environmental measures or policies just as a top-down style.

Organisational reforms

According to the manager of the environmental-supervising section, former self-evaluation was that the machine industry was thought as comparatively environmentally friendly and as an industry that was less involved in pollution and environmental problems. However, it had changed into a leading "Total Environmental Business Company". There were a couple of reasons. First, they found the some environmental problems in the production process, which should be improved in a sense of social responsibility. Second,

15 It was established more than 70 years ago. At a point of time in 1995, the work force was around 5,000, total capital was 25.7 trillion yen and sales volume was 347 trillion yen. The head office is located in Tokyo. It has 45 branch companies in Japan and 34 in overseas.

16 Now, he is a chairman of the board of directors. He is also insisting on the principle of "Zero Emission".

they thought it was no good to be indifferent to the environment, even if the point was not serious. Last, since the chemical industry had joined in this group, their profile as a machine industry was disappearing.

With this corporate principle, Company A had started reforming itself. First of all, an environmental committee, which integrates all sites under common environmental measures, was organised in 1992. In the next year, the environmental committee became the environmental-supervising committee, which controls the measures to step up, and the environmental-supervising section was newly set up under this committee.

After having this committee and section, integration is a key factor to analyse the change. Before there were complaints that reports of environmental measures from each site and factory were often hard to compare, because they were written in different standards. For solving these difficulties, the environmental-supervising section tried to integrate the style of report into the same way and to infiltrate it into every site and factory.

These experiences show that they achieved not only success in environmental measures but also an accomplishment of a new management system. As we already argued, it is thought that introducing environmental measures improves the management itself. Fortunately, the union and most of the employees did not have an opposite opinion, because of the good image to do environmental measures in their company. Furthermore, it became famous as a good company especially for the younger generation. According to the manager of personnel affairs, many students, who were motivated by working in this environmental friendly company, are applying to enter in every year.

"NOA" as a Promoting Circle

At the same time, "NOA"[17], which promotes environmental concern to all employees and their family members, formed in 1992, supported by younger employees. Around 30 members, who are mostly 20–39 years old, attended NOA voluntarily, also with the agreement by the environmental-supervising section.

It is interesting to see that NOA is an officially approved circle by Company A and it has an important function to connect employees' daily working life with environmental activities. For instance, they are publishing an information paper, organising a recycle market, planning nature-watching

17 NOA was named after Noah's ark. Its proper name is "Kankyo Hozen Suishin Iinkai (Committee for Promoting Environmental Preservation)".

tours and promoting an environmental household accounting book etc. These activities mean everyone can hit on but everyone cannot do it alone.

From the company side, there are some reasons to approve NOA's activities. Indeed, environmental activities in Company A, which decided to be an environmental friendly enterprise, are a good image to be recognised in the market and society. And, for advancing the bottom-up stream, managers do not stop employees' voluntary activities since the beginning of reforms with environmental measures.

Through these activities, most members have had or enhanced their environmental concern. To give an example, two members' testimonies are introduced:

1) He attended to NOA because it was offering many chances to have contacts with other company members. He did not always want to do environmental activities. However, he had an environmentally close mind since he was a university student, he recalls it was much more important to engage and to do something in NOA. Also, he wants that the activities have an effect on other company members (voice from leader of NOA).
2) He strongly expected to get in touch with environmental activities after joining Company A. Because he had an own image of environmental conservation, he wants to realise it. He felt the importance of NOA, as a circle of discussion with members of the same generation. He still constantly has an awareness of environmental issues. Of course, he said he is very happy to be active (voice from sub-leader of NOA).

These testimonies show that it is also significant to have some chance to be in touch with environmental activities at the work place. For employees, the work place is offering much information and many activities as a kind of daily life. Fortunately, Company A has an environmental circle NOA. Actually we can find results caused by NOA's promotion, which are improving employees' environmental minds. For example, according to the investigation[18] of Company A, more than 90% of the employees answered that their environmental interests were improving. Needless to say, top managers are pleased of these tendencies. For its reason, stimulus activities help to improve employees' positiveness in regard to their job.

---

18 This investigation is conducted once in 3 years for all employees.

The Environmental-supervising Section as an Intermediate Institution

According to Iijima (1995: 102–131), there are big differences in regard to environmental protection between two types of enterprise. There are these enterprises that have an initiative section for environment, and the enterprises that have none. So, establishing an environmental section is the first step to start environmental management.

The necessities of an environmental section for this enterprise are concluded like this: As mentioned above, the Voluntary Action Plan of MITI recommended setting up both an environmental policy and an environmental section. Another one was needs from Company A inside. This section, which supervises the daily work, was essential to promote those measures for everybody.

It is thought that to fix environmental concern needs daily, easy and understandable activities for all employees. In this company, the environmental-supervising section is advising waste separation, reusing copy-paper and saving energy etc. Moreover, they think that checking these data and the announcement to all employees and top management are other important roles. In short, they are connecting the policies from top management with the opinions of labour. So, it is thought that a two-way communication in Company A is accomplished. In this way, as this section is offering detailed data which are mostly related with the daily job, they can expect employees' understanding and reaching both management and environmental activities.

Evaluation of this Case

From these backgrounds, Company A had succeeded to certify ISO 14001 just after introducing it in Japan.[19] And they could keep many environmental technologies and an image of an environmental friendly enterprise. For the sake of it, Company A took merits of cutting down costs,[20] keeping value,[21] establishing an environmental education system and developing a business area which is consulting business using its matured technologies.

From business administration studies, three changes were pointed out (Yamada 1996: 22–25):

---

19 Company A was a first enterprise certified by ISO among environmental machinery industry (Adachi 1998: 49).

20 For instance, due to the replacement of electric pumps, they achieved 35% energy saving.

21 According to "DKM Management Report" (1996: 17), comparing with another enterprise, Company A was especially eager about environmental measures.

(1) fixing an environmental management policy,
(2) setting up an environmental section,
(3) initiative by top management.

In the case of Company A, the president had proposed the basic idea for an environmental policy and finally fixed it. Also, the environmental-supervising section had been arranged. Although it is said that realising top-down decision-making is the very form for a Japanese enterprise (Nogami 1997: 172), the above three points were all achieved in Company A including top-down. In respect to this success, it was natural that Company A had business merits.

Furthermore, as two NOA members' evidences are showing, it is interesting to see that the person who has environmental concern, however, it might be not so many, are appearing through the education system and NOA's activities. Of course, NOA is psychologically near to the employees, and it is planning easy daily activities and promoting environmental concern from experience-based activities. Almost all activities are having face-to-face communication. This example explains that it is meaningful to have some supportive activities, which make employees environmentally socialised on the enterprise level.

One manager of this company said that new problems like environmental conservation helpfully stimulate engineers. For instance, engineers who are engaging in a fully matured industry are less motivated, because there are very few occasions to give full play to their abilities. However, new tasks, which are an energy-saving function or environmental conservation technology, expand the fields where engineers can play an active part.

As the last one, the environmental-supervising section tries to develop many communication patterns, for example, between labour (bottom-up) and top management (top-down) or for employees and their family.

The Meaning of Environmental Measures

Four Significant Points

As we discussed above, it is possible to define the meaning of environmental measures as the following four points:

(1) Environmental Measures help to integrate or re-integrate an organisation, with a reformed information stream.

(2) It promotes employees' motivation. That means a kind of morale up through encouraging environmentally oriented activities, technologies and contrivances.
(3) In cope with ISO 14000,[22] environmental measures produce business chances, and usually include the merits inside of the organisation.
(4) We should not forget the functions of environmental measures, which are socialisation and environmental education of employees.[23] It is enough to be understandable, but not too much "everyday practices".[24]

Besides is positive meaning, it is thought that we have to refine these models on the base of Japanese management.

Environmental Measures and Japanese Management

The Japanese style of management is explained as the life-long employment, the seniority system and an enterprise union. In this management style, the employee education system is also defined as a special way. As mentioned above, however, there are some factors which are changing Japanese management style,[25] it might be still rather active in many enterprises.

These tendencies show that Japanese employees' talent improving system is strongly connecting with specified abilities those are gathering and managing special information in the enterprise. In other words, training in Japanese enterprises is strongly requested to be organised within the enterprise. In this structure, employees can easily develop their career and abilities. These abilities with special career experiences in the same enterprise are called "Organizational Specified" Ability (Yamagishi 2002: 92–101).

Improving this ability has two types of characters. From the employees' point of view, it guarantees their career. As employees are usually engaged in one enterprise for their whole life, they have to get the only ability to control

---

22 Particularly, ISO 14001 assists to accomplish environmental measures, because it is certifying each environmental measure in the enterprise. If some enterprise broke its promised measure, it will not be able to certify again until those measures will be recovered again. In this sense, regulations to enterprises become important to show the minimum line of environmental measures.

23 With this chance, an enterprise can contribute to environmental conservation by motivating employees. This educational function is a kind of a new role for the enterprise, which is connecting with social responsibility.

24 These strongest points are easy, daily and small activities in his/her work place, which have a chance to make their environmental consciousness and recognition.

25 For instance, those are meritocracy, mobility and diversity in employment etc.

a specific enterprise knowledge and information.[26] On the other hand, from management side, they can achieve cost reduction and avoid risk taking. As long as the specific organisational ability is useful in the labour market, top management can concentrate their attention on the training to manage employees without any failures.[27] More over, managers do not have to take risks to recruit well-trained persons from the labour market. For this reason, specific organisational abilities are helpful for both labour and management in the Japanese labour market. If that is the case, needless to say, environmental education is very effective in each enterprise.

Although, it must be said that the Japanese management system had already considerably declined. In any case, employees are always requested to produce "Employability"[28] (Miyamoto 2002: 179–187). On that particular occasion of thinking about the socialisation mechanism, employability also demands some abilities in accordance with enterprise's requests. The difference is that, where socialising employees takes place, it shifts from one enterprise to the whole labour market. However, it is not meant that the importance of character formation through the work is weakened.[29]

Are Environmental Measures Good for Environmental Protection?

As Simon (et al. 1950/1977: 168–192) said, we can also conceptualise the "authority" in enterprises. It means employees usually accept almost all decisions in the enterprise, if the concept of authority is well functioning enough. This also means that each person accepts directions stopping or slackening any critical abilities (Haneda 1993: 116). Well-organising these characteristics, it is thought that the enterprise can easily mobilise employees for environmental measures. When we analyse this situation from the outside, we can

---

26 It is explained that this regulated ability can allow employees to do many multi-section jobs in the enterprise. As a merit, without thinking to move to another organisation, they do not have to be trained widely.

27 Because they can believe that all employees will stay at the enterprise, it is able to make good use of training costs. If there are many employees who want to move, the result will be totally different. In these situations, many employees are sometimes requested to improve their specific abilities.

28 Different from specific organisational abilities, employability has a universal application, which is useful in any organisation.

29 This tendency is also explained by the fact that work still and forever exists as an important component of contemporary society. The reasons are (1) all people having a cash income by the expansion of monetary and market relations, (2) various phenomena are connected with the growing desire, and (3) many labourers promoting industrialisation and modernisation in many countries and continents (Beaud 1997 and 2000/2002: 278–279).

see a society, which is constructed by highly environmentally ethical individuals (Iwata 1997: 81). Only in the result, this is the same as the enterprises' contribution for environmental protection.

It is said that training in enterprises is not always effective outside the enterprise, but only at the work place. However, as Brender and Aglietta (1984/1990) argued, we can find that employees are able to establish their environmental awareness through socialisation mechanisms at the work place. As conditions for its success, there is a place where employees can get in touch with the daily, easy and understandable activities during their working time. Furthermore, a two-way information system between labour and management is helpful for this purpose. In short, it can be said that it is one of the keys to establish and fix own environmental measures.

Summary

When enterprises want to introduce and fix environmental measures, a promoter is always essential. Following his leadership, employees are integrated into measures and policies. In spite of the big effects of the educational system, it had been missed so far to examine the micro-level outcomes. The reason might be explained as follows: Because environmental measures were a comprehensive guide to establish a new management style, discussion with the outline or contents went ahead of those functions. Generally speaking, it is a necessary and sufficient condition for an environmentally friendly enterprise to support employees' awareness and to create environmental activities.

Indeed, industrial composition and labour problems are transformed on the grounds of economic conditions, the stage of the enterprise's development and various other factors. However, both the industrial composition and work are constantly having an influence on the individual, who is a social being. If all enterprises had well-organised environmental measures, we would have a chance to reach almost 70 million people, this is more than half of the population: So, employees can take over the environmental education in Japan. Furthermore, it shows another chance, which is the alternative way to pursue the change to an environmentally friendly society. That is not relying on an easygoing independent mind, and an unconditional following of market mechanisms or the absolutism of regulations.

References

Adachi, T. (1997), 'Strategies for Environmental Preservation in Contemporary Enterprise in Japan', in Natsume and Mishima (ed.), *Corporate Strategies in the Age of Globalization*, Yachiyo Shuppan.

Adachi, T. (1998), 'Contemporary Enterprise and Environment', in Adachi and Ito (ed.), *Fundamental Problems of Contemporary Enterprise*, Zeimukeirikyokai, Yachiyo Shuppan, Tokyo, pp. 231–254.

Ando, K. and Tagusagawa, R. (1990), *Sociology of Industry and Organization*, Gakubunsha, Tokyo.

Akimoto, R., Ishikawa, A., Haneda, A. and Sodei, T. (1990), *Introduction to Sociology*, Yuhikaku, Tokyo.

Beaud, M. (1997 and 2000), *Le Basculement Du Monde*, = 2002 (Japanese Translation), Fujiwarashoten, Tokyo.

Bell, D., *The Impact of Intellectual Society*, = 1995 (Japanese Translation), TBS Britannica. Tokyo.

Brender, A., and Aglietta, M. (1984), *Les Metamorphoses de la Sociétè Salariale*, = 1990 (Japanese Translation), Nihonhyoronsha, Tokyo.

DMK Management Report (1996), *Thinking of Environmental ISO from Environmental Activities in Company A*, vol. October.

EARC (Employment Advance Research Center) (2001), *Administration System in the New Century*, EARC, Tokyo.

EARC (2002), *Guide-book for Employee Administration in Out-sourcing Business of the General affairs, Personnel affairs and Training section*, EARC, Tokyo.

Fujii, Y. (2002), 'Developing the Environmental Protection Technologies and Industrial Organization – dynamic Process of Environmental Regulations and Industrial Technologies in Japanese Experiences', in Terao and Otsuka (ed.), *Policy Making and Dynamism of Development and Environment – Japanese Experiences, East-Asian Task*, Series of the Institute of Developing Economies (IDE) No. 527, IDE-JETRO, Chiba, pp. 79–106.

Haneda, A. (1993), *The Problems of Industrial Sociology*, Zeimukeirikyokai, Tokyo.

Hannigan, J.A. (1995), *Environmental Sociology – A Social Constructionist Perspective*, Routledge, London.

Hasegawa, K. (2001), 'Environmental Movement and Environmental Policy', in Hasegawa (ed.), *Environmental Sociology in Japan*, vol. 4, *Dynamism of Environmental Movement and Policy*, Yuhikaku, Tokyo, pp. 1–34.

Hiraoka, Y. (1993), 'Environmental Problems in Developing Countries', in Iijima (ed.), *Environmental Sociology*, Yuhikaku, Tokyo, pp. 169–192.

Iijima, N. (ed.) (1993), *Environmental Sociology*, Yuhikaku, Tokyo.

Iijima, N. (1995), *A Suggestion of Environmental Sociology*, Maruzen, Tokyo.

Ishii, M. (1999), *Introducing the Environmental Management System with Voluntary Enterprise's Action – Case Study from ISO14001 Certification in Japan*, Paper for the 72nd Conference of Japanese Sociological Association.

Ishikawa, A. (ed.) (1988), *Industrial Sociology*, Saiensusha, Tokyo.

Ishikawa, A. and Tajima, H. (eds) (1999), *Changing Organizations and Life in the Work Place*, Gakubunsha, Tokyo.

Iwata, K. (1997), 'A Criticism of an 'Ism' of Environmental Ethics', in Society for Environmental Economics and Policy Studies (ed.), *Environmental Ethics and Market Economy*, Toyokeizai Shinposha. Tokyo, pp. 69–81.

Japanese Ministry of Environment (2002), *White Paper of Environment 2002,* Gyosei, Tokyo.
Japanese Ministry of Health, Labour and Welfare (2002), *White Paper of Health, Labour and Welfare 2002,* Gyosei, Tokyo.
Japanese Ministry of Internal Trade and Industry (1994), *Plans for Environmental Activities in Enterprise – An Interim Report on Voluntary Plan in Each Enterprise for the Environmental Conservation,* Nihonkogyoshinbunsha, Tokyo.
Kato, H. (1991), *A Suggestion of Environmental Ethics,* Maruzen, Tokyo.
Mitsuhashi, T. (1997), *Zero Emission and Japanese Economy,* Iwanamishoten, Tokyo.
Miyamoto, M. (2002), 'Traps of the Safety Net', in Saeki and Matsubara (ed.), *A Plan for New Society in Market Mechanism,* Shinseisha, Tokyo, pp. 147–194.
Nogami, H. (1997), 'Environmental preservation and Enterprise Management – from an aspect of Market Mechanism and Environmental Ethics', in Society for Environmental Economics and Policy Studies (ed.), *Environmental Ethics and Market Economy,* Toyokeizai Shinposha, Tokyo, pp. 167–177.
Saeki and Matsubara (ed.) (2002), *A Plan for New Society in Market Mechanism,* Shinseisha, Tokyo.
Sakoi, I. (1997), 'Development of International Standardization on the Environment and Environmental Business in Japanese Enterprise', in Natsume and Mishima (ed.), *Corporate Strategies in the Age of Globalization,* Yachiyo Shuppan, Tokyo, pp. 209–230.
Simon, H.A. et al. (1950), *Public Administration,* [1977 (Japanese Translation), Diamondsha, Tokyo].
Széll, G. (1996), *Democracy and Sustainable Development. A Contribution to Environmental Sociology in Europe,* Paper for Conference at the Institute for Social Research at Chuo University, Tokyo.
Széll, G. (2002), *Environmental Kaizen: A German-Japanese Comparison,* [in this volume].
Tajima, H. (1999), 'Change of the Administration and Working Life in the Middle or High aged employees', in A. Ishikawa and H. Tajima (eds), *Changing Organizations and Life in the Work Place,* Gakubunsha, Tokyo, pp. 107–126.
Torigoe, H. (1998), 'Environment in the Daily Life and Lifestyle', in Hunabashi and Iijima (ed.), *Lectures on Sociology,* vol. 12, *Environment,* Tokyo University, pp. 75–102.
Umeda, M. (1997), 'Grapples for Environmental Problems in Enterprise Level – Case of Company A', *IE Review,* vol. 38, No. 5, pp. 37–42.
Umeda, M. (1998), 'Backstage of the ISO14001 Certifying', *Sangyokikai,* vol. March. pp. 45–49.
Yamada, K. (1996), 'A Study of Environmental Problems in Organizational Level', *Organizational Science,* vol. 30, No. 1, Hakuto Shobo, pp. 14–26.
Yamada, S. (1997), 'Decision-Making on Environmental Management in Japanese Enterprise', *Bulletin of Graduate School of Chuo University,* Vol. 29, pp. 71–83.
Yamada, S. (2000a), 'Action of Companies Toward the Acquisition of ISO14001: The Environmental Policy of Companies in Japan (5)', *Comprehensive Urban Studies,* Vol. 73, Tokyo Metropolitan University Center for Urban Studies, pp. 53–68.
Yamada, S. (2000b), 'The Influence of the Environmental Policy on the Eco-mind – Ecological Socialization in Enterprise', *Journal of the Faculty of Literature,* Vol. 10, Chuo University, pp. 169–198.
Yamada, S. (2001a), 'Waste and Environmental Management in Enterprises', in Iijima (ed.), *Environmental Sociological Studies on Waste Problems,* Series of Center for Urban Studies, No. 19, Tokyo Metropolitan University Press, pp. 145–176.

Yamada, S. (2001b), 'Environmental Problems And Environmental Measures in Enterprise', in Ishikawa and Tajima (ed.), *Changing Organizations and Life in the Work Place*, Gakubunsha, Tokyo, pp. 69–85.

Yamada, S. (2003a), 'Environmental Problems and Industrial Sociology – How Can We Connect?', *Bulletin of the Faculty of International Studies*, vol. 13, No. 2, Bunkyo University, pp. 75–88.

Yamada, S. (2003b), 'Case of the Environmental Machinery Business', *Expectancy to New Business Market and The Problems of Supply Side – Case Study of Health-Welfare, Medical and Environmental Market*, Japan Society for The Promotion of Machine Industry Economic Research Institute, pp. 69–71.

Yamagishi, T. (2002), *Big-Hearted Japanese: Illusion of Cultural Collectivism*, Nippon Keizai Shinbunsha, Tokyo.

18 Environmental Business Associations in Germany

GYÖRGY SZÉLL AND WIKING EHLERT

Preremarks

Thanks to the environmental movements but also scientific activities ecology has become one of the megathemes in the seventies of the last century, which culminated in the so-called Brundtland-report for the United Nations 'Our Common Future' in 1987 and the UNED-Rio-Conference 'Environment and Development' in 1992.

In this context in Germany Business Associations for the Environment (EBAs) had constituted themselves – like B.A.U.M.,[1] future, Modell Hohenlohe and the environmental initiative of the Chamber of Commerce and Industry in East-Westphalia –, which could each win hundreds of members. Two of them – namely B.A.U.M. and future – received in 1995 the environmental award by the German Federal Environmental Foundation, which has its seat in Osnabrück. This award is the highest in Europe with some 500.000 €. On the international level the Swiss entrepreneur Stephan Schmidheiny created the Business Council for Sustainable Development which assembled some of the leading companies in the world – the so-called global players like Shell, Dow Chemical etc. In 1992, just in time for the Rio-Conference, he published his much respected and translated book *Changing Course – A Global Perspective on Development and the Environment*. Quasi in parallel – certainly no hazard – with the end of the really existing socialism in Europe and with it the end of the East-West-conflict new perspectives of a peace dividend – as the late former German chancellor, Willy Brandt, dreamed of – seemed to take shape, which could be an ecological dividend (Széll 1990).

But in times of globalisation, in which shareholder-value dominates above all, the EBAs have difficulties to make themselves audible. Since the end of history, proclaimed as well in 1992 by the Japano-American Francis Fukuyama, which meant the final victory of capitalism, the liberal market-

---

1 Means "tree".

economy, every activity is oriented towards more profits – profit rates of 15 % are regarded as the minimum. Companies or parts of companies, which do not fulfil these criteria, are closed without mercy and respect of the interests of the employees and the respective regions. They publish quarterly their balances as the American standards ask for.[2] The share-values react frenetically according to these expectations and rhythms. The former president of the German Social Democratic Party and Minister of Finance, Oskar Lafontaine, and his wife, Christa Müller, published in 1998 their book 'Keine Angst vor der Globalisierung' (no fear of globalisation). In the meantime they got afraid and he stepped down in 1999 from all his offices.

Under these conditions the question of an 'ecologisation' of the economy becomes absolutely secondary. Though the red-green German government has introduced an eco-tax on energy, the intended effect of a sustained reduction of energy-consumption has not been realised – in spite of the three-litre car. The conflict between the Federal government and the energy conglomerates around the exit out of the nuclear energy has ended with an armistice, which gives the conglomerates thirty years to step out of this form of energy production. For the environment these are not good prospects.

But we should not despair too quickly. Political conjunctures may marginalise a topic; however, the question of sustainable development is a permanent topic, for which humanity has to find a solution. Although the forecasts from 1972 by the Club of Rome of the limits to growth have fortunately not materialised so far. And substantial progress for a sustainable development has been realised by companies, e.g. through the introduction of ISO 14.000 on a global level as well as the Eco-audit on the European level. Even against the opposition by the USA the Kyoto-protocol has already brought some progress as well. And the environmental summit in The Hague in November 2000 has been a disaster too. If the Johannesburg summit to celebrate the tenth anniversary of the Rio-Conference will be much better is very doubtful.

Our research-team from the Association Work-Environment-Society-Technology (VAUST) works together with the Department of Social Sciences of the University of Osnabrück in Germany since many years to realise since more than ten years a sustainable development on the regional and local level according to the slogan 'Think globally, act locally!'. We cooperate closely especially with Prof. Dr. Udo E. Simonis, from the Social

2 As we learnt recently from the scandals like ENRON, Worldcom, AOL and others this does not prevent from falsifications, on the contrary seems to encourage them even.

Science Centre Berlin, and Prof. Dr. Mitsuda Hisayoshi, from Bukkyo-University in Kyoto.

The project relates to a number of projects, which have to be mentioned in this context. At the first place this is the project 'Erfolgreicher als andere. Wissens- und Technologietransfer in deutschen und japanischen Unternehmen' (More successful than others. Knowledge- and technology transfer in German and Japanese companies), which has been conducted with five companies in each country between October 1995 and April 1997 together with ECOS Consult Osnabrück. As a result we coined in this project the term 'environmental Kaizen'.[3]

Partly in parallel we managed the project 'Ermittlung der Umweltschutzpotentiale der OBE-Region' (Analysis of the environmental protection potentials in the Region of Osnabrück-Bentheim-Emsland), together with the University of Applied Sciences Osnabrück.[4]

As a further praxis-related initiative we organised the first European Zero Emission Congress – together with University for Further Education Osnabrück and B.A.U.M. – in Osnabrück in October 1997 with more than 100 participants of leading German companies and associations like VW, BMW, the Federation of the German Chemical Industry, the German Federal Environmental Agency (Kühr and Széll 1997).

From October 1998 until September 2001 Olaf Winkelmann conducted his Ph.D. research with the support of the science foundation of the German Trade Union Confederation, the Hans Böckler Stiftung, on 'Potentials for a socially adapted strategy for regional sustainability at the example of the furniture-industry in the region East-Westphalia-Lippe' (SORENA). (Winkelmann 2002)

Finally I would like to make a special reference to another Ph.D. research by Klaus Tischler on 'Company management as learning process. Theory and praxis of environmental training as a contribution to the increase of the future orientation of companies', which may be regarded as a very successful example of combination between theory and praxis (Tischler 1998).

We started the project in autumn 1997 together with our partners from the praxis – three Environmental Business Associations and one foundation in creation. It was finished with a workshop in August 2000. The common application was preceded by a one-year preparatory phase, in which we con-

---

3 This project was co-financed with 50 % by the German Federal Environmental Foundation (DBU).

4 Again the DBU co-financed the project substantially.

ducted talks with potential partners. By the way, it was the first time that these associations sat around the same table.

The project 'Functioning and scope of action for multiplicators (associations, organisations) of company environmental protection in Germany' should determine the situation for the organised protection of the environment in structural as well as procedural perspective. The target was to submit to the actors proposals, how they could improve their organisation. To be sure that translation of sociological know-how into the economic world functions well, we subcontracted the consulting agency *ecco* at the University of Oldenburg with an adaptation study. With accompanying scientific expertise of the organisational improvements within the EBAs should be implemented. To summarise, the project was designed in this way to bring together as much analytical social science as necessary with as much as possible practical economy and environmental activities.[5]

Introduction

Environmental Business Associations (EBA) have been formed during the eighties and nineties of the last century with the perspective to become active for the environment within the companies and beyond them. The self-perception was to be an association of companies for mutual help in environmental issues. At the beginning all had the assumption that after a given period all environmental problems would be solved by the management of the companies, now it is clear for all concerned actors that they have committed themselves permanently for the environment. This change from a scheduled project to a sustained task became evident for these associations at about the same time with the understanding that they cannot convince the companies for the environment and has the easy success of the early years. And besides that due to the enlargement and deepening of the environmental debate in politics, the economy and society the tasks grew so enormously to which the associations had to take position. This dimension was not so easily

---

5 Fifty per cent of the project costs were subsidised by the DBU getragen. At this place we havealso to thank a number of persons for their support, who were active during different phases of the project: Thomas Claes, Mirko Ehlert, Pia M. Holle, Katharina Lippert, Martina Mrotzek, Peter Tänzer and Olaf Winkelmann. We have also to mention our consulting board: Prof. Dr. Winfried Anton (formerly University of Applied Sciences Osnabrück), Prof. Dr. Reinhold Pfriem (Carl von Ossietzky University of Oldenburg and ecco), Prof. Dr. Arno Rolf (University of Hamburg) and Prof. Dr. Eberhard Schmidt (Carl von Ossietzky University of Oldenburg).

to cope with the resources, which were available to the associations. In short, the environmental business associations were under pressure from many sides to be as effective for the economy, management and the companies as in the beginning.

Out of this perception, the Environmental Business Associations B.A.U.M., future, the Model Hohenlohe and the Umweltvorsorge (in creation)[6] have addressed themselves together with our research team to the German Environmental Foundation (DBU)[7] with the demand to analyse first of all the situation between the EBAs and their members, and then to prepare proposals as well as to test them, how the activities for the environment by the associations may be improved.[8]

In the following presentation the analysis of the activities of the EBAs and their optimisation shall be in the foreground. At the end we will enumerate the conditions, which all actors – irrespective of how they are organised – have to consider, if they still want to realise progress in the protection of the environment in the future.

Associations Associate

Associations have first of all to associate – irrespectively if they have economic, social or environmental tasks. Whereas the classical associations have become since long over the last couple of decades pillars of the economic and social self-managed order, new associations are far away from the integration into the structures of the German model of the social market economy. Therefore they should be understood less than associations with a given membership, but more as associations with a specific task and always-fluctuating membership and open future. At least the other perspective is named, into which newly formed interest associations may develop. This special role is expressed on the one hand in the difficulty to guarantee financial stability in the long run. On the other hand the continuous change of the topics makes it difficult to look for new and permanent members. Their not existing integration into the structure of the existing economic and social order is to be seen in the limited contribution, which the EBAs make as a

6 This association did not reach until the end of the project a legal status, so we do not refer to it in the further analysis.

7 The biggest of its kind in Europe.

8 Cf. for our detailed analysis the scientific report for the Deutsche Bundesstiftung Umwelt: Széll and Ehlert 2002, as well as Ehlert 2002.

pressure group for politics: The German Green Party is trying to establish its own environmental business association, UnternehmensGrün.

Seen from here it is not surprising that the EBAs live to a large part from subsidies, for which they have to apply for. They complain that only about one third of their membership is active themselves. Their charm as also their delimitation to the club of established associations is at the end too, to have not many tasks, but practically just one. They are a single purpose organisation. And as a single purpose organisation they are not taken for serious without support by the State. The supposition goes to be just an association of environmental protagonists, of which the political usefulness has not yet been established. So, the mostly self-nominated and not yet functional EBAs are not part of the German Associational State (Eschenburg). Their own role and understanding is largely not shared in the public and in the majority of the companies.[9] The biggest step they still have to realise. Though it is not yet clear where they will end. It is evident that they are at risk today from many and surprising sides. They face a difficult future.

From Political Mission to Changing Environmental Tasks

The creation of the EBAs[10] is not yet history for themselves. Two of the associations were founded by just a few people; among them was an outstanding personality and entrepreneur. In the whole economy the management and the organisation of environmental protection for the economy was conceived as a top-down process. At the beginning the environmental know-how had only to be transferred from the environmental movements to the managers. Today this advantage has disappeared at least by half, because the companies demand a more diversified answer to their problems and a specified adaptation to their resources, production and marketing. That for them leadership in environmental competence has become difficult may be interpreted as a contribution to their political normalisation. Great visions are transformed via history to business realities.

One association grew practically out of an environmental movement of SMEs in a socio-cultural homogenous region at the periphery of a metropolitan area (Stuttgart). There it started with a bottom-up process against a waste incineration factory. They continued on accumulating experiences and to look for solutions for environmental problems. They kept the successful

9 Results of the evaluation in control groups.

10 Summary of the history of the EBAs; cf. Széll and Ehlert 2000: pp. 56ff.

strategy to combine the environmental movement within the association for the companies. Working groups fit with small enterprises; one may already englobe them. For big enterprises they only fit for detailed and decentralised solutions. Here the realities of the economy hinder the extension of an idea and the appropriate strategy, which was simply 'good' on a small level.

For sociologists it is now self-speaking that the access to knowledge- and competence-transfer between the EBAs and their members resp. the economy and the public at large had to produce completely different results. And in effect today's EBAs are so different in size, and they are distributed in space only sporadically or regionally. In regard to the structure it is even more important that the relatively more top-down orientated EBAs, and against all proclamations have as the main focus the general. This happens, as a first view into their magazines shows, through a more general approach to environmental problems and their business solutions and the insistence of the own public presence (in the economy) mainly via congresses, workshops, campaigns etc. In contrast to that the other EBA has deficiencies when it has to demonstrate, what it can offer besides concrete support for companies for the environment or the economy as such. Its publications are specialised magazines, which can only reach specialists – may it be persons in specialised businesses or departments. The early success relates to the bottom-up principle as malediction of a gone past of concernedness and commitment.

These remarks should not be misinterpreted that the orientation towards general issues as well as the specialisation incorporate advantages and disadvantages for the EBAs on the company level. On the one side they are experienced as a chance to become more 'effective' through notoriousness (we still have to settle what this may mean), and to get more easily subsidies – mostly from the government. The great disadvantage for them, however, is that you have to demand eco-leadership in times where – according to interviewed experts – nothing new comes 'at the end of the pipe'.[11] Co-operation with science as a strategy is decreasing according to the active members of the EBAs.[12]

For the third EBA new environmental technologies and their development is not the problem. When they are treated in the SMEs in day-to-day practice there are always already experiences from other places available. Nothing has to be invented, but everything has to be rediscovered. The active and co-operating members themselves determine what an environmental

---

11 That is the outcome of interviews with experts for the future and the environment from the corresponding institutes.

12 According to the active members of the EBAs they rank in the last third in regard to their relevance (see also table 2).

problem is in form and content. Working groups are only constituted on demand. The advantage of this EBA in regard to topics, their specialisation and the level of treatment are its enormous flexibility. In reality the members take 'ready-made' solutions from the EBA home. The disadvantage lies in the lack of publicity in the public as well as in the difficulty not to reinvent the wheel every time. Two times the transfer of knowledge is questioned here. Where the future of this EBA lies is pre-determined by the other competing EBAs. The big ones are dominating the small ones – as often in life.

Naturally it would be wrong to suppose that the here – analytically presented – clear differentiation between top-down and bottom-up approach is to be found in every regard in the daily work of the EBAs. In the EBAs everything is possible – but some activities are practised and used today more or less successfully. The future will see the strengthening of the supposedly successful. And still the amount of possible members within the enterprises is so large that we cannot really speak of competition between the EBAs. The existence and the procedures are not really at stake; the only competition is around public subsidies.

To get them, both types of EBAs have problems. The first have the advantage of public recognition. This apparently honoured by the agencies, however, they ask also for the effectiveness of projects. The third one is very efficient in regard to implementation, whereas public recognition has still to be reached. But public subsidies are in face of the always increasing number of institutions and their demand, which all – and even it is only peripherally – want to promote a better environment,[13] are for the growth of the EBAs only one and steadily reduced source of income.

Consulting has been discovered since long as an other source of income. It has to be financed in first place by the receiving enterprise. This raises for the members of the EBAs the question why the have to pay for a service, which has been granted until then for free. In the meantime all EBAs have strengthened their consulting capacities. This was reached, however, to fulfil the expected demand by members and to circumvent fiscal problems as an association for general interest. For that purpose consulting agencies have been outsourced. What has been so far legally an association in public interest, has been slowly transformed functionally into a network of profit- and non-profit-organisations.

In regard to their finances the (functional) EBAs live today from their membership fees, public subsidies, public orders for projects and enterprise consulting. As more the associations grow, and as more they grow above the

---

13 Cf. for a general overview Schulz (1998).

possibilities of their membership to consult others, as more precarious becomes their financial situation. Limited project funds replace more and more yearly membership fees, and also the chances for reduced consulting; co-operative use of money for determined targets disappears finally in the financial mass of heteronomously determined projects and consulting activities of the in the meantime relatively successful EBAs. From the financial side the EBAs are reluctantly forced to change. The donors of subsidies determine the direction, which has to be followed, if one wants to continue to exist as an organisation and continue to grow. And the public donors have preferably general interests in mind.

For the environment and the solution of problems on the enterprise level this change of organisational forms on the long line may be of not much impact, but not for the very raison d'être of the EBAs. Projects and above all consulting activities open new income possibilities; but they also loosen the links between the association and its members. Projects and consulting activities become ever more important; it is not a good sign to refuse income and with it to rely only to the mutual help principle. The members and the association are reduced to publicity related appendixes; they are good as an area, out of which without problem new members may be recruited, which then may become customers of the consulting activities.

The result of this process is a kind of neutral distance from the members to their association. They leave the association with reference to the high membership fees.[14] On the other hand it is a kind of preventive deregulation on the micro level, which in regard to the sustainability of the association will leave behind as well the common association. The relativisation of the membership is in regard to the participation of companies at the classical associations no regress, but more a productive overcoming of historically obsolete relations of co-operation, to master economic-political and here of environmental problems in the sense of a further strengthened market-system.

The target in the future of the EBAs is already clear today. Even if the EBAs know about they could not propagate this future openly. Two sources out of three of their momentous income would disappear right away or at least would be obtained with much more difficulties. So, they move slowly on a way into a market dependency, which the classical (i.e. Chamber of Commerce and Industry etc.) associations are still afraid of. What has been named in former times association and had an organisation as its centre, mutes into a network of consulting agencies, in which the image and the

---

14 Results to the question in regard to the reason for exit within the control group.

economic success of the bought consultings will determine the relevance of the environmental actors and institutions. Mutual help will be exchanged to monetary relations between competitors.

What the Membership may Attend from their EBA

It is well known that motivation is not immediately followed by action. And in market economies the principle is valid too that the ongoing action can only approach the result of an invisible development, which – if recognised – determines what happens in and between the companies. The last target, i.e. the long-term development, finally realises itself behind the back of the actors. How much more difficult is it for the companies to decipher what the mid- or long-term future will bring for them and their environment. Settled between economic rationalisation options and/or ethical commitment, the market produces only unclear orientation for the fundamentally difficult action of EBA-members.

For the companies one way out of the multi-layer and multi-facetted insecurity is the reliance on the eco-leadership, which seems to lie in good hands at the EBAs. However, these associations are more and more questioned themselves in this role, but the top-down model of communication between association and members is still accepted in principle.

The other way of the companies is, to start themselves in one way or the other to find solutions for the optimisation of the environment, which may serve others as an example. Bottom-up processes of transfer seem to be appropriate. However, the companies tend more and more to leave to the EBAs the transfer and implementation of new insights and competencies around the environment to other companies. Still to be the leader in environmental affairs is determining the actors' conception. The transfer of knowledge becomes secondary. The 'business environmental movement' has passed already its climax. Targets, producers, users of environmental services have themselves separated. What links them becomes systemic and ephemere and enforced at the same time.

Apparently companies address themselves to others and to organisations. They become members of associations if they think the get the best answers to their problems. However, this hypothesis is only partly convincing.

Against it there are some structural reasons. Companies with more than 600 employees already have their own specialists for the environment who – when they have their own staff – think they are more capable to cope with environmental problems than the EBA, to which their company belongs to.[15] The more general environmental focus does not serve immediately the specific problems of the company – so their vote.

And then other entrance orientations are added. [16] About one third of the companies have become members of EBAs because they wanted to help other companies to solve their environmental problems. And at least for half of the active members their primary concern when entering to realise potential rationalisations by environmental protection; so the driving force was more profit than the environment. Whereas there are only a few – and only in second or third priority – which became members out of a political concern to save the environment. It is only at maximum one third of the companies, which in first priority and explicitly defends the environment, and which can be counted among the active EBA-members. So at the end it is only about 10 % of all EBA-members who are driven by this environmental motivation.

When today more than half of all active EBA-members are convinced that the environment is a permanent task for the companies, so in general the companies name only two problems, which are really relevant for them; this number even decreases in average just one in the mid-term perspective. And out of the great choice of possible environmental areas (from societal to public business debates as well as the prior problems, which create environmental problems) remain first of all end of the pipe-issues (from waste to recycling) and the contributions to publicly discussed topics which are mentioned. Environmental issues within the production sphere do not seem to be of much relevance for them.

Public debates and solutions in this already highly generalised environmental sphere are not expected in the future by the EBAs, the consultants or scientists, but by the owners of companies, politics and at large the debates within the environmental movements. Who today presents himself at a company as somebody who wants to help or even as a saviour of the environment has to count with much scepticism, distance or even refusal. That also the environmental consultants only cook with water has been spread. And even proper water remains water.

Finally it is still completely uncertain, what from the services of the EBAs will be implemented within the companies. Even the environmental specialist in the executive board in an ecologically active company can expect only in about half of the cases understanding by his colleagues; as more these cases are technologically focussed, as less support he gets. And this

15 Results of the accompanying and framework interviews to the empirical research.

16 Results of the the study of the active EBA-members.

environmental management understanding decreases more down over the hierarchy of the company. At the end committed individuals are more respected in their struggle for a better environment than parts of a company. Departments mostly do not bother what the environmental specialists up there in the hierarchy may tell them. They do business as usual.

Although the composition of companies within each EBA is rather heterogeneous, the history of the EBAs made it that there has been a mutual attraction effect between specific and appropriate companies. The founding history led to different types: we still can discern the top-down and bottom-up associations. But a strong will to pursue their initial mission is not to be found anymore. The EBAs – lost in the fog of lacking vision – are swimming in a lake, which has to be ecologically cleaned and with public access, which has unknown contours and depths. Really safe and environmental-friendly boats and ports are not yet discernible for the actors, the companies and the economy, not yet in sight.

However, this situation is no reason to despair. Really in danger are only small and medium-sized companies with up to 600 employees. They do not have capacities for the environment. They only treat very few environmental issues, which also have been enforced on them. They do not have the resources to find for them ecologically reasonable and economic solutions. They are the main target group for the EBAs.

The difficulties, however, for the EBAs that these SMEs have not been reached by the two main EBAs. And with their modest membership fees they could not sustain the work of the EBAs. The general demand to protect the environment and for ethical responsibility marginalise in the tendency just those companies, which are not able to helm themselves. The globalisation of the ecological demands (Agenda 21) will drive the SMEs even more to the margin of public awareness of the public support structures.

The bigger and really big companies are not so much interested in technical solutions. They already have the personnel to solve their own environmental problems by themselves. They are more specialised than the employees of the EBAs are. They have less problems of implementation, and they combine activities and experience. Learning for the environment is an integral part of their company. The big companies consider in the long run the (political) responsibility as the relevant criteria for environmental efforts of the associations.

The problem for the EBAs is their relatively modest size (just a few hundred members) to put real pressure on the different state levels for a better environment. The traditional associations do this still better – when they do it. Just one regional Chamber of Commerce and Industry has more members than a nation-wide acting EBA. So, for the EBAs remains the task to be an early warning system.

The balance on the side of structural determinants for environmental action is devastating for the companies and the EBAs. Those asking for support and those offering it go different directions materially, just when they tend ideally to look into the same direction to save the environment. Only if you do not look into the details, you talk of success on the general level, which actually did not happen on the enterprise level. The ideal capital of ethical responsibility for the environment, which has led to the existence of the EBAs, it is enough if it is perceived in public relations. The implementation remains secondary. This remains the task of every single company. The appropriate environmental learning process is guided by the market, and therefore nearly invisible.

Only Performance Counts

As already mentioned earlier, the further development perspectives for the EBAs and their members remain foggy. What we could win out of interviews with the actors was their opinion about the really existing EBA and the options, which they assume reasonable. What the EBAs offer as service, does not count bin the same way for the members. Service does not equal service.

Naturally the EBA-members welcome the services. But they do not expect and use them in the same way in amount and distribution, as it is conceived by the ideal and real work of the three EBAs. The association ideally expects that the members take part at the 'programme' of the EBA; even more, as they expect that through mutual aid they may help other companies to be successful for the environment. It is part of the minimal programme to use the services, which are at the core of the relationship between addressees and the environment: environmental management in the largest sense (top-down). The maximal programme includes substantial increase in amount, extension, largeness and depth of the efforts for the environment, and the active participation within the activities of the EBAs: Production of environmental know-how, its distribution (bottom-up) and implementation within the company. This core programme is most easily described and analysed as between Is and Should, when we differentiate between the transfer-forms of services and the services linked to specially offered and used services.

Table 1: Expected and used transfer-forms by the companies [17]

Expected			**Transfer-forms**	**Used**			**Diff.***		
TD1	**TD2**	**BU**		**TD1**	**TD2**	**BU**	**TD1**	**TD2**	**BU**
68,4	40,9	**57,9**	Leaflets, general info.	**84,2**	50,0	**65,8**	**16**	9	8
56,8	54,5	52,6	Guidelines/Handb., spec. info.	53,7	**86,4**	42,1	-3	**32**	-11
50,5	*36,4*	**55,3**	Info-leafl. sectors/ areas	44,2	*22,7*	**68,4**	-15	**-14**	**13**
66,3	**59,1**	42,1	Congresses/lectures	51,6	63,6	36,8	-15	-5	-5
63,2	45,5	42,1	Seminars/training	42,1	*40,9*	36,8	**-21**	-5	-5
35,8	45,5	*26,3*	Moderat./seminars/ workshops	*20,0*	63,3	*21,1*	-16	18	-5
24,2	50,0	39,5	Working groups	*12,6*	59,1	50,0	-12	9	11
42,1	*27,3*	28,9	Individual consultation	35,8	54,5	26,3	-6	27	11
56,8	**59,1**	*15,8*	EBA magazine	**66,3**	**81,8**	26,3	10	22	11
43,2	50,0	31,6	Public relations	34,7	50,0	*15,8*	-9	0	**-16**
5,3	4,5	2,6	Other	3,2	4,5	2,6	-2	0	0

TD1 Top-down association 1; TD2 Top-down association 2; BU Bottom-up association
TD1 – total 95 = 100 % TD2 – total 22 = 100 % BU – total 38 = 100 %
* in %; multiple nomination was possible

In regard to the expectations concerning the use of the different transfer-forms today written information is preferred to oral, and general knowledge and competencies more than specific, what may induce to a more precise knowledge on the side of the companies through the co-operation with the EBAs. The relative difference of EBA TD2 is to be explained by its specific membership-structure. They are mostly SMEs with a special commitment of their managers for the environment. Out of this mutual socio-cultural cohesion they seem to prefer to represent their EBA publicly in good shape. The relative difference of the EBA BU is also related to the small company size of its members, but also due to the selective services of this EBA, e.g. not like to others to offer their own magazine. But here too the soft factors of cultural cohesiveness play a great role. Instead of well-known personalities the specific region gives binding and attraction for the members.

17 Newly arranged results. Cf. Széll and Ehlert 2000.

The difference between top-down and bottom-up reappears only again within the working groups. Whereas the first prefer the traditional transferforms, the others are with the offers of teaching-/learning materials up-to-date. Pure top-down processes are not any more advantaged; bottom-up seems to be old-fashioned.

The difference between expectations and use shows, how important the EBA is for its members at the end. Also as an active EBA-member you may not participate in it programmatically. The interest to the EBA-magazine is explicit in this regard. To be informed what stands in the foreground in the public environmental debates, and what happens within the EBA seems to be enough. In regard to all other issues there is some distance. Modern moderation technologies are only welcome in one single case. Individual solutions are on a low level of some interest. Public relations seem to be less important as has been supposed in all debates around environmental management today.

Surprisingly is the large distance between the members and their EBA. It is less visible in the lack of consensus to the whole EBA, but more in the lack of the use of the offers and the lack of participation in the EBA-activities.

Naturally services, i.e. the content of the transferforms, incorporate first of all results, which give the possibility of a more precise insight into the relationship between the membership and the EBAs. Again it becomes evident that written and general information on the environment is preferred above all, what is more precise and demands personal presence. Services are particularly honoured, if they are targeted on the management and the environmental specialists. The promotion of co-operation between companies is well seen too, and everything which is concerned with the State (administration, subsidies etc.). Whereas it is relatively less desired or even undesired what in the business environmental debate is proposed as solutions to the companies by the EBAs. Environmental management techniques and everything, which treats with the production of goods and services, receives at utmost only half the attention. Here again the companies divert, which out of reason of size do not have their own environmental management kennen or where the actors in management have not – due to socio-cultural reasons – developed the conscience or do not have specialised environmental personnel. The originary task of an association – i.e. to influence politics – is regarded as secondary by them.

The distance of the membership is not only been visible in relation to general vs specific tasks, but expresses itself in the programme of the EBAs in regard to everything, which should be of relevance for the management: the knowledge of strategic-tactical measures of the implementation of environmental know-how for and in the enterprise. The spectrum of services of the EBAs is not honoured to the degree by the companies as it is offered to them by the EBAs in the context of the general environmental debate.

Table 2: Expected and used results by enterprises [18]

Expected			**EBA A**	**Used**			**Diff.***		
TD1	**TD2**	**BU**	**Services**	**TD1**	**TD2**	**BU**	**TD1**	**TD2**	**BU**
71,6	**54,5**	**55,3**	Information – general	**84,2**	**72,7**	**65,8**	**12**	18	**11**
44,2	36,4	**52,6**	Information – special	46,3	54,5	**60,5**	2	18	8
35,8	**54,5**	15,8	Competence/motiv. manag.	35,8	59,0	26,3	0	5	**11**
33,7	45,5	18,4	Competence/motiv. employ.	25,3	72,7	18,4	-8	27	0
49,5	**59,1**	26,3	Competence/motiv. eco-pers.	**48,4**	**77,3**	34,2	-1	18	8
26,3	31,8	23,7	Creation Eco-manag.-system	16,8	45,5	21,1	-10	14	-3
28,4	36,4	15,8	Help eco-cost calculation	14,7	36,4	18,4	-14	0	3
22,1	18,2	*10,5*	Help audit./certification	9,5	22,7	13,2	-13	5	3
17,9	*4,5*	*10,5*	Help market/distribution	12,6	9,0	13,2	-5	5	3
21,1	*4,5*	*7,9*	Help pack./logis./transp.	12,6	0,0	10,5	-9	-5	3
20,0	18,2	21,1	Help preventive prod.	11,6	9,0	18,4	-8	-10	-3
22,1	18,2	18,4	Help end of the pipe-solution	15,8	27,3	15,8	-6	10	-3
21,1	18,2	23,7	Help product-recycling	15,8	18,2	28,9	-5	0	5
36,8	13,6	15,8	Help env.-prot. management	36,8	31,8	10,5	0	18	*-5*
28,4	18,2	28,9	Help Energie-H2O-Manag.	20,0	13,6	39,5	-8	-5	**11**
10,5	13,6	18,4	Help work/health	11,6	18,2	15,8	1	5	-3
27,4	**54,5**	13,2	Business targets/policies	28,4	86,4	15,8	1	**32**	3
52,6	**54,5**	39,5	Co-op.-support companies	42,1	68,2	42,1	-11	14	3
34,7	40,9	18,4	Co-op.-support science	17,9	45,5	13,2	-17	5	*-5*
44,2	36,4	13,2	Info. Governm. subsidies	26,3	45,5	21,1	*-18*	9	8
29,5	36,4	26,3	Influence on law-making	14,7	9,0	21,1	-15	*-27*	-5
38,9	36,4	23,7	Help PR	30,5	63,6	21,1	-8	28	-3
6,6	4,5	2,6	Other	5,5	4,4	2,6	-1	0	0

TD1 Top-Down Association 1; TD2 Top-Down Association 1; BU Bottom-up Association
TD1 – Total 95 = 100 % TD2 – Total 22 = 100 % BU – Total 38 = 100 %
* Figures in full per cent; multiple answers were possible

[18] Newly arranged results. Cf. Széll and Ehlert 2000.

The Optimisation of the Efforts by Actors and EBAs

Out of the empirical findings we may conclude two things. The distance we stated between the membership and the EBAs as well as the weak acceptance of their core business may derive from the wrong orientation and/or the lacking quality of the services by the EBAs. But this conclusion is not convincing, as the active members have confirmed that the EBAs have in general done a good job. They also do not have addressed any specific wishes to EBAs for the future, which go beyond generalities like "more innovation".[19]

The distance, however, may come from the sphere of structural conditions for the EBAs out of a market society. Then they are out of reach from the individual actors as well as organisations, and they have their practical effect within them and on all and everything concerning the environment. The result would be that the EBAs lived more from the ideal capital of the companies and the responsibility for the environment felt there than the could practice really and materially for the companies. With the vanishing of the environmental topic in the public of the economy the benefits of the ideal capital for the EBAs would disappear too.

This second conclusion is not as relevant for the past of the EBAs, however, much more so for the presence and the future. In the presence the EBAs recognise ever more the three-phase model of participation of the membership: first interest, then commitment, and finally passive membership or exit. For the companies already the first phase helps them to clarify what they really want and with what the EBA may support them. If there is clarity, there is half retreat in form of an ideal support or complete retreat. The number of 'dead' or short-term members is surprising.[20]

In the future, when consulting will largely replace mutual help for the environment – which was at the core of the idea of an EBA – the insight in these relationship will disappear more and more. In many consultings only the money-relationship remains. The amount expresses the mutual respect and attention For the effect on the environment is not discernibly any more in the centre of the EBAs, but in the well-being of a great number of actors of a commerce amongst others in a market economy, it is not to be said or even proven that consulting may be the more efficient model for the future of the EBAs. More than to hope is today impossible; any other certainty demands new scientific research.

19 Cf. Széll and Ehlert 2000

20 Would we include all exits of passive members the percentage of active members would decrease dramatically. The EBAs would statistically become nearly insignificant.

When the actors in the EBAs as well as in the companies continue to optimise their services the will have less success in the future in the way they are organised as a solidaristic association. Scientifically we can enumerate up to six target areas[21] with each up to 22 topics,[22] in which chances for success exist, when the EBAs improve their services. Offer-demand as well as rating differences structures this potential. But in every of the pilot-projects we proposed,[23] and which have been conducted by the EBAs, the wish for optimisation encountered implementation difficulties, which the actors had not expected and they could not well control. I.e., the 'places' became clearer, at which the market-society restrictions for action emerged; the deficits were emphasised, which have to be confronted further and in a more concentrated way. And the actors within the EBAs understood better the optimisation projects as a developmental process, which in their time-scope go beyond the limits of an accompanying research project. The basic problems lying behind remained for all actors until now 'systematically-ununderstandable' or even invisible.

Real success was realised, where practically all elements of the pilot-projects remained in the hands of the involved actors, may it be associations or companies. Association-internal personnel- and organisation development are to be named as well, as the forced extension of the EBAs. The largest deficits lay in regard to knowledge and the capacities of the EBAs, to transform the commitment of the companies into services, where the companies had the impression that the service arrived as well, as they had hoped for. Here it is not just a problem of simple 'communication', where it is supposed that the sent was received (Shannon/Weaver). But it was the problem of unequal preconditions of mutually similarly understood tasks, services and results. The first were not well understood, because not explicitly formulated, kept by oneself and/or even unknown; the latter seemed only similar to the actors, as they abstractified from the preconditions, and common-sense conceptions followed a slogan-like environmental language. Those projects remained open, which at the beginning had been designed as an offer, but where the demand can only articulate itself much later.

---

21 They are determined on the one side by the possibilities for action for the EBAs andf the members, on then other hand by the levels of action, i.e. in other words to be able to become active through the chances and risks today, on mid-term and in the future on the background of changing structural conditions.

22 As topics the service types were used, which are the base for table 2.

23 The EBAs could chose the pilot-project out of the proposal by the scientists. The pilot-projects which were realised concerned the proper development of the EBAs, their relation of the companies to the public administration, and the options of the Internets for the work of the EBAs. (For more details cf. Ehlert 2002)

As the diversely conceived the pilot-projects were concerned with the question of competencies of mutually structured learning between the environment, the EBAs, companies, employees and a future for the companies, which may help to include and reach more and sustainable environmental concern.

Summary

The optimisation of the work of the EBAs can only produce limited results. Areas and topics, in which and on which can be worked, are to be named. Because the companies have gone more in distance to the EBAs as those had expected, they cannot win enough insight into the implementation problems of the environmental-friendly company. The company-related environmental know-how and the corresponding capacities, which they should have, to be able to accomplish their task really and well, is not developing. The EBAs dispose at the moment nor of the respective human resources nor the programme. As 'old-fashioned' associations they do not seem to be able to cope with these challenges; as consulting associations their qualification profile is still in the making.

EBAs, which want to survive in a market of environmental offers, have to look for new target groups, which necessarily will focus in the first place SMEs. Where there are no resources to develop own competencies, the demand for support is most easily to be seen, to grasp, to transform into action and to conclude with success for the companies. It is difficult only to address these companies, and the way how to win them for the EBAs, as well as any form of financial support. This demands a completely new approach for the EBAs.

So, the EBAs of the future will act more as private corporations with a public appendix. There, where they can consult, the general approach to speak for the whole environment disappears behind single consulting projects, for which it has to be paid. The amount guarantees the reciprocal seriousness. Consulting does not mean to take the responsibility for the implementation; this is the business of the company only. The attractiveness of the means will overcome the limited possibilities of action. It is more quantity than environmental quality, which had been the main target of the EBAs before.

As perfectly the consultants behave and what ever they may do, the big companies have completely different interests in regard to the environment. Their political ambitions imply that in the long run they will only co-operate with the old type of EBAs and the political administration. In the moment the EBAs will change into consulting agencies, they will lose the socially recognised competence as a co-operative institutions for politically organised interests.

The performances of EBAs of the old and the new type have to develop – where the do not just practice consulting – an other understanding of eco-management, and then to implement it into the companies anew. In place of learning-aids (top-down) has to come organised self-aid (bottom-up). It has to transfer the capacity to elaborate own business environmental compatibility into the company itself.

Evidently successful campaigns and publicly efficient environmental ratings of companies and environmental awards for personalities will accompany the activities of the EBAs. However, in face of all public success it has not be forgotten that all these activities do not reach the average company nor address the companies directly. These public relations have effect – if at all – only indirectly in a hectic media world. In the day-to-day environmental work within the companies these activities remain more or less marginal. The aura of the outstanding environmental managers is not so brightly shining anymore in the average company, because their management had to learn to cope with the daily hard labour for the environment. The way is the target.

References

Brickwedde, F. (2000), 'Environment and work – innovative environmental technology as a chance for economy and employment', in K. Busch et al. (eds), *Ways to social peace in Europe,* secolo, Osnabrück, pp. 128–132.

Burschel, C. (1995), *Umweltschutz als sozialer Prozeß,* Westdeutscher Verlag, Opladen and Wiesbaden.

Ehlert, W. (2002), *Unternehmen und betriebliche Umweltverbände,* Peter Lang, Frankfurt [u.a.].

Fukuyama, F.(1992), *The End of History,* The Free Press, New York.

Hans-Böckler-Stiftung (2000), *Arbeit und Ökologie*, Düsseldorf.

Hildebrandt, E. and Schmidt, E. (2001), 'The Ecological Challenge', in G. Széll (ed.), *European Labour Relations,* Gower/Ashgate, Aldershot, vol. 1, pp. 116–136.

Hoffmann, J.; Matthies, H. and Mückenberger, U. (eds) (1992), *Der Betrieb als Ort ökologischer Politik,* Verlag Westfälisches Dampfboot, Münster.

Kühr, R. and Széll, G. (eds) (1997), *Zero Emissions Research Initiative Kongreß Osnabrück. Dokumentation,* Universität Osnabrück.

Lafontaine, O. and Müller, C. (1998), *Keine Angst vor der Globalisierung,* Dietz, Bonn.

Schulz, W. (1998), 'Das Leistungsspektrum der Verbände', *BJU-Umweltschutz-Berater,* Frankfurt.

Széll, G. (1990), 'Möglichkeiten partizipatorischer Konversion auf der Basis der Mitbestimmung', in L. Köllner and B.J. Huck (eds), *Abrüstung und Konversion. Politische Voraussetzungen und wirtschaftliche Folgen in der Bundesrepublik*, Campus, Frankfurt/New York, pp. 491–510.

Széll, G. (1992b), 'Environment', in G. Széll (ed.), *Concise Encyclopaedia of Participation and Co-Management*, de Gruyter, Berlin/New York, pp. 318–325.

Széll, G. et al. (1998), *Unternehmenserfolg durch Umweltschutz. Herausforderung Japan: Umweltschutz als kontinuierlicher Prozeß – vom Qualitätsmanagement zum Umweltmanagement*, Forschungsstelle Japan (# 6), Universität Osnabrück.

Széll, G. (2001), 'To learn from Japan?', in G. Széll (ed.), *European Labour Relations*, Gower/Ashgate, Aldershot, vol. 2, pp. 191–205.

Széll, G. and Ehlert, W. (2000), *Wirkungsweisen und Handlungsspielräume von Multiplikatoren für Umweltschutz in Unternehmen in der Bundesrepublik Deutschland*, V.A.U.S.T., Juni, Osnabrück.

Tischler, K. (1998), *Betriebliches Management als Lernprozess. Theorie und Praxis organisationsorientierter Umweltbildung als Beitrag zur Steigerung der Fortschrittsfähigkeit der Unternehmung*, Peter Lang Verlag, Frankfurt a.M. [u.a.].

Winter, G. (1993), *Das umweltbewußte Unternehmen*, Beck, München.

Acknowledgements

I thank Prof. Dr. Ken'ichi Tominaga and his staff for the excellent organisation of the Congress in October 2002 at the Musashi Institute for Technology. We thank the Musashi Institute for Technology for the generous hosting, and the Deutsche Forschungsgemeinschaft which helped by a number of travel grants that so many German scientists could participate.

For putting this book into good shape, it was again as for many others before, done by Gabriele Parlmeyer from the Department of Social Sciences of the University of Osnabrück. Felix Mengel corrected the references. We thank both of them most sincerely.

Osnabrück, February 2004 György Széll

The Authors

Dr. Wiking EHLERT, born 1942 in Berlin, Professor for sociology at the University of Osnabrück, he studied political science and sociology in Berlin and Georgetown University, Washington D.C., main research areas are environmental and organisational sociology.
wiking-ehlert@t-online.de

Dr. Gesine FOLJANTY-JOST, Professor for Japanese Studies at Martin-Luther-University Halle-Wittenberg. She is specialised in Japanese politics, and is concentrating her research since years on Japanese environmental policy-making. <barth@japanologie.uni-halle.de>

Dr. Friedrich FÜRSTENBERG, born 1930 in Berlin, Professor Emeritus for sociology at Bonn University. He has done extensive empirical research on industrial organisations and labour markets both in Germany and Japan. His numerous publications also include books on social structure and mobility as well as the sociology of economics (translated in Japanese) and of religion. <f.fuerstenberg@uni-bonn.de>

Dr. Harutoshi FUNABASHI (1948), Professor, Faculty of Social Sciences, Hosei University, Ex-President of the Japanese Association for Environmental Sociology. <hfunabas@mtj.biglobe.ne.jp>

Dr. Koichi HASEGAWA, Professor of Sociology at Tohoku University and Chief of the Division of Studies of Minorities, The Center for the Study of Social Stratification and Inequality, Tohoku University, 21st Century Center of Excellence Program. He received his PhD from Tokyo University. His research mainly deals with environmental sociology, social movements and social change. His most recent book is Constructing Civil Society in Japan: Voices of Environmental Movements (Trans Pacific Press, 2004). <hasegawa@sal.tohoku.ac.jp>

Kazuo IWAMURA, born in1948 in Kobe, Japan. Graduated from Graduate School of Waseda University/M.E. in 1974. Architect (Council Member of UIA & JIA) at IWAMURA Atelier Co., Ltd. in Tokyo, and Professor at the Faculty of Environmental & Information Studies, Musashi Institute of Technology in Yokohama. <iwamura@yc.musashi-tech.ac.jp>

Dr. Makoto KOBAYASHI, Associate Professor for psychology at the Tamagawa University in Tokyo, College of Education. He studied philosophy and psychology at Keio University in Tokyo (B.A. and M.A.) and completed his doctoral study at the University of Konstanz, Germany, as a DAAD Scholarship fellow under the supervision of Prof. Dr. G. Trommsdorff. His main research areas are the self and moral development of youth from cross-cultural perspectives and the analysis of psychological preconditions for intercultural tolerance.
benedict@dream.ocn.ne.jp; makoto@edu.tamagawa.ac.jp

Rüdiger KÜHR, M.A. (1970), PhD Student University of Osnabrück; United Nations University, Zero Emissions Forum, European Focal Point (Stockholm/Hamburg), International co-ordinator and secretary "Alliance for Global Eco-Structuring" (AGES) under UNEP's Cleaner Production Network, freelance policy consultant to various international organisations, national government, and companies. <kuehr@online.de>

Dr. Volker LINNEWEBER, studied psychology, educational sciences and sociology at Münster University, (dissertation on aggression). Following postdoctoral studies at UC Irvine (US), he completed his habilitation at University of Saarland, worked as a research fellow at the Potsdam-Institute for Climate Impact Research, and now is a Professor for social psychology at Magdeburg University.
<Volker.Linneweber@gse-w.uni-magdeburg.de>

Helmut LOISKANDL, born in Vienna, Austria, in 1934, started his academic career at the University of Munich. After teaching at different places in the US and Australia he has held the chair for comparative sociology at Tokiwa University in Mito, Japan, since 1989, Visiting Professor at the University of Osnabrück since 2003.
<loisk@wc4.so-net.ne.jp>

Takashi NAMBA, is Associate Professor of urban sociology and computer sciences at Nagoya College. His main subject is urban renewal and gentrification in inner-city areas. <namba@nagoyacollege.ac.jp>

Seiichi NISHIKAWA, born in 1955, studied forestry, policy studies and sociology, and manages a private limited partnership named " Nishikawa Institute", while acting as a professional engineer in forestry and erosion control, and as an adjunct instructor at Bukkyo University.
<nekosoho@hi-net.zaq.ne.jp>

Dr. Yoshiki OTSUKA, received his PhD (Sociology) from University of Tsukuba, now teaches environmental sociology at Musashi Institute of Technology. Main research on "Socioeconomic considerations relevant to the sustainable development, use and control of genetically modified foods". <otsuka@yc.musashi-tech.ac.jp>

Dr. Dr. h.c. Erwin K. Scheuch, born 1928 in Cologne, died in 2003, Dipl.-Volksw., B.A. was Professor of sociology at the University of Cologne until 1993, many distinctions, amongst others President of the German Sociological Association and the International Institute of Sociology, main research areas empirical and comparative social research.

Dr. György SZÉLL, born 1941 in Budapest, Professor for sociology at the University of Osnabrück since 1973, he studied social sciences at the universities of Frankfurt, Münster, Aix-en-Provence and Algiers, main research areas: labour relations, international comparative social sciences and environmental sociology. <gszell@uni-osnabrueck.de>

Dr. Akira TOKUYASU, Professor of Sociology, Faculty of Social Sciences, Hosei University. His major research field is social systems theory. <atokuyas@mt.tama.hosei.ac.jp>

Dr. Dr. Ken'ichi TOMINAGA, born 1931 in Tokyo. Present Status: Professor Emeritus, the University of Tokyo; Dr. of Sociology, from the University of Tokyo, 1967; Dr. of Economics, from the University of Kyoto, 1998; Graduate Course of the University of Tokyo, 1959; Vice-President of the German Association of Social Sciences.
knch.tominaga@jcom.home.ne.jp

Shuji YAMADA, born in 1968, BA, MA, Associate Lecturer at the Faculty of International Studies Bunkyo University, studied Environmental Sociology and Industrial Sociology from 1987 to 1991, and from 1991–1997 at Chuo University Graduate School. <yamashu1968@hotmail.com>

WORK – TECHNOLOGY – ORGANIZATION – SOCIETY
ARBEIT – TECHNIK – ORGANISATION – SOZIALES

Edited by Wiking Ehlert, György Széll and Heinz Sünker
Herausgegeben von Wiking Ehlert, György Széll und Heinz Sünker

Band 1 György Széll/Wiking Ehlert (eds.): New Democracies and Old Societies in Europe. 2001.

Band 2 Norbert H. Pieper: Arbeitsmarktpolitik in Deutschland und Japan im Vergleich. 1996.

Band 3 Volker Telljohann: Globaler Wettbewerb – lokale Entwicklung. Gewerkschaftliche Konzepte innovativer Industriepolitik zur Nutzung endogener regionaler Entwicklungspotentiale am Beispiel der Erfahrungen im Bereich der Maschinenbauindustrie Bolognas. 1997.

Band 4 Roderick Martin/Akihiro Ishikawa/Csaba Makó/Francesco Consoli (eds.): Workers, Firms and Unions, Part 1. Industrial Relations in Transition. 1998.

Band 5 Martin Schwanholz: Informationstechnologien im Wandel. Marktstrukturen – Aufholprozesse – Wettbewerbschancen. 1998.

Band 6 Klaus Tischler: Betriebliches Umweltmanagement als Lernprozess. Theorie und Praxis organisationsorientierter Umweltbildung als Beitrag zur Steigerung der Fortschrittsfähigkeit der Unternehmung. 1998.

Band 7 Maria Deuling: Qualifikationsanforderungen in den kaufmännischen Berufen. Regionale Arbeitskräftebedarfsanalyse. 1999.

Band 8 Kotaro Oshige: Konvergenz der Interessenvertretungen durch Globalisierung? Ein Vergleich der Funktionsmechanismen der Arbeitnehmerinteressenvertretungssysteme in Deutschland und Japan am Beispiel der Elektroindustrie. 1999.

Band 9 Khaleda Akter Siddiqui: Urban Working Women in the Formal Sector in Bangladesh. 2000.

Band 10 Akihiro Ishikawa/Roderick Martin/Witold Morawski/Veljko Rus (eds.): Workers, Firms and Unions, Part 2. The Development of Dual Commitment. 2000.

Band 11 Karl Bruckmeier/Wiking Ehlert (eds.): The Agri-Environmental Policy of the European Union. The Implementation of the Agri-environmental Measures within the Common Agricultural Policy in France, Germany, and Portugal. 2002.

Band 12 Désirée M. Nana: L'éducation environnementale des enfants et jeunes au Sahel. L'exemple du Burkina Faso. 2001.

Band 13 Wiking Ehlert: Unternehmen und betriebliche Umweltschutzverbände. 2002.

Band 14 György Széll, Dasarath Chetty, Alain Chouraqui (eds.): Participation, Globalisation & Culture. International and South African Perspectives. 2002.

Band 15 Olaf Winkelmann: Entwicklung in der Mitbestimmung - Industrial Relations, Industrial Districts und Netzwerke. Das Beispiel *Betriebsräte-Netzwerk* im Industrial District Ostwestfalen-Lippe. 2002.

Band 16 Thomas Volkening: Ansätze zu organisatorischen Reformstrategien in einem Krankenhaus in kommunaler Trägerschaft. Eine Bestandsaufnahme. 2004.

Band 17 György Széll / Gian Primo Cella (eds.): The Injustice at Work. An International View on the World of Labour and Society. 2002.

Band 18 Lydia Kocar: Globale Ökonomie und nationale Politik. Die Forschungs- und Technologiepolitik der EU. 2003.

Band 19 Heinz Sünker: Politik, Bildung und soziale Gerechtigkeit. Perspektiven für eine demokratische Gesellschaft. 2003.

Band 20 Nicola Bücker: Historische Aspekte und Entwicklungstendenzen des japanischen Beschäftigungssystems. 2003.

Band 21 Faping He: Struktur und Funktionsweise des chinesischen Arbeitsmarktes. Der Gestaltungsprozess des Arbeitsmarktes unter dem wirtschaftlich-sozialen Strukturwandel in der VR China seit 1978. 2003.

Band 22 Thomas Swiderek: Kinderpolitik und Partizipation von Kindern. Im Spannungsfeld von Vergesellschaftung und der Möglichkeit größerer Selbstbestimmtheit, Mündigkeit und dem Erlernen von Demokratie. 2003.

Band 23 Heinz Sünker / Russell Farnen / György Széll (eds.): Political Socialisation, Participation and Education. Change of Epoch – Processes of Democratisation. 2003.

Band 24 Birgit Riße: Mündigkeit in der Postmoderne. Ansatzpunkte und Zielperspektiven von Erwachsenenbildung als Identitätsbildung im Rahmen der gegenwärtigen gesellschaftlichen Wandlungsprozesse. 2004.

Band 25 Francesco Garibaldo / Volker Telljohann (eds.): Globalisation, Company Strategies and Quality of Working Life in Europe. 2004.

Band 26 Serajul Hoque: Global Trade Liberalization. Impact on the Readymade Garments Industry in Bangladesh. 2004.

Band 27 György Széll / Ken'ichi Tominaga (eds.): The Environmental Challenges for Japan and Germany. Intercultural and Interdisciplinary Perspectives. 2004.

www.peterlang.de

FSC
www.fsc.org
MIX
Papier | Fördert gute Waldnutzung
FSC® C083411

Kolibri 360 GmbH
Ferdinand-Jühlke-Straße 7
99095 Erfurt, Deutschland
produktsicherheit@kolibri360.de

Druck:
CPI Druckdienstleistungen GmbH
im Auftrag der
Kolibri 360 GmbH
Ferdinand-Jühlke-Str. 7
99095 Erfurt

Neuroglia in the Aging Brain

Contemporary Neuroscience

Pathogenesis of Neurodegenerative Disorders, edited by **Mark P. Mattson,** 2001
Neurobiology of Spinal Cord Injury, edited by **Robert G. Kalb** and **Stephen M. Strittmatter,** 2000
Cerebral Signal Transduction: From First to Fourth Messengers, edited by **Maarten E. A. Reith,** 2000
Central Nervous System Diseases: Innovative Animal Models from Lab to Clinic, edited by **Dwaine F. Emerich, Reginald L. Dean, III, and Paul R. Sanberg,** 2000
Mitochondrial Inhibitors and Neurodegenerative Disorders, edited by **Paul R. Sanberg, Hitoo Nishino, and Cesario V. Borlongan,** 2000
Cerebral Ischemia: Molecular and Cellular Pathophysiology, edited by **Wolfgang Walz,** 1999
Cell Transplantation for Neurological Disorders, edited by **Thomas B. Freeman** and **Håkan Widner,** 1998
Gene Therapy for Neurological Disorders and Brain Tumors, edited by **E. Antonio Chiocca** and **Xandra O. Breakefield,** 1998
Highly Selective Neurotoxins: Basic and Clinical Applications, edited by **Richard M. Kostrzewa,** 1998
Neuroinflammation: Mechanisms and Management, edited by **Paul L. Wood,** 1998
Neuroprotective Signal Transduction, edited by **Mark P. Mattson,** 1998
Clinical Pharmacology of Cerebral Ischemia, edited by **Gert J. Ter Horst** and **Jakob Korf,** 1997
Molecular Mechanisms of Dementia, edited by **Wilma Wasco** and **Rudolph E. Tanzi,** 1997
Neurotransmitter Transporters: Structure, Function, and Regulation, edited by **Maarten E. A. Reith,** 1997
Motor Activity and Movement Disorders: Research Issues and Applications, edited by **Paul R. Sanberg, Klaus-Peter Ossenkopp,** and **Martin Kavaliers,** 1996
Neurotherapeutics: Emerging Strategies, edited by **Linda M. Pullan** and **Jitendra Patel,** 1996
Neuron–Glia Interrelations During Phylogeny: II. Plasticity and Regeneration, edited by **Antonia Vernadakis** and **Betty I. Roots,** 1995
Neuron–Glia Interrelations During Phylogeny: I. Phylogeny and Ontogeny of Glial Cells, edited by **Antonia Vernadakis** and **Betty I. Roots,** 1995
The Biology of Neuropeptide Y and Related Peptides, edited by **William F. Colmers** and **Claes Wahlestedt,** 1993
Psychoactive Drugs: Tolerance and Sensitization, edited by **A. J. Goudie** and **M. W. Emmett-Oglesby,** 1989
Experimental Psychopharmacology, edited by **Andrew J. Greenshaw** and **Colin T. Dourish,** 1987
Developmental Neurobiology of the Autonomic Nervous System, edited by **Phyllis M. Gootman,** 1986
The Auditory Midbrain, edited by **Lindsay Aitkin,** 1985
Neurobiology of the Trace Elements, edited by **Ivor E. Dreosti** and **Richard M. Smith:** Vol. 1: *Trace Element Neurobiology and Deficiencies,* 1983; Vol. 2: *Neurotoxicology and Neuropharmacology,* 1983

Neuroglia in the Aging Brain

Edited by

***Jean S. de Vellis,* PhD**

UCLA Mental Retardation Research Center,
Los Angeles, CA

Foreword by

***Paola S. Timiras,* MD, PhD**

University of California at Berkeley, Berkeley, CA

Springer Science+Business Media, LLC

Dedication

for Antonia Vernadakis

Originally published by Humana Press Inc in 2002
Softcover reprint of the hardcover 1st edition 2002
999 Riverview Drive, Suite 208
Totowa, New Jersey 07512

humanapress.com

This publication is printed on acid-free paper. ∞
ANSI Z39.48-1984 (American Standards Institute) Permanence of Paper for Printed Library Materials.

Production Editor: Mark J. Breaugh.

Cover design: Patricia Cleary.

For additional copies, pricing for bulk purchases, and/or information about other Humana titles, contact Humana at the above address or at any of the following numbers: Tel.: 973-256-1699; Fax: 973-256-8341; E-mail: humana@humanapr.com, Website:humana@humanapress.com

10 9 8 7 6 5 4 3 2 1

Library of Congress Cataloging in Publication Data

Neuroglia in the aging brain / edited by Jean S. de Vellis.
p. ; cm. -- (Contemporary neuroscience)
Includes bibliographical references and index.

1. Neuroglia. 2. Brain--Aging. 3. Brain--Aging--Molecular aspects. I. De Vellis, Jean
II. Series.
[DNLM: 1. Brain--cytology. 2. Neuroglia--physiology. 3. Aging--physiology. 4. Neurodegenerative Diseases. WL 300 N4938154 2001]
QP363.2 .N483 2001
612.8'2--dc21

2001026438

59-105-3

ISBN 978-1-61737-088-5 ISBN 978-1-59259-105-3 (eBook)

DOI 10.1007/978-1-59259-105-3

Foreword

As we begin the second millennium and evaluate human achievements, one of the greatest accomplishments of civilization is unquestionably the extraordinary increase of human life expectancy. In prehistoric and early historic times, the average span of life was as short as 25 years. In 1900, a child born in the United States might expect to live only about 50 years, given the mortality conditions of the time. Currently, in most developed countries, life expectancy has risen to around 75 to 80 years. Thus, about half of the remarkable increase in life expectancy has taken place during the 20th century.

Many factors have contributed to this rapid and unprecedented change, among which may be listed not only advances in biomedical sciences (e.g., discovery and use of antibiotics, an understanding of the structure of DNA and the gene), but also improvements in socioeconomic conditions (e.g., higher incomes, better access to medical/health care, better sanitation). One result has been the rapid aging of human populations. Now, greater numbers of individuals are surviving to become centenarians, with ages that approach or even exceed those thought previously to represent the maximal potential life span. Moreover, human life span appears still to continue to lengthen.

Given the rapidly increasing proportion of the aged in the present population, especially those 80 years and older, it is important to understand the physiopathology of aging to be able to improve and prolong functional competence and to prevent or treat the disabilities and diseases of old age. Despite numerous theories of aging at the molecular level [e.g., genetic mutations, DNA damage and faulty repair, expression of genes (gerontogenes, longevity assurance genes) promoting or inhibiting cell proliferation, and role of telomeres], at the cellular level (e.g., wear and tear, oxidative damage), and the organismic level (e.g., disruption of neuro-immuno-endocrine interactions), the cause(s) and nature of the aging process remain elusive.

In complex organisms (mammals and especially humans), the central nervous system, closely related to the immune and endocrine systems, plays a key role in regulating the various stages of the life-span from early development to childhood, adulthood, and old age. Thus, alterations in the nervous system associated with age, such as may occur in neurodegenerative diseases, may impair the capacity of the affected individual to adapt to environmental demands, weaken physiologic competence, and induce death. Aging-associated changes in neural structures and metabolism, synapses and neurotransmission, cell number and regeneration, growth-promoting or inhibiting factors, all play a crucial role in endowing the nervous system with a certain degree of plasticity. Plasticity in this sense is the capacity to show compensatory/adaptive responses on demand (e.g., generated by environmental stimuli/damage). This property, until 20–30 years ago, considered a privilege of the developing nervous system, is now viewed as potentially extended to include the adult and even the old central nervous system.

Indeed, the study of the aging nervous system and, particularly, of neurons, is one of the most active areas of research in the field of gerontology and geriatrics. Less well-known are changes with aging in glial cells and the factors that may modulate aging of these cells. However, it is expected that knowledge of the importance of glial cells in their relation to neuronal metabolism and transport (astrocytes), to myelin formation (oligodendrocytes), and as representative of the immune system (microglia) will become considerably enhanced in the near future. With increasing progress in central nervous system rehabilitation, the restoration of neuronal plasticity after damage caused by trauma, disease, or aging is becoming a reachable reality. Although adult neurons usually do not proliferate, this does not necessarily signify that they have lost the capacity to divide; perhaps inhibitory factors derived from the matrix or even the glial cells prevent adult neurons from manifesting this property. Such inhibition of cell division would ensure neuronal stability desirable for optimal nervous system function. However, under appropriate endogenous influences of their microenvironment, neurons may regenerate, not only in rodents and primates, but also in human brains. Because glial cells with the intercellular matrix and the blood vessels comprise this microenvironment, it may be argued that the study of factors that influence glial function represents a promising, albeit indirect, approach to eventual therapeutic manipulation of neuronal potential for self-renewal. Keeping this important role of neuroglia in mind, *Neuroglia in the Aging Brain* has been organized around six major topics. First to be examined are the cellular and molecular changes that occur with aging, especially in astrocytes, the aging-associated gliosis and its relation to neuronal injury and repair. This is followed by a discussion of neuron–glia intercommunication and of how glial signals may be modified/modulated by neurohormones, hormones, extra- and intracellular metabolism and transport, as well as aging of the blood–brain barrier. The last chapters examine the role of neuroglia, especially that of astrocytes, in the etiology of a number of neurodegenerative diseases and evaluate possible therapeutic interventions, specifically on glial responses, but also, indirectly on neurons.

Planning for this book was envisioned first by Dr. Antonia Vernadakis to whom the book is dedicated. She outlined the major topics to be considered, she enlisted the majority of the contributors and secured a publisher. However, her death in 1998, brought the preparation of the book to a halt and threatened its future completion. The book contributors, well aware of the basic and practical significant contribution of glial development, function and aging to CNS plasticity, were eager to see the book published. Therefore, I know that I express the feelings of all contributors, including my own, when I gratefully acknowledge the willingness of Dr. Jean De Vellis to take over the task of editing the book, which he has done with great expertise. We are also grateful to Elyse O'Grady from Humana Press for her steadfast support of the book publication.

Paola S. Timiras

Preface

The study of the neuroscience and neurobiology of aging continues to be an intensely active area of research. However, most of the research has been conducted on neurons. *Neuroglia in the Aging Brain* reviews the current knowledge of the supporting structure of the nervous system, the neuroglia, in aging.

Neuroglia in the Aging Brain discusses the role of glial cells in normal aging and in pathological aging, i.e., neurodegenerative diseases, such as Alzheimer's and Parkinson's diseases, because a great deal of knowledge can be gleaned from the study of these more severe forms of aging. The book is divided into six sections: cellular and molecular changes of aged and reactive astrocytes; neuron–glia intercommunication; neurotrophins, growth factors, and neurohormones in aging and regeneration; metabolic changes; astrocytes and the blood–brain barrier in aging; and astrocytes in neurodegenerative diseases.

It is my hope that all of those scientists engaged in this interesting area of research will find this book to be useful. Many thanks to all of the authors for their fine contributions.

Jean S. de Vellis

Contents

Contributors

THIERRY AMÉDÉE • *Neurobiologie Integrative, Bordeaux, France*

MICHAEL ASCHNER • *Department of Physiology and Pharmacology, Wake Forest University School of Medicine, Winston-Salem, NC*

THOMAS G. BEACH • *Civin Laboratory of Neuropathology, Sun Health Research Institute, Sun City, AZ*

MARIA T. BERCIANO • *Departamento de Anatomia y Biologia Celular, Facultad de Medicina, University of Cantabira, Santander, Spain*

ANGÉLIQUE BORDEY • *Department of Neurobiology, The University of Alabama at Birmingham, Birmingham, AL*

JESÚS BOYA • *Department of Histology, Faculty of Medicine, Complutense University, Madrid, Spain*

NEVILLE BROOKES • *Department of Pharmacology and Experimental Therapeutics, University of Maryland School of Medicine, Baltimore, MD*

KURT R. BRUNDEN • *Drug Discovery, Athersys Inc., Cleveland, OH*

J. L. CALVO • *Department of Histology, Faculty of Medicine, Complutense University, Madrid, Spain*

JONATHAN A. COLES • *INSERM U438, Grenoble, France*

AURORE COLOMAR • *INSERM U394, Neurobiologie Integrative, Bordeaux, France*

ROLF DERMIETZEL • *Department of Neuroanatomy and Molecular Brain Research, Ruhr-University, Bochum, Germany*

CHERYL F. DREYFUS • *UMDNJ/Robert Wood Johnson Medical School, Piscataway, NJ*

FRANÇOISE ECLANCHER • *Centre de Neurochimie, Strasbourg, France*

PEDRO M. FAUSTMANN • *Department of Neuroanatomy and Molecular Brain Research, Ruhr University, Bochum, Germany*

ROBERT FERN • *Department of Neurology, University of Washington, Seattle, WA*

ROBERT C. A. FREDERICKSON • *Discovery International Venture Catalysts, Victoria, British Columbia, Canada*

JOSÉ E. GARCÍA-MAURIÑO • *Department of Histology, Faculty of Medicine, Complutense University, Madrid, Spain*

CHRISTIAN GIAUME • *INSERM U114, College de France, Paris, France*

JACQUES GLOWINSKI • *INSERM U114, College de France, Paris, France*

ZEZONG GU • *Department of Human Biological Chemistry and Genetics, The University of Texas Medical Branch at Galveston, Galveston, TX*

BRIGITTE HAMON • *INSERM U114, College de France, Paris, France*

KEVIN HIGASHIGAWA • *Department of Molecular and Cell Biology, University of California, Berkeley, CA*

ELIZABETH A. HOWES • *The Babraham Institute, Babraham, Cambridge, United Kingdom*

CONCEPCION IRIBAR • *Department of Biochemistry, F. Oloriz Institute of Neurosciences, University of Granada, Granada, Spain*

DOROTHEE KRAUSE • *Department of Neuroanatomy and Molecular Brain Research, Ruhr-University, Bochum, Germany*

GERARD LABOURDETTE • *Centre de Neurochimie, Strasbourg, France*

JUDITH LACKLAND • *UMDNJ/Robert Wood Johnson Medical School, Piscataway, NJ*

MIGUEL LAFARGA • *Departamento de Anatomica y Biologia Celular, Facultad de Medicina, University of Cantabria, Santander, Spain*

RICHARD M. LOPACHIN • *University Hospital for Albert Einstein College of Medicine, Bronx, NY*

ANGEL LÓPEZ-CARBONELL • *Department of Histology, Faculty of Medicine, Complutense University, Madrid, Spain*

LATHA MALAIYANDI • *Department of Molecular and Cell Biology, University of California, Berkeley, CA*

SUDARSHAN K. MALHOTRA • *University of Alberta, Edmonton, Canada*

LEE J. MARTIN • *Department of Pathology, Johns Hopkins University School of Medicine, Baltimore, MD*

MANUEL MARTINEZ • *Department of Experimental Biology, University of Jaen, Jaen, Spain*

KEN D. MCCARTHY • *Department of Pharmacology, University of North Carolina, Chapel Hill, NC*

CATHERINE MYTILINEOU • *Department of Neurology, Mount Sinai School of Medicine, New York, NY*

MICHAEL D. NORENBERG • *Department of Pathology, University of Miami School of Medicine; Miami Veterans' Affairs Medical Center, Miami, FL*

JUAN ANGEL PEDROSA • *Department of Experimental Biology, University of Jaen, Jaen, Spain*

JOSE MARIA PEINADO • *Department of Biochemistry, F. Oloriz Institute of Neurosciences, University of Granada, Granada, Spain*

MARIA ANGELES PEINADO • *Department of Experimental Biology, University of Jaen, Jaen, Spain*

J. REGINO PEREZ-POLO • *Department of Human Biological Chemistry and Genetics, The University of Texas Medical Branch at Galveston, Galveston, TX*

ADORACION QUESADA • *Department of Experimental Biology, University of Jaen, Jaen, Spain*

MARIA JESUS RAMIREZ • *Department of Biology, University of Jaen, Jaen, Spain*

GESA RASCHER • *University of Tübingen, Tübingen, Germany*

JOHN F. REILLY • *Neurome, La Jolla, CA*

LUISA RONCALI • *Dipartimento di Anatomia Umana e Istologia, Facoltá di Medicina e Chirurgia, Universitá di Bari, Italy*

ALISA SEO • *Department of Molecular and Cell Biology, University of California, Berkeley, CA*

MARILEE K. SHELTON • *Department of Pharmacology, University of North Carolina, Chapel Hill, NC*

NAYAN SHETH • *Department of Molecular and Cell Biology, University of California, Berkeley, CA*

THEODOR K. SHNITKA • *University of Alberta, Edmonton, Alberta, Canada*

HOGAN SHY • *Department of Molecular and Cell Biology, University of California, Berkeley, CA*

PETER J. S. SMITH • *BioCurrents Research Center, Marine Biological Laboratory, Woods Hole, MA*

HARALD SONTHEIMER • *Department of Neurobiology, The University of Alabama at Birmingham, Birmingham, AL*

EVA SYKOVÁ • *Department of Neuroscience, Institute of Experimental Medicine ASCR, Prague, Czech Republic*

PAOLA S. TIMIRAS • *Department of Molecular and Cell Biology, University of California, Berkeley, CA*

GIORGIOS TSIANOS • *Department of Molecular and Cell Biology, University of California, Berkeley, CA*

NURIA T. VILLAGRA • *Departamento de Anatomia y Biologia Celular, Facultad de Medicina, University of Cantabira, Santander, Spain*

DOUGLAS G. WALKER • *Civin Laboratory of Neuropathology, Sun Health Research Institute, Sun City, AZ*

HARTWIG WOLBURG • *Institute of Pathology, University of Tübingen, Tübingen, Germany*

I

Cellular and Molecular Changes of Aged and Reactive Astrocytes

1

Neuromorphological Changes in Neuronal and Neuroglial Populations of the Cerebral Cortex in the Aging Rat

Neurochemical Correlations

Maria Angeles Peinado, Manuel Martinez, Maria Jesus Ramirez, Adoracion Quesada, Juan Angel Pedrosa, Concepcion Iribar, and Jose Maria Peinado

1. INTRODUCTION

The cerebral cortex is one of the most important anatomic foundations of cognitive and memory functions; both functions are affected by aging, particularly when this process is associated with dementias such as the Alzheimer type *(1,2)*. Nevertheless, available knowledge about the cytoarchitecture and quantitative structural changes in normal brain aging, particularly in the cortex are still scarce and even contradictory *(3,4)*. The heterogeneity of the studies, as well as the fact that the effects of aging differ in cerebral zones, animal models, and individuals, are the main causes of these discrepancies *(2,5)*.

In this chapter we examine the effects of aging on the cerebral cortex from two different but complementary views: The neuromorphological and the neurochemical. The neuromorphological studies reviewed here involve quantitative and cytoarchitectonic analyses of neuronal and neuroglial populations. Neurochemical studies evaluate endogenous levels of cytotoxic amino acids such as glutamate and aspartate, which may act as neurodegenerative agents, as well as determine some aminopeptidase activities involved in neuropeptidergic metabolism, that may affect the free amino acids pool.

2. MACROSCOPIC CHANGES

The first studies of cerebral aging were done by pathologists at the beginning of the twentieth century. They detected morphological differences between sections from young and aging brains, and showed that the nerve cells of aged brains had many deposits of a brown pigment called lipofuscin *(6)*. Today it is known that the senescent human brain weighs an average of 7–8% less, at 65 of age, than a mid-life adult's brain. Magnetic resonance studies suggest that this decrease is greater in white than in gray matter *(7)*, although there is considerable variation between different brain regions *(8)*. In the cerebral cortex the overall reduction in volume is about 5%, with losses close to

From: *Neuroglia in the Aging Brain*
Edited by: Jean S. de Vellis © Humana Press Inc., Totowa, NJ

10% in the frontal lobe, but only negligible losses in the parietal and occipital lobes *(9)*. Our volumetric studies of the frontal and parietal cortex of the rat indicate stability with aging. No changes with age were detected in any of the different layers (I, II–IV, V, and VI) of these cortical areas *(10–13)*.

3. NEURONAL ATROPHY AND DEATH

Volumetric studies are incomplete if nerve cell populations are not evaluated. In 1984, Haug and coworkers studied 76 human brains and found decreases in volume, stability in the total number of neurons, and therefore increases in neuronal density. These findings were explained as a progressive loss of neuropil but not neuronal loss. The shrinkage of the extracellular space would compensate neuronal swelling and gliosis *(14)*. Our studies in the rat cortex also included quantitative analyses of neuronal density and the total number of neurons; we did not detect significant neuronal losses, although there were some cytomorphometric changes consisting of small but significant decrease in the size of the neuronal nucleus in the frontal cortex *(10)*, as well as in the size of the neuronal soma and nucleus in the parietal cortex *(11)*. These results, in addition to the detection of some dark neuron profiles and gliosis in the aged rats (see below), suggested that aging produced a slight but real neurodegenerative damage in the cerebral cortex, which was particularlly evident in rats up to 32-mo-old. This is in agreement with other studies that described neuronal atrophy and the development of age-related neurodegenerative changes. In this sense, a progresive regression of the dendritic tree and neuronal atrophy have been reported in the aging frontal and temporal cortices *(14,15)*. This progressive loss of dendrites leads to partial deafferentation of the neurons *(16)* and to a decrease in the number of synapses *(17–19)*. The presynaptic bouton's decline in the neocortex of the aging rat seems to be more intense in deeper cortical lamina *(20)*. Although atrophic cells can remain metabolically active during aging *(21)*, the neurodegenerative process may eventually lead to the death of some neurons. However, in contrast to early morphological studies that supported the idea of major age-related neuronal loss *(22–24)*, current studies *(25)* accept that neuronal death probably occurs only in some specific areas *(26)* such as the substantia nigra *(27)*, hippocampus *(26)*, and nucleus centralis superior *(3)*. In most regions of the cortex, the total number of neurons, neuronal density, and the percentage of cells per area remain unchanged in aging humans *(2,28)*, monkeys, *(29–31)* and rodents *(10,11,32)*. In the frontal cortex synaptic density gradually decreases with aging, although synaptic vesicle density remained unchanged *(33)*. These data support the idea of synaptic plasticity in the "functioning" neurons.

We used differents histological, quantitative and image analysis techniques *(10–13)* to evaluate the dynamic changes in neuronal and neuroglial populations in the frontal and parietal cortex during aging. Studies of 10 μm thick histological sections stained with cresyl-fast violet indicated stability in the number of neurons but changes in neurocytomorphometric parameters *(10,11)*. In a separate set of experiments, we used 1 μm sections stained with toluidine blue as well as ultrathin sections. In this new material we detected the appearance of dark neurons profiles (DN) as shown in Figure 1. The results of DN quantification are shown in Figure 2. Only aged rats had a significant increased number of DN in both the frontal and parietal cortex; however, the number of normal neurons (nondark neurons: NDN) decreased with age in comparison to young control animals. In this group the number of DN was imperceptible. Figure 2 also shows that there was no change in the sum of DN plus NDN, total neurons.

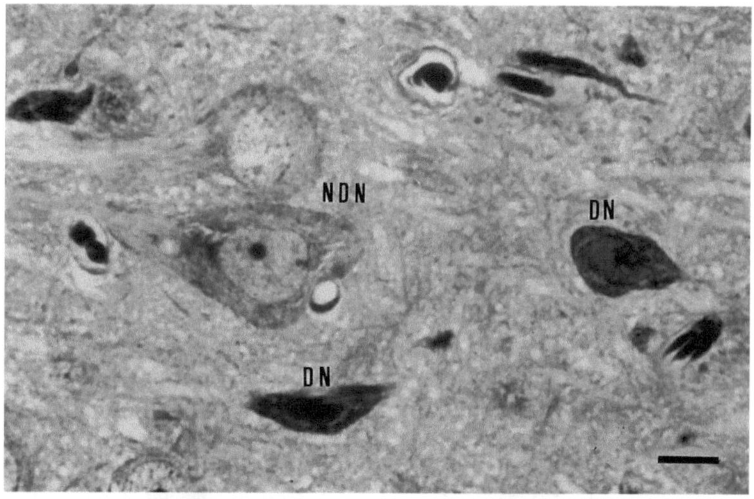

Fig. 1. Toluidine blue-stained semithin section of the parietal cortex of an aged rat showing several DN next to normal NDN (Bar: 10 μm).

In addition, we analyzed cytomorphometric variables, such as mean neuronal soma and nucleus areas, in a significant sample of DN and NDN in the frontal cortex. Samples were obtained by proportional affixation in different strata throughout the different cortical layers in the frontal and parietal cortices of aged and young rats *(12)*. Figure 3 shows that soma and nucleus area clearly decrease in DN. Considering the scarce number of DN in young rats, cytomorphometric evaluation of the total neuronal population (Fig. 4) (DN plus NDN = TN), indicated only a small decrease in soma and nucleus areas in aged animals as was shown in earlier work *(10,11)*.

DN are characterized by their high affinity for basic stains, their soma, nucleus, and nucleolus are strongly shrunk *(34–36)*. Ultrastructurally they showed microvacuolar degeneration *(37)*, lipofushin accumulation *(38)*, disaggregated polyribosome and disorganized endoplasmic reticulum and mitochondria *(37)*. Their plasma membrane is irregular and is surrounded by many glial processes *(34)*.

Dark neurons have been well characterized in ischemia *(39)*, glucose deprivation *(40)*, injury *(34)*, transitory breakdown of the blood-brain barrier *(41)*, poisoning with toxins such as kainic acid *(42)*, and elevated atmospheric pressure *(43)*. To explain the formation of these neurons Gallyas et al. *(35)* proposed a common generative mechanism independently of the cause: They suggested that energy deregulation might affect cytoskeletal neurofilaments *(34,43,36)*.

The detection of DN profiles in histological sections has been interpreted as a sign of neuronal atrophy, probably representing a degenerative step previous to cell death, which occurs later by necrotic or apoptotic mechanisms. Necrosis is the main consequence of mechanical, ischemic or toxic lesions, whereas apoptosis is a type of programmed cell death which normally occurs during development. In addition, necrosis involves an inflammatory response which does not occur in apoptosis. Recent studies describe apoptosis as a common mechanism of cell death in aging, in which intrinsic signals (cytokines, neurotransmitters, growth factors, and so on) as well as extrinsic (mediated by free radicals) signals would participate *(44,45)*. Regardless of the under-

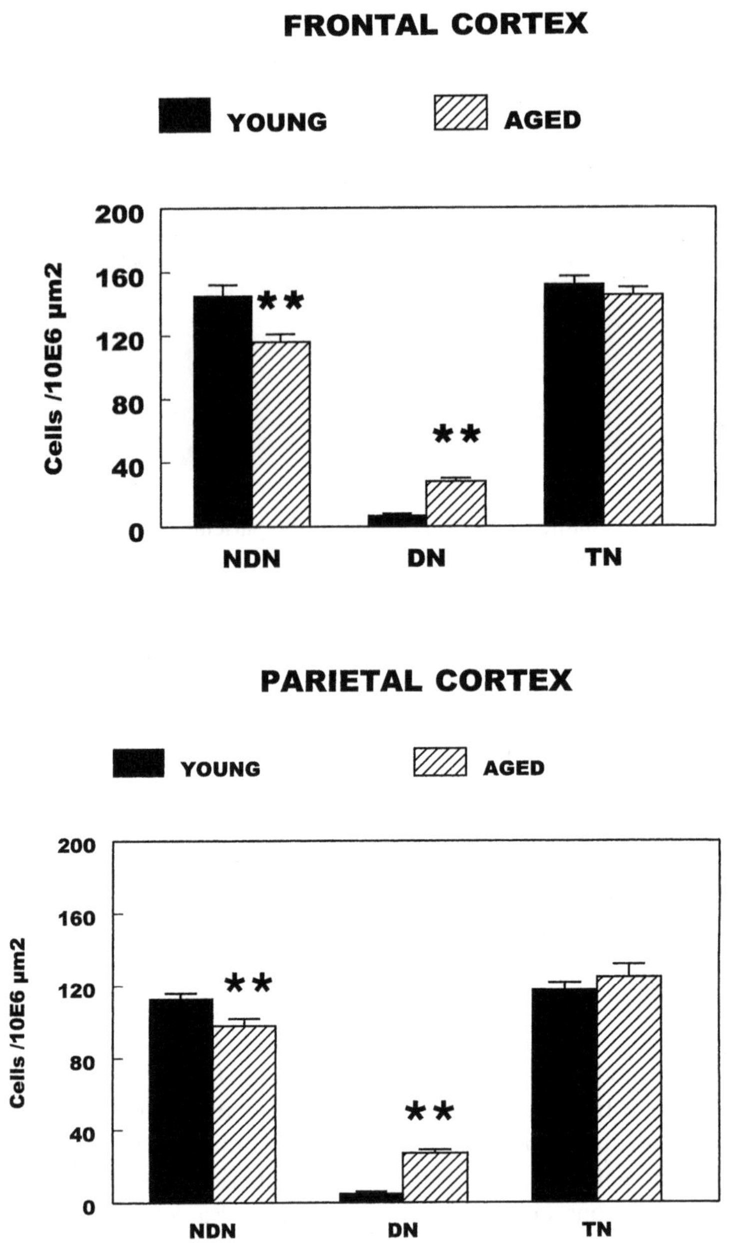

Fig. 2. Neuronal density (mean ± SEM in 10^6 µm^2 of nervous tissue) of NDN, DN and total neurons (TN) quantified in the frontal and parietal cortex of young and aged rats. (Frontal young: n=271; Frontal aged: n = 261; Parietal young: n = 246; Parietal aged: n = 225.) (For details of the quantitative method see *(13)*.)

lying mechanisms of cell death, we speculate that during aging, DN represent senescent neurons. In fact, atrophic neurons have been described in early aging studies as cells that showed decreases in the size of the pericaryon, nucleus, nucleolus, and dendritic tree *(46,47)*. This general shrinkage may reflect loss of the ability to maintain homeostasis, with gross dehydration and consequently an increased affinity for stains.

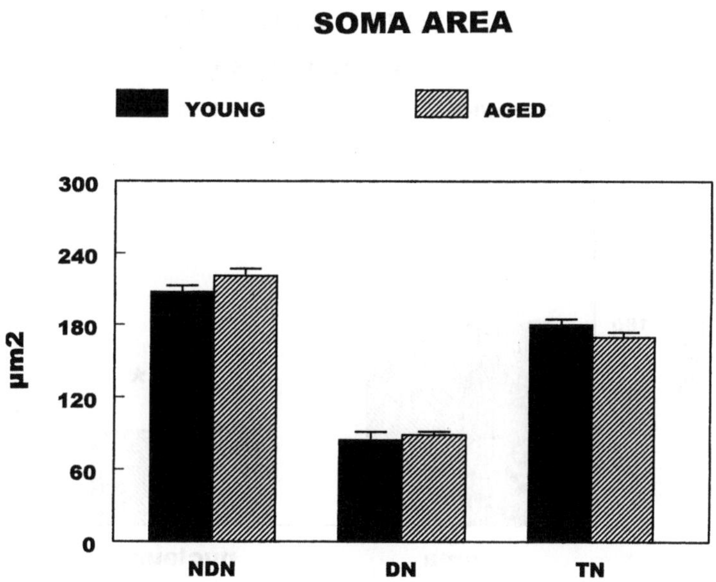

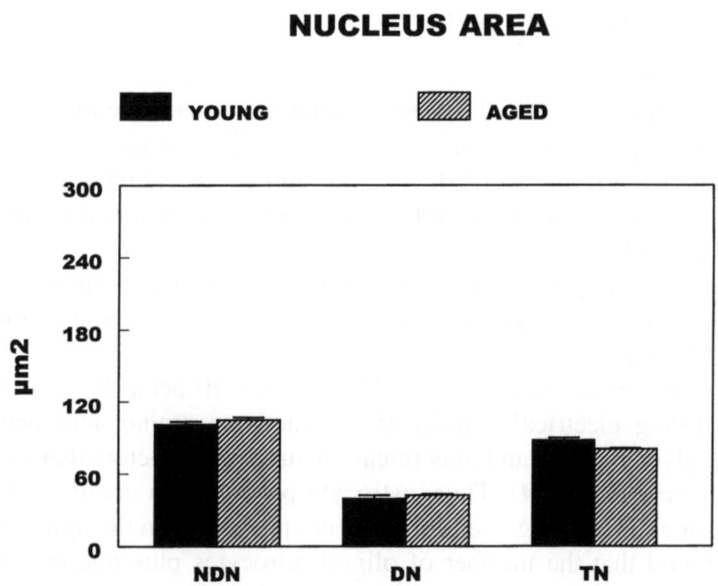

Fig. 3. Neuronal soma and nucleus areas (mean ± SEM expressed in μm^2) of NDN and DN in the frontal cortex of young and aged rats. (NDN young: n = 720; NDN aged: n = 600; DN young: n = 30; DN aged: n = 150.) (For details of the cytomorphometric method see [*10,11,13*].)

4. REACTIVE GLIOSIS

An other important issue in aging is the behavior of the glial population. Most studies report increases in glial density in parallel with neuronal degeneration *(48–50)*. In morphological studies we detected gliosis in both the frontal and parietal cortex *(10–13)*. Moreover, when we quantified astrocytes and oligodendrocytes-plus-microglial

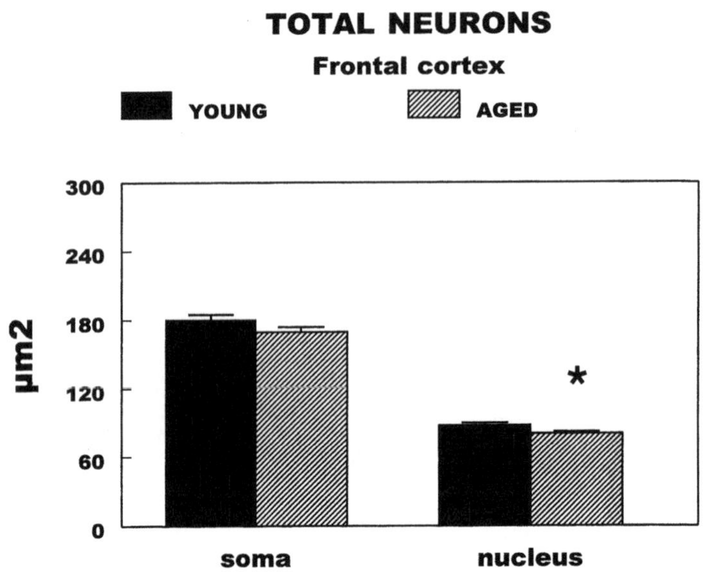

Fig. 4. Soma and nucleus areas (mean ± SEM expressed in µm^2) of the total neuronal population in the frontal cortex of young and aged rats. (Young: n = 750; Aged: n = 750.) (For details of the cytomorphometric method see [*10,11,13*].)

cells, we found the same results in the frontal and parietal cortex: Only astrocytes showed significant age-related increases (Fig. 5). The increases in astrocyte led to hyperplasia and hypertrophy, as shown by the immunodetection of glial fibrillar acidic protein (GFAP) (Fig. 6). These results agree with those of similar studies in humans *(51)* and rats *(52–54).*

The changes in glial population may correlate with increases in glucose uptake *(55)* as well as increases in the activity of enzymes such as glutamine synthetase *(56–58)* marker of glial fuction.

Astrocytes may protect neurons *(59)*. These glial cells act as buffers of ions released by neurons during electrical activity *(60),* inactivate amino acid neurotransmitters *(58,61),* store glycogen *(62)* and may release neurotrophic factors that support neuronal function and survival *(63,64).* Thus, astrocyte proliferation could be interpreted as a consequence of an increasing need for neuronal protection in the aging brain.

We have found that the number of oligodendrocytes plus microglial cells, remain stable with age in the different cortical layers and in the total thickness of the cortex (Fig. 5). In contrast to our results, studies in monkeys *(30,65)* showed an increase with age in oligodendrocytes and microglia, but not in astrocytes. Other studies, however, have reported no change in the microglial population with age in the mouse brain *(66).* In addition, Ogura et al. *(67),* using OX-42 and OX-6 antibodies, also failed to observe changes in the number of microglia, although aging induced their transformation into reactive cells. This was also found by Perry et al. *(68),* who detected MCH-II antigen in some microglial cells in the brains of aged rodents.

At the ultrastructural level, glial population also undergoes changes with age. These changes include the appearance of foamy inclusions and osmiophilic materials in all glial cell types *(30,69).* Similarly, some types of abnormal argyrophilic structures or

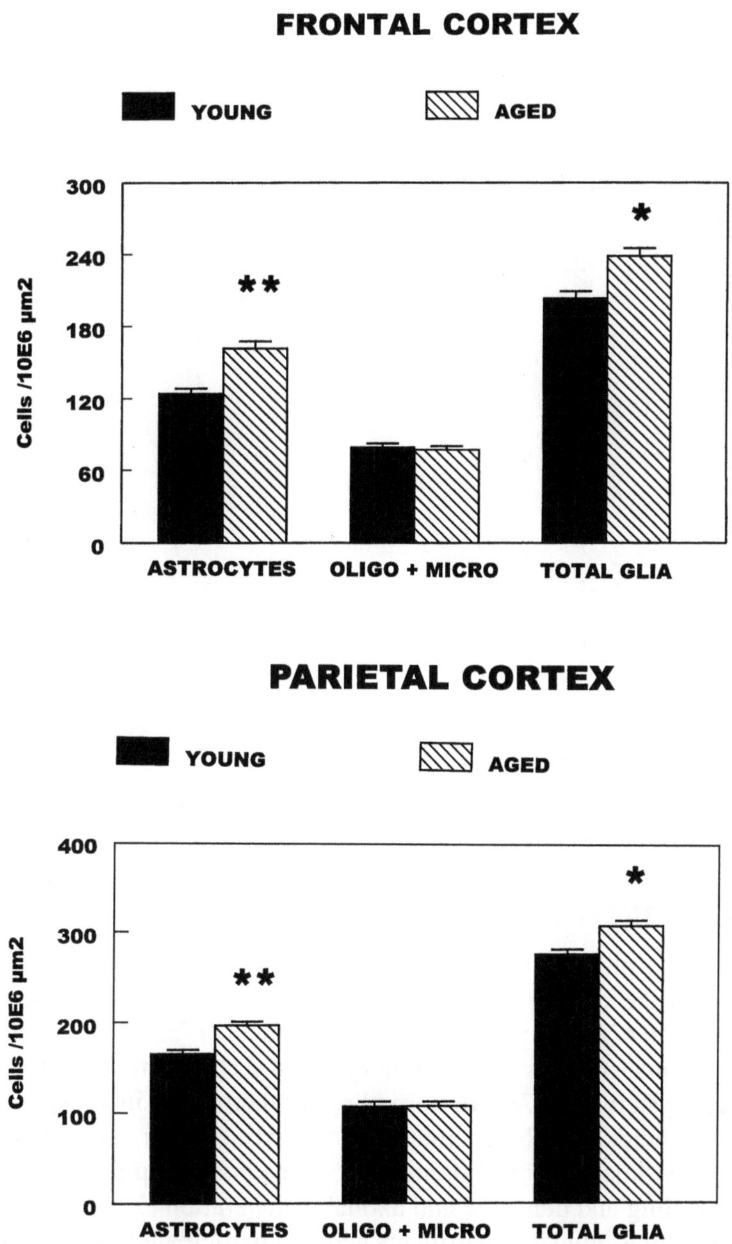

Fig. 5. Density (mean ± SEM in 10^6 μm^2 of nervous tissue) of astrocytes, oligodendrocytes plus microglia and total glial cells quantified in the frontal and parietal cortex of young and aged rats. (Frontal young: n = 500; Frontal aged: n = 500; Parietal young: n = 300; Parietal aged: n = 300.) (For details of the quantitative method *see [13].*)

inclusions have been recognized in senescent astrocytes and oligodendroglial cells; immunocytochemical techniques have revealed that phosphorylated *tau,* an abnormal protein located in neurofibrillary tangles, is also present in these cells, although the precise significance of this is still unknown *(70).*

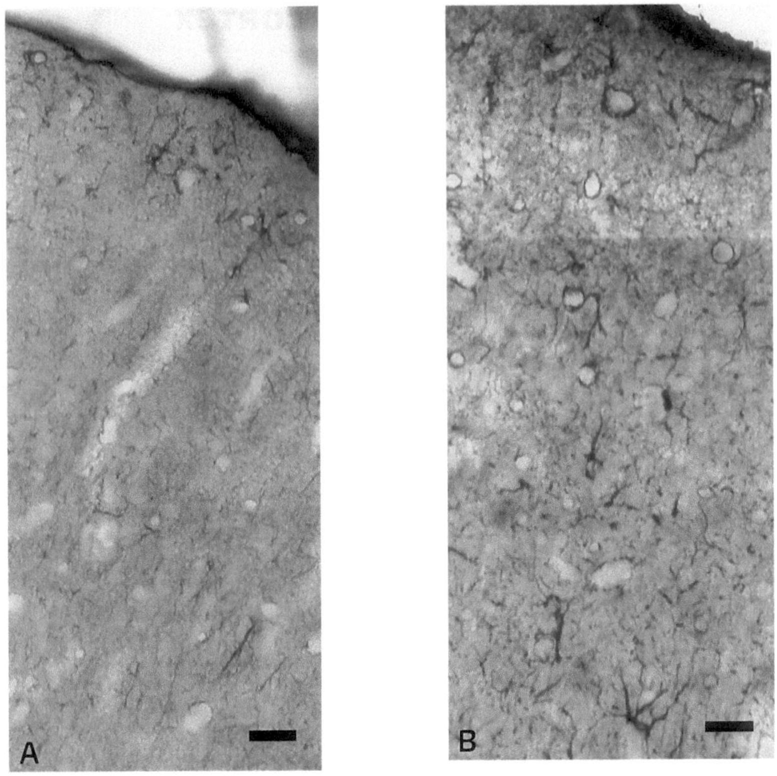

Fig. 6. Inmunostaining for GFAP in sections of the frontal cortex of young **(A)** and aged **(B)** rats. Note the increase in immunoreaction in the aged rat (Bar: 20 μm).

5. NEUROCHEMICAL FINDING

All the neuromorphological changes described in aging cells and organism should be interpreted as the consequence of metabolic alterations, including deficiences in energy metabolism, *(71)*, the production of free radicals *(72,73)*, and changes in neurotransmitter systems such as the cholinergic *(74,75)*, serotoninergic, *(76)* and the aminoacidergic systems *(77,78)*. In normal aging there is an increase in the release of excitatory amino acids, which accumulate in the extracellular space, where they act as potent toxins. The injection of glutamate produces postsynaptic lesions characterized by neuronal swelling and dendritic and axonal degeneration *(77,78)*.

Neuronal as well as glial compartments actively participate in glutamate metabolism *(79)*. The glutamate release from presyhaptic neurons interacts with different receptors that can be classified as ionotropic and metabotropic *(80)*. The N-methyl-D-aspartate receptor (NMDA) is a channel that allows the inward flow of Na^+ and Ca^{2+}. This ionotropic receptor has been implicated in the neuronal plasticity and memory funtions which are impaired during aging *(81,82)*.

Neurons and glia show mechanisms for the uptake of glutamate which regulate the extracellular concentration of this neurotransmitter, preventing excessive stimulation of glutamate receptors *(78)*. In glial cells glutamine synthetase converts glutamate to inactive glutamine, which returns to the neuronal terminal where it is transformed into glutamate *(61)*. An excess of extracellular glutamate can overstimulate the neurons and

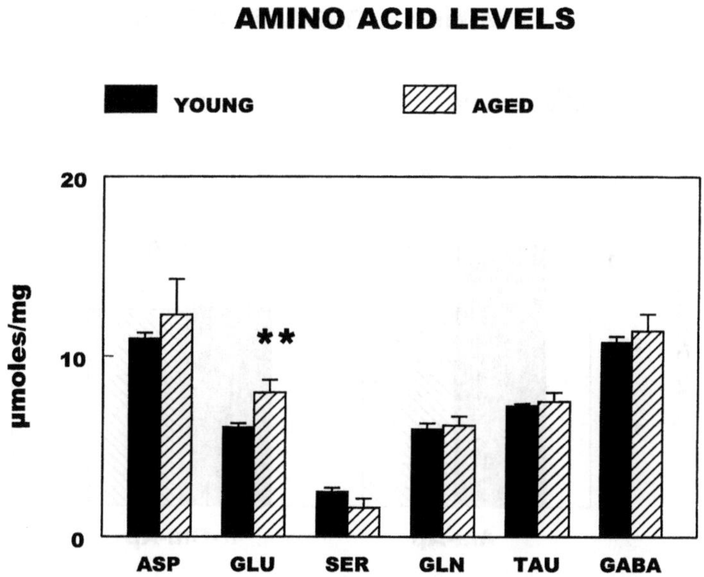

Fig. 7. Levels (mean ± SEM) of aspartate (Asp), glutamate (Glu), serine (Ser), glycine (Gly), taurine (Tau) and γ-aminobutiric acid (GABA) obtained in the frontal cortex of young and aged rats. Results are expressed as μmoles/mg of nervous tissue (n= 14; ** $p<0.01$).

lead to neurotoxicity. Two mechanism for these neurotoxic effects have been described. Permanent activation of NMDA produces short-term osmotic damage due to influx of ions/water into the cell, being characterized by neuronal swollen *(77,78,83)*. The long term mechanism is mainly based on Ca^{2+} permeability. The increase in cytosolic Ca^{2+} activates kinases, proteases and lipases and also generates free radicals. All of these activities can cause neuronal damage *(77,78)*.

We evaluated the levels of aspartate, glutamate, γ-amino-butiric acid (GABA), serine, glicine and taurine in the frontal and parietal cortex of the rat (Fig. 7). The results obtained indicated that only glutamate increased significantly in aged rats. This increase may be at least partly responsible for the neurodegenerative changes observed, and consequently for the memory and cognitive impairments that occur during aging. Furthermore, because astrocytes are to a great extend responsible for glutamate uptake and inactivation, metabolic deregulation of this cytotoxic amino acid activates these cells, which then may proliferate or become hypertrophic, as we found in aged rats.

The uptake mechanism may be inactivated by oxygen free radicals as have been demonstrated in astrocyte cultures *(72)*. In addition, Ca^{2+} through the activation of NMDA receptors, induces nitric oxide (NO) sythesis *(81)*. Considering the role of NO in memory *(82)*, it may be suggested that different mechanism interact and reinforce the neurodegenerative aging process.

Other mechanism may also be involved in the deregulation of amino acid metabolism. Proteases participate in neuropeptide activation and inactivation, releasing amino acids and therefore increasing the free amino acid pool. We determined some aminopeptidase activities such as alanine-aminopeptidase (ala-A), α-glutamate-aminopeptidase (α-Glu-Ap), pyro-glutamyll-aminopeptidase (p-Glu-Ap) and aspartyl-aminopeptidase (Asp-Ap). The results obtained (Fig. 8) indicated that only p-Glu-Ap and Asp-Ap were significantly

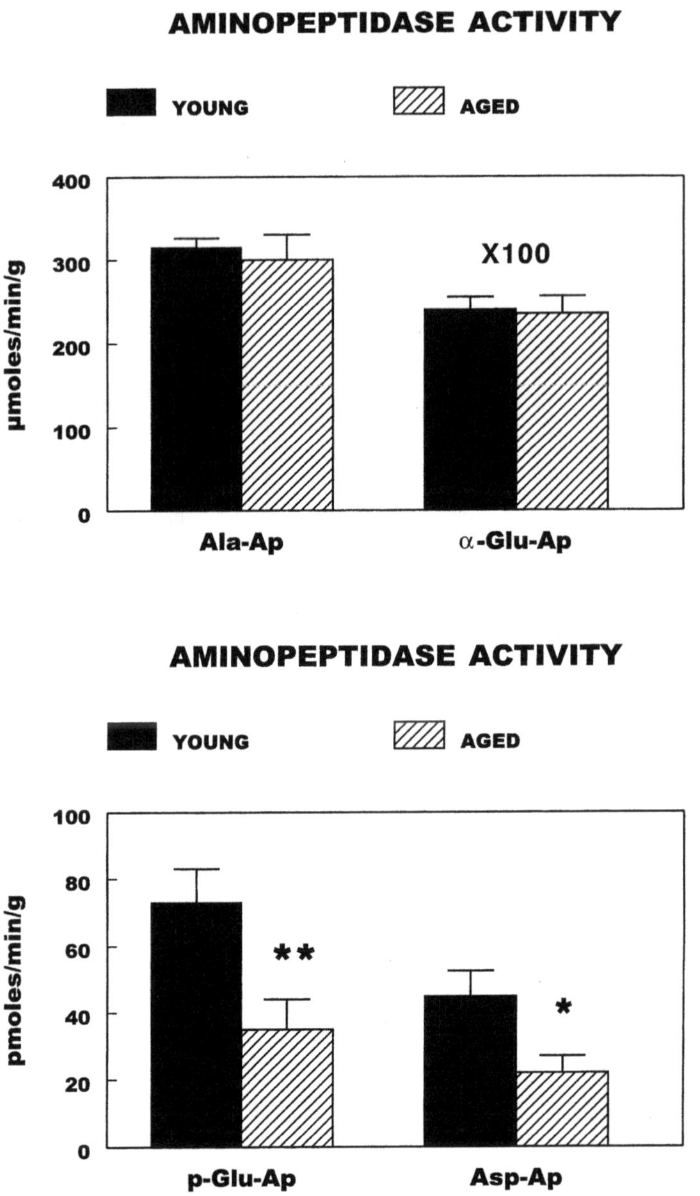

Fig. 8. Aminopeptidase activities obtained for alanine-aminopeptidase (ala-A), α-glutamate-aminopeptidase (α-Glu-Ap), piro-glutamate-aminopeptidase (p-Glu-Ap) and aspartate-amiopeptidase (Asp-Ap), in the frontal cortex of young and aged rats. Results are expressed as µmoles/min/g of nervous tissue or pmoles/min/g of nervous tissue. Values of α-Glu-Ap are multiplied by 100 (n = 14; $*p<0.05$; $*p<0.01$).

decreased in the cortex of aged rat *(84)*. A generalized decrease in protein degradation and RNA synthesis have been observed with aging in different tissues. In brain, the aged related accumulation of β-amyloid could reflect deficits in protease activity.

In summary, a series of changes occur in the neuronal and glial populations in the frontal and parietal cortex of the aging rat. The aging brain may have a greater suceptibility to functional challenges. Cognitive and memory impairments could be initially due to neuronal dysfunction rather than neuronal loss. This neurodegenerative process would lead finally to neuronal death and reactive gliosis. Changes in amino acid neurotransmitters metabolism could be underlying in this process.

ACKNOWLEDGMENTS

The authors are indebted to Dr. José Rodriguez Avi for his valuable help with the statistical study, and K Shashok for help with the language. This work was supported by grant DGICYT PM 90–0146 and by Junta de Andalucia PAI (CV-0184; CV-0164; group 3312).

REFERENCES

1. Coleman, P.D. and Flood, D.G. (1987) Neuron numbers and dendritic extent in normal aging and Alzheimer's disease. *Neurobiol. Aging* **8,** 521–545.
2. Haug, H. (1997) The aging human cerebral cortex: Morphometry of areal differences and their functional meaning. In: Principles of Neuronal Aging. (S.U. Dani, A. Hori, and G.F. Walter, eds.), Elsevier, Amsterdan, pp. 247–261.
3. Kemper, T.L., Moss, M.B., Rosene, D.L., and Killiany, R.J. (1997) Age-related neuronal loss in the nucleus centralis superior of the rhesus-monkey. *Acta Neuropathol.* **94,** 124–130.
4. Haug, H., Kühl, S., Mecke, E., Sass, N.L., and Wasner, K. (1984) The significance of morphometric procedures in the investigation of age changes in cytoarchitectonic structures of the human brain. *Z. Hirnforsch.* **25,** 353–374.
5. Begega, A., Cuesta, M., Santin, L.J., Rubio, S., Astudillo, A., and Arias, J.L. (1999) Unbiased estimation of the total number of nervous cells and volume of medial mammillary nucleus in humans. *Exp. Gerontol.* **34,** 771–782.
6. Stubel, H. (1911) Die Fluoreszenz tierscher Gewebe in ultravioletten Licht. Pfluegers Arch. Gesamte *Physiol.* **142,** 1–14.
7. Wickelgren, I. (1996) For the cortex, neuron loss may be less than thought. *Science* **273,** 48–50.
8. Insausti, R., Insausti, A.M., Sobreviela, M.T. Salinas, A., and Martinez- Peñuela J.M. (1998) Human medial temporal lobe in aging: Anatomical basis of memory preservation. *Microsc. Res. Tech.* **43,** 8–15.
9. Haug, H. and Eggers, R. (1991) Morphometry of the human cortex cerebri and corpus striatum during aging. *Neurobiol. Aging* **12,** 336–338.
10. Peinado, M.A., Martinez, M., Pedrosa, J.A., Quesada, A., and Peinado, J.M. (1993) Quantitative morphological changes in neurons and glia in the frontal lobe of the aging rat. *Anat. Rec.* **237,** 104–108.
11. Peinado, M.A., Quesada, A., Pedrosa, J.A., et al. (1997) Light microscopic quantification of morphological changes during aging in neurons and glia of the rat parietal cortex. *Anat. Rec.* **247,** 420–425.
12. Peinado, M.A., Ramirez, M.J., Pedrosa, J.A., et al. (1998a) Effects of contralateral lesions and aging on the neuronal and glial population of the cerebral cortex of the rat. In: Understanding Glia Cells. (B. Castellano, B. Gonzalez, and M. Nieto-Sanpedro, eds.), Kruwer Academic Publishers, Boston, pp: 297–317.
13. Peinado, M.A., Quesada A., Pedrosa J.A., et al. (1998b) Quantitative and ultrastructural changes in glia and pericytes in the parietal cortex of the aging rat. *Microsc. Res. Tech.* **43,** 34–42.
14. Nakamura, S., Akiguchi, I., Kameyama, M., and Mizuno, N. (1985) Age-related changes of pyramidal cell basal dendrites in layers III and V of human motor cortex: a quantitative Golgi study. *Acta Neuropathol.* **65,** 281–284.
14. Sykova, E., Mazel, T., and Simonova, Z. (1998) Diffusion constraints and neuron-glia interaction during aging. *Exp. Gerontol.* **33,** 837–851.

15. Anderson, B. and Rutledge, V. (1996) Age and hemisphere effects on dendritic structure. *Brain* **119,** 1983–1990.
16. Haug, H. (1985) Are neurons of the human cerebral cortex really lost during aging? In: Senile Dementia of the Alzheimer Type. J. Traber and V. H. Gispen, eds. Springer, Berlin- Heidelberg, pp. 150–163.
17. Jones, D.G. and Calverley, R.K.S. (1992) Changes of synapses with age: The respective contributions of non-perforated and perforated synapses in neocortex. In: Development and Involution of Neurones. (K. Fujisawa and Y. Morimatsu, eds.), Japan Scientific Societies Press, Tokyo. pp. 249–262.
18. Peters, A., Sethares, C., and Moss, B. (1998) The effects of aging on layer-1 in area-46 of prefrontal cortex in the Rhesus-Monkey. *Cer. Cortex.* **8,** 671–684.
19. Huang, C.M., Brown, N., and Huang, R.H. (1999) Age-related-changes in the cerebellum – parallel fibers. *Brain Res.* **840,** 148–152.
20. Wong, T.P., Campbell, P.M., Ribeirodasilva, A., and Cuello, A.C. (1998) Synaptic Numbers Across Cortical Laminae and Cognitive Performance of the Rat During Aging. *Neuroscience* **84,** 403–412.
21. Finch, C.E. (1993) Neuron atrophy during aging: Programmed or sporadic? *Trends Neurosci.* **16,** 104–110.
22. Brody, H. (1955) Organization of the cerebral cortex III: A study of aging in the human cerebral cortex. *J. Comp. Neurol.* **102,** 511–556.
23. Devany, K.O. and Johnson, H.A. (1980) Neuron loss in the aging visual cortex of man. *J. Gerontol.* **35,** 836–841.
24. Henderson, G., Tomlinson, B.E., and Gibson, P.H. (1980) Cell counts in human cerebral cortex in normal adults throughout life using an image analysing computer. *J. Neurol. Sci.* **46,** 113–136.
25. Calhoun, M.E., Kurth D., Phinney, A.L., et al. (1998) Hippocampal neuron and synaptophysin-positive bouton number in aging C57BL/6 mice. *Neurobiol. Aging* **19,** 599–606.
26. West, M.J. (1993) Regionally specific loss of neurons in the aging human hippocampus. *Neurobiol. Aging* **14,** 287–293.
27. McGeer, P.L., McGeer, E.G., and Suzuki, J.S. (1977) Aging and extrapyramidal function. *Arch. Neurol.* **34,** 33–35.
28. Terry, R.D., DeTeresa, R., and Hansen, L.A. (1987) Neocortical cell counts in normal human adult aging. *Ann. Neurol.* **21,** 530–539.
29. Vincent, S.L., Peters, A., and Tigges, J. (1989) Effects of aging on the neurons within area 17 of rhesus monkey cerebral cortex. *Anat. Rec.* **223,** 329–341.
30. Peters, A., Leahu, D., Moss, M.B., and McNally, K.J. (1994) The effects of aging on area-46 of the frontal-cortex of the Rhesus Monkey. *Cerebr. Cortex* **6,** 621–635.
31. Tigges, J., Herndon, J.G., and Rosene, D.L. (1996) Preservation into old-age of synaptic number and size in the supragranular layer of the dentate gyrus in Rhesus-Monkeys. *Acta Anatomica.* **157,** 63–72.
32. Curcio, C.A. and Coleman, P.D. (1982) Stability of neuron number in cortical barrels of aging mice. *J. Comp. Neurol.* **212,** 158–172.
33. Nakamura, H., Kobayashi, S., Ohashi, Y., and Ando, S. (1999) Age-Changes of Brain Synapses and Synaptic Plasticity in Response to an Enriched Environment. *J. Neurosci. Res.* **56,** 307–315.
34. Gallyas, F., Zoltay, G., and Balas I. (1992a) An immediate light microscopic response of neuronal somata, dendrites and axons to contusing concussive head injury in the rat. *Acta Neuropathol.* **83,** 394–401.
35. Gallyas, F., Zoltay, G., and Dames, W. (1992b) Formation of dark (argyrophilic) neurons of various origin proceeds with a common mechanism of biophysical nature (a novel hypothesis). *Acta Neuropathol.* **83,** 504–509.
36. Liposits, Z., Kallo, I., Hrabovszky, E., and Gallyas, F. (1997) Ultrastructural pathology of degenerating dark granule cells in the hippocampal dentate gyrus of adrenalectomized rats. *Acta Biologica Hungarica.* **48,** 173–187.

37. Tomimoto, H. and Yanahiara, T. (1992) Electron microscopic investigation of the cerebral cortex after cerebral ischemia and reperfusion in the gerbil. *Brain Res.* **598,** 87–97.
38. Ong, W.Y. and Garey, L.J. (1993) Ultrastructural features of biopsied temporopolar cortex (area 38) in a case of schizophrenia. *Schizophr. Res.* **10,** 15–27.
39. Czurko, A. and Nishino, H. (1993) Collapsed (argyrophilic, dark) neurons in rat model of transient focal cerebral ischemia. *Neurosci. Lett.* **162,** 71–74.
40. Tong, L. and Perezpolo, R. (1998) Brain-derived neurotrophic factor (bdnf) protects cultured rat cerebellar granule neurons against glucose deprivation-induced apoptosis. *J. Neural Transm.* **105,** 905–914.
41. Solohudding, T.S., Kalimo, H., Johansson, B.A., and Olsson, Y. (1998) Observations on exudation of fibronectin, fibrinogen and albumin in the brain after carotid infusion of hyperosmolar solutions. An immunohistochemical study in the rat indicating long-lasting changes in the brain microenviroment and multifocal nerve cell injuries. *Acta Neuropathol. (Berl)* **76,** 1–10.
42. Kiernan, J.A., Macpherson, C.M., Price, A., and Sun, T. (1998) A histochemical examination of the staining of kainate-induced neuronal degeneration by anionic dyes. *Biotechnic. Histochem.* **73,** 244–254.
43. Mennel, H.D., Stumm, G., And Wenzel, J. (1997) Early morphological findings in experimental high-pressure neurological syndrome. *Exp. Toxicol. Pathol.* **49,** 425–432.
44. Linden, R. (1997) Neuron death: A developmental perspective. In: Principles of Neuronal Aging. (S.U. Dani, A. Hori, and G.F. Walter, eds.), Elsevier, Amsterdan, pp. 229–246.
45. Tong, L.Q., Toliverkinsky, T., Taglialatela, G., Werrbachperez, K., Wood, T., and Perezpolo, J.R. (1998) Signal-transduction in neuronal death. *J. Neurochem.* **71,** 447–459.
46. Scheibel, A.B. (1978) Structural aspects of the aging brain: spine systems and the dendritic arbor. In: Aging. Alzheimer disease, senile dementia and related disorder. (R. Katzman, R.D. Terry, and K.L. Bick, eds.), Plenum Press, New York, pp. 352–273.
47. Henrique, R.M.F., Monteiro, R.A.F., Rocha, E., and Mariniabreu, M.M. (1997) A stereological study on the nuclear volume of cerebellar granule cells in aging rats. *Neurobiol. Aging* **18,** 199–203.
48. Landfield, P.W., Rose, G., Sandles, L., Wohstadter, T.C., and Lynch, G. (1977) Patterns of astroglial hypertrophy and neuronal degeneration in the hippocampus of aged memory deficient rats. *J. Gerontol.* **32,** 3–12.
49. Vaughan, D.W. and Peters, A. (1974) Neuroglial cells in the cerebral cortex of rats from young adulthood to old age: an electron microscope study. *J. Neurocytol.* **3,** 405–429.
50. Unger, I.W. (1998) Glial reaction in aging and alzheimer's disease. *Microsc. Res. Tech.* **43,** 24–28.
51. Hansen, L.A., Armstrong, D.M., and Terry, R.D. (1987) An immunohistochemical quantifica-tion of fibrous astrocytes in the aging human cerebral cortex. *Neurobiol. Aging* **8,** 1–6.
52. Lolova, I. 1991 Qualitative and quantitative glial changes in the hippocam-pus of aged rats. *Anat. Anz.* **172,** 263–271.
53. Amenta, F., Bograni, S., Cadel, S., Ferrante, F., Valsecchi, B., and Vega, J.A. (1994) Microanatomical changes in the frontal cortex of aged rats: Effect of L-Deprenyl treatment. *Brain Res. Bull.* **34,** 125–131.
54. Amenta, F., Bronzetti, E., Sabbatini, M., and Vega J.A. (1998) Astrocyte changes in aging cerebral cortex and hippocampus: A quantitative immunohistochemical study. *Microsc. Res. Tech.* **43,** 29–33.
55. Marrif, H. and Juurlink, B.H.J. (1999). Astrocytes respond to hypoxia by increasing glycolytic capacity. *J. Neurosci. Res.* **57,** 255–260.
56. Cao, D.H., Strolin, M., and Dos-tert, P. (1985) Age-related changes in glutamine synthetase activity of rat brain, liver and heart. *Gerontol.* **31,** 95–100.
57. Peinado, J.M. and R.D. Myers (1988) Cortical amino acid neuro-transmitter release is altered by CCK perfused in frontal region of unrestrained aged rat. *Peptides* **9,** 631–636.
58. Kanai, Y., Smith, C.P., and Hediger, M.A. (1994) A new family of neurotransmit-ter transporters: The high-affinity glutama-te transporters. *FASEB J.* **8,** 1450–1459.

59. Norenberg, M.D. (1994) Astrocyte responses to CNS injury. J. Neuropath. Exper. Neurol. **53,** 213–220.
60. Tardy, M. (1991) Astrocyte et homéostasie. *Médecine/Sciences* **8/7,** 799–804.
61. Worral, M. and Williams, C. (1994) Sodium ion-dependent transpor-ters for neurotrans-mitters: A review of recent developments. *Biochem. J.* **297,** 425–436.
62. Magistretti, P.J. (1988) Regulation of glycogenolysis by neurotransmitters in the central nervous system. *Diabet. Metab.* **14,** 237–246.
63. Hatten, M.E. (1990) Riding the glial monorail: A common mechanism for glial-guided neuronal migration in different regions of the developing mammaliam brain. *Trends Neurosci.* **13,** 179–184.
64. Shao, Y. and McCarthy, K.D. (1994) Plasticity of astrocytes. *Glia* **11,** 147–155.
65. Peters, A., Josephson, K., and Vincent, S.L. (1991) Effects of aging on the neuroglial cells and pericytes within area 17 of the rhesus monkeys cerebral cortex. *Anat. Rec.* **229,** 384–398.
66. Lawson, L.J., Perry, V.H., and Gordon, S. (1992) Turnover of resident microglia in the normal adult mouse brain. *Neurosci.* **48,** 505–415.
67. Ogura, K., Ogawa, M., and Yoshida, M. (1994) Effects of aging on microglia in the normal rat brain: Immunohistochemical observations. Neuroreport **5,** 1224–1226.
68. Perry, U.H., Matiszak, M.K., and Feurn, S. (1993) Altered antigen expression of microglia in the aged rodent CNS. *Glia* **7,** 60–67.
69. Peinado M.A. (1998) Histology and histochemistry of the aging cerebral cortex: An overview. *Microsc. Res. Tech.* **43,** 1–7.
70. Ikeda, K. (1997) Glial degeneration. In: Principles of Neuronal Aging. (S. U. Dani, A. Hori and G.F. Walter, eds.), Elsevier, Amsterdan, pp. 343–357.
71. Pedersen, J.Z., Bernardi, G., Centonze, D., et al. (1998). Hypoglycemia, hypoxia, and ischemia in a corticostriatal slice preparation-electrophysiologic changes and ascorbyl radical formation. *J. Cer. Blood Flow Metabol.* **18,** 868–875.
72. Sohal, R.S. (1993) The free radical hypothesis of aging: an appraisal of the current status. *Aging* **5,** 3–17.
73. Knight, J.A. (1995) The process and theories of aging. *Ann. Clin. Lab. Sci.* **25,** 1–12.
74. Taylor, L. and Griffith, W.H. (1993) Age-related decline in cholinergic synaptic transmission in hippocampus. *Neurobiol. Aging* **14,** 509–515.
76. Venero, J.L., Machado, A., and Cano, J. (1993) Effect of aging on monoamine turnover in the prefrontal cortex of rats. *Mech. Aging Dev.* **72,** 105–118.
75. Muir, J.L. (1997) Acetylcholine, Aging, and Alzheimers-disease. *Pharmacol. Biochem. Behav.* **56,** 687–696.
77. Choy, D.W (1992) Excitotoxic cell death. *J. Neurobiol.* **23,** 1261–1276.
78. Michaelis, E.K. (1998) Molecular-biology of glutamate-receptors in the central-nervous-system and their role in excitotoxicity, oxidative stress and aging. *Prog. Neurobiol.* **54,** 369–415.
79. Najlerhim, A., Francis, P.T., and Bowen, D.M. (1990) Aged related-alteration in excitatory amino acid neurotransmission in rat brain. *Neurobiol. Aging* **11,** 155–8.
80. Nakanishi, S. (1992) Molecular diversity of glutamate receptors and implications for brain funtion. *Science* **258,** 597–603.
81. O'Dell, T.J., Huang, P.L., Dawson, T.M., et al. (1994) Endothelial NOS and the blockade of LTP by NOS inhibitors in mice lacking neuronal NOS. *Science* **265,** 542–546.
82. Yamada, K. and Nabeshima, T. (1998) Changes in NMDA receptors nitric oxide signaling pathway in the brain with aging. *Microsc. Res. Tech.* **43,** 68–74.
83. Hablitz, J.J. and Langmoen, I.A. (1982) Excitation of hippocampal pyramidal cells by glutamate in the guinea pig and rat. *J. Physiol.* **325,** 317–331.
84. Iribar, M.C., Esteban, M.J., Martinez, J.M., and Peinado, J.M. (1995) Decrease in cytosoloc aspartyl-aminopeptidase but not in alanyl-aminopeptidase in thefrontal cortex of the aged rat. *Brain Res.* **687,** 211–213.

2
Diversity in Reactive Astrocytes

Sudarshan K. Malhotra and Theodor K. Shnitka

1. INTRODUCTION

Normal astrocytes in the adult undergo hypertrophy and proliferation and transform into reactive astrocytes following many types of central nervous system (CNS) injury *(1–5)*. This process is termed astrogliosis and may result in the formation of a glial scar.

Morphological studies on astrogliosis by neuroanatomists and pathologists at the beginning of this century focused on the most florid examples encountered in the immediate vicinity of destructive lesions of the CNS, such as lacerations, infarcts, abscesses, and multiple sclerosis plaques *(6)*. Accordingly, the criteria which became standard for defining the reactive state were astrocytic hypertrophy and mild to moderate proliferation, an elaboration of long, thick cytoplasmic processes and an increase in glial filaments composed of glial febrillary acidic protein (GFAP) *(7)*.

Duchesne et al. *(8,9)* were the first to draw attention to biochemical diversity in reactive astrocytes in four different models of CNS parenchymal injury. After a hiatus of a decade, there is now growing awareness that reactive astrocytes in different categories of CNS lesions are biochemically heterogeneous, largely as a result of detailed studies of the astroglial reaction in different experimental models of CNS injury *(1)*, combined with the use of a panel of methods to detect GFAP and a number of other conventional and novel "astrocyte-specific" and companion biochemical markers *(2,10)*. From these investigations it is evident that astrocytes do not respond in a stereotypic fashion to all forms of CNS insult, but rather are capable of a variety of types of response, as defined by qualitative, quantitative, temporal and spatial differences in the patterns of molecules which they elaborate in different types of parenchymal injury. This biochemical diversity in reactive astrocytes appears to largely depend on the nature of the CNS injury and the microenvironment of the injury site. Thus, in cellular and molecular terms it is no longer appropriate to hold to a single general definition of astrogliosis.

In this chapter, we consider the likely combinations of "damage signals" and factors in different CNS pathologies that act to induce subtypes of reactive astrocytes. Finally, attention is directed to the future use for mechanistic studies of new and evolving cell culture models of reactive astrogliosis.

From: *Neuroglia in the Aging Brain*
Edited by: Jean S. de Vellis © Humana Press Inc., Totowa, NJ

2. SUBTYPES OF REACTIVE ASTROCYTES

2.1. *Proximal and Distal Reactive Astrocytes in Trauma Models*

Proximal reactive astrocytes develop in the immediate vicinity of a destructive lesion of the adult CNS in which there is cellular necrosis and disruption of the blood brain barrier. Distal reactive astrocytes develop at a distance from such a lesion, in a much less disturbed microenvironment *(9)*. The "local astrocytic response" *(11)* and the "remote astrocytic response" *(12)* are alternative terms used to describe the foregoing two topographical and intensity-graded patterns of astroglial reactivity.

Proximal reactive astrocytes are the subtype which has been studied in the greatest detail *(2–4)*. Stab wounds of the rat cerebrum provide a convenient and reproducible model system for their induction *(1,13)*. The center of the lesion is necrotic and hemorrhagic. Astrocytes become reactive and undergo hypertrophy within 3 h postinjury in a 1 to 2 mm surrounding zone. The astroglial response spreads through the ipsilateral cortex and into subcortical white matter, reaching a peak between 3 and 7 d. The contralateral cerebral hemisphere also may be remotely affected. The margins of the wound are infiltrated by inflammatory mononuclear cells (i.e., lymphocytes, blood-derived macrophages, and intrinsic brain microglial cells). Between 6 and 12 h postinjury, there is a rapid increase in GFAP mRNA in proximal reactive astrocytes which is followed 2 d later by an increase in their GFAP content *(11,14,15)*. Both GFAP mRNA and GFAP decline to near normal levels by 21 d. Between 3 and 6 d postinjury, approx 13% of the proximal reactive astrocytes are the progeny of cells which have divided *(16)*. Thus, the majority of proximal reactive astrocytes arise from normal astrocytes or their precursors in the region. Over a period of several weeks proximal reactive astrocytes participate in the formation of a persistent astroglial scar (anisomorphic gliosis) whereas distal reactive astrocytes revert to normal (isomorphic gliosis) *(6,17)*. Allowing for differences in the design and execution of experiments, comparable findings to those obtained with cerebral stab wounds have been observed following a cryogenic injury, focal X-irradiation, laser-irradiation, and focal ischemia *(2)*.

Spinal cord transection in adult rats produces an astroglial reaction which is maximal at 14 d. Concurrently, there is both rostal and caudal spread of the reaction *(18)*. In neonatal rats, however, astrogliosis remains limited to the site of spinal cord injury. Barrett et al., *(19)* suggested that the remote astrocytic reaction is due to degeneration of the long ascending and descending fiber tracts which are myelinated in the adult but not in the neonate. Gliosis is more severe in the lacerated spinal cord than in the lacerated cerebrum, probably because of the greater amount of initial tissue necrosis at the former site. An immunohistochemical study of lacerated adult rat spinal cord by Predy et al., *(20)*, employing a monoclonal antibody (Mab J1–31) raised against multiple sclerosis plaque tissue by means of hybridoma technology *(21)*, revealed that J1–31 antigen is a more intense marker for proximal reactive astrocytes than GFAP, but is a less intense marker for distal reactive astrocytes than GFAP in the rat spinal cord *(22)* (Figs. 1 and 2).

Many antigens are either unregulated or are expressed *de novo* in proximal reactive astrocytes. GFAP, an intermediate filament protein found only in astrocytes, has become the classical marker for reactive astrocytes in general, and for proximal reactive astrocytes in particular *(7)*. Other molecules which are nonspecifically increased in

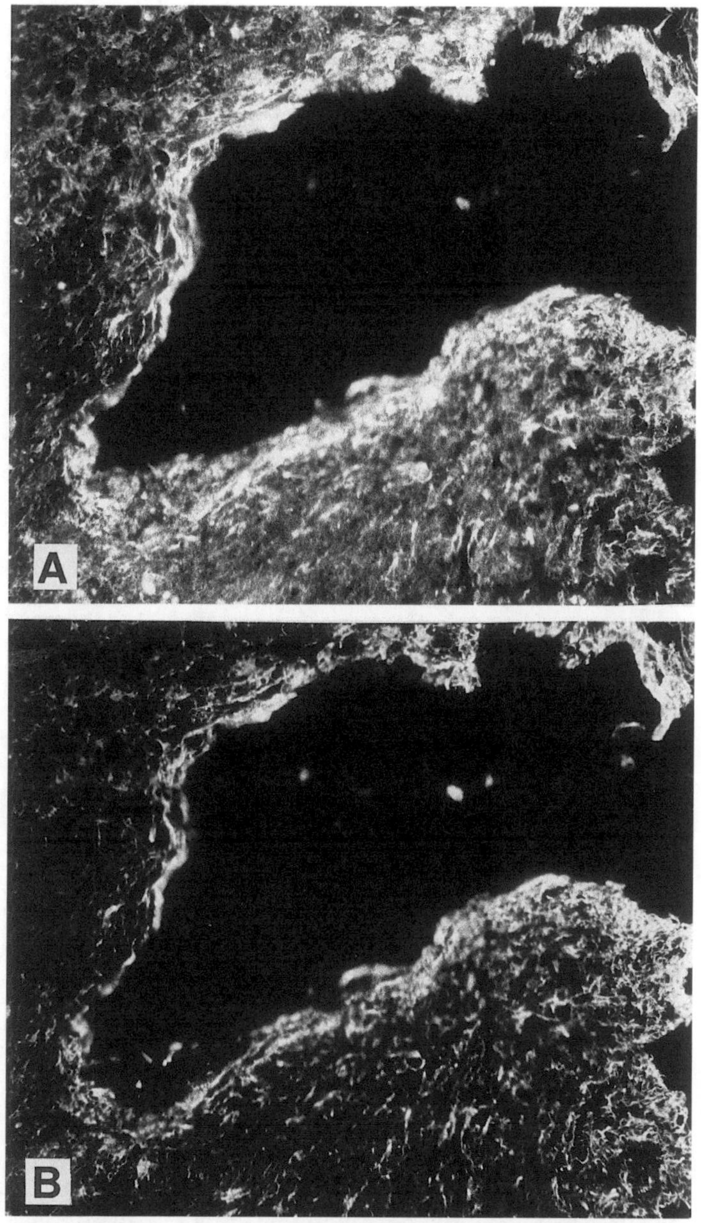

Fig. 1. Sectioned lacerated rat spinal cord showing more intense staining of proximal reactive astrocytes with Mab J1–31 (**A**) than with anti-GFAP (**B**) in the vicinity of the wound. Details of protocols for spinal cord laceration and immunofluorescence staining are given in Predy et al., *(20)* ×106. (Reproduced with permission from *(2)*.)

proximal reactive astrocytes include S-100β protein, vimentin; NSE, GS and GAP-43 protein *(2,4,5,23)*.

In recent years, hybridoma technology has fostered the production of some new monoclonal antibodies which recognize specific markers for proximal reactive astrocytes. These include J1–31 antigen *(21,22)*, 6.17 antigen and M22 antigen *(5)*, and

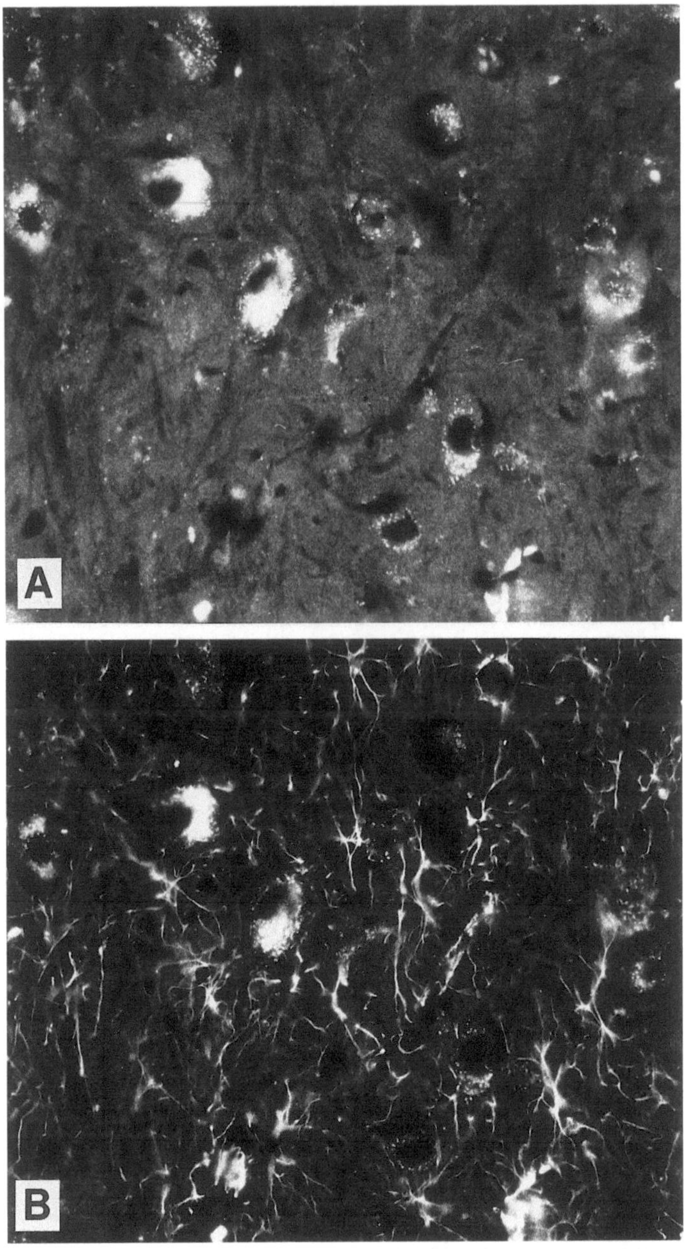

Fig. 2. Sectioned lacerated rat spinal cord in a region at a distance from the wound, showing a lack of astroglial staining with Mab J1–31 **(A)** in contrast to intense staining with anti-GFAP **(B)**, in a double-labeled preparation. Autofluorescence is apparent in the soma of neurons in both photographs (arrows). Details of protocols for spinal cord laceration and immunofluorescence staining are given in Predy et al. *(20)*. (Reproduced with permission from *(2)*.)

13All and 01E4 epitopes *(24,25)*. The foregoing require further characterization, however. Table 1 indicates that although proximal and distal reactive astrocytes share some biochemical characteristics in common, they also differ significantly with regard to other characteristics. To date, distal reactive astrocytes have been less thoroughly stud-

Table 1
Comparison of the biochemical characteristics of normal astrocytes, proximal reactive astrocytes and distal reactive astrocytes[2].

Biochemical characteristics Category[1]	Normal Astrocytes	Proximal Reactive Astrocytes (PRA's)	Distal Reactive Astrocytes (DRA's)
Intermediate filaments	GFAP ↑ GFAP content ↑ Vimentin-neg	GFAP immunoreactivity ↑↑↑ GFAP content ↑↑ Vimentin ↑ (10% of PRA's)	GFAP immunoreactivity ↑↑ GFAP content ↑ Vimentin ±
Gene expression	GFAP mRNA±	GFAP mRNA ↑↑	GFAP mRNA ↑↑
Enzymes	Oxidoreductive enzyme activities: succinate, glucose-6-phosphate, lactate and glutamate dehydrogenases are at normal levels	Oxidoreductive enzyme activities: succinate, glucose-6-phosphate ↑ and glutamate dehyrdrogenases ↑ Enzymes of glycolysis ↑↑↑ Enzymes of the hexose monophosphate shunt (LDH) ↑↑↑	Oxidoreductive enzyme activities: succinate, glucose-6-phosphate, lactate and glutamate dehydrogenases, remain at normal levels
Growth factors	βFGF-neg	βFGF ↑↑	bFGF-neg
Other biologically active proteins	S-100β protein ↑	S-100β protein ↑↑	S-100β protein ↑
Gangliosides	GD_3-neg	GD_3 ↑	GD_3-neg
Other marker epitopes	J1–31-neg 6.17-neg M22-neg	J1–31 ↑ 6.17 ↑↑ M22 ↑↑	J1–31-neg 6.17 ↑↑ M22 ↑

[1] The selection of marker molecules shown above is based on the availability of published comparative biochemical data for normal astrocytes and sub-types of reactive astrocytes under consideration.

[2] For sources of data see reviews by Eddleston and Mucke *(21)*; Malhotra and Shnitka *(2)*; Ridet, et al., *(3)*; Ridet and Privat *(4)*.

ied than proximal reactive astrocytes. Hence the imbalance in available data concerning these two sub-types of reactive astrocytes *(4,23)*.

2.2. Mechanisms of Activation of Proximal Reactive Astrocytes

The induction of proximal reactive astroctyes in the immediate vicinity of a traumatic lesion of the CNS is a highly complex process involving a multiplicity of "damage signals" and factors and severe alterations in local microenvironmental conditions. The details are far from clear, and remain to be fully elucidated. From pathological studies, it is obvious that mechanical trauma produces local destruction of neural and glial cells and their connections, disruption of blood vessels, and an escape of the cellular and fluid components of blood into the injury site. This complex environment contains serum proteins, blood platelets, eicosanoids, biologically active peptides, myelin breakdown products, fibrin split products and purine nucelosides and nucleotides, and so on, which can activate astrocytes and/or function as trophic factors for mononuclear inflammatory cell *(2,4,26)*. Soon after injury, activated astrocytes and mononuclear inflammatory cells (i.e., lymphocytes, blood-derived macrophages and intrinsic brain microglial cells) and activated endothelial cells elaborate many cytokines (such as ciliary neurotrophic factor [CNTF], nerve growth factor [NGF], basic fibroblast growth factor [bFGF], and insulin growth factor-1 [IGF-1]), which act over short distances in an autocrine or paracrine fashion, to produce a broad range of synergistic or antagonistic effects in different cell types *(5,27)*.

Although the intact normal brain is regarded as an immunoprivileged organ, it does contain resident astrocytes and microglia, which when activated produce a number of potent immune molecules, e.g., TNF-α, TGFβ, IL-1, IL-6 and IFN-γ *(5,27)*. TNF-α, IFN-γ and IL-6 have multiple effects in the CNS in controlling glial and neuronal activation, proliferation and survival, thus influencing both degenerative and repair processes. Some of the beneficial and detrimental actions induced by TNF-α and IFN-γ are largely mediated by nitric oxide synthase (NOS)-derived NO production *(5)*. Indeed, NO is involved not only in cytotoxic reactions, but also in the survival and differentiation of neurons. In general, high concentrations are neuroprotective. TNF-α and IFN-γ also upregulate the expression of ICAM-VCAM surface adhesion molecules by astrocytes and brain endothelial cells, thereby favoring astroglial migration, adhesion and anchoring, and neuronal differentiation, during the repair of CNS damage and commencing astrogliosis *(27)*. From the foregoing, it is clear that bidirectional communication exists between resident parenchymal cells (neurons and astrocytes), and resident and infiltrating cells of the immune system in the CNS. It logically follows that *in vitro* studies require *in vivo* confirmation, because indirect synergistic effects predominate *in vivo,* rather than single direct effects *(1)*. The challenge for the future will be to investigate in detail how different factors interact and transform normal astrocytes into proximal reactive astrocytes *in vivo*. The transcriptional mechanisms that are involved in the activation of astrocytes are still a matter of conjecture. There is abundant evidence from *in vitro* studies to indicate that glial cells receive numerous signals which employ cAMP as the second messenger *(28,29)*. Also, signaling through the protein kinase C pathway may play a major role in mediating the developmental proliferation of astrocytes and their activation in astrogliosis in the adult brain *(30)*.

2.3. *Mechanisms of Activation of Distal Reactive Astrocytes*

The origin of distal reactive astrocytes in both the ipsilateral and contralateral cerebral hemispheres following a cerebral laceration has been the subject of speculation and experimentation *(1,13,31,32)*. Several possible mechanisms have been suggested to explain the phenomenon:

1. The release and widespread diffusion from the site of injury of cytokines and growth factors induces astrocyte reactivity at a distance *(31)*.
2. Neuronal degeneration at the site of CNS injury causes anterograde and retrograde fibre degeneration and includes remote astroglial response in the corresponding projection territories *(12)*.
3. Proximal reactive astrocytes migrate away from the site of injury to remote regions *(33)*.
4. Injury results in decreased gap junctional ativity which may be propagated through the astroglial syncytical network *(34)*.

To specifically address the origin of contralateral (remote) astrogliosis, Moumdjian et al., *(31)* performed a callosotomy in rats which also had received a cerebral stab wound, in order to prevent the migration of astrocytes via the corpus callosum. Cerebral callosotomy also served to transect all associative fibers, and thereby enhance neuronal degeneration in both hemispheres. The results of this spatiotemporal study strongly implicated diffusable substances in the induction of the remote astorglial respone, rather than the other possibilities listed above. The lack of major local cell destruction, integrity of the blood-brain barrier, absence of blood components, and the pattern of the microglial/macrophage response in the remote astroglial reaction may be additional determinants of its special characteristics.

2.4. *Reactive Astrocytes in Axotomy Models*

Traumatic injury to a peripheral motor or sensory nerve or the sectioning of a CNS fiber tract (axotomy), induces in the corresponding projection territory, a series of structural and metabolic changes in neuronal cell bodies, dendrites and presynaptic terminals (axon reaction), which are rapidly followed by an astrocytic response (isomorphic gliosis) in the immediate vicinity *(35)*. Details are provided in published descriptions of the axotomy reaction after transection of various motor nerves in the rat *(36–41)*.

After facial nerve axotomy, protoplasmic astrocytes in the facial nucleus undergo hypertrophy and transform into fibrous astrocytes. Increased GFAP synthesis is detectable at 24 h and peaks by 3 d *(40)*. In these motor nerve axotomy models, reactive astrocytes do not proliferate *(39)*, or express vimentin *(40)*, or immunostain with Mab J1–31, (*22, see* Table 2 and Fig. 3). In the latter context, they resemble the distal reactive astrocytes observed in lacerated rat spinal cord *(22)*.

Until recently, the widely held view has been that "damage signals" emanating from axotomized neurons are responsible for the reactive changes that are observed in neighboring astrocytes. Candidates for the intercellular signaling process have included neurotransmitters (for which astroglial cells possess receptors), or ions particularly K^+, which is released by active neurons and is taken up by astrocytes *(42)*. Excitatory amino acids (particularly glutamate) may also play a role, because excitotoxicity has been implicated in a number of acute and chronic neurological diseases *(43)*. Also,

Table 2
Comparison of proximal reactive astrocytes induced by cerebral laceration vs reactive astrocytes induced by axotomy

Subtype of reactive astrocyte	Proximal reactive astrocyte	Axotomy-induced reactive astrocyte	References
Model system	Cerebral laceration	Axotomy	
Pattern of glial scar	Anisomorphic gliosis	Isomorphic gliosis	*(17,35)*
Blood brain barrier	Disrupted	Intact	*(71)*
Hyaluronate binding protein	Negative	Positive	*(72)*
Axonal growth	Permissive	Nonpermissive	*(72)*
Microglial proliferation	Immediate	Delayed	*(17)*
Astroglial proliferation	Yes	No	*(38)*
Vimentin	In about 10% of reactive astrocytes	Negative	*(16,40)*
J1–31 antigen	Postive	Negative	*(22)*

Reproduced with permission from *(2)*.

reactive changes in astroglia may be related to the synaptic reorganization that follows neuronal injury *(42)*.

However, evidence presented by Svensson et al., *(44)*, strongly suggests that the activation of astrocytes following hypoglossal nerve transection is mediated by factors (such as IL-1) released by reactive microglial cells. Their experimental approach was to block the usual axotomy-induced proliferation of indigenous microglial cells in rat brain by the intraventricular infusion of cytosine-arabinoside (ARA-c) *(44)*. Subsequently, astrocytes in the projection territory of the axotomized hypoglossal nerve failed to show expected increases in GFAP and GFAP mRNA, leading to the conclusion that axotomy-induced astrogliosis is mediated indirectly via the microglial reaction.

2.5. *Reactive Astrocytes in Cortical Epileptogenic Foci*

Temporal lobe seizures are the most common type of active epilepsy in adults, comprising about 40% of all cases *(45)*. A broad range of pathological lesions have been identified in the temporal lobe in association with intractable complex partial seizures, including mild to severe gliosis (Ammon's horn sclerosis), malformation (mild cortical dysplasia, microdysgenesis, tuberous sclerosis, or angiomatous malformations), neoplasms (gliomas or mixed tumors of the CNS), and inflammatory scars from infection or infarcts *(46–48)*.

Ammon's horn sclerosis (AHS) is the most common of the foregoing lesions among patients treated surgically for intractable temporal lobe epilepsy. Pathologically, AHS is characterized by atrophy of the hippocampal formation, loss of neurons and gliosis in CA1 and CA4 and in the dentate nucleus *(47)*. Several theories have been proposed to explain the pathogenesis of AHS, including abnormal circuitry in the hippocampal formation due to a developmental error, hypoxic-ischemic damage (due to birth injury or to post-natal trauma), or that epileptic seizures can cause cortical gliosis *(47,49)*. The currently favored view is that seizures provoke the death of neurons as a result of

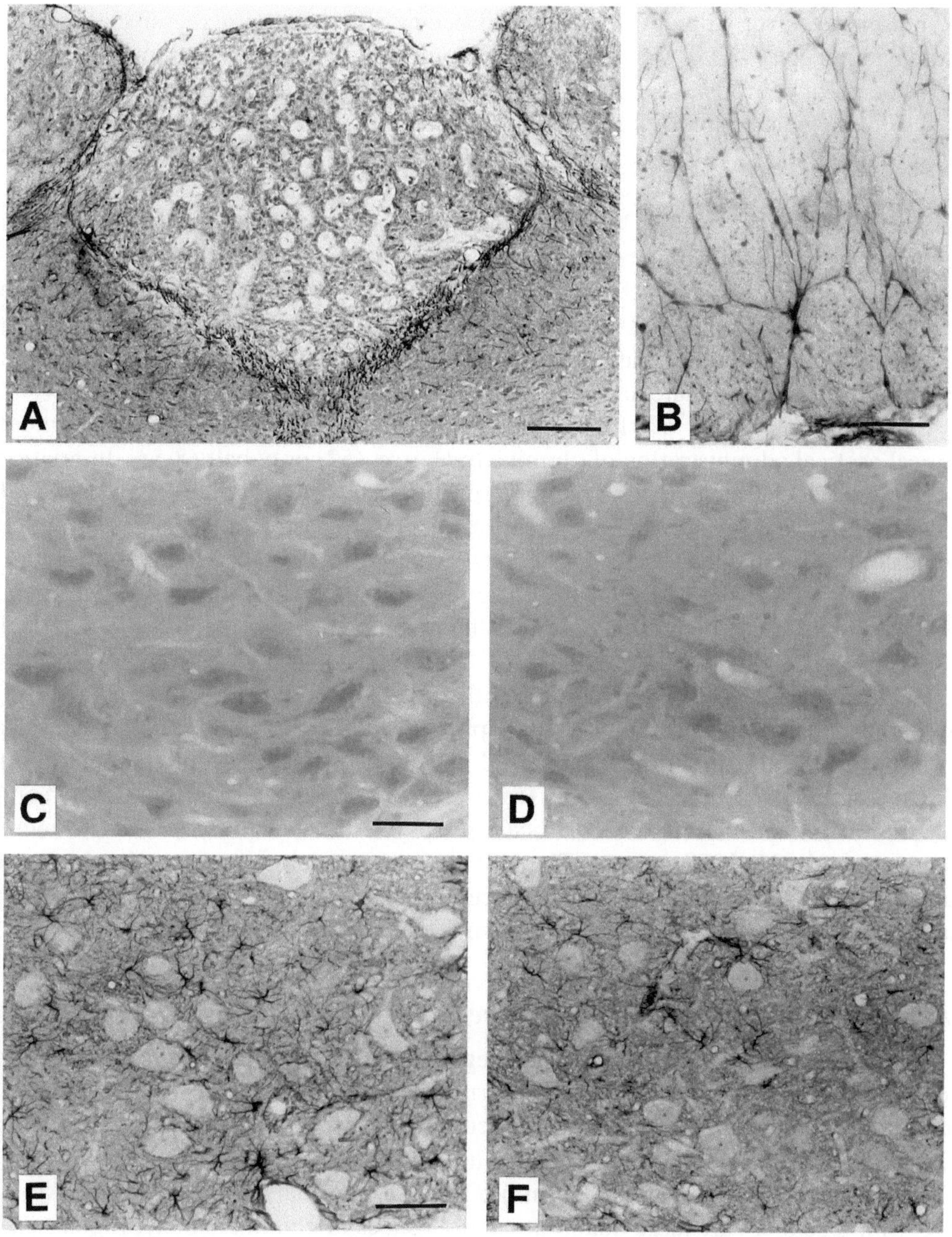

Fig. 3. **(A-D)** J1–31 immunolabeling of rat brain stem, 1 wk after transection of the right hypoglossal nerve. Immunoreactive processes surround the area postrema and extend toward the central canal **(A)**. Subpial astrocytes and their processes are likewise positive for J1–31 **(B)**. However, no specific immunoreactivity is observed in the right hypoglossal nucleus **(C,D)**. Bar 100 μm **(A)**, 50 μm **(B-D)**. **(E-F)** GFAP immunolabeling in the right hypoglossal nucleus 1 week after transection of the right hypoglossal nerve. Note the increased intensity of immunoreactivity on the side of the operation **(E,F)**. Bar 50 μm. (Reproduced with permission from *(22)*.)

excitotoxicity. Over-stimulation of neurons occurs when there is excessive activation of glutamate receptors *(45)*. According to this hypothesis, gliosis is a secondary response to the death of neurons. Also, neuronal cell death may be followed by the formation of new synaptic connections with abnormal hyperexcitablity.

Steward et al. *(42)* have shown experimentally that when mild electrical stimulation of the rat hippocampus was employed to elicit neuronal activity and acute-onset seizures, GFAP and mRNA levels increased rapidly and dramatically at the sites of stimulation, as well as in areas that were synaptically activated by the seizures. These early-stage increases in GFAP mRNA did not appear to be related to the presence of necrotic neurons. Candidates for signaling mechanisms included an upregulation of several "immediate early genes" (such as c-fos and c-jun), and an increased release by active neurons of glutamate and K^+ *(50)*.

Experimental models of chronic focal epilepsy have involved the intracortical or topical cortical administration of alumina *(51)*, iron *(52)* or cobalt *(8,53)*. In relation to the topic of biochemical diversity among reactive astrocytes, Brotchi et al. *(54)* applied the term "activated astrocytes" to a subset of reactive astrocytes in human epileptogenic cortex and in cobalt-treated rat cortex, which display elevated levels of glutamate dehydrogenase (GS), glucose-6-phosphate dehydrogenase, and lactate dehydrogenase. Glucose 6-phosphate dehydrogenase was the first enzyme to rise and the last to decline to normal *(9)*.

Hamberger et al. *(48)* have monitored the levels of neuron specific enolase (NSE), two glial cell specific proteins (S-100β protein and GFAP), and CNAM neural cell adhesion molecule) in surgically excised specimens of human epileptogenic cortex. Gliosis varied from mild to severe. However, no correlation was found between the GFAP content and the duration of focal epilepsy. There was only a 30–40% increase above normal of S-100β protein. The NSE values were close to normal, which correlated with the observation that the concentration of neurons in the cortex of most patients remains relatively unaffected, even after years of seizures.

2.6. Regional Differences in Resident Normal Astrocytes also May Influence the Development of Sub-Types of Reactive Astrocytes

Controversy exists as to whether the region-specific properties of normal astroglial cells are the result of intrinsic coded information within the glial cell, or whether neurons in different brain regions induce uncommitted astroglial cells to express specific proteins in support of a particular type of neuron-glial interaction *(55,56)*.

In different regions of the intact rat brain, not all normal astrocytes are GFAP-positive *(57)*; also there are large quantitative differences in glutamate dehydrogenase (GS) activity, and S100β protein in astrocytes in different brain regions *(58,59)*.

Reviews by Hatten et al. *(56)* and Norton et al. *(1)* give credence to the suggestion that in certain instances phenotypic diversity in reactive astrocytes may in part be related to the diversity of normal astrocytes in the region. We previously collected together some reports of experiments by others which could fit this pattern of reactivity *(2)*. Two examples suffice here. Alonso and Privat *(60)* studied the fine structural organization and immunostaining characteristics of reactive astrocytes in glial scars produced in adult rats by two surgical stab wounds in two close-by locations of the same brain area. With a lesion of the ventral hypothalamus, axonal regeneration occurred and was associated with GFAP-, vimentin-, laminin-, and PSA-NCAM-positive reactive astrocytes. On the other hand, in the dorsal hpothalamus, the same type of lesion resulted in a nonpermissive scar, which exhibited only slight PSA-NCAM and laminin staining. Electron microscopy disclosed that the nonpermissive scar in the dorsal hypothalamus contained significantly more gap junctions than the permissive gliotic scar in the ventral hypothalamus *(4)*.

Another example of regional diversity in reactive astrocytes is described by Fernaud-Espinosa, et al. *(17)* in a study on the differential activation of microglia and astrocytes in aniso- and isomorphic glial scars. Microglial and macroglial responses were compared after two different types of brain damage in two distinct regions in rat brain (i.e., the cerebral cortex and the hippocampus). Each site was subjected to trauma resulting in anisomorphic gliosis. The microglial response seemed to be only linked to the type of lesion inflicted on the CNS, whereas the astrocytic response also appered to depend on the region of the brain that was damaged. Reactive astrocytes in the hippocampus expressed β-amyloid precursor protein (β-APP) immunoreactivity after both aniso- and isomorphic gliosis, but this marker protein was not evident in reactive astrocytes in the cerebral cortex, under the same experimental conditions.

2.7. Models of Reactive Astrogliosis In Vitro

The complexity of the cellular and molecular events occurring in animal models of brain injury has prompted efforts to develops convenient, reliable model systems using astroglial cell cultures, for controlled experiments on astrogliosis *in vitro*. Wu and Schwartz *(25)* have reviewed the current range of cell culture models which are available to biochemically characterize reactive astrocytes, i.e., primary cultures of neonatal astrocytes, co-cultures of astrocytes with either neurons or microglia, and organ cultures. Each of the foregoing addresses a different set of questions. The *in vitro* models if employed as a panel, may help to identify the various patterns of "damage signals" and factors responsible for the biochemical diversity observed in reactive astrocytes *in vivo* in different categories of CNS lesions.

Yu et al. *(61,62)* have described a "mechanical injury model of reactive gliosis *in vitro*" evoked by scratching with a plastic pipet tip, confluent primary cultures of astrocytes prepared from the cerebral cortex of 7- to 10-d-old rat pups. Injured astrocytes along the scratch track started to swell within 6 h of injury, and between 1 and 3 d displayed stellation, an increased content of GFAP, and larger and more organized assemblies of cytoplasmic filaments.

Eng et al. *(63)* quantitated the changes in gene expression which occurred after a scratch wound of cultured rat astrocytes; *c-fos* mRNA increased 30-fold and heat shock protein (Hsp) mRNA increased four-fold within 60 min. Both returned to low levels by 12 h postinjury. The early genes which are activated may influence cell division, protect against further injury, and induce later genes (such as metallothionein) through the induction of transcription factors.

Two key nuclear signal transduction mechanisms in astroglia appear to be the basis for a flexible genomic switch which allows extracellular signals to change genetic programs and protein expression patterns:

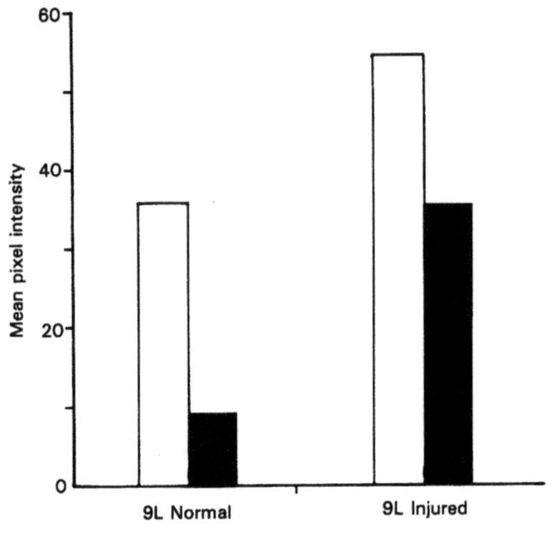

Fig. 4. Astrogliosis in vitro in 9L rat glioma cells induced by mechanical injury ("scratch wound" model). Histogram comparing the relative intensities of the fluorescent signals from double-labeled images for GFAP (solid columns) and J1–31 antigen (open columns) of normal and mechanically injured 9L cells. (Reproduced with permission from *(65).*)

1. Specific transmitter receptor stimulation by extracellular signals can produce a highly integrated nuclear third messenger (NTM) response with coordinated assembly at the level of NTM transcription and translation;
2. Multiple signals can dynamically change the expression of specific target genes via negative or positive cooperative interactions among various classes of NTMs at the genomic level *(64).*

In 1995, Malhotra et al. *(65)* showed that 9L rat glioma cells grown on coverslips, like primary cultures of astrocytes, undergo astrogliosis when subjected to mechanical trauma from a "scratch wound." Cells along the scratch track display mild hypertrophy and increases in GFAP and J1-31 antigen immunoreactivities, from trace to moderate levels in the case of GFAP, and from moderate to high levels in the case of the J1-31 antigen (Fig. 4). Reaction to mechanical injury by the 9L cells occurred without interactions with microglia, neurons or oligodendroglia. In the same report, we reviewed the literature on the origin and characteristics of the 9L rat glioma cell line and listed the advantages of using cultures of 9L cells, rather than primary cultures of astrocytes, for studies on astrogliosis in vitro. Space limitations preclude the presentation of full details here.

9L rat glioma cells *(66),* and primary cultures of astrocytes from neonatal rat brain *(67)* also undergo astrogliosis after exposure to low levels of cadmium chloride (a prototype neurotoxicant [*68*]). Moreover, when cultures of 9L cells were subjected to mechanical injury and then exposed to cadmium chloride, there was a marked increase in the expression of the astrocytic marker, J1-31 antigen. This was due to a summation of stimulatory effects from these two injurious agents. A moderate coordinated response was detected for the expression of the classical marker, GFAP, and cell hypertrophy was only slightly increased over that produced by either injurious agent alone (Fig. 5). The foregoing findings suggest that more than one transcription mechanism

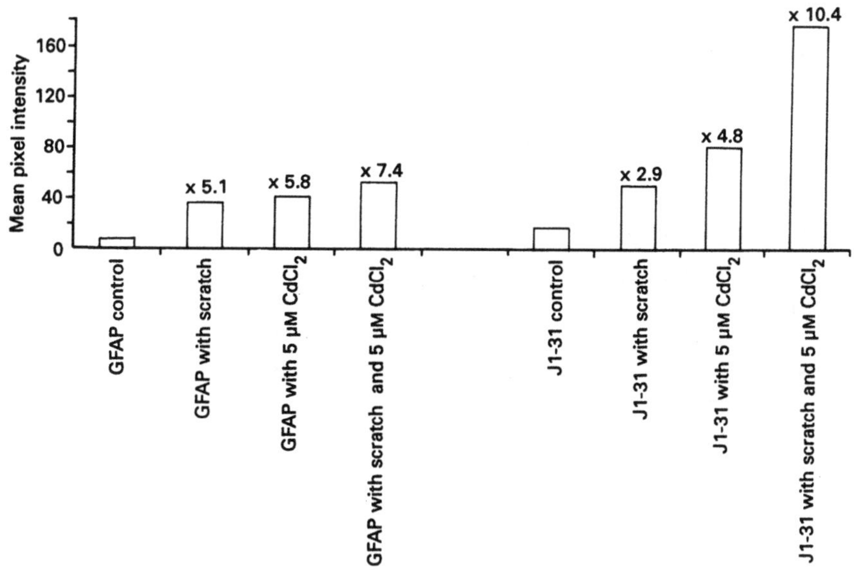

Fig. 5. Coordinated mechanical and chemical injuries upregulate astrogliosis in vitro in 9L rat glioma cells. Histogram comparing the relative intensities of the fluorescence signals from immunolabeled images from 9L cells for GFAP and J1–31 antigen respectively, in normal cells, mechanically injured cells, $CdCl_2$-chemically injured cells, and in cells with coordinated injuries from both agents. Increases shown above columns. Controls, without scratch or $CdCl_2$. (Reproduced with permission from *(66)*.)

is involved in the activation of astroglia, and that mechanical and $CdCl_2$-induced injuries probably respectively affect different receptors and second- and third-messenger pathways. Thus, in a relatively simple in vitro model system, biochemical diversity in reactive astrocytes could be demonstrated by altering the conditions of astroglial cell injury.

3. CONCLUDING REMARKS

Recent reports on a wide range of experimental animal models of CNS injury, combined with the application of immunostaining methods for a considerably expanded panel of conventional and novel astrocyte-specific markers, indicate that reactive astrocytes are capable of graded levels and different types or responses, as defined by the molecular they elaborate in different categories of local pathology *(1,2,4)*. Biochemical heterogeneity and functional diversity are also prominent among other activated types of specialized cells such as microglia *(69)* and fibroblasts *(70)*. This adaptive plasticity of reactive astrocytes largely appears to be a modulated response to the different microenvironmental conditions, particularly the combinations of immune molecules, cytokines, eicosinoids, serum factors, peptides, nucleotides, and adhesion molecules that exist in different categories of CNS lesions. Regional biochemical differences in the normal astrocytes from which reactive astrocytes arise, as well as local nonglial influences, may be contributory factors in certain brain regions.

Future research, no doubt, will expand on the lines of investigation already established. New and evolving cell culture models of reactive astrogliosis should facilitate mechanistic studies. The challenge then will be to determine both the common and the diverse signaling mechanism which underlie the conversion of normal astrocytes into their respective reactive subtypes. The foregoing should lead to a better understanding of the special roles played by the biochemical subtypes of reactive astrocytes which are present in different categories of CNS pathology.

ACKNOWLEDGMENTS

This work has been supported by a grant from the NSERC of Canada. Dr. K.B. Newbound graciously provided partial financial support for this project. We are grateful to Mr. Le Thoung Luong for his technical assistance and to Mr. Rakesh Bhatnagar for his expertise in the use of the Confocal Microscope. Ms. Brenda Metherell rendered invaluable secretarial help with the manuscript.

REFERENCES

1. Norton, W.T., Aquino, D.A., Hozumi, I., Chiu, F-C., and Brosnan, C.F. (1992) Quantitative aspects of reactive gliosis: a review. *Neurochem. Res.* **17,** 877–885.
2. Malhotra, S.K. and Shnitka, T.K. (1994) Adaptive plasticity and diversity among reactive astrocytes in central nervous system lesions. *Biomedical Letters* **49,** 273–302.
3. Montgomery, D.L. (1994) Astrocytes: form, functions, and roles in disease. *Vet. Path.* **31,** 145–167.
4. Ridet, J.L., Malhotra, S.K., Privat, A., and Gage, F.H. (1997) Reactive astrocytes: cellular and molecular cues to biological function. *Trends Neurosci.* **20,** 570–577.
5. Ridet, J.-L. and Privat, A. (1999) Reactive astrocytes, their roles in CNS injury and repair mechanisms, in *Advances in Structural Biology* (Malhotra, S.K. ed.) Vol. 6 (in press). Jai Press Inc. Stamford, CT, USA.
6. Greenfield, J.G. (1958) General pathology of nerve cell and neuroglia, in *Neuropathology* (Greenfield, J.G., Blackwood, W., Meyer, A., McMenemey, W.H. and Norman, R.M., eds.) Edward Arnold Ltd. London, pp. 1–66.
7. Eng, L.F. and Ghirnikar, R.S. (1994) GFAP and astrogliosis. *Brain Pathol.* **4,** 229–237.
8. Duchense, P.Y., Ghuens, J., Brotchi, J., and Gerebtzoff, M.A. (1979) Normal and reactive astrocytes: a comparative study by immunohistochemistry and by a classical histological technique. *Cell Molec. Biol.* **24,** 237–239.
9. Duchesne, P.Y., Gerebtzoff, M.A., and Brotchi, J. (1981) Four types of reactive astrocytes. *Biblthca Anat.* **19,** 313–316.
10. Malhotra, S.K., Shnitka, T.K., and Elbrink, J. (1990) Reactive astrocytes – a review. *Cytobios* **61,** 133–160.
11. Condorelli, D.F., Dell'Albani, P., Kaczmarek, L., Messina, L., Spampinato, G., Avola, R., Messina, A., and Giuffrida Stella, A.M. (1990) Glial fibrillary acidic protein messenger RNA and glutamine synthetase activity after nervous system injury. *J. Neurosci. Res.* **26,** 251–257.
12. Hajós, F., Kálmán, M., Zilles, K., Schleicher, A., and Sotonyi, P. (1990) Remote astrocytic response as demonstrated by glial fibrillary acidic protein immunohistochemistry in the visual cortex of dorsal lateral geniculate nucleus lesioned rats. *Glia* **3,** 301–310.
13. Mathewson, A.J. and Berry, M. (1985) Observations on the astrocyte response to a cerebral stab wound in adult rats. *Brain Res.* **327,** 61–69.
14. Hozumi, I., Chiu, F-C., and Norton, W.T. (1990a) Biochemical and immunocytochemical changes in glial fibrillary acidic protein after stab wounds. *Brain Res.* **524,** 64–71.
15. Hozumi, I., Aquino, D.A., and Norton, W.T. (1990b) GFAP mRNA levels following stab wounds in rat brain. *Brain Res.* **534,** 291–294.

16. Miyake, T., Hattori, T., Masaru, F., Kitamura, T., and Fujita, S. (1988) Quantitative studies on proliferative changes of ractive astrocytes in mouse cerebral cortex. *Brain Res.* **451,** 133–138.
17. Fernaud-Espinosa, I., Nieto-Sampedro, M., and Bovolenta, P. (1993) Differential activation of microglia and astrocytes in aniso- and isomorphic gliotic tissue. *Glia* **8,** 277–291.
18. Barrett, C.P., Guth, L., Donait, E.J., and Krikorian, J.G. (1981) Astroglial reaction in the gray matter of lumbar segments after midthoracic transection in the adult rat spinal cord. *Expl. Neurol.* **73,** 365–375.
19. Barrett, C.P., Donati, E.J., and Guth, L. (1984) Differences between adult and neonatal rats in their astroglial response to spinal injury. *Expl. Neurol.* **84,** 374–385.
20. Predy, R., Malhotra, S.K., and Das, G.D. (1988) Enhanced expression of a protein antigen (J1–31 antigen, 30 kilodaltons) by reactive astrocytes in lacerated spinal cord. *J. Neurosci. Res.* **19,** 397–404.
21. Malhotra, S.K., Wong, F., Cumming, P., Ross, S.D., Shnitka, T.K., Manickavel, V., Warren, K.G., and Jeffrey, V. (1984) A monoclonal antibody for cytoskeletal antigenic determinant(s) distinguishable from glial fibrillary acidic protein in astrocytes. *Microbios Letters* **26,** 151–157.
22. Malhotra, S.K., Svensson, M., Aldskogius, H., Bhatnagar, R., Das, G.D., and Shnitka, T.K. (1993) Diversity among reactive astrocytes: proximal reactive astrocytes in lacerated spinal cord preferentially react with monoclonal antibody J1–31. *Brain Res. Bull.* **30,** 395–404.
23. Eddleston, M. and Mucke, L. (1993) Molecular profile of reactive astrocytes – implications for their role in neurologic disease. *Neuroscience* **54,** 15–36.
24. Welter, E., Bolesta, M.J., and Landis, D.M.D. (1993) Monoclonal antibodies which bind to reactive astrocytes. *Soc. Neurosci. Abstr.* **19,** 60.
25. Wu, V.W. and Schwartz, J.P. (1998) Cell culture models for reactive gliosis: New perspectives. *J. Neurosci. Res.* **51,** 675–681.
26. McMillian, M.K., Thai, L., Hong, J-S., O'Callaghan, J.P., and Pennypacker, K.R. (1994) Brain injury in a dish: a model for reactive gliosis. *Trends Neurosci.* **17,** 138–142.
27. Muñoz-Fernńdez, M.A. and Fresno, M. (1998) The role of tumor necrosis factor, interleukin 6, interferon-γ and inducible nitric oxide synthease in the development and pathology of the nervous system. *Prog. Neurobiol.* **56,** 307–340.
28. Hansson, E. (1989) Co-existence between receptors, carriers, and second messengers on astrocytes grown in primary cultures. *Neurochem. Res.* **14,** 811–819.
29. Melcangi, R.C., Celotti, F., Castano, P., and Martini, L. (1992) Intracellular signalling sytems controlling the 5α-reductase in glial cell cultures. *Brain Res.* **585,** 411–415.
30. Yong, V.W. (1992) Proliferation of human and mouse astrocytes *in vitro:* signalling through the protein kinase C pathway. *J. Neurol Sci.* **111,** 92–103.
31. Moumdjian, R.A., Antel, J.P., and Yong, V.W. (1991) Origin of contralateral reactive gliosis in surgically injured rat cerebral cortex. *Brain Res.* **547,** 223–228.
32. Takamiya, Y., Kohsaka, S., Toya, S., Otani, M., and Tsukada, Y. (1988) Immunohistochemical studies on the proliferation of reactive astrocytes and the expression of cytoskeletal proteins following brain injury in rats. *Rev. Brain Res.* **38,** 201–210.
33. Janeczko, K. (1989) Spatiotemporal patterns of the astroglial proliferation in rat brain injured at the postmitotic stage of postnatal development: a combined immunocytochemical and autoradiographic study. *Brain Res.* **485,** 236–243.
34. Anders, J.J., Niedermair, S., Ellis, E., and Salopek, M. (1990) Response of rat cerebral cortical astrocytes to free- or cobalt induced injury: an immunocytochemical and Gap-FRAP study. *Glia* **3,** 476–486.
35. Aldskogius, H. and Svensson, M.A. (1993) Neuronal and glial cell responses to axon injury, in *Advances in Structural Biology, Vol. 2,* (Malhotra, S.K., ed) pp. 191–223. Jai Press.
36. Reisert, I., Wildermann, G., Grab, D., and Pilgrim, C.H. (1984) The glial reaction in the course of axon regeneration: a stereological study of the rat hypoglossal nucleus. *J. Comp. Neurol.* **229,** 121–128.

37. Cova, J.L. and Aldskogius, H. (1985) A morphological study of glial cells in the hypoglossal nucleus of the cat during nerve regeneration. *J. Comp. Neuro.* **233,** 421–428.
38. Cova, J.L. and Aldskogius, H. (1986) Effect of axotomy on perineuronal glial cells in the hypoglossal and dorsal motor vagal nuclei of the cat. *Expl. Neuro.* **93,** 662–667.
39. Graeber, M.B. and Kreutzberg, G.W. (1988) Delayed astrocyte reaction following facial nerve axotomy. *J. Neurocytol.* **17,** 209–220.
40. Tetzlaff, W., Graeber, M.B., Bisby, M.A., and Kreutzberg, G.W. (1988) Increased glial fibrillary acidic protein synthesis in astrocytes during retrograde reaction of the rat facial nucleus. *Glia,* **1,** 90–95.
41. Gilmore, S.A., Sims, T.J., and Leiting, J.E. (1990) Astrocytic reactions in spinal gray matter following sciatic axotomy. *Glia* **3,** 342–349.
42. Steward, O., Torre, E.R., Tomasulo, R., and Lothman, E. (1991) Neuronal activity up-regulates astroglial gene expression. *Proc. Natn. Acad. Sci. USA* **88,** 6819–6823.
43. Whetsell, W.O. Jr. and Shapira, N.A. (1993) Biology of disease. Neuroexcitation, excitotoxicity and human neurological disease. *Lab. Invest.* **68,** 372–387.
44. Svensson, M. and Aldskogius, H. (1993) Evidence for activation of astrocytes via reactive microglial cells following hypoglossal nerve transection. *J. Neurosci. Res.* **35,** 373–381.
45. McNamara, J.O. (1992) The neurobiological basis of epilepsy. *Trends Neurosci.* **15,** 357–359.
46. Paul, L.W. and Scheibel, A.B. (1986) Structural substrates of epilepsy. *Adv. Neurol* **44,** 775–786.
47. Armstrong, D.D. (1993) The neuropathology of temporal lobe epilepsy. *J. Neuropath. Exp. Neurol.* **52,** 433–443.
48. Hamberger, A., Bock, E., Nordborg, C., et al. (1993) Biochemical correlates to cortical dysplasia, gliosis, and astrocytoma infiltration in human epileptogenic cortex. *Neurochem. Res.* **18,** 511–518.
49. Mathieson, G. (1975) Pathological aspects of epilepsy with special reference to the surgical pathology of focal seizures. *Adv. Neurol.* **8,** 108–137.
50. Torre, E.R., Lothman, E., and Steward, O. (1993) Glial response to neuronal activity: GFAP-mRNA and protein levels are transiently increased in the hippocampus after seizures. *Brain Res.* **631,** 256–264.
51. Harris, A.B. (1975) Cortical neuroglia in experimental epilepsy. *Expl. Neurol.* **49,** 691–715.
52. Hammond, E.J., Ramsey, R.E., Villareal, H.J., and Wilder, G.J. (1980) Effects of intracortical injection of blood and blood components on the electrocortigram. *Epilepsia* **21,** 3–14.
53. Fisher, J., Holubar, J., and Malik, V. (1968) Neurohistochemical study of the development of experimental epileptogenic cortical cobalt-gelatine foci in rats and their correlation with the onset of epileptogenic electrical activity. *Acta Neuropathol.* **11,** 45–54.
54. Brotchi, J., Tanaka, T., and Leviel, V. (1978) Lack of activated astrocytes in the kindling phenomena. *Expl. Neurol.* **58,** 119–125.
55. Wilkin, G.P., Marriott, D.R., and Cholewinski, A.J. (1990) Astrocyte heterogeneity. *Trends Neurosci.* **13,** 43–46.
56. Hatten, M.E., Liem, R.K.H., Shelanski, M.L., and Mason, C.A. (1991) Astroglia in CNS injury. *Glia* **4,** 233–243.
57. Hajós, F. and Kálmán, M. (1989) Distribution of glial fibrillary acidic protein (GFAP)-immunoreactive astrocytes in the rat brain. II. Mesencephalon, rhombencephalon and spinal cord. *Expl. Brain Res.* **78,** 164–173.
58. Patel, A.J., Weir, M.D., Hunt, A., Tahourdin, C.S.M., and Thomas, M.G.T. (1985) Distribution of glutamine synthetase and glial fibrillary acidic protein and correlation of glutamine synthetase with glutamate decarboxylase in different regions of the rat central nervous system. *Brain Res.* **331,** 1–9.
59. Didier, M., Harandi, M., Aquera, M., et al. (1986) Differential immunocytochemical staining for glial fibrillary acidic (GFA) protein, S-100 protein and glutamine synthesis in the rat subcom-

misural organ, nonspecialized ventricular ependyma and adjacent neuropil. *Cell Tissue Res.* **245,** 343–351.

60. Alonso, G. and Privat, A. (1993) Reactive astrocytes involved in the formation of lesional scars differ in the mediobasal hypothalamus and in other forebrain regions. *J. Neurosci. Res.* **34,** 523–538.
61. Yu, A.C.H., Kwan, H.H., Lee, Y.L., and Eng, L.F. (1993a) Morphologic changes in mechanically damaged astrocytes. *Trans. Am. Soc. Neurochem.* **24,** 242.
62. Yu, A.C.H., Lee, Y.L., and Eng, L.F. (1993b) Astrogliosis in culture. 1. The model and the effect of antisense oligonucleotides on glial fibrillary acidic protein synthesis. *J. Neurosci. Res.* **34,** 295–303.
63. Eng, L.F., Lee, L.Y., Murphy, G.M., and Yu, A.C.H. (1995) A RT-PCR study of gene expression in a mechanical injury model. *Prog. Brain Res.* **105,** 219–229.
64. Szekely, A.M., Grayson, D., and Costa, E. (1993) Nuclear signal transduction via immediate early genes in neurons and glia. *Biochem. Soc. Trans.* **21,** 61–65.
65. Malhotra, S.K., Bhatnagar, R., Shnitka, T.K., Herrcra, J.J., Koke, J.R., and Singh, M.V. (1995) Rat glioma cell line as a model for astrogliosis. *Cytobios.* **82,** 39–51.
66. Malhotra, S.K., Luong, L.T., Bhatnagar, R., and Shnitka, T.K. (1997) Up-regulation of reactive astrogliosis in the rat glioma 9L cell line by combined mechanical and chemical injuries. *Cytobios.* **89,** 115–134.
67. Rising, L., Vitarella, D., Kimelberg, H.K., and Aschner, M. (1995) Cadmium chloride (CdC12)-induced metallothionein (MT) expression in neonatal rat primary astrocyte cultures. *Brain Res.* **678,** 91–98.
68. O'Callaghan, J.P. (1993) Quantitative features of reactive gliosis following toxicant-induced damage of the CNS. *Ann. N.Y. Acad. Sci.* **679,** 195–210.
69. Flaris, N.A., Densmore, T.L., Molleston, M.C., and Hickey, W.F. (1993) Characterization of microglia and macrophages in the central nervous system of rats: definition of the differential expression of molecules using standard and novel monoclonal antibodies in normal CNS and in four models of parenchymal reaction. *Glia* **7,** 34–40.
70. Sappino, A.P., Schurch, W., and Gabbiani, G. (1990) Differentiation repertoire of fibroblastic cells: expression of cytoskeletal proteins as marker of phenotypic modulations. *Lab. Invest.* **63,** 144–161.

3

Astrocytic Reaction After Traumatic Brain Injury

Jesús Boya, J. L. Calvo, Angel López-Carbonell
and José E. García-Mauriño

1. INTRODUCTION

Astrocytes and microglia respond to a great variety of lesions in the nervous tissue of the central nervous system (CNS). In this sense, although astroglial cells react in a relatively constant manner, the intensity and time course of the changes are largely dependent upon the type of lesion involved.

The present review fundamentally addresses traumatic lesions affecting the CNS. This type of lesion has been extensively studied from the experimental perspective and constitutes one of the most reliable and easily reproducible models for analyzing the reactive changes taking place in the glial component of nervous tissue.

One of the earliest morphological studies of glial reaction to traumatic injury of the CNS was carried out by Wilson in 1926 *(1)*. This author employed conventional histological techniques (cresyl violet and trichromic stains) to study the trajectory of therapeutic punctions made in two human brains. In this context, a glial reaction was observed, with connective tissue participation in the repair processes. However, the application of metallic impregnation techniques allowed more detailed research of the glial reaction in these lesions, and afforded a clear distinction between cell types such as astrocytes and microglia. In this sense, the classical studies of Penfield et al. were of great interest *(2–4)*. These authors, by producing punction lesions in the brain, detected a series of cellular changes in the astrocytes that comprised early edema, astrocyte proliferation, fibrous morphological transformation of the astrocytes, astrocyte hypertrophy, and participation of the connective tissue in the glial scar. In addition, the intensity of the glial reaction was related to the amount of necrotic tissue produced.

2. INTERMEDIATE FILAMENT PROTEIN EXPRESSION IN ASTROCYTES

Both glial fibrillary acidic protein (GFAP) and the protein vimentin (VIM) have been described in astrocytes. GFAP is considered to be characteristic of mature astrocytes *(5)*, whereas VIM—initially described in mesenchymal cells—has been detected in radial glia and in mature astrocytes *(6,7)*. Changes in the expression of both glial antigens are thought to occur in the course of normal astroglial development. Thus, in several mammal

From: *Neuroglia in the Aging Brain*
Edited by: Jean S. de Vellis © Humana Press Inc., Totowa, NJ

species, VIM-GFAP transition takes place in the first weeks of life after birth, as a result of which GFAP and VIM may be temporarily coexpressed by most astrocytes during such transition periods *(7–11)*. Moreover, biochemical studies have revealed changes in the presence of messenger RNA (mRNA) encoding for both GFAP and VIM during this period *(12)*.

Once development has been completed, astrocytes express variable amounts of GFAP. Whereas the concentration of the latter is high in fibrous astrocytes of the white matter *(13)*, it may be very low in certain regions of the brain cortex—where protoplasmic astrocytes (at least in the rat) cannot be detected by the habitual immunohistochemical techniques *(14)*—with the exception of layer I astrocytes that form the *glia limitans,* which are clearly GFAP-positive. However, astrocytes in special locations such as the cerebellum, retina, optic nerve, and major tracts of white matter continue to coexpress GFAP and VIM even in the adult animal *(7,15–17)*. In vitro immunoelectron microscopic studies of GFAP and VIM location by Abd-El-Basset et al. *(18)* have shown that in both astrocytes and their precursors, VIM and GFAP copolymerize in the same individual intermediate filament; as a result, the GFAP/VIM ratio present in these intermediate filaments reflects the degree of differentiation and functional status of these cells.

Gliosis is the typical astroglial response to CNS lesions. The term encompasses two phenomena: Astroglial proliferation and hypertrophy *(14)*. Astroglial hypertrophy (proliferation will be addressed in the corresponding section below) is the most important and easily detectable characteristic of reactive astrocytes. It may be demonstrated by argentic impregnation techniques (Cajal's gold sublimate, for example), though at present the immunohistochemical demonstration of increased GFAP (and occasionally VIM) expression is clearly the most reliable parameter.

2.1. Time Course

The early character of astrocyte reaction seems to be a constant finding in traumatic damage of the CNS—regardless of the lesion model employed. Thus, in experimental cerebral wounds, increased GFAP immunopositivity becomes detectable a few hours to two days after producing the lesion—with a maximum intensity peak that rarely exceeds 7 d *(18a–27)*. In relation to this phenomenon, the findings have been similar to those reported for other experimental models of traumatic injury of the CNS *(28–34)*.

It is interesting to point out that the early onset of astrocyte reaction (expressed by an increase in GFAP positivity) is very similar to that observed in other types of nervous tissue lesion *(35–59)*. In different types of lesion, the data obtained from biochemical studies—both as regards GFAP quantification *(20,36,55,59–62)* and GFAP mRNA assay *(46,55,60,62,63)*—clearly confirm the immunohistochemical findings. Likewise, GFAP mRNA *in situ* hybridization studies *(55,64)* have shown the astrocytes to be responsible for GFAP synthesis.

Based on electron microscopic studies of the first evolutive stages of experimentally induced cerebral lesions, we have found (nonpublished personal observations) reactive astrocytes to exhibit a manifest edematous appearance, with a very electron-transparent hyaloplasm and very scarce glial filaments (though the latter subsequently increase greatly in number). In this sense, the ultrastructural image is scantly compatible with the intense GFAP immunopositivity detected in light microscopic immunohistochemical studies. Considering that the intermediate filaments are highly dynamic structures *(65)*, a pool of kinetically active disassembled subfilamentous units would be expected to be in dynamic equilibrium with assembled filaments *(66)*. The surprising increase in astrocyte GFAP immunoreactivity so shortly after injury, and the apparently paradoxical ultrastructural absence of glial filaments, could be accounted for by a rapid increase in GFAP synthesis, which would expand the pool of subfilamentous units and thus increase the immunopositivity of the reactive astrocytes that nevertheless still require a period of time to complete filament assembly.

As seen above, the constant pattern of findings has led to widespread consensus on the rapid establishment of reactive astrocyte changes in highly diverse CNS lesions. However, variable results have been obtained as regards the permanent or transient nature of such changes, depending on the experimental model involved. An added difficulty is the tendency to carry out studies that conclude too early—thus frequently preventing the recording of reliable data on the true duration of the glial reaction.

Cerebral lesions exhibit a gradual decrease in astrocyte reaction over time, reflected by a drop in GFAP immunoreactivity. In this sense, the reaction is seen to have clearly decreased after approximately three weeks *(20)*. After this period of time, only a narrow band of immunopositive astrocytes remains in the cortex surrounding the lesion *(22)*. Biochemical studies have shown the amount of both GFAP *(67)* and GFAP mRNA *(68)* to return to values very close to normal after these three weeks. Likewise, in mild contusive brain lesions, immunopositivity practically returns to normal 30 d after injury *(28)*.

The duration of the glial reaction is much more controversial in other types of CNS lesion involving astrocyte response. Thus, in ischemic lesions the reaction is seen to remain stable after three months *(49)*, though it may descend to negligible levels after six months *(44)*. In turn, deafferentiation lesions appear to induce a transient astroglial reaction. As an example, deafferentiation caused by the injection of a neurotoxic agent (ibotenic acid) induces a glial reaction in the projection areas of the injected zone that disappears 5–6 mo later *(45)*. The astrocytes of a given region undoubtedly respond to lesions in a subcortical nucleus projecting fibers to that region. In this sense, astrocytes have been shown to respond to cholinergic but not to dopaminergic deafferentiation *(36)*. Additional examples of transient reactions are the glial response produced in the olfactory tract by sectioning the olfactory nerve—with regression occurring after one month *(35)*—and the glial reaction in the facial nucleus after compression of the facial nerve (with regression after 40 d) *(57)*.

However, other lesion models appear to induce a permanent glial response. Thus, in Wallerian degeneration of the optic nerve, the astrocyte reaction seems to persist for at least one year and apparently indefinitely *(69,70)*. Likewise, in unilateral sections of the subcommissural fornix, glial response persists for at least one year *(56)*. Toxic lesions *(40,57,71)* in turn lead to a practically permanent presence of reactive astrocytes.

In other situations, the greater or lesser persistence of the glial reaction is related to the severity of the damage produced. This has been observed following the induction of variable intensity brain ischemia *(53)*, and in toxic lesions *(44)*, where the astrocyte reaction persists after four months in certain areas of the hippocampus and fades in others—in close correlation to the degree of neuronal degeneration produced.

2.1.1. Other Intermediate Filament Proteins in Astrocytes

Although much less frequently than GFAP, VIM has also been used as a reactive astrocyte marker in traumatic lesions of the CNS. In cerebral injuries after 5 d of evolution, our group *(19)* has detected a reactive astrocyte band measuring 300–350 μm in thickness surrounding the punction trajectory. By applying techniques for the detection of VIM and GFAP, we were able to show that 60% of the reactive astrocytes coexpressed VIM and GFAP—the remaining astrocytes only being positive for GFAP. In no case did we observe simultaneously VIM-positive and GFAP-negative astrocytes. Other authors have also reported the coexpression of VIM and GFAP in reactive astrocytes *(50,53)*. By using this same lesion model, Takamiya et al. *(25)* have detected reactive astrocytes that transiently express VIM from 2–10 d post-lesion, with a maximum expression peak after 3–5 d. This latter observation has also been reported in other types of acute necrotic lesions *(53)*.

The expression of VIM has also been described in reactive astrocytes in noninvasive traumatic injuries, such as brain contusions *(28,31)* or spinal cord compression *(29)*. VIM expression has likewise been recorded in ischemic lesions *(44,49,52,54)*, toxic lesions *(44)*, in lesions involving retrograde degeneration *(42,59)*, or in Wallerian degeneration processes of the CNS *(50,56)*.

As pointed out at the start of this study, VIM is regarded as an immature astrocyte marker in the course of astroglial development; consequently, VIM expression by reactive astrocytes could reflect a temporal regression of these cells to immature states. In this sense, Abd-El-Basset et al. *(18)* have observed in vitro that highly mobile astrocyte precursors (proastroblasts and young astroblasts) possess intermediate filaments exclusively composed of VIM (proastroblasts) or composed of a heteropolymer in which the GFAP/VIM ratio is low (young astroblasts). In contrast, in more mature (and thus less mobile) astroblasts, and in immobile astrocytes, the intermediate filaments are composed of heteropolymers with a high GFAP/VIM ratio. These authors suggest that heteropolymer formation allows the astroglia to regulate the "stiffness" of its intermediate filament network by increasing or decreasing the GFAP/VIM ratio. The reactive astroglia could behave in the same manner, thus responding to different functional situations that may require increased cell mobility.

The idea that reactive astroglia undergoes a certain regression toward more immature states has been supported by new evidence in recent years. Thus, nestin (a type of intermediate filament expressed by neuroepithelial stem cells of the embryonic CNS) *(72)* is known to be expressed by reactive astrocytes (along with GFAP) in lesions of the CNS *(73–75)*.

In punction-induced cerebral lesions, reactive astrocytes *(27)* have been found to coexpress GFAP and a certain intermediate filament associated protein (IFAP) found in radial glia and derived elements, but not in the adult CNS.

2.2. *Spatial Spread*

In experimental cerebral wounds, astrocyte reaction is frequently not limited only to the proximity of the lesion, but extends through the damaged hemisphere and even to the opposite hemisphere (i.e., theoretically unaffected by the traumatism). The degree of homolateral spread and the presence or absence of contralateral involvement seems to depend upon the size of the lesion. Thus, in small cerebral wounds *(18,22,23,25)*, the astroglial reaction is circumscribed to 1 mm of tissue surrounding the lesion, or alternatively extends throughout the ipsilateral cerebral cortex *(18,22,25)* but without reaching the contralateral hemisphere. A similar situation is observed in percussion-induced contusive brain lesions *(33)*. It is interesting to note that the astroglial reaction in the damaged hemisphere is later in developing with a slower and more persistent spread in subcortical regions than in the cortex *(18a,22)*. In contrast, when the amount of damaged tissue is greater *(24),* the reaction likewise extends to the cortex opposite to the side of the lesion, albeit with less intensity. Unpublished observations by our group corroborate this finding. In addition, we have established that VIM is only expressed by the reactive astrocytes immediately adjacent to the punction trajectory, as a result of which this protein clearly cannot be employed for studying more distant glial responses. In nontraumatic brain lesions that nevertheless involve abundant damaged nervous tissue, as in experimental laser-induced injuries *(53),* astroglial reaction in the contralateral hemisphere can also be detected. According to these authors, the expression of VIM by the reactive astrocytes is restricted to the areas in proximity to the lesion.

A phenomenon requiring more detailed analysis is the presence of astrocyte reaction in the undamaged hemisphere. According to Moumdjian et al. *(24),* a number of causes may be postulated:

1. Distant tissue diffusion of soluble factors that induce astrocyte activation and which are initially released at the site of the lesion;
2. Wallerian degeneration necrosis of nerve fibers with cortico-cortical projections that traverse the *corpus callosum,* thereby triggering an astroglial reaction contralateral to the lesion that is in turn incremented by the astrocyte reaction produced by deafferentiation itself;
3. Migration of reactive astrocytes away from the wound site, through the *corpus callosum,* with secondary colonization of the contralateral hemisphere.

In this sense, it has been shown that transplanted astrocytes are able to migrate *(77–82)*—an important migration route being the parallel tracts of myelinated nerve fibers. However, native astrocytes do not appear to possess such mobility *(83).*

In order to evaluate the contribution of each of the mechanisms proposed above in accounting for the spread of astrocyte reaction to the contralateral hemisphere *(24),* a number of surgical procedures have yielded the following results:

1. Callosotomy alone induces mild gliosis in both hemispheres;
2. Unilateral brain lesion only produces severe ipsilateral and moderate contralateral gliosis;
3. Callosotomy with unilateral brain lesion induces effects similar to those of the second procedure above.

According to these authors, if astrocyte migration through the *corpus callosum* were a factor to be taken into account, then the contralateral gliosis seen in rats subjected to lesion plus callosotomy (which thus prevents migration) would have to be less than that recorded in rats with lesion only. However, no differences are observed between the two experimental groups. On the other hand, if Wallerian degeneration of axons projecting to the contralateral hemisphere—with the resulting deafferentiation—were an important consideration in the genesis of contralateral gliosis, then animals belonging to the lesion plus callosotomy group would be expected to develop more contralateral reactive gliosis than the group subjected to lesion only. Not only is this not the case, but callosotomy alone moreover induces less gliosis than the contralateral reaction of the brain lesion alone. As a result, Moumdjian et al. *(24)* clearly favor soluble factors diffusing from the lesion site as the basic mechanism underlying the observed contralateral glial reaction.

An additional factor must be considered in the origin of distant (and even contralateral) astrocyte reactions. In effect, it is well established that astrocytes are interlinked by gap junctions *(84).* These cells therefore form a type of "astroglial syncytium" throughout the nervous parenchyma. In experimental astroglial reactions, the immunohistochemical expression of connexin-43 (a predominant protein in interastrocyte gap junctions) is found to be increased *(85,86)*. This suggests that transformation to reactive astrocytes implies an increase in gap junctions and thus in the efficacy of the "astroglial syncytium." In this way, certain intracellular signals that trigger gliosis could pass from one cell to another—covering considerable distances in a relatively brief period of time. However, the fact that VIM expression by reactive astrocytes never extends beyond the injury zone could be in conflict with the idea of possible intercellular signals that diffuse across gap junctions to play an important role in the genesis of distant astroglial reaction.

3. ASTROCYTE PROLIFERATION

The use of double labeling techniques (GFAP and thymidine or similar compounds such as bromodeoxyuridine, widely employed in recent years) is essential for studying astrocyte division in CNS lesions. By applying these methods to nervous tissue lesions with scant production of necrotic material and an intact blood-brain barrier, most studies have shown that the astrocytes do not divide *(51,87,88)*. In contrast, deafferentiation lesions do seem to show astrocyte proliferation *(89)*.

In experimental traumatic lesions of the CNS *(23,25,90–99)*, astrocyte proliferation is generally an early and very brief phenomenon that only takes place between 24 h and 8 d after lesion induction. There is no evidence of astrocyte proliferation in later stages of the evolutive course, and the maximum mitotic peak is reached 3–4 d after injury. The labeling index is low and rarely exceeds 10%. Similar results have been obtained in other traumatic lesion models *(28,100)*.

Regional differences in astrocyte proliferation have been detected. Thus, Topp et al. *(26)* and Garcia-Estrada et al. *(91)* have found the labeling index to be higher in the cortex than in the hippocampus, whereas Janeczko *(93)* has reported greater proliferation in the white matter and deep-lying regions of the lesion. Although the proliferative phenomena appear to be limited to the damaged hemisphere *(23,25,93)*, other researchers have found a slight spread of astrocyte division towards the contralateral hemisphere *(101,102)*.

Thus, it seems clear that although many nervous tissue lesions exhibit an important number of GFAP-positive reactive astrocytes, the low cell proliferation figures recorded indicate that most such cells are preexisting GFAP-negative astrocytes that have become GFAP-positive.

4. PHAGOCYTIC ROLE OF THE ASTROGLIA

Although the microglia/macrophages are the principal cells in charge of eliminating foreign, nocive, or degenerated elements in nervous tissue injuries, evidence suggests that astrocytes also possess a certain phagocytic capacity.

In vitro research has shown that astrocytes in neonatal or early postnatal rats uptake polystyrene spheres *(79)* within lysosomal structures, where they are retained for several weeks. These cells are also able to engulf yeast cells *(103)* and latex particles *(104)*. Ronnevi *(105,106)* has in turn described astrocyte phagocytosis of synaptic buttons in the neonatal cat.

On the other hand, the phagocytosis of degenerative tissue debris by astrocytes has been described under pathological conditions such as nerve fiber injury and degeneration *(63,107–116)*, toxic lesions *(76,117,118)*, experimental allergic encephalitis or traumatic brain injury *(64)*.

However, ultrastructural studies carried out by our group involving experimental cerebral wounds and Wallerian degeneration of the optic nerve have failed to provide evidence of phagocytosis by astrocytes. In the case of Wallerian degeneration of the optic nerve, we have seen *(69)* an increase in lipid droplets and dense bodies within astrocytes. Although the dense bodies often adopt peculiar morphologies, we have never observed unequivocal images of astrocyte phagocytic activity. We consider that no clear phagocytic role can be attributed to these cells—at least in the experimental setting involved. In contrast, astrocytes may be implicated in the metabolic processing of certain products originating from intense tissue degeneration—a phenomenon that could explain the presence of lipid droplets.

An interesting experimental approach to the problem of a possible phagocytic role for the astroglia involves the *in situ* administration of particulate material—most of which will obviously be engulfed by the microglia/macrophages. A number of authors have applied this experimental design, using different materials. Thus, it has been found that astrocytes in the neonatal rat phagocytose exogenously injected carbon particles *(119),* and that astrocytes in adult rats are able to capture latex particles *(83).*

In a recent study in adult rats subjected to the local injection of colloidal carbon in cerebral wounds, Al-Ali and Al-Hussain *(120)* have observed that astrocytes are able to phagocytose carbon particles. However, this phenomenon was only recorded in astrocytes in brains subjected to two successive injections of colloidal carbon spaced one week apart; no uptake was seen in brains subjected to a single injection. The evident conclusion of this study is that astrocytes only act as phagocytes in the event of "professional" macrophage saturation, i.e., astrocytes would appear to function as a second line of defense.

5. NEWLY-FORMED *GLIA LIMITANS*

In extensive traumatic lesions of the CNS, meningeal cells of fibroblastic appearance (meningocytes) penetrate deeply within the cavity of the lesion and establish close relations with the astroglial cells. The repair process leads to the appearance of a newly-formed *glia limitans,* which closely resembles the normal *glia limitans.* This normal component of the surface of the nervous organs is known to be composed of a series of astrocytic prolongations externally lined by a continuous basal membrane, beyond which the connective elements of the leptomeningeal territory are found *(84).*

Our group has conducted structural studies of the meningeal regeneration process and the peculiar meningo-astroglial relations that are established in the course of the repair of cerebral lesions in rats *(121).* These experiments show that the appearance of a basal lamina limiting the lesion cavity is a crucial event, for it establishes a clear separation between the mesodermal or neuroectodermal elements. According to our results, patches of electron-dense material similar to the basal lamina are already apparent 10 d after inducing the lesion. After 14 d, practically the entire wound cavity is delimited by a typical basal lamina intimately attached to the underlying nervous parenchyma. This basal lamina always rests upon glial filament-rich cell prolongations identifiable as corresponding to astrocytes. Collagen microfibrils—either isolatedly or forming small bundles—as well as macrophages or regenerative meningocytic elements habitually appear in proximity to the astrocytic prolongations—though always covered by the basal lamina. The *glia limitans* thus regenerated is clearly identifiable 14 d after lesion induction. A feature of this newly-formed *glia limitans* is the marked irregularity of its surface *(121);* as a result, the meningeal territory frequently presents long solid glial cords, always totally enveloped by basal lamina, and incompletely covered by thin and delicate meningocytic prolongations (and even somata). Whole astrocyte somata are occasionally seen to form part of the glial cords, and always associated to the corresponding basal lamina cover. This irregularity of the surface of the newly-formed *glia limitans* has been confirmed in different types of lesion *(69,122–125)*. A more recent study by our group *(121a)* has shown the glial cords to coexpress GFAP and VIM in the meningeal territory.

As regards the appearance of the basal lamina during the lesion repair process, our findings basically coincide with the ultrastructural observations of other authors, though with minor chronological differences *(180,127,142,143)*.

Thus, the glial reaction leading to the appearance of a newly-formed *glia limitans* seems to be similar in response to different types of lesion. As a result of their plasticity and repair capacity, the subpial reactive astrocytes are able to extend their prolongations (and even somata) to the subarachnoid space.

Although ultrastructural observations suggest meningeal elements to induce basal lamina formation by astrocytes and the development of newly-formed *glia limitans,* evidence from experimental models also strongly supports this hypothesis. In astrocyte and meningocyte cultures *(144,145),* structures similar to *glia limitans* and even irregular deposits of electron-dense material analogous to the basal lamina are seen to form in the contact zones between both types of cell. Experiments have shown that meningocytes may even be essential for *glia limitans* formation: The destruction of meningocytes with 6-hydroxydopamine (6-OHDA) causes disorganization of the *glia limitans* and covering basal lamina *(146–149)*. Meningocyte destruction with 6-OHDA implies a decrease in the concentration of fibrilar collagen types I, III and VI, together with laminin, fibronectin and type IV collagen in relation to the *glia limitans (135)*. Moreover, steroids topically applied to cerebral wounds, which would cause a decrease in the proliferation of meningeal elements, has been found to induce a lesser organization of the astrocyte prolongations forming the *glia limitans,* with the appearance of laminin-negative zones at this level *(126)*.

Biochemical studies support the idea that the basal lamina is synthesized by astrocytes. These cells have been shown to produce molecules that form part of the basal lamina, including laminin *(127–131),* fibronectin *(131,132)* and proteoglycans *(133,134)*. On the other hand, the meningeal cells are able to produce *(135)* fibrilar collagen types I, III and VI, fibronectin, laminin, type IV collagen, and heparan-sulfate type proteoglycans. Thus, it seems possible that meningeal cells not only regulate astrocyte synthesis of the basal lamina but also contribute in part to production of the latter. In addition, astrocytes would be able to interact with the extracellular matrix by means of adhesion molecules such as the neural cellular adhesion molecule (NCAM) *(136),* or tenascin *(137–139)*—which are likewise expressed by these cells.

6. GROWTH FACTORS AND CYTOKINES

Nonreactive astrocytes possess few growth factors. However, when activated they are known to express multiple factors, cytokines and other substances *(140,141)*. Astrocyte expression of these diffusible compounds varies according to the type of lesion and region involved. Likewise, astrocyte reaction differs between zones close to the lesion and areas located at a distance from the injury site *(141)*.

This section deals with the cytokines and growth factors predominantly implicated in reactive gliosis—fundamentally in traumatic lesions of the CNS.

Three steps may be contemplated in the expression of diffusible factors in post-lesion gliosis, with reference to astrocytes: (a) The origin of the factors that activate astrocytes; (b) The factors expressed by astrocytes in response to injury; and (c) Actions of the diffusible factors secreted by reactive astrocytes upon neighboring structures.

(a) Astrocyte reaction to injury is a response to different soluble activating factors that in turn induce astrocytes to produce growth factors and cytokines. The source of these triggering substances is the nervous tissue affected by the lesion. Injury itself causes the destruction of many cells, including neurons which release factors that are normally found dissolved in the cytoplasm but are not habitually secreted. An example of such behavior is afforded by the basic fibroblast growth factor (bFGF) *(150)* and the ciliary neurotrophic factor (CNTF) *(151)*. These two factors, and particularly bFGF, seem to play a special role in astrocyte activation. In addition to destruction of nervous tissue, injury causes vascular rupture, with the subsequent loss of blood-brain-barrier integrity. Additional diffusible factors and cytokines from the extravasated blood plasma (thrombin, platelet-derived growth factor [PDGF], epidermal growth factor [EGF] and insulin, and so on) are thus added and contribute to activate the astrocytes. The extravasated mononuclear cells, together with the microglia, participate in the inflammatory reaction secondary to injury not only by removing the altered tissue debris but also by releasing cytokines such as interleukins (IL) IL-1 and IL-6, interferon-gamma (IFN-γ), tumor necrosis factor-alpha (TNF-α) and growth factors such as transforming growth factor-beta (TGF-β) *(152–156)* and bFGF *(157)*. All these agents induce the expression of specific reactive astrocyte receptors such as bFGFr *(158)*, CNTFr *(151)*, and EGFr *(159)*. The binding of these factors to their corresponding receptors located on the astrocyte membrane, induces the synthesis of further diffusible factors, and thus stimulates and modulates astrocyte response.

(b) The first sign of astrocyte activation corresponds to the biochemical or *in situ* hybridization detection within the cytoplasm of the mRNA encoding for the different soluble factors. However, astrocyte response does not become effective until the corresponding proteins are synthesized and secreted to the exterior, where they in turn act upon the specific target cells. Once the astrocytes have been activated, a factor synthesis cascade is triggered in sequence over time. In this sense, Cook et al. *(157)* inform of the existence of both early genes (i.e., which express their mRNA in the first 24 h following injury) and late genes (those with mRNA expression after 72 h).

    The early activation of certain genes encoding for soluble factors points to the existence of a rapid astrocyte response, though the first soluble factors to be demonstrated by immunohistochemical techniques appear around three days after the lesion is induced. These factors are bFGF *(102)*, CNTF *(151)*, vascular endothelial growth factor (VEGF) *(104)*, and TGF-β *(160–162)*. Their expression is maintained for prolonged periods of time. Posteriorly, about 5–6 d after injury, the astrocytes express other factors such as insulin-like growth factor-2 (IGF-2) *(161)*, IL-1, IL-6 and TNF-β *(154)*, and nerve growth factor (NGF) *(163,164)*. This latter factor has been detected by Goss et al. *(165)* after 24 h.

(c) The factors synthesized by the reactive astrocytes are secreted to the exterior, and act in paracrine fashion upon the neighboring cells (neurons, oligodendroglia or its precursors, endothelium, meningeal cells in the event that destruction affects the surface of the nervous organ, and macrophages located within the lesion site). They also interact with the secreting astrocytes themselves, in an autocrine manner. The action of the diffusible factors always takes place from the exterior; to this effect, it is essential for the factors to be secreted to the exterior, where they in turn bind to specific receptors located on the target cell surface membrane. The fact that astrocytes secrete numerous factors suggests that they play a special role in controlling the nerve tissue repair process, though response varies according to the region where the lesion is produced. The factors secreted by activated astrocytes act upon many different cells. In this sense, astrocytes may be regarded as the neuralgic center of the repair process, for their activation triggers "explosive" action upon many structures that intervene in the nerve tissue repair process.

These factors exert a protective effect upon the neurons. In this sense, bFGF *(157,166)*, NGF *(167)*, IGF-1 *(168)* and the leukemia inhibitor factor (LIF) *(169)* diminish neuronal death in the zones close to the lesion site. However, reactive astrocytes stimulated by IL-1 may exert a neurotoxic effect by producing nitric oxide *(170)*.

Astrocytes are important target cells of factors secreted by their own activated astrocytic counterparts. These factors exert the following effects:

Astrocyte proliferation is stimulated by bFGF *(171,172)*, EGF *(173,174)* and PDGF *(67)*. Fibronectin, expressed by reactive astrocytes, is a potent stimulator of astrocyte division *(175,176)*. Many factors also inhibit astrocyte proliferation, including IFN-γ *(177–179)*, and TGF-α 1, which inhibits the mitotic effects of FGF and EGF *(180)*. This stimulating and inhibiting behavior of factors produced by the astrocytes themselves supports the idea of their regulatory influence upon gliosis.

Astrocyte hypertrophy, with increases in GFAP and VIM synthesis. These actions are performed by bFGF *(171,181–183)*. However, Reilly et al. *(184)* observed that bFGF produces a decrease in GFAP mRNA in astrocytes in vitro, whereas TGF-α 1 increases it. Logan et al. *(155)* described a TGF-α 1 stimulating effect upon GFAP synthesis. CNTF is also a potent inducer of astrocyte hypertrophy *(67)*, stimulating the synthesis of GFAP and VIM—actions that are in turn enhanced by TNF-γ *(64,101)*. According to Kahn et al. *(185)*, CNTF appears to specifically increase astrocyte GFAP synthesis, though without affecting VIM levels.

According to Ridet et al. *(141)*, these soluble factors induce astrocyte synthesis of surface molecules and extracellular matrix. Astrocyte expression of intercellular adhesion molecules (ICAM-1) is stimulated by different growth factors and cytokines such as IL-1 *(186)*. NGF *(163)* stimulates the synthesis of neural cellular adhesion molecules (NCAM), although interleukin-beta 1 *(186)* and IFN-γ *(140)* stimulate ICAM production. Furthermore, NCAM and components of the extracellular matrix such as fibronectin, tenascin or heparan sulfate favor axon growth of the neurons damaged by the wound. However, when astrocytes synthesize other proteoglycans such as keratan sulfate or chondroitin sulfate, such action is effectively inhibited *(187,188)*. Other proteoglycan actions have been reported, including the ability of heparan sulfate to bind to FGF type 4 receptors and stimulate the cell in the absence of growth factor.

Activated astrocyte factors induce astrocyte production of other growth factors and of an increased number of receptors for their own secretory products thereby incrementing reactive astrocyte response to certain trophic factors and cytokines. bFGF induces NGF expression *(150,157)* and increases the number of bFGF receptors present both in astrocytes and in other cells *(157,158,172)*. Although EGF is not synthesized by reactive astrocytes, the latter do exhibit receptors for this factor *(159)*; as a result, these cells become susceptible to EGF action. In vitro studies have shown EGF to stimulate astrocyte production of bFGF, TGF-β 1 and NGF *(140)*. In turn, TGF-β 1 also induces NGF synthesis in astrocytes *(163,189)*. Interleukin-1 is synthesized by reactive astrocytes *(154)*, and it has been shown to stimulate astrocyte synthesis of NGF *(190,191)*. IGF-1 also stimulates the production of its own receptors on hypertrophic astrocytes *(102)*.

Many of the growth factors synthesized by reactive astrocytes induce these same cells to produce NGF—one of the main functions of which is the induction of a trophic effect upon the damaged neurons, thereby contributing to their survival. In addition, NGF stimulates the synthesis of its own receptors by the same reactive astrocytes *(192)*.

Not all growth factors and cytokines exert a stimulating effect upon astrocytes. Certain diffusible substances synthesized by inflammatory cells and by the reactive astrocytes themselves tend to buffer glial reaction. In this sense, IFN-γ inhibits astrocyte proliferation and the synthesis of extracellular matrix molecules *(191)*. In turn, interleukin-10 (produced by microglia/macrophages) buffers astrocyte reaction *(193)*, with a decrease in bFGF mRNA levels within the astrocytes *(194)*.

Certain growth factors secreted by reactive astrocytes stimulate the proliferation of oligodendroglia precursors in adults *(102,157,162)*. In contrast, Amur-Umarjee et al. *(195)*. have observed an inhibitory effect of astrocytes upon oligodendrocyte remyelinization capacity in vitro.

Other growth factors in turn induce vascular proliferation and restoration of the blood-brain-barrier. In this context, bFGF appears to intervene in vessel neoformation *(182)*. VEGF (synthesized by astrocytes and inflammatory cells) seems to stimulate vascular formation following injury *(196)*, while TGF-α 1 would favor reconstitution of the blood-brain barrier *(155)*.

When the lesion affects the surface of the nervous organ, these factors induce neoformation of the *glia limitans*. Logan et al. *(155)* detected TGF-α 1 in astrocytes, and TGF-α 1 mRNA in meningeal cells.

Astrocytes release factors that stimulate microglial proliferation; these cells concentrate in the lesion site, and in turn increment astrocyte reactivity. This is the case of TGF-α 1 *(197)* and bFGF *(198)*. In this way a mutual stimulation circuit is established based on the positive feedback principle, between astrocytes and microglia/blood mononuclear cells, thus favoring and reinforcing astrocyte activity in gliosis.

Other elements have also been implicated in the regulation of glial response, including norepinephrine via beta-adrenergic receptor action *(91,199)*, and corticoids *(192)*.

To summarize, the reactive astrocyte plays a prominent role in the regulation and modulation of the scar tissue resulting from damage of the nervous system. This action is largely, though not entirely, mediated by the cascade activation of different growth factors and cytokines, which in turn interact with each other upon the surrounding cells and structures.

7. GLIAL REACTION IN RELATION TO AGE

Most authors consider that astrocyte reaction to lesions of the CNS is less important during the fetal and neonatal period than in adults. In general, immature astrocyte response to injury is similar though less intense than mature astrocyte reaction in the adult *(70,152,193,200–202)*. An important difference in astrocyte response to lesion between the fetal and adult stages is the absence of glial scar formation following CNS injury in the neonatal period *(18,21,203,204)*. However, according to some authors, the response is more dependent upon the type and/or location of the lesion than on astrocyte immaturity. In this sense, glial scars have been described following neonatal injury *(70,152,193,200–203)*.

According to most researchers, the absence of glial reaction in neonatal lesions is a consequence of astrocyte immaturity *(205,206)*. However, Smith et al. *(207,208)*, are of the opinion that the absence of a glial scar is due to some specific property of the immature astrocytes, and is not a consequence of any cellular incapacity. According to Berry et al. *(18a)*, the absence of a glial reaction could be attributed to the microglia/macrophages rather than to the astrocytes as such. In neonatal animal models, debris elimination is faster and more effective thereby avoiding persistent macrophage presence within the lesion site *(152,193,202,206)*.

A critical period seems to exist in the early neonatal stages, after which the capacity to develop an "adult"-like glial response to CNS injury is acquired. A number of authors have pointed out the coincidence in time between the critical period in which the "adult" response to lesions is acquired and the myelinization phase *(168,207,208)*. This coincidence has led to the idea that the presence of myelin and/or oligodendrocyte degradation products determines the formation of a glial scar. However, as has been demonstrated by Berry et al. *(18a)*, lesions are followed by normal healing in myelin-deficient mutants.

The existence of changes in the astrocyte population (particularly hypertrophy, with increased GFAP expression) associated to advanced age is generally accepted. However, unlike in the neonatal period, very few studies have been carried out in individuals of advanced age, and the results obtained are moreover contradictory. According to some authors, astrocyte response in old animals is more intense, extensive, and prolonged in time *(209)*, and proliferation is both greater and earlier *(26)*. In contrast, Kane et al. *(46)* have observed no increase in astrocyte hypertrophic response in such individuals. In this sense, the conflicting results obtained may be partly attributed to the experimental methodology employed.

REFERENCES

1. Wilson, R.B. (1926) Brain repair. *Arch. Neurol. Psychiat.* **15,** 75–84.
2. Penfield, W. (1927) The mechanism of cicatricial contraction in the brain. *Brain* **50,** 499–517.
3. Penfield, W. and Buckley, R.C. (1928) Punctures of the brain: the factors concerned in gliosis and in cicatricial contraction. *Arch. Neurol. Psychiat.* **20,** 1–13.
4. Río-Hortega, P. and Penfield, W. (1927) Cerebral cicatrix. The reaction of neuroglia and microglia to brain wounds. *Bull. John Hopkins Hosp.* **41,** 278–303.
5. Bignami, A., Dahl, D., and Rueger, D.C. Glial fibrillary acidic (GFA) protein in normal neural cells and in pathological conditions. In S. Fedoroff and L. Hertz (Eds.), Advances in Cellular Neurobiology, Vol. 1. Academic Press, New York, 1980, pp. 285–310.
6. Bignami, A., Raju, R., and Dahl, D. (1982) Localization of vimentin, the nonspecific intermediate filament protein in embryonal glia and in early differentiating neurons. *Dev. Biol.* **91,** 286–295.
7. Schnitzer, J., Franke, W., and Schachner, M. (1981) Immunocytochemical demonstration of vimentin in astrocytes and ependymal cells of developing and adult mouse nervous system. *J. Cell Biol.* **90,** 435–447.
8. Ghooray, G.T. and Martin, G.F. (1993a) Development of radial glia and astrocytes in the spinal cord of the North American opossum (Didelphis virginiana): an immunohistochemical study using anti-vimentin and anti-glial fibrillary acidic protein. *Glia* **9,** 1–9.
9. Oudega, M. and Marani, E. (1991) Expression of vimentin and glial fibrillary acidic protein in the developing rat spinal cord: an immunocytochemical study of the spinal cord glial system. *J. Anat.* **179,** 97–114.
10. Schiffer, D., Giordana, M.T., Cavalla, P., Vigliani, M.C., and Attanasio, A. (1993) Immunohistochemistry of glial reaction after injury in the rat: double stainings and markers of cell proliferation. *Int. J. Dev. Neurosci.* **11,** 269–280.
11. Voigt, T. (1989) Development of glial cells in the cerebral wall of ferrets: direct tracing of their transformation from radial glia into astrocytes. *J. Comp. Neurol.* **289,** 74–88.

12. Kost, S.A., Chacko, K., and Oblinger, M.M. (1992) Developmental patterns of intermediate filament gene expression in the normal hamster brain. *Brain Res.* **95,** 270–280.
13. Duffy, P.E. (1983) Astrocytes: normal, reactive, and neoplastic. Raven Press, New York.
14. Malhotra, S.K., Shnitka, T.K., and Elbrink, J. (1990) Reactive astrocytes. A review. *Cytobios* **61,** 133–160.
15. Calvo, J.L., Carbonell, A.L., and Boya, J. (1990) Coexpression of vimentin and glial fibrillary acidic protein in astrocytes of the adult rat optic nerve. *Brain Res.* **532,** 355–357.
16. Pixley, S.K.R. and De Vellis, J. (1984) Transition between radial glia and mature astrocytes studied with a monoclonal antibody to vimentin. *Dev. Brain Res.* **15,** 201–209.
17. Shaw, G., Osborn, M., and Weber, K. (1981) An immunofluorescence microscopical study of the neurofilament triplets proteins, vimentin and glial fibrillary acidic protein within the adult rat brain. *Eur. J. Cell Biol.* **26,** 68–82.
18. Abd-EI-Basset, E.M., Ahmed, I., Kalnins, V.I., and Fedoroff, S. (1992) Immuno-electron microscopical localization of vimentin and glial fibrillary acidic protein in mouse astrocytes and their precursor cells in culture. *Glia* **6,** 149–153.

18a. Berry, M., Maxwell, W.L., Logan, A., Mathewson, A., McConnell, P., Ashhurst, D.E., and Thomas, G.H. (1983) Deposition of scar tissue in the central nervous system. *Acta Neurochir.* **Suppl. 32,** 31–53.

19. Calvo, J.L., Carbonell, A.L., and Boya, J. (1991) Co-expression of glial fibrillary acidic protein and vimentin in reactive astrocytes following brain injury in rats. *Brain Res.* **566,** 333–336.
20. Hozumi, I., Chiu, F.-C., and Norton, W.T. (1990b) Biochemical and immunocytochemical changes in glial fibrillary acidic protein after stab wounds. *Brain Res.* **524,** 64–71.
21. Ludwin, S.K. (1985) Reaction of oligodendrocytes and astrocytes to trauma and implantation. A combined autoradiographic and immunohistochemical study. *Lab. Invest.* **52,** 20–30.
22. Mathewson, A.J. and Berry, M. (1985) Observations on the astrocyte response to a cerebral stab wound in adult rats. *Brain Res.* **327,** 61–69.
23. Miyake, T., Hattori, T., Fukuda, M., Kitamura, T., and Fujita, S. (1988) Quantitative studies on proliferative changes of reactive astrocytes in mouse cerebral cortex. *Brain Res.* **451,** 133–138.
24. Moumdjian, R.A., Antel, J.P., and Wee Yong, V. (1991) Origin of contralateral reactive gliosis in surgically injured rat cerebral cortex. *Brain Res.* **547,** 223–228.
25. Takamiya, Y., Kohsaka, S., Toya, S., Otani, M., and Tsukada, Y. (1988) Immunohistochemical studies on the proliferation of reactive astrocytes and the expression of cytoskeletal proteins following brain injury in rats. *Dev. Brain Res.* **38,** 201–210.
26. Topp, K.S., Faddis, B.T., and Vijayan, V.K. (1989) Trauma-induced proliferation of astrocytes in the brains of young and aged rats. *Glia* **2,** 201–211.
27. Yang, H.I., Lieska, N., Kriho, V., Wu, C.M., and Pappas, G.D. (1997) A subpopulation of reactive astrocytes at the immediate site of cerebral cortical injury. *Exp. Neurol.* **146,** 199–205.
28. Baldwin, S.A. and Scheff, S.W. (1996) Intermediate filament change in astrocytes following mild cortical contusion. *Glia* **16,** 266–275.
29. Farooque, M., Badonic, T., Olsson, Y., and Holtz, A. (1995) Astrocytic reaction after graded spinal cord compression in rats: immunohistochemical studies on glial fibrillary acidic protein and vimentin. *J. Neurotrauma* **12,** 41–52.
30. Hadley, S.D. and Goshgarian, H.G. (1997) Altered immunoreactivity for glial fibrillary acidic protein in astrocytes within 1 h after cervical spinal cord injury. *Exp. Neurol.* **146,** 380–387.
31. Hill, S.J., Barbarese, E., and Mclntosh, T.K. (1996) Regional heterogeneity in the response of astrocytes following traumatic brain injury in the adult rat. *J. Neuropathol. Exp. Neurol.* **55,** 1221–1229.
32. Hinkle, D.A., Baldwin, S.A., Scheff, S.W., and Wise, P.M. (1997) GFAP and S100beta expression in the cortex and hippocampus in response to mild cortical contusion. *J. Neurotrauma* **14,** 729–738.
33. Okimura, Y., Tanno, H., Fukuda, K., Ohga, M., Nakamura, M., Aihara, N., and Yamaura, A. (1996) Reactive astrocytes in acute stage after experimental brain injury: relationship to extravasated plasma protein and expression of heat shock protein. *J. Neurotrauma* **13,** 385–393.

34. Theriault, E., Frankenstein, U.N., Hertzberg, E.L., and Nagy, J.I. (1997) Connexin 43 and astrocytic gap junctions in the rat spinal cord after acute compression injury. *J. Comp. Neurol.* **382,** 199–214.
35. Anders, J.J. and Johnson, J.A. (1990) Transection of the rat olfactory nerve increases glial fibrillary acidic protein immunoreactivity from the olfactory bulb to the piriform cortex. *Glia* **3,** 17–25.
36. Anezaki, T., Yanagisawa, K., Takahashi, H., Nakajima, T., Miyashita, K., Ishikawa, A., Ikuta, F., and Miyatake, T. (1992) Remote astrocytic response of prefrontal cortex is caused by the lesions in the nucleus basalis of Meynert, but not in the ventral tegmental area. *Brain Res.* **574,** 63–69.
37. Barron, K.D., Marciano, F.F., Amundson, R., and Mankes, R. (1990) Perineuronal glial responses after axotomy of central and peripheral axons. A comparison. *Brain Res.* **523,** 219–229.
38. Cancilla, P.A., Bready, J., Berliner, J., Sharifi-nia, H., Toga, A.W., Santori, E.M., Scully, S., and De Vellis, J. (1992) Expression of mRNA for glial fibrillary acidic protein after experimental cerebral injury. *J. Neuropathol. Exp. Neurol.* **51,** 560–565.
39. Cheng, H.W., Jiang, T., Brown, S.A., Pasinetti, G.M., Finch, C.E., and McNeill, T.H. (1994) Response of striatal astrocytes to neuronal deafferentation: an immunocytochemical and ultrastructural study. *Neuroscience* **62,** 425–439.
40. Dusart, I., Marty, S., and Peschanski, M. (1991) Glial changes following an excitotoxic lesion in the CNS. II. *Astrocytes. Neuroscience* **45,** 541–549.
41. Garrison, C.J., Dougherty, P.M., Kajander, K.C., and Carlton, S.M. (1991) Staining of glial fibrillary acidic protein (GFAP) in lumbar spinal cord increases following a sciatic nerve constriction injury. *Brain Res.* **565,** 1–7.
42. Gilmore, S.A., Sims, T.J. and Leiting, J.E. (1990) Astrocytic reactions in spinal gray matter following sciatic axotomy. *Glia* **3,** 342–349.
43. Herrera, D.G. and Cuello, A.C. (1992) Glial fibrillary acidic protein immunoreactivity following cortical devascularizing lesion. *Neuroscience* **49,** 781–791.
44. Jorgensen, M.B., Finsen, B.R., Jensen, M.B., Castellano, B., Diemer, N.H., and Zimmer, J. (1993) Microglial and astroglial reactions to ischemic and kainic acid-induced lesions of the adult rat hippocampus. *Exp. Neurol.* **120,** 70–88.
45. Kálmán, M., Csillag, A., Schleicher, A., Rind, C., Hajós, F., and Zilles, K. (1993) Long-term effects of anterograde degeneration on astroglial reaction in the rat geniculo-cortical system as revealed by computerized image analysis. *Anat. Embryol.* **187,** 1–7.
46. Kane, C.J.M., Sims, T.J., and Gilmore, S.A. (1997) Astrocytes in the aged rat spinal cord fail to increase GFAP mRNA following sciatic nerve axotomy. *Brain Res.* **759,** 163–165.
47. Kumar, K., Wu, X., and Evans, A.T. (1996) GFAP-immunoreactivity following hypothermic forebrain ischemia. *Metab. Brain. Dis.* **12,** 21–27.
48. Li, Y., Chopp, M., Zhang, Z.G., and Zhang, R.L. (1995) Expression of glial fibrillary acidic protein in areas of focal cerebral ischemia accompanies neuronal expression of 72-kDa heat shock protein. *J. Neurol. Sci.* **128,** 134–142.
49. Lin, R.C.S., Polsky, K., and Matesic, D.F. (1993) Expression of γ-aminobutyric acid immunoreactivity in reactive astrocytes after ischemia-induced injury in the adult forebrain. *Brain Res.* **600,** 1–8.
50. Mikucki, S.A. and Oblinger, M.M. (1991) Vimentin mRNA expression increases after corticospinal axotomy in the adult hanster. *Metabol. Brain Dis.* **6,** 33–49.
51. Mitchell, J., Sundstrom, L.E., and Wheal, H.V. (1993) Microglial and astrocytic cell responses in the rat hippocampus after an intracerebroventricular kainic acid injection. *Exp. Neurol.* **121,** 224–230.
52. Petito, C.K. and Halaby, I.A. (1993) Relationship between ischemia and ischemic neuronal necrosis to astrocyte expression of glial fibrillary acidic protein. *Int. J. Dev. Neurosci.* **11,** 239–247.
53. Schiffer, D., Giordana, M.T., Migheli, A., Giaccone, G., Pezzotta, S., and Mauro, A. (1986) Glial fibrillary acidic protein and vimentin in the experimental glial reaction of the rat brain. *Brain Res.* **374,** 110–118.

54. Schmidt-Kastner, R., Szymas, J., and Hossmann, K.-A. (1990) Immunohistochemical study of glial reaction and serum-protein extravasation in relation to neuronal damage in rat hippocampus after ischemia. *Neuroscience* **38,** 527–540.
55. Steward, O., Kelley, M.S., and Torre, E.R. (1993) The process of reinnervation in the dentate gyrus of adult rats: temporal relationship between changes in the levels of glial fibrillary acidic protein (GFAP) and GFAP mRNA in reactive astrocytes. *Exp. Neurol.* **124,** 167–183.
56. Stichel, C.C. and Müller, H.-W. (1994) Extensive and long-lasting changes of glial cells following transection of the postcommisural fornix in the adult rat. *Glia* **10,** 89–100.
57. Streit, W.J. and Kreutzberg, G.W. (1988) Response of endogenous glial cells to motor neuron degeneration induced by toxic ricin. *J. Comp. Neurol.* **268,** 248–263.
58. Tanaka, H., Araki, M., and Masuzawa, T. (1992) Reaction of astrocytes in the gerbil hippocampus following transient ischemia: immunohistochemical observations with antibodies against glial fibrillary acidic protein, glutamine synthetase and S-100 protein. *Exp. Neurol.* **116,** 264–274.
59. Tetzlaff, W., Graeber, M.B., Bisby, M.A., and Kreutzberg, G.W. (1988) Increased glial fibrillary acidic protein synthesis in astrocytes during retrograde reaction of the rat facial nucleus. *Glia* **1,** 90–95.
60. Kindy, M.S., Bhat, A.N., and Bhat, N.R. (1992) Transient ischemia stimulates glial fibrillary acidic protein and vimentin gene expression in the gerbil neocortex, striatum and hippocampus. *Molec. Brain Res.* **13,** 199–206.
61. O'Callaghan, J.P., Miller, D.B., and Reinhard, J.F. (1990) Characterization of the origins of astrocyte response to injury using the dopaminergic neurotoxicant, 1-methyl-4-phenyl-1,2,3,6-tetrahydropyridine. *Brain Res.* **521,** 73–80.
62. Torre, E.R., Lothman, E., and Steward, O. (1993) Glial response to neuronal activity: GFAP-mRNA and protein levels are transiently increased in the hippocampus after seizures. *Brain Res.* **631,** 256–264.
63. Fulcrand, J. and Privat, A. (1977) Neuroglial reactions secondary to Wallerian degeneration in the optic nerve of the postnatal rat: ultrastructural and quantitative study. *J. Comp. Neurol.* **176,** 189–224.
64. Castejon, O.J. (1998) Electron microscopic analysis of cortical biopsies in patients with traumatic brain injuries and dysfunction of neurobehavioural system. *J. Submicrosc. Cytol. Pathol.* **30,** 145–156.
65. Steinert, P.M. and Liem, R.K.H. (1990) Intermediate filament dynamics. *Cell* **60,** 521–523.
66. Nakamura, Y., Takeda, M., Angelides, K.J., Tada, K., Hariguchi, S., and Nishimura, T. (1991) Assembly, disassembly, and exchange of glial fibrillary acidic protein. *Glia* **4,** 101–110.
67. Hudgins, S.N. and Levison, S.W. (1998) Ciliary neurotrophic factor stimulates astroglial hypertrophy in vivo and in vitro. *Exp. Neurol.* **150,** 171–182.
68. Hozumi, I., Aquino, D.A., and Norton, W.T. (1990a) GFAP mRNA levels following stab wounds in rat brain. *Brain Res.* **534,** 291–294.
69. Carbonell, A.L., Boya, J., Calvo, J.L., and Marin, J.F. (1991) Ultrastructural study of the neuroglial and macrophagic reaction in Wallerian degeneration of the adult rat optic nerve. *Histol. Histopathol.* **6,** 443–451.
70. Trimmer, P.A. and Wunderlich, R.E. (1990) Changes in astroglial scar formation in rat optic nerve as a function of development. *J. Comp. Neurol.* **296,** 359–378.
71. Isacson, O., Fischer, W., Wictorin, K., Dawbarn, D., and Björklund, A. (1987) Astroglial response in the excitotoxically lesioned neostriatum and its projection areas in the rat. *Neuroscience* **20,** 1043–1056.
72. Lendahl, U., Zimmerman, L., and McKay, R. (1990) CNS stem cells express a new class of intermediate filament protein. *Cell* **60,** 584–595.
73. Duggal, N., Schmidt-Kastner, R., and Hakim, A.M. (1997) Nestin expression in reactive astrocytes following focal cerebral ischemia in rats. *Brain Res.* **768,** 1–9.

74. Frisen, J., Johansson, C.B., Torok, C., Risling, M., and Lendahl, U. (1995) Rapid, widespread, and longlasting induction of nestin contributes to the generation of glial scar tissue after CNS injury. *J. Cell Biol.* **131,** 453–464.
75. Holmin, S., Almqvist, P., Lendahl, U., and Mathiesen, T. (1997a) Adult nestin-expressing subependymal cells differentiate to astrocytes in response to brain injury. *Eur. J. Neurosci.* **9,** 65–75.
76. Murabe, Y., Ibata, Y., and Sano, Y. (1981a) Morphological studies on neuroglia. II.-Response of glial cells to kainic acid-induced lesions. *Cell Tissue Res.* **216,** 569–580.
77. Andersson, C., Tytell, M., and Brunso-Bechtold, J. (1993) Transplantation of cultured type 1 astrocyte cell suspensions into young, adult and aged rat cortex: cell migration and survival. Int. *J. Dev. Neurosci.* **11,** 555–568.
78. Booss, J., Solly, K.S., Collins, P.V., and Jacque, C. (1991) Migration of xenogenic astrocytes in myelinated tracts: a nobel probe for immune responses in white matter. *Acta Neuropathol.* **82,** 172–177.
79. Emmett, C.J., Lawrence, J.M., and Seeley, P.J. (1988) Visualization of migration of transplanted astrocytes using polystyrene microspheres. *Brain Res.* **447,** 223–233.
80. Emmett, C.J., Lawrence, J.M., Raisman, G., and Seeley, P.J. (1991) Cultured epithelioid astrocytes migrate after transplantation into the adult rat brain. *J. Comp. Neurol.* **310,** 330–341.
81. Goldberg, W.J. and Bemstein, J.J. (1988) Fetal cortical astrocytes migrate from cortical homografts throughout the
82. Hatton, J.D., Nguyen, M.H., and Sang, U.H. (1993b) Differential migration of astrocytes grafted into the developing rat brain. *Glia* **9,** 113–119.
83. Hatton, J.D., Finkelstein, J.P., and Sang, U.H. (1993a) Native astrocytes do not migrate de novo or after local trauma. *Glia* **9,** 18–24.
84. Peters, A., Palay, S.L., and Webster, H.F. (1991) The fine structure of the nervous system. Neurons and their supporting cells. Oxford University Press, New York.
85. Aldskogius, H. and Kozlova, E.N. (1998) Central neuron-glial and glial-glial interactions following axon injury. *Prog. Neurobiol.* **55,** 1–26.
86. Rohlmann, A., Laskawi, R., Hofer, A., Dermietzel, R., and Wolff, J.R. (1994) Astrocytes as rapid sensors of peripheral axotomy in the facial nucleus of rats. *Neuroreport* **5,** 409–412.
87. Matsumoto, Y., Ohmori, K., and Fujiwara, M. (1992) Microglial and astroglial reactions to inflammatory lesions of experimental autoimmune encephalomyelitis in the rat central nervous system. *J. Neuroimmunol.* **37,** 23–33.
88. Svensson, M., Mattsson, P., and Aldskogius, H. (1994) A bromodeoxyuridine labelling study of proliferating cells in the brainstem following hipoglossal nerve transection. *J. Anat.* **185,** 537–542.
89. Lurie, D.I. and Rubel, E.W. (1994) Astrocyte proliferation in the chick auditory brainstem following cochlea removal. *J. Comp. Neurol.* **346,** 276–288.
90. Amat, J.A., Ishiguro, H., Nakamura, K., and Norton, W.T. (1996) Phenotypic diversity and kinetics of proliferating microglia and astrocytes following cortical stab wounds. *Glia* **16,** 368–382.
91. García-Estrada, J., Del Río, J.A., Luquin, S., Soriano, E., and García-Segura, L.M. (1993) Gonadal hormones down-regulate reactive gliosis and astrocyte proliferation after a penetrating brain injury. *Brain Res.* **628,** 271–278.
92. Giordana, M.T., Attanasio, A., Cavalla, P., Migheli, A., Vigliani, M.C., and Schiffer, D. (1994) Reactive cell proliferation and microglia following injury to the rat brain. *Neuropathol. Appl. Neurobiol.* **20,** 163–174.
93. Janeczko, K. (1989) Spatiotemporal patterns of the astroglial proliferation in rat brain injured at the postmitotic stage of postnatal development: a combined immunocytochemical and autoradiographic study. *Brain Res.* **485,** 236–243.
94. Janeczko, K. (1991) The proliferative response of S-100 protein-positive glial cells to injury in the neonatal rat brain. *Brain Res.* **564,** 86–90.

95. Janeczko, K. (1992) A comparison of the proliferative activity of astrocytes and astrocyte-like cells expressing vimentin in the injured mouse cerebral hemisphere. *Folia Histochem. Cytobiol.* **30,** 27–34.
96. Janeczko, K. (1993) Co-expression of GFAP and vimentin in astrocytes proliferating in response to injury in the mouse cerebral hemisphere. A combined autoradiographic and double immunocytochemical study. *Int. J. Dev. Neurosci.* **11,** 139–147.
97. Latov, N., Nilaver, G., Zimmerman, E.A., Johnson, W.G., Silverman, A.-J., Defendini, R., and Cote, L. (1979) Fibrillary astrocytes proliferate in response to brain injury. A study combining immunoperoxidase technique for glial fibrillary acidic protein and radioautography of tritiated thymidine. *Dev. Biol.* **72,** 381–384.
98. Miyake, T., Hattori, T., Fukuda, M., and Kitamura, T. (1989) Reactions of S-100-positive glia after injury of mouse cerebral cortex. *Brain Res.* **489,** 31–40.
99. Miyake, T., Okada, M., and Kitamura, T. (1992) Reactive proliferation of astrocytes studied by immunohistochemistry for proliferating cell nuclear antigen. *Brain Res.* **590,** 300–302.
100. Du Bois, M., Bowman, P.D., and Goldstein, G.W. (1985) Cell proliferation after ischemic injury in gerbil brain. An immunocytochemical and autoradiographic study. *Cell Tissue Res.* **242,** 17–23.
101. Hodges-Savola, C., Rogers, S.A., Ghilardi, J.R., Timm, D.R., and Manthy, P.W. (1996) Beta-adrenergic receptors regulate astrogliosis and cell proliferation in the central nervous system in vivo. *Glia* **17,** 52–62.
102. Liu, H.M. and Chen, H.H. (1994) Correlation between fibroblast growth factor expression and cell proliferation in experimental brain infarct studied with proliferating cell nuclear antigen immunohistochemistry. *J. Neuropathol. Exp. Neurol.* **53,** 118–126.
103. Iacono, R.F., Berría, M.I., and Lascano, E.F. (1991) A triple staining procedure to evaluate phagocytic role of differentiated astrocytes. *J. Neurosci. Meth.* **39,** 225–230.
104. Roldan, A., Gogg, S., Ferrini, M., Schillaci, R., and De Nicola, A.F. (1997) Glucocorticoid regulation of in vitro astrocyte phagocytosis. *Biocell.* **21,** 83–89.
105. Ronnevi, L.O. (1977) Spontaneous phagocytosis of boutons on spinal motoneurons during early postnatal development. An electron microscopical study in the cat. *J. Neurocytol.* **6,** 487–504.
106. Ronnevi, L.O. (1978) Origin of the glial processes responsible for the spontaneous postnatal phagocytosis of boutons on cat spinal motoneurons. *Cell Tissue Res.* **189,** 203–217.
107. Bechmann, I. and Nitsch, R. (1997) Astrocytes and microglial cells incorporate degenerating fibers following entorhinal lesion: a light, confocal, and electron microscopical study using a phagocytosis-dependent labeling technique. *Glia* **20,** 145–154.
108. Blanco, R.E. and Orkand, P.M. (1996) Astrocytes and regenerating axons at the proximal stump of the severed frog optic nerve. *Cell Tissue Res.* **286,** 337–345.
109. Frank, M. and Wolburg, H. (1996) Cellular reactions at the lesion site after crushing of the rat optic nerve. *Glia* **16,** 227–240.
110. Lassmann, H., Ammerer, H.P., and Kulnig, W. (1978) Ultrastructural sequence of myelin degradation. I.-Wallerian degeneration in the rat optic nerve. *Acta Neuropathol.* **44,** 91–102.
111. Nathaniel, E.J.H. and Nathaniel, D.R. (1981) The reactive astrocyte. In Fedoroff, S. & Hertz, L. (Eds.), Advances in Cellular Neurobiology, Vol. 2. Academic Press, New York, pp. 249–301.
112. Shen, C.-L. and Liu, K.-M. (1984) Neuroglia of the adult rat optic nerve in the course of Wallerian degeneration. *Proc. Natl. Sci. Council (Taiwan). Part B: Life Sci.* **8,** 324–334.
113. Skoff, R.P. (1975) The fine structure of pulse labeled (^{3}H-Thymidine cells) in degenerating rat optic nerve. *J. Comp. Neurol.* **161,** 595–612.
114. Vaughn, J.E. and Pease, D.C. (1970) Electron microscopic studies on Wallerian degeneration in rat optic nerves. II.-Astrocytes, oligodendrocytes and adventitial cells. *J. Comp. Neurol.* **140,** 207–226.
115. Vaughn, J.E. and Skoff, R.P. Neuroglia in experimentally altered central nervous system. In Bourne, G.H. (Ed.), The structure and function of nervous tissue, Vol.5. Academic Press, New York, 1972. pp. 39–72.

116. Wender, M., Kozik, M., Goncerzewicz, A., and Mularek, O. (1980) Neuroglia of the developing optic nerve in the course of Wallerian degeneration. *J. Hirnforsch.* **21,** 417–428.
117. Lemkey-Johnston, N., Butler, V., and Reynolds, W.A. (1976) Glial changes in the progress of a chemical lesion. An electron microscopic study. *J. Comp. Neurol.* **167,** 481–502.
118. Murabe, Y., Ibata, Y., and Sano, Y. (1981b) Morphological studies on neuroglia. III.-Macrophage response and "microgliocytosis" in kainic-acid induced lesions. *Cell Tissue Res.* **218,** 75–86.
119. Al-Ali, S.Y., Al-Zuhair, A.G.H., and Dawod, B. (1988) Ultrastructural study of phagocytic activities of young astrocytes in injured neonatal rat brain following intracerebral injection of colloidal carbon. *Glia* **1,** 211–218.
120. Al-Ali, S.Y. and Al-Hussain, S.M. (1996) An ultrastructural study of the phagocytic activity of astrocytes in adult rat brain. *J. Anat.* **188,** 257–262.
121. Carbonell, A.L. and Boya, J. (1988) Ultrastructural study on meningeal regeneration and meningo-glial relationships after cerebral stab wound in the adult rat. *Brain Res.* **439,** 337–344.
121a. Carbonell, A.L., Boya, J., and Calvo, J.L. (1994) Coexpression of glial fibrillary acidic protein and vimentin in astrocytes within the newly formed subarachnoid space after an experimental brain wound. *Biomed. Res.* **5,** 9–13.
122. Moore, G.R.W. and Raine, C.S. (1986) Leptomeningeal and adventitial gliosis as a consequence of chronic inflammation. *Neuropathol. Appl. Neurobiol.* **12,** 371–378.
123. Nathaniel, E.J.H. and Nathaniel, D.R. (1977) Astroglial response to degeneration of dorsal root fibres in adult rat spinal cord. *Exp. Neurol.* **54,** 60–76.
124. Ramsey, H.J. (1965) Fine structure of the surface of the cerebral cortex of human brain. *J. Cell Biol.* **26,** 323–333.
125. Sims, T.J. and Gilmore, S.A. (1992) Glial response to dorsal root lesion in the irradiated spinal cord. *Glia* **6,** 96–107.
126. Li, M.S. and David, S. (1996) Topical glucocorticoids modulate the lesion interface after cerebral cortical stab wounds in adult rats. *Glia* **18,** 306–318.
127. Bernstein, J.J., Getz, R., Jefferson, M., and Kelemen, M. (1985) Astrocytes secrete basal lamina after hemisection of rat spinal cord. *Brain Res.* **327,** 135–141.
128. Chiu, A.Y., Espinosa de los Monteros, A., Cole, R.A., Loera, S., and De Vellis, J. (1991) Laminin and s-laminin are produced and released by astrocytes, Schwann cells, and Schwannomas in culture. *Glia* **4,** 11–24.
129. Liesi, P. (1985) Laminin-immunoreactive glia distinguish regenerative adult CNS systems from non-regenerative ones. *EMBO J.* **4,** 2505–2511.
130. Liesi, P., Dahl, D., and Vaheri, A. (1983) Laminin is produced by early rat astrocytes in primary culture. *J. Cell Biol.* **96,** 920–924.
131. Liesi, P., Kirkwood, T., and Vaheri, A. (1986) Fibronectin is expressed by astrocytes cultured from embryonic and early postnatal rat brain. *Exp. Cell Res.* **163,** 175–185.
132. Price, J. and Hynes, D.D. (1985) Astrocytes in culture synthesize and secrete a variant form of fibronectin. *J. Neurosci.* **5,** 2205–2211.
133. Ard, M.D. and Bunge, R.P. (1988) Heparan sulphate proteoglycan and laminin immunoreactivity on cultured astrocytes: relationship to differentiation and neurite growth. *J. Neurosci.* **8,** 2844–2858.
134. Johnson-Green, P.C., Dow, K.E., and Riopelle, R.J. (1991) Characterization of glycosaminoglycans produced by primary astrocytes in vitro. *Glia* **4,** 314–321.
135. Sievers, J., Pehlemann, F.W., Gude, S., and Berry, M. (1994) Meningeal cells organize the superficial glia limitans of the cerebellum and produce components of both the interstitial matrix and the basement membrane. *J. Neurocytol.* **23,** 135–149.
136. Le Gal La Salle, G., Rougon, G., and Valin, A. (1992) The embryonic form of neural cell surface molecule (E-NCAM) in the rat hippocampus and its reexpression on glial cells following kainic acid-induced status epilepticus. *J. Neurosci.* **12,** 872–882.

137. Ajemian, A., Ness, R., and David, S. (1994) Tenascin in the injured rat optic nerve and in non-neuronal cells in vitro: potential role in neural repair. *J. Comp. Neurol.* **340,** 233–242.
138. Laywell, E.D., Dorries, U., Bartsch, U., Faissner, A., Schachner, M., and Steindler, D.A. (1992) Enhanced expression of the developmentally regulated extracellular matrix molecule tenascin following adult brain injury. *Proc. Natl. Acad. Sci. USA* **89,** 2634–2638.
139. McKeon, R.J., Schreiber, R.C., Rudge, J.S., and Silver, J. (1991) Reduction of neurite outgrowth in a model of glial scarring following CNS injury is correlated with the expression of inhibitory molecules on reactive astrocytes. *J. Neurosci.* **11,** 3398–3411.
140. Eddleston, M. and Mucke, L. (1993) Molecular profile of reactive astrocytes implications for their role in neurologic disease. *Neuroscience* **54,** 15–36.
141. Ridet, J.L., Malhotra, S.K., Privat, A., and Gage, F.H. (1997) Reactive astrocytes: cellular and molecular cues to biological function. *Trend Neurosci.* **20,** 570–577.
142. Feringa, E.R., Vahlsing, H.L., and Woodward, M. (1984) Basal lamina at the site of spinal cord transection in the rat: an ultrastructural study. *Neurosci. Lett.* **51,** 303–308.
143. Suzuki, M. and Choi, B.H. (1990) The behavior of the extracellular matrix and the basal lamina during the repair of cryogenic injury in the adult rat cerebral cortex. *Acta Neuropathol.* **80,** 355–361.
144. Abnet, K., Fawcett, J.W., and Dunnett, S.B. (1991) Interactions between meningeal cells and astrocytes in vivo and in vitro. *Dev. Brain Res.* **59,** 187–196.
145. Moore, I.E., Buontempo, J.M., and Weller, R.O. (1987) Response of fetal and neonatal rat brain to injury. *Neuropathol. Appl. Neurobiol.* **13,** 219–228.
145. Struckhoff, G. (1995) Cocultures of meningeal and astrocytic cells. A model for the formation of the glial limiting membrane. *Int. J. Dev. Neurosci.* **13,** 595–606.
146. Sievers, J., Mangold, U., Berry, M., Allen, C., and Schlossberger, H.G. (1981) Experimental studies on cerebellar foliation. I. A qualitative morphological analysis of cerebellar foliation defects after neonatal treatment with 6-OHDA in the rat. *J. Comp. Neurol.* **203,** 751–769.
147. Sievers, J., Mangold, U., and Berry, M. (1983) 6-OHDA-induced ectopia of external granule cells in the subarachnoid space covering the cerebellum. Genesis and topography. *Cell Tissue Res.* **230,** 309–336.
148. Sievers, J., Pehlemann, F.W., and Berry, M. (1986) Influences of meningeal cells on brain development: findings and hypothesis. *Naturwissenschaften* **73,** 188–194.
149. Sievers, J., Hartmann, D., Gude, S., Pehlemann, F.W., and Berry, M. (1987) Influences of meningeal cells on the development of the brain. In Wolff, J.R., Sievers, J. and Berry, M. (Eds.), Mesenchymal-epithelial interactions in neural development. Springer, Berlin, pp. 171–188.
150. Yoshida, K. and Gage, F.H. (1991) Fibroblast growth factors stimulate nerve growth factor synthesis and secretion by astrocytes. *Brain Res.* **538,** 118–126.
151. Lee, M.Y., Deller, T., Kirsch, M., Frotscher, M., and Hofmann, H.D. (1997) Differential regulation of ciliary neurotrophic factor (CNTF) and CNTF receptor alpha expression in astrocytes and neurons of the fascia dentata after entorhinal cortex lesion. *J. Neurosci.* **17,** 1137–1146.
152. Balasingam, V., Tejada-Berges, T., Wright, E., Bouchova, R., and Yong, V.W. (1994) Reactive astrogliosis in the neonatal mouse brain and its modulation by cytokines. *J. Neurosci.* **14,** 846–856.
153. Chiang, C.S., Stalder, A., Samimi, A., and Campbell, I.L. (1994) Reactive gliosis as a consecuence of interleukine-6 expression in the brain: studies in transgenic mice. *Dev. Neurosci.* **16,** 212–221.
154. Holmin, S., Schwalling, M., Hijerberg, B., Nordqvist, A.C.S., Skeftruna, A.K., and Mathiesen, T. (1997b) Delayed cytokine expression in the rat brain following experimental contusion. *J. Neurosurgery* **86,** 493–504.
155. Logan, A., Frautschy, S.A., Gonzalez, A.M., Sporn, M.B., and Baird, A. (1992) Enhaced expression of transforming growth factor β1 in the rat brain after a localized cerebral injury. *Brain Res.* **587,** 216–225.

156. Uno, H., Matsuyama, T., Akita, H., Nishimura, H., and Sugita, M. (1997) Induction of tumor necrosis factor alpha in the mouse hippocampus following transient forebrain ischemia. *J. Cerebral Blood-Flow & Metabolism* **17,** 491–499.
157. Cook, J.L., Marchaselli, V., Alam, J., Deininger, P.L., and Bazan, N.G. (1998) Temporal changes in gene expression following cryogenic rat brain injury. *Mol. Brain Res.* **55,** 9–19.
158. Reilly, J.F. and Kumari, V.G. (1996) Alterations in fibroblast growth factor receptor expression following brain injury. *Exp. Neurol.* **140,** 139–150.
159. Nieto-Sampedro, M., Gomez-Pinilla, F., Knaver, D.J., and Broderick, J.T. (1988) Epidermal growth factor receptor immunoreactivity in rat brain astrocytes. Response to injury. *Neurosci. Lett.* **91,** 276–282.
160. Knuckey, N.W., Finch, P., Palm, D.E., Primiano, M.J., Johanson, C.E., Flanders, K.C., and Thompson, N.L. (1996) Differential neuronal and astrocytic expression of transforming growth factor beta isomorfism in the rat hippocampus following transient forebrain ischemia. *Mol. Brain Res.* **40,** 1–14.
161. Williams, C., Guan, j., Miller, O., Beilharz, E., McNeill, H., Sirimanne, E., and Gluckman, P. (1995) The role of the growth factor IGF-1 and TGFβ1 after hypoxic-ischemic brain injury. *Ann. N.Y. Acad. Sci.* **765,** 306–307.
162. Yao, D.L., West, N.R., Bondy, C.A., Brenner, M., Hudson, J.D., Zhou, J., Collins, G.H., and Webster, H.D. (1995) Cryogenic spinal cord injury induces astrocytic gene expression of insulin-like growth factor binding protein 2 during myelin regeneration. *J. Neurosci. Res.* **40,** 647–659.
163. Aubert, I., Ridet, J.L., and Gage, F.H. (1995) Regeneration in the adult mammalian CNS: guied by development. *Curr. Opin. Neurobiol.* **5,** 625–635.
164. Strauss, S., Otten, U., Joggerst, B., Plüss, K., and Volk, B. (1994) Increased levels of nerve growth factor (NGF) protein and mRNA and reactive gliosis following kainic acid injection into the rat striatum. *Neurosci. Lett.* **168,** 193–196.
165. Goss, J.R., O'Malley, M.E., Zou, L.L., Styren, S.D., Kochanek, P.M., and Dakosky, S.T. (1998) Astrocytes are the major source of nerve growth factor upregulation following traumatic brain injury in the rat. *Exp. Neurol.* **149,** 301–309.
166. Frautschy, S.A., Walicke, P.A., and Baird, A. (1991) Localization of basic fibroblast growth factor and its mRNA after CNS injury. *Brain Res.* **553,** 291–299.
167. Longo, F.M., Holtzman, D.M., Grimes, M.L., and Mobley, W.C. (1993) Nerve growth factor: Actions in peripheral and central nervous system. In: Neurotrophic Factors (Loughlin, S.E. & Fallon, I.H., eds.). Academic Press, San Diego, pp. 209–256.
168. García-Estrada, J., García-Segura, L.M., and Torres-Alemán, I. (1992) Expression of insulin-growth factor I by astrocytes in response to injury. *Brain Res.* **59,** 343–347.
169. Branner, L.R., Moayer, N.N., and Patterson, P.H. (1997) Leukemia inhibitor factor is expressed in astrocytes following cortical brain injury. *Exp. Neurol.* **147,** 1–9.
170. Chao, C.C., Sheng, W.S., Bu, D., Bukrinski, M.I., and Peterson, P.K. (1996) Cytokine-stimulated astrocytes damage human neurons via a nitric oxide mechanism. *Glia* **16,** 276–284.
171. Eclancher, F., Labourdette, G., and Sensebrenner, M. (1996) Basic fibroblast growth factor (bFGF) injection activates the glial reaction in the injured adult rat brain. *Brain Res.* **737,** 201–214.
172. Gomez-Pinilla, F., Vu, L., and Cotman, C.W. (1995) Regulation of astrocyte proliferation by FGF-2 and heparan sulfate in vivo. *J. Neurosci.* **15,** 2021–2029.
173. Morrison, R. (1993) Epidermal growth factor: structure, expression, and functions in the central nervous system. In: Neurotrophic Factors (Loughlin, S.E. & Fallon, I.H., eds.). Academic Press, San Diego, pp. 339–357.
174. Planas, A.M., Justicia, C., Soriano, M.A., and Ferrer, I. (1998) Epidermal growth factor receptor in proliferating reactive glia following transient focal ischemia in the rat brain. *Glia* **23,** 120–129.
175. Goetschy, J.F., Ulrich, G., Aunis, A., and Ciesielski-Treska, J. (1987) Fibronectin and collagens modulate the proliferation and morpholgy of astroglial cells in culture. *Int. J. Dev. Neuroscience* **5,** 63–70.

176. Niquet, J., Jorquera, I., Ben-Ari, Y., and Represa, A. (1994) Proloiferative astrocytes may express fibronectin-like protein in the hippocampus of epileptic rats. *Neurosci. Lett.* **180,** 13–16.
177. DiProspero, N.A., Meiners, S., and Geller, H.M. (1997) Inflamatory cytokines interact to modulate extracellular matrix and astrocytic support of neurite outgrowth. *Exp. Neurol.* **148,** 628–639.
178. Pawlinski, R. and Janezko, K. (1997a) Intracerebral injection of interferon-gamma inhibits the astrocyte proliferation following the brain injury in the 6-day-old rat. *J. Neurosci. Res.* **50,** 1018–1022.
179. Pawlinski, R. and Janezko, K. (1997b) An inhibitory effect of interferon gamma on the injury-induced astrocyte proliferation in the postmitotic rat brain. *Brain Res.* **773,** 231–234.
180. Lindholm, D., Castren, E., Kiefer, R., Zafra, F., and Thoenen, H. (1992) Transforming growth factor-beta 1 in the rat brain: increase after injury and inhibition of astrocyte proliferation. *J. Cell Biol.* **117,** 395–400.
181. Hou, Y.J., Yu, A.C., García, J.M., Aotaki-Keen, A., Lee, Y.L., Hjelmeland, L.J., and Menon, V.K. (1995) Astrogliosis in culture. IV. Effects of basic fibroblast growth factor. *J. Neurosci Res.* **40,** 359–370.
182. Lippoldt, A., Andjer, B., Gerst, H., Ganten, D., and Fuxe, K. (1996) Basic fibroblast growth factor expression and tenascin C immunoreactivity after partial unilateral hemitransection of the rat brain. *Brain Res.* **730,** 1–16.
183. Rowntree, S. and Kolb, B. (1997) Blockade of basic fibroblast growth factor retards recovery from motor cortex injury in rats. *Eur. J. Neurosci.* **9,** 2432–2441.
184. Reilly, J.F., Maher, P.A., and Kumari, V.G. (1998) Regulation of astrocyte GFAP expression by TGF-beta 1 and FGF-2. *Glia* **22,** 202–210.
185. Kahn, M.A., Ellison, J.A., Ghang, R.P., Speight, G.J., and Devellis, J. (1997) CNTF induces GFAP in a S-100 alpha brain cell population: The pattern of CNTF-alpha R suggest an indirect mode of action. *Dev. Brain Res.* **98,** 221–223.
186. Rudge, J.S., Smith, G.M., and Silverm, J. (1989) An in vitro model of wound healing in the CNS: analysis of cell reaction and interaction at different ages. *Exp. Neurol.* **103,** 1–16.
186. Shibayama, M., Kuchiwaki, H., Inao, S., Yoshida, K., and Ito, M. (1996) Intercellular adhesion molecule-1 expression on glia following brain injury: Participation of interleukin-1 beta. *J. Neurotrauma* **13,** 801–808.
187. McKeon, R.I. and Silver, J. (1995) Functional significance of glial-derived matrix during development and regeneration. In Neuroglia (Kettenman, H. & Ranson, B.R. eds.) Oxford University Press, pp. 398–410.
188. Powell, E.M., Meiners, S., DiProspero, N.A., and Geller, H.M. (1997) Mechanism of astrocyte-directed neurite guidance. *Cell Tissue Res.* **290,** 385–393.
189. Labourdette, G. and Sensenbrenner, M. (1995) Growth factors and their receptors in the central nervous system. In Neuroglia (Kettenman, H & Ranson, B.R. eds.) Oxford University Press, Oxford, pp. 441–459.
190. Carman-Krzan, M., Vigé, X., and Wise, B.C. (1991) Regulation by interleukin-1 of nerve growth factor secretion and nerve growth factor mRNA expression in rat primary astroglial cultures. *J. Neurochem.* **56,** 636–643.
191. Dekosky, S.T., Styren, S.D., O'Malley, M.E., Goss, J.R., Kochanec, P., Marion, D., Evans, C.H., and Robbins, P.D. (1996) Interleukin-1 receptor antagonist suppresses neurotrophin response in injured rat brain. *Ann. Neurol.* **39,** 123–127.
192. Riva, M.A., Fumagalli, F., and Racagni, G. (1995) Opposite regulation of fibroblast growth factor and nerve growth factor gene expression in rat cortical astrocytes following dexametasone treatment. *J. Neurochem.* **64,** 2526–2533.
193. Balasingam, V., Dickson, K., Brade, A., and Yong, V.W. (1996) Astrocyte reactivity in neonatal mice: apparent dependence on the presence of reactive microglia/macrofages. *Glia* **18,** 11–26.

194. Zocchia, C., Spiga, G., Rabin, S.J., Grekova, M., Richert, J., Chemyshev, D., Colton, C., and Mochetti, I. (1997) Biological activity of interleukin-10 in the central nervous system. *Neurochem. Int.* **30,** 433–439.
195. AmurUmarjee, S., Schonmann, V., and Campagnon, A.T (1997) Neuronal regulation of myelin basic protein mRNA translocation in oligodendrocytes is mediated by platelet-derived growth factor. *Dev. Neurosci.* **19,** 143–151.
196. Nag, S., Takahashi, J.L., and Kilty, D.W. (1997) Role of vascular endothelial growth factor in blood-brain barrier breakdown and angiogenesis in brain trauma. *J. Neuropath. Exp. Neurol.* **56,** 912–921.
197. Yao, J., Harvarth, L., Gilbert, D.L., and Colton, C.A. (1990) Chemotaxis by a CNS macrophage, the microglia. *J. Neurosci. Res.* **27,** 36–42.
198. Menon, V.K. and Landerholm, T.E. (1994) Intralesion injection of basic fibroblast growth factor alters glial reactivity to neural trauma. *Exp. Neurol.* **129,** 142–154.
199. Schwartz, J.P. and Nishiyama, N. (1994) Neurotrophic factor gene expression in astrocytes during development and following injury. *Brain Res. Bull.* **35,** 402–407.
201. Roessman U. and Gambetti, P. (1986) Pathological reaction of astrocytes in perinatal brain injury. Immunohistochemical study. *Acta Neuropathol.* **70,** 302–307.
202. Rostworowski, M., Balasingam, V., Chabot-Owens, T., and Yong, V.W. (1997) Astrogliosis in the neonatal and adult murine brain post-trauma: elevation of inflammatory cytokines and the lack of requeriment for endogenous intergeron-gamma. *J. Neurosci.* **17,** 3664–3674.
203. Barret, C.P., Donati, E.J., and Guth L. (1984) Differences between adult and neonatal rats in their astroglial response to spinal injury. *Exp. Neurol.* **84,** 374–385.
204. Sijbesma, H. and Leonard. C.M. (1986) Developmental changes in the astrocytic response to lateral olfactory tract section. *Anat. Rec.* **215,** 374–382.
205. Ghooray, G.T. and Martin, G.F. (1993b) Development of an astrocytic response to lesions of the spinal cord in the North American opossum: an immunohistochemical study using antiglial fibrillary acidic protein. *Glia* **9,** 10–17.
206. Sumi S.M. and Hager, H. (1968) Electron microscopic study of the reaction of the newborn rat brain to injury. *Acta Neuropathol.* **10,** 324–335.
207. Smith. G.M. and Miller, R.H. (1991) Immature type-1 astrocytes suppress glial scar formation, are motile and interact with blood vessels. *Brain Res.* **543,** 111–122.
208. Smith, G.M., Miller, R.H., and Silver, J. (1986) Changing role of forebrain astrocytes during development, regenerative failure, and induced regeneration upon transplantation. *J. Comp. Neurol.* **251,** 23–43.
209. Gordon, M.N., Schreier, W.A., Ou, X., Holcomb, L.A., and Morgan, D.G. (1997) Exaggerated astrocyte reactivity after nigrostriatal deafferentiation in the aged rat. *J. Comp. Neurol.* **388,** 106–119.
210. Suzuki, M., Iwasaki, Y., Umezawa, K., Motohashi, O., and Shida, N. (1995) Distribution of extravasated serum protein after cryoinjury in neonatal and adult rat brains. *Acta Neuropathol.* **89,** 532–536.

II

Neuron–Glia Intercommunication

4
Astrocytes *In Situ* Exhibit Functional Neurotransmitter Receptors

Marilee K. Shelton and Ken D. McCarthy

1. INTRODUCTION

Astrocytes constitute a major portion of brain cells and envelop most neuronal elements. The intimate association of astrocytes and neurons led early anatomists to speculate that astrocytes interact with neurons *(1–4)*. Work beginning in the 1970s determined that astroglia (astrocytes in culture) exhibit a wide variety of neurotransmitter receptors that regulate both second messenger systems and ion channels. These findings suggested that astrocytes in vivo might have neurotransmitter receptors enabling them to respond to neuronal activity. The importance, however, of studying astrocytes without culturing them was underscored by reports indicating that the neuroligand responsiveness of astroglia changes in culture *(5–7)*. That placing astroglia in culture altered their phenotype was not surprising given that these cells are generally isolated from their normal cellular and chemical milieu at an early developmental stage and placed into an artificial environment that almost certainly lacks critical developmental cues. Unfortunately, the complex morphology of astrocytes together with their inability to propagate action potentials makes it difficult to study these cells in vivo with methods that were so powerful in elucidating the neuronal signaling systems in vivo.

Over the past decade, a number of laboratories have carried out studies to determine if astrocytes *in situ* exhibit the plethora of neuroligand receptors expressed by astroglia in culture. Although the results from *in situ* experiments are very limited relative to in vitro studies, it has become clear that astrocytes *in situ* exhibit functional neuroligand receptors. Approaches used to study the expression of astrocytic signaling systems *in situ* include immunological staining with receptor-specific antibodies *(8)*, mRNA detection *(9)*, electrophysiological recording of neuroligand responses *(10)*, and analysis of neuroligand-mediated changes in astrocytic Ca^{2+} using Ca^{2+}-indicator dyes *(11)*. Each of these approaches has advantages and disadvantages and they frequently compliment one another. Here we have attempted to summarize the evidence for the expression of "functional" astrocytic receptors *in situ;* by functional, we are referring to receptors shown to regulate either ion channels or second messenger systems. One caveat in the study of functional astrocytic receptors *in situ* is that astrocytic responses are generally measured in astrocyte perikarya. *In situ* analysis, therefore, may not

From: *Neuroglia in the Aging Brain*
Edited by: Jean S. de Vellis © Humana Press Inc., Totowa, NJ

detect signaling in astrocyte processes where morphological evidence suggests neuronal-astrocytic interactions may occur. Other caveats in the analysis of functional astrocyte receptors *in situ* are that, generally, few cells are analyzed; and negative findings should be given less value since technical issues with brain slices might prevent detection of functional receptors. In the following sections, we have attempted to summarize the results of studies that focused on the expression of functional astrocyte receptors *in situ.*

2. EVIDENCE FOR FUNCTIONAL GLUTAMATERGIC RECEPTORS ON ASTROCYTES *IN SITU*

Glutamate is the major excitatory neurotransmitter in the brain. It stimulates the ionotropic glutamate receptors (iGluRs), including NMDA and AMPA/kainate receptor subtypes, that gate ion channels. Glutamate also stimulates the metabotropic glutamate receptors (mGluRs) that are coupled to second messenger systems through G proteins. Substantial evidence suggests that astrocytes *in situ* exhibit functional iGluRs and mGluRs.

The first evidence for functional mGluRs on astrocytes *in situ* came from the work of Porter and McCarthy *(12)*. Astrocytes were found to respond to mGluR agonist 1-aminocyclopentane-*trans*-1,3-dicarboxylic acid (t-ACPD) with increases in intracellular Ca^{2+} when studied in the CA1 stratum radiatum region of hippocampal slices prepared from 9–13-day-old (P9-P13) rats *(12)*. Furthermore, responses to t-ACPD were blocked by the mGluR antagonist antagonist α-methyl-4-carboxyphenylglycine (MCPG). Studies from other laboratories indicate that treatment with t-ACPD elicits Ca^{2+} increases in astrocytes acutely isolated from the cerebral cortex *(7,13)* and from astrocytes *in situ* in the visual cortex *(14)*.

An early report indicated that astrocytes in P21–P42 hippocampal slices did not exhibit mGluRs *(15)*. This finding suggested that hippocampal astrocytic mGluRs may serve a developmental role but not be important in the mature hippocampus. We have since demonstrated, however, that mature astrocytes (P30–P35) within hippocampal brain slices have mGluRs coupled to Ca^{2+} increases (*16;* Fig. 1). tACPD also increased Ca^{2+} levels in GFAP/S100^{+} astrocytes within a ~P66–P71 hippocampal slice (Fig. 2), suggesting that mGluRs exist well into adulthood. Consistent with the detection of functional astrocytic mGluRs *in situ,* immunocytochemical staining suggests that astrocytes within this region express mGluR5, an mGluR subunit coupled to Ca^{2+} increases *(17–20)*. Antibodies for mGluR5 lightly stained astrocytes in adult hippocampus and cortex *(19)*. Interestingly, mGluR5-immunoreactive astrocytes in the hypothalamus envelop presynaptic boutons, where they might bind neuronally-released glutamate *(17)*.

The CA1 stratum radiatum region of the hippocampus receives excitatory glutamatergic input from the pyramidal neurons of CA3 via Schaffer collateral processes. The Schaffer collateral-CA1 pyramidal neuron (SC-CA1) synapse is one of the most studied synapses in brain. Experiments carried out by Porter and McCarthy *(21)* indicate that stimulation of this pathway leads to astrocytic Ca^{2+} responses that are blocked by

1. tetrodotoxin (which blocks action potential propagation),
2. conotoxin MVIIC (which blocks neurotransmitter release in this region) and
3. the mGluR antagonist, MCPG.

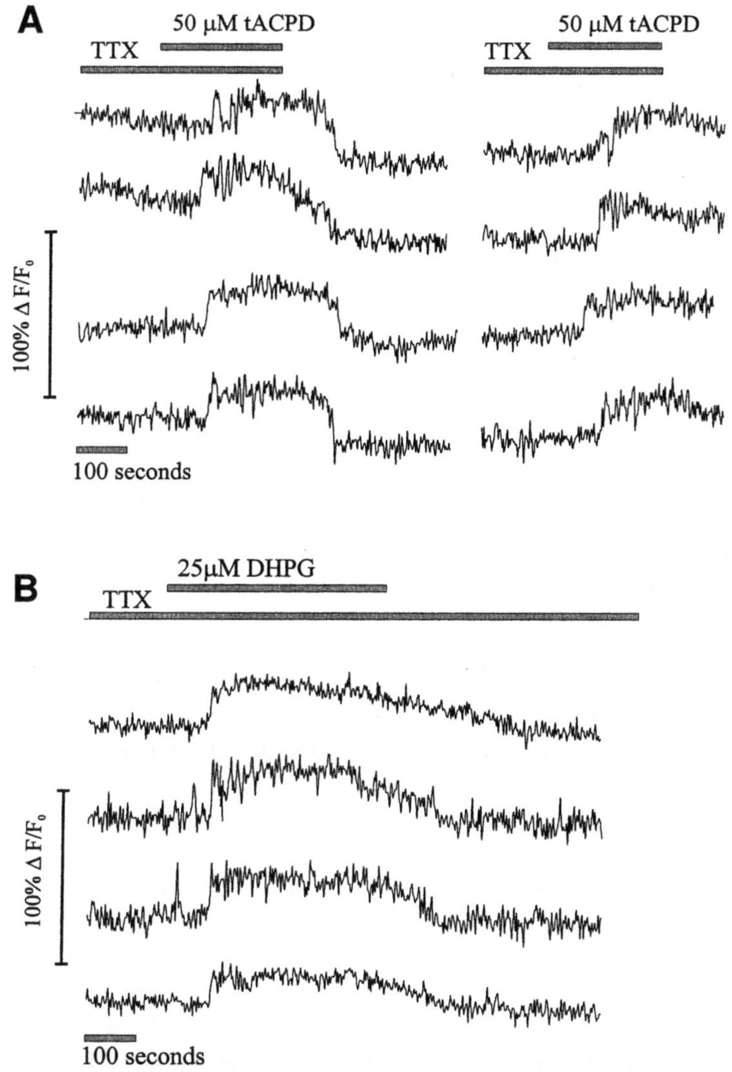

Fig. 1. Mature astrocytes in the CA1 stratum radiatum region of acutely-isolated hippocampal brain slices respond to metabotropic glutamate agonists with calcium increases. Fluorescence increases represent increases in intracellular calcium, as measured after loading cells with the Ca^{2+} sensitive dye Calcium Green-1 AM. Traces shown are from cells that were immunologically identified as astrocytes by immunocytochemically staining slices for astrocyte markers GFAP and S100. Traces are arbitrarily separated along the y-axis. Slices were pretreated with 1 μM tetrodotoxin (TTX) to prevent stimulation of action potentials. **(A)** P31 astrocytes exhibit calcium increases in response to metabotropic glutamate receptor agonist 1-aminocyclopentane-trans-1,3-dicarboxylic acid (t-ACPD). **(B)** P32 astrocytes exhibit calcium increases in response to the metabotropic glutamate receptor agonist dihydrophenylglycine (DHPG). Shelton, M.K. and McCarthy, K.D. (1999) Mature hippocampal astrocytes exhibit functional metabotropic and ionotropic glutamate receptors in situ. From Shelton and McCarthy 1999, *(16)* with permission.

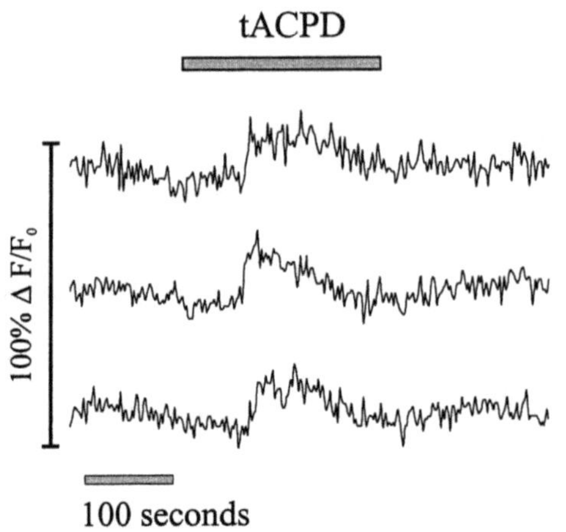

Fig. 2. Mature astrocytes, in the CA1 stratum radiatum region of ~P66–P71 acutely-isolated hippocampal brain slices, respond to the metabotropic glutamate receptor agonist tACPD with calcium increases. Shelton and McCarthy, unpublished results.

Other investigators have subsequently verified and extended these observations by suggesting that increasing Ca^{2+} in these astrocytes triggers them to release glutamate that subsequently increases Ca^{2+} in CA1 pyramidal neurons *(14)*. Overall, it is clear from these studies that hippocampal astrocytes express functional mGluRs that respond to applied mGluR agonists as well as glutamate released by synaptic activation.

There is also convincing evidence that astrocytes exhibit functional iGluRs. The AMPA/Kainate iGluRs (AMPA/KA-Rs) consist of the AMPA-preferring (AMPA-Rs) and kainate-preferring receptors (KA-Rs). AMPA-Rs and KA-Rs are collectively referred to as AMPA/KA-Rs, or non-NMDA receptors, since many reports do not distinguish between these two receptor types. Reports demonstrate that AMPA/KA-Rs stimulate depolarization and/or Ca^{2+} elevations in freshly-isolated and *in situ* astrocytes, as well as presumed astrocyte progenitors *(10,12,22–27)*. As with mGluRs, an initial report suggested that mature (P21–P42) hippocampal astrocytes lack AMPA/KA-Rs *(15)*. Subsequent studies from two separate laboratories, however, indicate that mature astrocytes express functional AMPA/KA-Rs coupled to depolarization *(28)* and Ca^{2+} increases (*16*, Fig. 3). AMPA/KA-R stimulation affected astrocytes both when they were isolated from slices and when neuronal action potentials were blocked, indicating that AMPA/KA-R agonists were directly affecting astrocytes *(10,12,28)* (Shelton and McCarthy, Neurochemistry). When AMPA/KA-R responses have been closely examined, they appear to occur via AMPA-Rs in hippocampal cells presumed to be astrocyte progenitors *(10,28)*.

In addition to the physiological evidence for AMPA/KA-Rs, there is also evidence that astrocytes express AMPA-R and KA-R subunits. AMPA-Rs are composed of four subunits named GluR1–4, and KA-Rs are composed of subunits GluR5–7 and KA1–2. GluR4 and GluR1 receptors or mRNA have been localized to astrocytes from cerebellum, hippocampus, cerebral cortex, and several other brain regions *(9,30–43)*. Bergmann glia, a type of specialized astrocyte, do not appear to express GluR2 mRNA *(33,44)*. This is in contrast to the detection of GluR2 transcripts in each presumed astrocyte progenitor examined from the hippocampus *(9)*. The GluR2 expression is likely to control Ca^{2+}

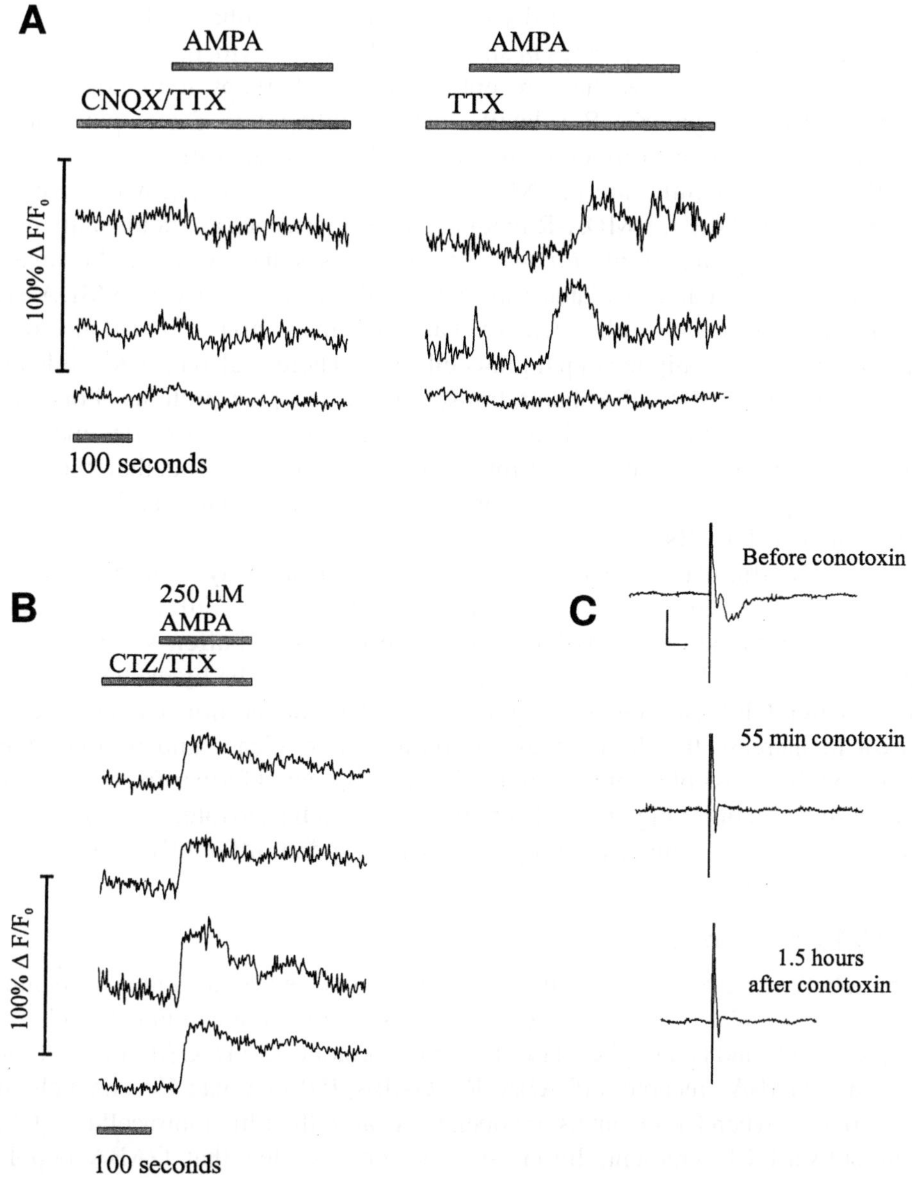

Fig. 3. Mature astrocytes in the CA1 stratum radiatum region of acutely-isolated hippocampal brain slices respond to the non-NMDA receptor agonist AMPA with calcium increases. (*see* Fig. 1) Slices were treated with cyclothiazide (CTZ) to prevent desensitization of α-amino-3-hydroxy-5-methylisoxazole-4-propionic acid (AMPA)-preferring non-NMDA receptors. **(A)** Astrocytes do not respond to AMPA in the presence of the non-NMDA receptor antagonist 6-cyano-7-nitroquinoxaline-2,3-dione (CNQX). Following CNQX washout, AMPA elicits Ca^{2+} responses. **(B)** P35 astrocytes respond to AMPA after treatment with 5 μM conotoxin MVIIC to prevent neurotransmitter release. **(C)** Conotoxin blocks postsynaptic field potentials, indicating that neuronal release is inhibited by conotoxin treatment. From *(16)*, with permission.

influx through these receptors, as it does in neurons *(45)*. Although functional KA-Rs have not been detected on astrocytes *in situ,* limited immunocytochemical evidence suggests that astrocytes express these receptors *(42,46)*. Perhaps immunocytochemical staining detects astrocyte KA-R subunits that are difficult to detect physiologically because they rapidly desensitize or are masked by AMPA-R activation.

The third type of iGluRs are the NMDA receptors (NMDA-Rs), which are permeable to Na^+, K^+, and Ca^{2+}. NMDA-R mediated Ca^{2+} elevations are not usually detected on astrocytes in culture or *in situ (47–50)*. There is some evidence, however, that NMDA depolarizes acutely-isolated astrocytes and astrocytes *in situ*. NMDA elicited small, atypical currents in Muller cells *(51)* and Bergmann glia *(47),* and also depolarized a subpopulation of hippocampal glial cells *(25)*. There is also limited evidence that astrocytes express NMDA-R subunit transcripts or receptor proteins. For example, Muller cells, Bergmann glia, cortical astrocytes, and thalamic astrocytes may express one or more NMDA-R subunit transcripts or receptors *(52–56)*. Together, the data suggest that subpopulations of astrocytes may express NMDA-Rs but that these may differ from neuronal NMDA-Rs.

Interestingly, stimulation of the Schaffer collateral pathway at levels that consistently activate astrocytic mGluRs, fails to elevate Ca^{2+} via iGluRs *(21)*. In contrast, when 4-aminopyridine (4-AP) is included in the perfusion buffer, stimulation of the Schaffer collateral pathway leads to responses that can only be blocked by the combination of both mGluR and iGluR antagonists. 4-AP is a potassium channel blocker that is known to increase the duration of neurotransmitter release and prolong synaptic transmission. It is possible that astrocytic iGluRs are located farther from sites of glutamate release and require higher levels of stimulation to be activated. This is interesting and may reflect a mechanism whereby astrocytes respond differently to varying levels of neuronal activity.

2.1. GABA Receptors

γ-aminobutyric acid (GABA) is the major inhibitory neurotransmitter in mammalian brain. GABA stimulates $GABA_B$ receptors ($GABA_BRs$) that are coupled to Ca^{2+} channels, K^+ channels and adenylyl cyclase through G proteins *(56A)*. GABA also stimulates $GABA_A$ and $GABA_C$ receptors ($GABA_ARs$, $GABA_CRs$) that gate Cl^- channels. Astrocytes depolarize when Cl^- channels are opened because their high intracellular $[Cl^-]$ creates an outward Cl^- gradient. Initial studies demonstrated that GABA depolarizes cultured astroglia *(57)*. $GABA_ARs$ have since been reported to depolarize acutely isolated and *in situ* astrocytes from hippocampus, cerebellum, and pituitary *(58–60)*. GABA not only depolarizes astrocytes, but may also elevate Ca^{2+} if astrocytes are sufficiently depolaried to open voltage-dependent Ca^{2+} channels (VDCC). For example, GABA induced $[Ca^{2+}]_i$ elevations that were depressed by VDCC antagonists in hippocampal astrocytes *(60)*. It should be mentioned, however, that data from another laboratory suggests that hippocampal astrocytes do not exhibit VDCC *(15)*. Interestingly, the function of $GABA_ARs$ may be developmentally regulated. This is suggested by observations that $GABA_ARs$ depolarize optic nerve and cerebellar astrocytes less as animals mature *(59,61)*. In addition to functional evidence for $GABA_AR$-mediated signaling in astrocytes, there is also evidence that astrocytes express several of the $GABA_AR$ α, β, γ, and δ subunits. Hippocampal astrocytes were labeled by antibodies for α_1 and β_1 subunits *(60)*.

Interestingly, α_2, α_3, and δ subunits are transiently expressed at about the same age that $GABA_ARs$ mediate large depolarizations in Bergmann glia *(59)*. Subunit expression and GABA-mediated depolarizations both decreased with maturation *(59)*. It may be that $GABA_ARs$ are only transiently expressed because they are important in development.

Importantly, $GABA_AR$ channels may be opened by neuronally-released GABA. Electrically stimulating stalk neurons in isolated pituitaries evoked depolarizations in stellate pituicytes (a specialized type of astrocyte; *(58)*). Blocking action potential propagation or $GABA_A$ receptors with TTX or bicuculline, respectively, blocked electrically-induced astrocytic depolarizations, suggesting that they depend on neuronal activity and $GABA_ARs$.

In addition to ionotropic $GABA_ARs$, evidence also suggests that neuronal activity stimulates astrocytic $GABA_ARs$, For example, direct stimulation of inhibitory neurons leads to Ca^{2+} elevations in neighboring hippocampal astrocytes *(62)*. These Ca^{2+} elevations were blocked by the potent $GABA_BR$ antagonist CGP55845A *(62,63)*, suggesting that they occur through $GABA_BRs$. Furthermore, GABA and $GABA_BR$ agonist baclofen elicited astrocyte Ca^{2+} responses that were blocked by CGP55845A but not by the $GABA_AR$ antagonist bicuculline *(62)*. These Ca^{2+} responses to baclofen and GABA depended on extracellular Ca^{2+} suggesting that a Ca^{2+} influx was necessary for $GABA_B$-mediated increases in astrocytic Ca^{2+}. Together, these data suggest that astrocyte $GABA_BRs$ mediate Ca^{2+} elevations that depend on Ca^{2+} influx. A puzzling aspect of these findings is that $GABA_B$ receptors are not typically associated with increased Ca^{2+} influx. Another report, however, suggests that cortical astroglia may also exhibit $GABA_BR$-mediated Ca^{2+} responses that depend on extracellular Ca^{2+} *(64)*. Perhaps astrocytes have a unique GABA receptor that is mediating these responses.

2.2. *Adrenergic Receptors*

Norepinephrine and epinephrine are important modulatory neurotransmitters within the central nervous system (CNS) that stimulate α and β adrenergic receptors (αARs and βARs). It has long been known that cultured astroglia express both α_1- and α_2-adrenergic receptors *(65–67)* There are several lines of evidence suggesting that astrocytes *in situ* also express αARs. For example, norepinephrine or the α_1AR-selective agonist phenylephrine evoked Ca^{2+} elevations in Bergmann glia and hippocampal astrocytes of acutely isolated brain slices *(15,50,68)*. These Ca^{2+} elevations were inhibited by the α_1AR selective antagonist prazosin, but they were not affected by α_2AR or βAR ligands *(15,68)*. Astrocyte responses were not secondary to transmitter release from neurons, since neither isolating astrocytes away from intact brain tissue, nor treating slices with tetrodotoxin (TTX) prevented α_1AR-mediated responses *(15,50,68)*. α_1Rs were also localized to acutely-isolated astrocytes with a radiolabeled α_1AR antagonist *(69,70)*.

In addition to α_1AR, two reports suggest that α_2-ARs are present on astrocytes *in situ*. Antibodies for α_2ARs reportedly bound astrocytes in hippocampus *(71)* and locus ceruleus *(72)*. Interestingly, these immunoreactive astrocytes were observed near dendrites and terminals of catecholamine neurons *(71,72)*, where they might bind released norepinephrine.

A number of ligand binding and immunocytochemical reports also localize βARs to astrocytes of different brain regions *(8,27,70,73–76A)*. For example, radiolabeled βAR ligands bound to protoplasmic and fibrous astrocytes that were acutely-isolated from cerebral cortex, striatum, cerebellum, and trigeminal motor nucleus *(70,73)*. Aoki and colleagues combined βAR immunostaining with electron microscopy to assess whether astrocytes express βARS. They concluded that βARs are expressed by astrocytes in visual cortex *(74,76A)* and cerebral cortex *(8)*. They also observed βAR positive astrocytes closely associating with axon terminals, dendrites, and synaptic junctions of catecholamine neurons *(74)*. In summary, astrocytes in various brain regions appear to express α_1, α_2 and β-adrenergic receptors. These receptors have been observed where they are likely to be exposed to neuronally-released norepinephrine *(71,72,74)*.

2.3. *Acetylcholine Receptors*

There is also evidence that astrocytes have G protein-coupled muscarinic cholinergic receptors (mAChRs). mAChR stimulation increases Ca^{2+} in immunologically-identified astrocytes in acutely-isolated hippocampal slices *(77a)*. These Ca^{2+} increases were inhibited by the mAChR antagonist pirenzepine, suggesting that the M1 mAChR subtype may be involved. In addition to physiological studies *in situ,* hippocampal and cortical astrocytes are labeled by an antibody that does not distinguish among mAChR subtypes *(77–80)*. Interestingly, this immunoreactivity increased in the hippocampus when seizure-like activity was induced *(78),* hinting that neuronal activity might alter astrocyte mAChR expression. Finally, the number of astrocytes stained by this M35 antibody markedly increased in aged rats, compared to young rats *(80,81)*. It is not clear if this reflects astrocyte proliferation/gliosis with aging, increases in the number of receptors per astrocyte, or changes in the availability of the epitope recognized by the antibody. The same antibody reportedly labeled gray matter astrocytes intensely in some Alzheimer's diseased brain *(82)*. It is interesting to speculate that intense mAChR immunoreactivity in astrocytes from Alzheimer's brain reflects an attempt on the part of the astrocytes to compensate for cholinergic neuron loss associated with this disease. In summary, it seems quite clear that mammalian astrocytes in vivo express mAChRs.

2.4. *Histamine Receptors*

Histamine is a modulatory neurotransmitter that influences physiological states such as wakefulness and behaviors such as learning and aggression *(83)*. H1 and H2 receptors are associated with cultured astroglia *(84,85)*. *In situ,* histamine elicited Ca^{2+} increases in Bergmann glia, immunologically-identified hippocampal astrocytes, and glial precursor cells in the corpus callosum (*68,77a,86*). H1 receptors probably mediate hippocampal astrocyte and Bergmann glia Ca^{2+} responses, as shown by their sensitivity to H1-selective antagonist chlorpheniramine *(68),* Shelton & McCarthy, submitted). Bergmann glia responses were dependent on intracellular Ca^{2+} release and were not mimicked by H2 or H3 agonists. *(68)*. In corpus callosum, histamine elicited Ca^{2+} elevations in P3–P7 cell bodies that were probably glial precursors *(86)*. Histamine, however, rarely elicited responses in P11–P18 corpus callosum cell bodies that are primarily glial precursors and astrocytes *(86)*. In summary, hippocampal astrocytes and Bergmann glia *In situ* express histamine receptors coupled to Ca^{2+} increases.

2.5. Adenosine and P2 Receptors

A variety of studies suggest that ATP and adenosine can act as traditional neurotransmitters that are released from neurons and activate signaling systems through surface receptors *(87,88)*. Adenosine stimulates a family of receptors known as P1 receptors, whereas ATP and related substances stimulate a family of receptors known as P2 receptors. Adenosine stimulates Ca^{2+} elevations in hippocampal astrocytes *in situ* and freshly isolated cortical astrocytes *(11,89)*. Pharmacological sensitivities suggest that Ca^{2+} elevations in hippocampal and cortical astrocytes occur by activation of A_{2B} adenosine receptors *(11,89)*. Adenosine activates Ca^{2+} release through intracellular stores, since responses depend on phospholipase C activity but not on extracellular Ca^{2+} *(11,89)*. As responses were observed either with TTX or in isolated astrocytes, it is unlikely that they are secondary to neuronal stimulation *(11,89)*. A_{2B} adenosine receptors, however, may not be the only P1 receptor subtype on astrocytes *in situ*. Some hippocampal astrocytes responded to an agonist that can activate $A_{2A}$and A_3 receptors, suggesting one or both of these receptors may also be on astrocytes *(11)*.

In addition to adenosine receptors, astrocytes also have P2 receptors that are activated by ATP and other nucleotides. Both ionotropic and metabotropic P2 receptors have been observed on cultured astroglia *(85)*. Studies suggest that astrocytes *in situ* may also express P2 receptors. For example, ATP elicited Ca^{2+} increases that were inhibited by the P2 antagonist suramin in Bergmann glia. These Ca^{2+} elevations were not accompanied by conductance changes, suggesting that P2Y metabotropic receptors were involved *(90)*. Hippocampal astrocytes also appear to have P2 receptors since the $P2Y_1$ agonist, ADPβS, elicited Ca^{2+} elevations in about 20% of astrocytes examined *(11)*. ADPβS evoked responses in the presence of TTX, suggesting that ATP directly stimulates astrocyte receptors *(11)*. In contrast to P2Y receptors, ionotropic P2X receptors have not been directly detected on astrocytes *in situ*. For example, ATP did not elicit conductance changes in electrically recorded hippocampal astrocytes *(91)* or Bergmann glian *in situ* *(90)*. Neal et al., however, suggest that P2X receptors might modulate GABA uptake into retinal glial cells *(92)*. In summary, ATP, adenosine and other nucleotides can signal to astrocytes *in situ* through P2 and A2 adenosine receptors.

2.6. Serotonin Receptors

Seven classes of serotonin receptors (5HT) have been identified. Cultured astroglia appear to have $5HT_2$ receptors coupled to Ca^{2+} *(93,94)*. In contrast to astroglia, there is little evidence for functional 5HT receptors on astrocytes *in situ*. Serotonin did not elicit Ca^{2+} elevations in Bergmann glia *in situ* or in acutely isolated cortical or hippocampal astrocytes *(6,7,13,68,95)*. 5HT receptors were expressed when acutely-isolated astrocytes were cultured in serum containing medium, suggesting that culturing 5HT receptors astroglia may induce 5HT receptor expression *(6,7)*. Serotonin did elicit a small number of responses in corpus callosum glial precursors *in situ;* however, these responses occurred less frequently as the precursors develop into astrocytes and oligodendrocytes *(86)*. Although astrocytes do not appear to express the Ca^{2+}-elevating $5HT_2$ receptors until they are cultured, they may express $5HT_{1A}$ receptors. Azmitia et al. report that immunostaining colocalizes $5HT_{1A}$ receptors and GFAP in septum, hippocampus, cerebral cortex, and other regions *(95,96)*. In these studies, 5HT immunostaining varied substantially in the different brain regions examined *(95)*. In summary, there is little evidence to date that astrocytes *in situ* express 5HT receptors coupled to Ca^{2+} increases; however, they may express other types of 5HT receptors.

2.7. Opioid Receptors

Opioid receptors δ, κ, and μ are activated by certain endogenous peptides such as β-endorphin, enkephalins, endomorphins, and/or dynorphin A. In general, opioid receptors modulate adenylate cyclase activity, increase K^+ channel activity, and inhibit Ca^{2+} channel activity through G proteins *(97)*. There is limited evidence that opioid receptors are on astrocytes *in situ*. For example, the δ receptor agonist [D-Pen5, D-Pen5]-enkephalin elevated Ca^{2+} in several acutely-isolated cells presumed to be astrocytes, and δ receptor antagonist ICI 174.864 blocked these Ca^{2+} elevations *(13)*. As the Ca^{2+} source was not investigated, the receptor signaling mechanism responsible for Ca^{2+} increases in these studies remains uncertain *(13)*. Ligand binding studies suggest pituicytes *in situ* may express κ opioid receptors. In these studies, neuronal innervation was removed from the pituitary by transecting pituitary stalks. Opioid binding sites that remained after neuronal degeneration were attributed to the expression of these receptors by pituicytes *(98,99)*. In summary, it seems that certain populations of astrocytes express opioid receptors *in situ*. Additional studies, however, are required to verify that nonspecialized astrocytes express opioid receptors.

3. ROLE OF ASTROCYTE NEUROTRANSMITTER RECEPTORS

The results of in vitro studies carried out over the past three decades show that astrocytes have the potential to express most neurotransmitter receptors. Unfortunately, it has been far more difficult to establish the repertoire of neurotransmitter receptors expressed by astrocytes *in situ*. To date, only a small percentage of the receptors shown to be expressed by astroglia in culture are known to be expressed by astrocytes *in situ*. Furthermore, we know little about the spatial distribution of astrocytic receptors or the effects of development and aging on the expression and distribution of astrocytic signaling systems. Less is known about astrocytes *in situ* primarily because it is more difficult to study astrocyte signaling systems *in situ*. It is clear, however, that astrocytes *in situ* express functional receptors that are linked to second messenger systems and ion channels. Importantly, in certain situations, it has been possible to demonstrate that astrocytic receptors are activated during neuronal activity. The emerging story suggests that astrocytes are likely to express a similar array of neurotransmitter receptors as neurons. The expression and activation of these receptors may vary both spatially and temporally throughout brain. It is possible that functional microdomains of astrocytic syncytia interact with local neuronal elements without involving perikarya where astrocyte responses *in situ* are monitored. If this occurs, it is conceivable that individual astrocytes exhibit marked receptor heterogeneity among their microdomains and that the complement of receptors expressed in each domain depends on the types of neurotransmitters released in that specific location. Developing new methods to study astrocytic signaling within astrocytic microdomains may be critical for us to understand the significance of bidirectional communication between neurons and astrocytes in the developing, mature, and aging brain.

The prevailing view is that the activation of astrocytic signaling systems may support and modulate neuronal signaling and function. A few of the different ways that astrocytic signaling systems may influence neurotransmission are discussed below.

Although we are a very long way from understanding the role of astrocytes in brain function, there is a growing consensus among neurobiologists that astrocytes are likely to play a much greater role in neurophysiology than imagined only a few years ago.

Astrocytes *in situ* are known to express receptors linked to most of the known second messenger systems including cyclic AMP, cyclic GMP, Ca^{2+}, and NO. Furthermore, astrocytes exhibit a number of receptors that directly or indirectly regulate the opening and closing of ion channels. Collectively, these different signaling systems affect a wide variety of processes that could markedly influence the composition of the extracellular space and in turn, neurotransmission. Receptor stimulation, for example, changes astroglial volume in vitro. In vivo, such an astrocyte volume change could alter the extracellular volume and thereby influence local ion and neurotransmitter concentrations, as well as the distances between neurotransmitter release sites and target receptors. Since modulating K^+, Ca^{2+}, or neuroligand concentrations affects neuronal activity, these volume changes could affect neurotransmission. Astrocytes might also modulate $[K^+]_o$ by changing their capacity to buffer $[K^+]_o$. Indeed, the conductance of the astrocytic K^+ channels and gap junctions thought to be involved in K^+ buffering can be regulated in vitro through signaling pathways stimulated by neuroligands *(100,101)*. If astrocyte microdomains respond to neurotransmitters independently, they could selectively alter the neuronal environment at particular spines.

In addition to influencing the neuronal environment by changing their volume and K^+ buffering, astrocyte receptor stimulation may also affect the uptake of neuroligands released during synaptic activity. It has become clear over the past several years that astrocytes also exhibit transport systems that are important in removing neurotransmitters *(102–105)*. For example, astrocytes express glutamate transporters that may participate in terminating the effects of synaptically released glutamate. Barbour et al. and Yu et al. have reported that the uptake of glutamate into astrocytes is affected by arachidonic acid *(106,107)*, a product generated following the activation of certain astrocytic glutamate receptors *(108)*. It is easy to imagine that regulating the removal of glutamate by astrocytes could markedly affect neuronal activity in a spatially restricted manner.

Astrocytes could also affect neuronal activity by slowly changing neuronal morphology and connections. Studies suggest that neuroligands stimulate astrocytes in vitro to synthesize neurotrophic factors *(109,110)*. Perhaps neuronal plasticity involves the activation of astrocytic receptors that regulate the synthesis and release of neurotrophic factors. The release of these neurotrophic factors could induce dendritic remodeling and long term changes in neuronal connections. Again, this could occur in a spatially-restricted manner.

In addition to ways that astrocytes might indirectly modulate neuronal communication, recent findings suggests that astrocytes could affect neuronal activity rather directly. Several laboratories have presented data suggesting that astrocytes release glutamate in a Ca^{2+}-dependent manner. Ca^{2+}-triggered neurotransmitter release from astrocytes was first demonstrated in cocultures of neurons and astroglia. In these experiments, astroglia released glutamate in response to several stimuli that elevated $[Ca^{2+}]_i$. This glutamate release stimulated Ca^{2+} in neighboring neurons and affected their activity through neuronal GluRs *(111–113)*.

Experiments also suggest that astrocytes *in situ* release glutamate in response to neuroligands that increase $[Ca^{2+}]_i$. To date, astrocyte Ca^{2+} responses in hippocampus, visual cortex and retina have been reported to affect neurons through what appears to be a glutamate release. In one study, the mGluR agonist tACPD elicited neuronal Ca^{2+} elevations that were dependent on the activation of iGluRs *(14)*. Through a series of experiments, the investigators made a convincing case that the neuronal iGluR-mediated Ca^{2+} responses were due to glutamate release from surrounding astrocytes *(14)*. Kang et al. also reported that increasing astrocyte Ca^{2+} influenced pyramidal neurons *(62)*. In this study, electrically stimulating astrocytes increased the frequency of miniature inhibitory postsynaptic currents (mIPSCs) in pyramidal neurons. Additional experiments were consistent with the conclusion that Ca^{2+} elevations in astrocytes triggered them to release glutamate that stimulates iGluRs on GABAergic neurons and that GABAergic neurons then increase their release of GABA onto pyramidal neurons *(62)*. Retinal astrocytes also appear to modulate synaptic activity *in situ (114)*. In these experiments, glial Ca^{2+} waves often decreased evoked neuronal spiking. As above, it appeared that increasing astrocytic Ca^{2+} led to the release of glutamate which then stimulated inhibitory neurons and decreased the evoked spiking of other neurons *(114)*. Collectively, the findings suggest that astrocytes release glutamate that modulates neuronal activity. Furthermore, it appears that astrocytic glutamate release can enhance or inhibit synaptic transmission, depending on which neurons are stimulated. These findings suggest that astrocytes may modulate synaptic transmission *in situ*. The impact of the release of neuroactive molecules by astrocytes on neurotransmission will be a very exciting area of investigation for some time to come.

Very little is known about the role of astrocyte receptor signaling during development and aging. It seems likely that expression and astrocyte receptor signaling does change throughout life. It is easy to imagine a role for astrocytic receptor signaling during development, where receptor stimulation could affect the secretion of neurotrophic factors or expression of adhesion molecules important in neuronal migration and differentiation. In the adult brain, it may be that astrocytic receptor signaling systems serve primarily to subtly modulate neurotransmission. It seems premature to speculate about how the process of aging affects astrocytic receptor signaling until we know more about the expression and role of astrocytic receptors in adult brain. However, given that the morphology and number of astrocytes change during aging, it seems likely that the impact of astrocytic receptor signaling systems will also change.

4. CONCLUSION

To conclude, it appears that astrocytes express a variety of neuroligand receptors whose activation directly affects neuronal excitability. The challenge will be to develop new methods for determining what conditions allow astrocytic receptor activation and for determining the functional outcome of activating astrocytic receptors. It seems likely that both high resolution imaging of astrocytic-neuronal interactions at the synapse and molecular disruption of specific astrocytic signaling molecules will be required to sort out the role of astrocytes in neurophysiology and brain function.

REFERENCES

1. Golgi, G. (1885) *Sulla fina anatomia della sistema nervosa* Riv. Sper. Freniatr.
2. His, W. (1889) Die Neuroblasten und deren Entstehungen im embryonalen Mark. *Arch. Anat. u. Phys.* **5,** 249–300.
3. Lugaro, E. (1907) Sulle funzioni della nevroglia. *Riv. Pat. Nerv. Ment.* **12,** 225–233.

4. Somjen, G.G. (1988) Nervenkitt: Notes on the history of the concept of neuroglia. *Glia* **1,** 2–9.
5. Shao, Y. and McCarthy, K.D. (1993) Regulation of astroglial responsiveness to neuroligands in primary culture. *Neuroscience* **55,** 991–1001.
6. Cai, Z. and Kimelberg, H.K. (1997) Glutamate receptor-mediated calcium responses in acutely isolated hippocampal astrocytes. *Glia* **21,** 380–389.
7. Kimelberg, H.K., Cai, Z., Rastogi, P., et al. (1997) Transmitter-induced calcium responses differ in astrocytes acutely isolated from rat brain and in culture. *J. Neurochem.* **68,** 1088–1098.
8. Aoki, C., Joh, T.H., and Pickel, V.M. (1987) Ultrastructural localization of beta-adrenergic receptor-like immunoreactivity in the cortex and neostriatum of rat brain. *Brain Research* **437,** 264–282.
9. Seifert, G., Rehn, L., Weber, M., and Steinhauser, C. (1997b) AMPA receptor subunits expressed by single astrocytes in the juvenile mouse hippocampus. *Brain Res. Mol. Brain Res.* **47,** 286–294.
10. Rothstein, J.D., Dykes-Hoberg, M., Pardo, C.A., et al, (1996) Knockout of glutamate transporters reveals a major role for astroglial transport in excitotoxicity and clearance of glutamate. *Neuron* **16,** 675–686.
10. Seifert, G. and Steinhauser, C. (1995) Glial cells in the mouse hippocampus express AMPA receptors with an intermediate Ca^{2+} permeability. *Eur. J. Neurosci.* **7,** 1872–1881.
11. Porter, J.T. and McCarthy, K.D. (1995a) Adenosine receptors modulate [Ca^{2+}] in hippocampal astrocytes *in situ. Journal of Neurochemistry* **65,** 1515–1523.
12. Porter, J.T. and McCarthy, K.D. (1995b) GFAP-positive hippocampal astrocytes *in situ* respond to glutamatergic neuroligands with increases in $[Ca^{2+}]_i$. *Glia* **13,** 101–112.
13. Thorlin, T., Eriksson, P.S., Ronnback, L., and Hansson, E. (1998) Receptor-activated Ca^{2+} increases in vibrodissociated cortical astrocytes: a nonenzymatic method for acute isolation of astrocytes. *J. Neurosci. Res.* **54,** 390–401.
14. Pasti, L., Volterra, A., Pozzan, T., and Carmignoto, G. (1997) Intracellular calcium oscillations in astrocytes: a highly plastic, bidirectional form of communication between neurons and astrocytes in situ. *J. Neurosci.* **17,** 7817–7830.
15. Duffy, S. and Mac Vicar, B.A. (1995) Adrenergic Calcium Signaling in Astrocyte Networks within the Hippocampal Slice. *J. Neurosci.* **15,** 5535–5550.
16. Shelton, M.K. and McCarthy, K.D. (1999) Mature hippocampal astrocytes exhibit functional metabotropic and ionotropic glutamate receptors in situ. *Glia* **26,** 1–11.
17. van den Pol, A.N., Romano, C., and Ghosh, P. (1995) Metabotropic glutamate receptor mGluR5 subcellular distribution and developmental expression in hypothalamus. *Journal of Comparative Neurology* **362,** 134–150.
18. Liu, X.B., Munoz, A., and Jones, E.G. (1998) Changes in subcellular localization of metabotropic glutamate receptor subtypes during postnatal development of mouse thalamus. *J. Comp. Neurol.* **395,** 450–465.
19. Romano, C., Seema, M.A., McDonald, C.T., O'Malley, K., van den Pol, A.N., and Olney, J.W. (1995) Distribution of metabotropic glutamate receptor mGluR5 immunoreactivity in rat brain. *Journal of Comparative Neurology* **355,** 455–469.
20. van den Pol, A.N. (1995) Presynaptic metabotropic glutamate receptors in adult and developing neurons: autoexcitation in the olfactory bulb. *J. Comp. Neurol.* **359,** 253–271.
21. Porter, J.T. and McCarthy, K.D. (1996) Hippocampal astrocytes *in situ* respond to glutamate released from synaptic terminals. *Journal of Neuroscience* **16,** 5073–5081.
22. Muller, T., Moller, T., Berger, T., Schnitzer, J., and Kettenmann, H. (1992) Calcium entry through kainate-receptors and resulting potassium-channel blockade in Bergmann glial cells. *Science* **256,** 1563–1566.
23. Clark, B. and Mobbs, P. (1992) Transmitter-operated channels in rabbit retinal astrocytes studies in situ by whole cell patch clamp. *Journal of Neuroscience* **12,** 664–673.

24. Jabs, R., Kirchhoff, F., Aronica, E.M., and Steinhauser, C. (1994) Kainate activates Ca^{2+}-permeable glutamate receptors and blocks voltage-gated K^+ currents in glial cells of mouse hippocampal slices. *Pfluggers Archives European Journal of Physiology* **426,** 310–319.
25. Steinhauser, C., Jabs, R., and Kettenmann, H. (1994) Properties of GABA and glutamate responses in identified glial cells of the mouse hippocampal slice. *Hippocampus* **4,** 19–35.
26. Backus, K.H. and Berger, T. (1995) Developmental variation of the permeability to Ca^{2+} of AMPA receptors in presumed hilar glial precursor cells. *Pflugers Arch.* **431,** 244–252.
27. Akopian, G., Kuprijanova, E., Kressin, K., and Steinhuser, C. (1997) Analysis of ion channel expression by astrocytes in red nucleus brain stem slices of the rat. *Glia* **19,** 234–246.
28. Seifert, G., Zhou, M., and Steinhauser, C. (1997a) Analysis of AMPA Receptor Properties During Postnatal Development of Mouse Hippocampal Astrocytes. *J. Neurophysiol.* **78,** 2916–2923.
29. Blackstone, C.D., Moss, S.J., Martin, L.J., Levey, A.I., Price, D.L., and Huganir, R.L. (1992) Biochemical characterization and localization of a non-N-methyl-D-aspartate glutamate receptor in rat brain. *J. Neurochem.* **58,** 1118–1126.
30. Petralia, R.S. and Wenthold, R.J. (1992) Light and electron immunocytochemical localization of AMPA-selective glutamate receptors in the rat brain. *J. of Comparative Neurology* **318,** 329–354.
31. Matute, C. and Miledi, R. (1993) Neurotransmitter receptors and voltage-dependent Ca^{2+} channels encoded by mRNA from the adult corpus callosum. *Proceedings of the National Academy of Sciences* **90,** 3270–3274.
32. Martin, L.J., Blackstone, C.D., Levey, A.I., Huganir, R.L., and Price, D.L. (1993) AMPA glutamate receptor subunits are differentially distributed in rat brain. *Neuroscience* **53,** 327–358.
33. Sato, K., Kiyama, H., and Tohyama, M. (1993) The differential expression patterns of messenger RNAs encoding non-N-methyl-D-aspartate glutamate receptor subunits (GluR1–4) in the rat brain. *Neuroscience* **52,** 515–539.
34. Molnar, E., Baude, A., Richmond, S.A., Patel, P.B., Somogyi, P., and McIlhinney, R.A. (1993) Biochemical and immunocytochemical characterization of antipeptide antibodies to a cloned GluR1 glutamate receptor subunit: cellular and subcellular distribution in the rat forebrain. *Neuroscience* **53,** 307–326.
35. Conti, F., Minelli, A., and Brecha, N.C. (1994) Cellular localization and laminar distribution of AMPA glutamate receptor subunits, mRNAs and proteins in the rat cerebral cortex. *Journal of Comparative Neurology* 241–259.
36. Matute, C., Gutierrez-Igarza, K., Rio, C., and Miledi, R. (1994) Glutamate receptors in astrocytic end-feet. *Neuroreport* **5,** 1205–1208.
37. Spreafico, R., Frassoni, C., Arcelli, P., Battaglia, G., Wenthold, R.J., and De Biasi, S. (1994) Distribution of AMPA selective glutamate receptors in the thalamus of adult rats and during postnatal development. A light and ultrastructural immunocytochemical study. *Brain Res. Dev. Brain Res.* **82,** 231–244.
38. Baude, A., Molnar, E., Latawiec, D., McIlhinney, R.A., and Somogyi, P. (1994) Synaptic and nonsynaptic localization of the GluR1 subunit of the AMPA-type excitatory amino acid receptor in the rat cerebellum. *J. Neurosci.* **14,** 2830–2843.
39. Day, N.C., Williams, T.L., Ince, P.G., Kamboj, R.K., Lodge, D., and Shaw, P.J. (1995) Distribution of AMPA-selective glutamate receptor subunits in the human hippocampus and cerebellum. *Molecular Brain Research* **31,** 17–32.
40. Mick, G. (1995) Non-N-methyl-D-aspartate glutamate receptors in glial cells and neurons of the pineal gland in a higher primate. *Neuroendocrinology* **61,** 256–264.
41. Peng, Y.W., Blackstone, C.D., Huganir, R.L., and Yau, K.W. (1995) Distribution of glutamate receptor subtypes in the vertebrate retina. *Neuroscience* **66,** 483–497.
42. Garcia-Barcina, J.M. and Matute, C. (1996) Expression of kainate-selective glutamate receptor subunits in glial cells of the adult bovine white matter. *European Journal of Neuroscience* **8,** 2379–2387.

43. Petralia, R.S., Wang, Y.X., Zhao, H.M., and Wenthold, R.J. (1996) Ionotropic and metabotropic glutamate receptors show unique postsynaptic, presynaptic, and glial localizations in the dorsal cochlear nucleus. *J. Comp. Neurol.* **372,** 356–383.
44. Burnashev, N., Khodorova, P., Jonas, P., et al. (1992) Calcium-permeable AMPA-kainate receptors in fusiform cerebellar glial cells. *Science* **256,** 1566–1570.
45. Geiger, J.R., Melcher, T., Koh, D.S., et al. (1995) Relative abundance of subunit mRNAs determines gating and Ca^{2+} permeability of AMPA receptors in principal neurons and interneurons in rat CNS. *Neuron* **15,** 193–204.
46. Petralia, R., Wang, Y.-X., and Wenthold, R.J. (1994) Histological and ultrastructural localization of the kainate receptor subunits, KA2 and GluR6/7, in the rat nervous system using selective antipeptide antibodies. *J. Compar. Neurol.* **349,** 85–110.
47. Muller, T., Grosche, J., Ohlemeyer, C., and Aronica, E.M. (1993) NMDA-activated currents in Bergmann glial cells. *Neuroreport* **4,** 671–674.
48. Wakakura, M. and Yamamoto, N. (1994) Cytosolic calcium transient increase through the AMPA/kainate receptor in cultured Muller cells. *Vision Res.* **34,** 1105–1109.
49. Holzwarth, J.A., Gibbons, S.J., Brorson, J.R., Phillipson, L.H., and Miller, R.J. (1994) Glutamate receptor agonists stimulate diverse calcium responses in different types of cultured rat cortical glial cells. *J. Neurosci.* **14,** 1879–1891.
50. Shao, Y. and McCarthy, K.D. (1997) Responses of Bergmann glia and granule neurons in situ to N-methyl-D-aspartate, norepinephrine, and high potassium. *J. Neurochem.* **68,** 2405–2411.
51. Puro, D.G., Yuan, J.P., and Sucher, N.J. (1996) Activation of NMDA receptor-channels in human retinal Muller glial cells inhibits inward-rectifying potassium currents. *Vis. Neurosci.* **13,** 319–326.
52. Aoki, C., Venkatesan, C., Go, C.-G., Mong, J.A., and Dawson, T.M. (1994) Celllular and subcellular localization of NMDA-R1 subunit immunoreactivity in the visual cortex of adult and neonatal rats. *J. Neurosci.* **14,** 5202–5222.
53. Luque, J.M. and Richards, J.G. (1995) Expression of NMDA 2B receptor subunit mRNA in Bergmann glia. *Glia* **13,** 228–232.
54. Conti, F., DeBiasi, S., Minelli, A., and Melone, M. (1996) Expression of NR1 and NR2A/B subunits of the NMDA receptor in cortical astrocytes. *Glia* **17,** 254–258.
55. Jones, E.G., Tighilet, B., Tran, B.V., and Huntsman, M.M. (1998) Nucleus- and cell-specific expression of NMDA and non-NMDA receptor subunits in monkey thalamus. *J. Comp. Neurol.* **397,** 371–393.
56. Goebel, D.J., Aurelia, J.L., Tai, Q., Jojich, L., and Poosch, M.S. (1998) Immunocytochemical localization of the NMDA-R2A receptor subunit in the cat retina. *Brain Res.* **808,** 141–154.
56a. Bettler, B., Kaupmann, K., and Bowery, N. (1998) $GABA_B$ receptors: drugs meet clones. *Current Opinion in Neurobiology* **8,** 345–350.
57. Kettenmann, H., Backus, K.H., and Schachner, M. (1984) Aspartate, glutamate and gamma-aminobutyric acid depolarize cultured astrocytes. *Neurosci. Lett.* **52,** 25–29.
58. Mudrick-Donnon, L.A., Williams, P.J., Pittman, Q.J., and Mac Vicar, B.A. (1993) Postsynaptic potentials mediated by GABA and dopamine evoked in stellate glial cells of the pituitary pars intermedia. *Journal Neuroscience* **13,** 4660–4668.
59. Muller, T., Fritschy, J.M., Grosche, J., Pratt, G.D., Mohler, H., and Kettenmann, H. (1994) Developmental regulation of voltage-gated K^+ channel and GABAA receptor expression in Bergmann glial cells. *J. Neurosci.* **14,** 2503–2514.
60. Fraser, D.D., Duffy, S., Angelides, K.J., Perez-Velazquez, J.L., Kettenmann, H., and Mac Vicar, B.A. (1995) $GABA_A$/benzodiazepine receptors in acutely isolated hippocampal astrocytes. *J. Neurosci.* **15,** 2720–2732.
61. Butt, A.M. and Jennings, J. (1994) The astrocyte response to gamma-aminobutyric acid attenuates with age in the rat optic nerve. *Proc. R. Soc. Lond. B. Biol. Sci.* **258,** 9–15.

62. Kang, J., Jiang, L., Goldman, S.A., and Nedergaard, M. (1998) Astrocyte-mediated potentiation of inhibitory synaptic transmission. *Nature Neuroscience* **1,** 683–692.
63. Blake, J.F., Cao, C.Q., Headley, P.M., Collingridge, G.L., Brugger, F., and Evans, R.H. (1993) Antagonism of baclofen-induced depression of whole-cell synaptic currents in spinal dorsal horn neurons by the potent $GABA_B$ antagonist CGP55845A. *Neuropharmacology* **32,** 1437–1440.
64. Nilsson, M., Eriksson, P.S., Ronnback, L., and Hansson, E. (1993) GABA induces Ca^{2+} transients in astrocytes. *Neuroscience* **54,** 605–614.
65. Lerea, L.S. and McCarthy, K.D. (1990) Neuron-associated astroglial cells express beta- and alpha 1-adrenergic receptors in vitro. *Brain Res.* **521,** 7–14.
66. Lerea, L.S. and McCarthy, K.D. (1989) Astroglial cells in vitro are heterogenous with respect to expression of the alpha 1-adrenergic receptor. *Glia* **2,** 135–147.
67. McCarthy, K.D., Enkvist, K., and Shao, Y. (1995) Astroglial adrenergic receptors: expression and function. In: *Neuroglia,* edited by Kettenmann, H. & Ransom, B.R. Oxford, New York, pp. 354–366.
68. Kirischuk, S., Tuschick, S., Verkhratsky, A., and Kettenmann, H. (1996) Calcium signalling in mouse Bergmann glial cells mediated by alpha1-adrenoreceptors and H1 histamine receptors. *Eur. J. Neurosci.* **8,** 1198–1208.
69. Sutin, J. and Shao, Y. (1992) Resting and reactive astrocytes express adrenergic receptors in the adult rat brain. *Brain Res. Bull.* **29,** 277–284.
70. Shao, Y. and Sutin, J. (1992) Expression of adrenergic receptors in individual astrocytes and motor neurons isolated from the adult rat brain. *Glia* **6,** 108–117.
71. Milner, T.A., Lee, A., Aicher, S.A., and Rosin, D.L. (1998) Hippocampal alpha2a-adrenergic receptors are located predominantly presynaptically but are also found postsynaptically and in selective astrocytes. *J. Comp. Neurol.* **395,** 310–327.
72. Lee, A., Rosin, D.L., and Van Bockstaele, E.J. (1998) alpha2A-adrenergic receptors in the rat nucleus locus coeruleus: subcellular localization in catecholaminergic dendrites, astrocytes, and presynaptic axon terminals. *Brain Res.* **795,** 157–169.
73. Salm, A.K. and McCarthy, K.D. (1989) Expression of beta-adrenergic receptors by astrocytes isolated from adult rat cortex. *Glia* **2,** 346–352.
74. Aoki, C. (1992) Beta-adrenergic receptors: astrocytic localization in the adult visual cortex and their relation to catecholamine axon terminals as revealed by electron microscopic immunocytochemistry. *J. Neurosci.* **12,** 781–792.
75. Aoki, C. and Pickel, V.M. (1992) C-terminal tail of beta-adrenergic receptors: immunocytochemical localization within astrocytes and their relation to catecholaminergic neurons in N. tractus solitarii and area postrema. *Brain Res.* **571,** 35–49.
76. Mantyh, P.W., Rogers, S.D., Allen, C.J., et al. (1995) Beta 2-adrenergic receptors are expressed by glia in vivo in the normal and injured central nervous system in the rat, rabbit, and human. *J. Neurosci.* **15,** 152–164.
76a. Aoki, C. (1997) Differential timing for the appearance of neuronal and astrocytic beta-adrenergic receptors in the developing rat visual cortex as revealed by light and electron-microscopic immunocytochemistry. *Vis. Neurosci.* **14,** 1129–1142.
77. Van der Zee, E.A., Matsuyama, T., Strosberg, A.D., Traber, J., and Luiten, P.G. (1989) Demonstration of muscarinic acetylcholine receptor-like immunoreactivity in the rat forebrain and upper brainstem. *Histochemistry* **92,** 475–485.
77a. Shelton, M.K. and McCarthy, K.D. (2000) Hippocampal astrocytes in situ exhibit functional M1 muscarinic acetylcholine and H1 histamine receptors in situ. *J. Neurochem.* **79,** 555–563.
78. Beldhuis, H.J., Everts, H.G., Van der Zee, E.A., Luiten, P.G., and Bohus, B. (1992) Amygdala kindling-induced seizures selectively impair spatial memory. 2. Effects on hippocampal neuronal and glial muscarinic acetylcholine receptor. *Hippocampus* **2,** 411–419.

79. Van der Zee, E.A., Streefland, C., Strosberg, A.D., Schroder, H., and Luiten, P.G. (1992) Visualization of cholinoceptive neurons in the rat neocortex: colocalization of muscarinic and nicotinic acetylcholine receptors. *Brain Res. Mol. Brain Res.* **14,** 326–336.
80. Van der Zee, E.A., De Jong, G.I., Strosberg, A.D., and Luiten, P.G. (1993) Muscarinic acetylcholine receptor-expression in astrocytes in the cortex of young and aged rats. *Glia* **8,** 42–50.
81. Van der Zee, E.A., Streefland, C., Strosberg, A.D., Schroder, H., and Luiten, P.G. (1991) Colocalization of muscarinic and nicotinic receptors in cholinoceptive neurons of the suprachiasmatic region in young and aged rats. *Brain Res.* **542,** 348–352.
82. Messamore, E., Bogdanovich, N., Schroder, H., and Winblad, B. (1994) Astrocytes associated with senile plaques possess muscarinic acetylcholine receptors. *Neuroreport* **5,** 1473–1476.
83. Schwartz, J.C., Arrang, J.M., Garbarg, M., Pollard, H., and Ruat, M. (1991) Histaminergic transmission in the mammalian brain. *Physiol. Rev.* **71,** 1–51.
84. Inagaki, N. and Wada, H. (1994) Histamine and prostanoid receptors on glial cells. *Glia* **11,** 102–109.
85. Verkhratsky, A., Orkand, R.K., and Kettenmann, H. (1998) Glial calcium: homeostasis and signaling function. *Physiol. Rev.* **78,** 99–141.
86. Bernstein, M., Lyons, S.A., Moller, T., and Kettenmann, H. (1996) Receptor-mediated calcium signalling in glial cells from mouse corpus callosum slices. *J. Neurosci. Res.* **46,** 152–163.
87. Maire, J.C., Medilanski, J., and Straub, R.W. (1984) Release of adenosine, inosine and hypoxanthine from rabbit non-myelinated nerve fibres at rest and during activity. *J. Physiol. (Lond)* **357,** 67–77.
88. Edwards, F.A., Gibb, A.J., and Colquhoun, D. (1992) ATP receptor-mediated synaptic currents in the central nervous system. *Nature* **359,** 144–147.
89. Pilitsis, J.G. and Kimelberg, H.K. (1998) Adenosine receptor mediated stimulation of intracellular calcium in acutely isolated astrocytes. *Brain Research* **798,** 294–303.
90. Kirischuk, S., Moller, T., Voitenko, N., Kettenmann, H., and Verkhratsky, A. (1995) ATP-induced cytoplasmic calcium mobilization in Bergmann glial cells. *J. Neurosci.* **15,** 7861–7871.
91. Jabs, R., Paterson, I.A., and Walz, W. (1997) Qualitative analysis of membrane currents in glial cells from normal and gliotic tissue in situ: down-regulation of Na^+ current and lack of P2 purinergic responses. *Neuroscience* **81,** 847–860.
92. Neal, M.J., Cunningham, J.R., and Dent, Z. (1998) Modulation of extracellular GABA levels in the retina by activation of glial P2X-purinoceptors. *Br. J. Pharmacol.* **124,** 317–322.
93. Nilsson, M., Hansson, E., and Ronnback, L. (1991) Heterogeneity among astroglial cells with respect to 5HT-evoked cytosolic Ca^{2+} responses. A microspectrofluorimetric study on single cells in primary culture. *Life Sciences* **49,** 1339–1350.
94. Deecher, D.C., Wilcox, B.D., Dave, V., Rossman, P.A., and Kimelberg, H.K. (1993) Detection of 5-hydroxytryptamine2 receptors by radioligand binding, northern blot analysis, and Ca^{2+}responses in rat primary astrocyte cultures. *J. Neurosci. Res.* **35,** 246–256.
95. Jalonen, T.O., Margraf, R.R., Wielt, D.B., Charniga, C.J., Linne, M.L., and Kimelberg, H.K. (1997) Serotonin induces inward potassium and calcium currents in rat cortical astrocytes. *Brain Res.* **758,** 69–82.
96. Azmitia, E.C., Gannon, P.J., Kheck, N.M., and Whitaker-Azmitia, P.M. (1996) Cellular localization of the 5-HT1A receptor in primate brain neurons and glial cells. *Neuropsychopharmacology* **14,** 35–46.
96. Whitaker-Azmitia, P.M., Clarke, C., and Azmitia, E.C. (1993) Localization of 5-HT1A receptors to astroglial cells in adult rats: implications for neuronal-glial interactions and psychoactive drug mechanism of action. *Synapse* **14,** 201–205.
97. Dhawan, B.N., Raghubir, C.R., Reisine, T., Bradley, P.B., Portoghese, P.S.A., and Hamon, M. (1996) International union of pharmacology. XII, Classification of opioid receptors. *Pharmacolog. Reviews* **48,** 567–592.

98. Lightman, S.L., Ninkovic, M., Hunt, S.P., and Iversen, L.L. (1983) Evidence for opiate receptors on pituicytes. *Nature* **305,** 235–237.
99. Bunn, S.J., Hanley, M.R., and Wilkin, G.P. (1985) Evidence for a kappa-opioid receptor on pituitary astrocytes: an autoradiographic study. *Neurosci. Lett.* **55,** 317–323.
100. Enkvist, M.O., Holopainen, I., and Akerman, K.E. (1989) Alpha-receptor and cholinergic receptor-linked changes in cytosolic Ca^{2+} and membrane potential in primary rat astrocytes. *Brain Res.* **500,** 46–54.
101. Enkvist, M.O. and McCarthy, K.D. (1992) Activation of protein kinase C blocks astroglial gap junction communication and inhibits the spread of calcium waves. *J. Neurochemistry* **59,** 519–526.
102. Rothstein, J.D., Martin, L., Levey, A.I., et al. (1994) Localization of neuronal and glial glutamate transporters. *Neuron* **13,** 713–725.
103. Lehre, K.P., Levy, L.M., Otterson, O.P., Storm-Mathisen, J., and Danbolt, N.C. (1995) Differential expression of two glial glutamate transporters in the rat brain: quantitative and immunocytochemical observations. *J. Neurosci.* **15,** 1835–1853.
105. Tanaka, K., Watase, K., Manage, T., et al. (1997) Epilepsy and exacerbation of brain injury in mice lacking the glutamate transporter GLT-1. *Science* **276,** 1699–1702.
106. Yu, A.C., Chan, P.H., and Fishman, R.A. (1986) Effects of arachidonic acid on glutamate and gamma-aminobutyric acid uptake in primary cultures of rat cerebral cortical astrocytes and neurons. *J. Neurochem.* **47,** 1181–1189.

5
Glia and Extracellular Space Diffusion Parameters in the Injured and Aging Brain

Eva Syková

1. INTRODUCTION

The extracellular space (ECS) is the microenvironment of the nerve cells and an important communication channel *(1–5)*. It includes ions, transmitters, metabolites, peptides, neurohormones, other neuroactive substances and molecules of the extracellular matrix, and directly or indirectly affects neuronal and glial cell functions. Populations of neurons can interact both by synapses and by the diffusion of ions and neurotransmitters in the ECS Since glial cells do not have synapses, their communication with neurons is mediated by the diffusion of ions and neuroactive substances in the ECS.

Neurons and glia release ions, transmitters, and various other neuroactive substances into the ECS. Substances which are released nonsynaptically diffuse through the ECS and bind to extrasynaptic, usually high-affinity, binding sites located on neurons, axons, and glial cells. This type of extrasynaptic transmission is also called "diffusion transmission" (neuroactive substances diffuse through the ECS) or "volume transmission" (neuroactive substances move through the volume of the ECS) *(4–7)*. Neuroactive substances can diffuse through the ECS to target neurons, glial cells or capillaries. This mode of communication without synapses can function between neurons (even those far distant from release sites), as well as between neurons and glial cells, and may provide a mechanism of long-range information processing in functions such as vigilance, sleep, chronic pain, hunger, depression or plastic changes and memory formation. The size and irregular geometry of diffusion channels in the ECS (tissue tortuosity and anisotropy, *see* **Subheading 2.**) substantially affect and/or direct the movement of various neuroactive substances in the CNS (Fig. 1) and thereby modulate neuronal signaling and neuron-glia communication. Changes in the ECS diffusion parameters may, therefore, result in impairment of the signal transmission and contribute to functional deficits and to neuronal damage. Dynamic changes in ECS ionic composition, volume, and geometry accompany neuronal activity, glial development and proliferation, aging, and some brain pathological states.

2. *ECS Composition*

Cellular elements and blood vessels fill about 80% of the total CNS tissue volume and the remaining portion (15–25 %) is the ECS. ECS ionic changes resulting from

From: *Neuroglia in the Aging Brain*
Edited by: Jean S. de Vellis © Humana Press Inc., Totowa, NJ

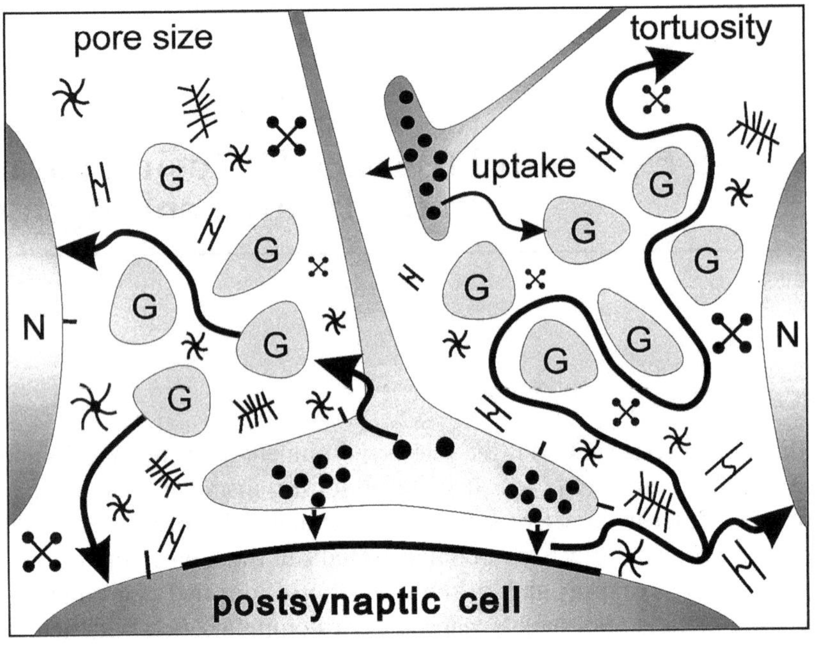

Fig. 1. Schematic of CNS architecture. The CNS architecture is composed of neurons (N), axons, glial cells (G), cellular processes, molecules of the extracellular matrix and intercellular channels between the cells. The architecture affects the movement (diffusion) of substances in the brain, which is critically dependent on channel size, extracelluar space tortuosity and cellular uptake.

transmembrane ionic shifts during neuronal activity depolarize neighboring neurons and glial cells, enhance or depress their excitability, and affect ion channel permeability *(1,3,4,8,9,10)*. These ionic changes may also lead to the synchronization of neuronal activity and stimulate glial cell function.

In the mammalian CNS, the average ionic constituents of the ECS are basically the same as in the cerebrospinal fluid: About 141 m*M* Na^+, 124 m*M* Cl^-, 3 m*M* K^+, 121 m*M* HCO_3^-, 1.2 m*M* Ca^{2+} and about 2.5 m*M* Mg^{2+}. However, in vivo measurements with ion-selective microelectrodes have revealed local changes in ECS ionic composition resulting from neuronal activity. The local changes in ion activity are localized to areas of high spontaneous activity (Fig. 2) *(11,12)*, or in areas being activated by adequate stimuli, e.g., tactile, visual, auditory, taste, aversive and painful stimuli *(13–17)*.

An activity-related increase in extracellular K^+ activity ($[K^+]_e$), alkaline and acid shifts in extracellular pH (pH_e), and a decrease in extracellular Ca^+ concentration ($[Ca^{2+}]_e$) have been found to accompany neuronal activity in a variety of animals and brain regions in vivo, as well as in vitro *(1,3,4,9)*. After sustained adequate stimulation or after repetitive electrical stimulation of the afferent input, the ionic changes reach a certain steadystate, the so-called "ceiling" level (Fig. 2). The K^+ ceiling level in mammalian brain or spinal cord is 7–8 m*M* K^+; the alkaline shifts do not exceed 0.02 pH units and the acid shifts 0.25 pH units. There is now convincing evidence for the neuronal origin of the extracellular alkaline shift and the glial origin of the activity-related acid shift. After tetanic stimulation of the sciatic nerve (30–100 Hz), the $[K^+]_e$ ceiling

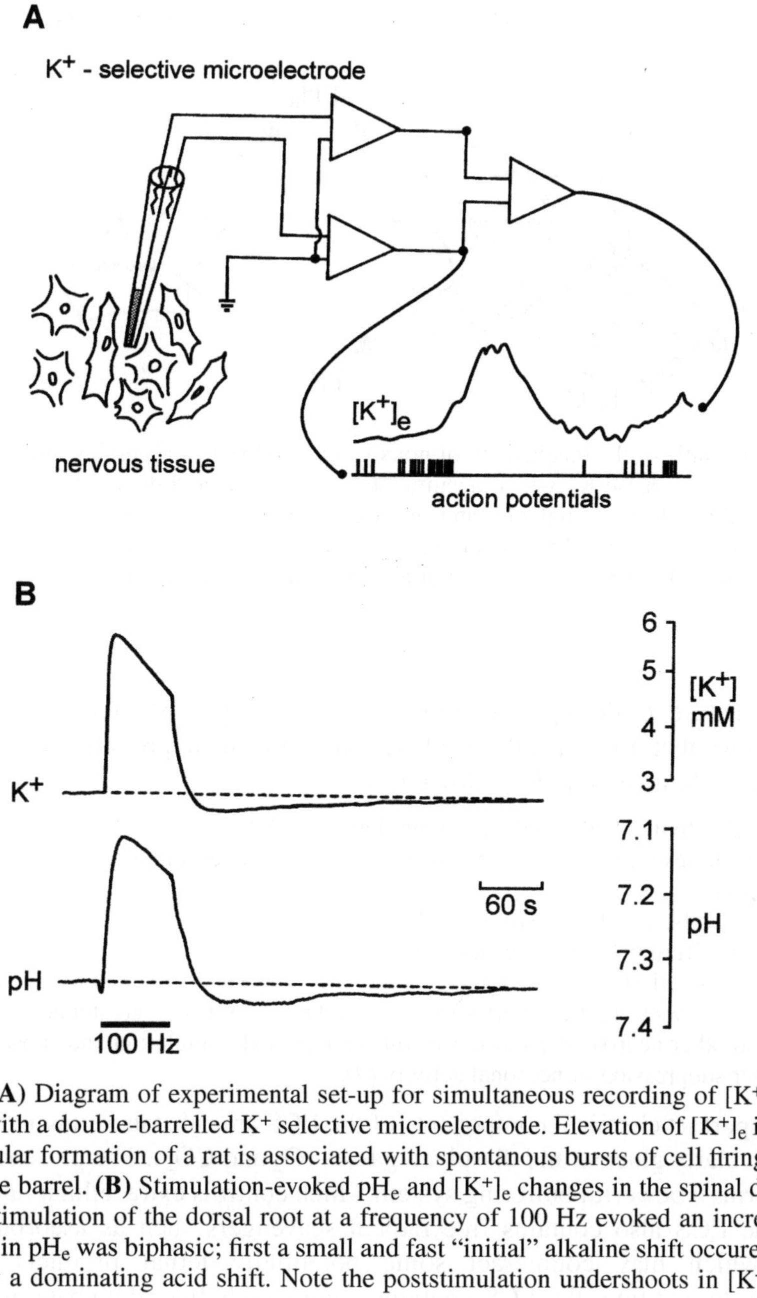

Fig. 2. **(A)** Diagram of experimental set-up for simultaneous recording of $[K^+]_e$ and action potentials with a double-barrelled K^+ selective microelectrode. Elevation of $[K^+]_e$ in the unstimulated reticular formation of a rat is associated with spontanous bursts of cell firing recorded by the reference barrel. **(B)** Stimulation-evoked pH_e and $[K^+]_e$ changes in the spinal dorsal horn of a rat. The stimulation of the dorsal root at a frequency of 100 Hz evoked an increase in $[K^+]_e$. The change in pH_e was biphasic; first a small and fast "initial" alkaline shift occured, which was followed by a dominating acid shift. Note the poststimulation undershoots in $[K^+]_e$ as well as in pH_e.

level in the adult rat is attained in 5–8 s, whereas the ceiling level of the acid shift is reached in 10–20 s. When stimulation is continued beyond this, a gradual decrease of both transients, $[K^+]_e$ and pH_e, occur after the ceiling levels are reached because of homeostatic mechanisms in neurons and glia (Fig. 2). Extra- and intracellular K^+ and pH homeostesis is ensured by a number of mechanisms in adult brain *(1,3,8,9,10)*. However, homeostatsis is not so efficiently controlled during development, when glial

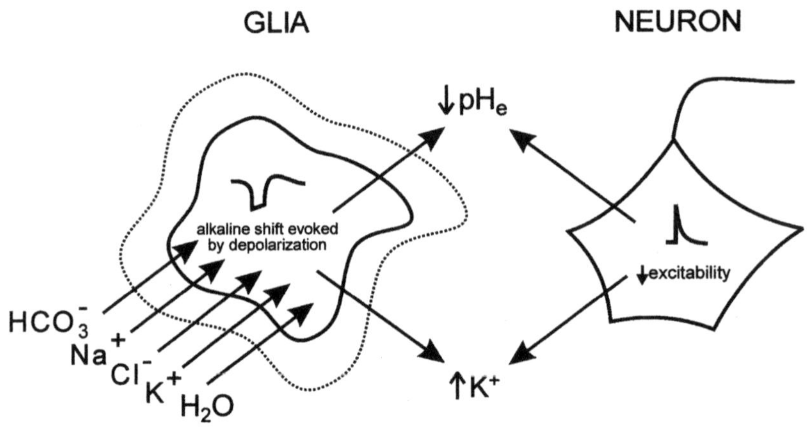

Fig. 3. Schematic of the mechanism of nonspecific feedback suppressing neuronal excitability. Active neurons release K^+ which accumulates in the ECS and depolarizes glial cells. This causes an alkaline shift in glial pH_i and an acid shift in pH_e. Extracellular acidosis further supresses neuronal activity. Transmembrane ionic movements result in glial swelling, ECS volume decrease and therefore in the greater accumulation of ions and neuroactive substances in the ECS.

cells are immature, or during pathological states when glia becomes reactive *(18–20)*. Figure 3 shows that a nonspecific feedback mechanism suppressing neuronal activity exists in the CNS, involving the following steps:

1. Neuronal activity results in the accummulation of $[K^+]_e$;
2. K^+ depolarizes glial cells, and this depolarization induces an alkaline shift in the glial cytoplasm;
3. The glial cells, therefore, extrude acid;
4. The acid shift in pH_e decreases the neuronal excitability;
5. Glial cells swell, because the ionic changes are always accompanied by water shifts;
6. The ECS shows a compensatory shrinkage, and this results in a greater accumulation of ions and other neuroactive substances during repeated stimulation and subsequently in a further suppression of neuronal activity *(4)*.

Other important chemical components of the ECS are substances involved in metabolism, particularly glucose and dissolved gasses (O_2 and CO_2). The presence of HCO_3^- and CO_2 forms a powerful buffering system which controls extracellular and intracellular pH. The ECS also contains free radical scavengers such as ascorbate and glutathione, which may counteract some potentially lethal products of oxygen metabolism. In addition, the ECS contains amino acids like glutamate and aspartate, catecholamines, indolamines such as dopamine and serotonin, various opioid peptides, nitric oxide (NO) and growth hormones. Transmitters in the ECS can bind to extrasynaptically located high affinity binding sites on neurons and glia.

The solution in the ECS is not a simple salt solution. It has become apparent that long-chain polyelectrolytes, either attached or unattached to cellular membranes, are present in ECS. The ECS also contains a number of glycosaminoglycans (e.g., hyaluronate), glycoproteins, and proteoglycans that constitute the extracellular matrix (ECM). Various ECM molecules and adhesion molecules have also been described, e.g., fibronectin, tanescin, laminin, and so on, *(21,22)*, the content of which can dynamically change during development, aging, wound healing, and many pathological processes. ECM molecules are produced by both neurons and glia. These molecules have been suggested to cordon off distinct functional units in the CNS (groups of neurons, axon tracts, and nuclear groups). As shown in figure 1, these large molecules can slow down the movement (diffusion) of various neuroactive substances through the ECS. More importantly, these molecules can hinder the diffusion of molecules so that they are confined to certain places, whereas the diffusion to other brain regions will be facilitated.

Glial cells maintain not only ECS ionic homeostasis, but also ECS volume homeostasis (by swelling and shrinking during ionic shifts); they produce various extracellular matrix molecules, and when hypertrophied or proliferating form diffusion barriers *(4)*. In this way glial cells can critically affect the permissiveness of the tissue, synaptic as well as extrasynaptic transmission, activity-dependent synaptic plasticity, neurogenesis, and regeneration.

3. ECS VOLUME AND GEOMETRY: DIFFUSION PARAMETERS

The diffusion of substances released from neurons or glia into the narrow ECS is hindered by the size of the channels between cells, the presence of membranes, fine processes, macromolecules of the extracellular matrix, charged molecules, and also the cellular uptake or degradation of these substances by enzymes (Fig. 1). Diffusion in the ECS obeys Fick's law, albeit subject to important modifications. First, diffusion is constrained by the restricted volume of the tissue available for diffusing particles, i.e., by the extracellular space volume fraction (α), which is the ratio between the volume of the ECS and total tissue volume. It is now clear that the ECS in adult brain amounts to about 20% of the total brain volume, i.e. $\alpha = 0.2$. The second modification to Fick's law is that the free diffusion coefficient *(D)* in the brain is reduced by the tortuosity factor (λ). ECS tortuosity is defined as $\lambda = (D/ADC)^{0.5}$, where *ADC* is the apparent diffusion coefficient in the brain and *D* is a free diffusion coefficient. As a result of tortuosity, *D* is reduced to an apparent diffusion coefficient $ADC = D/\lambda^2$. Thus, any substance diffusing in the ECS is hindered by membrane obstructions, glycoproteins, macromolecules of the ECM, charged molecules, and fine neuronal and glial cell processes. Third, substances released into the ECS are transported across membranes by nonspecific concentration-dependent uptake *(k')*. In many cases however, these substances are transported by energy-dependent uptake systems that obey nonlinear kinetics *(23)*. When these three factors (α, λ and k') are incorporated into Fick's law, diffusion in the CNS is described fairly satisfactorily *(24)*. To describe the diffusion constraints on neuroactive substances and their dynamic changes in vivo, the diffusion parameters of the ECS are studied using the real-time ionotophoretic method *(4,5)*. This method allows us to follow the diffusion of extracellular markers applied by iontophoresis (Fig. 4). The absolute values of the ECS volume, *ADCs,* tortuosity as well as nonspecific cellular uptake can be obtained both in vivo or in vitro, during physiological as well as in pathological states. These three parameters are extracted by a nonlinear curve-fitting simplex algorithm operating on the diffusion curve described by Equation [1] below, which represents the behavior of TMA^+, assuming that it spreads out with spherical symmetry, when the iontophoresis current is applied for duration *S.*

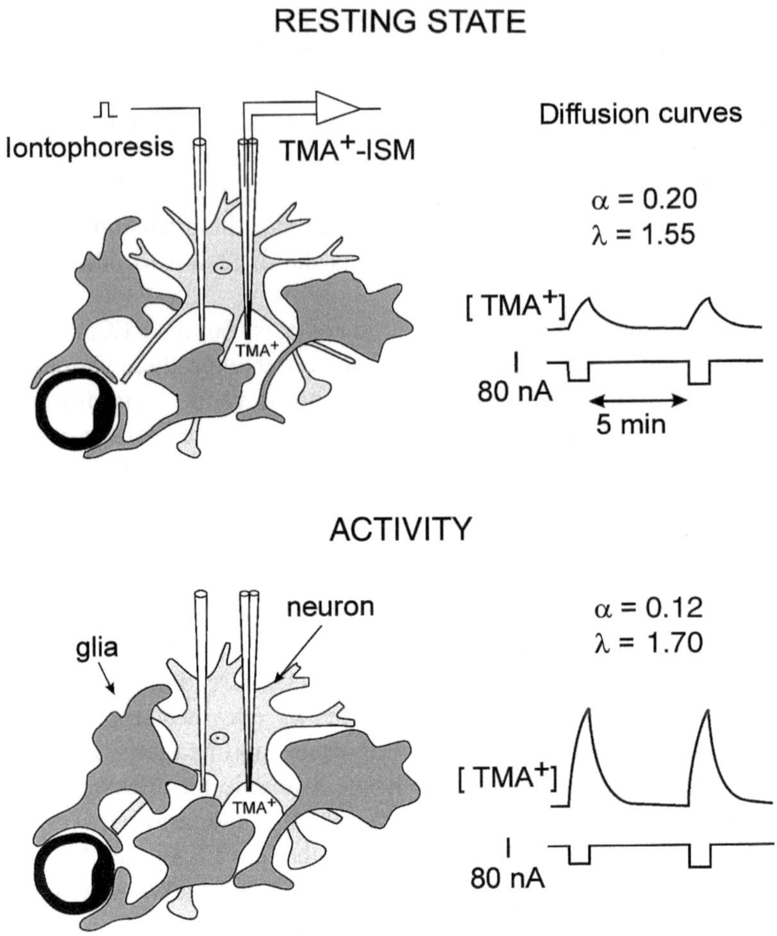

Fig. 4. Experimental set-up, TMA^+ diffusion curves and typical ECS diffusion parameters α (volume fraction) and λ (tortuosity) in the CNS. Left: Schema of the experimental arrangement. A TMA^+ selective double-barrelled ion-selective microelectrode (ISM) was glued to a bent iontophoresis microelectrode. The separation between electrode tips was 130–200 μm. Right: Typical TMA^+ diffusion curves obtained in unstimulated brain (resting state) and during stimulation (activity), evoked by the same iontophoretic current of 80 nA. ECS in unstimulated brain is 20% (volume fraction $\alpha = 0.20$) and tortuosity is about 1.55. When the ECS is smaller due to cell swelling during activity, the diffusion curves are bigger. ECS volume may decrease to about 12% ($\alpha = 0.12$) although tortuosity increases ($\lambda = 1.70$).

In this expression, C is the concentration of the ion at time t and distance r. The equation governing the diffusion in brain tissue is:

$$C = G(t) \qquad t < S, \text{ for the rising phase of the curve}$$

$$C = G(t) - G(t - S) \qquad t < S, \text{ for the falling phase of the curve.}$$

The function $G(u)$ is evaluated by substituting t or $t - S$ for u in the following equation *(24):*

$$G(u) = (Q\lambda^2/8\pi D\alpha r)\ \{\exp[r\lambda(k'/D)^{1/2}]\ \text{erfc}\ [r\lambda/2(Du)^{1/2} + (k'u)^{1/2}] + \exp[-r\lambda(k'/D)^{1/2}]\ \text{erfc}[r\lambda/2(Du)^{1/2} - (k'u)^{1/2}]\} \quad [1]$$

The quantity of TMA^+ or TEA^+ delivered to the tissue per second is $Q = In/zF$, where I is the step increase in current applied to the iontophoresis electrode, n is the transport number, z is the number of charges associated with the substance iontophoresed (+ for TMA^+ or TEA^+), and F is Faraday's electrochemical equivalent. The function "erfc" is the complementary error function. When the experimental medium is agar, by definition, $\alpha=1=\lambda$ and $k' = 0$, and the parameters n and D are extracted by the curve fitting. Knowing n and D, the parameters α, λ, and k' can be obtained when the experiment is repeated in neural tissue.

The other methods used so far to study ECS volume and geometry in vivo have been less comprehensive, since either they can only give information about relative changes in ECS volume, or these changes are only partly related to cell swelling and ECS shrinkage and some other, often unknown mechanisms, can contribute to these signals. These methods include light scattering *(25),* measurements of tissue resistance *(26–28),* measurement of the *ADCs* of molecules tagged with fluorescent dye and followed by optical imaging *(29)* or changes in the *ADC* of water (ADC_W) measured by diffusion-weighted NMR *(30–33).* The optical methods, light reflectance and light transmittance, used in vitro, particularly on brain slices, are believed to reflect changes in ECS volume, however direct evidence is missing. Recently, we found that changes in ECS volume in spinal cord slices measured by the "TMA^+ method" have a different time course than those revealed by light scattering (Syková et al., unpublished data). Diffusion-weighted NMR methods give information about the water diffusion coefficient, but the relationship between water movement, diffusion ADC_W maps and the changes in cell volume and ECS diffusion parameters (ECS volume fraction and tortuosity) is not yet well understood.

Neuroactive substances released constantly into the ECS will accumulate in this limited volume more rapidly than in free solution. Tortuosity (which is absent in a free medium) also causes a greater and more rapid accumulation of released substances. CNS tortuosity reduces the diffusion coefficient for small molecules by a factor of about 2.5 in many CNS regions. Larger molecules (with a relative molecular mass above 10 kDa), have a smaller diffusion coefficient than small molecules and are significantly more hindered in their diffusion, and therefore exhibit larger tortuosity *(34,35).* Even large proteins, e.g., negatively charged globular proteins such as bovine serum albumin (66 kDa), or dextrans of 70 kDa, still migrate through the narrow interstices of brain slices *(29,36).* Recently, the diffusion properties of two types of rather large copolymers of N-(2-hydroxypropyl)methacrylamide (HPMA), developed as water-soluble anti-cancer drug carriers, were studied in rat cortical slices—long-chain HPMA with M_r = 220 kDa and globular (bulky) HPMA, containing either albumin (179 kDa) or immunoglobulin (IgG) (319 kDa) in the center with HPMA side branches. Using the integrative optical imaging method and pressure microinjection of fluorescein-tagged polymers, the apparent diffusion coefficients *(ADC)* were obtained in rat cortical slices *(37).* The tortuosity for long-chain HPMA was found to be smaller than the tortuosity for globular copolymers *(38).* These data show that rather than M_r, the shape of the substance is the limiting factor in its movement through the extracellular space.

4. DIFFUSION INHOMOGENEITY AND ANISOTROPY

ECS diffusion parameters are different in different parts of the CNS. For example, it has been recognized that the TMA^+ diffusion parameters in the sensorimotor cortex of young adult rats in vivo are heterogeneous *(39)*. The mean volume fraction gradually increases from $\alpha = 0.19$ in cortical layer II to $\alpha = 0.23$ in cortical layer VI. These typical differences are apparent in each individual animal. In subcortical white matter (corpus callosum) the volume fraction is always lower than in cortical layer VI, often between 0.19–0.20 *(40)*. The mean tortuosity values are typically in the range of 1.51–1.65, and k' values vary between $3.3–6.3 \times 10^{-3}\ s^{-1}$. There is also a heterogeneity in the spinal cord, the mean values of the volume fraction being highest in the ventral horn ($\alpha = 0.23$) and lowest in the white matter ($\alpha = 0.18$) *(41–43)*. Similar α values ($\alpha = 0.21–0.22$) have been found throughout the rat hippocampus in vivo *(44,45)*, as well as in hippocampal slices *(46)*.

Significant differences in tortuosity have been found in various brain regions, showing that the local architecture is significantly different. There is increasing evidence that diffusion in brain tissue is anisotropic. Isotropy is defined as the state of constant λ in any direction from a point source, whereas anisotropy indicates a difference in λ along different axes. To test for anisotropy, the ECS diffusion parameters are measured in three orthogonal axes *x, y* and *z*. Anisotropic diffusion, mediated by the structure of neurons, dendrites, axons and glial processes, can channel the migration of substances in the ECS (preferred diffusion in one direction, e.g., along the axons) and may, therefore, account for a certain degree of specificity in diffusion transmission. Indeed, anisotropic diffusion was described using the "TMA^+ method" in the white matter of the corpus callosum and spinal cord *(40,43)* as well as in gray matter in the molecular layer of the cerebellum *(47)*, in hippocampus *(44)* and in the auditory but not in the somatosensory cortex *(48)*. Using MRI, evidence of anisotropic diffusion in white matter was found in cat brain *(49)* as well as in human brain *(50)*. Therefore, not only the diffusion of large molecules, such as TMA^+ or dextrans, but even the diffusion of water, are prevented by the presence of myelin sheaths. Because of the distinct diffusion characteristics, the extracellular molecular traffic will be different in various brain regions. The anisotropy of the white and gray matter could enable different modes of diffusion transmission in these regions.

4.1. Role of Glia in ECS Volume and Geometry Changes

Ions as well as neurotransmitters released into the ECS during neuronal activity or pathological states interact not only with the postsynaptic and presynaptic membranes, but also with extrasynaptic receptors, including those on glial cells. Stimulation of glial cells leads to the activation of ion channels, second messengers and intracellular metabolic pathways, and to changes in glial volume that are accompanied by dynamic variations in ECS volume, particularly the swelling and possible rearrangement of glial processes. In addition to their role in the maintenance of extracellular ionic homeostasis, glial cells may thus, by regulating their volume, influence the extracellular pathways for neuroactive substances.

Many pathological processes in the CNS are accompanied by a loss of cells or neuronal processes, astrogliosis, demyelination, and changes in the extracellular matrix, all of which may affect the apparent diffusion coefficients of neuroactive substances. Several animal models have been developed to study changes in ECS diffusion parameters. Brain injury of any kind elicits reactive gliosis, involving both hyperplasia and hypertrophy of astrocytes, which show intense staining for glial fibrillary acidic protein (GFAP) *(51)*. Astrogliosis is also a typical characteristic of cortical stab wounds in rodents *(51)*. The lesion is typically accompanied by an ECS volume increase and a substantial tortuosity increase to mean values of α of about 0.26 and λ of about 1.77 (Fig. 5A) *(52)*.

Similarly, both the size of the ECS (α) and, surprisingly, also λ, were significantly higher in cortical grafts than in host cortex, about 0.35 and 1.79, respectively, as is also the case in gliotic cortex after stab wounds. Both α and λ were increased in cortical grafts of fetal tissue transplanted to the midbrain, where severe astrogliosis compared to host cortex was found, but not in fetal grafts placed into a cavity in the cortex, where only mild astrogliosis occurred *(53)*. Another characteristic feature of cortical grafts into midbrain was the variability of α and λ. The different values found at various depths of the grafts correlated with the morphological heterogeneity of the graft neuropil. These measurements show that even when the ECS in gliotic tissue or in cortical grafts is larger than in normal cortex, the tortuosity is still higher, and the diffusion of chemical signals in such tissue may be hindered. Limited diffusion may also have a negative impact on the viability of grafts in host brains. Compared to host cortex, immunohistochemistry showed myelinated patches and a larger number of hypertrophic astrocytes in areas of high λ values, suggesting that more numerous and/or thicker glial cell processes might be a cause of the increased tortuosity.

Cell swelling and astrogliosis (manifested as an increase in GFAP) were also evoked in isolated rat spinal cords of 4–21-d-old rats by incubation in either 50 m*M* K^+ or hypotonic solution (235 mosmol kg^{-1}). Application of K^+ and hypotonic solution resulted at first in a decrease in the ECS volume fraction and in an increase in tortuosity in spinal gray matter (Fig. 5B). These changes resulted from cell swelling, since the total water content (TW) in spinal cord was unchanged and the changes were blocked in Cl^--free solution and slowed down by furosemide and bumetanide. During a continuous 45 min application, α and λ started to return toward control values, apparently due to the shrinkage of previously swollen cells since TW remained unchanged. This return was blocked by fluoroacetate, suggesting that most of the changes were due to the swelling of glia. A 45 min application of 50 m*M* K^+ and, to a lesser degree, of hypotonic solution evoked astrogliosis, which persisted after washing out these solutions with physiological saline. During astrogliosis λ increased again to values as high as 2.0, whereas α either returned to or increased above control values. This persistant increase in λ after washout was also found in white matter. These data show that glial swelling and astrogliosis are associated with a persistant increase in ECS diffusion barriers.

4.2. Diffusion Parameters and Extracellular Matrix

ECM molecules and other large molecules can also affect the tortuosity of the ECS. Their possible effect on changes in TMA^+ diffusion parameters has been studied in rat cortical slices *(34,36,37)* and in isolated rat spinal cord *(54)*. Superfusion of the slice or spinal cord with a solution containing either 40-kDa or 70-kDa dextran or hyaluronic acid (HA) resulted in a significant increase in λ. In standard physiological solution, λ was about 1.57, whereas in a 1% or 2% solution of 40-kDA or 70-kDa dex-

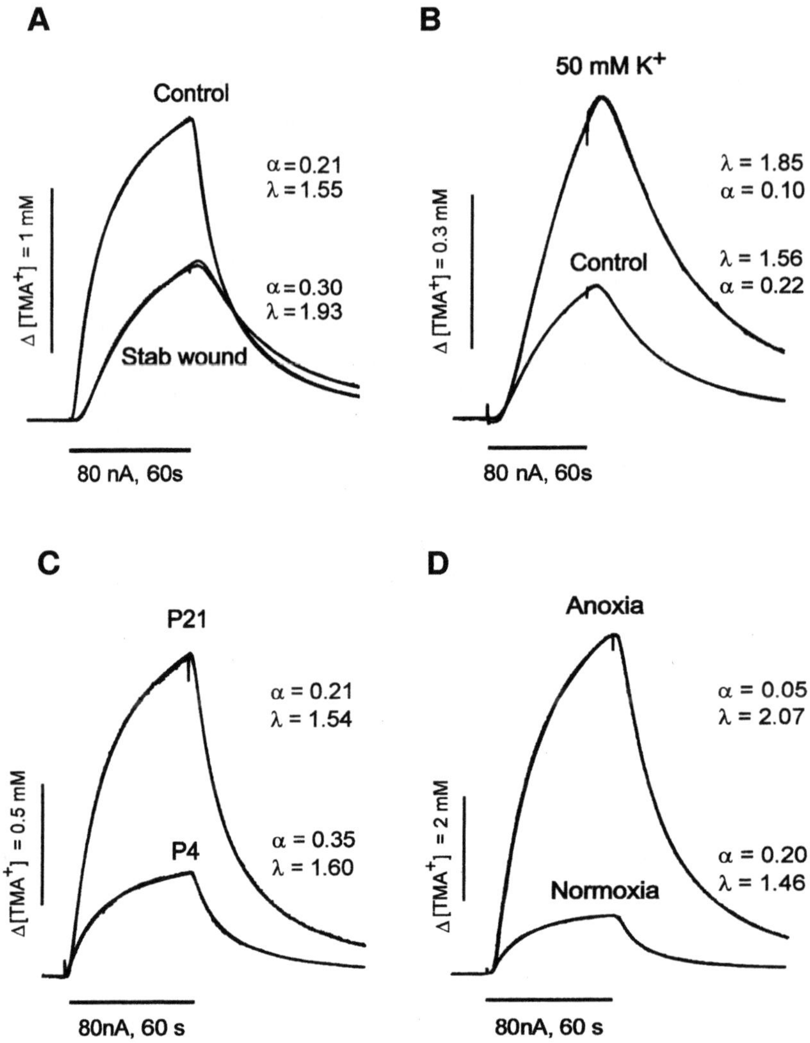

Fig. 5. TMA^+ diffusion curves under different experimental conditions. For each curve, the ECS diffusion parameters α (volume fraction) and λ (tortuosity) were extracted by appropriate nonlinear curve fitting. Experimental and theoretical curves are superimposed in each case. For each figure the concentration scale is linear. **(A)** Typical recordings obtained in rat cortex (control). Values of α and λ are incresed in the gliotic cortex around a stab wound. Note that the larger the curve, the smaller the value of α; a slower rise and decay indicate higher tortuosity. **(B)** Typical recordings in isolated rat spinal cord of a 13-d-old rat before (control) and during application of 50 m*M* K^+. Cellular, particularly glial, swelling results in an ECS volume decrease and a tortuosity increase. **(C)** Typical recordings obtained in rat cortex at postnatal days (P) 4 and 21. Note the dramatic decrease in the ECS volume and the increased tortuosity during maturation. **(D)** Typical recordings obtained in adult rat cortex (lamina V) during normoxia and in the same animal about 10 min after cardiac arrest (anoxia).

tran, λ increased to about 1.72–1.77. Application of a 0.1% solution of HA (1.6×10^6Da) resulted in an increase in λ to about 1.75. The α was unchanged in these experiments, suggesting that these substances had no effect on cell volume and the viability of the preparation.

A decrease in tortuosity and a loss of anisotropy have also been found during aging. This decrease correlates with the disapperance of fibronectin and chondroitin sulphate proteoglycans, forming perineuronal nets around granular and pyramidal cells in the hippocampus of young adult rats. These results suggest that bigger molecules such as 40- and 70-kDa dextran, hyaluronic acid, and molecules of the extracellular matrix may slow down the diffusion of small molecules such as TMA^+ (74 Da) in the ECS.

5. ECS DIFFUSION PARAMETERS DURING DEVELOPMENT

Compared to healthy adults, ECS diffusion parameters significantly differ during postnatal development *(39,40,43,55)*. The ECS volume in the cortex is about twice as large (α= 0.36–0.46) in the newborn rat as in the adult rat (α=0.21–0.23), while the tortuosity increases with age (Fig. 5C). The reduction in the ECS volume fraction correlates well with the growth of blood vessels. In rat spinal cord gray matter, α decreases with neuronal development and gliogenesis from postnatal day 4 to 12 by about 15%, while λ significantly increases, showing that the diffusion of molecules becomes more hindered with age. The large ECS channels during development may allow the migration of larger substances (e.g., growth factors) and better conditions for cell migration during development. On the other hand, the large ECS in the neonatal brain could significantly dilute ions, metabolites, and neuroactive substances released from cells, relative to release in adults, and may be a factor in the prevention of anoxic injury, seizure and spreading depression in young individuals. The diffusion parameters could also play an important role in the developmental process itself. Diffusion parameters are substantially different in myelinated and unmyelinated white matter *(40,43)*. Isotropic diffusion was found in corpus callosum and spinal cord white matter of young rats with incomplete myelination. In myelinated spinal cord and corpus callosum, the tortuosity is higher (the apparent diffusion coefficient is lower) when $TMA^{\%}$ diffuses across the axons than when it diffuses along the fibers (Fig. 6).

6. ECS IN PATHOLOGICAL STATES

Pathological states, e.g., anoxia/ischemia, are accompanied by a lack of energy, seizure activity, the excessive release of transmitters and neuroactive substances, neuronal death, glial cell loss or proliferation, glial swelling, the production of damaging metabolites including free radicals and the loss of ionic homeostasis. Others are characterized by inflammation, edema or demyelination. It is therefore evident that they will be accompanied not only by substantial changes in ECS ionic composition *(1,3)* but also by various changes in ECS diffusion parameters according to the different functional and anatomical changes.

5.1. ECS Ionic Composition and Volume during Anoxia/Ischemia

Dramatic K^+ and pH_e changes occur in the brain and spinal cord during anoxia and/or ischemia *(3,41,56)*. Within 2 min after respiratory arrest in adult rats, blood

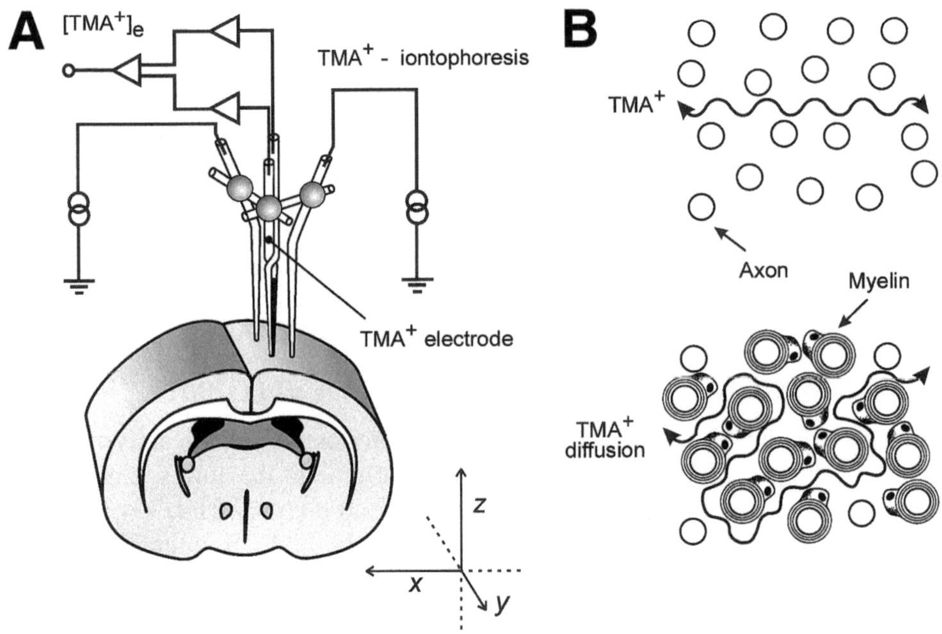

Fig. 6. **(A)** Schema of experimental arrangement to study diffusion anisotropy. A TMA^+ selective double-barrelled ion-selective microelectrode (ISM) was glued to two bent iontophoresis microelectrodes. Separation between electrode tips was 110–180 μm. Tips of the 3 pipets formed a 90° horizontal angle for simultaneous measurements along the *x*- and *y*-axes. Similarly, for measurements along the *z*-axis, 1 iontophoresis pipet tip was lowered 110–180 μm below the tip of the ISM. **(B)** Extracellular diffusion in the direction perpendicular to the orientation of the axons, i.e., around axons, is compromised by the number of myelin sheaths, the number of myelinated axons, and the length of the myelin shealths along axons. The scheme demonstrates the increased anisotropy as myelination progresses.

pressure begins to increase and pH_e begins to decrease (by about 0.1 pH unit), although $[K^+]_e$ remains unchanged. With the subsequent blood pressure decrease, the pH_e decreases by 0.6–0.8 pH units to pH 6.4–6.6. This pH_e decrease is accompanied by a steep rise in $[K^+]_e$ to about 50–70 m*M;* decreases in $[Na^+]_e$ to 48–59 m*M,* $[Cl^-]_e$ to 70–75 m*M,* $[Ca^{2+}]_e$ to 0.06–0.08 m*M,* and pH_e to 6.1–6.8; an accumulation of excitatory amino acids; a negative DC slow potential shift; and a decrease in ECS volume fraction to 0.04–0.07. The ECS volume starts to decrease when the blood pressure drops below 80 mm Hg and $[K^+]_e$ rises above 6 m*M (41).*

Figure 5D shows that during hypoxia and terminal anoxia, the ECS volume fraction in rat cortex or spinal cord decreases from about 0.20 to about 0.05, although tortuosity increases from 1.5 to about 2.1 *(41,57).* The same ultimate changes were found in all neonatal, adult and aged rats, in grey and white matter, in the cortex, corpus callosum and spinal cord. However, the time course in white matter was significantly slower than in gray matter; and the time course in neonatal rats was about 10 times slower than in adults *(55).* Linear regression analysis revealed a positive correlation between the normoxic size of the ECS volume and the time course of the changes. This corresponds to the well-known resistance of the immature CNS and the greater susceptibility of the aged brain to anoxia.

5.2. Changes in Extracellular Space Volume and Geometry during X-irradiation Injury

Radiation therapy is an effective treatment for some human cancers. However, this therapy is limited by radiation injury. Clinical correlates of demyelination and necrosis, which occur early, include somnolence, changes in intellect, radiation myelopathies, and leucoencepathies. The responses of the normal tissue immediately surrounding the tumor to its unavoidable irradiation are therefore of considerable interest. It is generally accepted that radiation injury is caused by mitotic death and the depletion of the various cell populations. The main symptoms of radiation injury in the CNS are very similar in rodents and humans. The immature CNS is more sensitive to radiation than the adult nervous system, apparently due to the proliferative potential and increased radiation sensitivity of glial and/or vascular endothelial cells in the developing nervous system *(58)*.

In experiments on the somatosensory neocortex and subcortical white matter of one-day-old (P1) rats, X-irradiation with a single dose of 40 Gy resulted in radiation necrosis with typical early morphological changes in the tissue, namely cell death, DNA fragmentation, extensive neuronal loss, blood-brain-barrier (BBB) damage, activated macrophages, astrogliosis, an increase in extracellular fibronectin, and concomitant changes in all three diffusion parameters. The changes were observed as early as 48 h postirradiation and persisted at P21 *(59)*.

In the nonirradiated cortex, the volume fraction, α, of the ECS is large in newborn rats and diminishes with age *(39)*. X-irradiation with a single dose of 40 Gy blocked the normal pattern of volume fraction decrease during postnatal development and, in fact, brought about a significant increase *(59)*. At P4–P5, α in both cortex and corpus callosum increased to about 0.5. The large increase in α persisted at 3 wk after X-irradiation. Tortuosity, λ, and nonspecific uptake, k', significantly decreased at P2–P5; at P8–P9 they were not significantly different from those of control animals, although they significantly increased at P10–P21. This means that in chronic lesions, e.g., those occuring 1–3 wk after X-irradiation and/or in gliotic tissue, the volume fraction remains elevated, but tortuosity increases. Interestingly, X-irradiation with a single dose of 20 Gy, which resulted in relatively light neuronal damage and loss and BBB damage, did not produce changes in diffusion parameters significantly different from those found with 40 Gy. Less pronounced but still significant changes in diffusion parameters were also found in areas of the ipsilateral and contralateral hemispheres adjacent to the directly X-irradiated cortex *(59)*.

The observed increase in the extracellular volume fraction in the X-irradiated tissue could also contribute to the impairment of signal transmission frequently observed after X-irradiation employed in the clinical treatment of various tumors, e.g., by diluting ions and neuroactive substances released from cells, and may play an important role in functional deficits as well as in the impairment of the developmental processes. Moreover, the increase in tortuosity, inferred from the decrease in the ADC_{TMA},, in the X-irradiated cortex and in the contralateral hemisphere, suggests that even long after mild irradiation the diffusion of substances can be substantially hindered. The observed increase in tortuosity seems to be related to astrogliosis, but changes in adhesion molecules or extracellular matrix molecules could also account for it.

5.3. ECS Volume and Geometry During Inflammation and Demyelination

Changes in ECS diffusion parameters can be expected during inflammation during which brain edema may develop. In an experimental model, inflammation was evoked by an intracerebral inoculation of a weakly pathogenic strain of *Staphylococcus aureus (60)*. Acute inflammation and an increase in BBB permeability in the abscess region resulted in rather mild changes in the ECS diffusion parameters, i.e., volume fraction tended to be somewhat larger and the tortuosity somewhat smaller.

Dramatic changes in the ECS diffusion parameters were found in the spinal cord of rats during experimental autoimmune encephalomyelitis (EAE), an experimental model of multiple sclerosis *(42)*. EAE, which was induced by the injection of guinea-pig myelin basic protein (MBP), resulted in typical morphological changes in the CNS tissue, namely demyelination, an inflammatory reaction, astrogliosis, BBB damage, and paraparesis, at 14–17 d postinjection. Paraparesis was accompanied by increases in α in the dorsal horn, in the intermediate region, in the ventral horn and in white matter from about 0.18 to about 0.30. There were significant decreases in λ in the dorsal horn and the intermediate region and decreases in k' in the intermediate region and the ventral horn *(42)*. There was a close correlation between the changes in ECS diffusion parameters and the manifestation of neurological abnormalities.

These results suggest that the expansion of the ECS alters diffusion parameters in inflammatory and demyelinating diseases and may affect the accumulation and movement of ions, neurotransmitters, neuromodulators and metabolites in the CNS in these disorders, possibly by interfering with axonal conduction.

6. ECS DIFFUSION PARAMETERS AND AGING

Aging, Alzheimer's disease, and many degenerative diseases are accompanied by serious cognitive deficits, particularly impaired learning and memory loss. This decline in old age is a consequence of changes in brain anatomy, morphology, volume, and functional deficits. Nervous tissue, particularly in the hippocampus and cortex, is subject to various degenerative processes including a decreased number and efficacy of synapses, a decrease in transmitter release, neuronal loss, astrogliosis, demyelination, deposits of beta amyloid, changes in extracellular matrix proteins, and so on. These and other changes not only affect the efficacy of signal transmission at synapses, but could also affect extrasynaptic ("volume") transmission mediated by the diffusion of transmitters as well as other substances through the volume of the extracellular space (ECS). Many mediators, including glutamate and GABA, bind to high affinity binding sites located at nonsynaptic parts of the membranes of neurones and glia. Mediators that escape from the synaptic clefts at an activated synapse, particularly following repetitive stimulation, diffuse in the ECS and can crossreact with receptors in nearby synapses. This phenomenon, called "cross-talk" between synapses by the "spillover" of a transmitter (e.g., glutamate, GABA, glycine), has been proposed to account for LTP and LTD in the rat hippocampus *(61,62)*. The cross-talk between synapses, and the efficacy and directionality of volume transmission, could be critically dependent on the diffusion properties of the ECS.

The ECS diffusion parameters α, $\lambda_{x,y,z}$ and k' were measured in the cortex, corpus callosum, and hippocampus (CA1, CA3 and in dentate gyrus). If diffusion in a particular brain region is anisotropic, then the correct value of the ECS volume fraction cannot be calculated from measurements done only in one direction. For anisotropic diffusion, the diagonal components of the tortuosity tensor are not equal, and generally its nondiagonal components need not be zero. Neverthless, if a suitable referential frame is chosen (i.e., if we measure in three privileged orthogonal directions), neglecting the nondiagonal components becomes possible, and the correct value of the ECS size can thus be determined *(44,47)*. Therefore, TMA^+ diffusion was measured in the ECS independently along three orthogonal axes (x – transversal, y – sagital, z – vertical). In all three regions—cortex, corpus callosum and hippocampus—the mean ECS volume fraction α was significantly lower in aged rats (26–32-mo-old), ranging from 0.17 to 0.19, than in young adults (3–4-mo-old) in which α ranged from 0.21 to 0.23 (Fig. 7). Nonspecific uptake k' was also significantly lower in aged rats. From Fig. 7 it is evident that the diffusion curves for the hippocampus are larger in the aged rat than in the young one, i.e., the space available for TMA^+ diffusion is smaller. Although the mean tortuosity values along the x-axis are not significantly different between young and aged rats, the values are significantly lower in aged rats along the y- and z-axes (Fig. 7), and thus the values along all three axes become the same. This means that there is a loss of anisotropy in the aging hippocampus, particularly in the CA3 region and the dentate gyrus *(45)*.

The three-dimensional pattern of diffusion away from a point source can be illustrated by constructing iso-concentration spheres (isotropic diffusion) or ellipsoids (anisotropic diffusion) for extracellular TMA^+ concentration. The surfaces in figure 7 represent the locations where TMA^+ concentration first reached 1 m*M*, 60s after its application in the center. The ellipsoid in the hippocampus of the young adult rat reflects the different abilities of substances to diffuse along the x-, y- and z-axes, whereas the sphere in the hippocampus of the aged rat shows isotropic diffusion (Fig. 7). The smaller ECS volume fraction in aged rats is reflected in the sphere's being larger than the ellipsoid.

Morphological changes during aging include cell loss, loss of dendritic processes, demyelination, astrogliosis, swollen astrocytic processes, and changes in the extracellular matrix. It is reasonable to assume that there is a significant decrease in the *ADC* of many neuroactive substances in the aging brain, which accompanies astrogliosis and changes in the extracellular matrix. One of the explanations why α in the cortex, corpus callosum, and hippocampus of senescent rats is significantly lower than in young adults could be astrogliosis in the aged brain. Increased GFAP staining and an increase in the size and fibrous character of astrocytes have been found in the cortex, corpus callosum and hippocampus of senescent rats, which may account for changes in the ECS volume fraction *(45)*. Other changes could account for the decreases in λ values and for the disruption of tissue anisotropy. In the hippocampus in CA1, CA3, as well as in the dentate gyrus, we observed changes in the arrangement of fine astrocytic processes. These are normally organized in parallel in the x-y plane (Fig. 8A,B), and this organization totally disappears during aging. Moreover, the decreased staining for chondroitin sulfate proteoglycans and for fibronectin (Fig. 8C,D) suggests a loss of extracellular matrix macromolecules.

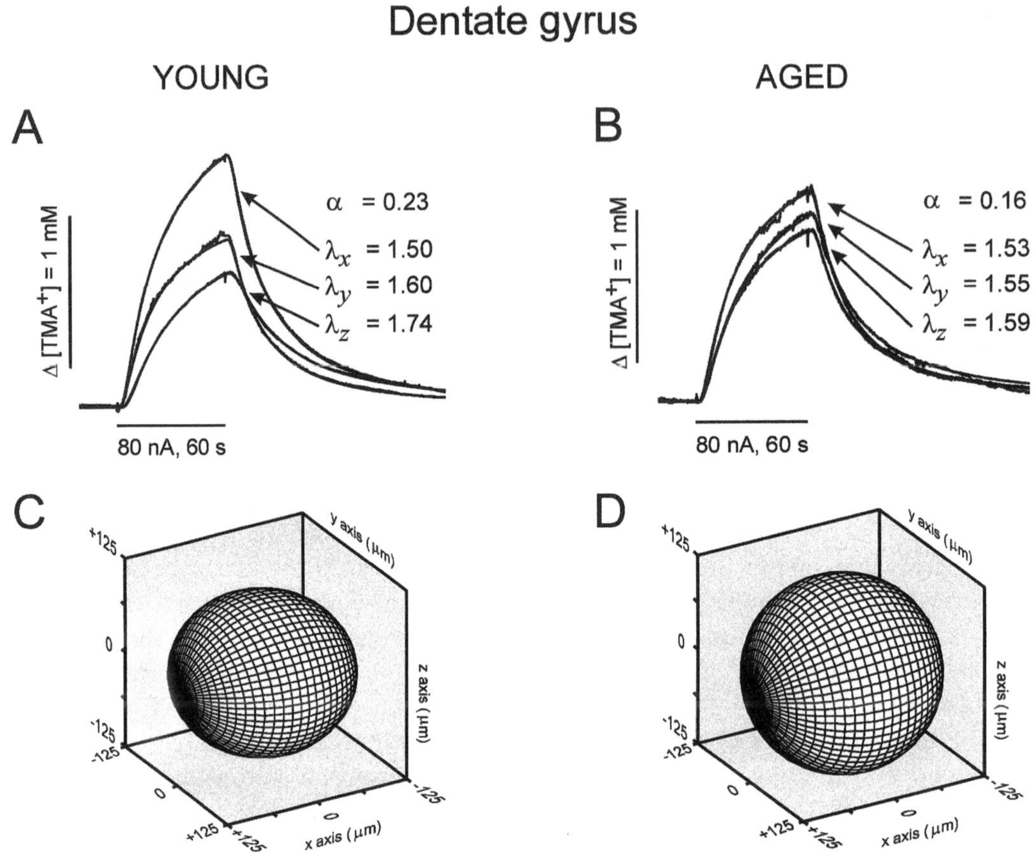

Fig. 7. Diffusion parameters in the hippocampus dentate gyrus of a young adult and an aged rat. **(A)** Anisotropic diffusion in the dentate gyrus of a young adult rat. TMA^+ diffusion curves (concentration-time profiles) were measured along three orthogonal axes (*x*-mediolateral, *y*-rostrocaudal, *z*-dorsoventral). The slower rise in the *z* than in the *y* direction and in the *y* than in the *x* direction indicates a higher tortuosity and more restricted diffusion. The amplitude of the curves shows that TMA^+ concentration, at approximately the same distance from the tip of the iontophoresis electrode, is much higher along the *x*-axis than along the *y*-axis and even higher than along the *z*-axis (λ_x, λ_x, λ_z). This can be explained if we realize that TMA^+ concentration decreases with the "diffusion distance" from the iontophoretic micropipet and that the real "diffusion distance" is not *r* but λr. Note that the actual ECS volume fraction α is about 0.2 and can be calculated only when measurements are done along the *x*-, *y*- and *z*-axes. **(B)** Anisotropy is almost lost in an aged rat; diffusion curves are higher, showing that α is smaller. C and D: Iso-concentration surfaces for a 1.0 m*M* TMA^+ concentration contour 60s after the onset of a 80 nA iontophoretic pulse. The surfaces were generated using mean values of volume fraction and tortuosity. **(C)** The ellipsoid represent anisotropic diffusion in a young adult rat. **(D)** The larger sphere in an aged rat, corresponding to isotropic diffusion and to a lower ECS volume fraction, demonstrates that diffusion from any given source will lead to a higher concentration of substances in the surrounding tissue and a larger action radius in aged rats than in young adults.

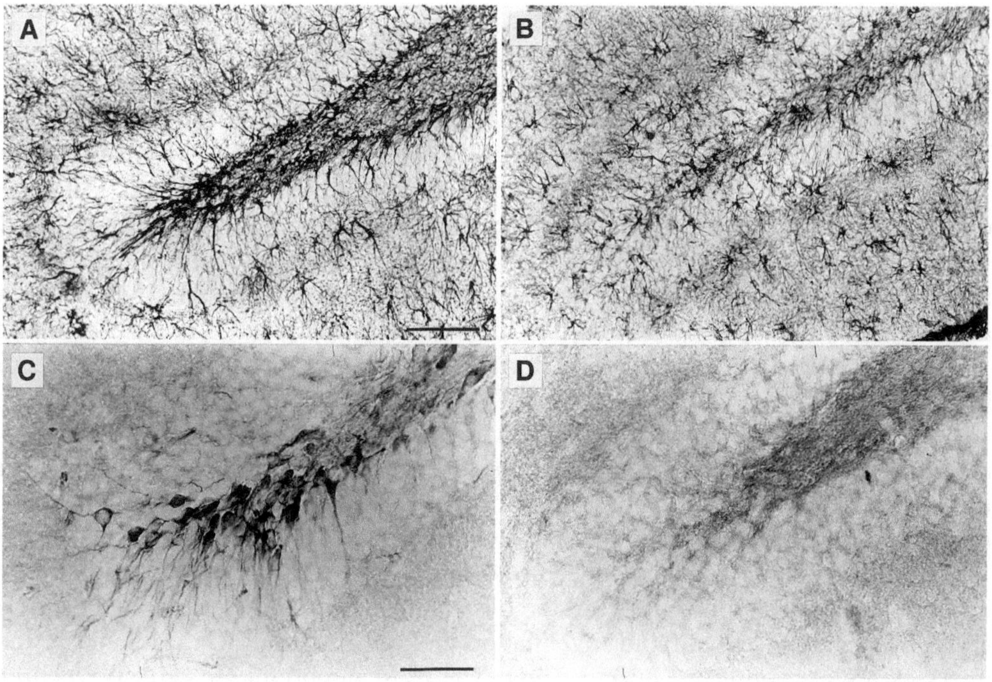

Fig. 8. Structural changes in the hippocampus gyrus dentatus region of aged rats. **(A)** Astrocytes stained for GFAP in a young adult rat; note the radial organization of the astrocytic processes between the pyramidal cells (not stained). **(B)** In an aged rat the radial organization of the astrocytic processes is lost. **(C)** Staining for fibronectin in a young adult rat shows densely stained cells, apparently due to perineuronal staining around granular cells. **(D)** In an aged rat the fibronectin staining is lost. Scale bar: in **A** and **B** = 100 μm, **C** and **D** = 50 μm.

Because α is lower in aging rats, we expected some differences in the ECS diffusion parameter changes during ischemia in senescent rats. Our study revealed that the final values of α, λ and *k′* induced by cardiac arrest are not significantly different between young and aged rats; however, the time course of all the changes is faster in aged animals *(45)*. The first peak in TMA^+ baseline changes, indicating an ECS volume decrease and a K^+ increase, was reached in young adult rats at about 3 min, whereas in senescent rats the time to peak was significantly shorter, about 1.5 min. The second and final peak in the TMA^+ baseline, associated with an extracellular acid shift, was reached in young adult rats at about 20 min, although in senescent rats this time to peak was again significantly shorter, about 12 min. The faster changes in extracellular space volume fraction and tortuosity in nervous tissue during aging can contribute to a faster impairment of signal transmission, e.g., because of a faster accumulation of ions and neuroactive substances released from cells and their slower diffusion away from the hypoxic/ischemic area in the more compacted ECS.

Our recent study also revealed that the degree of learning deficit during aging correlates with the changes in ECS volume, tortuosity and nonspecific uptake *(63)*. The hippocampus is well-known for its role in memory formation, especially the declarative memory. It is therefore reasonable to assume that diffusion anisotropy, which leads to a certain degree of specificity in extrasynaptic communication, may play an important role in memory formation. There was a significant difference between mildly and severely behaviorally impaired rats (rats were tested in a Morris water maze), which was particularly apparent in the hippocampus. The ECS in the dentate gyrus of severely impaired rats was significantly smaller than in mildly impaired rats. Also, anisotropy in the hippocampus of severely impaired rats, particularly in the dentate gyrus, was much reduced, although a substantial degree of anisotropy was still present in aged rats with a better learning performance. Anisotropy might be important for extrasynaptic transmission by channeling the flux of substances in a preferential direction. Its loss may severely disrupt extrasynaptic communication in the CNS, which has been suggested to play an important role in memory formation *(5,45)*.

Volume fraction is thus decreasing during the entire postnatal life with the steepest decrease in early postnatal development *(39,40)*. The larger ECS (30–45%) in the first days of postnatal development in the rat can be attributed to incomplete neuronal migration, gliogenesis, angiogenesis, and to the presence of large extracellular matrix proteoglycans, particularly hyaluronic acid, which because of the mutual repulsion of its highly negatively charged branches, occupies a great deal of space and holds cells apart from each other. The ensuing decrease in the ECS size could be explained by the disappearance of a significant part of the ECS matrix, neuron migration, and the development of dendritic trees, rapid myelination and proliferation of glia. Some of these processes are also observed during aging. The most important are probably neuronal degeneration, a further loss of extracellular matrix and astrogliosis. Indeed, we observed a decrease of fibronectin and chondroitin sulfate proteoglycan staining in the hippocampus of mildly impaired aged rats and almost a complete loss of staining in severely impaired aged rats. Chondroitin sulfate proteoglycans participate in multiple cellular processes (64,65). These include axonal outgrowth, axonal branching and synaptogenesis, which are important for the formation of memory traces.

One of the most striking changes we observed was the reduction of nonspecific TMA^+ uptake in aged rats. The underlying processes for TMA^+ uptake are so far unknown. These may include transfer into cells or binding to cellular surfaces or to negatively charged molecules of the extracellular matrix. All of these may be reduced during aging. Transfer into cells might decrease due to reduced pinocytosis (stiffer membranes due to a higher proportion of cholesterol), binding to cellular surfaces because of reduced membrane potential and binding to the extracellular matrix because of its loss. It is also interesting that uptake was reduced mainly in cell-body-rich areas and not in the corpus callosum.

What is the functional significance of the observed changes in ECS diffusion parameters during aging? Anisotropy, which, particularly in the hippocampus and corpus callosum, may help to facilitate the diffusion of neurotransmitters and neuromodulators to regions occupied by their high affinity extrasynaptical receptors, might have crucial importance for the specificity of signal transmission. The importance of anisotropy for the "spill-over of glutamate," "cross-talk" between synapses, and for LTP and LTD has been proposed *(61,62)*. The observed loss of anisotropy in senescent rats could therefore lead to impaired cortical and, particularly, hippocampal function. The decrease in ECS size could be responsible for the greater susceptibility of the aged brain to pathological events (particularly ischemia) *(45)*, the poorer outcome of clinical therapy and the more limited recovery of affected tissue after insult.

7. CONCLUSION

There is increasing evidence that long-term changes in the physical and chemical parameters of the ECS accompany many physiological and pathological states including CNS trauma and aging. The "acute" or relatively fast changes in the size of the intercellular channels are apparently a consequence of cellular (particularly glial) swelling. Abrupt ECS volume decrease may cause "molecular crowding" which can lead to an acute increase in tortuosity. Long-term changes in diffusion would require changes in ECS composition, either permanent changes in the size of the intercellular channels, changes in extracellular matrix molecules, or in the number and thickness of cellular (glial) processes. Available data suggest that in some pathophysiological states α and λ behave as independent variables. A persistent increase in λ (without a decrease in ECS volume) is always found during astrogliosis and in myelinated tissue, suggesting that glial cells can form diffusion barriers, make the nervous tissue less permissive and play an important role in signal transmission, tissue regeneration, and pathological states. This observation has important implications for our understanding of the function of glial cells. The extracellular matrix apparently also contributes to diffusion barriers and to diffusional anisotropy, because its loss, e.g., during aging, correlates with a tortuosity decrease and loss of anisotropy.

In general, changes in ECS ionic composition, size and geometry may affect:

1. Synaptic transmission (width of synaptic clefts, permeability of ionic channels, concentration of transmitters, dendritic length constant, and so on).
2. Extrasynaptic "volume" transmission by diffusion (diffusion of factors such as ions, NO, transmitters, neuropeptides, neurohormones, growth factors and metabolites).
3. Neuronal interaction and synchronization.
4. Neuron-glia communication.
5. ECS ionic homeostasis and glial function.
6. Clearance of metabolites and toxic products.
7. Permeability of ionic channels.
8. Regeneration processes.
9. Long-term changes in local CNS architecture including ECS volume and tortuosity, which apparently can influence plastic changes, LTP or LTD, changes in behaviour and memory formation.

Hopefully, we will more fully understand all of those consequences of an altered ECS in the near future.

ACKNOWLEDGMENTS

Dr. Syková's research is supported by grants VS96-130, GAČR 307/96/K226, GA ČR 309/97/K048, and GA ČR 309/99/0657.

REFERENCES

1. Syková, E. (1983) Extracellular K^+ accumulation in the central nervous system. *Prog. Biophys. Molec. Biol.* **42,** 135–189.
2. Fuxe, K. and Agnati, L.F., eds. (1991) *Volume Transmission in the Brain. Novel Mechanisms for Neural Transmission.* Raven Press, New York.
3. Syková, E. (1992) Ionic and volume changes in the microenvironment of nerve and receptor cells, in *Progress in Sensory Physiology* (Ottoson, D., ed.), Springer-Verlag, Heidelberg, pp. 1–167.

4. Syková, E. (1997) The extracellular space in the CNS: Its regulation, volume and geometry in normal and pathological neuronal function. *Neuroscientist* **3,** 28–41.
5. Nicholson, C. and Syková, E. (1998) Extracellular space structure revealed by diffusion analysis. *Trends Neurosci.* **21,** 207–215.
6. Agnati, L.F., Zoli, M., Strömberg, I., and Fuxe, K. (1995) Intercellular communication in the brain: Wiring versus volume transmission. *Neuroscience* **69,** 711–726.
7. Zoli, M., Jansson, A., Syková, E., Agnati, L.F., and Fuxe, K. (1999) Volume transmission in the CNS and its relevance for neuropsychopharmacology. *Trends Pharmacol. Sci.* **20,** 142–150.
8. Walz, W. (1989) Role of glial cells in the regulation of the brain ion microenvironment. *Prog. Neurobiol.* **33,** 309–333.
9. Chesler, M. (1990) The regulation and modulation of pH in the nervous system. *Prog. Neurobiol.* **34,** 401–427.
10. Deitmer, J.W. and Rose, C.R. (1996) pH regulation and proton signalling by glial cells. *Prog. Neurobiol.* **48,** 73–103.
11. Syková, E., Rothenberg, S., and Krekule, I. (1974) Changes of extracellular potassium concentration during spontaneous activity in the mesencephalic reticular formation ot the rat. *Brain Res.* **79,** 333–337.
12. Syková, E., Kříž, N., and Preis, P. (1983) Elevated extracellular potassium concentration in unstimulated spinal dorsal horns of frogs. *Neurosci. Lett.* **43,** 293–298.
13. Singer, W. and Lux, H.D. (1975) Extracellular potassium gradients and visual receptive fields in the cat striate cortex. *Brain Res.* **96,** 378–383.
14. Svoboda, J., Motin, V., Hájek, I., and Syková, E. (1988) Increase in extracellular potassium level in rat spinal dorsal horn induced by noxious stimulation and peripheral injury. *Brain Res.* **458,** 97–105.
15. Johnston, B.M., Patuzzi, R., Syka, J., and Syková, E. (1989) Stimulus-related potassium changes in organ of Corti of guinea-pig. *J. Physiol. (Lond).* **408,** 77–92.
16. Syková, E., and Svoboda, J. (1990) Extracellular alkaline-acid-alkaline transients in the rat spinal cord evoked by peripheral stimulation. *Brain Res.* **512,** 181–189.
17. Syková, E., Jendelová, P., Svoboda, J., Sedman, G., and Ng, K.T. (1990) Activity-related rise in extracellular potassium concentration in the brain of 1–3-day-old chicks. *Brain Res. Bull.* **24,** 569–575.
18. Jendelová, P. and Syková, E. (1991) Role of glia in $K^+$and pH homeostasis in the neonatal rat spinal cord. *Glia* **4,** 56–63.
19. Syková, E., Jendelová, P., Šimonová, Z., and Chvátal, A. (1992) $K^+$and pH homeostasis in the developing rat spinal cord is impaired by early postnatal X-irradiation. *Brain Res.* **594,** 19–30.
20. Syková, E., Mazel, T., and Šimonová, Z. (1998) Diffusion constraints and neuron-glia interaction during aging. *Exp. Gerontol.* **33,** 837–851.
21. Thomas, L.B. and Steindler, D.A. (1995) Glial boundaries and scars: programs for normal development and wound healing in the brain. *Neuroscientist* **1,** 142–154.
22. Celio, M.R., Spreafico, R., De Biasi, S., and Vitellare-Zuccarello, L. (1998) Perineuronal nets: past and present. *Trends Neurosci.* **21,** 510–515.
23. Nicholson, C. (1995) Interaction between diffusion and Michaelis-Menten uptake of dopamine after iontophoresis in striatum. *Biophys. J.* **68,** 1699–1715.
24. Nicholson, C. and Phillips, J.M. (1981) Ion diffusion modified by tortuosity and volume fraction in the extracellular microenvironment of the rat cerebellum. *J. Physiol.(Lond).* **321,** 225–257.
25. Andrew, R.D. and MacVicar, B.A. (1994) Imaging cell volume changes and neuronal excitation in the hippocampal slice. *Neuroscience* **62,** 371–383.
26. Van Harreveld, A., Dafny, N., and Khattab, F.I. (1971) Effects of calcium on electrical resistance and the extracellular space of cerebral cortex. *Exp. Neurol.* **31,** 358–367.
27. Matsuoka, Y. and Hossmann, K.A. (1982) Cortical impedance and extracellular volume changes following middle cerebral artery occlusion in cats. *J. Cereb. Blood Flow Metab.* **2,** 466–474.

28. Korf, J., Klein, H.C., and Postrema, F. (1988) Increases in striatal and hippocampal impedance and extracellular levels of amino acids by cardiac arrest in freely moving rats. *J. Neurochem.* **50,** 1087–1096.
29. Nicholson, C. and Tao, L. (1993) Hindered diffusion of high molecular weight compounds in brain extracellular microenvironment measured with integrative optical imaging. *Biophys. J.* **65,** 2277–2290.
30. Benveniste, H., Hedlund, L.W., and Johnson, G.A. (1992) Mechanism of detection of acute cerebral ischemia in rats by diffusion-weighted magnetic resonance microscopy. *Stroke* **23,** 746–754.
31. Latour, L.L., Svoboda, K., Mitra, P.P., and Sotak, C.H. (1994) Time-dependent diffusion of water in a biological model system. *Proc. Natl. Acad. Sci. USA.* **91,** 1229–1233.
32. Norris, D.G., Niendorf, T., and Leibfritz, D. (1994) Healthy and infarcted brain tissues studied at short diffusion times: The origins of apparent restriction and the reduction in apparent diffusion coefficient. *NMR in Biomedicine* **7,** 304–310.
33. Van der Toorn, A., Syková, E., Dijkhuizen, R.M., et al. (1996) Dynamic changes in water ADC, energy metabolism, extracellular space volume, and tortuosity in neonatal rat brain during global ischemia. *Magn. Reson. Med.* **36,** 52–60.
34. Tao, L., Vořísek, I., Lehmenkuhler, A., Syková, E., and Nicholson, C. (1995) Comparison of extracellular tortuosity derived from diffusion of 3 kDa dextran and TMA^+ in rat cortical slices. *Soc. Neurosci. Abstr.* **21,** 604.
35. Križaj, D., Rice, M.E., Wardle, R.A., and Nicholson, C. (1996) Water compartmentalization and extracellular tortuosity after osmotic changes in cerebellum of *Trachemys scripta. J. Physiol.(Lond).* **492,** 887–896.
36. Tao, L. and Nicholson, C. (1996) Diffusion of albumins in rat cortical slices and relevance to volume transmission. *Neuroscienc.* **75,** 839–847.
37. Vargová, L., Tao, L., Syková, E., Ulbrich, K., Šubr, V., and Nicholson, C. (1998) Diffusion of large polymers in rat cortical slices measured by integrative optical imaging. *J. Physiol. (Lond).* **511,** 16P.
38. Prokopová-Kubinová, Š., Vargová, L., Tao, L., Ulbrich, K., Šubr, V., Syková, E., and Nicholson, C. (2001) Poly[N-(2-hydroxypropy1) methacrylamide] polymers diffuse in brain extracellular space with same tortuosity as small molecules. *Biophys. J.* **80,** 542–548.
39. Lehmenkühler, A., Syková, E., Svoboda, J., Zilles, K., and Nicholson, C. (1993) Extracellular space parameters in the rat neocortex and subcortical white matter during postnatal development determined by diffusion analysis. *Neuroscience* **55,** 339–351.
40. Vořísek, I. and Syková, E. (1997) Evolution of anisotropic diffusion in the developing rat corpus callosum. *J. Neurophysiol.* **78,** 912–919.
41. Syková, E., Svoboda, J., Polák, J., and Chvátal, A. (1994) Extracellular volume fraction and diffusion characteristics during progressive ischemia and terminal anoxia in the spinal cord of the rat. *J. Cereb. Blood Flow Metab.* **14,** 301–311.
42. Šimonová, Z., Svoboda, J., Orkand, R., Bernard, C.C.A., Lassmann, H., and Syková, E. (1996) Changes of extracellular space volume and tortuosity in the spinal cord of Lewis rats with experimental autoimmune encephalomyelitis. *Physiol. Res.* **45,** 11–22.
43. Prokopová, Š., Vargová, L., and Syková, E. (1997) Heterogeneous and anisotropic diffusion in the developing rat spinal cord. *Neuroreport* **8,** 3527–3532.
44. Mazel, T, Šimonová, Z., and Syková, E. (1998) Diffusion heterogeneity and anisotropy in rat hippocampus. *Neuroreport* **9,** 1299–1304.
45. Syková, E., Mazel, T., and Šimonová, Z. (1998) Diffusion constraints and neuron-glia interaction during aging. *Exp. Gerontol.* **33,** 837–851.
46. Pérez-Pinzon, M.A., Tao, L., and Nicholson, C. (1995) Extracellular potassium, volume fraction, and tortuosity in rat hippocampal CA1, CA3 and cortical slices during ischemia. *J. Neurophysiol.* **74,** 565–573.

47. Rice, M.E., Okada, Y., and Nicholson, C. (1993) Anisotropic and heterogeneous diffusion in the turtle cerebellum. *J. Neurophysiol.* **70,** 2035–2044.
48. Syková, E., Mazel, T., Roitbak, T., and Šimonová, Z. (1999) Morphological changes and diffusion barriers in auditory cortex and hippocampus of aged rats. *Assoc. Res. Otolaryngol. Abs.* **22,** 117.
49. Moseley, M.E., Cohen, Y., Mintorovitch, L., et al. (1990) Early detection of regional cerebral ischemia in cats: comparison of diffusion and T2-weighted MRI and spectroscopy. *Magn. Reson. Med.* **14,** 330–346.
50. Le Bihan, D., Turner, R., and Douek, P. (1993) Is water diffusion restricted in human brain white matter? An echoplanar NMR imaging study. *Neuroreport* **4,** 887–890.
51. Norton, W.T., Aquino, D.A., Hosumi, I., Chiu, F.C., and Brosnan, C.F. (1992) Quantitative aspects of reactive gliosis: a review. *Neurochem. Res.* **17,** 877–885.
52. Roitbak, T. and Syková, E. (1999) Diffusion barriers evoked in the rat cortex by reactive astrogliosis. *Glia* **28,** 40–48.
53. Syková, E., Roitbak, T., Mazel, T., Šimonová, Z., and Harvey, AR. (1999) Astrocytes, oligodendroglia, extracellular space volume and geometry in rat fetal brain grafts. *Neuroscience* **91,** 783–798.
54. Prokopová, Š., Nicholson, C., and Syková, E. (1996) The effect of 40-kDa or 70-kDa dextran and hyaluronic acid solution on extracelluar space tortuosity in isolated rat spinal cord. *Physiol. Res.* **45,** P28.
55. Voříšek, I. and Syková, E. (1997) Ischemia-induced changes in the extracellular space diffusion parameters, K^+ and pH in the developing rat cortex and corpus callosum. *J. Cereb. Blood Flow Metab.* **17,** 191–203.
56. Xie, Y., Zacharias, E., Hoff, P., and Tegtmeier, F. (1995) Ion channel involvment in anoxic depolarisation indused by cardiac arrest in rat brain. *J. Cereb. Blood Flow Metab.* **15,** 587–594.
57. Lundbaek, J.A. and Hansen, A.J. (1992) Brain interstitial volume fraction and tortuosity in anoxia. Evaluation of the ion-selective microelectrode method. *Acta Physiol. Scand.* **146,** 473–484.
58. Gutin, P.H., Leibel, S.A., and Sheline, G.E., eds. (1991) *Radiation Injury to the nervous system.* Raven Press, New York.
59. Syková, E., Svoboda, J., Šimonová, Z., Lehmenkühler, A., and Lassmann, H. (1996) X-irradiation-induced changes in the diffusion parameters of the developing rat brain. *Neuroscience* **70,** 597–612.
60. Lo, W.D., Wolny, A.C., Timan, C., Shin, D., and Hinkle, G.H. (1993) Blood-brain barrier permeability and the brain extracellular space in acute cerebral inflammation. *J. Neurol. Sci.* **118,** 188–193.
61. Asztely, F., Erdemli, G., and Kullmann, D.M. (1997) Extrasynaptic glutamate spillover in the hippocampus: dependence on temperature and the role of active glutamate uptake. *Neuron* **18,** 281–293.
62. Kullmann, D.M., Erdemli, G., and Asztely, F. (1996) LTP of AMPA and NMDA receptor-mediated signals: evidence for presynaptic expression and extrasynaptic glutamate spill-over. *Neuron* **17,** 461–474.
63. Syková, E., Mazel, T., Hasenöhrl, R.U., Harvey, A.R., Šimonová, Z., Mulders, W.H.A.M., and Huston, J.P. Learning deficits in aged rats related to decrease in extracellular volume and loss of diffusion anisotropy in hippocampus. *Hippocampus,* In press.
64. Hardington, T.E. and Fosang, A.J. (1992) Proteoglycans: many forms and many functions. *FASEB J.* **6,** 861–870.
65. Margolis, R.K. and Margolis, R.U. (1993) Nervous tissue proteoglycans. Experientia **49,** 429–466.

6

Intercellular Diffusional Coupling between Glial Cells in Slices from the Striatum

Brigitte Hamon, Jacques Glowinski, and Christian Giaume

1. INTRODUCTION

In the central nervous system (CNS), macroglial cells (astrocytes and oligodendrocytes) represent the main cell population coupled by gap junctions. Indeed, a much higher level of gap junctional communication is observed in macroglial cells than in neurons and this communication process persists up to the adult stage in these cells *(1)*. Gap junctions are characterized by clusters of intercellular channels connecting the cytoplasm of adjacent cells. These channels are constituted by two hemichannels, the connexons, each of them being formed by six structural subunit proteins, the connexins (Cxs). Cxs are organized around a relatively large hydrated pore which allows the diffusion of ions and small molecules up to 1–1.2 kDa, providing thus the morphological basis for electrical and metabolic coupling *(2)*. Although Cxs are expressed in neurons as well as in glial and ependymal cells *(3)*, glial cells and astrocytes, particularly, contain the highest level of Cxs from the late embryo to the adult *(4)*. Cx43, which is prevalent in astrocytes, is detected early during development, peaks at birth, and remains constant throughout the adult life *(4)*. Cx30 is also found in mature astrocytes and its distribution resembles that of Cx43 *(5)*. Experiments performed with low molecular weight tracers (Lucifer yellow, biocytin, neurobiotin) have indicated that populations of 10 to 100 astrocytes can communicate through gap junction channels in either cortical, hippocampal, or cerebellar slices *(6–8)*. However, although Cx43 was also detected in striatal astrocytes, as reported in a previous study, the intercellular diffusion of the low molecular weight fluorescent dye Lucifer yellow was not detected in astrocytes injected with this dye in striatal slices *(9,10)*. This is surprising because functional studies performed on astrocytic primary cultures from different brain structures, including the striatum, have demonstrated that these cells are highly coupled by gap junction channels *(3,11)*. A marked regional heterogeneity was found in these cell culture studies *(12,13)* but astrocytes from either the cerebral cortex, the hippocampus, the brain stem, the cerebellum or the striatum were all shown to express Cx43 and to exchange fluorescent dyes, calcium signaling molecules, glucose, and glutamate *(14)*.

From: *Neuroglia in the Aging Brain*
Edited by: Jean S. de Vellis © Humana Press Inc., Totowa, NJ

Due to the occurrence of cytoplasmic exchanges through gap junction channels, the contribution of astrocytes to several brain functions should not only be considered as resulting from individual elements, but also from groups of communicating and coordinated cells. This is likely to be the case either for intracellular and extracellular ionic homeostasis *(15,16)*, trafficking of metabolic substrates *(17)* or for the implication of astrocytes in neuronal survival and toxicity *(18,19)*.

In the present study, acute slices were used to determine whether gap junction-mediated intercellular communication occurs between striatal glial cells. Intracellular injections of Lucifer yellow or neurobiotin were performed thanks to the patch-clamp technique using the whole-cell configuration. In agreement with previous findings made on primary cultures, striatal glial cells identified as putative astrocytes were found to be coupled by gap junction channels. In addition, several electrophysiological passive properties of these coupled glial cells were found to be modified by treatments known to inhibit junctional permeability. According to these observations, astrocytes in striatal slices are connected by functional gap junctions and regulations of these intercellular channels may affect the basic electrophysiological properties of these cells.

2. EXPERIMENTAL PROCEDURES

2.1. *Slice Preparation*

OFA rats (8- to 14-d-old) were decapitated and their brain was quickly removed and stored in cold (4°C) standard bath solution containing (in m*M*) NaCl 125, KCl 2.5, $NaHCO_3$ 22, KH_2PO_4 1.15, 5 N-2-hydroxyethylpiperazine-N′-2-ethanesulfonic acid (HEPES) 5, $MgCl_2$ 1.5, $CaCl_2$ 1.5, D-glucose 10, oxygenated with 95% O2–5% CO_2. Coronal slices from the striatum (300 μm thick) were cut using a microslicer (DTK 1000, DSK). After 45 min at least of recovery in a storage beaker at room temperature (20–25°C), a slice was transferred to the recording chamber mounted on an upright microscope (Axioskop, Zeiss) and perfused with standard bath solution at the rate of 2.5–3 mL per min.

2.2. *Electrophysiology*

Cells were visualized using differential interference contrast (DIC) microscopy, equipped with a 40X water immersion objective and an infrared-sensitive video camera (C2400, Hamamatsu) in combination with a video graphic printer (UP 890-CE, Sony). Patch-clamp recordings were performed in whole-cell configuration with an Axopatch-lD amplifier (Axon Instruments). Electrodes were filled with a buffer solution containing (in m*M*): KCl 120, $MgCl_2$ 2, $CaCl_2$ 0.5, ethylene glycol-bis (-aminoethyl ether)-N,N,N′,N′-tetraacetic acid (EGTA) 0.5, HEPES (Na salt) 10, pH adjusted at 7.2 with KOH. The patch pipet resistance was in the range of 3–5 MΩ. The capacity transient minimization and series resistance compensation were performed in response to 10 mV and 10 ms hyperpolarizing voltage steps. After compensation to 50–80%, series resistances ranged from 4 to 9 MΩ. Similar voltage steps were used to determine the cell input membrane resistance. For each depolarizing response, the leak current was determined in response to an hyperpolarizing step of the same amplitude and subtracted off-line. Signals were filtered at 5 kHz, digitized with an ACQUIS1 program (G. Sadok, Paris) for storage and off-line analysis.

2.3. *Study of Intercellular Diffusional Coupling*

In preliminary experiments, gap junction-mediated coupling was studied by adding 0.1% Lucifer yellow CH (dipotassium salt, Sigma) to the pipet solution. However, most results were obtained using pipets filled with 0.5% neurobiotin (Vector Labs), the KCl concentration of the internal solution being lowered to 105 m*M* for compensation of the hyperosmolarity. In all cases, only one cell was stained per slice to avoid misinterpretation in the field of diffusional coupling. Cells were uncoupled by adding halothane (diluted in absolute ethanol, final concentration 2 m*M*) in the bathing solution. Slices were continuously superfused with halothane 5 min before recording and remained under this condition until tissue fixation. After removal of the electrode from the recorded cell, neurobiotin was allowed to spread for 15 min into neighboring cells before final histological procedure. Briefly, tissues were fixed overnight in 4% paraformaldehyde at 4°C. After three 5 min washes in 0.1 *M* phosphate-buffered saline (PBS,pH 7.3), slices were treated for 30 min with 1% hydrogen peroxyde and 10% methanol to block endogenous peroxydases. Slices were then washed three times in 0.1 *M* PBS, incubated overnight in avidin-biotin horseradish peroxidase (ABC kit, Vectastain Elite, Vector Labs) and 0.1% Triton X-100. Finally, slices were exposed to 0.05% diaminobenzidine tetrahydrochloride used as the chromophore, dried for several hours, mounted on slides and coverslipped in Eukitt for permanent storage and microscopic examination.

2.4. *Statistical Analysis*

Data were expressed as means ± standard error of the mean and statistical significance was established using the Student's *t*-test or Mann-Whitney test in case of significantly different standard deviations.

3. RESULTS

3.1. *Visual Selection of Cells*

In preliminary experiments, the morphology of each cell visualized under DIC microscopy during patch-clamp recording was compared with the shape of the cell staining on-line with Lucifer yellow. Indeed, after achievement of the whole-cell configuration, Lucifer yellow was observed to fill almost instantaneously the recorded cell. Then, the dye fluorescence decreased rapidly within the target cell, this decline likely corresponding to a fast diffusion of the dye into neighboring cells (see below). In addition, because of fluorescence fading, morphological and coupling observations were found to be unstable within the 45 min duration of the experiment. Therefore, neurobiotin was routinely used for further analysis of recorded cells and study of intercellular communication. Comparison of DIC on-line microphotographs with neurobiotin staining after histological treatment and electrophysiological criteria of cell identification indicated that glial cells could be generally preselected for recording. This preselection was mainly based on the size of the cell. Thus, at all postnatal ages between PN8 and PN14, glial cell bodies were generally oval and significantly smaller than those of neurons (Fig. 1A, *see also* morphological data).

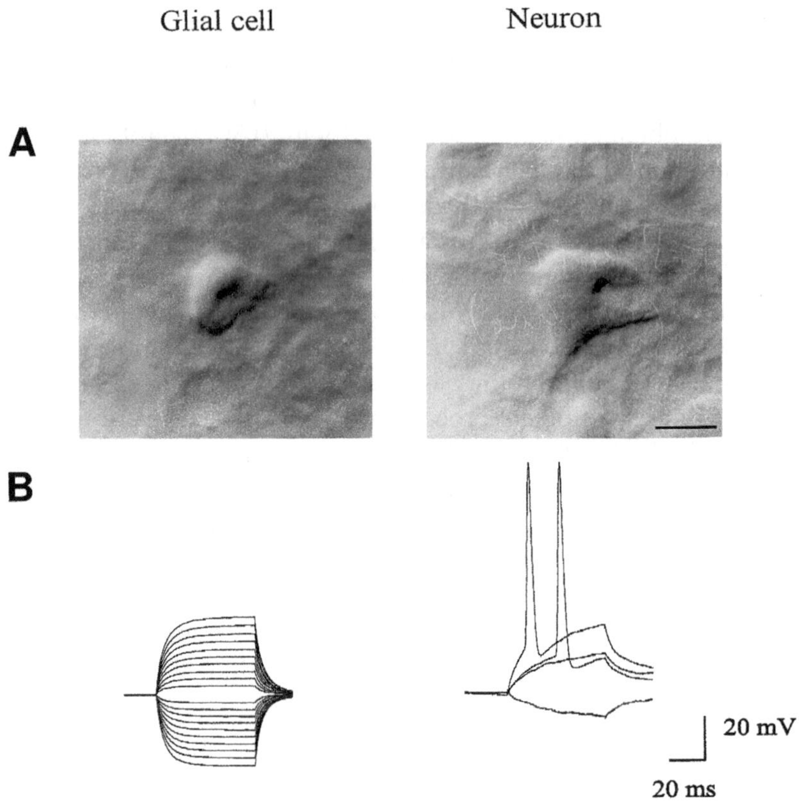

Fig. 1. Representative glial cells and neurons studied in striatal slices. **(A)** Differential interference contrast microphotographs showing that glial cells can be distinguished from neurons by their small cell body. Left panel: a glial cell. Right panel: a neuron. Scale bar, 10 µm. **(B)** Glial cells and neurons were also characterized by their electrophysiological properties examined in the whole-cell current-clamp mode. Left panel: responses of a glial cell to the injection of incremental current steps. No action potential was evoked by depolarizing steps. Right panel: traces from a neuron: a spike was emitted during depolarizing steps. Resting membrane potentials were –78 and –65 mV in the glial cell and the neuron, respectively.

3.2. Electrical Properties of Striatal Glial Cells: Current-Clamp Recordings

All patch-clamp recordings were performed within the striatum of animals aged from 8 to 14 d after birth. No difference was observed in the rate of success to establish the whole-cell configuration with different pipet solutions: standard solution alone or with the addition of Lucifer yellow or neurobiotin. Cell excitability was first examined in the current-clamp mode by analysis of voltage responses to the injection of several depolarizing and hyperpolarizing currents. (Fig. 1B). In general, the electrophysiological identification of the recorded cell, i.e., glia vs neuron, confirmed the selection achieved under DIC and infrared microscopy. Indeed, glial cells selected according to the small size of their soma never fired action potentials during the injection of large depolarizing currents (Fig. 1B, left panel), nor showed spontaneous potential changes in membrane potential (not shown). These data were obtained with either standard pipet solution alone ($n = 11$) or pipet solutions containing a staining marker (Lucifer yellow: $n = 14$; neurobiotin: $n = 29$). No significant difference in the electrophysiological properties of the cells were observed with these different solutions. Therefore, all data obtained under current-clamp were pooled. Cells selected as neurons were recorded with pipets containing neurobiotin. As expected, these cells emitted typical action potentials in response to depolarization ($n = 8$; Fig. 1B, right panel). In addition, in 5 out of 8 cases, spontaneous spikes or spontaneous postsynaptic potentials were recorded at resting membrane potential (not shown).

When glial cells were recorded, in few cases (16% of the recorded cells, $n = 38$), the membrane potential shifted toward values more negative than –95 mV 15 min after the beginning of recordings. Thus, in all cells, the resting membrane potential (E_m) recorded during the first 2 min was only taken into account and found to be –82±2 mV ($n = 38$). In addition, when recordings were performed with pipet containing neurobiotin, a shift of the resting membrane potential toward more positive values was observed (standard with or without Lucifer yellow: –87±2 mV, n = 13; neurobiotin –79±2 mV, $n = 25$, $p < 0.0001$).

When neurons were recorded, the resting membrane potential was more positive: –56±5 mV ($n = 8$, $p<0.0001$). Some neurons had a membrane potential of –50 mV or less and did not emit spikes in response to depolarizing stimulations even when prehyperpolarized with sustained current injection. Consequently, these cells often associated with unstable whole-cell recordings were not taken into account. The neuronal input resistance did not significantly differ from that of glial cells when recordings were performed with either a standard pipet solution or a pipet solution containing Lucifer yellow (126±31 MΩ, $n = 10$). In contrast, when cells were recorded with neurobiotin in the patch-pipet, the neuronal input resistance (151±16 MΩ, $n = 8$) was higher than that of glial cells (89 ± 10 MΩ, $n = 24$, $p < 0.02$).

3.3. Voltage-Clamp Recordings

The voltage-dependent membrane properties of striatal glial cells (n=38) was investigated under voltage-clamp by stepping the membrane potential from –100 to +90 mV at a holding potential of –70 mV. Since similar observations were made with different pipet media, all data were pooled. Currents elicited in response to this family of voltage steps revealed different types of conductance profiles. "Passive" and "complex" cells were distinguished according to a previous classification originally proposed by Steinhaüser et al. *(20,21)*. "Passive" profiles corresponded to a nearly ohmic and symmetrical behavior of the membrane currents in response to depolarizing and hyperpolarizing voltage steps (n=24). These "passive" cells were also characterized by a lack or a very slow decay time (several hundred milliseconds). In 16 out of 24 "passive" cells, pulse responses were followed by small currents at the end of the voltage step (Fig. 2A). In the eight remaining "passive" cells, currents recorded in response to voltage steps were associated with large, slowly decaying offset currents (Fig. 2B). In contrast, "complex" profiles consisted in voltage-activated currents superimposed on voltage-independent currents ($n = 14$, Fig. 2C). This observation was further confirmed in 14 out of 38 cells by a point-by-point analysis performed by adding positive and negative current traces (Fig. 2C, insert). In 12 out of 14 cases, "complex" cells were typically associated with small offset currents, which in few cases were found to be more pronounced in

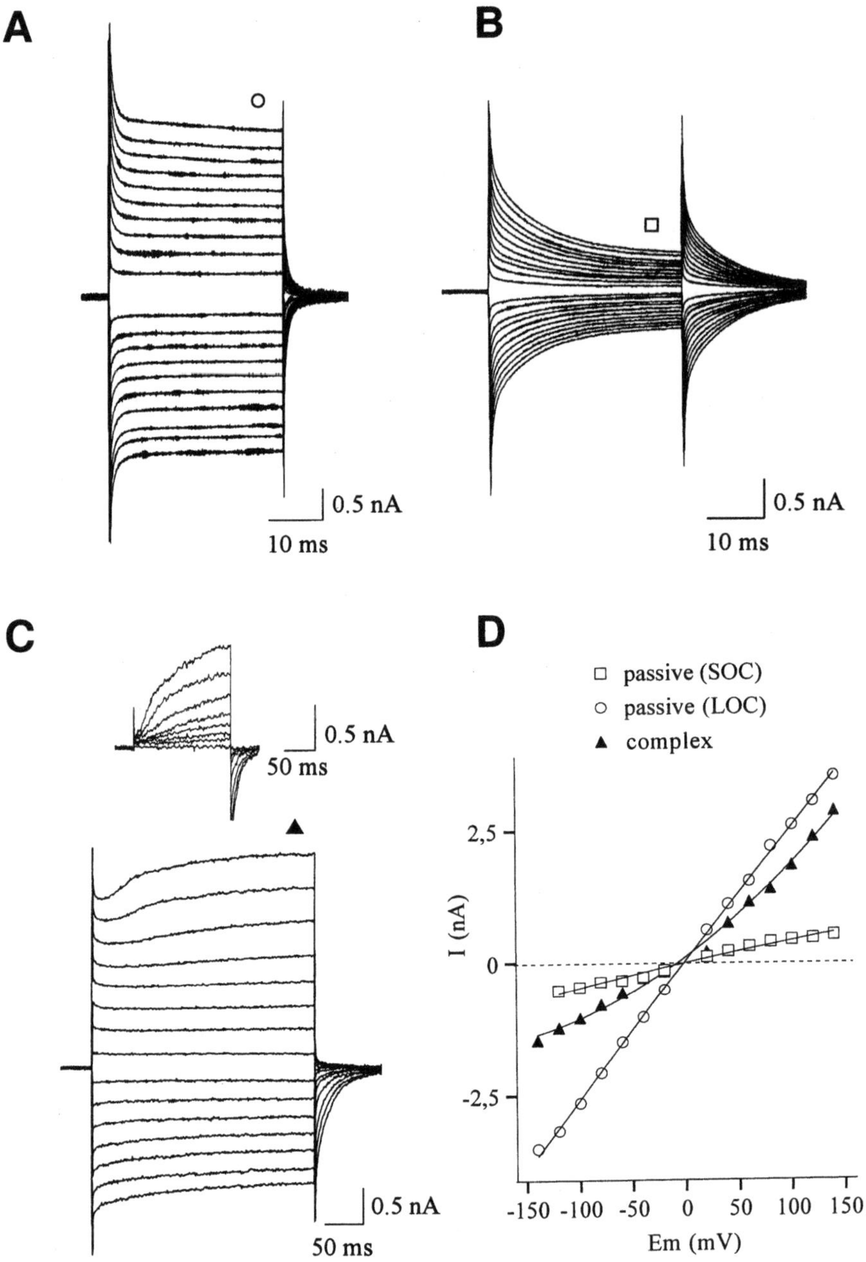

Fig. 2. Whole-cell current patterns of glial cells recorded at a holding potential of –70 mV, in response to de- and hyperpolarizing voltage steps ranging from –160 to +40 mV. **(A)** A "passive" cell with symmetrical currents showing neither voltage-, nor time-dependence. Small offset currents (SOC) are observed at the end of the stimulation pulses. **(B)** A "passive" cell expressing large offset currents (LOC) at the end of the stimulation. **(C)** A "complex" cell, endowed with voltage-dependent currents. Insert: the point-by-point addition of inward to outward currents reveals slowly activating outward currents of the delayed rectifier type. **(D)** I–V curves drawn from traces illustrated in **A** (white circles), **B** (white squares) and **C** (black triangles). Resting membrane potentials of these cells were –73 **(A)**, –69 **(B)** and –77 mV **(C)**.

Table 1
Different types of currents evoked in striatal glial cells by voltage steps between PN8 and PN14

	Cells with SOC		Cells with LOC	
	Symmetrical	Voltage-dependent	Small ratio	Large ratio
Time-independent	12	–	1	2
Slowly inactivating	10	–	3	2
Delayed rectifier(I_{DR})	–	2	–	–
Inward rectifier I_{DR})	–	2	–	–
I_A	–	5	–	2
$I_{Na}{}^+$ and I_A	–	3	–	–

Numbers represent the different types of voltage-gated and voltage-independent currents in response to voltage steps ranging from –100 to +90 mV from a holding potential of –70 mV. SOC, small offset currents; LOC, large offset currents. Voltage-dependent conductances were in most cases superimposed on symmetrical, ohmic conductances.

response to hyperpolarization (Fig. 2C). In 2 out of 14 neurons, "complex" cells were also characterized by large offset currents (*see* Table 1).

As previously reported, large offset currents are typically found in oligodendrocyte precursors and mature oligodendrocytes as identified by immunocytochemistry *(22)*. Profiles of conductances in striatal glial cells were thus classified on the basis of their small (SOC) or large (LOC) offset currents. Decay time constants of offset currents were analyzed in current responses induced by a 50 mV step and calculated in the absence of de activation of voltage-dependent currents. These decay times, often well fitted with a simple exponential function, were significantly larger in LOC than in SOC cells (9.2±2.6 ms, n = 7, and 2.4±0.8 ms, n = 11, respectively, $p < 0.05$). LOC cells were morphologically characterized by their parallel processes as revealed by microscopic analysis of neurobiotin injected cells. From these observations, LOC and SOC cells were considered to correspond to oligodendrocytes and astrocytes, respectively.

Further electrophysiological properties of "passive" and "complex" glial cells recorded from striatal slices were determined. Current-voltage (I/V) relationships established for responses measured at 50 ms before the end of the voltage pulse (250 ms duration) are illustrated in figure 2D for SOC and LOC "passive" cells and in figure 2C for "complex" cells. As expected from their ohmic properties, the I/V relationships were linear in SOC and LOC cells. In contrast, in "complex" cells, I/V curves were not linear. Indeed, in 9 out of 14 "complex" cells, outward currents were mainly observed whereas inward currents were also found in the five remaining cells. In two glial cells, outward currents were characterized by a delayed activation and no inactivation during 300–500 ms long depolarizing steps (not shown), reminiscent of that classically observed for delayed rectifying potassium currents. In 5 other glial cells, outward currents inactivated with a mean time constant of 17.2±4.2 ms for depolarizing steps from –80 to –0 mV. The amplitude of these currents increased with conditioning hyperpolarizing steps more negative than –110 mV, which fits with the I_A type of potassium current. Furthermore, in 3 out of these 5 cells, these I_A-like outward current responses were superimposed with inward sodium currents. Finally, 2 other glial cells were endowed with net inward currents inactivating at potentials more negative than –140 mV, reminiscent of an inward rectifying potassium conductance (*see* Table 1).

3.4. Morphological Analysis

Cell-filling with neurobiotin lasted from 3 to 30 min (n=21). The recording duration affected neither the staining intensity of somata and processes nor the tracer diffusion in the processes (n =12). Therefore, as initially observed *on line* with Lucifer yellow, neurobiotin appears to rapidly and completely fill the recorded cell after the membrane rupture.

Neurobiotin filling was considered to be adequate when no evidence of neurobiotin leakage into the extracellular space was found. Successful stainings were found in 18 glial cells and 3 neurons from a total of 29 control striatal cells recorded with neurobiotin. Average sizes of glial cell bodies were 6.5±0.4 µm broad, 11.4±0.8 µm long (n= 18) whereas those of neurones were 10±2 µm broad, 16.8±1 µm long (n = 3). Glial cell somata was elongated in most cases (6.2±0.4 µm broad, 12.5±0.8 µm long, n = 14) and occasionally round (7.5±1.1 µm diameter, n = 4). Glial cells were also classified according to the morphological aspects of their processes. Several patterns were distinguished:

1. Branched processes orientated along one or two predominant directions (length range: 7 to 95 µm, n = 9),
2. scarcely branched processes (maximal length range: 5 to 36 µm, n = 6), and finally,
3. few short processes (length range: 1 to 3 µm, n = 3).

In 7 out of 9 cells, either parallel elements elongating symetrically from a thick bifurcating primary process were observed (total length range: 35 to 95 µm; n = 3) or a few shorter elements directly originating from the soma (maximal length range: from 11 to 55 µm; n = 4) were observed.

3.5. Intercellular Diffusion of Neurobiotin

Lucifer yellow was initially used for on-line estimation of gap junction-mediated coupling between striatal glial cells. Coupling was observed in 48% of the trials (23 experiments), the number of coupled cells ranging from 1 to 9. Halothane (2 m*M*), an inhibitor of gap junction channels in cultured striatal astrocytes (Mantz et al., 1993) was used to demonstrate that the staining of surrounding cells was mediated by gap junction channels. This uncoupling agent was applied 5 min before patch-clamp recording. As expected no dye diffusion was detected in the three experiments made in these uncoupling conditions.

Neurobiotin was then used instead of Lucifer yellow. When more than one cell was stained, the staining intensity was highly similar among stained cells, except at the cluster periphery where a weaker staining was detected (Fig. 3A). The recorded cell was identified according to its morphometric and electrophysiological characteristics (*see* above). When these identification criteria were not satisfied, the cell was eliminated from the coupling analysis. Cells were recorded in either the absence (control, n = 18) or presence of halothane (2 m*M*, n = 15), the mean recording duration being 16±2 min (range from 3 to 30 min). In the absence of halothane, single stained cells were observed in 3 cases, and double or multiple staining in 15 other cases (Fig. 3A, left panel). In contrast, in the presence of halothane, 6 out of 15 cells were found to be

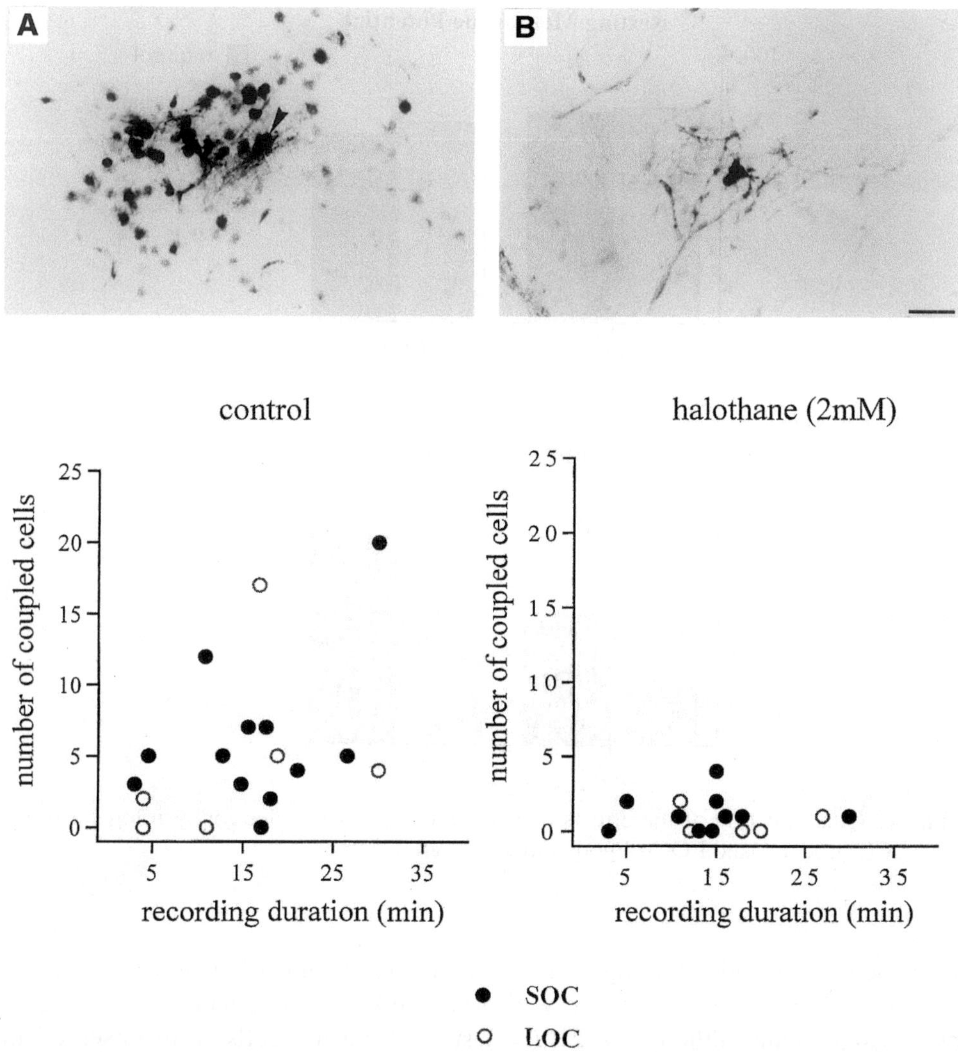

Fig. 3. Dye coupling of striatal glial cells and uncoupling effect of halothane. Upper part: microphotographs showing stained glial cells in the absence **(A)** or presence of halothane **(B).** Target cells (arrowhead) were recorded using a patch-pipet solution containing 0.5% neurobiotin in either the absence (control, left panel) or presence of 2 m*M* halothane (right panel). Lower part: quantitative representation of coupling under control conditions (left panel) and halothane perfusion (right panel). Number of coupled cells against recording duration is plotted for cells with small offset currents (SOC, black circles) and large offset currents (LOC, white circles).

single-stained (Fig. 3A, right panel) while the number of surrounding stained cells decreased from 7±2 ($n = 14$) to 2 ($n = 9$) for control and halothane-treated cells, respectively ($p < 0.0005$, Mann-Whitney test).

Attempts were then made to look for a different sensitivity to halothane of the SOC and LOC glial cells. In SOC cells (classified as astrocytes), the number of associated stained cells decreased from 6±2 ($n = 11$) to 1 ($n = 10$) ($p < 0.002$, Mann-Whitney test)

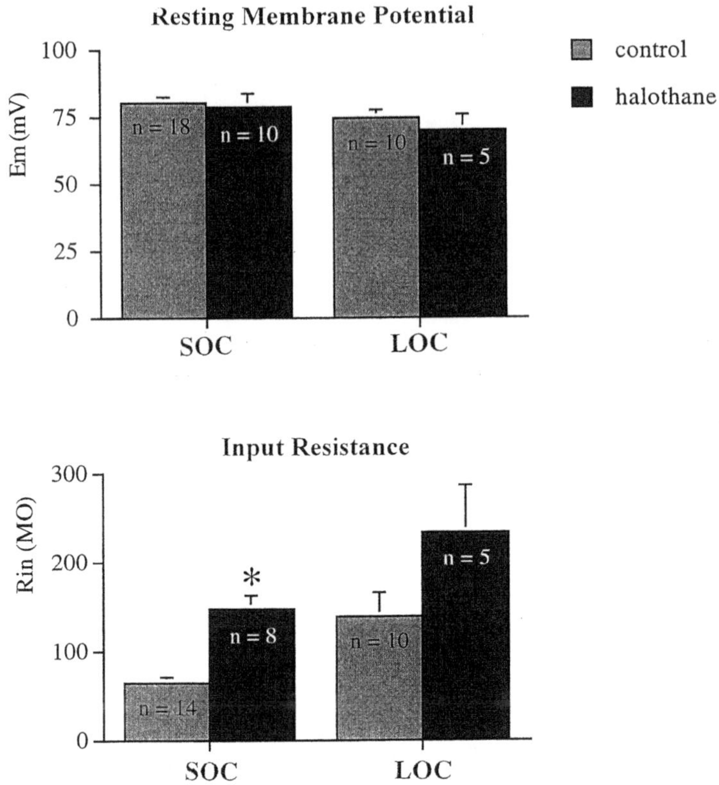

Fig. 4. Quantification of the effects of halothane on passive electrophysiological properties of gl SOC(astrocytes)and LOC(oligodendrocytes) glial cells.

the incidence of single staining being of 1 out of 11 and 3 out of 10 in control and halothane-treated slices, respectively. In LOC cells, the halothane uncoupling action was not significant, although 4±2 (n=7) stained partner cells were observed in the absence of the uncoupling agent and 1 (n=5) stained neighbor cell under halothane (the incidence of single cell staining being 2 out of 7, and 3 out of 5 in control and halothane conditions, respectively). Altogether, these results indicate that multiple neurobiotin staining can be observed between striatal glial cells and that this intercellular diffusion is prevented, or at least reduced, in the presence of the blocker of gap junction channels.

3.6. *Effects of Halothane on Passive Properties of Glial Cells*

The effect of halothane on the passive membrane properties of glial cells were also tested in cells recorded with neurobiotin in the patch-pipette. In SOC cells (astrocytes), the membrane potential remained unchanged in the presence of halothane (–80±2 mV, n = 18 and –79±5mV, n = 10, in control and halothane treated cells, respectively). Halothane treatment was also without significant effect on the resting membrane potential of LOC cells (oligodendrocytes) (–75±3 mV, n = 10, and –70 ±6 m V n = 5, in the absence and presence of halothane, respectively) (Fig. 4, upper panel). In contrast, halothane treatment increased the membrane input resistance in astrocytes (SOC cells) from 65±7 MΩ ($n = 14$) to 148±15 MΩ (n = 10), in control and halothane treated cells, resepectively but not in oligodendrocytes (LOC cells) (139±27 MΩ, $n = 8$, and 233±53 MΩ, $n = 5$, in the absence and presence of halothane, respectively) (Fig. 4, middle panel). These observations agree with the greater uncoupling efficacy of halothane in astrocytes than in oligodendrocytes (see above). Finally, halothane decreased the amplitude of passive currents in astrocytes (1.9±0.2 nA, $n = 18$, to 0.8±0.1 nA, $n = 11$ in absence and presence of halothane, respectively) but not in oligodendrocytes (0.9±0.2 nA, $n = 8$, and 0.5±0.1 nA, $n = 6$, in absence and presence of halothane, respectively).

4. DISCUSSION

The present study demonstrates that in acute slices of the rat striatum, glial cells are connected by functional gap junction channels. This agrees with results of previous studies that indicated the presence of mRNAs coding for CX43, the astrocytic gap junction protein, and the expression of this protein within the striatum *(9,10,24)* Our conclusion based on results obtained both with Lucifer Yellow and neurobiotin thus differs from that of Rufer et al. *(9),* who failed to detect astrocytic dye-coupling in striatal slices using Lucifer yellow. This functional study completes our previous investigations which indicated that cultured striatal astrocytes are coupled by functional junctional channels mainly constituted of Cx43 *(25).* The present results obtained in the striatum provide further support for the existence of extensive glial communication through gap junctions,in several brain regions in vivo or in acute slices as already shown in the neocortex *(26),* the visual cortex *(27),* the hippocampus *(6,5,8),* and the cerebellum *(7).* Nevertheless, the extent of dye and tracer diffusion between striatal astrocytes appears rather moderate when compared with that observed in other structures. This was also found in our study in parallel experiments performed on cerebral cortical slices (data not shown). Such a difference agrees with results of comparative regional investigations on the distribution of Cx43 immunoreactive sites and mRNA *(10,24),* and with functional observations performed on cultured astrocytes from different brain structures *(12,13).*

One of our main concerns was to identify the recorded cells initially stained with the tracer. This identification was mainly based on electrophysiological and morphological criteria previously established in studies performed on the spinal cord, the red nucleus or the hippocampus *(8,22,28,29).* The absence of action potentials in response to injections of depolariziting currents allowed to conclude that recordings were made from glial cells instead of neurons. Oligodendrocytes were characterized by their long processes oriented in a parallel fashion to each other. Putative astrocytes were identified by their small and in most cases oval somata with short processes emerging in all directions. The morphological distinction between the two main types of glial cells was correlated in most cases with the presence of LOC and SOC, two electrophysiological properties previously proposed for the discrimination of mature oligodendrocytes and astrocytes, respectively *(22).* In addition, "passive" currents were mainly observed in oligodendrocytes while "complex" current/voltage relationships were found in astrocytes in agreement with previous published observations *(8,22,28,29).* Finally, the main conductances (I_A, I_{DR}, I_{IR}, I_{Na+}) known to be expressed in astrocytes and oligodendrocytes were also recorded from striatal glial cells.

Table 2
Morphological analysis of neurobiotin injected cells

Size of somata (mean ± s.e.m., μm)			Size of processes (range, μm)	
Round, *n* = 4 (diameter)	Elongated, *n* = 14 (bride/length)	Oriented, *n* = 9	Diffuse, *n* = 6	Scarce, *n* = 3
7.5 ± 1.1	6.2 ± 0.4/12.5 ± 0.8	7–95	5–36	1–3

In agreement with previous observations made on cultured glial cells, in striatal slices, astrocytes were more frequently and extensively coupled to neighboring cells than oligodendrocytes *(30,31)*. The identity of the cells coupled with recorded glial cells remains to be established. Likely, this should be solved in the future by combining staining with the intercellular tracer and immunocytochemistry with either GFAP antibodies or oligodendrocytic specific makers. Presently, this identification was achieved on the basis of the size of the soma and the pattern of emerging processes of the coupled cells labeled by the tracer. According to these parameters and in agreement with previous observations made on cultured glial cells *(31)*, the occurrence of homotypic (astrocyte-astrocyte or oligodendrocyte-oligodendrocyte) and heterotypic (astrocyte-oligodendrocyte) couplings could be expected. However, the existence of gap junction-mediated communication between a recorded glial cell with surrounding neurons cannot be excluded. Indeed, junctional coupling between astrocytes and neurons has been reported *(32,33)* and Cx43 immunoreactive staining between these two cell types has been detected *in situ* in the cerebral cortex *(34)*.

Finally, the occurrence of functional gap junction channels between glial cells was also found to modify the basic electrophysiological properties of the recorded glial cells. Indeed, the input resistance was increased by about 50% and the amplitude of passive currents was reduced by 50% when junctional communication was inhibited by halothane. These observations demonstrate that electrical coupling through open gap junction channels may contribute to the passive electrophysiological properties of the astrocytes *in situ (34)*. Accordingly, the presence of these intercellular communication processes should be taken into account when glial cells, and astrocytes, particularly, are recorded using the patch-clamp technique. This has been taken into account by D'Ambrosio et al. *(8)* who performed whole-cell recordings in hippocampal astrocytes by omitting ATP in their pipet solution, a procedure which depletes intracellular ATP and thus is expected to block gap junction channels *(38)*.

AKNOWLEDGMENTS

The authors wish to thank A.M. Godeheu for excellent technical assistance during the completion of this work.

REFERENCES

1. Dermietzel, R. and Spray, D.C. (1998) From Neuro-Glue (Nervenkitt) to glia: a prologue. *Glia* **24,** 1–7.

2. Bruzzone, R., White, T., and Paul, D. (1996) Connections with connexins: The molecular basis of direct intercellular signalling. *Eur. J. Biochem.* **238,** 1–27.
3. Dermietzel, R. and Spray, D.C. (1993) Gap junctions in the brain: where, what type and why? *TINS* **16,** 186–192.
4. Dermietzel, R., Traub, O., Hwang, T.K., Beyer, E., Bennett, M.V.L., and Spray, D.C. (1989) Differential expression of three gap junction proteins in developing and mature brain tissues. *P.N.A.S.* **87,** 1328–1331.
5. Kunzelmann, P., Schroder, W, Traub, O., Steinhauser, C., Dermietzel, R., and Willecke, K. (1999) Late onset and increasing expression of the gap junction protein connexin 30 in adult murine brain and long-term cultured astrocytes. *Glia* **15,** 111–119.
6. Konietzko, U. and Müller, C. (1994) Astrocytic dye coupling in rat hippocampus: topography, developmental onset, and modulation by protein kinase C. *Hippocampus* **4,** 297–306.
7. Müller, T., Möller, T., Neuhaus, J., and Kettenmann, H. (1996) Electrical coupling among Bergmann glial cells and its modulation by glutamate receptor activation. *Glia* **17,** 274–284.
8. D'Ambrosio, R., Wenzel, J., Schwartzkroin, P.A., McKkhann II G.M., and Janigro, D. (1998) Functional specialization and topographic segregation of hippocampal astrocytes. *J. Neuroscience* **18,** 4425–4438.
9. Rufer, M., Wirth, S.B., Hofer, A., et al. (1996) Regulation of connexin-43, GFAP, and FGF-2 is not accompanied by changes in astroglial coupling in MPTP-lesioned, FGF-2-treated parkinsonian mice. *J. Neurosci. Res.* **46,** 606–617.
10. Yamamoto, T., Ochalski, A., Hertzberg, E.L., and Nagy, J.I. (1990) On the organization of astrocyte gap junctions in rat brain as suggested by LM and EM immunohistochemistry of connexin-43 expression. *J. Comp. Neurol.* **302,** 853–883.
11. Giaume, C. and McCarthy, K.D. (1996) Control of gap-junctional communication in astrocytic networks. *TINS* **19,** 319–325.
12. Batter, D.K., Corpina, R.A., Roy, C., Spray, D.C., Hertzberg, E.L. and Kessler, J.A. (1992) Heterogeneity in gap junction expression in astrocytes cultured from different brain regions. *Glia* **6,** 213–221.
13. Blomstrand, F., Aberg, N.D., Erikson, P.S., Hanson, E. and Rönnbäck, L. (1999) Extent of intercellular calcium wave propagation is related to gap junction permeability and level of connexin-43 expression in astrocytes in primary cultures from four brain regions. *Neuroscience* **92,** 255–265.
14. Ransom, B.R. (1995) Gap junctions In "Neuroglia" Eds. Kettenmann H. and Ransom B.R., pp, 299–318.
15. Rose, C.R. and Ransom, B.R. (1997) Gap junctions equalize intracellular Na+ concentration in astrocytes. *Glia* **20,** 299–307.
16. Venance, L., Prémont, J., Glowinski, J. and Giaume, C. (1998) Gap junctional permeability and pharmacological heterogeneity in astrocytes cultured from the rat striatum. *J. Physiol. (London)* **510,** 429–440.
17. Giaume, C. Tabernero, A. and Medina, J.M. (1997) Metabolic trafficking through astrocytic gap junctions. *Glia,* **21,** 114–123.
18. Blanc, E.M., Bruce-Keller, A.J., and Mattson, M.P. (1998) Astrocytic gap junctional communication decreases neuronal vulnerability to oxidative stress-induced disruption of Ca^{2+} homeostasis and cell death. *J. Neurochem.* **70,** 958–970.
19. Lin, J. H.-C., Weigel, H., Cotrina, M.L., et al. (1998) Gap-junction-mediated propagation and amplification of cell injury. *Nature Neurosci.* **1,** 494–500.
20. Steinhäuser, C., Berger, T., Frotscher, M. and Kettenmann, H. (1992) Heterogeneity in membrane current pattern of identified glial cells in the hippocampal slice. *Eur. J. Neurosci.* **4,** 472–484.
21. Sontheimer, H. and Waxman, S.G. (1993) Expression of voltage-activated ionic channels by astrocytes and oligodendrocytes in the hippocampus slice. *J. Neurophysiol.* **70,** 1863–1873.

22. Chvatal, A., Pastor, A., Mauch, M., Sykova, E., and Kettenmann, (1995) Distinct populations of identified glial cells in the developing rat spinal cord slice: ion channels properties and cell physiology. *Eur. J. Neurosci.* **7,** 129–142.
23. Mantz, J., Cordier, J. and Giaume, C. (1993) Effects of general anesthetics on intercellular communication mediated by gap junctions between astrocytes in primary culture. *Anesthesiology* **78,** 892–901.
24. Micevytch, P.E. and Abelson, L. (1991) Distribution of mRNAs coding for liver and heart gap junction proteins in the rat central nervous system. *J. Comp. Neurol.* 305, 96–118
25. Giaume, C., Fromaget, C., El Aoumari, A., Cordier, J., Glowinski, J. and Gros, D. (1991) Gap junctions in cultured astrocytes: single-channel currents and characterization of channel-forming protein. *Neuron* **6,** 133–143.
26. Connors, B.W., Bernardo, L.S. and Prince, D.A. (1984) Carbon dioxide sensitivity of dye coupling amoung glia and neurons. *J. Neurosci.* **4,** 1324–1330.
27. Binmöller, F.J. and Müller, C.M. (1992) Postnatal development of dye-coupling among astrocytes in the rat visual cortex. *Glia* **6,** 127–137.
28. Akopian, G., Kuprijanova, E., Kressin, K. and Steinhäuser, C. (1997) Analysis of ion channels exppression by astrocytes in red nucleus brain stem slices of the rat. *Glia* **19,** 234–246.
29. Bordey, A. and Sontheimer, H. (1997) Postnatal development of ionic currents in rat hippocampal astrocytes In Situ. *J. Neurophysiol.* **78,** 461–477.
30. Ransom, B. and Kettenmann, H. (1990) Electrical coupling, without dye coupling, between mammalian astrocytes and oligodendrocytes in cell culture. *Glia* **3,** 258–266.
31. Venance, L., Cordier, J., Monge, M., Zalc, B., Glowinski, J. and Giaume, C. (1995) Homotypic and heterotypic coupling mediated by gap junctions during glial differentiation in vitro. *Eur. J. Neurosci.* **7,** 451–461.
32. Nedergaard, M (1994) Direct signaling from astrocytes to neurons in cultures of mammalian brain cells. *Science* **263,** 1768–1771.
33. Fróes, M.M., Corriea, H.P, Garcia-Abreu, J., Spray, D.C., Campos de Carvalho, A.C. and Moura Neto, V. (1998) Gap junction coupling between neurons and astrocytes in primary central nervous system. *P.N.A.S. (USA)* **96,** 7541–7546.
34. Nadarajah, B. and Parnavelas, J.G. (1999) Gap junction-mediated communication in the developing and adult cerebral cortex. In "Gap junction-mediated intercellular signalling in health and disease" Novartis Foundation Symposium, Wiley Editors. pp: 157–170.
37. Kressin, K., Kuprijanova, E., Jabs, R., Seifert, G. and Steinhäuser, C. (1995) Developmental regulation of Na+ and K+ conductances in glial cells of mouse hippocampal brain slices. *Glia* **15,** 173–187.
38. Vera, B., Sanchez-Abarca, L.I., Bolanos, J.P. and Medina, J.M. (1996) Inhibition of astrocyte gap junctional communication by ATP depletion is reversed by calcium sequestration. *FEBS Lett.* **392,** 225–228.

7

Glial Cell Involvement in Brain Repair and the Effects of Aging

Elizabeth A. Howes and Peter J. S. Smith

1. INTRODUCTION

The integrity of the central nervous system (CNS) is essential for the normal functioning of all aspects of animal behavior from the coordination of segmental muscle contractions in annelid locomotion to the development and maintenance of synaptic pathways involved in human memory and thought. The effects of brain injury are likely, therefore, to be devastating. It follows that repair of damage caused by disease or accident is of enormous importance to the well-being of the individual affected and yet the adult mammalian brain has poor powers of recovery from injury. If nerve tracts in the peripheral nervous system are damaged, it does not follow that the region of injury will not recover its full range of movement since axons in the periphery are able to regrow and reestablish innervation. A similar injury within the CNS, however, will not be repaired in a way that will lead to functional recovery despite the fact that CNS axons are intrinsically capable of growth and that sprouting and extension of neurites are initiated by injury. The main factor that controls the different outcomes in these two cases is the glial environment, which is permissive of axonal growth in the peripheral nervous system but checks growth within the CNS *(1)*.

Glial cells were once regarded as mere supporting tissue for the neurones, a kind of "sponge" that held the neurones in place, provided nutrients, and mopped up waste products of neural activity. Thanks to the work of many researchers over the last three decades a much more realistic view of glial function has emerged and glial cells are now known to be crucially involved in the development and effective functioning of the nervous system throughout life. It follows, therefore, that perturbation of glial cell function is likely to have severe consequences: A view borne out by the links that have been established between glia and many neurological diseases including Parkinson's disease, multiple sclerosis, progressive supranucleur palsy, and Alzheimer's disease.

Although gross damage to CNS nerve tracts cannot be reversed, mechanisms exist for limiting and repairing damage caused by infection, oxidative stress, and mechanical lesions to the brain. Glial cells are intimately involved in these processes: They can proliferate in response to injury, act as controllers of inflammation, phagocytose damaged tissue, secrete neuroprotective and neurotrophic molecules, form boundaries to isolate regions of damage and help to maintain ionic homeostasis within damaged

From: *Neuroglia in the Aging Brain*
Edited by: Jean S. de Vellis © Humana Press Inc., Totowa, NJ

regions of the brain. However, they can also promote dangerous levels of inflammation, secrete neurotoxic molecules, inhibit neurite growth, and promote demyelination. In this chapter we will describe injury-induced cellular responses associated with glial activation and consider both the balance that exists between beneficial and harmful effects of activation and the effects of aging in weighting this balance.

2. BRAIN INJURY AND ITS IMMEDIATE EFFECTS

The main experimental technique used to induce in vivo CNS injury in vertebrates is occlusion of cerebral arteries for shorter or longer periods thus causing different levels of ischemia throughout the brain and mimicking the effects of stroke injury. Some studies have also been carried out using mechanical means, e.g., stab-wounds to the brain and the severing of specific nerve tracts, or by chemical lesioning, e.g., injection of kainic acid or 3-nitroproprionic acid into the brain. Selective lesioning of glial elements has also been carried out in invertebrates where the layout of the nervous system enables localized destruction of glia through the application of ethidium bromide to regions lacking nerve cell nuclei. In this chapter, we will largely consider the sequence of repair that follows ischemia but will include data from studies of other types of brain injury where these shed light on additional repair processes.

The immediate effect of restricting the flow of blood in the carotid arteries is to reduce oxygen levels throughout the brain. The resulting anoxia causes a rapid decline in normal energy supplies and a switch to anerobic metabolism in which lactate is produced as a by-product. These events lead to failure of the Na^+K^+-ATPase exchanger in neuronal membranes and an accumulation of axoplasmic Na^+ via voltage-gated Na^+ channels. The combination of high internal concentrations of Na^+ and a concomitant depletion of K^+ depolarizes the neuronal membrane. This stimulates a reverse N^+-Ca^{2+} exchange, and the resulting sharp rise in Ca^{2+} levels within the neurone activates Ca^{2+}-dependent enzymes such as phospholipases, calpain, and protein kinase C that can cause irreversible damage and death of the affected neurones *(2,3)*. Not all regions of the brain are equally susceptible to ischemia: CA1 hippocampal neurones in adult rat brains are killed by a 10 min exposure to ischemia but there is little or no damage to neurones in other regions of the brain *(4)*. Similarly, in organotypic cultures, CA1 hippocampal neurones are damaged by short (7 min) exposure to ischemia while CA3 neurones and the most susceptible neocortical neurones withstand longer periods (up to 30 min) of anoxia. Dentate gyrus and some neocortical neurones are even more resistant and are only damaged after 60 min exposure to anoxic conditions *(5)*.

Glial cells are considered to be more robust than neurones in their response to ischemia but, following even brief periods of global ischemia, some damaged glial cells appear in the cortex and thalamus. Most of these are oligodendrocytes, possibly because they are susceptible to glutamate released from damaged neurones. In the case of prolonged ischemia or mechanical injury, external levels of glutamate may be augmented by release from astrocytes at the site of injury given that a gradual loss of their transmembrane Na^+-gradient would lead to conditions favoring carrier reversal and the extrusion of glutamate from the cells *(6)*. Since glutamate is also a ligand of the *N*-methyl-D-aspartate receptor on neurones, these elevated levels of glutamate would have the effect of inducing further excito-toxic neuronal injury and death, thus extending the effect of the initial lesion *(7)*

Mechanical injuries such as stab wounds destroy neural tissues directly and introduce breaches in the blood-brain-barrier that threaten the homeostasis of the brain. Death of neurones and glial cells is augmented as blood leaks into the damaged region changing the ionic environment around undamaged cells and inducing fluctuations in membrane potential and a cycle of events resembling that outlined above. Uncontrolled entry of circulating lymphocytes and other cells of the immune system to previously protected neural tissue may mount an autoimmune response in addition to the inflammatory reaction set in motion by the resident immune system of the brain.

Chemical lesions are designed to cause specific types of injury e.g., the use of serotonin neurotoxin to cause death of serotonergic neurones without gross damage to other cells or kainic acid injection to cause selective damage to hippocampal CA3 pyramidal cells. Similarly, axotomy of specific nerves, such as the facial nerve has been used to produce limited damage in the region of the brain to which the damaged axons project. Localized application of ethidium bromide to interganglionic connectives within the insect CNS enables selective destruction of glial cells and allows glial response and repair to be studied in the absence of neuronal damage. All types of injury appear to initiate a similar chain of glial response.

3. GLIAL RESPONSES TO INJURY

3.1. *Glial Types*

In discussing glial responses to damage it is necessary to differentiate between the three major types of glia recognized in the brain:

1. Microglial cells form between 5 and 20% of the neuroglial population and differ from other glial cell types in the brain in being derived from mesenchymal tissues. The relationship between microglia and circulating macrophages has been discussed by Perry and Gordon *(8)* and it is clear that microglial cells form the main element in the immune system of the brain. They are particularly responsive to small changes in their environment, carry macrophage markers on their cell surface and respond rapidly to damage. They are also able to change their morphology, act as phagocytes and secrete a range of cytokines and growth factors that modulate neural repair *(9)*.
2. Astrocytes in the intact brain are largely responsible for maintaining its homeostasis both by forming the blood-brain-barrier and by uptake and secretion of molecules in the extracellular fluid e.g., astrocytes metabolize glutamate released at nerve endings before it can reach cytotoxic levels and they contain glycogen stores that can be released to supply neuronal energy demands. In addition, astrocytes have been shown to release neurotrophic factors that support neuronal survival and sprouting, which may be of major importance during both development and neural repair.
3. Oligodendrocytes are the glial cells most intimately associated with neurones within the brain, they are responsible for myelination, provide metabolic support for neurones and are closely involved in signaling between neurones and glia via voltage- and ligand-gated ion channels *(10,11)* They do not appear to play a role in the glial reactivity that follows brain injury, but are a factor in preventing repair in the CNS since myelin-associated proteins can inhibit neurite growth. As described above, they are the glial cell type most susceptible to injury by anoxia but such damage can be repaired by recruitment from a limited pool of progenitor cells resident within the brain *(12)*

3.2. Pattern of Glial Cell Response to Injury

The general response of glial cells to neural tissue damage is similar over a range of different injuries although the level of reactivity depends on the severity of the injury. There is an initial activation of resting microglial cells, which involves a change in cell shape and accumulation of reactive microglia in and around the site of injury. This is followed by a period of astrogliosis during which increasing numbers of activated astrocyes appear within the brain. The majority of these cells are associated with the damaged tissues, and their most obvious activity is the formation of a glial scar. The recruitment of cells in these two processes is a prerequisite for the repair processes that are set in train by cell damage and death within the brain. The initial inflammatory reaction in which they are involved is an important event since the level of this response may determine the success of any future repair. A combination of upregulation of signaling proteins, cytokines, neurotrophic proteins, and other proteins involved in the cycle of events eliciting brain repair is accompanied by apoptotic cell death and the induction of phagocytotic activity to clear debris within the lesion. Tight control of these processes is essential to prevent the inflammatory process itself causing further and more widespread cellular damage. Consequently, a considerable amount of research has been directed toward understanding the processes of microglial and astrocytic activation, the results of which we summarize here.

3.3. Activation of Microglial Cells

Activation can occur extremely rapidly and is often detected before damage to neurones is visible; Ivacko et al., *(13)* using Griffonia simplicifolia B4-isolectin to identify microglial cells were able to determine subtle changes in their appearance within 10 min of administering hypoxic injury to young rats. Four hours after injury, activated microglial cells could be identified in the lesion zone and the number of microglial cells present continued to rise with numbers peaking 2 to 4 d after injury. In a similar study of brief ischaemic injury in rat brain, proliferation of microglial cells was reported to peak seven days after injury *(14)*. Intraventricular kainic acid injection also results in the accumulation of microglial cells around damaged pyramidal neurones within 3 h and, prior to this, alterations in the shape of the microglial cells to an amoeboid form are detectable *(15)*. The fact that microglial cells play a continuing role in brain repair processes is demonstrated by a study in which kainic acid lesion-induced changes in the morphology and number of microglial cells in the mouse brain were monitored over a period ranging from 12 h to 81 d postlesion *(16)*. This showed a decrease in the number of cell processes and an accompanying rounding of the microglial cell 12 h post injury. Short, thick processes unlike those of resting microglia began to re-extend from the cell 48 h postinjury and very gradually changed to give a return to the normal, resting morphology after approximately one month. Over the same period, changes in the number of microglial cells in regions of the brain associated with damaged neurones were apparent. Cell counts around the lesion remained stable for two days and then increased from approx 100 cells/mm^2 to 500 cells/mm^2 3 d postlesion. The number of microglial cells present continued to rise for a further two days and remained high for approximately a month before gradually falling to their original level.

Observations such as these raise the question of whether this increase in reactive microglial cell numbers is the result of migration to the site of injury from undamaged regions of the brain, of proliferation of resident microglia or of the recruitment of monocytes across the blood-brain-barrier and their transformation into microglial-like cells. It seems likely that in the case of noninvasive brain injury the initial accumulation of cells in and around the lesion is due to microglial cell migration since Akiyama et al. *(15)* noted a reduction in the number of microglial cells in adjacent regions concomitant with the appearance of cells in the lesion zone. In addition, the ability of activated microglial cells to migrate toward injured neurones has been demonstrated in vitro in hippocampal slice cultures *(17)*. Proliferation of microglial cells has also been identified at sites of injury in rat brain *(18,19)*. Following cortical stab wounds maximal microglial proliferation is reported to occur between 2 and 3 d postlesion *(20)*, coinciding with the timing of the increase in cell numbers described by Andersson et al. *(16)*. Although such studies indicate that cell division contributes to the long-term increase in microglial cell numbers, they do not rule out the possibility that blood monocytes may also be involved in the secondary increase in microglial cell numbers seen between 2 and 5 d postlesion. Even in the case of treatments that do not compromise the blood-brain-barrier, monocytes are recruited into the brain within 2 d of injury and subsequently transform to macrophages *(21)*. More recently Akiyama et al. *(15)* have shown that vascular endothelial cells begin to express ICAM-1 approximately six h after ischemic injury with levels gradually increasing over the next 42 h. As a ligand for leucocyte function associated antigen (LFA)-1, the expression of ICAM-1 precedes the accumulation of leucocytes in the brain vasculature. These then pass through the blood-brain-barrier to infiltrate the brain, where they may assume some of the functions of reactive microglia. A similar involvement of blood cells in repair of selective glial lesions to cockroach CNS was demonstrated by Treherne et al., *(22)* who showed that hemocytes pass through the intact neural lamella, accumulating in the site of injury where they transform into granule-containing cells that play a crucial role in subsequent repair processes.

3.3.1. Upregulation of Microglial Proteins

The transformation from resting to reactive state is accompanied by upregulation of a number of proteins, some of which are specific to microglia, although others are also expressed by reactive astrocytes (Table 1). In keeping with their immuno-protective role, many of the proteins on reactive microglial cells are associated with aspects of the immune reaction. Thus, macrophage scavenger receptors appear on microglial cells within 24 h of kainic acid injection into rat brain, consistent with the view that microglial cells take on a phagocytic role and assist in the removal of debris from damaged cells *(23)*. The class II major histocompatibility complex (MHC class II), which is expressed at low levels on at least some resting microglia, increases in response to infection or injury. The degree of upregulation is controlled so that although, for example, a viral infection may induce almost all microglia to become MHC class II positive, limited brain damage following peripheral axotomy results in upregulation confined to the region of the brain into which the damaged axons project. The role of MHC class II in the peripheral immune system is to present antigen to helper T-lymphocytes, but it is a matter of current debate whether it has this function in the CNS. In inflammatory

Table 1
Glial factors upregulated in response to brain lesion

	Factor	Function
Activated Microglia		
cytokines:	IL-1	– enhances dopaminergic neuron sprouting
	IL-1α	– activates astrocytes
		– upregulates β-amyloid precursor and S110β protiens
	IL-1β	– proinflammatory
		– inhibits Ca^{2+} influx, protein kinase activity, ACh and glutamate release from neurons
		– induces reactive oxygen species formation
ICE:		– appears prior to apoptosis; possible mediator of inflammation
	IL-6	– increases blood brain barrier permeability
		– proinflammatory
		– role in migration and differentiation of oligodendrocyte precursors
		– neuroprotective
	IGF-I, II	– neuroprotective
		– neurotrophic
	TGFβ-1	– immunosuppresive
		– inhibits ICAM-1 expression
		– chemotactic for astocytes and microglia
		– inhibits astrocyte proliferation
		– induces differentiation of oligodendrocyte precursors
		– upregulates GFAP expression
		– neuroprotective
	β-chemokines	– monocyte chemoattractant
		– macrophage inflammatory protein

Table 1
(Continued)

	Factor	Function
		– upregulated by LPS
cell adhesion molecules:	ICAM-1	– important for lymphocyte activity in CNS
macrophage-associated antigen:	MHC I	– antigen-presenting molecule
	MHC II	– antigen presentation
	complement type-e receptor	
	leucocyte common antigen	
	macrophage scavenger receptor	
extracellular matrix molecules:	laminin	– enhances neurite outgrowth
	fibronectin	– promotes oligodendrocyte migration
	osteopontin	– glial scar formation
others:	cytosolic phospholipase A	– breakdown of phospholipids
	HSp 70	– neuroprotective
	thromboxane	
	ATP-hydrolysing enzyme	– neuroprotective via release of adenosine
	amyloid precursor protein	
	cystatin C	– cysteine protease inhibitor
	prostanoids	– neuroprotective
	NO synthase	– upregulates nitric oxide, a cytotoxic signalling molecule
	CCAAT-enhancer binding protein α	– sequence-specific binding protein, possible regulator of gene expression
Activated Astrocytes		
cytokines:	bFGF	– trophic for neurons, glia and endothelial cells
		– induces activated morphology in astrocytes
		– supports oligodendrocyte growth
		– protective against reactive oxygen species
	FGFR	
	EGFR	

(continues)

Table 1
(Continued)

	Factor	Function
Activated Astrocytes		
	BDNF	– supports axonal regrowth
	NGF	– enhances oligodendrocyte survival
	CNTF	– oligodendrocyte differentiation, astrocyte activation
	NT-3	– neurotrophic
	IGF-1	– neuroprotective
		– neurotrophic
	IGF-II	
chemokines:	MCP-1	– recruitment of blood-derived inflammatory cells
	IP-10	– recruitment of blood-derived inflammatory cells
cell adhesion molecules:	VCAM-1	– aids leukocyte entry through blood brain barrier
	ICAM-1	
intermediate filament proteins:	GFAP	– morphological changes
	nestin	– morphological changes
	vimentin	– morphological changes
extracellular martix molecules:	vitronectin	– supports astrocyte migration
	laminin	
	fibronectin	– promotes oligodendrocyte migration
	tenascin-C	– inhibits oligodendrocyte migration
antioxidants:	(mSOD	
	glutathione peroxidase	– elimination of free radicals
	catalase	– protection against oxidative stress
	glutathione	– neuroprotective
	hemeoxygenase 1	
	NO synthase	– upregulates NO production
	NO synthase)	– signalling molecule
		– cytotoxic

cytosolic phospholipase A	– phospholipid breakdown
substance P	– regulator of inflammation, wound healing and immune response
HSp 70	– neuroprotective
calbindin D28K	– neuroprotective
	– Ca^{2+} homeostasis
L type Ca^{2+} channels	– Ca^{2+} homeostasis
	– Ca^{2+} signalling
endothelin β receptor	– mitogenic to astrocytes
	– mediates actin reorganization
NF_kB	– gene transcription factor

Abbreviations:

ACh	acetylcholine	IL	interleukin
BDNF	brain-derived neurotrophic factor	IP-10	IFN-γ-inducible protein
CNTF	ciliary neurotrophic factor	LPS	lipopolysaccharide
ICAM-1	intracellular cell adhesion molecule 1	MCP-1	monocyte chemoattractant peptide-1 GFAP
ECM	extracellular matrix	NGF	nerve growth factor
EGFR	epidermal growth factor receptor	NO	nitric oxide
bFGF	basic fibroblast growth factor receptor	NT-3	neurotrophin 3
GFAP	glial fibrillary acidic protein	MHC	major histocompatibility complex
HSp 70	heat shock protein	SOD	superoxide dismutase
ICE	interleukin-1 β converting enzyme	NFkB	nuclear factor kappa B
IGF	insulin-like growth factor		

brain lesions, the apoptosis of T cells that occurs in neuroectodermal parenchyma does not depend on antigen presentation by glial cells and the T cells that infiltrate the brain are destroyed whatever their antigen specificity or level of activation *(24)*.

A number of macrophage cytokines are expressed by microglia and the timecourse over which they are upregulated in vivo is consistent with the view that they are involved at early stages of glial activation. Transforming growth factor-β1 (TGF-β1) mRNA is expressed by activated microglial cells within 6 h of ischaemia and since TGF-β1 has an inhibitory effect on astrocyte proliferation *(25)*, it may have a role in limiting potentially damaging inflammatory responses after injury. The continued expression of TGF-β1 by both microglial cells and macrophages for periods of up to 3 mo suggests that it also has a role in long-term repair and remodeling of damaged neurones. Cytokines also have mitogenic, neuroprotective, and neurotoxic properties when tested on cultured cells but it is difficult to predict how the effects of different cytokines will be modulated within the complex environment of the brain. There are increasing numbers of in vivo studies, however, to support the view that cytokines play an important role in CNS repair. For example, the finding that injection of interleukin-6 (IL-6) into rat brain reduces the level of neuronal damage produced by prolonged ischemia coupled with the fact that increased levels of endogenous IL-6 activity are measurable in ischemic regions of the brain within 2 h of damage *(26)*, suggest that upregulation of IL-6 in activated microglia has a physiological role in vivo. Similarly, macrophage colony stimulating factor (M-CSF) has been shown to have a mitogenic and morphological effect on microglial cells in vitro, although in vivo infusion of a related colony stimulating factor, GM-CSF, results in an increase in microglial cells in the treated area *(27)*. The finding that a mouse strain carrying a mutation silencing the gene encoding M-CSF fails to show the usual reactive microglial proliferation following facial nerve axotomy *(28)* also indicates that colony stimulating factors are important for microglial responsiveness.

Activated microglia in vivo can secrete growth factors including insulin-like growth factor (IGF-1), epidermal growth factor (EGF) and nerve growth factor, and it is likely that these have a neurotrophic role. A recent paper shows that 3 d after lesion mRNA for IGF-1 is present in microglial cells in regions of delayed neuronal cell death. A concomitant increase in IGF-binding protein in reactive astrocytes closely associated with surviving neurones suggests that microglial IGF-1 is transferred to these astrocytes and may be used by them for the continued support of the neurones *(29)*.

Other proteins expressed by activated microglial cells may be involved in limiting neuronal damage by other means. An example of this is the proposed role of microglial cells in the generation of adenosine from ATP. Adenosine has neuroprotective properties and increased levels of adenosine are found in damaged areas of ischaemic brains. Although these may partly result from release by damaged cells, it has also been shown that upregulation of ectoapyrase and ecto-5′-nucleotidase, which together can hydrolyze nucleoside 5′-mono-, di- and triphosphates, occurs within 2 d of ischaemia. They are still present 28 d post-lesion, particularly in the vulnerable CA1 hippocampal region, and their pattern of expression coincides with the distribution of markers for reactive glia, especially those associated with microglia *(30)*. Extracellular matrix (ECM) molecules such as laminin and fibronectin are expressed by microglial cells after CNS injury and are known to influence neuronal growth in vitro. Their potential role in vivo is discussed in relation to astrocyte activation in a subsequent section. Another ECM molecule, osteopontin, provides an example of the inter-relationship between microglial and astroglial cells following injury. Osteopontin mRNA is expressed *de novo* by microglial cells at the margin of the damaged region within 3 h of ischaemic lesion. The protein itself subsequently occurs in this same area within 24 h and is later also expressed within the lesion zone. The integrin receptor for osteopontin, α(v)β(3), is expressed by astrocytes 5–15 d post-lesion, initially at a distance from the osteopontin-expressing microglia but by day 15 coincident with them *(31)*. This observation suggests that activated microglia may initially be responsible for delineating the area of the lesion by laying down extracellular material with which migratory reactive astrocytes will then interact during glial scar formation.

3.4. Activation of Astrocytes

Reactive astrocytes begin to appear in the brain shortly after the initial activation of microglial cells *(32)*. Morphologically these astrocytes are characterized by an overall increase in size which includes lengthening and thickening of cellular processes. In addition, glial fibrillary acid protein (GFAP) expression is upregulated, the number and size of the mitochondria present increases and the endoplasmic reticulum becomes enlarged. When rats are subjected to brief ischaemic injury, reactive astrocytes appear at the site of injury within a week of damage, with numbers peaking at around 2 wk *(33)*, which is in the same range as the glial repair processes that follow glial lesioning in insect CNS. An increase in perineurial glial cells is apparent in the cockroach within four days of chemical lesion with numbers peaking after 6 d, at which time there is a further recruitment of subperineurial glia in the lesion zone, thus extending the period of gliosis to approximately 2 wk *(34)*. Much more rapid induction of astrocytic gliosis occurred in hippocampal and neocortical organotypic cultures subjected to histotoxic ischaemia where gliosis could be detected within 30 min of injury *(5)*.

The extent of astrogliosis within the brain varies with the type and severity of injury and tends to be more widespread than the distribution of activated microglia. In general, astroglyosis following ischaemic injury is largely restricted to the area of the lesion but, following a stab wound to the cerebral cortex, astrogliosis may occur over a widespread area with GFAP-positive astrocytes even present in the hemisphere contralateral to the wound *(35)*. Such distant responses may depend on the extensive communication system, based on changes in intracellular free Ca^{2+} concentrations transmitted from cell to cell through gap juctions, that exists in astrocytes *(36)*. Gap junctions can carry apoptotic signals from damaged cells to adjacent healthy cells causing widespread death at sites distant from the original lesion *(33)*, but the importance of such signaling in extending and controlling cell damage within the CNS is not known. During the early stages of neuronal degeneration following ischemic injury neurones can trigger Ca^{2+} waves in adjacent astrocytes via gap junctions or glutamate release *(37)* and although these signals may initiate cell death it is also possible that they trigger astrogliosis. In this context, it may be relevant to note that, a calcium-binding protein, calbindin-D28K, normally present in neurones is also found in reactive astrocytes in the CA1 subfield after ischaemia *(38)*.

Whatever the signal pathways involved, the appearance of large numbers of activated astrocytes within a lesion must involve transformation of a preexisting population of astrocytes and/or glial cell proliferation. As with microglia, it appears that both processes occur but the balance between them varies in different situations. Eclancher et al., *(39)* reported that some 22% of GFAP-positive astrocytes incorporated ^{3}H-thymidine following an electrolytic lesion of rat brain. Similar evidence has been obtained for astrocyte division following ischemia, mechanical wounding and chemical lesioning. In other studies, however, attempts to label astrocytes with ^{3}H-thymidine have shown little or no incorporation *(40,41)*.

3.4.1. Upregulation of Astrocyte Proteins

In addition to upregulation of GFAP, which is the most commonly used marker of astrocyte activation, a number of other proteins are upregulated or newly expressed (Table 1). The functions of these proteins fall into four broad areas:

1. Containment of the site of injury,
2. Regulation of inflammation,
3. Promotion of neuronal survival,
4. Reduction of oxidative stress.

Morphological changes, cell migration to lesion sites, cell proliferation, and production of extracellular matrix molecules are all likely to be involved in restricting the site of lesion by forming a glial scar. The expression of both GFAP and another intermediate filament protein, nestin, is linked with the change in shape that accompanies activation. This process must involve reorganization of other cytoskeletal elements by changes in the disposition of actin and actin-binding proteins, for example, and the effects of this on surface membrane protein activity have yet to be investigated. A recent in vitro study has, however, demonstrated a role for the small GTP-binding protein Rho in modulating changes in astrocyte shape *(42)*. Rho is involved in the regulation of F-actin in some cell types, where its activity is modulated by components of the extracellular matrix. The production of ECM molecules may thus have an effect on both the morphology and the migration of astrocytes since, in addition to the Rho pathway, ECM molecules that bind to integrins on the cell surface influence cell shape in some systems, while vitronectin has been shown to support migration of astrocytes in vitro in the presence of TGFβ1. Secretion of ECM by activated astrocytes will also influence other aspects of the repair process: Different ECM proteins can exert modulatory effects on neuronal and oligodendrocyte cell extension and affect the migration of oligodendrocyte precursor cells. Laminin, which promotes neurite outgrowth in developing brain, is transiently expressed by activated astrocytes *(43)* and there may be a causal relationship between the secretion of laminin by astrocytes and the short-lived neurite sprouting seen after CNS damage. It may be significant that in species such as frog and goldfish where the adult CNS is permissive of axonal growth laminin is expressed in the brain throughout life *(44)*. Laminin, together with fibronectin also supports the migration of oligodendrocyte precursor cells, however, a third ECM molecule, tenascin-C, inhibits this migration *(45)* and its presence in the glial scar may prevent remyelination of axons within the lesion.

The effects of the extracellular matrix are modulated by cytokines, a number of which have been implicated in the processes accompanying astrocyte activation. For example, bFGF enhances the ability of astrocyte-secreted ECM to support oligodendrocyte outgrowth in vitro *(46),* whereas TNFα blocks astrocyte migration, even in the presence of a migration-inducing cytokine, TGFβ, unless vitronectin is also present *(47).* Cytokines are involved in many other aspects of the response of the brain to lesion including control of inflammation, effects on proliferation and migration of all glial cell classes and production of nitric oxide and reactive oxygen species *(48).* The diversity of cytokine effects is demonstrated by the fact that a single cytokine, bFGF, is reported to protect neurones from glutamate toxicity, upregulate antioxidant defenses via glutathione, induce an activated morphology in astrocytes and have a potent trophic effect on neurons, glia, and endothelial cells.

A variety of other astrocyte-associated proteins affect these same processes: Substance P modulates inflammation, heat shock proteins protect neuronal function, and heme oxygenase-1 protects against oxidative stress, whereas the upregulation of L-type Ca^+ channels may regulate homeostasis and enhance release of neurotrophic agents as well as being involved in signaling in astrocytic networks.

As the above description of microglial and astrocyte activation indicates, glial cells mount a robust reponse to injury and are able to control and reverse at least some of the effects of brain injury. The level of permanent damage inflicted by a brain lesion depends upon the number of neurones that are killed; astrocytes and microglia are capable of proliferation, oligodendrocytes can be replaced by differentiating precursor cells but neurones cannot be replaced. In addition, the more widespread the damage to neurones and myelin-producing cells, the more intense the inflammatory response is likely to be, culminating in the destruction of previously undamaged neurones as activated microglial cells attack their myelin sheaths. The major role of glia may, therefore, be to control inflammation by rapidly isolating the region of damage from the rest of the brain thus restricting cytokine activation to a limited area and enabling a balanced process of tissue destruction and repair to proceed. The effects of aging on the maintenance of this balance are likely to be variable and complex.

4. CHANGES IN THE AGING BRAIN THAT MAY AFFECT REPAIR PROCESSES

Studies of the ways in which the CNS changes during normal aging are less common than studies of the effects of diseases of the brain and there have been few direct studies of the effects of age on responses to lesions. However, by evaluating results from studies that have clarified aspects of the aging processes in nondiseased brains, it is possible to select age-related changes that are likely to have an impact on glial responses to brain injury. Although there is some work on aging in the CNS of normal human subjects most studies in this area use rodents as models. Work with rats and mice of different ages, as well as studies on senescence-accelerated mice, have provided much information on the effects of age on brain morphology and function, whereas in vitro studies have enabled examination of the effects of cell "age" on cell-cell interactions. It is worth noting that rodents are often maintained to advanced ages by strictly controlling their food intake, but since recent work has shown that dietary restriction influences various aspects of brain aging through reduction of oxidative stress *(49),* such animals are not good models of normal aging processes.

4.1. Morphological Changes in Aging Brain

Overall changes in brain volume as well as more specific changes in the number and morphology of brain cells have been reported from studies of aging subjects but the extent of such changes is unclear. There is a general consensus that as the brain ages there is a gradual decrease in neuronal cell number, especially among cholinergic neurones, probably accompanied by a reduction in dendritic spine density and an increase in glial cell numbers. Changes in oligodendrocytes have not been widely reported although some vacuolation and demyelination may take place in older brains. With the development of new technologies including improved methods for estimating cell numbers and the availability of specific glial markers, more accurate studies of neural cells have become possible, and it is instructive to look at some of the most recent examinations of changes in cell types with aging.

A study in rats revealed no differences in cortical volume or neuronal density between young (4–6-mo-old) and aged (30–32-mo-old) rats but a decrease in the area occupied by neuronal somata was identified in layers II-VI. This was accompanied by changes in neuronal shape in these same layers and a decrease in neuronal nuclear area in layer VI. Glial density increased by ~17% throughout the cortical layers *(50).* This increase was subsequently shown to result from changes in the number of astrocytes present, appearing as astrocytic clusters in older brains, whereas the numbers of microglia and oligodendrocytes remained constant between the two groups. However, changes in the morphology of microglial cells were noted, the significantly larger numbers of cytoplasmic inclusions present in older animals presumably reflecting increased phagocytotic activity with age *(51).* A similar result was obtained by Amenta *(52),* who examined the number and morphology of GFAP-positive astrocytes in the frontal cortex and CA1 hippocampal subfield of adult (12-mo-old) and aged (24-mo-old) rats. An age-related increase in the number and size of GFAP-positive astrocytes was found in the two regions but they differed in that, although increases in cell numbers was more apparent in the CA1 subfield, increased astrocytic size was more noticeable in the frontal cortex. Although comparable results have been found in a number of studies, they are by no means a universal finding. An examination of the supraoptic nucleus region in groups of rats ranging in age from 5 to 24 mo failed to find any significant changes in the number of neurons or astrocytes between groups, although some degenerating neurons were observed in older animals. This same study does, however, describe age-related changes in the morphology of astrocytes including an increase in nuclear size associated with the appearance of nuclear bodies, clear patches within the nucleoplasm and changes in the organization of the nucleolus. The cytoplasm of the astrocytes also showed changes including hypertrophy of intermediate filaments and the appearance of an extensive tubular network linking stacks of Golgi cisternae. These morphological changes were accompanied by an increase in GFAP and vimentin expression *(53).* A stereological analysis of three different regions of the hippocampus in mice ranging in age from 4 to 28 mo found no statistical differences in the numbers of astrocytes or microglia in any of the three areas studied *(54).* The varying responses obtained in these studies undoubtedly reflect variability in the cellular responses to aging in different regions of the brain and may also indicate that not all species show identical responses. They also point to differences both between individuals and between groups of animals from different backgrounds.

Similar studies of brain tissue from humans with no signs of neurological disease suggest that only limited changes in cell numbers occur with age. Neuronal cell death is not widespread *(55,56),* though some regions of the brain such as the hippocampus may undergo greater loss of neurones than other less vulnerable regions *(57)* and neuron shrinkage and synaptic loss may also occur *(58).* These changes can, however, be partly offset by growth and elaboration of dendrites in affected areas and such compensatory growth occurs even in aging (>70-y-old) humans *(59).* Changes in glial cell numbers are similarly limited. There appears to be no overall increase in glial proliferation with age but some reactive gliosis is associated with those regions of the brain where synaptic modification occurs in response to the aging process.

There is no reason to believe that these modest morphological responses to aging would have an effect on responses to brain injury were they not accompanied by a notable upregulation of both astrocyte and microglial cell activation. Studies designed to identify activation of astrocytes during aging show an increase in expression of GFAP with age, which is unrelated to cell numbers *(60)* but depends upon increased levels of GFAP gene transcription *(61,62).* Thus, in neurologically normal humans significant correlations between GFAP levels and age have been identified. A study by David et al. *(63)* showed increased expression of GFAP with age, visible first in the hippocampus, and subsequently, in the entorhinal cortex and isocortex. Levels of expression did not show a steady rise throughout life but increased sharply at ~65-yr-of-age. Similar increases have been shown in other species, for example, in the hippocampus, septum and corpus callosum of the rat *(64).* Astrocyte activation in aging is accompanied by upregulation of a number of proteins including GFAP, MHC class II, S100β, monoamine oxidase, superoxide dismutase and αβ-crystallin. Upregulation does not occur globally throughout the brain: Levels of S100β increased most markedly in the hippocampus of aging senescence-accelerated mice and paralleled the observed increase in the number of astrocytes expressing GFAP *(65).*

Microglial activation is also recognized as a common feature of aging brains. A study of differences in antigen expression by microglial cells from young (3–6-mo-old) and aged (24–30-mo-old) rats showed upregulation of complement type-3 receptor, MHC class II, leucocyte common antigen, CD4, and a macrophage antigen recognized by a monoclonal antibody, ED1 *(66).* A similar result is obtained in monkeys, with microglial cells showing an age-related increase in MHC II expression *(67).* Both studies showed that enhancement of MHC II expression is largely associated with microglial cells in the white matter. Signals from degenerating white matter induce expression of MHC Class II on microglial cells *(68)* and thus the upregulation of this molecule in aging brains may be a marker of degenerative changes taking place within the myelinated axon bundles. The upregulation of MHC Class II molecules occuring on both microglia and astrocytes in aged brain may also reflect the integrity of adjacent neurones since, in rat hippocampal slice cultures, interferon-γ only induced MHC class II in regions where there was substantial neuronal damage. Intact neurones prevented astrocyte expression unless channel blockers were used to suppress spontaneous neuronal activity *(69).*

Increased production of cytokines by activated microglia may also play an important role in the ability of the aging brain to respond to stress or injury. A study of age-related interleukin-1α-immunoreactivity (IL-1α+) in mesial temporal lobe microglial cells of neurologically normal humans indicated that total numbers of activated IL-1α+ cells increased with age. Three types of IL-1α+ cells were identified: Primed, enlarged, and phagocytic, and these were taken to represent three progressive stages of activation. The proportion of enlarged IL-1α+ cells increased threefold whereas phagocytic IL-1α+ cells increased 11-fold in individuals over 60 yr of age. At the same time, high levels of IL-1α mRNA were found in tissue extracts *(70)*. The effects of IL-1α include upregulation of neuronal β-amyloid prescursor protein expression, activation of astrocytes and promotion of S110β expression. A similar increase in expression of interleukin-1β has been found in the hippocampus of aged rats *(71)*. This cytokine has the potential to influence repair processes at a number of levels since, when applied exogenously to the hippocampus, IL-1β inhibits Ca-influx, protein kinase C and the release of acetylcholine and glutamate. It has also been shown to inhibit long-term potentiation, possibly through the production of reactive oxygen species, which enhance lipid peroxidation and cause neuronal membrane damage *(72)*. Another inflammatory cytokine, IL-6, is overexpressed by brain microglia in the cerebellum, cerebral cortex and hippocampus, but not the hypothalamus, of aged mice *(73)*.

The underlying causes of increased glial cell activation in aging brain are not well understood but since glial cells are known to be particularly responsive to changes in their environment they may include factors that affect brain homeostasis. Among these is a reported reduction in intracellular pH with aging *(74)* and a reduction in Ca^{2+} channel antagonist receptors on neuronal membranes leading to increased intracellular concentrations of Ca^{2+} *(75)*. The possibility of age-related modulation of blood-brain-barrier permeability suggested by the increased rates of transfer of macromolecules from the blood into the hippocampus and dorsal thalamus that occur in SAM-P8 mice *(76)* is supported by the observation that specific carrier-mediated transport systems in the blood-brain-barrier of humans, including choline and glucose transfer systems, alter with aging *(77)*. Such changes, coupled with the increased levels of free radicals associated with aging, could impose further strains on the mechanisms responsible for maintaining brain homeostasis.

5. GLIAL RESPONSES TO INJURY IN AGING CENTRAL NERVOUS SYSTEMS

There are strong parallels between glial responses to injury and to aging (Table 2) and it appears that aging induces a constant level of raised reactivity in both microglia and astrocytes. The effect of this on glial responses to lesion are not clear, although a few studies have now been carried out and these suggest an impairment of responses to injury in the CNS of aged, but otherwise healthy individuals. In one such study, the levels of GFAP mRNA were measured in the spinal cord of young (2-mo-old) and older (8 to 17-mo-old) rats. The level increased with age, being 0.4-fold higher in the older group. Sciatic nerve axotomy caused an increase in mRNA levels for GFAP in the young group indicating that astrogliosis was initiated but no response occurred in the older group *(78)*. This suggests that the increased astrocyte reactivity indicated by high levels of GFAP mRNA in older animal prevents further upregulation of GFAP in response to a lesion. Does this occur because age-induced upregulation of GFAP reaches a maximum level above which no further increase in GFAP production is pos-

Table 2
A comparison between the responses evoked by injury and by normal aging in brain tissues

Effects of injury	Eeffects of age
perturbation of brain homeostasis: changes in ionic flux across cell membranes	changes in brain microenvironment: lowering of intracellular pH
neuronal damage: debris removed by phagocytic activity of microglial cells and invading macrophages	some neuronal loss by apoptosis: sequestration of waste products in microglial cell cytoplasm
release of potentially toxic materials from damaged cells and activated glia	build up of potentially damaging reactive oxygen species and advanced glycation end products
oligodendrocyte death: demyelination may be enhanced by reactive microglia	some loss of oligodendrocytes
rapid activation of microglia and induction of inflammation	increasing numbers of activated microglia present in some brain regions
activation of astrocytes: upregulation of GFAP and immunomodulatory cytokines	increased levels of astrocyte activation marked by increased level of GFAP and morphological changes
upregulation of MHC class II and other immune system antigens	increased expression of MHC class II and other immune system antigens
secretion of IL-6 by microglia: proinflammatory	increased levels of IL-6
TGF-β-1 secreted by microglia cells has immunosuppressive effect	loss of sensitivity to regulatory effects of TGFβ-1

sible, or does the upregulation render the CNS less responsive to signals from damaged neurons, particularly in cases like sciatic nerve axotomy where neuronal damage is restricted? It appears that microglial cells do react to this type of sciatic nerve injury in older rats and only the astrocyte response fails *(79),* indicating that signals from damaged neurones are not suppressed but may be insufficient to activate further astrogliosis against a background of permanently enhanced astrocyte activity.

The fact that in aged animals microglia can continue to upregulate their activity in response to injury although astrocytes cannot could have serious consequences for repair processes. The inflammatory activity of the microglia, which in younger animals is modulated by anti-inflammatory cytokines from reactive astrocytes, will escalate in the absence of astrocyte upregulation. The resulting uncontrolled inflammation will cause increased neuronal loss, demyelination, and oligodendrocyte death. In addition, glial scar formation to isolate the lesion site will be impeded, although ECM molecules and cytokines inducing oligodendrocyte precursor differentiation and migration will not be upregulated. In fact, the number of precursor cells available to replace damaged oligodendrocytes will decline throughout life as cells are withdrawn from the limited pool of progenitors to effect repair. Thus, depending on the previous history of the individual, oligodendrocyte replacement may be doubly challenged by aging.

An in vitro study comparing the behavior of glial cells derived from the brains of young (3-mo-old) and old rats (30-mo-old) shows that both astrocytes and microglial cells from older animals proliferate more readily in culture and that they are less responsive to controlling factors such as GM-CSF and TGF-β1 *(80)* However, some astrocyte responses appear to be less affected by aging; in cultured cells derived from aged rats, astrocytes are responsive to neurotrophic signals for many passages although a stage is eventually reached (after ~ 46 passages) when responsiveness decreases and cell homeostasis is impaired *(81)*.

Even in cases where glial cells respond to injury, the responses may be reduced in older animals. A proliferative glial response to deafferentation of the dentate gyrus occurs in both young and aging rats but the older animals (12–26-mo-old) showed a much lower level of increased insulin-like growth factor (IGF-1) mRNA expression in response to the injury. and a reduced capacity for neuronal sprouting *(82)*. The increases in IGF-1 mRNA seen after deafferentation in young rats are associated with microglial cells, so it appears that although microglia in aging individuals can still proliferate in response to injury they may lose their ability to upregulate neurotrophic molecules such as IGF-1. Alternatively, the signaling events that lead to upregulation of IGF-1 in younger animals may be impaired with aging.

The failure of the astrocyte reaction in aged rats is not confined to that species. A study of astrocyte activation in response to ischemic stroke in middle-aged (40–65-yr-old) and elderly (80–101-yr-old) humans showed that reactive astrocytes appeared in the lesion more slowly in the elderly group (5 d post-lesion compared to 3 d) and that the reaction was both less pronounced and of shorter duration *(83)*. Again it is possible to speculate that the reasons for this may depend upon a decreased responsivity of the aging brain to proliferative and migration-inducing signals from cytokines or a decreased sensitivity to neuronal damage. An age-related decrease in sensitivity to extracellular signals might result from a change in the glial cells themselves or to changes in the brain environment rendering the glial cells less responsive. More research is needed to establish the relative importance of these different elements in the impairment of glial responses to damage that occurs with age but, given the rapid advances that have been made in elucidating glial cell function, the prospects for unraveling the relationships between glial cells, neurons, and brain function in normal aging have never looked more hopeful.

REFERENCES

1. Aguayo, A.J., David, S., and Bray, G.M. (1981) Influences of the glial environment on the elongation of axons after injury: transplantation studies in adult rodents *J. Exp. Biol.* **95,** 231–240.
2. Lehning, E.J., Doshi, R., Isaksson, N., Stys, P.K., and LoPachin, R.M. (1996) Mechanisms of injury-induced calcium entry into peripheral nerve myelinated axons: Role of reverse sodium-calcium exchange. *J. Neurochem.* **66,** 493–500.
3. Stys, P.K. (1998) Anoxic and ischemic injury of myelinated axons in CNS white matter. From mechanistic concepts to therapeutics. *J. Cerebral Blood Flow & Metabolism* **18,** 2–25.
4. Petito, C.K., Olarte, J.P., Roberts, B., Nowak, T.S., and Pulsinelli, W.A. (1998) Selective glial vulnerability following transient global ischemia in rat brain. *J. Neuropathol. Exp. Neurol.* **57,** 231–238.

5. Bernaudin, M., Nouvelot, Mackenzie, E.T., and Petit, E. (1998) Selective neuronal vulnerability and specific glial reactions in hippocampal and neocortical organotypic cultures submitted to ischemia. *Experimental Neurol.* **150,** 30–39.
6. Rose, C.R., Waxman, S.G., and Ransom, B.R. (1998) Effects of glucose deprivation, chemical hypoxia, and simulated ischemia on Na^+ homeostasis in rat spinal cord astrocytes. *J. Neurosci.* **18,** 3554–3562.
7. Vornow, J.J. (1995) Toxic NMDA-receptor activation occurs during recovery in a tissue culture model of ischemia. *J. Neurochem.* **65,** 1681–1691.
8. Perry, V.H. and Gordon, S. (1991) Macrophages and the nervous system. *Int. Rev. Cytol.* **125,** 203–244.
9. Barron, K.D. (1995) The microglial cell. A historical review. *J. Neurol. Sci.* **134,** (Supplement) 57–68.
10. Barres, B.A. (1991) New roles for glia. *J. Neurosci* **11,** 3683–3694.
11. He M.H., Howe, D.G., and McCarthy, K.D. (1996) Oligodendroglial signal transduction systems are regulated by neuronal contact. *J. Neurochem.* **67,** 1491–1499.
12. Compston, A., Zajicek, J., Sussman, J. et al. (1997) Glial lineages and myelination in the central nervous system. *J. Anat.* **190,** 161–200.
13. Ivacko, J.A., Sun, R., and Silverstein, F.S. (1996) Hypoxic-ischemic brain injury induces an acute microglial reaction in perinatal rats. *Pediatric Res.* **39,** 39–47.
14. Lin, B.W., Ginsberg, M.D., Busto, R., and Dietrich, W.D. (1988) Sequential analysis of subacute and chronic neuronal, astrocytic and microglial alterations after transient global ischemia in rats. *Acta Neuropathologica* **95,** 511–523.
15. Akiyama, H., Tooyama, I., Kondo, H. et al, (1994) Early responses of brain resident microglia to kainic acid-induced hippocampal lesions *Brain Res.* **635,** 257–268.
16. Andersson, P.-B., Perry, V.H., and Gordon, S. (1991) The kinetics and morphological characteristics of the macrophage-microglial response to kainic acid-induced neuronal degeneration. *Neuroscience* **42,** 201–214.
17. Heppner, F.L., Skutella, T., Hailer, N.P., Haas, D., and Nitsch, R. (1998) Activated microglial cells migrate towards sites of exocitotoxic neuronal injury inside organotypic hippocampal slice cultures. *Eur. J. Neurosci.* **10,** 3284–3290.
18. Graeber, M.B., Streit, W.J., and Kreutzberg, G.W. (1988). Axotomy of the rat facial nerve leads to CR3 complement receptor expression by activated microglial cells. *J. Neurosci. Res.* **21,** 18–24.
19. Streit, W.J. and Kreutzberg, G.W. (1988) Response of endogenous glial cells to motor neuron degeneration induced by toxic ricin *J. Comp. Neurol.* **268,** 248–263.
20. Amat, J.A., Ishiguro, H., Nakamura, K., and Norton, W.T. (1996) Phenotypic diversity and kinetics of proliferating microglia and astrocytes following cortical stab wounds. *Glia* **16,** 368–382.
21. du Bois, M., Bowman, P.D., and Goldstein, G.W. (1985) Cell proliferation after ischemic injury in gerbil brain. An immunocytochemical and autoradiographic study. *Cell Tissue Res.* **242,** 17–23.
22. Treherne, J.E., Smith, P.J.S., and Howes, E.A. (1988) Cell recruitment during glial repair: the role of exogenous cells. *Cell Tiss. Res.* **251,** 339–343.
23. Bell, M.D., Lopez Gonzales, R., Lawson, L. et al. (1994) Upregulation of the macrophage scavenger receptor in response to different forms of injury in the CNS. *J. Neurocytol.* **23,** 605–613.
24. Bauer, J., Bradl, M., Hickey, W.F. et al. (1998) T-cell apoptosis in inflammatory brain lesions – Destruction of T cells does not depend on antigen recognition. *Amer. J. Pathol.* **153,** 715–724.
25. Lindholm, D., Castren, E., Kiefer, R., Zafra, F., and Thoenen, H. (1992) Transforming growth factor β1 in the rat brain: Increases after injury and inhibition of astrocytes proliferation. *Biol.* **117,** 395–400.

26. Loddick, S.A., Turnbull, A.V., and Rothwell, N.J. (1998) Cerebral interleukin-6 is neuroprotective during permanent focal cerebral ischemia in the rat. *J. Cerebral Blood Flow and Metabolism* **18,** 176–179.
27. Giulian, D. and Ingeman, J.E. (1988) Colony stimulating factors as promoters of ameboid microglia. *J. Neurosci.* **8,** 4707–4717.
28. Raivich, G., Moreno Flores, M.T., Moller, J.C., and Kreutzberg, G.W. (1994) Inhibition of post-traumatic microglial proliferation in a genetic model of macrophage colony-stimulating factor deficiency in the mouse. *Eur. J. Neurosci.* **6,** 1615–1618.
29. Beilharz, E.J., Russo, V.C., Butler, G. et al. (1998) Co-ordinated and cellular specific induction of the components of the IGF/IGFBP axis in the rat brain following hypoxic-ischemic injury. *Mol. Brain Res.* **59,** 119–134.
30. Braun, N., Zhu, Y., Kreglstein, J., Culmsee, C., and Zimmermann, H. (1998) Upregulation of the enzyme chain hydrolysing extracellular ATP after transient forebrain ischemia in the rat. *J. Neurosci.* **18,** 4891–4900.
31. Ellison, J.A., Velier, J.J., Spera, P. et al. (1998) Osteopontin and its integrin receptor alpha(v)beta(3) are upregulated during formation of the glial scar after focal stroke. *Stroke* **29,** 1698–1706.
32. Hatten, M.E., Liem, R.K.H., Shelanski, M.L., and Mason, C.A. (1991) Astroglia in CNS injury. *Glia* **4,** 233–243.
33. Lin, J.H.-C, Weigel, H., Cotrina, M.L. et al. (1998) Gap-junction-mediated propagation and amplification of cell injury. *Nature Neurosci.* **1,** 494–500.
34. Smith, P.J.S., Howes, E.A., and Treherne, J.E. (1990) Cell proliferation in the repairing adult insect central nervous system: incorporation of the thymidine analogue 5-bromo-2-deoxyuridine in vivo. *J. Cell Sci.* **95,** 599–604.
35. Condorelli, D.F., Dell'Albani, P., Kaczmarek L. et al. (1990) Glial fibrillary acidic protein messenger RNA and glutamine synthetase activity after nervous system injury. *J. Neurosci. Res.* **26,** 251–257.
36. Finkbeiner, S. (1992) Calcium waves in astrocytes. Filling in the gaps. *Neuron.* **8,** 1101–1108.
37. Budd, S.L. and Lipton, S.A. (1998) Calcium tsunamis: do astrocytes transmit cell death messages via gap junctions during ischemia? *Nature Neurosci.* **1,** 431–432.
38. Toyoshima, T., Yamagami, S., Ahmed, B.Y. et al. (1996) Expression of calbindin-D28K by reactive astrocytes in gerbil hippocampus after ischaemia. *Neuroreport* **7,** 2087–2091.
39. Eclancher, F., Kehri, P., Labourdette, G., and Sensenbrenner, M. (1996) Basic fibroblast growth factor (bFGF) injection activates the glial reaction in the injured rat brain. *Brain Res.* **737,** 201–214.
40. Gall, C., Rose, G., and Lynch, G. (1979) Proliferative and migratory activity of glial cells in the partially deafferented hippocampus. *J. Comp. Neurol.* **183,** 539–550.
41. Miyake, T., Hattori, T., Fukuda, M., Kitamura, T., and Fujita, S. (1988) Quantitative studies on proliferative changes in reactive astrocytes in mouse cerebral cortex. *Brain Res.* **451,** 133–138.
42. Suidan, H.S., Nobes, C.D., Hall, A., and Monard, D. (1997) Astrocyte spreading in response to thrombin and lysophosphatidic acid is dependent on the Rho GTPase. *Glia* **21,** 244–252.
43. Liesi, P., Kaakkola, S., Dahl, D., and Vaheri, A. (1984) Laminin is induced in astrocytes of adult rat brain by injury. *EMBO J.* **3,** 683–686.
44. Liesi, P. (1985) Laminin-immunoreactive glia distinguish regenerative adult CNS systems from non-regenerative ones. *EMBO J.* **4,** 2505–2511.
45. Frost, E., Kiernan, B.W., Faissner, A., and ffrench Constant, C. (1996) Regulation of oligodendrocyte precursor migration by extracellular matrix: Evidence for substrate-specific inhibition of migration by tenascin-C. *Dev. Neurosci.* **18,** 266–273.
46. Oh, L.Y.S. and Yong, V.W. (1996) Astrocytes promote process outgrowth by adult human oligodendrocytes in vitro through interaction between bFGF and astrocyte extracellular matrix. *Glia* **17,** 237–253.

47. FaberElman, A., Lavie, V., Schvartz, I., Shatiel, S., and Schwartz, M.V. (1995) Vitronectin overides a negative effect of TNF-alpha on astrocyte migration. *Faseb. J.* **15,** 1605–1613.
48. Merrill, J.E. and Benveniste, E.N. (1996) Cytokines in inflammatory brain lesions: helpful and harmful. *Trends Neurosci.* **19,** 331–338.
49. Morgan, T.E., Rozovsky, I., Goldsmith, S.K., Stone, D.J., Yoshida, T., and Finch, C.E. (1997) Increased transcription of the astrocyte gene GFAP during middle-age is attenuated by food restriction: Implications for the role of oxidative stress. *Free Radical Biol. and Med.* **23,** 524–528.
50. Peinado, M.A., Quesada, A., Pedrosa, J.A., et al. (1997) Light microscopic quantification of morphological changes during aging in neurons and glia of the rat parietal cortex. *Anat. Rec.* **247,** 420–425.
51. Peinado, M.A., Quesada, A., Pedrosa, J.A., et al. Quantitative and ultrastructural changes in glia and pericytes in the parietal cortex of the aging rat. *Microscop. Res. & Tech.* **43,** 34–42.
52. Amenta, F., Bronzetti, E., Sabbatini, M., and Vega, J.A. (1998) Astrocyte changes in aging cerebral cortex and hippocampus: a quantitative immunohistochemical study. *Microscopy Res. Tech. 43,* 29–33.
53. Berciano, M.T., Andres, M.A., Calle, E., and Lafarga, M., (1995) Age-induced hypertrophy of astrocytes in rat supraoptic nucleus – a cytological, morphometric, and immunocytochemical study. *Anat. Rec.* **243,** 129–144.
54. Long, J.M., Kalehua, A.N., Muth, N.J., et al. (1998) Stereological analysis of astrocyte and microglia in aging mouse hippocampus. *Neurobiol. Aging* **19,** 495–503.
55. Anderton, B.H. (1997) Changes in the ageing brain in health and disease. *Phil. Trans. Roy Soc. Lond. (Biol.)* **352,** 1781–92
56. Morrison, J.H. and Hof. P.R. (1997) Life and death of neurons in the aging brain. *Science* **278,** 412–419.
57. West, M.J. (1993) Regionally specific loss of neurones in the aging human hippocampus. *Neurobiol. Aging* **14,** 287–293.
58. Masliah, E., Mallory, M., Hansen, I., DeTeresa, R., and Terry, R.D. (1993) Quantitative synaptic alterations in the human neocortex during normal aging. *Neurology* **43,** 192–197.
59. Coleman, P.D. and Flood, D.G. (1987) Neuron numbers and dendrite extent in normal aging and Alzheimer's disease. *Neurobiol. Aging* **8,** 521–545.
60. Gordon, M.N. and Morgan, D.G. (1991) Increased GFAP expression in the aged rat brain does not result from increased astrocyte density. *Soc. Neurosci. Abstracts* **17,** 53.
61. Krekowski, G.A., Parhad, I.M., Fung, T.S., and Clark, A.W. (1996) Aging is associated with divergent effects on Nf-L and GFAP transcription in rat brain. *Neurobiol. Aging* **17,** 833–841.
62. Yoshida, T., Goldsmith, S.K., Morgan, T.E., Stone, D.J., and Finch, C.E. (1966) Transcription supports age-related increase of GFAP gene expression in the male rat brain. *Neurosci. Lett.* **215,** 107–110.
63. David, J.P., Ghozali, F., FalletBianco, C. et al. (1997) Glial reaction in the hippocampal formation is highly correlated with aging in human brain. *Neurosci. Lett.* **237,** 53–56.
64. Jeglinski, W., Pepeu, G., and OderfeldNowak, B. (1997) Differential susceptibility of senile and lesion-induced astrogliosis to phosphatidylserine. *Neurobiology of Aging* **18,** 81–86.
65. Griffin, W.S.T., Sheng, J.G., and Mrak, R.E. (1998) Senescence-accelerated overexpression of S100β in brain of SAMP6 mice. *Neurobiol. Aging* **19,** 71–76.
66. Perry, V.H., Matyszak, M.K., and Fearn, S. (1993) Altered antigen expression of microglia in the aged rodent CNS. *Glia* **7,** 60–67.
67. Sheffield, L.G. and Berman, N.E.J. (1998) Microglial expression of MHC class II increases in normal aging of nonhuman primates. *Neurobiology of Aging* **19,** 47–55.
68. Rao, K. and Lund, R.D. (1989) Degeneration of optic axons induces the expression of major histocompatibility antigens. *Brain Res.* **488,** 332–335.

69. Neumann, H., Boucraut, J., Hahnel, C., Misgeld, T., and Wekerle, H. (1996) Neuronal control of MHC class II inducibility in rat astrocytes and microglia. *Eur. J. Neurosci.* **8,** 2582–2590.
70. Sheng, J.G., Mrak, R.E., and Griffin, W.S.T. (1998) Enlarged and phagocytic, but not primed, interleukin-1α-immunoreactive microglia increase with age in normal human brain. *Acta Neuropathol* **95,** 229–234.
71. Murray, C. and Lynch, M.A. (1997) Impaired ability of aged animals to sustain long-term potentiation may result from increased hippocampal expression of interleukin-1β. *J. Physiol (Lond)* **501,** 87P.
72. Murray, C.A. and Lynch, M.A. (1998) Evidence that increased hippocampal expression of the cytokine interleukin-1β is a common trigger for age- and stress-induced impairments in long-term potentiation. *J. Neurosci.* **18,** 2974–2981.
73. Ye, S.M. and Johnson, R.W. (1999) Increased interleukin-6 expression by microglia from brain of aged mice. *J. Neuroimmunol.* **93,** 139–148.
74. Roberts, E.L. and Sick, T.J. (1996) Aging impairs regulation of intracellular pH in rat hippocampal slices. *Brain Res.* **735,** 339–342.
75. Kawamata, T., Akiguchi, I., Maeda, K. et al. (1998) Age-related changes in the brains of senescence-accelerated mice (SAM): Association with glial and endothelial reactions. *Microscop. Res. and Tech.* **43,** 59–67.
76. Ueno, M., Akiguchi, I., Yagi, H., et al. (1993) Age-related changes in barrier function in mouse brain.I. Accelerated age-related increase of brain transfer of serum albumin in accelerated senescence prone SAM-P/8 mice with deficits in learning and memory. *Arch Gerentol Geriatr* **16,** 233–248.
77. Shah, G.N. and Mooradian, A.D. (1997) Age-related changes in the blood-brain barrier. *Exp. Gerontol.* **32,** 501–519.
78. Kane, C.J., Sims, T.J., and Gilmore, S.A. (1997) Astrocytes in the aged rat spinal cord fail to increase GFAP mRNA following sciatic nerve axotomy. *Brain Res.* **759,** 163–165.
79. Gilmore, S.A. and Kane, C.J.M. (1998) Microglia, but not astrocytes, react to sciatic nerve injury in aging rats. *Brain Res.* **806,** 113–116.
80. Rozovsky, I., Finch, C.E., and Morgan, T.E. (1998) Age-related activation of microglia and astrocytes: In vitro studies show persistent phenotypes of aging, increased proliferation, and resistance to down-regulation. *Neurobiol. Aging* **19,** 97–103.
81. Grove, J., Gomez, J., Kentroti, S., and Vernadakis, A. (1996) Plasticity of astrocytes derived from aged mouse cerebral hemispheres: Changes with cell passage and immortalization. *Brain Res. Bull.* **39,** 211–217.
82. Woods, A.G., Guthrie, K.M., Kurlawalla, M.A., and Gall, C.M. (1998) Deafferentation-induced increases in hippocampal insulin-like growth factor-1 messenger RNA expression are severely attenuated in middle aged and aged rats. *Neurosci.* **83,** 663–668.
83. Dziewulska, D. (1997) Age-dependent changes in astroglial reactivity in human ischemic stroke. Immunohistochemical study. *Folia Neuropath* **35,** 99–106.

8
ATP Signaling in Schwann Cells

Thierry Amédée, Aurore Colomar, and Jonathan A. Coles

1. INTRODUCTION

Since the early 1970s, strong evidence has been provided that adenosine 5′-triphosphate (ATP) acts as a potent extracellular chemical messenger in cell of many types. It is now well-known that ATP acts as a neurotransmitter at synapses in both the central and peripheral nervous systems. In addition of being a neurotransmitter, ATP and its products of degradation (ADP, AMP, and adenosine) are also known to act as trophic factors by promoting cell proliferation and/or differentiation, stimulation of synthesis and/or release of neurotrophic factors both under physiological and pathological conditions. All these diverse biological effects are mediated via cell surface receptors termed purinoceptors. Purinoceptors of the P_2 type, for which ATP is the physiological ligand, are subdivided into receptors that directly gate ion channels (P_{2X}) and G protein-coupled receptors (P_{2Y}).

Within the peripheral nervous system (PNS), Schwann cells establish and maintain a very intimate relationship with neurons. As ATP is released from virtually all peripheral nerves, neighboring Schwann cells are natural candidates for being modulated by extracellular purines. The recent discovery of the expression of both types of P_2 purinoceptors by Schwann cells brings new insights to neurone-Schwann cell interactions, but how and what for ATP is signaling to Schwann cells is still poorly understood.

This chapter will mainly focus on release of ATP in the PNS and its actions on different P_2 purinoceptors on Schwann cells. We will also discuss putative roles of ATP signaling to Schwann cells.

2. ATP IS RELEASED BY NEURONS OF THE PNS

2.1. *ATP is Released from Peripheral Nerve Terminals*

Holton and Holton *(1)* and Holton *(2)* were the first to show that ATP is released during peripheral nerve stimulation. They reported that electrical stimulation of sensory nerves in the rabbit ear evoked release of large amounts of nucleotides in the bloodstream which caused vasodilatation, and that ATP was the nucleotide that most potently induced vasodilatation.

From: *Neuroglia in the Aging Brain*
Edited by: Jean S. de Vellis © Humana Press Inc., Totowa, NJ

Since this pioneer work, it has become apparent that ATP signaling is ubiquitous within the PNS, ATP being costored and coreleased with other neurotransmitters (see below) by a large variety of peripheral nerves. It has been shown that ATP is released from motor nerves terminals, from some sensory-motor nerves, and from sensory nerves specific for a number of different modalities including mechanoreception, chemoreception, and nociception *(3)*. ATP is involved in the nonadrenergic noncholinergic (NANC) neurotransmission which mediates in part the vagal innervation of the heart *(4)* and also regulates gastro-intestinal motility *(5)*.

ATP storage and release have also been reported for cultured PNS neurons. This has been shown in chick sympathetic neurons *(6)*, rat superior cervical neurons *(7)*, and rat dorsal root ganglion neurons *(8)*.

2.2. Vesicular Release of ATP

The nucleotide contents of synaptic vesicles of a large variety of peripheral synapses have been closely examined. ATP does not appear to be stored on its own in vesicles, but is costored and coreleased together with other neurotransmitters *(9)*. The nature of the co-transmitters varies considerably with location and species *(10)* but a general pattern is emerging. ATP is usually costored and coreleased with acetylcholine in motor nerves and parasympathetic nerves. In sympathetic nerves supplying vascular and visceral musculature, ATP is costored and coreleased with noradrenaline. Last, in some sensory-motor nerves ATP co-exists with calcitonin gene related peptide (CGRP) and substance P. Uptake of ATP into vesicles is achieved by a vesicular nucleotide transporter of low affinity *(11)* which stores additional nucleotides like UTP and GTP.

Elegant studies done at the frog neuromuscular junction showed that motor nerve stimulation *in situ* induced an increase in intracellular calcium in perisynaptic Schwann cells *(12)* that are mimicked by local application of acetylcholine and ATP *(13)*. Then Robitaille *(14)* established that neuronal ATP released in the synaptic cleft during neurotransmission was activating two types of purinoceptors (P_{2X}-like and P_{2Y}-like, *see* **Subheadings 3.1, and 3.2**) expressed by perisynaptic Schwann cells and linked to intracellular calcium signaling.

The mechanism of release of ATP follows some of the general criteria of neurotransmitter release. ATP release is triggered by evoked action potentials, abolished by blockers of voltage-dependent calcium and sodium channels and by withdrawal of calcium from the extracellular medium in cultured chick postganglionic sympathetic neurons *(6)*. In the same study, the authors reported that electrical activity increased the basal efflux of ATP fivefold (respectively 0.15 ± 0.04 pmol ATP mg^{-1} protein min^{-1} and 0.82 ± 0.15 pmol ATP mg^{-1} protein min^{-1}) and that the ATP/noradrenaline ratio in storage vesicles was 1:5.8. ATP is released in quanta from peripheral adrenergic varicosities *(15)*.

Estimations of the intra-vesicle concentration of ATP reveal considerable variance, probably due to methodological limitations, but it is generally accepted that ATP is stored in the hundred-millimolar range *(16)* although the concentration of the co-transmitter is always several fold higher.

Taken together, all these data are consistent with vesicular release of ATP. However some other results suggest different mechanisms. For example, synaptosomal release of acetylcholine and ATP from peripheral electromotor synapses of the Torpedo electric organ is differentially blocked by Ω-conotoxin, a voltage-dependent calcium channel blocker *(17)*. Release of ATP and noradrenaline from sympathetic nerves is differentially affected by the electrical pattern of stimulation *(6)*. These data suggest that ATP may be released from neuronal cytoplasmic sources other than synaptic vesicles, although preferential release of one neurotransmitter from a mixture in a vesicle seems to be not necessarily impossible.

2.3. *Possible Mechanisms of Nonvesicular Release of ATP*

Nonvesicular release of ATP along axons has been postulated by Lyons et al. *(18)* and more recently substantiated by Stevens and Fields *(18a)*. As a matter of fact, nonvesicular release of other neurotransmitters along central and peripheral axons have been reported in batrachians and mammalians. Electrical stimulation triggered release of glutamate and aspartate from peripheral and central frog nerve trunks *(19)* and release of a neurotransmitter (possibly glutamate or adenosine) in rat optic nerve *(20)*. Acetylcholine is released by depolarization along preganglionic axons of sympathetic nerve trunk of the cat *(21)*. Since axons lack vesicular means of releasing neurotransmitters *(22)*, release mechanisms based on transporters or others transmembrane pathways have to be considered. In **Subheading 2.3.1**, we will present recent data on different possible routes for ATP release in the extracellular medium.

2.3.1. *ATP-Binding Cassette Family of Transporters*

Members of the ATP-binding cassette (ABC) (*see* ref. *(23)*) family of transporters are thought to be able to conduct large anions such as ATP. Among the ABC transporter family, the cystic fibrosis transmembrane conductance regulator (CFTR) and the P-glycoprotein have been studied in detail. A new pathway, (NPPB pathway) probably also mediated by an ABC transporter has been very recently identified and will also be presented.

2.3.1.1. CFTR

The CFTR belongs to a superfamily of proteins conducting ions, proteins, and hydrophobic substances. Recent studies have shown that CFTR functions as a protein kinase A-sensitive chloride channel regulated by intracellular ATP *(24,25)*. Loss of CFTR-mediated chloride conductance from the apical plasma membrane of epithelial cells is a primary physiological lesion in cystic fibrosis *(26)*. In addition to its well-known chloride conductance activity, work on human epithelial CFTR, either expressed in mouse mammary carcinoma cells *(27)* or incorporated into a lipid bilayer *(28)*, has been recently shown that it can mediate release of ATP to the extracellular medium. Whether or not CFTR does possess intrinsic ATP channel activity or triggers release of ATP indirectly remains controversial *(25)*.

2.3.1.2. P-Glycoprotein

The multidrug resistance (mdrl) gene product, P-glycoprotein is responsible for the ATP-dependent extrusion of a large variety of compounds from cells including multiple cytotoxic drugs. How this efflux transporter functions and what is its physiological relevance remains to be determined *(29)*. P-glycoprotein also regulates a volume-sensitive chloride channel *(30)* via a protein kinase C-mediated mechanism *(31)* and functions as an ATP channel *(32)*.

2.3.1.3. NPPB Pathway

In a very recent study, Mitchell et al. *(33)* have identified another route for release of ATP in cultured ciliary epithelial cells. Release of ATP is triggered by the activation by extracellular hypotonicity of a chloride conductance and blocked by the chloride channel blocker 5-nitro-2-(3-phenylpropylamino) benzoic acid (NPPB). The pharmacological profile of the NPPB pathway excludes CFTR and P-glycoprotein but does not rule out the involvement of other members of the ABC transporter family *(34)*.

2.3.2. Connexins

Gap junctions proteins (connexins) are specialized channels formed between cells or different compartments of individual cells which permit the passage of ions, amino acids, second messengers molecules and small metabolites *(35)*. Connexins span two plasma membranes and result from the association of two connexin-hemichannels (connexons). Cotrina et al. *(36)* recently reported that ATP release by different cell lines (C6 glioma, HeLa, and U373 glioblastoma) was tightly linked to connexin expression (Cx43, Cx32, Cx26). Furthermore, ATP release was facilitated in calcium-free medium which is consistent with the known regulation of the connexin permeability by calcium *(37)*. Then Cotrina et al. *(36)* suggested that ATP could be released in the extracellular medium through connexons.

2.3.3. Do Such Pathways Exist Within the PNS?

To our knowledge, the detection of members of the ABC transporter family in the PNS has not been reported in the literature, in contrast to the CNS where they are widely expressed by neurons throughout the rat and mouse brain *(38)* and by astrocytes *(39)*. In contrast to the CNS where connexins are expressed both by neurons and glial cells, connexins in the PNS seem to be only expressed by Schwann cells *(40)*. However a recent study by Dezawa et al. *(41)* suggest that Cx32 is located between regenerating axons and Schwann cells. It is therefore conceivable that axonal ATP might be released by this pathway during regeneration and could act on neighboring Schwann cells (*see* **Subheading 5.1.**).

2.4. Fate of Extracellular ATP

By being released either in the synaptic cleft or possibly in the periaxonal space, one can expect, because of the narrowness of these extracellular compartments, that the concentration of ATP could rise up to millimolar levels. However, when present in the extracellular space, ATP is degraded by sequential removal of phosphate groups to give ADP, AMP, and finally adenosine. This processus is catalyzed by a series of enzymes (ecto-nucleotidases) located on the extracellular surface of plasma membrane *(42)*.

Two ecto-nucleotidases have been studied in details: The ecto-ATPase which hydrolyses ATP in ADP and the ecto-5′-nucleotidase which hydrolyses AMP giving adenosine. The localization of ecto-nucleotidases has been investigated using enzyme cytochemical techniques *(16)*. In the PNS, ecto-ATPase is mainly localized on the plasma membrane of unmyelinated fibers *(43)* and also on the surface of nerve terminals of the myenteric plexus *(44)*. The ecto-5′-nucleotidase is found on glial cells (Schwann cells and satellite cells) and also on capillary endothelial cells *(16)* but has not been detected in peripheral neurons *(45,44)*.

Because biochemical studies of ecto-nucleotidases and determination of their kinetic constants (K_m and V_{max}) have been carried out on synaptosomal preparations or on *Xenopus laevis* oocytes, very little is known about their direct functional role *in situ* in nucleotide degradation at the synaptic compartment. However, the higher potency to ATP itself of weakly hydrolyzable analogs of ATP (e.g., α,β-methylene ATP or ATP-γ-S) in causing contraction of various visceral smooth muscle tissues *(46)* suggests that endogenous ecto-ATPases are functional *in situ.* By using suramin (a P_2 purinoceptor antagonist which also inhibits ecto-ATPase activity) in the rabbit ear artery, Crack et al. *(47)* have concluded that the potency of ATP for P_{2X} purinoceptors would be 400-fold higher in the absence of endogenous ecto-ATPase.

At the neuromuscular junction in rat tail artery, the time for clearance by ecto-ATPase of a single quantum of ATP released by a single pulse or short trains (*10* pulses) at 50 Hz has been estimated to be less than 100 msec *(48).* Same experimental measurements of the time of clearance of ATP have not been done for the periaxonal space. However, if we make the assumption than the ecto-ATPase expressed by Schwann cell plasma membranes display the same surface density and kinetic constants than the ecto-ATPase expressed by *Xenopus laevis (49),* it is possible to estimate the time of clearance of ATP in the periaxonal space. Considering that the diameter of an oocyte is 1 mm, its area will be 3.14 10^6 μm^2. The V_{max} of the ecto-ATPase is 15.8 pmoles P_i min^{-1} $cell^{-1}$ which recalculated and expressed by surface unit gives 8.38 10^{-8} pmoles P_i sec^{-1} $(\mu m^2)^{-1}$. Assuming that the width of the periaxonal space is about 10 nm *(50),* to a Schwann cell membrane area of 1 μm^2 will correspond a volume of periaxonal space of 0.01 μm^3. If the concentration of ATP released in the periaxonal space reaches 1 mM, then the quantity of ATP in 0.01 μm^3 will be 0.01 10^{-6} pmoles. Therefore the time necessary for the ecto-ATPase to hydrolyse ATP will be around 120 msec.

3. ATP ACTS ON SCHWANN CELLS VIA DIFFERENT TYPES OF P_2 PURINOCEPTORS

The classification and the distribution of purinoceptors has been widely and extensively reviewed in the past few years *(9,10,51–53).* Basically there are two main families of purinoceptors, P_1 purinoceptors for which adenosine is the natural ligand and P_2 purinoceptors for which ATP is the natural ligand. ATP responses reported in Schwann cells mainly involved P_2 purinoceptors and we will therefore focus on this family (*see* **Subheading 3.1.1.**). The P_2 purinoceptor family is divided into two main subtypes based on their receptor signal transduction mechanisms: P_{2X} purinoceptors are ligand-gated ion channels (ionotropic purinoceptors) while P_{2Y} purinoceptors are G-protein coupled receptors (metabotropic purinoceptors). Schwann cells express purinoceptors of both subtypes.

3.1. P_{2X} Purinoceptors

3.1.1. General Features

Up to now, seven P_{2X} purinoceptors ($P_{2X1\,...\,2X7}$) have been cloned and expressed in Xenopus oocytes *(53).* P_{2X1} to P_{2X6} purinoceptors are ATP-gated ion channels selectively permeable to cations (Na^+, K^+ and Ca^{2+}). Because ATP directly gates the cationic channel, the onset of the current is rapid (within 10 msec). The transmembrane influx of cations results in a membrane depolarization and an increase in the intracellular calcium concentration ($[Ca^{2+}]_i$). The membrane depolarization leads to the secondary activation of voltage-dependent calcium channels which reinforces the intracellular calcium signal. The P_{2X7} purinoceptor is somewhat special as it has been identified as a cytolytic purinoceptor *(54)*. Depending on the concentration of extracellular ATP and the length of stimulation, P_{2X7} purinoceptors form a nonselective cationic pore. These pores carry molecules with a molecular mass up to 900 Da, and may lead to necrosis or apoptosis of the cell *(55)*.

3.1.2. P_{2X} Purinoceptors Expressed by Schwann Cells

Robitaille *(14)* was the first to report that Schwann cells expressed P_{2X} purinoceptors. He studied perisynaptic Schwann cells at the frog neuromuscular junction which possess a P_{2X} purinoceptor activated by ATP released during synaptic transmission. Its activation by ATP or L-AMP-PCP (a nonhydrolyzable ATP analog with a high affinity for P_{2X} purinoceptors) triggers an increase in $[Ca^{2+}]_i$ which is dependent on external calcium and strongly reduced by L-type calcium channel blockers. Robitaille *(14)* concluded that the membrane depolarization resulting from a modest influx of calcium through the P_{2X} purinoceptors activated L-type calcium channels and thereby produced a large calcium entry. The pharmacological profile of this response has been not investigated further so the subtype of P_{2X} purinoceptor involved remains to be identified.

Amédée and Despeyroux *(55a)* first reported the existence of an unusually low affinity (K_d = 8.4 mM) ATP-activated mixed current composed of cationic and anionic conductances in Schwann cells cultured from dorsal root ganglia of the mouse. Later using potent agonists (BzATP and ATP^{4-}) and antagonist (oATP) together with a polyclonal antibody raised against P_{2X7} purinoceptors, Colomar and Amédée *(56)* clearly demonstrated the expression of a P_{2X7} purinoceptor in mouse Schwann cells. In addition to the well documented formation of a non selective cationic pore, the stimulation of the mouse P_{2X7} purinoceptor leads to the activation of two other associated ionic conductances, that is a Ca^{2+}-dependent K^+ conductance and Cl^- conductance. An ionotropic purinoceptor with electrophysiological and pharmacological characteristics similar to the P_{2X7} subtype has also been identified in the paranodal membrane of rat Schwann cell *(57)*. ATP and BzATP activate a nonspecific cation current and confocal calcium imaging showed a sustained rise in $[Ca^{2+}]_i$ which is exclusively dependent on external calcium.

3.2. P_{2Y} Purinoceptors

3.2.1. General Features

To date five P_{2Y} purinoceptors (P_{2Y1}, P_{2Y2}, P_{2Y4}, P_{2Y6} and P_{2Y11}) have been cloned and firmly identified as members of the P_2 purinoceptor family *(53)*. P_{2Y} purinoceptors belong to the superfamily of receptors that have seven transmembrane domain and are coupled to G-proteins. Binding of ATP on P_{2Y} purinoceptors activates phospholipase C and the formation of inositol 1,4,5-trisphosphate (IP_3) which releases calcium from intracellular stores. Ion channels either regulated by G-proteins or by an increase in the $[Ca^{2+}]_i$ can also be activated following P_{2Y} stimulation. Because of the involvement of second-messenger systems, the delay between the stimulation of the receptor and the onset of the cellular response is longer (> 100 msec) than that of P_{2X} purinoceptors.

3.2.2. P_{2Y} Purinoceptors Expressed by Schwann Cells

Lyons et al. *(58)* were the first to report the expression of P_{2Y} purinoceptors by Schwann cells cultured from rat sciatic nerves. ATP induced a transient increase in $[Ca^{2+}]_i$. This response disappeared within 4 d in vitro, but was restored by culturing the cells in the presence of cAMP analogs. In the subsequent work, Lyons et al. *(18)* reported that the ATP-mediated calcium response of the Schwann cells could also be maintained or restored by contact with neurons. According to Ralevic and Burnstock *(53),* the sequence of relative agonist potency (ATP=2MeSATP=ADPβS; adenosine and UTP ineffective) of this receptor was similar to those described for the cloned P_{2Y1} purinoceptors.

Since then, ATP-responses mediated by P_{2Y} purinoceptors have been reported in other Schwann cells. P_{2Y} purinoceptors responses have been reported in cultured Schwann cells from rat and rabbit sciatic nerves *(59),* mouse dorsal root ganglia *(60),* rat dorsal root ganglia *(61)* and on Schwann cells acutely dissociated from the neuro-electrocyte junction in skate *(62).* In all these studies, the ATP signaling was invariably associated with release of intracellular calcium from IP_3 sensitive calcium stores.

In most cases an adequate pharmacological characterization and/or molecular identification of the P_{2Y} purinoceptors subtype involved is lacking even though it seems to be an important issue for axon-Schwann cell interactions. For example, the Schwann cell phenotype (i.e., myelinating vs nonmyelinating) and the molecular identity of the P_{2Y} subtype are linked. Myelinating Schwann cells from human sural nerves, rat ventral roots, and vagus nerves express a P_{2Y2}-like purinoceptor, whereas nonmyelinating Schwann cells express a P_{2Y1}-like purinoceptor *(63).* Moreover, the P_{2Y2}-like purinoceptor is preferentially expressed in the paranodal region of myelinating Schwann cells *(64).*

4. EFFECTS OF EXTRACELLULAR ATP ON SCHWANN CELLS

4.1. Change of Membrane Potential

Modulation by ATP of membrane potential may involve different subtypes of P_2 purinoceptors. Basically ATP can either gate directly nonselective cationic channels (P_{2X}-like purinoceptors) and therefore allow the influx of Na^+, K^+, and Ca^{2+} which will depolarize the cell, or by acting on G-protein coupled P_{2Y}-like purinoceptors, release calcium from intracellular stores which may in turn trigger the activation of calcium-dependent ion channels. The resulting change in membrane potentiel may lead to the secondary activation of voltage-dependent ion channels.

Such mechanisms or at least part of these sequences of events are known for neurons and glial cells of the CNS *(65,66)* but are still hypothetical for Schwann cells.

Amédée and Despeyroux *(55a)* reported that in Schwann cells cultured from dorsal root ganglia of the mouse, ATP induced a mixed current composed of a cationic (K^+ and anionic (Cl^-) conductances. This current by depolarizing the Schwann cell up to –40 mV may activate voltage-dependent Ca^{2+} channels *(67).*

ATP (1 mM) induces a nonspecific cation current in cultured rat Schwann cells from spinal roots *(57)* via an ionotropic purinoceptor of characteristics similar to those of P_{2X7}-like purinoceptors.

4.2. Modulation of Intracellular Calcium

Calcium ions plays an important role in regulating very many glial functions both on short time scale (release of neuroactive molecules, ion regulation…) through the activation of protein kinases, protein phosphatases, phospholipases, adenylate cyclases, and gating of ion channels, and on a longer time scale (cell cycle, proliferation, and differentiation…) through gene expression *(68–70)*.

Intracellular free calcium concentration whatever the cell type, is tightly regulated by several mechanisms. It is sequestered in intracellular stores and complexed to calcium binding proteins. Calcium fluxes occur through different pathways (voltage-dependent calcium channels, ionotropic receptors, store-operated channels…) and its extrusion is carried by specialized proteins (Ca-ATPase, Na/Ca exchanger). In Schwann cells, resting $[Ca^{2+}]_i$ varies from 35 to 100 nM *(59,62,71,72)*, a value within the range reported for glial cells of the CNS (30–40 to 200–400 nM,*[73]*).

Modulation of $[Ca^{2+}]_i$ triggered by ATP stimulation is dependent on intracellular and/or extracellular sources of calcium in Schwann cells. The ratio of both sources varies among the species. In cultured adult rat and rabbit Schwann cells, ATP (100 μM) triggers a fast monophasic transient increase in $[Ca^{2+}]_i$ which is independent of the presence of external calcium and fades rapidly upon repeated stimulation *(59)*.

In contrast, in cultured mouse Schwann cells, ATP (1 nM–100μM) triggers a biphasic increase in $[Ca^{2+}]_i$: an early fast transient increase followed by a plateau phase which slowly declined to resting level *(60)*. The plateau phase is suppressed by removal of external calcium, whereas the early fast transient increase is unaffected by such removal. Similar biphasic ATP responses have been reported in perisynaptic Schwann cells at the neuroelectrocyte junction in skate *(62)*.

The amplitude of the $[Ca^{2+}]_i$ increase varies with the concentration of ATP used and the species. In cultured rodent Schwann cells, ATP (100 μM) increases $[Ca^{2+}]_i$ two- to fourfold *(59)* in skate perisynaptic Schwann cells, ATP (50 μM) increases $[Ca^{2+}]_i$ sevenfold *(62)*.

From all these different studies, it can be concluded that the first phase is due to the release of calcium from intracellular calcium stores following stimulation of P_{2Y}-like purinoceptors leading to the generation of IP_3. The secondary plateau phase probably involves extracellular calcium influx activated by the depletion of intracellular stores ([capacitative calcium entry],*[74]*). This influx is thought to be necessary for the refilling and/or maintenance of IP_3 mobilized intracellular calcium stores.

4.3. Release of Neurotransmitters

The release of neurotransmitters by glial cells of the CNS is well-known to be a calcium-dependent process *(75)*. Because Schwann cell calcium levels are modulated by ATP and other neuroligands, this has prompted studies to investigate the possibility of a neuroligand evoked calcium-dependent release of neurotransmitters by Schwann cells.

Acetylcholine was probably the first neurotransmitter found to be released by Schwann cells. Katz and Miledi *(76)* reported the occurrence of spontaneous miniature end-plate potentials (min.e.p.p.s) after complete denervation of amphibian end-plates and suggested that these potentials were due to acetylcholine release by Schwann cells. Later, Dennis and Miledi *(77)* showed that electrically stimulated denervated Schwann cells released acetylcholine in a nonquantal form. In giant axons of the squid, Lieberman et al. *(78)* have reported that nerve stimulation caused Schwann cell hyperpolarization. They suggested that glutamate released by axons caused Schwann cell depolarization which in turn triggered the release of acetylcholine and caused by an autocrine loop Schwann cell hyperpolarization. Although these observations have not been substantiated in other molluscan species and need further confirmation, the calcium permeability of NMDA receptors and their presence in squid Schwann cells *(79)* make possible that a glutamate-mediated $[Ca^{2+}]_i$ increase in Schwann cells mediates acetylcholine release.

In mammalian Schwann cells, recent work has clearly established the existence of a neurotransmitter-evoked calcium dependent release of neuroligands. Bradykinin is released in the PNS in response to trauma *(80)*. In rat Schwann cells, bradykinin (10 nM) increased $[Ca^{2+}]_i$ sevenfold (from 102 to 747 nM) which caused the release of glutamate and aspartate *(71)*. In the same preparation, ATP (100 μM), acting on P_{2Y} purinoceptors, also released glutamate and aspartate *(61)*. Glutamate and aspartate release was due to mobilization of IP_3 sensitive calcium stores. Furosemide (5 mM), a chloride cotransport inhibitor blocked ATP-induced glutamate release. The calcium-dependent release mechanisms remain to be determined, but Schwann cells do not express synaptic proteins (synaptophysin and synaptotagmin) characteristic of neuronal transmitter release *(71)* which suggest that the release mechanism is nonvesicular.

4.4. Cell Death by Apoptosis

It has been well-known for almost 20 yr that high concentrations of extracellular ATP cause cell death in many different cell types including glial cells *(81)*. However, it is only very recently that the membrane receptors involved in cell necrosis and/or apoptosis have been characterized and cloned. Surprenant et al. *(54)* cloned from rat brain the P_{2X7} purinoceptor and identified it as a cytolytic purinoceptor for extracellular ATP. P_{2X7} purinoceptors have been reported to mediate apoptosis of activated mouse thymocytes *(82)*. The intracellular pathways activated following P_{2X7} purinoceptor stimulation are still poorly known, but there is no doubt that the major pertubation in the intracellular ion homeostasis resulting from the gating of the nonselective pore plays an important role. There is no doubt that the activation by ATP of different caspases lead to cell apoptosis. ATP activates caspase 1 (IL-1β converting enzyme) in mouse microglial cells *(83)*, caspase 3 and caspase 8, both leading to the apoptosis of microglial cells from the mouse cell line N13 *(84)*. Interestingly, the activation of caspase 1 triggers apoptosis in fibroblasts *(85)*.

The potent ligand for P_{2X7} purinoceptor is known to be the tetraanionic form of ATP (ATP^{4-}). As in physiological solutions, ATP is almost exclusively coordinated to calcium and/or magnesium *(86)*, ATP^{4-} represents only 1–5% of total ATP which may explain the high concentrations of ATP (mM to ten-mM range) needed to activate the receptor. However, bearing in mind that cytoplasmic ATP concentration is in the same range (5–10 mM, *[81]*), it is plausible that neural ATP released during pathological conditions could reach concentrations high enough to induce necrosis and/or apoptosis of surrounding Schwann cells.

5. PUTATIVE ROLES OF ATP SIGNALING IN SCHWANN CELLS

5.1. *Trophic Actions of Extracellular ATP on Schwann Cells*

The trophic actions of extracellular ATP in the PNS are poorly known. However, some of the purinoceptors expressed by Schwann cells have been reported to mediate mitogenic and morphogenic effects of ATP on astrocytes in the CNS *(87,88)*. This is the case for P_{2Y1}-like and P_{2Y2}-like purinoceptors which are linked to the activation of a mitogen-activated protein kinase (MAPK) in rat cortical astrocytes *(89)*. Activation of the MAPK pathway leads to the phosphorylation of transcription factors which bind to the serum responsive element (SRE) sequence on DNA and trigger changes in gene expression *(90)*.

The activation by extracellular ATP of the MAPK pathway has not been shown directly in Schwann cells. However, Mutoh et al. *(91)* reported that the MAPK pathway in Schwann cells was activated by changes in the intracellular cAMP levels. Application of 8-bromo cAMP (a cell permeable derivative of cAMP) had effects that depended on the concentration and the length of the stimulation: Schwann cells were driven to a proliferative state by brief stimulation with low concentrations or to a differentiated state by prolonged stimulation with higher concentration. Hence since the P_{2Y1}-like purinoceptor and to a lesser extent the P_{2Y2}-like purinoceptor decrease cAMP levels in rat Schwann cells *(72)*, activation of these receptors is expected to control the switching of the cells between proliferation and differentiation.

5.2. *Schwann Cell Apoptosis During Physiopathological Conditions*

Apoptosis, also known as programmed cell death, is a highly orchestrated cell suicide involving specific cysteine proteases (caspases) which lead to nuclear fragmentation and formation of membrane-packaged cellular debris (apoptotic bodies) that are phagocytosed by macrophages or other surrounding cells *(92)*. Apoptosis plays a major role not only in the developing nervous system but also occurs at adult stages of cell life *(93)*. Neuronal cell apoptosis has been extensively studied during the last decade and it is generally believed that supernumery neurons which are initially generated, die by apoptosis during normal development to match their number to the requirements of the neuronal circuitry.

Because glial cells are intimately associated with neuronal cells and participate actively in the development of the nervous system, it was therefore of interest to study glial cell apoptosis. Apoptosis has been reported for rat oligodendrocytes in the developing optic nerve *(94)* and rat astrocytes during the development of cerebellum *(95)*.

5.2.1. Schwann Cell Apoptosis During Development

Schwann cells die by apoptosis during normal development, of ventral and dorsal nerve roots in the chick embryo *(96)* and rat sciatic nerve *(97)*. The mechanisms responsible for the developmental apoptosis of Schwann cells are not completely clear, but neuronal derived trophic factors of the neuregulin family *(98)* seem to reduce, at least in part, the Schwann cell apoptosis. This has been shown in the developing rat sciatic nerve, where Schwann cell apoptosis is markedly reduced in vivo by injection of neuregulin in newborn rats nerves *(97)*.

A contribution by purinoceptors to the regulation of apoptosis of Schwann cells during peripheral nerve development has not yet been reported but appears plausible. The growth cones of developing neurons are known to release neurotransmitters. This has been shown in vitro for mammalian CNS *(99)* and PNS *(100)* neurons. It is therefore conceivable that like other transmitters, ATP could be released from growth cones and/or along axons of peripheral neurons during development and modulates apoptosis of surrounding Schwann cells.

5.2.2. Schwann Cell Apoptosis During Pathological Conditions

During the Wallerian degeneration which follows experimental or traumatic peripheral nerve transection, the distal stump (transected axon and surrounding Schwann cells) undergo major cellular events *(101)*. Axonal cytoskeleton and axoplasm disintegrate and are converted into granular and amorphous debris, while the vast majority of Schwann cells throughout the distal stump rapidly (in a few hours) undergo apoptosis. This has been shown in the chick embryo *(96)*, rat sciatic nerve *(97,102)*, and rat neuromuscular junction *(103)*.

Because ATP is massively released when cells are damaged *(82)* it is very likely that it could trigger apoptosis of surrounding Schwann cells. However, in such pathological conditions, loss of neuronal trophic factors also favours Schwann cell apoptosis. For example, Schwann cells at the developing rat neuromuscular junction are rescued in vivo from axotomy-induced apoptosis by glial growth factor *(103)* one of the members of the neuregulin family *(98)*.

5.3. Schwann Cell Modulation of Neuronal Functioning

Extracellular ATP induces a calcium-dependent release of aspartate and glutamate from cultured Schwann cells *(61)*. In the CNS of mammals, glutamate is the major excitatory neurotransmitter and mediates neuron-glial calcium signaling *(20,104,105)*. One consequence of neuron-glial calcium signaling is the increase or the decrease of the neuronal synaptic efficacy *(106)*. For example, an increase in the synaptic efficacy may be achieved by the calcium dependent activation of astrocytic phospholipase A_2 which produces arachidonic acid and inhibits glutamate uptake.

As peripheral synapses are devoid of glutamate receptors, the existence of this modulatory mechanism is very unlikely in the PNS. However, cytotoxic effects of glutamate and aspartate have been reported in rat sciatic nerve *(107)* which suggest that at least some peripheral axons can respond to extracellular glutamate.

As in astrocytes, where glutamate induces the release of another neurotransmitter (GABA) through an autocrine pathway *(108)*, glutamate/aspartate release from Schwann cells could stimulate the release of another neurotransmitter which could act on neurons. A glutamate/acetylcholine autocrine/paracrine pathway has been suggested for Schwann cells of the giant axon of the squid (*see* **Subheading 4.3.**). Figure 1 summarizes some of the possible actions of ATP on Schwann cells in physiological and pathological conditions.

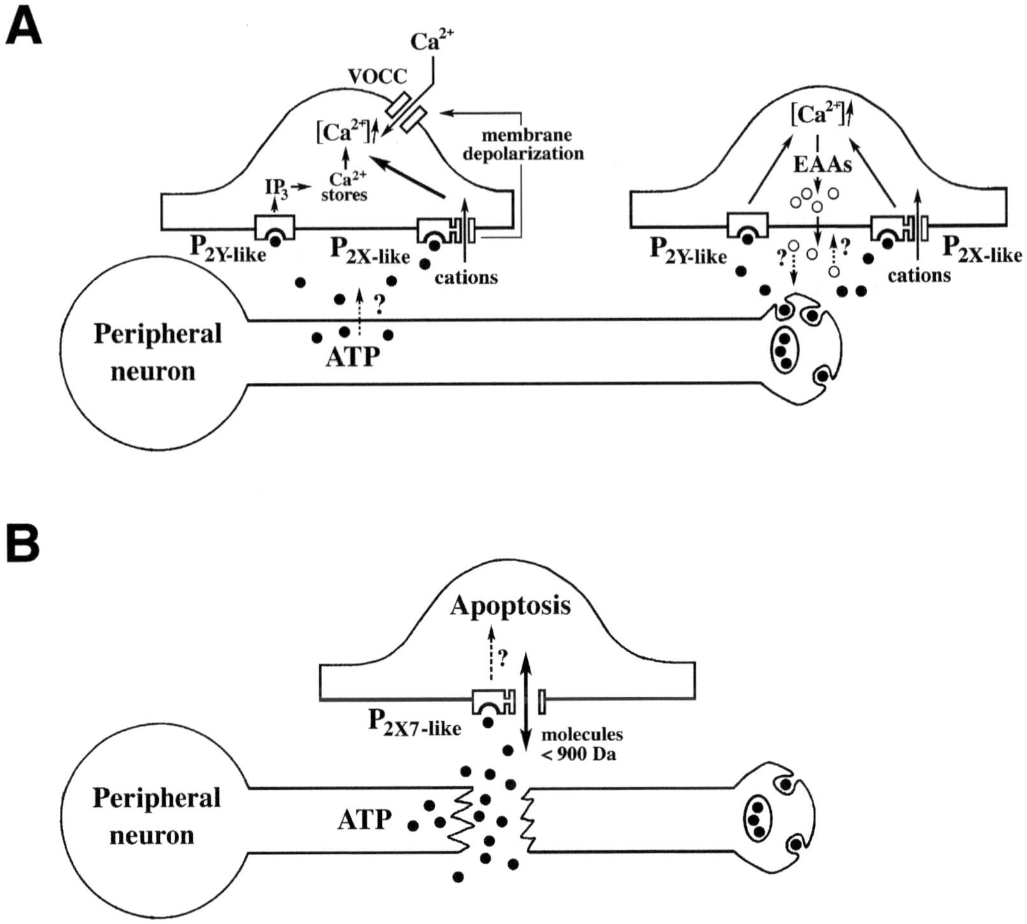

Fig. 1. Scheme of ATP signaling in Schwann cells under physiopathological conditions. **A.** In physiological conditions, ATP is released at synapses and may be released along the axon. For sake of clarity, membrane and intracellular mechanisms leading to an increase in $[Ca^{2+}]_i$ have been detailed in the left hand Schwann cell and calcium-dependent release of EAAs in the right hand Schwann cell. **B.** In pathological conditions, ATP is massively released by lesioned axon and by activating P_{2X7}-like purinoceptor may trigger Schwann cell apoptosis. VOOC: voltage operated calcium channel; EAAs: excitatory amino acids.

5.4. Secretion of Cytokines

Cytokines are pleiotropic signaling peptides which act through paracrine and autocrine networks. They are secreted not only by inflammatory and immune system cells, but also by accessory immune cells such as glial cells of the CNS (astrocytes, oligodendrocytes, and microglia) *(109)*.

Schwann cells have cellular and humoral immune functions *(110)*. They can process exogenous antigens and express MHC molecules (cellular immune functions) but also they synthesize and secrete a number of proinflammatory cytokines including IL-1, IL-6, IFN-α and TNF-α as well as immunoregulatory cytokines including IL-10 or TGF-β *(111)*. In the case of mouse microglial cells *(84)* and human macrophages *(112)* release of IL-1β is stimulated by ATP. The ATP binds to a P_{2X7}-like purinoceptor and activates IL-1β converting enzyme (ICE) which converts IL-1β from its inactive precursor to its biologically active form.

For those Schwann cells which express P_{2X7}-like purinoceptors, their activation may lead to the activation of ICE. This could be a pathway leading to the synthesis and/or secretion of IL-1β. The consequences are probably quite complex since the same cytokine, i.e., IL-1β, can stimulate either the immune response or serve as growth or protective factors. For example, IL-1β induces synthesis of nerve growth factor by Schwann cells *(113)* which stimulates neuronal repair.

6. CONCLUSION

In this chapter we have highlighted ATP as a neural signaling molecule in the PNS by examining neuronal sites of storage, pathways of release, as well as known and putative effects on surrounding Schwann cells. ATP has a wide range of effects on Schwann cell physiology. It may affect the Schwann cell differentiated/proliferative state by interacting with gene expression through cAMP level and MAPK pathway and/or the synthesis and release of cytokines. By evoking release of excitatory amino acids from Schwann cells, ATP signals to Schwann cells which in turn could signal to neurons (paracrine loop) and/or to Schwann cell itself (autocrine loop).

Most of the time, the response to ATP involves changes in the Schwann cell $[Ca^{2+}]_i$. How does ATP induce such diverse responses in Schwann cells? Some specificity is given by ATP causing either an inhibition of adenylate cyclase via the stimulation of P_{2Y1}-like purinoceptors or an increase in $[Ca^{2+}]_i$. In the latter case, further specificity probably results from the spatiotemporal pattern of the increase in $[Ca^{2+}]_i$. For example the increase in $[Ca^{2+}]_i$ may be fast and transient when it is only due to release of calcium from intracellular stores, while it will be slow and sustained when capacitative calcium entry is involved. Last, because ATP is released under different circumstances and in different extracellular spaces, it is likely that the extracellular concentration varies and therefore the subtype of purinoceptor activated may be different. For example, the activation of P_{2X7}-like purinoceptors will require higher concentrations than activation of P_{2Y}-like purinoceptors. Such high concentrations of ATP are likely to be reached during pathological conditions.

By the diversity of its actions, ATP appears to be one of the most powerful signaling molecules involved in neuro-glial interactions in the PNS.

REFERENCES

1. Holton, F.A. and Holton, P. (1953) The possibility that ATP is a transmitter at sensory nerve endings. *J. Physiol. London* **119,** 50–51P.
2. Holton, P. (1959) The liberation of adenosine triphosphate on antidromic stimulation of sensory nerves *J. Physiol. London* **145,** 494–504.
3. Thorne, P.R. and Housley, G.D. (1996) Purinergic signalling in sensory systems. In: *Seminars in The Neurosciences* **8,** 233–246.
4. Fieber, L.A. and Adams, D.J. (1991) Adenosine triphoshate-evoked currents in cultured neurons dissociated from rat parasympathetic cardiac ganglia. *J. Physiol. London* **434,** 239–256.
5. Boeckxstaens, G.E., Pelckmans, P.A., Rampart, M., Verbeuren, T.J., Herman, A.G., and Van Maercke, Y.M. (1990) Evidence against ATP being the inhibitory transmitter released by nonadrenergic noncholinergic nerves in the canine ileocolonic junction. *Pharmacol Exp Ther.* **254(2),** 659–63.

6. Von Kügelgen, I., Allgaier, C., Schobert, A., and Starke, K. (1994) Co-release of noradrenaline and ATP from cultured sympathetic neurons. *Neuroscience* **61(2),** 199–202.
7. Wolinsky, E.J. and Patterson, P.H. (1985) Potassium-stimulated purine release by cultured sympathetic neurons. *J. Neurosci.* **5(7),** 1680–1687.
8. Jahr, C.E. and Jessell, J.M. (1983) ATP excites a subpopulation of rat dorsal horn neurones. *Nature* **304,** 730–733.
9. Burnstock, G. (1996) P2 purinoceptors: historical perspective and classification. In *P2-purinoceptors: localization, function and transduction mechanisms,* Wiley, Chichester (Ciba Foundation Symposium **198**), 1–34.
10. Burnstock, G. (1997) The past, present and future of purine nucleotides as signalling molecules. *Neuropharmac.* **36,** 1127–1139.
11. Gualix, J., Abal, M., Pintor, J., Garcia-Carmona, F., and Miras-Portugal, M.T. (1996) Nucleotide vesicular transporter of bovine chromaffin granules. Evidence for a mnemonic regulation. *J. Biol. Chem.* **271(4),** 1957–65.
12. Reist, N.E. and Smith, S.J. (1992) Neurally evoked calcium transients in terminal Schwann cells at the neuromuscular junction. *Proc. Natl. Acad. Sci. USA* **89,** 7625–7629.
13. Jahromi, B.S., Robitaille, R., and Charlton, M.P. (1992) Transmitter release increases intracellular calcium in perisynaptic Schwann cells in situ. *Neuron* **8,** 1069–1077.
14. Robitaille, R. (1995) Purinergic receptors and their activation by endogenous purines at perisynaptic glial cells of the frog neuromuscular junction. *J. Neurosci.* **15,** 7121–7131.
15. Stjärne, L., Astrand, P., Bao, J.X., Gonon, F., Msghina, M., and Stjarne, E. (1994) Spatiotemporal pattern of quantal release of ATP and noradrenaline from sympathetic nerves: consequences for neuromuscular transmission. *Adv. Second. Messenger Phosphoprotein Res.* **29,** 461–496.
16. Zimmermann, H. (1996) Biochemistry, localization and functional roles of ecto-nucleotidases in the nervous system. *Progress Neurobiol.* **49,** 589–618.
17. Farinas, I., Solsona, C., and Marsal, J. (1992) Omega-conotoxin differentially blocks acetylcholine and adenosine triphosphate releases from Torpedo synaptosomes. *Neuroscience* **47(3),** 641–648.
18. Lyons, S.A., Morell, P., and McCarthy, K.D. (1995) Schwann cell ATP-mediated calcium increases in vitro and in situ are dependent on contact with neurons. *Glia* **13,** 27–38.

18a. Stevens, B. and Fields, R.D. (2000) Response of Schwann cells to action potentials in development. *Science* **287,** 2267–2271.

19. Weinreich, D. and Hammerschlag, R. (1975) Nerve impulse-enhanced release of amino acids from non-synaptic regions of peripheral and central nerve trunks of bullfrog. *Brain Res.* **84(1),** 137–142.
20. Kriegler, S. and Chiu, S.Y. (1993) Calcium signaling of glial cells along mammalian axons. *J. Neurosci.* **13(10),** 4229–4245.
21. Vizi, E.S., Gyires, K., Somogyi, G.T., and Ungvary, G. (1983) Evidence that transmitter can be released from regions of the nerve cell other than presynaptic axon terminal: axonal release of acetylcholine without modulation. *Neuroscience* **10(3),** 967–972.
22. Martin, D.L. (1992) Synthesis and release of neuroactive substances by glial cells. *Glia* **5(2),** 81–94.
23. Schwiebert, E.M. (1999) ABC transporter-facilitated ATP conductive transport. *Am. J. Physiol.* **276,** C1–C8.
24. Reisin, I.L., Prat, A.G., Abraham, E.H., Amara, J.F., Gregory, R.J., Ausiello, D.A., and Cantiello, H.F. (1994) The cystic fibrosis transmembrane conductance regulator is a dual ATP and chloride channel. *J. Biol. Chem.* **269(32),** 20584–20591.
25. Li, C., Ramjeesingh, M., and Bear, C.E. (1996) Purified cystic fibrosis transmembrane conductance regulator (CFTR) does not function as an ATP channel. *J. Biol. Chem.* **271(20),** 11623–11626.

26. Reddy, M.M., Quinton, P.M., Haws, C., et al. (1996) Failure of the cystic fibrosis transmembrane conductance regulator to conduct ATP. *Science* **271,** 1876–1879.
27. Prat, A.G., Reisin, I.L., Ausiello, D.A., and Cantiello, H.F. (1996) Cellular ATP release by the cystic fibrosis transmembrane conductance regulator. *Am. J. Physiol.* **270,** C538–C545.
28. Cantiello, H.F., Jackson, G.R., Grosman, C.F., et al. (1998) Electrodiffusional ATP movement through the cystic fibrosis transmembrane conductance regulator. *Am. J. Physiol.* **274,** C799–C809.
29. Gottesman, M.M., Pastan, I., and Ambudkar, S.V. (1996) P-glycoprotein and multidrug resistance. *Curr. Opin. Genet. Dev.* **6(5),** 610–617.
30. Gill, D.R., Hyde, S.C., Higgins, C.F., Valverde, M.A., Mintening, G.M., and Sepulveda, F.V. (1992) Separation of drug transport and chloride channel functions of the human multidrug resistance P-glycoprotein. *Cell* **71,** 23–32.
31. Hardy, S.P., Goodfellow, H.R., Valverde, M.A., Gill, D.R., Sepulveda, V., and Higgins, C.F. (1995) Protein kinase C-mediated phosphorylation of the human multidrug resistance P-glycoprotein regulates cell volume-activated chloride channels. *EMBO J.* **14(1),** 68–75.
32. Abraham, E.H., Prat, A.G., Gerweck, L., et al. (1993) The multidrug resistance (mdr1) gene product functions as an ATP channel. *Proc. Natl. Acad. Sci. USA* **90(1),** 312–316.
33. Mitchell, C.H., Carré, D.A., McGlinn, A.M., Stone, R.A., and Civan, M.M. (1998) A release mechanism for stored ATP in ocular ciliary epithelial cells. *Proc. Natl. Acad. Sci. USA* **95,** 7174–7178.
34. Allikmets, R., Gerrard, B., Hutchinson, A., and Dean, M. (1996) Characterization of the human ABC superfamily: isolation and mapping of 21 new genes using the expressed sequence tags database. *Hum. Mol. Genet.* **5(10),** 1649–1655.
35. Bruzzone, R. and Ressot, C. (1997) Connexins, gap junctions and cell-cell signalling in the nervous system. *Eur. J. Neurosci.* **9,** 1–6.
36. Cotrina, M.L., Lin, J.H., Alves-Rodrigues, A., et al. (1998) Connexins regulate calcium signaling by controlling ATP release. *Proc. Natl. Acad. Sci. USA* **95(26),** 15735–15740.
37. Giaume, C. and Venance, L. (1998) Intercellular calcium signalling and gap junctional communication in astrocytes. *Glia* **24,** 50–64.
38. Karschin, C., Ecke, C., Ashcrift, F.M., and Karschin, A. (1997) Overlapping distribution of K(ATP) channel-forming Kir6.2 subunit and the sulfonylurea receptor SUR1 in rodent brain. *FEBS Lett.* **401(1),** 59–64.
39. Rathbone, M.P., Middlemiss, P., Andrew, C., et al. (1998) The trophic effects of purines and purinergic signaling in pathologic reactions of astrocytes. *Alzheimer Dis. Assoc. Disord.* **12,** S36–S45.
40. Chandross, K.J. (1998) Nerve injury and inflammatory cytokines modulate gap junctions in the peipheral nervous system. *Glia* **24,** 21–31.
41. Dezawa, M., Mutoh, T., Dezawa, A., and Adachi-Usami, E. (1998) Putative gap junctional communication between axon and regenerating Schwann cells during mammalian peipheral nerve regeneration. *Neuroscience* **85,** 663–667.
42. Ziganshin, A.U., Hoyle, C.H.V., and Burnstock, G. (1994) Ecto-enzymes and metabolism of extracellular ATP. *Drug Dev. Res.* **32,** 134–146.
43. Mata, M., Staple, J., and Fink, D.J. (1988) Cytochemical localization of Ca^{2+}-ATPase activity in peripheral nerve. *Brain Res.* **445(1),** 47–54.
44. Nitahara, K., Kittel, A., Liang, S.D., and Vizi, E.S. (1995) A1-receptor-mediated effect of adenosine on the release of acetylcholine from the myenteric plexus. Role and localization of ecto-ATPase and 5′-nucleotidase. *Neuroscience* **67,** 159–168.
45. Grondal, E.J.M., Janetzko, A., and Zimmermann, H. (1988) Monospecific antiserum against 5′-nucleotidase from *Torpedo* electric organ. Immunocytochemical distribution of the enzyme and its association with Schwann cell membrane. *Neuroscience* **24,** 351–363.

46. Hourani, S.M.O., Bailey, S.J., Nicholls, J., and Kitchen, I. (1991) Direct effects of adenylyl 5′-(β,γ-methyle)diphosphonate a stable ATP analogue, on relaxant P1-purinoceptors in smooth muscle. *Br. J. Pharmac.* **104,** 685–690.
47. Crack, B.E., Beukers, M.W., McKechnie, K.C.W., Ijzerman, A.P., and Leff, P. (1984) Pharmacological analysis of ecto-ATPase inhibition: evidence for combined enzyme inhibition and receptor antagonism in P_{2X}-purinoceptor ligands. *Br. J. Pharmac.* **113,** 1432–1438.
48. Bao, J.X. (1993) Sympathetic neuromuscular transmission in rat tail artery: a study based on electrochemical, electrophysiological and mechanical recording. *Acta Physiol. Scand. Suppl.* **610,** 1–58.
49. Ziganshin, A.U., Ziganshin, L.E., King, B.F., and Burnstock, G. (1995) Characteristics of ecto-ATPase of *Xenopus* oocytes and the inhibitory actions of suramin on ATP breakdown. *Pflüeg Archiv.* **429,** 412–418.
50. Brown, E.R. and Abbott, N.J. (1993) Ultrastructure and permeability of the Schwann cell layer surrounding the giant axon of the squid. *J. Neurocytol.* **22,** 283–298.
51. Dalziel, H.H. and Westfall, D.P. (1994) Receptors for adenine nucleotides and nucleosides: subclassification, distribution and molecular characterization. *Pharmacol. Rev.* **446,** 449–466.
52. Bhagwatt and Williams, M. (1997) P_2 purine and pyrimidine receptors: emerging superfamilies of G-protein-coupled and ligand-gated ion channel receptors. *Eur. J. Med. Chem.* **32,** 183–193.
53. Ralevic, V. and Burnstock, G. (1998) Receptors for purines and pyrimidines. *Pharmacol. Rev.* **50,** 413–492.
54. Surprenant, A., Rassendren, F., Kawashima, E., North, R.A., and Buell, G. (1996) The cytolytic P_{2Z} receptor for extracellular ATP identified as a P_{2X} receptor (P_{2X7}). *Science* **272,** 735–738.
55. DiVirgilio, F., Chiozzi, P., Falzoni, S., et al. (1998) Cytolytic P_{2x} purinoceptors. *Cell death and differentiation* **5,** 191–199.
55a. Amédée, T. and Despeyroux, S. (1995) ATP activates cationic and anionic conductances in Schwann cells cultured from dorsal root ganglia of the mouse. *Proc. R. Soc. Lond. B. Biol. Sci* **259,** 277–284.
56. Colomar, A., and Amédée, T. (2001) ATP stimulation of $P2X_7$ receptors activates three different ionic conductances on cultured mouse Schwann cells. *Eur. J. Neurosci.* (in press).
57. Grafe, P., Mayer, C., Takigawa T., Kamleiter M., and Sanchez-Brandelik R. (1999) Confocal calcium imaging reveals an ionotropic P2 nucleotide receptor in the paranodal membrane of rat Schwann cells. *J. Physiol. London* **515.2.,** 377–383.
58. Lyons, S.A., Morell, P., and McCarthy, K.D. (1994) Schwann cells exhibit P_{2Y} purinergic receptors that regulate intracellular calcium and are up-regulated by cyclic AMP analogues. *J. Neurochem.* **63,** 552–560.
59. Ansselin, A.D., Davey, D.F., and Allen, D.G. (1997) Extracellular ATP increases intracellular calcium in cultured adult Schwann cells. *Neuroscience* **76,** 947–955.
60. Amédée T., Cantereau A., Beaudu-Lange.C., Georgescauld D. and Coles J.A.(1998) High affinity and low affinity purinergic receptors on mouse Schwann cells. In *Dialogue between glia and neurons* (Third European Meeting on glial cell function in health and disease), Abstract 49.
61. Jeftinija S.D. and Jeftinija K.V. (1998) ATP stimulates release of excitatory amino acids from cultured Schwann cells. *Neuroscience* **82,** 927–934.
62. Green, A.C., Dowdall, M.J., and Richardson, C.M. (1997) ATP acting on P_{2Y} receptors triggers calcium mobilization in Schwann cells at the neuroelectrocyte junction in skate. *Neuroscience* **80,** 635–651.
63. Mayer C., Quasthoff S. and Grafe P. (1998) Differences in the sensitivity to purinergic stimulation of myelinating and non-myelinating Schwann cells in peripheral human and rat nerve. *Glia,* **23,** 374–382.

64. Mayer, C., Wächtler, J., Kamleiter, M. and Grafe, P. (1997) Intracellular calcium transients mediated by P_2 receptors in the paranodal Schwann cell region of myelinated rat spinal root axons. *Neurosci. Lett.* **224,** 49–52.
65. Zimmermann, H. (1994) Signalling via ATP in the nervous system. *Trends Neurosci.* **17,** 420–426.
66. Illes P. and Nörenberg W. (1993) Neuronal ATP receptors and their mechanism of action. *Trends Pharmacol. Sci.* **14,** 50–54.
67. Amédée, T., Ellie, E., Dupouy, B., and Vincent, J.D. (1991) Voltage-dependent calcium and potassium channels in Schwann cells cultured from dorsal root ganglia of the mouse. *J. Physiol. London.* **441,** 35–56.
68. Clapham, D.E. (1995) Calcium signalling. *Cell* **80,** 259–268.
69. Lovisolo, D.E., Distasi, C., Antoniotti, S., and Munaron, L. (1997) Mitogens and calcium channels. *News Physiol. Sci.* **12,** 279–285.
70. Berridge, M.J. (1998) Neuronal calcium signaling. *Neuron* **21,** 13–26.
71. Parpura, V., Liu, F., Jeftinija, K.V., Haydon, P.G., and Jeftinija, S.D. (1995) Neuroligand-evoked calcium-dependent release of excitatory amino acids from Schwann cells. *J. Neurosci.* **15,** 5831–5839.
72. Berti-Mattera, L.N., Wilkins, P.L., Madhun, Z., and Suchovsky, D. (1996) P_2-purinergic receptors regulate phospholipase C and adenylate cyclase activities in immortalized Schwann cells. *Biochem. J.* **314,** 555–561.
73. Verkhratsky, A., Orkand, R.K., and Kettenmann, H. (1998) Glial calcium: Homeostasis and signalling function. *Physiol. Rev.* **78,** 99–141.
74. Putney, J.W. (1986) A model for receptor-regulated calcium entry. *Cell Calcium* **7,** 1–12.
75. Jeftinija, S.D., Jeftinija, K.V., Stefanovic, G., and Liu, F. (1996) Neuroligand-evoked calcium-dependent release of excitatory aminoacids from cultured astrocytes. *J. Neurochem.* **58,** 1277–1284.
76. Katz, B. and Miledi, R. (1959) Spontaneous subthreshold activity at denervated amphibian endplates. *J. Physiol. London* **146,** 44–45P.
77. Dennis, M.J. and Miledi, R. (1974) Electrically induced release of acetylcholine from denervated Schwann cells. *J. Physiol. London* **237,** 431–452.
78. Lieberman, E.M., Abbott, N.J. and Hassan, S. (1989) Evidence that glutamate mediates axon-to-Schwann cell signaling in the squid. *Glia* **2,** 94–102.
79. Evans, P.D., Reale, V., Merzon, R.M., and Villegas, J. (1992) N-methyl-D-aspartate (NMDA) and non-NMDA (metabotropic) type glutamate receptors modulate the membrane potential of the Schwann celll of the squid giant nerve fibre. *J. Exp. Biol.* **173,** 229–249.
80. Dray, A. and Perkins, M. (1993) Bradykinin and inflammatory pain. *Trends Neurosci.* **16,** 99–104.
81. Ferrari, D., Chiozzi, P., Falzoni, S., et al. (1997) ATP-mediated cytotoxicity in microglial cells. *Neuropharmacol.* **36,** 1295–1301.
82. Chow, S.C., Kass, G.E.N. and Orrenius, S. (1997) Purines and their roles in apoptosis. *Neuropharmac.* **36,** 1149–1156.
83. Ferrari, D., Villaba, M., Chiozzi, P., Falzoni, S., Ricciardi-Castagnoli, P., and Di Virgilio, F. (1996) Mouse microglial cells express a plasma membrane pore gated by extracellular ATP. *J. Immunol.* **156,** 1531–1539.
84. Ferrari, D., Los, M., Bauer, M., Vandenabeele, P., Wesselborg, S., and Schulze-Osthoff, K. (1999) P2Z purinoceptor ligation induces activation of caspases with distinct roles in apoptotic and necrotic alterations of cell death. *FEBS Letters* **447,** 71–75.
85. Miura, M., Zhu, H., Rotello, R., Hartweig, E.A., and Yuan, J. (1993) Induction of apoptosis in fibroblasts by IL-1β – Converting Enzyme, a mammalian homolog of the C.elegans cell death gene ced-3. *Cell,* **75,** 653–660.

86. Cohn, M. (1990) Structural and chemical properties of ATP and its metal complexes in solution. *Ann. N. Y. Acad. Sci.* **603,** 151–164.
87. Neary, J.T., Rathbone, M.P., Cattabeni, F., Abbracchio, M.P., and Burnstock, G. (1996) Trophic actions of extracellular nucleotides and nucleosides on glial and neuronal cells. *Trends Neurosci.* **19,** 13–18.
88. Neary, J.T. (1996) Trophic actions of extracellular ATP on astrocytes, synergistic interactions with fibroblast growth factors and underlying signal transduction mechanisms. In *P2-purinoceptors: localization, function and transduction mechanisms,* Wiley, Chichester (Ciba Foundation Symposium **198**), 130–139.
89. Neary, J.T. and Zhu, Q. (1994) Signaling by ATP receptors in astrocytes. *Neuroreport,* **5(13),** 1617–1620.
90. Seger, R. and Krebs, E.G. (1995) The MAPK signaling cascade. *FASEB J.* **9,** 726–735.
91. Mutoh, T., Li, M., Yamamoto, M., Mitsuma, T., and Sobue, G. (1998) Differential signaling cascade of MAP kinase and S6 kinase depends on 3′,5′-monophosphate concentration in Schwann cells: correlation to cellular differentiation and proliferation. *Brain Res.* **810(1–2),** 274–278.
92. Thornberry, N.A. and Lazebnik, Y. (1998) Caspases: enemies within. *Science* **281,** 1312–1316.
93. Oppenheim, R.W. (1991) Cell death during development of the nervous system. *Ann. Rev. Neurosci.* **14,** 453–501.
94. Barres, B.A. and Raff, M.C. (1994) Control of oligodendrocyte number in the developing rat optic nerve. *Neuron.* **12(5),** 935–942.
95. Krueger, B.K., Burne, J.F. and Raff, M.C. (1995) Evidence for large-scale astrocyte death in the developing cerebellum. *Neuroscience* **15,** 3366–3374.
96. Ciutat, D., Caldero, J., Oppenheim, R.W., and Esquerda, J.E. (1996) Schwann cell apoptosis during normal development and after axonal degeneration induced by neurotoxins in the chick embryo. *J. Neurosci.* **16,** 3979–3990.
97. Grinspan, J.B., Marchionni, M.A., Reeves, M., Coulaglou, M., and Scherer, S.S. (1996) Axonal interactions regulate Schwann cell apoptosis in developing peripheral nerve: Neuregulin receptors and the role of neuregulins. *J. Neurosci.* **16,** 6107–6118.
98. Jessen, K.R. and Mirsky, R. (1997) Embryonnic Schwann cell development: the biology of Schwann cell precursors and early Schwann cells. *J. Anat.* **191,** 501–505.
99. Tatsumi, H., Tsuji, S., Anglade, P., Motelica-Heino, I., Soeda, H., and Katayama, Y. (1995) Synthesis, storage and release of acetylcholine at and from growth cones of rat central cholinergic neurons in culture. *Neurosci. Lett.* **202,** 25–28.
100. Soeda, H., Tatsumi, H., and Katayama, Y. (1997) Neurotransmitter release from growth cones of rat dorsal root ganglion neurons in culture. *Neuroscience* **77,** 1187–1199.
101. Griffin, J.W. and Hoffman, P.N. (1993) Degeneration and regeneration in the peripheral nervous system. In *Peripheral neuropathy* (Griffin, J.W., Low, P.A., and Poduslo, J.F., eds) Saunders, Philadelphia, pp 361–376.
102. Whiteside, G., Doyle, C.A., Hunt S.P., and Munglani, R. (1998) Differential time course of neuronal and glial apoptosis in neonatal rat dorsal root ganglia after sciatic nerve axotomy. *Eur. J. Neurosci.* **10,** 3400–3408.
103. Trachtenberg, J.T. and Thompson W.J. (1996) Schwann cell apoptosis at developing neuromuscular junctions is regulated by glial growth factor. *Nature* **379,** 174–177.
104. Parpura, V., Basarsky, T.A., Liu, F., Jeftinija, K.V., Jeftinija, S.D., and Haydon P.G. (1994) Glutamate-mediated astrocyte-neuron signaling. *Nature* **369,** 744–747.
105. Vernadakis, A. (1996) Glia-neuron intercommunications and synaptic plasticity. *Progress Neurobiol.* **49,** 185–214.
106. Finkbeiner, S.M. (1993) Glial calcium. *Glia* **9,** 83–104.
107. Hugon, J., Vallat, J.M., and Leboutet M.J. (1987) Cytotoxic properties of glutamate and aspartate in rat peripheral nerves: histological findings. *Neurosci. Lett.* **81,** 1–6.

108. Gallo, V., Patrizio, M., and Levi, G. (1991) GABA release triggered by the activation of neuron-like non-NMDA receptors in cultured type 2 astrocytes is carrier-mediated. *Glia* **4**(3), 245–255.
109. Bennveniste, E. (1992) Inflammatory cytokines within the central nervous system: sources, function and mechanism of action. *Am. J. Physiol.* **269,** C1–C6.
110. Gold, R., Archelos, J.J., and Hartung, H-P. (1999) Mechanisms of immune regulation in the peripheral nervous system. *Brain Pathol.* **9,** 343–360.
111. Lisak, R.P., Skundric, D., Bealmear, B., and Ragheb, S. (1997) The role of cytokines in Schwann cell damage, protection and repair. *J. Infect. Dis.* **176,** S173–S179.
112. Ferrari, D., Chiozzi, P., Falzoni S., et al. (1997) Extracellular ATP triggers IL-1β release by activating the purinergic P2Z receptor of human macrophages. *J. Immunol.* **159,** 1451–1458.
113. Lindholm, D., Heumann, R., Meyer, M., and Thoenen H. (1987) Interleukin regulates synthesis of nerve growth factor in no-neuronal cells of rat sciatic nerve. *Nature* **330,** 658–659.

III

Neurotrophins, Growth Factors, and Neurohormones in Aging and Regeneration

9
Gliosis Growth Factors in the Adult and Aging Rat Brain

Gérard Labourdette and Françoise Eclancher

1. INTRODUCTION

Gliosis is a response of glial cells (consisting usually in hypertrophy, migration, proliferation, and phenotypic changes) to various types of neural tissue perturbations. This glial response occurs naturally in aged animals. Notwithstanding the nature of the trauma, the glial response is more or less stereotypical. A reproducible succession of cellular and molecular events develops, coordinated by intercellular signaling in which growth factors and extracellular matrix seem to play a major role. Gliosis can occur in any part of the central nervous system (CNS). Prominent responses that follow brain trauma include the activation of microglia, recruitment of blood-derived macrophages, and astroglial reactivity. Occasionally, depending on the zone of gliosis, other cell types can be activated like ependymal cells, endothelial cells, meningeal cells, and Schwann cells. The main aspect of localized gliosis is the elaboration of a glial scar which forms a physical and biochemical barrier preventing axonal regeneration and remyelination. In addition to this physical repair, many data suggest that gliosis contributes to nerve survival and to immunological protection.

2. GLIOSIS RESULTING FROM BRAIN INJURY IN THE ADULT RAT

2.1. *Triggering of Gliosis*

Gliosis can be triggered by a great variety of perturbations of the nervous system:

1. Physical injury which can be a stab-wound, possibly associated with heating, freezing or chemical attack, or a physical shock without wound;
2. Infection by external organisms;
3. Various diseases which can affect the CNS such as ischemia, seizures, neurodegenerative diseases, autoimmune diseases or tumors. It seems that the primary event triggering gliosis is a destruction of neurons, but also of oligodendrocytes, astrocytes or other cells. Demyelination is also able to induce gliosis. In order to understand the mechanisms of gliosis or of CNS degenerative diseases, and possibly to prevent them or attenuate their effects, the level of various molecules can be modulated in the CNS or in some parts of it. Some of them will trigger gliosis.

From: *Neuroglia in the Aging Brain*
Edited by: Jean S. de Vellis © Humana Press Inc., Totowa, NJ

4. The two main ways to achieve this task are: Injection of substances, locally, or systemically when they are able to pass the blood-brain-barrier or to break it, such as growth factors, and cytotoxic molecules;
5. Overproduction or defect of molecules normally present in the CNS, for example by the use of transgenic mice, of mRNA antisense strategies or of immunochemical methods.

2.2. Cells Involved in Gliosis

Gliosis was first defined by the activation of astrocytes. However, it appears that microglia are necessary for astrogliosis to develop *(1)*. Hypertrophy of microglia starts 24 h after injury, and after 2–3 d, they begin to proliferate and to migrate toward the site of the injury. These cells secrete cytokines and factors which can be involved in inflammation, immune defense, neuron protection or regulation of astrocyte activation *(2)*. Within 24 h after cerebral injury, the level of glial fibrillary acidic protein (GFAP) and vimentin *(3)* increases in astrocytes with an increase of their volume and of the number and length of their processes. GFAP is involved in the mechanism of this activation. Later, they migrate and proliferate, produce extracellular matrix and form a scar composed of cells associated to this matrix. The expression of many proteins is upregulated; nestin, various cytokines, growth factors, carbonic anhydrase, and L-type Ca^{2+} channels. The Ca^{2+} S-100 protein present mainly in astrocytes is increased after aspiration of the occipital cortex and participates in the trophism and plasticity of the injured visual pathway *(4)*. Astrocytes can elicit other effects through the release of growth factors and particularly neurotrophic factors which will help the survival of the damaged neurons.

Recent data suggest that oligodendrocytes are involved in gliosis. There is a swelling of their nucleus and cytoplasm and an increase in number and size of their processes. An increase in the number of oligodendrocytes and an enhanced expression of 2′, 3′-cyclic nucleotide phosphohydrolase (CNP) and myelin basic protein (MBP) has also been reported. Ependymal cells can also be involved in gliosis. In response to spinal cord injury, their proliferation increases dramatically to generate migratory cells that differentiate into astrocytes and participate in scar formation *(5)*. The migration and proliferation of ependymal cells appear to depend on growth factors like EGF, PDGF, TGF beta, and thrombin *(6)*. Meningeal cells may invade the lesion and contribute also to the formation of scar.

3. GROWTH FACTORS INVOLVED IN ADULT BRAIN GLIOSIS

This chapter describes which factors are released in gliosis, and when known, in which cells they are produced, on which cells they elicit their main effects, what are these effects, how their expression is regulated, and what are their functions.

3.1. Fibroblast Growth Factor-2 (FGF-2)

FGF-2 is an ubiquitous factor found at a high concentration in adult brain. It elicits pleiotrophic effects on all the nerve cells. In normal rat brain, it is localized in astrocytes, in selected neuronal populations, and occasionally in microglial cells. FGF-2 protein and mRNA are found increased in astrocytes at the site of focal brain wounds *(7)* and after ischemia, convulsive seizures or kainic acid injection *(8)*, and sometimes, to less extent, in neurons and microglia *(9)*. Reactive gliosis can be stimulated by injection of FGF-2 in injured rat brain *(10–12)*. Expression of FGF receptor-1 (FGFR-1) increases on astrocytes adjacent to the wound cavity by day 2–7 postlesion until day 10 and decreases to control values by day 28. Thus, through autocrine or paracrine mechanisms FGF-2 could be an early response of reactive astrocytes and it may act on this reactive astrogliosis by stimulating the proliferation, hypertrophy, and migration of astrocytes. FGF-2 also promotes the synthesis and release of extracellular matrix (ECM) (tenascin C) and growth factors, such as, neurotrophic factors (NTFs). In the early phase following brain injury, cytokine-activated astrocytes can rescue the damaged neurons and contribute to the process of axonal outgrowth and synaptogenesis via NTFs and other molecules. After a moderate contusion of rat spinal cord, administration of FGF-2 enhances functional recovery and tissue sparing. After 6 wk, astroglial and microglial cell immunoreactivity was as in controls *(13)*. Since FGF-2 activates quiescent microglia and enhances their proliferation in vitro, FGF-2 released by astrocytes should contribute to this activation. In astrocytes, FGF-2 and FGFR can be stimulated by interleukine-1 (IL-1) secreted by activated microglia and by free radicals produced by invading neutrophils. FGF-2 elicits neuroprotective or neurotrophic effects on neurons, directly or indirectly by inducing the expression of neurotrophic factors in astrocytes.

Fibroblasts genetically engineered to produce FGF-2 and implanted in the striatum can protect the nigrostriatal dopaminergic system and may be useful in the treatment of Parkinson's disease *(14)*.

3.2. *Platelet-Derived Growth Factor (PDGF)*

In normal rats, PDGF B-chain is preferentially expressed within neural cell bodies in the cortex, hippocampus, and cerebellum. An increase in PDGF expression has been observed in various traumatic conditions in the CNS such as after human brain abcess or in proliferative retinopathies *(6)* or after mechanical injury in rat brain *(15)*. In vitro, astrocytes produce A and B chains. Moreover, TGF-β 1, TNF-α, and IL-1 β are able to stimulate astrocyte PDGF B production. In human brain abcess, PDGF immunoreactivity was found in glia, and also in endothelial cells and fibroblasts *(16)*. An enhanced expression of PDGF was also found in macrophages on days 3 and 4 after mechanical lesion *(15)*. PDGF is a strong mitogen for astrocytes in vitro. PDGF-AA increases the number of retinal astrocytes and these cells promote the survival of endothelial cells as well as their expression of barrier characteristics *(17)*. PDGF-BB has been shown to exert neuroprotective action on various types of neurons like GABAergic interneurons.

3.3. *Transforming Growth Factor-β (TGF-β)*

There are three forms of TGF-β (1, 2, and 3) and two types of receptors. All are present in the CNS. Although in vitro astrocytes display these three isoforms, neurons express only TGF-β 2. Both type I and type II TGF-β receptors are present in cortical neurons and astrocytes in vitro. Thus, TGF-β may act as autocrine and paracrine signals between neurons and astrocytes in the CNS. TGF-β 1 in the rat brain increases after injury and inhibits astrocyte proliferation *(18)*. At days 1–3 following ischemia TGF-β and their receptors are upregulated in the perifocal neurons, reactive astroglial cells, endothelial cells and macrophages *(19)*. Astroglial overproduction of TGF-β 1 in transgenic mice triggers a pathogenic cascade leading to AD-like cerebrovascular amyloidosis, microvascular degeneration, and local alterations in brain metabolic activities *(20)*. The production of TGF-β 1 is stimulated by IL-1 in astrocytes in vitro. TGF-β inhibit the proliferation of microglia. In vitro, TGF-β 1 promotes survival of rat spinal cord motoneurons and dopaminergic neurons. It protects cultured neurons from damage. The neuroprotective action of TGF-β 1 is mediated by the protease inhibitor PAI-1 produced by astrocytes and possibly by NGF in vivo. Treatment of cerebral wounds with anti-TGF-β 2 antibody leads to a marked attenuation of all aspects of CNS scarring *(21)*.

3.4. Epidermal Growth Factor (EGF) Family

Three members of the EGF family are present in the nervous system, EGF, TGF-α and heparin binding (HB)-EGF. They share the same receptor (EGFR). TGF-α and EGFR produced by neurons and glia, play an important role in the development of the nervous system. Following different CNS traumas, EGFR appears mostly within and around the damaged areas, in reactive astrocytes *(22)* as well as in reactive microglial cells and released after injury. EGF is not a mitogen for microglia, but rather a chemoattractant, so it may serve to direct microglial cells to the lesion site. Moreover, since EGF is secreted by activated microglia themselves in vivo, it may act as an autocrine modulator of microglial cell function *(23)*. Synthesis of TGF-α, HB-EGF and EGFR in reactive astrocytes following CNS traumas suggests that they play a role in the development of astrogliosis. They are involved in mediating glial-neuronal and axonal-glial interactions and participate in injury associated astrocytic gliosis *(24)*.

The effect of TGF-α has been examined by the use of transgenic mice bearing the human TGF-α cDNA. These mice display enhanced GFAP and vimentin protein levels in brain. Thus enhanced TGF-α synthesis is sufficient to trigger astrogliosis, whereas microglia are unaffected *(22)*. Moreover, a soluble astrocyte mitogen inhibitor is reduced after injury. Intracerebral injection of antibody against it causes the appearance of reactive astrocytes. So, this inhibitor may play a key role in the control of glial cell division in both normal and injured brain. In addition to their role in gliosis, factors of the EGF family also exert neurotrophic activity. EGF extends the survival of cultured neurons, facilitates neurite outgrowth, and prevents neuronal damage caused by various cytotoxic compounds.

3.5. Insulin-Like Growth Factors (IGFs)

This is a family of three factors, insulin, IGF-1, and IGF-2. Six specific IGF binding proteins (IGFBPs) have been identified, which are able to transport and/or modulate the action of IGFs. IGF-1 receptor is ubiquitous. IGF-1 is present mainly in neurons and stimulates the proliferation and survival of neurons and oligodendrocytes. Intact animals showed no detectable IGF-1 immunoreactivity in astrocytes. IGF-2 mRNA expression is observed in the choroid plexus, meningeal membranes, and in blood vessel endothelium. Several observations suggest a role for IGFBPs in targeting the neuroprotective actions of IGF-1. Following electrolytical lesion in rat hippocampus or unilateral hypoxic-ischemic injury, there is an increase in secretion of IGF-1 at the lesioned site *(25)*. IGF-1 may have a role as a neuroprotectant for surviving neurons and a signal for local neuronal sprouting, as well as in reactive astrogliosis. After partial deafferentation of the cerebellar cortex by 3-acetylpyridine injection, continuous infusion of IGF-1 significantly decreases reactive gliosis while IGFR1 antagonist or IGF-1 antisense increase it. Several authors have shown that IGF-1 is a survival factor, particularly after ischemia, for hippocampal and cortical neurons. Some data provide experimental evidences which suggest that IGF and BDNF may exert their potential neuroprotective effects via regulation of NOS activity *(26)*. IGF-2 has been shown to be secreted within the injured rat brain. During the acute phase, the secretion of IGF-2 from the choroid plexus into the CSF is upregulated and IGF-2 is complexed to IGFBP-2 to be transported to the wound. In the chronic phase of the injury response, IGF-2 reasserts itself to a predominantly autocrine/paracrine role restricted to the mesenchymal support structures, including the glia limitans, which may help to reestablish and maintain tissue homeostasis. Reactive IGF-2 is found in microglia and astrocytes *(27)*. IGF-II receptors were found widely in all neuronal regions and plexus choroids but were not detected in any of the astroglial plaques *(26,28)*.

3.6. Neurotrophins (NGF, BDNF, NT-3, GDNF)

3.6.1. Nerve Growth Factor (NGF)

NGF is a neurotrophic factor for sympathetic and sensory neurons as well as for basal forebrain cholinergic neurons in the brain. In the mature mammalian brain, NGF expression is restricted to neurons. However, astrocytes activated by various cytokines released after brain injury, produce a significant amount of NGF in vitro and in vivo *(29)*. In the early phase following brain injury, these cytokine-activated astrocytes rescue the damaged neurons via NGF and other biologically active molecules *(30)*. The increase in NGF after CNS trauma is directly mediated through IL-1 beta and IL-6. Other cytokines like IL-4, IL-10, and IFN gamma induced the secretion of NGF by activated astrocytes and IL-1 and IL-4 synergized with lipopolysaccharide (LPS) and TNF-α in this effect.

3.6.2. Brain-Derived Growth Factor (BDNF)

BDNF exerts neuroprotective as well as neurotrophic roles for the survival and differentiation of dopaminergic and spinal cord neurons and for certain subpopulations of hippocampal neurons. In the CNS, the highest levels of BDNF mRNA are found in the hippocampus. As seen for other factors, BDNF level is rapidly increased in brain after various types of gliosis-promoting damages induced by physical lesion, spinal cord crushing, kainic injection, seizures or ischemia *(31)*. Several authors have reported modulation of the expression of a truncated form of the BDNF receptor TrkB after injury *(32)*. Expression of this truncated form, which lacks the catalytic tyrosine kinase (TK) domain, increases and stays elevated for at least 2 mo. This suggests that BDNF and TrkBTK- play a role in dopaminergic regeneration and repair. The truncated TrkB which is expressed in reactive astrocytes may act as negative regulators of neurite growth in damaged regions *(33)*. After injury, BDNF can prevent neuronal loss and promote some regeneration and this is also observed in vitro. Transplants of fibroblasts genetically modified to express BDNF promote regeneration of adult rat rubrospinal and spinal cord after injury in the adult rat.

3.6.3. Neurotrophin-3 (NT-3)

NT-3 is a neurotrophic factor which prevents the death of adult central noradrenergic neurons in vivo, but it also induces microglia proliferation and phagocytic activity, suggesting that the factor plays a role in cellular activation. After crushing of the spinal cord, an increased level of BDNF and NT3 mRNAs is first observed at 6 h and the labeling is enhanced at 24 h and 72 h *(34)*. This NT-3 promotes growth of lesioned dorsal column axons with an abundance of fiber sprouting apparent at the lesion site, and many fibers extending into and beyond the lesion epicenter. After brain injury, a moderate increase in BDNF mRNA expression is observed bilaterally in the CA3 region of the hippocampus at 1,3, and 6 h, but expression declined to control levels by 24 h. Conversely, NT-3 mRNA is significantly decreased in the dentate gyrus *(35)*. Topical application of NT-3 attenuates ischemic brain injury after transient middle cerebral artery occlusion in rats.

3.6.4. Glia Cell Line-Derived Neurotrophic Factor (GDNF)

After a wire knife in adult mouse striatum GDNF mRNA expression increases within 6 h, doubles after 1 wk and remains elevated for at least 1 mo. Most GDNF expression is associated with brain macrophages. The dopaminergic sprouting that accompanies striatal injury appears to result from neurotrophic factor secretion by activated microglia at the wound site. An increase in GDNF receptors occurs in neurons after ischemia or facial nerve injury *(36)*. The main known effect of GDNF is to protect nigral dopaminergic neurons involved in Parkinson's disease. GDNF also protects motoneurons and calbindin neurons. In addition, the increase in nitric oxide that accompanies ischemia and subsequent reperfusion is blocked almost completely by this protein. Thus, GDNF may be efficient in the treatment of cerebrovascular occlusive disease.

3.7. Cytokines (Interleukins, Tumor Necrosis Factor-α, Interferon-γ)

Traumatic injury to the CNS initiates inflammatory processes that are implicated in secondary tissue damage. These processes include the synthesis of pro- and antiinflammatory cytokines, leukocyte extravasation, vasogenic edema, and blood-brain-barrier breakdown. IL-1, IL-6, TNF and IL-8, which are increased after head injury, play a role in the cellular cascade of injury *(37)*. Many cytokines (IL-1, IL-2, IL-6, TNF-α and IFN-γ) induce a significant increase of astrogliosis.

3.7.1. Interleukin-1 (IL-1)

IL-1 β is a proinflammatory cytokine, produced by blood-borne and resident brain inflammatory cells. IL-1 level is regulated through the activity of the specific protease IL-1 β converting enzyme. This enzyme dependence is suggested by its ischemia-induction and by the reduced brain reaction when it is inhibited. IL-1 is produced by astrocytes *(38)*, by microglia *(39)*, by some neurons and by endothelial cells. IL-1 β is synthesized early after ischemia or after other brain damages and may play a role in postlesion recovery. After injury, IL-1 receptors are located on neurons and, at the site of gliosis, on astrocytes and microglia *(37a)*.

IL-1 is a potent mitogen for astrocytes in which it stimulates the production of cytokines such as NGF, PDGF, TNF-α and IL-6, which, in turn, will act on neurons and on microglial cells. Increase of IL-1 reduces ischemic and excitotoxic brain, however, deleterious effects have also been reported like exacerbation of ischemic brain damage, and blocking of IL-1 activity can be beneficial for ischemic brain edema. Thus, excessive production of IL-1 appears to mediate experimentally induced neurodegeneration in vivo, whereas neuroprotective effects of low concentrations of the cytokine suggest a dual role for IL-1 in neuronal survival *(40)*. An endogenous IL-1 receptor antagonist (IL-1ra) prevents binding of IL-1β to the signaling receptor. It was found to attenuate or inhibit neuronal damage in several types of injury. IL-1ra also suppresses NGF production in injured rat brain. The fact that IL-1ra inhibits neuronal damage in the rat confirms that endogenous IL-1 is a mediator of this type of damage. Thus, inhibitors of IL-1 action may be of therapeutic value in the treatment of acute or chronic neuronal death.

3.7.2. Interleukin-2 (IL-2)

IL-2 is an immunoregulatory cytokine, initially discovered for its mitogenic activity on T cells. It also acts on monocytes, resulting in the activation of cytokine production, superoxide production, and tumoricidal activity. IL-2 is present in injured rat brain, mainly in the lesion site, reaching a maximal activity at 10 d postlesion. LPS- or IL-2-activated microglia express IL-2 receptor beta-chain protein in culture *(41)*. IL-2 treatment results in an intracranial accumulation of T and B lymphocytes within the infused brain hemisphere. Adjacent brain regions were characterized by reactive astrogliosis, microglial activation, endothelial upregulation of adhesion molecules, myelin damage and neuronal loss. Severe brain damage is observed in IL-2-transgenic mice but IL-2 elicits some neurotrophic effects *(42)*.

3.7.3. Interleukin-3, -4, and -5 (IL-3, IL-4, IL-5)

IL-3 (multi-CSF) is a hematopoietic and immunomodulatory cytokine. It stimulates the proliferation and activation of microglia and can enhance differentiation of cholinergic and sensory neurons. Chronic CNS production of low levels of IL-3 promotes the recruitment, proliferation and activation of macrophage/microglial cells in white matter regions with consequent primary demyelination and motor disease *(43)*. Microglia produce IL-3 constitutively with the probable activation of an autocrine loop. Astrocytes too, synthesize IL-3. This factor prevents delayed neuronal death in the hippocampal CA1 field.

IL-4 is a cytokine which plays an important role in the function of various immunocompetent cells as well as in the pathophysiology of various CNS disorders. It inhibits astrocyte activation and induces NGF secretion *(44)*. IL-4 induces proliferation and activation of microglia.

IL-5 is a microglia mitogen which is secreted by both microglia and astrocytes in response to inflammatory stimuli. Thus, IL-5 may be involved in the cytokine-immune cascades leading to microglia proliferation in damaged areas.

3.7.4. Interleukin-6 (IL-6)

IL-6 is a proinflammatory cytokine with neuroprotective properties. In the intact brain, IL-6 is mainly localized in cholinergic and GABAergic neurons of the basal forebrain. After brain damage, there is a rapid and concomitant increased expression of IL-1 β *(45)*, IL-6 and their receptors *(46)*. Their predominantly neuronal localization in rat brain suggests a role for IL-6 in activating micro- and astroglial cells in response to degenerating cholinergic neurons. However, it has been reported that after traumatic injury, IL-1 and TNF-α induce IL-6 production by astrocytes *(47)*. IL-6 is mitogenic for astrocytes and it induces NGF production in these cells, suggesting that the IL-6 producing brain injury may likely contribute to the release of neurotrophic factors by astrocytes. Overexpression of IL-6 in astrocytes or in neurons (IL-6 transgenic mice) *(48)*, or of IL-6/soluble IL-6R alpha *(49)*, induce massive reactive astrogliosis and microgliosis *(48)*. Accelerated regeneration of the axotomized nerve observed in these transgenic mice suggests that the IL-6 signal may play an important role in nerve regeneration after trauma. In IL-6 knock-out mice, the initial inflammatory response to cortical injury is diminished *(50)*. These results emphasize the role of IL-6 in the global regulation of neurons, astrocytes, and microglia and their activation in the injured nervous system.

3.7.5. Interleukin-8, -12 (IL-8, IL-12)

IL-8, a neutrophil-activating cytokine that promotes invasion of these leukocytes into brain parenchyma, is not expressed in normal brains. It is expressed after reperfusion in brain ischemia and contributes to edema. IL-8 can be induced in astrocytes surrounding tumor lesion *(51)*, after acidosis, and in microglia, after LPS treatment *(52)*. IL-12, a proinflammatory cytokine which activates cellular immune response mechanisms, has a significantly elevated level in all patients in the course of 14 d after severe head trauma *(53)*. The upregulation of IL-12 in activated microglia depends on an autocrine loop of TNF-α.

3.7.6. Interleukin-10 (IL-10)

IL-10, a cytokine with antiinflammatory properties, negatively modulates proinflammatory cascades at multiple levels. It is a potent inhibitor of cytokine synthesis by microglia. After traumas, IL-10 can be found in the CNS and in the CSF *(54)*. Rapid monocytic IL-10 release after sympathetic activation may represent a common pathway for immunodepression induced by stress and injury. Acute stroke significantly increases CSF levels of IL-8 and IL-10 *(55)*. IL-10 and its receptor mRNA are enhanced in both microglia and astrocytes after stimulation of these cells with LPS. After traumatic brain injury, IL-10 treatment improves neurological recovery and significantly reduces TNF expression. This improvement may relate, in part, to reductions in proinflammatory cytokine synthesis like TNF-α, IL-12, and NO production by microglia *(56)*. Four days after local application of IL-10 to the site of corticectomy in adult mice, the number of reactive astrocytes and their state of hypertrophy were reduced (by 60%). In control mice, IL-10 induced a dose-dependent increase of NGF secretion in astrocytes.

3.7.7. TNF-α

The cytokine tumor necrosis factor (TNF-α) is a pleiotrophic polypeptide that plays a significant role in brain immune and inflammatory activities. TNF-α is produced in the brain in response to various pathological processes such as infectious agents, ischemia, and trauma *(37)*. TNF-α is present early in neuronal cells, in and around the ischemic tissue, and later in macrophages and in astrocytes *(57)*. Exposure of glial cells to the cerebrospinal fluid of AD patients was followed by TNF-α release *(58)*. TNF-α activates glial cells, thereby regulating tissue remodeling, gliosis, and scar formation. IL-1 β and TNF-α synergistically stimulate NGF release from cultured rat astrocytes. They downregulate the expression of GFAP. TNF-α, IL-1 and, to a lesser degree, IL-6, are mitogenic for astrocytes in vitro. However, TNF-α inhibits proliferation induced by other mitogens. Thus, TNF-α may be important in regulating the proliferative response of astrocytes during reactive astrogliosis in vivo *(59)*. Tumor necrosis factors protect neurons against metabolic-excitotoxic insults and promote maintenance of calcium homeostasis.

3.7.8. Interferon-gamma (IFN-γ)

IFN-γ, appear to have complex and broad-ranging actions in the CNS that may result in protection or injury. It is a potent microglia/macrophage activator. Although IFN-γ is induced after injury, it does not appear as perequisite for gliosis induction since in IFN-γ knockout adult mice, astrogliosis and increases of IL-1 α and TNF-α levels are induced rapidly by injury *(60)*. However, IFN-γ can induce gliosis and CNS demyelination *(61)*. IFN-γ specific transcripts are detected in rat astrocytes and are upregulated after treatment with IFN-γ itself. IFN-γ increases synthesis of NGF in reactive astrocytes but not in the normal adult astrocytes. It has been reported to inhibit the proliferation of rat cortical astrocytes and to reduce the levels of the ECM molecules. The production of IFN-γ in the CNS can be a two-edged sword that on the one hand confers protection against a lethal viral infection but on the other hand causes significant gliosis to the brain.

3.8. Leukemia Inhibitory Factor (LIF) and Oncostatin M (OSM)

LIF mRNA is increased 30-fold 6 h after surgical lesion of the cortex in adult rat brain, it reaches a peak at 24 h and returns to baseline after 7 d. LIF mRNA is induced in GFAP-positive astrocytes as well as in a small number of microglial cells. The striking induction of LIF transcripts in glia suggests that this cytokine may play a key injury-response role in the CNS as it does in the PNS, where LIF has been demonstrated to regulate neuropeptide expression both in vivo and in vitro *(62)*. In LIF knock-out mice, the microglial and astroglial responses to surgical injury of the cortex are slower compared to wild-type mice. LIF is a key regulator of neural injury in vivo, where it is produced by glia and can act directly on neurons, glia, and inflammatory cell *(50)*. Levels of OSM, LIF, and CNTF were all increased in the hippocampus after seizure, to different extents and with different time courses. The majority of LIF+ cells were astrocytes, while the majority of OSM+ cells had the morphology of interneurons *(63)*.

3.9. Ciliary Neurotrophic Factor (CNTF)

CNTF elicits trophic effects on ciliary, motor sympathetic, sensory, retinal and hippocampal neurons. It ameliorates axotomy-induced degeneration of CNS neurons and is upregulated at wound sites in the brain *(64)*. The increased level of CNTF mRNA in lesioned hippocampus is maximal by 3 d and is sustained for up to 20 d. Increased CNTF level in CNS recapitulates the glial response to CNS lesion with astrogliosis and appearance of activated microglia *(65)*. These data are consistent with a model whereby CNTF (which is synthesized and stored by astrocytes) would be released when the integrity of the astrocyte membrane is compromised, whereupon it would elicit an inflammatory response. CNTF has neurotrophic and neuroprotective effects, but more limited than those of other neurotrophic factors. It prevents medial septum neuron degeneration and promotes low affinity NGF receptor expression in the adult rat CNS. CNTF stimulates astroglial hypertrophy in vivo and in vitro and GFAP expression *(66,67)*.

3.10. Hepatocyte Growth Factor (HGF/SF)

In brain, HGF/SF is found in astrocytes and microglia, as well as in rare cortical neurons. Expression of both HGF and c-Met/HGF receptor was markedly induced in response to cerebral ischemic injury *(68)*. HGF prolonged survival of embryonic hippocampal neurons in primary culture. In vivo, it has a profound neuroprotective effect against postischemic delayed neuronal death in the hippocampus.

3.11. Thrombin

Thrombin is a protease/growth factor the effects of which are mediated by the protease activated receptors (PAR-1, 2, 3). This factor is liberated in the site of injury by the blood clotting process but it is also synthetized by nerve cells *(69)*. Its involvement in gliosis is suggested by various observations. Infusion in rat caudate nucleus caused infiltration of inflammatory cells, proliferation of mesenchymal cells, induction of angiogenesis, and an increase in vimentin-positive reactive astrocytes *(70)*. Thrombin inhibitors ameliorate secondary damage in rat brain injury by suppressing inflammatory cells and vimentin-positive astrocytes *(71,72)*. Thrombin activates cultured rodent microglia *(73)* and activates intracellular death protease pathways inducing apoptosis in model motor neurons *(74)*. It is an endogenous mediator of hippocampal neuroprotection against ischemia at low concentrations but causes degeneration at high concentrations *(75)*.

4. GROWTH FACTORS AND GLIOSIS IN THE AGING BRAIN

4.1. Alterations Linked to Naturally Occurring Gliosis

In aged rats, the volume of hippocampus was higher while that of frontal cortex was unchanged. Various brain tissue alterations have been detected by the use of magnetic resonance imaging (MRI). With advancing age, the periventricular and subcortical white matter becomes susceptible to a heterogeneous assortment of such alterations that cannot be easily categorized in terms of traditionally defined neuropathologic disease. Frequently observed microscopic changes include dilated perivascular (Virchow-Robin) spaces, mild demyelination, gliosis, and diffuse regions of neuropil vacuolation. Associated clinical abnormalities are usually confined to deficits of attention, mental processing speed, and psychomotor control. Occasionally, histologically severe white matter lesions may occur that result in dementia and focal neurologic impairment. These lesions are characterized by extensive arteriosclerosis, diffuse white matter necrosis, and lacunar infarction *(76)*. Some results suggest that in aging, gliosis is secondary to the neurodegenerative changes. Synaptic loss is likely to be a common pathogenetic feature of aging and neurodegenerative disorders and a likely cause of clinical symptoms *(77)*.

4.2. Microglia

With aging, microglial cells grow more numerous, become activated and may enter the phagocytic or reactive stage. They also become richer in iron and ferritin and exhibit phenotypic alteration, e. g., the expression of MHC class II antigens that are not ordinarily demonstrable immunohistochemically in the resting state *(78)* and an increase in TGF-β 1 level. Microglial cells are prominently involved in such pathologic processes as the acquired immunodeficiency syndrome, multiple sclerosis, prion diseases, and the degenerative disorders, e.g., Alzheimer's disease and Parkinson's disease.

4.3. *Astrocytes*

The aging process increases astrocyte and pericyte populations in the rat parietal cortex *(79)*. The predominant change is glial activation characterized first by GFAP expression. In human, GFAP level increases dramatically after the age of 65 yr, and more especially in the hippocampal formation. This glial reaction was observed in aged patients that do not show cognitive impairment and the neuropathological hallmarks of Alzheimer's disease *(80)*. GFAP is a unique example of a gene that shows increased expression during aging. Astrocytosis and production of APP-derived fragments occur markedly in senescence-accelerated mice (SAMP-8) brains. During aging, astrocytes also exhibit signs of metabolic activation, an hypertrophy of their cell body and processes and the extension of these astrocytic processes around neurons. Tissue levels of S100 β as well as the number of S100 β-immunoreactive astrocytes, increased with advancing age. In hippocampus, astrocytes loose their radial organization.

Age has a differential effect on astrocytic and microglial hyperactivity in gray vs white matter areas. This mosaic of glial aging suggests that multiple mechanisms are at work during aging *(81)*. The regional heterogeneity of aging changes has been reported by many authors. For example, glial expression of GFAP, apolipoprotein E, apolipoprotein J (clusterin), heme oxygenase-1, complement 3 receptor (OX42), MHC class II and TGF-β 1 were elevated in the corpus callosum during aging but other regions showed marked dissociation of the extent and direction of changes *(81)*. The change in glial volume and ECM observed in old animals, results in a decrease in the apparent diffusion coefficient of neuroactive substances. This could lead to the impairment of the diffusion of neuroactive substances, extrasynaptic transmission, "crosstalk" between synapses and neuron-glia communication.

4.4. *Origin of Age-Related Gliosis*

The origin of age-related gliosis is not known but since gliosis is supposed to be triggered by neural lesions, the presence of such lesions in patients over 60 yr has been investigated by using MRI and microscopic observation *(82)*. Periventricular lesions graded as moderate or severe were found in 10% of the patients in the age group between 60 and 69 yr, and in 50% between 80 and 89 yr. The presence or absence of periventricular lesions on MRI correlated well with the severity of demyelination and astrocytic gliosis. Demyelination was always associated with an increased ratio between wall thickness and external diameter of arterioles (up to 150 microns). A variable degree of axonal loss in Bodian-stained sections was present in the white matter of all brains with demyelination. Dilated perivascular spaces were found; their presence correlated strongly with corrected brain weight, but incompletely with demyelination and arteriolosclerosis. The observations suggest that arteriolosclerosis is the primary factor in the pathogenesis of diffuse white matter lesions in the elderly. This is soon followed by demyelination and loss of axons, and only later by dilatation of perivascular spaces *(82)*. Some other histological changes appear to be universal in aged human brains. These include increasing numbers of corpora amylacea within astrocytic processes near blood-brain or cerebrospinal fluid-brain interfaces, accumulation of the "aging" pigment lipofuscin in all brain regions, and appearance of neurofibrillary tangles (but not necessarily amyloid plaques) in mesial temporal structures.

An increasing body of evidence indicates that aging-related impairments of nervous functions are caused by loss of neurons. It has been found that about half of the motoneuron somata examined by electron microscopy, from aged Sprague-Dawley rats, had a reduced (50%) bouton coverage, which seemed to be caused by a smaller number of axosomatic bouton profiles. Long stretches of the cell body plasma membrane were apposed by palc processes, and immunolabeling for GFAP disclosed that a number of the aged motoneurons appeared embedded in GFAP immunopositive processes. Astrogliosis is also observed in the spinal cord of old rats *(83)*.

Two mutant mice have been compared, SAM-P8 characterized as mice in which aging is accelerated with occurrence of memory disturbances and the senescence-resistant (SAM-R1) mice. The concentrations of the β-subunit of NGF in the hippocampus were reduced at 8-mo-of-age in both mutants but the decrease was more conspicuous in SAM-R1. In the olfactory bulbs, the β-NGF level was already higher in SAM-P8 than in SAM-R1, at 2-mo-of-age. The acceleration of age-related increase was apparent in the levels of S100 β and Mn-SOD in the cerebral cortex from SAM-P8. In these last mutants, a fibrous gliosis was present at quite an early age in selected regions of the brain *(84)*. In contrast, in aged rats, there is a reduced content of NGF in the frontal and parietal cortices. Repeated oral administration of the stimulators of the NGF synthesis, idebenone and propentofylline, produced a significant recovery of the reduced NGF content *(85)*.

4.5. Inhibition of Gliosis

In enriched environment conditions, there are less astrocytes and they are less hypertrophied than in naive animals, GFAP level is also lowered *(86)*. In general, the effects of steroids oppose the effects of aging. Recent data indicate that steroid treatment can decrease the expression of GFAP in the aged brain, yet GFAP is resistant to downregulation by endogenous glucocorticoids *(87)*.

4.6. Gliosis and Release of Trophic Factors after Brain Lesion in the Aging Rat

To our knowledge there are few data about gliosis following damage in aged animals. As seen before, diffuse glial reaction occurs naturally in aging brain, but most studies deal with gliosis which is associated to age-related diseases or which is artificially induced.

4.6.1. Lesions Induced in Normal Animals

In the deafferented neostriatum produced by unilateral 6-hydroxydopamine injection in 6-, 15-, or 24-mo-old-rats, the time-dependent induction of GFAP was larger and persisted longer in the aged rats. The response of middle-aged rats was intermediate. After stab wounds in the neostriatum, there were substantially larger GFAP inductions than after deafferentation, but fewer effects of age. In both lesion paradigms, GFAP staining increased in the contralateral striatum of old rats, but not in young rats.

The consistency of the exaggerated glial reactivity in the aged brain after modest injury involving either the nigrostriatal system or the septohippocampal one suggests some remarks. It appears that aged astrocytes are more sensitive to gliotrophic factors released by terminal degeneration. Larger quantities of such factors are produced after injury and their clearance is delayed in old rodents and/or aged astrocytes are less able to terminate GFAP inductions after activation *(88)*.

The Cotman group performed deafferentation of the hippocampus by ablation of the entorhinal/occipital cortex in young and aged rats. These last rats in which reactive sprouting is deficient showed no increase in neurite-promoting activity by 12 d after hippocampal deafferentation while young animals had a several fold increase of this activity between 9 and 15 d postlesion *(89)*. In middle-aged and aged-rats, deafferentation of dentate gyrus induces reactive axonal growth by surviving afferent systems, which is delayed and reduced relative to young adults *(90)*. The cause for this age-related decline may reflect a decrement in trophic signals which initiate sprouting. IGF-1 may play a role in sprouting because it promotes axonal growth and it is expressed at elevated levels by microglia just prior to sprouting onset. It is also expressed with better spatiotemporal correspondence to hippocampal sprouting than other trophic factors. Messenger RNA levels of IGF-1 were markedly elevated 4 d after an entorhinal cortex lesion performed at 3-mo-of-age and there was only a modest increase 8 d after the lesion performed at 18–26-mo-of-age. These results support the association between IGF-1 expression impairment in microglia and reduced axonal plasticity with age.

Following a unilateral ibotenic acid lesion of the nucleus magnocellularis (NBM) in 3- and 15-mo-old rats, there was a loss of NBM choline acetyltransferase-(ChAT)-positive cells, a decrease in cortical ChAT activity and impairment of the acquisition of avoidance response. When NGF was administered with GM1 ganglioside immediately after surgery, there was an attenuation of the lesion-induced changes, however, GM1 had no effect on ChAT activity decrease and behavioral impairment in 15-mo-old rats. The age effect on gliosis resulting from brain lesion has characteristics which depend on the type of lesion (mechanical or chemical) and on the cerebral structure concerned. After total immunolesion resulting from intraventricular injections of the immunotoxin, 192 IgG-saporin in aged rats, there was an increase of microglia but not astrocytes in the subnuclei of basal forebrain. This immunotoxin was effective in young and aged rats, in producing cholinergic lesions. NGF protein levels were significantly increased in the hippocampus, cortex, and olfactory bulb of the young rats but not of the aged rats, except for small increases in the olfactory bulb. The total immunolesions had no effects on NGF and BDNF mRNA levels in the hippocampus and cortex *(91)*. No difference was found between normal young adult (2–5 mo) and aged (24 mo) rats in terms of hippocampal NGF levels. However, following medial septal lesions, only young rats demonstrated significant increases in hippocampal NGF-like activity *(92)*.

4.6.2. Lesions Occurring in Transgenic Animals

Homozygous beta amyloid precursor protein (APP)-deficient mice are viable and fertile. However, the mutant animals weighed 15%–20% less than age-matched wild-type controls. Neurological evaluation showed that the APP-deficient mice exhibited a decreased locomotor activity and forelimb grip strength, indicating a compromised neuronal or muscular function. In addition, four out of six homozygous mice showed reactive gliosis at 14-wk-of-age *(93)*. These null mice also had marked reactive gliosis in many areas, especially in cortex and hippocampus throughout the CA1 region *(11)*. It is not known if the impaired synaptic plasticity and long term potentiation are related to gliosis. Normal β-APP may serve an essential role in the maintenance of synaptic function during ageing. Alteration of this function may contribute to the progression of the memory decline and the neurodegenerative changes seen in Alzheimer's disease. Elevated GFAP levels have been reported in transgenic mice expressing the gene for bovine growth hormone (bGH) with increased body size, reduced reproductive capacity, and high basal levels of several hormones including corticosterone *(94)*. This may reflect increased neural damage due to accelerated aging processes or damage associated with high circulating levels of bGH or corticosterone.

Alternatively, this increased expression of GFAP may reflect altered regulation of GFAP rather than an increased signal by neural damage. Young glycosyl asparaginase-deficient mice demonstrate many pathological changes found in human aspartylglycosaminuria patients. After the age of 10 mo, the general condition of null mutant mice gradually deteriorate, with a progressive motoric impairment and impaired bladder function until premature death. A widespread lysosomal hypertrophy in CNS was detected. The oldest animals (20-mo-old) display a clear neuronal loss and gliosis, particularly in brain regions with the most severe vacuolation. The severe ataxic gait of the older mice is likely due to the dramatic loss of Purkinje cells, intensive astrogliosis, and vacuolation of neurons in the deep cerebellar nuclei, and the severe vacuolation of the cells in vestibular and cochlear nuclei *(95)*.

4.6.3. Administration of Growth Factors

There is a reduced content of NGF in the frontal and parietal cortices of aged rats. Repeated oral administration of the stimulators of NGF synthesis, idebenone, and propentofylline, produced a significant recovery of the reduced NGF content *(85)*. With trimethylquinone derivative, they induced an increase in NGF protein and mRNA, and in choline acetyltransferase activity, in basal forebrain-lesioned and aged rats, but not in intact young rats. Many authors reported the reparative effects of NGF with survival of the cholinergic neurons and regrowth of damaged fibers after fimbria fornix lesion in aged rats. Atrophic cholinergic neurons in aged animals were similarly responsive to NGF treatment, like those in the young animals. The responses of basal forebrain cholinergic neurons to NGF treatment varied with time after the lesion and imply that NGF administration could promote the collateral sprouting from spared cholinergic fibers after the lesion in the septohippocampal system and forebrain. A role for glial cells was proposed to clarify how NGF availability may be regulated during the degenerative and regenerative events. Concerning the neostriatum, it has been reported that neostriatal NGF binding sites and intrinsic cholinergic neurons are co-localized and are lost to an equal extent following age- or injury-induced loss of neostriatal neurons *(95A)*. It appears that NGF has potential use as a cholinergic "neurotrophic-factor therapy." Stimulation of glucocorticoid or beta-adrenergic receptors by dexamethasone and clenbuterol, respectively, has been shown to increase NGF mRNA levels in adult rat brain. A more modest increase was observed for the aged rats. The variation throughout life, of induction of NGF expression by neurotransmitter/hormone receptor activation suggests that pharmacological agents might be useful tools to enhance trophic support in aging. Concerning the NGF-receptor, a reduced immunoreactivity was found within the nucleus basalis of Meynert (NBM) neurons in aging (18–24 mo) rats *(96)*.

4.7. Growth Factors in Gliosis Resulting from Age-Related Diseases

Aging is accompanied by increases in glial cell activation, in oxidative damage to proteins and lipids, in irreversible protein glycation, and in damage to DNA, and such changes may underlie in part the age-associated increasing incidence of degenerative conditions such as Alzheimer's disease (AD) and Parkinson's disease. Then the degenerative diseases can induce gliosis. Aging is also characterized by an accumulation of lipofuscin pigments, formation of paired filaments and increased immunoreactivity to microtubule-assembling protein tau occurring in paired filaments of neurofibrillary tangles, beta-amyloid proteins, and ubiquitin *(97)*.

Astrogliosis is observed in AD *(98)* and it appears that the neurons and their processes although not exclusively may be the site of EGF-R immunoreactivity. An EGF/EGF-R system within the CNS may play an important part in scar formation in response to neuronal injury and death occurring in AD and also in aging. It may also function as a trophic factor important in axonal or dendritic sprouting. FGF-2 is increased in the brains of AD patients. Furthermore, FGF-2 is not distributed in an identical fashion to normal and AD brains, but is found in association with the lesions that characterize this disease. PDGF is associated with neuronal and glial alterations of AD. Antibodies against PDGF-BB (but not PDGF-AA) recognized the neurofibrillary alterations of this disease. The levels of PDGF-BB correlated with the patterns of synaptic loss and sprouting while PDGF-AA immunostaining of the vessels was correlated with glial proliferation. PDGF protein was mildly increased in AD. These data suggest that PDGF, as well as other neurotrophic factors, play an important role in the mechanisms of neurofibrillary pathology. TGF-β 1 is in senile plaques in AD and limited to neuritic profiles within them. TGF-β 2 is seen in neuronal neurofibrillary tangles, plaque neurites, microglia, astrocytes and macrophages. TGF-β 3 is restricted to Hirano bodies. Such expression is not observed in every gliosis since in infarction, reactive astrocytes and macrophages express the three isoforms. The localization of TGF-β isotypes to the lesions of AD supports the hypothesis that these cytokines may influence lesion expression. Overexpression of TGF-β 1 may initiate or promote amyloidogenesis in AD *(99)* and could be related to its capacity to induce ECM component production which could promote the formation of amyloid plaques *(100)*. It has been shown that TGF-β can increase amyloid-beta protein immunoreactive plaque-like deposits in rat brain. This effect could be exerted by astrocytes in which TGF-β 1 elevates APP mRNA level, and also increases the half-life of APP message by at least fivefold *(101)*. Finally, induction of TGF-β 2 which occurs in various degenerative diseases, may be an intrinsic part of the processes that induce neurofibrillar tangles formation and reactive gliosis.

Gliosis is also detected in vascular dementia. In Creutzfeldt-Jakob disease, there is a diffuse astrogliosis determined by cell hypertrophy *(102)* and elevation of GFAP in the later stages of disease. Microgliosis is also observed in the grey matter where astrogliosis and prion protein (PrP) deposits were prominent. Activated microglia may contribute to neuronal damage due to their cytotoxic potential *(103)*. In areas where spongiform degeneration affected the entire depth of the cortex, microglia exhibited atypical, tortuous cell processes, and occasionally intracytoplasmic vacuoles, suggesting that microglia themselves may become a disease target.

Following bacterial or viral infections, toxic products such as bacterial LPS circulate to the brain, where they induce IL-1 and iNOS mRNA synthesis. After several hours delay, massive quantities of NO are released. The large amounts of NO released by iNOS may well produce death of not only neurons but also glial cells. NO stimulates glutamic acid release, which by N-methyl-D- aspartate (NMDA) receptors causes release of NO leading to neuronal cell death. Repeated bouts of systemic infection could result in induction of iNOS in the CNS and lead to large fall out of neurons in hippocampus to impair mcmory, in hypothalamus to decrease fever, and neuroendocrine response to infection, and could play a role in the pathogenesis of degenerative neuronal diseases of aging, such as AD. The damage to the pituitary could also impair responses to stress and infection, and the release of NO during infection could be responsible for the degenerative changes in the pineal and diminished release of melatonin, an antioxident, and consequently, an antiaging hormone, that occur with age*(104)*.

REFERENCES

1. Balasingam, V., Dickson, K., Brade, A., and Yong, V.W. (1996) Astrocyte reactivity in neonatal mice: apparent dependence on the presence of reactive microglia/macrophages. *Glia* **18,** 11–26.
2. Streit, W.J. (2000) Microglial response to brain injury: a brief synopsis. *Toxicol Pathol.* **28,** 28–30.
3. Herpers, M.J., Ramaekers, F.C., Aldeweireldt, J., Moesker, O., and Slooff, J. (1986) Co-expression of glial fibrillary acidic protein- and vimentin-type intermediate filaments in human astrocytomas. *Acta Neuropathol.* **70,** 333–339.
4. Cerutti, S.M. and Chadi, G. (2000) S100 immunoreactivity is increased in reactive astrocytes of the visual pathways following a mechanical lesion of the rat occipital cortex. *Cell Biol. Int.* **24,** 35–49.
5. Johansson, C.B., Momma, S., Clarke, D.L., Risling, M., Lendahl, U., and Frisen, J. (1999) Identification of a neural stem cell in the adult mammalian central nervous system. *Cell* **96,** 25–34.
6. Seo, M.S., Okamoto, N., Vinores, M.A., et al. (2000) Photoreceptor-specific expression of platelet-derived growth factor-B results in traction retinal detachment. *Am. J. Pathol.* **157,** 995–1005.
7. Eriksson, C., Winblad, B., and Schultzberg, M. (1998) Kainic acid induced expression of interleukin-1 receptor antagonist mRNA in the rat brain. *Mol. Brain Res.* **58,** 195–208.
8. Ballabriga, J., Pozas, E., Planas, A.M., and Ferrer, I. (1997) bFGF and FGFR-3 immunoreactivity in the rat brain following systemic kainic acid administration at convulsant doses: localization of bFGF and FGFR-3 in reactive astrocytes, and FGFR-3 in reactive microglia. *Brain Res.* **752,** 315–318.
9. Logan, A., Frautschy, S.A., Gonzalez, A.M., and Baird, A. (1992) A time course for the focal elevation of synthesis of basic fibroblast growth factor and one of its high-affinity receptors (flg) following a localized cortical brain injury. *J. Neurosci.* **12,** 3828–3837.
10. Barotte, C., Eclancher, F., Ebel, A., Labourdette, G., Sensenbrenner, M., and Will, B. (1989) Effects of basic fibroblast growth factor (bFGF) on choline acetyltransferase activity and astroglial reaction in adult rats after partial fimbria transection. *Neurosci. Lett.* **101,** 197–202.
11. Dawson, G.R., Seabrook, G.R., Zheng, H., et al. (1999) Age-related cognitive deficits, impaired long-term potentiation and reduction in synaptic marker density in mice lacking the beta-amyloid precursor protein. *Neuroscience* **90,** 1–13.

11a. Eclancher, F., Perraud, F., Faltin, J., Labourdette, G., and Sensenbrenner, M. (1990) Reactive astrogliosis after basic fibroblast growth factor (bFGF) injection in injured neonatal rat brain. *Glia* **3,** 502–509.

12. Eclancher, F., Kehrli, P., Labourdette, G., and Sensenbrenner, M. (1996) Basic fibroblast growth factor (bFGF) injection activates the glial reaction in the injured adult rat brain. *Brain Res.* **737,** 201–214.

13. Rabchevsky, A.G., Fugaccia, I., Turner, A.F., Blades, D.A., Mattson, M.P., and Scheff, S.W. (2000) Basic fibroblast growth factor (bFGF) enhances functional recovery following severe spinal cord injury to the rat. *Exp. Neurol.* **164,** 280–291.

14. Shults, C.W., Ray, J., Tsuboi, K., and Gage, F.H. (2000) Fibroblast growth factor-2-producing fibroblasts protect the nigrostriatal dopaminergic system from 6-hydroxydopamine. *Brain Res.* **883,** 192–204.

15. Takayama, S., Sasahara, M., Iihara, K., Handa, J., and Hazama, F. (1994) Platelet-derived growth factor B-chain-like immunoreactivity in injured rat brain. *Brain Res.* **653,** 131–140.

16. Liu, H.M., Yang, H.B., and Chen, R.M. (1994) Expression of basic fibroblast growth factor, nerve growth factor, platelet-derived growth factor and transforming growth factor-beta in human brain abscess. *Acta Neuropathol.* **88,** 143–150.

17. Yamada, H., Yamada, E., Ando, A., et al. (2000) Platelet-derived growth factor-A-induced retinal gliosis protects against ischemic retinopathy. *Am. J. Pathol.* **156,** 477–487.

18. Lindholm, D., Castren, E., Kiefer, R., Zafra, F., and Thoenen, H. (1992) Transforming growth factor-beta 1 in the rat brain: increase after injury and inhibition of astrocyte proliferation. *J. Cell. Biol.* **117,** 395–400.

19. Ata, K.A., Lennmyr, F., Funa, K., Olsson, Y., and Terent, A. (1999) Expression of transforming growth factor-betal, 2, 3 isoforms and type I and II receptors in acute focal cerebral ischemia: an immunohistochemical study in rat after transient and permanent occlusion of middle cerebral artery. *Acta Neuropathol. (Berl)* **97,** 447–455.

20. Wyss-Coray, T., Lin, C., von Euw, D., Masliah, E., Mucke, L., and Lacombe, P. (2000) Alzheimer's disease-like cerebrovascular pathology in transforming growth factor-beta 1 transgenic mice and functional metabolic correlates. *Ann N Y Acad Sci* **903,** 317–323.

21. Logan, A., Green, J., Hunter, A., Jackson, R., and Berry, M. (1999) Inhibition of glial scarring in the injured rat brain by a recombinant human monoclonal antibody to transforming growth factor-beta 2. *Eur. J. Neurosci.* **11,** 2367–2374.

22. Rabchevsky, A.G., Weinitz, J.M., Coulpier, M., Fages, C., Tinel, M., and Junier, M.P. (1998) A role for transforming growth factor alpha as an inducer of astrogliosis. *J. Neurosci.* **18,** 10541–10552.

23. Nolte, C., Kirchhoff, F., and Kettenmann, H. (1997) Epidermal growth factor is a motility factor for microglial cells in vitro: evidence for EGF receptor expression. *Eur. J. Neurosci.* **9,** 1690–1698.

24. Xian, C.J. and Zhou, X.F. (2000) Roles of transforming growth factor-alpha and related molecules in the nervous system. *Mol Neurobiol* **20,** 157–183.

25. Gluckman, P., Klempt, N., Guan, J., et al. (1992) A role for IGF-1 in the rescue of CNS neurons following hypoxic- ischemic injury. *Biochem. Biophys Res. Commun.* **182,** 593–599.

26. Sharma, H.S., Nyberg, F., Westman, J., Alm, P., Gordh, T., and Lindholm, D. (1998) Brain derived neurotrophic factor and insulin like growth factor-1 attenuate up-regulation of nitric oxide synthase and cell injury following trauma to the spinal cord. An immunohistochemical study in the rat. *Amino Acids* **14,** 121–129.

27. Beilharz, E.J., Bassett, N.S., Sirimanne, E.S., Williams, C.E., and Gluckman, P.D. (1995) Insulin-like growth factor II is induced during wound repair following hypoxic-ischemic injury in the developing rat brain. *Mol. Brain Res.* **29,** 81–91.

28. Wilczak, N., De Bleser, P., Luiten, P., Geerts, A., Teelken, A., and De Keyser, J. (2000) Insulin-like growth factor II receptors in human brain and their absence in astrogliotic plaques in multiple sclerosis. *Brain Res* **863,** 282–8.
29. Ridet, J.L., Malhotra, S.K., Privat, A., and Gage, F.H. (1997) Reactive astrocytes: cellular and molecular cues to biological function [published erratum appears in Trends Neurosci 1998 Feb;21(2):80]. *Trends Neurosci* **20,** 570–577.
30. Gottlieb, M. and Matute, C. (1999) Expression of nerve growth factor in astrocytes of the hippocampal CA1 area following transient forebrain ischemia. *Neuroscience* **91,** 1027–1034.
31. Oyesiku, N.M., Evans, C.O., Houston, S., et al. (1999) Regional changes in the expression of neurotrophic factors and their receptors following acute traumatic brain injury in the adult rat brain. *Brain Res.* **833,** 161–172.
32. Hicks, R.R., Li, C., Zhang, L., Dhillon, H.S., Prasad, M.R., and Seroogy, K.B. (1999) Alterations in BDNF and trkB mRNA levels in the cerebral cortex following experimental brain trauma in rats. *J. Neurotrauma* **16,** 501–510.
33. Goutan, E., Marti, E. and Ferrer, I. (1998) BDNF, and full length and truncated TrkB expression in the hippocampus of the rat following kainic acid excitotoxic damage. Evidence of complex time-dependent and cell-specific responses. *Mol. Brain Res.* **59,** 154–164.
34. Hayashi, M., Ueyama, T., Tamaki, T., and Senba, E. (1997) [Expression of neurotrophin and IL-1 beta mRNAs following spinal cord injury and the effects of methylprednisolone treatment]. *Kaibogaku Zasshi* **72,** 209–213.
35. Hicks, R.R., Numan, S., Dhillon, H.S., Prasad, M.R., and Seroogy, K.B. (1997) Alterations in BDNF and NT-3 mRNAs in rat hippocampus after experimental brain trauma. *Mol. Brain Res.* **48,** 401–406.
36. Kitagawa, H., Sasaki, C., Zhang, W.R., et al. (1999) Induction of glial cell line-derived neurotrophic factor receptor proteins in cerebral cortex and striatum after permanent middle cerebral artery occlusion in rats. *Brain Res.* **834,** 190–195.
37. Balasingam, V., Tejada-Berges, T., Wright, E., Bouckova, R., and Yong, V.W. (1994) Reactive astrogliosis in the neonatal mouse brain and its modulation by cytokines. *J. Neurosci.* **14,** 846–856.
37a. Finklestein, S.P., Apostolides, P.J., Caday, C.G., Prosser, J., Philips, M.F., and Klagsbrun, M. (1988) Increased basic fibroblast growth factor (bFGF) immunoreactivity at the site of focal brain wounds. *Brain Res.* **460,** 253–259.
38. Nieto-Sampedro, M. and Berman, M.A. (1987) Interleukin-1-like activity in rat brain: sources, targets, and effect of injury. *J. Neurosci. Res.* **17,** 214–219.
39. Woodroofe, M.N., Sarna, G.S., Wadhwa, M., et al. (1991) Detection of interleukin-1 and interleukin-6 in adult rat brain, following mechanical injury, by in vivo microdialysis: evidence of a role for microglia in cytokine production. *J Neuroimmunol* **33,** 227–236.
40. Rothwell, N.J. and Hopkins, S.J. (1995) Cytokines and the nervous system II: Actions and mechanisms of action [see comments]. *Trends Neurosci.* **18,** 130–136.
41. Sawada, M., Suzumura, A., and Marunouchi, T. (1995) Induction of functional interleukin-2 receptor in mouse microglia. *J. Neurochem.* **64,** 1973–1979.
42. Sarder, M., Abe, K., Saito, H., and Nishiyama, N. (1996) Comparative effect of IL-2 and IL-6 on morphology of cultured hippocampal neurons from fetal rat brain. *Brain Res.* **715,** 9–16.
43. Chiang, C.S., Powell, H.C., Gold, L.H., Samimi, A., and Campbell, I.L. (1996) Macrophage/microglial-mediated primary demyelination and motor disease induced by the central nervous system production of interleukin-3 in transgenic mice. *J. Clin. Invest.* **97,** 1512–1524.
44. Brodie, C., Goldreich, N., Haiman, T., and Kazimirsky, G. (1998) Functional IL-4 receptors on mouse astrocytes: IL-4 inhibits astrocyte activation and induces NGF secretion. *J. Neuroimmunol* **81,** 20–30.

45. Streit, W.J., Hurley, S.D., McGraw, T.S., and Semple-Rowland, S.L. (2000) Comparative evaluation of cytokine profiles and reactive gliosis supports a critical role for interleukin-6 in neuron-glia signaling during regeneration. *J. Neurosci. Res.* **61,** 10–20.
46. Loddick, S.A., Turnbull, A.V., and Rothwell, N.J. (1998) Cerebral interleukin-6 is neuroprotective during permanent focal cerebral ischemia in the rat. *J. Cereb. Blood Flow Metab.* **18,** 176–179.
47. Hariri, R.J., Chang, V.A., Barie, P.S., Wang, R.S., Sharif, S.F., and Ghajar, J.B. (1994) Traumatic injury induces interleukin-6 production by human astrocytes. *Brain Res.* **636,** 139–142.
48. Castelnau, P.A., Garrett, R.S., Palinski, W., Witztum, J.L., Campbell, I.L., and Powell, H.C. (1998) Abnormal iron deposition associated with lipid peroxidation in transgenic mice expressing interleukin-6 in the brain. *J. Neuropathol. Exp. Neurol.* **57,** 268–382.
49. Brunello, A.G., Weissenberger, J., Kappeler, A., et al. (2000) Astrocytic alterations in interleukin-6/Soluble interleukin-6 receptor alpha double-transgenic mice. *Am. J. Pathol.* **157,** 1485–1493.
50. Sugiura, S., Lahav, R., Han, J., et al. (2000) Leukaemia inhibitory factor is required for normal inflammatory responses to injury in the peripheral and central nervous systems in vivo and is chemotactic for macrophages in vitro. *Eur. J. Neurosci.* **12,** 457–466.
51. Nitta, T., Allegretta, M., Okumura, K., Sato, K., and Steinman, L. (1992) Neoplastic and reactive human astrocytes express interleukin-8 gene. *Neurosurg. Rev.* **15,** 203–207.
52. Ehrlich, L.C., Hu, S., Sheng, W.S., et al. (1998) Cytokine regulation of human microglial cell IL-8 production. *J. Immunol.* **160,** 1944–1948.
53. Stahel, P.F., Kossmann, T., Joller, H., Trentz, O., and Morganti-Kossmann, M.C. (1998) Increased interleukin-12 levels in human cerebrospinal fluid following severe head trauma. *Neurosci. Lett.* **249,** 123–126.
54. Bell, M.J., Kochanek, P.M., Doughty, L.A., et al. (1997) Interleukin-6 and interleukin-10 in cerebrospinal fluid after severe traumatic brain injury in children. *J. Neurotrauma* **14,** 451–457.
55. Tarkowski, E., Rosengren, L., Blomstrand, C., et al., (1997) Intrathecal release of pro- and anti-inflammatory cytokines during stroke. *Clin. Exp. Immunol.* **110,** 492–499.
56. Knoblach, S.M. and Faden, A.I. (1998) Interleukin-10 improves outcome and alters proinflammatory cytokine expression after experimental traumatic brain injury. *Exp. Neurol.* **153,** 143–151.
57. Chung, I.Y. and Benveniste, E.N. (1990) Tumor necrosis factor-alpha production by astrocytes. Induction by lipopolysaccharide, IFN-gamma, and IL-1 beta. *J. Immunol.* **144,** 2999–3007.
58. Viviani, B., Corsini, E., Galli, C.L., Padovani, A., Ciusani, E., and Marinovich, M. (2000) Dying neural cells activate glia through the release of a protease product. *Glia* **32,** 84–90.
59. Tejada-Berges, T. and Yong, V.W. (1994) The astrocyte mitogen, tumor necrosis factor-alpha, inhibits the proliferative effect of more potent adult human astrocyte mitogens, gamma-interferon and activated T-lymphocyte supernatants. *Brain Res.* **653,** 297–304.
60. Bruccoleri, A., Brown, H., and Harry, G.J. (1998) Cellular localization and temporal elevation of tumor necrosis factor-alpha, interleukin-1 alpha, and transforming growth factor-beta 1 mRNA in hippocampal injury response induced by trimethyltin. *J. Neurochem.* **71,** 1577–1587.
61. Horwitz, M.S., Evans, C.F., McGavern, D.B., Rodriguez, M., and Oldstone, M.B. (1997) Primary demyelination in transgenic mice expressing interferon-gamma. *Nat. Med.* **3,** 1037–1041.
62. Banner, L.R., Moayeri, N.N., and Patterson, P.H. (1997) Leukemia inhibitory factor is expressed in astrocytes following cortical brain injury. *Exp. Neurol.* **147,** 1–9.
63. Jankowsky, J.L. and Patterson, P.H. (1999) Differential regulation of cytokine expression following pilocarpine-induced seizure. *Exp. Neurol.* **159,** 333–346.
64. Lin, T.N., Wang, P.Y., Chi, S.I., and Kuo, J.S. (1998) Differential regulation of ciliary neurotrophic factor (CNTF) and CNTF receptor alpha (CNTFR alpha) expression following focal cerebral ischemia. *Brain Res. Mol. Brain Res.* **55,** 71–80.

65. Kahn, M.A., Ellison, J.A., Speight, G.J., and de Vellis, J. (1995) CNTF regulation of astrogliosis and the activation of microglia in the developing rat central nervous system. *Brain Res.* **685,** 55–67.
66. Kahn, M.A., Ellison, J.A., Chang, R.P., Speight, G.J., and de Vellis, J. (1997) CNTF induces GFAP in a S-100 alpha brain cell population: the pattern of CNTF-alpha R suggests an indirect mode of action. *Dev. Brain Res.* **98,** 221–233.
67. Kahn, M.A., Huang, C.J., Caruso, A., et al. (1997) Ciliary neurotrophic factor activates JAK/Stat signal transduction cascade and induces transcriptional expression of glial fibrillary acidic protein in glial cells. *J. Neurochem.* **68,** 1413–1423.
68. Honda, S., Kagoshima, M., Wanaka, A., Tohyama, M., Matsumoto, K., and Nakamura, T. (1995) Localization and functional coupling of HGF and c-Met/HGF receptor in rat brain: implication as neurotrophic factor. *Mol. Brain Res.* **32,** 197–210.
69. Dihanich, M., Kaser, M., Reinhard, E., Cunningham, D., and Monard, D. (1991) Prothrombin mRNA is expressed by cells of the nervous system. *Neuron* **6,** 575–581.
70. Nishino, A., Suzuki, M., Ohtani, H., et al. (1993) Thrombin may contribute to the pathophysiology of central nervous system injury. *J. Neurotrauma* **10,** 167–179.
71. Kubo, Y., Suzuki, M., Kudo, A., et al. (2000) Thrombin inhibitor ameliorates secondary damage in rat brain injury: suppression of inflammatory cells and vimentin-positive astrocytes. *J. Neurotrauma* **17,** 163–172.
72. Motohashi, O., Suzuki, M., Shida, N., Umezawa, K., Sugai, K., and Yoshimoto, T. (1997) Hirudin suppresses the invasion of inflammatory cells and the appearance of vimentin-positive astrocytes in the rat cerebral ablation model. *J. Neurotrauma* **14,** 747–754.
73. Moller, T., Hanisch, U.K., and Ransom, B.R. (2000) Thrombin-induced activation of cultured rodent microglia. *J. Neurochem.* **75,** 1539–1547.
74. Smirnova, I.V., Zhang, S.X., Citron, B.A., Arnold, P.M., and Festoff, B.W. (1998) Thrombin is an extracellular signal that activates intracellular death protease pathways inducing apoptosis in model motor neurons. *J. Neurobiol.* **36,** 64–80.
75. Striggow, F., Riek, M., Breder, J., Henrich-Noack, P., Reymann, K.G., and Reiser, G. (2000) The protease thrombin is an endogenous mediator of hippocampal neuroprotection against ischemia at low concentrations but causes degeneration at high concentrations. *Proc. Natl. Acad. Sci. USA* **97,** 2264–2269.
76. Golomb, J., Kluger, A., Gianutsos, J., Ferris, S.H., de Leon, M.J., and George, A.E. (1995) Nonspecific leukoencephalopathy associated with aging. *Neuroimaging Clin. N. Am.* **5,** 33–44.
77. Liu, X., Erikson, C., and Brun, A. (1996) Cortical synaptic changes and gliosis in normal aging, Alzheimer's disease and frontal lobe degeneration. *Dementia* **7,** 128–134.
78. Barron, K.D. (1995) The microglial cell. A historical review. *J Neurol Sci* **134 Suppl,** 57–68.
79. Peinado, M.A., Quesada, A., Pedrosa, J.A., et al. (1998) Quantitative and ultrastructural changes in glia and pericytes in the parietal cortex of the aging rat. *Microsc. Res. Tech.* **43,** 34–42.
80. David, J.P., Ghozali, F., Fallet-Bianco, C., et al. (1997) Glial reaction in the hippocampal formation is highly correlated with aging in human brain. *Neurosci. Lett.* **235,** 53–56.
81. Morgan, T.E., Xie, Z., Goldsmith, S., et al. (1999) The mosaic of brain glial hyperactivity during normal ageing and its attenuation by food restriction. *Neuroscience* **89,** 687–699.
82. van Swieten, J.C., van den Hout, J.H., van Ketel, B.A., Hijdra, A., Wokke, J.H., and van Gijn, J. (1991) Periventricular lesions in the white matter on magnetic resonance imaging in the elderly. A morphometric correlation with arteriolosclerosis and dilated perivascular spaces. *Brain* **114,** 761–774.
83. Kullberg, S., Ramirez-Leon, V., Johnson, H., and Ulfhake, B. (1998) Decreased axosomatic input to motoneurons and astrogliosis in the spinal cord of aged rats. *J. Gerontol. A. Biol. Sci. Med. Sci.* **53,** B369–379.

84. Katoh-Semba, R. and Kato, K. (1994) Age-related changes in levels of the beta-subunit of nerve growth factor in selected regions of the brain: comparison between senescence-accelerated (SAM-P8) and senescence-resistant (SAM-R1) mice. *Neurosci. Res.* **20,** 251–256.
85. Nabeshima, T., Nitta, A., Fuji, K., Kameyama, T., and Hasegawa, T. (1994) Oral administration of NGF synthesis stimulators recovers reduced brain NGF-content in aged rats and cognitive dysfunction in basal-forebrain- lesioned rats. *Gerontol.* **40,** 46–56.
86. Soffie, M., Hahn, K., Terao, E., and Eclancher, F. (1999) Behavioural and glial changes in old rats following environmental enrichment. *Behav. Brain Res.* **101,** 37–49.
87. Nichols, N.R. (1999) Glial responses to steroids as markers of brain aging [Review]. *J. Neurobiol.* **40,** 585–601.
88. Gordon, M.N., Schreier, W.A., Ou, X., Holcomb, L.A., and Morgan, D.G. (1997) Exaggerated astrocyte reactivity after nigrostriatal deafferentation in the aged rat. *J. Comp. Neurol.* **388,** 106–119.
89. Needels, D.L., Nieto-Sampedro, M., and Cotman, C.W. (1986) Induction of a neurite-promoting factor in rat brain following injury or deafferentation. *Neuroscience* **18,** 517–526.
90. Woods, A.G., Guthrie, K.M., Kurlawalla, M.A., and Gall, C.M. (1998) Deafferentation-induced increases in hippocampal insulin-like growth factor-1 messenger RNA expression are severely attenuated in middle aged and aged rats. *Neuroscience* **83,** 663–668.
91. Gu, Z., Yu, J., and Perez-Polo, J.R. (1998) Responses in the aged rat brain after total immunolesion. *J. Neurosci. Res.* **54,** 7–16.
92. Scott, S.A., Liang, S., Weingartner, J.A., and Crutcher, K.A. (1994) Increased NGF-like activity in young but not aged rat hippocampus after septal lesions. *Neurobiol. Aging* **15,** 337–346.
93. Zheng, H., Jiang, M., Trumbauer, M.E., et al. (1995) beta-Amyloid precursor protein-deficient mice show reactive gliosis and decreased locomotor activity. *Cell* **81,** 525–531.
94. Miller, D.B., Bartke, A., and O'Callaghan, J.P. (1995) Increased glial fibrillary acidic protein (GFAP) levels in the brains of transgenic mice expressing the bovine growth hormone (bGH) gene. *Exp. Gerontol.* **30,** 383–400.
95. Gonzalez-Gomez, I., Mononen, I., Heisterkamp, N., Groffen, J., and Kaartinen, V. (1998) Progressive neurodegeneration in aspartylglycosaminuria mice. *Am. J. Pathol.* **153,** 1293–1300.
95a. Altar, C.A. (1991) Nerve growth factor and the neostriatum. *Prog Neuropsychopharmacol Biol Psychiatry* **15,** 157–169.
96. Unger, J.W. and Schmidt, Y. (1992) Quisqualic acid-induced lesion of the nucleus basalis of Meynert in young and aging rats: plasticity of surviving NGF receptor-positive cholinergic neurons. *Exp. Neurol.* **117,** 269–277.
97. Isaeff, M., Goya, L., and Timiras, P.S. (1993) Alterations in the growth and protein content of human neuroblastoma cells in vitro induced by thyroid hormones, stress and ageing. *J. Reprod. Fertil. Suppl* **46,** 21–33.
98. Kalanj, S., Kracun, I., Rosner, H., and Cosovic, C. (1991) Regional distribution of brain gangliosides in Alzheimer's disease. *Neurol. Croat.* **40,** 269–281.
99. Wyss-Coray, T., Masliah, E., Mallory, M., et al. (1997) Amyloidogenic role of cytokine TGF-beta 1 in transgenic mice and in Alzheimer's disease. *Nature* **389,** 603–606.
100. Wyss-Coray, T., Feng, L., Masliah, E., et al. (1995) Increased central nervous system production of extracellular matrix components and development of hydrocephalus in transgenic mice over-expressing transforming growth factor-beta 1. *Am J Pathol* **147,** 53–67.
101. Amara, F.M., Junaid, A., Clough, R.R., and Liang, B. (1999) TGF-beta(1), regulation of alzheimer amyloid precursor protein mRNA expression in a normal human astrocyte cell line: mRNA stabilization. *Mol. Brain Res.* **71,** 42–49.

102. Gertz, H.J., Stoltenburg, G., Cruz-Sanchez, F., Lafuente, J., and Schopol, R. (1988) [Panencephalopathy of the Creutzfeldt-Jakob disease type]. *Nervenarzt* **59,** 110–112.
103. Muhleisen, H., Gehrmann, J., and Meyermann, R. (1995) Reactive microglia in Creutzfeldt-Jakob disease. *Neuropathol. Appl. Neurobiol.* **21,** 505–517.
104. McCann, S.M. (1997) The nitric oxide hypothesis of brain aging. *Exp. Gerontol.* **32,** 431–440.

10
Role of Fibroblast Growth Factor-2 in Astrogliosis

John F. Reilly

1. INTRODUCTION

Astrocytes comprise the most numerous cellular element of the brain. Though once thought to merely provide structural support for neurons, astrocytes are now known to participate in many aspects of normal nervous system development and function, including neuronal migration, synaptogenesis, induction of the blood-brain-barrier, metabolic support of neurons, and regulation of extracellular fluid composition. Astrocytes throughout the brain are electrically coupled, forming a glial syncytium that may play a substantial role in signaling within the nervous system *(1)*. In addition to functions in the developing and adult nervous systems, astrocytes respond to changes in the brain caused by aging and injury.

2. ASTROGLIOSIS

The responses of astrocytes to aging and injury are termed astrogliosis, and can include proliferation, an increase in cell number through mitosis, and hypertrophy, an increase in cell size and process outgrowth. Many ultrastructural changes accompany astrogliosis, including enlargement of the nucleus, increased numbers of mitochondria and lysosomes, and increased glycogen. A diverse array of molecules, including various cell adhesion proteins, major histocompatibility complex class I and II proteins, early-response gene products, cytokines and growth factors, proteases, and protease inhibitors are upregulated in astrocytes following injury *(2)*. However, the most striking ultrastructural feature of astrocyte hypertrophy is the increased cytoplasmic accumulation of glial fibrillary acidic protein (GFAP). GFAP is the principal intermediate filament of the astrocyte cytoskeleton *(3)*, and immunostaining for GFAP is commonly used both to identify astrocytes and to assess their morphological changes.

2.1. *Age-Related Astrogliosis*

Ultrastructural studies have demonstrated hypertrophy of astrocytes in the aging brain *(4)*. The increases in astrocyte cell size with advancing age are not accompanied by an overall increase in the number of astrocytes identified by morphological criteria, suggesting that age-related astrogliosis is characterized primarily by hypertrophy, rather than proliferation. GFAP protein levels have also been shown to increase with

From: *Neuroglia in the Aging Brain*
Edited by: Jean S. de Vellis © Humana Press Inc., Totowa, NJ

age in the mouse, rat, and human brain *(5–8)*. This increase occurs in both gray and white matter in several brain regions. Though an increase in the number of GFAP-positive astrocytes has been observed with aging *(9,10)*, this most likely results from enhanced immunodetection due to increased GFAP expression. Astrocytes in the intact central nervous system (CNS), especially protoplasmic astrocytes in the gray matter, normally express very low levels of GFAP protein, which can be below the threshold for routine immunohistochemical stains *(11)*.

The increases in GFAP protein content result from an increase in the amount of GFAP mRNA in astrocytes *(6,12,13)*. As with the increase in GFAP protein expression, this increase occurs in many brain regions, and progresses with advancing age. However, the increased expression of GFAP mRNA does not appear to result from increased transcription *(14)*. This suggests that age-induced astrogliosis is associated with increased stability of GFAP mRNA, protein, or both.

2.2. *Injury-Induced Astrogliosis*

Astrocytes exhibit one of the earliest and most remarkable cellular responses following a variety of insults to the CNS. Both proliferation and hypertrophy of astrocytes occur, and it is likely that these events are independently regulated. It has been shown that in response to facial nerve transection, astrocytes in the rat facial nucleus exhibit hypertrophy in the absence of an appreciable proliferative response *(15,16)*. This phenomenon has also been demonstrated in the rat visual cortex following optic nerve transection *(17)*. Following a traumatic injury to the CNS, only a small proportion of astrocytes undergo cell division, suggesting that astrocyte hypertrophy is the major determinant of trauma-induced astrogliosis *(18–20)*.

Very few studies of astrocyte hypertrophy have relied on morphological criteria; in general, astrocyte hypertrophy is evaluated using GFAP immunostaining. Astrocyte hypertrophy and the concomitant increase in expression of GFAP mRNA and protein are detectable less than 24 h after a traumatic injury *(21–23)*. The response is likely to be more rapid than these data suggest, since expression of a reporter gene driven by the GFAP promoter is detectable as early as one hour following traumatic injury in a transgenic mouse model *(24)*. Expression of GFAP mRNA and protein reach a peak between seven and 14 d following injury, and persist at elevated levels beyond 30 d. Although the time course varies with the nature and severity of the injury, expression of GFAP is increased in a wide variety of experimental models, including ischemia, systemic administration of kainic acid, and deafferentation.

Astrocytes in the aged brain respond differently to injury than those in the neonatal or adult brain. In deafferentation models of injury, astrocytes in the aged brain demonstrate increases in GFAP mRNA and protein that are greater in magnitude than in adult rats *(25,26)*. These increases appear to develop more slowly in the aged rodent brain, and persist longer. The delayed but stronger response to injury may reflect a difference in the astrocytes of the aged brain, or may result from alterations in the factors that regulate astrogliosis.

3. FIBROBLAST GROWTH FACTOR-2

The response of astrocytes to injury, including alterations in the expression of GFAP, is thought to be initiated and regulated by local chemical mediators released at the site

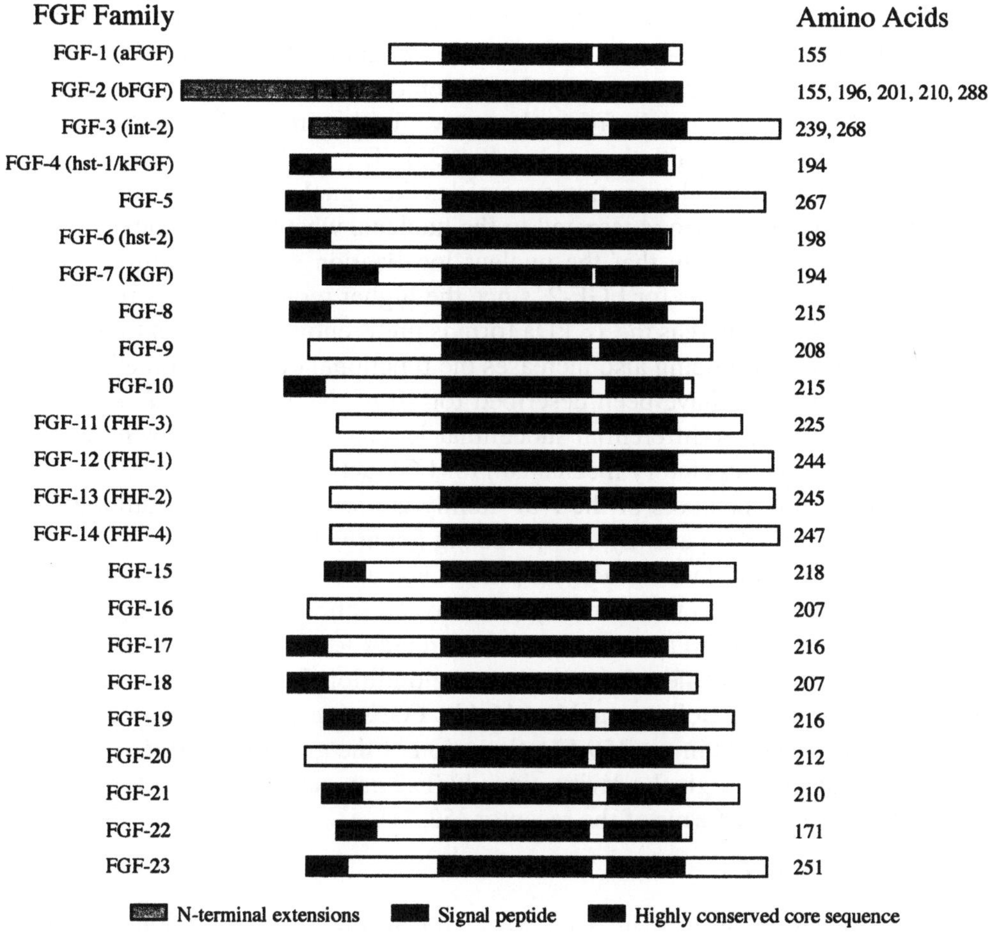

Fig. 1. Structural features of the FGF family. The core structure contains the regions of greatest sequence identity between family members. A hydrophobic signal peptide is present on several of the FGFs. N-terminal extensions of FGF-2 and -3 represent alternate initiation sites, and contain nuclear localization sequences. References: FGF-1 *(29–31),* FGF-2 *(28,34,35),* FGF-3 *(33,133),* FGF-4 *(134,135),* FGF-5 *(136),* FGF-6 *(137),* FGF-7 *(138),* FGF-8 *(139),* FGF-9 *(46),* FGF-10 *(140),* FGF-11 and -12 *(141),* FGF-13 and -14 *(142),* FGF-15 *(143),* FGF-16 *(45),* FGF-17 *(144,145),* FGF-18 *(146,147),* FGF-19 *(148),* FGF-20 *(47),* FGF-21 *(149),* FGF-22 (GenBank AB021925), FGF-23 *(150).*

of injury. Many growth factors and cytokines have been identified in the wounded rat brain, and some or all may be involved in the activation of glial cells. Substantial evidence suggests that basic fibroblast growth factor (FGF-2) plays a significant role in the response of astrocytes to injury.

3.1. *FGF Gene Family*

The FGF family comprises a rapidly expanding group of polypeptide growth factors with 23 members identified to date (Fig. 1), which share 25–70% sequence homology. This family of proteins is mitogenic and morphogenic for a variety of cell types *(27).* The first two members of the family to be purified and characterized were acidic FGF

(FGF-1) and basic FGF *(28–31)*. The core structure of the FGF family is represented by the 18 kDa FGF-2 polypeptide. This structure contains the regions of the molecule that are required for binding to cell surface receptors *(32)*. The N-terminal extensions found on FGF-2 and-3 result from alternative initiation at CUG codons located upstream from the AUG initiation codon *(33–35)*. The nuclear localization sequence contained in the N-terminal extension supersedes the signal peptide in FGF-3, since the higher molecular mass form is targeted to the nucleus, whereas the nonextended form is secreted *(33)*. It is likely that the nuclear localization sequence is dominant over other targeting mechanisms for FGF-2, since the higher molecular mass forms are targeted to the nucleus, whereas the 18 kDa form is preferentially exported to the cell surface *(36)*. Alternative splicing also increases the functional diversity of the FGF family. To date, splice variants have been described for 10 of the 23 FGFs *(37–43)*. In many cases, splicing results in differential subcellular localization of the growth factor, suggesting an additional regulatory mechanism for FGF action.

Although several members of the FGF family are secreted by the traditional endoplasmic reticulum-Golgi pathway, an interesting feature of FGF-1, -2, -9, -11 through -14, -16, and -20 is the lack of a hydrophobic signal sequence in the N-terminus to target the growth factor to the secretory pathway. Despite the lack of a signal peptide, FGF-1, -2, -9, -16, and -20 are known to be secreted *(44–48)*. Secretion of FGF-2 into the extracellular compartment occurs via a nontraditional pathway distinct from the endoplasmic reticulum-Golgi complex *(44,49)*. FGFs can also be released from cells due to lethal or sublethal injury *(50,51)*. Secretion of the growth factor may not be necessary for all of its biological activity, since FGF-2 can act in an intracrine manner by binding to intracellular forms of the receptor *(52,53)*.

3.2. Expression of FGF-2

FGF-2 mRNA is present in astrocytes and several subpopulations of neurons *(54,55)*. FGF-2 protein is also present in astrocytes and neurons in vivo, although the mRNA does not always colocalize with protein expression *(56)*. A comprehensive review of the expression of FGF-2 in the rat brain has been published *(56)*. Astrocytes in culture have also been shown to express FGF-2 mRNA and protein *(57,58)*.

Local expression of FGF-2 mRNA and protein are increased following traumatic brain injury *(59,60)*. Ischemia, neuronal degeneration, stimulant drugs, and excitotoxic injury also result in enhanced expression of FGF-2 *(61–64)*. An increase in FGF-2 mRNA is detected as early as 4 h after injury, and persists for at least 14 d, with a peak at around 7 d post-lesion. Expression of FGF-2 protein follows a similar time course. Although FGF-2 is expressed in neurons, microglia, and macrophages, the majority of FGF-2 mRNA and protein is expressed by astrocytes *(60)*. The peak of FGF-2 expression also coincides with the peak of the astrogliotic response.

Expression of FGF-2 is altered with aging. FGF-2 immunoreactivity decreases with advancing age in several regions of the rat brain *(65–67)*. This decrease is not associated with a reduction in the number of GFAP-positive astrocytes. Despite a decrease in the level of FGF-2 expression in the intact brain, systemic administration of glucocorticoids or β-adrenergic agonists results in an increase in FGF-2 mRNA comparable to that seen in young animals, indicating that induction of FGF-2 can occur in the aged brain *(68)*. However, FGF-2 expression following injury to the aged brain has not been examined.

3.2. FGF Receptors

Three broad classes of receptors for FGFs have been identified. High affinity FGF receptors (FGFR) are cell surface receptor tyrosine kinases that bind FGFs with affinities of 10–500 pM. Low affinity FGF receptors consist of cell surface heparan sulfate proteoglycans (HSPG), which bind FGFs with affinities of 2–10 nM. A cysteine-rich FGF receptor (CFR) has also been identified which does not exhibit significant sequence identity with FGFR, and binds several of the FGFs with high affinity *(69)*. The CFR is localized to the Golgi apparatus and plays a role in the intracellular trafficking of FGFs *(70)*. FGFR represents the primary signal transducing receptor for the FGFs.

To date, four distinct FGFR genes have been identified, sharing 55–72% amino acid sequence identity *(71,72)*. Full-length FGFR consists of an extracellular domain with three immunoglobulin-like domains, a hydrophobic transmembrane domain, and a cytoplasmic domain containing a tyrosine kinase domain split by a short insert. All four FGFRs are expressed in multiple forms resulting from alternative splicing *(72,73)*. Although alternative splicing is observed in other growth factor receptor systems, the FGFR system is unique in the vast array of splice variants identified. All four of the FGFRs are known to be activated by FGF-2. Interestingly, the FGFs are not the only ligands capable of binding to the FGFRs. A yeast expression screening method has been used to identify two additional ligands which bind to and activate FGFR1 *(74)*. These ligands are distantly related to the epidermal growth factor and angiogenin families, and show no homology to the FGFs. In addition to soluble factors, cell adhesion molecules (CAM) may bind to and activate FGFR. FGFR1 contains a CAM-homology domain immediately downstream of the acid box *(75)*, and interaction with N-CAM, N-cadherin, and L1 at this site may be responsible for CAM-induced neurite outgrowth *(76)*.

3.3. Expression of FGFR

Early studies suggested that FGFR1 was expressed specifically by neurons, FGFR2 and 3 were expressed by glia, and FGFR4 was not expressed in the CNS *(77)*. However, this localization was based primarily upon the detection of FGFR1 mRNA in gray matter and FGFR2 and 3 mRNA in white matter. More recent studies have demonstrated the in vivo expression of FGFR1 mRNA in neurons and astrocytes *(56,60)*, FGFR2 mRNA in neurons, astrocytes, and oligodendrocytes *(78,79)*, FGFR3 mRNA in astrocytes *(80)*, and FGFR4 mRNA in a small population of cholinergic neurons in the medial habenular nucleus *(81,82)*.

Increased astrocyte expression of FGFR1 mRNA has been observed in response to traumatic brain injury *(60)*, and increases in FGFR1 and FGFR2 mRNA expression by astrocytes have been reported in response to transient ischemia *(79)*. Expression of FGFR1 protein in astrocytes is also increased following traumatic injury (Fig. 2). This increase is seen as early as one day following injury and further increases are seen for at least 10 d *(83)*. By 28 d following a traumatic injury, expression of FGFR1 has returned to control levels, except for a sustained upregulation in the glia limitans at the edge of the lesion. The temporal pattern of FGFR1 expression following injury correlates well with GFAP expression, suggesting a potential role for FGFR1 upregulation in astrocyte hypertrophy.

Expression of FGFR1 in the brain, both before and after injury, is regionally heterogeneous. For example, astrocytes in the mid-cortex express lower levels of FGFR1 than

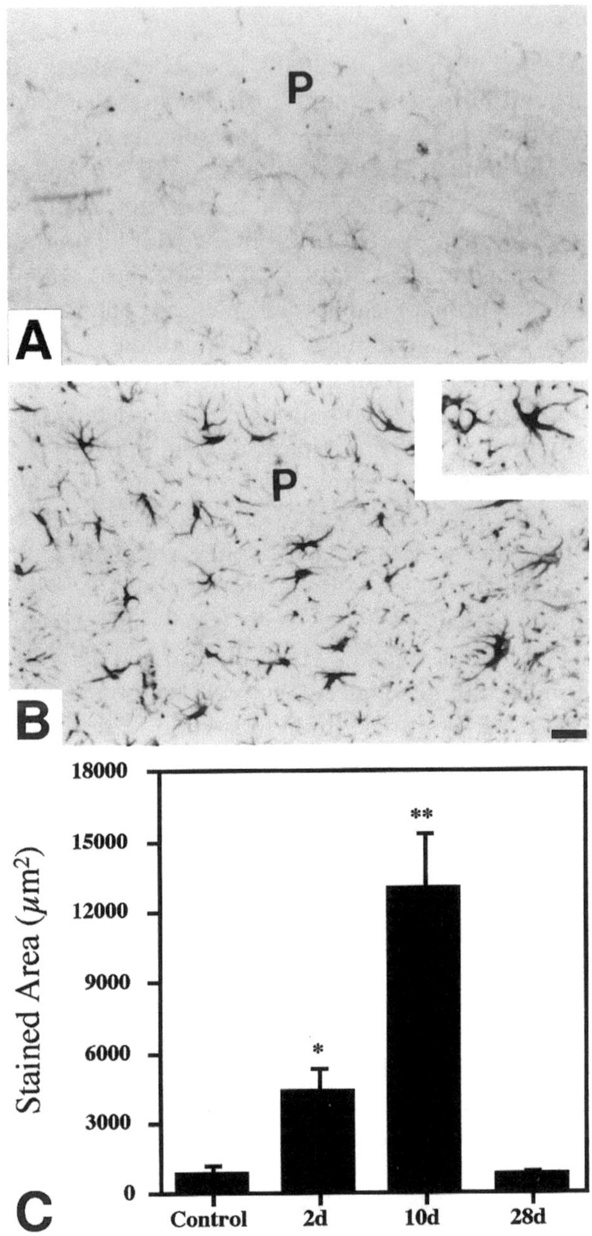

Fig. 2. Hippocampal FGFR1 expression following traumatic brain injury. In nonlesioned animals **(A),** expression is low. At seven days post-lesion **(B),** a substantial increase over nonlesioned controls is evident. A high magnification view is shown (inset). Wound edge is to the right of the fields shown; the stratum pyramidale (P) is indicated. Scale bar, 20 μm. Quantitative image analysis **(C)** of FGFR1 expression after injury demonstrates a 16-fold increase by 10 d post-lesion, with a return to control levels by day 28. ($*p < 0.05$ vs nonlesioned; $**p < 0.005$ vs non-lesioned). Modified from *(83).*

those in the hippocampus in the intact brain, but display a substantially greater relative increase following injury *(83)*. This heterogeneity is consistent with differences in astrocyte expression of other proteins *(84)*. These differences may have functional consequences for wound healing, regenerative efforts and the potential for functional recovery. It has been shown that reactive astrocytes can, in certain locations in vivo, permit axonal regrowth following a traumatic brain injury *(85)*.

4. EFFECTS OF FGF-2 ON ASTROCYTE HYPERTROPHY

4.1. *In Vitro Models*

Much of the work on the effects of FGF-2 on astrocytes has been performed using in vitro models. Treatment of cultured astrocytes with FGF-2 enhances cell proliferation in a dose-dependent manner *(86,87)*. Cultured astrocytes also undergo morphological differentiation with FGF-2 treatment, changing from a flat, polygonal morphology to a stellate morphology and elaborating long processes *(86,88)*. In addition to mitogenic and morphogenic effects on astrocytes in culture, FGF-2 enhances the expression of several gene products, including immediate early genes *(89)*, glutamine synthetase *(87)*, and extracellular matrix (ECM) molecules *(90,91)*.

4.2. *Intralesion Injection of FGF-2*

In vivo studies on the effects of FGF-2 on astrocytes have focused primarily on the modification of the astrocyte response to injury. Several studies have demonstrated that FGF-2 enhances trauma-induced astrogliosis *(20,92,93)*. The effects of FGF-2 on astrogliosis primarily represent an increase in astrocyte hypertrophy, as opposed to proliferation. Increased numbers of GFAP-positive cells following traumatic injury are sometimes attributed to astrocyte proliferation. However, studies using ^{3}H-thymidine or bromodeoxyuridine uptake to directly assess proliferation have shown that cell division represents only a minor component of astrogliosis *(18,19)*, and that the increase in astrocyte cell number represents increased detection of astrocytes previously expressing GFAP at subdetectable levels. Direct injection of FGF-2 into a brain lesion increases astrocyte hypertrophy in a dose-dependent manner *(20)*. This increase is detectable as early as two days following injection. Astrocytes in the vicinity of the injection demonstrate enhanced cell size and ramification of processes compared to diluent-injected controls.

In addition to effects on astrocyte hypertrophy, intralesion injection of FGF-2 has effects on other glial cells at the site of injury (Fig. 3). Microglial cells proliferate extensively in response to injection of FGF-2 *(20)*. Although microglia express all four FGFRs *(94,95)*, it is possible that this proliferation represents an indirect response to FGF-2. Pure cultures of microglia do not show a substantial proliferative response to FGF-2 *(96)*, but do proliferate extensively in response to granulocyte-macrophage colony stimulating factor (GM-CSF; *97,98*). GM-CSF is produced by astrocytes *(99)*, and is released in response to injury *(100)*, and this may be further enhanced by FGF-2 injection.

4.3. *Co-administration of Heparan Sulfate*

Heparan sulfate (HS) represents the binding component of the low-affinity receptors for FGFs. HSPG is expressed in the ECM of the cortex and hippocampus, and in the

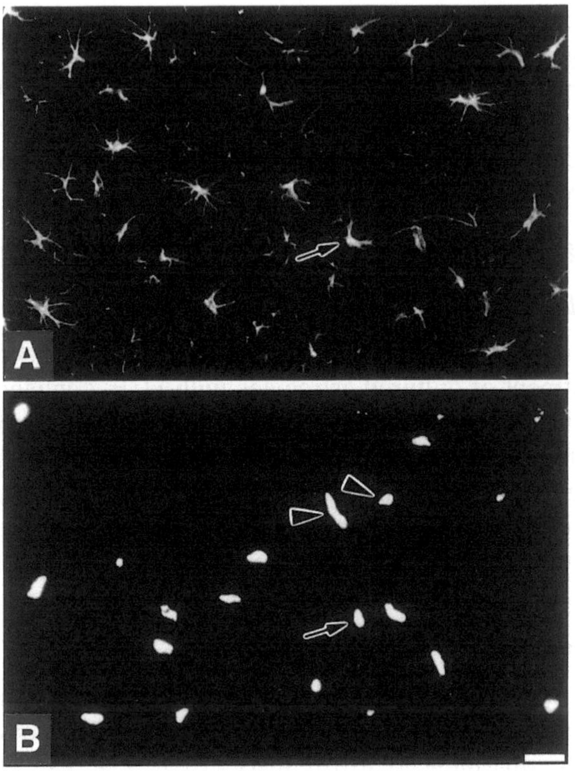

Fig. 3. Immunofluorescent double-labeling for GFAP and bromodeoxyuridine (BrdU) in the cortex of animals injected with FGF-2. GFAP immunoreactivity **(A)** is present in the cytoplasm and processes of astrocytes. BrdU staining **(B)** of nuclei of dividing cells in the same field. Arrows indicate astrocyte processes **(A)** and nucleus **(B).** Nuclei of other dividing cells, primarily of macrophage-microglial phenotype, are also indicated (arrowheads, **B**). Scale bar, 30 μm. From *(106)*.

basal laminae of capillaries throughout the brain *(101,102)*. Although there has been debate in the literature about an absolute requirement for HS participation in the high-affinity binding of FGF to FGFR, HS does seem to facilitate this interaction, perhaps by inducing a conformational change in FGF, or by promoting dimerization of FGF leading to dimerization of FGFR *(32,103)*. In addition to participation in FGF binding to FGFR, HS may serve as a sink for FGF, sequestering free growth factor. Intracerebrally injected FGF-2 exhibits only limited diffusion into the brain parenchyma, as determined by studies using iodinated FGF-2 *(104)*. The growth factor remains bound to HSPG adjacent to blood vessels, and is available only to cells in the immediate vicinity of the injection site. In the injured brain, these low-affinity binding sites are likely to prevent lateral diffusion of FGF released from injured cells or secreted by intact cells.

Several studies have demonstrated that coinjection of HS with FGF-2 into a lesion enhances the glial response *(93,105,106)*. The combination of FGF-2 and HS results in greater astrocyte hypertrophy in the vicinity of the lesion compared to FGF-2 alone. Another important effect of coadministration of free HS is an increase in the radius of growth factor diffusion. In vitro studies have demonstrated that heparin, and to a lesser

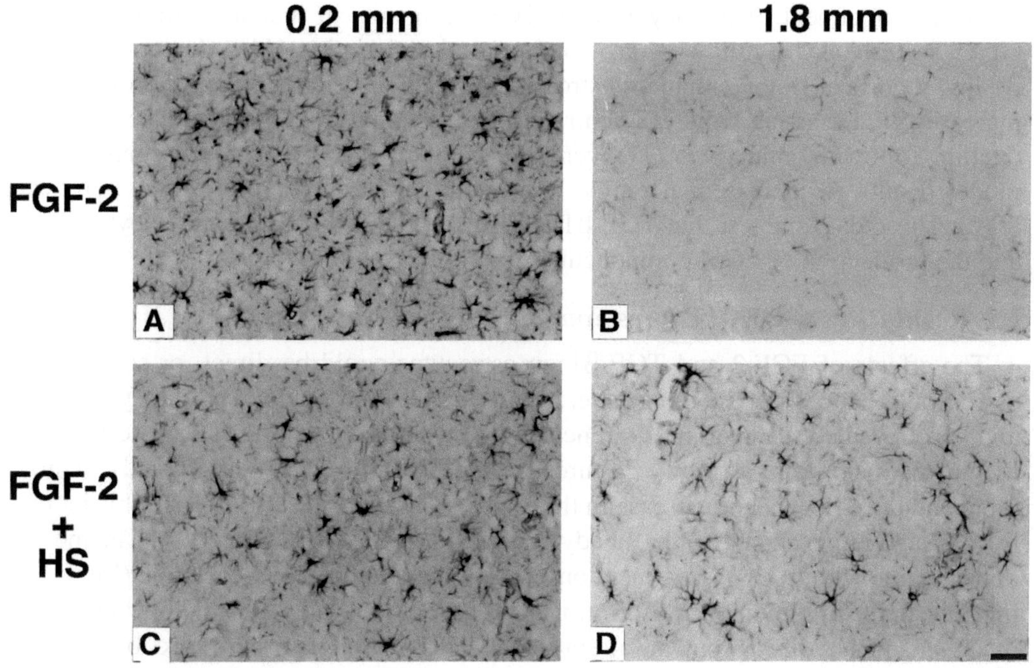

Fig. 4. Immunohistochemical staining for GFAP in following intralesion injection of FGF-2 (**A, B**) or FGF-2 + HS (**C, D**). Subfields are centered 0.2 mm (**A, C**) or 1.8 mm (**B, D**) from the lesion. Astrocyte hypertrophy is evident at a substantially greater distance from the lesion with coadministration of HS. Scale bar, 30 µm. Modified from *(106)*.

extent HS, stimulate morphological changes in a monolayer of endothelial cells at a 10-fold greater radius than FGF-2 alone *(107)*. When FGF-2 is injected into a cortical lesion, astrocyte hypertrophy is significantly enhanced within 0.8 mm of the injection (Fig. 4). The combination of FGF-2 and HS produces an increase in astrocyte hypertrophy to a distance of at least 2.8 mm from the injection site *(106)*. This effect may result from HS binding to the low affinity receptor binding sites on FGF-2, allowing the complex to bypass low affinity receptors in the injured tissue, as suggested by Schlessinger et al. *(103)*.

5. EFFECTS OF FGF-2 AND TGF-β1 ON ASTROCYTE GFAP EXPRESSION

5.1. Transforming Growth Factor-β

In addition to FGF-2, other growth factors and cytokines play a role in the regulation of astrogliosis, including transforming growth factor-β (TGF-β). To date, three TGF-β isoforms (TGF-β1, -β2, and -β3) have been reported in mammals *(108)*. TGF-β1 is now known to be representative of a large family of growth factors, whose members share 25–90% sequence identity. This family comprises over 40 members, including the inhibins and activins, the bone morphogenetic proteins, decapentaplegic and other Drosophila proteins, Müllerian inhibiting substance, the growth-differentiation factors, and glial cell-line derived neurotrophic factor. Members of the TGF-β superfamily exert two major types of effects on most cells: they inhibit proliferation and they strongly activate the production of ECM proteins *(109)*.

In vivo treatment of astrocytes with TGF-β1 results in increased transcription of GFAP as early as 4 h following growth factor administration *(110)*. Injection of TGF-β1 into a cortical lesion enhances astrocyte production of laminin and fibronectin and increases glial scarring, whereas neutralizing antibodies to TGF-β1 reduce ECM production following injury *(111)*. Overexpression of TGF-β1 in a transgenic mouse model results in increased astrocyte expression of GFAP, laminin, and fibronectin *(112)*. These data suggest that TGF-β1 is a key regulator of the astrocyte cytoskeleton and the production of ECM components.

5.2. Regulation of GFAP Expression and Process Outgrowth

The effects of FGF-2 and TGF-β1 on astrocytes could be direct, or secondary to effects on other cells at the injury site. In vitro studies have helped to clarify this issue, and have revealed substantial differences in the response to short- vs long-term growth factor treatment. Pure astrocyte cultures treated for up to 24 h with TGF-β1 exhibit a rapid and dose-dependent increase in the expression of GFAP mRNA, and this increase is translated into a change in the steady-state level of GFAP protein *(113)*. The increase in detectable within 3 h of treatment onset, and maintained for at least 24 h (Fig. 5). An increase in the initiation of transcription is at least partially responsible for the increase in GFAP mRNA *(110)*, but an alteration in mRNA stability may also be involved. Treatment of astrocytes with FGF-2 results in a rapid and dose-dependent decrease in GFAP expression over a time course similar to that observed with TGF-β1 (Fig. 5). Treatment with a combination of FGF-2 and TGF-β1 also results in decreased GFAP expression, suggesting that FGF-2 can inhibit the TGF-β1-mediated increase in GFAP synthesis.

Although short-term TGF-β1 treatment results in increased expression of GFAP mRNA and protein, the growth factor has only minor effects on astrocyte morphology *(113,114)*. Conversely, FGF-2 treatment decreases expression of GFAP up to 24 h following administration, yet produces a dramatic morphological change to a process-bearing phenotype *(86,88,113)*. These data demonstrate that astrocyte GFAP expression is uncoupled from process formation. It is likely that the initial events of process formation represent an actin-dependent process *(115)*.

When applied for longer time periods, FGF-2 and TGF-β1 have effects on GFAP expression very different from those in the short term. Treatment of cultured astrocytes with FGF-2 for 72 h results in a significant increase in the expression of GFAP, while TGF-β1 treatment has no effect on GFAP expression (Fig. 5C). These effects highlight the complexity of GFAP regulation; following 1 day of in vitro treatment, FGF-2 inhibits a TGF-β1-mediated increase in astrocyte GFAP expression, while after 3 days of treatment, TGF-β1 potentiates an FGF-2-induced increase. The data also suggest that the early increase in GFAP expression caused by TGF-β1 is not sufficient to produce a long-term change in GFAP protein levels or astrocyte morphology. FGF-2 treatment, which induces a rapid morphological response in the absence of increased GFAP expression, also results in a delayed increase in GFAP expression. This suggests that the increase in GFAP expression seen at later time points is necessary to stabilize astrocyte processes elaborated due to FGF-2-induced rearrangements of the actin cytoskeletal network.

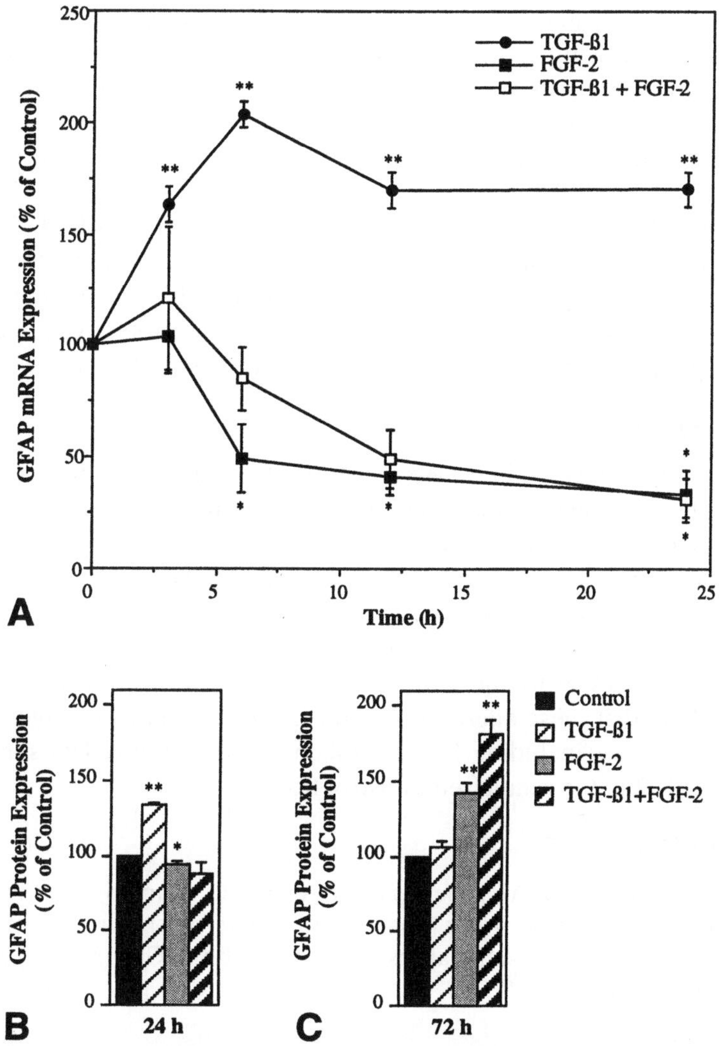

Fig. 5. Effects of TGF-β1 and FGF-2 on GFAP expression. GFAP mRNA expression (**A**) following 3, 6, 12, or 24 h treatment with TGF-β1 (2 ng/mL), FGF-2 (20 ng/mL), or a combination of the two is increased with TGF-β1 treatment and decreased with FGF-2 or the combination of the two growth factors. Changes in GFAP protein expression with 24 h treatment (**B**) follow a similar pattern. After treatment for 72 h (**C**), FGF-2 and TGF-β1 +FGF-2-treated cultures show increases in the expression of GFAP protein, while GFAP protein expression in TGF-β1-treated cultures has returned to control levels. All values are expressed as the percent of control, and represent the mean ± standard error of three independent cultures. * $p < 0.05$ vs control, ** $p < 0.005$ vs control. **A** and **B** modified from *(113)*.

6. ASTROCYTE-DERIVED FGF-2

As discussed in **Subheading 5.2.**, astrocytes express FGF-2, and this expression is upregulated following injury. There is also evidence that FGF-2 induces its own sythesis *(116)*. In addition to effects on astrocytes, FGF-2 is available to act on other cells in the injured area, such as microglia and macrophages. FGF-2 is also a survival factor for neurons *(117–121)*. Furthermore, FGF-2 has been used successfully in vivo as a neuroprotective agent following traumatic, ischemic, neurotoxic, and excitotoxic injuries *(122–124)*, and has been shown to promote functional recovery in the injured spinal cord *(125)*. Taken together, these data indicate that the role of FGF-2 in astrogliosis extends beyond effects on astrocytes themselves. The upregulation of FGF-2 expression in reactive astrocytes at the site of injury is likely to be part of a generalized trophic response for support of injured neurons. In addition to FGF-2, astrocytes in vivo produce a variety of other growth factors following injury, including nerve growth factor *(126)*, insulin-like growth factor I *(127)*, TGF-β1 *(128)*, and ciliary neurotrophic factor *(129)*. These growth factors, like FGF-2, have neuroprotective effects *(130–132)*, and their upregulation represents a beneficial aspect of astrogliosis.

7. CONCLUSIONS

Astrogliosis represents a coordinated response to a complex set of stimuli. Injury to the CNS results in changes in the microenvironment including ionic and pH perturbations, breakdown of the blood-brain-barrier, entry of nonneural cells, and the release of a variety of cytokines, growth factors, and other local chemical mediators. The interplay of these factors is a dynamic process, having a variety of effects on all cells in the injured area. FGF-2 induces astrocyte hypertrophy in vivo, and modifies the biochemical responses of astrocytes to injury. In the aged brain, astrocyte responses to injury are altered. It is likely that these alterations result from age-related changes in the expression of growth factors, including FGF-2. Taken together, these data support a significant role for FGF-2 in the regulation of astrogliosis.

ACKNOWLEDGMENTS

I thank Ms. Shawndra D. Martinez and Dr. Vijaya G. Kumari for providing valuable comments on the manuscript. Work was carried out in the laboratory of Dr. Vijaya G. Kumari and supported by the National Institutes of Health (AG06159) and the Department of Veterans Affairs (640-D56325).

REFERENCES

1. Dermietzel, R. and Spray, D.C. (1998) From neuro-glue ('Nervenkitt') to glia: a prologue. *Glia* **24,** 1–7.
2. Eddleston, M. and Mucke, L. (1993) Molecular profile of reactive astrocytes – implications for their role in neurologic disease. *Neuroscience* **54,** 15–36.
3. Eng, L.F. and Ghirnikar, R.S. (1994) GFAP and astrogliosis. *Brain Pathol.* **4,** 229–237.
4. Berciano, M.T., Andres, M.A., Calle, E., and Lafarga, M. (1995) Age-induced hypertrophy of astrocytes in rat supraoptic nucleus: a cytological, morphometric, and immunocytochemical study. *Anat. Rec.* **243,** 129–144.
5. David, J.-P., Ghozali, F., Fallet-Bianco, C., et al. (1997) Glial reaction in the hippocampal formation is highly correlated with aging in human brain. *Neurosci. Lett.* **235,** 53–56.
6. Kohama, S.G., Goss, J.R., Finch, C.E., and McNeill, T.H. (1995) Increases of glial fibrillary acidic protein in the aging female mouse brain. *Neurobiol. Aging* **16,** 59–67.
7. Bronson, R.T., Lipman, R.D., and Harrison, D.E. (1993) Age-related gliosis in the white matter of mice. *Brain Res.* **609,** 124–128.
8. Day, J.R., Frank, A.T., O'Callaghan, J.P., Jones, B.C., and Anderson, J.E. (1998) The effect of age and testosterone on the expression of glial fibrillary acidic protein in the rat cerebellum. *Exp. Neurol.* **151,** 343–346.

9. Amenta, F., Bronzetti, E., Sabbatini, M., and Vega, J.A. (1998) Astrocyte changes in aging cerebral cortex and hippocampus: a quantitative immunohistochemical study. *Microsc. Res. Tech.* **43,** 29–33.
10. Björklund, H., Eriksdotter-Nilsson, M., Dahl, D., Rose, G., Hoffer, B., and Olson, L. (1985) Image analysis of GFA-positive astrocytes from adolescence to senescence. *Exp. Brain Res.* **58,** 163–170.
11. Eng, L.F. and Lee, Y.-L. (1995) Intermediate filaments in astrocytes. In: *Neuroglia,* Kettenmann, H. and Ransom, B.R., ed. Oxford University Press, New York, pp. 650–667.
12. Nichols, N.R., Day, J.R., Laping, N.J., Johnson, S.A., and Finch, C.E. (1993) GFAP mRNA increases with age in rat and human brain. *Neurobiol. Aging* **14,** 421–429.
13. Goss, J.R., Finch, C.E., and Morgan, D.G. (1991) Age-related changes in glial fibrillary acidic protein mRNA in the mouse brain. *Neurobiol. Aging* **12,** 165–170.
14. Laping, N.J., Teter, B., and Anderson, C.P. (1994) Age-related increases in glial fibrillary acidic protein do not show proportionate changes in transcription rates or DNA methylation in the cerebral cortex and hippocampus of male rats. *J. Neurosci. Res.* **39,** 710–717.
15. Graeber, M.B., Tetzlaff, W., Streit, W.J., and Kreutzberg, G.W. (1988) Microglial cells but not astrocytes undergo mitosis following rat facial nerve axotomy. *Neurosci. Lett.* **85,** 317–321.
16. Tetzlaff, W., Graeber, M.B., Bisby, M.A., and Kreutzberg, G.W. (1988) Increased glial fibrillary acidic protein synthesis in astrocytes during retrograde reaction of the rat facial nucleus. *Glia* **1,** 90–95.
17. Hajós, F., Gerics, B., and Turai, É. (1993) Astroglial reaction following Wallerian degeneration in the rat visual cortex: proliferation or hypertrophy? *Neurobiology* **1,** 123–131.
18. Janeczko, K. (1989) Spatiotemporal patterns of the astroglial proliferation in rat brain injured at the postmitotic stage of postnatal development: a combined immunocytochemical and autoradiographic study. *Brain Res.* **485,** 236–243.
19. Miyaki, T., Hattori, T., Fukuda, M., Kitamura, T., and Fujita, S. (1988) Quantitative studies on proliferative changes of reactive astrocytes in mouse cerebral cortex. *Brain Res.* **451,** 133–138.
20. Menon, V.K. and Landerholm, T.E. (1994) Intralesion injection of basic fibroblast growth factor alters glial reactivity to neural trauma. *Exp. Neurol.* **129,** 142–154.
21. Hozumi, I., Chiu, F.-C., and Norton, W.T. (1990) Biochemical and immunocytochemical changes in glial fibrillary acidic protein after stab wounds. *Brain Res.* **524,** 64–71.
22. Hozumi, I., Aquino, D.A., and Norton, W.T. (1990) GFAP mRNA levels following stab wounds in rat brain. *Brain Res.* **534,** 291–294.
23. Vijayan, V.K., Lee, Y.-L., and Eng, L.F. (1990) Increase in glial fibrillary acidic protein following neural trauma. *Mol. Chem. Neuropathol.* **13,** 107–118.
24. Mucke, L., Oldstone, M.B.A., Morris, J.C., and Nerenberg, M.I. (1991) Rapid activation of astrocyte-specific expression of GFAP-lacZ transgene by focal injury. *New Biologist* **3,** 465–474.
25. Gordon, M.N., Schreier, W.A., Ou, X., Holcomb, L.A., and Morgan, D.G. (1997) Exaggerated astrocyte reactivity after nigrostriatal deafferentation in the aged rat. *J. Comp. Neurol.* **388,** 106–119.
26. Goss, J.R. and Morgan, D.G. (1995) Enhanced glial fibrillary acidic protein RNA response to fornix transection in aged mice. *J. Neurochem.* **64,** 1351–1360.
27. Baird, A. (1994) Fibroblast growth factors: activities and significance of non-neurotrophin neurotrophic growth factors. *Curr. Opin. Neurobiol.* **4,** 78–86.
28. Esch, F., Baird, A., Ling, N., et al. (1985) Primary structure of bovine pituitary basic fibroblast growth factor (FGF) and comparison with the amino-terminal sequence of bovine brain acidic FGF. *Proc. Natl. Acad. Sci. USA* **82,** 6507–6511.
29. Thomas, K.A., Rios-Candelore, M., and Fitzpatrick, S. (1984) Purification and characterization of acidic fibroblast growth factor from bovine brain. *Proc. Natl. Acad. Sci. USA* **81,** 357–361.

30. Bohlen, P., Baird, A., Esch, F., Ling, N., and Gospodarowicz, D. (1984) Isolation and partial molecular characterization of pituitary fibroblast growth factor. *Proc. Natl. Acad. Sci. USA* **81,** 5364–5368.
31. Jaye, M., Howk, R., Burgess, W., et al. (1986) Human endothelial cell growth factor: cloning, nucleotide sequence, and chromosome localization. *Science* **233,** 541–545.
32. Fernig, D.G. and Gallagher, J.T. (1994) Fibroblast growth factors and their receptors: an information network controlling tissue growth, morphogenesis and repair. *Prog. Growth Factor Res.* **5,** 353–377.
33. Acland, P., Dixon, M., Peters, G., and Dickson, C. (1990) Subcellular fate of the *int*-2 oncoprotein is determined by choice of initiation codon. *Nature* **343,** 662–665.
34. Arnaud, E., Touriol, C., Boutonnet, C., et al. (1999) A new 34-kilodalton isoform of human fibroblast growth factor 2 is cap dependently synthesized by using a non-AUG start codon and behaves as a survival factor. *Mol. Cell. Biol.* **19,** 505–514.
35. Florkiewicz, R.Z. and Sommer, A. (1989) Human basic fibroblast growth factor gene encodes four polypeptides: three initiate translation from non-AUG codons. *Proc. Natl. Acad. Sci. USA* **86,** 3978–3981.
36. Florkiewicz, R.Z., Baird, A., and Gonzalez, A.-M. (1991) Multiple forms of bFGF: differential nuclear and cell surface localization. *Growth Factors* **4,** 265–275.
37. Myers, R.L., Payson, R.A., Chotani, M.A., Deaven, L.L., and Chiu, I.M. (1993) Gene structure and differential expression of acidic fibroblast growth factor mRNA: identification and distribution of four different transcripts. *Oncogene* **8,** 341–349.
38. Zuniga Mejia Borja, A., Murphy, C., and Zeller, R. (1996) *alt*FGF-2, a novel ER-associated FGF-2 protein isoform: its embryonic distribution and functional analysis during neural tube development. *Dev. Biol.* **180,** 680–692.
39. Mansour, S.L. (1994) Targeted disruption of int-2 (fgf-3) causes developmental defects in the tail and inner ear. *Mol. Reprod. Dev.* **39,** 62–68.
40. Ozawa, K., Suzuki, S., Asada, M., et al. (1998) An alternatively spliced fibroblast growth factor (FGF)-5 mRNA is abundant in brain and translates into a partial agonist/antagonist for FGF-5 neurotrophic activity. *J. Biol. Chem.* **273,** 29262–29271.
41. Ghosh, A.K., Shankar, D.B., Shackleford, G.M., et al. (1996) Molecular cloning and characterization of human FGF8 alternative messenger RNA forms. *Cell Growth Differ.* **7,** 1425–1434.
42. Xu, J., Lawshe, A., MacArthur, C.A., and Ornitz, D.M. (1999) Genomic structure, mapping, activity and expression of fibroblast growth factor 17. *Mech. Dev.* **83,** 165–178.
43. Munoz-Sanjuan, I., Smallwood, P.M., and Nathans, J. (2000) Isoform diversity among fibroblast growth factor homologous factors is generated by alternative promoter usage and differential splicing. *J. Biol. Chem.* **275,** 2589–2597.
44. Mignatti, P., Morimoto, T., and Rifkin, D.B. (1992) Basic fibroblast growth factor, a protein devoid of secretory signal sequence, is released by cells via a pathway independent of the endoplasmic reticulum-Golgi complex. *J. Cell. Physiol.* **151,** 81–93.
45. Miyake, A., Konishi, M., Martin, F.H., et al. (1998) Structure and expression of a novel member, FGF-16, of the fibroblast growth factor family. *Biochem. Biophys. Res. Commun.* **243,** 148–152.
46. Miyamoto, M., Naruo, K., Seko, C., Matsumoto, S., Kondo, T., and Kurokawa, T. (1993) Molecular cloning of a novel cytokine cDNA encoding the ninth member of the fibroblast growth factor family, which has a unique secretion property. *Mol. Cell. Biol.* **13,** 4251–4259.
47. Ohmachi, S., Watanabe, Y., Mikami, T., et al. (2000) FGF-20, a novel neurotrophic factor, preferentially expressed in the substantia nigra pars compacta of rat brain. *Biochem. Biophys. Res. Commun.* **277,** 355–360.
48. Shin, J.T., Opalenik, S.R., and Wehby, J.N. (1996) Serum-starvation induces the extracellular appearance of FGF-1. *Biochim. Biophys. Acta.* **1312,** 27–38.

49. Florkiewicz, R.Z., Majack, R.A., Buechler, R.D., and Florkiewicz, E. (1995) Quantitative export of FGF-2 occurs through an alternative, energy-dependent, non-ER/Golgi pathway. *J. Cell. Physiol.* **162,** 388–399.
50. McNeil, P.L., Muthukrishnan, L., Warder, E., and D'Amore, P.A. (1989) Growth factors are released by mechanically wounded endothelial cells. *J. Cell Biol.* **109,** 811–822.
51. D'Amore, P.A. (1990) Modes of FGF release *in vivo* and *in vitro. Cancer Metastasis Rev.* **9,** 227–239.
52. Bikfalvi, A., Klein, S., Pintucci, G., Quarto, N., Mignatti, P., and Rifkin, D.B. (1995) Differential modulation of cell phenotype by different molecular weight forms of basic fibroblast growth factor: possible intracellular signaling by the high molecular weight forms. *J. Cell Biol.* **129,** 233–243.
53. Sherman, L., Stocker, K.M., Morrison, R., and Ciment, G. (1993) Basic fibroblast growth factor (bFGF) acts intracellularly to cause the transdifferentiation of avian neural crest-derived Schwann cell precursors into melanocytes. *Development* **118,** 1313–1326.
54. Powell, P.P., Finklestein, S.P., Dionne, C., Jaye, M., and Klagsbrun, M. (1991) Temporal, differential and regional expression of mRNA for basic fibroblast growth factor in the developing and adult rat brain. *Mol. Brain Res.* **11,** 71–77.
55. Emoto, N., Gonzalez, A.M., Walicke, P.A., et al. (1989) Basic fibroblast growth factor (FGF) in the central nervous system: identification of specific loci of basic FGF expression in the rat brain. *Growth Factors* **2,** 21–29.
56. Gonzalez, A.M., Berry, M., Maher, P.A., Logan, A., and Baird, A. (1995) A comprehensive analysis of the distribution of FGF-2 and FGFR1 in the rat brain. *Brain Res.* **701,** 201–226.
57. Ferrara, N., Ousley, F. and Gospodarowicz, D. (1988) Bovine brain astrocytes express basic fibroblast growth factor, a neurotropic and angiogenic mitogen. *Brain Res.* **462,** 223–232.
58. Vijayan, V.K., Lee, Y.L. and Eng, L.F. (1993) Immunohistochemical localization of basic fibroblast growth factor in cultured rat astrocytes and oligodendrocytes. *Int. J. Dev. Neurosci.* **11,** 257–267.
59. Frautschy, S.A., Walicke, P.A. and Baird, A. (1991) Localization of basic fibroblast growth factor and its mRNA after CNS injury. *Brain Res.* **553,** 291–299.
60. Logan, A., Frautschy, S.A., Gonzalez, A.-M., and Baird, A. (1992) A time course for the focal elevation of synthesis of basic fibroblast growth factor and one of its high-affinity receptors *(flg)* following a localized cortical brain injury. *J. Neurosci.* **12,** 3828–3837.
61. Chadi, G., Cao, Y., Pettersson, R.F., and Fuxe, K. (1994) Temporal and spatial increase of astroglial basic fibroblast growth factor synthesis after 6-hydroxydopamine-induced degeneration of the nigrostriatal dopamine neurons. *Neuroscience* **61,** 891–910.
62. Flores, C., Rodaros, D. and Stewart, J. (1998) Long-lasting induction of astrocytic basic fibroblast growth factor by repeated injections of amphetamine: blockade by concurrent treatment with a glutamate antagonist. *J. Neurosci.* **18,** 9547–9555.
63. Riva, M.A., Donati, E., Tascedda, F., Zolli, M., and Racagni, G. (1994) Short- and long-term induction of basic fibroblast growth factor gene expression in rat central nervous system following kainate injection. *Neuroscience* **59,** 55–65.
64. Speliotes, E.K., Caday, C.G., Do, T., Weise, J., Kowall, N.W., and Finklestein, S.P. (1996) Increased expression of basic fibroblast growth factor (bFGF) following focal cerebral infarction in the rat. *Mol. Brain Res.* **39,** 31–42.
65. Bhatnagar, M., Cintra, A., Chadi, G., et al. (1997) Neurochemical changes in the hippocampus of the brown Norway rat during aging. *Neurobiol. Aging* **18,** 319–327.
66. Cintra, A., Lindberg, J., Chadi, G., et al. (1994) Basic fibroblast growth factor and steroid receptors in the aging hippocampus of the brown Norway rat: immunocytochemical analysis in combination with stereology. *Neurochem. Int.* **25,** 39–45.

67. Lolova, I.S. and Lolov, S.R. (1995) Age-related changes in basic fibroblast growth factor-immunoreactive cells of rat substantia nigra. *Mech. Ageing Dev.* **82,** 73–89.
68. Colangelo, A.M., Follesa, P. and Mocchetti, I. (1998) Differential induction of nerve growth factor and basic fibroblast growth factor mRNA in neonatal and aged rat brain. *Mol. Brain Res.* **53,** 218–225.
69. Burrus, L.W., Zuber, M.E., Lueddecke, B.A., and Olwin, B.B. (1992) Identification of a cysteine-rich receptor for fibroblast growth factors. *Mol. Cell. Biol.* **12,** 5600–5609.
70. Zuber, M.E., Zhou, Z., Burrus, L.W., and Olwin, B.B. (1997) Cysteine-rich FGF receptor regulates intracellular FGF-1 and FGF-2 levels. *J. Cell. Physiol.* **170,** 217–227.
71. Jaye, M., Schlessinger, J. and Dionne, C.A. (1992) Fibroblast growth factor receptor tyrosine kinases: molecular analysis and signal transduction. *Biochim. Biophys. Acta.* **1135,** 185–199.
72. Johnson, D.E. and Williams, L.T. (1993) Structural and functional diversity in the FGF receptor multigene family. *Adv. Cancer Res.* **60,** 1–41.
73. Takaishi, S., Sawada, M., Morita, Y., Seno, H., Fukuzawa, H., and Chiba, T. (2000) Identification of a novel alternative splicing of human FGF receptor 4: Soluble-form splice variant expressed in human gastrointestinal epithelial cells. *Biochem. Biophys. Res. Commun.* **267,** 658–662.
74. Kinoshita, N., Minshull, J. and Kirschner, M.W. (1995) The identification of two novel ligands of the FGF receptor by a yeast screening method and their activity in Xenopus development. *Cell* **83,** 621–630.
75. Williams, E.J., Furness, J., Walsh, F.S., and Doherty, P. (1994) Activation of the FGF receptor underlies neurite outgrowth stimulated by L1, N-CAM, and N-cadherin. *Neuron.* **13,** 583–594.
76. Doherty, P. and Walsh, F.S. (1996) CAM-FGF receptor interactions: a model for axonal growth. *Mol. Cell. Neurosci.* **8,** 99–111.
77. Wanaka, A., Johnson, Jr, E.M. and Milbrandt, J. (1990) Localization of FGF receptor mRNA in the adult rat central nervous system by in situ hybridization. *Neuron.* **5,** 267–281.
78. Asai, T., Wanaka, A., Kato, H., Masana, Y., Seo, M., and Tohyama, M. (1993) Differential expression of two members of the FGF receptor gene family, FGFR-1 and FGFR-2, in the adult rat central nervous system. *Mol. Brain Res.* **17,** 174–178.
79. Takami, K., Kiyota, Y., Iwane, M., et al. (1993) Upregulation of fibroblast growth factor-receptor messenger RNA expression in rat brain following transient forebrain ischemia. *Exp. Brain Res.* **97,** 185–194.
80. Yazaki, N., Hosoi, Y., Kawabata, K., et al. (1994) Differential expression patterns of mRNAs for members of the fibroblast growth factor receptor family, FGFR-1-FGFR-4, in rat brain. *J. Neurosci. Res.* **37,** 445–452.
81. Itoh, N., Yazaki, N., Tagashira, S., et al. (1994) Rat FGF receptor-4 mRNA in the brain is expressed preferentially in the medial habenular nucleus. *Mol. Brain Res.* **21,** 344–348.
82. Miyake, A., and Itoh, N. (1996) Rat fibroblast growth factor receptor-4 mRNA in the brain is preferentially expressed in cholinergic neurons in the medial habenular nucleus. *Neurosci. Lett.* **203,** 101–104.
83. Reilly, J.F., and Kumari, V.G. (1996) Alterations in fibroblast growth factor receptor expression following traumatic brain injury. *Exp. Neurol.* **140,** 139–150.
84. Montgomery, D.L. (1994) Astrocytes: form, functions, and roles in disease. *Vet. Pathol.* **31,** 145–167.
85. Alonso, G., and Privat, A. (1993) Reactive astrocytes involved in the formation of lesional scars differ in the mediobasal hypothalamus and in other forebrain regions. *J. Neurosci. Res.* **34,** 523–538.
86. Hou, Y.-J., Yu, A.C.H., Garcia, J.M.R.Z., et al. (1995) Astrogliosis in culture. IV. Effects of basic fibroblast growth factor. *J. Neurosci. Res.* **40,** 359–370.
87. Perraud, F., Besnard, F., Pettmann, B., Sensenbrenner, M., and Labourdette, G. (1988) Effects of acidic and basic fibroblast growth factors (aFGF and bFGF) on the proliferation and the glutamine synthetase expression of rat astroblasts in culture. *Glia* **1,** 124–131.

88. Petroski, R.E., Grierson, J.P. Choi-Kwon, S. and Geller, H.M. (1991) Basic fibroblast growth factor regulates the ability of astrocytes to support hypothalamic neuronal survival in vitro. *Dev. Biol.* **147,** 1–13.
89. Simpson, C.S., and Morris, B.J. (1994) Basic fibroblast growth factor induces c-fos expression in primary cultures of rat striatum. *Neurosci. Lett.* **170,** 281–285.
90. Flanders, K.C., Lüdecke, G. Renzing, R. Hamm, C. Cissel, D.S. and Unsicker K. (1993) Effects of TGF-βs and bFGF on astroglial cell growth and gene expression in vitro. *Mol. Cell. Neurosci.* **4,** 406–417.
91. Meiners, S., Marone, M. Rittenhouse, J.L. and Geller, H.M. (1993) Regulation of astrocytic tenascin by basic fibroblast growth factor. *Dev. Biol.* **160,** 480–493.
92. Eclancher, F., Kehrli, P. Labourdette, G. and Sensenbrenner, M. (1996) Basic fibroblast growth factor (bFGF) injection activates the glial reaction in the injured adult rat brain. *Brain Res.* **737,** 201–214.
93. Gómez-Pinilla, F., Vu, L. and Cotman, C.W. (1995) Regulation of astrocyte proliferation by FGF-2 and heparan sulfate in vivo. *J. Neurosci.* **15,** 2021–2029.
94. Presta, M., Urbinati, C. Dell'Era, P. et al. (1995) Expression of basic fibroblast growth factor and its receptors in human fetal microglia cells. *Int. J. Dev. Neurosci.* **13,** 29–39.
95. Balaci, L., Presta, M. Ennas, M.G. et al. Differential expression of fibroblast growth factor receptors by human neurones, astrocytes and microglia. *Neuroreport* **6,** 197–200.
96. Araujo, D.M., and Cotman, C.W. (1992) Basic FGF in astroglial, microglial, and neuronal cultures: characterization of binding sites and modulation of release by lymphokines and trophic factors. *J. Neurosci.* **12,** 1668–1678.
97. Lee, S.C., Liu, W. Brosnan, C.F. and Dickson, D.W. (1994) GM-CSF promotes proliferation of human fetal and adult microglia in primary cultures. *Glia* **12,** 309–318.
98. Giulian, D., and Ingeman J.E. (1988) Colony-stimulating factors as promoters of ameboid microglia. *J. Neurosci.* **8,** 4707–4717.
99. Ohno, K., Suzumura, A., Sawada, M., and Marunouchi, T. (1990) Production of granulocyte/macrophage colony-stimulating factor by cultured astrocytes. *Biochem. Biophys. Res. Commun.* **169,** 719–724.
100. Giulian, D., Johnson, B., Krebs, J.F., George, J.K. and Tapscott, M. (1991) Microglial mitogens are produced in the developing and injured mammalian brain. *J. Cell Biol.* **112,** 323–333.
101. Fuxe, K., Chadi, G., Tinner, B., Agnati, L.F., Pettersson, R., and David, G. (1994) On the regional distribution of heparan sulfate proteoglycan immunoreactivity in the rat brain. *Brain Res.* **636,** 131–138.
102. Lin, W.-L. (1990) Immunogold localization of basal laminar heparan sulfate proteoglycan in rat brain and retinal capillaries. *Brain Res. Bull.* **24,** 533–536.
103. Schlessinger, J., Lax, I., and Lemmon, M. (1995) Regulation of growth factor activation by proteoglycans: what is the role of the low affinity receptors? *Cell* **83,** 357–360.
104. Gonzalez, A.M., Carman, L.S., Ong, M., et al. (1994) Storage, metabolism, and processing of ^{125}I-fibroblast growth factor-2 after intracerebral injection. *Brain Res.* **665,** 285–292.
105. Gómez-Pinilla, F., Miller, S., Choi, J., and Cotman, C.W. (1997) Heparan sulfate potentiates the autocrine action of basic fibroblast growth factor in astrocytes: an in vivo and in vitro study. *Neuroscience* **76,** 137–145.
106. Reilly, J.F., Bair, L., and Kumari, V.G. (1997) Heparan sulfate modifies the effects of basic fibroblast growth factor on glial reactivity. *Brain Res.* **759,** 277–284.
107. Flaumenhaft, R., Moscatelli, D., and Rifkin, D.B. (1990) Heparin and heparan sulfate increase the radius of diffusion and action of basic fibroblast growth factor. *J. Cell Biol.* **111,** 1651–1659.
108. Massagué, J. (1990) The transforming growth factor-β family. *Annu. Rev. Cell Biol.* **6,** 597–641.
109. Lawrence, D.A. (1996) Transforming growth factor-β: a general review. *Eur. Cytokine Netw.* **7,** 363–374.

110. Laping, N.J., Morgan, T.E., Nichols, N.R., et al. (1994) Transforming growth factor-β1 induces neuronal and astrocyte genes: tubulin α1, glial fibrillary acidic protein and clusterin. *Neuroscience* **58,** 563–572.
111. Logan, A., Berry, M., Gonzalez, A.M., Frautschy, S.A., Sporn, M.B., and Baird, A. (1994) Effects of transforming growth factor β1 on scar production in the injured central nervous system of the rat. *Eur. J. Neurosci.* **6,** 355–363.
112. Wyss-Coray, T., Feng, L., Masliah, E., Ruppe, M.D., Lee, H.S., Toggas, S.M., Rockenstein, E.M., and Mucke, L. (1995) Increased central nervous system production of extracellular matrix components and development of hydrocephalus in transgenic mice overexpressing transforming growth factor-β1. *Am. J. Pathol.* **147,** 53–67.
113. Reilly, J.F., Maher, P.A., and Kumari, V.G. (1998) Regulation of astrocyte GFAP expression by TGF-β1 and FGF-2. *Glia,* **22,** 202–210.
114. Toru-Delbauffe, D., Baghdassarian-Chalaye, D., Gavaret, J.M., Courtin, F., Pomerance, M., and Pierre, M. (1990) Effects of transforming growth factor β1 on astroglial cells in culture. *J. Neurochem.* **54,** 1056–1061.
115. Baorto, D.M., Mellado, W., and Shelanski, M.L. (1992) Astrocyte process growth induction by actin breakdown. *J. Cell Biol.* **117,** 357–367.
116. Moffett, J., Kratz, E., Myers, J., Stachowiak, E.K., Florkiewicz, R.Z., and Stachowiak, M.K. (1998) Transcriptional regulation of fibroblast growth factor-2 expression in human astrocytes: implications for cell plasticity. *Mol. Biol. Cell* **9,** 2269–2285.
117. Charon, I., Zuin-Kornmann, G., Bataille, S., and Schorderet, M., (1998) Protective effect of neurotrophic factors, neuropoietic cytokines and dibutyryl cyclic AMP on hydrogen peroxide-induced cytotoxicity on PC12 cells: a possible link with the state of differentiation. *Neurochem. Int.* **33,** 503–511.
118. Schubert, D., Ling, N., and Baird, A. (1987) Multiple influences of a heparin-binding growth factor on neuronal development. *J. Cell Biol.* **104,** 635–643.
119. Unsicker, K., Reichert-Preibsch, H., and Wewetzer, K. (1992) Stimulation of neuron survival by basic FGF and CNTF is a direct effect and not mediated by non-neuronal cells: evidence from single cell cultures. *Dev. Brain Res.* **65,** 285–288.
120. Vaca, K., and Wendt, E. (1992) Divergent effects of astroglial and microglial secretions on neuron growth and survival. *Exp. Neurol.* **118,** 62–72.
121. Westermann, R., Grothe, C., and Unsicker, K. (1990) Basic fibroblast growth factor (bFGF), a multifunctional growth factor for neuroectodermal cells. *J. Cell Sci. Suppl.* **13,** 97–117.
122. Otto, D., and Unsicker, K. (1994) FGF-2 in the MPTP model of Parkinson's disease: effects on astroglial cells. *Glia,* **11,** 47–56.
123. Dietrich, W.D., Alonso, O. Busto, R. and Finklestein, S.P. (1996) Posttreatment with intravenous basic fibroblast growth factor reduces histopathological damage following fluid-percussion brain injury in rats. *J. Neurotrauma* **13,** 309–316.
124. Dreyfus, H., Sahel, J., Heidinger, V. et al. (1998) Gangliosides and neurotrophic growth factors in the retina. Molecular interactions and applications as neuroprotective agents. *Ann. New York Acad. Sci.* **845,** 240–252.
125. Rabchevsky, A.G., Fugaccia, I., Fletcher-Turner, A., Blades, D.A., Mattson, M.P., and Scheff, S.W. (1999) Basic fibroblast growth factor (bFGF) enhances tissue sparing and functional recovery following moderate spinal cord injury. *J. Neurotrauma* **16,** 817–830.
126. Goss, J.R., O'Malley, M.E., Zou, L., Styren, S.D., Kochanek, P.M., and DeKosky, S.T. (1998) Astrocytes are the major source of nerve growth factor upregulation following traumatic brain injury in the rat. *Exp. Neurol.* **149,** 301–309.
127. Garcia-Estrada, J., Garcia-Segura, L.M., and Torres-Aleman, I. (1992) Expression of insulin-like growth factor I by astrocytes in response to injury. *Brain Res.* **592,** 343–347.

128. De Groot, C.J.A., Montagne, L., Barten, A.D., Sminia, P., and Van Der Valk, P. (1999) Expression of transforming growth factor (TGF)-β1, -β2, and β3 isoforms and TGF-β type I and type II receptors in multiple sclerosis lesions and human adult astrocyte cultures. *J. Neuropathol. Exp. Neurol.* **58,** 174–187.
129. Lee, M.Y., Kim, C.J., Shin, S.L., Moon, S.H., and Chun, M.H. (1998) Increased ciliary neurotrophic factor expression in reactive astrocytes following spinal cord injury in the rat. *Neurosci. Lett.* **255,** 79–82.
130. Gluckman, P., Klempt, N., Guan, J., et al. (1992) A role for IGF-1 in the rescue of CNS neurons following hypoxic-ischemic injury. *Biochem. Biophys. Res. Commun.* **182,** 593–599.
131. Henrich-Noack, P., Prehn, J.H., and Krieglstein, J. (1996) TGF-beta 1 protects hippocampal neurons against degeneration caused by transient global ischemia. Dose-response relationship and potential neuroprotective mechanisms. *Stroke* **27,** 1609–1615.
132. Korsching, S. (1993) The neurotrophic factor concept: a reexamination. *J. Neurosci.* **13,** 2739–2748.
133. Smith, R., Peters, G., and Dickson, C. (1988) Multiple RNAs expressed from the int-2 gene in mouse embryonal carcinoma cell lines encode a protein with homology to fibroblast growth factors. *EMBO J.* **7,** 1013–1022.
134. Taira, M., Yoshida, T., Miyagawa, K., Sakamoto, H., Terada, M., and Sugimura, T. (1987) cDNA sequence of human transforming gene hst and identification of the coding sequence required for transforming activity. *Proc. Natl. Acad. Sci. USA* **84,** 2980–2984.
135. Delli-Bovi, P., Curatola, A.M., Kern, F.G., Greco, A., Ittmann, M., and Basilico, C. (1987) An oncogene isolated by transfection of Kaposi's sarcoma DNA encodes a growth factor that is a member of the FGF family. *Cell* **50,** 729–737.
136. Zhan, X., Bates, B., Hu, X.G., and Goldfarb, M. (1988) The human FGF-5 oncogene encodes a novel protein related to fibroblast growth factors. *Mol. Cell Biol.* **8,** 3487–3495.
137. Marics, I., Adelaide, J., Raybaud, F., et al. (1989) Characterization of the HST-related FGF.6 gene, a new member of the fibroblast growth factor gene family. *Oncogene* **4,** 335–340.
138. Finch, P.W., Rubin, J.S., Miki, T., Ron, D., and Aaronson, S.A. (1989) Human KGF is FGF-related with properties of a paracrine effector of epithelial cell growth. *Science* **245,** 752–755.
139. Tanaka, A., Miyamoto, K., Minamino, N. (1992) Cloning and characterization of an androgen-induced growth factor essential for the androgen-dependent growth of mouse mammary carcinoma cells. *Proc. Natl. Acad. Sci. USA* **89,** 8928–8932.
140. Yamasaki, M., Miyake, A., Tagashira, S., and Itoh, N. (1996) Structure and expression of the rat mRNA encoding a novel member of the fibroblast growth factor family. *J. Biol. Chem.* **271,** 15918–15921.
141. Coulier, F., Pontarotti, P., Roubin, R., Hartung, H., Goldfarb, M., and Birnbaum, D. (1997) Of worms and men: an evolutionary perspective on the fibroblast growth factor (FGF) and FGF receptor families. *J. Mol. Evol.* **44,** 43–56.
142. Smallwood, P.M., Munoz-Sanjuan, I., Tong, P., et al. (1996) Fibroblast growth factor (FGF) homologous factors: new members of the FGF family implicated in nervous system development. *Proc. Natl. Acad. Sci. USA* **93,** 9850–9857.
143. McWhirter, J.R., Goulding, M., Weiner, J.A., Chun, J., and Murre, C. (1997) A novel fibroblast growth factor gene expressed in the developing nervous system is a downstream target of the chimeric homeodomain oncoprotein E2A-Pbx1. *Development* **124,** 3221–3232.
144. Greene, J.M., Li, Y.L., Yourey, P.A., et al. (1998) Identification and characterization of a novel member of the fibroblast growth factor family. *Eur. J. Neurosci.* **10,** 1911–1925.
145. Hoshikawa, M., Ohbayashi, N., Yonamine, A., et al. (1998) Structure and expression of a novel fibroblast growth factor, FGF-17, preferentially expressed in the embryonic brain. *Biochem. Biophys. Res. Commun.* **244,** 187–191.

146. Hu, M.C.-T., Qiu, W.R., Wang, Y.-P., et al. (1998) FGF-18, a novel member of the fibroblast growth factor family, stimulates hepatic and intestinal proliferation. *Mol. Cell. Biol.* **18,** 6063–6074.
147. Ohbayashi, N., Hoshikawa, M., Kimura, S., Yamasaki, M., Fukui, S., and Itoh, N. (1998) Structure and expression of the mRNA encoding a novel fibroblast growth factor, FGF-18. *J. Biol. Chem.* **273,** 18161–18164.
148. Nishimura, T., Utsunomiya, Y., Hoshikawa, M., Ohuchi, H., and Itoh, N. (1999) Structure and expression of a novel human FGF, FGF-19, expressed in the fetal brain. *Biochim. Biophys. Acta* **1444,** 148–151.
149. Nishimura, T., Nakatake, Y., Konishi, M., and Itoh, N. (2000) Identification of a novel FGF, FGF-21, preferentially expressed in the liver. *Biochim. Biophys. Acta.* **1492,** 203–206.
150. Yamashita, T., Yoshioka, M., and Itoh, N. (2000) Identification of a novel fibroblast growth factor, FGF-23, preferentially expressed in the ventrolateral thalamic nucleus of the brain. *Biochem. Biophys. Res. Commun.* **277,** 494–498.

11
Trophins as Mediators of Astrocyte Effects in the Aging and Regenerating Brain

Judith Lackland and Cheryl F. Dreyfus

1. INTRODUCTION

In their role as cells that provide multiple forms of support to the central nervous system (CNS), astrocytes are beginning to be appreciated as suppliers of critical survival and differentiation factors to neurons and other glia. Current work suggests that astrocytes may express these trophic factors throughout life, not only in the developing fetus, but also in the adult and aged individual. Moreover, studies on brain injury suggest that expression of astrocyte-derived factors is plastic and may be regulated by molecules elicited by injury, such as cytokines, hormones, and growth factors, as well as neurotransmitters. The hope is that as a result of these studies these regulatory molecules may be utilized to enhance astrocyte trophin function and thereby support the aging and injured brain. To examine these contentions, this chapter

1. Defines expression of growth factors by adult astrocytes,
2. Considers the regulation of this expression by injury,
3. Details effects of injury-induced molecules on growth factor expression and
4. Discusses the possible implications of alterations in astrocyte-derived trophic support on learning, memory, and the vulnerability of the CNS to aging and injury.

We focus in this review on astrocyte expression of trophic factors, bearing in mind that astrocytes express a host of other molecules that enhance or diminish neuron function *(1,2)*. In addition, we have limited this review to astrocytes, but it should be appreciated that other glial subtypes, i.e., oligodendrocytes *(3)* and microglia *(4)* also express survival factors and they too undoubtedly contribute to the support of neurons, as well as astrocytes, in CNS function.

2. EXPRESSION OF GROWTH FACTORS BY ASTROCYTES IN THE ADULT BRAIN

2.1. The NGF Gene Family

That trophic factors are expressed by astrocytes was first suggested 20 yr ago by identification of nerve growth factor (NGF) in astrocytes in culture. In these studies, conditioned medium (CM), derived from lesioned adult rat brain astrocytes, was found

From: *Neuroglia in the Aging Brain*
Edited by: Jean S. de Vellis © Humana Press Inc., Totowa, NJ

to enhance survival of NGF-responsive, as well as NGF-insensitive, peripheral neurons *(5)*. Effects of CM on NGF-dependent neurons were blocked by neutralizing antibodies to NGF, supporting the claim that at least some of these effects could be attributed to NGF. These observations were confirmed by others who found NGF protein and mRNA were expressed by cultured neonatal mouse and rat brain astrocytes and that NGF was in CM derived from these cells *(6–13)*. Not only was NGF present in these astrocytes, the expression appeared to be regulated by growth status; proliferating cells expressed more NGF than quiescent ones *(7)*.

These findings prompted other studies that investigated glial NGF mRNA expression in vivo and evaluated how expression changes as the brain ages. The optic nerve provided a glial population for such in vivo examination. Optic nerves from neonatal and adult rats were monitored for NGF mRNA using nuclease protection assays. NGF mRNA was found to be decreased in the optic nerve from adult as compared to neonatal animals, suggesting that, as in culture, NGF mRNA in vivo may be highly expressed in young animals in which glial cells are known to be proliferating. When glia become quiescent, such as in older animals, NGF mRNA decreases *(14)*.

However, these studies of optic glia, although exciting, were limited. The optic nerve is composed of multiple glial subtypes, including oligodendrocytes as well as astrocytes. Study of this model did not address the question of NGF expression specifically in astrocytes. To do so, it was necessary to evaluate relative levels of NGF mRNA or protein in glial fibrillary acidic protein-positive (GFAP+) astrocyte cells. Therefore, *in situ* hybridization studies were performed that examined coexpression of NGF mRNA with GFAP. The findings indicated that NGF mRNA was present at low levels in adult astrocytes of the basal forebrain, neocortex, amygdala, cerebellar cortex, and hypothalamus in vivo *(15)*.

Subsequent work examined how such expression of NGF may be regulated during maturation. In particular, cultures were utilized in which cortical astrocytes derived from fetal, neonatal or adult animals were isolated and examined for NGF mRNA. The astrocytes were found to express progressively lower levels of NGF mRNA and protein as the age of the animal from which the astrocytes were derived increased *(11)*. As the brain ages, then, NGF in astrocytes appears to decline.

NGF is a member of a gene family, known as the neurotrophins that includes brain-derived neurotrophic factor (BDNF), neurotrophin-3 (NT-3) and neurotrophin-4 (NT-4). Although little is known about NT-4 in astrocytes, BDNF and NT-3 have been found to be expressed by astrocytes. For example, BDNF and NT3 mRNAs were found to be expressed in astrocytes in culture *(10,13,16–18)* and BDNF, as well as NT3 proteins were localized to astrocyctes and astrocyte-like cells in vivo *(19–21)*. Moreover, neurotrophin expression was found to be decreased during brain maturation. When astrocytes from the cortex were dissociated from brains of animals of increasing ages and grown in culture, BDNF mRNA expression decreased as animals matured from fetus to adult (personal observation). As a group, then, NGF, BDNF, and NT-3 are expressed by astrocytes and are downregulated during brain maturation.

2.2. Other Growth Factors

Although the neurotrophins have been examined most extensively, other trophic factors have now been found to be produced by astrocytes. These include ciliary neurotrophic factor (CNTF) *(10,22,23),* fibroblast growth factor-2 (FGF-2) *(24–26),* enkephalin, somatostatin *(27),* and glial cell line-derived neurotrophic factor (GDNF) *(28,29).* Interestingly, in vivo studies indicated that while expression of some of these growth factors, for example, FGF-2, decreased in vivo with maturity *(30),* expression of others, such as CNTF, increased with age *(31).* Moreover, culture studies indicated that neuropeptides were also regulated in an age-dependent manner. For example, cerebellar astrocytes exhibited an increase in expression of proenkephalin and a decrease in somatostatin when derived from progressively older animals *(32).* Astrocytes, then, appear not only to express multiple factors, but also to alter their roles as trophin providers during maturity by exhibiting selective regulation of the different factors. Relative levels of individual neurotrophins, as well as FGF-2, CNTF, somatostatin and proenkephalin change as aging occurs.

3. EFFECTS OF INJURY ON ASTROCYTIC EXPRESSION OF GROWTH FACTORS

A new literature has indicated that the in vivo expression of astrocyte-derived molecules, as defined in the CNS above, is plastic and can be regulated by environmental signals. This has been most clearly demonstrated by studies examining the effects of injury on profiles of growth factor expression. For example, in the adult, a lesion of the optic nerve resulted in enhancement of NGF mRNA expression *(14),* and chronic neurotoxic damage to the basal forebrain and hippocampus *(15),* or trauma to the cortex *(33)* led to an increase in NGF mRNA in GFAP+ cells. Ischemia also led to an increase in NGF protein expression in astrocytes in the hippocampus *(34).* Studies in culture complemented these in vivo findings. When compared with astrocytes from unlesioned animals, NGF protein and mRNA were markedly elevated in cultures of striatal astrocytes derived from brain lesioned by treatment with 6-hydroxydopamine (6HD) or 1-methyl-4-phenyl-1,2,3,6-tetrahydropyridine (MPTP). These astrocytes from the lesioned brain morphologically resembled reactive astrocytes in vivo. GFAP mRNA and protein were elevated as a result of the lesion along with vimentin and two epitopes that are only expressed in vivo on reactive astrocytes *(35,36).*

Injury could similarly affect other growth factors. For example, FGF-2 mRNA and protein were found to be increased in GFAP+ astrocytes *(37–39)* or in astrocyte-like cells *(40,41)* following damage. Increases in CNTF mRNA, in concert with increases in GFAP+ astrocytes *(22,42),* as well as increases in IGF-1 mRNA in GFAP+ astrocytes *(43)* have also been reported.

Importantly, not all astrocyte-derived factors are elevated by injury. For example, in vivo studies that indicated that CNTF was increased, demonstrated that BDNF and NT-3 were slightly decreased after injury *(22)* and culture studies that utilized lesioned astrocytes, showed that NGF mRNA was elevated although proenkephalin was decreased *(35).* Thus, trophin production in astrocytes is plastic and can be altered by lesions in a selective manner.

The studies of injury have prompted analyses of the molecules that are regulated by injury. These experiments have suggested signals that may underlie plasticity in the aging brain. These include cytokines, hormones, neuronal growth factors, and possibly neurotransmitters (Fig. 1). New studies have begun to explore the effects of these molecules on astrocytes.

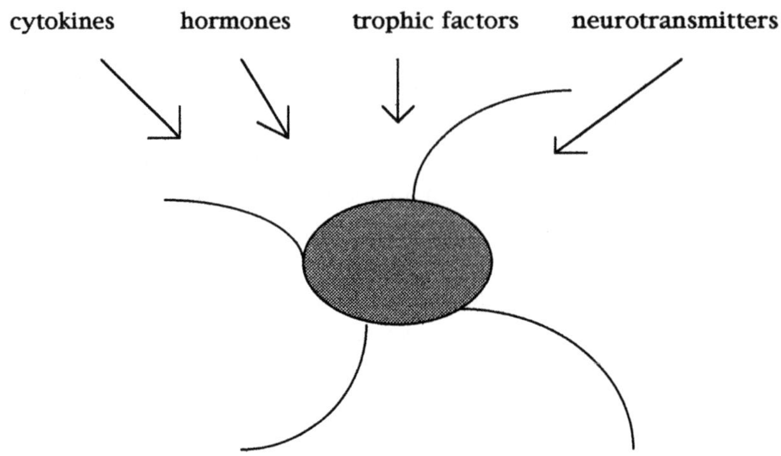

Fig. 1. In the maturing brain astrocytes express receptors for, and respond to, cytokines, hormones, trophic factors and neurotransmitters.

4. REGULATION OF ASTROCYTE GROWTH FACTORS BY MOLECULES ELEVATED FOLLOWING LESIONS

4.1. Cytokines

To begin to explore injury-induced molecules, cytokines, such as interferon-gamma (IFN-γ), interleukin 1β (IL-Iβ), and transforming growth factor-β1 (TGF-β1) are being investigated *(44–46)*. These molecules originate in the brain from microglia and astrocytes, or they enter the brain through a disrupted blood-brain-barrier. TGF-β1, for example, is abundant in platelets that enter the brain after specific lesions *(47)*.

What are the effects of these cytokines on growth factor expression in maturing astrocytes? To address this issue a growing literature has emerged primarily based on studies not of the aging or adult brain, but rather of cultures of early postnatal astrocytes. It has been relatively easy to look at direct effects of cytokines on expression of specific trophins in such neonatal models. For example, neonatal astrocytes have been shown to exhibit increased levels of NGF mRNA in response to bacterial lipopolysaccharide, a potent activator of the inflammatory response *(48,49)* and NGF mRNA was also found to be elevated in response to multiple cytokines, including IFN-γ *(11)*, TGF-β1 *(13,50–53)* and IL-Iβ *(11,13,52,53)*. The mechanisms underlying these responses are being elucidated *(54,55)*.

By extension of the work on early postnatal cells, new studies also have begun to evaluate adult astrocytes. Notably, adult cells in culture have been found to respond differently to cytokines than do neonatal cells. For example, adult astrocytes responded only minimally to IL-1β and IFNγ, whereas neonatal cells exhibited dramatic increases in NGF mRNA expression *(36)*. Lesions, however, modulated responsivity in the adult cells. Cultured adult astrocytes, derived from 6-HD lesioned brains, exhibited a marked elevation of NGF mRNA in response to cytokines *(36)*. These observations suggest that injury not only elicits an increase in molecules that regulate neurotrophin expression, but also sensitizes the cells to cytokines, thus augmenting the response.

As is the case with injury in general, all astrocyte-derived growth factors are not similarly regulated by cytokines. For example, although NGF, as well as FGF-2 expression was elevated by inflammatory molecules *(25)*, BDNF, NT-3, NT-4, GDNF, and CNTF expression was not *(13,25)*. Thus, in response to a lesion, cytokines elicit elevation of a precise combination of growth factors that may be released into the microenvironment to influence nearby neurons and other glial cells. These cytokines are prime candidates for further examination of molecules that regulate astrocytes in the aging brain.

4.2. *Glucocorticoids*

In addition to altering cytokine levels, brain damage elicits marked increases in glucocorticoid hormones *(56,57)*. Interestingly, these increases also occur during aging *(58)*. The elevated glucocorticoids may have direct effects on astrocytes. Glucocorticoid receptors are expressed in these cells *(59,60)*. Studies of postnatal hippocampal astrocytes revealed that dexamethasone regulated NGF mRNA expression *(51,61)*. Moreover, influences of glucocorticoids on individual growth factors have been found to differ; for example glucocorticoids increase FGF-2 mRNA in the same region where NGF mRNA is decreased *(61)*.

Elevation of multiple factors as a result of injury leads to interaction among these factors that influence astrocyte function. Thus, glucocorticoids have been shown to interact with cytokines, leading to a greater susceptibility to stress. For example, corticosterone inhibited the stimulatory effects of TGF-β1 and IL-1β on NGF expression in an astrocyte cell line *(62)*. Interestingly, another steroid, 1, 25-dihydroxyvitamin D3 (calcitriol), potentially produced by microglia *(63)*, stimulated NGF mRNA expression in an astrocyte cell line *(62)* and in cultured astrocytes *(64)*. Moreover, it augmented the effects of IL-1β and TGF-β1 *(62)*. Thus, immune responses to injury are complex and involve interactions of disparate molecules that have the potential to elevate as well as reduce trophin expression. Knowledge of these interactions is important in identifying signals that regulate aging astrocytes, even in the absence of injury.

4.3. *Gonadal Steroids*

Recent observations have predicted that another class of steroids, the estrogens, also affect growth factor expression in astrocytes. Estrogen response elements have been reported to be present in NGF and BDNF genes *(65,66)* and these genes have been shown to be regulated by estrogen in neurons in vivo *(65–68)*. Moreover, estrogen receptors have been identified in astrocytes and estrogen effects on morphology of postnatal cells in culture and adult cells in vivo have been described *(69–71)*. Current work, therefore, has examined effects of estrogen on astrocytes derived from postnatal and adult animals. When estradiol was added to enriched cortical astrocyte cultures, it elevated expression of BDNF mRNA in both adult and neonatal cells. In contrast, NT-3 was elevated in adult, but not neonatal cells *(72)*. Molecules other than neurotrophins, moreover, may be regulated by estrogens. For example, IGF-I in astrocytes appeared to be affected by estradiol in vivo *(73)*.

It remains to be determined how responsivity to estrogen may be affected by injury and how estrogen may interact with glucocorticoids or cytokines to influence astrocyte function. However, it is interesting to note that in astrocytes the estrogen synthetic enzyme, aromatase, and aromatase activity were upregulated following a lesion *(74)*. These cells normally produce little or none of this protein *(75)*. Therefore, CNS injury may result in an increase in the local astrocyte concentration of estrogen in the damaged region.

4.4. Effects of Astrocyte-Derived Growth Factors

As noted in **Subheading 4.3.,** expression of astrocyte growth factors (neurotrophins, FGF-2, and CNTF) decreased or increased as the brain aged or following a lesion. This, potentially resulted in a change in the trophic factor milieu. Effects of such factors on neurons have been examined in some detail. In contrast, the effects on astrocytes have been relatively unknown, but are beginning to be explored.

4.4.1. The Role of Neurotrophins on Astrocytes

In the case of the neurotrophins, data supporting a role for these factors on astrocytes is sparse. This is due in part to the lack of evidence that functional neurotrophin receptors are expressed on astrocytes and the expectation, therefore, that neurotrophins would play no role. Recent studies, however, have begun to force a reexamination of this issue.

Neurotrophin receptors consist of two transmembrane glycoproteins, the p75 neurotrophin receptor and the trk receptor tyrosine kinases *(76,77)*. The p75 receptor has been known as the common neurotrophin receptor because it binds all neurotrophins with equal affinity *(78)*. Astrocytes express the p75 receptor in culture *(79–81)* and in the adult brain in vivo *(81,82)* and this expression has been reported to be elevated following a lesion *(83)*. However, the role of the p75 receptor on astrocytes is unclear, particularly in response to damage or aging.

Trk receptors, in contrast to p75, are relatively specific in action. NGF binds most specifically to trkA, BDNF and NT-4 to trkB and NT3 to trkC. These receptors have been reported to be expressed on astrocytes in both full-length and truncated form *(79–81)*. The predominant form, however, is the truncated one, considered to be unable to transduce a neurotrophin signal as measured by immediate early gene induction or by protein tyrosine phosphorylation *(81)*. Recent studies, however, have suggested that these receptors are functional. Truncated trkB receptors were found to bind and endocytose BDNF which could be released as a bioactive molecule to influence neuron survival *(84)*. Moreover, truncated trkB, transfected into a specific cell line, mediated effects of BDNF on the rate of acidic metabolite release *(85)*. These findings have been complemented by observations that the truncated receptor was elevated in response to IFN-γ *(84)*, a molecule upregulated by injury, suggesting that the effects of neurotrophins on astrocytes may be enhanced following injury.

Supplementing the observation that astrocytes express neurotrophin receptors, are the demonstrations that neurotrophins affect astrocytes. For example, it had been shown that NGF regulated levels of p75 *(79,80,86)* and trkB mRNA *(80)* in cultured cells. NGF also elicited dramatic increases in astrocyte cell numbers in embryonic septum (Fig. 2) and substantia nigra, as well as alterations in astrocyte morphology (Fig. 3) *(87) see also (86)*. The findings that both receptors and ligands are present in the same cells in the adult brain suggests that astrocytes may, through neurotrophins, signal to themselves or neighboring cells to enhance function, not only during development, but also in maturity.

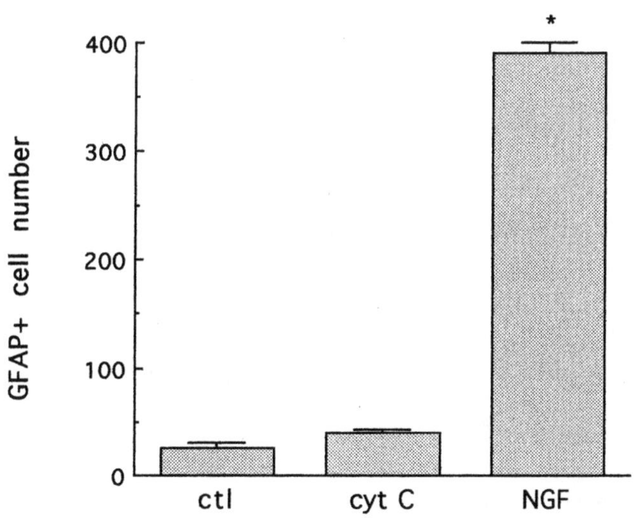

Fig. 2. Effect of NGF on GFAP-positive cell numbers in embryonic septal culture. Cultures were grown for 7 days in the presence (NGF;1-2ug /ml) or absence (ctl) of NGF or cytochrome C (cyt C; 1-2ug/ml), a compound similar in structure and physicochemical properties to NGF. NGF was prepared from adult mouse salivary glands. NGF from a different preparation (Boehringer Mannheim) also increased astrocytic cell number, indicating that the effect was not due to a potential contaminant in any single preparation. **p* <0.0001, ctl vs NGF and cyt C vs NGF. Reprinted with permission *(87)*.

4.4.2. The Role of Other Trophic Factors on Astrocytes

An emerging literature suggests that other trophic factors expressed by astrocytes also affect astrocytes. F6Fs, for example, have been reported to elevate expression of NGF in astrocytes *(46,47)*. Moreover, lesions may enhance this FGF action. For example, expression of the FGF receptor *flg,* present on adult astrocytes, was found to be upregulated following injury *(35,83)*, In addition, cytokines, that are increased by a lesion, enhanced actions of FGF-2 *(47)*, supporting the concept that lesions elicit enhanced responsiveness. CNTF has also been reported to affect astrocyte function. For example, it increases GFAP and induces hypertrophy *(84)*. Here, too, injury appears to enhance responsivity. For example, receptors for CNTF were upregulated on astrocytes by damage, in some cases from almost undetectable levels *(25,81)*. This suggests that following injury astrocytes became further sensitized to the very molecules they themselves express.

4.5. Other Factors that Influence Astrocyte Function: The Role of Neuronal Signals

Increasing reports have indicated that astrocytes express a variety of K^+ channels *(90)*. Moreover, they exhibit α- *(91)* and β-*(92)* noradrenergic receptors, dopaminergic receptors *(93)*, muscarinic *(94,95)*, and nicotinic *(96)* cholinergic receptors, non-NMDA ionotropic and metabotropic glutamatergic receptors *(97–100)*, and peptidergic receptors *(93,101)*. The findings that many of these are present on astrocytes throughout life *(95,102–104)* and that they are, in some cases, upregulated following injury

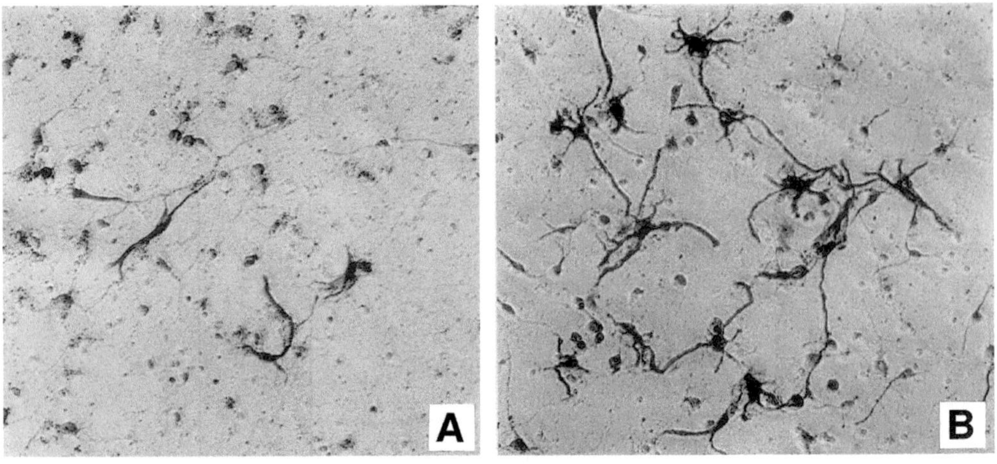

Fig. 3. Astrocytes visualized in dissociated septal cultures using GFAP immunocytochemical procedures. Astrocytes were detected in cultures grown in the absence (**A**) and presence (**B**) of NGF (1 ug/mL). Reprinted with permission *(87)*.

(105), indicate that neuronal signals may regulate astrocytes into adulthood and old age. Moreover, astrocytes have been shown to express specific receptors in close apposition to appropriate axons in vivo *(92)*, and K^+ is released from axons at locations in close proximity to astrocytes *(106,107)*, suggesting that astrocytes are in position to respond to neuronal signals. These observations have been complemented by the finding that treatment of astrocytes with appropriate ligands *(108–112)*, or with K^+ *(18,113)* elicited increases in $[Ca^{2+}]_i$ and in some cases altered astrocyte morphology.

To begin to determine whether astrocyte response to neuronal signals may extend to effects on trophin expression, effects of norepinephrine and other β-adrenergic agonists on NGF have been evaluated. Norepinephrine and other β-adrenergic agonists found to alter astrocyte expression of NGF *(114–116)* and BDNF *(13)*, and increase NGF secretion in whole brain astrocytes *(114)*. Dopamine and epinephrine also were to increase NGF and BDNF secretion *(8)*. Moreover, effects of β-adrenergic agonists on astrocytes were not limited to those on neurotrophins. Expression of proenkephalin, somatostatin *(11,115)* and FGF-2 *(117)* mRNAs and proteins was enhanced by these ligands. It should be noted, moreover, that all catecholamine effects were not stimulatory. CNTF mRNA and protein as well as FGF-1 mRNA were downregulated in astrocytes that exhibited enhanced expression of NGF or FGF-2 mRNAs and proteins *(116–118)*.

Glutamate also was found to influence growth factor expression. For example, glutamate enhanced expression of BDNF and NGF in the basal forebrain and cortex *(17,119)*, as well as FGF-2 in cortical astrocytes *(17)*, and GDNF in striatal astrocytes *(120)*.

What is the meaning of these responses to neuronal signals for the aging and lesioned brain? To the best of our knowledge, little is known of how neurotransmitter actions on astrocytes may be modified during development and after damage. It is clear that specific neurotransmitter receptors are located on adult astrocytes, suggesting that astrocytes are sensitive to this type of stimulus throughout life. Furthermore, astrocytes of damaged brain also have been reported to express neurotransmitter receptors *(95)* that may be upregulated in comparison to unlesioned brain *(105)*. However, undoubtedly the modulation of sensitivity to neuronal signals with aging and in response to lesion again depends on the specific stimulus, the brain region being examined and the responding growth factor being assessed. We have, for example, grown astrocytes from postnatal basal forebrain in culture and compared these to astrocytes derived from the adult. Although neonatal astrocytes responded to K^+ and glutamate by increasing expression of neurotrophins, adult basal forebrain cells did not, suggesting that responsivity to these signals decreases as aging proceeds *(121)*. However, it should be remembered that astrocytes may modulate responsivity to one set of stimuli in the presence of another set of stimuli. It remains to be determined whether the adult basal forebrain astrocytes increase responsivity to neuronal signals in the presence of the other categories of molecule associated with lesions.

5. CONSEQUENCES OF AGING ON LEARNING, MEMORY, AND VULNERABILITY TO LESION: THE POTENTIAL ROLE OF ASTROCYTES

5.1. Effects of Aging on Brain Function and Trophin Expression

As an animal ages, well documented alterations have been reported to occur in the morphology of neurons and astrocytes. For example, a loss of neurons and dendritic processes has been described that was accompanied by astrogliosis, characterized by elevated GFAP staining, astrocyte hypertrophy, and an increase in the fibrous character of astrocytes *(122–124)*. As a result of these changes, it has been suggested that there is an alteration in the neurotransmitter composition of the brain and that this may underlie decreases in cognition *(125)*. It has further been suggested that neuronal and glial alterations are associated with a decrease in trophic factor expression *(126–129)*. However, when examined, these changes in trophin expression have been found to be variable. Such modulation of trophins has been evaluated most extensively in the basal forebrain-hippocampal system where, in some cases neurotrophins *(129)* and their receptors *(128)* were reported to be decreased with aging. In contrast, other studies have found no change in expression *(126–128)*, or an increase in expression *(129)*. Interestingly, in spite of the variability in reports of effects of aging on neurotrophin expression, administration of exogenous NGF resulted in an improvement in cognitive function *(130–133)* and neuronal function, suggesting that some deficit in trophin expression was indeed present in aging that was, at least partly, corrected.

In an analogous situation, deficits in behavior and neuronal function in aged animals were associated with the nigro-striatal system. Levels of GDNF mRNA were not altered in the striatum of the aged animal *(134)*. However, as with the basal forebrain-hippocampus, both behavioral and neuronal deficits were improved by administration of GDNF *(135,136)*.

The deficits associated with aging extend to responsivity to injury and stress. For example, aged animals exposed to partial hippocampal lesions, exhibited greater decreases in cholinergic neuron size and choline acetyltransferase staining than did young rats *(137)*. This increase in vulnerability may have been due to a deficit in trophin response to the injury. When gelfoam was used to capture trophic activity that accumulated following lesion of the striatum, trophic activity derived from aged animals exhibited reduced capacity to support neurite outgrowth in culture than did trophic activity accumulated from younger animals *(138)* In other studies, trophic activity in lesioned aged brains took longer to attain maximal activity when evaluated in culture than did activity in young brain *(139)*.

Stressed rats similary appear deficient in trophin response. Changes in BDNF expression in response to stress were attenuated in the aged animal. Moreover, these changes were specific and were not noted with respect to NGF or NT-3 *(140)*.

The molecules underlying these alterations in responsivity have begun to be delineated. In the denervated striatum, for example, interleukin 1 expression has been reported to be markedly reduced in older rats as compared to younger animals *(141)*. The literature reviewed in this chapter gives hints as to what other molecules may be altered as aging progresses.

5.2. Involvement of Astrocytes in Deficits Associated with Aging

Data on the aging brain imply that aging may be accompanied by deficits in trophin expression that lead to increased neuronal vulnerability. What role do astrocytes play in this deficit and can we ultimately make use of these cells to counteract loss?

An emerging literature suggests that astrocytes from the aging brain do indeed exhibit impairments in function. For example, astrocytes grown for prolonged time in culture were reported to be deficient in their ability to respond to oxidative injury *(142)*. In another model of aging, astrocytes passaged for 46–51 passages exhibited diminished ability to respond to microenvironmental signals *(143)*. We *(121)* and others *(36)* have found that astrocytes derived from adult brain expressed reduced levels of mRNA encoding trophic molecules and a diminished response to neurotransmitters or cytokines. In vivo, glia of aging animals demonstrated reduced expression of specific trophic molecules *(14)*.

This chapter indicates that astrocytes express multiple trophic factors that may contribute to the milieu of the aging and injured brain. The expression of these trophic molecules, moreover, is plastic, and is regulated by cytokines, hormones, growth factors and neurotransmitters. Following a lesion, and as aging progresses, astrocytes have been widely reported to increase in number and size relative to neurons and they emerge as prime candidates to target in order to reverse deficits in trophin production that may be associated with aging and responsivity to injury.

Future studies must make use of the current culture models, complemented by in vivo studies, to further determine events modulation astrocyte function in aging and injured animals, The excitement of the field is the hope that with extension of the information currently available, we may be able to manipulate the complex variety of factors that are produced by astrocytes to ameliorate the deficits associated with aging and disease.

ACKNOWLEDGMENTS

The authors gratefully acknowledge the valuable discussion of this work with Dr. Wilma Friedman of the Department of Pathology, Columbia College of Physicians and Surgeons, New York, New York and Dr. Karen Dougherty and Tanya Zaremba of the Department of Neuroscience and Cell Biology, UMDNJ/Robert Wood Johnson Medical School, Piscataway, New Jersey. The work of Dr. Dreyfus et al. presented in this chapter was supported by NIH grants (HD 23315 and NS 10259)

REFERENCES

1. Tacconi, M.T. (1998) Neuronal death: is there a role for astrocytes. *Neurochem. Res.* **23,** 759–765.
2. Montgomery, D.L. (1994) Astrocytes: form, function, and roles in disease. *Veterinary Pathology* **31,** 145–167.
3. Byravan, S., Foster, L.M., Phan, T., Verity, A.N., and Campagnoni, A.T. (1994) Murine oligodendroglial cells express nerve growth factor. *Proc. Natl. Acad. Sci. USA* **91,** 8812–8816.
4. Elkabes, S., DiCicco-Bloom, E.M., and Black, I.B. (1996) Brain microglia/macrophages express neurotrophins that selectively regulate microglial proliferation and function. *J. Neurosci.* **16,** 2508–2521
5. Lindsay, R.M. (1979) Adult rat brain astrocytes support survival of both NGF-dependent and NGF-insensitive neurones. *Nature* **282,** 80–82.
6. Furukawa, S., Furukawa, Y., Satoyoshi, E., and Hayashi, K. (1986) Synthesis and secretion of nerve growth factor by mouse astroglial cells in culture. *Biochem. Biophys. Res. Commun* **136,** 57–63
7. Furukawa, S., Furukawa, Y., Satoyoshi, E., and Hayashi, K. (1987) Synthesis/secretion of nerve growth factor is associated with cell growth in cultured mouse astroglial cells. *Biochemical and Biophysical Research Communications* **142,** 395–402.
8. Inoue, S., Susukida, M., Ikeda, K., Murase, K., and Hayashi, K. (1997) Dopaminergic transmitter up-regulation of brain-derived neurotrophic fact (BDNF) and nerve growth factor (NGF) synthesis in mouse astrocytes in culture. *Biochem. Biophys Res Commun.* **238,** 468–472.
9. Lindholm, D., Castren, E., Berzaghi, M., Blochl, A., and Thoenen, H. (1994) Activity-dependent and hormonal regulation of neurotrophin mRNA levels in the brain-implications for neuronal plasticity. *J. Neurobiol.* **25,** 1362–1372.
10. Rudge, J.S., Alderson, R.F., Pasnikowski, E., McClain, J., Ip, N.Y., and Lindsay, R.M. (1992) Expression of ciliary neurotrophic factor and the neurotrophins- nerve growth factor, brain-derived neurotrophic factor and neurotrophin 3 -in cultured rat hippocampal astrocytes. *Eur. J. Neurosci.* **4,** 459–471.
11. Schwartz, J.P. and Nishiyama, N. (1994) Neurotropic factor gene expression in astrocytes during development and following injury. *Brain Res. Bull.* **35(5/6),** 403–407.
12. Wu, H., Qu, P., Friedman, W.J., Black, I.B., and Dreyfus, C.F. (1995) Differential regulation of neurotrophin expression in basal forebrain astrocytes by neuronal activity related signals. *Soc.Neurosci. Abstr.* **21,** 304.
13. Zafra, F., Lindholm, D., Castren, E., Hartikka, J., and Thoenen, H. (1992) Regulation of brain-derived neurotrophic factor and nerve growth factor mRNA in primary cultures of hippocampal neurons and astrocytes. *J. Neurosci.* **12(12),** 4793–4799.
14. Lu, B., Yokoyama, M., Dreyfus, C.F., and Black, I.B. (1991) NGF gene expression in actively growing brain glia. *J. Neurosci.* **11,** 318–326.
15. Arendt, T., Bruckner, M.K., Krell, T., Pagliusi, S., Kruska, L., and Heumann, R. (1995) Degeneration of rat cholinergic basal forebrain neurons and reactive changes in nerve growth factor expression after chronic neurotoxic injury-II. Reactive expression of the nerve growth factor gene in astrocytes. *Neuroscience* **65(3),** 647–659.
16. Condorelli, D.F., Salin, T., Dell'Albani, P., et al. (1996) Neurotrophins and their trk receptors in cultured cells of the glial lineage and in white matter of the central nervous system. *Journal Molec. Neurosci.* **6,** 237–248.
17. Pechan, P.A., Chowdhury, K., Gerdes, W., and Seifert, W. (1993) Glutamate induces the growth factors NGF, bFGF, and the receptor FGF-R1 and c-fos mRNA expression in rat astrocyte culture. *Neuroscience* **153,** 111–114.

18. Wu, H., Qu, P., Black, I.B., and Dreyfus, C.F. (1994) The regulation of astrocyte function by depolarizing signals. *Soc. Neurosci. Abstr.* **20,** 1314.
19. Zhou, X.-F. and Rush, R.A. (1994) Localization of neurotrophin-3-like immunoreactivity in the rat central nervous system. *Brain Res.* **1994,** 162–172.
20. Friedman, W.J., Black, I.B., and Kaplan, D.R. (1998) Distribution of the neurotrophins brain-derived neurotrophic factor, neurotrophin-3, and neurotrophin-4/5 in the postnatal rat brain: an immunocytochemical study. *Neurosci.* **84,** 101–114.
21. Dreyfus, C.F., Dai, X., Lercher, L.D., Racey, B.R., Friedman, W.J., and Black, I.B. (1999) Expression of neurotrophins in the adult spinal cord in vivo. *J. Neurosci. Res.* **56,** 1–7.
22. Ip, N.Y., Wiegand, S.J., Morse, J., and Rudge, J.S. (1993) Injury-induced regulation of ciliary neurotrophic factor mRNA in the adult rat brain. *Eur. J. Neurosci.* **5,** 25–33.
23. Kamiguchi, H., Yoshida, K., Sagoh, M. (1995) Release of ciliary neurotrophic factor from cultured astrocytes and its modulation by cytokines. *Neurochemical Research* **20(10),** 1187–1193.
24. Ferrara, N., Ousley, F., and Gaspodarowicz, D. (1988) Bovine brain astrocytes express basic fibroblast growth factor, a neurotrophic and angiogenic mitogen. *Brain Research* **462,** 223–232.
25. Yoshida, K. and Toya, S. (1997) Neurotrophic activity in cytokine-activated astrocytes. *Keio J. Medicine* **46(2),** 55–60.
26. Hatten, M.E., Lynch, M., Rydel, R.E., et al. (1988) In vitro neurite expression by granule neurons is dependent upon astroglial-derived fibroblast growth factor. *Developmental Biology* **125,** 280–289.
27. Schwartz, J.P. (1994) Neuropeptide synthesis in astrocytes: Possible trophic roles, in *Trophic regulation of the basal ganglia: Focus on dopamine neurons* (K. Fuxe, ed.) Pergamon Press, New York, pp. 317–327.
28. Schaar, D.G., Sieber, B.-A., Sherwood, A.C., et al. (1994) Multiple astrocytes transcripts encode nigral trophic factors in rat and human. *Experimental Neurol.* **130,** 387–393.
29. Moretto, G., Walker, D.G., Lanteri, P., et al. (1996) Expression and regulation of glial-cell-line-derived neurotrophic factor (GDNF) mRNA in human astrocytes in vitro. *Cell Tissue Reseach* **286,** 257–262.
30. Cintra, A., Lindberg, J., Chadi, G., et al. (1994) Basic fibroblast growth factor and steroid receptors in the aging hippocampus of the brown Norway rat: immunocytochemical analysis in combination with stereology. *Neurochem. Int.* **25,** 39–45.
31. Stockli, K.A., Lillien, L.E., Naher-Noe, M., et al. (1991) Regional distribution, developmental changes, and cellular localization of CNTF-mRNA and protein in the rat brain. *J. Cell Biol.* **115,** 447–459.
32. Shinoda, H., Marini, A.M., and Schwartz, J.P. (1992) Developmental expression of the proenkephalin and prosomatostatin genes in cultured cortical and cerebellar astrocytes. *Dev. Brain Res.* **67,** 205–210.
33. Goss, J.R., O'Malley, M.E., Zou, L., Styren, S.D., Kochanek, P.M., and DeKosky, S.T. (1997) Astrocytes are the major source of nerve growth factor upregulation following traumatic brain injury in the rat. *Experimental Neurology* **149,** 301–309.
34. Lee, T.-H., Kato, H., Chen, S.-T., Kogure, K., and Itoyama, Y. (1998) Expression of nerve growth factor and trk A after transient focal cerebral ischemia in rats. *Stroke* **29,** 1687–1697.
35. Schwartz, J.P., Sheng, J.G., Mitsuo, K., Shirabe, S., and Nishiyama, N. (1993) Trophic factor production by reactive astrocytes in injured brain, in *Annals NYAS*New York Academy of Sciences, New York.
36. Wu, V.W., Nishiyama, N., and Schwartz, J.P. (1998) A culture model of reactive astrocytes: increased nerve growth factor synthesis and reexpression of cytokine responsiveness. *J. Neurochem.* **71,** 749–756.

37. Logan, A., Frautschy, S.A., Gonzalez, A.-M., and Baird, A. (1992) A time course for the focal elevation of synthesis of basic fibroblast growth factor and one of its high-affinity receptors (flg) following a localized cortical brain injury. *J. Neurosci.* **12,** 3828–3387.
38. Humpel, C., Chadi, G., Lippoldt, A., Ganten, D., Fuxe, K., and Olson, L. (1994) Increase of basic fibroblast growth factor (bFGF, FGF-2) messenger RNA and protein following implantation of a microdialysis probe into rat hippocampus. *Exp. Brain Res.* **98,** 229–237.
39. Gomez-Pinilla, F., Lee, J.W.-K., and Cotman, C. (1992) Basic FGF in adult rat brain: cellular distribution and response to entorhinal lesion and fimbria-fornix transection. *J. Neurosci.* **12,** 345–355.
40. Finklestein, S.P., Apostolides, P.J., Caday, C.G., Prosser, J., Philips, M.F., and Klagsbrun, M. (1988) Increased basic fibroblast growth factor (bFGF) immunoreactivity at the site of focal brain wounds. *Brain Res.* **460,** 253–259.
41. Takami, K., Iwane, M., Kiyota, Y., Miyamoto, M., Tsukuda, R., and Shiosaka, S. (1992) Increase of basic fibroblast growth factor immunoreactivity and its mRNA level in rat brain following transient forebrain ischemia. *Exp. Brain Res.* **90,** 1–10.
41. Otero, G.C. and Merrill, J.E. (1994) Cytokine receptors on glial cells. *Glia* **11,** 117–124.
42. Asada, H., Ip, N.Y., L.Pan, Razack, N., Parfitt, M.M., and Plunkett, R.J. (1995) Time course of ciliary neurotrophic factor mRNA expression is coincident with the presence of protoplasmic astrocytes in traumatized rat striatum. *J. Neurosci. Res.* **40,** 22–30.
43. Lee, W.-H., Clemens, J.-A., and Bondy, C.A. (1992) Insulin-like growth factors in the response to cerebral ischemia. *Mol. & Cell. Neurosci.* **3,** 36–43.
44. Hetier, E., Ayala, J., Denefle, P., Bousseau, A., Rouget, P., Mallat, M., and Prochiantz, A. (1988) Brain macrophages synthesize interleukin-1 mRNAs in vitro. *Eur. J. Neurosci.* **2,** 762–768.
45. Krupinski, J., Kumar, P., Kumar, S., and Kaluza, J. (1995) Increased expression of TGF-b1 in brain tissue after ischemic stroke in humans. *Stroke.* **27(5),** 852–857.
46. Lindholm, D., Castren, E., Kiefer, R., Zafra, F., and Thoenen, H. (1992) Transforming growth factor-B1 in the rat brain: increase after injury and inhibition of astrocyte proliferation. *J. Cell Biol.* **117,** 395–400.
47. Assoian, R.K., Komoriya, A., Meyers, C.A., Miller, D.M., and Sporn, M.B. (1983) Transforming growth factor-beta in human platelets. Identification of a major storage site, purification, and characterization. *J. Biol. Chem.* **258,** 7155–7160.
48. Galve-Roperh, I., Malpartida, J.M., Haro, A., Brachet, P., and Diaz-Laviada, I. (1997) Regulation of nerve growth factor secretion and mRNA expression by bacterial lipopolysaccharide in primary cultures of rat astrocytes. *Journal of Neuroscience Reseach* **49,** 569–575.
49. Friedman, W.J., Larkfors, L., Ayer-LeLievre, C., Ebendal, T., Olson, L., and Persson, H. (1990) Regulation of β-nerve growth factor expression by inflammatory mediators in hippocampal cultures. *Journal of Neuroscience Research* **27,** 374–382.
50. Lindholm, D., Hengerer, B., Zafra, F., and Thoenen H. (1990) Transforming growth factor-beta 1 stimulates expression of nerve growth factor in the rat CNS. *Neuroreport* **1,** 9–12.
51. Lindholm, D., et.al. (1992) Differential regulation of nerve growth factor (NGF) synthesis in neurons and astrocytes by glucocorticoid hormones. *European Journal of Neuroscience* **4,** 404–410.
52. Yoshida, K. and Gage, F.H. (1991) Fibroblast growth factors stimulate nerve growth factor synthesis and secretion by astrocytes. *Brain Research* **538,** 118–126.
53. Yoshida, K. and Gage, F.H. (1992) Cooperative regulation of nerve growth factor synthesis and secretion in fibroblasts and astrocytes by fibroblast growth factor and other cytokines. *Brain Research* **569,** 14–25.
54. Friedman, W.J., Thakur, S., Seidman, L., and Rabson, A.B. (1996) Regulation of nerve growth factor mRNA by interleukin-1 in rat hippocampal astrocytes is mediated by NFκB. *Journal of Biological Chemistry* **271(49),** 31115–31120.

55. Pshenichkin, S.P. and Wise, B.C. (1997) Okadaic acid stimulates nerve growth factor production via an induction of interleukin-1 in primary cultures of cortical astroglial cells. *Neurochemistry International* **30(4/5),** 507–514.
56. deKloet, E.R. (1992) Corticosteroids, stress, and aging, in *Aging and Cellular Defense Mechanisms* (C. Franceschi, et al., eds), New York Academy of Sciences, New York.
57. Sapolsky, R.M. (1998) Deleterious and salutary effects of steroid hormones in the nervous system, in *Neuroprotective signal transduction* (M.P. Mattson, ed.) Humana Press, Inc., Totowa, NJ.
58. Sapolsky, R.M. (1992) Do glucocorticoid concentrations rise with age in the rat? *Neurobiol. Aging* **13,** 171–174.
59. Bohn, M.C., O'Banion, M.K., Young, D.A., et al. (1994) In vitro studies of glucocorticoid effects on neurons and astrocytes, in *Brain Corticosteroid Receptors* (E.R. deKloet, E.C. Azmitia, and P.W. Landfield, eds.), New York Academy of Sciences, New York.
60. Cintra, A., Bhatnagar, M., Chadi, G., et al. (1994) Glial and neuronal glucocorticoid receptor immunoreactive cell populations in developing, adult, and aging brain, in *Brain Corticosteroid Receptors* (E.R. deKloet, E.C. Azmitia, and P.W. Landfield, eds), New York Academy of Sciences, New York.
61. Niu, H., Hinkle, D.A., and Wise, P.M. (1997) Dexamethasone regulates basal fibroblast growth factor, nerve growth factor and S100B expression in cultured hippocampal astrocytes. *Molecular Brain Research* **51,** 97–105.
62. Hahn, M., Lorez, H., and Fischer, G. (1997) Effect of Calcitriol in combination with corticosterone, interleukin-1β, and transforming growth factor-β on nerve growth factor secretion in an astroglial cell line. *Journal of Neurochemistry* **69,** 102–109.
63. Neveu, I., Naveilhan, P., Menaa, C., Wion, D., Brachet, P., and Garabedian, M. (1994) Synthesis of 1,25-dihydroxyvitamin D3 by rat brain macrophages in vitro. *Journal of Neuroscience Research* **38,** 214–220.
64. Neveu, I., Naveilhan, P., Jehan, F., et al. (1994) 1,25-dihydroxyvitamin D3 regulates the synthesis of nerve growth factor in primary cultures of glial cells. *Mol. Brain Res.* **24,** 70–76.
65. Sohrabji, F., Miranda, R.C.G., and Toran-Allerand, C.D. (1995) Identification of a putative estrogen response element in the gene encoding brain-derived neurotrophic factor. *Proceedings of the National Academy of Science USA* **92,** 11110–11114.
66. Toran-Allerand, C.D. (1996) Mechanisms of estrogen action during neural development: mediation by interactions with the neurotrophins and their receptors? *J. Steroid Biochem. and Molec. Biol.* **56(1–6),** 169–178.
67. Gibbs, R.B., Wu, D., Hersh, L.B., and Pfaff, D.W., (1994) Effects of estrogen replacement on the relative levels of choline acetyltransferase, trk A, and nerve growth factor messenger RNAs in the basal forebrain and hippocampal formation of adult rats. *Experimental Neurology* **129,** 70–80.
68. Singh, M., Meyer, E.M., and Simpkins, J.W. (1995) The effect of ovariectomy and estradiol replacement on brain-derived neurotrophic factor messenger ribonucleic acid expression in cortical and hippocampal brain regions of female Sprague-Dawley rats. *Endocrinology* **136,** 2320–2324.
69. Santagati, S., Melcangi, R.C., Celotti, F., Martini, L., and Maggi, A. (1994) Estrogen receptor is expressed in different types of glial cells in culture. *Journal of Neurochemistry* **63,** 2058–2064.
70. Garcia-Segura, L.M., Torres-Aleman, I., and Naftolin, F. (1989) Astrocytic shape and glial fibrillary acidic protein immunoreactivity are modified by estradiol in primary rat hypothalamic cultures. *Developmental Brain Research* **47,** 289–302.
71. Garcia-Segura, L.M., Luquin, S., Parducz, A., and Naftolin, F. (1994) Gonadal hormone regulation of glial fibrillary acidic protein immunoreactivity and glial ultrastructure in the rat neuroendocrine hypothalamus. *Glia* **10,** 59–69.

72. Lackland, J., Wu, H., Black, I.B., and Dreyfus, C.F. (1997) Differential regulation of neurotrophin expression in adult and neonatal cortical astrocytes by estrogen. *Soc. Neurosci. Abstr.* **23,** 49.
73. Duenas, M., Luquin, S., Chowen, J.A., I.Torres-Aleman, Naftolin, F., and Garcia-Segura, L.M. (1994) Gonadal hormone regulation of insulin-like growth factor-1-like immunoreactivity in hypothalamic astroglia of developing and adult rats. *Neuroendocrinology* **59,** 528–538.
74. Garcia-Segura, L.M., Wozniak, A., Azcoitia, I., Rodriguez, J.R., Hutchison, R.E., and Hutchison, J.B. (1999) Aromatase expression by astrocytes after brain injury: implications for local estrogen formation in brain repair. *Neuroscience* **89,** 567–578.
75. Negri-Cesi, P.N., Melcangi, R.C., Celotti, F., and Martini, L. (1992) Aromatase activity in cultured brain cells: difference between neurons and glia. *Brain Research* **589,** 327–332.
76. Barbacid, M. (1994) The trk family of neurotrophin receptors. *J. Neurobiol.* **25,** 1386–1403.
77. Chao, M.V. (1992) Neurotrophin receptors: a window into neuronal differentiation. *Neuron* **9,** 583–593.
78. Dechant, G., Rodriguez-Tebar, A., and Barde, Y.-A. (1994) Neurotrophin receptors. *Prog. Neurobiol.* 347–352.
79. Hutton, L.A., de Vellis, J., and Perez-Polo, J.R. (1992) Expression of $p75^{NGFR}$ trkA, and trkB mRNA in rat C6 glioma and type I astrocytes cultures. *Journal of Neuroscience Research* **32,** 375–383.
80. Kumar, S., Pena, L.A., and de Vellis, J. (1993) CNS glial cells express neurotrophin receptors whose levels are regulated by NGF. *Molecular Brain Research* **17,** 163–168.
81. Rudge, J.S., Li, Y., and Pasnikowski, E.M. (1994) Neurotrophic factor receptors and their signal transduction capabilities in rat astrocytes. *European Journal of Neuroscience* **6,** 693–705.
82. Dougherty, K.D. and Milner, T.A. (1999) p75NTR immunoreactivity in the rat dentate gyrus is mostly within presynaptic profiles but is also found in some astrocytic and postsynaptic profiles. *J. Comp. Neurol.* **407,** 77–91.
83. Junier, M.P., Suzuki, F., Onteniente, B., and Peschanski, M. (1994) Target-deprived CNS neurons express the NGF gene while reactive glia around their axonal terminals contain low and high affinity NGF receptors. *Molecular Brain Research* **24,** 247–260.
84. Rubio, N. (1997) Mouse Astrocytes store and deliver brain-derived neurotrophic factor using non-catalytic $pg95^{trk\ B}$ receptor. *Euro. J. Neurosci.* **9,** 1847–1853.
85. Baxter, G.T., Radeke, M.J., Kuo, R.C., et al. (1997) Signal transduction mediated by the truncated trk B receptor isoforms, trk B.T1 and trk B.T2. *J. Neurosci.* **17(8),** 2683–2690.
86. Hutton, L.A. and Perez-Polo, J.R. (1995) In vitro glial responses to nerve growth factor. *J. Neurosci. Res.* **41,** 185–196.
87. Yokoyama, M., Black, I.B., and Dreyfus, C.F. (1993) NGF increases brain astrocyte number in culture. *Experimental Neurology* **124,** 377–380.
88. Gomez-Pinilla, F., Vu, L., and Cotman, C.W. (1995) Regulation of astrocyte proliferation by FGF-2 and heparan sulfate in vivo. *J. Neurosci.* **15,** 2021–2029.
89. Levison, S., Hudgins, S., and Crawford, J. (1998) Ciliary neurotrophic factor stimulates nuclear hypertrophy and increases the GFAP content of cultured astrocytes. *Brain Res.* **803,** 189–193.
90. Duffy, S. and MacVicar, B.A. (1994) Potassium-dependent calcium influx in acutely isolated hippocampal astrocytes. *Neuroscience* **61(1),** 51–61.
91. Lerea, L.S. and McCarthy, K.D. (1989) Astroglial cells in vitro are heterogeneous with respect to expression of the alpha(1)-adrenergic receptor. *Glia* **2,** 135–147.
92. Aoki, C. (1992) Beta-adrenergic receptors: Astrocytic localization in the adult visual cortex and their relation to catecholamine axon terminals as revealed by electron microscopic immunocytochemistry. *Journal of Neuroscience* **12(3),** 781–792.
93. Murphy, S. and Pearce, B. (1987) Functional receptors for neurotransmitters on astroglial cells. *Neuroscience* **22(2),** 381–394.

94. Ashkenazi, A., Ramachandran, J., and Capon, D.J. (1989) Acetylcholine analogue stimulates DNA synthesis in brain-derived cells via specific muscarinic receptor subtypes. *Nature* **340,** 146–150.
95. Messamore, E., Bogdanovich, N., Schroder, H., and Winblad, B. (1994) Astrocytes associated with senile plaques possess muscarinic acetylcholine receptors. *NeuroReport* **5,** 1473–1476.
96. Hosli, L., Hosli, E., Winter, T., and Stauffer, S. (1994) Coexistence of cholinergic and somatostatin receptors on astrocytes of rat CNS. *NeuroReport* **5,** 1469–1472.
97. Backus, K.H., Kettenmann, H., and Schachner, M. (1989) Pharmacological characterization of the glutamate receptor in cultured astrocytes. *J. Neurosci. Res.* **22,** 274–282.
98. Sontheimer, H., Kettenmann, H., Backus, K.H., and Schacher, M. (1988) Glutamate opens Na^+/K^+ channels in cultured astrocytes. *Glia* **1,** 328–336.
99. Usowicz, M.M., Gallo, V., and Cull-Candy, S.G. (1989) Multiple conductance channels in type 2 cerebellar astrocytes activated by excitatory amino acids. *Nature* **339,** 380–383.
100. von Blankenfeld, G. and Kettenmann, H. (1991) Glutamate and GABA receptors in vertebrate glial cells. *Molec. Neurobiol.* **5,** 31–45.
101. Sumners, C., Tang, W., Paulding, W., and Raizada, M.K. (1994) Peptide receptors in astroglia: Focus on angiotensin II and atrial natiuretic peptide. *Glia* **11,** 110–116.
102. Matute, C. and Miledi, R. (1993) Neurotransmitter receptors and voltage-dependent Ca2+ channels encoded by mRNA from the adult corpus callosum. *Proc. Nat. Acad. Sci. USA* **90,** 3270–3274.
103. Matute, C., Garcia-Barcina, J.M., and Miledi, R. (1994) Expression of neurotransmitter receptors and Ca^{2+} channels in the adult fornix and optic nerve. *Neuro Report* **5,** 1457–1460.
104. Salm, A.K. and McCarthy, K.D. (1989) Expression of beta-adrenergic receptors by astrocytes isolated from adult rat cortex. *GLIA* **2,** 346–352.
105. Shao, Y. and McCarthy, K.D. (1994) Plasticity of astrocytes. *Glia* **11,** 147–155.
106. Kuffler, S.W. (1967) Neuroglial cells: physiological properties and a potassium mediated effect of neuronal activity on the glial membrane potential. *Proceedings of the Royal Society of Britain* **168,** 1–21.
107. Sykova, E., Jendelova, P., Svoboda, J., and Chvatal, A. (1992) Extracellular K^+, pH, and volume changes in spinal cord of adult rats and during postnatal development. *Cancer Journal of Physiology and Pharmacology* **70,** S301–S309.
108. McCarthy, K., Enkvist, K., and Shao, Y. (1995) Astroglial adrenergic receptors: expression and function, in *Neuroglia* (H. Kettenmann and B.R. Ranson, eds.), Oxford University Press, New York.
109. de Barry, J., Ogura, A., and Kudo, Y. (1991) Ca2+ mobilization in cultured rat cerebellar cells: Astrocytes are activated by t-ACPD. *Eur. J. Neurosci.* **3,** 1146–1154.
110. Kim W. T., Rioult, M.G., and Cornell-Bell, A.H. (1994) Glutamate-induced calcium signaling in astrocytes. *Glia,* **11,** 173–184.
111. Porter, J.T. and McCarthy, K. D. (1995) GFAP-positive hippocampal astrocytes in situ respond to glutamatergic neuroligands with increases in $[Ca^{2+}]i$. *Glia* **13,** 101–112.
112. Cornell-Bell, A.H., Thomas, P.G., and Smith, S.J. (1990) The excitatory neurotransmitter glutamate causes filopodia formation in cultured hippocampal astrocytes. *Glia* **3,** 322–334.
113. Canaday, K.S. and Rube., E.W. (1992) Rapid and reversible astrocytic reaction to afferent activity blockade in chick cochlear nucleus. *J. Neurosci.* **12,** 1001–1009.
114. Furukawa, S., Furukawa, Y., Satoyoshi, E., and Hayashi, K. (1987) Regulation of nerve growth factor synthesis/secretion by catecholamine in cultured mouse astroglial cells. *Biochem. Biophys. Res. Comm.* **147,** 1048–1054.
115. Schwartz, J.P., Nishiyama, N., Wilson, D., and Taniwaki, T. (1994) Receptor-mediated regulation of neuropeptide gene expression in astrocytes. *Glia* **11,** 185–190.
116. Rudge, J.S., Morrissey, D., Lindsay, R.M., and Pasnikowski, E.M. (1994) Regulation of ciliary neurotrophic factor in cultured rat hippocampal astrocytes. *Eur. J. Neurosci.* **6,** 218–229.117.

117. Riva, M.A., Molteni, R., Lovati, E., Fumagalli, f., Resnati, M., and Racagni, G. (1996) Cyclic AMP-dependent regulation of fibroblast growth factor-2 messenger RNA levels in rat ccortical astrocytes: comparison with fibroblast growth factor-1 and ciliary neurotrophic factor. *Mol. Pharmacol.* **49(4),** 699–706.
118. Nagao, H., Matsuoka, I., and Kurihara, K. (1995) Effects of adenylyl cyclase-linked neuropeptides on the expression of ciliary neurotrophic factor mRNA in cultured astrocytes. *FEBS Lett.* **362,** 75–79.
119. Wu, H., Black, I.B., and Dreyfus, C.F. (1996) Differential regulation of neurotrophin expression in basal forebrain astrocytes. *Soc. Neurosci. Abstr.* **22,** 548.
120. Ho, A., Gore, A.C., Weickert, C.S., and Blum, M. (1995) Glutamate regulation of GDNF gene expression in the striatum and primary striatal astrocytes. *Neuroreport,* **6,** 1454–1458.
121. Lackland, J., Campos, S.J., Black, I.B., and Dreyfus, C.F. (1999) Basal forebrain (BF) astrocyte function is developmentally regulated. *Soc. Neurosci. Abstr.* **25,** 1787.
122. Goss, J.R., Finch, C.F., and Morgan, D.G. (1991) Age-related changes in glial fibrillary acidic protein mRNA in the mouse brain. *Neurobiol Aging.* **12(2),** 165–170.
123. O'Callaghan, J.P. and Miller, D.B. (1991) The concentration of glial fibrillary acidic protein increases with age in the mouse and rat brain. *Neurobiol Aging* **12,** 171–174.
124. Nichols, N.R., Day, J.R., Laping, N.J., Johnson, S.A., and Finch, C.F. (1993) GFAP mRNA incrases with age in rat and human brain. *Neurobiol. Aging* **14(5),** 421–429.
125. Fischer, W., Chen, K.S., Gage, F.H., and Bjorklund, A. (1992) Progressive decline in spatial learning and integrity of forebrain cholinergic neurons in rats during aging. *Neurobiol. Aging* **13(1),** 9–23.
126. Lapchak, P.A., Araujo, D.M., Beck, K.D., Finch, C.E., Johnson, S.A., and Hefti, F. (1993) BDNF and trkB mRNA expression in the hippocampal formation of aging rats. *Neurobiol. Aging* **14,** 121–126.
127. Taglialatela, G., Robinson, R., Gegg, M., and perez-Polo, J.R. (1997) Nerve growth factor, central nervous system apoptosis, and adrenocortical activity in aged Fischer-344/Brown Norway F1 hybrid rats. *Brain Res. Bull* **43,** 229–233.
128. Hesenohrl, R.U., Soderstrom, S., Mohammed, A.H., Ebendal, T., and Huston, J.P. (1997) Reciprocal changes in expression of mRNA for nerve growth factor and its receptors TrkA and LNGFR in brain of aged rats in relation to maze learning deficits. *Exp. Brain Res.* **114,** 205–213.
129. Narisawa-Saito, M. and Nawa, H. (1996) Differential regulation of hippocampal neurotrophins during aging in rats. *J. Neurochem.* **67,** 1124–1131.
130. fischer, W., Wictorin, K., Bjorklund, A., Williams, L.R., Varon, S., and Gage, F.H. (1987) Amelioration of cholinergic neuron atrophy and spatial memory impairment in aged rats by nerve growth factor. *Nature* **329,** 65–68.
131. Pelleymounter, M.A., Cullen, M.J., Baker, M.B., Gollub, M., and Wellman, C. (1996) The effects of intrahippocampal BDNF and NGF on spatial learning in aged Long Evans rats. *Molec. & Chem. Neuropathol.* **29,** 211–226.
132. Bergado, J.A., Fernandez, C.I., Gomez-Soria, A., and Gonzalez, O. (1997) Chronic intraventricular infusion with NGF improves LTP in old cognitively-impaired rats. *Brain Res.* **770,** 1–9.
133. Markowska, A.L., Price, D., and Koliatsos, V.E. (1996) Selective effects of nerve growth factor on spatial recent memory as assessed by a delayed nonmatching-to-position task in the water maze. *J. Neurosci.* **16,** 3541–3548.
134. Blum, M. and Weickert, C.S. (1995) GDNF mRNA expression in normal postnatal development, aging, and in Weaver mutant mice. *Neurobiol. Aging,* **16,** 925–929.
135. Lapchak, P.A., Miller, P.J., and Jiao, S. (1997) Glial cell line-derived neurotrophic factor induces the dopaminergic and cholinergic phenotype and increases locomotor activity in aged Fischer 344 rats. *Neurosci.* **77,** 745–752.

136. Hebert, M.A. and Gerhardt, G.A. (1997) Behavioral and neurochemical effects of intranigral administration of glial cell line-derived neurotrophic factor on aged Fischer 344 rats. *J.P.F.T.* **282,** 760–768.
137. Cooper, J.D. and Sofroniew, V. (1996) Increased volnerability of septal cholinergic neurons to partial loss of target neurons in aged rats. *Neurosci.* **75,** 29–35.
138. Kascloo, P.A., Lis, A., Asada, H., Barone, T.A., and Plunkett, R.J. (1996) In vitro assessment of neurotrophic activity from the striatum of aging rats. *Neurosci. Lett.* **218,** 157–160.
139. Whittemore, S.R., Nieto-Sampedro, M., Needels, D.L., and Cotman, C.W. (1985) Neuronotrophic factors for mammalian brain neurons: injury induction in neonatal, adualt, and aged rat brain. *Dev. Brain Res.* **20,** 169–178.
140. Smith, M.A. and Cizza, G. (1996) Stress-induced changes in brain-derived neurotrophic factor expression are attenuated in aged Fischer 344/N rats. *Neurobiol. Aging* 17, 859–864.
141. Ho, A. and Blum, M. (1998) Induction of interleukin-1 associated with compensatory dopaminergic sprouting in the denervated stiatum of young mice: model of aging and neurodegenerative disease. *J. Neurosci.* **18,** 5614–5629.
142. Papadopoulos, M.D., Koumenis, I.L. Yuan, T.Y., and Giffard, R.G. (1998) Increasing vulnerability of astrocytes to oxidative injury with age despite constant antioxidant defenses. *Neurosci.* **82,** 915–925.
143. Grove, J., Gomez, J., Kendroti, S., and Vernadakis, A. (1996) Plasticity of astrocytes derived from aged mouse cerbral hemispheres: changes with cell passage and immortalization. *Brain Res. Bull.* **39,** 211–217.

12

Responses in the Basal Forebrain Cholinergic System to Aging

Zezong Gu and J. Regino Perez-Polo

1. INTRODUCTION

Documentation of age-related mental deficiency, described as the decline and decay of the human body and regression of mental capacities by 63- and 81-yr-of-age, dates to the description of the life cycle by Pythagoras, a Greek physician of the 7th century B.C. He described the late stage of life where "*the scene of mortal existence closes, after a great length of time, ... The system returns to the imbecility of the first epoch of the infancy.*" The more modern description of senile dementia dates to Alois Alzheimer's description in 1906 of the first documented case of Alzheimer's disease (AD) *(1)*. During the last quarter of this century there has been significant progress in the fields of neuropathology, biochemistry, and genetics, as well as some pharmatherapeutic interventions aimed at this dementing disorder. However, there is no definitive characterization of the mechanisms of this disease *(2)*.

Alzheimer's disease, the most common dementing disorder of late life and a major cause of disability and death in the elderly, is manifested by behavior deficits involving memory and cognition, accompanied by personality changes that increase in frequency with advancing years. Alzheimer's disease is associated with an excessive loss of synapses and mainly cholinergic neurons in specific brain regions, deficits in neurotransmitter function, and marked extracellular deposition of amyloid deposits that form senile plaques and the prominent appearance of intracellular neurofibrilllary tangles [for reviews, see *(3,4,5)*.] Although to a lesser extent, there are similar changes evident in the healthy aged brain *(6)*. Also, neurotransmitter- and/or amyloid precursor protein (APP)-mediated disruptions of signal transduction cascades may exacerbate AD pathology *(7)*.

The appearance of a link between cholinergic dysfunction and AD has focused scientific efforts on developing tools that elucidate the regulatory events controlling cholinergic system function with the goal of developing therapeutic interventions for the disorder. An important step in understanding the mechanisms underlying cholinergic dysfunction has been the development of in vivo rodent models that mimic some features of AD. Using such in vivo rodent models, several cholinergic enhancement strategies have been tested that have proven to be somewhat effective in alleviating

From: *Neuroglia in the Aging Brain*
Edited by: Jean S. de Vellis © Humana Press Inc., Totowa, NJ

lesion-induced cognitive deficits *(8)*. These include acetylcholine precursor loading approaches (e.g., lecithin); treatment with postsynaptic agonists, acetylcholinesterase inhibitors (e.g., tacrine and donepezil), neurotrophic factors (e.g., nerve growth factor); and transplantation of cholinergic-enriched fetal grafts.

2. CHOLINERGIC DEFICITS IN THE AGED AND IN ALZHEIMER'S DISEASE

2.1. The Cholinergic Basal Forebrain

Cholinergic neurons are those that produce, store, and release acetylcholine (ACh), the first substance recognized to be a neurotransmitter. Choline acetyltransferase (ChAT) and acetylcholinesterase (AChE), are the two rate-limiting enzymes that regulate together with choline uptake, ACh synthesis, and degradation. Thus, measurement of the specific activities of ChAT and AChE are useful markers for cholinergic function. Cholinergic neurons are medium to large sized cells (range, 18–43 µm in maximum soma extent) with long axons that can be identified by histochemical staining with AChE, immunocytochemistry for ChAT, and *in situ* hybridization for ChAT mRNA *(9,10,11,12)*. In the brain, cholinergic neurons are mostly localized within the basal forebrain and to a lesser extent in the brainstem and cerebellum.

The basal forebrain cholinergic system in the adult contains three major subnuclei of projection neurons within the medial septum (MS), the diagonal band of Broca (DB) and the nucleus basalis of Meynert (NBM), along with interneurons of the striatum. Anatomical tracking studies *(13)* have shown that cholinergic basal forebrain neurons (CBFNs) give rise to a dense network of cholinergic fiber projections. The basal forebrain cholinergic neurons project from:

1. the NBM to the cerebral cortex and amygdala, bundling as the basalo-cortical and basalo-amygdalar afferents;
2. the MS and vertical limb of DB to the hippocampus via the fimbria-fornix forming the septo-hippocampal afferents;
3. the lateral portion of the horizontal limb of DB to the olfactory bulb and associated nuclei *(9)* (Fig. 1).

There are also noncholinergic neurons intermingled with the CBFNs within or surrounding the basal forebrain, some of which are GABAergic septal neurons that contain parvalbumin *(14,15)*; and substantia inominata neurons that contain calbindin-D_{28K} or NADPH-diaphorase *(16)*. CBFNs with extensive dendrites also intersect with other cholinergic and noncholinergic neurons within the basal forebrain extending into and intertwining with heavily myelinated fibers. In the hippocampus and cortex, pyramidal neurons appear to constitute the primary synaptic target for CBFNs *(12)*.

That there is a relationship between cholinergic function and memory and learning was initially ascertained from observations of aged and demented patients in their responses to various pharmacological manipulations. Application of the muscarinic receptor antagonist, scopolamine, to young normal subjects produces deficits in cognition and memory that are similar to those seen in aged subjects. Additionally, administration of the cholinergic agonist, physostigmine, to aged human subjects results in an improvement in memory function *(17,3,18)*. Neuropsychological, biochemical, and

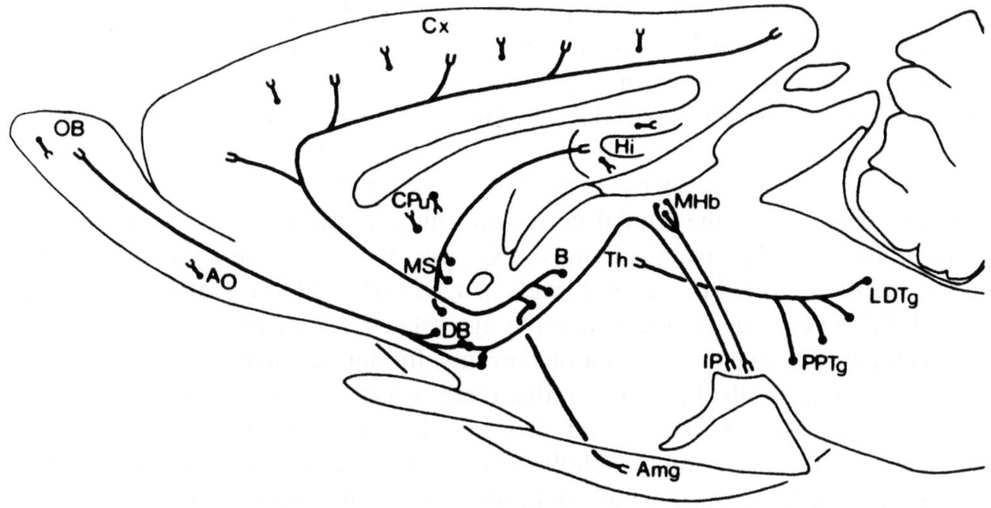

Fig. 1. Schematic illustration of the horizontal plane of the major cholinergic systems in the rat brain. The basal forebrain cholinergic system contains three major subnuclei of projection neurons within the medial septum (MS), the diagonal band of Broca (DB) and the nucleus basalis of Meynert (B), along with interneurons of the striatum. The basal forebrain cholinergic neurons (CBFNs) project from: **(1)** the nucleus basalis of Meynert to the cerebral cortex (Cx) and amygdala (Amg), bundling as the basalo-cortical and basalo-amyadalar afferents; **(2)** the MS and vertical limb of DB to the hippocampus (Hi) via the fimbria-fornix forming the septo-hippocampal afferents; and **(3)** the lateral portion of the horizontal limb of DB to the olfactory bulb (OB) and associated nuclei. Modified from Butcher, et al. *(10)*.

pharmacological evidence further supports the notion of a significant role for cholinergic function in age-related memory disturbance leading to a proposal of a cholinergic hypothesis of dementia in aging and age-related disorders.

2.2. Cholinergic Deficits in Aging and Alzheimer's Disease

The presence of cholinergic deficits and the loss of CBFNs in the aged and AD brains have been well documented *(19,20,21)*. Postmortem studies have shown that there is a profound reduction in cortical presynaptic cholinergic markers in patients with AD and senile dementia of the Alzheimer's type. For example, ChAT activity decreases 60–90% in the cerebral cortices and hippocampi of AD patients. The cholinergic neurons of the NBM, the major source of cortical cholinergic innervation, undergo a profound (greater than 75%) and selective degeneration in these patients and the degree of loss correlates with the severity of the observed cognitive impairments *(21,22)*. In addition, an *in situ* hybridization study *(23)* has shown decreased ChAT mRNA expression in the NBM in AD patients, suggesting that expression of ChAT mRNA might be downregulated in surviving cholinergic neurons. Reductions are also observed in high-affinity choline uptake (HACU) *(24)*, ACh and AChE levels in cortex *(25)* and cerebrospinal fluid (CSF) *(26)*. Nerve growth factor (NGF) receptor and ChAT remain colocalized in the NBM in AD patients *(27)*. Apart from the presence of cholinergic dysfunction, there are *(28)* moderate increases in NGF-like activity throughout the brain of AD coupled with significant declines in NBM cell numbers compared to aged healthy individuals. Although the mechanisms that lead to the degeneration of cholinergic neurons in the NBM are unknown, it has been speculated that neuronal death may result in part due to a failure of neurotrophic support for the maintenance of oxidant-antioxidant and glutathione peroxidase homeostasis *(29,30,31)*.

There is a substantial decrease in ChAT activity in the striatum, and there are decreases in HACU in the frontal cortex and hippocampus, of 24-mo-old rats when compared to 4-mo-old rats *(32)*. However, Ogawa, et al. *(33)* found that reduced ChAT activity and muscarinic M1 receptor levels in aged Fisher 344 rat brains did not parallel their mRNA levels, suggesting that some age-related impairments of the cholinergic system may be due to posttranscriptional events. Immunocytochemistry and retrograde transport labeling results have shown that there is a decline in the number of neurons retrogradely transporting tracers and also, that there is a significant shrinkage in cell surface area in the basal forebrain cholinergic system of aged rats, consistent with there being atrophy of cholinergic basal forebrain neurons and impairment of uptake or retrograde transport mechanisms in the aged brain *(34)*.

2.3. *In Vivo Immunolesion Model of a Cholinergic Deficit*

In order to understand the mechanisms responsible for age- and AD-associated cholinergic deficits, it is necessary to develop animal models that display specific cholinergic deficits in vivo and allow for the evaluation of the biochemical, neuropathological, and behavioral consequences of the lesion model. Models must also be amenable to the study of lesion-induced responses and the repair mechanisms that are triggered by the cholinergic deficits.

At present there is no adequate animal model available which can mimic all of the behavioral, biochemical, and histopathological abnormalities observed in AD patients. However, partial clues can be obtained from a number of cholinergic lesion paradigms that have been used to produce cholinergic hypofunction in vivo. These include fimbria-fornix transections, mechanical lesions with radiofrequency and electrolysis, systemic or intracerebral injections of excitotoxins, which are analog of the excitatory amino acid neurotransmitter glutamate (e.g., kainic acid, ibotenic acid, quisqualic acid, NMDA, and AMPA), high-affinity choline transport inhibitors (ethylcholine mustard aziridinium ion, AF64A), or murine anti-AChE monoclonal antibodies. The limitations of these lesion paradigms are that they not only cause cholinergic deafferentation, but that they also deplete noncholinergic projections, such as GABAergic, serotonergic, noradrenergic, and dopaminergic innervations. Thus, the differential affinities of the different excitotoxins to distinct glutamate receptor subtypes may partly explain the differential cytotoxic effects of each glutamate analogue on different regions of the brain. Selectively destroying cholinergic neurons, although sparing other cell types, would be useful to characterize forebrain cholinergic deficits without the limitations mentioned above *(35)*.

Most of the CBFNs possess NGF receptors in contrast to the other neurons in this region, including cholinergic cells present in the nearby striatum, that do not express detectable levels of NGF receptors *(36,37,38)*. It has been shown that a well-characterized monoclonal antibody to the low-affinity neurotrophin receptor—$p75^{NTR}$, 192 IgG, accumulates bilaterally and selectively in CBFNs following intracerebral ventricle (i.c.v.) administration. Crosslinking 192 IgG via a disulfide bond to the ribosomal inactivating protein, saporin, generates an immunotoxin (IT), the 192 IgG-saporin complex, that is also taken up via p75^{NTR} receptors and is a specific and selective tool for producing lesions to CBFNs *(39)*. Thus, after an i.c.v. injection of 192 IgG-saporin, the IT is specifically internalized by the terminals of p75^{NTR}-bearing CBFNs, retrogradely transported, and accumulated in the cell bodies of CBFNs. Intraventricular administration of IT produces a selective loss of p75^{NTR}-bearing cholinergic neurons that results in substantial reductions in ChAT activity in a time and dose-dependent manner in the rat basal forebrain and their neocortical and hippocampal afferents *(40,41,16,14,42,43,44,45,39,46,47)*, whereas cholinergic interneurons in the striatum, noncholinergic septal neurons containing parvalbumin *(48,14,42)* and substantia inominata GABAergic neurons containing calbindin-D28K or NADPH-diaphorase *(16)* as well as tyrosine hydroxylase-positive neurons *(39)*, glutamic acid dehydrogenase-positive neurons *(44)*, galanin-, neuropeptide Y- and neurotensin B-positive neurons *(49,50)* are not affected. Not surprisingly, IT also can affect NGF-reactive cerebellar Purkinje neurons at higher doses *(16)*. Prelabeling cortical projecting neurons in NBM with fluoro-gold shows that only those neurons that are also labeled for ChAT are destroyed in the IT-treated animals, suggesting that IT is lethal to cholinergic cells, rather than suppressing cholinergic expression in existing cells *(51)*. Thus, this model allows for dissociation between cholinergic neuronal loss and the attenuation of cholinergic function in spared neurons, a critical distinction if we are to dissect the causes for cholinergic dysfunction in the aged and AD brain.

An early study *(52)* showed that partial deafferentation of the cholinergic and adrenergic innervations in the hippocampus obtained by destruction of the dorsal routes (through the fimbria-fornix and the supracallosal striae) results in a delayed, but significant recovery of the original cholinergic and adrenergic innervation patterns by remaining cholinergic and adrenergic inputs that reach the target via the ventral route. Because of its slow and protracted time-course and its ability to reestablish innervation in initially denervated areas, this compensatory collateral sprouting phenomenon may be of particular interest for the development of long-term, protracted functional recovery as documented for both experimental brain lesions and accidental severe brain injuries. A partial degeneration to basal forebrain cholinergic projections can also be induced by the administration of sub-maximal doses of IT *(53,54,45)*. These latter partial lesions are comparable to the situation described for AD, in that there is a graded behavioral and biochemical change suitable for the study of cholinergic functional recovery and reorganization in the cholinoceptive target areas.

Behavioral studies have shown that using either i.c.v. or parenchymal delivery into NBM or cortex of 192 IgG-saporin *(55,56,57,14,58)*, impairs performance in learning and memory tasks in a manner consistent with the extensive loss of CBFNs *(59)*.

3. TROPHIC ACTIONS OF NGF ON PLASTICITY FOR CHOLINERGIC NEURONS

The neurotrophic hypothesis, that the survival of neurons depends on their competition for neurotrophic factors synthesized in limiting amounts by their target fields, provides the best available explanation for how target fields influence the neuronal populations that innervate them and the subsequent processes associated with neuronal plasticity *(60,61)*.

3.1. Nerve Growth Factor and Its Receptors

Nerve growth factor (NGF) is the first characterized member of the neurotrophin (NT) family, which also includes brain-derived neurotrophic factor (BDNF), neurotrophin-3 (NT-3), neurotrophin-4/5 (NT-4/5), and neurotrophin-6 (NT-6) (for reviews, *see 62,63,64,65,66,67*). NTs are highly basic proteins (pI 9–10.5) each composed of approx 120 amino acids with three intrachain disulfide bonds. Although sharing 50–60% amino acid homology, NTs display distinct, yet overlapping, regional distribution and biological effects that regulate the survival, differentiation and phenotypic maintenance of specific neuronal populations and also play a less well understood role in neuronal plasticity. The biologically active form for NGF is made from a homodimer with two identical beta-subunits. In the central nervous system (CNS), NGF is expressed in different brain areas with the highest levels present in the hippocampus, cerebral cortex, and olfactory bulb, the principal target areas of CBFNs *(61,68,69)* and affects a variety of cholinergic populations of the forebrain, including those of the medial spetum, nuceus basalis of Meynerts, substantia innominata and striatum.

Neurotrophin action is via a receptor-signaling system that is composed of two transmembrane receptor proteins, a $p140^{trk}$ family of the high affinity neurotrophin receptors, and the low affinity neurotrophin receptor, $p75^{NTR}$ *(70–73)*. Most of the NGF receptors inthe CNS are synthesized in the CBFNs and anterogradely transported to hippocampal and cortical axon terminals that innervate NGF-producing neurons. NGF interaction with its receptors initiates a series of signal transduction events that begin with the binding of neurotrophins to receptors on the membrane of CBFN terminals in the hippocampus and cortex, internalization of a neurotrophin-receptor complex, and the retrograde transport of the neurotrophin to the neuronal soma.

The $p75^{NTR}$ receptor is a transmembrane glycoprotein which belongs to a family of diverse cell surface proteins that include tumor necrosis factor (TNF) receptors, Fas antigen, and CD40 *(74,75)*. Although $p75^{NTR}$ was originally characterized as the only NGF receptor, it has subsequently been shown to bind the neurotrophins BDNF and NT-3 with a similar K_d ($10^{-9}M$). Immunocytochemical and *in situ* hybridization studies have demonstrated the presence of $p75^{NTR}$ within several distinct populations of neurons in the brains of adult rats and the normal aged human brain, with the highest expression (greater than 95%), of both protein and mRNA, in the ChAT- or AChE-positive cholinergic projection neurons of the basal forebrain, and moderate levels present in cerebellar Purkinje cells, hypothalamus, and brain stem *(9,76,77,38)*.

The $p140^{trk}$ receptor is a *trk*-related proto-oncogene glycoprotein that contains an intracellular tyrosine kinase domain. There are three specific genes forming the *trk* family in mammals, *trkA, trkB,* and *trkC,* whose gene products bind NGF, BDNF, and NT-3 respectively *(78,79,80)*. *TrkA* expression is primarily colocalized with $p75_{NTR}$ in CBFNs, although there is no such coexpression in striatal cholinergic interneurons *(81,39)*. Coexpression of the two genes for $p75^{NTR}$ and *trkA* can potentially lead to greater responsiveness to NGF *(71)* since the $p75^{NTR}$ receptor modulates *trk* receptor function as a result of direct physical interaction between the two receptors *(82–84)*.

The expression of $p75^{NTR}$ and $p140^{trk}$ mRNA in the basal forebrain neurons is also induced by NGF *(85)*.

The $p75^{NTR}$ receptor is involved in the internalization and retrograde transport of all neurotrophins *(86)*. The $p75^{NTR}$ receptor also triggers some cellular responses independent from *trk* receptors by a signal transduction cascade through ceramide that ultimately activates NF-kappaB *(87,88)* and deters apoptotic cell commitment *(89,90,91)*.

3.1. Neurotrophic Actions on CBFNs

The viability of CBFNs is dependent in part on the neurotrophic support provided by NGF during early development and in adulthood [for reviews, *see (92,93,94)*]. In vitro studies *(95,96)* have shown that NGF promotes ChAT activity and increases ChAT-immunoreactivity (IR) of neurons exhibiting dendritic and axonal processes as demonstrated using organotypic cultures of fetal rat striatum, although the action of NGF in these cultures may be dependent on ambient conditions *(97,98)*.

NGF effectively prevents the degeneration of axotomized CBFNs in both young and aged animals *(99–105)*. Furthermore, the atrophy of CBFNs and the cognitive deficits displayed by aged rats can be reversed by NGF *(106,107)*. Chronic i.c.v. injections of NGF elevate hippocampal ChAT activity in adult rats after partial septohippocampal lesions *(108)*, induce synaptogenesis and hypertrophy after decortication of adult rats *(109)*, and ameliorate cholinergic neuronal atrophy and behavioral impairment after brain injury or extreme aging *(110,109,111)*.

Hellweg, et al. *(112)* reported that moderate lesions of the septohippocampal pathway by intraventricular infusions of ethylcholine aziridinium (AF64A) induces a dose-dependent decrease of hippocampal ChAT activity, which returns to control levels within five weeks of treatment. This cholinergic deficit is associated with an increase in hippocampal NGF mRNA, but not BDNF- or NT-3 mRNA. The content of NGF protein transiently increases in the hippocampus before returning to control levels by five weeks after AF64A infusion. At that time, however, NGF content as well as ChAT activity, are significantly increased in the septum, suggesting an increased uptake of NGF by the remaining spared cholinergic neurons. These data provide correlative evidence for a critical role for endogenous NGF in neuroregeneration and plasticity of the cholinergic basal forebrain after incipient damage. Results by Lapchak and Hefti *(113)* have also shown that chronic or repeated administration of NGF during the onset of degenerative events is necessary for the stimulation of presynaptic cholinergic function in the hippocampus of adult rats with partial fimbrial transections. Another study *(114)* reveals that cultured basal forebrain from neonatal rats shows both overlapping and nonoverlapping responses to treatment with different neurotrophins. Nonner, et al. *(115)* demonstrated that NGF and BDNF enhance ChAT activity and survival of cholinergic neurons after hypoglycemic stress. Studies *(116,117,118)* showed that both NGF and BDNF preserve ChAT activity after lesions, but that only NGF treatment upregulates ChAT or *trkA* expression in individual cells surviving the lesion and induces hypertrophy of the remaining neurons in the lesioned NBM.

Several studies *(119,120,54,121)* have shown that NGF treatment, by means of NGF infusion or gene transfer, promotes long term CNS neuronal rescue and functional recovery from cognitive impairments associated with aging, lesions of the septohippocampal projection, or partial immunolesions to CBFNs. Dekker, et al. *(119)* also reported that grafting of NGF-producing fibroblasts reduces behavioral deficits in rats with lesions of the NBM. NGF-producing fibroblasts, but not beta-galactosidase-producing fibroblasts, ameliorate impairments demonstrated in a Morris water-maze task assay. In addition, spatial acuity is improved to near-normal levels by the NGF-producing graft-treatment. However, ChAT activity in cortical areas and hippocampus is not affected by the NGF-producing grafts. Both experimental and sham grafted groups show a similar reduction in the level of dopamine, but not homovanillic acid or 3-methoxytyramine, in the frontal cortex. Levels of norepinephrine, epinephrine, and serotonin and their metabolites in the neocortex and hippocampus are also not affected by the lesion or the grafts. NGF-producing grafts increase the size of remaining $p75^{NTR}$-positive neurons in the NBM by 25% NBM lesions reduce the levels of integrated optic density seen after ChAT-positive fiber staining in the ventral neocortex by 46%, whereas NGF-producing grafts restore staining in this area to 86% of controls. These data suggest that NGF-producing grafts may improve behavior through the partial restoration of the lesioned projections from the NBM to the neocortex.

One study *(121)* has shown that grafts of substrates favorable axonal growth combined with transient NGF infusions promote morphological and functional recovery in the adult rat brain after lesions of the septohippocampal projection. Long-term septal cholinergic neuronal rescue and partial hippocampal reinnervation are achieved, resulting in partial functional recovery as measured by a simple task that assesses habituation but not a more complex task that assesses spatial reference memory indicating that

(i) partial recovery from central nervous system injury may be stimulated both by preventing host neuronal losses and by promoting host axonal regrowth and
(ii) long-term neuronal losses can also be prevented with transient NGF infusions. Others *(122,123)* have shown that there are long-term protective effects after treatment with human recombinant NGF on primate nucleus basalis cholinergic neurons after neocortical infarction or a unilateral transection of the fornix.

3.2. NGF-mediated Gene Expression

In situ hybridization results *(124)* relying on use of ^{35}S-labeled antisense oligonuceotide probes show that NGF specifically increases the transcription of the ChAT and m2 muscarinic receptor genes. Chronic infusion of NGF into the forebrain of the adult rat also increases NGF receptor mRNA, ChAT mRNA, and neuronal hypertrophy *(125)*. Since *trkA* mRNA and protein are present in basal forebrain neurons during the entire postnatal period and their distribution is identical to that for ChAT mRNA, *trkA* gene expression also colocalizes to BFCNs *(126)*, where NGF stimulates *trkA* and ChAT mRNA and anti-NGF suppresses the expression of both genes *(118)*. These results suggest that endogenous NGF regulates the expression of *trkA* and ChAT. Finally, whereas NGF infusion increases the size of developing CBFNs, NGF antibody inhibits their normal developmental growth. These results are evidence that endogenous NGF acts on developing CBFNs to enhance gene expression and cellular differentiation. Venero, et al. *(127)* also reported that intrastriatal NGF infusion after quinolinic acid lesions prevents lesion-induced decreases in ChAT and *trkA* mRNA, but has no effect on glutamate decarboxylase mRNA expression.

Stressful conditions in the CNS, such as ischemia, hypoxia, hemorrhage, trauma, and aging, are known to induce complex responses in neuronal gene expression with

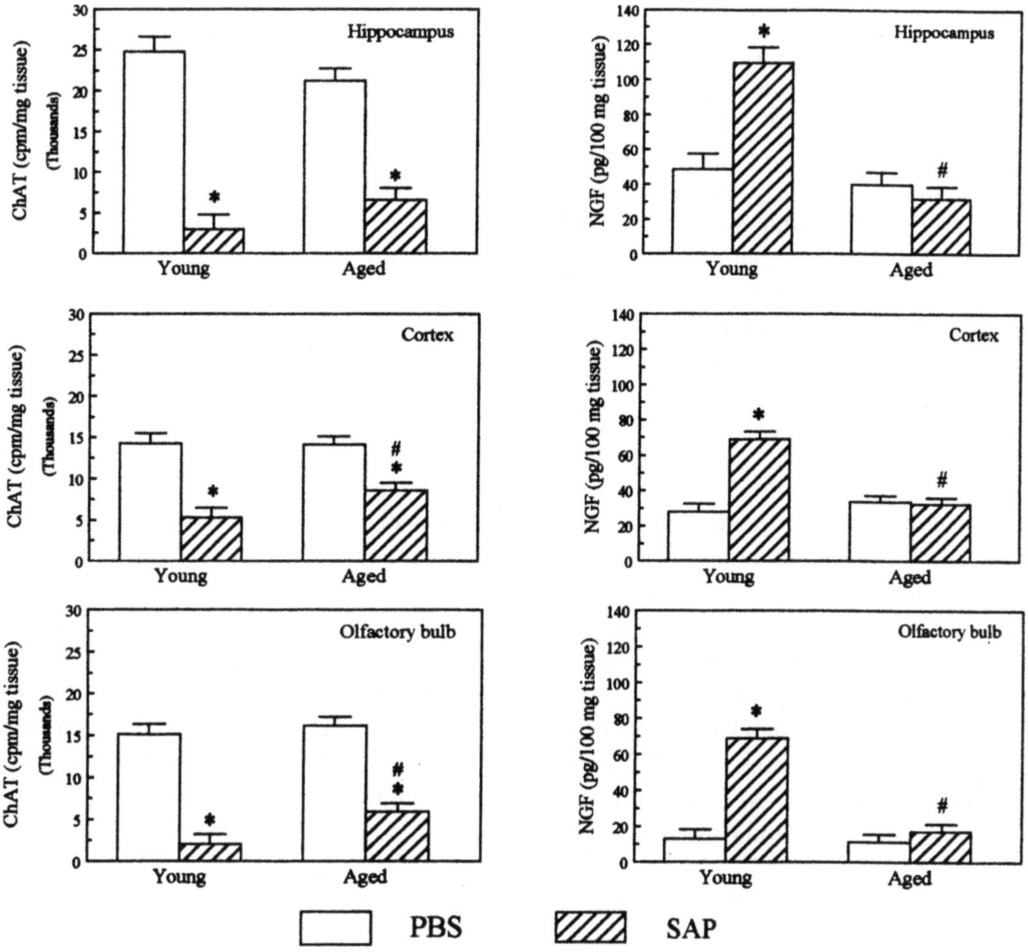

Fig. 2. Effect of total immunolesions on brain ChAT activity and NGF protein levels in the CBFN target areas of young and aged Fisher 344xBrown Norway hybrid rats at 4 week postlesion. Values are mean ± S.E.M. of 4 young control, 4 young lesioned, 8 aged control, and 7 aged lesioned animals. *$p < 0.05$, significantly different from young control rats; # $p<0.05$, significantly different from young lesioned rats, two-way ANOVA with post-hoc Fisher's LSD analysis. Adapted from Gu, et al. (131).

delayed effects on neurotransmission and eventual neuronal commitment to apoptosis, two outcome measures with important physiological consequences. There are fewer cholinergic neurons in the aged basal forebrain, and they express lower levels of ChAT and ACh uptake than their counterparts in young rats. These differences are consistent with both age-associated increases in proapototic markers and decreases in ChAT activity *(3,17,19,34,60,107,128,129,130)*. Also, cholinergic deafferenting insults to the brain (e.g., fimbria fornix transections or immunolesions) typically result in increases in neurotrophin protein in the brains of young rats, but not their aged counterparts *(107,130,131,132)* (Fig. 2). These observations are consistent with the hypothesis that there are age-associated impairments of stress response signaling pathways that regulate recovery of neurotransmitter function and abrogate neuronal commitment to apoptosis after injury. Although we have information as to the general nature of cholinergic regulation and the more prevalent signals regulating apoptosis in the CNS, the mechanisms that regulate the cholinergic cell loss and the downregulation of ChAT activity in the injured aged or AD brain, the major focus of this project, are less well understood. We hypothesize that the activities of transcription factors that regulate gene expression responsible for neurotransmitter function and apoptotic responses to oxidative stress and trauma are perturbed in the aged brain, in part as a response to chronic stressful events associated with aging-related pathology, some of which is likely to involve oxidative stress. We submit that an understanding of the concerted action of multiple regulatory factors is necessary to understand age-associated changes in gene expression.

4. INTRACELLULAR REGULATION OF CBFNS

4.1. Programmed Cell Death

Programmed cell death (PCD) is a fundamental phenomenon for development and homeostasis maintenance in biological organisms to describe spatially and temporally predictable cell loss. Homeostatic control of cell number results from a dynamic balance between cell proliferation and cell death. It occurs as a physiological process during organogenesis in embryos, as well as in adult cell turnover and differentiation and as a pathological process in response to various injuries [for reviews, see *(133,134)*]. During the development of the nervous system, neuronal cell death has been considered as a prime example of PCD, which primarily serves to match neuronal number with target innervation. Competition for limiting amounts of neurotrophic factors that are secreted by neuronal targets determines which and how many neurons ultimately survive *(135)*.

Apoptosis, the best characterized form of PCD, is a morphological term coined by Kerr, et al. *(136)* from ancient Greek for "falling of leaves," in recognition of its significance in tissue homeostasis. In contrast to necrosis, a term for the features of "accidental" cell death, in which cells undergo swelling, loss of membrane integrity, release of lysosomal contents, and which ends with total cell lysis, apoptosis is characterized by a process beginning with cytoplasmic shrinkage and nuclear condensation, followed by loss of the nuclear membrane, fragmentation of the nuclear chromatin, and subsequent partition of condensed nuclear material and cytoplasm into membrane bound-vesicles defined as apoptotic bodies. These apoptotic bodies are then engulfed by adjacent cells through endocytosis. An inflammatory reaction with infiltration of leukocytes that have typically seen in necrosis is absent.

Apoptosis involves a primary perturbation of nuclear chromatin in which activation or *de novo* synthesis of endonuclease(s) results in cleavage of DNA at linker regions between nucleosomes to form multiple units of chromatin pieces with about 180 base pairs (bp) fragments of double-stranded DNA, which present as a DNA "ladder" by electrophoresis *(137,138,139)*. Appearance of DNA ladders has been reported in many apoptotic processes, and is widely used as a marker of apoptosis with the exception that some necrotic cells also show formation of DNA ladders *(140)*. Furthermore, apoptotic endonucleases also generate free 3′-OH DNA double-strand breaks with the sticky ends, which can be identified using the template-dependent *in situ* end-labeled translation staining with Klenow DNA polymerase, where the terminal deoxy-transferase dUTP-nick end labeling (TUNEL) detects both sticky and blunt ends *(141,142)*.

Neuronal cell death occurs not only during development but also as a consequence of acute traumatic events or chronic degenerative disorders that disturb cellular homeostasis, such as trauma, ischemia, and aging *(143,144,135,136)*. For example, prolonged ischemia results in varying degrees of neuronal cell death in the region outside the stroke's core, which is affected by a decrease in tissue perfusion that results in an inadequate supply of oxygen, glucose, and other metabolites *(137)*. Although the machinery leading to cell death is not fully understood, some important aspects of the cell death process show the characteristics of apoptosis, such as DNA fragmentation, chromatin condensation, apoptotic bodies, and cycloheximidie reduction of ischemic damage *(138–145)*. In addition, studies *(146,147,148,149)* show that experimental cerebral ischemia induces activation and cleavage of caspase-3 in apoptotic cortex, delayed activation of caspase prolongs ischemic brain protection, elevation of neuronal expression of neuronal apoptosis inhibitory protein (NAIP) reduces ischemic damage in the rat hippocampus.

A wide variety of neurodegenerative diseases such as AD, Parkinson's disease (PD) and amyotrophic lateral sclerosis (ALS) are characterized by a gradual loss of specific neuronal populations. LeBlanc *(149a)* reported that in serum-deprived cultured neurons, metabolism of amyloid precursor proteins (APPs) through the non-amyloidogenic secretory pathway is decreased by 20% to 40% in control cultures whereas 4KD amyloid-beta (A-beta) peptide is increased one- to four-fold suggesting that human neurons undergoing apoptosis generate excess A-beta. Zhao, et al. *(150)* provided evidence that expression of mutant APPs in differentiated PC12 cells induces cell death via an apoptotic pathway. Observations of age-dependent damage, deterioration of respiratory enzyme activities, and susceptibility of mitochondrial DNA to oxidative stress have suggested that the delayed onset and age dependence of neurodegenerative diseases may result from the deterioration of stress response mechanisms *(151)*. A unified model for aging mechanisms encompassing intrinsic mediators and extrinsic agents proposes that altered regulation of gene expression and signal transduction-redox pathways is the basis for age-associated neurodegenerative diseases *(152)* (Fig. 3). Apoptotic events reflect an interplay between intrinsic signaling events that rely on cytokines, neurotransmitters, and growth factors and responses to extrinsic events that increase levels of reactive oxygen species (ROS). Both intrinsically and extrinsically driven signal-transduction pathways act via transcription factors that regulate the coordinated timely expression of stress-response genes as part of a decision-making process that can commit cells to apoptosis or survival *(153)*.

Neuronal cells, like most other cell types, contain endogenous suicide mechanisms that are activated when specific cell numbers are to be reduced during development, or in response to genotoxic signals *(134,154)*. A report by Sinson, et al. *(155)* indicates that NGF administration beginning 24 h after fluid-percussion brain injury improves cognitive outcome and decreases cholinergic neuronal cell loss and apoptotic cell death. Another study *(156)* also shows that NGF prevents cell death in lesioned central cholinergic neurons. NGF also protects hippocampal and cortical neurons against iron-induced degeneration via generation of free radicals *(157)*.

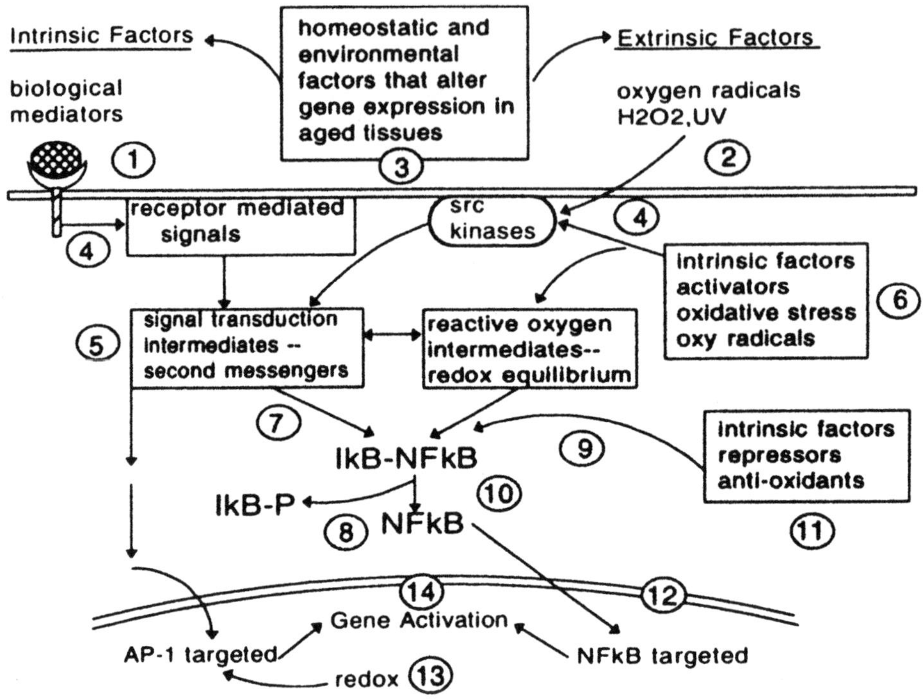

Fig. 3. Model outlining the general interactions of the signal transduction-redox pathways that regulate trans-acting factors of the stress response genes. **(1)** Biological mediators activate stress response genes via a receptor-ligand complex that transmits signals to the cytoplasm. **(2)** Extrinsic agents exert their effects upon the early response cytoplamic intermediate. **(3)** Homeostatic and environmental changes of intrinsic and/or extrinsic factors could result in altered gene expression and regulation in aged. **(4)** Membrane-bound *src* kinases are the earliest cytoplasmic factors. **(5)** Signal transduction intermediates conduct a cascade of modifications that lead to gene regulation. **(6)** Oxidative metabolism is a source of free radicals that can act as second messengers to stimulate signal transduction intermediates and affect the redox equilibrium. **(7–8)** Activation of NF-κB requires its release from I-kappaB. **(12)** Nuclear translocation is mediated by a specific carrier protein. **(13–14)** Activation of NF-κB results in gene activation. Modified from Papaconstantinou (152).

Oxidative stress has been linked to neuronal cell death resulting from stressful conditions such as ischemia, trauma, excitotoxicity, and neurodegenerative diseases *(158,159,160)*. As oxidative stimuli, ROS are generated as byproducts of normal and aberrant metabolic processes using molecular oxygen (O_2) as substrate. Cellular defense systems against ROS include enzymes that convert ROS to less reactive species and antitoxidants that quench the different ROS directly. Due to the high oxygen consumption and high concentration of oxidizable substances such as catecholamines, ascorbic acid, and lipids in brain, imbalances in oxidative homeostasis result in oxidative stress leading to neuronal damage *(161,162)*. Both transcription factors activator protein-1 (AP-1) and nuclear factor-kappaB (NF-kappaB) play roles in the signal transduction pathways potentially associated with both trkA- and $p75^{NTR}$-mediated regulation of apoptosis *(163,91,164,165)*. Serum and NGF deprivation, oxidative stress, and glucose deprivation induce apoptosis in PC12 cells. AP-1 activity changes are associated with both induced apoptosis and NGF-mediated cell rescue in PC12 cells. The survival of PC12 cells is significantly reduced by the inhibition of NF-κB activation in the presence or absence of NGF. Apoptosis is induced via the stimulation of several different cell surface receptors in association with caspase activation. Carter, et al. *(87)* has shown that NGF binding to $p75^{NTR}$ activates the transcription factor NF-kappaB in rat Schwann cells. This activation is not observed in Schwann cells isolated from mice that lacked $p75^{NTR}$. Selective for NGF, NF-κB is not activated by BDNF or NT-3.

4.2. Nuclear Factor-κB

4.2.1. Structure, Regulation and Function

Although the transcription factor NF-κB was first described as a nuclear factor found in B cells which binds to the immunoglobulin kappa light chain enhancer *(166),* it is widely expressed outide the immune system. NF-κB belongs to a family of homo- and hetero-dimeric proteins related by a conserved ~300 amino acid residue NH_2-terminal Rel/homology domain; members of this family include the proteolytically processed p65 (also referred to as RelA), p50 and p49 (also referred to as p52) as well as RelB and c-Rel. The dimeric subunits of NF-κB can recognize and potentially bind to a 10-bp generic DNA consensus sequence 5′-GGGRNNYYCC-3′ (G=guanine, R=purine, N=any nucleotide, Y=pyrimidine, C=cytosine). The most commonly described dimeric subunit combination of NF-κB is the p65/p50 heterodimer found in the cytoplasm bound to an inhibitory subunit, I-κB, in the absence of stimulation. Both the p65 and p50 subunits of NF-κB are regulated by redox modification *(166,167,168,169)* and manipulations of intracellular glutathione (GSH) and cysteine have significant consequences for NF-κB activity *(170).* A role for ROS as second messengers in NF-κB regulation is suggested by the activation of NF-κB by H_2O_2 and its inhibition by antioxidants, such as N-acetylcysteine (NAC) and thioredoxin *(171,172,173,174,175).* Upon stimulation, I-κB is phosyphorylated and degraded *(176,177),* and nuclear translocation (NLS) domains on NF-κB proteins are exposed, which facilitate translocation to the nucleus for DNA-binding. Heterodimerization of NF-κB subunits produces species with various intrinsic DNA-binding specificities, transactivation properties and subcellular localization [for reviews, see *(178,179)*].

NF-κB is ubiquitously expressed in most cell types and tissues, although its activation process has been mostly studied in lymphoid cells, epithelial cells, and fibroblasts. NF-κB is an inducible transcription factor that is responsive to a broad range of stimuli *(178,180),* such as interleukin-1 (IL-1) and other cytokines, tumor necrosis factor (TNF), bacterial lipopolysaccharide (LPS), human immunodeficiency type-I virus (HIV-1), T-cell mitogens (e.g., lectins, phorbol esters), UVA radiation *(181),* hydrogen peroxide *(182),* and hypoxia *(183).* The converging event for different stimuli appears to be the removal of I-κB proteins from a cytoplasmic complex with NF-κB via phosphorylation of I-κB and subsequent ubiquitination and degradation by proteasomes *(184) (185,186).* Several kinases, such as protein kinase C (PKC), cAMP-dependent protein kinase and casein kinase II may mediate I-κB phosphorylation and subsequent dissociation *(187).* The signal pathway for TNF-α is mediated via the sphingomyelin pathway with the production of ceramide and the activation of PKCζ *(188–193).*

Since NF-κB is a preformed transcriptional factor with regulatory activity, it can be rapidly activated by mechanisms that do not require *de novo* protein synthesis in contrast to the immediate-early transcription factor AP-1 family, whose activity is regulated via prompt, robust and transient gene induction. Activation of NF-κB by various stimuli leads to subsequent transcriptional activation of many target genes (for reviews, *see 187,7*), including proinflammatory cytokines such as TNF, IL-1, IL-6, interferon-; inducible nitric oxidase synthase (iNOS), manganese superoxide dismutase (Mn-SOD), cyclooxygenase-2 (COX-2); major histocompatibility complex Class-I (MHC-I), vascular cell adhesion molecule-1 (VCAM-1); as well as the neuropeptide dynorphin, and viral HIV-1 gene.

4.2.2. NF-κB Role in CNS

Although early work was confined to gene activation during the inflammatory and immune response in lymphocytes, epithelial cells and fibroblasts, there is evidence that NF-κB has a unique role in neurons and astroglia in CNS, where NF-κB acts in both normal function and acute responses to injury, as well as chronic neurodegenerative disorders—such as AD and PD (for reviews, *see (7,184,187)*).

There is constitutive NF-κB activity present in the hippocampus, cortex and basal forebrain *(194,195,171,173,196)*. In addition to NF-κB in neuronal cell bodies, both I-κ B-α and NF-κB immunoreactivity are present in the postsynaptic densities (PSDs) in the rat hippocampus and cortex *(195)*. Increased NF-κB has been identified in neurons and astrocytes of brain sections from AD patients *(197,198)*. NF-κB is also present in the microglia and macrophages in both parenchymal and perivascular areas of multiple sclerosis patients *(199)*.

In AD brains, the p65 subunit is most abundant in the particulate fraction of the temporal cortex *(200)* where the proportion of large—probably cholinergic—neurons with nuclear NF-κB staining is significantly increased *(197,198)*. There is increased NF-κB activity—assayed with an activity-specific monoclonal antibody (mAb) to the p65 subunit—in neurons and astroglia of brain sections from AD patients, where increased NF-κB activation is restricted to cells in close proximity to early plaques. Also, the neurotoxic peptide A-β is a potent NF-κB activator in primary neuronal cultures. This activation requires reactive oxygen intermediates as messengers because an antioxidant, pyrrolidine dithiocarbamate, prevents A-β induced NF-κB activation *(172)*. In PD patients, the proportion of dopaminergic neurons with immunoreactive NF-κB in their nuclei is more than 70-fold higher than that in control subjects *(201,202)*. In vitro treatment of primary cultures of rat mesencephalon with C2-ceramide activates the sphingomyelin-dependent signaling pathway resulting in translocation of NF-κB and the transient production of free radicals that induce apoptosis, indicating a possible relationship between the nuclear localization of NF-κB in mesencephalic neurons of PD patients and oxidative stress in such neurons.

4.2.3. Delayed NF-κB Alteration

The outcome of NF-κB activation may depend on its time course. For example, the transient activation of NF-κB observed at 24 h after ischemia may be responsible for the induction of protective factors—an antioxidant (LY231617) present in neurons that survive the ischemic insult, whereas the persistent activation of NF-κB in hippocampal neurons after global ischemia could be responsible for the induction of proteins that result in CA1 neuronal death *(203)*. Interestingly, this is the opposite of AP-1 where transient increases in activity are associated with apoptosis and persistent increases in activity are associated with neuronal rescue by NGF *(165)*. Thus, NGF treatment of serum-deprived PC12 cells results in a temporally delayed decrease in NF-κB DNA-binding activity that contrasts with the prompt and robust increases in AP-1 binding activity associated with serum deprivation and are further enhanced and prolonged by NGF treatment. In vivo, transient focal ischemia-reperfusion in the cerebral cortex causes regional alteration of DNA-binding activity of transcription factors AP-1, CREB, Sp-1, and NF-κB in a time-dependent fashion. There is an increase in AP-1 activity surrounding the ischemic cortex during early reperfusion that is, followed by an increase in the CREB, Sp-1, and NF-κB, but not AP-1, binding activity in the ischemic cortex itself, and to a lesser extent of Sp-1 and NF-κB binding activity in the surrounding region, after five ds of reperfusion *(204)*. Traumatic brain injury (TBI) also increases NF-κB DNA-binding activity in cerebral cortex ipsilateral to the injury site after three days *(205)*. After injection with the toxic glutamate receptor agonist, kainic acid (KA), p65-IR is markedly increased in reactive astrocytes after 7–10 days while binding activity to the kappa light chain enhancer consensus sequence increases two days after KA treatment *(206)*. In spinal cord, there is increased NF-κB activation that persists for at least 72 h after a contusion injury *(207)*. This injury-induced increase in activated NF-κB in spinal cord co-localizes with inducible nitric oxide synthase (iNOS) protein, a NF-κB dependent gene product. These studies suggest that there are critical differences between transient and prolonged activation of transcription factors that determine apoptotic outcomes.

4.2.4. Role of NF-κB in Cell Death

Results of Bales, et al. *(208)* show that A-β induced toxicity of cultured fetal rat cortical neurons is associated with internucleosomal DNA fragmentation early after exposure to A-β. There are increased I-κB-alpha mRNA and its protein that may be responsible for the retention of NF-κB in the cytoplasm. In contrast, exposure of rat primary astrocyte cultures to A-β results in the activation of NF-κB and the subsequent stimulation of IL-1β and IL-6 levels, suggesting that perturbations in NF-κB-induced gene expression may contribute to the neurodegenerative and inflammatory response reported for AD. Grilli, et al. *(209)* have reported that aspirin, and its metabolite sodium salicylate, protect against neurotoxicity in glutamate treated rat primary neuronal cultures and hippocampal slices through inhibition of glutamate-mediated induction of NF-κB. Ko, et al. *(210)* has also shown that aspirin exerts its neuroprotective action against NMDA. Also, by blocking the NMDA-induced activation of NF-κB and JNK. Mattson, et al. *(211)* have shown that the induction of Mn-SOD by TNF-α and C2-ceramide treatment, and the suppression of peroxynitrite formation and membrane lipid peroxidation by the peroxynitrite scavenger uric acid, is via increased NF-κB activation. Thus, NF-κB plays an antiapoptotic role under neurodegenerative conditions resulting from metabolic and oxidative insults due to FeSO4 and amyloid β-peptide. Furukawa and Mattson *(212)* have reported that NF-κB mediates increased calcium currents and decreased NMDA- and AMPA/KA-induced currents in hippocampal neurons treated with TNF-α. Pahan, et al. *(213)* has shown that the sphingomyelin-ceramide signaling pathway stimulates the expression of iNOS via LPS- or cytokine-mediated activation of NF-κB in astrocytes. However, Lezoualc'h, et al. *(214)* have reported that high constitutive NF-κB activity mediates resistance to oxidative stress in neuronal cell populations as NF-κB regulates the transcription of a wide variety of genes that contribute to different physiological outcomes.

To identify whether NF-κB regulates the transcription of the synthesis of ACh genes and whether this regulation is affected in the aged CNS, it is necessary to assess NF-κB responses in terms of its binding to the NF-κB site on the ChAT promoter. The ChAT promoter contains NGF inducible elements and consensus sequences for several transcription factors, including those of the basic helix-loop-helix family *(215)*. The mouse ChAT gene promoter region contains a potential NF-κB binding site as well as sites that bind the serum response element (SRE), the cAMP response element (CRE), and the activator protein-1 (AP-1) *(216,217,218)* have shown that nuclear protein extracts from rat basal forebrain and PC12 cultures bind with more affinity to the specific NF-κB consensus sequence present in the ChAT promoter as compared to the kappa light chain enhancer NF-κB binding sequence. A supershift reaction using purified NF-κB proteins has shown that p49 and p65 bind to the ChAT NF-κB consensus sequence, but not the p50 subunit, in agreement with the hypothesis that different NF-κB consensus sequences bind to different NF-κB subunits in a tissue- or cell type-specific manner. Quantitative determination of binding affinities of different NF-κB proteins to the different consensus sequences (64 different combinations) may provide strategies for selective intervention approaches that may have therapeutic potential.

ACKNOWLEDGMENTS

This work was supported in part by NINDS NS33288, a grant from the Sealy Center on Aging, and grants from the Spinal Cord Research Foundation and TIRR.

REFERENCES

1. Bick, K., Amaducci, L. and Pepeu, G. (1987) *The Early Story of Alzheimer's Disease. Translation of the historical papers.* Liviana Press: Padova, Italy.
2. Berchtold, N.C. and Cotman, C.W. (1998) Evolution in the conceptualization of dementia and Alzheimer's disease: Greco-Roman period to the 1960s. *Neurobiol. Aging* **19,** 173–189.
3. Coyle, J.T., Price, D.L., and DeLong, M.R. (1983) Alzheimer's disease: a disorder of cortical cholinergic innervation. *Science* **219,** 1184–1190.
4. Whitehouse, P.J., Struble, R.G., Hedreen, J.C., Clark, A.W., and Price, D.L. (1985) Alzheimer's disease and related dementias: selective involvement of specific neuronal systems. *CRC Critical Reviews Clin. Neurobiol.* **1,** 319–339.
5. Yankner, B.A. (1996) Mechanisms of neuronal degeneration in Alzheimer's disease. *Neuron* **16,** 921–932.
6. Johnson, S.A. and Finch, C.E. (1996) Changes in gene expression during brain againg: a survey. In: *Handbook of the Biology of Aging,* pp. 300–27. Eds. E.L. Schneider and J.W. Rowe. Academic Press: San Diego.
7. O'Neill, L.A. and Kaltschmidt, C. (1997) NF-kappa B: a crucial transcription factor for glial and neuronal cell function [see comments]. *Trends in Neurosci.* **20,** 252–258.
8. Winkler, J., Thal, L.J., Gage, F.H., and Fisher, L.J. (1998) Cholinergic strategies for Alzheimer's disease. *Journal of Molecular Medicine* **76,** 555–567.

9. Butcher, L.L., Oh, J.D. and Woolf, N.J. (1993) Cholinergic neurons identified by in situ hybridization histochemistry. *Progress in Brain Research* **98,** 1–8.
10. Butcher, L.L. and Woolf, N.J. (1984) Histochemical distribution of acetylcholinesterase in the central nerve system: Clues to the localization of cholinergic neurons. In: *Handbook of Chemical Neuroanatomy Vol. 3,* pp. 1–50. Elsevier: Amsterdam.
11. Oh, J.D., Woolf, N.J., Roghani, A., Edwards, R.H., and Butcher, L.L. (1992) Cholinergic neurons in the rat central nervous system demonstrated by in situ hybridization of choline acetyltransferase mRNA. *Neuroscience* **47,** 807–822.
12. Woolf, N.J. (1991) Cholinergic systems in mammalian brain and spinal cord. *Progress in Neurobiology* **37,** 475–524.
13. Bigl, V., Woolf, N.J. and Butcher, L.L. (1982) Cholinergic projections from the basal forebrain to frontal, parietal, temporal, occipital, and cingulate cortices: a combined fluorescent tracer and acetylcholinesterase analysis. *Brain Res. Bull.* **8,** 727–749.
14. Leanza, G., Nilsson, O.G., Wiley, R.G., and Bjorklund, A. (1995) Selective lesioning of the basal forebrain cholinergic system by intraventricular 192 IgG-saporin: behavioural, biochemical and stereological studies in the rat. *European Journal of Neuroscience* **7,** 329–343.
15. Rossner, S., Schliebs, R., and Bigl, V. (1995b) 192IgG-saporin-induced immunotoxic lesions of cholinergic basal forebrain system differentially affect glutamatergic and GABAergic markers in cortical rat brain regions. *Brain Research* **696,** 165–176.
16. Heckers, S., Ohtake, T., Wiley, R.G., Lappi, D.A., Geula, C., and Mesulam, M.M. (1994) Complete and selective cholinergic denervation of rat neocortex and hippocampus but not amygdala by an immunotoxin against the p75 NGF receptor. *Journal of Neuroscience* **14,** 1271–1289.
17. Bartus, R.T., Dean, R.L.d., Beer, B., and Lippa, A.S. (1982) The cholinergic hypothesis of geriatric memory dysfunction. *Science* **217,** 408–414.
18. Drachman, D.A. (1977) Memory and cognitive function in man: does the cholinergic system have a specific role? *Neurology* **27,** 783–790.
19. Davies, P. and Maloney, A.J. (1976) Selective loss of central cholinergic neurons in Alzheimer's disease [letter]. *Lancet* **2,** 1403.
20. Perry, E.K., Gibson, P.H., Blessed, G., Perry, R.H., and Tomlinson, B.E. (1977) Neurotransmitter enzyme abnormalities in senile dementia. Choline acetyltransferase and glutamic acid decarboxylase activities in necropsy brain tissue. *Journal of the Neurological Sciences* **34,** 247–265.
21. Whitehouse, P.J., Price, D.L., Struble, R.G., Clark, A.W., Coyle, J.T., and Delon, M.R. (1982) Alzheimer's disease and senile dementia: loss of neurons in the basal forebrain. *Science* **215,** 1237–1239.
22. Perry, E.K., Tomlinson, B.E., Blessed, G., Bergmann, K., Gibson, P.H., and Perry, R.H. (1978) Correlation of cholinergic abnormalities with senile plaques and mental test scores in senile dementia. *British Medical Journal* **2,** 1457–1459.
23. Strada, O., Vyas, S., Hirsch, E.C. et al. (1992) Decreased choline acetyltransferase mRNA expression in the nucleus basalis of Meynert in Alzheimer disease: an in situ hybridization study. *Proceedings of the National Academy of Sciences of the United States of America* **89,** 9549–9553.
24. Rylett, R.J., Ball, M.J., and Colhoun, E.H. (1983) Evidence for high affinity choline transport in synaptosomes prepared from hippocampus and neocortex of patients with Alzheimer's disease. *Brain Research* **289,** 169–175.
25. Richter, J.A., Perry, E.K., and Tomlinson, B.E. (1980) Acetylcholine and choline levels in postmortem human brain tissue: preliminary observations in Alzheimer's disease. *Life Sciences* **26,** 1683–1689.
26. Elble, R., Giacobini, E., and Higgins, C. (1989) Choline levels are increased in cerebrospinal fluid of Alzheimer patients. *Neurobiology of Aging* **10,** 45–50.

27. Kordower, J.H., Gash, D.M., Bothwell, M., Hersh, L., and Mufson, E.J. (1989) Nerve growth factor receptor and choline acetyltransferase remain colocalized in the nucleus basalis (Ch4) of Alzheimer's patients. *Neurobiol. Aging* **10,** 67–74.
28. Scott, S.A., Mufson, E.J., Weingartner, J.A., Skau, K.A., and Crutcher, K.A. (1995) Nerve growth factor in Alzheimer's disease: increased levels throughout the brain coupled with declines in nucleus basalis. *Journal of Neuroscience* **15,** 6213–6221.
29. Jackson, G.R., Werrbach-Perez, K., Pan, Z., Sampath, D., and Perez-Polo, J.R. (1994) Neurotrophin regulation of energy homeostasis in the central nervous system. *Developmental Neuroscience* **16,** 285–290.
30. Pan, Z., Sampath, D., Jackson, G., Werrbach-Perez, K., and Perez-Polo, R. (1997) Nerve growth factor and oxidative stress in the nervous system. *Advances in Experimental Medicine & Biology* **429,** 173–193.
31. Perez-Polo, J.R., Foreman, P.J., Jackson, G.R. et al. (1990) Nerve growth factor and neuronal cell death. *Molecular Neurobiology* **4,** 57–91.
32. Williams, L.R. and Rylett, R.J. (1990) Exogenous nerve growth factor increases the activity of high-affinity choline uptake and choline acetyltransferase in brain of Fisher 344 male rats. *Journal of Neurochemistry* **55,** 1042–1049.
33. Ogawa, N., Asanuma, M., Kondo, Y., Nishibayashi, S., and Mori, A. (1994) Reduced choline acetyltransferase activity and muscarinic M1 receptor levels in aged Fisher 344 rat brains did not parallel their respective mRNA levels. *Brain Research* **658,** 87–92.
34. De Lacalle, S., Cooper, J.D., Svendsen, C.N., Dunnett, S.B., and Sofroniew, M.V. (1996) Reduced retrograde labelling with fluorescent tracer accompanies neuronal atrophy of basal forebrain cholinergic neurons in aged rats. *Neuroscience* **75,** 19–27.
35. Schliebs, R., Rossner, S., and Bigl, V. (1996) Immunolesion by 192IgG-saporin of rat basal forebrain cholinergic system: a useful tool to produce cortical cholinergic dysfunction. *Progress in Brain Research* **109,** 253–264.
36. Gage, F.H., Batchelor, P., Chen, K.S. et al. (1989) NGF receptor reexpression and NGF-mediated cholinergic neuronal hypertrophy in the damaged adult neostriatum. *Neuron* **2,** 1177–1184.
37. Steininger, T.L., Wainer, B.H., Klein, R., Barbacid, M., and Palfrey, H.C. (1993) High-affinity nerve growth factor receptor (Trk) immunoreactivity is localized in cholinergic neurons of the basal forebrain and striatum in the adult rat brain. *Brain Research* **612,** 330–335.
38. Yan, Q. and Johnson, E.M., Jr. (1989) Immunohistochemical localization and biochemical characterization of nerve growth factor receptor in adult rat brain. *Journal of Comparative Neurology* **290,** 585–598.
39. Wiley, R.G., Oeltmann, T.N., and Lappi, D.A. (1991) Immunolesioning: selective destruction of neurons using immunotoxin to rat NGF receptor. *Brain Research* **562,** 149–153.
40. Wiley R.G. (1992) Neural lesioning with ribosome-inactivating proteins: suicide transport and immunoleioning. *Trends in Neurosciences* **15,** 285–290.
41. Book, A.A., Wiley, R.G. and Schweitzer, J.B. (1992) Specificity of 192 IgG-saporin for NGF receptor-positive cholinergic basal forebrain neurons in the rat. *Brain Research* **590,** 350–355.
42. Rossner, S., Hartig, W., Schliebs, R. et al. (1995a) 192IgG-saporin immunotoxin-induced loss of cholinergic cells differentially activates microglia in rat basal forebrain nuclei. *Journal of Neuroscience Research* **41,** 335–346.
43. Rossner, S., Schliebs, R., Hartig, W., and Bigl, V. (1995c) 192IGG-saporin-induced selective lesion of cholinergic basal forebrain system: neurochemical effects on cholinergic neurotransmission in rat cerebral cortex and hippocampus. *Brain Research Bulletin* **38,** 371–381.
44. Torres, E.M., Perry, T.A., Blockland, A. (1994) Behavioural, histochemical and biochemical consequences of selective immunolesions in discrete regions of the basal forebrain cholinergic system. *Neuroscience* **63,** 95–122.

45. Waite, J.J., Chen, A.D., Wardlow, M.L., Wiley, R.G., Lappi, D.A., and Thal, L.J. (1995) 192 immunoglobulin G-saporin produces graded behavioral and biochemical changes accompanying the loss of cholinergic neurons of the basal forebrain and cerebellar Purkinje cells. *Neuroscience* **65,** 463–476.
46. Yu, J., Pizzo, D.P., Hutton, L.A., and Perez-Polo, J.R. (1995) Role of the cholinergic system in the regulation of neurotrophin synthesis. *Brain Research* **705,** 247–252.
47. Yu, J., Wiley, R.G., and Perez-Polo, R.J. (1996) Altered NGF protein levels in different brain areas after immunolesion. *Journal of Neuroscience Research* **43,** 213–223.
48. Leanza, G., Nilsson, O.G., Nikkhah, G., Wiley, R.G., and Bjorklund, A. (1996c) Effects of neonatal lesions of the basal forebrain cholinergic system by 192 immunoglobulin G-saporin: biochemical, behavioural and morphological characterization. *Neuroscience* **74,** 119–141.
49. Stoehr, J.D., Mobley, S.L., Roice, D. et al. (1997) The effects of selective cholinergic basal forebrain lesions and aging upon expectancy in the rat. *Neurobiology of Learning & Memory* **67,** 214–227.
50. Wenk, G.L., Stoehr, J.D., Quintana, G., Mobley, S., and Wiley, R.G. (1994) Behavioral, biochemical, histological, and electrophysiological effects of 192 IgG-saporin injections into the basal forebrain of rats. *Journal of Neuroscience* **14,** 5986–5995.
51. Book, A.A., Wiley, R.G. and Schweitzer, J.B. (1994) 192 IgG-saporin: I. Specific lethality for cholinergic neurons in the basal forebrain of the rat. *J. Neuropath. & Experimen. Neurol.* **53,** 95–102.
52. Gage, F.H., Bjorklund, A., and Stenevi, U. (1983) Reinnervation of the partially deafferented hippocampus by compensatory collateral sprouting from spared cholinergic and noradrenergic afferents. *Brain Research* **268,** 27–37.
53. Leanza, G., Nikkhah, G., Nilsson, O.G., Wiley, R.G., and Bjorklund, A. (1996b) Extensive reinnervation of the hippocampus by embryonic basal forebrain cholinergic neurons grafted into the septum of neonatal rats with selective cholinergic lesions. *Journal of Comparative Neurology* **373,** 355–357.
54. Rossner, S., Yu, J., Pizzo, D. et al. (1996) Effects of intraventricular transplantation of NGF-secreting cells on cholinergic basal forebrain neurons after partial immunolesion. *Journal of Neuroscience Research* **45,** 40–56.
55. Baxter, M.G., Bucci, D.J., Gorman, L.K., Wiley, R.G., and Gallagher, M. (1995) Selective immunotoxic lesions of basal forebrain cholinergic cells: effects on learning and memory in rats. *Behav. Neurosci.* **109,** 714–722.
56. Berger-Sweeney, J., Heckers, S., Mesulam, M.M., Wiley, R.G., Lappi, D.A., and Sharma, M. (1994) Differential effects on spatial navigation of immunotoxin-induced cholinergic lesions of the medial septal area and nucleus basalis magnocellularis. *J. Neurosci.* **14,** 4507–4519.
57. Leanza, G., Muir, J., Nilsson, O.G., Wiley, R.G., Dunnett, S.B., and Bjorklund, A. (1996a) Selective immunolesioning of the basal forebrain cholinergic system disrupts short-term memory in rats. *European Journal of Neuroscience* **8,** 1535–1544.
58. Nilsson, O.G., Leanza, G., Rosenblad, C., Lappi, D.A., Wiley, R.G., and Bjorklund, A. (1992) Spatial learning impairments in rats with selective immunolesion of the forebrain cholinergic system. *Neuroreport* **3,** 1005–1008.
59. Wiley, R.G. (1997) Findings about the cholinergic basal forebrain using immunotoxin to the nerve growth factor receptor. *Annals of the New York Academy of Sciences* **835,** 20–29.
60. Davies, P. (1979) Neurotransmitter-related enzymes in senile dementia of the Alzheimer type. *Brain Research* **171,** 319–327.
61. Thoenen, H. (1995) Neurotrophins and neuronal plasticity. *Science* **270,** 593–598.
62. Bothwell, M. (1995) Functional interactions of neurotrophins and neurotrophin receptors. *Annual Review of Neuroscience* **18,** 223–253.

63. Hefti, F. (1997) Pharmacology of neurotrophic factors. *Annual Review of Pharmacology & Toxicology* **37,** 239–267.
64. Ibanez, C.F. (1998) Emerging themes in structural biology of neurotrophic factors. *Trends in Neurosciences* **21,** 438–444.
65. Lindsay, R.M. (1994b) Neurotrophins and receptors. *Progress in Brain Research* **103,** 3–14.
66. Thal, L.J. (1996) Neurotrophic factors. *Progress in Brain Research* **109,** 327–330.
67. Thoenen, H. (1991) The changing scene of neurotrophic factors. *Trends in Neurosciences* **14,** 165–170.
68. Thoenen, H., Bandtlow, C., and Heumann, R. (1987) The physiological function of nerve growth factor in the central nervous system: comparison with the periphery. *Reviews of Physiology Biochemistry & Pharmacology* **109,** 145–178.
69. Whittemore, S.R. and Seiger, A. (1987) The expression, localization and functional significance of beta-nerve growth factor in the central nervous system. *Brain Research* **434,** 439–464.
70. Chao, M., Casaccia-Bonnefil, P., Carter, B., Chittka, A., Kong, H., and Yoon, S.O. (1998) Neurotrophin receptors: mediators of life and death. *Brain Research – Brain Research Reviews* **26,** 295–301.
71. Chao, M.V. and Hempstead, B.L. (1995) p75 and Trk: a two-receptor system. *Trends in Neurosci.* **18,** 321–326.
72. Frade, J.M. and Barde, Y.A. (1998) Nerve growth factor: two receptors, multiple functions. *Bioessays* **20,** 137–145.
73. Kaplan, D.R. and Miller, F.D. (1997) Signal transduction by the neurotrophin receptors. *Current Opinion in Cell Biology* **9,** 213–221.
74. Bothwell, M. (1996) p75NTR: a receptor after all [see comments]. *Science* **272,** 506–507.
75. Smith, C.A., Farrah, T., and Goodwin, R.G. (1994) The TNF receptor superfamily of cellular and viral proteins: activation, costimulation, and death. *Cell* **76,** 959–962.
76. Mesulam, M.M., Mufson, E.J., Wainer, B.H., and Levey, A.I. (1983) Central cholinergic pathways in the rat: an overview based on an alternative nomenclature (Chl-Ch6). *Neuroscience* **10,** 1185–1201.
77. Mufson, E.J., Bothwell, M., Hersh, L.B., and Kordower, J.H. (1989) Nerve growth factor receptor immunoreactive profiles in the normal, aged human basal forebrain: colocalization with cholinergic neurons. *Journal of Comparative Neurology* **285,** 196–217.
78. Kaplan, D.R., Martin-Zanca, D., and Parada, L.F. (1991) Tyrosine phosphorylation and tyrosine kinase activity of the trk proto-oncogene product induced by NGF. *Nature* **350,** 158–160.
79. Klein, R., Nanduri, V., Jing, S.A. (1991) The trkB tyrosine protein kinase is a receptor for brain-derived neurotrophic factor and neurotrophin-3. *Cell* **66,** 395–403.
80. Lamballe, F., Klein, R., and Barbacid, M. (1991) trkC, a new member of the trk family of tyrosine protein kinases, is a receptor for neurotrophin-3. *Cell* **66,** 967–979.
81. Gibbs, R.B. and Pfaff, D.W. (1994) In situ hybridization detection of trkA mRNA in brain: distribution, colocalization with p75NGFR and up-regulation by nerve growth factor. *J. Compara. Neurol.* **341,** 324–339.
82. Barker, P.A. and Shooter, E.M. (1994) Disruption of NGF binding to the low affinity neurotrophin receptor p75LNTR reduces NGF binding to TrkA on PC12 cells. *Neuron* **13,** 203–215.
83. Verdi, J.M., Birren, S.J., Ibanez, C.F. et al. (1994) p75LNGFR regulates Trk signal transduction and NGF-induced neuronal differentiation in MAH cells. *Neuron* **12,** 733–745.
84. Wolf, D.E., McKinnon, C.A., Daou, M.C., Stephens, R.M., Kaplan, D.R., and Ross, A.H. (1995) Interaction with TrkA immobilizes gp75 in the high affinity nerve growth factor receptor complex. *Journal of Biological Chemistry* **270,** 2133–2138.

85. Holtzman, D.M., Li, Y., Parada, L.F., Kinsman, S., Chen, C.K., Valletta, J.S., Zhou, J., Long, J.B., and Mobley, W.C. (1992) p140trk mRNA marks NGF-responsive forebrain neurons: evidence that trk gene expression is induced by NGF. *Neuron* **9,** 465–478.
86. Johnson, E.M., Jr., Taniuchi, M., Clark, H.B. (1987) Demonstration of the retrograde transport of nerve growth factor receptor in the peripheral and central nervous system. *J. Neurosci.* **7,** 923–929.
87. Carter, B.D., Kaltschmidt, C., Kaltschmidt, B., Offenhauser, N., Bohm-Matthaei, R., Baeuerle, P.A., and Barde, Y.A. (1996) Selective activation of NF-kappa B by nerve growth factor through the neurotrophin receptor p75 [see comments]. *Science* **272,** 542–545.
88. Taglialatela, G., Hibbert, C.J., Hutton, L.A., Werrbach-Perez, K., and Perez-Polo, J.R. (1996) Suppression of p140trkA does not abolish nerve growth factor-mediated rescue of serum-free PC12 cells. *J. Neurochem.* **66,** 1826–1835.
89. Casaccia-Bonnefil, P., Carter, B.D., Dobrowsky, R.T., and Chao, M.V. (1996) Death of oligodendrocytes mediated by the interaction of nerve growth factor with its receptor p75. *Nature* **383,** 716–719.
90. Rabizadeh, S., Oh, J., Zhong, L.T., Yang, J., Bitler, C.M., Butcher, L.L., and Bredesen, D.E. (1993) Induction of apoptosis by the low-affinity NGF receptor. *Science* **261,** 345–348.
91. Taglialatela, G., Robinson, R., and Perez-Polo, J.R. (1997) Inhibition of nuclear factor kappa B (NFkappaB) activity induces nerve growth factor-resistant apoptosis in PC12 cells. *J. Neurosci. Research* **47,** 155–162.
92. Connor, B. and Dragunow, M. (1998) The role of neuronal growth factors in neurodegenerative disorders of the human brain. *Brain Research – Brain Research Reviews* **27,** 1–39.
93. Cuello, A.C. (1996) Effects of trophic factors on the CNS cholinergic phenotype. *Progress in Brain Research* **109,** 347–358.
94. Lindsay, R.M. (1994a) Neurotrophic growth factors and neurodegenerative diseases: therapeutic potential of the neurotrophins and ciliary neurotrophic factor. *Neurobiology of Aging* **15,** 249–251.
95. Martinez, H.J., Dreyfus, C.F., Jonakait, G.M., and Black, I.B. (1985) Nerve growth factor promotes cholinergic development in brain striatal cultures. *Proceed. Nation. Acad. Scien. USA* **82,** 7777–7781.
96. Mobley, W.C., Rutkowski, J.L., Tennekoon, G.I., Buchanan, K., and Johnston, M.V. (1985) Choline acetyltransferase activity in striatum of neonatal rats increased by nerve growth factor. *Science* **229,** 284–287.
97. Hefti, F., Hartikka, J., Eckenstein, F., Gnahn, H., Heumann, R., and Schwab, M. (1985) Nerve growth factor increases choline acetyltransferase but not survival or fiber outgrowth of cultured fetal septal cholinergic neurons. *Neuroscience* **14,** 55–68.
98. Nonner, D., Brass, B.J., Barrett, E.F., and Barrett, J.N. (1993) Reversibility of nerve growth factor's enhancement of choline acetyltransferase activity in cultured embryonic rat septum. *Experimental Neurology* **122,** 196–208.
99. Gage, F.H., Armstrong, D.M., Williams, L.R., and Varon, S. (1988) Morphological response of axotomized septal neurons to nerve growth factor. *Journal of Comparative Neurology* **269,** 147–155.
100. Hefti, F. (1986) Nerve growth factor promotes survival of septal cholinergic neurons after fimbrial transections. *Journal of Neuroscience* **6,** 2155–2162.
101. Koliatsos, V.E., Clatterbuck, R.E., Gouras, G.K., and Price, D.L. (1991a) Biologic effects of nerve growth factor on lesioned basal forebrain neurons. *Annals of the New York Academy of Sciences* **640,** 102–109.
102. Kordower, J.H., Winn, S.R., Liu, Y.T. et al. (1994) The aged monkey basal forebrain: rescue and sprouting of axotomized basal forebrain neurons after grafts of encapsulated cells secreting human nerve growth factor. *Proceed. Nation. Acad. Scien. USA* **91,** 10898–10902.

103. Tuszynski, M.H., Buzsaki, G., and Gage, F.H. (1990) Nerve growth factor infusions combined with fetal hippocampal grafts enhance reconstruction of the lesioned septohippocampal projection. *Neuroscience* **36,** 33–44.
104. Tuszynski, M.H., Sang, H., Yoshida, K., and Gage, F.H. (1991) Recombinant human nerve growth factor infusions prevent cholinergic neuronal degeneration in the adult primate brain. *Annals of Neurology* **30,** 625–636.
105. Williams, L.R., Varon, S., Peterson, G.M. et al. (1986) Continuous infusion of nerve growth factor prevents basal forebrain neuronal death after fimbria fornix transection. *Proceedings of the National Academy of Sciences of the United States of America* **83,** 9231–9235.
106. Fischer, W., Bjorklund, A., Chen, K., and Gage, F.H. (1991) NGF improves spatial memory in aged rodents as a function of age. *J. Neurosci.* **11,** 1889–1906.
107. Fischer, W., Wictorin, K., Bjorklund, A., Williams, L.R., Varon, S., and Gage, F.H. (1987) Amelioration of cholinergic neuron atrophy and spatial memory impairment in aged rats by nerve growth factor. *Nature* **329,** 65–68.
108. Hefti, F., Dravid, A., and Hartikka, J. (1984) Chronic intraventricular injections of nerve growth factor elevate hippocampal choline acetyltransferase activity in adult rats with partial septo-hippocampal lesions. *Brain Research* **293,** 305–311.
109. Garofalo, L., Ribeiro-da-Silva, A., and Cuello, A.C. (1992) Nerve growth factor-induced synaptogenesis and hypertrophy of cortical cholinergic terminals. *Proceedings of the National Academy of Sciences of the United States of America* **89,** 2639–2643.
110. Chen, K.S. and Gage, F.H. (1995) Somatic gene transfer of NGF to the aged brain: behavioral and morphological amelioration. *J. Neurosci.* **15,** 2819–2825.
111. Martinez-Serrano, A. and Bjorklund, A. (1998) Ex vivo nerve growth factor gene transfer to the basal forebrain in presymptomatic middle-aged rats prevents the development of cholinergic neuron atrophy and cognitive impairment during aging. *Proceed. Nation Acad. Scien. USA* **95,** 1858–1863.
112. Hellweg, R., Humpel, C., Lowe, A., and Hortnagl, H. (1997) Moderate lesion of the rat cholinergic septohippocampal pathway increases hippocampal nerve growth factor synthesis: evidence for long-term compensatory changes? *Brain Research. Molecular Brain Research* **45,** 177–181.
113. Lapchak, P.A. and Hefti, F. (1991) Effect of recombinant human nerve growth factor on presynaptic cholinergic function in rat hippocampal slices following partial septohippocampal lesions: measures of [3H]acetylcholine synthesis, [3H]acetylcholine release and choline acetyltransferase activity. *Neuroscience* **42,** 639–649.
114. Nonomura, T., Nishio, C., Lindsay, R.M., and Hatanaka, H. (1995) Cultured basal forebrain cholinergic neurons from postnatal rats show both overlapping and non-overlapping responses to the neurotrophins. *Brain Research* **683,** 129–139.
115. Nonner, D., Barrett, E.F., and Barrett, J.N. (1996) Neurotrophin effects on survival and expression of cholinergic properties in cultured rat septal neurons under normal and stress conditions. *J. Neurosci.* **16,** 6665–6675.
116. Dekker, A.J., Fagan, A.M., Gage, F.H., and Thal, L.J. (1994a) Effects of brain-derived neurotrophic factor and nerve growth factor on remaining neurons in the lesioned nucleus basalis magnocellularis. *Brain Research* **639,** 149–155.
117. Hefti, F., Knusel, B., and Lapchak, P.A. (1993) Protective effects of nerve growth factor and brain-derived neurotrophic factor on basal forebrain cholinergic neurons in adult rats with partial fimbrial transections. *Progress in Brain Research* **98,** 257–263.
118. Venero, J.L., Knusel, B., Beck, K.D., and Hefti, F. (1994b) Expression of neurotrophin and trk receptor genes in adult rats with fimbria transections: effect of intraventricular nerve growth factor and brain-derived neurotrophic factor administration. *Neuroscience* **59,** 797–815.

119. Dekker, A.J., Winkler, J., Ray, J., Thal, L.J., and Gage, F.H. (1994b) Grafting of nerve growth factor-producing fibroblasts reduces behavioral deficits in rats with lesions of the nucleus basalis magnocellularis. *Neuroscience* **60,** 299–309.
120. Martinez-Serrano, A., Fischer, W., and Bjorklund, A. (1995) Reversal of age-dependent cognitive impairments and cholinergic neuron atrophy by NGF-secreting neural progenitors grafted to the basal forebrain. *Neuron* **15,** 473–484.
121. Tuszynski, M.H. and Gage, F.H. (1995) Bridging grafts and transient nerve growth factor infusions promote long-term central nervous system neuronal rescue and partial functional recovery. *Proceed. Nation. Acad. Sci. USA* **92,** 4621–4625.
122. Koliatsos, V.E., Clatterbuck, R.E., Nauta, H.J. et al. (1991b) Human nerve growth factor prevents degeneration of basal forebrain cholinergic neurons in primates. *Annals of Neurol.* **30,** 831–840.
123. Liberini, P., Pioro, E.P., Maysinger, D., Ervin, F.R., and Cuello, A.C. (1993) Long-term protective effects of human recombinant nerve growth factor and monosialoganglioside GM1 treatment on primate nucleus basalis cholinergic neurons after neocortical infarction. *Neuroscience* **53,** 625–637.
124. Ebstein, R.P., Bennett, E.R., Sokoloff, M., and Shoham, S. (1993) The effect of nerve growth factor on cholinergic cells in primary fetal striatal cultures: characterization by in situ hybridization. *Brain Research. Developmental Brain Research* **73,** 165–172.
125. Higgins, G.A., Koh, S., Chen, K.S., and Gage, F.H. (1989) NGF induction of NGF receptor gene expression and cholinergic neuronal hypertrophy within the basal forebrain of the adult rat. *Neuron* **3,** 247–256.
126. Li, Y., Holtzman, D.M., Kromer, L.F. (1995c) Regulation of TrkA and ChAT expression in developing rat basal forebrain: evidence that both exogenous and endogenous NGF regulate differentiation of cholinergic neurons. *J. Neurosci.* **15,** 2888–2905.
127. Venero, J.L., Beck, K.D., and Hefti, F. (1994a) Intrastriatal infusion of nerve growth factor after quinolinic acid prevents reduction of cellular expression of choline acetyltransferase messenger RNA and trkA messenger RNA, but not glutamate decarboxylase messenger RNA. *Neuroscience* **61,** 257–268.
128. Drachman, D.A. and Leavitt, J. (1974) Human memory and the cholinergic system. A relationship to aging? *Archives of Neurology* **30,** 113–121.
129. Drachman, D.A. and Sahakian, B.J. (1980) Memory and cognitive function in the elderly. A preliminary trial of physostigmine. *Archives of Neurology* **37,** 674–675.
130. Williams, L.R., Rylett, R.J., Ingram, D.K. et al. (1993) Nerve growth factor affects the cholinergic neurochemistry and behavior of aged rats. *Progress in Brain Research* **98,** 251–256.
131. Gu, Z.Z., Yu, J.A., and Perez-Polo, J.R. (1998) Responses in the aged rat brain after total immunolesion. *Journal of Neuroscience Research* **54,** 7–16.
132. Scott, S.A., Liang, S., Weingartner, J.A., and Crutcher, K.A. (1994) Increased NGF-like activity in young but not aged rat hippocampus after septal lesions. *Neurobiol. Aging* **15,** 337–346.
133. Jacobson, M.D., Weil, M., and Raff, M.C. (1997) Programmed cell death in animal development. *Cell* **88,** 347–354.
134. Steller, H. (1995) Mechanisms and genes of cellular suicide. *Science* **267,** 1445–1449.
135. Oppenheim, R.W. (1991) Cell death during development of the nervous system. *Annual Review of Neuroscience* **14,** 453–501.
136. Kerr, J.F., Wyllie, A.H., and Currie, A.R. (1972) Apoptosis: a basic biological phenomenon with wide-ranging implications in tissue kinetics. *British Journal of Cancer* **26,** 239–257.
137. Gerschenson, L.E. and Rotello, R.J. (1992) Apoptosis: a different type of cell death. *FASEB Journal* **6,** 2450–2455.
137. Thompson, C.B. (1995) Apoptosis in the pathogenesis and treatment of disease. *Science* **267,** 1456–1462.

138. Du, C., Hu, R., Csernansky, C.A., Hsu, C.Y., and Choi, D.W. (1996a) Very delayed infarction after mild focal cerebral ischemia: a role for apoptosis? *Journal of Cerebral Blood Flow & Metabolism* **16,** 195–201.

138. Schwartzman, R.A. and Cidlowski, J.A. (1993) Apoptosis: the biochemistry and molecular biology of programmed cell death. *Endo. Reviews* **14,** 133–151.

139. Du, C., Hu, R., Csernansky, C.A., Liu, X.Z., Hsu, C.Y., and Choi, D.W. (1996b) Additive neuroprotective effects of dextrorphan and cycloheximide in rats subjected to transient focal cerebral ischemia. *Brain Research* **718,** 233–236.

139. Wyllie, A.H., Kerr, J.F., and Currie, A.R. (1980) Cell death: the significance of apoptosis. *International Review of Cytology* **68,** 251–306.

140. Fukuda, K., Kojiro, M., and Chiu, J.F. (1993) Demonstration of extensive chromatin cleavage in transplanted Morris hepatoma 7777 tissue: apoptosis or necrosis? *American Journal of Pathology* **142,** 935–946.

140. Li, Y., Chopp, M., Jiang, N., and Zaloga, C. (1995a) In situ detection of DNA fragmentation after focal cerebral ischemia in mice. *Brain Res. Mole. Brain Res.* **28,** 164–168.

141. Gavrieli, Y., Sherman, Y., and Ben-Sasson, S.A. (1992) Identification of programmed cell death in situ via specific labeling of nuclear DNA fragmentation. *Journal of Cell Biology* **119,** 493–501.

141. Li, Y., Chopp, M., Jiang, N., Zhang, Z.G., and Zaloga, C. (1995b) Induction of DNA fragmentation after 10 to 120 minutes of focal cerebral ischemia in rats. *Stroke* **26,** 1252–1257. discussion 1257–1258.

142. Li, Y., Sharov, V.G., Jiang, N., Zaloga, C., Sabbah, H.N., and Chopp, M. (1995d) Ultrastructural and light microscopic evidence of apoptosis after middle cerebral artery occlusion in the rat. *American Journal of Pathology* **146,** 1045–1051.

142. Meyaard, L., Otto, S.A., Jonker, R.R., Mijnster, M.J., Keet, R.P., and Miedema, F. (1992) Programmed death of T cells in HIV-1 infection. *Science* **257,** 217–219.

143. Cotman, C.W., Whittemore, E.R., Watt, J.A., Anderson, A.J., and Loo, D.T. (1994) Possible role of apoptosis in Alzheimer's disease. *Annals of the NY Acad. of Sci.* **747,** 36–49.

143. Linnik, M.D., Miller, J.A., Sprinkle-Cavallo, J. et al. (1995) Apoptotic DNA fragmentation in the rat cerebral cortex induced by permanent middle cerebral artery occlusion. *Brain Research. Molecular Brain Research* **32,** 116–124.

144. Ellis, R.E., Yuan, J.Y., and Horvitz, H.R. (1991) Mechanisms and functions of cell death. *Annual Review of Cell Biology* **7,** 663–698.

145. Liu, P.K., Hsu, C.Y., Dizdaroglu, M. et al. (1996) Damage, repair, and mutagenesis in nuclear genes after mouse forebrain ischemia-reperfusion. *J. Neurosci.* **16,** 6795–6806.

146. Cheng, Y., Deshmukh, M., D'Costa, A., et al. (1998) Caspase inhibitor affords neuroprotection with delayed administration in a rat model of neonatal hypoxic-ischemic brain injury [see comments.] *J. Clin. Invest.* **101,** 1992–1999.

147. Fink, K., Zhu, J., Namura, S. et al. (1998) Prolonged therapeutic window for ischemic brain damage caused by delayed caspase activation. *Journal of Cerebral Blood Flow & Metabolism* **18,** 1071–1076.

148. Namura, S., Zhu, J., Fink, K. et al. (1998) Activation and cleavage of caspase-3 in apoptosis induced by experimental cerebral ischemia. *J. Neurosci.* **18,** 3659–3668.

149. Xu, D.G., Crocker, S.J., Doucet, J.P. et al. (1997) Elevation of neuronal expression of NAIP reduces ischemic damage in the rat hippocampus. *Nature Medic.* **3,** 997–1004.

149A.LeBlanc, A. (1995) Increased production of 4 kDa amyloid beta peptide in serum deprived human primary neuron cultures: possible involvement of apoptosis. *J. Neurosci.* **15,** 7837–7846.

150. Zhao, B., Chrest, F.J., Horton, W.E., Jr., Sisodia, S.S. and Kusiak, J.W. (1997) Expression of mutant amyloid precursor proteins induces apoptosis in PC12 cells. *Journal of Neuroscience Research* **47,** 253–263.

151. Beal, M.F. (1995) Aging, energy, and oxidative stress in neurodegenerative diseases. *Annals of Neurology* **38,** 357–366.
152. Papaconstantinou, J. (1994) Unifying model of the programmed (intrinsic) and stochastic (extrinsic) theories of aging. The stress response genes, signal transduction-redox pathways and aging. *Annals of the New York Academy of Sciences* **719,** 195–211.
153. Tong, L., Toliver-Kinsky, T., Taglialatela, G., Werrbach-Perez, K., Wood, T., and Perez-Polo, J.R. (1998) Signal transduction in neuronal death. *Journal of Neurochemistry* **71,** 447–459.
154. Vaux, D.L. (1993) Toward an understanding of the molecular mechanisms of physiological cell death. *Proceedings of the National Academy of Sciences of the United States of America* **90,** 786–789.
155. Sinson, G., Perri, B.R., Trojanowski, J.Q., Flamm, E.S., and McIntosh, T.K. (1997) Improvement of cognitive deficits and decreased cholinergic neuronal cell loss and apoptotic cell death following neurotrophin infusion after experimental traumatic brain injury. *Journal of Neurosurgery* **86,** 511–518.
156. Wilcox, B.J., Applegate, M.D., Portera-Cailliau, C., and Koliatsos, V.E. (1995) Nerve growth factor prevents apoptotic cell death in injured central cholinergic neurons. *Journal of Comparative Neurology* **359,** 573–585.
157. Fischer W., Wictorin K., Bjorklund A., Williams L.R., Varon S., and Gage F.H. (1987) Amelioration of cholineergic neuron atropy and spatial memory impairment in aged rats by nerve growth factor. *Nature* **329,** 65–68.
158. Coyle, J.T. and Puttfarcken, P. (1993) Oxidative stress, glutamate, and neurodegenerative disorders. *Science* **262,** 689–695.
159. Olanow, C.W. (1993) A radical hypothesis for neurodegeneration [see comments]. *Trends in Neurosciences* **16,** 439–444.
159. Zhang, Y., Tatsuno, T., Carney, J.M., and Mattson, M.P. (1993) Basic FGF, NGF, and IGFs protect hippocampal and cortical neurons against iron-induced degeneration. *Journal of Cerebral Blood Flow & Metabolism* **13,** 378–388.
160. Siesjo, B.K., Agardh, C.D., and Bengtsson, F. (1989) Free radicals and brain damage. *Cerebrovascular & Brain Metabolism Reviews* **1,** 165–211.
161. Basaga, H.S. (1990) Biochemical aspects of free radicals. *Biochemistry & Cell Biology* **68,** 989–998.
162. Graham, D.G. (1978) Oxidative pathways for catecholamines in the genesis of neuromelanin and cytotoxic quinones. *Molecular Pharmacology* **14,** 633–643.
163. Cui, J.K., Hsu, C.Y., and Liu, P.K. (1999) Suppression of postischemic hippocampal nerve growth factor expression by a c-fos antisense oligodeoxynucleotide. *Journal of Neuroscience* **19,** 1335–1344.
164. Tong, L. and Perez-Polo, J.R. (1995) Transcription factor DNA binding activity in PC12 cells undergoing apoptosis after glucose deprivation. *Neuroscience Letters* **191,** 137–140.
165. Tong, L. and Perez-Polo, J.R. (1996) Effect of nerve growth factor on AP-1, NF-kappa B, and Oct DNA binding activity in apoptotic PC12 cells: extrinsic and intrinsic elements. *Journal of Neuroscience Research* **45,** 1–12.
166. Hayashi, T., Ueno, Y., and Okamoto, T. (1993) Oxidoreductive regulation of nuclear factor kappa B. Involvement of a cellular reducing catalyst thioredoxin. *Journal of Biological Chemistry* **268,** 11380–11388.
167. Matthews, J.R., Wakasugi, N., Virelizier, J.L., Yodoi, J., and Hay, R.T. (1992) Thioredoxin regulates the DNA binding activity of NF-kappa B by reduction of a disulphide bond involving cysteine 62. *Nucleic Acids Research* **20,** 3821–3830.
168. Meyer, M., Pahl, H.L., and Baeuerle, P.A. (1994) Regulation of the transcription factors NF-kappa B and AP-1 by redox changes. *Chemico-Biological Interactions* **91,** 91–100.

169. Mosialos, G., Hamer, P., Capobianco, A.J., Laursen, R.A., and Gilmore, T.D. (1991) A protein kinase-A recognition sequence is structurally linked to transformation by p59v-rel and cytoplasmic retention of p68c-rel. *Molecular & Cellular Biology* **11,** 5867–5877.
170. Mihm, S., Galter, D., and Droge, W. (1995) Modulation of transcription factor NF kappa B activity by intracellular glutathione levels and by variations of the extracellular cysteine supply. *FASEB Journal* **9,** 246–252.
171. Kaltschmidt, B., Baeuerle, P.A., and Kaltschmidt, C. (1993) Potential involvement of the transcription factor NF-kappa B in neurological disorders. *Molecular Aspects of Medicine* **14,** 171–190.
172. Kaltschmidt, C., Kaltschmidt, B., and Baeuerle, P. A. (1995) Stimulation of ionotropic glutamate receptors activates transcription factor NF-kappa B in primary neurons. *Proceed Nation Acad Scien USA* **92,** 9618–9622.
173. Kaltschmidt, C., Kaltschmidt, B., Neumann, H., Wekerle, H., and Baeuerle, P.A. (1994) Constitutive NF-kappa B activity in neurons. *Molecular & Cellular Biology* **14,** 3981–3992.
174. Meyer, M., Schreck, R., and Baeuerle, P.A. (1993) H2O2 and antioxidants have opposite effects on activation of NF-kappa B and AP-1 in intact cells: AP-1 as secondary antioxidant-responsive factor. *EMBO Journal* **12,** 2005–2015.
175. Schenk, H., Klein, M., Erdbrugger, W., Droge, W., and Schulze-Osthoff, K. (1994) Distinct effects of thioredoxin and antioxidants on the activation of transcription factors NF-kappa B and AP-1. *Proceedings of the National Academy of Sciences of the United States of America* **91,** 1672–1676.
176. Ghosh, S. and Baltimore, D. (1990) Activation in vitro of NF-kappa B by phosphorylation of its inhibitor I kappa B. *Nature* **344,** 678–682.
177. Lin, Y.C., Brown, K., and Siebenlist, U. (1995) Activation of NF-kappa B requires proteolysis of the inhibitor I kappa B-alpha: signal-induced phosphorylation of I kappa B-alpha alone does not release active NF-kappa B. *Proceed. Nation. Acad. Scien. USA* **92,** 552–556.
178. Baeuerle, P.A. (1991) The inducible transcription activator NF-kappa B: regulation by distinct protein subunits. *Biochimica et Biophysica Acta* **1072,** 63–80.
179. Siebenlist, U., Franzoso, G., and Brown, K. (1994) Structure, regulation and function of NF-kappa B. *Annual Review of Cell Biology* **10,** 405–455.
180. Baeuerle, P.A. and Baltimore, D. (1988) Activation of DNA-binding activity in an apparently cytoplasmic precursor of the NF-kappa B transcription factor. *Cell* **53,** 211–217.
181. Vile, G.F., Tanew-Ilitschew, A., and Tyrrell, R.M. (1995) Activation of NF-kappa B in human skin fibroblasts by the oxidative stress generated by UVA radiation. *Photochemistry & Photobiology* **62,** 463–468.
182. Schmidt, K.N., Amstad, P., Cerutti, P., and Baeuerle, P.A. (1995) The roles of hydrogen peroxide and superoxide as messengers in the activation of transcription factor NF-kappa B. *Chemistry & Biology* **2,** 13–22.
183. Koong, A.C., Chen, E.Y., and Giaccia, A.J. (1994) Hypoxia causes the activation of nuclear factor kappa B through the phosphorylation of I kappa B alpha on tyrosine residues. *Cancer Research* **54,** 1425–1430.
184. Baeuerle, P.A. and Baltimore, D. (1996) NF-kappa B: ten years after. *Cell* **87,** 13–20.
185. Woronicz, J.D., Gao, X., Cao, Z., Rothe, M., and Goeddel, D.V. (1997) IkappaB kinase-beta: NF-kappaB activation and complex formation with IkappaB kinase-alpha and NIK [see comments]. *Science* **278,** 866–869.
186. Zandi, E., Rothwarf, D.M., Delhase, M., Hayakawa, M., and Karin, M. (1997) The IkappaB kinase complex (IKK) contains two kinase subunits, IKKalpha and IKKbeta, necessary for IkappaB phosphorylation and NF-kappaB activation. *Cell* **91,** 243–252.
187. Baeuerle, P.A. and Henkel, T. (1994) Function and activation of NF-kappa B in the immune system. *Annual Review of Immunology* **12,** 141–179.

188. Diaz-Meco, M.T., Dominguez, I., Sanz, L. et al. (1994) zeta PKC induces phosphorylation and inactivation of I kappa B-alpha in vitro. *EMBO Journal* **13,** 2842–2848.
189. Dominguez, I., Sanz, L., Arenzana-Seisdedos, F., Diaz-Meco, M.T., Virelizier, J.L., and Moscat, J. (1993) Inhibition of protein kinase C zeta subspecies blocks the activation of an NF-kappa B-like activity in Xenopus laevis oocytes. *Molecular & Cellular Biology* **13,** 1290–1295.
190. Kolesnick, R. and Golde, D.W. (1994) The sphingomyelin pathway in tumor necrosis factor and interleukin-1 signaling. *Cell* **77,** 325–328.
191. Lozano, J., Berra, E., Municio, M.M. (1994) Protein kinase C zeta isoform is critical for kappa B-dependent promoter activation by sphingomyelinase. *J. Biolog. Chem.* **269,** 19200–19202.
192. Machleidt, T., Wiegmann, K., Henkel, T., Schutze, S., Baeuerle, P., and Kronke, M. (1994) Sphingomyelinase activates proteolytic I kappa B-alpha degradation in a cell-free system. *J. Biolog. Chemis.* **269,** 13760–13765.
193. Schutze, S., Potthoff, K., Machleidt, T., Berkovic, D., Wiegmann, K., and Kronke, M. (1992) TNF activates NF-kappa B by phosphatidylcholine-specific phospholipase C-induced "acidic" sphingomyelin breakdown. *Cell* **71,** 765–776.
194. Helenius, M., Hanninen, M., Lehtinen, S.K., and Salminen, A. (1996) Changes associated with aging and replicative senescence in the regulation of transcription factor nuclear factor-kappa B. *Biochem. J.* **318,** 603–608.
195. Suzuki, T., Mitake, S., Okumura-Noji, K., Yang, J.P., Fujii, T., and Okamoto, T. (1997) Presence of NF-kappaB-like and IkappaB-like immunoreactivities in postsynaptic densities. *Neuroreport* **8,** 2931–2935.
196. Toliver-Kinsky, T., Papaconstantinou, J., and Perez-Polo, J.R. (1997) Age-associated alterations in hippocampal and basal forebrain nuclear factor kappa B activity. *Journal of Neuroscience Research* **48,** 580–587.
197. Boissiere, F., Hunot, S., Faucheux, B., Duyckaerts, C., Hauw, J.J., Agid, Y., and Hirsch, E.C. (1997) Nuclear translocation of NF-kappaB in cholinergic neurons of patients with Alzheimer's disease. *Neuroreport* **8,** 2849–2852.
198. Kaltschmidt, B., Uherek, M., Volk, B., Baeuerle, P.A., and Kaltschmidt, C. (1997) Transcription factor NF-kappaB is activated in primary neurons by amyloid beta peptides and in neurons surrounding early plaques from patients with Alzheimer disease. *Proceed. Nation. Acad. Scien. USA* **94,** 2642–2647.
199. Gveric, D., Kaltschmidt, C., Cuzner, M.L., and Newcombe, J. (1998) Transcription factor NF-kappaB and inhibitor I kappaBalpha are localized in macrophages in active multiple sclerosis lesions. *Journal of Neuropathology & Experimental Neurology* **57,** 168–178.
200. Kitamura, Y., Shimohama, S., Ota, T., Matsuoka, Y., Nomura, Y., and Taniguchi, T. (1997) Alteration of transcription factors NF-kappaB and STAT1 in Alzheimer's disease brains. *Neuroscience Letters* **237,** 17–20.
201. France-Lanord, V., Brugg, B., Michel, P.P., Agid, Y., and Ruberg, M. (1997) Mitochondrial free radical signal in ceramide-dependent apoptosis: a putative mechanism for neuronal death in Parkinson's disease. *Journal of Neurochemistry* **69,** 1612–1621.
202. Hunot, S., Brugg, B., Ricard, D. et al. (1997) Nuclear translocation of NF-kappaB is increased in dopaminergic neurons of patients with parkinson disease. *Proceedings of the National Academy of Sciences of the United States of America* **94,** 7531–7536.
203. Clemens, J.A., Stephenson, D.T., Yin, T., Smalstig, E.B., Panetta, J.A., and Little, S.P. (1998) Drug-induced neuroprotection from global ischemia is associated with prevention of persistent but not transient activation of nuclear factor-kappaB in rats. *Stroke* **29,** 677–682.
204. Salminen, A., Liu, P.K., and Hsu, C.Y. (1995) Alteration of transcription factor binding activities in the ischemic rat brain. *Biochemical & Biophysical Research Communications* **212,** 939–944.

205. Yang, K., Mu, X.S., and Hayes, R.L. (1995) Increased cortical nuclear factor-kappa B (NF-kappa B) DNA binding activity after traumatic brain injury in rats. *Neuroscience Letters* **197,** 101–104.
206. Perez-Otano, I., McMillian, M.K., Chen, J., Bing, G., Hong, J.S., and Pennypacker, K.R. (1996) Induction of NF-kB-like transcription factors in brain areas susceptible to kainate toxicity. *Glia* **16,** 306–315.
207. Bethea, J.R., Castro, M., Keane, R.W., Lee, T.T., Dietrich, W.D., and Yezierski, R.P. (1998) Traumatic spinal cord injury induces nuclear factor-kappaB activation. *J. Neurosci.* **18,** 3251–3260.
208. Bales, K.R., Du, Y., Dodel, R.C., Yan, G.M., Hamilton-Byrd, E., and Paul, S.M. (1998) The NF-kappaB/Rel family of proteins mediates Abeta-induced neurotoxicity and glial activation. *Brain Research. Molecular Brain Research* **57,** 63–72.
209. Grilli, M., Pizzi, M., Memo, M., and Spano, P. (1996) Neuroprotection by aspirin and sodium salicylate through blockade of NF-kappaB activation. *Science* **274,** 1383–1385.
210. Ko, H.W., Park, K.Y., Kim, H. (1998) Ca2+-mediated activation of c-Jun N-terminal kinase and nuclear factor kappa B by NMDA in cortical cell cultures. *J. Neurochem.* **71,** 1390–1395.
211. Mattson, M.P., Goodman, Y., Luo, H., Fu, W., and Furukawa, K. (1997) Activation of NF-kappaB protects hippocampal neurons against oxidative stress-induced apoptosis: evidence for induction of manganese superoxide dismutase and suppression of peroxynitrite production and protein tyrosine nitration. *Journal of Neuroscience Research* **49,** 681–697.
212. Furukawa, K. and Mattson, M.P. (1998) The transcription factor NF-kappaB mediates increases in calcium currents and decreases in NMDA- and AMPA/kainate-induced currents induced by tumor necrosis factor-alpha in hippocampal neurons. *Journal of Neurochemistry* **70,** 1876–1886.
213. Pahan, K., Sheikh, F.G., Khan, M., Namboodiri, A.M., and Singh, I. (1998) Sphingomyelinase and ceramide stimulate the expression of inducible nitric-oxide synthase in rat primary astrocytes. *Journal of Biological Chemistry* **273,** 2591–2600.
214. Lezoualc'h, F., Sagara, Y., Holsboer, F., and Behl, C. (1998) High constitutive NF-kappaB activity mediates resistance to oxidative stress in neuronal cells. *J. Neurosci.* **18,** 3224–3232.
215. Bejanin, S., Habert, E., Berrard, S., Edwards, J.B., Loeffler, J.P., and Mallet, J. (1992) Promoter elements of the rat choline acetyltransferase gene allowing nerve growth factor inducibility in transfected primary cultured cells. *J. Neurochem.* **58,** 1580–1583.
216. Misawa, H., Ishii, K., and Deguchi, T. (1992) Gene expression of mouse choline acetyltransferase. Alternative splicing and identification of a highly active promoter region. *J. Biolog. Chemis.* **267,** 20392–20399.
217. Pu, H.F., Zhai, P., and Gurney, M. (1993) Enhancer, silencer, and growth factor responsive regulatory sequences in the promoter for the mouse choline acetyltransferase gene. *Molecular and Cellular Neurosciences* **4,** 131–142.
218. Toliver-Kinsky, T., Rassin, D., and Perez-Polo, J.R. (1998) Negative regulation of choline acetyltransferase expression by nuclear factor kappa B. *Society for Neuroscience 28th Annaul Meeting Abstract* **24,** 1565.

13

Effects of Estrogens and Thyroid Hormone on Development and Aging of Astrocytes and Oligodendrocytes

Kevin Higashigawa, Alisa Seo, Nayan Sheth, Giorgios Tsianos, Hogan Shy, Latha Malaiyandi, and Paola S. Timiras

1. INTRODUCTION

The regulatory role of hormones and growth factors on glial cell development has been demonstrated in several studies *(1–3)*. Hormones also influence the aging of glial cells and their role in the aging brain *(4)*. In early reports, the gliosis that occurred in the brain of aged animals consisted primarily of astrocytes and oligodendrocytes and was interpreted as a compensatory response to the diminishing structural integrity and functional competence of neurons and their increasing pathology *(5,6)*. The stimulatory effects of thyroid hormones on development and adult function of glial cells are numerous and well established *(4,7)*. Endocrine-glia interactions may support normal function of neurons and repair damage of central nervous system (CNS) regions. The current view is that neurons may regenerate not only during development but also in adulthood, and not only in laboratory animals but also in humans *(8–10)*. The "post-developmental" CNS may not be "as hostile an environment for the regeneration of neuronal networks as once believed" *(10)* and it may be induced to regenerate by internal influences *(8)*. As glial cells are an important component of the neuronal microenvironment, they may represent a key factor for CNS repair after injury or in prevention and treatment of neurodegenerative diseases.

Glial cells have multiple actions on CNS repair that may be either antagonistic (e.g., formation of a "glial scar" that prevents cell migration to the site of injury) *(11,12)* or facilitatory (e.g., release of growth factors and axon-guidance and adhesion molecules providing trophic and directional support for regenerating neurons) *(12–14)*. The well-known Spanish neuroscientist Ramon y Cajal had advocated, in 1913, the concept of an "immutable, fixed and rigid CNS" *(15)*. However, Cajal also prophezied that future science may well replace this early concept with that of a more flexible, "plastic" CNS which would respond to environmental and endogenous stimuli capable of modulating neural cells growth, differentiation, and function *(15)*.

Topics in the present chapter focus on the possible role of glial cells as mediators of neural plasticity and the role of hormones thereon. The glial cells studied here

From: *Neuroglia in the Aging Brain*
Edited by: Jean S. de Vellis © Humana Press Inc., Totowa, NJ

derive from a rat glioma cell line that consists of a mixture of astrocytes and oligodendrocytes, with the proportion of each cell type depending on several factors including the number of cell passages and the type and concentration of the hormone in the culture medium *(16–18)*. C-6 glioma cells provide a useful model to study glial cell properties, glial factors, and sensitivity of glial cells to various substances, including hormones. Among the many hormones/factors significant in glia-endocrine interrelationships, the effects of estrogens and thyroid hormone were studied because of their significance in regulating several important CNS functions, their efficacy in providing reliable treatment of clinical disorders, and the many years they have been investigated in our laboratory.

2. EFFECTS OF ESTROGENS ON PROLIFERATION AND METABOLISM OF GLIAL CELLS

2.1. *Estradiol and Tamoxifen Protect Astrocytes and Oligodendrocytes from Glutamate Toxicity*

Estrogen actions on neurons have been extensively studied since the 1960s and have been reported to induce maturation, differentiation, and excitation as well as protection of neurons *(19–21)*. By contrast, estrogen actions on glial cells are little and only recently, known *(2–4,22–27)*. In experiments from our laboratory, the ovarian steroid, 17β-estradiol, and the "designer," nonsteroidal estrogen, Tamoxifen (triphenylethylene), effectively protected cultured glial cells from the cytotoxicity of the excitatory neurotransmitter, glutamate *(28)*. One of the purposes of these experiments was to compare the effects of 17β-estradiol, the most biologically potent ovarian steroid, with those of tamoxifen, one of the nonsteroidal estrogens frequently used for treatment *(29)* and prevention *(30)* of breast cancer. Traditionally, tamoxifen is considered as an antiestrogen that prevents the actions of estradiol by competitively binding to the two estrogen receptors (ERαβ) *(23,26,31,32)*. This competitive binding results in apparently opposite actions, with Tamoxifen being an effective estrogen antagonist in breast tissue but an estrogen agonist in bone, liver, and uterus.

In our experiments, exposure of C-6 glioma cells to 10–20 m*M* glutamate concentration induced 61–78% cell death. Pretreatment of the cells with 0.01 m*M* estradiol or with 2 μM tamoxifen significantly reduced the glutamate-induced cell death, estradiol being the most effective in this regard (Table 1). These estrogen concentrations were selected because previous concentration/effect studies with graded concentrations of estrogens had shown their effectiveness *(27,28)*. Thus, estrogens often used in therapy (estradiol as replacement after menopause and tamoxifen for treatment/prevention of breast cancer) significantly protect glial cells in vitro against glutamate toxicity *(28)*.

Although the mechanism of this protective action may be quite complex, it is known that glutamate is removed from the synapse into astrocytes where it is transformed into the nontoxic glutamine by the enzyme glutamine synthetase abundantly present in astrocytes *(33–36)*. As shown in previous studies *(27)* and further discussed in subheading 2.2., both estradiol and tamoxifen significantly stimulate glutamine synthetase activity. Thus, one of the mechanisms of neural cell protection against glutamate toxicity by estrogens may be an accelerated conversion of glutamate to glutamine in astrocytes.

Table 1
Comparison of the protective action of 17β-estradiol and Tamoxifen against glutamate toxicity in glial cells[a]

	% cell death		
Glutamate concentration m*M*	Control without estrogens	17β-estradiol concentration 0.01 m*M*	Tamoxifen concentration 2μ*M*
10	60.60 ± 9.98[b]	2.82 ± 0.59	18.06 ± 0.99
15	68.32 ± 8.92	6.25 ± 1.15	25.65 ± 2.13
20	72.75 ± 10.20	19.95 ± 0.98	29.67 ± 1.97

[a] C-6 rat glioma 2B-clone cells (50–60 passage cells) were grown in Dulbecco's Modified Eagle Medium (DMEM) supplemented with 10% fetal bovine serum (FBS), and 1% penicillin/streptomycin/fungizone in 95% humidified air with 5% CO_2 and at 37°C. Concentration of 17β-estradiol (0.01 m*M*) and tamoxifen (2 μM) were those proven effective in previous dose-response experiments. Glutamate toxicity was measured by cell counts (with a Coulter Counter ZM) and by the activity of the enzyme lactate dehydrogenase which is released from lysed cells and, therefore, taken as an index of cell death *(30,38)*. Three increasing concentrations of glutamate, 10, 15, and 20 m*M*, were used.

[b] Means ± SE were analyzed by Student's *t*-test. Cell deaths were statistically ($p \leq 0.05$) higher in controls than in cells treated with both estrogens, and higher in tamoxifen-treated than estradiol-treated cells.

2.2. *Estradiol and Tamoxifen may Affect Differently Glial Cell Proliferation and Maturation Depending on Cell Age (Number of Passages) and Type of Estrogen*

Estrogens have opposite actions on cell proliferation of early passage cells as compared to late passage cells. In young (10–30 passage) glial cells, 17β-estradiol promotes cell growth in a concentration-dependent manner *(37)*. However, in studies with older (50–60 passage) cells *(27)*, the response induced by 17β-estradiol on cell proliferation is inhibitory: the 0.01 m*M* concentration of 17β-estradiol that stimulated proliferation in younger cells, reduced cell number by 50% after 8 d in culture in later passage cells; likewise, tamoxifen reduced cell growth in late passage cells, although inhibition was less pronounced than with estradiol (Fig. 1). However, in late passage cells, despite the reduction in cell number, the amount of protein per cell is significantly increased above controls with tamoxifen (by 23%) and even more with estradiol (by 86%) (Table 2). The apparent discrepancy between decreased growth rate and increased protein/cell content suggests that glial cells cultured with estradiol or tamoxifen mature more rapidly than the controls (without estrogen) and thereby become more specialized and lose or reduce their ability to proliferate.

Estradiol stimulates activity of marker enzymes: Glutamine synthethase (GS) for astrocytes and 2′3′-cyclic-nucleotide 3′-phospho-hydrolase (CNP) for oligodendrocytes (involved in the production of myelin). In young glial cells, the increased activity was proportional to the concentration of the hormone. For the 0.01 m*M* estrogen concentration, GS activity was increased by 65% whereas CNP activity was increased by 20% over controls *(37)*. The stimulation of glutamine synthetase by estradiol is particularly significant in these early passage glial cells; young cultures contain a higher proportion of oligodendrocytes than astrocytes while older cultures contain a higher

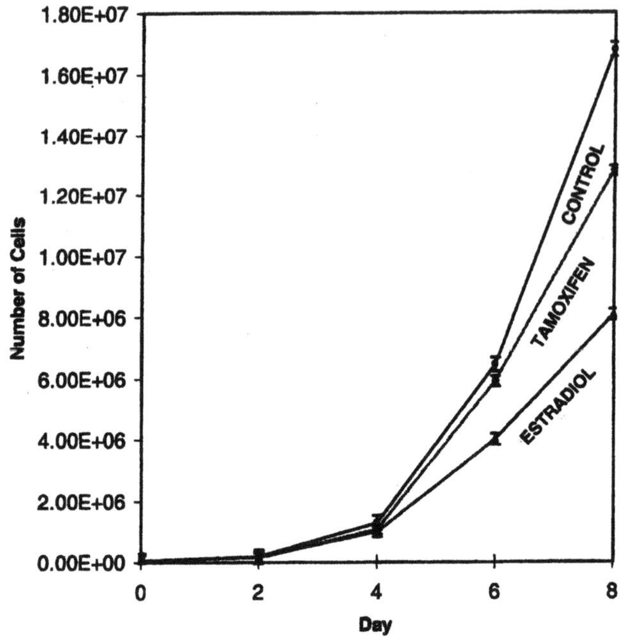

Fig. 1. Comparison of the effects of 17β-estradiol and tamoxifen on proliferation of 50–60 passage glial cells. Cells (C-6 rat glioma 2B-clone, 50–60 passages) were grown in Dulbecco's Modified Eagle Medium (DMEM) supplemented with 10% fetal bovine serum (FBS) and 1% penicillin/streptomycin/fungizone in 95% humidified air with 5% CO_2 and at 37°C. Concentration of 17β-estradiol and tamoxifen were 0.01 m*M* and 2 μ*M*, respectively and were chosen for their effectiveness in previous dose-response experiments. Dots and bracketed lines represent the mean ± SE.

proportion of astrocytes *(16)*. Therefore, GS activity would be expected to be lower in earlier than in later passage cells where astrocytes predominate. Indeed, in later passage cells cultured in 0.01 m*M* estradiol concentration, the GS activity undergoes a marked increase over controls of 136% whereas CNP activity is increased by 10% only. Conversely, Tamoxifen significantly increases CNP activity of late passage cells (by 83%) but stimulates GS activity to a much lesser degree than estradiol (Table 2).

With respect to the actions of estrogens on glial cell maturation of late passage cells, histologic examination indicates that both estrogens promote cell maturation as indicated by the more elongated and stellar shape of cell body and the longer and more abundant neurites than in control cells (without estrogens) (Fig. 2). This more advanced maturation is associated with a slower growth but with increased cell protein content and GS and CNP activity.

It may be noted that, whereas estradiol appears to promote astrocytic differentiation, tamoxifen may preferentially stimulate olidogendrocytic differentiation (Fig. 2). The lineage of glial cells, originally thought to be totally independent, may in fact have a common origin. Changes in environmental conditions, such as the addition of specific hormones, may direct the differentiation of glial precursors into cells exhibiting astrocytic or oligodendrocytic phenotypes and, conversely, astrocytes and oligodendrocytes may revert into precursor cells *(38)*. The differential action of the two

Table 2
Comparison of stimulation of total protein/cell and glutamine synthetase (GS) and 2′3′ cyclic nucleotide 3′ phosphohydrolase (CNP) activity by 17β-estradiol and tamoxifen in glial cells[a]

	% change from control (without estrogens)		
Estrogen type	Protein/cell[b]	GS[c]	CNP[d]
17B-estradiol 0.01 m*M*	86.27 ± 3.31[e]	136.33 ± 6.18	9.85 ± 1.5
Tamoxifen 2μ*M*	23.33 ± 1.55	82.13 ± 4.17	83.46 ± 2.87

[a] Glial cells (50–60 passages) were grown without (controls) and with estrogens under the same culture conditions described in Table 1.

[b] Protein content was measured spectrophotometrically using Coomassie blue dye, by the method of Holbrook and Leaver (1976) *Analyt. Biochem.*, **75,** 634.

[c] Glutamine synthetase (GS) activity was measured spectrophotometrically by the method of Rowe et al. (1970) *Meth. Enzymol.*, **17,** 900 and taken as an astrocytic marker.

[d] 2′3′-cyclic nucleotide 3-phosphohydrolase (CNP) activity was measured by the method of Prohaska et al. (1973) *Analyt. Biochem.*, **56,** 275 and taken as an oligodendrocytic marker.

[e] Means ± SE were analyzed by Student's *t*-test. Increase in protein/cell was statistically ($p \leq 0.05$) greater after estradiol and tamoxifen treatment than in controls, with estradiol being greater. Enzyme activity was significantly ($p \leq 0.05$) higher for both GS and CNP activity after 17β-estradiol and tamoxifen treatment. However, GS activity was highest after estradiol and CNP after tamoxifen.

estrogens *(27)* may correspond to a selective binding of each compound to either of the two ERs and the formation of two distinct, very different estrogen receptor-ligand complexes *(39).* The different ER-ligand complexes may undergo varying levels of activation, bind to separate DNA-enhancer elements and both functions could lead to differential gene activations. Tamoxifen also has nongenomic, non-ER related actions: It is a potent protein kinase-C inhibitor and at high doses may induce apoptosis (5–15 μM) *(40–42).*

The demonstration of the potential ability of tamoxifen to stimulate the differentiation of precursor glial cells into oligodendrocytes, is in agreement with in vivo studies in which estrogens stimulated myelinogenesis in rat brain and spinal cord *(43).* In demyelinating diseases such as multiple sclerosis, the early active lesions of the disease may undergo an initial stimulation of oligodendrocytic proliferation followed by a decline *(44).* Estrogens, such as tamoxifen, may stimulate the differentiation of glial cells into oligodendrocytes leading to subsequent remyelination and thereby opening new avenues of therapy for diseases characterized by deficits of myelin formation. Data presented here suggest that the two estrogens studied may influence differentially glial cell maturation: Estradiol would promote the development of the astrocytic expression and tamoxifen would promote the oligodendrocytic phenotype. So far, such a differential maturation is tentatively based on the increased activity of specific marker enzymes (GS or CNP) and cell morphology. Further studies are now ongoing in our laboratory to immunologically identify specific marker proteins to definitely confirm whether each of the two estrogens preferentially stimulates the astrocytic or oligodendrocytic phenotype.

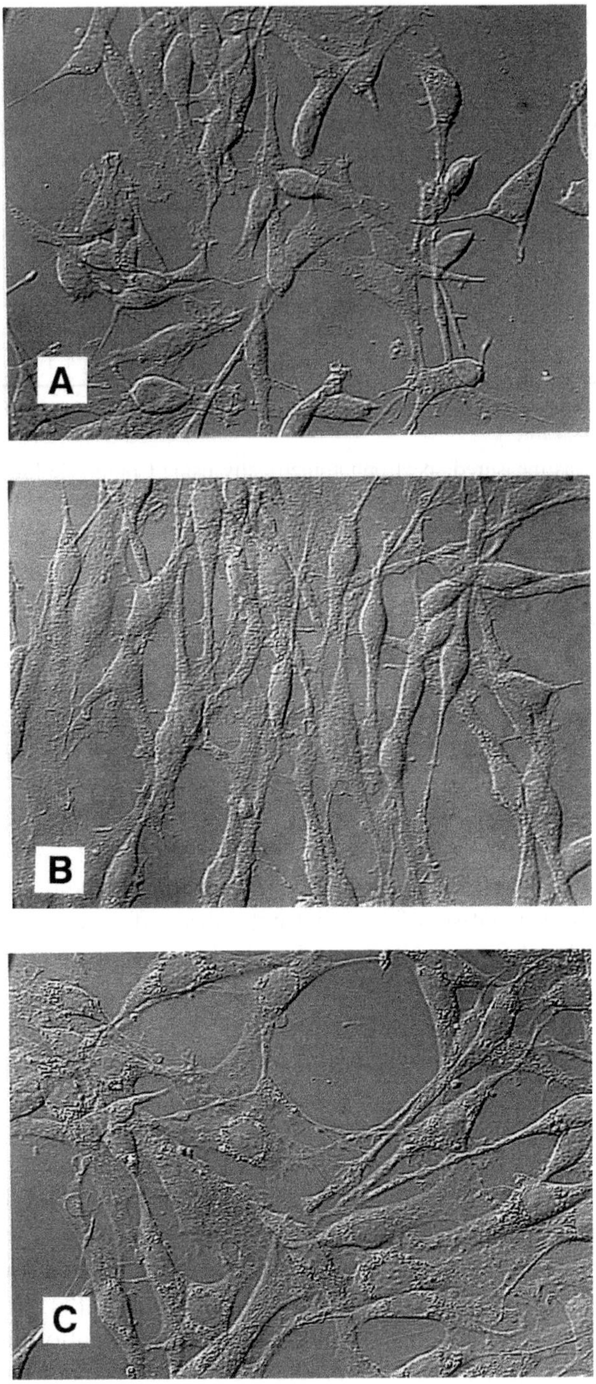

Fig. 2. Estrogens stimulate glial cell differentiation. Cells and experimental conditions were similar to those described in Fig. 1. Cells were grown to 85–90% confluency, washed, trypsinized, resuspended in 10 mL of medium, grown to a concentration of 10^6 mL and then photographed on an Olympus BH2 RFCA inverted microscope with a Technical Instruments COSTAR video camera CV 730 YCB, magnification is ×400 **(A)** Control cells present a rounded or oblong shape with processes of various lengths. **(B)** 7-d pretreatment with 17β-estradiol induces maturational changes characterized by more stellate (astrocytic) cell shape, cells oriented in vertical rows, and with longer processes. **(C)** 7-d pretreatment with tamoxifen also induces maturational changes characterized by a flat, rounded and enlarged cell shape with longer processes forming an extensive network.

3. EFFECTS OF THE THYROID HORMONE, TRIIODOTHYRONINE (T3) ON PROLIFERATION AND METABOLISM OF GLIAL CELLS

The importance of thyroid hormones, particularly the most biologically active hormone, triiodothyronine (T3), in brain development is well documented *(4,7)*. During mammalian CNS development, thyroid hormones are necessary, pre- and neonatally, for cell growth and proliferation of cerebral and cerebellar neurons, proliferation of dendrites, establishment of synapses, synthesis of neurotransmitters, maturation of enzymes for transport and cellular metabolism; during postnatal development, thyroid hormones promote glial cell differentiation, stimulate myelin formation, and local growth factors secretion (e.g., nerve growth factor, NGF). In pre- and neonatal hypothyroidism (i.e., T3 absence or deficit), the severe impairment of brain development results in a profound retardation of neurologic and mental development that leads to "cretinism," a severe form of cognitive deficiency. If, however, replacement therapy with T3 is initiated early during the neonatal period, most of the neurologic and mental damage may be prevented and brain development and maturation essentially proceed as in euthyroid infants. In the adult brain, normal thyroid function is necessary for optimal neurotransmission and for neurologic and behavioral competence: Deficiency or excess of thyroid hormones results in neurologic and mental disorders that may be routinely corrected by replacement therapy (in hypothyroidism) or the use of inhibitors of thyroid hormone biosynthesis (goitrogens) or by surgical or radiological removal of the hypersecreting thyroid tumors.

The presence of T3 receptors has been demonstrated in cultured glial cells *(45)*. These T3 glial receptors, as those on neurons, mediate the major T3 actions both during development and in aging. Although three T3 receptor genes are expressed in several human tissues including the brain, the significance of the expression of multiple genes remains unclear *(46)*. In culture, the number of T3 receptors is significantly greater in neuroblastoma (N2A and N18) than in glioma cells and increases in both cell types when grown in a medium deficient in T3 *(45)*.

3.1. Differential Effects of T3 on Glial Cell Proliferation in Early and Late Passage Glial Cells

In a previous study in young (10 passages) cells cultured with T3, proliferation was stimulated in a concentration-dependent manner: proliferation was unchanged from controls when cells were incubated with the lowest (0.001 m*M*) T3 concentration; with the two higher concentrations (0.01 and 0.1 m*M*), proliferation was decreased initially but significantly increased above controls on day 6 in culture, with the highest T3 concentration showing the most growth *(47)*.

Whereas cell proliferation is stimulated in young cells, in old (passage 85) cells, the same T3 concentrations of 0.01 and 0.1 m*M* inhibited cell proliferation after 6 d in culture, with the highest T3 concentration of 0.1 m*M* inhibiting cell proliferation the most. Cell proliferation was unaffected by the lowest T3 concentration (0.001 m*M*) (Fig. 3). As in previous studies, the lowest T3 concentration had no effect on GS and CNP activities, in neither young nor old cells. The intermediate T3 concentration was significantly less effective in the old than the young cells in stimulating both GS and CNP activities (Table 3).

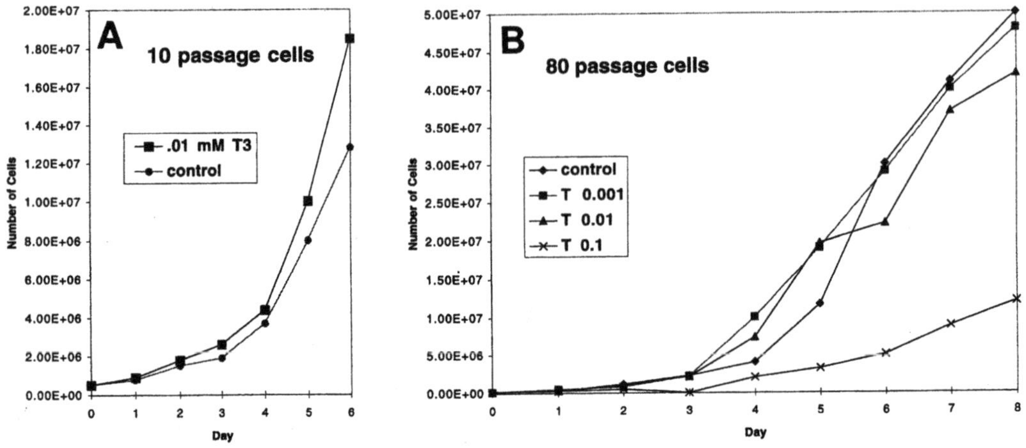

Fig. 3. Comparison of the effects of T3 on glial cell proliferation at early *(10)* and late (85) cell passages. Cells and culture conditions were similar to those described in Fig. 1. T3 is 3,5,3′-triiodo-L-thyronine.

An overall evaluation of the effects of T3 on GS activity shows that the greatest increase in GS activity (337%) occurs in the old (85 passage) cells with the highest (0.1 m*M*) T3 concentration. As already mentioned, early passage cultures consist predominately of oligodendrocytes and fewer astrocytes and the reverse is true for the late passage cultures *(16,38)*. The differential proportion of the two types of glial cells with the number of cell passages is reflected in our findings that stimulation of GS activity by T3 is greatest in late passages cultures rich in astrocytes. The observation of an accelerated maturation of astrocytes after thyroid hormone treatment reported several years ago *(1)*, is supported by the observations presented here, and by other recent studies demonstrating an increased GS expression and activity *(3)* as well as the increased expression of the glial fibrillary acidic protein (GFAP), another astrocyte marker, in various areas of the brain and in cultured cerebellar astrocytes *(48,49)*.

CNP stimulation is greatest (458%) in the early passage cultures rich in oligodendrocytes (Table 3). The stimulatory action of T3 on oligodendrocytes is in agreement with previous studies that demonstrate a regulatory role of thyroid hormones on myelinogenesis in vivo *(50,51)*.

3.2. Comparison of Estrogen and Thyroid Hormone Actions on Glial Cells: Role of Cell Age and Phenotype and of Hormone Type and Concentration

The observations presented here suggest that estrogens and T3 may influence proliferation, differentiation, and maturation of astrocytes and oligodendrocytes. Which of these actions will be operative at any given time will depend on the hormone itself (estradiol, tamoxifen, or T3) and on the nature (astrocytes vs oligodendrocytes), age (early vs late passage cells), and degree of maturation (immature/progenitor vs mature/differentiated) of the glial cells. Thus, while in early passage cells, both estrogens and T3 stimulate cell proliferation, in late passage cells they inhibit proliferation (Figs. 1 and 2). This age-related differential action on cell proliferation is not unique

Table 3
Comparison of the effects of three T3 concentrations on glutamine synthetase and 2′3′ cyclic nucleotide 3′ phosphohydrolase activity in young (10 passages) and old (85 passages) glial cells[a]

	% change from control (without T3)	
T3 concentration (m*M*)	Glutamine synthetase	2′3′ cyclic nucleotide 3′ phosphohydrolase
	Young (10 passage) cells	
0.001	1.5 ± 0.04[b]	10.5 ± 1.6
0.01	171.0 ± 10.6	204 ± 9.8
0.1	277.0 ± 21.5	458 ± 16.5
	Old (85 passage) cells	
0.001	4.00 ± 0.23	0.2 ± 0.00
0.01	21.8 ± 0.45	4.6 ± 0.52
0.1	337.8 ± 2.12	244.6 ± 1.82

[a] Cells were grown without (controls) and with T3 under the same culture conditions as described in Table 1. Assays for GS and CNP were the same as indicated in Table 2 (c and d). Three graded T3 concentrations were tested.

[b] Means ± SE were analyzed by Student's *t*-test. In both young and old cells, enzyme activity was slightly increased above that of controls with the lowest T3 concentration but was significantly ($p \leq 0.05$) higher than than controls with the two higher doses. Peak activity for GS was reached with the highest T3 concentration in old (85 passages) cells and peak activity for CNP was also reached with the highest T3 concentration but in young (10 passages) cells.

to estrogens and T3 but has been demonstrated also in other hormones, such as insulin, and growth factors, such as platelet-derived growth factor (PDGF), each with a specific timetable of action: Insulin and PDGF stimulate growth in late passage cells only *(52)*.

Estrogens and T3 appear to stimulate glial cell maturation overall, as indicated by the increased cell protein content and activity/cell of the marker enzymes, GS for astrocytes and CNP for oligodendrocytes, as compared to controls (cells cultured in medium without the hormones). However, there are some subtle differences in hormone action. Estradiol appears to promote astrocytic expression, whereas tamoxifen seems to selectively promote oligodendrocytic expression (Table 2). T3 stimulation was less effective in stimulating marker enzymes in old (85 passage) cells than in young (10 passage) cells, especially when the hormone concentration was low (Table 3); with the highest T3 concentration, GS activity was greatest in the old passage cells (which are prevalently astrocytic), whereas CNP activity was highest in the early passage cells (which are prevalently oligodendrocytic) (Table 3). In addition to these effects on astrocytes and oligodendrocytes, estrogens may also stimulate microglia to exert both immune stimulatory and antiinflammatory actions and to display neuroprotective actions *(53,54)*. Thus, the availability of a large repertoire of glial responses induced by hormones such as estrogens and T3 reflects the potential significance of hormonal regulation of the neuronal microenvironment.

With increasing progress in CNS rehabilitation, the perspective of restoring neuronal plasticity after damage because of trauma, disease, or aging is becoming a reachable reality. Perhaps, in the adult CNS, inhibition of neuronal replication is the trade-off for neuronal stability necessary for optimal function. However, under appropriate endogenous influences of their microenvironment, neurons may regenerate in animal, and human brains *(8,10)*. Since glial cells, the extracellular matrix, and blood vessels are part of this microenvironment, it may be argued that the study of hormonal influences on glial function represents an indirect but promising approach to eventual therapeutic manipulation of neuronal potential for self-renewal.

ACKNOWLEDGEMENTS

This research was supported in part by a donation from BioTime Inc., Berkeley, CA.

REFERENCES

1. Aizenman, Y. and De Vellis, J. (1987) Synergistic action of thyroid hormone, insulin and hydrocortisone on astrocyte differentiation. *Brain Res.* **414,** 301–308.
2. Jung-Testas, I., Schumacher, M., Robel, P., and Baulieu, E.E. (1994) Actions of steroid hormones and growth factors on glial cells of the central and peripheral nervous system. *J. Steroid Biochem. Mol. Biol.* **48,** 145–154.
3. Baas, D., Fressinaud, C., Vitkovic, L., and Sarlieve, L.L. (1998) Glutamine synthetase expression and activity are regulated by 3,5,3'-triiodo-L-thyronine and hydrocortisone in rat oligodendrocyte cultures. *Int. J. Devl. Neuroscience* **16,** 333–340.
4. Vernadakis, A. (1995) Effects of hormones on neural tissue: In vivo and in vitro studies, in *Hormones and Aging* (Timiras, P.S., Quay, W.L. and Vernadakis, A., eds.), CRC Press, Boca Raton, FL, pp. 291–314.
5. Brizzee, K.R., Sherwood, N., and Timiras, P.S. (1968) A comparison of cell populations at various depth levels in cerebral cortex of young and aged Long-Evans rats. *J. Gerontol.* **23,** 289–297.
6. Brizzee, K.R., Cancilla, P.A., Sherwood, N., and Timiras, P.S. (1969) The amount and distribution of pigments in neurons and glia of the cerebral cortex. Autofluorescent and ultrastructural studies. *J. Gerontol.* **24,** 127–135.
7. Timiras, P.S. (1988) Thyroid hormones and the developing brain, in Handbook of Human Growth and Developmental Biology, Vol.1, Part C, (Meisami, E. and Timiras, P.S., eds), CRC Press, Boca Raton, FL, pp. 59–82.
8. Eriksson, P.S., et al. (1998) Neurogenesis in the adult human hippocampus. *Nature Medicine* **4,** 1313–1317.
9. Gould, E., Tanapat, P., McEwen, B.S., Fluegge, G., and Fuchs, E. (1998) Proliferation of granule cell precursors in the dentate gyrus of adult monkeys is diminished by stress. *Proc. Natl. Acad. Sci. USA* **95,** 3168–3171.
10. Lowenstein, D.H. and Parent, J.M. (1999) Brain, heal thyself. *Science* **283,** 1126–1127.
11. Reier, P.J. (1986) Gliosis following CNS injury: The anatomy of astrocytic scars and their influences on axonal elongation, in *Astrocytes,* volume 3 (Fedoroff, S. and Vernadakis, A., eds.) Academic Press, New York, NY, pp. 263–324.
12. Ide, C., et al. (1996) Cellular and molecular correlates to plasticity during recovery from injury in the developing mammalian brain. *Progr. Brain Res.* **108,** 365–377.
13. Johansson, C.B., et al. (1999) Identification of a neural stem cell in the adult mammalian central nervous system. *Cell* **96,** 25–34.

14. Powell, E.M., Meiners, S., Di Prospero, N.A., and Geller, H.M. (1997) Mechanisms of astrocyte-directed neurite guidance. *Cell & Tissue Res.* **290,** 385–393.
15. y Cajal, S.R., and May, R.T., eds. (1959) *Degeneration and Regeneration of the Nervous System.* Hafner, N.Y., pp. 750.
16. Parker, K.P., Norenberg, M.D., and Vernadakis, A. (1980) Transdifferentiation of C6 glial cells in culture. *Science* **208,** 179–181.
17. Vernadakis, A. (1988) Neuron-glia interactions, *Int. Neurobiol. Rev.* **30,** 149–223.
18. Cole, R. and De Vellis, J. (1997) Astrocyte and oligodendrocyte cultures, in *Protocols for Neural Cell Culture,* 2nd edition (Fedoroff, S. and Richardson, A., eds), Humana, Totowa, N.J., pp. 117–130.
19. Timiras, P.S. (1982) The timing of hormone signals in the orchestration of brain development, in *The Development of Attachment and Affiliative Systems, Topics in Developmental Psychobiology* (Emde, R.N. and Harmon, R.J., eds), Plenum Press, New York, NY, pp. 47–63.
20. McEwen, B.S. (1995) Oestrogens and the structural and functional plasticity of neurons: Implications for memory, aging and neurodegenerative processes, in *CIBA Foundation Symposium, Non-reproductive actions of sex steroids.* (Bock, G.J. and Goode, J.A., eds.), John Wiley and Sons, Ltd., New York, NY, pp. 52–73.
21. Chang, D., Kwan, J., and Timiras, P.S. (1997) Estrogens influence growth, maturation and amyloid β-peptide production in neuroblastoma cells and in β-APP transfected kidney 293 cell line, in *Brain Plasticity.* (Filogamo, G., Vernadakis, A., Gremo, F., Privat, A.M., and Timirs, P.S., eds.), Plenum Press, New York, NY, pp. 261–271.
22. Bishop, J. and Simpkins, J.W. (1994) Estradiol treatment increases viability of glioma and neuroblastoma cells in vitro. *Mol. Cell Neurosci.* **5,** 303–308.
23. Santagati, S., Melcangi, R.C., Celotti, F., Martini, L., and Maggi, A. (1994) Estrogen receptor is expressed in different types of glial cells in culture. *J. Neurochem.* **63,** 2058–2064.
24. Drekic, D., Mallobabic, S., Gledic, D., and Cvetkovic, D. (1995) Different neuronal and glial cell groups in corticomedial amygdala react differently to neonatally administered estrogen. *Neuroscience* **66,** 475–481.
25. Garcia-Segura, L.M., Chowen, J.A., and Naftolin, F. (1996) Endocrine glia: Roles of glial cells in the brain actions of steroid and thyroid hormones and in the regulation of hormone secretion. *Frontiers Neuroendocrinol.* **17,** 180–211.
26. Jung-Testas, I. and Baulieu, E.E. (1998) Steroid hormone receptors and steroid action in rat glial cells of the central and peripheral nervous system. *J. Steroid Biochem. Mol. Biol.* **65,** 243–251.
27. Higashigawa, K.H., Vorgas, D., and Timiras, P.S. (1998) Effects of estradiol and selective estrogen receptor modulators on proliferation and enzyme activity of glial cells. *Exp. Gerontol.* **33,** 913.
28. Shy, H., Malaiyandi, L., and Timiras, P.S. (2000) Protective action of 17β-estradiol and tamoxifen on glutamate toxicity in glial cells. *Int. J. Devl. Neuroscience* **18,** 289–297.
29. DeGregorio, M.W. and Wiebe, V.J. (1996) *Tamoxifen and Breast Cancer.* Yale University press, New Haven, CT.
30. Fischer, B., et al. (1998) Tamoxifen for prevention of breast cancer: Report of the National Surgical Adjuvant Breast and Bowel Project P-1 study. *J. Natl. Cancer Inst.* **90,** 1371–1388.
31. Paech, K., et al. (1997) Differential ligand activation of estrogen receptors ERα and ERβ at AP1 sites. *Science* **277,** 1508–1510.
32. MacGregor, J.I, Jordan, V.C. (1998) Basic guide to the mechanisms of antiestrogen action. *Pharmacol. Rev.* **50,** 151–196.
33. Schousboe, A., et al. (1993) Glutamate and glutamine metabolism and compartmentation in astrocytes. *Dev. Neurosci.* **15,** 359–366.
34. Dugan, L.L., Bruno, V.M.G., Amagasu, S.M., and Giffard, R.G. (1995) Glia modulate the response of murine cortical neurons to excitotoxicity: Glia exacerbate AMPA neurotoxicity. *J. Neurosci.* **15,** 4545–4555.

35. Goodman, Y., Bruce, A.J., Cheng, B., and Mattson, M.P. (1996) Estrogens attenuate and corticosterone exacerbates excitotoxicity, oxidative injury and amyloid β-petide toxicity in hippocampal neurons. *J. Neurochem.* **66,** 1836–1844.
36. Singer, C.A., Rogers, K.L., Strickland, T.M., and Dorsa, D.M. (1996) Estrogen protects primary cortical neurons from glutamate toxicity. *Neurosci. Lett.* **212,** 13–16.
37. Kapur, R. and Timiras, P.S. (1997) Estrogen-brain relationship during development and aging: Effects on glial cells. *Neuroendocrinol. Lett.* **8,** 215–220.
38. Vernadakis, A., Kentroti, S., Brodie, C., Mangoura, D., and Sakellaridis, N. (1991). C-6 glioma cells of early passage have progenitor properties in culture, in *Plasticity and Regeneration of the Nervous System,* (Timiras, P.S., Privat, A., Giacobini, E., Lauder, J., and Vernadakis, A., eds.) Plenum Press, New York, NY, pp. 181–195.
39. Navarro, D., et al. (1998) The two native estrogen receptor forms of 8s and 4s present in cytosol from human uterine tissue display opposite reactivities with the antiestrogen tamoxifen aziridine and the estrogen responsive element. *J. Steroid Biochem. Mol. Biol.* **64,** 49–58.
40. Couldwell, W.T., et al. (1994) Protein kinase C inhibitors induce apoptosis in human malignant glioma cell lines. *FEBS Lett.* **345,** 43–46.
41. Wiseman, H. (1994) Tamoxifen: New membrane-mediated mechanisms of action and therapeutic advances. *TiPS.* **15,** 83–89.
42. Hashimoto, M., Inoue, S., Muramatsu, M., and Masliah, E. (1997) Estrogen stimulate tamoxifen-induced neuronal cell apoptosis *in vitro:* A possible nongenomic action. *Biochem. Biophys. Res. Commun.* **240,** 464–470.
43. Casper, R., Vernadakis, A., and Timiras, P.S. (1967) Influence of estradiol and cortisol in lipids and cerebrosides in the developing brain and spinal cordof the rat. *Brain Res.* **5,** 524–526.
44. Schoenrock, L.M., Kuhlmann, T., Adler, S., Bitsch, A., and Brueck, W. (1998) Identification of glial cell proliferation in early multiple sclerosis lesions. *Neuropathol. Appl. Neurobiol.* **24,** 320–330.
45. Draves, D.J., Manley, N.B., and Timiras, P.S. (1986) Glia hormone receptors: Thyroid hormones and microtubules in gliomas and neuroblastomas, in *Astrocytes: Cell Biology and Pathology of Astrocytes,* vol. 3, (Fedoroff, S. and Vernadakis, A., eds) Academic Press, New York, NY, pp. 183–201.
46. Sakurai, A., Nakai, A., and Degroot, L.J. (1989) Expression of three forms of thyroid hormone receptor in human tissues. *Mol. Endocrin.* **3,** 392–399.
47. Mozaffarieh, N., Ghafouri, P., and Timiras, P.S. (1998) Stimulatory action of triiodothyronine on proliferation and enzyme activity of glial cells. *FASEB J,* **12,** Abstracts, Part II, p. A757.
48. Lima, F.R.S., Trentin, A.G., Rosenthal, D., Chagas, C., and V. Moura Neto. (1997) Thyroid hormone induces protein secretion and morphological changes in astroglial cells with an increase in expression of glial fibrillary acidic protein. *J. Endocrinol.* **154,** 167–175.
49. Lima, F.R.S., et al. (1998) Thyroid hormone action on astroglial cells from distinct brain regions during development. *Int. J. Devl. Neuroscience* **16,** 19–27.
50. Dalal, K.B., Valcana, T., Timiras, P.S., and Einstein, E.R. (1971) Regulatory role of thyroxine on myelinogenesis in the developing rat. *Neurobiology* **1,** 211–224.
51. Valcana, T., Einstein, E.R., Csejtey, J., Dalal, K.B., and Timiras, P.S. (1975) Influence of thyroid hormones on myelin proteins in the developing rat brain. *J. Neurol. Sci.* **25,** 19–27.
52. Goya, L., Feng, P.T., Aliabadi, S., and Timiras, P.S. (1996) Effect of growth factors on the in vitro growth and differentation of early and late passage C6 glioma cells. *Int. J. Devl. Neuroscience* **14,** 409–417.
53. Ganter, S., Northoff, H., Mannel, D., and Genicke-Harter, P.J. (1992) Growth control of cultured microglia. *J. Neurosci. Res.* **33,** 218–230.
54. Harris-White, M.E., Chu, T., Miller, S.A., Simmons M., Nash, D., Teter, B., Cole, G.M., and Frautschy, S.A. (2001) Estrogen (E2) and glucocorticoid (Gc) effects on μglia and Aβ clearance in vitro and in vivo. *Neurochem. Int.* **39,** 435–448.

IV Metabolic Changes

14
Neurotoxic Injury and Astrocytes

Michael Aschner and Richard M. LoPachin

1. INTRODUCTION

The unique functions of the nervous system are largely attributable to the properties of its electrically excitable cells, the neurons. However, there is a more abundant class of nonexcitable cells, collectively referred to as the neuroglia. Within the central nervous system (CNS) they comprise the astrocytes, oligodendrocytes, and microglia. Progress in the understanding of neuroglial function was originally based upon the pioneering histological staining developed by Golgi and Ramon y Cajal, around 1870 and 1890, respectively. The term neuroglia was derived from the essentially erroneous concept of the German pathologist Virchow *(1)*, who postulated that neurons were embedded in a connective tissue to which he coined the name neuroglia, or nerve glue. Although erroneous, it has persisted as the preferred and generic term for these cells, or in its shortened form—"Glia."

Originally, astrocytes were viewed as mere passive support cells for neurons. However, modern experimental techniques have provided ample evidence that astrocytes serve in numerous additional capacities to maintain an optimal environment for neuronal function. Direct contact between astrocytes and neurons determines the morphological and functional differentiation of the latter. It is now well established that the role of astrocytes extends well beyond passive structural support and sensitivity to axon commands. In fact, astrocytes and neurons establish a highly dynamic reciprocal relationship that influences growth, morphology, behavior, and repair within the CNS. Astrocyte interactions with neuronal and nonneuronal cells (oligodendrocytes, microglia, and endothelial cells), and between themselves and the complexity of these interactions provide numerous strategic sites for neurotoxic action. This chapter will provide examples of astrocytic modulation of neurotoxicity. Examples include selective astrocytic toxicants, parent compounds that are metabolized within astrocytes to reactive intermediates with subsequent propensity to selectively damage neurons, as well as toxicants and pathophysiological conditions that affect astrocytic function and lead to altered extracellular fluid composition and secondary neuronal dysfunction.

From: *Neuroglia in the Aging Brain*
Edited by: Jean S. de Vellis © Humana Press Inc., Totowa, NJ

2. EXCITATORY AMINO ACIDS: THE ASTROCYTIC POOL

Astrocytes occupy about 25% of the brain volume *(1A),* and their processes are found around synapses and in close association with nodes of Ranvier, axon tracts, and blood vessels. In addition to their structural support for neurons, a partial list of astrocyte functions includes secretion of neurotrophic factors, K^+ buffering, control of extracellular fluid pH, inactivation of extracellular glutamate, glycogen storage, synaptic remodeling, and uptake and metabolism of neurotransmitters. During development, astrocytes prominently function in guiding neurons to their final target. The physiological functions of astrocytes in the developing and mature CNS extend beyond the scope of this review. Hence, they will not be discussed herein, but the reader is referred to a number of recent publications that provide extensive reviews of astrocytic functions *(2–7).*

CNS damage in a number of pathological states (e.g., hypoxia, seizures, hypoglycemia, and hepatic encephalopathy), neurodegenerative disorders (Parkinson's disease and Huntington's disease), and aging is thought to be partly due to excessive stimulation of neuronal glutamate-gated ion channels (reviewed in *8*). The origin of glutamate (and its analog, aspartate) has been tacitly assumed to be presynaptic nerve endings. However, it is known that astrocytes remove extracellular glutamate by a Na^+-dependent mechanism *(9).* This transport has a likely stoichiometry of 1 glutamate and 3 Na^+ transported inwardly, and 1 K^+ transported outwardly to offset the negative charge of glutamate. In the presence of ammonia, glutamate is metabolized to glutamine by the astrocyte-specific enzyme glutamine synthetase (GS); *(10,11),* maintaining $[glutamate]_o$ at 0.3 μ*M* *(12,13).* This represents a 10,000-fold gradient vs $[glutamate]_i$ (3 m*M*). This glutamate-glutamine pathway constitutes the pool of brain glutamate originally described by Berl *(14).* Astrocytes also efficiently remove extracellular taurine by a Na^+-dependent mechanism *(15).* This transport system generates and maintains a $[taurine]/[taurine]_o$ of about 10,000, and has a likely stoichiometry of 1 taurine and 2 Na^{++} ions transported inwardly, to generate and maintain the observed taurine gradient *(15,16).* Release from both the glutamate and taurine pools occurs as a result of astrocytic swelling *(2,17).* Astrocytic swelling is seen as an early event (within an hour of injury), followed by regulatory volume decrease (RVD). RVD is characterized by astrocyte reestablishment of preswelling volume, a process involving the extrusion of ions such as K^+ and Cl^-, and compensatory organic osmolytes (e.g., taurine, myoinositol). To a lesser extent, glutamate and aspartate *(17,18)* are also released by swollen astrocytes. The prominence of astrocytic swelling in various diseases, the rapidity of the astroglial response and its evolutionary conservation indicate that astrocytic swelling may fulfill a number of important functions. Unlike gliosis it occurs rapidly *(19),* and may reverse slowly with time. The consequences and the mechanisms of astrocytic swelling are as yet not fully defined, but are beginning to yield to experimental analysis in vitro. Astrocytic swelling *in situ* is routinely found to be associated with early pathological states affecting the CNS. Ultrastructural features of head injury suggest that astrocytic swelling precedes neuronal damage *(20).* A combined magnetic resonance and histochemical study suggests that brain injury after acute cerebral hypoxia is also secondary to astrocytic swelling *(21). In situ,* astrocytes are also known to swell more readily than neurons in response to lactic acidosis and elevated extracellular K^+, glutamate, and other monoamine transmitters *(22,23).* Similar findings of rapid and extensive astrocytic swelling have been reported in vivo after cerebral ischemia *(23)*. Mechanistically, astrocytic swelling appears to be a complex phenomenon, with several potential causes and consequences. For example, swelling may occur by simultaneous operation of Cl^-/HCO_3^- and Na^+/H^+ exchange transporters, with H^+ and HCO_3^- cycling from the intra- to extracellular spaces via membrane-permeant CO_2 when the increased intracellular NaCl cannot be pumped out. Acidosis occurring with increased lactate increases tissue CO_2 but this will by itself only lead to swelling if $pH_i << pH_o$, increasing $[HCO_3^-]_i$ relative to $[HCO_3^-]_o$. However, when extracellular Na^+ exchanges for intracellular H^+, and extracellular Cl^- for intracellular HCO_3^-, Na^+ and Cl^- replaces intracellular H^+ and HCO_3^- which can then continue to cycle inside and outside the cell via membrane-permeant CO_2 bringing in one Na^+ and one Cl^- for each turn of the cycle.

Astrocytic swelling and the associated release of excitatory amino acids is likely suspect in mediating neuronal cytotoxicity. In addition, the observed swelling of astrocytes is associated with a close to 50% decrease in the average capillary lumen, likely resulting in decreased blood flow because the flow of red blood cells is impeded. Furthermore, perivascular astrocytic swelling potentially can increase diffusion distances for substrates and waste products to blood vessels that would not otherwise be affected by the primary occlusion.

3. ASTROCYTIC MODULATION OF NEUROTOXICITY

3.1. Methylmercury (MeHg)

MeHg is a particular threat to the CNS, as evidenced by MeHg poisoning in Japan *(24)*, and Iraq *(25)*. Because methylation of inorganic mercury to MeHg by microorganisms is known to take place in waterways *(26,27)*, resulting in its accumulation in the food chain, any source of environmental mercury represents a potential source for MeHg poisoning. Industrial sources *(28)* of mercury culminating in the acidification of freshwater streams and lakes, and the impoundment of water for large hydroelectric schemes *(29)* have led to increases in MeHg concentrations in fish, posing increasingly greater risk to human populations.

In the mid 1950s, a chemical plant near Minamata Bay in Japan, discharged mercury into the bay as part of waste sludge. The inorganic mercury was methylated to the organic species, MeHg, and fish and shellfish became contaminated. Consumption of MeHg-adulterated fish by the local population led to an epidemic of MeHg poisoning and severe neurotoxicological and developmental effects *(24,30)*. Following a major drought in Iraq in 1971, the local government opted to switch to a resilient variety of wheat. The order was placed with the government of Mexico. However, a single letter typographical order was made in the name of the fungicide the wheat was to be treated with. Thus, the wheat was treated with MeHg instead of a relatively harmless mercury containing fungicide. The wheat had arrived in Iraq too late to be planted and was used instead to make the traditional pita bread. The farmers were unaware of the significance of the labeling (skull and crossbones poison designation), nor the pink dye additive that was added to warn them of the poisonous nature of the wheat. Weeks later the effects of MeHg intoxication started to appear, leading to mass poisoning epidemic with more than 450 reported deaths *(25)*.

That astrocytes are involved in the etiology of MeHg neurotoxicity is consistent with a number of observations.

1. In vivo, MeHg preferentially accumulates in astrocytes, both in humans and nonhuman primates *(31–34)*.
2. In vitro, MeHg inhibits glutamate uptake in astrocytes *(35)*. Other transport systems examined are 2–5-fold less sensitive to inhibition by MeHg *(36)*.
3. In the absence of extracellular glutamate, cultured neurons are unaffected by acute exposure to mercury *(37)*.
4. Exposure to MeHg is associated with astrocytic swelling in vivo *(31,33,34)* and in vitro, inhibition of regulatory volume decrease (RVD), as well as increased release of endogenous excitatory amino acids, such as glutamate and aspartate *(38)*.

Several reported actions of MeHg on membrane transporters can lead to astrocytic swelling and release of endogenous glutamate. Based on our, and other investigators' data *(39–42)* the following model was recently proposed for MeHg-induced astrocytic swelling: MeHg rapidly activates an anion exchange system (HCO_3^-/Cl^-) leading to unidirectional Na^+ influx and osmotically obligated influx of water. The cellular swelling leads to diminished K^+ clearance from the extracellular fluid either by diminished spatial buffering or reduced KCl uptake, due to the depolarization and appearance of other significant ion conductances, some of which can also lead to a net release of K^+ and Cl^-. Because the astrocytic glutamate carrier is both voltage- and ion gradient-dependent, swelling also reduces glutamate uptake (due to membrane depolarization and increased $[K^+]_o$). Swollen astrocytes also represent a source for the release of $[\text{glutamate}]_i$ presumably via the activation of leak pathways *(38,43)*. This sequence of events leads to elevated concentrations of excitatory amino acids (EAA) in the extracellular fluid, activating *N*-methyl-D-asparate (NMDA) receptors and damaging neurons *en masse*.

3.2. *Ammonia*

Hepatic encephalopathy (HE) or congenital and acquired hyperammonemia result in excessive ammonia (ammonium, NH_4^+) accumulation within the CNS. The condition is due to liver failure. Experimental studies in vivo have shown that the effects of ammonia on the CNS vary with its concentration. At high concentrations of ammonia within the CNS it produced seizures, resulting from its depolarizing action on cell membranes, whereas, at lower concentrations, ammonia produced stupor and coma, consistent with its hyperpolarizing effects.

Ammonia intoxication is commonly associated with astrocytic swelling. In addition, astrocytes undergo morphological changes upon chronic exposure to ammonia, yielding the so-called Alzheimer type II astrocytes common to most hyperammonemic conditions. Notably, the astrocytic changes precede any other morphological change in the CNS *(44–46)*. As alluded to earlier, the exclusive site for the detoxification of glutamate to glutamine occurs within the astrocytes. This process requires ATP-dependent amidation of glutamate to glutamine, a process mediated by the astrocyte-specific enzyme, glutamine synthetase (GS; *(47)*. In vivo chronic exposure to ammonia leads to diminished glutamine metabolism within the astrocytes, as well as impairment of astrocytic energy metabolism *(48,49)*. In addition, it has been reported that the reduced astrocytic capacity to metabolize ammonia leads to ammonia-induced cytotoxicity in juxtaposed neurons, promoting accumulation of glutamine. The latter, in turn, leads to decreased cerebral glucose consumption and amino acid imbalances *(50,51)*. Increased intracellular ammonia concentrations have also been implicated in the inhibition of neuronal glutamate precursor synthesis, resulting in diminished glutamatergic neurotransmission, changes in neurotransmitter uptake (glutamate), and changes in receptor-mediated metabolic responses of astrocytes to neuronal signals (reviewed in *49*). For additional details see chapters 15 and 26 in this volume.

3.3. 1-Methyl-4-Phenyl-1,2,3,6-Tetrahydropyridine (MPTP)

1-methyl-4-phenyl-1,2,3,6-tetrahydropyridine, or MPTP, is an analog of the opiate analgesic meperidine (Demerol). In 1982, this compound surfaced on the illegal market in the San Francisco Bay area as a "synthetic heroin." Heroin addicts (generally in their 20s and 30s) exposed to MPTP produced in an unregulated "clandestine" laboratory, rapidly developed a permanent disorder, clinically indistinguishable from nonidiopathic Parkinson's disease *(52)*. Subsequently, MPTP was shown to specifically damage nigrostriatal dopaminergic neurons. The discovery of MPTP led to resurgence of research on the etiology of Parkinson's disease, as well as its treatment, for it afforded an experimental model for the disease. When pieced together, the results from many studies suggest a pivotal role for astrocytes in the underlying neurodegenerative changes.

A classical compound that is metabolized within astrocytes to a reactive neurotoxic intermediate is MPTP. The parent compound is metabolized to 1-methyl-4-phenylpyridinium ion (MPP^+) that selectively destroys nigrostriatal dopaminergic neurons *(53–55)*. The mechanisms of MPTP neurotoxicity, although not fully understood *(56,57)*, suggest that the oxidized pyridinium metabolite (MPP^+) is the primary mediator of MPTP neurotoxicity. MPP^+ is apparently able to damage neuronal cells after being formed within and released from astrocytes *(57)*. Because of their ability to accumulate MPP^+ (via dopamine uptake) and to retain it for a prolonged period of time, dopaminergic neurons are particularly vulnerable to MPTP toxicity. Two pathways of MPP^+ formation have been identified within astrocytes. The first of these is dependent upon the activity of monoamine oxidase (MAO) and the other is related to the presence of transition metals *(57)*, such as iron. Increased glutamatergic drive to basal ganglia output nuclei has also been postulated to contribute to the pathogenesis of MPTP neurotoxicity. Since astrocytes possess efficient transport mechanisms for both MPTP and glutamate uptake, and increased excitatory tone may be related to aberrant glutamate uptake, recent studies by Hazell et al. *(58)* have examined the effect of MPTP on astrocytic D-aspartate uptake. Their studies corroborate that MPTP reversibly compromises glutamate uptake in cultured astrocytes, and that this effect is dependent on the conversion of MPTP to MPP^+. Such findings suggest that the glutamate transporter in astrocytes may play an important role in MPTP-induced neurotoxicity *(58)*. Another recent study investigated the cellular changes within astroglial cells in the MPTP-lesioned striatum. Specifically, striatal expression and regulation of connexin-43 (cx43), the principal gap junction protein of astroglial cells were evaluated *(59)*. These studies confirm that MPTP is cytotoxic altering the expression and protein levels of astrocytic cx43, this providing for another possible mechanism for MPTP-induced neurotoxicity.

That astrocytes mediate an essential step in MPTP toxicity is also supported by in vivo studies with the astroglial-selective toxicant, α-aminoadipic acid (α-AA; *see also* below) *(60)*. When MPTP is directly injected into the substantia nigra, a loss of nigral neurons is observed as revealed by fluorescent retrograde axonal tracing. In contrast, co-injection of MPTP plus α-AA into the substantia nigra is associated with reduced neuronal degeneration, presumably due to the initial destruction of resident astrocytes, and reduced conversion of MPTP to MPP^+. The protective effect of α-AA is curtailed, however, if reactive astrocytosis is allowed to proceed and MPTP is injected into the substantia nigra one week after the initial injection of α-AA. Thus, once repopulated with astrocytes (astrocytic scar), MPTP is again oxidized to MPP^+, leading to nigral damage and loss of dopaminergic neurons. Furthermore, because of the increased number of astrocytes upon the formation of an astrocytic scar (gliosis), MPTP-damage is increased compared to the experimental paradigm where it is co-injected with α-AA, reflecting increased metabolic conversion of MPTP to MPP^+. It is noteworthy, however, that other researchers were unable to reproduce the gliotoxic effects of α-AA *(61)*. Although the two groups used very different methods to assess the degree of cell death, the reason for the different outcomes of these studies remains unclear. Clearly, these studies point out the importance of convincingly deciphering the modality and permanence of the effects of α-AA within the context of astrocytic functions, and provide added impetus for future studies to outline the mechanistic link between MPTP and its cytotoxicity.

3.4. Methionine Sulfoximine

Methionine sulfoximine (MSO) is an irreversible inhibitor of the astrocyte-specific enzyme, GS *(62)*. This enzyme in the presence of ammonia catalyzes the conversion of glutamate to glutamine. When administered to animals, MSO leads to rapid convulsions (see below). Although at present, there appears to be no known environmental exposure to MSO, the literature is replete with examples of accidental poisonings with this compound. For example, wheat flour that had been bleached with agene (containing nitrogen trichloride) and accidentally consumed by dogs led to "running fits", "canine hysteria", convulsions, and anoxia *(63–65)*. Oxidation of methionine residues during the bleaching process of the wheat proteins is believed to have led to the formation of the toxic species, MSO.

At the morphological level, ingestion of large amounts of adulterated flour resulted in neuronal cell loss in the hippocampal fascia dentata and pyramidal cell layer, in the short association fibers and lower layers of the cerebral cortex, and in cerebellar Purkinje cells. MSO leads to large increases of glycogen levels *(66)*, primarily within astrocytic cell bodies, but not in other neuroglial cells (oligodendrocytes and microglia) or neurons *(67)*. In astrocytic cell bodies and in subpial, pericapillary and perineuronal astrocyte processes the glycogen often completely fills the cytoplasm, crowding the remaining organelles and inclusions. MSO treatment is also associated with swollen and damaged astrocytic mitochondria *(68)*. An ultrastructural study of cerebral cortex following administration of MSO revealed morphologic changes in astrocytes, consisting of cytoplasmic enlargement, mitochondrial and rough endoplasmic reticulum proliferation, development of cisternal and saccular smooth endoplasmic reticulum, nuclear chromatin clumping, and hydropic degenerative changes *(69,70)*. As noted by these authors, these changes are similar to those seen in experimental ammonia encephalopathy, suggesting perhaps an important role of ammonia in the evolution of these morphologic changes. Prolonged exposure to MSO was recently shown to lead to the appearance of swollen astrocytes, with watery nuclei reminiscent of Alzheimer type II glia, primarily in the neocortex, hippocampus, and lateral thalamus *(71)*. In agreement with earlier studies *(66)*, these changes were accompanied by simultaneous accumulation of glycogen in the superficial three layers of the neocortex, hippocampus, and pyriform cortex. GS immunoreactivity appeared enhanced in the cortex, hippocampus and lateral thalamus with a parallel increase in GFAP immunoreactivity. It is noteworthy, that the area of glycogen accumulation coincided with the known distribution of *N*-methy-D-aspartate (NMDA) and glutamate receptors *(71)*. This suggests that GS may play an important role in NMDA receptor-mediated glutamate metabolism.

Although it is generally accepted that MSO inhibits GS, it remains unclear whether this inhibition represents the primary mechanism of MSO neurotoxicity. Studies in cultured astrocytes incubated in the presence of [^{3}H]guanosine and MSO have shown that three of the four [^{3}H]methyl guanines formed were more highly labeled in the [^{3}H]tRNA of the MSO-exposed cells, relative to that of the control cells *(72)*. These findings suggest a stimulatory effect of MSO on the methylation of neural tRNA guanines, which was previously observed both in vitro using [^{14}C]S-adenosyl-L-methionine and in vivo using [methyl-^{3}H]L-methionine. These authors suggest that the effect of MSO is mediated via an effect on methyltransferase enzymes *(73)*. The same group has also reported that following intraperitoneal administration of MSO, S-adenosyl-L-methionine levels in brain are maximally reduced at the same time that MSO reaches its highest CNS concentration. Unfortunately, however, a mechanistic link between an effect on methyltransferase enzymes and cellular morphology and specificity was not established.

The relationship between inhibition of GS by MSO and seizure generation is also not well understood. Recent hypotheses have focused on the role of glutamate and GABA in seizure generation, since glutamine provides the precursor for these neurotransmitters. Rothstein and Tabakoff *(74,75)* have demonstrated that the calcium-dependent, potassium-stimulated release of glutamate and aspartate was inhibited in striatal tissue after intracerebroventricular injection of MSO. Addition of glutamine to the perfusion medium could reverse this effect. Furthermore, MSO has been shown to reduce the synthesis and stimulation-induced release of GABA and glutamate both in brain slices and in vivo, pointing to the importance of glutamine synthesis for neurotransmitter amino acid synthesis *(76,77)*. In other studies MSO increased the rate of efflux of newly loaded radiolabeled glutamine from rat cortical astrocytes in primary culture to more than 400% of the basal efflux, but it did not affect the efflux of the neurotransmitter amino acids, GABA or D-aspartate, under the same experimental conditions. It was suggested by these authors, that MSO-induced overflow with glutamine, which strongly interacts with the NMDA receptor complex, may therefore contribute to the convulsive action of MSO.

3.5. α-*Aminoadipic Acid* (α-*AA*)

A six-carbon homologue of the excitatory amino acids (EAA), glutamate and aspartate, α-aminoadipic acid (α-AA) is naturally produced in small amounts within the CNS as an intermediate of the degradation of L-lysine *(78)*. An inborn error of metabolism where the oxidative degradation of α-ketoadipic acid is depressed (produced by the transamination of α-AA) is associated with abnormally large urinary excretion of α-AA. These same individuals also have increased plasma levels of α-AA *(79,80)*.

Toxicity associated with the racemic mixture of α-AA (D,L-α-AA) was first uncovered in the mouse model *(81)*. D,L-α-AA was shown to affect ependymal and glial cells in the hippocampal arcuate nucleus, as well as Muller cells of the retina *(81)*. Additional studies have shown that enantiomorph type (D- vs L-), the temporal profile of exposure, as well as the tissue or cell type involved, determine the cytotoxic outcome. Other factors are also known to determine cell susceptibility to α-AA or its racemic mixture (D,L-α-AA). For example, when D,L-α-AA is applied to C6 neurogliomas they readily degenerate. However, if the cells are pretreated with dibutyryl cyclic AMP or sodium butyrate prior to D,L-α-AA exposure the cells appear insensitive to the racemic mixture *(82)*.

As alluded to in the section on MPTP neurotoxicity the ability of α-AA to selectively destroy astrocytes is somewhat controversial. The precise reason for different outcomes of astrocytic degeneration upon microinjection of α-AA in different experimental approaches remains unclear. The mechanism by which α-AA exerts its cytotoxicity is also a subject for debate. It is generally accepted that this effect is not attributable to activation of the EAA receptors, because specific ligands of these receptors (NMDA, kainate, AMPA) do not induce gliotoxicity *(83–86)*. The uptake of α-AA was reported to be astrocyte-specific, and it was purported to be Na^+-dependent and insensitive to tetrodotoxin *(87,88)*. A number of potential mechanisms for the astrocyte-specific effects of α-AA were postulated. These include:

1. a reduction of intracellular glutathione levels and subsequent oxidative damage *(82,89)* and
2. a rapid increase in the transient of intracellular Ca^{2+} *(90)*. In the absence of mechanistic understanding on α-AA neurotoxicity, it is rather surprising that there are no published papers on the effect of α-AA in astrocytes since 1993.

3.6. Fluoroacetate And Fluorocitrate

The Krebs cycle inhibitor fluorocitrate (FC) and its precursor fluoroacetate (FA) are taken up in brain preferentially by glia. FA occurs naturally in a number of plants in the southern hemisphere, and is available commercially as a rodenticide (Compound 1080). It is prevalent in the South African plant *Dichapetalum cymosum,* commonly referred to as the Gifblaar plant. Exposure to FA may also occur via exposure to the anticancer drug 5-fluorouracil *(91)*. Ingestion of large amounts of FA results in ionic convulsions within 30–60 min. Animals consuming FA commonly seize within minutes of exposure, and those surviving these episodes frequently die later on due to respiratory arrest or heart failure. Ruminants are particularly sensitive to FA, likely because during digestion in the prestomach they rely on formed acetate as an energy source *(86)*.

The actions of FC and FA have been attributed to both the disruption of carbon flux through the Krebs cycle and to impairment of ATP production *(92)*. FA can be metabolized to fluoroacetyl CoA, followed by condensation with oxaloacetate to form FC by citrate synthase *(93)*. A second hypothesis implies that FA toxicity is associated with the inhibition of a bi-directional citrate carrier in mitochondrial membranes *(94)*, which would also be expected to lead to elevated intramitochondrial citrate and could affect citrate-dependent ATP synthesis *(95)*. Finally, it has been suggested that elevated citrate, secondary to inhibition of aconitase, is associated with the cytotoxicity of these compounds. The latter catalyzes the reactions involved in mitochondrial energy production.

Both FA and FC have been shown to reduce the incorporation of radioactive label from several substrates into glutamine to a much greater extent than into glutamate *(96)*. Since glutamine synthesis occurs exclusively in astrocytes, it was suggested that the reduction of glutamine labeling is related to a direct inhibitory effect of FA and FC on GS. However, this does not appear to be the case. A predominant theory attributes the reduction in astrocytic glutamine to impaired uptake by astrocytes of neuronally released glutamate by FA *(97)*. Other studies have demonstrated that FA applied by microdialysis acted locally on astrocytes and, therefore, impaired astrocytic function was hypothesized to contribute to the development of hepatic encephalopathy by facilitating the entry of ammonia into the brain *(98)*. Inhibition of excitatory synaptic transmission by elevated brain ammonia has been suggested by the same authors as a potential mechanism for CNS depression in hepatic encephalopathy. Other findings strongly suggest that endogenous citrate released specifically from astrocytes into the extracellular space in the brain may function to modulate NMDA receptor activity *(99)*.

These findings, as well those by Hassel at al.*(100)* collectively suggest that FA lowers the level of glutamine, and inhibits glutamine formation in the brain in vivo. Furthermore, this occurs not by depletion of glial cells ATP, but by causing a rerouting of 2-oxoglutarate from glutamine synthesis into the TCA cycle during inhibition of aconitase *(100)*. After the inhibition of aconitase, citrate accumulates, whereas the levels of isocitrate, and α-ketoglutarate decrease. The reversible enzyme glutamate dehydrogenase begins to work in the opposite direction feeding more α-ketoglutarate into the TCA cycle *(86)*.

3.7. Potential Astrocytic Involvement in Neurodegenerative Disorders

Mucke and Eddleston *(101)*, Eddleston and Mucke *(102)*, Benveniste *(103)*, and Norenberg *(104)* recently reviewed astrocyte roles in CNS immune responses. The prevailing theories about CNS immune responses suggest that astrocytes, in concert with microglial cells recruit and activate infiltrating hematogenous cells and that they regulate blood-CNS interfaces. Reactive astrocytes produce and secrete a host of cytokines, proteases, protease inhibitors, adhesion molecules, and extracellular matrix components all of which mediate immune and inflammatory responses. Astrocytes have been previously implicated as the cells that are responsible for the initiation of inflammatory demyelinating disease by presenting CNS antigens to autoreactive immune cells. However, more recent studies discount such a role for astrocytes, and favor resident microglia as the CNS antigen presenting cells (APC) (reviewed in *(102,103)*).

In reviewing the role of glial cytokines in CNS repair, Mrak et al. *(105)*, have identified the following cascade as potentially playing an etiologic role in Alzheimer's disease. The microglial immune response-generated cytokine, interleukin-1 (IL-1), upregulates both the expression and processing of β-amyloid precursor proteins (β-APPs), hence favoring β-amyloid deposition, the morphological hallmark of Alzheimer's disease. Proliferation and activation of astrocytes are also promoted by β-APP, in turn, up regulating S100 β protein synthesis and secretion by astrocytes.

Increased concentrations of extracellular S100 β increase the intracellular concentrations of Ca^{2+} within neurons, stimulating neurite growth and neuritic plaque formation. Direct β-amyloid activation and/or Ca^{2+}-induced neuronal injury and death further stimulate microglial release of IL-1, an added stimulus for astrocytic activation. Accordingly, chronic, self-propagating, cytokine-mediated molecular and cellular reactions are invoked to explain the progressive neurodegeneration and dementia of Alzheimer's disease *(105)*. A similar mechanism has been postulated in the etiology of temporal lobe epilepsy *(106)* and HIV *(107)*. Interestingly, both CNS IL-1, as well as S100 β immunoreactivity are reported to be increased in Alzheimer's disease *(108)*. In temporal lobe neocortical tissue surgically resected from patients with intractable epilepsy, astrocytic S100 β-immunoreactivity is also 3–5 times higher than in control patients *(106)*. Stanley et al. *(107)* have recently noted an increase in the number of activated astroglia expressing elevated levels of S100 β in HIV-infected patients. Although at this point these findings are viewed as correlative, an assessment of the relative contribution of astrocytes to neurodegenerative disorders should be a particularly fruitful subject for future studies.

Granulocyte/macrophage-colony stimulating factor (GM-CSF) and granulocyte-colony stimulating factor (G-CSF) are cytokines necessary for growth and differentiation of macrophages. Both cytokines have been reported to lead to an accumulation of macrophages at the site of inflammatory lesions. Furthermore, both GM-CSF and G-CSF enhance a number of functional activities of mature macrophages, such as their phagocytic, cytotoxic, and microbicidal activities (reviewed in *(103)*. GM-CSF and G-CSF produced locally by astrocytes are an essential element for the recruitment and activation of hematogenous cells. Secretion of G-CSF and GM-CSF by astrocytes would, therefore, be expected to increase granulocyte and macrophage survival within the CNS and augment their activity against invading microbes. However, because viral replication in cultured HIV-infected monocytes is increased by GM-CSF, it has been suggested that astrocytic release of GM-CSF may augment viral production in monocytes and microglia, thus potentially worsening the spread of the infection within the CNS *(101,109)*. That astrocyte cytokine production, namely transforming growth factor-β (TGFβ), may lead to the recruitment and spread of cell-borne virus in HIV-1 infection was also postulated by Wahl et al. *(110)*. However, as suggested by Eddleston and Mucke "many cytokines appear to fulfill a multitude of functions and their effects in the intact adult CNS are only now beginning to be defined. It is therefore, perhaps not too surprising that the effects of cytokines in specific neurologic diseases have been difficult to predict." Exposure of primary cultured rat astrocytes to the major HIV envelop glycoprotein, gp120, and also leads to alterations of ion and solute transport *(111)*. It is yet to be explained how these changes contribute to neuronal cell injury associated with AIDS dementia.

4. SUMMARY

The foregoing Subheadings have provided a brief discussion on the potential involvement of astrocytes in neurotoxicity. Both in vivo and in vitro studies corroborate that, in principle, astrocytes are capable of damaging neurons in response to chemical exposure. Both direct impairment of astrocyte function by parent compounds or metabolism of a parent compound to an intermediate metabolite can play an active role in assaulting healthy neurons of the CNS. The relatively sparse amount of information currently available on the effects of neurotoxic compounds on astrocytes makes the expansion of such studies timely and worthwhile. Expanded investigations on astrocytic involvement in neurotoxicity is clearly warranted, and as new experimental tools are developed it is likely that further strides will be made in understanding astrocytic-mediated mechanisms of neurotoxicity and neurodegeneration.

ACKNOWLEDGMENTS

This review was partially supported by grants from PHS, ES 07331 (MA), and ES 03830 (RML).

REFERENCES

1. Virchow, R. (1946) Ueber das granulierte Ansehsn der Wandungen der Gehirnventrikel. *Allg Z Psychiatrie* **3,** 242–250.
1A. Pope, A. (1977) Neuroglia: Quantitative aspects. In *Dynamic Aspects of Glia Cells,* ed. E. Schofferiels, L., Hertz, and D.B. Tower, pp. 13–20. Pergamon Press, London, UK.
2. Kimelberg, H.K. and Norenberg, M.D. (1989) Astrocytes. *Scientific American* **260,** 66–76.
3. Kimelberg, H.K. and Aschner, M. (1994) Astrocytes and their functions, past and present. In *National Institute on Alcohol Abuse and Alcoholism Research Monograph, Alcohol and Glial Cells.* NIH Publication No. 94-3742, Bethesda, MD, Monograph 27, pp. 1–40.
4. LoPachin, R. and Aschner, M. (1993) Glial-neuronal interactions: the relevance to neurotoxic mechanisms. *Toxicol. Appl. Pharmacol.* **118,** 141–158.
5. Murphy, S., ed. (1994) In *Astrocytes: Pharmacology and Function,* Academic Press, New York.
6. Kettenmann, H., and Ransom, B.R., ed. (1995) In *Neuroglia.* Oxford University Press, New York, NY.
7. Aschner, M. and Kimelberg, H.K., eds. (1996) *The Role of Glia in Neurotoxicity.* CRC Press, Boca Raton.
8. Coyle, J.T., and Puttfarcken, P. (1994) Oxidative stress, glutamate, and neurodegenerative disorders. *Science* **262,** 689–695.
9. Hertz, L. (1979) Functional interactions between neurons and astrocytes I. Turnover and metabolism of putative amino acid transmitters. In *Progress in Neurobiology* vol. **13,** pp. 277–323, Pergamon, Oxford.
10. Martinez-Hernandez, A., Bell, K.P., and Norenberg, M.D. (1977) Glutamine synthetase: Glial localization in brain. *Science* **195,** 1356–1358.
11. Norenberg, M.D. and Martinez-Hernandez, A. (1979) Fine structural localization of glutamine synthetase in astrocytes of rat brain. *Brain Research* **161,** 303–310.
12. Schousboe, A. and Divac, I. (1979) Differences in glutamate uptake in astrocytes cultured from different brain regions. *Brain Research* **177,** 407–409.
13. Waniewski, R.A. and Martin, D.L. (1986) Exogenous glutamate is metabolized to glutamine and exported by primary astrocyte cultures. *Journal of Neurochemistry* **47,** 304–313.
14. Berl, S., Lajtha, A., and Waelsch, A. (1961) Amino acid and protein metabolism. VI. Cerebral compartments of glutamic acid metabolism. *J. Neurochem.* **7,** 186–197.
15. Martin, D.L. (1992) Synthesis and release of neuroactive substances by glial cells. *Glia* **5,** 81–94.
16. Shain, W. and Martin, D.L. (1990) Uptake and release of taurine – an overview. In *Taurine, Functional Neurochemistry, Physiology And Cardiology,* H. Pasantes-Morales, D.L. Martin, W. Shain, and R. del Rioeds., pp. 243–252. Wiley-Liss, Inc., New York.
17. Vitarella, D., DiRisio, D.J., Kimelberg, H.K., and Aschner, M. (1994) Potassium and taurine release are highly correlated with regulatory volume decrease in neonatal primary rat astrocyte cultures. *Journal of Neurochemistry.* **63,** 1143–1149.

18. Gilles, R., Hoffman, E.K., and Bolis, L. eds. (1991) In *Advances in Comparative and Environmental Physiology: Volume and Osmolality Control in Animal Cells.* Springer Verlag, Berlin.
19. Whetsell, W.O., Jr., C., Köhler, and R., Schwarcz, A. (1988) In *The Biochemical Pathology of Astrocytes* ed. M.D., Norenberg, L., Hertz, and A., Schousboe, pp. 191–202. Alan R. Liss, New York.
20. Bullock, R., Maxwell, W.L., Graham, D.I., Teasdale, G.M., and Adams, J.H. (1991) Glial swelling following human cerebral contusion: An ultrastructural study. *Journal of Neurology, Neurosurgery and Psychiatry* **54,** 427–434.
21. Rumpel, H., Nedelcu, J., Aguzzi, A., and Martin, E. (1997) Late glial swelling after acute cerebral hypoxia-ischemia in the neonatal rat: A combined magnetic resonance and histochemical study. *Pediatric Research* **42,** 54–59.
22. Kimelberg, H.K., Sankar, P., O'Connor, E.R., Jalonen, T., and Goderie, S.K. (1992) Functional consequences of astrocytic swelling. *Progress in Brain Research* **94,** 57–68.
23. Kempski, O., Staub, F., Schneider, H.-H., Weigt, H., and Baethmann, A. (1992) Swelling of C6 glioma cells and glial cells from glutamate, high K^+ concentrations or acidosis. *Progress in Brain Research* **94,** 69–76.
24. Takeuchi, T., Morikawa, N., Matsumoto, H., and Shiraishi, A. (1962) A pathological study of Minamata disease. *Acta Neuropathologica* (Berlin) **2,** 40–57.
25. Bakir, F., Damluji, S.F., Amin-Zaki, L., et al. (1973) Methylmercury poisoning in Iraq. *Science* **181,** 230–242.
26. Wood, J.M., Kennedy, F.S., and Rosen, C.E. (1968) Synthesis of methylmercury compounds by extracts of methanogenic bacterium. *Nature* **220,** 173–174.
27. Jensen, S. and Jernelov, A. (1969) Biological methylation of mercury in aquatic organisms. *Nature* **223,** 753–754.
28. Veiga, M.M., Meech, J.A., and Onate, N. (1994) Mercury pollution from deforestation. *Nature* **368,** 816–817.
29. Stokes, P.M. and Wren, C.D. (1987) Bioaccumulation of mercury by aquatic biota in hydroelectric reservoirs: a review and consideration of mechanisms. In *Lead, Mercury, and Arsenic in the Environment,* ed. T.C. Hutchinson and K.M. Meema, pp. 255–278. John Wiley and Sons, New York.
30. Takeuchi, T. (1972) Biological reactions and pathological changes in human beings and animals caused by organic mercury contamination. In *Environmental Mercury Contamination,* ed. R. Hartung, and B.D. Dinman, pp. 247–289. Ann Arbor Science, Ann Arbor.
31. Oyake, Y., Tanaka, M., Kubo, H., and Cichibu, H. (1966) Neuropathological studies on organic mercury poisoning with special reference to the staining and distribution of mercury granules. *Advances in Neurolog. Sci.* **10,** 744–750.
32. Garman, R.H., Weiss B., and Evans, H.L. (1975) Alkylmercurial encephalopathy in the monkey; a histopathologic and autoradiographic study. *Acta Neuropathologica* (Berlin) **32,** 61–74.
33. Charleston, J.S., Bolender, R.P., Mottet, N.K., Body, R.L., Vahter, M.E., and Burbacher, T.M. (1994) Increases in the number of reactive glia in the visual cortex of *Macaca fascicularis* following subclinical long-term methyl mercury exposure. *Toxicology and Applied Pharmacology* **129,** 196–206.
34. Vahter, M., Mottet, N.K., Friberg, L., Lind, B., Shen, D.D., and Burbacher, T. (1994) Speciation of mercury in primate blood and brain following long-term exposure to methyl mercury. *Toxicology and Applied Pharmacology* **124,** 221–229.
35. Albrecht, J., Talbot, M., and Kimelberg, H.K., and Aschner, M. (1993) The role of sulfhydryl groups and calcium in the mercuric chloride-induced inhibition of glutamate uptake in rat primary astrocyte cultures. *Brain Research* **607,** 249–254.
36. Brookes, N. and Kristt, D.A. (1989) Inhibition of amino acid transport and protein synthesis by $HgCl_2$ and methylmercury in astrocytes: Selectivity and reversibility. *J. Neurochem.* **53,** 1228–1237.

37. Brookes, N. (1992) In vitro evidence for the role of glutamate in the CNS toxicity of mercury. *Toxicology* **76,** 245–256.
38. Aschner, M., Du, Y.-L., Gannon, M., and Kimelberg, H.K. (1993) Methylmercury-induced alterations in excitatory amino acid efflux from rat primary astrocyte cultures. *Brain Res.* **602,** 181–186.
39. Rothstein, A. and Mack, E. (1992) Volume-activated calcium uptake: its role in cell volume regulation of Madin-Darby canine kidney cells. *American Journal of Physiology* **262,** C339–C347.
40. Jensen, B.S., Kramhoft, B., Jessen, F., Lambert, I.H., and Hoffman, E.K. (1993) $HgCl_2$-induced ion transport in Ehrlich ascites tumor cells. *Cellular Physiology and Biochemistry* **3,** 97–110.
41. Aschner, M., Vitarella, D., Allen, J.W., Conklin, D.R., and Cowan, K.S. (1998) Methylmercury-induced astrocytic swelling is associated with activation of the Na^+/H^+ antiporter, and is fully reversed by amiloride. *Brain Research* **799,** 207–214.
42. Aschner, M., Vitarella, D., Allen, J.W., Conklin, D.R., and Cowan, K.S. (1998) Methylmercury-induced inhibition of regulatory volume decrease in Astrocytes: Characterization of osmoregulator efflux and its reversal by amiloride. *Brain Research.* **811,** 133–142.
43. Aschner, M., Eberle, N.B., Miller, K., and Kimelberg, H.K. (1990) Interactions of methylmercury with rat primary astrocyte cultures: Inhibition of rubidium and glutamate uptake and induction of swelling. *Brain Research* **530,** 245–250.
44. Martin, H., Voss, K., Hufnagl, P., Wack, R., and Wassilew, G. (1987) Morphometric and densitometric investigations of protoplasmic astrocytes and neurons in human hepatic encephalopathy. *Experimental Pathology* **32,** 241–250.
45. Mossakowski, M.J., Renkawek, K., Krasnicka, Z., Smialek, M., and Pronaszko, A. (1970) Morphology and histochemistry of Wilsonian and hepatic gliopathy in tissue culture. *Acta Neuropathologica* (Berlin) **16,** 1–16.
46. M.D. Norenberg, M.D. (1981) In *Advances in Cellular Neurology,* Vol. 2, ed. S. Fedoroff and L. Hertz, pp. 304–338. Academic Press, New York.
47. Norenberg, M.D. (1986) In *Astrocytes: Cell Biology and Pathology of Astrocytes,* Vol. 3, ed. S. Fedoroff and A. Vernadakis, pp. 425–460. Academic Press, New York, 1986.
48. Albrecht, J., Hilgier, W., Lazarewicz, J.W., RafaAEowska, U., and Wysmyk-Cybula, U. (1988) In *Biochemical Pathology of Astrocytes,* ed. M.D. Norenberg, L. Hertz and A. Schousboe, pp. 465–476. Alan R. Liss, New York.
49. Albrecht, J. (1996) Astrocytes and ammonia neurotoxicity. *In The Role of Glia in Neurotoxicity,* M. Aschner, and H.K. Kimelberg, eds., pp. 137–153. CRC Press, Boca Raton.
50. Hawkins, R.A., and Jessy, J. (1991) Hyperammonemia does not impair brain function in the absence of net glutamine synthesis. *Biochemical Journal,* **277,** 697–703.
51. Hawkins, R.A., Jessy, J., Mans, A.M., and De Joseph, M.R. (1993) Effect of reducing brain glutamine synthesis on metabolic symptoms of hepatic encephalopathy. *J. Neurochem.* **60,** 1000–1006.
52. Langston, J.W., Ballard, P., Tetrud, J.W., and Irwin, I. (1982) Chronic parkinsonism in humans due to a product of Meperidine-analog synthesis. *Science* **219,** 979–980.
53. Burns, R.S., Chiueh, C.C., Markey, S.P., Ebert, M.H., Jacobowitz, D.M., and Kopin, I.J. (1983) A primate model of parkinsonism: Selective destruction of dopaminergic neurons in the pars compacta of the substantia nigra by N-methyl-4-phenyl-1,2,3,6-tetrahydropyridine. *Proceedings of the National Academy of Sciences,* USA, **80,** 4546–4550.
54. Heikkila, R.E., Hess, A., and Duvoisin, R.C. (1984) Dopaminergic neurotoxicity of 1-methyl-4-phenyl-1,2,5,6-tetrahydropyridine in mice. *Science* **224,** 1451–1453.
55. Langston, J.W., Forno, L.S., Rebert, C.S., and Irwin, I. (1984) Selective nigral toxicity after systemic administration of 1-methyl-4-phenyl-1,2,3,5,6-tetrahydropyridine (MPTP) in the squirrel monkey. *Brain Research* **292,** 390–394.

56. Marini, A.M., Schwartz, J.P., and Kopin, I.J. (1989) The neurotoxicity of 1-methyl-4-phenylpyridinium in cultured cerebellar granule cells. *Journal of Neuroscience* **9,** 3665–3672.
57. Di Monte, D.A., Royland, J.E., Irwin, I., and Langston, J.W. (1996) Astrocytes as the site for bioactivation of neurotoxins. *Neurotoxicology* **17,** 697–703.
58. Hazell, A.S., Itzhak, Y., Liu, H., and Norenberg, M.D. (1997). 1-Methyl-4-phenyl-1,2,3,6-tetrahydropyridine (MPTP) decreases glutamate uptake in cultured astrocytes. *Journal of Neurochemistry* **68,** 2216–2219.
59. Rufer, M., Wirth, S.B., Hofer, A., et al. (1996) Regulation of connexin-43, GFAP, and FGF-2 is not accompanied by changes in astroglial coupling in MPTP-lesioned, FGF-2-treated parkinsonian mice. *Journal of Neuroscience Research* **46,** 606–617.
60. Takada, M., Li, Z.K., and Hattori, T. (1990) Astroglial ablation prevents MPTP-induced nigrostriatal neuronal death. *Brain Research* **509,** 55–61.
61. Saffran, B.N. and Crutcher, K.A. (1987) Putative gliotoxin, α-aminoadipic acid, fails to kill hippocampal astrocytes in vivo. *Neuroscience Letters* **81,** 215–220.
62. Albrecht, J., Norenberg, M.D. (1990) L-methionine-DL-sulfoximine induces massive efflux of glutamine from cortical astrocytes in primary culture. *Euro. J. Pharmacol.* **182,** 587–599.
63. Lewey, F.H. (1950) Neuropathological changes in nitrogen trichloride intoxication of dogs. *Journal of Neuropathology and Experimental Neurology* **9,** 396.
64. Innes, J.R.M., and Saunders, L.Z. (1962) In *Comparative Neuropathology,* Academic Press, New York, 1962.
66. Folbergrova, J. (1973) Glycogen and glycogen phosphorylase in the cerebral cortex of mice under the influence of methionine sulfoximine. *Journal of Neurochemistry* **20,** 547–557.
67. Phelps, C.H. (1975) An ultrastructural study of methionine sulphoximine-induced glycogen accumulation in astrocytes of the mouse cerebral cortex. *Journal of Neurocytology* **4,** 479–490.
68. Hevor, T.K. (1994) Some aspects of carbohydrate metabolism in the brain. *Biochimie* **76,** 111–120.
69. Gutierrez, J.A., and Norenberg, M.D. (1975) Alzheimer II astrocytosis following methionine sulfoximine. *Archives of Neurology* **32,** 123–126.
70. Gutierrez, J.A., and Norenberg, M.D. (1977) Ultrastructural study of methionine sulfoximine-induced Alzheimer type II astrocytosis. *American Journal of Pathology* **86,** 285–300.
71. Yamamoto, T., Iwasaki, Y., Sato, Y., Yamamoto, H., and Konno, H. (1989) Astrocytic pathology of methionine sulfoximine-induced encephalopathy. Acta *Neuropathologica* (Berlin) **77,** 357–368.
72. Sellinger, O.Z. and Der, O. (1981) Effect of methionine sulfoximine on methylation of guanine residues in astroglial transfer ribonucleic acids. *Neurochemical Research* **6,** 153–162.
73. Sellinger, O.Z., Schatz, R.A., and Gregor, P. (1986) α-methylations in epileptogenesis. In *Advances in Neurology,* Vol 44, A.V. Delgado-Escueta, A.A. Ward, D.M. Woodbury, and R.J. Porter eds. Raven Press, New York.
74. Rothstein, J.D. and Tabakoff, B. (1984) Alteration of striatal glutamate release after glutamine synthetase. *J. Neurochem.* **43,** 1438–1446.
75. Rothstein, J.D. and Tabakoff, B. (1986) Regulation of neurotransmitter aspartate metabolism by glial glutamine synthetase. *Journal of Neurochemistry* **46,** 1923–1928.
76. Paulsen, R.E., Contestabile, A., Villani, A., and Fonnum, F. (1987) An in vivo model for studying function of brain tissue temporarily devoid of glial cell metabolism; the use of fluorocitrate *Journal of Neurochemistry* **48,** 1377–1385.
77. Paulsen, R.E. and Fonnum, F. (1989) Role of glial cells for the basal and $Ca2^+$-dependent K^+ evoked release of transmitter amino acids investigated by microdialysis. *Journal of Neurochemistry* **52,** 1823–1829.
78. Charles, K.A., and Chang, Y.-F. (1981) Uptake, release, and metabolism of D- and L-α-aminoadipate by rat cerebral cortex. *Journal of Neurochemistry* **36,** 1127–1136.

79. Fischer, M.H., Gerritsen, T., and Opitz, J.M. (1974) α-aminoadipic aciduria, a non-deleterious inborn metabolic defect. *Humangenetik* **24,** 265–270.
80. Jakobs, C., and de Grauw, A.J. (1992) A fatal case of 2-keto-,2-hydroxy-, and 2-aminoadipic aciduria: relation of organic aciduria to phenotype. *Journal of Inherited Metabolic Disease* **15,** 279–280.
81. Olney, J.W., Ho, O.L., and Rhee, V. (1971) Cytotoxic effects of acidic and sulfur-containing amino acids on the infant mouse central nervous system. *Experimental Brain Research* **14,** 61–76.
82. Kato, S., Higashida, H., Higuchi, Y., Hakatenaka, S., and Negeshi, K. (1984) Sensitive and insensitive states of cultured glioma cells to glutamate damage. *Brain Research* **303,** 365–373.
83. Olney, J.W. (1982) The toxic effects of glutamate and related compounds in the retina and the brain. *Retina* **2,** 341–359.
84. Bridges, R.J., Hatalski, C.G., Shim, S.N., Cummings, B.J., Vijayan, V., Kundi, A., and Cotman, C.W. (1992) Gliotoxic actions of excitatory amino acids. *Neuropharmacology* **32,** 899–907.
85. Yamada, K., Goto, S., Oyama, T., Inoue, N., Nagahiro, S., and Ushio, Y. (1994) Elevated immunoreactivity for glutamic acid deacrboxylase in the rat cerebral cortex following transient middle cerebral artery occlusion. *Acta Neuropathologica* (Berlin) **88,** 553–557.
86. Martin, D.L. and Waniewski, R.A. (1996) Precursor synthesis and neurotransmitter uptake by astrocytes as targets of neurotoxicants. In *The Role of Glia in Neurotoxicity,* ed. M. Aschner and H.K. Kimelberg, pp. 335–357. CRC Press, Boca Raton, 1996.
87. Huck, S., Grass, F., and Hatten, M.E. (1984a) Gliotoxic effect of α-aminoadipic acid on monolayer cultures of dissociated postnatal mouse cerebellum. *Neuroscience* **12,** 783–791.
88. Huck, S., Grass, F., and Hortnagl, H. (1984b) The glutamate analog α-aminoadipic acid is taken up by astrocytes before exerting its gliotoxic effect *in vitro. Journal of Neuroscience* **4,** 2650–2657.
89. Kato, S., Ishita, K., Sugawara, K., and Mawatari, K. (1993) Cystine/glutamate antiporter expression in retinal Muller glial cell: implications for DL-alpha-aminoadipate toxicity. *Neuroscience,* **57,** 473–482.
90. Wakakura, M. and Yamamoto, N. (1992) Rapid increase of intracellular Ca^{2+} concentration caused by aminoadipic acid enantiomers in retinal Muller cells and neurons in vitro. *Documenta Ophthalmologica.* **80,** 385–392.
91. Okeda, R., Shibutani, M., Matsuo, T., Kuroiwa, T., Shimokawa, R., and Tajima, T. (1990) Experimental neurotoxicity of 5-fluorouracil and its derivatives is due to poisoning by monofluorinated organic metabolites, monofluoroacetic acid and alpha-fluoro-beta-alanine. *Acta Neuropathologica* (Berlin) **81,** 66–73.
92. Swanson, R.A. and Graham, S.H. (1994) Fluorocitrate and fluoroacetate effects on astrocyte metabolism in vitro. *Brain Research* **664,** 94–100.
93. Brady, R.O. (1955) Fluoroacetyl coenzyme A. *J. Biolog. Chemis.* **217,** 213.
94. Kun, E., Kirsten, E., and Sharma, M. (1977) Enzymatic formation of glutathione-citryl thioester by mitochondrial system and its inhibition by (–)erythrofluorocitrate. *Proceedings of the National Academy of Sciences of the USA* **74,** 4942–4946.
95. Kirsten, E., Sharma, M.L., and Kun, E. (1978) Molecular toxicology of (–)erythro-fluorocitrate: Selective inhibition of citrate transport I mitochondria and the binding of fluorocitrate to mitochondrial proteins. *Molecular Pharmacology* **14,** 172–184.
96. Clarke, D.D., Nicklas, W.J., and Berl, S. (1970) Tricarboxylic acid-cycle metabolism in brain: Effect of fluoroacetate and fluorocitrate on the labeling of glutamate, aspartate, glutamine, and γ-aminobutyrate. *Biochem. J.* **120,** 345–351.
97. Szerb, J.C. and Issekutz, B. (1987) Increase in the stimulation-induced overflow of glutamate by fluoroacetate, a selective inhibitor of the glial tricarboxylic cycle. *Brain Research* **410,** 116–120.

98. Szerb, J.C. and Redondo, I.M. (1993) Astrocytes and the entry of circulating ammonia into the brain: effect of fluoroacetate. *Metabolic Brain Disease* **8,** 217–234.
99. Westergaard, N., Banke, T., Wahl, P., Sonnewald, U., and Schousboe, A. (1995) Citrate modulates the regulation by Zn2+ of N-methyl-D-aspartate receptor-mediated channel current and neurotransmitter release. *Proceedings of the National Academy of Sciences of the USA* **92,** 3367–3370.
100. Hassel, B., Sonnewald, U., Unsgard, G., and Fonnum F. (1994) NMR spectroscopy of cultured astrocytes: effects of glutamine and the gliotoxin fluorocitrate. *Journal of Neurochemistry* **62,** 2187–2194.
101. Mucke, L. and Eddleston, M. (1993) Astrocytes in infectious and immune-mediated diseases of the central nervous system. *FASEB Journal* **7,** 1226–1232.
102. Eddleston, M., and Mucke, L. (1993) Molecular profile of reactive astrocytes – implications for their role in neurologic disease. *Neuroscience* **54,** 15–36.
103. Benveniste, E.N. (1995) The role of cytokines in multiple sclerosis/autoimmune encephalitis and other neurological disorders. In *Human Cytokines, Their Role in Research and Therapy,* ed. B. Agrawal, and R. Puri, pp. 195–216. Blackwell Science Publications, Boston, MA.
104. Norenberg, M.D. (1996) Reactive astrocytosis. In *The Role of Glia in Neurotoxicity,* ed. M. Aschner, and H.K. Kimelberg, pp. 93–107. CRC Press, Boca Raton, FL.
105. Mrak, R.E., Sheng, J.G., and Griffin, W.S. (1995) Glial cytokines in Alzheimer's disease: review and pathogenic implications. *Human Pathology* **26,** 816–823.
106. Griffin, W.S., Yeralan, O., Sheng, J.G., et al, (1995) Overexpression of the neurotrophic cytokine S100 beta in human temporal lobe epilepsy. *Journal of Neurochemistry* **65,** 228–233.
107. Stanley, L.C., Mrak, R.E., Woody, R.C., et al. (1994) Glial cytokines as neuropathogenic factors in HIV infection: pathogenic similarities to Alzheimer's disease. *Journal Neuropathology and Experimental Neurology* **53,** 231–238.
108. Griffin, W.S.T., Stanley, L.C., Ling, C., et al. (1989) Brain interleukin 1 and S-100 immunoreactivity are elevated in Down syndrome and Alzheimer disease. *Proceedings of the National Academy of Sciences, USA* **86,** 7611–7615.
109. Tweardy, D.J., Mott, P.L., and Glazer, E.W. (1990) Monokine modulation of human astroglial cell production of granulocyte colony-stimulating factor and granulocyte-macrophage colony stimulating factor. 1. Effects of IL-1 alpha and IL-1 beta. *Journal of Immunology* **144,** 2233–2241.
110. Wahl, S.M., Allen, J.B., McCartney-Francis, N., et al. Mergenhagen, S.E. and Orenstein, J.M. (1991) Macrophage- and astrocyte-derived transforming growth factor beta as a mediator of central nervous system acquired immune deficiency syndrome. *Journal of Experimental Medicine* **173,** 981–991.
111. Benos, D.J., Hahn, B.H., Shaw, G.M., Bubien, J.K., and Benveniste, E.N. (1994) gp120-mediated alterations in astrocyte ion transport. *Advances in Neuroimmunol.* **4,** 175–179.

15

Ammonium Ion Transport in Astrocytes

Functional Implications

Neville Brookes

1. INTRODUCTION

1.1. Brain-to-Blood Distribution of Ammonia

The gross compartmentation of ammonia in the brain appears as if consistent with the commonly held assumption that the uncharged base (NH_3) is much more permeant than its conjugate cation (NH_4^+). For example, the normal brain-to-blood concentration ratio of ammonia is in the range 1.5–3 *(1),* where "ammonia" is defined as the total of NH_3 plus NH_4^+. This is the predicted distribution when the concentrations of NH_3 in blood and brain are nearly equal, and the differences in NH_4^+ levels are determined simply by the pH of each compartment and the pK_a of ammonia, which is approx 9.2 in plasma. An estimated intracellular pH of 7.1 in brain, as compared with 7.4 in blood, accounts for a twofold excess of NH_4^+ in the brain. This distribution of NH_4^+ approximates the distribution of total ammonia because more than 98% of ammonia exists as NH_4^+ at pH 7.4 or below.

1.2. Effect of Ammonia on Intracellular pH

The rapid intracellular alkalinization usually observed when neural cells are exposed to ammonium chloride is further evidence of the relative permeability of cell membranes to NH_3 *(2).* Alkalinization occurs because NH_3, entering more rapidly than NH_4^+, associates with cytosolic H^+. Boron and DeWeer *(2)* observed that the NH_4Cl-induced alkalinization of squid giant axon was followed by a slowly acidifying "plateau phase" (similar to the methylamine tracing shown in Fig. 1). They hypothesized that a slow influx of NH_4^+ contributes to this gradual decline of pH_i toward baseline. Once a steady-state distribution of NH_3 is reached, entering NH_4^+ now dissociates to form H^+ and NH_3, the latter diffusing out of the axon. Ammonium ions enter because an electrochemical gradient for NH_4^+ entry continues to exist even when the intracellular and extracellular concentrations of NH_3 equalize at steady-state *(3).*

The basis for this continuing inward electrochemical gradient of NH_4^+ is as follows. Assuming that the value of pK_a (= $[H^+][B]/[BH^+]$) is not different inside the cell than

From: *Neuroglia in the Aging Brain*
Edited by: Jean S. de Vellis © Humana Press Inc., Totowa, NJ

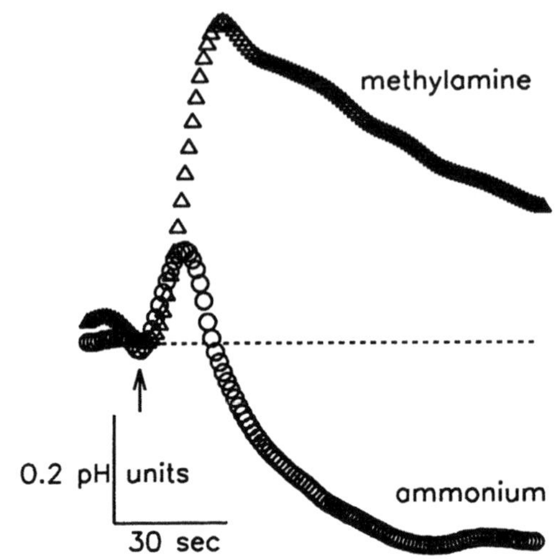

Fig. 1. Effect of ammonium (O) contrasted with methylamine (Δ) on intracellular pH (pH_i) in mouse cerebral astrocytes. At the arrow, each weak base (10 m*M*) was added in nominally bicarbonate-free solution at pH 7.4. Upward deflection of the superimposed fluorometric recordings indicates alkalinization. The dashed horizontal indicates resting pH_i (~7.1). Recordings are smoothed and replotted from Fig. 2 of *(4)*.

outside, it may be stated that when $[B]_i = [B]_o$, then $[H^+]_i/[H^+]_o = [BH^+]_i/[BH^+]_o$ (where B is base, BH^+ is conjugate cation, and subscripts i and o indicate inside or outside of the cell). Stated in other terms, the equilibrium potential for NH_4^+ adopts the same value as the equilibrium potential for H^+. Because active extrusion of acid by pH-regulatory transport makes the H^+ equilibrium potential considerably less negative than the membrane potential in most cells, thus creating an electrochemical gradient for H^+ influx, a similar gradient will drive NH_4^+ entry. Boron and DeWeer *(2)* found that their quantitative model of the response of intracellular pH (pH_i) to NH_4Cl best fit their data when the permeability of the membrane to NH_4^+ was 10^5 times lower than the permeability to NH_3 (that is, $P_B/P_{BH^+} = 10^5$)

1.3. Reevaluating Ammonia Fluxes in Astrocytes

An effect of NH_4Cl on pH_i resembling that described by Boron and DeWeer *(2)* has been observed in a variety of neurons and glia. However, more recent findings in glia indicate that the initial alkalinization may be absent or greatly curtailed, whereas the subsequent acidification seen in the continued presence of NH_4Cl is greatly intensified (4–6). This behavior suggests rapid transmembrane fluxes of NH_4^+, and small P_B/P_{BH^+} ratios of less than 10^3. It is argued here that these low values of P_B/P_{BH^+} are likely to be associated with the accumulation of ammonia in astrocytes. Existing evidence is compatible with a role for astrocytic accumulation of ammonia. For example, an overall brain-to-blood ratio of 3 allows for the possibility of severalfold regional variation in ratios among different intracellular compartments. When blood ammonia increases, the brain-to-blood ratio is observed to rise transiently to ~8 or more *(1)*.

The purpose of this chapter is to summarize the evidence that fluxes of NH_4^+ across the cell membranes of some mammalian astrocytes are more rapid than previously suspected, and to explore functional implications relating to the distribution of ammonia in the brain. Because NH_4^+ fluxes are primarily via transport pathways shared by K^+, and neuronal activity raises extracellular K^+ concentration ($[K^+]_o$) *(7),* it is predictable that the interaction between NH_4^+ and K^+ fluxes will be activity-dependent. A "potassium-ammonium countercurrent" model is proposed that hypothesizes opposing fluxes through the astrocytic syncytium, with NH_4^+ migrating toward, as K^+ migrates away from, regions of increased neuronal activity *(4,8).* The reactive astrocytes observed to proliferate in the aging brain, and in neurodegenerative disease, express an altered K^+-transport phenotype *(9)* that could disrupt this hypothesized potassium-ammonium countercurrent.

2. AMMONIUM ION TRANSPORT AND DISTRIBUTION

2.1. Mammalian Astrocytes

Figure 1 illustrates the markedly differing effects of methylamine hydrochloride and NH_4Cl on pH_i in mouse astrocyte cultures *(4).* Methylamine hydrochloride elicits the classic response to a permeant weak base with a relatively impermeant conjugate cation. In this astrocytic response, the decline of pH_i during the plateau phase following methylamine-induced alkalinization is likely to involve pH-regulatory transport (for example, Cl^-/HCO_3^- exchange) and metabolic production of acid *(10).* By contrast, NH_4Cl elicits a brief alkalinization succeeded rapidly by a marked and sustained acidification below resting pH_i. Figure 2 shows that the initial velocity of this acidification, expressed in terms of NH_4^+ influx, does not approach saturation in the concentration range 0.25–20 m*M* NH_4Cl, suggestive of channel-mediated NH_4^+ flux.

Low concentrations of Ba^{2+} inhibited up to 81% of the acidifying NH_4^+ influx produced by 1 m*M* NH_4Cl (*see* Fig. 3), implicating inwardly rectifying potassium (Kir) channels as the major path for NH_4^+ entry. Kir channels are the predominant K^+ channels determining the membrane potential of astrocytes, and they are characteristically sensitive to Ba^{2+} *(11,12).* Although the initial velocity of acidification did not saturate, the steady-state acidification did so at concentrations above 5 m*M* NH_4Cl. Steady-state acidification saturates because as pH_i falls, the velocity of pH-regulatory acid extrusion increases and the electrochemical gradient driving NH_4^+ influx declines, limiting the maximum extent of acidification.

Astrocytes also transport NH_4^+ actively *(4,6).* Bumetanide (100 μM) blocked 34% of the acidifying influx elicited by 1 m*M* NH_4Cl, indicating that NH_4^+ is taken up via cation-chloride cotransport *(13).* In glia acutely isolated from the retina of the bee, cation-chloride cotransport of NH_4^+ accounted entirely for an acidifying response to NH_4Cl *(6).* However, in mouse astrocytes, the combination of Ba^{2+} plus bumetanide was required to prevent net acidification in response to 1 m*M* NH_4Cl (*see* Fig. 3).

A well-established precedent for rapid, acidifying uptake of NH_4^+ by active and passive routes exists in the thick ascending limb of Henle's loop *(14).* In this nephron segment, the primary routes of NH_4^+ flux across the apical membrane similarly are Ba^{2+}-inhibitable K^+ channels and bumetanide-inhibitable Na^+-K^+-$2Cl^-$ cotransporters carrying NH_4^+ in place of K^+. It is noteworthy that the estimated value of P_B/P_{BH^+} in

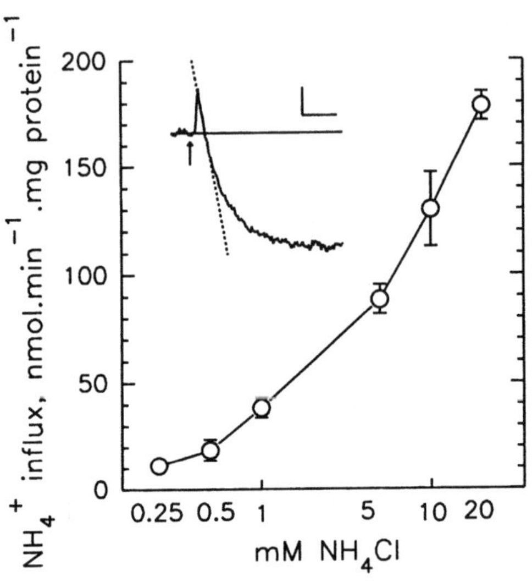

Fig. 2. The initial velocity of intracellular acidification (expressed here as equivalent influx of NH_4^+) in mouse cerebral astrocytes does not saturate with increasing NH_4Cl concentration. The plot was generated from data (means ± SEM, n = 3–6) shown in Fig. 3B of *(4)*, using estimates of intracellular buffering power (15.8 mmol.l^{-1}.pH unit^{-1}) and intracellular volume (4 µL/mg protein) *(23)*. **Inset:** Typical response of pH_i to 5 m*M* NH_4Cl (added at the arrow) in nominally bicarbonate-free solution (calibration, 0.1 pH u/30 s). The initial velocity of intracellular acidification is given by the slope of the dashed line.

rabbit thick ascending limb was as low as 20 *(15)*. In other words, the apical membrane may be only 20-fold more permeable to NH_3 than to NH_4^+.

2.2. Role of Inwardly-Rectifying Potassium Channels

The cloned inwardly rectifying K^+ channel Kir4.1 is expressed predominantly in mammalian glia *(16)*. The astrocytic property of high membrane K^+ conductance, resulting in a membrane potential close to the K^+ equilibrium potential, is likely to be in some measure a function of Kir4.1 expression *(17)*. Consequently, Kir4.1 channels, which are blocked by 3–30 µM Ba^{2+} *(16)*, are also a probable route of Ba^{2+} sensitive, passive NH_4^+ entry in astrocytes. The findings on the regulation and modulation of Kir4.1 channels suggest a basis for the variable permeability of glia to NH_4^+, and point to limitations of astrocyte culture as a model system. Expression of Kir4.1 disappeared when glia isolated from mammalian retina were maintained in culture for 4 d *(17)*. By use of the appropriate attachment factor and hormone supplementation during culture, either diffuse or clustered Kir4.1 expression could be restored to the cell membrane. Moreover, coexpression of specific membrane-associated anchoring proteins triggered both the clustering of Kir4.1 and marked enhancement of the Kir4.1-mediated current *(18)*.

Functional studies in astrocyte cultures show that Kir conductance is more evident in stellate, process-bearing cells than in morphologically-undifferentiated, proliferating astrocytes *(12,19)*. It is well documented that K^+ fluxes in astrocytes cultured from mouse cerebrum exceed those measured in rat astrocytes cultured under similar conditions by at least 10-fold, suggesting that the expression of Kir conductance, in cultured astrocytes also varies with species *(20)*. Thus the evidence strongly suggests that high

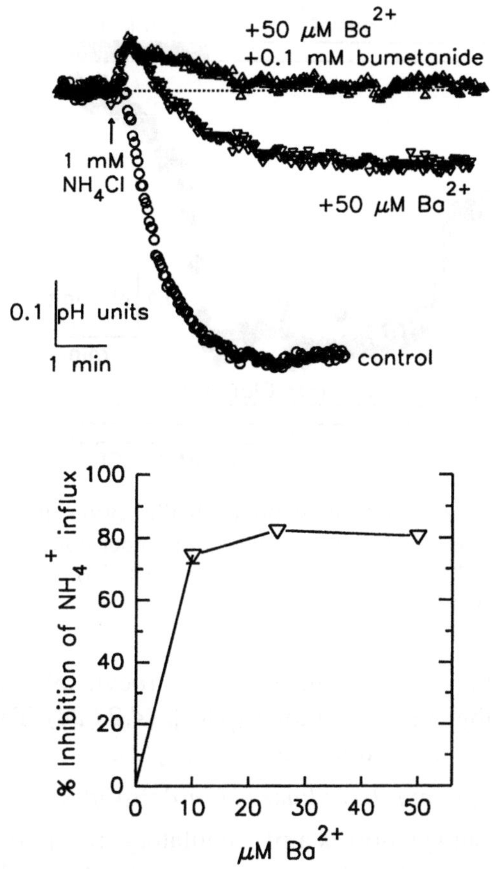

Fig. 3. Ammonium-induced intracellular acidification in mouse cerebral astrocytes is blocked by inhibitors of Kir channels (10–50 μ*M* Ba^{2+}) and cation-chloride cotransport (0.1 m*M* bumetanide). **Top:** Cumulative inhibition by Ba^{2+} and bumetanide. Recordings of fluorometric responses to 1 m*M* NH_4Cl in the presence and absence of inhibitors are shown superimposed on a dashed horizontal indicating resting pH_i **Bottom:** Inhibition of initial acidification rate, expressed as NH_4^+ influx elicited by 1 m*M* NH_4Cl, saturates at low [Ba^{2+}] (means ± SEM, n = 7–11). The data are replotted from Figs. 7A and 6B of *(4)*.

Kir conductance, together with an associated high permeability to NH_4^+, is a feature of the postmitotic astrocyte phenotype whose expression is variably subject to impairment in cell culture. It is reasonable to conclude that the use of astrocyte cultures can lead to an underestimate of the extent of NH_4^+ permeation in vivo. This is not to imply uniformly high Kir conductance in astrocytes in vivo. There is clear evidence of nonuniform distribution of Kir conductance not only within the cell membranes of single glia *(21)*, but also between regionally distinct glial populations *(22)*.

2.3. *Effects of Extracellular Potassium*

The extracellular potassium concentration in the brain is basally 2–3 m*M*, rising maximally to ~12 m*M* in areas of intense neuronal activity, or higher when neural

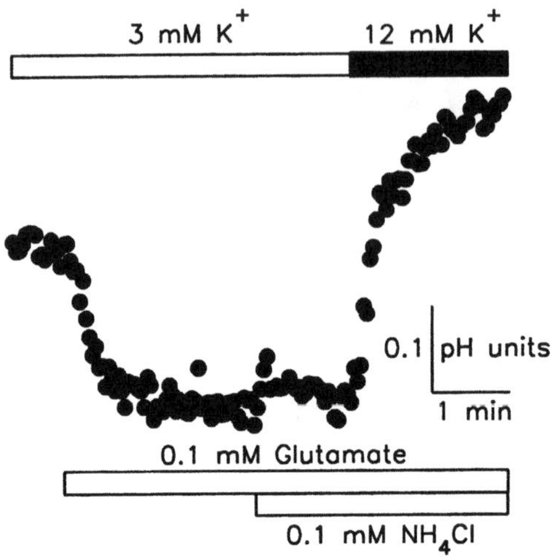

Fig. 4. Dominance of K^+-induced intracellular alkalinization in mouse cerebral astrocytes exposed to 0.1 m*M* glutamate plus 0.1 m*M* ammonium, under conditions of physiologic bicarbonate buffering (26 m*M* HCO_3^-, gassed with 5% CO_2/95% O_2). The fluorometric recording is replotted from Fig. 2A of *(31)*.

depolarization is pathologically sustained *(7)*. Astrocyte pH_i, extracellular pH, and NH_4^+ transport are all subject to regulation by $[K^+]_o$ *(4,23,24)*. Thus $[K^+]_o$ is a key variable affecting the distribution of ammonia in astrocytes.

2.3.1. Potassium-Dependence of Intracellular pH in Astrocytes

The sodium-bicarbonate symport is a pH-regulatory transporter found characteristically in glia, and not detected so far in neurons *(10)*. Inwardly transported HCO_3^- increases pH_i by associating with H^+. Carbonic anhydrase activity catalyzes the conversion of carbonic acid thus formed to membrane permeant CO_2 and water. Transport of each Na^+ ion is coupled to the transport of either 2 or 3 HCO_3^- ions, resulting in a net movement of charge and a consequent dependence of transport on membrane potential *(25)*. Potassium-induced depolarization of the astrocyte membrane accelerates influx of HCO_3^- via the symport, thus alkalinizing the cytosol, the loss of HCO_3^- to the intracellular compartment causes extracellular carbonic acid to dissociate and acidify extracellular pH *(24)*.

Potassium- and ammonia-induced pH changes are affected very differently by physiologic bicarbonate buffering. Because the effect of $[K^+]_o$ on pH_i is mediated by HCO_3^- transport, this effect is fully expressed in physiologic bicarbonate buffer (~26 m*M* HCO_3^-, *see* Fig. 4) and markedly diminished in HCO_3^--depleted solution *(23)*. In a sense, $[K^+]_o$ regulates the "set-point" at which pH-regulatory transport seeks to maintain pH_i. By contrast, the direct effects of NH_3 and NH_4^+ on pH_i are attenuated by bicarbonate buffering, as well as by the intrinsic buffering capacity of the cytosol and by pH-regulatory transport *(4)*. Normal extracellular levels of ammonia (0.1–0.2 m*M*) show little effect on astrocyte pH_i when applied in bicarbonate-buffered solution in vitro (*see* Fig. 4).

It is noteworthy that K^+-induced alkalinization is the dominant effect of neuronal stimulation in vivo on astrocyte pH_i *(26)*, surmounting the acidifying action of glutamate (*see*

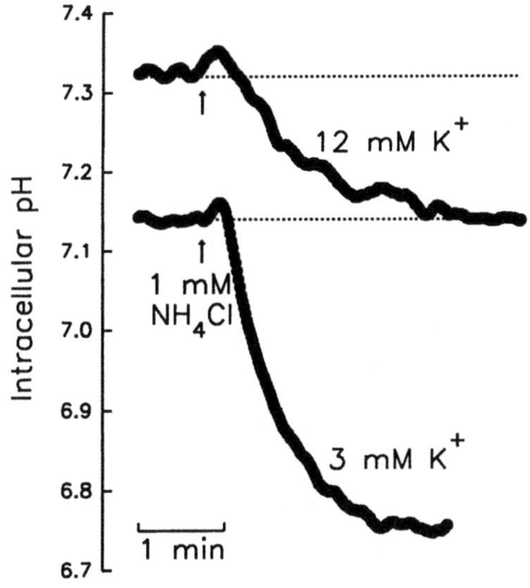

Fig. 5. Potassium inhibits ammonium-induced intracellular acidification in mouse cerebral astrocytes. The fluorometric recordings, replotted from Fig. 5D of *(4)*, show responses of pH_i to 1 m*M* NH_4Cl in nominally bicarbonate-free solution containing 3 m*M* or 12 m*M* K^+. The upper recording is displaced by the alkalinizing effect of increased $[K^+]_o$.

Fig. 4) and other mediators and metabolites released by neuronal activity. A large physiologic increase in $[K^+]_o$ from 3 m*M* to 12 m*M* in vivo can reverse the transmembrane gradient of pH in astrocytes. This would eliminate the normal concentration gradient of NH_4^+ between the astrocyte compartment and blood, assuming that P_B/P_{BH^+} is high. However, it will be argued here that the effect of $[K^+]_o$ on ammonia distribution is amplified further when P_B/P_{BH^+} is basally low and rises with increasing $[K^+]_o$.

2.3.2. Effects of Potassium on Ammonium Transport

Studies in mouse astrocyte cultures show that increasing $[K^+]_o$ in the physiological range markedly slows the acidifying influx of NH_4^+ (*see* Fig. 5). A depolarization-induced decline in the inward electrochemical gradient of NH_4^+ is unlikely to account for this effect of K^+ because it is offset by an increase in this electrochemical gradient caused by K^+-induced alkalinization of pH_i. Further, this inhibitory effect of K^+ on NH_4^+-induced acidification is too large to be attributable to reduced cation-chloride cotransport of NH_4^+ *(4)*. The remaining possibility is that K^+ inhibits permeation of NH_4^+ via Kir channels. However, raising $[K^+]_o$ is well known to *increase* the K^+ conductance of Kir channels in astrocytes *(12)*. Thus inhibition of NH_4^+ permeation presumably involves some form of competition between K^+ and NH_4^+ for transport via these channels *(27)*.

2.4. Dependence of Ammonium Ion Distribution on Permeability

When NH_3 is the permeating species and NH_4^+ is assumed relatively impermeant, the distribution of ammonia is independent of cell membrane potential and is dependent only on the pH of the intracellular and extracellular compartments, as outlined in the Introduction. The opposite condition of very low P_B/P_{BH^+} results in a distribution

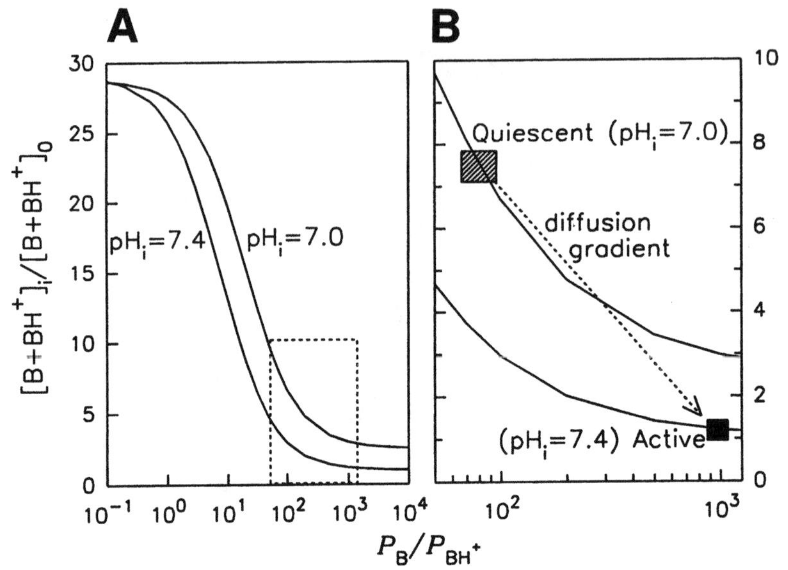

Fig. 6. **(A)** Distribution ratio of total ammonia (B + BH^+) as a function of the relative permeability of astrocyte membrane to the conjugate species (P_B/P_{BH^+}), plotted using the following expression (Eqn. 3 of Ref. 29):

$$\frac{[H^+]_i + K_a}{[H^+]_o + K_a} \cdot \frac{(P_B/P_{BH}+) - \{f\,[H^+]_o/K_a\,(1 - e^f)\}}{(P_B/P_{BH}+) - \{f\,[H^+]_i\,e^{>f}/K_a\,(1 - e^f)\}}$$

where $f = FV_m/RT$ (V_m is membrane potential, and *F, R* and *T* have their usual meaning). Plots were constructed for the putative limiting physiologic values of astrocyte pH_i shown (pH_o = 7.4, pK_a = 9.2, V_m = –90 mV, and temperature = 37°C). **(B)** An expansion of the area of Graph A enclosed by the dashed line. See **Subheadings 2.4.** and **3.2.** of the text for discussion.

of the conjugate cation determined entirely by the membrane potential, and independent of pH. Roos and Boron *(28,29)* explored quantitatively how the distribution of weak bases relates to intermediate values of P_B/P_{BH^+} between these two extremes. The expression developed by Roos *(28)* is used in figure 6 to model the transmembrane distribution of total ammonia in astrocytes as a function of P_B/P_{BH^+}. This relationship is plotted for two different values of pH_i, representing approximate limits of pH_i variation in response to the physiologic range of $[K^+]_o$. Significant voltage-dependent intracellular accumulation of NH_4^+ is seen when the value of P_B/P_{BH^+} falls below 10^3. Observations of a rapid acidifying influx of NH_4^+ in astrocytes justify considering the functional implications of P_B/P_{BH^+} values in this low range.

3. POTASSIUM-AMMONIUM COUNTERCURRENT

3.1. Astrocyte Regulation of Extracellular Potassium

The maintenance of neuronal excitability by buffering $[K^+]_o$ is a glial function that has been explored extensively *(11)*. The expression of Kir conductance, and perhaps Kir4.1 channels specifically, in astrocytes is thought to be critical for this function *(17)*.

Proposed mechanisms of K^+ homeostasis rely on a high-conductance pathway for K^+ influx which remains open throughout the physiologic range of $[K^+]_o$. The "spatial buffering" model proposes that K^+ enters astrocytes in regions where neuronal activity raises $[K^+]_o$, and exits at a distance in electrically more quiescent regions of lower $[K^+]_o$. A second proposal is that astrocytes transiently accumulate K^+, together with chloride and water, when $[K^+]_o$ rises in electrically active regions. Both mechanisms create an intracellular concentration gradient of K^+ within astrocytes, or within a syncytium of astrocytes coupled by gap junctions, such that K^+ flows away from active zones towards more quiescent zones. The following consideration of the effect of $[K^+]_o$ on ammonia distribution leads to the conclusion that an opposing gradient of intracellular NH_4^+ concentration exists, causing NH_4 to migrate away from quiescent zones towards active zones.

3.2. An Intracellular Gradient of Ammonium Concentration

As already discussed, pH_i and P_B/P_{BH}^+ are major factors governing the intracellular concentration of NH_4^+. In astrocytes, the evidence suggests that both of these factors are regulated by $[K^+]_o$. A localized elevation of $[K^+]_o$ will elicit a localized depolarization-induced alkalinization when the space constant of the astrocyte cell membrane is small compared to the dimensions of the cell or syncytium. Given the physiological range of $[K^+]_o$ in the brain, reasonable limits of pH_i are 7.4 in an alkalinized active zone and 7.0 in a quiescent zone. From the curves shown in figure 6, assuming for the moment that P_B/P_{BH} is high ($>10^3$) throughout the membrane, these values of pH_i translate into an approximately threefold concentration gradient of intracellular NH_4^+, causing NH_4^+ to migrate *toward* the active zone. Note that the curves in figure 6 are based on a constant extracellular pH of 7.4, whereas variations in $[K^+]_o$ regulate extracellular pH as well as pH_i. Thus it is the distribution of ammonia between astrocytic cytosol and blood that is modeled here, rather than the distribution between astrocytic cytosol and the extracellular compartment of the brain.

When the evidence of rapid NH_4^+ permeation in astrocytes is taken into account, allowing that P_B/P_{BH}^+ may fall below 10^3, figure 6 shows how the intracellular gradient of NH_4^+ concentration is further amplified. Because the apparent permeability of NH_4^+ declines in the presence of increased $[K^+]_o$, it follows that $P_B/P_{BH}+$ is higher in the active zone. However, in the absence of direct measurements, the range of values of P_B/P_{BH}^+ selected in figure 6B represents no more than a partially informed guess. It should be noted that the time course of K^+-induced effects on pH_i and P_B/P_{BH}^+ differ markedly. Responses of pH_i to changes in $[K^+]_o$ are slow, with a half-time of ~0.5 min in vitro *(23)*, whereas direct effects of $[K^+]_o$ on P_B/P_{BH}^+ presumably are very rapid.

The dependence on pH of the velocity of ammonia metabolism by glutamine synthase is an additional factor that potentially can amplify the intracellular gradient of NH_4^+ concentration, and speed the migration of NH_4^+ toward an active zone. Glutamine synthesis from ammonia and glutamate is maximal near the pH optimum for glutamine synthase activity, which coincides with an active zone pH_i of 7.4 *(30,31)*.

Figure 7 is an attempt to summarize in a simple scheme some of the factors that generate a potassium-ammonium countercurrent within a syncytium of astrocytes. The arguments supporting this scheme suggest that normal extracellular ammonia concen-

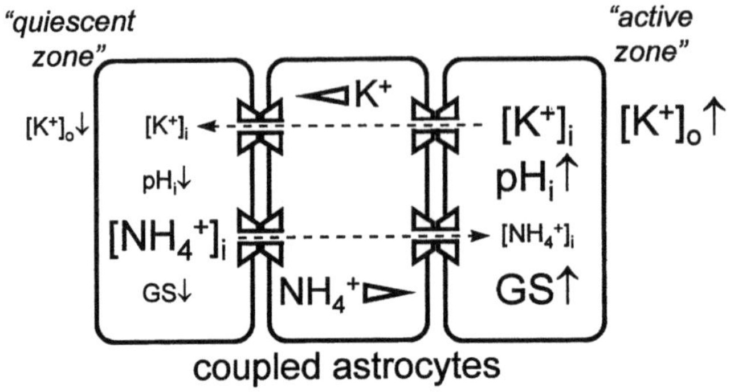

Fig. 7. Hypothetical intracellular "potassium-ammonium countercurrent" in a glial syncytium, between neurally active and neurally quiescent zones. In the active zone, $[K^+]_i$, pH_i and glutamine synthase activity (GS) increase, and $[NH_4^+]_i$ decreases, in response to rising $[K^+]_o$. Opposite changes occur in the quiescent zone. See **Subheading 3.2.** of the text for discussion.

trations of 0.1–0.2 m*M* *(1)* result in intracellular NH_4^+ concentrations of 1 m*M* or more in astrocyte quiescent zones, and 0.1–0.2 m*M* or below in active zones. Assuming that quiescent zones are normally more extensive than active zones, the intracellular flux of NH_4^+ produced by this gradient could balance much or all of the opposing flux of K^+ away from active zones. It should be noted that the short space constant characteristic of high-conductance glial cell membranes favors this scheme, although presenting difficulties for the spatial buffering model *(11)*. Nevertheless, a potassium-ammonium countercurrent presumably would complement other mechanisms of K^+ homeostasis rather than represent an alternative to them.

4. HYPERAMMONEMIA AND AMMONIUM ION TRANSPORT

Measurements of astrocyte pH_i and ammonia distribution in animal models of human hyperammonemia appear at first sight to challenge the interpretations of in vitro data offered here, but further consideration reveals a basis for consistency. Specifically, two studies of hyperammonemic rats report *elevations* of glial pH_i in vivo relative to controls *(32,33)*, in contrast to the acidifying responses of astrocytes in vitro. However, high ammonia concentrations cause a loss of intracellular K^+ that raises $[K^+]_o$ markedly in vivo *(34,35)*, whereas the volume bathing astrocyte cultures is too large for $[K^+]_o$ to be affected by such K^+ loss. As noted in **Subheading 2.3.1.**, K^+-induced alkalinization is a dominant regulator of pH_i in vivo. For example, the recordings in Fig. 5 show that, even in a nominally bicarbonate-free solution, raising $[K^+]_o$ from 3 m*M* to 12 m*M* alkalinizes pH_i and diminishes the acidifying effect of 1 m*M* NH_4Cl sufficiently that steady-state pH_i in the presence of 1 m*M* NH_4Cl does not differ from control pH_i at $[K^+]_o$ = 3 m*M*. In physiologically bicarbonate-buffered solution, the K^+-induced alkalinization is amplified, and the NH_4^+-induced acidification reduced by >70% *(4)*, such that the combined response of pH_i to 1 m*M* NH_4Cl and elevated $[K^+]_o$ will be net alkalinization in vitro as in vivo. Interestingly, 1 m*M* NH_4Cl *acidified* pH_i in brain slices *(36)*, perhaps because the short diffusion path to the bathing solution diminishes the ability of K^+ to accumulate extracellularly in the slice.

Similar arguments can be invoked to account for the transience of the rise in the brain-to-blood ratio to ~8 or more observed after an increase in blood ammonia concentration *(1)*. Ammonia-induced membrane depolarization and loss of intracellular K^+ occurs slowly, in a time frame of tens of minutes *(37)*. This allows astrocytes to accumulate ammonia initially, with a subsequent decline in distribution ratio induces pH_i as rising $[K^+]_o$ to alkalinize and P_B/P_{BH^+} to increase. However, these factors are not sufficient to account for the inability of brain slices to accumulate ammonia when exposed to ≥ 1 m*M* NH_4Cl in vitro *(38)*. The evidence suggests that these high concentrations of NH_4C1 only partially depolarize astrocytes *(37)*, have little effect on astrocyte Kir conductance *(34,37)*, and acidify pH_i in brain slices *(36)*. Some other, as yet undetermined, mechanism must prevent ammonia accumulation under these conditions in vitro.

A related question is why, if the brain-to-blood ratio can rise transiently to ~8 at the onset of hyperammonemia, is a similarly high brain-to-blood ratio not found at normal blood ammonia concentrations which do not cause a loss of intracellular K^+? A possible reason is suggested by the observation that methionine sulfoximine, an inhibitor of glutamine synthase activity, increases the normal ratio by approximately threefold *(1)*. Ammonia accumulation in astrocytes may therefore be limited normally by a velocity of ammonia fixation exceeding NH_4^+ influx. Similar reasoning suggests that diminished glutamine synthase activity in vitro may contribute to the large distribution ratio of ~19 measured in brain slices when the ammonia concentration in the bathing solution is not raised above the normal physiologic range for blood *(38)*.

5. REACTIVE ASTROCYTES

Glial proliferation occurs at loci of neuronal degeneration in the normal aging brain, also early in the formation of the senile plaques of Alzheimer's disease, and commonly in other neurodegenerative disorders *(39)*. The membrane properties of proliferating "reactive astrocytes" revert to a phenotype expressed during gliogenesis, and evident also in a variety of other proliferating cell types *(9)*. There is a marked decrease in Kir conductance and an increase in outwardly rectifying K^+ conductance. This signature change is detectable in gliotic tissue of patients with recurrent seizures, as well as in the proliferating astrocytes of a cell culture model of gliosis *(9)*. Given the key role of Kir channels in K^+ homeostasis, impairment of this astrocytic function is expected in gliosis, and is in fact observable in the gliotic tissue surrounding an experimental epileptic focus *(40)*. Accordingly, the loss of astrocyte Kir conductance, leading to high P_B/P_{BH^+} values for ammonia and diminished modulation of pH_i by $[K^+]_o$, will impair also the potassium-ammonium countercurrent mechanism proposed here. It would be valuable to assess the nature and extent of the decrements in CNS function attributable to impairment of astrocyte-mediated K^+ homeostasis both in hyperammonemic syndromes and in the aging brain. The isolation of glia-specific Kir channels *(16)* brings this goal closer to attainment.

6. CONCLUSIONS

Evidence of rapid permeation of NH_4^+ ions in astrocytes in vitro raises the possibility that the distribution of ammonia in the brain is determined not only by the pH of intracellular and extracellular compartments, but also by membrane potential-dependent accumulation of NH_4^+ in astrocytes. Based on this premise, intracellular NH_4^+ concentrations in astrocytes *in situ* should not be uniform, but should vary as a function of local neuronal activity. This linkage derives from activity-related fluctuation of $[K^+]_o$, together with the role of $[K^+]_o$ as a regulator of (a) $NH_4{}^+$ permeation, (b) compartmental pH, and (c) NH_4^+ metabolism by glutamine synthase. In quiescent regions of faster NH_4^+ permeation, decreased pH_i, and decreased glutamine synthase activity, the local intracellular NH_4^+ concentration in astrocytes rises. In active zones, where $[K^+]_o$ is high, opposite trends decrease intracellular NH_4^+ concentration. Consequently, intracellular NH_4^+ in the astrocyte syncytium migrates towards zones of neuronal activity as intracellular K^+ migrates away. The molecular identification of the Kir channels that are a probable major pathway for NH_4^+ permeation in astrocytes will facilitate a more direct examination of the premises upon which this concept of a potassium-ammonium countercurrent is based, and permit an evaluation of impaired K^+ homeostasis associated with reactive astrogliosis in aging and disease.

ACKNOWLEDGMENT

The author acknowledges the support of US Public Health Service grant AG16951 from the National Institute on Aging.

REFERENCES

1. Cooper, A.J.L. and Plum, F. (1987) Biochemistry and physiology of brain ammonia. *Physiol. Rev.* **67,** 440–519.
2. Boron, W.F. and De Weer, P. (1976) Intracellular pH transients in squid giant axons caused by CO_2, NH_3, and metabolic inhibitors. *J. Gen. Physiol.* **67,** 91–112.
3. Putnam, R.W. (1988) Basic Principles of pH Regulation, in *Na^+/H^+ Exchange* (Grinstein, S., ed.), CRC Press, Boca Raton, FL, pp. 139–153.
4. Nagaraja, T.N. and Brookes, N. (1998) Intracellular acidification induced by passive and active transport of ammonium ions in astrocytes. *Am. J. Physiol.* **274,** C883–C891.
5. Norenberg, M.D. (1998) Astroglial dysfunction in hepatic encephalopathy. *Metab. Brain. Dis.* **13,** 319–335.
6. Marcaggi, P., Thwaites, D.T., Deitmer, J.W., and Coles, J.A. (1999) Chloride-dependent transport of NH_4^+ into bee retinal glial cells. *Eur.J. Neurosci.* **11,** 167–177.
7. Somjen, G.G. (1979) Extracellular potassium in the mammalian central nervous system. *Annu. Rev. Physiol.* **41,** 159–177.
8. Brookes, N. (2000) Functional integration of the transport of ammonium, glutamate and glutamine in astrocytes. *Neurochem. Int.* **37,** 121–129.
9. MacFarlane, S.N. and Sontheimer, H. (1997) Electrophysiological changes that accompany reactive gliosis in vitro. *J. Neurosci.* **17,** 7316–7329.
10. Deitmer, J.W. and Rose, C.R. (1996) pH regulation and proton signalling by glial cells. *Progr. Neurobiol.* **48,** 73–103.
11. Barres, B.A., Chun, L.L.Y., and Corey, D.P. (1990) Ion channels in vertebrate glia. *Annu. Rev. Neurosci.* **13,** 441–474.
12. Ransom, C.B. and Sontheimer, H. (1995) Biophysical and pharmacological characterization of inwardly rectifying K^+ currents in rat spinal cord astrocytes. *J. Neurophysiol.* **73,** 333–346.
13. Moore-Hoon, M.L. and Turner, R.J. (1998) Molecular characterization of the cation-chloride cotransporter family. *Eur. J. Morphol.* **36,** Suppl., 137–41.
14. Good, D.W. (1994) Ammonium transport by the thick ascending limb of Henle's loop. *Annu. Rev. Physiol.* **56,** 623–647.

15. Garvin, J.L., Burg, M.B., and Knepper, M.A. (1988) Active NH_4^+ absorption by the thick ascending limb. *Am. J. Physiol.* **255,** F57–F65.
16. Takumi, T., Ishii, T., Horio, Y., et al. (1995) A novel ATP-dependent inward rectifier potassium channel expressed predominantly in glial cells. *J. Biol. Chem.* **270,** 16339–16346.
17. Ishii, M., Horio, Y., Tada, Y., et al. (1997) Expression and clustered distribution of an inwardly rectifying potassium channel, K-AB-2/Kir4.1, on mammalian retinal Muller cell membrane: Their regulation by insulin and laminin signals. *J. Neurosci.* **17,** 7725–7735.
18. Horio, Y., Hibino, H., Inanobe, A., et al. (1997) Clustering and enhanced activity of an inwardly rectifying potassium channel, Kir4.1, by an anchoring protein, PSD-95/SAP90. *J. Biol. Chem.* **272,** 12885–12888.
19. Ferroni, S., Marchini, C., Schubert, P., and Rapisarda, C. (1995) Two distinct inwardly rectifying conductances are expressed in long term dibutyryl-cyclic-AMP treated rat cultured cortical astrocytes. *FEBS Letters* **367,** 319–325.
20. Walz, W. and Kimelberg, H.K. (1985) Differences in cation transport properties of primary astrocyte cultures from mouse and rat brain. *Brain Res.* **340,** 333–340.
21. Newman, E.A. (1986) High potassium conductance in astrocyte end-feet. *Science* **233,** 453–454.
22. D'Ambrosio, R., Wenzel, J., Schwartzkroin, P.A., McKhann II, G.M., and Janigro, D. (1998) Functional specialization and topographic segregation of hippocampal astrocytes. *J. Neurosci.* **18,** 4425–4438.
23. Brookes, N. and Turner, R.J. (1994) K^+-induced alkalinization in mouse cerebral astrocytes mediated by reversal of electrogenic Na^+-HCO_3^- cotransport. *Am. J. Physiol.* **267,** C1633–C1640.
24. Grichtchenko, I.I. and Chesler, M. (1994) Depolarization-induced acid secretion in gliotic hippocampal slices. *Neuroscience* **62,** 1057–1070.
25. Bevensee, M.O., Apkon, M., and Boron, W.F. (1997) Intracellular pH regulation in cultured astrocytes from rat hippocampus. 2. Electrogenic Na/HCO_3 cotransport. *J. Gen. Physiol.* **110,** 467–483.
26. Chesler, M. and Kraig, R.P. (1989) Intracellular pH transients of mammalian astrocytes. *J. Neurosci.* **9,** 2011–2019.
27. Moroni, A., Bardella, L., and Thiel, G. (1998) The impermeant ion methylammonium blocks K^+ and NH_4^+ currents through KAT1 channel differently: evidence for ion interaction in channel permeation. *J. Membr. Biol.* **163,** 25–35.
28. Roos, A. (1975) Intracellular pH and distribution of weak acids across cell membranes. A study of D- and L-lactate and of DMO in rat diaphragm. *J. Physiol. (Lond.)* **249,** 1–25.
29. Boron, W.F. and Roos, A. (1976) Comparison of microelectrode, DMO, and methylamine methods for measuring intracellular pH. *Am. J. Physiol.* **231,** 799–809.
30. Brookes, N. (1992) Effect of pH on glutamine content derived from exogenous glutamate in astrocytes. *J. Neurochem.* **59,** 1017–1023.
31. Brookes, N. and Turner, R.J. (1993) Extracellular potassium regulates the glutamine content of astrocytes: mediation by intracellular pH. *Neurosci. Lett.* **160,** 73–76.
32. Swain, M.S., Blei, A.T., Butterworth, R.F., and Kraig, R.P. (1991) Intracellular pH rises and astrocytes swell after portacaval anastomosis in rats. *Am. J. Physiol.* **261,** R1491–R1496.
33. Kanamori, K. and Ross, B.D. (1997) Glial alkalinization detected in vivo by ^1H-^{15}N heteronuclear multiple-quantum coherence-transfer NMR in severely hyperammonemic rat. *J. Neurochem.* **68,** 1209–1220.
34. Szerb, J.C. and Butterworth, R.F. (1992) Effect of ammonium ions on synaptic transmission in the mammalian central nervous system. *Progr. Neurobiol.* **39,** 135–153.
35. Sugimoto, H., Koehler, R.C., Wilson, D.A., Brusilow, S.W., and Traystman, R.J. (1997) Methionine sulfoximine, a glutamine synthetase inhibitor, attenuates increased extracellular potassium activity during acute hyperammonemia. *J. Cereb. Blood Flow Metab.* **17,** 44–49.

36. Brooks, K.J., Kauppinen, R.A., Williams, S.R., Bachelard, H.S., Bates, T.E., and Gadian, D.G. (1989) Ammonia causes a drop in intracellular pH in metabolizing cortical brain slices. A [^{31}P]- and [^{1}H] nuclear magnetic resonance study. *Neuroscience* **33,** 185–192.
37. Allert, N., Koller, H., and Siebler, M. (1998) Ammonia-induced depolarization of cultured rat cortical astrocytes. *Brain Res.* **782,** 261–270.
38. Benjamin, A.M., Okamoto, K., and Quastel, J.H. (1978) Effects of ammonium ions on spontaneous action potentials and on contents of sodium, potassium, ammonium and chloride ions in brain *in vitro. J. Neurochem.* **30,** 131–143.
39. Pike, C.J., Cummings, B.J., and Cotman, C.W. (1995) Early association of reactive astrocytes with senile plaques in Alzheimer's disease. *Exp. Neurol.* **132,** 172–179.
40. Lewis, D.V., Mutsuga, N., Schuette, W.H., and Van Buren, J. (1977) Potassium clearence and reactive gliosis in the alumina gel lesion. *Epilepsia* **18,** 499–506.

V

Astrocytes and the Blood-Brain Barrier in Aging

16
Molecular Anatomy of the Blood-Brain Barrier in Development and Aging

Dorothee Krause, Pedro M. Faustmann and Rolf Dermietzel

1. INTRODUCTION

The blood-brain barrier (BBB) is characterized by bidirectional exchange mechanisms which occur at the morphological interfaces separating the intravascular compartment from the brain parenchyma.

The cerebral endothelium plays a crucial role in maintaining the BBB properties. The morphological equivalents of the endothelial barrier are represented by two structural components: nonfenestrated endothelial cells, which are sealed by impermeable tight junctions *(1,2)* and a low appearance of endocytotic vesicles within the endothelium. In addition to the cerebral endothelium further structural components serve a function as a secondary barrier:

1. Pericytes, which cover the endothelial cells separated by a perivascular basement membrane from the endothelium.
2. The basement membrane itself, which surrounds both, the pericytes and the endothelial cells. The basement membrane plays a protective role and functions as an electrostatic selective filter for charged macromolecules *(3)*.
3. Astroglial endfeet, which are actively involved in maintaining the BBB properties.

In addition to the structural components, various enzymes, transporters, and receptors within the cerebral microvasculatur provide the biochemical and metabolic constituents of the BBB, which regulate the transport of various biological substrates, such as ions, nucleotides, amino acids, glucose, peptides, and hormones *(4)*.

2. DEVELOPMENT AND AGING IS ASSOCIATED WITH CHANGES IN THE BBB PHENOTYPE

In fetuses and newborns the BBB is immature, which means that it lacks structural and functional properties prevalent in the mature brain.

At embryonic day 11 (Ell) of rat, the first blood vessels invade the brain anlagen from the perineural plexus and at embryonic day 12–14 (E12–14) the first intracerebral blood vessels become visible. At this time perivascular glia is undifferentiated. At embryonic day 15 the closure of the BBB tight junctions occurs and thereafter, intracerebral angiogenesis proceeds within the differentiating brain tissue concurrent with the increase of volume of brain parenchyma.

From: *Neuroglia in the Aging Brain*
Edited by: Jean S. de Vellis © Humana Press Inc., Totowa, NJ

The formation of endothelial tubes is accompanied by association of locally-derived mesenchymal cells that are prone to either differentiate into pericytes (in capillaries and postcapillary venules) or smooth muscle cells (in larger vessels).

The ensheathment of cerebral microvessels by perivascular astrocytic endfeet is a late event in BBB development occuring postnatally around week 2–3 *(5)*, followed by completion of the basement membrane (week 3–4 postnatally).

This brief description of the schedule of BBB differentiation shows that maturation of the BBB is a sequential process that follows a well-defined morphogenetic program.

Our review is aimed to summarize relevent data on the morphology and function of the BBB in the developing and aging brain.

3. AGE-RELATED STRUCTURAL CHANGES OF THE BLOOD-BRAIN BARRIER (BBB)

Age-related changes of structural components of the BBB include alterations of the microvascular and perivascular complexes of the BBB (for extensive reviews *see 6,7,8,9*).

Throughout aging and senescence the number of endothelial cells is steadily reduced, the endothelium is elongated *(10,11)*, and the thickness of the subendothelial basement membrane is increased *(12)*. The development of tight junctions can be monitored by the expression of tight junction associated protein (ZO-1). ZO-1 is a peripheral membrane protein and is concentrated at the cytoplasmatic surface of tight junctions *(13)*. It is a reliable indicator for the presence of tight junctions between cerebral endothelial cells *(14)*. Occludin, an integral membrane protein, has been identified as a further component of tight junction strands *(15)*. The expression of occludin is developmentally regulated. It is low in rat brain endothelium at postnatal day 8 but clearly detectable in adults *(16)*. Recently, the claudin gene family was described to be directly involved in the barrier function of tight junctions *(17)* but no data on its expression in cerebral blood vessels have been published thus far.

An interesting aspect of BBB maturation is that the susceptibility of the BBB to pathologic agents differs with age. For instance, susceptibility to mumps-associated hydrocephalus is age-dependent and considered to be related to the maturity of the BBB. In hamster 2-d-of-age occurrence of pathologic signs are associated with the immaturity and fragility of endothelial tight junctions as evidenced by the determination of endothelial ZO-1 immunoreactivity *(18)*.

In addition, besides changes in the classical features of the BBB complex (i.e., endothelial tightness and integrity of the metabolic complement) alterations of secondary structures like pericytic engulfment and the organization of perivascular astrocytes are essential for determining BBB integrity.

During aging the number of pericytes is increased and the cells show signs of degeneration with a high rate of inclusion bodies and vacuoles *(19,20)*.

Concomitant, the number of astrocytes is increased *(20)* and the glial fibrillary astrocytic protein (GFAP) reveals enhanced expression *(21)* accompanied by a reduction of polarity of astrocytes *(22)*. The functional meaning of these changes is essentially unknown, but studies on null mutations (i.e., GFAP –/– mice, *(23)*) will help to clarify the role which the diverse structural components play in the concert of BBB preservation.

Table 1
Age-related structural changes of the blood-brain barrier

Component	Age related changes	Species-Region	Reference
Microvascular part of BBB			
Endothelial cell	Number is reduced, cells are elongated	Rat cortex	*(10)*
		Human cortex	*(11)*
Basement membrane	Increased thickness	Rat hippocampus	*(12)*
Pericyte	Degeneration	Rat	*(19)*
	Increased number	Rat	*(20)*
Perivascular part of BBB			
perivascular space	Formation of lacunae	Human	*(25)*
Astrocyte	Increased number	Rat	*(20)*
	Increased GFAP expression	Human	*(21)*
	Reduced polairty	Rat optic nerve	*(22)*
Microglia	Increased number of activated microglia	Human	*(24)*

The microglia shows no changes of its gross morphology. But there were significant age-associated increases in the total numbers of activated microglial cells, as evidenced by interleucin-1 alpha-immunoreactivity *(24)*. A summary of the age-related structural changes of the BBB is given in Table 1. In the context of structural alterations it is important to emphasize that these changes of the BBB in aging are region- and species-specific.

4. AGE-RELATED FUNCTIONAL CHANGES OF THE BBB

4.1. Cerebrospinal Fluid and Serum

By comparing the protein-concentrations in the human cerebrospinal fluid (CSF) and blood serum, age-related changes in the function of the BBB can be assessed. Data from Garton et al. *(25)* and Kleine et al. *(26)* indicate that in human from 12th yr of age to adulthood the CSF/serum ratio for prealbumin, albumin, immunoglobulin G (IgG), and α2-macroglobulin increases. It was taken into account, however, that the age-related changes of the CSF/serum quotients, with exception of α2-macroglobulin, may also result from the decrease of blood serum protein concentrations. Therefore it was suggested that α2-macroglobulin is the most sensitive indicator of age-related changes of BBB permeability *(25)*. In children the CSF-concentration of albumin and IgG decreases from birth until 7-mo-of-age subsequently, thereafter a steady increase can be observed until 3-yr-of-age. Transthyretin and transferrin show an increased CSF/serum ratio until about 3-yr-of-age and decrease from then on *(27)*. These differences in CSF protein concentration may reflect a phasic evolvement of BBB permeability including differential expression of receptor complements. A summary of age-related functional changes of the blood-brain-barrier indicated through CSF/serum analysis is given in Table 2.

Table 2
Functional changes in the BBB with aging. Permeability of the BBB as determined by CSF/serum ratio of individual proteins

Protein	Age related change	Reference
Prealbumin Albumin IgG α2-Macroglobulin	Increased CSF/serum ratio correlate with age (adults aged 18–89 years)	*(25)*
Albumin α2-Macroglobulin	Increased CSF/serum ratio correlate with age (child/adults aged 12–90 years)	*(26)*
Albumin IgG	Decreased CSF concentration from birth until 7 mo age	*(27)*
Albumin α2-Macroglobulin Transthyretin	Increased (slightly) CSF concentration from 7 mo until 3 yr age	*(27)*
IgG	Increased (strongly) CSF concentration from 7 mo until 3 yr age	*(27)*
Transthyretin Transferrin	Increased CSF/serum ratio until about 3 yr and decreased from then on	*(27)*

4.2. Cerebral Endothelium

Substrate movement through the lipid bilayer of the cerebral endothelium is mediated by four mechanisms: Passive diffusion, facilitated diffusion, active transport, and receptor-mediated endocytosis accomplished by transport systems. In the following a brief account of the developmental- and age-related differentiation of these four mechanisms is given.

4.2.1. Transport- and Carrier-Systems

4.2.1.1. Glucose Transporter Most BBB transport systems appear to be operational at birth *(28)*. One specific example is the glucose transport system. The brain endothelial glucose transporter 1 (Glut 1), in its 55 kDa vascular isoform, is a well characterized intrinsic endothelial membrane glycoprotein which mediates the translocation of glucose across the plasma membranes of the cerebral endothelium. The expression of the glucose transport system is developmentally regulated and coincides with the maturation of tightness of cerebral microvessels *(29,30,31)*.

4.2.1.1.1. Expression of Glut 1 in the Immature State During early angiogenesis a phasic expression of glucose transporter was observed in brain. While the neuroepithelium in rats at embryonic day 12 and 13 shows a high Glut 1 concentration, successive vascularization of the brain anlage and development of vascular tightness is accompanied by a significant reduction of Glut 1 expression in neuroepithelial cells and confinement of Glut 1 expression to the cerebral endothelium.

The level of Glut 1 remains unchanged during the first postnatal week but increases during the next two wk in postnatal development to attain adult levels at P21. Interestingly, levels of Glut 1 mRNA remain unchanged *(30)*. However, the length of the poly(A) tail of Glut 1 mRNA, a determinant of mRNA translatability, decreases with age *(32)* indicative for a translational regulation of Glut 1 expression and alteration of Glut 1 protein turnover with age. Transport kinetics in rabbits have been shown to exhibit a lower capacity (V_{max}), but no significant changes in the half-saturation constant (K_m) of the Glut 1 in newborns as compared to the 28 day animals have been found *(33)*. The age related increase in V_{max} is a function of increased numbers of transporters on the endothelial plasma membranes. This period correlates well with the increase in brain activity and metabolism between birth and P21–P30. Furthermore, the temporal profile of Glut 1 expression during the first postnatal week follows the time course of the most active angiogenetic period in rat brain development which occurs between P3 and P5 *(34)*.

These findings indicate that the abundance of Glutl transporter correlates with the development of cerebral blood vessels.

4.2.1.1.2. Expression of Glut 1 in the Mature State and Senescence The amount of glucose transporter does not change with senescence, whereas the brain uptake of glucose decreases with age. This suggests that the reduced BBB transport of glucose is due to reduction in kinetics and/or affinity of the transport system to glucose rather than to age-related endothelial cell drop-outs *(35)*.

4.2.1.3. TRANSPORTERS FOR OTHER SUBSTRATES Nine separate transport systems exist in cerebral capillaries for amino acid transport *(36)*.

The cerebral endothelium is further furnished with a set of transporter systems for ions such as sodium, potassium and phosphate, for cytokines, transferrin, several neurotransmitters and opiate peptides, as well as hormones *(37)*. The brain substitutes its iron through transcytosis of iron-loaded transferrin across the brain microvessels. The transferrin receptor is a well characterized receptor which is localized at the luminal and abluminal plasma membrane of the cerebral endothelium *(38)*. Immunocytochemical localization of transferrin and its receptor in the developing chicken brain shows immunoreactive neurons at embryonic day 10. Thereafter the immunoreactivity of the neuronal transferrin receptor becomes weaker around day 11–16 whereas the BBB vessels show increased positivity *(39)*. In the adult brain neurons are negative for transferrin and its receptor *(40)*. This schedule is similar to the phasic expression of the Glut 1 transporter system and speaks in favor of a general morphogenetic program that governs transporter expression during BBB maturation.

P-glycoprotein (PGP) is a product of the multidrug resistance 1 (mdr 1) gene which is believed to act as an active efflux system for endothelial cells. It effectively blocks the penetration of lipid-soluble drugs into the brain *(41)*. PGP is highly enriched in adult brain vessels. Mice exhibiting a null mutation of the mdr 1 gene (–/–) show an increase of BBB permeability *(42)*. In rat brain PGP shows a late onset of expression. In prenatal brains and in early postnatal stages of development PGP is undetectable. It is first detectable around P 7 and increases gradually to reach plateau levels *(42)*. The lack of PGP in the embryonic brain is of considerable clinical importance, explicitely when the intake of drugs which bypass the maternal-embryonic placenta barrier is considered. Apparently, the embryo and developing fetus is unable to eliminate cytotoxic drugs and to prevent them from entering the cerebral parenchyma.

4.2.2. Enzymes

One example of a blood-brain barrier specific enzyme is y-glutamyl-transpeptidase (y-GT) which serves a major function in the transport system for large neutral amino acids in cerebral endothelium. The presence of astroglial cells is required for the expression of y-GT in cerebral endothelial cells. Around the time of BBB closure y-GT appears in rat brain capillaries *(43)*. The first postnatal period (day 2–12) is characterized by a significant decrease in y-GT activity. During the second phase of postnatal period (day 12–21) a fast increase in y-GT activity can be measured. Between postnatal day 21 and 60 the enzyme activity acquires the adult level *(44)*.

Several peptidases control the exchange of neuropeptides at the blood-brain or the brain-blood interfaces. A similar time course as described for postnatal activity of y-GT is observed for aminopeptidase A, whereas alkaline phosphatase, a nonspecific phosphomonoesterase, dipeptidyl peptidase and aminopeptidase N increase steadily during the first 2 wk of postnatal development *(45)*.

Monoamine oxidase (MAO) impedes the traffic of monoamine precursors into the brain by enzymatic breakdown of biogenic amines, thus preventing uptake of monoaminergic peripheral neurotransmitters into the interstitial cerebral environment. MAO is associated with endothelial cells. MAO activity also increases postnatally and reaches a peak at 3-wk-of-age *(46)*. In aged rats alkaline phosphatase, monoaminoxidase B, and tau-glutamyltranspeptidase increase. On the contrary, lactate dehydrogenase activity and monoamine oxidase A decrease with age. No noticeable changes can be seen in acetyl-cholinesterase activity during aging.

A general concept of the pattern of enzyme activity at the BBB interface with respect to age-related changes cannot be drawn, in particular if interspecies differences are taken into account.

4.2.3. Receptors

A number of biologically active proteins, including immunoproteins such as IgG, hormones like insulin and N-tyrosinated peptides are actively transcytosed through cerebral endothelium via receptor/carrier mechanisms *(47)*.

Insulin for example has limited access to the adult brain but has increased availability to the newborn brain where it may function as a neonatal brain growth promotor *(48)*. The binding site for insulin at the BBB is the insulin receptor on the capillary endothelium. Whether the binding capacity of the receptor or the receptor density is responsible for the lower amount of insulin in adult brain is still unknown.

Aging in mammals is characterized by a decline in plasma levels of insulin-like growth factor 1. The cerebral vasculature is an important source of insulin-like growth factor 1 for the brain. With age there are no changes in mRNA expression but the protein level decreases. There is also no change in type 1 insulin-like growth factor receptor mRNA with age but the receptor decrease in several brain regions *(49)*. As already indicated for Glut 1 expression age-related adapation of BBB protein levels seems to be primarily regulated through translational or posttranslational regulation.

4.2.4. Cell Adhesion Molecules

Besides the classical receptor/enzyme complement of the BBB some additional functional moieties like cell adhesion molecules exhibit age-related expression patterns. For instance, the platelet endothelial cell adhesion molecule-1 (PECAM-1) plays an important role in immune responses of the brain. In mouse BBB PECAM-1 is initially upregulated prior to structural maturity of the BBB with an increase up to day 7–10 postpartum and then decreases two weeks after birth. These findings suggest that the development of an "immune" BBB manifests prior to anatomical closure of the BBB *(50).*

The adhesion molecule ICAM-1 shows the same upregulation prior to the structural maturity of the BBB on the luminal and abluminal endothelial surface and a downregulation two weeks after birth *(51).*

An endothelial specific antigen with unknown function is the MECA-32 (mouse endothelial cell antigen). It is expressed during embryogenesis in the mouse brain microvasculature. This antigen is lost during the differentiation of the BBB, but remains present on the brain vasculature that does not exhibit barrier characteristics, i.e., choroid plexus and circumventricular organs. These findings suggest that MECA-32 is involved in maintaining the fenestrated phenotype in leaky brain capillaries *(52).*

4.2.5. Nonendothelial Components Associated with the BBB

The cerebral vascular bed contains additional structural components, which serve complementary functions in maintaining the metabolic BBB complex.

A predominant role among these additive components is played by perivascular cells (pericytes) which are distinctively shaped forming numerous cytoplasmatic processes that encircle endothelial cells. Pericytes are separated from the cerebral endothelium by a common basement membrane. They take up blood-borne substances through phagocytosis *(67)* and they are identified as immunocompetent antigen-presenting cells *(68),* which participate in angiogenesis and are suggested to play a regulatory function in the control of capillary growth *(69).*

Cerebral pericytes constitute a significant secondary element in metabolic BBB functions *(70).* Immunocytochemical studies provide evidence for the existence of a BBB-specific enzymatic machinery in pericytes which exhibits y-glutamyltranspeptidase, alkaline phosphatase, aminopeptidase A, and aminopeptidase N. Pericytes reveal a developmentally regulated pattern of their enzymatic complement similar to the differentiation of endothelial BBB constituents during development. The onset of yGT-expression in embryonic rat brain occurs on E15 accompanied by closure of the BBB whereas the first noticable expression of pericytic aminopeptidase N (pAPN) occurs on day E18. A steady-state level is reached at day 6–8 postnatally *(71).*

In summary, the described developmental changes in BBB phenotype are subject of morphogenetic control mechanisms that regulate the maturation of cerebral blood vessels. These mechanisms are prevalent from early embryonic angiogenesis to postnatal stages and exhibit a phasic schedule. Susceptibility of the developing BBB to pathogenic agents is regarded to be phase specific. Alterations of the aging BBB are manifold including increase in serum proteins as well as changes in the kinetics and affinity of transporters to their substrates. However, the molecular mechanisms which underly the functional deficits of the aging BBB are essentially unknown and remain to be eludicated by a cooperative structural/molecular approach.

Table 3
Summarizes the systems governing the blood-brain exchange and shows the effects of development and aging on the exchange rate

Transport System	Development and age related change	Species	Reference
Transporters			
Glucose Transporter 1	decreased (day 1–14)	Rabbit	*(53)*
	increased (day 14–70)	Rabbit	*(33)*
	increased (day 1–30)	Rat	*(30)*
Glucose Transporter 1 mRNA	unchanged	Rabbit	*(53)*
		Rat	*(32)*
Neutral Amino Acids	unchanged	Human	*(54)*
Tryptophan	decreased	Rat	*(55)*
Arginine	decreased	Rabbit	*(56)*
Adenine	decreased	Rabbit	*(56)*
P-Glycoprotein	increased	Rat	*(42)*
Enzymes			
Adenylate cyclase	decreased	Mouse	*(57)*
Aminopeptidase A	decreased (week 1–2) increased (week 2–8)	Rat	*(45)*
Aminopeptidase M (N)	increased (week 1–8)	Rat	*(45)*
Alkaline Phosphatase	increased (week 1–2)	Rat	*(45)*
	increased	Rat	*(58)*
	increased	Mouse	*(59)*
Dipeptidyl-Aminopeptidase IV	increased (week 1–8)	Rat	*(45)*
Tau-Glutamyl-Transpeptidase	increased	Rat	*(58)*
Gamma-Glutamyl-Transpeptidase	increased (week 1–8) (fourfold)	Rat	*(45)*
	increased (day 10–180) decreased (day 180–360)	Rat	*(59)*
	decreased (day 2–12) increased (day 12–21)	Rat	*(44)*
	increased (day 7–month 12) decreased (month 12–)	Mouse	*(60)*
Lactate Dehydrogenase	decreased	Rat	*(58)*
Monoaminooxidase	increased (day 1–14) decreased (day 14–)	Rat	*(46)*
Monoaminooxidase A	decreased	Rat	*(58)*
Monoaminooxidase B	increased	Rat	*(58)*
Small Peptides	increased	Rat	*(61)*
Choline	decreased	Human	*(62)*
Cholinesterase	increased	Rat	*(63)*
Protein Kinase C β-Isoform	unchanged	Rodent, Rat, Human	*(64)*
	decreased	Human-Alzheimer Disease	*(64)*

(continues)

Table 3
(Continued)

Transport System	Development and age related change	Species	Reference
Transferrin	Shift from neurons to blood vessels	Chicken	*(40)*
Receptors			
β-Adrenergic	decreased (mo 3–24)	Rat	*(57)*
Insulin-like growth factor 1 receptor mRNA	Unchanged	Rat	*(49)*
Insulin-like growth factor 1 Protein concentration	decreased		
Insulin-line growth factor 1 receptor level	decreased (in hippocampus and cortical layers II/III and V/VI		
Antigens			
Neurothelin (HT7)	Shift from neuroblasts to endothelial cells	Chicken	*(65)*
Intercellular-adhesion-molecule (ICAM-1) (CD54)	increased (day 2–7), decreased (day 7–14), no expression (day 14–)	Mouse	*(51)*
Platelet-endothelial-cell-adhesion-molecule (PECAM-1) (CD31)	See ICAM-1	Mouse	*(50)*
(neural)-N-Cadherin	decreased (with the onset of BBB differentiation)	Chicken	*(66)*
Occludin	Low level (day 8), increase (day 70)	Rat	*(16)*
MECA-32	Until E16, decreased (E17–)	Mouse	*(52)*

REFERENCES

1. Reese, T. S. and Karnovsky, M. J. (1967) Fine structural localization of a blood-brain barrier to exogenous peroxidase. *J. Cell Biol.* **34,** 207–217.
2. Dermietzel, R. (1975) Junctions in the central nervous system of the cat. IV. Interendothelial junctions of cerebral blood vessels from selected areas of the brain. *Cell Tissue Res.* **164,** 45–62.
3. Pardridge, W. E. (1986) Receptor-mediated peptide transport through the blood-brain barrier. *Endocr. Rev.* **7,** 314–330.
4. Dermietzel, R. and Krause, D. (1991) Molecular anatomy of the blood-brain barrier as defined by immunohistochemistry. *Int. Rev. Cytol.* **127,** 57–109.
5. Bär, Th. and Wolff, J. R. (1976) Development and adult variations of the wall of brain capillaries in the neocortex of rat and cat. in *The cerebral vessel wall* (Cervos-Navarro, J. et al., eds.), Raven Press, New York, pp. 1–6.
6. Kalaria, R. N. (1996) Cerebral vessels in ageing and Alzheimer's disease. *Pharmacol. Ther.* **72,** 193–214.

7. Shah, G. N. and Mooradian, A. D. (1997) Age-related changes in the blood-brain barrier. *Exp. Gerontol.* **32,** 501–519.
8. De Jong, G. I., De Vos, R. A., Steur, E. N. and Luiten P. G. (1997) Cerebrovascular hypoperfusion: a risk factor for Alzheimer's disease? Animal model and postmortem human studies. *Ann. N. Y. Acad. Sci.* **826,** 56–74.
9. Mrak, R. E., Griffin, S. T. and Graham, D. I. (1997) Aging-associated changes in human brain. *J. Neuropathol. Exp. Neurol.* **56,** 1269–1275.
10. Barr, T. (1978) Morphological evolution of capillaries in different laminae of rat cerebral cortex by autonomic image analysis: Changing during development and aging. *Adv. Neurol.* **20,** 1–9.
11. Hunzicker, O., Al Abdel, S. and Schultz, U. (1979) The aging human cerebral cortex: A steriological characterization of changes in the capillary net. *J. Gerontol.* **34,** 345–350.
12. Topple, A., Fifkova, E., Baumgardner, D. and Cullen-Dockstader, K. (1991) Effect of age on blood vessels and neurovascular appositions in the CA1 region of the hippocampus. *Neurobiol. Aging* **12,** 211–217.
13. Fanning, A. S., Jameson, B. J., Jesaitis, L. A. and Anderson, J. M. (1998) The tight junction protein ZO-1 establishes a link between the transmembrane protein occludin and the actin cytoskeleton. *J. Biol. Chem.* **273,** 29745–29753.
14. Krause, D., Mischeck, U., Galla, H. J. and Dermietzel, R. (1991) Correlation of zonula occludens ZO-1 antigen expression and transendothelial resistance in porcine and rat cultured cerebral endothelial cells. *Neuroscience Lett.* **128,** 301–304.
15. Tsukita, S. and Furuse, M. (1999) Occludin and claudins in tight-junction stands: leading or supporting players? *Trends Cell Biol.* **9,** 268–273.
16. Hirase, T., Staddon, J. M., Saitou, M., Ando-Akatsuka, Y., Itoh, M., Furuse, M., Fujimoto, K., Tsukita, S. and Rubin, L. L. (1997) Occludin as a possible determinant of tight junction permeability in endothelial cells. *J. Cell Sci.* **110,** 1603–1613.
17. Morita, K., Furuse, M., Fujimoto, K. and Tsukita, S. (1999) Claudin multigene family encoding four-transmembrane domain protein components of tight junction stands. *Proc. Natl. Acad. Sci. USA* **96,** 511–516.
18. Uno, M., Takano, T., Yamano, T. and Shimada, M. (1997) Age-dependent susceptibility in mumps-associated hydrocephalus: neuropathologic features and brain barriers. *Acta Neuropathol.* **94,** 207–215.
19. De Jong, G. I., Jansen, A. S., Horvath, E., Gispen, W. H. and Luiten, P. G. (1992) Nimodipine effects on cerebral microvessels and sciatic nerve in aging rats. *Neurobiol. Aging* **13,** 73–81.
20. Peinado, M. A., Quesada, A., Pedrosa, J. A., Torres, M. I., Martinez, M., Esteban, F. J., Del Moral, M. L., Hernandez, R., Rodrigo, J. and Peinado, J. M. (1998) Quantitative and ultrastructural changes in glia and pericytes in the parietal cortex of the aging rat. *Microsc. Res. Techn.* **43,** 34–42.
21. Hansen, L. A., Armstrong, D. M. and Terry, R. D. (1987) An immunohistochemical quantification of fibrous astrocytes in the aging human cerebral cortex. *Neurobiol. Aging* **8,** 1–6.
22. Wolburg, H. and Buerle, C. (1993) Astrocytes in the lamina cribrosa of the rat optic nerve: are their morphological peculiarities involved in an altered blood-barrier? *J. Hirnforsch.* **34,** 445–459.
23. Liedtke, W., Edelmann, W., Bieri, P. L., Chiu, F. C., Cowan, N. J., Kucherlapati, R. and Raine, C. S. (1996) GFAP is necessary for the integrity of CNS white matter architecture and long-term maintenance of myelination. *Neuron* **17,** 607–615.
24. Sheng, J., Mrak, R. E. and Griffin, W. S. T. (1998) Enlarged and phagocytic, but not primed, interleukin-alpha immunoreactive microglia increase with age in normal human brain. *Acta Neuropathol.* **95,** 229–234.
25. Dozono, K., Ishii, N., Nishihara, Y. and Horie, A. (1991) An autopsy study of the incidence of lacunes in relation to age, hypertension, and arteriosclerosis. *Stroke* **22,** 993–996.

25. Garton, M. J., Keir, G., Vijaya Lakshmi, M. and Thompson, E. J. (1991) Age-related changes in cerebrospinal fluid protein concentrations. *J. Neurol. Sci.* **104,** 74–80.
26. Kleine, T. O., Hackler, R. and Zofel, P. (1993) Age-related alterations of the blood-brain-barrier (BBB) permeability to protein molecules of different size. *Z. Gerontol.* **26,** 256–259.
27. Barnard, K., Herold, R., Siemes, H. and Siegert, T. (1998) Quantification of cerebrospinal fluid proteins in children by high-resolution agarose gel electrophoresis. *J. Child. Neurol.* **13,** 51–58.
28. Cornford, E. M. and Cornford, M. E. (1986) Nutrient transport and the blood-brain barrier in developing animals. *Fed. Proc.* **45,** 2065–2072.
29. Dermietzel, R., Krause, D., Kremer, M., Wang, C. and Stevenson, B. (1992) Pattern of the glucose transporter (GLUT1) expression in embryonic brain is related to maturation of blood-brain barrier tightness. *Dev. Dyn.* **193,** 152–163.
30. Vannucci, S. J. (1994) Developmental expression of GLUT1 and GLUT3 glucose transporters in rat brain. *J. Neurochem.* **62,** 240–246.
31. Bolz, S., Farrell, C. L., Dietz, K. and Wolburg, H. (1996) Subcellular distribution of glucose transporter (Glut-1) during development of the blood-brain barrier in rats. *Cell Tissue Res.* **284,** 355–365.
32. Mooradian, A. D. and Shah, G. N. (1997) Age-related changes in glucose transporter-one mRNA structure and function. *Proc. Soc. Exp. Biol. Med.* **216,** 380–385.
33. Cornford, E. M., Hyman, S. and Landaw, E. M. (1994) Developmental modulation of the blood-brain-barrier glucose transport in the rabbit. *Brain Res.* **663,** 7–18.
34. Robertson, P. L., Dubois, M., Bowman, P. D. and Goldstein, G. W. (1985) Angiogenesis in developing rat brain: an in vivo and in vitro study. *Dev. Brain Res.* **23,** 219–223.
35. Mooradian, A. D., Morin, A. M., Cipp, L. J. and Haspel, H. C. (1991) Glucose transport is reduced in the blood-brain barrier of aged rats. *Brain Res.* **551,** 145–149.
36. Smith, Q. R. and Stoll, J. (1999) Molecular characterization of amino acid transporters at the blood-brain barrier, in *Brain Barrier Systems* (Paulson O., Knudsen, G. M., Moos, T. eds.) Alfred Benzon Symposium 45, Munksgaard, Copenhagen, pp. 303–317.
37. Betz, A. L. (1992) An overview of the multiple functions of the blood-brain barrier. *Nida. Res. Monogr.* **120,** 54–72.
38. Pardridge, W. E. (1999) A morphological approach to the analysis of blood-brain barrier transport function. in *Brain Barrier Systems* (Paulson, O., Knudsen, G. M., Moos, T. eds.), Alfred Benzon Symposium 45, Munksgaard, Copenhagen, pp. 19–42.
39. Moos, T., Oates, P. S. and Morgan, E. H. (1998) Expression of the neuronal transferrin receptor is age dependent and susceptible to iron deficiency. *J. Comp. Neurol.* **398,** 420–430.
40. Oh, T. H., Markelonis, G. J., Royal, G. M. and Bregman, B. S. (1986) Immunocytochemical distribution of transferrin and its receptor in the developing chicken nervous system. *Brain Res.* **395,** 207–220.
41. Tatsuta, T., Naito, M., Ohara, T., Sugawara, I. and Tsururuo, T. (1992) Functional involvement of P-glycoprotein in blood-brain barrier. *J. Biol. Chem.* **28,** 20383–20391.
42. Matsuoka, Y., Okazzaki, M., Kitamura, Y. and Taniguchi, T. (1999) Developmental expression of P-glycoprotein (multidrug resistance gene product) in the rat brain. *J. Neurobiol.* **39,** 383–392.
43. Risau, W., Hallman, R. and Albrecht, U. (1986) Differentiation-dependent expression of proteins in brain endothelium during development of the blood-brain barrier. *Dev. Biol.* **117,** 537–545.
44. Budi Santoso, A. W. and Bar, T. (1986) Postnatal development of gamma-GT activity in rat brain microvessels corresponds to capillary growth and differentiation. *Int. J. Dev. Neurosci.* **4,** 503–511.
45. Brust, P., Bech, A., Kretschmar, R. and Bergmann, R. (1994) Developmental changes of enzymes involved in peptide degradation in isolated rat brain microvessels. *Peptides* **15,** 1085–1088.
46. Kalaria, R. N. and Harik, S. I. (1987) Differential postnatal development of monoamine oxidase A and B in the blood-brain barrier of the rat. *J. Neurochem.* **49,** 1589–1594.

47. Poduslo, J. F., Curran, G. L. and Berg, C. T. (1994) Macromolecular permeability across the blood-nerve and blood-brain barriers. *Proc. Natl. Acad. Sci. USA* **91,** 5705–5709
48. Frank, H. J., Jankovic-Vokes, T., Pardridge, W. M. and Morris, W. L. (1985) Enhanced insulin binding to blood-brain barrier in vivo and to brain microvessels in vitro in newborn rabbits. *Diabetes* **34,** 728–733.
49. Sonntag, W. E., Lynch, C. D., Bennett, S. A., Khan, A. S., Thornton, P. L., Cooney, P. T., Ingram, R. L., McShane, T. and Brunso-Bechtold, J. K. (1999) Alterations in insulin-like growth factor-1 gene and protein expression and type 1 insulin-like growth factor receptors in the brains of ageing rats. *Neurosci.* **88,** 269–279.
50. Lossinsky, A. S., Wisniewski, H. M., Dambska, M. and Mossakowski, M. J. (1997) Ultrastructural studies of PECAM-1/CD31 expression in the developing mouse blood-brain barrier with the application of a pre-embedding technique. *Folia Neuropathol.* **35,** 163–170.
51. Lossinsky, A. S. and Wisniewski, H. M. (1998) Immunoultrastructural expression of ICAM-1 and PECAM-1 occurs prior to structural maturaty of the murine blood-brain barrier. *Dev. Neurosci.* **20,** 518–524.
52. Hallmann, R., Mayer, D., Broermann, R., Berg, E. and Butcher, E. C. (1995) Novel endothelial cell marker is suppressed during formation of the blood brain barrier. *Dev. Dyn.* **202,** 325–332.
53. Dwyer, K. J. and Pardridge, W. M. (1993) Developmental modulation of blood-brain barrier and choroid plexus GLUT1 glucose transporter messenger ribonucleic acid and immunoreactive protein in rabbits. *Endocrinology* **132,** 558–565.
54. Mooradian, A. D. (1994) Potential mechanisms of the age-related changes in the blood-brain barrier. *Neurobiol. Aging* **15,** 751–755.
55. Tang, J. P. and Melethil, S. (1995) Effect of aging on the kinetics of blood-brain barrier uptake of tryptophan in rats. *Pharm. Res.* **12,** 1085–1091.
56. Braun, L. D., Cornford, E. M. and Oldendorf, W. H. (1980) Newborn rabbit blood-brain barrier is selectively permeable and differs substantially from the adult. *J. Neurochem.* **34,** 147–152.
57. Mooradian, A. D. and Scarpace, P. J. (1991) Beta-adrenergic receptor activity of cerebral microvessels is reduced in aged rats. *Neurochem. Res.* **16,** 447–451.
58. Agrawal, A., Shukla, R., Tripathi, L. M., Pandey, V. C. and Srimal, R. C. (1996) Permeability function related to cerebral microvessel enzymes during ageing in rats. *Int. J. Dev. Neurosci.* **14,** 87–91.
59. Rao, V. L. and Murhy, C. R. (1991) Variations in the effects of L-methionine-DL-sulfoximine on the activity of cerebral gamma-glutamyltranspeptidase in rats as a function of age. *Neurosci. Lett.* **126,** 13–17.
59. Vorbrodt, A. W., Lossinsky, A. S. and Wisniewski, H. M. (1986) Localization of alkaline phosphatase activity in the endothelia of developing and mature mouse blood-brain barrier. *Dev. Neurosci.* **8,** 1–13.
60. Lisy, V., Stastny, F. and Lodin, Z. (1983) Gamma-glutamyl transpeptidase activity in the isolated cells and in various regions of the mouse brain during early development and aging. *Physiol. Bohemoslov* **32,** 1–9.
61. Banks, W. A. and Kastin, A. J. (1985) Aging and the blood-brain barrier: changes in the carrier-mediated transport of peptides in rats. *Neurosci. Lett.* **61,** 171–175.
62. Cohen, B. M., Renshaw, P. F., Stoll, A. L., Wurtman, R. J., Yurgelun-Todd, D. and Babb, S. M. (1995) Decreased brain choline uptake in older adults. An in vivo proton magnetic resonance spectroscopy study. *JAMA* **271,** 902–907.
63. Panula, P. and Rechardt, L. (1978) Age-dependent increase in the non-specific cholinesterase activity of the capillaries in the rat neostriatum. *Histochemistry* **55,** 49–54.
64. Grammas, P., Moore, P., Botchelet, T., Hanson-Painton, O., Cooper, D. R., Ball, M. J. and Roher, A. (1995) Cerebral microvessels in Alzheimer's have reduced protein kinase C activity. *Neurobiol. Aging* **16,** 563–569.

65. Schlosshauer, B., Bauch, H. and Frank, R. (1995) Neutothelin: amino acid sequence, cell surface dynamics and actin cololacization. *Eur. J. Cell Biol.* **68,** 159–166.
66. Gerhardt, H., Liebner, S., Redies and C., Wolburg, H. (1999) N-cadherin expression in endothelial cells during early angiogenesis in the eye and brain of the chicken: relation to blood-retina and blood-brain barrier development. *Eur. J. Neurosci.* **11,** 1191–1201.
67. Broadwell, R. D., Balin, B. J. and Salcman, M. (1988) Transcytosis of blood-borne protein through the blood-brain barrier. *Proc. Natl. Acad. Sci. U.S.A.* **85,** 632–636.
68. Hickey, W. F. and Kimura, H. (1988) Perivascular microglial cells of the CNS are bone marrow-derived and present antigen in vivo. *Science* **239,** 290–292
69. Hirschi, K. K. and D'Amore, P. A. (1996) Pericytes in the microvasculature. *Cardiovasc. Res.* **32,** 687–698.
70. Kunz, J., Krause, D., Kremer, M. and Dermietzel, R. (1994) The 140-kDa protein of blood-brain barrier-associated pericytes is identical to aminopeptidase N. *J. Neurochem.* **62,** 2375–2386.
71. Krause, D., Vatter, B. and Dermietzel, R. (1988) Immunochemical and immunocytochemical characterization of a novel monoclonal antibody recognizing a 140 kDa protein in cerebral pericytes of the rat. *Cell Tissue Res.* **252,** 543–555.

17
The Blood-Brain Barrier in the Aging Brain

Gesa Rascher and Hartwig Wolburg

1. INTRODUCTION

Homeostasis of the extracellular microenvironment in the neural tissue of the brain as well as its protection against neurotoxic compounds and variations in the composition of the blood are important for normal function of the neurons. It is warranted by a structure formed between blood and brain which is therefore called the blood-brain barrier (BBB). This barrier has been postulated in earlier decades by experiments using dyes which directly visualized both the protection of the brain if injected into the vasculature, and free access of the brain if injected into the cerebrospinal fluid. Since in the first experiment some areas of the brain around the ventricle were stained and in the second experiment the identical areas were not, the barrier necessarily had to be separated into a BBB and a blood-cerebrospinal fluid barrier.

In most vertebrates, the BBB is located in the brain capillary endothelial cells (*1;* for a review, *see* ref. *2*). Where the endothelial BBB is leaky in order to gain access for neurosecretory neurons to the vasculature, a barrier has to be formed to avoid free substance exchange between blood and cerebrospinal fluid. These barrier cells in the so-called circumventricular organs are glial tanycytes, which are in line with the periventricular ependymal cells *(3)*. As well, the plexus epithelial cells of the choroid plexus are destined to produce the cerebrospinal fluid from the blood. The blood vessels in these areas are fenestrated *(4)*, and the fenestration is now known to be induced by the vascular permeability factor (VPF) which is identical with the vascular endothelial growth factor (VEGF) secreted by, for example, the choroid plexus epithelial cells *(5)*. At the surface of the brain, there is a third type of barrier, the outer blood-cerebrospinal fluid barrier which protects the brain against blood-borne influences from leaky vessels of the dura mater. This outer blood-cerebrospinal fluid barrier is located in the meningeal cells *(6, 7)*.

The endothelial, glial, and meningeal subtypes of the plasma-interstitial fluid barrier, which together form the BBB in a wider sense, have a common feature: A dense meshwork of intramembranous strands occluding the intercellular cleft. These tight junctions (TJs), together with a low number of endocytotic vesicles, are generally believed to represent the basis of low transcellular permeability *(1,8)*. If considering the BBB properties of the aging brain we have to look both at those structures which are responsible for the low permeability and at those cells which are suggested to

From: *Neuroglia in the Aging Brain*
Edited by: Jean S. de Vellis © Humana Press Inc., Totowa, NJ

induce it. In a first part, we will briefly describe BBB properties of the normal brain vasculature including endothelial TJs, receptors, transporters, and enzymes as well as the capillary-associated cells,and in a second part, we will summarize alterations of the BBB during aging.

2. PROPERTIES OF THE BLOOD-BRAIN BARRIER

2.1 *The Endothelial Barrier*

Mature BBB capillaries in the mammalian brain are mainly characterized by the small height of endothelial cells, the complex interendothelial TJs, the small number of caveolae at the luminal surface of the cell *(4),* and the high number of endothelial mitochondria *(9)*. In addition, subendothelial pericytes which are completely surrounded by a basal lamina, phagocytic perivascular cells and astrocytic processes belong to the set of elements directly adjacent to the cerebral vasculature *(10,11)*

The microvascular endothelial cells are doubtless most important in the determination of the BBB-related permeability. From transplantation experiments it became evident that BBB characteristics are determined to a great extent by intrinsic factors of the brain (as summarized in ref. *12*). Initially, proliferating and migrating endothelial cells are suggested to be committed to express the BBB phenotype by the interaction with neuroectodermal cells *(13)*. The expression of N-cadherin on endothelial cells invading the neuroectoderm has recently been proposed to represent an initial and transient signal involved in this commitment *(14)*. Consecutively, astrocytes would stabilize and maintain the acquired barrier properties. This proposal of BBB development is supported by in vitro-experiments showing that factors released from astrocytes seem to be necessary to induce BBB characteristics *(15)*. However, they do not seem to do so in a quantitatively sufficient manner *(16,17)*. Nature and mechanism of the putative BBB-inducing factors remain elusive.

The morphology of TJs as the elements restricting the paracellular permeability still is most suitably investigated by means of the freeze-fracture technique. In epithelia, the TJ strands are almost completely associated with the protoplasmatic fracture face (P-face). On the external fracture face (E-face), the grooves, which are mainly free of particles, may represent imprints caused by the P-face-associated strands. It is conspicuous that in rat brain endothelium *in situ,* the P-face association of TJ particles is significantly lower than in epithelia and higher than in non-BBB endothelia *(15)*. Since in cultured BBB endothelial cells both the electrical resistance and the P-face association of the tight junction particles are reduced, it was speculated that the extent of the P-face association may inversely correlate with the transendothelial permeability *(17,18)*.

Regarding the molecular composition of TJs, there are two competing hypotheses: the lipid model proposes inverse lipidic micelles to represent the TJ strands *(19,20)* and implies the continuity of the external leaflets of two adjacent cells. The protein model suggests the insertion of proteins into the membrane occluding the intercellular cleft *(21)*. Indeed, four-transmembrane domain protein components of the TJs, occludin and the claudins, were identified and characterized so far *(18,22–25)*. Transfected cells expressing claudin-1 form TJs with a high P-face association resembling those in epithelial cells, whereas transfectants expressing claudin-2 develop TJs with E-face-associated chains of particles resembling TJs in peripheral endothelial cells or BBB endothelial cells in vitro *(23)*. In the central nervous system (CNS), the choroid plexus was the only site where claudin-2 was detected *(26)*. The claudins found to be expressed in the CNS thus were claudin-1, claudin-2, claudin-5 and claudin-11. Claudin-11 was identified as the oligodendroglia-specific protein OSP *(27)*; consequently, the claudins 1 and 5 are specific for brain endothelial cells. Indeed, Morita et al. *(28)* were able to characterize claudin-5 as endothelial-specific. Furthermore, the authors transfected fibroblasts with the cDNA of claudin-5 and found that the particles of the TJs that were formed in these transfected cells were associated with the E-face. Since the BBB endothelial cells show variable degrees of the association with the P- or E-face under different conditions, we are tempted to speculate that the permeability of the BBB TJs may be regulated via the stochiometry of endothelial claudins 1 and 5. This view is supported by the observation that in human astrocytomas the BBB is dysregulated, the endothelial TJs are predominantly associated with the E-face, and the expression of claudin-1 is downregulated *(29)*.

In the last years, several TJ-associated proteins were identified such as ZO-1, ZO-2, ZO-3, 7H6 and cingulin. All these proteins are under the control of different kinases, and their phosphorylation state correlates with barrier function (as summarized in ref. *15*). In addition, there exists an interplay between tight and adhesion junction components which is not well understood so far.

The TJ-based paracellular impermeability of the brain capillary endothelial cells implicates that hydrophilic substances that are essentially needed for the metabolism of the brain require transport through the endothelial wall. Several independent carrier systems for the transport of hexoses (glucose, galactose), neutral, basic and acidic amino acids, monocarboxylic acids (lactate, pyruvate, ketone bodies), purines (adenine, guanine), nucleosides (adenosine, guanosine, uridine), amines (choline), and ions have been described (for details, *see 12,30*). Among these transporters, the glucose transporter is of special importance due to the fact that glucose is the main energy source of the brain. The 55kDa form of the glucose transporter isoform Glut1 as one of five members of a supergene familiy of sodium-independent glucose transporters is highly restricted to the capillary endothelial cells in the brain *(31)*. This restriction to endothelial cells *(32)* and the asymmetrical distribution in the endothelial cell membranes (the density of transporter molecules is four times higher in the ablumenal than in the lumenal membrane) are believed to reflect the onset of BBB function during development *(33,34)*.

In addition to the carrier-mediated transport from blood to brain, the receptor-mediated transport has also been investigated intensively. In BBB endothelial cells, receptors for the endocytosis of transferrin, LDL, immunoglobulin G, insulin and insulin-like growth factor have been described *(12,35–38)*. Furthermore, a large number of receptors which are by no means specific for brain endothelial cells (for example for adrenalin [α and β], histamin, endothelin, adenosine as well as for angiogenic growth factors such as fibroblast growth factor, vascular endothelial growth factor or angiopoietin) has been described to participate partly in the regulation of BBB-specific enzymes *(30)* or to be responsible for the induction of angiogenesis *(39–41)*. Enzymes associated with the BBB such as the adenylate cyclase, the alkaline phosphatase, the nonspecific phosphomonoesterase, the Na^+/K^+-ATPase, the NO-synthase, the guanylate cyclase or the cholinesterase have been suggested to play an important role in the regulation of transendothelial permeability or glio-vascular interrelationships *(30,42)*, whereas the DOPA-decarboxylase and the monoaminooxidase are connected to the so-called metabolic BBB which restricts the transfer of L-DOPA between brain and blood by enzymatic conversion within the endothelial cell *(43)*. Other enzymes such as the γ-glutamyl-transpeptidase and antioxidant enzymes (glutathione peroxidase, glutathione reductase, catalase, and superoxid dismutase) are involved in the transfer of γ-glutamyl residues of glutathione to amino acids or the cleavage of the reduced form of glutathione *(44)* and the protection of the BBB against free radical peroxidation *(45)*. Some of these enzyme activities decline during aging (*see* **Subheading 3.1.**).

There are still other antigens associated with the BBB. For example, in order to restrict free diffusion also of lipophilic compounds from blood to brain, a control mechanism is required to export them effectively out of the brain endothelium. In this context, the P-glycoprotein which is a transmembrane protein that confers multidrug resistance is of special importance. P-glycoprotein *(46)* as well as the multidrug resistance-associated protein Mrp1 *(47)* have been identified in the brain endothelial cell membranes. Moreover, an important class of surface molecules mediating the interaction between blood cells and endothelium represent the adhesion molecules. Whereas the function of neurothelin/HT7, a member of the immunoglobulin superfamily *(48,49)*, is still unclear, the expression and upregulation of the intercellular and vascular cell adhesion molecules (ICAM-1 and VCAM-1, respectively) have been identified as essential steps for lymphocyte recruitment during inflammation *(50,51)*.

2.2. Interactions Between Endothelial Cells and Glial Cells or Pericytes

The extracellular space between the vascular complex and the astrocytes is extremely narrow *(52)*, and the membrane of astrocytes directly adjacent to the blood vessel is crowded with a high density of so-called orthogonal arrays of particles *(53)*. Recently, it was found that the water channel-protein aquaporin-4 is a principal constitutent of these arrays *(54,55)*. A typical feature of astrocytes *in situ* is the high polarity of astrocytes: It is characterized by the polar distribution of both OAP and aquaporin-4 and has been proposed to be an essential constituent of the BBB in vivo *(56)*.

The endothelial basal lamina is believed to play a crucial role for the maintenance of the BBB. The composition of the basal lamina includes many compounds of the extracellular matrix (ECM) such as fibronectin, laminin, type IV collagen and glycosaminoglycans. Since the ECM of cerebral microvascular vessels changes dramatically during aging, it is important to test the influence of single ECM-components on the BBB directly. Nevertheless, detailed information about that are still lacking so far. Agrin which is a heparan sulfate proteoglycan originally discovered at the neuromuscular junction and more recently shown to be committed to multiple functions in the CNS (survey in ref. *57*,) was found to be accumulated in the brain microvascular basal lamina. Because of the parallel onset of agrin expression and BBB development, and the strict localization with BBB endothelial cells, it was suggested to be important for its formation and maintenance of the barrier *(58)*.

The role of other vessel-associated cells such as pericytes, perivascular cells or microglial cells in the formation and maintenance of the BBB is unknown or still circumstantial. The pericytes seem to control the microvasculature by expressing recep-

Table 1
Age-related changes in permeability

	Human	Monkey, Rat, Mouse
Changes in permeability in aging	*(111,112)* ↔	*(99,108,114)*↔ *(108) (in gray matter);* ↑ *(in white matter)*
Changes in permeability in aging- associated diseases	*(112,113)* *Alzheimer's disease* ↑ *Vascular dementia* ↑↑	*Hypertension:* *(99,115,116)*↑ *Ischemia:* *(66,117)*↑

↔ no change; ↑, increase; ↓, decrease;

tors for catecholamines, endothelin-1, vasoactive intestinal peptide, vasopressin and angiotensin, and they may play an important role in the process of inflammation *(11)*. BBB pericytes have been regarded as a second line of defense *(59)* which is reflected under pathological conditions by the increase of pinocytosis and of enzyme activities which normally are characteristic of endothelial cells *(11)*. The pericyte:endothelial cell ratio is four times higher in the retina than in the brain *(60)*, but generally higher in the CNS when compared with other organs *(11)*. In platelet-derived growth factor (PDGF)-B-deficient mice, Lindahl et al. *(61)* reported the loss of pericytes and the formation of microaneurysms. This can be explained by PDGF-receptors expressed by pericytes. If the close interplay between pericytes and endothelial cells which is—for example—represented by the expression of angiopoietin-1 by pericytes and its receptor tie-2 by endothelial cells, is disturbed by tie-2- or angiopoietin-1-null mutation, the mice fail to develop a functional integrity of blood vessels *(62,63)*.

3. ALTERATIONS OF THE BLOOD-BRAIN BARRIER DURING AGING

After the description of the morphological, molecular, and functional properties of the BBB of the adult brain, we now will have a look at the alterations taking place during aging. First of all, we have to state that the permeability of the BBB in the intact aged brain seems not to be increased. However, several authors claimed that the susceptibility or vulnerability of the BBB to factors compromising the cellular integrity of the vascular wall is increased during aging. For example, Sankar et al. *(64)* could show that amphetamine induces an increase in permeability to albumin in aging rats compared to younger control animals suggesting an age-related deficiency to compensate environmental stimuli. Under pathological conditions, such as hypertension, ischemia, Alzheimer's disease, an increase of the BBB permeability has been described which could lead to severe brain edema (for literature, see Table 1). The BBB breakdown may be due to a higher rate of transcytotic vesicular transport of high molecular weight proteins *(65)* which could be induced by the release of serotonin from blood platelets (*(66)*; compare also the serotonin-dependent extravasation of ferritin across venules via formation of vesiculo-vacuolar organelles, as described by Feng et al. *(67)*). A destruc-

Table 2
Age-related changes in CNS microvasculature

Components	Human	Monkey	Rat
Cross-sectional area of capillary wall	*(118)* ↑	*(76)* ↓	*Hippocampus:* *(100)* ↓ *Cortex:* *(99,100)* ↓ *(120)* ↓ *(76)* ↔
Number of endothelial cells	*(118)* ↓	*(94)* ↓	*(120)* ↓
Numner of mitochondria	*(68)* ↔	*(76,94)* ↓	*(120)* ↔ *(121)* ↑
Gliofibrillary proliferation	*(119)* ↑	NR	NR
Thickness of basement membrane	*(119)* ↑	*(76)* ↑	*(74,92)* *(76,94,100,103, 104)* ↑
Changes in number of pericytes	*(68)* ↓	*(98)* ↔	*(74,104)* ↑ *(103)* ↓

↔ no changes; ↑, increase; ↓, decrease; NR, not reported

tion of TJs could neither be observed nor ruled out. Thus, an increased BBB permeability is not a necessary consequence of aging, but seems to reflect a restricted ability to react on unfavorable conditions such as inflammation, hypertension or high concentration of any neurotoxic compounds in the blood. There are many details known about alterations of structures or enzymatic activities in the different cells of the BBB in the aged brain *(30,68–74)* but essentially less is known about which change is causally connected to which functional inferiority. Investigations have been carried out not only in normally aged animals or human beings, but frequently in the senescence-accelerated mouse strain (SAM) which was inaugurated as a model system by Takeda et al. (*75;* for a recent review, *see* ref. 73).

3.1. Age-Related Alterations of the Endothelial Cells

There exists an enormous body of literature on age-related morphological changes of the CNS microvasculature including perivascular cells. Briefly, most authors decribed changes in the cross-sectional area of the capillary wall, the intercapillary distance, the number of endothelial cells, the number and density of mitochondria per cell, and the thickness of the subendothelial basal lamina *(68,71,76–78). see* Table 2 and Fig. 1. That the information about these parameters frequently are controversial, may be due to differences among the species (mostly human, monkey, and rat), the neuroanatomical site looked at, in whether venules, arterioles or capillaries were considered, and finally in the methodology. The mostly described morphological age-related change in the CNS vasculature concerns the thickening of the subendothelial basal

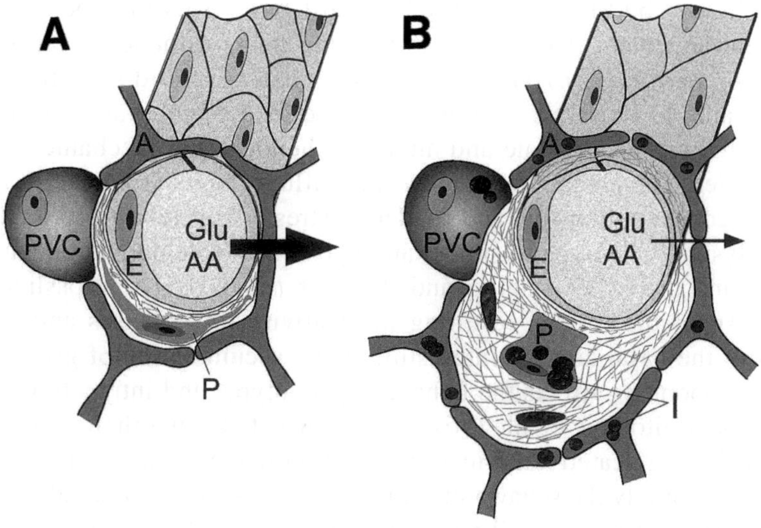

Fig. 1. Schematic view of a normal **(A)** and a brain capillary showing age-related changes **(B).** The most conspicuous alterations are the attenuation and elongation of the endothelial cells (E), the tickening of the basal lamina, the vacuolization of pericytes (P), perivascular cells (PVC) and astrocytic endfeet (A) and the appearance of inclusion bodies (I), and the decline of the transendothelial glucose (Glu) and amino acid (AA) transport (thin vs thick arrows; *see* ref. *(30)* 1997). Changes of the content of endothelial mitochondria was not referred to because of controversial data in the literature (*see* Table 2).

lamina (Table 2, Fig. 1). This increment might have consequences for the barrier regulation of endothelial cells. For example, Tilling et al. *(79)* reported on a decrease of the transendothelial permeability in cultured endothelial cells when grown on laminin, type IV collagen or fibronectin, when compared to rat tail collagen. The actual composition of the ECM, in particular the basal lamina, is dependent on matrix metalloproteinases, serine proteinases and their inhibitors which are normally present in the brain and differentially activated. MunBryce and Rosenberg *(80)* were able to show that after an intracerebral injection of type IV collagenase the vascular basal lamina was disrupted and the BBB opened. A growing body of literature presently describes the involvement of matrix metalloproteinases in the disruption of the BBB *(81)* suggesting that the degradation of the basal lamina and a breakdown of the BBB are causally related processes. If so, integrins could play a major role in the signaling pathway from the ECM components to the mechanisms which are involved in triggering the BBB permeability *(82)*.

The effect of basal lamina thickening on endothelial permeability has been studied in organs other than the brain, but it cannot be excluded that in BBB endothelial cells, too, these results may be valid. For example, the nonenzymatic glycation of collagen probably is in part responsible for the age-related thickening of basal lamina and far-reaching pathological effects on the microvasculature. The first step of this reaction is the nonenzymatic condensation of glucose and the epsilon amino group of lysine, which is followed by the formation of a Schiff base. This may be stabilized by an Amadori rearrangement from which in turn the nonenzymatic glycation is initiated (for a review, *see* ref. *(83)*.) The resulting so-called advanced glycation endproducts (AGEs) are able to crosslink proteins. Because of the long half-life of many ECM proteins, AGEs can accumulate and alter the chemical and mechanical properties of these molecules. The glycation pathways are influenced by factors such as phosphate and glucose concentration, pH and oxidative stress. Well investigated AGEs are carboximethyllysine (CLM), pentosidine, and pyrraline. Protein glycation and crosslinking of proteins increase in aging and diabetes *(83–85)*. The crosslinked collagen might be responsible for the resulting insensitivity to proteases and the following thickening of the basal lamina. Interestingly, the accumulation of glycation products on matrix-components affects the behavior, phenotype, and intracellular signaling of resident cells. Haitoglou et al. *(86)* have shown that endothelial cells, grown on nonenzymatically glycated laminin and type IV collagen, reduced their spreading and adhesion. Accordingly, Hasegawa et al. *(87)* observed in endothelial cells, plated on glycated basement membrane-like matrix, both a reduced intracellular tyrosine phosphorylation of paxillin, a member of focal adhesion proteins, and a distribution of paxillin and actin filaments which were different to control cells. We now know that the matrix glycation affects the cells via receptors which bind AGEs and were identified by Vlassara et al. *(88)*. This AGE-receptor is distinct from the receptor for AGE that was detected by the Stern group and called RAGE *(89,90)*. Ligand binding by RAGE resulted in the induction of oxidant stress in endothelial cells and in increased permeability *(91)*. The activation of these receptors can also result in the synthesis and release of cytokines such as IL-1 and TNA-α *(88)*. However, it should be held in mind that most of these results were gained by investigating other than BBB endothelial cells. Nevertheless, Mooradian and Meredith *(92)* reported that the content of AGEs, namely pentosidine, in cerebral microvessels did not change with age. Therefore, the correlation of AGE production with thickening of the basal lamina and the increased permeability of the BBB should carefully be examined. Handa et al. *(84)* showed that the AGE pentosidine is effective in inducing the platelet-derived growth factor PDGF in retinal pigment epithelial cells in vitro. The authors speculate that AGE-induced changes in Bruch's membrane could alter the outer blood retinal barrier function. The β-amyloid protein which is deposited during normal aging around brain microvessels and increasingly in Alzheimer's disease has been found to bind to RAGE and to be involved in transendothelial migration of monocytes *(93)*. Thus, the roles of glycation products and RAGE for the maintenance of the integrity of brain microvessels seem to be more relevant than thought previously, and the investigation of their pathological consequences in the context of BBB aging will be a challenge for the future.

The attenuation of endothelial cells is another parameter frequently described in the literature on aging. In an attenuated endothelial cell, the mitochondrial distribution and density obviously change, thus rendering it difficult to determine the true shift in the number of mitochondria per cell. Nevertheless, it seems generally accepted that mitochondrial content in brain capillary endothelial cells declines during aging *(71,76,94)*. Possibly, this decline of mitochondrial content is causally related to an increased risk of the cells to be impaired by free radicals and lipid peroxides. Mooradian and Uko-eninn *(95)* reported on an increase of lipid peroxidation by-products, such as malondialdehyde, in cerebral microvessels as a consequence of an age-related decrease of the antioxidative capacity. On the other side, Tayarani et al. *(96)* investigated antioxidative enzymes in aging brain capillaries and did not find a consistent decline of glutathione peroxidase and reductase activities. The activity of the superoxide dismutase was even increased with age; only the activity of catalase was reduced.

3.2. *Age-Related Alterations of the Pericytes*

During aging, pericytes show ultrastructural changes such as vesicular and lipofuscin-like inclusions *(74,97–99);* (Fig. 1), an increased size of mitochondria *(100)* and a foamy transformation *(101)*. Ueno et al. *(102)* reported on membranous inclusions within the basal lamina or the ECM of the microvessels and suggested the degeneration of pericytes (Fig. 1). De Jong et al. *(103)* in the aging rat and Stewart et al. *(68)* in the aging human brain showed the degeneration and loss of pericytes. Burns et al. *(76,94)* described the reduction of the cross-sectional area of pericytes in aging monkey and rat brain. However, Peters et al. *(98)* found no change in the number of pericytes in monkey, and Heinsen and Heinsen *(104)* and Peinado et al. *(74)* showed an increase in the number of pericytes in the rat aging brain. The obvious discrepancies in the literature could be explained by considering the degeneration and the proliferation of pericytes as distinct age-related mechanisms. Accordingly, Scheibel and Fried *(105)* could show that pericytes in aging brain begin to loose their contact to the endothelial cell and begin to migrate. This behavior seems to be a response of the pericytes to a yet unidentified age-related effect. The PDGF released by endothelial cells is known to bind to the PDGF receptor expressed by pericytes and to attract pericytes to endothelial cells *(106,107)*. If this ligand-receptor system is disturbed, the functional integrity of the vessel is impaired *(61,107)*. With age, this system could deteriorate as well, but data are not available as yet.

4. PERSPECTIVES

The huge number of information on the normal and developing BBB comprising morphological, physiological, biochemical, immunological, pharmacological, and clinical data are far from being integrated into a commonly accepted scientific concept. Considering the data on the BBB in the aging brain, they are predominantly descriptive and less functional. They mainly concern morphological and enzymatic alterations and are considerably controversial. It would be very attractive to compare age-related changes with those observed in diabetes, gliomas or neurodegenerative disorders; however, due to the space available, this comparison has to be limited. In the first part, we gave a survey of the cellular and molecular components of the BBB and described, in the second part, some elected alterations in the aging brain. Finally, we wish to briefly delineate some strains of research which seem to us to merit interest in the future. Again, we are not able to cover all age-related changes of endothelial physiology and pharmacology. Namely, we did not refer to changes of transporters and receptors in aging, which have been reviewed by Shah and Mooradian *(30)* previously (but see Fig. 1).

First of all, although the knowledge of the molecular constitution of TJs as well as of interactions and interdependencies of junctional components connected to the barrier formation in endothelial cells has considerably advanced, the characterization of these components in the aging brain still is difficult and equivocal. The permeability appears not to be substantially altered in aging brain, at least if considering high molecular weight compounds; regarding the permeability related to low molecular weight compounds, it seems to be slightly increased, at least in the white matter *(108)*. The morphological parameters of endothelial TJs could be expected to change with age in a similar way as observed in glioma vessels or under pathological conditions (unpublished observations): This alteration of the TJ freeze-fracture morphology concerns the switch of the particles from the protoplasmic to the E-face, and this could mean an altered linkage to the cytoskeleton or a switch from claudin 1 to claudin 5 as has been already discussed by Tsukita and Furuse *(18)* and by Liebner et al. *(29)*. Furthermore, the altered ECM (the thickening of the basal lamina, for example) and/or the nonenzymatic glycation of collagen are candidate factors influencing the permeability of endothelial cells *(79,83,92)*. In particular, agrin was found to be widely expressed in senile plaques and neurofibrillary tangles in Alzheimer's disease *(109)*. Interestingly, Donahue et al. *(57)* reported that in some cases of Alzheimer's disease the immunostain of agrin around vessels had a ragged profile and that senile plaques were seen frequently in the vicinity of capillaries. A large fraction of the agrin in Alzheimer's diseased brains changed biochemically into becoming unsoluble in 1% SDS *(57)*. The mechanisms by which agrin is involved in the maintenance of the BBB and by which the biochemical modification of agrin is involved in the breakdown of the BBB under pathological conditions such as Alzheimer's disease or tumor remain to be elucidated.

Although the activities of a number of endothelial enzymes are increased or decreased with age *(30)*, no clear-cut concept on functional implications of these alterations has been proposed.

Similarly, the implications of the alterations of the pericytes are far from being elucidated. The understanding of the complex interplay between endothelial cells and pericytes is actually still at its beginning *(11,14,61)*. Also, the role of astrocytes and the significance of the specific distribution of their orthogonal arrays of particles (corresponding to water channels which consists of aquaporin 4 *(54,55,110)* in the membranes of perivascular astrocytic endfeet is not elucidated. This distribution is a typical property of the mature BBB and lost or reduced in culture, after injury, in a glioma *(53,56)* and possibly during aging.

Since insight into regulatory mechanisms of the BBB are of primary significance for the development of new therapeutic strategies in neurology and neurosurgery, for instance for the treatment of gliomas or neurodegenerative disorders, the molecular analysis of these mechanisms in the developing adult and aging brain is highly important. The transfer of results from basic research on angiogenesis, adhesion molecules, junctional components, and the ECM to the field of research on aging will improve understanding age-related processes in the brain.

ACKNOWLEDGMENT

G.R. is supported by a grant of the Bundesministerium fr Bildung, Wissenschaft, Forschung und Technologie (01 KS 9602). We are grateful to Stefan Liebner for critical reading the manuscript and for help with figure 1.

REFERENCES

1. Brightman, M.W., and Reese, T.S. (1969) Junctions between intimately apposed cell membranes in the vertebrate brain. *J. Cell Biol.* **40,** 648–677.
2. Risau, W. and Wolburg, H. (1990) Development of the blood-brain barrier. *Trends Neurosci.* **13,** 174–178.
3. Leonhardt, H. (1980) Ependym und Cirkumventrikuläre Organe. *Handb. mikr. Anat. Mensch.* **4,** 177–666.
4. Peters, A., Palay, S.L., and Webster, H., deF. (1991b) The fine structure of the nervous system: neurons and their supporting cells. Oxford University Press, New York.
5. Breier, G.U., Albrecht, U., Sterrer, S., and Risau, W. (1992) Expression of vascular endothelial growth factor during embryonic angiogenesis and endothelial cell differentiation. *Development* **114,** 521–532.
6. Nabeshima, S., Reese, T.S., Landis, D.M., and Brightman, M.W. (1975) Junctions in the meninges and marginal glia. *J. Comp. Neurol.* **164,** 127–170.
7. Rascher, G. and Wolburg, H. (1997) The tight junctions of the leptomeningeal blood-cerebrospinal fluid barrier during development. *J. Brain Res.* **38,** 525–540.
8. Reese, T.S. and Karnovsky, M.J. (1967) Fine structural localization of a blood-brain barrier to exogenous peroxidase. *J. Cell Biol.* **34,** 207–217.
9. Coomber, B.L., and Stewart, P.A. (1985) Morphometric analysis of CNS microvascular endothelium. *Microvasc. Res.* **30,** 99–115.
10. Angelov, D.N., Walther, M., Streppel, M. Guntinas-Lichius, O., and Neiss, W.F. (1998) The cerebral perivascular cells. *Adv. Anat. Embroyl. Cell Biol.* **147,** 1–90.
11. Balabanov, R. and Dore-Duffy, P. (1998) Role of the CNS microvascular pericyte in the blood-brain barrier. *J. Neurosci. Res.* **53,** 637–644.
12. Wolburg, H. and Risau, W. (1995) Formation of the blood-brain barrier. In: Neuroglia, (H., Kettenmann, B.R., Ransom, eds.) Oxford University Press, New York, Oxford, pp. 763–776.
13. Engelhardt, B., Risau W.(1995) Development of the blood-brain barrier. In New concepts of a blood-brain barrier (Greenwood JEA, ed), Plenum, New York, pp. 11–31.
14. Gerhardt, H., Liebner, S., Redies, C., and Wolburg, H. (1999) N-cadherin expression in endothelial cells during early angiogenesis in the eye and brain of the chicken: relation to blood-retina and blood-brain barrier development. *Europ J Neurosci* **11,** 1191–1201.
15. Kniesel, U. and Wolburg, H. (2000) The tight junctions of the blood-brain barrier. *Cell Mol. Neurobiol.* **20,** 57–76.
16. Rubin, L.L., Hall, D.E., Porter, S., et al. (1991) A cell culture model of the blood-brain barrier. *J. Cell Biol.* **115,** 1725–1736.
17. Wolburg, H., Neuhaus, J., Kniesel, U., et al. (1994) Modulation of tight junction structure in blood-brain barrier ECs. Effects of tissue culture, second messengers and cocultured astrocytes. *J. Cell Sci.* **107,** 1347–1357.
18. Tsukita, S. and Furuse, M. (1999) Occludin and claudins in tight-junction strands, leading or supporting players? *Trends Cell Biol.* **9,** 268–273.
19. Kachar, B. and Reese, T.S. (1982) Evidence for the lipidic nature of tight junction strands. *Nature* **296,** 464–466.
20. Grebenkämper, K., and Galla, H-J. (1994) Translational diffusion measurements of a fluorescent phospholipid between MDCK-1 cells support the lipid model of the tight junctions. *Chem. Phys. Lipids* **71,** 133–143.
21. Caldéron, V., Lázaro, A., Contreras, R.G., et al. (1998) Tight junctions and the experimental modifications of lipid content. *J. Membrane Biol.* **164,** 59–69.
22. Furuse, M., Hirase, T., Itoh, M., Nagafuchi, A., Yonemura, S., Tsukita, S. (1993) Occludin: a novel integral membrane protein localizing at tight junctions. *J. Cell Biol.* **123,** 1777–1788.

23. Furuse, M., Sasaki, H., Fujimoto, K., Tsukita, S. (1998) A single gene product, claudin-1 or—2, reconstitutes tight junction strands and recruits occludin in fibroblasts. *J. Cell Biol.* **143,** 391–401.
24. Hirase, T., Staddon, J.M., Ando-Akatsuka, Y., et al. (1997) Occludin as a possible determinant of tight junction permeability in endothelial cells. *J. Cell Sci.* **110,** 1603–1613.
25. Morita, K., Furuse, M., Fujimoto, K., and Tsukita, S. (1999a) Claudin multigene family encoding four-transmembrane domain protein components of tight junction strands. *Proc. Natl. Acad. Sci. USA* **96,** 511–516.
26. Lippoldt, A., Liebner, S., Andbjer, B., et al. (2000) Organization of choroid plexus epithelial and endothelial cell tight junctions and regulation of claudin-1, -2 and -5 expressions by protein kinase C. *NeuroReport* **11,** 1427–1431.
27. Morita, K., Sasaki, H., Fujimoto, K., Furuse, M., and Tsukita, S. (1999b) Claudin-11/OSP-based tight junctions of myelin sheaths in brain and Sertoli cells in testis. *J. Cell Biol.* **145,** 579–588.
28. Morita, K., Sasaki, H., Furuse, M., and Tsukita, S. (1999c) Endothelial claudin: claudin-5/TMVCF constitutes tight junction strands in endothelial cells. *J. Cell Biol.* **147,** 185–194.
29. Liebner, S., Fischmann, A., Rascher, G., et al. (2000) Claudin-1 and claudin-5 expression and tight junction morphology are altered in blood vessels of human *glioblastoma multiforme. Acta Neuropathol.* **100,** 323–331.
30. Shah, G.N. and Mooradian, A.D. (1997) Age-related changes in the blood-brain barrier. *Exp. Gerontol.* **32,** 501–519.
31. Maher, F., Vannucci, S.J., and Simpson, J.A. (1994) Glucose transporter proteins in brain. *FASEB J.* **8,** 207–212.
32. Bauer, H., Sonnleitner, U., Lametschwandtner, A., Steiner, M., Adam, H., and Bauer, H.C. (1995) Ontogenetic expression of the erythroid-type glucose transporter (Glut-1) in the telencephalon of the mouse: correlation to the tightening of the blood-brain barrier. *Develop. Brain Res.* **86,** 317–325.
33. Farrell, C.L. and Pardridge, W.M. (1991) Blood-brain barrier glucose transporter is asymmetrically distributed on brain capillary endothelial lumenal and ablumenal membranes. An electron microscopic immunogold study. *Proc. Nat. Acad. Sci. USA* **88,** 5779–5783.
34. Bolz, S., Farrell, C.L., Dietz, K., and Wolburg, H. (1996) Subcellular distribution of glucose transporter (GLUT-1) during development of the blood-brain barrier in rats. *Cell Tissue Res.* **284,** 355–365.
35. Jefferies, W.A., Brandon, M.R., Hunt, S.V., Williams, A.F., Gatter, K.C., and Mason, D.Y. (1984) Transferrin receptor on endothelium of brain capillaries. *Nature* **312,** 162–163.
36. Méresse, S., Delbart, J.C., Fruchart, J.C., and Cechelli, R. (1989) Low-density lipoprotein receptor on endothelium of brain capillaries. *J. Neurochem.* **53,** 340–345.
37. Zlokovic, B.V., Shundric, D.S., Segal, M.B., Lipovac, M.V., Mackic, J.B., and Davson, H. (1990) A saturable mechanism for transport of immunoglobulin G across the blood-brain barrier of the guinea pig. *Exp. Neurol.* **107,** 263–290.
38. Bradbury, M.W.B. (1997) Trnsport of iron in the blood-brain-cerebrospinal fluid system. *J. Neurochem.* **69,** 443–454.
39. Hanahan, D. (1997) Signalling vascular morphogenesis and maintenance. *Science* **277,** 48–50.
40. Risau, W. (1997) Mechanisms of angiogenesis. *Nature* **386,** 671–674.
41. Yancopoulos, G.D., Klagsbrun, M., and Folkman, J. (1998) Vasculogenesis, angiogenesis and growth factors: ephrins enter the fray at the border. *Cell* **93,** 661–664.
42. Murphy, S., Simmons, M.L., Agullo, L., et al. (1993) Synthesis of nitric oxide in CNS glial cells. *Trends Neurosci.* **16,** 323–328.
43. Goldstein, G.W., and Betz, A.L. (1986) The blood-brain barrier. *Sci Am* **255,** 70–79.

44. Dringen, R., Pfeiffer, B., and Hamprecht, B. (1999) Synthesis of the antioxidant glutathione in neurons: supply by astrocytes of CysGly as precursor for neuronal glutathione. *J. Neurosci.* **19,** 562–569.
45. Nagashima, T., Wu, S., Yamaguchi, M., and Tamaki, N. (1999) Reoxygenation injury of human brain capillary endothelial cells. *Cell Mol. Neurobiol.* **19,** 151–161.
46. Cordon-Cardo, C. O'Brien, J.P., Casals, D., et al. (1989) Multidrug-resistance gene (P-glycoprotein) is expressed by endothelial cells at blood-brain barrier sites. *Proc. Natl. Acad. Sci. USA* **86,** 695–698.
47. Regina, A., Koman, A., Piciotti, M., El. Hafny, B., Center, M.S., Bergmann, R., Couraud, P.O., and Roux, F. (1998) Mrp1 multidrug resistance-associated protein and P-glycoprotein expression in rat brain microvessel endothelial cells. *J. Neurochem.* **71,** 705–715.
48. Schlosshauer, B. and Herzog, K.-H. (1990) Neurothelin: an inducible cell surface glycoprotein of blood-brain barrier-specific endothelial cells and distinct neurons. *J. Cell Biol.* **110,** 1261–1274.
49. Seulberger, H., Lottspeich, F., and Risau, W. (1990) The inducible blood-brain barrier specific molecule HT7 is a novel immunoglobuline-like cell surface glycoprotein. *EMBO J.* **9,** 2151–2158.
50. Weller, R.O., Engelhardt, B., and Phillips, M.J. (1996) Lymphocyte targeting of the central nervous system: a review of afferent and efferent CNS-immune pathways. *Brain Pathol.* **6,** 275–288.
51. Engelhardt, B. (1997) The blood-brain barrier. In Molecular biology of multiple sclerosis (Russel WC, ed.) John Wiley & Sons, pp. 137–160.
52. Dermietzel, R., Krause, D. (1991) Molecular anatomy of the blood-brain barrier as defined by immunocytochemistry. *Int. Rev. Cytol.* **127,** 57–109.
53. Wolburg, H. (1995a) Orthogonal arrays of intramembranous particles. A review with special reference to astrocytes. *J. Brain Res.* 239–258.
54. Verbavatz, J.-M., Ma, T., Gobin, R., and Verkman, A.S. (1997) Absence of orthogonal arrays in kidney, brain and muscle from transgenic knockout mice lacking water channel aquaporin-4. *J. Cell Sci.* **110,** 2855–2860.
55. Rash, J.E., Yasumura, T., Hudson, C.S., Agre, P., and Nielsen, S. (1998) Direct immunogold labeling of aquaporin-4 in square arrays of astrocyte and ependymocyte plasma membranes in rat brain and spinal cord. *Proc. Natl. Acad. Sci. USA* **95,** 11981–11986.
56. Wolburg, H. (1995b) Glia-neuronal and glia-vascular interrelations in blood-brain barrier formation and axon regeneration in vertebrates. In: Neuron-Glia interrelations during phylogeny. II Plasticity and regeneration (Vernadakis, A., Roots, B.I., eds.) Humana Press, Totowa, pp. 479–510.
57. Donahue, J.E., Berzin, T.M., Rafii, M.S., et al. Agrin in Alzheimer's disease: Altered solubility and abnormal distribution within microvasculature and brain parenchyme. *Proc. Natl. Acad. Sci.* **96,** 6468–6472.
58. Barber, A.J. and Lieth, E. (1997) Agrin accumulates in the brain microvascular basal lamina during development of the blood-brain barrier. *Dev. Dyn.* **208,** 62–74.
59. Broadwell, R.D., Salcman, M. (1981) Expanding the definition of the blood-brain barrier to protein. *Proc. Natl. Acad. Sci. USA* **78,** 7820–7824.
60. Stewart, P.A. and Tuor, U.I. (1994) Blood-eye barriers in the rat: correlation of ultrastructure with function. *J. Comp. Neurol.* **340,** 566–576.
61. Lindahl, P., Johansson, B.R., Levéen, P., and Betsholtz. (1997) Pericyte loss and microaneurysm formation in PDGF-B-deficient mice. *Science* **277,** 242–245.
62. Sato, T.N., Tozawa, Y., Deutsch, U., et al. (1995) Distinct roles of the receptor tyrosine kinases tie-1 and tie-2 in blood vessel formation. *Nature* **376,** 70–74.

63. Suri, C., Jones, P.F. Patan, S., et al. (1996) Requisite role of angiopoietin-1, a ligand for the tie2 receptor, during embryonic angiogenesis. *Cell* **87,** 1171–1180.
64. Sankar, R., Blossom, E., Clemons, K., and Charles, P. (1983) Afe-associated changes in the effects pf amphetamine on the blood-brain barrier of rats. *Neurobiol. Aging* **4,** 65–68.
65. Westergaard, E. (1977) The blood-brain barrier to horseradish peroxidase under normal and experimental conditions. *Acta Neuropathol. (Berl)* **39,** 181–187.
66. Sage, J.I., van Uitert, R.L., and Duffy T.E. (1984) Early changes in blood-brain barrier permeability to small molecules after transient cerebral ischemia. *Stroke* **15,** 46–50.
67. Feng, D., Nagy, J.A., Hipp, J., Dvorak, H.F., and Dvorak, A.M. (1996) Vesiculo-vacuolar organelles and the regulation of venule permeability to macromolecules by vascular permeability factor, histamine, and serotonin. *J. Exp. Med.* **183,** 1981–1986.
68. Stewart, P.A., Magliocco M., Hayakawa K., et al. (1987) A quantitative analysis of blood-brain barrier ultrastructure in the aging human. *Microvasc. Res.* **33,** 270–282.
69. Mooradian, A.D. (1988) Effect of aging on the blood-brain barrier. *Neurobiol. Aging* **9,** 31–39.
70. Mooradian, A.D. (1994) Potential mechanisms of the age-related changes in the blood-brain barrier. *Neurobiol. Aging* **15,** 751–755.
71. Kalaria, R.N. (1996) Cerebral vessels in ageing and Alzheimer s disease. *Pharmacol. Ther.* **72,** 193–214.
72. Unger, J.W., (1998) Glial reaction in aging and Alzheimer's disease. *Micr. Res. Techn.* **43,** 24–28.
73. Kawamata, T., Akiguchi, I., Maeda, K., et al. (1998) Age-related changes in the brains of senescence-accelerated mice (SAM): association with glial and endothelial reactions. *Micr. Res. Techn.* **43,** 59–67.
74. Peinado, M.A., Quesada, A., Pedrosa, J.A., et al. (1998) Quantitative and ultrastructural changes in glia and pericytes in the parietal cortex of the aging rat. *Microsc. Res. Tech.* **43,** 34–42.
75. Takeda, T., Hosokawa, M., Takeshita, S., et al. (1981)A new murine model of accelerated senescence. *Mech. Ageing. Dev.* **17,** 183–194.
76. Burns, E.M., Kruckeberg, T.W., and Gaetano, P.K. (1981) Changes with age in cerebral capillary morphology. *Neurobiol Aging* **2,** 285–291.
77. De la Torre, J.C., and Mussivand, T. (1993) Can disturbed microcirculation cause Alzheimer s disease? *Neurol. Res.* **15,** 146–153.
78. De Jong, G.I., De Vos, R.A., Steur, E.N., and Luiten, P.G. (1997) Cerebrovascular hyperfusion: a risk factor for Alzheimer disease? Animal model and postmortem human studies. *Ann. NY Acad. Sci.* **826,** 56–74.
79. Tilling, T., Korte, D., Hoheisel, D., and Galla, H.J. (1998) Basement membrane proteins influence brain capillary endothelial barrier function in vitro. *J. Neurochem.* **71,** 1151–1157.
80. MunBryce, S. and Rosenberg, G.A. (1998) Matrix metalloproteinases in cerebrovascular disease. *J. Cerebr. Blood Flow Metabol.* **18,** 1163–1172.
81. Fujimura, M., Gasche, Y., Morita-Fujimura, Y., Massengale, J., Kawase, M., and Chan, P.H. (1999) Early appearance of activated matrix metalloproteinase-9 and blood-brain-brarrier disruption in mice after focal cerebral cerebral ischemia and reperfusion. *Brain Res.* **842,** 92–100.
82. Gautam, N., Herwald, H., Hedqvist, P., and Lindbom, L. (2000) Signaling via β_2 integrins triggers neutrophil-dependent alteration in endothelial barrier function. *J. exp. Med.* **191,** 1829–1839.
83. Reiser, K. (1998) Nonenzymatic glycation of collagen in aging and diabetes. *Proc. Soc. Exp. Biol. Med.* **18,** 23–37.
84. Handa, J.T., Reiser, K.M., Matsunaga, H., and Hjelmeland, L.M. (1998) The advanced glycation endproduct pentosidine induces the expression of PDGF-B in human retinal pigment epithelial cells. *Exp. Eye. Res.* **66,** 411–419.

85. Farboud, B., Aotaki-Keen, A., Miyata, T., Hjelmeland, L.M., and Handa, J.T. (1999) Development of a polyclonal antibody with broad epitope specificity for advanced glycation endproducts and localization of these epitopes in Bruch s membrane of the aging eye. *Mol. Vis.* **5,** 11 (*http://*www.molvis.org/molvis/v5/p11).
86. Haitoglou, C.S., Tsilibary, E.C., Brownlee, M., and Charonis, A.S. (1992) Altered cellular interactions between endothelial cells and nonenzymatically glycosylated laminin/type IV collagen. *J. Biol. Chem.* **267,** 12404–12407.
87. Hasegawa, G., Hunter, A.J., and Charonis, A.S. (1995) Matrix nonenzymatic glycosylation leads to altered cellular phenotype and intracellular tyrosine phosphorylation. *J. Biol. Chem.* **270,** 3278–3283.
88. Vlassara, H., Brownlee, M., and Cerami, A. (1985) Recognition and uptake of human diabetic peripheral nerve myelin by macrophages. *Diabetes* **34,** 553–557.
89. Schmidt, A.M., Hasu, M., Popov, D., et al. (1994) Receptor for advanced glycation end products (AGEs) has a central role in vessel wall interactions and gene activation in response to circulating AGE proteins. *Proc. Natl. Acad. Sci. USA* **91,** 8807–8811.
90. Schmidt, A.M., Hori, O., Cao, R., et al. (1996) A novel cellular receptor for advanced glycation end products. *Diabetes* **45**(Suppl 3): S77–S80.
91. Wautier, J.L., Zoukourian, C., and Chappey, O., et al. (1996) Receptor-mediated endothelial cell dysfunction in diabetic vasculopathy: soluble receptor for advanced glycation end products blocks hyperpermeability in diabetic rats. *J. Clin. Invest.* **97,** 238–243.
92. Mooradian, A.D. and Meredith, K.E. (1992) The effect of age on protein composition of rat cerebral microvessels. *Neurochem. Res.* **17,** 665–670.
93. Giri, R., Shen, Y., Stins, M. et al. (2000) β-amyloid-induced migration of monocytes across human brain endothelial cells involves RAGE and PECAM-1.*Am J Physiol* **279,** C1772–C1781.
94. Burns, E.M., Kruckeberg, T.W., Gaetano, P.K., and Shulman, L.M. (1983) Morphological changes in cerebral capillaries with age, in Brain aging: Neuropathology and Neuropharmacology (Cerv s-Navarro, J, Sarkander H-I, eds), Raven Press, New York, 115–132.
95. Mooradian, A.D. and Uko-eninn, A. (1995) Age-related changes in the antioxidative potential of cerebral microvessels. *Brain Res.* **671,** 159–163.
96. Tayarani, I., Colez, I., Clement, M., and Bourre J.M., (1989) Antioxidant enzymes and related trace elements in aging brain capillaries and choroid plexus. *J. Neurochem.* **53,** 817–824.
97. Tigges, J., Herndon, J.G., and Rosene, D.L. (1995) Mild age-related changes in the dentate gyrus of adult rhesus monkeys. *Acta Anat. (Basel)* **153,** 39–48.
98. Peters, A., Josephson, K., and Vincent, S.L. (1991a) Effects of aging on the neuroglial cells and pericytes within area 17 of the rhesus monkey cerebral cortex. *Anat. Rec.* **229,** 384–398.
99. Knox, C.A., Yates, R.D., Chen, I., and Klara, P.M. (1980) Effects of aging on the structural and permeability characteristics of cerebrovasculature in normotensive and hypertensive strains of rats. *Acta Neuropathol. (Berlin)* **51,** 1–13.
100. Hicks, P., Rolsten, C., Brizzee, and D., Samorajski, T. (1983) Age-related changes in rat brain capillaries. *Neurobiol. Aging* **4,** 69–75.
101. Sturrock, R.R., (1980) A comparative and morphological study of ageing in the mouse neostriatum, indusium griseum and anterior commissure. *Neuropathol. Appl. Neurobiol.* **6,** 51–68.
102. Ueno, M., Akiguchi, I., Hosokawa, M., et al. (1998) Ultrastructural and permeability features of microvessels in the olfactory bulbs of SAM mice. *Acta Neuropathol* **96,** 261–270.
103. De Jong, G.I., Horvath, E., and Luiten, P.G. (1990) Effects of early onset of nimodipine treatment on microvascular integrity in the aging rat brain. *Stroke* **21 (Suppl 12),** 113–116.
104. Heinsen, H., and Heinsen, Y.L. (1983) Cerebellar capillaries. Qualitative and quantitative observations in young and senile rats. *Anat. Embryol.* **168,** 101–106.

105. Scheibel, A.B. and Fried, I. (1983) Age-related changes in the peri-capillary envirpment of the brain. In Aging of the brain, Vol 22, Raven, New York.
106. Li, J., Zhang, M., and Rui, Y.C. (1997) Tumor necrosis factor mediated release of platelet-derived growth factor from bovine cerebral microvascular endothelial cells. *Acta Pharmacol. Sin.* **18,** 133–136.
107. Benjamin, L.E., Hemo, I., and Keshet, E. (1998) A plasticity window for blood vessel remodelling is defined by pericyte coverage of the preformed endothelial network and is regulated by PDGF-B and VEGF. *Development* **125,** 1591–1598.
108. Rapoport, S.I., Ohno, K. and Pettigrew, K.D. (1979) Blood-brain barrier permeability in senescent rats. *J. Gerontol.* **34,** 162–169.
109. Verbeek, M.M., Otte-H ller, I., van den Born J., et al. (1999) Agrin is a major heparan sulfate proteoglycan accumulating in Alzheimers's disease brain. *Am. J. Pathol.* **155,** 2115–2125.
110. Venero, J.L., Vizuete, M.L., Machado, A. and Cano, J. (2001) Aquaporins in the central nervous system. *Progr. Neurobiol* **63,** 321–336.
111. Blennow, K., Wallin, A., Fredman, P., Karlsson, I., Gottfries, C.G., and Svennerholm, L. (1990) Blood-brain barrier disturbance in patients with Alzheimer's disease is related to vascular factors. *Acta Neurol. Scand.* **81,** 323–326.
112. Wada, H. (1998) Blood-brain barrier permeability of the demented elderly as studied by cerebrospinal fluid-serum albumin ratio. *Intern. Med.* **37,** 509–513.
113. Skoog, I., Wallin A., Fredman, P., et al. (1998) A population study on blood-brain barrier function in 85-year-olds: Relation to Alzheimer‘s disease and vascular dementia. *Neurology* **50,** 966–971.
114. Rudick, R.A. and Buell, S.J. (1983) Integrity of blood-brain barrier to horseradish peroxidase in senescent mice. *Neurobiol. Aging* **4,** 283–287.
115. Mayhan, W.G. (1990) Disruption of the blood-brain barrier during acute hypertension in adult and aged rats. *Am. J. Physiol.* **285,** H1735–H1738.
116. Westergaard, E.K., Go, G., Klatzo, I., and Spatz, M. (1976) Increased permeability of cerebral vessels to horseradish peroxidase induced by ischemia in Mongolian gerbils. *Acta Neuropathol.* **35,** 307–325.
117. Westergaard, E., Deurs, D., and Bronsted, H.E. (1977) Increased vesicular transfer of horseradish peroxidase across cerebral endothelium evoked by acute hypertension. *Acta Neuropathol.* **37,** 141–152.
118. Hunzicker, O., Abdel‘Al, S., Frey, H., Veteau, J., and Meier-Ruge, W. (1978) Quantitative studies in the cerebral cortex of aging humans. *Gerontology* **24,** 27–31.
119. Ravens, J.R. (1978) Vascular changes in the human senile brain. *Adv. Neurol.* **20,** 487–501.
120. Bär, T. (1978) Morphometric evaluation of capillaries in different laminae of rat cerebral cortex by automatic image analysis: changes during development and aging. *Adv. Neurol.* **20,** 1–9.
121. Topple, A., Fifkova, E., and Cullen-Dockstader, K. (1990) Effect of age on blood vessels and neurovascular appositions in the rat dentate fascia. *Neurobiol. Aging* **11,** 371–380.

18

Astrocytes and Barrier-provided Microvasculature in the Developing Brain

Luisa Roncali

1. INTRODUCTION

The morphofunctional specificity of the cerebral endothelia has been ascribed to the close association between these cells and the astrocytes, glial cells that in the central nervous system (CNS) microvasculature are arranged in an almost continuous perivascular sheath. The astrocytes are thought to play a role in inducing the tightness and other properties of the endothelial cells (the blood-brain barrier [BBB] that preserve the nervous tissue separate from the blood environment and guarantee the unique and privileged metabolism of the neuronal cells.

Many studies have been performed to investigate glial control on the differentiated status of the cerebral endothelia, starting from the suggestion by Davson and Oldendorf *(1)* that endothelial permeability via pinocytosis is restricted in the brain vessels by the perivascular glial action. In the cerebral endothelial cells, the low permeability, tightness of junctions, expression of specific enzymatic activities and transport systems, glucose uptake, as well as the cell membrane resistance and polarity, have been demonstrated to be influenced by astrocytes in various, mainly in vitro, BBB models *(2–16)*. Although an intimate astrocyte-endothelium relationship has been shown to characterize the barrier-provided CNS microvasculature, whereas this association is absent where the brain vessels are barrier-free, such as in the circumventricular organs and in pathological situations of the brain, the inducing role exerted on cerebral endothelium by astrocytes is still under discussion *(17–22)*.

Investigations carried out *in situ* on the developmental organization of the astrocyte-microvascular interface show that relationships between glial cells and blood vessels take place precociously in the CNS and that the first astroglial forms to send processes to the growing cerebral vessels are the radial glial cells. The radial glial cells, once referred to as matrix cells or spongioblasts, are transient cells committed to giving rise to multipolar astrocytes in the mature brain. Their body is located in the deep, ventricular layer of the nervous wall and they reach the external limiting membrane through a slender distal process that elongates and branches as the intermediate, mantle layer thickens. The radial glial cells are involved in several fundamental functions during CNS morphohistogenesis; among these they may provide a scaffolding in the nervous

From: *Neuroglia in the Aging Brain*
Edited by: Jean S. de Vellis © Humana Press Inc., Totowa, NJ

wall for both neuroblast migration and the spatial arrangement of the first CNS vessels that grow radially from the perineural plexuses *(23–26)*. Observations with both light and electron microscopy of cells of the glial lineage sending processes to the developing CNS vasculature date back to many years ago. Glial end feet attached to blood vessels are described at the 12th embryonic day in the chicken, at the 15th day in the mouse and rat fetuses, at the 48th embryonic day in the monkey and beginning from the 7th gestational week in the human brain *(27–32)*. As soon as differentiation occurs, firstly of astroblasts and later of astrocytes, their processes also contribute to the building of the vessel glial sheaths that progressively widen as the perivascular space disappearance. In the perinatal and adult brain microvasculature, the astrocytic coverage is almost complete (85% in the rat cerebral cortex; 96% in the optic *tectum* of hatching chickens; 86% in the human cerebral cortex), while other glial cells and neuron processes participate, at least to a small extent, in the composition of the perivascular sheath *(33–36)*.

2. ASTROGLIA AND MICROVASCULATURE IN EARLY NEURAL DEVELOPMENT

The close, anatomical relationships established between astroglia and microvessels during early neural development suggest that astrocytes, as well as their precursors, may play a role in forming the blood-brain-barrier. Nevertheless, the perivascular glia lack barrier properties and constitute a blood-brain-barrier only in some invertebrates. During evolution, the endothelial barrier replaces the glial barrier, although the glial cells seem to maintain some barrier competence even in the vertebrate brain *(37)*.

Developmental studies on the endothelial barrier have to be considered the most likely to provide information on the role of the glial cells in differentiation of the cerebral endothelium. In this context, it should be noted that whereas some investigations seem to indicate that the brain vessels are provided with barrier properties from the beginning of the neural angiogenesis, others suggest that the morphofunctional tightening of the endothelia takes place gradually during embryonic development *(38,39)*.

Detailed results of this process have been obtained in the brain microvasculature of the chick embryo. The endothelial cells undergo continuous molding processes, involving their thickness and surface features, pinocytotic structures, junctional systems, and biochemical and antigen apparatuses. Only by birth does the chick endothelia seem to have achieved almost complete differentiation. By this time they have become flattened and smooth, with only few endocytotic vesicles and vacuoles, and are sealed by extensive pentalaminar junctions that appear as networks of interconnected fibrils on the replicas of freeze-fractured vessels *(27,40–42)*. The vessel permeability to the classical markers, horseradish peroxidase and Evans blue, parallels the morphological changes in the endothelial cells since progressive restriction takes place that ends with complete blockage of vascular escape of the tracer by hatching time, for horseradish peroxidase, and by one month after hatching for the Evans blue marker *(43,44)*. The time course of the metabolic maturation of the endothelial barrier investigated on isolated microvessels of the chick embryo brain is the same as that required for the anatomical differentiation of the endothelial cells. In fact in the endothelial cells there is a progressive decline in the activity of the alkaline phosphatase enzyme and in the expression of the transport systems for neutral aminoacids and D-glucose during embryonic life. They are further reduced in the adult brain, in accordance with the increased tightness of the endothelial junctions *(45)*. Among the cerebral endothelium antigens, the HT7 glycoprotein was recognized to be specifically expressed in the chick brain endothelial cells, where it correlates with the glucose transporter activity and BBB functioning *(46)*. Like other barrier markers, HT7 expression modifies during chick embryo brain development, becoming specifically localized on the endothelial luminal and lateral surfaces at the late embryonic stages and in adulthood. The glycoconjugates of the endothelial cell plasma membranes also undergo changes during maturation of the BBB in developing chick brain vessels. N-acetylglucosamine and sialic residues, specifically located on the luminal surface of the endothelial cells and expressing their polarity, increase significantly in number parallel with morphofunctional differentiation of the endothelial cells *(47,48)*.

The main morphological changes undergone by the endothelial cells of the mouse embryo and fetus brains are similar to those demonstrated in the chick embryo brain endothelia, and develop according to the same time sequence, mouse gestation being quite as long as chick incubation. In addition, in the mouse fetal brain, ZO-1 phosphoprotein expression, which correlates with the maturity of the tight junctions, modifies during development and shows a selective localization in the endothelial cells only by the prenatal age *(49)*.

Recent studies carried out on the morphological characteristics of the microvasculature of the human fetus telencephalon further support the concept that the endothelial cells evolve toward a BBB status in the prenatal age (Fig. 1a, b) *(50,51)*. This possibility is substantiated by some aspects of endothelial cell function, such as glucose transport and permeability to endogenous albumin. The glucose uptake by the barrier-provided microvasculature is mediated by the glucose transporter isoform 1 (GLUT1) whose expression in the endothelial cells is developmentally regulated *(52,53)*. In the human telencephalon microvessels, GLUT1 antigen sites already show the specific localization on the abluminal and lateral endothelial plasma membrane by the 12th gestational week *(54–56)*. The prenatal GLUT1 expression correlates with the efficacy of the endothelial barrier to the endogenous albumin (Fig. 2a, b) and with the expression of the multidrug-resistance P-glycoprotein *(57,58)*.

Besides the endothelial cell characteristics, the cerebral microvessel specificity and BBB functioning is also because of the subendothelial pericytes and basal lamina. The CNS pericytes are contractile and phagocytic cells that express specific surface and cytoplasmic markers as well as enzymatic activities, and play vasodynamic, hemostatic and immunological roles *(59)*. Pericytes escort cerebral endothelia as from vessel growth into the embryonic brain, and undergo remarkable modifications in shape, extension, and surface features during the neural angiogenesis and BBB formation *(60,61)*. In the chick embryo optic *tectum,* these cells are irregularly surfaced in the proliferating vessels and cover the 97% of the vessel perimeter when the endothelial cell tight junctions begin to form and the vessel permeability decreases; they are flattened and smooth-surfaced, surrounded by the basal lamina, and almost completely encircled by astrocytes in the mature, barrier-provided vessels *(27,61)*. These findings together with observations that the pericyte investment constitutes a second, defense layer when the endothelial barrier is immature or damaged, and widens when the integrity of the perivascular astrocytic sheath fails, introduce quite rightly the pericytes among the cells having relationship in the BBB system *(43,59,62)*.

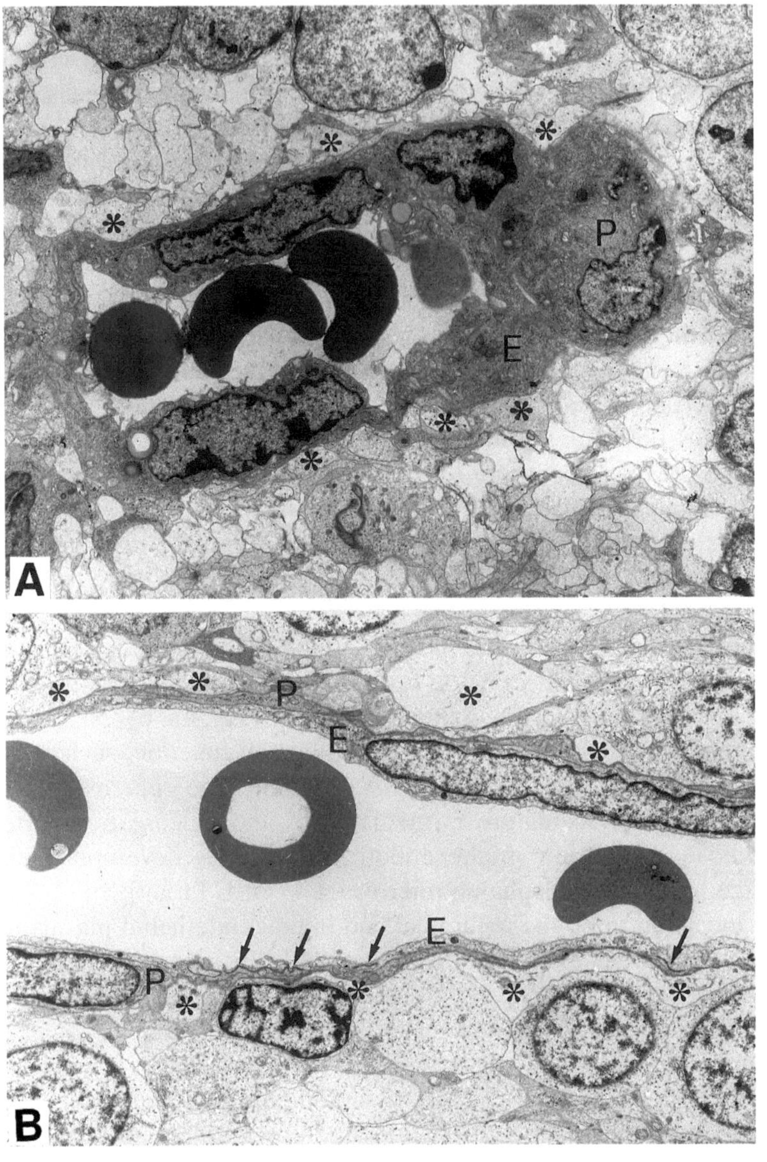

Fig. 1. Electron micrographs of radial microvessels of the human fetus telencephalon at the 12th (**A**) and 18th (**B**) gestational weeks. Differences are detectable between the microvessels in the two developmental stages. In comparison with the 12th wk, the 18th wk microvessel is characterized by reduced thickness and remarkable smoothness of the endothelial cells (E) that are sealed by extensive tight junctions (arrows); there is flattening of the pericyte body and processes (P) that are included in a continuous basal lamina. In both microvessels the endothelium-pericyte layer is closely surrounded by processes of astrocytes (asterisks). (**A**) ×6000; (**B**) ×5500.

In the adult brain microvessels, the basal lamina is thick and continuous under the endothelial cells and splits into two layers surrounding the pericytes interposed between the endothelial layer and the astrocytic perivascular sheath. The various basal lamina components, type IV collagen and noncollagenous glycoproteins such as

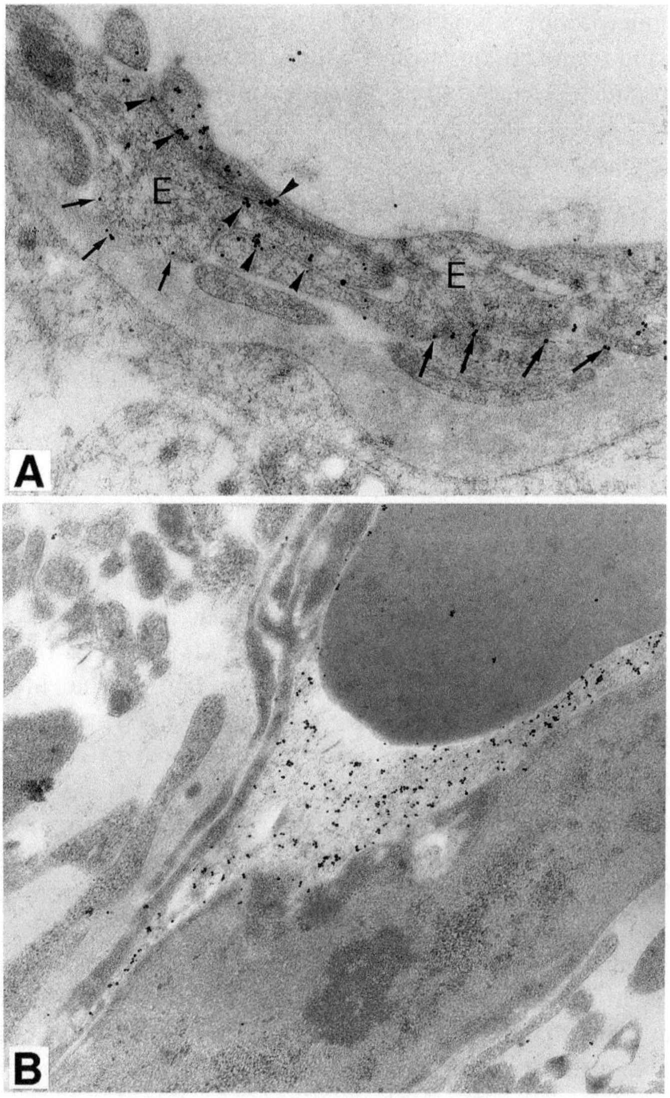

Fig. 2. **(A)** GLUT1 electron immunohistochemistry of a radial microvessel in the 12 wk human fetus telencephalon. The ends of two endothelial cells (E) welded by a tight junction are labeled by gold particles that are mainly distributed along the abluminal (arrows) and lateral (arrowheads) plasma membranes. **(B)** Albumin electron immunohistochemistry of a radial microvessel in the 12 wk human fetus telencephalon. The gold particles are localized exclusively inside the vessel lumen. **(A)** ×68000; **(B)** ×40000.

fibronectin, laminin, tenascin, as well as proteoglycans, are produced by both vascular cells and perivascular astrocytes and are differently expressed in the neural angiogenesis and BBB development, fibronectin and tenascin being more expressed in early stages *(26,63–67)*. A continuous basal lamina, rich in laminin and glycoconjugates containing β-D galactose residues, is contemporary with the endothelial growth slowing down and BBB maturation *(48,68)*. Despite the difficulties of immunodetection of the molecular components of the developing basal lamina and the discrepancies between light microscopy immunocytochemistry and electron microscopy observations, previous and recent studies strongly suggest that the basal lamina development is enhanced when and where astroglial processes make contact with the vessel surface and that the basal lamina maturation coincides with the organization of an astroglial perivascular sheath *(27,69,70)*.

3. CHANGES IN THE ASTROCYTIC PERIVASCULAR GLIA

Like the endothelial cells and basal lamina, the astrocytic perivascular glia also undergo remarkable changes during development. Progressively more differentiated types of cells of astroglial lineage establish relationships with the brain vessels. As stated above, their walls are contacted by processes of the radial glial cells at an early stage, and by processes of astroblasts and astrocytes later on. These processes contain both vimentin and glial fibrillary acidic protein cytoskeletal filaments, the vimentin expression prevailing in the early steps of gliogenesis *(26,71–73)*. Further developmental changes undergone by the astrocyte perivascular processes concern the distribution of the intramembrane particles (IMPs) on their joined plasma membranes. The changes consist of the formation of orthogonal assemblies of small intramembrane particles (OAPs) and of gap junctions that, together, characterize the plasma membranes of the astrocytic processes. The OAPs are accumulated in the plasma membrane of the perivascular astrocytic end feet facing the blood vessels and progressively increase in number and size during astrocyte maturation and BBB formation (Fig. 3a, b), presumably playing a physiological role in the water and ion transport in the mature form *(74–80)*. The OAPs are seen to be numerous where the astrocyte plasma membranes are surrounded by a basal lamina, as in the subpial astrocytic processes where these particle assemblies might share in membrane stabilization *(81)*. In the chick embryo brain vessels, the OAP arrangement is time-related to basal lamina formation and BBB differentiation; the possibility that OAPs may have a mechanical function in astrocyte membrane stabilization has therefore to be considered even in the setting up of the BBB. The gap junctions are assemblies of large IMPs that form channels for the diffusion of ions and appear closely packed in hexagonal arrays in the freeze-fractured vessels. In the adult human brain microvasculature, gap junctions are extensively distributed on the plasma membranes of adjacent perivascular astrocytic end feet *(82)*, but gap junctions are also seen by electron microscopy in developing cerebral vessels (Fig. 4a, b). It seems conceivable that the final, ordered alignment of astrocyte end feet close to the microvessel wall in the adult brain may depend on the attraction forces bringing astrocyte processes close together throughout the CNS *(83)* and is concluded with the establishment of the gap junctional systems between them.

4. GAP JUNCTIONS

The early formation of gap junctions in the microvessels of the prenatal brain is also documented by research carried out in the human fetus telencephalon with connexin 43 (Cx43) immunocytochemistry (Fig. 5a, b, c). Cx43 is a protein component of the gap junction units that constitute the intercellular channels coupling adjacent cells. Unlike other CNS connexins, Cx43 is detectable in the prenatal and postnatal brain and is specifically expressed by both astrocyte precursors (radial glia) and mature astrocytes

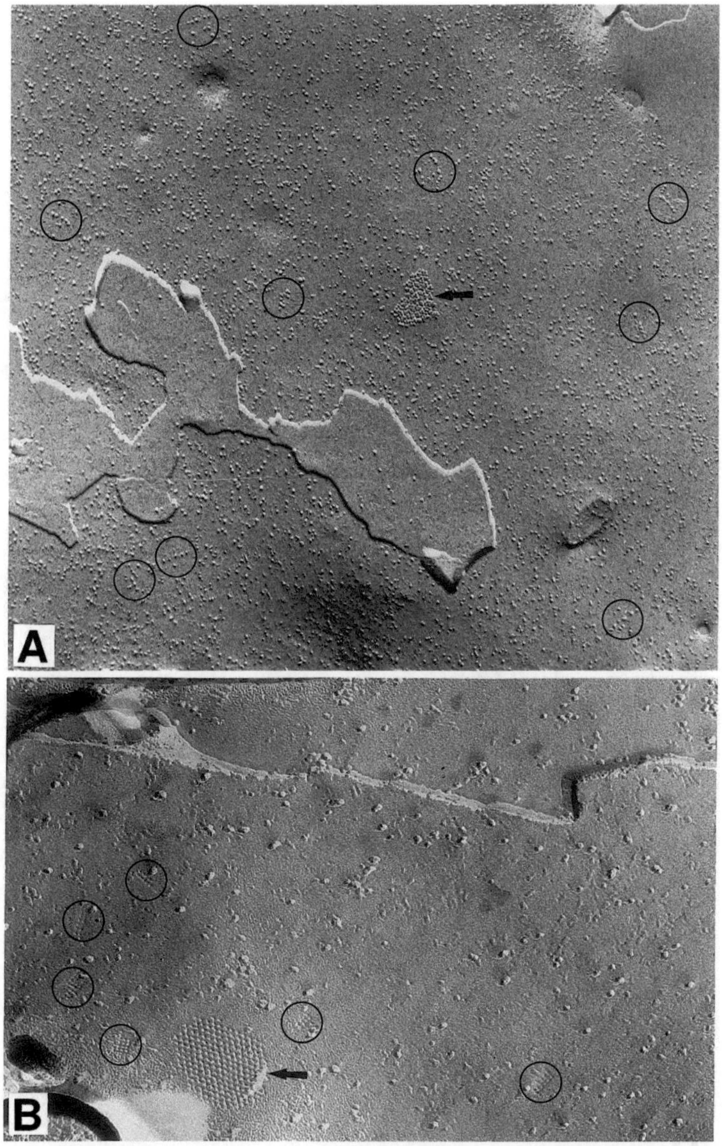

Fig. 3. Replicas from microvessels of 14 d (**A**) and 20 d (**B**) chick embryo optic *tecta* showing plasma membranes of perivascular astrocyte processes. (**A**) Single chains, and short double chains, of aligned individual IMPs (circles) are recognizable, together with a particle cluster identified as a primordial gap junction (arrow), on the P-face of the fractured plasma membrane. (**B**) The astrocytic plasma membrane E-face shows orthogonal assemblies, square or rectangular in shape and of varying sizes (circles); a gap junction plaque is also present (arrow). (**A**) ×76000; (**B**) ×155000.

(84,85). In the human telencephalon, Cx43 labels the radial glial cells and fibers, as well as bipolar and multipolar astroblast perikarya and processes. Cx43 is also detectable in the vascular compartment of the telencephalon when the corticogenesis is under way and radial vessels cross the cortical plate and the subcortical zones coming

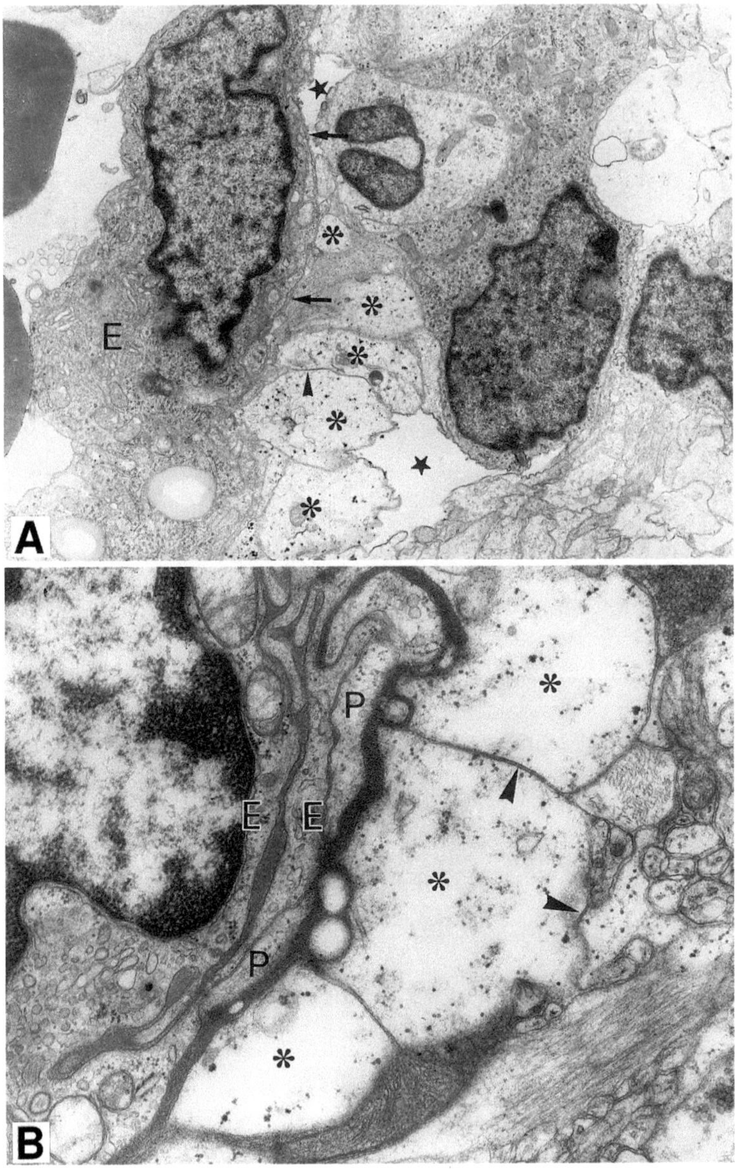

Fig. 4. (**A**) Electron micrograph of a radial microvessel of the human fetus telencephalon at the 12th gestational week. The abluminal side of a thick endothelial cell (E) is partly surrounded by a basal lamina (arrows) and by perivascular astrocyte processes (asterisks). A gap junction is detectable between adjacent glial end feet (arrowhead). Small spaces are observed in the perivascular tissue (stars). (**B**) Electron micrograph of a capillary of the adult human cerebral cortex. The thin vessel wall, built by endothelial cells (E) facing a very narrow lumen and by flattened pericytic processes (P) enclosed within the basal lamina, is surrounded by tightly packed astrocyte end feet (asterisks) and a compact neuropil. The arrowheads point to interastrocytic gap junctions. (**A**) ×14000; (**B**) ×42000.

together in a plexus located in the ventricular zone. In the ventricular and subcortical zones, the Cx43 immunoreactivity is mainly continuous in the endothelial cells, pericytes and surrounding glia, whereas in the cortical plate the immunoreactivity appears finely punctate (Fig. 5A,B,C) indicating the presence of differentiated vascular gap junctions. Gap junctions between endothelial cells have been already described in the embryonic cerebral microvessels and are transient structures decreasing in number as the endothelial tight junctions differentiate and BBB maturation progresses *(86)*. Gap junctions between endothelial cells and pericytes control the vessel growth processes in the embryonic CNS and persist in the adult brain, where the endothelium-pericyte coupling is involved in the regulation of the blood flow *(87,88)*. The punctate Cx43 immunolabeling shown in the cortical plate of the human telencephalon (Fig. 5c) indicates that gap junctions earlier establish in the wall of microvessels distributing to the most differentiated telencephalic layers. Furthermore, the developmental expression of the same connexin in endothelial cells, pericytes and perivascular astrocytes strongly suggests that communication by gap junction may be involved in the control of the BBB maturation during brain ontogenesis *(89)*.

5. CONCLUSION

In conclusion, many observations of vessel features and properties made in *in situ* studies during the brain embryonic and fetal development seem to indicate that the vascular and perivascular cells undergo differentiating changes and suggest that the BBB progressively matures during prenatal life. In addition, the results demonstrate that processes of astrocyte precursors or mature astrocytes have reached the vessel walls and are arranged close to their endothelium-pericyte layer by the time that the endothelial cells express some signs of morphofunctional tightness and biochemical specificity.

The parallel, prenatal "aging" involving both endothelial cells and perivascular astrocytes strongly substantiates the hypothesis that relationships established between them are involved in setting up the BBB. Such relationships could be mutual and imply not only the widely discussed inducing role that the astrocytes may play on endothelial cell differentiation but also, vice versa, a controlling role exerted by the endothelial cells on the astrocytes *(90–92)*.

ACKNOWLEDGMENTS

The author wishes to thank Drs. Mirella Bertossi, Beatrice Nico, Daniela Virgintino, for contributing electronmicroscopical and immunocytochemical micrographs, Dr. Mariella Errede for the skilful technical support and Mrs. Marisa Ambrosi for her expert photographic assistance (Dipartimento di Anatomia Umana e Istologia, Università di Bari, Italy). The author is also grateful to Drs. Dario Cantino (Dipartimento di Anatomia, Farmacologia e Medicina Legale, Università di Torino, Italy), Rainer Herken and Fabio Quondamatteo (Zentrum Anatomie, Abteilung Histologie, Universität Göttingen, Germany), Paul Monaghan and David Robertson (Institute of Cancer Research, Sutton, United Kingdom), for the joint collaboration in freeze-fracture, lectin histochemistry, and immunocytochemistry techniques.

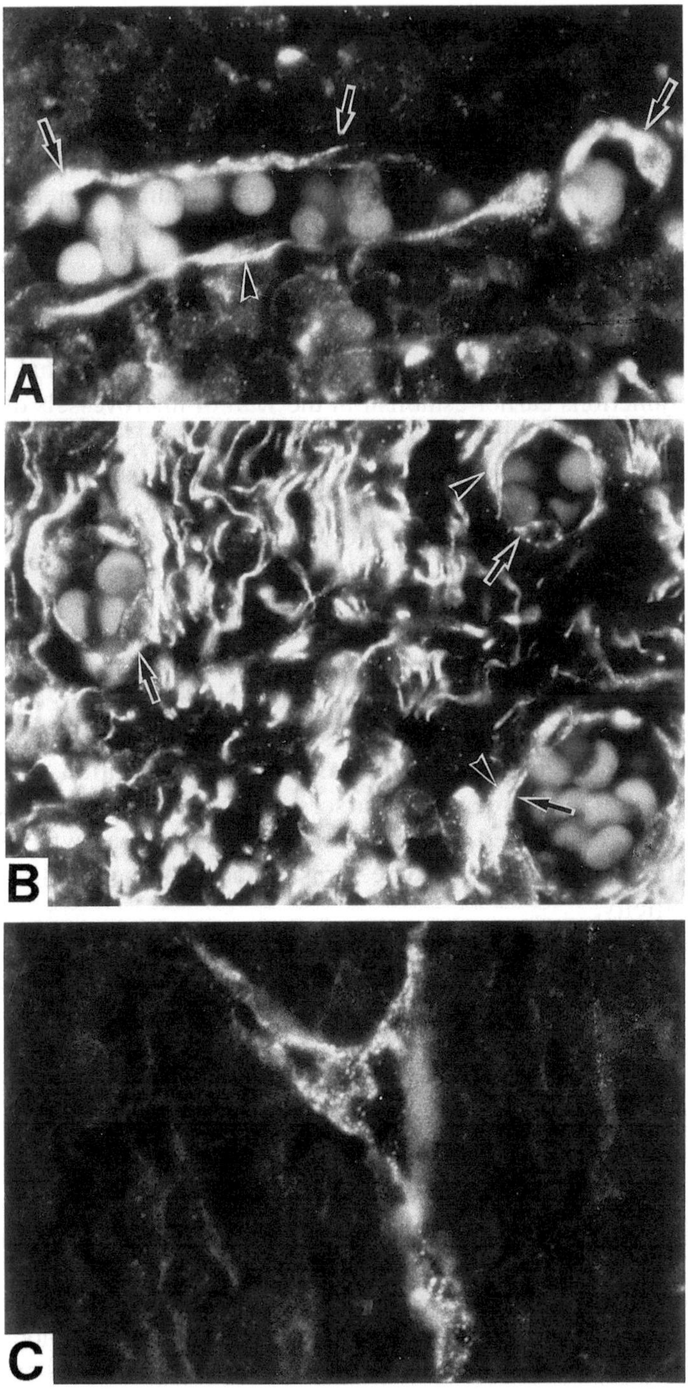

Fig. 5. **(A–C)** C×43 immunohistochemistry in the 18 wk human fetus telencephalon. **(A)** The microvessels of the ventricular plexus show a strong immunoreactivity of endothelial cells (arrows) and subjacent pericytes (arrowhead). **(B)** In the subcortical zone immunolabeled radial glial fibers are seen to establish close contacts with the vessel wall formed by heavily stained endothelial cells (arrows) and pericytes (arrowheads). **(C)** A radial vessel of the cortical plate shows a punctate C×43 immunolabeling. **(A,B)** ×1100; **(C)** ×1250.

REFERENCES

1. Davson, H. and Oldendorf, W. H. (1967) Symposium on membrane transport. Transport in the central nervous system. *Proc. R. Soc. Med.* **60,** 326–329.
2. De Bault, L. E. and Cancilla, P. A. (1980) γ-glutamyltranspeptidase in isolated brain endothelial cells: induction by glial cells *in vitro. Science* **207,** 653–655.
3. Beck, D. W., Vinters, H. V., Hart, M. N., and Cancilla, P. A. (1984) Glial cells influence polarity of the blood-brain barrier. *J. Neuropathol. Exp. Neurol.* **43,** 219–224.
4. Beck, D. W., Roberts, R. L. M., and Olson, J. J. (1986) Glial cells influence membrane-associated enzyme activity at the blood-brain barrier. *Brain Res.* **381,** 131–137.
5. Tao-Cheng, J., Nagy, Z., and Brightman, M. W. (1987) Tight junctions of brain endothelium *in vitro* are enhanced by astrocytes. *J. Neurosci.* **7,** 3293–3299.
6. Arthur, F. E., Shivers, R. R., and Bowman, P. D. (1987) Astrocyte-mediated induction of tight junctions in brain capillary endothelium: an efficient *in vitro* model. *Dev. Brain Res.* **36,** 155–159.
7. Janzer, R. C. and Raff, M. C. (1987) Astrocytes induce blood-brain barrier properties in endothelial cells. *Nature* **325,** 253–257.
8. Goldstein, G. W. (1988) Endothelial cell-astrocyte interactions. A cellular model of the blood-brain barrier. *Ann. N. Y. Acad. Biol. Sci.* **529,** 31–39.
9. Maxwell, K., Berliner, J. A., and Cancilla, P. A. (1989) Stimulation of glucose analogue uptake by cerebral microvessel endothelial cells by a product released by astrocytes. *J. Neuropathol. Exp. Neurol.* **48,** 69–80.
10. Takakura, Y., Trammel, A. M., Kuentzel, S. L. (1991) Hexose uptake in primary cultures of bovine brain microvessel endothelial cells. II. Effects of conditioned media from astroglial and glioma cells. *Biochem. Biophys. Acta* **1070,** 11–19.
11. Tontsch, U. and Bauer, H. (1991) Glial cells and neurons induce blood-brain barrier related enzymes in cultured cerebral endothelial cells. *Brain Res.* **539,** 247–253.
12. Lobrinus, J. A., Juillerat-Jeanneret, L., Darekar, P., Schlosshauer, B., and Janzer, R. C. (1992) Induction of the blood-brain barrier specific HT7 and neurothelin epitopes in endothelial cells of the chick chorioallantoic vessels by a soluble factor derived from astrocytes. *Brain Res. Dev. Brain Res.* **70,** 207–211.
13. Pákáshi, M. and Kása, P. (1992) Glial cells in coculture can increase the acetylcholinesterase activity in human brain endothelial cells. *Neurochem. Int.* **21,** 129–133.
14. Joó, F. (1995) Isolated brain microvessels and cultured cerebral endothelial cells in blood-brain barrier research: 20 years on, in: *New concepts of a blood-brain barrier* (Greenwood, J., Begley, D. J. and Segal, M. B., eds.), Plenum Press, New York, pp. 229–237.
15. Ceballos, G. and Rubio, R. (1995) Coculture of astroglial and vascular endothelial cells as apposing layers enhances the transcellular transport of hypoxanthine. *J. Neurochem.* **64,** 991–999.
16. Rao, J. S., Sawaya, R., Gokaslan, Z. L., Yung, W. K., Goldstein, G. W., and Laterra, J. (1996) Modulation of serine proteinases and metalloproteinases during morphogenic glial-endothelial interactions. *J. Neurochem.* **66,** 1657–1664.
17. Holash, J. A., Noden, D. M., and Stewart, P. A. (1993) Re-evaluating the role of astrocytes in blood-brain barrier induction. *Dev. Dyn.* **197,** 14–25.
18. Krum, J. M. and Rosenstein, J. M. (1993) Effect of astroglial degeneration on the blood-brain barrier to protein in neonatal rats. *Brain Res. Dev. Brain Res.* **74,** 41–50.
19. Hirano, A., Kawanami, T., and Llena, J. F. (1994) Electron microscopy of the blood-brain barrier in disease. *Microsc. Res. Technique* **27,** 543–556.
20. Krum, J. M. (1996) Effect of astrocyte degeneration on neonatal blood-brain barrier marker expression. *Exp. Neurol.* **142,** 29–35.

21. Felts, P. A. and Smith, K. J. (1996) Blood-brain barrier permeability in astrocyte-free regions of the central nervous system and demyelinated by Schwann cells. *Neuroscience* **75,** 643–655.
22. Baver H.C. and Baver, H. (2000) Neural induction of the blood-brain barrier: Still an enigma, *Cell Mol. Neurobiol.* **20,** 13–28.
23. Roncali, L., Ribatti, D., and Ambrosi, G. (1985) Vasculogenesis in the chick embryo optic tectum. *Acta Anat.* **122,** 229–234.
24. Norman, M. G. and O'Kusky, J. R. (1986) The growth and development of microvasculature in human cerebral cortex. *J. Neuropathol. Exp. Neurol.* **46,** 222–232.
25. Rakic, P. (1995) Radial glial cells: scaffolding for brain construction, in *Neuroglia* (Kettenmann, H. and Ransom, B. R., eds.), Oxford University Press, Oxford, pp. 746–762.
26. Virgintino, D., Maiorano, E., Errede, M., et al. (1998) Astroglia-microvessel relationship in the developing human telencephalon. *Int. J. Dev. Biol.* **42,** 1165–1168.
27. Roncali, L., Ribatti, D., and Ambrosi, G. (1985) Ultrastructural basis of the vessel wall differentiation in the chick optic tectum. *J. Submicrosc. Cytol. Pathol.* **17,** 83–88.
28. Phelps, C. H. (1972) The development of glio-vascular relationships in the rat spinal cord. *Z. Zellforsch.* **128,** 555–563.
29. Fedoroff, S. (1986) Prenatal ontogenesis of astrocytes, in *Astrocytes* (Fedoroff, S. and Vernadakis, A., eds.), Academic Press, London, pp. 35–74.
30. Schmechel, D. E. and Rakic, P. (1979) A Golgi study of radial glia cells in developing monkey telencephalon: morphogenesis and transformation into astrocytes. *Anat. Embryol* **156,** 115–152.
31. Povlishock, J. T., Martinez, A. J., and Moossy, J. (1977) The fine structure of blood vessels of the telencephalic germinal matrix in the human fetus. *Am. J. Anat.* **149,** 439–452.
32. Choi, B. H. and Lapham, L. W. (1978) Radial glia in the human fetal cerebrum: a combined Golgi, immunofluorescent and electron microscopic study. *Brain Res.* **148,** 295–311.
33. Maynard, E. A., Scultz, R. L., and Pease, D. C. (1957) Electron microscopy of the vascular bed of the rat cerebral cortex. *Am. J. Anat.* **100,** 409–422.
34. Bertossi, M., Ribatti, D., Virgintino, D., Mancini, L., and Roncali, L. (1989) Computerized three-dimensional reconstruction of the developing blood-brain barrier. *Acta Neuropathol.* **79,** 48–51.
35. Ambrosi, G., Virgintino, D., Benagiano, V., Maiorano, E., Bertossi, M., and Roncali, L. (1995) Glial cells and blood-brain barrier in the human cerebral cortex. *It. J. Anat. Embryol.* **100,** 177–184.
36. Virgintino, D., Monaghan, P., Robertson, D., Errede, M., Bertossi, M., Ambrosi, G., and Roncali, L. (1997) An immunohistochemical and morphometric study on astrocytes and microvasculature in the human cerebral cortex. *Histochem. J.* **29,** 655–660.
37. Abbott, N. J. (1987) Glia and the blood-brain barrier. *Nature* **325,** 195.
38. Saunders, N. R., Dziegielewska, K. M., and Møllgard, K. (1991) The importance of the blood-brain barrier in fetuses and embryos. *Trends Neurosci.* **14,** 14.
39. Risau, W. and Wolburg, H. (1991) The importance of the blood-brain barrier in fetuses and embryos. *Trends Neurosci.* **14,** 15.
40. Bertossi, M., Riva, A., Virgintino, D., and Roncali, L. (1992) A correlative SEM/TEM examination of the endothelium surface in neural capillaries. *J. Submicrosc. Cytol. Pathol.* **24,** 215–224.
41. Bertossi, M., Mancini, L., Favia, A. et. al. (1992) Permeability-related structures in developing and mature microvessels of the chicken optic tectum. *Biol. Struct. Morphol.* **4,** 144–152.
42. Nico, B., Cantino, D., Bertossi, M., Ribatti, D., Sassoé, M., and Roncali, L. (1992) Tight endothelial junctions in the developing microvasculature: a thin section and freeze-fracture study in the chick embryo optic tectum. *J. Submicrosc. Cytol. Pathol.* **24,** 85–95.
43. Roncali, L., Nico, B., Ribatti, D., Bertossi, M., and Mancini, L. (1986) Microscopical and ultrastructural investigations on the development of the blood-brain barrier in the chick embryo optic tectum. *Acta Neuropathol.* **70,** 193–201.

44. Ribatti, D., Nico, B., and Bertossi, M. (1993) The development of the blood-brain barrier in the chick. Studies with Evans blue and horseradish peroxidase. *Ann. Anat.* **175,** 85–88.
45. Nico, B., Cardelli, P., Fiori, A., et al. (1997) Developmental study of ultrastructural and biochemical changes in isolated chick brain microvessels. *Miscovasc. Res.* **53,** 79–91.
46. Albrecht, U., Seulberger, H., Schwarz, H., and Risau, W. (1990) Correlation of blood-brain barrier function and HT7 protein distribution in chick brain circumventricular organs. *Brain Res.* **535,** 49–61.
47. Vorbrodt, A. W. (1988) Ultrastructural cytochemistry of blood-brain barrier endothelia. *Progr. Histochem. Cytochem.* **18,** 1–99.
48. Nico, B., Quondamatteo, F., Ribatti, D., et al. (1998) Ultrastructural localization of lectin binding sites in the developing brain microvasculature. *Anat. Embryol.* **197,** 305–315.
49. Nico, B., Quondamatteo, F., Herken, R., et al. (1999) Developmental expression of ZO-1 antigen in the mouse blood-brain barrier. *Dev. Brain Res.* **114,** 161–169.
50. Bertossi, M., Virgintino, D., Errede, M., and Roncali, L. (1999) Immunohistochemical and ultrastructural characterization of cortical plate microvasculature in the human foetus telencephalon. *Microvasc. Res.* **58,** 49–61.
51. Saunders, N. R., Habgood, M. D., and Dziegelewska, K. M. (1999) Barrier mechanisms in the brain, II. Immature brain. *Clin. Exp. Pharmacol. Physiol.* **26,** 85–91.
52. Pardridge, W. M., Boado, R. J., and Farrell, C. R. (1990) Brain-type glucose transporter (GLUT-1) is selectively localized to the blood-brain barrier. *J. Biol. Chem.* **265,** 18035–18040.
53. Bauer, H., Sonnleitner, U., Lametschwandtner, A., Steiner, M., Adam, H., and Bauer, H. C. (1995) Ontogenic expression of the erythroid-type glucose transporter (Glut 1) in the telencephalon of the mouse: correlation to the tightening of the blood-brain barrier. *Dev. Brain Res.* **86,** 317–325.
54. Farrell, C. L. and Pardridge, W. M. (1991) Blood-brain barrier glucose transporter is asymmetrically distributed on brain capillary endothelial luminal and abluminal plasma membranes: an electron microscopic immunogold study. *Proc. Natl. Acad. Sci. USA* **88,** 5779–5783.
55. Virgintino, D., Robertson, D., Monaghan, P., et al. (1997) Glucose transporter GLUT1 in human brain microvessels revealed by ultrastructural immunocytochemistry. *J. Submicrosc. Cytol. Pathol.* **29,** 365–370.
56. Virgintino, D., Robertson, D., Monaghan, P. et al. (1998b) Glucose transporter GLUT1 localization in human foetus telencephalon. *Neurosci. Lett.* **256,** 147–150.
57. Schumacher, U. and Møllgärd, K. (1997) The multidrug-resistance P-glycoprotein (Pgly, MDR1) is an early marker of blood-brain barrier development in the microvessels of the developing human brain. *Histochem. Cell Biol.* **108,** 179–182.
58. Balslev, Y., Dziegielewska, K. M., and Møllgärd, K. (1997) Intercellular barriers to and transcellular transfer of albumin in the fetal sheep brain. *Anat. Embryol.* **195,** 229–236.
59. Balabanov, R. and Dore-Duffy, P. (1998) Role of the CNS microvascular pericyte in the blood-brain barrier. *J. Neurosci. Res.* **53,** 637–644.
60. Bauer, H. C., Bauer, H., Lemetschwandtner, A., Amberger, A., Ruiz, P., and Steiner, M. (1993) Neovascularization and the appearance of morphological characteristics of the blood-brain barrier in the embryonic mouse central nervous system. *Brain Res.* **75,** 269–278.
61. Bertossi, M., Riva, A., Congiu, T., Virgintino, D., Nico, B., and Roncali, L. (1995) A compared TEM/SEM investigation on the pericytic investment in developing microvasculature of the chick optic tectum. *J. Submicrosc. Cytol. Pathol.* **27,** 349–358.
62. Van Deurs, B. (1980) Structural aspects of brain barriers, with special reference to the permeability of the cerebral endothelium and choroidal epithelium. *Int. Rev. Cytol.* **65,** 117–191.
63. Hirshi, K. K. and D'Amore, P. (1997) Control of angiogenesis by pericytes: molecular mechanisms and significance. *EXS* **79,** 419–428.
64. Risau, W. and Lemmon, V. (1988) Changes in the vascular extracellular matrix during embryonic vasculogenesis and angiogenesis. *Dev. Biol.* **125,** 441–450.

65. Herken, R., Götz, W., and Thies, M. (1990) Appearance of laminin, heparan sulphate proteoglycan and collagen Type IV during initial stages of vascularization of the neuroepithelium of the mouse embryo. *J. Anat.* **169,** 189–195.
66. Chiu, A. Y., Espinosa de los Monteros, A., Cole, R. A., Loera, S., and de Vellis, J. (1991) Laminin and s-laminin are produced and released by astrocytes, Schwann cells, and schwannomas in culture. *Glia* **4,** 11–24.
67. van der Laan, L. J., Groot, C. J., and Dijkstra, C. D. (1997) Extracellular matrix proteins expressed by human adult astrocytes *in vivo* and *in vitro:* an astrocyte surface protein containing the CS1 domain contibutes to binding of lymphoblasts. *J. Neurosci. Res.* **15,** 539–548.
68. Bär, T. (1980) The vascular system of the cerebral cortex. *Adv. Anat. Embryol. Cell Biol.* **59,** 1–62.
69. Bär, T. and Wolff, J. R. (1972) The formation of capillary basement membranes during internal vascularization of the rat's cerebral cortex. *Z. Zellforsch.* **133,** 231–248.
70. Krum, J. M., More, N. S., and Rosenstein, J. M. (1991) Brain angiogenesis: variations in vascular basement membrane glycoprotein immunoreactivity. *Exp. Neurol.* **111,** 152–165.
71. Pixley, S. K. and de Vellis, J. (1984) Transition between radial glia and mature astrocytes studied with a monoclonal antibody to vimentin. *Dev. Brain Res.* **15,** 201–209.
72. Virgintino, D., Maiorano, E., Bertossi, M., Pollice, L., Ambrosi, G., and Roncali, L. (1993) Vimentin- and GFAP-immunoreactivity in developing and mature neural microvessels. Study in the chicken *tectum* and *cerebellum. Eur. J. Histochem.* **37,** 353–362.
73. Eng, L. F. E. and Lee, Y., L. (1995) Intermediate filaments in astrocytes, in *Neuroglia* (Kettenmann, H. and Ramson, B. R., eds.), Oxford University Press, Oxford, pp. 650–667.
74. Dermietzel, R. (1974) Junctions in the central nervous system of the cat. III. Gap junctions and membrane-associated orthogonal particle complexes in astrocytic membranes. *Cell Tissue Res.* **149,** 121–135.
75. Landis, D. M. D. and Reese, T. S. (1974) Arrays of particles in freeze-fractured astrocytic membranes. *J. Cell. Biol.* **60,** 316–320.
76. Anders, J. J. and Brightman, M. W. (1979) Assemblies of particles in the cell membranes of developing, mature and reactive astrocytes. *J. Neurocytol.* **8,** 777–795.
77. Landis, D. and Reese, T. S. (1981) Membrane structure in mammalian astrocytes: a review of freeze-fracture studies on adult, developing, reactive and cultured astrocytes. *J. Exp. Biol.* **95,** 35–48.
78. Neuhaus, J. (1990) Orthogonal arrays of particles in astroglial cells: quantitative analysis of their density, size, and correlation with intramembranous particles. *Glia* **3,** 241–251.
79. Nico, B., Cantino, D., Sassoé Pognetto, M., Bertossi, M., Ribatti, D., and Roncali, L. (1994) Orthogonal arrays of particles (OAPs) in perivascular astrocytes and tight junctions in endothelial cells. A comparative study in developing and adult brain microvessels. *J. Submicrosc. Cytol. Pathol.* **26,** 103–109.
80. Rash, J. E., Yasumara, T., Hudson, C. S., Agre, P. P., and Nielsen, S. (1998) Direct immunogold labeling of aquaporin-4 in square arrays of astrocyte and ependymocyte plasma membranes in rat brain and spinal cord. *Proc. Natl. Acad. Sci.* **95,** 11981–11986.
81. Gotow, T. and Hashimoto, P. H. (1989) Developmental alterations in membrane organization of rat subpial astrocytes. *J. Neurocytol.* **18,** 731–747.
82. Bertossi, M., Virgintino, D., Maiorano, E., Occhiogrosso, M., and Roncali, L. (1997) Ultrastructural and morphometric investigation of human brain capillaries in normal and peritumoral tissue. *Ultrastruct. Pathol.* **21,** 41–49.
83. Distler, C., Dreher, Z., and Stone, J. (1991) Contact spacing among astrocytes in the central nervous system: an hypothesis of their structural role. *Glia* **4,** 484–494.

84. Dermietzel, R., Traub, O., Hwang, T. K., et al. (1989) Differential expression of three gap junction proteins in the developing and mature brain tissues. *Proc. Natl. Acad. Sci. USA* **86,** 10148–10152.
85. Nadarajah, B., Jones, A. M., Evans, W. H., and Parnavelas, J. G. (1997) Differential expression of connexins during neocortical development and neuronal circuit formation. *J. Neurosci.* **17,** 3096–3111.
86. Delorme, P., Gayet, J., and Grignon, G. (1970) Ultrastructural study on transcapillary exchanges in the developing telencephalon of the chicken. *Brain Res.* **22,** 269–283.
87. Fujimoto, K. (1995) Pericyte-endothelial gap junctions in developing rat cerebral capillaries: a fine structural study. *Anat. Rec.* **242,** 562–565.
88. Cuevas, P., Gutierrez Diaz, J. A., Reimers, D., Dujovny, M., Diaz, F. G., and Ausman, J. I. (1984) Pericyte endothelial gap junctions in human cerebral capillaries. *Anat. Embryol.* **170,** 155–159.
89. Leybaert, L., Paemeleire, K., Strahonja, A., and Sanderson, M. J. (1998) Inositol-trisphosphate-dependent intercellular calcium signaling in and between astrocytes and endothelial cells. *Glia* **24,** 398–407.
90. Bertossi, M., Roncali, L., Nico, B., et al. (1993) Perivascular astrocytes and endothelium in the development of the blood-brain barrier in the optic tectum of the chick embryo. *Anat. Embryol.* **188,** 21–29.
91. Tao-Cheng, J., Nagy, Z., and Brightman, M. W. (1990) Astrocytic orthogonal arrays of intramembranous particle assemblies are modulated by brain endothelial cells *in vitro*. *J. Neurocytol.* **19,** 143–153.
92. Estrada, C., Bready, J. V., Berliner, J. A., Pardrige, W. M., and Cancilla, A. (1990) Astrocyte growth stimulation by a soluble factor produced by cerebral endothelial cells *in vitro*. *J. Neuropathol. Exp. Neurol.* **49,** 539–549.

VI Astrocytes in Neurodegenerative Diseases

19

Microglial and Astrocytic Reactions in Alzheimer's Disease

Douglas G. Walker and Thomas G. Beach

1. INTRODUCTION

In the last 20 years, there have been major advances in understanding the causes of Alzheimer's disease (AD) that have allowed the development of different treatment strategies. Progress has been particularly rapid since 1984, with the determination of the amino acid sequence of the amyloid beta (Aβ) peptide in cerebrovascular amyloid *(1)*, and plaque cores *(2)*. From this finding, the identification of the amyloid precursor protein (and its gene), *(3,4,5)* as being the protein from which plaque Aβ is derived, was possible.

There are now known to be gene defects on chromosome 1, 14 or 21 that can lead to early onset forms of AD *(6,7,8,)*. These cases are extremely rare, but have helped in the understanding of the possible mechanisms involved in AD cases that have no indentifiable genetic component (possibly 80%). The majority of AD cases can be considered late-onset, with significant incidence only being detectable after age 65. Although AD is primarily a disease of aging, a number of risk factors associated with this disease have been identified. The major one is possession of the apolipoprotein E4 (ApoE4) allele. This allele is possessed by approx 16% of the normal human population, whereas approx 50% of late-onset AD patients possess the ApoE4 allele *(9)*. This finding has been repeated in a large number of subsequent studies with similar results. It does mean however that approx 50% of AD patients *do not* have the ApoE4 allele. Other genetic markers, including one on chromosome 12 have been proposed, but confirmations are still required *(10–12)*. Environmental risk factors that have been linked to AD are head trauma *(13)*, old age *(14)*, and coronary artery and cerebrovascular disease *(15)*. A role for aluminum *(16)* or viruses *(17)* has also been proposed, however the evidence has never been convincing.

Common pathological features for all types of AD are the development of Aβ-containing plaques, neurofibrillary tangles, dystrophic neurites and also the development of a reactive glial response. The purpose of this article is to consider the possible role of the reactive glial response in the neurodegenerative changes and how it may be involved in the pathogenesis of AD.

From: *Neuroglia in the Aging Brain*
Edited by: Jean S. de Vellis © Humana Press Inc., Totowa, NJ

2. MICROGLIAL REACTIONS IN AD BRAINS

Microglia are considered to be a population of dendritic cells that have the same origin as monocytes and macrophages, but are normally resident in the brain. In comparison to the other glial cells, the oligodendrocytes and astrocytes, microglia are believed to have migrated into the brain tissue during early brain development, and are not derived from neuroectoderm cells. With the development of modern immunohistochemical techniques and monoclonal antibodies that recognize macrophage proteins, there has been a number of published reports describing the interactions of microglia with the hallmark pathological structures of AD (examples *(18–26,68,97,99,117,123,125,144)*. However it should be remembered that the early pioneers of neuroscience, including Hortega, *(27)* and Penfield *(28)* using classical histochemical staining methods, observed the microglia in the brain, and suggested their potential role in degenerative diseases such as AD.

The typical appearances of microglia that are closely associated with senile plaques are shown in Figure 1A. This section from the temporal cortex of an AD patient was stained with the antibody CR3/43 that recognizes the major histocompatibility complex class II proteins, particularly HLA-DR. In many immunohistochemical studies, this marker has been used to identify activated microglia within gray matter regions of affected brains. In the figure shown (Fig. 1A), the staining identifies a cluster of microglial cells with hypertrophied cell bodies and shortened-enlarged processes, which are located on and around an Aβ immunoreactive plaque. The morphology of reactive and also resting microglia is shown in figure 1B. This section was stained with an antibody to CD64 that reacts with all microglial cells. The resting cells have a smaller cell body with longer multipolar-ramified processes. The significance of MHC class II expression as a marker of microglial activation is complicated by the observations that robust expression of this marker can be observed on microglia in white matter regions of normal human brain. Figure 1C shows the margin between the white matter region of cortex (lower section) and gray matter (upper section) of an AD case. Although this case had extensive AD neuropathological changes in the gray matter, the strongest HLA-DR immunoreactive microglia were found throughout the white matter. Similarly, in non-AD cases, HLA-DR immunoreactive microglia are also found throughout the white matter *(21,29)*. Although the significance of increased microglial

Fig. 1. ***(Right)*** **(A).** Photomicrograph showing highly reactive microglia clustered on and around Aβ immunoreactive plaques. The section is from the inferior temporal gyrus of an AD case. The section was stained with the antibody CR3/43 that recognizes the MHC class II protein HLA-DR. **(B).** Photomicrograph showing a range of different morphologies of microglia in a section from an AD case. The section was stained with an antibody to CD64 (FcγRI). This antibody reacts with all microglia, but more strongly with those with a reactive morphology. **(C).** Photomicrograph showing the strong HLA-DR immunoreactivity of microglia in white matter. The section is from an AD case and shows the boundary between white matter (lower section) and gray matter (upper section). In the gray matter, the microglia are more associated with pathological structures, while in the white matter the staining is more diffuse and not apparently associated with pathology. In normal cases, HLA-DR immunoreactive microglia can be observed in white matter, but generally not in gray matter.

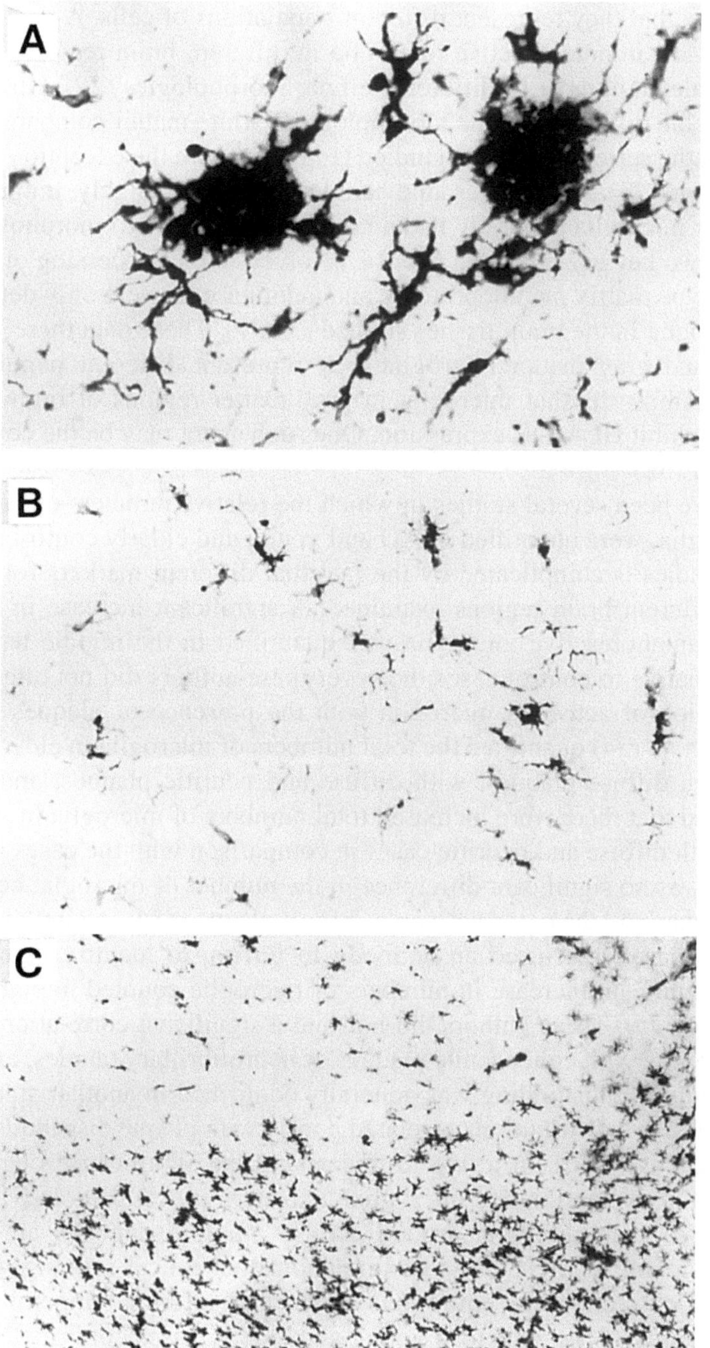

HLA-DR immunoreactivity in relation to defining their "activated" state still remains to be determined, it has proven to be the best marker for identifying activated microglia in areas where degenerative pathology exists, particularly in gray matter regions of the brain. The difference in reactivity between microglia cells in different brain regions could suggest that they represent different populations of cells. A detailed comparison of the HLA-DR immunoreactive microglia in different brain regions, including white matter, revealed a number of different cellular morphologies *(29)*. However, there was not a noticeable difference in the morphology of white matter compared to gray matter microglia in the series of sections studied (Fig. 1C), with the exception of the microglia associated with the Aβ plaques and tangles. Although highly immunoreactive, the white matter microglia generally had a ramified (unactivated) morphology. It has been shown that two key enzymes that may be involved in the processing of Aβ peptide, the membrane-type matrix metalloprotease and gelatinase A, were only detectable in white matter microglia in the brain tissues studied *(30,31)*. These data therefore also suggest that white and gray matter microglia may represent different populations of cells. Another possibility is that microglia in gray matter regions of brain are exposed to agents that inhibit HLA-DR expression. One such agent may be the colony stimulating factor CSF-1 *(32)*.

There have been several studies in which the relative numbers of microglia, or activated microglia, were quantified in AD and young and elderly control brains. Comparing these studies is complicated by the fact that different markers for microglia were used and different brain regions examined. A significant increase in total number of HLA-DR immunoreactive microglia was quantified in the middle temporal gyrus of AD cases relative to control cases, however these authors did not attempt to correlate the distribution of activated microglia with the presence of plaques or tangles *(33)*. Mackenzie et al. *(34)* quantified the total numbers of microglia in elderly cases with no plaques, with diffuse plaques, with diffuse and neuritic plaques, and with AD. This study showed that there were increased total numbers of microglia in AD cases, and in the cases with diffuse and neuritic cases in comparison with the cases with no plaques. Also, there was no significant difference in the number of microglia between the cases with no plaques and the cases with only diffuse plaques *(34)*. A similar study of young, old and AD cases, that used an antibody to ferritin to identify activated microglia, showed a significant increase in numbers of microglia counted in eight regions of the hippocampus *(35)*. These authors did not find a significant correlation of the localization of microglia with either senile plaques or neurofibrillary tangles, except for tangles in the subiculum. This finding was generally confirmed in another study, which found that microglial cell distribution correlated poorly with plaque distribution in the dentate gyrus of AD cases *(36)*. In trying to understand the significance of microglial MHC class II protein expression and microglial activation in AD pathology, it has generally been observed that microglia around diffuse amyloid deposits are not expressing increased amounts of MHC class II. Immunohistochemical staining with an antibody to HLA-DR of sections of entorhinal cortex and inferior temporal gyrus from the brains of a series of nondemented young and elderly cases confirmed this observation (Schwab et al., unpublished observations). There was a significant increase in numbers of HLA-DR immunoreactive microglia when comparing the elderly with the young cases. However, there was no significant difference in numbers of HLA-DR reactive microglia between the elderly cases without Aβ plaques and those that did have plaques. These cases primarily had diffuse plaques in the regions studied (Schwab et al., unpublished observations). This finding generally confirms the observations of others where there was not increased HLA-DR immunoreactivity associated with diffuse plaques in AD cases, in the cerebellum of control or AD cases, or diffuse Aβ plaques of cases of hereditary cerebral hemorrhage with amyloidosis (Dutch) *(20,21,23,26,37)*. Fukumoto et al. *(38)* have refined these observations with the use of antibodies that can distinguish between Aβ plaques ending at amino acid 40 and plaques ending at amino acid 42. A significantly higher proportion of Aβ-40 plaques were associated with activated microglia than Aβ-42 plaques. These results were similar in the AD cases and nondemented elderly cases. This result has been interpreted as evidence that the microglia around the plaques are processing the Aβ-42 to the Aβ-40 form, however it might also be interpreted to indicate that the Aβ-42 is deposited earlier than Aβ-40, and prior to microglial activation.

A series of studies by Griffin and coworkers have indicated that microglia around diffuse plaques may also be activated. In contrast to most studies, these authors have used increased expression of the cytokine interleukin 1α (IL-1α) as the marker for microglial activation *(57,131,134,136)*. IL-1α is a potent proinflammatory cytokine that is induced in macrophage cells by a range of proinflammatory stimuli. These authors observed the highest numbers of IL-1α immunoreactive microglia associated with uncored neuritic plaques *(39)*. In vitro studies have shown that the activation signals that cause induction of IL1α are not the same as those that activate the transcription of the MHC class II genes *(27,32)*.

A similar type of microglia response to Aβ has been observed in the brains of transgenic mice carrying copies of the human APP gene with AD-associated mutations, or rats whose brains have been infused with preparations of synthetic Aβ. Some of these animals develop AD-like neuritic plaques which have associated reactive microglia and astrocytes *(40–43)*. Evidence that the microglial reaction is in response to the Aβ plaques was shown by the fact that it was more pronounced in the transgenic mice that had the higher burden of amyloid plaques *(43)*.

There is still some controversy relating to whether there is proof that the activated microglia are contributing to the AD pathology or whether the observed activation is an epiphenomenon. Although in vitro studies indicate the potential for microglia to produce neurotoxic products, there is only indirect evidence that this is occurring in vitro. Morphometric studies that counted the number and distribution of activated microglia in AD and aging brains showed that there was not significant correlation between microglial cell density and plaque or tangle density *(35,36)*. Both of those studies utilized ferritin as the marker to identify activated microglia. Although, as shown, the most activated microglia appear to be arranged around dense Aβ plaques, overall the pattern of activation is quite diffuse throughout the neuropil of affected brain regions. This might indicate that the insoluble Aβ or tangles were not the primary activation signals. Microglial activation may also be in response to neuropil threads, synaptic loss, neurofibrillary tangle formation, axonal degeneration or other factors. Many studies have shown that microglial activation is significantly associated with neuritic plaques. It can thus be suggested that some soluble proteins released from damaged neurons may be the key factors involved in microglial activation *(44)*.

3. INCREASED ASTROCYTOSIS IN AD

It has been established that there are increased numbers of glial fibrillary acidic protein (GFAP)-immunoreactive astrocytes in the AD cerebral cortex. A qualitative report

Table 1
Biologically active compounds secreted by astrocytes and associated with AD pathogenesis

Cytokines
Interleukin 6 *(166)*
GM-CSF *(58)*
MCSF-1 *(32)*
Complement proteins
Complement C3 *(167)*
Complement C4 *(168)*
Complement C9 *(168)*
Complement inhibitor proteins
C1-inhibitor *(169,170)*
Clusterin *(169)*
Vitronectin *(171)*
Protease/Protease inhibitors
α-1 antichymotrypsin *(173)*
α-2-macroglobulin *(174)*
Lipid transport molecules
Apolipoprotein E *(175)*
Apolipoprotein J (clusterin) *(169)*
Free radicals
Nitric oxide (Nitric oxide synthase) *(93)*

by Duffy et al. *(45)* was followed by quantitative work by Schechter et al. *(46)* and Mancardi et al. *(47)*, which demonstrated fourfold or greater increases in GFAP-immunoreactive astrocytes. Mapping of the pattern of cerebral GFAP immunoreactivity in AD cases *(48)* found all cortical areas to be consistently and uniformly affected, with subcortical nuclei being frequently but inconsistently involved.

One of the established concepts that requires discussion is the idea that reactive astrocytes are playing a supportive role in trying to help "unhealthy" neurons to survive *(49)*. It has been documented that reactive astrocytes can express a range of neurotrophic agents including nerve growth factor (NGF), acidic and basic fibroblast growth factor (FGF), neurotrophin (NT)-3, brain-derived neurotrophic factor (BDNF), midkine and glial cell line-derived neurotrophic factor (GDNF) *(50–54)*. In addition, astrocytes have been shown to produce a range of antioxidants, protease, and complement inhibitors (see Table 1). However, evidence now exists that astrocytes could be contributing directly or indirectly to the pathological process in AD. This includes the possible production of neurotoxic substances, inhibition of microglial phagocytosis of amyloid plaques *(55,56)* and the production of pathological chaperone proteins *(57)*.

Staining of human AD tissue sections with an antibody to the glial fibrillary acidic protein (GFAP), the most widely used marker to identify astrocytes, shows that the most reactive astrocytes are arranged around the compact Aβ plaques (Fig. 2A and 2B).

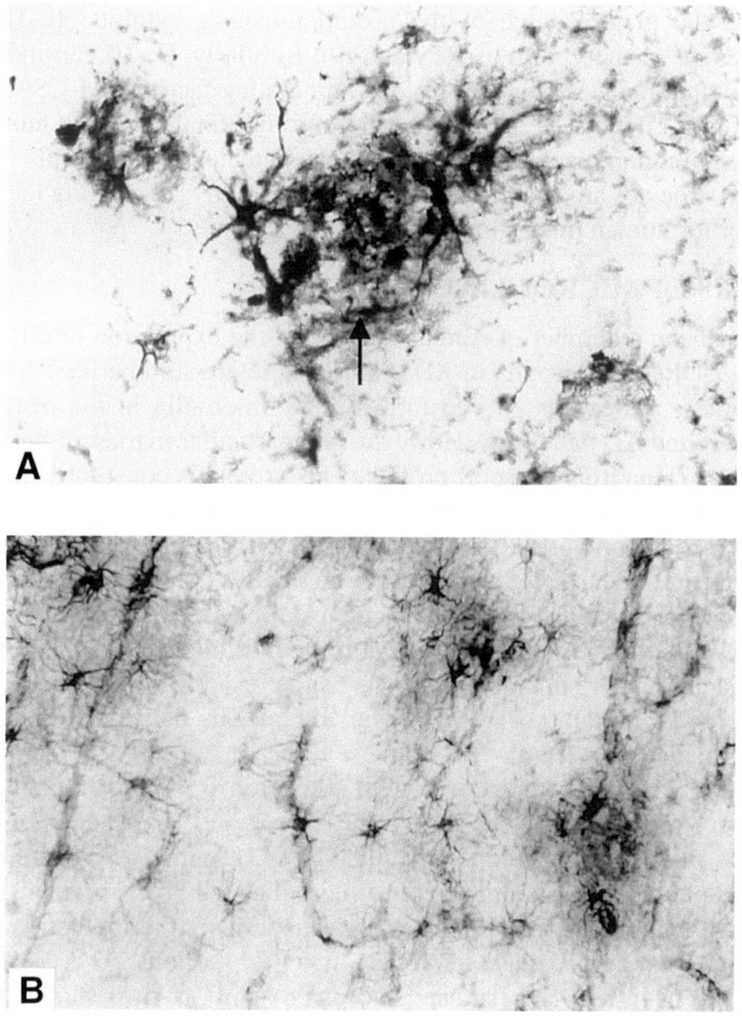

Fig. 2. (A). Photomicrograph of reactive astrocytes clustered around an Aβ-immunoreactive plaque (arrow). The section from an AD case was immunostained with an antibody to glial fibrillary acidic protein. In comparison to the microglia illustrated in Fig. 1A, reactive astrocytes tend to be found around the periphery of plaques. **(B).** Lower magnification photomicrograph of reactive astrocytes in the gray matter of an AD case. The immunoreactive cells are found associated with AD pathological structures. In human brains, gray matter astrocytes with a resting morphology exhibit weak or undetectable GFAP immunoreactivity.

This is in contrast to the clusters of highly reactive microglia that tend to be located on the plaque core (Fig. 1A). In the latter case it can be suggested that the microglia are attempting to phagocytose the extracellular Aβ peptide, however the function of the astrocytes in this process is not clear. Similar to what can be observed for HLA-DR in normal brain, there is stronger GFAP immunoreactivity in the white matter brain regions than the gray matter. One unique feature about the glial reaction in the AD brain may be the cooperation between the highly activated microglia and astrocytes located around the plaque. This could be exacerbating the glial response. Microglia appear to be the major source of the proinflammatory cytokine IL-1β, which can induce a reactive phenotype in astrocytes *(49)*. Similarly, IL-1β can induce astrocyte expression of granulocyte-monocyte colony stimulating factor (GM-CSF). This growth factor is the most effective in inducing proliferation of fetal and adult human microglia *(58)*. GM-CSF immunoreactive astrocytes can be detected around senile plaques in AD brains *(58)*. Astrocytes are also the major source of colony stimulating factor-1, another growth factor for human microglia *(58)*.

4. MARKERS OF MICROGLIAL ACTIVATION

There have been a number of studies describing the expression of different markers (besides HLA-DR) on microglia in AD brains. These are summarized in Table 2. Most of these markers are expressed constitutively by microglia in the resting state, and upregulated around AD pathology. However, one particular marker of interest is melanotransferrin (p97), an iron transport protein. This protein is constitutively expressed by vascular endothelial cells, but is also selectively expressed by a small subset of microglia in association with Aβ plaques *(59,60)*. Significantly elevated serum concentrations of this protein have been measured in AD cases compared to control cases, and it has been proposed as a potential diagnostic test for AD *(61)*. A number of other studies have shown that ferritin, an iron binding protein, is also a good marker for identifying microglia and under suitable conditions can be used to identify reactive microglia *(35,38)*. Microglial immunoreactivity for ferritin appears not to be as restricted as that observed for melanotransferrin.

Another potential marker for microglial activation is immunoreactivity with antibodies that recognize phosphotyrosine-modified proteins. There has been limited application of this marker for histological studies *(40,43)*, with only one report of results showing phosphotyrosine immunreactive microglia in human brain tissue from AD cases *(62)*. Although many cells possess tyrosine kinases, antibodies to phosphotyrosine tend to recognize activated microglia in tissue sections. The usefulness of this marker has not been fully investigated, however many in vitro studies with cultured macrophages or microglia show that increased tyrosine phosphorylation occurs rapidly as a consequence of many different stimuli that activate tyrosine kinases *(63,64)*. These studies generally showed that tyrosine phosphorylation was short lived as a result of the action of different cellular phosphatases. Immunoreactivity for phosphotyrosine would thus appear to identify microglia that were acutely activated at the time of death. It will be of interest to compare the correlation between microglial expression of HLA-DR and phosphotyrosine.

5. MARKERS OF ASTROCYTIC ACTIVATION

GFAP is a 57 kD intermediate filament protein that is only expressed by astrocytes. In the normal human brain, astrocytes in gray matter regions show weak or undetectable GFAP immunoreactivity. A noticeable increase in GFAP immunoreactivity becomes apparent in areas of degeneration or injury, along with an increase in the size of the astrocyte cell body and its processes. Similar to what has been described for HLA-DR, strong GFAP immunoreactivity is observable in white matter areas of human brains lacking pathology. The significance of increased GFAP expression by astrocytes

Table 2
Expression of markers on microglia in Alzheimer's Disease

Marker	Function	Expression in AD (ref)
MHC class I	Antigen presentation to CD8 positive T cells	Increased *(120)*
MHC class II	Antigen presentation to CD4 positive cells	Increased *(22)*
CD11a (LFA-1)	ICAM-1 and ICAM-11 receptor	Increased *(121,122)*
CD11b (CR3)	Complement C3bi receptor	Increased *(121,122)*
CD11c (CR4/p150)	Complement C3bi receptor	Increased *(121,122)*
CD18 (β2 integrin)	Subunit of CD11 complex	As CD11a, b, c
CD25 (Tac)	Interleukin-2 receptor	Increased *(22)*
CDw32 (FcγRII)	IgG Fc fragment receptor	Increased *(121)*
CD43 (leukosialin)	Antiadhesion molecule	Decreased *(123)*
CD45 (LCA)	Receptor with phosphotyrosine phosphatase activitiy	Increased *(124)* CD45RO
CD51 (αV integrin)	Vitronectin receptor	Increased *(122,125)*
CD64 (FcγRI)	IgG Fc fragment receptor	Increased *(121)*
CD68 (macrosialin)	Lysosome-associated membrane protein	Increased *(126)*
Interleukin-1α	Proinflammatory cytokine	Increased *(39)*
Interleukin-1β	Proinflammatory cytokine	Increased *(127)*
Interferon-α	Pro/anti-inflammatory cytokine	Increased *(128)*
Interferon-α receptor	Interferon-α receptor	Increased *(129)*
Monocyte chemoattractant protein-1	Chemotactic cytokine	Increased *(130)*
Transforming growth factor-β2	Anti-inflammatory cytokine	Increased *(28)*
Tumor necrosis factor α	Proinflammatory cytokine	Increased *(127)*
MCSF-receptor (c-fms)	Colony stimulating factor receptor	Increased *(131)*
Ferritin	Iron transport protein	Increased *(38)*
Melanotransferrin	Iron transport protein	Increased *(59)*
Phosphotyrosine-modified proteins	Product of tyrosine kinase action	Increased *(132)*
Macrophage scavenger receptor	Lipoprotein receptor	Increased *(133,134)*
VLDL-receptor	Lipoprotein receptor	Increased *(136)*
Vimentin	Cytoskeletal protein	Increased *(137)*
Plasminogen activator inhibitor	Serpin protease inhibitor	Increased *(138)*
Factor XIIIa	Coagulation factor	Increased *(139)*
Bcl-xl	Apoptosis inhibitor	Increased *(140)*
Tissue factor pathway inhibitor-1	Protease inhibitor	Increased *(141)*
Myeloperoxidase	Enzymes, generates free radicals	Increased *(142)*
CCR3, CCR5	Chemokine receptors	Increased *(143)*
C/EBP α	Transcription factor	Increased *(144)*

Table 3
Expression of markers on astrocytes in Alzheimer's Disease

Marker	Function	Expression in AD (ref)
GFAP	Intermediate filament	Increased *(48)*
S100 β	calcium binding protein	Increased *(69)*
TGF-β2	Anti-inflammatory cytokine	Increased *(145)*
Heat shock protein 27	Stress protein	Increased *(146)*
Basis fibroblast growth factor	Neurotrophic agent	Increased *(53)*
Acidic fibroblast growth factor	Neurotrophic agent	Increased *(54)*
Fibroblast growth factor-9	Neurotrophic agent	Increased *(147)*
FGF receptor 1	Fibroblast growth factor receptor	Increased *(148)*
TrkA receptor	Nerve growth factor receptor	Increased *(149)*
TrkB receptor	Nerve growth factor receptor	Increased *(150)*
MIP-1β	Chemokine	Increased *(143)*
Hepatocyte growth factor	Growth factor	Increased *(151,152)*
Metallothioneins I and II		Increased *(153)*
α-1 antichymotrypsin	Protease inhibitor	Increased *(154)*
Apolipoprotein E	Lipid transport protein	Increased *(37)*
Apolipoprotein J (clusterin)	Lipid transport protein	Increased *(155)*
PP-2A, PP-2B	Protein phosphatases	Increased *(156)*
Insulin growth factor-1	Growth factor	Increased *(157)*
Manganese superoxide dismutase	Free radical scavenger	Increased *(158)*
Bcl-xl	Apoptosis inhibitor	Increased *(140)*
Cytosolic phospholipase A2	Phospholipase	Increased *(159)*
Fas	Cell surface apoptosis initiator	Increased *(160)*
Nitric oxide synthase	Enzyme	Increased *(95)*
CD44	Receptor/adhesion molecule	Increased *(161)*
Intercellular adhesion molecule-1	Adhesion molecule	Increased *(162)*
Vimentin	Intermediate filament	Increased *(137)*
Vasoconstrictor endothelin-1	Vasoconstrictor peptide	Increased *(163)*

as they develop a reactive profile remains to be determined. There have been conflicting reports on this matter. Treatment of cultured astrocytes with antisense sequences of GFAP prevented them from developing a reactive profile *(65–67)*. However, on the other hand, mice with a deletion of the GFAP gene were able to develop a normal reactive gliosis response to an experimental insult *(66)*. However, a recent study showed that astrocytes from mice lacking expression of GFAP could not develop a reactive response to Aβ *(68)*.

Other markers for activated astrocytes have been described in AD brains. These are listed in Table 3. In contrast to GFAP, none of these are specific to astrocytes, but are more highly expressed by these cells compared to others.

Of the listed markers, S100β has been especially interesting as a marker for activated astrocytes. S100β is a calcium-binding protein secreted by glial cells in the central and peripheral nervous systems. S100β promotes neuronal differentiation and survival but may be detrimental to cells if overexpressed. It is also a mitogen for glial cells. The overexpression of S100 β in AD brains relative to control cases, correlated well with the pattern of regional involvement by neuritic plaques, with the largest increases being in the hippocampus and temporal cortex, and the smallest in the cerebellum *(19,69)*. S100β was shown to increase APP mRNA in neuronal cultures, and Aβ peptide increased S100β expression in astrocyte cultures *(70–73)*. It has been suggested that neurite-promoting properties of S100β may contribute to the formation of dystrophic neurites associated with certain classes of Aβ plaques *(74)*. The expression of S100β by astrocytes was increased by the cytokines IL-1α and IL-1β *(75.)* These data confirm the probable interaction between microglia and astrocytes contributing to AD neuropathology.

6. CONTRIBUTION OF GLIA TO AD PATHOLOGY

There has been much speculation as to the contribution of reactive glia to the formation of AD type pathology. Some researchers believe that it is the interaction of these glial cells with the Aβ peptide, rather than the direct toxic effect of Aβ on neurons, that is causing much of the AD neuropathology. There has been a range of glial-derived toxic agents suggested, including nitric oxide, reactive oxygen species, cytokines, glutamate, phenolic amine like compound and peptides *(76–84)*. We will not try and discuss them all, but concentrate on discoveries where the in vitro data is backed up with in vivo observations.

7. MICROGLIA-DERIVED NEUROTOXINS

Of particular note has been the observations on the production by cultured microglia of a putative small molecule phenolic-amine like toxin (designated NTox), when these cells interact with Aβ derived from plaque cores purified from AD brains, or synthetic Aβ peptides corresponding to the human, but not rodent, sequences *(85,86)*. This compound has not been fully characterized chemically, but is specifically induced by microglia upon interaction with Aβ peptides containing the 10–16 domain sequence and the sequences at the C terminal of the peptide (ending in 40 or 42) *(86)*. The data suggest that the Aβ region 13–16 is required for binding of the peptide to a putative microglia receptor, whereas the C-terminal sequence is necessary for the induction of the NTox. NMDA receptor antagonists could block the action of this toxin on neuronal cells *(86)*. The induction of toxin by microglia could be blocked by treatment of cells with the peptide HHQK, which corresponds to Aβ amino acids 13–16 *(87)*. Using synthetic Aβ peptides, there did not appear to be a significant difference between Aβ 1–40 and Aβ 1–42 in their efficacy in inducing the production of this toxin. This toxin was produced in a similar manner by Aβ-stimulated human or rodent microglia. A toxin that was isolated directly from AD brains behaved in the same neurotoxic manner as the toxin produced by Aβ-stimulated microglia *(85)*.

8. ASTROCYTE-DERIVED NEUROTOXINS

A possible involvement of toxic forms of nitric oxide (NO) in AD has been suggested by the demonstration of increased amounts of nitrated residues of tyrosine in neurofibrillary tangles, and in neurons showing evidence of DNA damage *(88,89)*. The presence of nitrotyrosine-modified proteins is direct evidence for the action of peroxynitrite. Increased production of NO, along with its interaction with superoxide, can result in the formation of peroxynitrite, a highly reactive product. This can oxidize proteins resulting in the formation of carbonyl, as well as modify tyrosine to form nitrotyrosine. Nitrotyrosine can be measured by high performance liquid chromatography techniques, along with electrochemical detection, whereas it can be localized by immunohistochemistry with nitrotyrosine specific antibodies *(90)*.

There are three different nitric oxide synthase (NOS) enzymes present in different cells of the brain. NOS type 1 is found in neurons, NOS type III is found mainly in endothelial cells, whereas NOS type II (also called inducible NOS) is induced by astrocytes and microglia in response to inflammatory stimuli. It is thus possible that the increased synthesis of NO could be derived from cells other than glia, however activated glial cells can produce large amounts of NO relative to neurons or endothelial cells. It has been demonstrated that microglia from rodents can be stimulated to produce large amount of NO as a result of stimulation of expression of the inducible NOS gene. However it appears that human microglia do not readily express inducible NOS *(26,64,80)*, though this is still a subject of controversy *(91)*.. Human astrocytes can be stimulated by inflammatory cytokines to produce NO *(92,93)*. NO derived from activated astrocytes can be neurotoxic to human neurons *(77,94)*. In vitro studies have shown that the astrocytic produced NO can induce apoptosis in cocultured neurons. The neurotoxic effect can be blocked by coincubation of stimulated astrocytes with a NOS inhibitor *(77,94)*. Astrocytes expressing inducible NO synthase have been observed associated with plaques in AD brains, using enzyme and immunohistochemistry techniques *(95)*. This in vivo observation provides evidence for the possible contribution of astrocyte derived NO to AD pathogenesis.

9. ASTROCYTE-DERIVED PATHOLOGICAL CHAPERONE PROTEINS

The contribution of the amyloid-associated proteins to the pathogenesis of the disease has been intensely studied. A large number of proteins have been shown to become associated with Aβ plaques and neurofibrillary tangles. Of particular interest are the complement proteins. The potential role of these in AD have been extensively reviewed *(96,97,98)*.

The possible microglial and astrocytic origins of some of these are listed in Tables 1 and 4. The proteins that have been most intensively studied have been apolipoprotein E (apoE) *(57)*, α-1-antichymotrypsin *(99)*, heparan sulfate proteoglycan *(100)*, apolipoprotein J (apoJ-clusterin) *(101)*, and complement C1q *(102)*. These compounds are of particular interest as they have been shown to modulate the rate of formation of Aβ peptide into a beta-pleated sheet configuration. A consequence of this would be either to transform soluble (diffuse) Aβ to the toxic fibril form, or to stabilize the plaques and tangles thus rendering them more resistant to degradation. ApoJ is also associated with Aβ plaques, however binding of this protein to Aβ appears to inhibit fibril formation *(101)*. Some studies have suggested that apoJ-Aβ complexes are also neurotoxic *(103)*. With the exception of C1q, which is derived from microglia or neurons, the other components are primarily derived from astrocytes. In vivo studies have demonstrated immunoreactivity or mRNA for apo E, apo J, α-1-antichymotrypsin and heparan sulfate proteoglycan in astrocytes, with increased expression being detectable in those cells closely associated with plaques.

Table 4
Biologically active compounds produced by microglia and associated with AD pathogenesis

Cytokines (107)
Interleukin-1α
Interleukin-1β
Interleukin-6
Tumor necrosis factor α
Complement proteins (107,126)
Complement C1q
Complement C3
Complement C4
Complement inhibitor proteins
C1-inhibitor (164)
C4 binding protein
Free radicals/small molecules
Reactive oxygen species *(165)*
NTox (phenolic amine) *(86)*
Glutamate *(80,81)*
Quinolinic acid *(48)*

It can be suggested that the astrocytic response in AD that appears to develop as a consequence of the microgliosis is likely to have a detrimental effect on the pathological process, rather than a reparative response.

10. MODELING GLIAL RESPONSES IN ALZHEIMER'S DISEASE

A number of research groups are discovering that glial cells derived from postmortem or biopsy samples of human brains can be prepared and used in experiments *(104,85,105,106,107,108)*. Techniques based on those developed for isolating human oligodendrocytes have proven suitable for preparing pure cultures of microglia from postmortem brains, as well as human biopsy material *(109)*. The availability of such cells is dependent on being able to obtain brain tissue from cases as soon after death as possible (ideally less than 4 h) *(106,107)*. In a series of glial cell isolations, using white matter tissue from cases that were obtained less than 5 h after death, we were able to obtain 3.02×10^6 viable cells/gram of tissue (standard deviation of 1.69×10^6, n=25 cases). In comparison to microglia derived from fetal brains, postmortem brain-derived microglia do not readily replicate so the numbers available for experimentation, even if tissue is available, remains limited. Their responses to stimuli may more accurately reflect microglial reactivity in elderly human brains than do those from fetal microglia. Differences between adult and fetal microglia have been identified, as have differences between microglia derived from young and aged rats. It was observed that human fetal microglia and astrocytes were more responsive to proinflammatory stimuli than were adult brain-derived glial cells *(105)*.

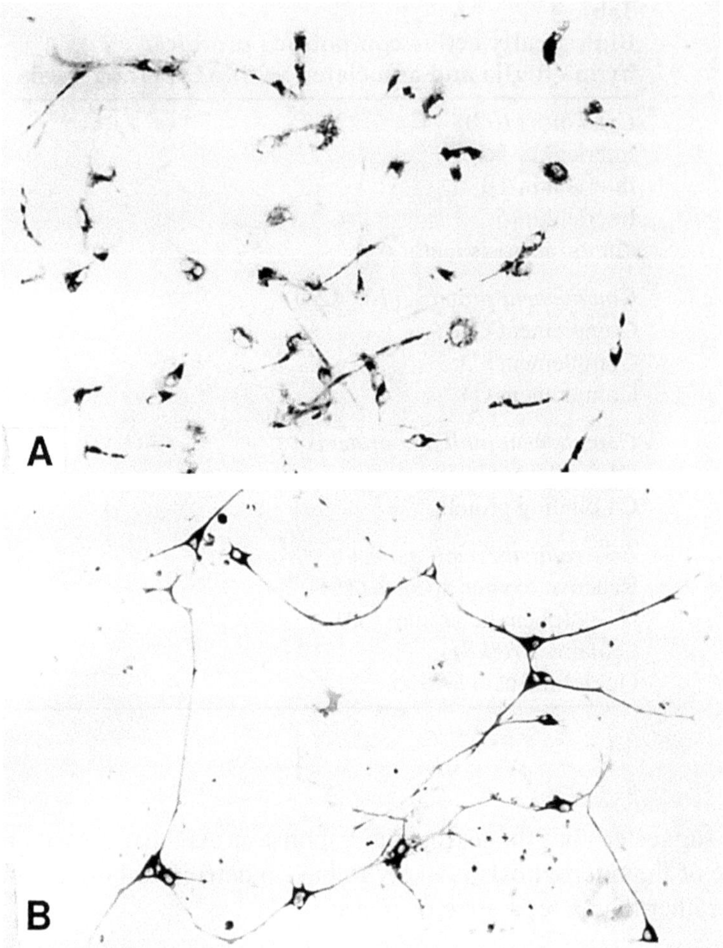

Fig. 3. (A). Photomicrograph of microglia derived from human postmortem brain tissue. The cells were cultured in vitro for three weeks after death prior to use in experiments. The cells were identified using an antibody that recognizes the macrophage specific marker CD68 (macrosialin). This protein is located on lysosomal membranes. Reactive microglia that have phagocytosed material show stronger CD68 immunoreactivity. **(B).** Photomicrograph of fibrous astrocytes-derived from human postmortem brain tissue. The cells were cultured in vitro for four weeks after death prior to use in experiments. These cells are separated from microglia. Cells were identified by immunostaining with an antibody to GFAP. These types of cells can be maintained up to eight weeks in culture. They do not exhibit evidence of cell division. In similar cultures, protoplasmic astrocytes can also be seen. These cells will divide slowly.

In contrast, microglia cultured from elderly rats appeared to have a higher degree of spontaneous reactivity, as measured by MHC class II expression, compared to cells from young rats *(110)*. In addition, the microglia from elderly rats appeared less response to the antiinflammatory effects of transforming growth factor β *(110)*.

Examples of glial cells isolated and grown from human postmortem brains are shown in Fig. 3.

11. DISCUSSION

There has been tremendous scientific investigation into the glial responses that are occurring in AD, and how they may be involved in the pathology of AD. It is possible to list all the potentially damaging products of these glial cells, and assume that they are doing damage in the human brain, because this can be demonstrated in isolated cell culture systems. However there is also evidence for increased expression of antiinflammatory cytokines, antioxidants, and protease and complement inhibitors by these same glial cells. In particular, it is likely that false conclusions will be drawn if emphasis is placed on the role of only one or two of these glial products. As shown there are now a large number of these products that interact with microglia, astrocytes, plaques or tangles. As AD tends to be a chronic disease, with the affected showing gradual cognitive decline, it can be suggested that the inhibitory proteins may gradually be overcome, leading to synaptic loss, cell death, and dementia.

Final interpretation of all the data, both from pathological studies and in vitro model system, indicate that it has not been definitively proven that glial cells are contributing to the pathology of AD, though the data strongly suggests this conclusion. It has been suggested that proof of microglial involvement in AD has come from the findings that antiinflammatory drugs are effective at slowing the progression of AD *(111)*. The problem with this interpretation is that the effective drugs share the common property of being cyclooxygenase (COX) inhibitors *(111)*. There are two different cyclooxygenases. COX-1 (constitutive cyclooxygenase) and COX-2 (inducible cyclooxygenase) exist in different cell populations. The effective drugs tend to have greater activity as COX-1 inhibitors, however, they can also inhibit COX-2 at higher concentrations. In the brain, COX-2 is readily detectable, although COX-1 is present at much lower abundance *(112)*. Also of interest is the fact that different populations of neurons express high levels of COX-2, whereas microglial expression of this protein has not been definitively demonstrated in vivo in the human brain *(110,113)*. Thus the "antiinflammatory" drugs may be acting directly on vulnerable neurons, not on the inflammatory associated cells *(114)*. Model systems for ischemia have shown that COX-2 is increased in neurons following toxic insult *(115,116)*. This increased neuronal COX-2 expression can lead directly to cell death.

Similarly, propentofylline has been proposed to be effective in AD *(117)* because of its antiinflammatory properties, possibly by inhibiting microglial superoxide production *(118)*. However this agent, an adenosine reuptake inhibitor, is also effective in stimulating increased nerve growth factor secretion by astrocytes *(119)*.

Considerable progress has been made in the last decade to describe the glial response that is occurring in the brains of AD cases. These observations have been combined with a range of experimental model systems that describe mechanisms that might be occurring in the AD brain. Continued detailed studies are required to confirm that inhibiting the glial response will be beneficial to the AD patient. This will be aided by the availability of agents that are specific in inhibiting the glial-associated response.

ACKNOWLEDGMENTS

The authors are supported by grants from the National Institutes of Health Alzheimer's Association and the Ronald and Nancy Reagan Research Institute. This article is dedicated to the memory of our colleague, Dr. W. H. Civin.

REFERENCES

1. Glenner, G.G. and Wong, C.W. (1984) Alzheimer's disease: initial report of the purification and characterization of a novel cerebrovascular amyloid protein. *Biochem. Biophys. Res. Commun.* **120,** 885–890.
2. Masters, C.L., Simms, G., Weinman, N.A., Multhaup, G., McDonald, B.L., and Beyreuther, K. (1985) Amyloid plaque core protein in Alzheimer disease and Down syndrome. *Proc. Natl. Acad. Sci. U.S.A.* **82,** 4245–4249.
3. Goldgaber, D., Lerman, M.I., McBride, O.W., Saffiotti, U., and Gajdusek, D.C. (1987) Characterization and chromosomal localization of a cDNA encoding brain amyloid of Alzheimer's disease, *Science* **235,** 877–880.
4. Kang, J., Lemaire, H.G., Unterbeck, A., et al. (1987) The precursor of Alzheimer's disease amyloid A4 protein resembles a cell-surface receptor. *Nature* **325,** 733–736.
5. Tanzi, R.E., Gusella, J.F., Watkins, P.C., Bruns, G.A., St George-Hyslop, P., Van Keuren, M.L., Patterson, D., Pagan, S., Kurnit, D.M. and Neve, R.L. (1987) Amyloid beta protein gene: cDNA, mRNA distribution, and genetic linkage near the Alzheimer locus, *Science* **235,** 880–884.
6. Goate, A., Chartier-Harlin, M.C., Mullan M., et al. (1991) Segregation of a missense mutation in the amyloid precursor protein gene with familial Alzheimer's disease. *Nature* **349,** 704–706.
7. Levy-Lahad, E., Wasco, W., Poorkaj, P., Romano, D.M., Oshima, J., Pettingell, W.H., Yu, C.E., Jondro, P.D., Schmidt, S.D., and Wang, K. (1995) Candidate gene for the chromosome 1 familial Alzheimer's disease locus. *Science* **269,** 973–977.
8. Sherrington, R., Rogaev, E.I., Liang, Y., Rogaeva, E.A., Levesque, G., Ikeda, M., Chi, H., Lin, C., Li, G. and Holman, K. (1995) Cloning of a gene bearing missense mutations in early-onset familial Alzheimer's disease, *Nature* **375,** 754–760.
9. Saunders, A.M., Strittmatter, W.J., Schmechel, D., George-Hyslop, P.H., Pericak-Vance, M.A., Joo, S.H., Rosi, B.L., Gusella, J.F., Crapper-MacLachlan, D.R. and Alberts, M.J. (1993) Association of apolipoprotein E allele epsilon 4 with late-onset familial and sporadic Alzheimer's disease [see comments], *Neurology* **43,** 1467–1472.
10. Beffert, U., Arguin, C., and Poirier, J. (1999) The polymorphism in exon 3 of the low density lipoprotein receptor-related protein gene is weakly associated with Alzheimer's disease. *Neurosci. Lett.* **259,** 29–32.
11. Scott, W.K., Grubber, J.M., Abou-Donia, S.M., Church, T.D., Saunders, A.M., Roses, A.D., Pericak-Vance, M.A., Conneally, P.M., Small, G.W. and Haines, J.L. (1999) Further evidence linking late-onset Alzheimer disease with chromosome 12 [letter], *JAMA* **281,** 513–514.
12. Wu, W.S., Holmans, P., Wavrant-DeVrieze, F., Shears, S., Kehoe, P., Crook, R., Booth, J., Williams, N., Perez-Tur, J., Roehl, K., Fenton, I., Chartier-Harlin, M.C., Lovestone, S., Williams, J., Hutton, M., Hardy, J., Owen, M.J. and Goate, A. (1998) Genetic studies on chromosome 12 in late-onset Alzheimer disease, *JAMA* **280,** 619–622.
13. Schofield, P.W., Tang, M., Marder, K., Bell, K., Dooneief, G., Chun, M., Sano, M., Stern, Y. and Mayeux, R., Alzheimer's disease after remote head injury: an incidence study, *J. Neurol. Neurosurg. Psychiatry* **62,** 119–124.
14. Launer, L.J., Andersen, K., Dewey, M.E., Letenneur, L., Ott, A., Amaducci, L.A., Brayne, C., Copeland, J.R., Dartigues, J.F., Kragh-Sorensen, P., Lobo, A., Martinez-Lage, J.M., Stijnen, T., and Hofman, A. (1999) Rates and risk factors for dementia and Alzheimer's disease: results from EURODEM pooled analyses. EURODEM Incidence Research Group and Work Groups. European Studies of Dementia. *Neurology* **52,** 78–84.
15. Sparks, D.L. (1997) Coronary artery disease, hypertension, ApoE, and cholesterol: a link to Alzheimer's disease?, *Ann. N.Y. Acad. Sci.* **826,** 128–146.
16. Martyn, C.N., Coggon, D.N., Inskip, H., Lacey, R.F., and Young, W.F. (1997) Aluminum concentrations in drinking water and risk of Alzheimer's disease. *Epidemiology.* **8,** 281–286.

17. Itzhaki, R.F., Lin, W.R., Wilcock, G.K., and Faragher, B. (1998) HSV-1 and risk of Alzheimer's disease [letter; comment]. *Lancet* **352,** 733–736.
18. Cras, P., Kawai, M., Siedlak, S., et al. (1990) Neuronal and microglial involvement in beta-amyloid protein deposition in Alzheimer's disease. *Am. J. Pathol.* **137,** 241–246.
19. Griffin, W.S., Stanley, L.C., Ling, C., White, L., MacLeod, V., Perrot, L.J., White, C.L., and Araoz, C. (1989) Brain interleukin 1 and S-100 immunoreactivity are elevated in Down syndrome and Alzheimer disease. *Proc. Natl. Acad. Sci. U.S.A.* **86,** 7611–7615.
20. Itagaki, S., McGeer, P.L., Akiyama, H., Zhu, S., and Selkoe, D. (1989) Relationship of microglia and astrocytes to amyloid deposits of Alzheimer disease. *J. Neuroimmunol.* **24,** 173–182.
21. Mattiace, L.A., Davies, P. and Dickson, D.W. (1990) Detection of HLA-DR on microglia in the human brain is a function of both clinical and technical factors. *Am. J. Pathol.* **136,** 1101–1114.
22. McGeer, P.L., Itagaki, S., Tago, H., and McGeer, E.G. (1987) Reactive microglia in patients with senile dementia of the Alzheimer type are positive for the histocompatibility glycoprotein HLA-DR. *Neurosci. Lett.* **79,** 195–200.
23. Perlmutter, L.S., Barron, E., and Chui, H.C. (1990) Morphologic association between microglia and senile plaque amyloid in Alzheimer's disease. *Neurosci. Lett.* **119,** 32–36.
24. Rogers, J., Luber-Narod, J., Styren, S.D. and Civin, W.H. (1988) Expression of immune system-associated antigens by cells of the human central nervous system: relationship to the pathology of Alzheimer's disease, *Neurobiol. Aging* **9,** 339–349.
25. Rozemuller, J.M., Eikelenboom, P., Stam, F.C., Beyreuther, K. and Masters, C.L. (1989) A4 protein in Alzheimer's disease: primary and secondary cellular events in extracellular amyloid deposition, *J. Neuropathol. Exp. Neurol.* **48,** 674–691.
26. Styren, S.D., Civin, W.H. and Rogers, J. (1990) Molecular, cellular, and pathologic characterization of HLA-DR immunoreactivity in normal elderly and Alzheimer's disease brain, *Exp. Neurol.* **110,** 93–104.
27. Colton, C.A., Yao, J., Keri, J.E., and Gilbert, D. (1992) Regulation of microglial function by interferons. *J. Neuroimmunol.* **40,** 89–98.
28. Peress, N.S. and Perillo, E. (1995) Differential expression of TGF-beta 1, 2 and 3 isotypes in Alzheimer's disease: a comparative immunohistochemical study with cerebral infarction, aged human and mouse control brains. *J. Neuropathol. Exp. Neurol.* **54,** 802–811.
29. Gehrmann, J., Banati, R.B., and Kreutzberg, G.W. (1993) Microglia in the immune surveillance of the brain: Human microglia constitutively express HLA-DR molecules. *J. Neuroimmunol.* **48,** 189–198.
30. Yamada, T., Miyazaki, K., Koshikawa, N., Takahashi, M., Akatsu, H. and Yamamoto, T. (1995) Selective localization of gelatinase A, an enzyme degrading beta- amyloid protein, in white matter microglia and in Schwann cells, *Acta Neuropathol. (Berl.)* **89,** 199–203.
31. Yamada, T., Yoshiyama, Y., Sato, H., Seiki, M., Shinagawa, A. and Takahashi, M. (1995) White matter microglia produce membrane-type matrix metalloprotease, an activator of gelatinase A, in human brain tissues, *Acta Neuropathol. (Berl.)* **90,** 421–424.
32. Lee, S.C., Liu, W., Roth, P., Dickson, D.W., Berman, J.W., and Brosnan, C.F. (1993) Macrophage colony-stimulating factor in human fetal astrocytes and microglia. Differential regulation by cytokines and lipopolysaccharide, and modulation of class II MHC on microglia. *J. Immunol.* **150,** 594–604.
33. Carpenter, A.F., Carpenter, P.W., and Markesbery, W.R. (1993) Morphometric analysis of microglia in Alzheimer's disease. *J. Neuropathol. Exp. Neurol.* **52,** 601–608.
34. Mackenzie, I.R., Hao, C. and Munoz, D.G. (1995) Role of microglia in senile plaque formation. *Neurobiol. Aging* **16,** 797–804.
35. DiPatre, P.L. and Gelman, B.B. (1997) Microglial cell activation in aging and Alzheimer Disease: Partial linkage with neurofibrillary tangle burden in the hippocampus. *J. Neuropathol. Exp. Neurol.* **56,** 143–149.

36. Roe, M.T., Dawson, D.V., Hulette, C.M., Einstein, G., and Crain, B.J. (1996) Microglia are not exclusively associated with plaque-rich regions of the dentate gyrus in Alzheimer's disease. *J. Neuropathol. Exp. Neurol.* **55,** 366–371.
37. Styren, S.D., Kamboh, M.I. and DeKosky, S.T. (1998) Expression of differential immune factors in temporal cortex and cerebellum: the role of alpha-1-antichymotrypsin, apolipoprotein E, and reactive glia in the progression of Alzheimer's disease, *J. Comp. Neurol.* **396,** 511–520.
38. Fukumoto, H., Asami-Odaka, A., Suzuki, N., and Iwatsubo, T. (1996) Association of Aβ-40-positive senile plaques with microglial cells in the brains of patients with Alzheimer's disease and in non-demented aged individuals. *Neurodegeneration* **5,** 13–17.
39. Griffin, W.S., Sheng, J.G., Roberts, G.W., and Mrak, R.E. (1995) Interleukin-1 expression in different plaque types in Alzheimer's disease: significance in plaque evolution. *J. Neuropathol. Exp. Neurol.* **54,** 276–281.
40. Frautschy, S.A., Horn, D.L., Sigel, J.J., Harris-White, M.E., Mendoza, J.J., Yang, F., Saido, T.C., and Cole, G.M. (1998) Protease Inhibitor Coinfusion with Amyloid β-Protein Results in Enhanced Deposition and Toxicity in Rat Brain, *J. Neurosci.* **18,** 8311–8321.
41. Frautschy, S.A., Yang, F., Irrizarry, M., Hyman, B., Saido, T.C., Hsiao, K., and Cole, G.M. (1998) Microglial response to amyloid plaques in APPsw transgenic mice, *Am. J. Pathol.* **152,** 307–317.
42. Irizarry, M.C., Soriano, F., McNamara, M., et al. (1997) A beta deposition is associated with neuropil changes, but not with overt neuronal loss in the human amyloid precursor protein V717F (PDAPP) transgenic mouse, *J. Neurosci.* **17,** 7053–7059.
43. Sturchler-Pierrat, C., Abramowski, D., Duke, M., Wiederhold, K.H., Mistl, C., Rothacher, S., Ledermann, B., Burki, K., Frey, P., Paganetti, P.A., Waridel, C., Calhoun, M.E., Jucker, M., Probst, A., Staufenbiel, M. and Sommer, B. (1997) Two amyloid precursor protein transgenic mouse models with Alzheimer disease-like pathology, *Proc. Natl. Acad. Sci. U.S.A.* **94,** 13287–13292.
44. Sudo, S., Tanaka, J., Toku, K., Desaki, J., Matsuda, S., Arai, T., Sakanaka, M. and Maeda, N. (1998) Neurons induce the activation of microglial cells in vitro, *Exp. Neurol.* **154,** 499–510.
45. Duffy, P.E., Rapport, M., and Graf, L. (1980) Glial fibrillary acidic protein and Alzheimer-type senile dementia. *Neurology* **30,** 778–782.
46. Schechter, R., Yen, S.-H.C. and Terry, R.D. (1981) Fibrous astrocytes in senile dementia of the Alzheimer type., *J. Neuropathol. Exp. Neurol.* **40,** 95–101.
47. Mancardi, G.L., Liwnicz, B.H. and Mangiarotti, F. (1983) Fibrous astrocytes in Alzheimer's disease and senile dementia of Alzheimer's type. *Acta. Neuropathol.* **61,** 76–80.
48. Beach, T.G., Walker, R., and McGeer, E.G. (1989) Patterns of gliosis in Alzheimer's disease and aging cerebrum. *Glia* **2,** 420–436.
48. Espey, M.G., Chernyshev, O.N., Reinhard, J.F.J., Namboodiri, M.A., and Colton, C.A. (1997) Activated human microglia produce the excitotoxin quinolinic acid, *Neuroreport.* **8,** 431–434.
49. Giulian, D. (1993) Reactive glia as rivals in regulating neuronal survival. *Glia* **7,** 102–110.
50. Moretto, G., Walker, D.G., Lanteri, P., Taioli, F., Zaffagnini, S., Xu, R.Y., and Rizzuto, N. (1996) Expression and regulation of glial-cell-line-derived neurotrophic factor (GDNF) mRNA in human astrocytes in vitro. *Cell Tissue Res.* **286,** 257–262.
51. Moretto, G., Xu, R.Y., Walker, D.G., and Kim, S.U. (1994) Co-expression of mRNA for neurotrophic factors in human neurons and glial cells in culture. *J. Neuropathol. Exp. Neurol.* **53,** 78–85.
52. Satoh, J., Muramatsu, H., Moretto, G., Muramatsu, T., Chang, H.J., Kim, S.T., Cho, J.M. and Kim, S.U. (1993) Midkine that promotes survival of fetal human neurons is produced by fetal human astrocytes in culture, *Brain Res. Dev. Brain Res.* **75,** 201–205.
53. Stopa, E.G., Gonzalez, A.M., Chorsky, R., Corona, R.J., Alvarez, J., Bird, E.D. and Baird, A. (1990) Basic fibroblast growth factor in Alzheimer's disease, *Biochem. Biophys. Res. Commun.* **171,** 690–696.

54. Tooyama, I., Akiyama, H., McGeer, P.L., Hara, Y., Yasuhara, O. and Kimura, H. (1991) Acidic fibroblast growth factor-like immunoreactivity in brain of Alzheimer patients, *Neurosci. Lett.* **121,** 155–158.
55. Brosnan, C.F., Battistini, L., Raine, C.S., Dickson, D.W., Casadevall, A., and Lee, S.C. (1994) Reactive nitrogen intermediates in human neuropathology: an overview. *Dev. Neurosci.* **16,** 152–161.
56. DeWitt, D.A., Perry, G., Cohen, M., Doller, C., and Silver, J. (1998) Astrocytes regulate microglial phagocytosis of senile plaque cores of Alzheimer's disease. *Exp. Neurol.* **149,** 329–340.
57. Castano, E.M., Prelli, F., Wisniewski, T., Golabek, A., Kumar, R.A., Soto, C., and Frangione, B. (1995) Fibrillogenesis in Alzheimer's disease of amyloid beta peptides and apolipoprotein E. *Biochem. J.* **306,** 599–604.
58. Lee, S.C., Liu, W., Brosnan, C.F., and Dickson, D.W. (1994) GM-CSF promotes proliferation of human fetal and adult microglia in primary cultures. *Glia* **12,** 309–318.
59. Jefferies, W.A., Food, M.R., Gabathuler, R., Rothenberger, S., Yamada, T., Yasuhara, O., and McGeer, P.L. (1996) Reactive microglia specifically associated with amyloid plaques in Alzheimer's disease brain tissue express melanotransferrin. *Brain Res.* **712,** 122–126.
60. Rothenberger, S., Food, M.R., Gabathuler, R., Kennard, M.L., Yamada, T., Yasuhara, O., McGeer, P.L. and Jefferies, W.A. (1996) Coincident expression and distribution of melanotransferrin and transferrin receptor in human brain capillary endothelium, *Brain Res.* **712,** 117–121.
61. Kennard, M.L., Feldman, H., Yamada, T., and Jefferies, W.A. (1996) Serum levels of the iron binding protein p97 are elevated in Alzheimer's disease. *Nat. Med.* **2,** 1230–1235.
62. Wood, J.G. and Zinsmeister, P. (1991) Tyrosine phosphorylation systems in Alzheimer's disease pathology, *Neurosci. Lett.* **121,** 12–16.
63. Combs, C.K., Johnson, D.E., Cannady, S.B., Lehman, T.M., and Landreth, G.E. (1999) Identification of microglial signal transduction pathways mediating a neurotoxic response to amyloidogenic fragments of beta-amyloid and prion proteins. *J. Neurosci.* **19,** 928–939.
63. Hortega, P.D.R. (1919) El tercer elemente de los centros nerviosis: Poder fagocitario y movilidad de la microglia. *Biol. Soc. esp. de biol.* **9,** 69–120.
64. Hu, S., Chao, C.C., Khanna, K.V., Gekker, G., Peterson, P.K., and Molitor, T.W. (1996) Cytokine and free radical production by porcine microglia. *Clin. Immunol. Immunopathol.* **78,** 93–96.
64. McDonald, D.R., Brunden, K.R. and Landreth, G.E. (1997) Amyloid fibrils activate tyrosine kinase-dependent signaling and superoxide production in microglia. *J. Neurosci.* **17,** 2284–2294.
65. Lefrancois, T., Fages, C., Peschanski, M., and Tardy, M. (1997) Neuritic outgrowth associated with astroglial phenotypic changes induced by antisense glial fibrillary acidic protein (GFAP) mRNA in injured neuron-astrocyte cocultures. *J. Neurosci.* **17,** 4121–4128.
66. Wu, V.W. and Schwartz, J.P. (1998) Cell culture models for reactive gliosis: new perspectives, *J. Neurosci. Res.* **51,** 675–681.
67. Yu, A.C., Lee, Y.L. and Eng, L.F. (1991) Inhibition of GFAP synthesis by antisense RNA in astrocytes, *J. Neurosci. Res.* **30,** 72–79.
68. Xu, K., Malouf, A.T., Messing, A. and Silver, J. (1999) Glial fibrillary acidic protein is necessary for mature astrocytes to react to beta-amyloid, *Glia* **25,** 390–403.
69. Marshak, D.R., Pesce, S.A., Stanley, L.C., and Griffin, W.S. (1992) Increased S100 beta neurotrophic activity in Alzheimer's disease temporal lobe. *Neurobiol. Aging* **13,** 1–7.
70. Li, Y., Wang, J., Sheng, J.G., Liu, L., Barger, S.W., Jones, R.A., Van Eldik, L.J., Mrak, R.E., and Griffin, W.S. (1998) S100 beta increases levels of beta-amyloid precursor protein and its encoding mRNA in rat neuronal cultures. *J. Neurochem.* **71,** 1421–1428.

71. Sheng, J.G., Mrak, R.E. and Griffin, W.S. (1994) S100 beta protein expression in Alzheimer disease: potential role in the pathogenesis of neuritic plaques, *J. Neurosci. Res.* **39,** 398–404.
72. Sheng, J.G., Mrak, R.E., Rovnaghi, C.R., Kozlowska, E., Van Eldik, L.J. and Griffin, W.S. (1996) Human brain S100 beta and S100 beta mRNA expression increases with age: pathogenic implications for Alzheimer's disease, *Neurobiol. Aging* **17,** 359–363.
73. Van Eldik, L.J. and Griffin, W.S. (1994) S100 beta expression in Alzheimer's disease: relation to neuropathology in brain regions, *Biochim. Biophys. Acta* **1223,** 398–403.
74. Mrak, R.E., Sheng, J.G. and Griffin, W.S. (1996) Correlation of astrocytic S100 beta expression with dystrophic neurites in amyloid plaques of Alzheimer's disease. *J. Neuropathol. Exp. Neurol.* **55,** 273–279.
75. Sheng, J.G., Ito, K., Skinner, R.D., Mrak, R.E., Rovnaghi, C.R., VanEldik, L.J. and Griffin, W.S. (1996) In vivo and in vitro evidence supporting a role for thc inflammatory cytokine interleukin-1 as a driving force in Alzheimer pathogenesis., *Neurobiol. Aging* **17,** 761–766.
76. Chao, C.C., Hu, S., and Peterson, P.K. (1996) Glia: the not so innocent bystanders. *J. Neurovirol.* **2,** 234–239.
77. Chao, C.C., Hu, S., Sheng, W.S., Bu, D., Bukrinsky, M.I., and Peterson, P.K. (1996) Cytokine-stimulated astrocytes damage human neurons via a nitric oxide mechanism. *Glia* **16,** 276–284.
79. Ii, M., Sunamoto, M., Ohnishi, K., and Ichimori, Y. (1996) beta-Amyloid protein-dependent nitric oxide production from microglial cells and neurotoxicity. *Brain Res.* **720,** 93–100.
80. Piani, D. and Fontana, A. (1994) Involvement of the cystine transport system xc- in the macrophage- induced glutamate-dependent cytotoxicity to neurons. *J. Immunol.* **152,** 3578–3585.
81. Piani, D., Spranger, M., Frei, K., Schaffner, A., and Fontana, A. (1992) Macrophage-induced cytotoxicity of N-methyl-D-aspartate receptor positive neurons involves excitatory amino acids rather than reactive oxygen intermediates and cytokines. *Eur. J. Immunol.* **22,** 2429–2436.
82. Tsirka, S.E. (1997) Clinical implications of the involvement of tPA in neuronal cell death, *J. Mol. Med.* **75,** 341–347.
83. Westmoreland, S.V., Kolson, D. and Gonzalez-Scarano, F. (1996) Toxicity of TNF alpha and platelet activating factor for human NT2N neurons: a tissue culture model for human immunodeficiency virus dementia, *J. Neurovirol.* **2,** 118–126.
84. Yoshida, T., Tanaka, M., Sotomatsu, A., Hirai, S. and Okamoto, K. (1998) Activated microglia cause iron-dependent lipid peroxidation in the presence of ferritin, *Neuroreport.* **9,** 1929–1933.
85. Giulian, D., Haverkamp, L.J., Li, J., Karshin, W.L., Yu, J., Tom, D., Li, X., and Kirkpatrick, J.B. (1995) Senile plaques stimulate microglia to release a neurotoxin found in Alzheimer brain. *Neurochem. Int.* **27,** 119–137.
86. Giulian, D., Haverkamp, L.J., Yu, J.H., Karshin, W., Tom, D., Li, J., Kirkpatrick, J., Kuo, Y.M., and Roher, A.E. (1996) Specific domains of beta-amyloid from Alzheimer plaque elicit neuron killing in human microglia. *J. Neurosci.* **16,** 6021–6037.
87. Giulian, D., Haverkamp, L.J., Yu, J., Karshin, W., Tom, D., Li, J., Kazanskaia, A., Kirkpatrick, J., and Roher, A.E. (1998) The HHQK domain of beta-amyloid provides a structural basis for the immunopathology of Alzheimer's disease. *J. Biol. Chem.* **273,** 29719–29726.
88. Good, P.F., Werner, P., Hsu, A., Olanow, C.W., and Perl, D.P. (1996) Evidence of neuronal oxidative damage in Alzheimer's disease. *Am. J. Pathol.* **149,** 21–28.
89. Su, J.H., Deng, G. and Cotman, C.W. (1997) Neuronal DNA damage precedes tangle formation and is associated with up- regulation of nitrotyrosine in Alzheimer's disease brain, *Brain Res.,* **774,** 193–199.
90. Hensley, K., Maidt, M.L., Yu, Z., Sang, H., Markesbery, W.R., and Floyd, R.A. (1998) Electrochemical analysis of protein nitrotyrosine and dityrosine in the Alzheimer brain indicates region-specific accumulation. *J. Neurosci.* **18,** 8126–8132.

91. Ding, M., St.Pierre, B.A., Parkinson, J.F., Medberry, P., Wong, J.L., Rogers, N.E., Ignarro, L.J., and Merrill, J.E. (1997) Inducible nitric-oxide synthase and nitric oxide production in human fetal astrocytes and microglia. A kinetic analysis. *J. Biol. Chem.* **272,** 11327–11335.
92. Lee, S.C. and Brosnan, C.F. (1996) Cytokine Regulation of iNOS Expression in Human Glial Cells. *Methods* **10,** 31–37.
93. Lee, S.C., Dickson, D.W., Liu, W., and Brosnan, C.F. (1993) Induction of nitric oxide synthase activity in human astrocytes by interleukin-1 beta and interferon-gamma. *J. Neuroimmunol.* **46,** 19–24.
94. Chao, C.C., Hu, S., Ehrlich, L., and Peterson, P.K. (1995) Interleukin-1 and tumor necrosis factor-alpha synergistically mediate neurotoxicity: involvement of nitric oxide and of N-methyl-D-aspartate receptors. *Brain Behav. Immun.* **9,** 355–365.
95. Wallace, M.N., Geddes, J.G., Farquhar, D.A. and Masson, M.R. (1997) Nitric oxide synthase in reactive astrocytes adjacent to beta-amyloid plaques, *Exp. Neurol.* **144,** 266–272.
96. Eikelenboom, P. and Veerhuis, R. (1996) The role of complement and activated microglia in the pathogenesis of Alzheimer's disease, *Neurobiol. Aging* **17,** 673–680.
97. McGeer, P.L. and McGeer, E.G. (1995) The inflammatory response system of brain: implications for therapy of Alzheimer and other neurodegenerative diseases. *Brain Res. Brain Res. Rev.* **21,** 195–218.
98. Pasinetti, G.M. (1996) Inflammatory mechanisms in neurodegeneration and Alzheimer's disease: the role of the complement system. *Neurobiol. Aging* **17,** 707–716.
99. Aksenov, M.Y., Aksenova, M.V., Carney, J.M., and Butterfield, D.A. (1996) Alpha 1-antichymotrypsin interaction with A beta (1–42) does not inhibit fibril formation but attenuates the peptide toxicity. *Neurosci. Lett.* **217,** 117–120.
100. Castillo, G.M., Ngo, C., Cummings, J., Wight, T.N., and Snow, A.D. (1997) Perlecan binds to the beta-amyloid proteins (A beta) of Alzheimer's disease, accelerates A beta fibril formation, and maintains A beta fibril stability. *J. Neurochem.* **69,** 2452–2465.
101. Choi-Miura, N.H. and Oda, T. (1996) Relationship between multifunctional protein "clusterin" and Alzheimer disease. *Neurobiol. Aging* **17,** 717–722.
102. Webster, S. and Rogers, J. (1996) Relative efficacies of amyloid beta peptide (A beta) binding proteins in A beta aggregation, *J. Neurosci. Res.* **46,** 58–66.
103. Oda, T., Wals, P., Osterburg, H.H., Johnson, S.A., Pasinetti, G.M., Morgan, T.E., Rozovsky, I., Stine, W.B., Snyder, S.W., and Holzman, T.F. (1995) Clusterin (apoJ) alters the aggregation of amyloid beta-peptide (A beta 1–42) and forms slowly sedimenting A beta complexes that cause oxidative stress. *Exp. Neurol.* **136,** 22–31.
105. Lafortune, L., Nalbantoglu, J. and Antel, J.P. (1996) Expression of tumor necrosis factor alpha (TNF alpha) and interleukin 6 (IL-6) mRNA in adult human astrocytes: comparison with adult microglia and fetal astrocytes. *J. Neuropathol. Exp. Neurol.* **55,** 515–521.
106. Lue, L.F., Brachova, L., Walker, D.G., and Rogers, J. (1996) Characterization of glial cultures from rapid autopsies of Alzheimer's and control patients. *Neurobiol. Aging* **17,** 421–429.
107. Walker, D.G., Kim, S.U. and McGeer, P.L. (1995) Complement and cytokine gene expression in cultured microglia derived from postmortem human brains, *J. Neurosci. Res.* **40,** 478–93.
108. Williams, K., Bar-Or, A., Ulvestad, E., Olivier, A., Antel, J.P. and Yong, V.W. (1992) Biology of adult human microglia in culture: comparisons with peripheral blood monocytes and astrocytes, *J. Neuropathol. Exp. Neurol.* **51,** 538–549.
109. Becher, B. and Antel, J.P. (1996) Comparison of phenotypic and functional properties of immediately ex vivo and cultured human adult microglia. *Glia* **18,** 1–10.
109. Kim, S.U., Sato, Y., Silberberg, D.H., Pleasure, D.E., and Rorke, L.B. (1983) Long-term culture of human oligodendrocytes. Isolation, growth and identification. *J. Neurol. Sci.* **62,** 295–301.
110. Pasinetti, G.M. and Aisen, P.S. (1998) Cyclooxygenase-2 expression is increased in frontal cortex of Alzheimer's disease brain. *Neuroscience* **87,** 319–324.

111. Stewart, W.F., Kawas, C., Corrada, M. and Metter, E.J. (1997) Risk of Alzheimer's disease and duration of NSAID use., *Neurology* **48,** 626–632.
112. Kitamura, Y., Shimohama, S., Koike, H., Kakimura, J., Matsuoka, Y., Nomura, Y., Gebicke-Haerter, P.J., and Taniguchi, T. (1999) Increased expression of cyclooxygenases and peroxisome proliferator- activated receptor-gamma in Alzheimer's disease brains. *Biochem. Biophys. Res. Commun.* **254,** 582–586.
113. Oka, A. and Takashima, S. (1997) Induction of cyclo-oxygenase 2 in brains of patients with Down's syndrome and dementia of Alzheimer type: specific localization in affected neurones and axons. *Neuroreport.* **8,** 1161–1164.
114. Pasinetti, G.M. (1998) Cyclooxygenase and inflammation in Alzheimer's disease: experimental approaches and clinical interventions. *J. Neurosci. Res.* **54,** 1–6.
115. Nakayama, M., Uchimura, K., Zhu, R.L., Nagayama, T., Rose, M.E., Stetler, R.A., Isakson, P.C., Chen, J., and Graham, S.H. (1998) Cyclooxygenase-2 inhibition prevents delayed death of CA1 hippocampal neurons following global ischemia, *Proc. Natl. Acad. Sci. U.S.A.* **95,** 10954–10959.
115. Penfield, W. (1932) Neuroglia and microglia. The interstitial tissue of the central nervous system. In E.V. Cowdry (Ed.) *Special cytology vol. III* Hoeber, New York, pp. 1445–1482.
116. Nogawa, S., Zhang, F., Ross, M.E., and Iadecola, C. (1997) Cyclo-oxygenase-2 gene expression in neurons contributes to ischemic brain damage. *J. Neurosci.* **17,** 2746–2755.
117. Kittner, B., Rossner, M. and Rother, M. (1997) Clinical trials in dementia with propentofylline. *Ann. N.Y. Acad. Sci.* **826,** 307–316.
118. Banati, R.B., Schubert, P., Rothe, G., Gehrmann, J., Rudolphi, K., Valet, G., and Kreutzberg, G.W. (1994) Modulation of intracellular formation of reactive oxygen intermediates in peritoneal macrophages and microglia/brain macrophages by propentofylline. *J. Cereb. Blood Flow Metab.* **14,** 145–149.
119. Yamada, K., Nitta, A., Hasegawa, T., Fuji, K., Hiramatsu, M., Kameyama, T., Furukawa, Y., Hayashi, K. and Nabeshima, T. (1997) Orally active NGF synthesis stimulators: potential therapeutic agents in Alzheimer's disease, *Behav. Brain Res.* **83,** 117–122.
120. Tooyama, I., Kimura, H., Akiyama, H. and McGeer, P.L. (1990) Reactive microglia express class I and class II major histocompatibility complex antigens in Alzheimer's disease, *Brain Res.* **523,** 273–280.
121. Akiyama, H. and McGeer, P.L. (1990) Brain microglia constitutively express beta-2 integrins. *J. Neuroimmunol.* **30,** 81–93.
122. Eikelenboom, P., Zhan, S.S., Kamphorst, W., van, d., V., and Rozemuller, J.M. (1994) Cellular and substrate adhesion molecules (integrins) and their ligands in cerebral amyloid plaques in Alzheimer's disease, *Virchows Arch.* **424,** 421–427.
123. Matsuo, A., Walker, D.G., Terai, K., and McGeer, P.L. (1996) Expression of CD43 in human microglia and its downregulation in Alzheimer's disease. *J. Neuroimmunol.* **71,** 81–86.
124. Akiyama, H., Ikeda, K., Katoh, M., McGeer, E.G., and McGeer, P.L. (1994) Expression of MRP14, 27E10, interferon-alpha and leukocyte common antigen by reactive microglia in postmortem human brain tissue. *J. Neuroimmunol.* **50,** 195–201.
125. Akiyama, H., Kawamata, T., Dedhar, S., and McGeer, P.L. (1991) Immunohistochemical localization of vitronectin, its receptor and beta- 3 integrin in Alzheimer brain tissue. *J. Neuroimmunol.* **32,** 19–28.
126. Walker, D.G. (1998) Expression and regulation of complement C1q by human THP-1-derived macrophages., *Mol. Chem. Neuropathol.* **34,** 197–218.
127. Dickson, D.W., Lee, S.C., Mattiace, L.A., Yen, S.H., and Brosnan, C. (1993) Microglia and cytokines in neurological disease, with special reference to AIDS and Alzheimer's disease. *Glia* **7,** 75–83.

128. Funato, H., Yoshimura, M., Yamazaki, T., et al. (1998) Astrocytes containing amyloid beta-protein (Abeta)-positive granules are associated with Abeta40-positive diffuse plaques in the aged human brain. *Am. J. Pathol.* **152,** 983–992.
129. Yamada, T. and Yamanaka, I., Microglial localization of alpha-interferon receptor in human brain tissues, *Neurosci. Lett.* **189,** 73–76.
130. Ishizuka, K., Kimura, T., Igata-yi, R., Katsuragi, S., Takamatsu, J., and Miyakawa, T. (1997) Identification of monocyte chemoattractant protein-1 in senile plaques and reactive microglia of Alzheimer's disease. *Psychiatry Clin. Neurosci.* **51,** 135–138.
131. Akiyama, H., Nishimura, T., Kondo, H., Ikeda, K., Hayashi, Y., and McGeer, P.L. (1994) Expression of the receptor for macrophage colony stimulating factor by brain microglia and its upregulation in brains of patients with Alzheimer's disease and amyotrophic lateral sclerosis. *Brain Res.* **639,** 171–174.
131. Sheng, J.G., Griffin, W.S., Royston, M.C. and Mrak, R.E. (1998) Distribution of interleukin-1-immunoreactive microglia in cerebral cortical layers: implications for neuritic plaque formation in Alzheimer's disease, *Neuropathol. Appl. Neurobiol.* **24,** 278–283.
132. Karp, H.L., Tillotson, M.L., Soria, J., Reich, C., and Wood, J.G. (1994) Microglial tyrosine phosphorylation systems in normal and degenerating brain. *Glia* **11,** 284–290.
133. Christie, R.H., Freeman, M., and Hyman, B.T. (1996) Expression of the macrophage scavenger receptor, a multifunctional lipoprotein receptor, in microglia associated with senile plaques in Alzheimer's disease. *Am. J. Pathol.* **148,** 399–403.
134. Honda, M., Akiyama, H., Yamada, Y., et al. (1998) Immunohistochemical evidence for a macrophage scavenger receptor in Mato cells and reactive microglia of ischemia and Alzheimer's disease. *Biochem. Biophys. Res. Commun.* **245,** 734–740.
134. Sheng, J.G., Mrak, R.E. and Griffin, W.S. (1995) Microglial interleukin-1 alpha expression in brain regions in Alzheimer's disease: correlation with neuritic plaque distribution, *Neuropathol. Appl. Neurobiol.* **21,** 290–301.
135. Sheng, J.G., Mrak, R.E. and Griffin, W.S. (1997) Neuritic plaque evolution in Alzheimer's disease is accompanied by transition of activated microglia from primed to enlarged to phagocytic forms, *Acta Neuropathol. (Berl.)* **94,** 1–5.
136. Christie, R.H., Chung, H., Rebeck, G.W., Strickland, D., and Hyman, B.T. (1996) Expression of the very low-density lipoprotein receptor (VLDL-r), an apolipoprotein-E receptor, in the central nervous system and in Alzheimer's disease. *J. Neuropathol. Exp. Neurol.* **55,** 491–498.
136. Sheng, J.G., Mrak, R.E. and Griffin, W.S. (1998) Enlarged and phagocytic, but not primed, interleukin-1 alpha-immunoreactive microglia increase with age in normal human brain, *Acta Neuropathol. (Berl.)* **95,** 229–234.
137. Yamada, T., Kawamata, T., Walker, D.G. and McGeer, P.L. (1992) Vimentin immunoreactivity in normal and pathological human brain tissue, *Acta Neuropathol. (Berl.)* **84,** 157–162.
138. Akiyama, H., Ikeda, K., Kondo, H., Kato, M., and McGeer, P.L. (1993) Microglia express the type 2 plasminogen activator inhibitor in the brain of control subjects and patients with Alzheimer's disease. *Neurosci. Lett.* **164,** 233–235.
139. Akiyama, H., Kondo, H., Ikeda, K., Arai, T., Kato, M., and McGleer, P.L. (1995) Immunohistochemical detection of coagulation factor XIIIa in postmortem human brain tissue. *Neurosci. Lett.* **202,** 29–32.
140. Drache, B., Diehl, G.E., Beyreuther, K., Perlmutter, L.S., and Konig, G. (1997) Bcl-xl-specific antibody labels activated microglia associated with Alzheimer's disease and other pathological states. *J. Neurosci. Res.* **47,** 98–108.
141. Hollister, R.D., Kisiel, W. and Hyman, B.T. (1996) Immunohistochemical localization of tissue factor pathway inhibitor-1 (TFPI-1), a Kunitz proteinase inhibitor, in Alzheimer's disease. *Brain Res.* **728,** 13–19.

142. Reynolds, W.F., Rhees, J., Maciejewski, D., Paladino, T., Sieburg, H., Maki, R.A., and Masliah, E. (1999) Myeloperoxidase polymorphism is associated with gender specific risk for Alzheimer's disease. *Exp. Neurol.* **155,** 31–41.
143. Xia, M.Q., Qin, S.X., Wu, L.J., Mackay, C.R. and Hyman, B.T. (1998) Immunohistochemical study of the beta-chemokine receptors CCR3 and CCR5 and their ligands in normal and Alzheimer's disease brains, *Am. J. Pathol.* **153,** 31–37.
144. Walton, M., Saura, J., Young, D., MacGibbon, G., Hansen, W., Lawlor, P., Sirimanne, E., Gluckman, P. and Dragunow, M. (1998) CCAAT-enhancer binding protein alpha is expressed in activated microglial cells after brain injury, *Brain Res. Mol. Brain Res.* **61,** 11–22.
145. Lippa, C.F., Smith, T.W. and Flanders, K.C. (1995) Transforming growth factor-beta: neuronal and glial expression in CNS degenerative diseases. *Neurodegeneration.* **4,** 425–432.
146. Renkawek, K., Bosman, G.J., and de Jong, W.W. (1994) Expression of small heat-shock protein hsp 27 in reactive gliosis in Alzheimer disease and other types of dementia. *Acta Neuropathol. (Berl.)* **87,** 511–519.
147. Nakamura, S., Arima, K., Haga, S., Aizawa, T., Motoi, Y., Otsuka, M., Ueki, A., and Ikeda, K. (1998) Fibroblast growth factor (FGF)-9 immunoreactivity in senile plaques [In Process Citation]. *Brain Res.* **8,** 222–225.
148. Takami, K., Matsuo, A., Terai, K., Walker, D.G., McGeer, E.G. and McGeer, P.L. (1998) Fibroblast growth factor receptor-1 expression in the cortex and hippocampus in Alzheimer's disease, *Brain Res.* **802,** 89–97.
149. Aguado, F., Ballabriga, J., Pozas, E., and Ferrer, I. (1998) TrkA immunoreactivity in reactive astrocytes in human neurodegenerative diseases and colchicine-treated rats. *Acta Neuropathol. (Berl.)* **96,** 495–501.
150. Connor, B., Young, D., Lawlor, P., Gai, W., Waldvogel, H., Faull, R.L., and Dragunow, M. (1996) Trk receptor alterations in Alzheimer's disease. *Brain Res. Mol. Brain Res.* **42,** 1–17.
151. Fenton, H., Finch, P.W., Rubin, J.S., Rosenberg, J.M., Taylor, W.G., Kuo-Leblanc, V., Rodriguez-Wolf, M., Baird, A., Schipper, H.M., and Stopa, E.G. (1998) Hepatocyte growth factor (HGF/SF) in Alzheimer's disease, *Brain Res.* **779,** 262–270.
152. Yamada, T., Yoshiyama, Y., Tsuboi, Y. and Shimomura, T. (1997) Astroglial expression of hepatocyte growth factor and hepatocyte growth factor activator in human brain tissues, *Brain Res.* **762,** 251–255.
153. Zambenedetti, P., Giordano, R. and Zatta, P. (1998) Metallothioneins are highly expressed in astrocytes and microcapillaries in Alzheimer's disease, *J. Chem. Neuroanat.* **15,** 21–26.
154. Pasternack, J.M., Abraham, C.R., Van Dyke, B.J., Potter, H., and Younkin, S.G. (1989) Astrocytes in Alzheimer's disease gray matter express alpha 1- antichymotrypsin mRNA, *Am. J. Pathol.* **135,** 827–834.
155. Lidstrom, A.M., Bogdanovic, N., Hesse, C., Volkman, I., Davidsson, P., and Blennow, K. (1998) Clusterin (apolipoprotein J) protein levels are increased in hippocampus and in frontal cortex in Alzheimer's disease. *Exp. Neurol.* **154,** 511–521.
156. Pei, J.J., Grundke-Iqbal, I., Iqbal, K., Bogdanovic, N., Winblad, B., and Cowburn, R.F. (1997) Elevated protein levels of protein phosphatases PP-2A and PP-2B in astrocytes of Alzheimer's disease temporal cortex. *J. Neural Transm.* **104,** 1329–1338.
157. Connor, B., Beilharz, E.J., Williams, C., Synek, B., Gluckman, P.D., Faull, R.L., and Dragunow, M. (1997) Insulin-like growth factor-I (IGF-I) immunoreactivity in the Alzheimer's disease temporal cortex and hippocampus. *Brain Res. Mol. Brain Res.* **49,** 283–290.
158. Maeda, M., Takagi, H., Hattori, H., and Matsuzaki, T. (1997) Localization of manganese superoxide dismutase in the cerebral cortex and hippocampus of Alzheimer-type senile dementia. *Osaka. City. Med. J.* **43,** 1–5.
159. Stephenson, D.T., Lemere, C.A., Selkoe, D.J. and Clemens, J.A. (1996) Cytosolic phospholipase A2 (cPLA2) immunoreactivity is elevated in Alzheimer's disease brain, *Neurobiol. Dis.* **3,** 51–63.

160. Nishimura, T., Akiyama, H., Yonehara, S., Kondo, H., Ikeda, K., Kato, M., Iseki, E., and Kosaka, K. (1995) Fas antigen expression in brains of patients with Alzheimer-type dementia, *Brain Res.* **695,** 137–145.
161. Akiyama, H., Tooyama, I., Kawamata, T., Ikeda, K., and McGeer, P.L. (1993) Morphological diversities of CD44 positive astrocytes in the cerebral cortex of normal subjects and patients with Alzheimer's disease. *Brain Res.* **632,** 249–259.
162. Akiyama, H., Kawamata, T., Yamada, T., Tooyama, I., Ishii, T., and McGeer, P.L. (1993) Expression of intercellular adhesion molecule (ICAM)-1 by a subset of astrocytes in Alzheimer disease and some other degenerative neurological disorders. *Acta Neuropathol. (Berl.)* **85,** 628–634.
163. Zhang, W.W., Badonic, T., Hoog, A., Jiang, M.H., Ma, K.C., Nie, X.J. and Olsson, Y. (1994) Astrocytes in Alzheimer's disease express immunoreactivity to the vaso- constrictor endothelin-1. *J. Neurol. Sci.* **122,** 90–96.
164. Walker, D.G., Yasuhara, O., Patston, P.A., McGeer, E.G. and McGeer, P.L., Complement C1 inhibitor is produced by brain tissue and is cleaved in Alzheimer disease, *Brain Res.* **675,** 75–82.
165. Colton, C., Wilt, S., Gilbert, D., Chernyshev, O., Snell, J., and Dubois-Dalcq, M. (1996) Species differences in the generation of reactive oxygen species by microglia. *Mol. Chem. Neuropathol.* **28,** 15–20.
166. Lee, S.C., Liu, W., Dickson, D.W., Brosnan, C.F., and Berman, J.W. (1993) Cytokine production by human fetal microglia and astrocytes. Differential induction by lipopolysaccharide and IL-1 beta. *J. Immunol.* **150,** 2659–2667.
167. Barnum, S.R. and Jones, J.L. (1994) Transforming growth factor-beta 1 inhibits inflammatory cytokine- induced C3 gene expression in astrocytes. *J. Immunol.* **152,** 765–773.
168. Walker, D.G., Kim, S.U. and McGeer, P.L. (1998) Expression of complement C4 and C9 genes by human astrocytes, *Brain Res.* **809,** 31–38.
169. Gasque, P., Fontaine, M., and Morgan, B.P. (1995) Complement expression in human brain. Biosynthesis of terminal pathway components and regulators in human glial cells and cell lines, *Journal of Immunology* **154,** 4726–33.
170. Veerhuis, R., Janssen, I., Hoozemans, J.J., De Groot, C.J., Hack, C.E. and Eikelenboom, P. (1998) Complement Cl-inhibitor expression in Alzheimer's disease., *Acta Neuropathol. (Berl.)* **96,** 287–296.
171. Walker, D.G. and McGeer, P.L. (1998) Vitronectin expression in Purkinje cells in the human cerebellum., *Neurosci. Lett.* **251,** 109–112.
173. Das, S. and Potter, H. (1995) Expression of the Alzheimer amyloid-promoting factor antichymotrypsin is induced in human astrocytes by IL-1. *Neuron.* **14,** 447–456.
174. Higuchi, M., Ito, T., Imai, Y., et al. (1994) Expression of the alpha 2-macroglobulin-encoding gene in rat brain and cultured astrocytes. *Gene* **141,** 155–162.
175. Zarow, C. and Victoroff, J. (1998) Increased apolipoprotein E mRNA in the hippocampus in Alzheimer disease and in rats after entorhinal cortex lesioning, *Exp. Neurol.* **149,** 79–86.

20
Activated Neuroglia in Alzheimer's Disease

Kurt R. Brunden and Robert C. A. Frederickson

1. INTRODUCTION

The distinguishing pathological hallmark of Alzheimer's disease (AD) is the presence of proteinaceous deposits, referred to as senile plaques, within the higher learning centers of the brain *(1)*. These plaques are comprised predominantly of multimeric fibrils of the amyloid β (Aβ) peptide *(2)*, a 40–43 amino acid long proteolytic fragment that is cleaved from the larger amyloid precursor protein (APP) *(3)*. Genetic data suggest that senile plaques are key causative agents in AD, as all known mutations that cause early-onset familial AD *(4–8)* result in an increased production of the amyloidogenic $A\beta_{1-42}$ isoform *(9–13)*. Although the deposition of multimeric Aβ fibrils into plaques is likely to be a requisite step in AD onset, there is still uncertainty as to how Aβ and neuritic plaques might cause the neuropathology that leads to the characteristic dementia of this disease. One compelling hypothesis that is supported by substantial experimental data postulates that senile plaques cause glial cell activation, resulting in the release of a host of glial-derived molecules that contribute to disease progression.

Immunohistochemical analyses reveal a distinct and robust glial response in the immediate vicinity of senile plaques. Notably, activated microglia *(14,15)* that express surface major histocompatability antigens *(16)* are found associated with the plaque core. Moreover, glial fibrillary acidic protein (GFAP)-positive reactive astrocytes circumscribe senile plaques *(17)*. The localization of activated glia is relatively specific to the area near the senile plaques, and it is particularly noteworthy that activated microglia and astrocytes are not typically located at diffuse plaques *(18)*. These deposits contain amorphous Aβ aggregates and are generally devoid of neuritic damage. This suggests that the glial cells respond specifically to Aβ fibrils and subsequently promote neuritic damage. This premise is supported by the observation that there is a positive correlation between activated glia and the presence of tau-2-positive neuritic plaques *(19)*.

Based on the above mentioned findings, as well as many further reports on amyloid and glial phenomena in AD, we and others have proposed hypotheses for the possible role of activated astrocytes and microglia in the etiology of AD *(20–24)*. Hypotheses are of little benefit unless they can be tested and discarded or refined. Therefore, we will try to review the progress made over the last few years in testing and evaluating a number of these hypotheses. It is beyond the scope of this chapter to review the

From: *Neuroglia in the Aging Brain*
Edited by: Jean S. de Vellis © Humana Press Inc., Totowa, NJ

progress of all hypotheses concerning the role of neuroglia in AD and we will thus focus on two promising concepts which have the benefit of being testable and which have received much confirmatory attention over the last half-decade. The first is the modified excitotoxicity hypothesis in which β-amyloid is proposed to induce its neurotoxicity indirectly by an inhibitory action on the uptake of synaptic glutamate by astrocytes *(21)*. The second is the immuno-inflammatory hypothesis, which postulates that activated microglia and their various secreted products are the intermediaries of the Aβ-induced neurotoxicity presumed to lead to the dementia of AD *(22,23)*. These two hypotheses are not mutually exclusive, nor do they exclude other contributions to the pathology of AD, but both provide very promising approaches to providing new means of treatment for the various pathologies of AD.

The remainder of this chapter summarizes and discusses the body of literature that addresses these possible roles of glial cells in the pathology of AD. There are compelling data to support the concept that therapeutic strategies aimed at reducing the extent of glial activity in the AD brain should be tested in controlled clinical trials.

2. EXCITOTOXICITY MEDIATED BY AMYLOID-COMPROMISED ASTROCYTES IN AD BRAIN

It was proposed almost 8 years ago that Aβ may induce glutamate excitotoxicity in AD brain by an indirect mechanism. This hypothesis *(21)* suggested that neuronal cell death in AD may be secondary to an effect of Aβ on astrocytes in AD brain, impairing their ability to remove glutamate from neuronal synapses. Much circumstantial evidence supported this premise, but the critical experimental test of whether or not Aβ could indeed impair glutamate uptake by astrocytes had not yet been demonstrated. In subsequent years, a number of groups have demonstrated that Aβ does indeed impair the uptake of glutamate by astrocytes in culture *(25–27)*, thus strengthening this hypothesis. Although the mechanism of the amyloid-induced impairment of astrocytic glutamate uptake has not been definitively established, it has been suggested that it may be secondary to the action of free radicals, nitric oxide (NO) or arachidonic acid *(23,25)*, the production of which by activated neuroglia in AD brain is discussed later in this chapter.

The observations of Aβ-induced attenuation of glutamate transport seem particularly pertinent when combined with the understanding that glutamate is the major excitatory neurotransmitter in brain regions such as hippocampus and frontal cortex that experience significant neuronal cell loss in AD *(28,29)*. Moreover, it has been reported that glutamate uptake is decreased in the brains of AD patients *(30)*.

Klegeris et al. *(31)*, furthermore, report that Aβ*(1–40)* significantly increased the extracellular glutamate levels in co-cultures of THP-1 monocytes and U-373 MG astrocytoma cells whereas the antiinflammatory drug dexamethasone significantly decreased these levels. The authors conclude that inflammatory stimuli may increase extracellular glutamate whereas antiinflammatory drugs decrease it. This group had previously reported that another antiinflammatory agent, indomethacin, significantly reduced the progression of the symptomatology of AD *(32)* and a mechanism has been suggested whereby this may have been due to a reduction in glutamate excitotoxicity *(23)*. An immuno-inflammatory contribution to the pathology of AD has gained strong support and the studies above suggest that the excitotoxin, glutamate, could be a significant contributor to the inflammatory pathology in AD. Other factors contributing to the immunoinflammatory pathology are discussed in the following sections.

3. IMMUNO-INFLAMMATORY PATHOLOGY IN AD

3.1. Glial-Derived Proinflammatory Molecules in AD

3.1.1. Cytokines

Activated microglia within the AD brain have a morphology that is similar to peripheral macrophage. The prevailing view is that microglia are cells of the monocyte/macrophage lineage that migrate to the brain during early development, prior to formation of an intact blood-brain-barrier *(33)*. Like macrophage, microglia are capable of expressing the proinflammatory cytokines interleukin-1α (IL-1α), IL-1β, IL-6, and tumor necrosis factor α (TNFα). Griffin and colleagues *(34)* reported a 30-fold increase in the number of IL-1-positive microglia in the AD brain relative to age-matched control tissue, suggestive of a chronic inflammatory condition. There is also evidence that the levels of IL-1β within cerebrospinal fluid may be increased in individuals with AD *(35,36)*. Further proof of an inflammatory component in AD is provided by the observation that IL-6 levels appear to be increased in the AD brain *(35)*. Finally, it has been reported that serum TNFα levels are elevated in AD patients *(37)*.

As noted in the introduction, the specific activation of microglia and astrocytes near senile plaques implies that these cells are responding to one or more plaque components. Several in vitro studies have revealed that Aβ peptide is capable of eliciting a robust upregulation of cytokine expression by microglia and macrophage. Exposing mouse microglia to Aβ results in increased mitosis and an associated potentiation of IL-1 release *(38)*. More recent studies have found that either nonstimulated mouse microglia *(39)* or those treated with γ-interferon *(40)* respond to Aβ fibrils by increasing their release of TNFα. It appears that the response of microglia and macrophage to Aβ is specific to the fibrillar form of the peptide. For example, human THP-1 monocytes that are differentiated to a macrophage phenotype increase their release of IL-1β when treated with Aβ fibrils *(41)*. However, no cytokine elevation was observed upon addition of nonfibrillar Aβ to the THP-1 cells. Recent data reveal that these cells also increase their release of certain inflammation-associated chemokines in response to Aβ fibrils *(42)*.

A localized, chronic increase of IL-1 in the AD brain is likely to lead to changes in astrocyte expression of IL-6, as evidenced by the significant release of IL-6 by these cells in culture after addition of IL-1 *(43)*. Aβ fibrils further enhance IL-1-induced IL-6 release from astrocytes *(44)*. Thus, it is possible that the reported IL-6 elevation in AD *(35)* is due to both microglial-derived IL-1 and plaque-associated Aβ fibrils.

Astrocytes also respond to IL-1 by increasing the expression of S100β *(19)*, a protein with growth-promoting activity that has been suggested to contribute to the formation of dystrophic neurites around senile plaques. The reactive astrocytes found in the vicinity of plaques overexpress S100β *(34)*, and levels of S100β were found to be significantly correlated with the number of tau-2-positive neuritic plaques *(45)*. Furthermore, transgenic mice that overexpress S100β show an age-dependent increase in cytoskeletal abnormalities that may be related to those observed in AD *(46)*. Potential neuropathological effects of S100β may result not only from its ability to provoke neuritic growth, but also from a potentiation of astrocyte NO production *(47)*.

IL-1 may further contribute to AD pathology by causing increased production of astrocyte-derived α1-antichymotrypsin (ACT) *(48)*, a protein that may enhance the assembly of Aβ fibrils *(49)*. In addition, several studies have demonstrated that IL-1 can increase APP expression *(50,51)*. The release of IL-1 by activated microglia in AD could thus initiate a vicious cycle in which additional Aβ peptide is formed from APP, with ACT potentially contributing to greater plaque deposition and resulting glial activation.

Finally, an elevation of proinflammatory cytokines in the AD brain could lead directly to neuropathology. The combination of TNFα and IL-1 causes neuronal death in culture *(52)* and transgenic mice that express elevated levels of IL-6 in their brain show severe neuron loss as they age *(53)*. In summary, a chronic release of inflammatory cytokines by glia in the AD brain could readily contribute to the neuronal damage and death that leads to the dementia of this disease. This concept has recently been bolstered by the finding that a genetic variation in the IL-6 gene reduces the risk of AD *(54)*.

3.1.2. Arachidonic Acid and Eicosanoids

Among the variety of potential detrimental effects triggered by IL-1 elevation in the AD brain may be the production of arachidonic acid and its metabolites. Astrocyte cultures that are treated with IL-1 increase their release of prostaglandin E2 (PGE2) *(55)*. In fact, IL-1-induced IL-6 release by astrocytes has been suggested to be mediated by PGE2 *(56)*. An IL-1-initiated elevation of PGE2 may result from an upregulation of phospholipase A2 (PLA2), the enzyme responsible for the release of arachidonic acid, since both IL-1 and TNFα cause increased expression of PLA2 by astrocytes *(57–59)*.

In addition to serving as a prostaglandin precursor, arachidonic acid could contribute to AD neuropathology by additional mechanisms. This fatty acid has been implicated as a direct neurotoxin *(60)* and is an effective inhibitor of high-affinity glutamate uptake by astrocytes *(61,62)*. As noted above, because astrocytes are the primary regulator of glutamate concentrations near NMDA and AMPA receptors, a reduction in their ability to remove glutamate could lead to excitotoxic neuronal damage in the cognitive regions of the brain *(21,23)*.

Consistent with the possibility of microglial IL-1 release leading to enhanced PLA2 activity and consequent arachidonic acid release is the finding that astrocyte PLA2 immunoreactivity is increased around senile plaques in AD *(63)*. There is also evidence of cyclooxygenase-2 elevation in neurons in AD *(64)*, leading to the possibility of astrocyte-derived arachidonic acid being converted to eicosanoids by neurons in the AD brain.

3.2. Glial-Derived NO and Free Radicals

Glial cells may further contribute to neuropathology in AD through their release of agents that lead to free radical damage. There is evidence of oxidative damage in the vicinity of senile plaques in AD *(65,66)* and microglia are known to release increased levels of superoxide anion into culture medium after exposure to Aβ fibrils *(40,67)*. This suggests that the microglia that are in intimate contact with senile plaques might release superoxide and other free radicals that are detrimental to neurons.

An additional glial-derived agent that can result in oxidative damage is NO, which can form the highly toxic peroxynitrite ion upon interaction with superoxide *(68)*. It has recently been revealed that there is an upregulation of inducible NO synthase (iNOS) in the reactive astrocytes that are adjacent to Aβ-containing plaques in AD *(69)*. This elevation in astrocyte iNOS could result from the actions of cytokines and/or S100β which, as discussed under **Subheading 3.1.1.**, are elevated in AD. In vitro studies have documented that the combination of IL-1 and TNFα result in enhanced NO production by astrocytes *(70)*. Likewise, treatment of astrocyte cultures with S100β results in iNOS activation and NO release *(47)*. Because astrocyte synthesis of S100β in AD may be triggered by microglial-derived IL-1, a compelling argument can be made that Aβ activation of microglia and a resulting release of IL-1 and TNFα may start a cascade of events that ultimately leads to the production of additional neurotoxic agents such as S100β, IL-6, arachidonic acid, superoxide anion, and NO.

3.3. Glia as Potential Sources of Complement Proteins

The early activation products of the classical complement pathway (i.e., C1 and C4) are found associated with senile plaques in AD brain *(71,72)*. Moreover, the terminal complement activation product, membrane-attack complex (MAC or C5b-9), appears to be localized to dystrophic neurites found in the vicinity of senile plaques *(73,74)*. These observations have resulted in the hypothesis that complement activation may contribute to neuronal damage in AD.

Although the classical complement cascade may progress to completion around senile plaques, it appears that the pathway only proceeds through C3 in diffuse plaques *(71)*. The apparent absence of later complement components, such as MAC, at diffuse plaques suggests that there may be relatively low levels of one or more of the C5–C9 precursors in the region of these deposits. The complement seen in AD brain is likely synthesized by resident central nervous system cells since there does not appear to be a general disruption of the blood-brain-barrier *(75)*. *In situ* hybridization studies indicate that microglia and neurons of the AD brain contain elevated amounts of C1q and C4 mRNA, respectively *(76)*. Both astrocytes and microglia can synthesize most of the classical pathway proteins in vitro *(77,78)*. It is therefore possible that the activated glia found at senile plaques synthesize key complement proteins, and the absence of these cells in the vicinity of diffuse plaques may account for the observed shortage of later complement proteins and activation products. If this is true, then agents that suppress plaque-associated glial activation should prove useful in attenuating any damage associated with complement activation in the AD brain.

4. CONCLUDING REMARKS

The findings discussed above suggest a number of potential approaches that might be employed to minimize the contribution of glia to neuronal damage in AD. Since several reports indicate that the interaction of Aβ fibrils with microglia triggers cellular activation, it would seem desirable to block this amyloid-associated event. It has been suggested that Aβ can bind to specific receptors on microglia, including the scavenger receptor *(67)* and the receptor for advanced glycation end-products (RAGE) *(39)*. Hence, molecules that prevent amyloid peptide binding to one or both of these receptors may reduce microglial activation at senile plaques and consequently be of therapeutic value. However, there is some question as to whether either of these receptor systems are truly responsible for initiating the immuno-inflammatory events described in this chapter *(79,80)*, and the potential effectiveness of RAGE or scavenger receptor antagonists in AD awaits further testing.

Since many of the changes in glial cell protein expression in AD may result from an initial release of IL-1 and other proinflammatory cytokines, compounds that reduce inflammation would seem to be reasonable choices as potential drug candidates. In fact, there are several epidemiological studies which conclude that classical antiinflammatory drug treatments reduce the onset of AD (reviewed in *(81,82)*). A confounding factor in utilizing existing steroidal or nonsteroidal antiinflammatory drugs (NSAIDs) in AD is the general intolerance of the elderly to these compounds. Thus, newer drugs without the side-effect profiles of typical antiinflammatory treatments are needed to serve the AD population. Furthermore, current NSAIDs are primarily cyclooxygenase (COX) inhibitors that affect cytokine expression only at very high concentrations *(83)*. Although COX inhibition would protect against any detrimental effects related to arachidonic acid metabolites in the AD brain (*see* **Subheading 3.1.2.**), NSAIDs would be predicted to be relatively ineffective at attenuating glial cytokine release. Accordingly, specific inhibitors of proinflammatory cytokine expression may prove to be more valuable in slowing AD progression.

In addition to cytokines, activated microglia and reactive astrocytes in the AD brain appear to release a variety of additional harmful molecules. It is difficult to predict the relative contribution of these various bioactive species to disease neuropathology. As discussed, free radicals, NO, glutamate, arachidonic acid and/or complement proteins may play a role in neuronal damage in AD. Thus, drugs directed toward one or more of these downstream targets may prove to be efficacious. For example, antioxidants, glutamate receptor antagonists, PLA2 inhibitors or inhibitors of complement activity could reasonably be postulated as possible drug candidates for AD. The next several years promise to be very interesting since it is likely that one or more of these approaches will be pursued clinically in the hopes of slowing the glial contribution to AD.

REFERENCES

1. Terry, R. and Katzman, R. (1983) Senile dementia of the Alzheimer type: defining a disease. In Neurology of Aging. (Katzman, R. and Terry, R. eds.) Davis, F.A. 51–84.
2. Glenner, G.G. and Wong, C.W. (1984) Alzheimer's disease and Down's syndrome: sharing of a unique cerebrovascular amyloid fibril protein. *Biochem. Biophys. Res. Comm.* **122,** 1131–1135.
3. Kang, J., Lemaire, H.G., Unterbeck, A., et al. (1987) The precursor of Alzheimer's disease amyloid A4 protein resembles a cell-surface receptor. *Nature* **325,** 733–736.
4. Chartier-Harlin, M.-C., Crawford, F., Houlden, H., et al. (1991) Early-onset Alzheimer's disease caused by mutations at codon 717 of the β-amyloid precursor protein gene. *Nature* **353,** 844–845.
5. Goate, A., Chartier-Harlin, M., Mullan, M., et al. (1991) Segregation of a missense mutation in the amyloid precursor protein gene with familial Alzheimer's disease. *Nature* **349,** 704–706.
6. Mullan, M., Crawford, F., Axelman, K., et al. (1992) A pathogenic mutation for probable Alzheimer's disease in the APP gene at the N-terminus of β-amyloid. *Nat. Genet.* **1,** 345–347.
7. Levy-Lahad, E., Wasco, W., Poorkaj, P., et al. (1995) Candidate gene for the chromosome 1 familial Alzheimer's disease locus. *Science* **269,** 973–977.
8. Sherrington, R., Rogaev, E.I., Liang, Y., et al. (1995) cloning of a gene bearing missense mutations in early-onset familial Alzheimer's disease. *Nature* **375,** 754–760.

9. Cai, X.-D., Golde, T.E., and Younkin, S.G. (1993) Release of excess amyloid β protein from a mutant amyloid β protein precursor. *Science* **259,** 514–516.
10. Citron, M., ltersdorf, T., Haass, C., et al. (1992) Mutation of the β-amyloid precursor protein in familial Alzheimer's disease increases β-protein production. *Nature* **360,** 672–674.
11. Suzuki, N., Cheung, T.T., Cai, X.-D., Odaka, A., Otvos, L., Eckman, C., Golde, T.E., and Younkin, S.G. (1994) An increased percentage of long amyloid β protein secreted by familial amyloid β protein precursor (βAPP717) mutants. *Science* **264,** 1336–1340.
12. Duff, K., Eckman, C., Zehr, C., et al. (1996) Increased amyloid-42(43) in brains of mice expressing mutant presenilin 1. *Nature* **383,** 710–713.
13. Scheuner, D., Eckman, C., Jensen, M., et al. (1996) Secreted amyloid β-protein similar to that in the senile plaques of Alzheimer's disease is increased in vivo by the presenilin 1 and 2 and APP mutations in familial Alzheimer's disease. *Nature Med.* **2,** 864–870.
14. Haga, S., Adai, K., and Ishii, T. (1989) Demonstration of microglial cells in and around senile (neuritic) plaques in the Alzheimer brain: an immunohistochemical study using a novel monoclonal antibody. *Acta Neuropathol.* **77,** 569–575.
15. McGeer, P.L., Itagaki, S., Tago, H., and McGeer, E.G. (1987) Reactive microglia in patients with senile dementia of the Alzheimer type are positive for the histocompatibility glycoprotein HLA-DR. *Neurosci. Lett.* **9,** 195–200.
16. Rogers, J., Luber-Narod, J., Styren, S.D., and Civin, W.H. (1988) Expression of immune system-associated antigens by cells of the human central nervous system: relationship to the pathology of Alzheimer's disease. *Neurobiol. Aging* **9,** 339–349.
17. Itagaki, S., McGeer, P.L., Akiyama, H., Zhu, S., and Selkoe, D.J. (1989) Relationship of microglia and astrocytes to amyloid deposits of Alzheimer's disease. *J. Neuroimmunol.* **24,** 173–182.
18. Dickson, D.W. (1997) The pathogenesis of senile plaques. *J. Neuropath. Exp. Neurol.* **56,** 321–339.
19. Sheng, J.G., Mrak, R.E., Rovnaghi, C.R., Kozlowska, E., van Eldik, L.J., and Griffin, W.S.T. (1996) Human brain S100 and S100 mRNA expression increases with age: pathogenic implications for Alzheimer's disease. *Neurobiol. Aging* **17,** 359–363.
20. McGeer, P.L. and Rogers, J. (1992) Anti-inflammatory agents as a therapeutic approach to Alzheimer's disease. *Neurology* **42,** 447–449.
21. Frederickson, R.C.A. (1992) Astroglia in Alzheimer's disease. *Neurobiol. Aging* **13,** 239–253.
22. McGeer, P.L., Rogers, J., and McGeer, E.G. (1994) Neuroimmune mechanisms in Alzheimer disease pathogenesis. *Alzheimer Dis. Assoc. Disorders* **8,** 149–158.
23. Frederickson, R.C.A. and Brunden, K.R. (1994) New opportunities in AD research-roles of immunoinflammatory responses and glia. *Alzheimer Diseases and Associated Disorders* **8,** 159–165.
24. Breitner, J.C. (1996) Inflammatory processes and anti-inflammatory drugs in Alzheimer's disease: a current appraisal. *Neurobiol. Aging* **17,** 789–794.
25. Harris, M.E., Carney, J.M., Cole, P.S., et al. (1995) β-Amyloid peptide-derived, oxygen-dependent free radicals inhibit glutamate uptake in cultured astrocytes: implications for Alzheimer's disease. *Neuroreport* **6,** 1875–1879.
26. Harris, M.E., Wang, Y., Pedigo, N.W., Hensley, K., Butterfield, D.A., and Carney, J.M. (1996) Amyloid beta peptide (25–35) inhibits Na^+-dependent glutamate uptake in rat hippocampal astrocyte cultures. *J. Neurochem.* **67,** 277–286.
27. Parpura-Gill, A., Beitz, D., and Uemura, E. (1996) The inhibitory effects of β-Amyloid on glutamate and glucose uptake by cultured rat astrocytes. *Brain Res.* **754,** 65–71.
28. Storm-Mathisen, J. (1977) Glutamic acid and excitatory nerve endings: reductions of glutamic acid uptake after axotomy.*Brain Res.* **120,** 379–386.
29. White, W.F., Nadler, J.V., Hamberger, A., Cotman, C.W., and Cummins, J.T. (1977) Glutamate as a transmitter of the hippocampal perforant path. *Nature* **270,** 356–367.

30. Kanai, Y., Smith, C.P., and Hediger, A. (1993) The elusive transporters with a high affnity for glutamate uptake. *Trends Neurosci.* **16,** 365–370.
31. Klegaris, A., Walker, D.G., and McGeer, P.L. (1997) Regulation of glutamate in cultures of human monocytic THP-1 and astrocytoma U-373 MG cells. *J. Neuroimmunol.* **78,** 152–161.
32. Rogers, J., Kirby, L.C., Hempelman, S.R., et al. (1993) Clinical trial of indomethacin in Alzheimer's disease. *Neurology* **43,** 1609–1611.
33. Ling, E.A. and Wong, W.C. (1993) The origin and nature of ramified and amoeboid microglia. *Glia* **7,** 9–18.
34. Griffin, W.S.T. Stanley, L.C., Ling, C., et al. (1989) Brain interleukin-1 and S-100 immunoreactivity are elevated in Down syndrome and Alzheimer disease. *Proc. Natl. Acad. Sci. USA* **86,** 7611–7615.
35. Blum-Degen, D., Muller, T. Kuhn, W., Gerlach, M., Przuntek, H., and Riederer, P. (1995) Interleukin-1β and interleukin-6 are elevated in the cerebrospinal fluid of Alzheimer's and de novo Parkinson's disease patients. *Neurosci. Lett.* **202,** 17–20.
36. Cacabelos, R., Alvarez, X.A., Fernandez-Novoa, L., et al. (1993) Brain interleukin-1β in Alzheimer's disease and vascular dementia. *Meth. Find. Exp. Clin. Pharmacol.* **16,** 141–151.
37. Fillit, H., Ding, W., Buee, L., et al. (1991) Elevated circulating tumor necrosis factor in Alzheimer's disease. *Neurosci. Lett.* **129,** 318–320.
38. Araujo, D.M. and Cotman, C.W. (1992) β-Amyloid stimulates glial cells in vitro to produce growth factors that accumulate in senile plaques in Alzheimer's disease. *Brain Res.* **569,** 141–145.
39. Yan, S.D., Chen, X., Fu, J., et al. (1996) RAGE and amyloid-peptide neurotoxicity in Alzheimer's disease. *Nature* **382,** 685–691.
40. Meda, L., Cassatella, M.A., Szendrel, G.I. (1995) Activation of microglial cells by β-amyloid protein and interferon-γ. *Nature* **374,** 647–650.
41. Lorton, D., Kocsis, J., King, L., Madden, K., and Brunden, K.R. (1996) β-Amyloid induces increased release of interleukin-1 from lipopolysaccharide-activated human monocytes. *J. Neuroimmunol.* **67,** 21.
42. Yates, S.L., Burgess, L.H., Kocsis-Angle, J., J.M., et al. (2000) Amyloid β and amylin induce increases in proinflammatory cytokine and chemokine production by THP-1 cells and murine microglia. *J. Neurochem.* **74,** 1017–1025.
43. Cadman, E.D., Witter, D.G., and Lee, C.-M. (1994) Regulation of the release of interleukin-6 from human astrocytoma cells. *J. Neurochem.* **63,** 980–987.
44. Gitter, B.C., Cox, L.M., Rydel, R.E., and May, P.C. (1995) Amyloid beta peptide potentiates cytokine secretion by interleukin-1 β-activated human astrocytoma cells. *Proc. Natl. Acad. Sci. USA* **92,** 10738–10741.
45. Sheng, J.G., Mrak, R.E., and Griffin, W.S.T. (1994) S100β protein expression in Alzheimer's disease: potential role in the pathogenesis of neuritic plaques. *J. Neurosci. Res.* **39,** 398–404.
46. Whitaker-Azmitia, P.M., Wingate, M., Borella, A., Gerlai, R., Roder, J., and Azmitia, E.C. (1997) Transgenic mice overexpressing the neurotrophic factor S100β show neuronal cytoskeletal and behavioral signs of altered aging processes: implications for Alzheimer's disease and Down's syndrome. *Brain Res.* **776,** 51–60.
47. Hu, J., Ferrira, A., and van Eldik, L.J. (1997) S100β induces neuronal cell death through nitric oxide release from astrocytes. *J. Neurochem.* **69,** 2294–2301.
48. Das, S. and Potter, H. (1995) Expression of the Alzheimer amyloid-promoting factor antichymotrypsin is induced in human astrocytes by IL-1. *Neuron* **14,** 447–456.
49. Ma, J., Yee, A., Brewer, H.B., Das, S., and Potter, H. (1994) Amyloid-associated proteins α1-antichymotrypsin and apolipoprotein E promote assembly of Alzheimer β-protein into filaments. *Nature* **372,** 92–94.

50. Goldgaber, D., Harris, H.W., Hla, T., et al. (1989) Interleukin-1 regulates synthesis of amyoid β-protein precursor mRNA in human endothelial cells. *Proc. Natl. Acad. Sci. USA* **86,** 7606–7610.
51. Forloni, G., Demicheli, F., Giorgi, S., Bendotti, C., and Angeretti, N. (1992) Expression of amyloid precursor protein mRNAs in endothelial, neuronal and glial cells: modulation by interleukin-1. *Mol. Brain Res.* **16,** 128–134.
52. Chao, C.C., Hu, S., Ehrlich, L., and Peterson, P.K. (1995) Interleukin-1 and tumor necrosis factor-α synergistically mediate neurotoxicity: involvement of nitric oxide and of N-methyl-D-aspartate receptors. *Brain, Behavior and Immunity* **9,** 355–365.
53. Heyser, C.J., Masliah, E., Samimi, A., Campbell, I.L., and Gold, L.H. (1997) Progressive decline in avoidance learning paralled by inflammatory neurodegeneration in transgenic mice expressing interleukin-6 in the brain. *Proc. Natl. Acad. Sci. USA* **94,** 1500–1505.
54. Papassotiropoulos, A., Bagli, M., Jessen, F., Bayer, T.A., Maier, W., Rao, M.L., and Heun, R. (1999) A genetic variation of the inflammatory cytokine interleukin-6 delays the initial onset and reduces the risk for sporadic Alzheimer's disease. *Ann. Neurology* **45,** 666–668.
55. Katsuura, G., Gottschall, P.E., Dahl, R.R., and Arimura, A. (1989) Interleukin-1β increases prostaglandin E2 in rat astrocyte cultures: modulatory effect of neuropeptides. *Endocrinology* **124,** 3125–3127.
56. Fiebich, B.L., Hull, M., Lieb, K., Gyufko, K., Berger, M., and Bauer, J. (1997) Prostaglandin E2 induces interleukin-6 synthesis in human astrocytoma cells. *J. Neurochem.* **68,** 704–709.
57. Oka, S., and Arita, H. (1991) Inflammatory factors stimulate expression of group II phospholipase A2 in rat cultured astrocytes. Two distinct pathways of gene expression. *J. Biol. Chem.* **266,** 9956–9960.
58. Ozaki, M., Morii, H., Qvist, R., and Watanbe, Y. (1994) Interleukin-1 beta induces cytosolic phospholipase A2 gene in rat glioma line. *Biochem. Biophys. Res. Comm.* **205,** 12–17.
59. Stella, N., Estelles, A., Siciliano, J., et al. (1997) Interleukin-1 enhances the ATP-evoked release of arachidonic acid from mouse astrocytes. *J. Neurosci.* **17,** 2939–2946.
60. Okuda, S., Saito, H., and Katsuki, H. (1994) Arachidonic acid: toxic and trophic effects on cultured hippocampal neurons. *Neuroscience* **63,** 691–699.
61. Barbour, B., Szatkowski, M., Ingledew, N., and Attwell, D. (1989) Arachidonic acid induces a prolonged inhibition of glutamate uptake in glial cells. *Nature* **342,** 918–920.
62. Yu, A.C.H., Chan, P.H., and Fishman, R.A. (1986) Effects of arachidonic acid on glutamate and γ-aminobutyric acid uptake in primary cultures of rat cerebral cortical astrocytes and neurons. *J. Neurochem.* **47,** 1181–1189.
63. Stephenson, D.T., Lemere, C.A., Selkoe, D.J., and Clemens, J.A. (1996) Cytosolic phospholipase A2 (cPLA2) immunoreactivity is elevated in Alzheimer's disease brain. *Neurobiol. Disease* **3,** 51–63.
64. Oka, A. and Takashima, S. (1997) Induction of cyclooxygenase 2 in brains of patients with Down's syndrome and dementia of Alzheimer type: specific localization in affected neurons and axons. *Neuroreport* **8,** 1161–1164.
65. Furuta, A., Price, D.L., Pardo, C.A., et al. (1995) Localization of superoxide dismutases in Alzheimer's disease and Down's syndrome neocortex and hippocampus. *Am. J. Pathol.* **146,** 357–367.
66. Papolla, M.A., Omar, R.A., Kim, K.S., and Robakis, N.K. (1992) Immunohistochemical evidence of antioxidant stress in Alzheimer's disease. *Am. J. Pathol.* **140,** 621–628.
67. El Khoury, J., Hickman, S.E., Thomas, C.A., Cao, L., Silverstein, S.C., and Loike, J.D. (1996) Scavenger receptor-mediated adhesion of microglia to β-amyloid fibrils. *Nature* **382,** 716–719.
68. Dawson, V.L. (1995) Nitric oxide: role in neurotoxicity. *Clin. Exp. Pharmacol. Physiol.* **22,** 305–308.

69. Wallace, M.N., Geddes, J.G., Farquhar, D.A., and Masson, M.R. (1997) Nitric oxide synthase in reactive astrocytes adjacent to beta-amyloid plaques. *Exp. Neurol.* **144,** 266–272.
70. Liu, J., Zhao, M.L., Brosnan, C.F., and Lee, S.C. (1996) Expression of type II nitric oxide synthase in primary human astrocytes and microglia: role of IL-1β and IL-1 receptor antagonist. *J. Immunol.* **157,** 3569–3576.
71. Lue, L.-F. and Rogers, J. (1992) Full complement activation fails in diffuse plaques of the Alzheimer's disease cerebellum. *Dementia* **3,** 308–313.
72. Veerhuis, R., Janssen, I., Hack, E., and Eikelenboom, P. (1996) Early complement components in Alzheimer's disease brains. *Acta Neuropathol.* **91,** 53–60.
73. Itagaki, S., Akiyama, H., Saito, H., and McGeer, P.L. (1994) Ultrastructural localization of complement membrane attack complex (MAC)-like immunoreactivity in brains of patients with Alzheimer's disease. *Brain Res.* **645,** 78–84.
74. McGeer, P.L., Akiyama, H., Itagaki, S., and McGeer, E.G. (1989) Immune system response in Alzheimer's disease. *Can. J. Neurol. Sci.* **16,** 516–527.
75. Rogers, J., Cooper, N.R., Webster, S., et al. (1992) Complement activation by β-amyloid in Alzheimer disease. *Proc. Natl. Acad. Sci. USA* **89,** 10016–10020.
76. Johnson, S.A., Lampert-Etchells, M., Pasinetti, G.M., Rozovsky, I., and Finch, C.E. (1992) Complement mRNA in the mammalian brain: responses to Alzheimer's disease and experimental brain lesioning. *Neurobiol. Aging* **13,** 641–648.
77. Walker, D.G., Kim, S.U., and McGeer, P.L. (1995) Complement and cytokine gene expression in cultured microglia derived from postmortem human brains. *J. Neurosci. Res.* **40,** 478–493.
78. Gasque, P., Fontaine, M., and Morgan, B.P. (1995) Complement expression in human brain: biosynthesis of terminal pathway components and regulators in human glial cells and cell lines. *J. Immunol.* **154,** 4726–4733.
79. Combs, C.K., Johnson, D.E., Cannady, S.B., Lehman, T.M., and Landreth, G.E. (1999) Identification of microglial signal transduction pathways mediating a neurotoxic response to amyloidogenic fragments of β-amyloid and prion proteins. *J. Neurosci.* **19,** 928–939.
80. McDonald, D.R., Brunden, K.R., and Landreth, G.E. (1997) Amyloid fibrils activate tyrosine kinase-dependent signaling and superoxide production in microglia. *J. Neurosci.* **17,** 2284–2294.
81. Breitner, J.C. (1996) The role of anti-inflammatory drugs in the prevention and treatment of Alzheimer's disease. *Annu. Rev. Med.* **47,** 401–411.
82. McGeer, P.L., Schulzer, M., and McGeer, E.G. (1996) Arthritis and anti-inflammatory agents as possible protective factors for Alzheimer's disease: a review of 17 epidemiologic studies. *Neurology* **47,** 425–432.
83. Jiang, C., Ting, A.T., and Seed, B. (1998) PPAR-γ agonists inhibit production of monocyte inflammatory cytokines. *Nature* **391,** 82–85.

21

Reactive Astroglia in the Ataxic Form of Creutzfeldt-Jakob Disease

Cytology and Organization in the Cerebellar Cortex

Miguel Lafarga, Nuria T. Villagra, and Maria T. Berciano

1. INTRODUCTION

Transmissible spongiform encephalopathies (TSE) are sub-acute neurodegenerative disorders which include scrapie in sheep, bovine spongiform encephalopathy in cows, and Creutzfeldt-Jakob disease (CJD), kuru, Gerstmann-Straussler syndrome, and fatal familial insomnia in humans. The essential pathogenic component of TSE is an abnormal isoform of the prion protein designated PrP^{Sc} (for reviews and discussions *see 1*). The naturally occurring, cellular prion protein, called PrP^{C}, encoded by the *Prnp* gene, is expressed in neurons *(2)* and glia *(3)*. PrP^{C} is glycosilated and anchored to the plasma membrane, but little is known about its cellular function (*see* references in *1*). During the disease process, PrP^{C} is converted into PrP^{Sc} in both neurons and astrocytes. The "converted" PrP^{Sc} is partly resistant to proteinase K digestion and has an altered conformational state whereby the amount α-helical structure decreases and the content of β-sheet increases (for review *see 1*). Prion-induced encephalopathies are characterized by intracerebral accumulation of PrP^{Sc} and deposition of PrP amyloid (for review *see 1,4*). The observation that *Prnp* knockout mice do not develop spontaneous scrapie *(5)* supports the view that the accumulation of PrP^{Sc} and PrP amyloid in the brain mediate neuronal toxicity in the central nervous system.

Common neuropathological hallmarks of prion diseases are neuronal degeneration, spongiform changes, formation of amyloid plaques, and astrogliosis (for review *see* refs. *6–8*). The reactive response of astrocytes includes cellular hypertrophy and hyperplasia *(9,10)*. Although the pathogenesis of the astrogliosis in TSE is still undetermined, some factors such as the neuronal damage *(9)* and the effect of the tumor necrosis factor a *(11)* and other factors released by microglia *(12)* seem to be involved in the astroglial reaction. In addition, several experimental studies have demonstrated in scrapie-infected mice and hamsters that PrP^{Sc} accumulates in astrocytes prior to the development of neuronal vacuolation and reactive gliosis *(13–16)*. Whereas the conversion of PrP^{C} to PrP^{Sc} in neurons leads to degeneration and neuronal death, in astrocytes it precedes upregulation of glial fibrillary acidic protein (GFAP), suggesting that hyper-

From: *Neuroglia in the Aging Brain*
Edited by: Jean S. de Vellis © Humana Press Inc., Totowa, NJ

trophy and proliferation of astroglia occurs as a response to the presence of PrP^{Sc} (for review *see 15,16*). In fact, recent studies in vitro have shown that PrP^{C}-expressing astrocytes proliferate in response to a neurotoxic peptide corresponding to residues 106–126 of human PrP (PrP 106–126), a sequence which is included in the protease-resistant fraction of the PrP^{Sc} *(12,17)*.

The current presentation will focus on the astrogliosis seen in the cerebellar cortex in three cases of the rare ataxic form of CJD with different degrees of neuronal loss. This ataxic form of CJD causes a progressive neurodegenerative disorder of the cerebellar cortex with loss of granule cells *(18),* diffuse isomorphic astrogliosis *(19–21),* upregulation of the GFAP gene expression *(9)* and deposition of PrP in the granular layer *(22)*. Apart from the possible direct effect of the PrP^{Sc} on the astroglial response, the ataxic form of CJD offers a useful pathological system of granule cell depletion to investigate the plasticity of neuron-astroglia interactions in response to acquired neuronal loss in a noninvasive injury of the cerebellar cortex. An additional advantage of this region is the geometrical cytoarchitecture of both neurons and astroglia. The latter includes two astroglial subtypes, Bergmann glia and granular layer astrocytes, which are segregated in different cortical layers and interact differently with each type of neuron. Thus, Bergmann glia establishes major interactions with Purkinje cells and granule cell axons, the parallel fibers, whereas astrocytes are involved in the clustering of granule cell perikarya and in the formation of perisynaptic glia around cerebellar glomeruli *(23)*.

2. NEURONAL DEATH AND NUMERICAL DENSITY OF NEURONS AND ASTROGLIAL CELLS IN THE CEREBELLAR CORTEX OF CJD

A neuropathological hallmark of the cerebellar cortex of the three cases studied of CJD is the severe loss of granule. Estimation of the numerical surface density of granule cells per mm^2 of granule cell layer has shown a range of reduction between 40% to 60% in CJD cerebella with respect to the controls *(18,21)*. The massive depletion of granule cells results in the virtual disappearance of the characteristic clusters or rosettes of granule cell perikarya (Figs. 1 and 2). Furthermore, numerous degenerating granule cells with pyknotic nuclei and spongiform changes in the neuropil are observed (Fig. 2). These findings indicate that granule cells are particularly vulnerable to the prions in the ataxic form of CJD, which is also supported by the PrP deposition in the granular layer *(22)*. The quantitative estimation of the astrocyte-granule cell ratio in the granular layer has registered a dramatic change from 1:98 in human controls vs 1:22 to 1:17 in the CJD cerebella *(18)*. This increased density of astrocytes may simply be due to the higher concentration of this glial population following granule cell loss and subsequent reduction in cerebellar size. However, our observation in all cases of CJD, but not in control cerebella, that a few reactive astrocytes express vimentin *(18,21),* a marker of immature differentiating astrocytes *(24,25),* suggests that a minor proliferation of astrocytes or their precursors takes place in the CJD cerebella. In agreement with this interpretation, Biernat et al. *(10)* have reported that a small fraction (less than 5%) of reactive astrocytes show proliferating cell nuclear antigen (PCNA) immunoreactivity in mice with experimental CJD. Furthermore, recent studies in vitro have also demonstrated that PrP^{c}-expressing astrocytes proliferate in response to the prion peptide PrP 106–126 *(12,17)*.

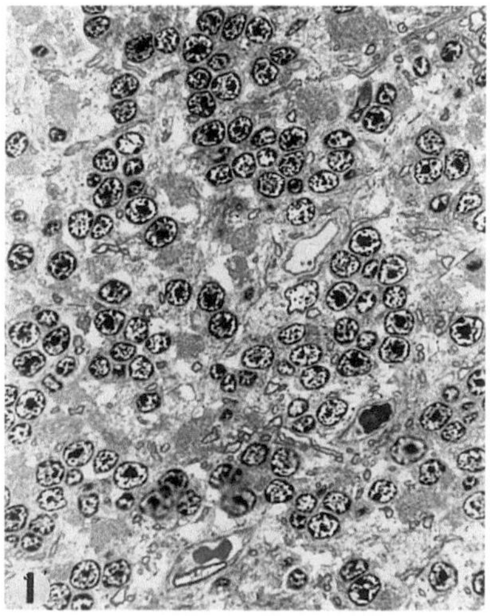

Fig. 1. Semithin section stained with toluidine blue of the granular layer from a control cerebellum showing the typical clusters of granule cell perikarya. ×900.

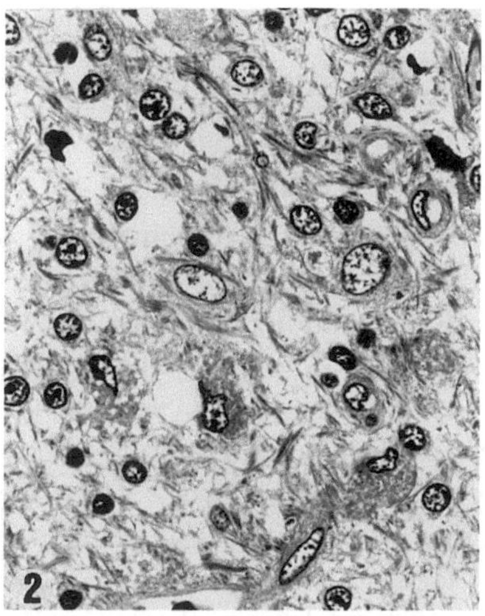

Fig. 2. Semithin section of the granular layer from a CJD cerebellum. Severe spongiosis and depletion of granule cells is seen. Numerous pyknotic nuclei of degenerating granule cells and two hypertrophied astrocytes are also observed. ×900.

We have also estimated the numerical density of Purkinje cells and Bergmann glial cells per mm of the Purkinje cell layer and their ratios in two cases of CJD (CJ-1 and CJ-2) and two controls. Although in the CJ-1 the number of Purkinje and Bergmann glial cells remained unchanged with respect to controls (Purkinje cell-Bergmann glia ratio 1:10), a significant reduction (about 32%) in Purkinje cell density accompanied by an increase (about 35%) in Bergmann glia density was detected in the CJ-2 (Purkinje cell-Bergmann glia ratio 1:20). This provides the opportunity to evaluate the reac-

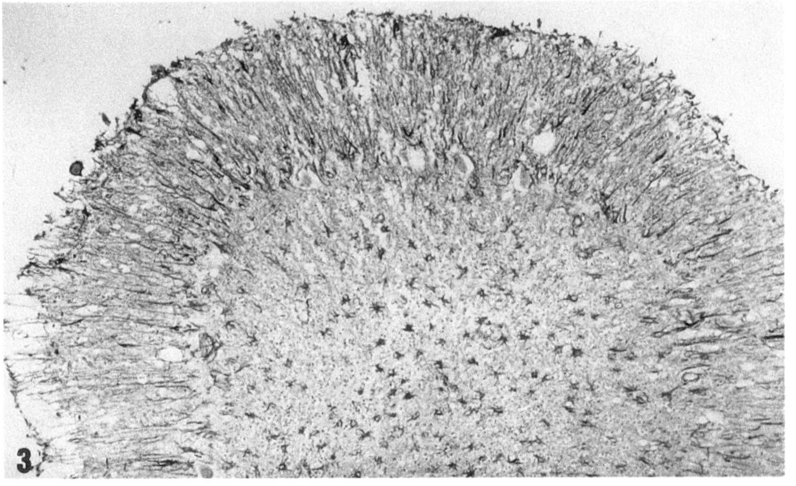

Fig. 3. GFAP immunocytochemistry of a sagittal section of the cerebellar cortex from a case of CJD. The molecular layer shows a hypertrophied palisade of normally oriented Bergmann fibers. In the granular layer the hypertrophied astrocytes appear regularly arranged. ×125.

tive response of Bergmann glia to the partial loss of their specific neuronal partner, the Purkinje cells (*see* **Subheadings 3–6**).

3. CYTOARCHITECTURE OF REACTIVE ASTROGLIA IN THE CEREBELLAR CORTEX OF CJD

In the normal cerebellar cortex, the cytoarchitecture of Bergmann glial cells and granule cell layer astrocytes is clearly demonstrated by using GFAP immunostaining. In all cases of ataxic form of CJD studied, we have found *(18,21)* a marked astroglial hypertrophy and overexpression of GFAP, but the normal geometry and spatial arrangement of both astroglial subtypes in the cortical layers appears strictly preserved (Figs. 3–5). Thus, in the molecular layer Bergmann fibers show a dense palisade organization of straight and hypertrophied processes ascending throughout this layer. The regular alignment of Bergmann fibers is clearly seen in semithin, 1μm thick, sagittal and horizontal cerebellar sections immunostained with the GFAP antibody (Figs. 4A,C). However, the comparative analysis of the cases CJ-1 and CJ-2 with preservation and loss of Purkinje cells, respectively, show that the degeneration and loss of Purkinje cells in CJ-2 results in a greater cellular hypertrophy of Bergmann glia with very robust glial processes (Fig.4B). Although in this case there was an increased numerical density of Bergmann glia, vimentin immunocytochemistry does not allow us to infer a possible hyperplasia because, unlike granule cell layer astrocytes, mature Bergmann glia constitutively express this cytoskeletal protein *(26)*. However, the normal numerical densities of both Purkinje cells and Bergmann glia found in the CJ-1 support that, at least in this pathological case, the astroglial reaction does not involve proliferation of Bergmann glia or their precursors *(18)* and may be partially induced by the granule cell loss and subsequent denervation of dendritic spines of Purkinje cells. In the CJD cerebella, the few Fañanas glial cells found in the upper portion of the molecular layer also exhibit a reactive appearance with prominent GFAP expression (Figs. 4D,E).

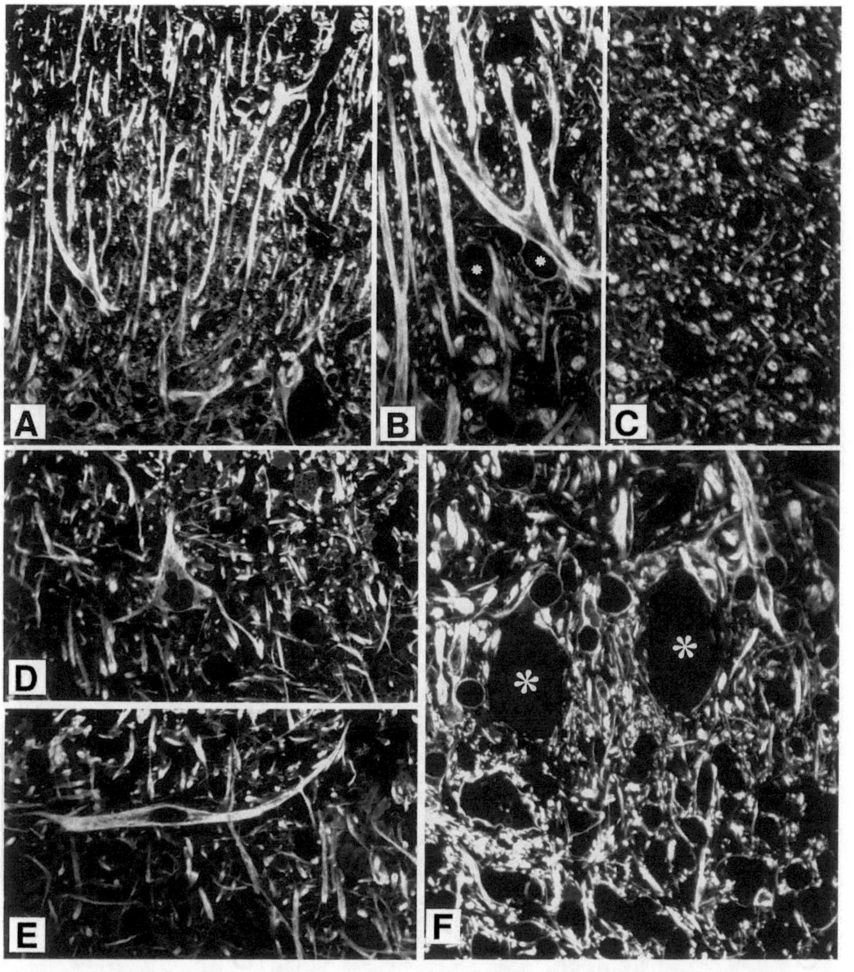

Fig. 4. GFAP immunofluorescence on 1 μm thick semithin sections of the cerebellar cortex from CJD cerebella. **(A)** and **(B)** Sagittal sections illustrating the dense palisade of straight, very thick Bergmann fibers. Note in **(B)** the hypertrophied Bergmann glia perikarya (asterisks). **(C)** Horizontal section of the molecular layer showing the regular arrangement of cross-sectioned GFAP positive Bergmann fibers. **(D)** and **(E)** Examples of hypertrophied GFAP positive Fañanas cells in the upper half of the molecular layer. **(F)** Purkinje cell layer showing a dense perineuronal net of immunofluorescent glial processes surrounding Purkinje cell perikarya (asterisks). In the underlying granule cell layer brilliant astrocytic processes delineate the unlabelled granule cell perikarya. **(A)**×750, **(B–F)**×1000.

Other interesting findings in the CJD are the reinforcement of the astroglial perineuronal nets that cover Purkinje cell perikarya (Fig. 4F) and the formation of a plexus of basal processes of Bergmann glia located at the interface between Purkinje and granular layers (Fig. 5B). This latter component of the glial response to neurodegeneration tends to create a barrier between these two cortical layers that may preserve the specific functional domains of Bergmann glia and astrocytes in molecular and granular layers, respectively.

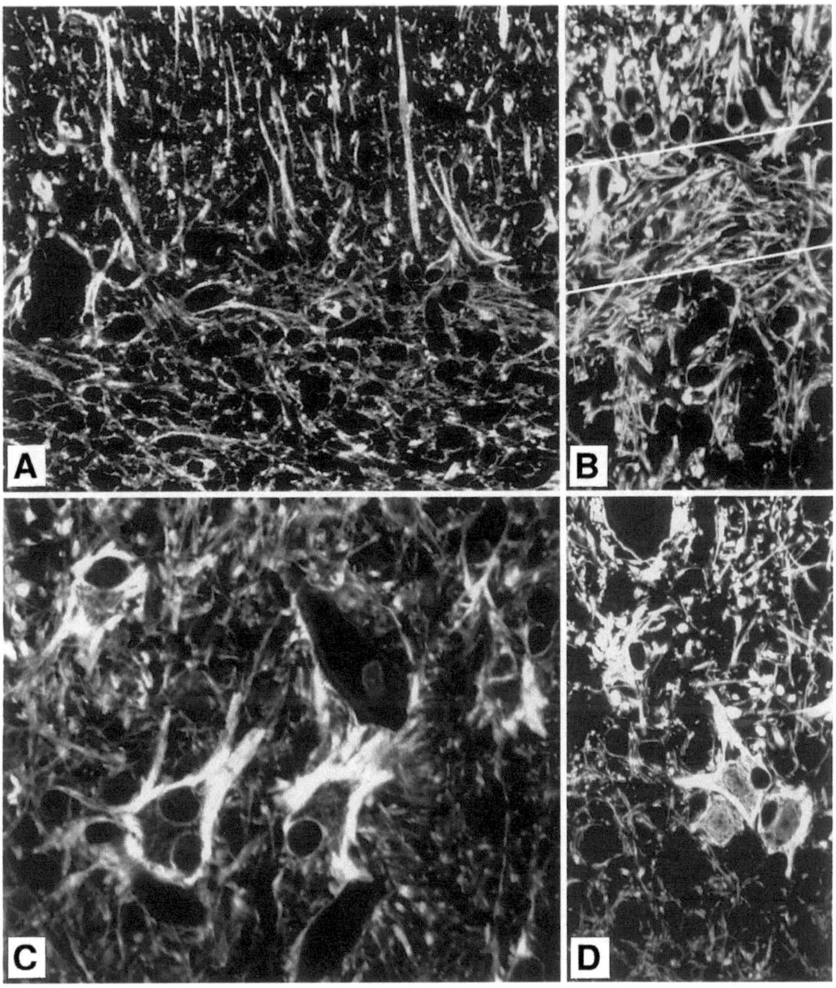

Fig. 5. Semithin sections of the cerebellar cortex immunostained with a GFAP antibody from CJD cerebella. (**A**) and (**B**) These pictures illustrate molecular-Purkinje-granule cell layers transition. The upper region of the granular layer seen in (**A**) shows a dense network of GFAP immunofluorescent astrocytic processes demarcating granule cell perikaryon compartments. (**B**) A dense basal plexus of Bergmann glia (within white lines) is observed at the transition between Purkinje cell and granule cell layers. (**C**) Detail of reactive astrocytes with perivascular endfeet in the granular layer. Note a binucleolated cell and the very intense GFAP staining of the marginal cytoplasm. (**D**) A cluster of hypertrophied astrocytes. A network of prominent profiles of reactive astrocytes is visible throughout the granule cell layer neuropil. (**A, B** and **D**), ×750, (**C**)×1000.

Regarding the granular layer astrocytes, GFAP immunostaining reveals a great hypertrophy of perikarya and cellular processes (Figs. 3, 5) which is clearly associated with the severe degeneration and loss of granule cells and their synaptic contacts with mossy fibers, the cerebellar glomeruli, in the CJD cerebella. In spite of the severe neuronal degeneration, reactive astrocytes appear regularly spaced (Fig. 3) and their hypertrophied radial processes form an extensive structural network of communication

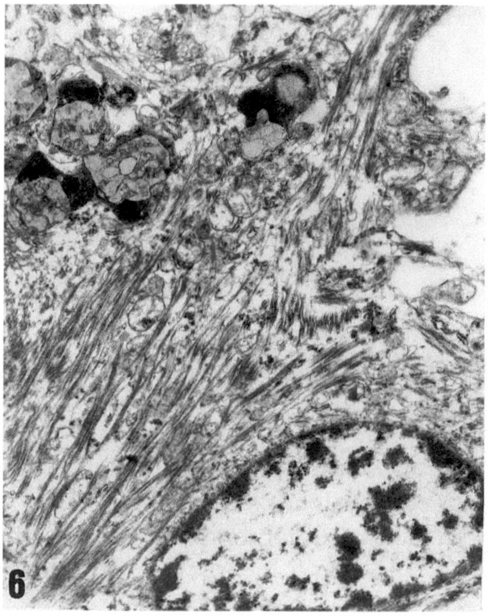

Fig. 6. Electron micrograph of the perikaryon of a reactive astrocyte with abundant bundles of glial filaments and some lipofucsin bodies. ×10,000.

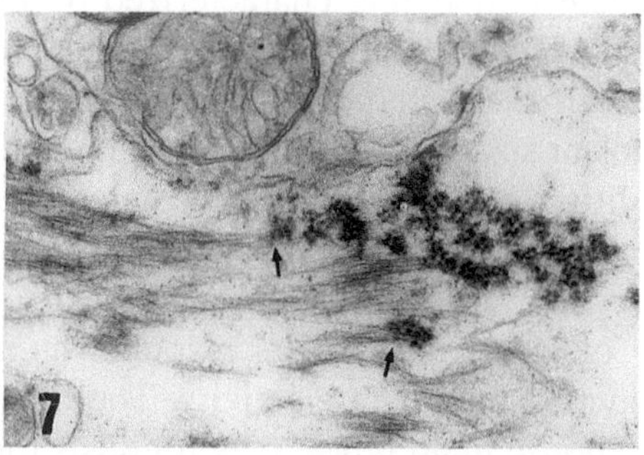

Fig. 7. Association of intermediate glial filaments with polyribosomes (arrows) in the cytoplasm of a reactive astrocyte. ×30,000.

among astrocytes which is closely associated with the surviving granule cells and blood vessels (Figs. 3, 4F, 5A). This reflects that "contact spacing" *(27)* behavior observed among normal astrocytes is also preserved among reactive astrocytes in the isomorphic form of CJD astrogliosis. The perikaryal cytoplasm of CJD reactive astrocytes shows intense GFAP immunostaining, particularly concentrated at the marginal and perinuclear cytoplasm. Some gemistocytic astrocytes with extensive cytoplasma are also observed (Fig. 5D). Electron microscopy studies in CJD cerebella *(21)* have revealed the characteristic hypertrophied bundles of glial filaments crossing the perikaryal cytoplasm and running in glial processes of reactive astrocytes (Figs. 6–8). In addition, we have observed close structural relationships between glial filaments and rosettes of polyribosomes (Fig. 7), suggesting that some mRNAs are associated with

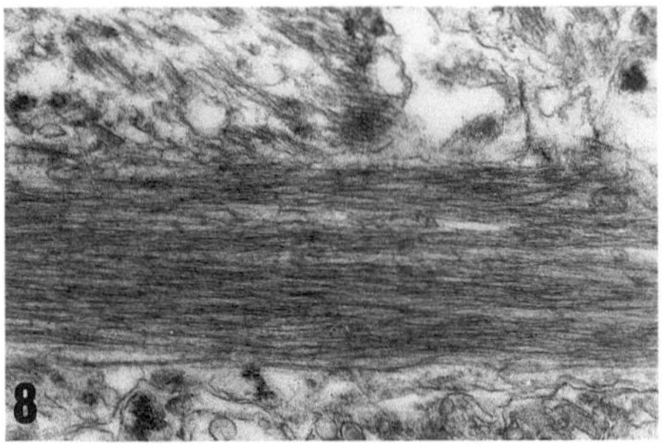

Fig. 8. Longitudinal section of a hypertrophic Bergmann fiber showing a prominent fascicle of densely packed glial filaments. ×30,000.

cytoskeletal elements of reactive astrocytes, as previously described in nonneural cells *(28)*. In this way, the intracellular localization of mRNAs is thought to play an important role in guiding protein toward appropriate cellular compartments with minimal transport costs *(29)*.

4. NUCLEAR AND NUCLEOLAR ORGANIZATION IN REACTIVE ASTROGLIA OF CJD CEREBELLA

The reactive response of astroglia to neurodegeneration in the CJD cerebellar cortex includes a profound nuclear remodeling characterized by the increase of nuclear and nucleolar size, chromatin decondensation and formation of nuclear bodies (NBs). Karyometric analysis performed in Bergmann glia and granule cell layer astrocytes has demonstrated greater nuclear size in both cell types in the CJD cerebella with respect to normal cerebella. In addition, the level of nuclear enlargement of Bergmann glia and granule cell layer astrocytes is positively related to the degree of neuronal loss of their specific neuronal partners, Purkinje cells and granule cells, respectively *(18,21)*. This highlights the plasticity of reactive astroglial response as a function of the neuronal loss and subsequent reduction of synaptic activity. The increase in the nuclear size of astrocytes is a normal response in all types of astrogliosis *(30)* associated with the cellular hypertrophy and is also observed in hypothalamic astrocytes of aged rats *(31)* and in reactive-like astrocytes induced in vitro by dibutyryl cyclic AMP treatment *(32)*. Nuclear enlargement usually reflects the activation of transcriptional activity which involves increased RNA synthesis, enhanced import of kayophilic proteins and chromatin unfolding *(33,34)*. Furthermore, the presence in the CJD of some reactive astroglial cells with very large nuclei or binucleolated (Fig. 5C) supports that cellular mechanisms of endoreduplication and endomitosis may occur resulting in an increase of ploidy.

The morphology of nuclei in reactive astroglial cells is characterized by extensive areas of dispersed chromatin (euchromatin) in the nuclear interior, the presence of small clusters of interchromatin granules, nuclear domains enriched in splicing factors for pre-mRNAs *(35)*, and the appearance of NBs (Figs. 9–11), all parameters related to activation of transcription and RNA processing *(35)*. NBs correspond to simple and complex nuclear bodies reported by Bouteille et al. *(36)* in several cell types. Although simple NBs are composed of an amorphous fibrillar material usually organized in con-

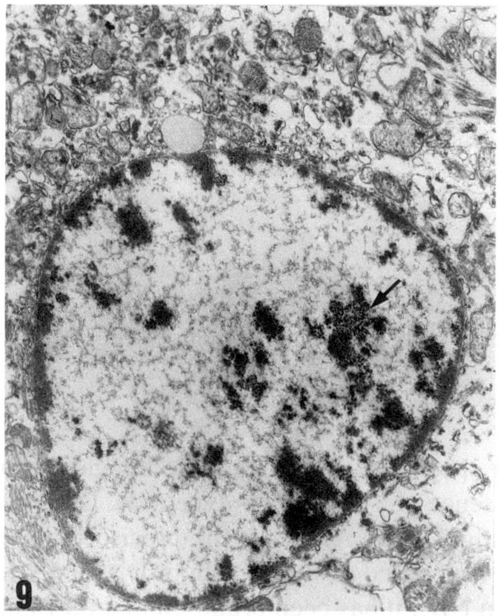

Fig. 9. A large cell nucleus of a reactive astrocyte with extensive domains of euchromatin and some small clusters of densely packed interchromatin granules (arrow). ×12,000.

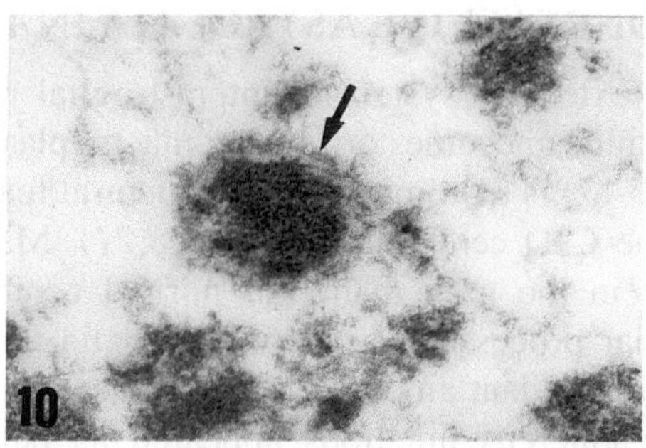

Fig. 10. Detail of a complex granular body (arrow) composed of a microfibrillar capsule and a dense granular core within the cell nucleus of a reactive astrocyte. ×30,000.

centric layers or spiral configuration (Fig. 11), the complex NBs exhibit a larger size and are formed by a fibrillar capsule that encloses a dense granular core (Fig. 10). In the CJD, both types of NBs have been found in reactive astrocytes from several brain regions *(21,37)*. They also appear in axotomy-induced reactive astrocytes *(38)*, hypothalamic astrocytes of aged rats *(31)* and in reactive Schwann cells observed in peripheral neuropathies such as the Guillain-Barré syndrome *(39)*. The formation of NBs in these glial cells seems to be related to an active metabolic state. This is consistent with several observations in nonneural cell types showing that the production of NBs or an increase in their frequency may be induced in experimental models of cellular activation, such as stimulated lymphocytes and estrogen-responsive cells (for review *see 34,40*). Cytochemical analysis has shown the proteinaceous nature of NBs with the additional presence of ribonucleoproteins in the granular core *(34,36,39,40)*, but little

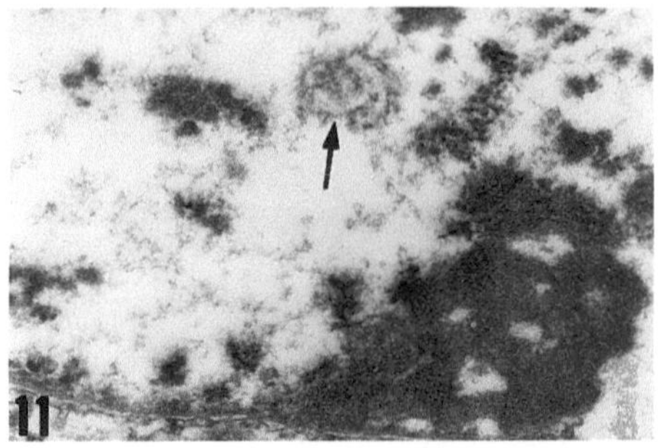

Fig. 11. Simple nuclear body formed by concentric shells of microfibrillar material **(arrow)**, ×30,000.

is known about their specific molecular nature. Recent immunocytochemical studies have detected the presence of a 126-kDa polypeptide, DNA polymerase α and proliferating cell nuclear antigen *(41)*. Although the precise functional significance of NBs remains to be established, based on the content of DNA polymerase α and proliferating cell nuclear antigen in NBs, it has been proposed that they play a role in DNA replication *(41)*. Further investigation will determine if the occurrence of NBs in some reactive astroglial cells is related to the induction of proliferation.

5. INTERASTROGLIAL JUNCTIONS AND PERIVASCULAR ORGANIZATION OF REACTIVE ASTROGLIA IN THE CJD CEREBELLA

The presence of an extensive system of interastroglial junctions, namely maculae adherens junctions, hemidesmosomes or focal contacts (plasmalemmal undercoat) and gap junctions (Figs. 12–17) is a prominent ultrastructural feature in the organization of reactive astroglia in the CJD cerebellar cortex *(18,21)*. Maculae adherens junctions, which are rarely seen in the gray matter of normal cerebella *(23)*, were regularly observed among cellular processes of reactive astroglia in all cases of CJD studied. They were particularly abundant among the lateral faces of perivascular endfeet (Fig. 12). Typical maculae adherens junctions appeared as buttonlike points of contact between glial processes that consist of dense cytoplasmic plaques linked to the cytosolic face of the plasma membranes of adjacent cells. Unlike hemidesmosomes, no intermediate glial filaments were anchored to the adhesion plaques. The proliferation of macular adherens junctions should increase the adhesion and structural integration among reactive astroglial cells. This may be essential to maintain the reinforced astroglial scaffold of the CJD cerebellar cortex in response to the alteration of neuron-astroglia interactions provoked by neurodegenerative changes *(21)*. Since the normal intercellular space is preserved, these junctions do not represent a barrier to extracellular diffusion of large molecules.

The organization of perivascular glia in the CJD cerebella is characterized by the formation of hypertrophied endfeet with increased GFAP immunoreactivity (Fig. 5c) and prominent intermediate filament bundles (Figs. 15, 16). Furthermore, it has been demonstrated that perivascular endfeet are associated with thickened basal lamina and exhibit conspicuous plasmalemmal undercoat on the cell membrane facing the perivascular space *(21)*. This undercoat, with an electron-dense and homogeneous

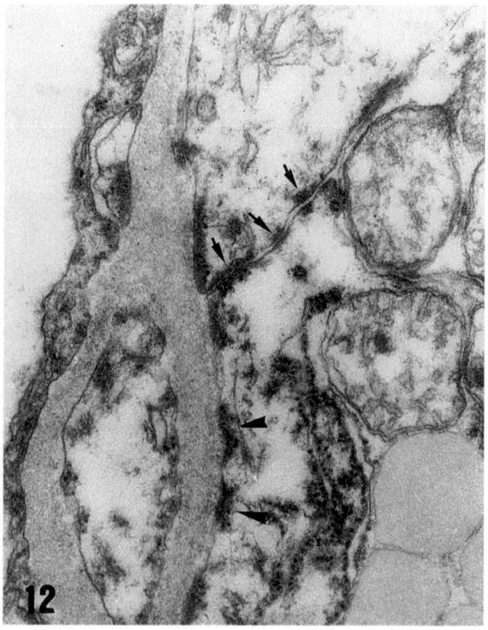

Fig. 12. Arrays of macular adherens junctions **(arrows)** and dense plaques of plasmalemmal undercoat **(arrowheads)** in perivascular endfeet of reactive astroglia. ×50,000.

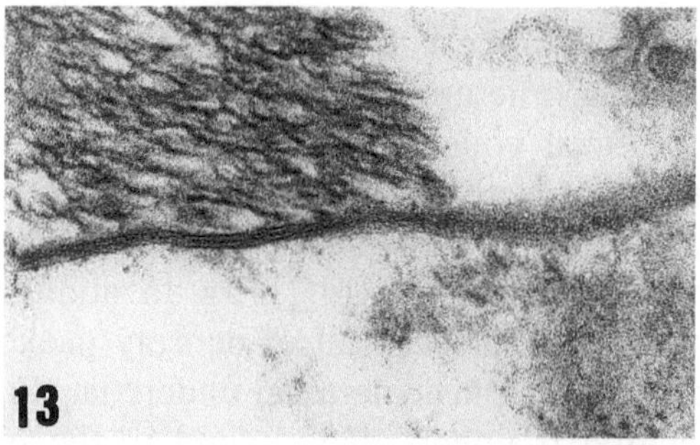

Fig. 13. A long gap junctional profile between reactive Bergmann glial processes. Note a fascicle of glial filaments in the vicinity of the junction. ×80,000.

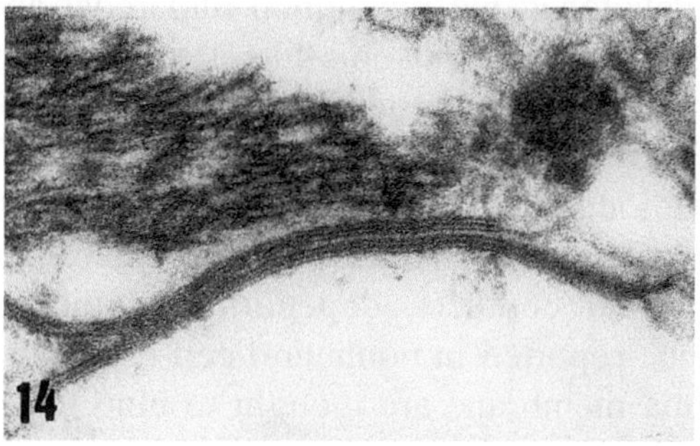

Fig. 14. Detail of paired gap junctions formed by lamellar processes of reactive astrocytes in the granule cell layer. ×70,000.

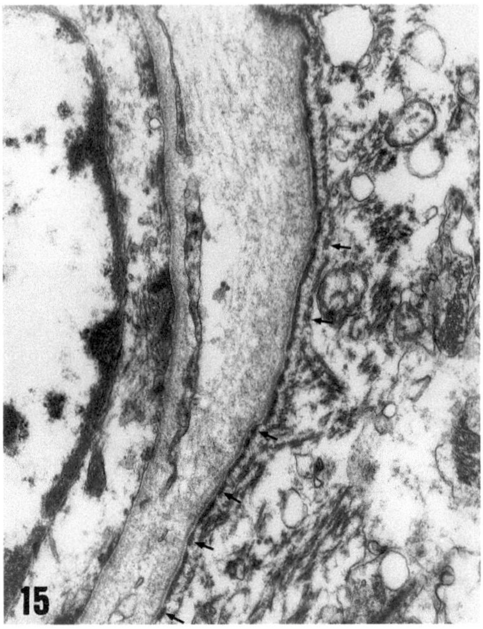

Fig. 15. Perivascular portion of the perikaryon of a reactive astrocyte around an arteriole. Note an almost continuous plasmalemal undercoat and the terminal web of glial filaments **(arrows)** running in parallel to the plasma membrane. ×20,000.

texture, is usually arranged as linear dense plaques of variable length in capillaries and venules (Figs. 12, 16), a structural configuration similar to those of dense plaques of hemidesmosomes or focal contacts, depending on the presence or absence of anchored glial filaments, respectively. In arterioles, the reinforcement of the plasmalemmal undercoat leads to dense plaques joining one another resulting in a continuous subplasmalemmal linear density (Fig. 15). In addition, perivascular endfeet frequently contain a terminal web or sheet of densely packed glial filaments which run parallel to the profile of the plasmalemmal undercoat (Figs. 15). From the terminal web, intermediate glial filaments may extend to the plasmalemmal undercoat which provides the attachment sites for this cytoskeletal component (Fig. 17). Under normal conditions, less developed subplasmalemmal densities have been observed on the pial astrocytic processes coated with the basal lamina *(42)*. Under neuropathological conditions, well developed plasmalemmal undercoat and terminal web of glial filaments, also referred to as desmosome-like structures and anchorage densities, respectively *(43)*, have also been reported in devastated cerebral regions with gliosis, indicating that these structural organizations are nonspecific products of reactive astroglial cells *(43)*. Although the molecular nature of the plasmalemmal undercoat in astroglial cells is unknown, its cytological organization clearly corresponds to hemidesmosomes or focal contacts, depending on the presence or absence of anchored glial filaments, reported in nonneural cell types. In both structures integrin molecules of the plasma membrane are thought to bind to proteins within the dense plaques and to extracellular matrix protein laminin, allowing the attachment of cells to the extracellular matrix *(44)*.

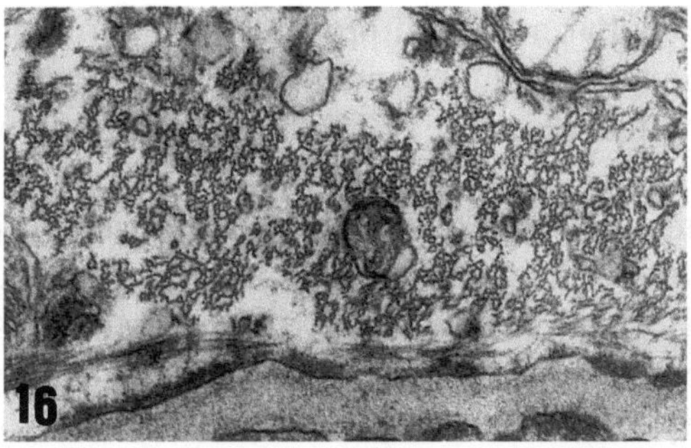

Fig. 16. Perivascular glial endfoot illustrating cross-sectioned bundles of glial filaments and several, hemidesmosome-like, dense plaques of the plasmalemmal undercoat **(arrows).** ×50,000.

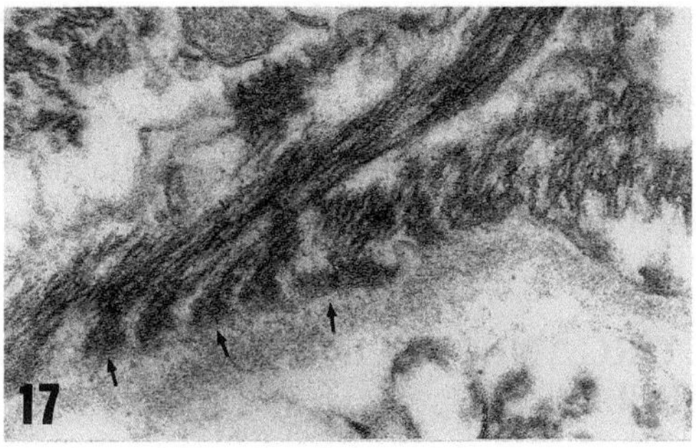

Fig. 17. Perivascular glial endfoot showing the anchorage of intermediate glial filaments to the dense plaques of the plasmalemma **(arrows).** ×60,000.

In this context, the plasmalemmal undercoat with the anchored glial filaments may increase the overall rigidity and mechanical support of perivascular domains of reactive astroglia *(21)*. Moreover, since astroglial cells induce blood-brain-barrier properties of brain endothelia *(45)*, the reinforcement of the perivascular processes of reactive astroglia observed in CJD may contribute to protect endothelial cells against the neurotoxic effect of the PrP^{Sc}. In fact, the structure of the interendothelial tight, junctions, a basic component of the blood-brain-barrier, is apparently preserved in the cerebellar cortex of CJD *(21)*. Finally, plasmalemmal undercoat, through molecular interactions with membrane proteins, may retain astrocyte-specific complexes of integral membrane proteins, such as orthogonal assemblies of intramembranous particles *(46)*, in the plasma membrane domains facing the perivascular space. In this way, previous freeze-fracture studies of reactive astrocytes in experimenatal gliosis *(47)*, including the scrapie-innoculated mice *(48)*, have shown a significant increase in the number of orthogonal assemblies in comparison with normal astrocytes.

Regarding interastroglial gap junctions, they are frequently found in both Bergmann glia and granule cell layer astrocytes in the cerebellar cortex of the CJD cerebella *(18,21)*. They usually appear as long linear profiles of adjacent plasma membranes both in astroglial processes (Fig. 13) and perikarya, being particularly abundant between tha lateral faces of perivascular glial endfeet. Paired gap junctions on opposite membranes of thin lamellar astroglial processes are commonly seen in the CJD cerebella (Fig. 14). Although gap junctions are occasionally found in ultrathin sections from normal human cerebellar cortex *(23)*, their frequent appearance in CJD cerebella suggests that the reactive state of astroglial cells induces the formation of these junctions. This is consistent with the behavior of astrocytic gap junctions in experimental models of gliosis. In the ischemia-induced brain injury, Hossain et al. *(49)* have shown that during the transformation to a reactive state, astrocytes undergo a gap junction remodeling with increased immunostaining for the astrocytic gap junctional protein connexin43. Similarly, reactive astrocytes involved in the formation of lesional scars are also connected by extended gap junctions in forebrain regions *(50)*. It is now well established that gap junctions are involved in metabolic and electrical coupling between normal astroglial cells (for review *see 51,52*). By means of exchange ions such as K^+ and second messengers such as Ca^{2+} and IP3, astrocytic gap junctions may actively participate in the redistribution of focal extracellular K^+ increases generated by synaptic activity *(51)*, and in the generation of intercellular calcium waves which constitute a long-range signaling system in the brain *(53,54)*. The proliferation of gap junctions among reactive astroglial cells in the CJD cerebella suggests that there are greater requirements for intercellular signaling and metabolic cooperation between these hypertrophic glial cells.

6. PERISYNAPTIC ORGANIZATION OF REACTIVE ASTROCYTES IN THE CJD CEREBELLA

As indicated previously, the pathological hallmark of the cerebellar cortex in the ataxic form of CJD is characterized by spongiform changes, degeneration of afferent mossy fibers, severe loss of granule cells, variable degeneration of Purkinje cells, and extensive astrogliosis *(19,20,55)*. Degeneration of mossy fibers and loss of granule cell produces a virtual disappearance of the typical rosettes of granule cell perikarya (Figs. 1, 2) accompanied by a dramatic reduction of synaptic glomeruli in the granule cell layer. In the molecular layer, a consequence of granule cell loss is the degeneration of parallel fibers resulting in a depletion of the synaptic contacts between these fibers and Purkinje cell spines *(55)*. Remaining synapses in the CJD cerebellar cortex appear intimately associated with perisynaptic processes of reactive astroglia, which usually have a watery cytoplasm (Fig. 18). It is noteworthy in the molecular layer the existence of a numerical mismatching between presynaptic parallel fibers and postsynatic Purkinje cell spines resulting in the frequent observation of vacant dendritic spines with preserved postsynaptic membrane densities (Fig. 19). They appear enclosed by simple or multilayered processes of reactive Bergmann glia *(18)*. Similar vacant dendritic spines of Purkinje cells have been reported in natural occurring granule cell ectopias *(56)* and in neurological mutant mice with granule cell death *(57)*, where they persist for long periods of time in the absence of synaptic transmission. This raises the speculation that the reactive Bergmann glia might maintain and signal postsynaptic sites potentially available for reinervation *(18)*.

Changes in synaptic activity can mediate other cellular events of the response of cerebellar astroglia to neurodegeneration in the CJD. Thus, our comparative analysis of the three cases of CJD studied has shown that the different degrees of granule cell loss and Purkinje cell degeneration, which indirectly reflect the availability of synapses in

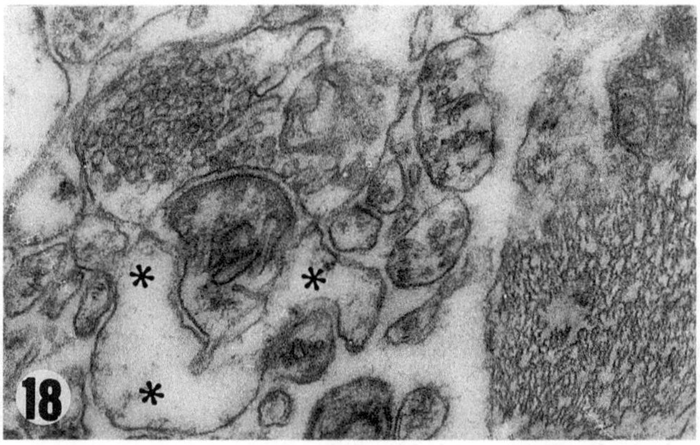

Fig. 18. A relatively well preserved synapse between a granule cell axon and dendritic spine of a Purlinje cell. Note the perisynaptic processes of Bergmann fibers **(asterisks)** and a prominent bundle of densely packed glial filaments. ×35,000.

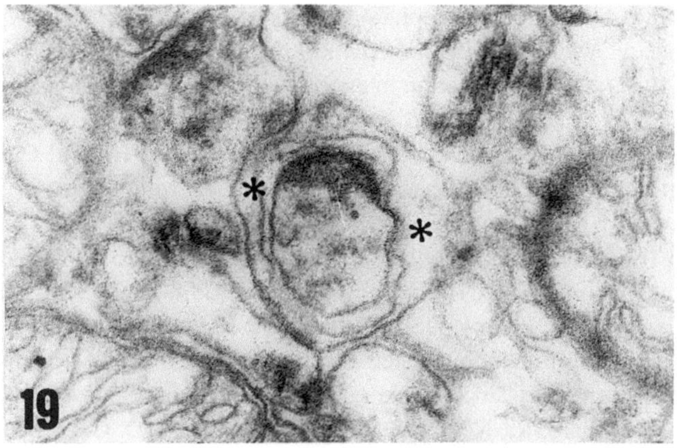

Fig. 19. A vacant dendritic spine with a prominent membrane density is surrounded by pale processes of Bergmann glia **(asterisks).** ×40,000.

the molecular layer, are closely related to the levels of hypertrophy and expression of GFAP in Bergmann glia *(18)*. In this way, a glial gene regulation by synaptic activity is an emerging concept of the last few years (for review *see 52*). Canady and Rubel *(58)* have demonstrated that the blockade with tetrodotoxin of afferent neural activity in chick cochlear nucleus produces a reversible astrocytic reaction with increased expression of GFAP. This effect is extensible to the model of nerve terminal and perisynaptic Schwann cell that covers the neuromuscular junction in which the deprivation of nerve activity rapidly upregulates GFAP expression in Schwann cells *(59)*. These findings suggest that the disruption of neuron-astroglia interactions or signaling at the synapses plays a striking role in the induction of the reactive astroglial response in CJD cerebella.

In conclusion, the pattern of astrogliosis in the CJD cerebella is characterized by a reinforced astroglial scaffold which may preserve the normal geometry, spatial arrangement, and specific astroglial associations with surviving neurons. Recent data in experimental scrapie and CJD supports that PrP may directly induce a reactive response in astrocytes (for review *see 15,16*). However, in the CJD cerebella studied here the level of the reactive astroglial response seems to be proportional to the degree of neurodegeneration and subsequent alteration of normal neuron-astroglia interactions.

REFERENCES

1. Prusiner, S. B. (1998) Prions. *Proc. Natl. Acad. Sci. USA.* **95,** 13363–13383.
2. Kretzschmar, H. A., Prusiner, S. B., Stowring, L. E., and DeArmond, S. J. (1986) Scrapie prion proteins are synthesized in neurons. *Am. J. Pathol.* **122,** 1–5.
3. Moser, M., Colello, R. J., Pott, U., and Oesch, B. (1995) Developmental expression of the prion protein gene in glial cells. *Neuron.* **14,** 509–517.
4. Prusiner, S. B. and DeArmond, S. J. (1994) Prion diseases and neurodegeneration. *Annu. Rev. Neurosci.* **17,** 311–339.
5. Büeler, H. R., Aguzzi, A., Sailer, A., Greiner, R. A., Autenried, P., Aguer, M., and Weissmann, C. (1993) Mice devoid of PrP are resistant to scrapie. *Cell.* **73,** 1339–1347.
6. Gray, E. G. (1986) Spongiform encephalopathy: a neurocytologist's viewpoint with a note on Alzheimer's disease. *Neuropathol. Appl. Neurobiol.* **12,** 149–172.
7. Liberski, P. P. (1987) Astrocyte reaction in experimental scrapie in hamsters. *J. Comp. Pathol.* **97,** 73–78.
8. Liberski, P. P., Guiroy, D. C., Yanaghiara, D. C., Wells, G. A. H., and Gajdusek, D. C. (1993) Astrocytic changes, in *Light and Electron Microscopic Neuropathology of Slow Virus Infections* (Liberski, P. P., ed.), CRC Press, Boca Raton, pp. 187–232.
9. Manuelidis, L., Tesin, D. M., Sklaviadis, T., and Manuelidis, E. E. (1987) Astrocyte gene expression in Creutzfeld-Jakob disease. *Proc. Natl. Acad. Sci. USA.* **84,** 5937–5941.
10. Biernat, W., Liberski, P. P., Guiroy, D. C., Yanahihara, R., and Gajdusek, D. C. (1995) Proliferating cell nuclear antigen immunohistochemistry in astrocytes in experimental Creutzfeldt-Jakob disease and Gerstmann-Strässler-Scheinker syndrome. *Neurodegeneration* **4,** 195–201.
11. Liberski, P. P., Yanagihara, R., Nerurkar, V. R., and Gajdusek, D. C. (1993) Tumor necrosis factors produces Creutzfeldt-Jakob disease-like lesions in vivo. *Neurodegeneration* **2,** 215–225.
12. Brown, D. R., Schmidt, B., and Kretzschmar, H. A. (1998) A prion protein fragment primes type 1 astrocytes to proliferation signals from microglia. *Neurobiol. Disease* **4,** 410–422.
13. Brown, H. R., Goller, N. L., Rudelli, R. D., et al. (1990) The mRNA encoding the scrapie agent protein is present in a variety of non-neuronal cells. *Acta. Neuropathol.* **80,** 1–6.
14. Diedrich, J. K., Bendheim, P. E., Kim, Y. S., Carp, R. I., and Haase, A. T. (1991) Scrapie-associated prion protein accumulates in astrocytes during scrapie infection. *Proc. Natl. Acad. Sci. USA.* **88,** 375–379.
15. DeArmond, S. J., Kristensson, K., and Bowler, R. P. (1992) PrP^{SC} Causes nerve cell death and stimulates astrocyte proliferation: a paradox, in *A.C.H.* (Yu, L., Hertz, M. D., Norenberg, E., Sykova, S. G., and Waxman, Eds.), Progress in Brain Res, Elsevier, Amsterdam, vol. 94, pp. 437–446.
16. Ye, X., Scallet, A. C., Kascsak, R. J., and Carp, R. I. (1998) Astrocytosis and amyloid deposition in scrapie-infected hamsters. *Brain Res.* **809,** 277–287.
17. Forloni, G., Del Bo, R., Angeretti, N., et al. (1994) A neurotoxic prion protein fragment induces rat astroglial proliferation and hypertrophy. *Eur. J. Neurosci.* **6,** 1415–1422.
18. Lafarga, M., Berciano, M. T., Suarez, I., Andres, M. A., and Berciano, J. (1993) Reactive astroglia-neuron relationships in the human cerebellar cortex: a quantitative, morphological and immunocytochemica study in Creutzfeldt-Jakob disease. *Int. J. Devl. Neurosci.* **11,** 199–213.

19. Gomori, A. J., Partnow, M. J., Horoupian, D. S., and Hirano, A. (1973) The ataxic form of Creutzfeldt-Jakob disease. *Arch. Neurol.* **29,** 318–323.
20. Jellinger, K., Heiss, W. D., and Deisenhammer, E. (1974) The ataxic (cerebellar) form of Creutzfeldt-Jakob disease. *J. Neurol.* **207,** 289–305.
21. Lafarga, M., Berciano, M. T., Suarez, I., Viadero, C. F., Andres, M. A., and Berciano, J. (1991) Cytology and organization of reactive astroglia in human cerebellar cortex with severe loss of granule cells: a study on the ataxic form of Creutfeldt-Jakob disease. *Neuroscience* **40,** 337–352.
22. Kretzschmar, H. A., Kitamoto, T., Doerr-Schott, J., Mehracin, P., and Tateishi, J. (1991) Diffuse deposition of immunohistochemically labeled prion protein in the granular layer of the cerebellum in a patient with Creutzfeldt-Jakob disease, *Acta. Neuropathol.* **82,** 536–540.
23. Palay, S. L. and Chan-Palay, V. (1974) *Cerebellar Cortex: Cytology and Organization.* Springer, New York.
24. Dahl, D., Reuger, D. C., and Bigniami, A. (1981) Vimentin, the 57000 molecular weight protein of fibroblast filaments, is the major cytoskeletal component of immature glia. *Eur. J. Cell. Biol.* **24,** 191–196.
25. Takamiya, Y., Kohsaka, S., Toya, S., Otani, M., and Tsukada, Y. (1988) Immunohistochemical studies on the proliferation of reactive astrocytes and the expression of cytoskeletal proteins following brain injury rats. *Dev. Brain Res.* **38,** 201–210.
26. Bovolenta, P., Liem, R. K. H., and Mason, C. A. (1984) Development of cerebellar astrogia: Transitions in form and cytoskeletal content. *Dev. Biol.* **102,** 248–259.
27. Chan-Ling, T. and Stone, J. (1991) Factors determining the morphology and distribution of astrocytes in the cat retina: a "contact-spacing" model of astrocyte interaction. *J. Comp. Neurol.* **303,** 387–399.
28. Hamill, D., Davis, J., Drawbridge, J., and Suprenant, K. A. (1994) Polyribosome targeting to microtubules: Enrichment of specific mRNA in a reconstituted microtubule preparation from sea urchin embryos. *J. Cell Biol.* **127,** 973–984.
29. St. Johnston, D. (1995) The intracellular localisation of messenger RNAs. *Cell.* **81,** 161–170.
30. Nathaniel, E. H. and Nathaniel, D. R. (1981) The reactive astrocytes, In *Advances in Cellular Neurobiology* (Fedoroff, S. and Hertz, L., eds.), Academic Press, New York.
31. Berciano, M. T., Andres, M. A., Calle, E., and Lafarga, M. (1995) Age-induced hypertrophy of astrocytes in rat supraoptic nucleus: a cytological and morphometric and immunocytochemical study. *Anat. Rec.* **243,** 129–144.
32. Fedoroff, S., McAuley, W. A., Houle, J. D., and Devon, R. M. (1984) Astrocyte cell lineage. V. Similarity of astrocytes that form in the presence of dBcAMP in cultures to reactive astrocytes in vivo. *J. Neurosci. Res.* **12,** 15–27.
33. Garcia-Segura, L. M., Berciano, M. T., and Lafarga, M. (1993) Nuclear compartmentalization in transcriptionally activated hypothalamic neurons. *Biol. Cell.* **77,** 143–154.
34. Brasch, K. and Ochs, R. L. (1995) Nuclear remodeling in response to steroid hormone action. *Int. Rev. Cytol.* **159,** 161–194.
35. Lamond, A. I. and Earnshaw, W. C. (1998) Structure and function in the nucleus. *Science.* **280,** 547–553.
36. Bouteille, M., Laval, M., and Dupuy-Coin, A. M. (1974) Localization of nuclear functions as revealed by ultrastructural autoradiography and cytochemistry. In *The Cell Nucleus* (Busch, H., ed.) vol. I, Academic Press, New York.
37. Payne, C. M. and Sibley, W. A. (1975) Intranuclear inclusion in a case of Creutzfeldt-Jakob disease. An ultrastructural study. *Acta. Neuropathol.* **31,** 353–361.
38. Uehara, M., Imagawa, T., and Kitagawa, H. (1998) Quantitative studies of nuclear bodies in astrocytes of chicken spinal cord following axotomy. *J. Vet. Med. Sci.* **60,** 773–775.

39. Berciano, M. T., Calle, E., Andres, M. A., and Berciano, J. (1996) Schwann cell nuclear remodelling and formation of nuclear and coiled bodies in Guillain-Barré syndrome. *Acta. Neuropathol.* **92,** 386–394.
40. Brasch, K. and Ochs, R. L. (1992) Nuclear Bodies (NBs): a newly "rediscovered" organelle. *Exp. Cell Res.* **202,** 211–223.
41. Hozák, P., Jackson, D. A., and Cook, P. R. (1994) Replication factories and nuclear bodies: the ultrastructural characterization of replication sites during the cell cycle. *J. Cell Sci.* **107,** 2191–2204.
42. Bondareff, W. and McLone, D. G. (1973) The external glial limiting membrane in Macaca: ultrastructure of a laminated glioepithelium. *Am. J. Anat.* **136,** 277–296.
43. Nakano, I., Kato, S., Yazawa, I., and Hirano, A. (1992) Anchorage densities associated with hemidesmosome-like structures in perivascular reactive astrocytes. *Acata Neuropathol.* **84,** 85–88.
44. Lodish, H., Baltimore, D., Berk, A., Zipursky, S. L., Matsudaira, P., and Darnell, J. (1995) *Molecular Cell Biology* W. H. Freeman and Company, New York.
45. Janzer, R. C. and Raff, M. C. (1987) Astrocytes induce blood-brain barrier properties in endothelial cells. *Nature* **325,** 235–256.
46. Landis, D. M. and Reese, T. S. (1974) Arrays of particles in freeze-fractured astrocytic membranes. *J. Cell Biol.* **60,** 316–320.
47. Anders, J. J. and Brightman, M. W. (1979) Assemblies of particles in the cell membranes of developing, mature and reactive astrocytes. *J. Neurocytol.* **8,** 777–795.
48. Dubois-Dalcq, M., Rodriguez, M., Reese, T. S., Gibbs, J., and Gajdusek, D. C. (1977) Search for a specific marker in the neural membranes of scrapie mice. A freeze-fracture study. *Lab. Invest.* **36,** 547–553.
49. Hossain, M. Z., Peeling, J., Sutherland, G. R., Hertzberg, E. L., and Nagy, J. L. (1994) Ischemia-induced cellular redistribution of the astrocytic gap junctional protein connexin43 in rat brain. *Brain Res.* **652,** 311–322.
50. Alonso, G. and Privat, A. (1993) Reactive astrocytes involved in the formation of lesional scars differ in the mediobasal hypothalamus and other forebrain regions. *J. Neurosci.* **34,** 523–538.
51. Walz, W. and Hertz, L. (1983) Functional interactions between neurons and astrocytes. II. Potassium homeostasis at the cellular level. *Progr. Neurobiol.* **20,** 133–183.
52. Vernadakis, A. (1996) Glia-neuron intercommunications and synaptic plasticity. *Progress Neurobiol.* **49,** 185–214.
53. Cornell-Bell, A. H., Finkbeiner, S. M., Cooper, M. S., and Smith, S. J. (1990) Glutamate induces calcium waves in cultured astrocytes: long-range glial signaling. *Science.* **247,** 470–473.
54. Giaume, C. and Venance, L. (1998) Intercellular calcium signaling and gap junctional communication in astrocytes. *Glia* **64,** 50–64.
55. Berciano, J., Berciano, M. T., Polo, J. M., Figols, J., Ciudad, J., and Lafarga, M. (1990) Creutzfeldt-Jakob disease with severe involvement of cerebral white matter and cerebellum. *Virchows Archiv. A.* **417,** 533–538.
56. Berciano, M. T., Conde B., and Lafarga M. (1990) Interactions between astroglia and ectopic granule cells in the cerebellar cortex of normal adult rats: a morphological and cytochemical study. *Exp. Brain Res.* **80,** 397–408.
57. Sotelo, C. (1973) Permanence and fate of paramembranous synaptic specializations in mutants and experimental animals. *Brain Res.* **62,** 345–351.
58. Canady, K. S. and Rubel, E. (1992) Rapid and reversible astrocytic reaction to afferent blockade in chick cochlear nucleus. *J. Neurosci.* **12,** 1001–1007.
59. Georgiou, J., Robitaille, R., Trimble, W. S., and Charlton, M. P. (1994) Synaptic regulation of glial protein expression in vivo. *Neuron.* **12,** 443–455.

22
Ischemic Injury, Astrocytes, and the Aging Brain

Robert Fern

1. INTRODUCTION

Astrocytes are the housekeeping cells of the central nervous system (CNS). Their primary function is to control the environment within which neurons operate. Stroke events are characterized by a breakdown in the normal environment of neurons, leading to loss of brain function and permanent injury. Astrocytes play a critical role in this process. They are involved in the unfolding events that initiate the cellular response to ischemia, they have a significant impact upon the interactions that determine the fate of individual neurons during ischemia, and they are of central importance in the events that follow a stroke which may result in the formation of glial scar tissue. In addition to the role astrocytes play in neuronal survival during stroke, astrocytes are themselves subject to ischemic injury via mechanisms that differ significantly from those seen in neurons or other types of glia.

Stroke injury is most common in the older CNS, and the role that astrocytes play in ischemia is of greatest importance in the aged brain. The majority of basic ischemia research, however, has been performed on cultured cells taken from juvenile or neonatal animals or upon isolated tissue harvested from young animals. Bearing this in mind, I will review what is known at the basic science level about the involvement of astrocytes in ischemic injury, paying particular attention to studies that extend to the aged CNS.

Astrocytes are involved in stroke at three levels:

1. During the initial insult, where astrocytes may be damaged or killed by ischemia;
2. In the injury process itself, where astrocytes affect the fate of other cell types both in the stroke core and in the penumbra region surrounding the core and;
3. In the postischemic period where the formation of a glial scar may affect cellular reorganization and restoration of function.

Gliosis is covered in detail in a previous chapter (*see* Part I: Cellular and Molecular Changes of Aged and Reactive Astrocytes) to which the reader is directed; only reactive gliosis following ischemia will be covered here.

2. ASTROCYTES ARE SUBJECT TO INJURY DURING ISCHEMIA

Following the onset of ischemia, interruption of glucose and oxygen supply places astrocytes in a metabolically precarious position. Cellular energy reserves become

From: *Neuroglia in the Aging Brain*
Edited by: Jean S. de Vellis © Humana Press Inc., Totowa, NJ

depleted and cytoplasmic ATP levels fall *(1)*. The cell loses the ability to maintain normal ion concentration gradients across the cell membrane as the Na^+-K^+ ATPase fails, resulting in membrane depolarization. This series of events, which will occur in all nucleated cells during ischemia, may proceed rather slowly in astrocytes due to the presence of significant stores of the glycolytic energy reserve glycogen *(2)*. For example, intracellular $[Na^+]$ increases rather slowly in cultured astrocytes exposed to simulated ischemia, rising to a concentration about 3 m*M* higher than resting after 60 min *(3)*. Other important cellular events overlie the gradual degradation of cell viability resulting from ion gradient breakdown due to energy depletion. For example, in cultured astrocytes complete metabolic arrest produces a membrane depolarization over a 45 min time course that is dependent upon the presence of extracellular Ca^{2+} and is associated with a large increase in membrane input resistance *(4)*. These changes in membrane properties are responsible for ischemic depolarization in these cells rather than a gross loss of ion gradients due to Na^+-K^+ ATPase failure *(4)*. The situation in vivo is further complicated by changes in the extracellular environment during ischemia. For example, elevated $[K^+]_o$ will depolarize astrocytes far more rapidly than either of the two mechanisms mentioned above *(5)*, possibly evoking cellular damage in vivo that is not seen in cultured cells (see below).

The metabolic rate of astrocytes in culture is not significantly different from that of neurons *(6,7)* yet ischemic conditions produce injury in cultured neurons far more rapidly then they do in cultured astrocytes. For example, Goldberg and Choi *(8)* have shown that withdrawal of glucose and oxygen for 70 min produces near-complete neuronal death (assessed 20–28 hr later), while 4–12 hr of glucose and oxygen withdrawal are required to produce significant death of astrocytes. The high sensitivity of neurons to ischemic conditions in vitro is due to the presence of NMDA-type glutamate receptors on the cells. These receptors are gated by an accumulation of glutamate in the bath solution during ischemia, resulting in Ca^{2+} influx and subsequent cellular injury *(8)*. Unlike neurons *(8)*, and oligodendrocytes *(9,10)* glutamate receptors do not seem to be important in astrocyte injury either in vitro or in vivo *(7,11–13)*

Cell culture studies such as those described above have left the impression that astrocytes are extremely resistant to ischemic injury. In vivo studies do not support this. Dead astrocytes appear in vivo after a period of ischemia as short as 30 min *(14)*, and almost all astrocytes within the central ischemic region are dead after 4hr *(14,15)*. At this time point the region of infarction is bordered by an area where astrocytes are actively losing their membrane integrity *(14,15)*. This high sensitivity of astrocytes to ischemic injury in vivo compared to in vitro is a matter of considerable importance. As has recently been pointed out *(16)*, a region of postischemic brain infarction is pannecrotic, with all cell types dead *(17)*. In areas where astrocytes survive ischemia but neurons die, a region of selective neuronal loss exists *(18)*. Some restoration of function may be possible in this latter case but functional recovery will not be possible in areas of infarction. The extent of astrocyte death therefore determines the size of a region of infarction and affects the extent of unrecoverable brain function.

When the effects of ischemia are examined in astrocytes *in situ*, rapid ischemic $[Ca^{2+}]_i$ rises are observed *(12,13)* (Fig 1). In CNS white matter, ischemic Ca^{2+} increases are associated with cell death, with 47% of astrocytes dying during an 80 min period of ischemia *(13)*. The cellular injury occurs as uncontrolled amounts of Ca^{2+}

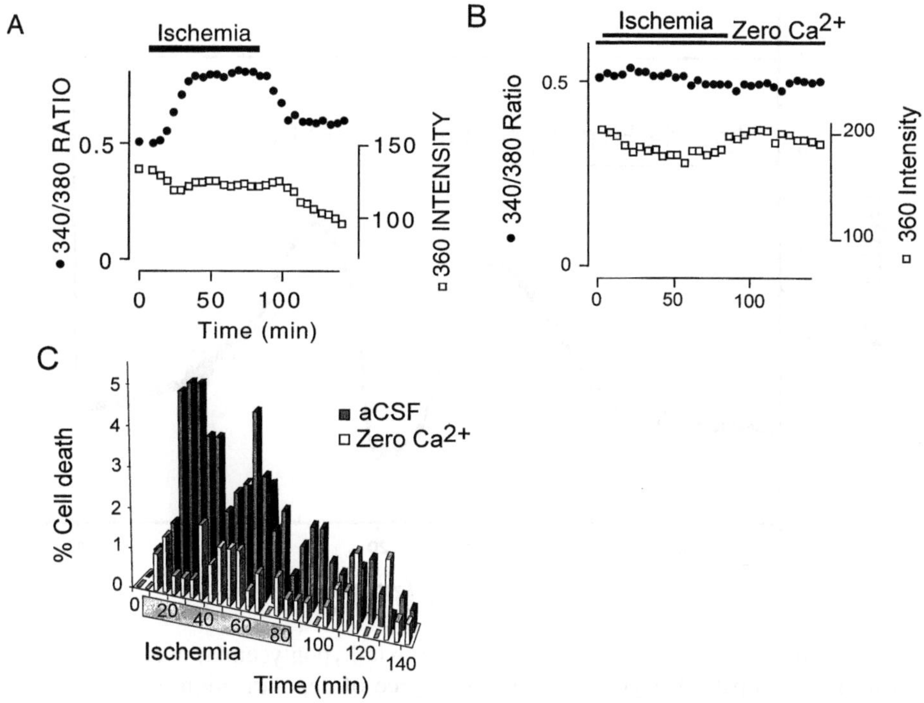

Fig. 1. Changes in $[Ca^{2+}]_i$ and cell death in astrocytes *in situ* during ischemia. Neonatal white matter astrocytes in rat optic nerve were loaded with the calcium indicator FURA-2 (*see 13*). **(A)** Following the withdrawal of oxygen and glucose ("ischemia"), the 340/380 ratio of the dye increases (filled circles), corresponding to an increase in $[Ca^{2+}]_i$. The 360 intensity of the dye (the calcium independent signal) drifts slowly lower, indicating continued cell viability in this cell. **(B)** In the absence of extracellular Ca^{2+} there is no increase in $[Ca^{2+}]_i$ during ischemia. **(C)** The occurrence of cell death (assessed as lose of 360 signal) in normal aCSF and in zero Ca^{2+} aCSF during and after an 80 min period of ischemia; demonstrating the rapid onset of cell death during ischemia, and the Ca^{2+} -dependence of the phenomena. Adapted from ref. *13* with permission.

enter the cell through voltage-gated calcium channels (VGCCs). Duffy and MacVicar *(12)* found that astrocytes in hippocampal slices respond to ischemia with a large rapid depolarization (about a 50 mV change after 10 min of ischemia; Fig 2). A corresponding rapid depolarization is not observed under similar conditions in cultured astrocytes *(4,12),* and is similar to the depolarization produced by elevating $[K^+]_o$ to 50 m*M* (12; Fig 2). It would appear that astrocytes in vivo respond to the early ischemic rise in $[K^+]_o$ with a large influx of Ca^{2+} mediated by membrane depolarization and activation of VGCCs *(12,13).* This phenomena may account for the much higher sensitivity of astrocytes in vivo *(14,15)* to ischemia than astrocytes in vitro *(8).* The mechanism of ischemic injury of astrocytes in vivo is shown in figure 3.

The mechanisms linking an ischemic rise in $[Ca^{2+}]_i$ with subsequent astrocyte death are incompletely understood, and may vary between brain regions and the time course and extent of the ischemic event. It is clear that Ca^{2+} influx is the trigger that leads to cell death since removing Ca^{2+} from the extracellular space is protective (*13;* Fig 1). High $[Ca^{2+}]_i$ will activate a number of destructive calcium-dependent lipases, calpains, and protieses that are harmful to cell viability *(19).* Generation of free radicals will also

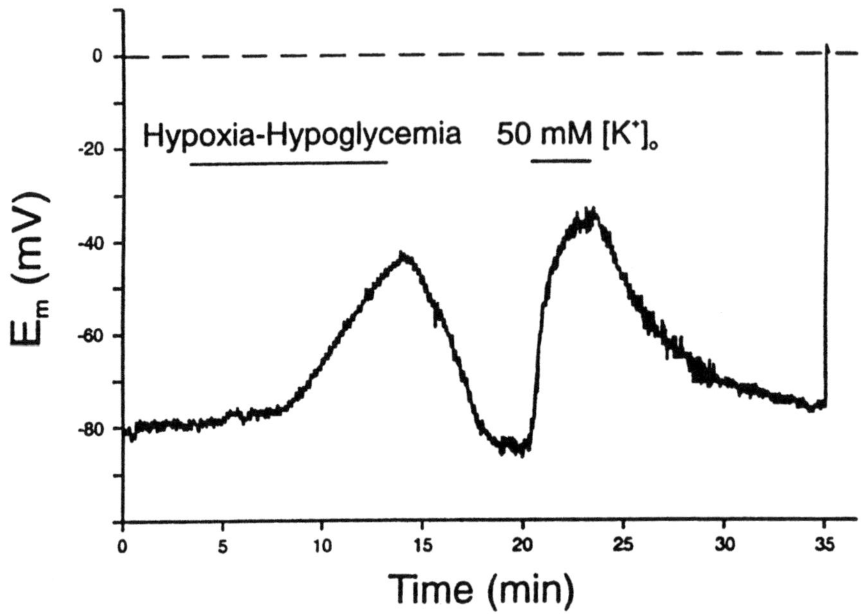

Fig. 2. Withdrawal of oxygen and glucose (hypoxia-hypoglycemia) evokes a rapid depolarization in hippocampal astrocytes in vivo. The degree of depolarization is similar to that produced by perfusion with 50 m*M* K^+. Adapted from Duffy and ref. *12* with permission.

occur and may contribute to cell injury *(20)*. Following the reinitiation of blood supply to an ischemic region, reperfusion phenomena occur in astrocytes that may also have important consequences for their survival. For example, removal of extracellular Ca^{2+}, as will occur during ischemia due to Ca^{2+} influx into the intracellular space, followed by return to normal extracellular Ca^{2+} gives rise to a significant Ca^{2+} influx into astrocytes *(21)*. This phenomena (a Ca^{2+} paradox) is selectively harmful to cultured astrocytes compared to other cell types *(21)*. A similar postischemic increase in intracellular Na^+ has been shown in astrocytes, of uncertain significance *(3)*. Under conditions that do not lead directly to necrotic astrocyte death, high $[Ca^{2+}]_i$ is linked to the onset of apoptotic cell death *(22)*. Apoptotic astrocytes are observed in the ischemic core and in the border zone between the core and the penumbra in ischemic lesions in mice *(22)*; and in man *(23)*. Apoptosis may be an important mechanism responsible for cell death in astrocytes that survive the initial wave of necrotic death.

3. ASTROCYTES IN THE INJURY PROCESS

3.1. The Initial Increase in $[K^+]_o$

Astrocytes are responsible for homeostatic control of the CNS extracellular space. Astrocytes maintain $[K^+]_o$ at around 3 m*M* under resting conditions, and prevent $[K^+]_o$ from exceeding about 12 m*M* even during the most intense periods of neuronal activity *(5)*. During the initial stages of ischemia $[K^+]_o$ rapidly elevates above this plateau level, reaching concentrations as high as 80 m*M* *(5)*. Loss of astrocyte regulation of $[K^+]_o$, and loss of the homeostatic function of these cells in general, is a key element in the onset of a stroke event. $[K^+]_o$ is likely to increase in astrocytes during ischemia *(24)*,

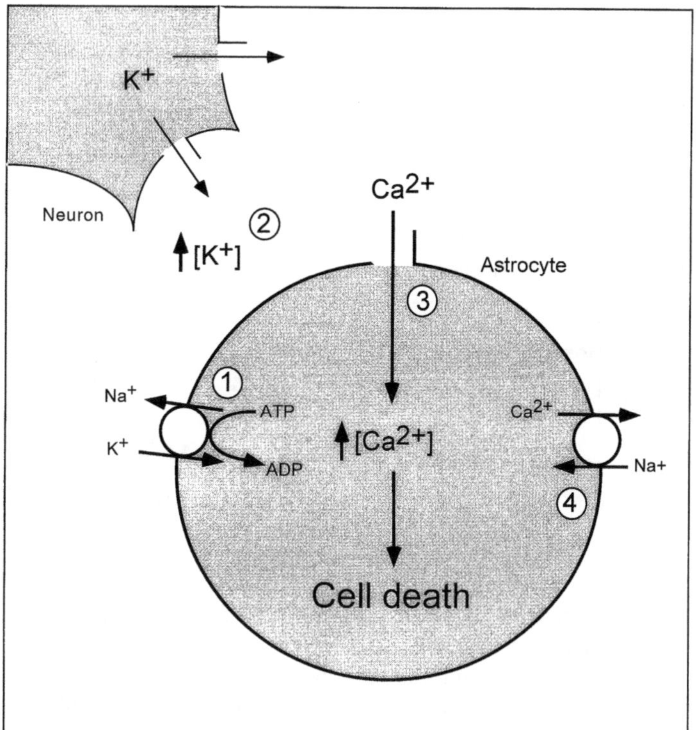

Fig. 3. A model of ischemic injury of astrocytes in vivo. During ischemia, the Na^+-K^+ -ATPase fails as the cytoplasmic ATP level falls ("1"). There is a rapid increase in extracellular K^+, largely as a result of efflux from neurons ("2"). Astrocyte membrane potential is depolarized as a result, leading to the opening of voltage-gated Ca^{2+} channels ("3"). The increase in $[Ca^{2+}]_i$ that results is lethal to cell viability. Astrocytes retain the ability to remove Ca^{2+} via the Na^+/Ca^{2+} – exchanger during ischemia ("4"). Adapted from ref. *13.* with permission.

and the accumulation of K^+ in the extracellular space is largely a product of K^+ efflux from neurons (*see* Fig. 3). The capacity of astrocytes to clear extracellular K^+ is limited during ischemia due to the compromised state of the cells. For example, ischemia reduces gap-junction coupling between astrocytes *(25),* which will limit the ability of astrocytes to remove extracellular K^+ via spatial buffering *(26).* The high rate of K^+ efflux into the extracellular space during ischemia rapidly swamps any remaining ability the astrocytes have to remove K^+, leading to high $[K^+]_o$, membrane depolarization, and release of neuroactive substances such as glutamate from neurons and glia (*see* **Subheading 3.4**).

3.2. Astrocyte Swelling

Astrocyte swelling is a prominent feature of CNS ischemia *(24);* it is the first morphological change observed *(18).* The mechanisms coupling these two events are complex and the situation in vivo is unclear. Movement of K^+, Cl^-, and H^+/HCO_3^- across the cell membrane, and the effects of high extracellular glutamate have all been implicated *(24).* Both the cell bodies and the processes of astrocytes swell during ischemia *(18,27),* leading to shrinkage of the extracellular space. Astrocyte swelling is associ-

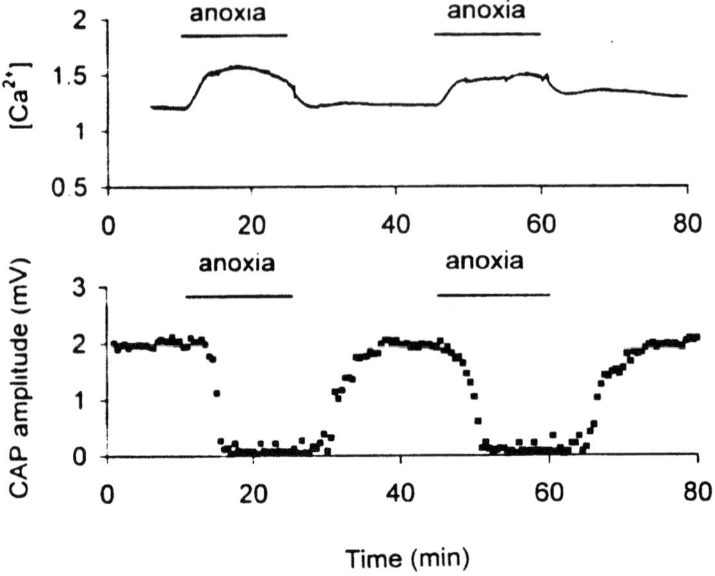

Fig. 4. Changes in extracellular $[Ca^{2+}]$ and compound action potential amplitude recorded from rat optic nerve (a CNS white matter tract). Extracellular $[Ca^{2+}]$ was measured with a Ca^{2+}–sensitive electrode, which shows a significant increase in extracellular $[Ca^{2+}]$ during anoxia. The increase in extracellular $[Ca^{2+}]$ is associated with shrinkage of extracellular space *(89)*. Such changes in extracellular $[Ca^{2+}]$ are not as apparent in CNS gray matter during anoxia/ischemia due to the rapid influx of Ca^{2+} into cells, a process that occurs far slower in the optic nerve. Adapted from ref. *29* with permission.

ated with intracellular acidosis *(28)*, and activation of ion channels which results in the net release of ions *(24)*. Cell swelling is potentially protective of the astrocytes, since an increase in water content will dilute the intracellular concentration of harmful Ca^{2+}. However, extracellular space shrinkage with concomitant concentration of extracellular ions is probably detrimental to neurons during ischemia. For example, in CNS white matter $[Ca^{2+}]_o$ can increase via this mechanism during relatively brief periods of anoxia (*29;* Fig 4), and extracellular space shrinkage will contribute to ischemic rises in $[K^+]_o$ throughout the CNS. Since neuronal and oligodendrocyte swelling is limited during ischemia *(24)*, such effects are primarily accountable to astrocytes. Astrocyte swelling is also associated with release of amino acids (including glutamate) into the extracellular space *(30,31)*, which have important consequences for neurons during ischemia (see **Subheading 3.1**).

3.3. pH_o Changes

Another major dysfunction of brain homeostasis during ischemia is a prominent extracellular acidosis associated with a build up of extracellular lactate *(5,32)*. Following the withdrawal of glucose and oxygen during ischemia, any glucose remaining in the cytoplasm together with the glycogen present in astrocytes is metabolized via glycolysis to lactate, yielding ATP *(7,33,34)*. The lactate produced will dissociate with a pKa of 3.84, generating a significant increase in hydrogen ion concentration. As a result, extracellular pH reaches values between 6.2 to 6.9 during ischemia *(16,32,35)*.

In addition to lactate production in the ischemic core, the capacity of astrocytes to utilize glycolytic respiration under conditions of low oxygen in the presence of glucose *(36,37),* will make these cells a major contributor to lactate production in the ischemic penumbra of a stroke where such conditions exist *(7,16).* In addition to lactate production, other sources of ischemic H^+ production include carbonic acid and the breakdown of ATP without corresponding ATP synthesis *(38).*

The precise consequences of ischemic acidosis are complex, and will relate to the extent and duration of the pH change. For example, H^+ accumulation will block NMDA-type glutamate receptors on neurons, limiting the conductance of this important pathway for ischemic Ca^{2+} influx *(39,40).* As a result, acid pH is protective against hypoxia or glutamate exposure in cultured neurons *(41,42).* The presence of lactate in the extracellular space may also be beneficial to neurons under conditions of low glucose in the presence of oxygen, since it will act as an energy substrate *(43,44).* However, acidosis can also be directly harmful to neurons *(45,46),* and accentuates injury during hypoxia in rat brain slices *(47,48).* Hyperglycemia can either exacerbate *(49,50),* or ameliorate *(51,52),* ischemic brain injury apparently apparently due to enhanced lactate release and acidosis; confirming the complexity of the effects of acidosis during brain ischemia *(7).*

Cultured astrocytes are more sensitive to the toxic effects of acidosis than are neurons *(32,46).* The importance of this finding is unclear, since acid levels at the extreme of those seen during ischemia are required before any significant astrocyte death is produced even if the acidosis is applied for long periods of time *(16,46).* The situation in vivo is also more complex than that in vitro. Swanson et al. *(16)* have shown significant synergy between the effects of hypoxia and acidosis upon cultured astrocyte survival (Fig 5). For example, acid pH (6.2) combined with hypoxia produce significant cell death after 5 h of exposure while under normoxic conditions this level of acidosis had no significant effect after 7 h exposure. The mechanism of this synergy may involve acid inhibition of glycolysis, upon which the cells depend during hypoxia *(16).* When acidosis was co-applied with high $[K^+]_o$ however, the effects of hypoxia were reduced. The interaction of environmental factors with acidosis during ischemia is not fully understood, and the importance of acidosis to astrocyte survival in vivo remains obscure. A further complication involves the potential accumulation of H^+ equivalents in astrocytes during ischemia. Kraig and colleagues *(32)* have provided evidence for a very low pH in astrocytes during ischemia. The consequences of this putative phenomena are not at all clear.

3.4. Extracellular Glutamate

High extracellular glutamate is a feature of brain ischemia *(53),* that has significant consequences for neurons (see above). As the major excitatory neurotransmitter of the CNS, glutamate is continually released from synaptice endings. Although some glutamate is taken back up by neurons, astrocytes play the major role in extracellular glutamate homeostasis *(54).* To fulfill this function, astrocytes express distinct glutamate uptake proteins in their cell membranes, including GLAST and GLT-1 *(55).* These proteins accrue intracellular glutamate with the co-transport of Na^+ and the counter-transport of K^+ and a net HCO_3^- *(56,57).* Membrane depolarization, high $[K^+]_o$ and high $[Na^+]_i$ during ischemia act to reduce the capacity of astrocytes to take up extracellular

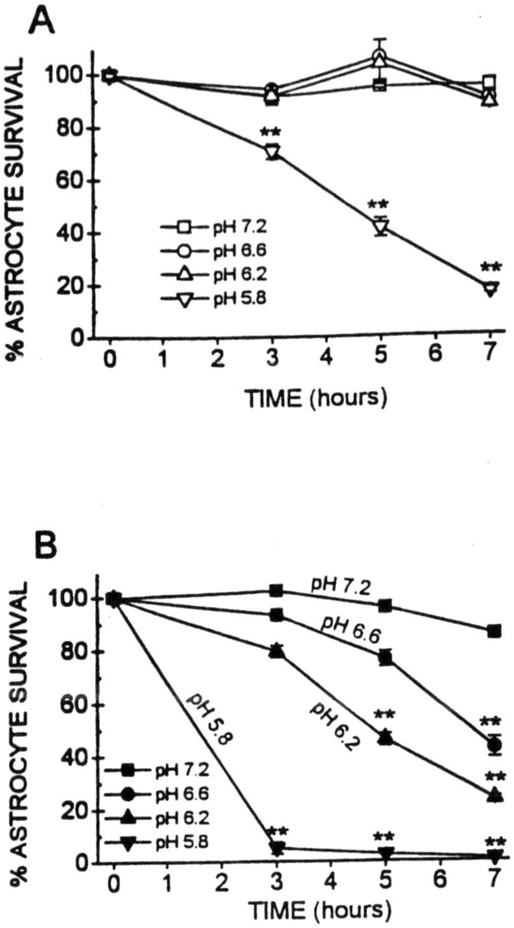

Fig. 5. Cultured astrocyte survival during acidosis is affected by hypoxia. **(A)** Astrocyte survival (assessed by lactate dehydrogenase release) over a 7-h period is affected only by very acid extracellular pH. **(B)** During hypoxia, cell death occurs at lower extracellular pH, within the range observed during ischemia in vivo. Adapted from ref. *16* with permission.

glutamate; and in fact these conditions may favor the net extrusion of glutamate from the cell since the uptake proteins are reversible *(58–60)*. In cultured astrocytes, ischemic-like conditions reduce the capacity for glutamate uptake *(11,16)* and cause glutamate release *(59,61)*. Glutamate release from cultured astrocytes is greater during ischemic conditions than that from cultured neurons *(61)*, and there is good evidence that ischemic increases in extracellular glutamate arise either partly or largely from nonvesicular release from astrocytes *(59,62)*.

In addition to glutamate, astrocytes play an important role in the uptake of other neurotransmitters. Many of the uptake proteins that perform this task are membrane potential and/or Na^+ dependent and have the potential to work in reverse during ischemia. Ischemia is accompanied by the extracellular accumulation of many neurotransmitters, such as GABA and glycine, that are normally regulated by astrocytes *(31,62)*. The role these neurotransmitters play during ischemia is currently unclear, and is likely to be both important and complex *(63,64)*. It has been argued that the ratio of excitatory neurotransmitter release to inhibitory neurotransmitter release during ischemia might be an important factor determining the toxicity of excitatory neurotransmitter build up. Furthermore, since binding of glycine to the NMDA-receptor enhances glutamate activation of the receptor *(65)*, glycine release may be an important determinant of the level of NMDA-receptor activation during ischemia. The role of astrocytes in the ischemic release of neurotransmitters other than glutamate may therefore be of some significance.

4. POSTISCHEMIC GLIOSIS

Following a 30 min period of ischemia induced by vessel occlusion, changes occur in astrocytes within 2 d both in regions of neuronal loss and in regions where neuronal loss is not significant *(66)*. These changes include an increase in GFAP content and an increase in cell size. These changes only persist in brain regions that suffer neuronal loss, where they are associated with an increase in astrocyte numbers *(66)*. In areas that do not suffer neuronal loss, reactive changes in astrocytes return to baseline. These three phenomena (increased GFAP expression, hypertrophy, and hyperplasia) are the cardinal features of reactive gliosis *(32,67)*, with the second form described by Petito et al *(66)* (absence of hyperplasia) forming a separate category *(32)*. The two types of reactive astrocytes described by Pitito et al *(66)* have been termed "anisomorphic" (in the immediate site of the lesion) and "somorphic" (distal from the lesion site). These terms refer to the distortion of CNS morphology produced by the generation of a glial scar at the lesion site *(67)*. The trigger for the reactive changes seen in astrocytes following ischemia are unclear, but high $[K^+]_o$ and high pH_i are features of CNS ischemia that induce reactive changes in astrocytes *(68)*.

The presence of a dense network of reactive astrocytes in regions of selective neurenal lose following ischemia is widely believed to be deleterious to functional recovery *(69)*. It has been argued that reactive astrocytes form a physical barrier to the regrowth of neuronal elements such as axons *(69,70)*. However, under some conditions, axon regrowth can be supported by reactive astrocytes *(71–74)* and it is possible that other factors present in the glial scar, such as oligodendrocyte or microglial cells, may be the major barrier to regrowth *(67,70)*.

The presence of an astrocyte barrier around a lesion may act to isolate the damaged region and protect the surrounding tissue *(67)*. Postischemic reactive astrocytes increase expression of a number of potentially protective proteins. These include metallothionein that may act as a free radical scavenger and metal ion chelator *(75)*; glutathione peroxidase which may also protect against free radical attack *(76)*; and ecto-5′-nucleotidase which may be protective since it increases production of extracellular adenosine *(77)*. Other upregulated proteins may be harmful. For example, nitric oxide generation has been implicated in cell damage after ischemia *(78)*, and a portion of the reactive astrocytes that form the postischemia glial scar are immunoposotive for inducible nitric oxide synthase *(79)*. In addition, postischemic astrocytes have elevated levels of PLA_2, which has been implicated in delayed neuronal injury *(88)*.

4.1. *The Aging Brain*

The aged brain appears to be more easily damaged during stroke *(81)*. However, it is unclear if this is due to cellular changes at the leavel of neurons and glia, or whether other age-related factors are responsible. The aged human brain exhibits slow and less profound reactive changes in astrocytes following ischemia compared to younger brain (*(82)*. This corresponds to a lower sensitivity to oxidative stress in a model of the senescent astrocyte compared to normal cells *(83)*. In the model used by Manganaro et al *(83)*, cysteamine is applied to cultured rat astrocytes resulting in cellular changes similar to those seen in a population of astrocytes in the aged CNS; including increased GFAP levels and the induction of cellular inclusions. The treated astrocytes resembled closely the gomori-posotive astrocytes that increase in number in the senescent brain *(84)*.

Despite the greater predilection to stroke injury, the aged brain has a lower innate sensitivity to excitotoxic injury, apparent when an excitotoxic agent is injected directly into the brain *(85,86)*. It is consistent with this observation that the aged brain can endure relatively high extracellular glutamate for extended periods without concomitant cell injury *(87)*. The greater resistance at the cellular level to excitotoxic damage and the apparent higher resistance of aged astrocytes to ischemic injury correlate to a slowing of brain metabolic rate as it ages *(88)*. A lower metabolic rate would in itself predispose the brain to a lower sensitivity to injury, although it is unclear whether this is an important factor or not. The consequences of the changes that occur in astrocytes as they age to the response of the brain to ischemia have clearly not been the subject of sufficient experimentation to date. The review of the literature presented above illustrates the important role of astrocytes in ischemic injury. Hopefully, future studies will concentrate on the role these cells play during stroke in the aging brain.

REFERENCES

1. Yager, J. Y., Brucklacher, R. M. and Vannucci, R. C. (1992) Cerebral energy metabolism during hypoxia-ischemia and early recovery in immature rats. *Amer. J. Physiol.* **262,** H672–H677.
2. Hamprecht, R. and Dringen, R. Energy metabolism. (1995) In: (Kettenmann H and Ransom BR, eds.) *Neuroglia.* New York: Oxford University Press, 919–936.
3. Rose, C. R., Waxman, S. G. and Ransom, B. R. (1998) Effects of glucose deprivation, chemical hypoxia, and simulated ischemia on Na+ homeostasis in rat spinal cord astrocytes. *J. Neurosci.* **18,** 3554–3562.
4. Walz, W. (1992) Coupling of metabolism and electrical activity in cortical astrocytes. *Can. J. Physiol. Pharmacol.* **70,** S176–S180.
5. Hansen, A. J. (1985) Effect of anoxia on ion distribution in the brain. *Physiol. Rev.* **65,** 101–148.
6. Hertz, L. and Peng, L. (1992) Energy metabolism at the cellular level of the CNS. *Can J. Physiol. Pharmacol* **70,** S145–157.
7. Ransom, B. R. and Fern, R. (1996) Anoxic-ischemic glial cell injury:Mechanisms and consequences. In: Haddad, G. G. and Lister, G. editors. Tissue oxygen deprivation. New York: Marcel Dekker, Inc., 617–652.
8. Goldberg, M. P. and Choi, D. W. (1993) Combined oxygen and glucose deprivation in cortical cell culture. *Journal of Neuroscience,* **13,** 3510–3524.
9. McDonald, J. W., Althomsons, S. P., Hyrc, K. L., Choi, D. W., and Goldberg, M. P. (1998) Oligodendrocytes from forebrain are highly vulnerable to AMPA/kainate receptor-mediated excitotoxicity. *Nat. Med.* **4,** 291–297.
10. Matute, C., Sanchez-Gomez, M. V., Martinez-Millan, L., and Miledi, R. (1997) Glutamate receptor-mediated toxicity in optic nerve oligodendrocytes. *Proc. Natl. Acad. Sci. USA* **94,** 8830–8835.
11. Yu, A. C., Gregory, G. A. and Chan, P. H. (1989) Hypoxia-induced dysfunctions and injury of astrocytes in primary cell cultures. *J. Cereb. Blood Flow Metab.* **9,** 20–28.

12. Duffy, S. and MacVicar, B. A. (1996) In vitro ischemia promotes calcium influx and intracellular calcium release in hippocampal astrocytes. *J. Neurosci.* **16,** 71–81.
13. Fern, R. (1998) Intracellular calcium and cell death during ischemia in neonatal rat white matter astrocytes in situ. *J. Neurosci.* **18,** 7232–7243.
14. Davies, C. A., Loddick, S. A., Stroemer, R. P., Hunt, J., and Rothwell, N. J. (1998) An integrated analysis of the progression of cell responses induced by permanent focal middle cerebral artery occlusion in the rat. *Exp. Neurol.* **154,** 199–212.
15. Schmidt-Kastner, R., Wietasch, K., Weigel, H., and Eysel, U. T. (1993) Immunohistochemical staining for glial fibrillary acidic protein (GFAP) after deafferentation or ischemic infarction in rat visual system: features of reactive and damaged astrocytes. *Int. J. Dev. Neurosci.* **11,** 157–174.
16. Swanson, R. A., Farrell, K. and Stein, B. A. (1997) Astrocyte energetics, function, and death under conditions of incomplete ischemia: a mechanism of glial death in the penumbra. *Glia* **21,** 142–153.
17. Brierly, J. B. and Graham, D. I. Hypoxia and vascular disorders of the nervous system. (1984) New York: John Wiley and Sons.
18. Petito, C. K. and Babiak, T. (1982) Early proliferative changes in astrocytes in postischemic noninfarcted rat brain. *Ann. Neurol.* **11,** 510–518.
19. Siesjo, B. K., Katsura, K., Zhao, Q., et al. (1995) Mechanisms of secondary brain damage in global and focal ischemia: a speculative synthesis. *Neurotrauma.* **12,** 943–956.
20. Papadopoulos, M. C., Koumenis, I. L., Dugan, L. L., and Giffard, R. G. (1997) Vulnerability to glucose deprivation injury correlates with glutathione levels in astrocytes. *Brain Res.* **748,** 151–156.
21. Kim-Lee, M. H., Stokes, B. T. and Yates, A. J. (1992) Reperfusion paradox: a novel mode of glial cell injury. *Glia* **5,** 56–64.
22. Li, Y., Chopp, M., Jiang, N., and Zaloga, C. (1995) In situ detection of DNA fragmentation after focal cerebral ischemia in mice. *Brain. Res. Mol. Brain Res.* **28,** 164–168.
23. Love, S., Barber, R. and Wilcock, G. K. (1998) Apoptosis and expression of DNA repair proteins in ischaemic brain injury in man. *Neuroreport* **9,** 955–959.
24. Kimelberg HK. Brain Edema. In: (Kettenmann, H. and Ransom, B. R., eds.) Neuroglia.(1995) New York: Oxford University Press, 919–936.
25. Cotrina, M. L., Kang, J., Lin, J. H., et al. (1998) Astrocytic gap junctions remain open during ischemic conditions. *J. Neurosci.* **18,** 2520–2537.
26. Newman EA. (1995) Glial cell regulation of extracellular potassium. In: Kettenmann H and Ransom BR, editors. Neuroglia. New York: Oxford University Press, 717–731.
27. Van Harreveld, A. (1982) Swelling of the Muller fibers in the chicken retina. *J. Neurobiol.* **13,** 519–536.
28. Volkl, H., Busch, G. L., Haussinger, D. and Lang, F. (1994) Alkalinization of acidic cellular compartments following cell swelling. *FEBS Lett.* **338,** 27–30.
29. Brown, A. M., Fern, R., Jarvinen, J. P., Kaila, K., and Ransom, B. R. (1998) Changes in [Ca2+]0 during anoxia in CNS white matter. *Neuroreport* **9,** 1997–2000.
30. Kimelberg, H. K., Goderie, S. K., Higman, S., Pang, S., and Waniewski, R. A. (1990) Swelling-induced release of glutamate, aspartate, and taurine from astrocyte cultures. *J. Neurosci.* **10,** 1583–1591.
31. Levi, G. and Gallo, V. (1995) Release of neuroactive amino acids from glia. In: (Kettenmann, H. and Ransom, B. R. eds.) Neuroglia. New York: Oxford University Press, 815–828.
32. Kraig, R. P., Lascola, C. D. and Caggiano, A. (1995) Glial response to brain ischemia. In: (Kettenmann, H. and Ransom, B. R., eds.) Neuroglia. New York: Oxford University Press, 964–976.
33. Dringen, R., Gebhardt, R., and Hamprecht, B. (1993) Glycogen in astrocytes: possible function as lactate supply for neighboring cells. *Brain Res.* **623,** 208–214.

34. Walz, W. and Mukerji, S. (1988) Lactate production and release in cultured astrocytes. *Neuroso. Lett.* **86,** 296–300.
35. Obrenovitch, T. P., Garofalo, O., Harris, R. J., et al. (1988) Brain tissue concentrations of ATP, phosphocreatine, lactate, and tissue pH in relation to reduced cerebral blood flow following experimental acute middle cerebral artery occlusion. *J. Cereb. Blood Flow Metab.* **8,** 866–874.
36. Callahan, D. J., Engle, M. J., and Volpe, J. J. (1990) Hypoxic injury to developing glial cells: protective effect of high glucose. *Pediatr. Res.* **27,** 186–190.
37. Kelleher, J. A., Chan, P. H., Chan, T. Y., and Gregory, G. A. (1993) Modification of hypoxia-induced injury in cultured rat astrocytes by high levels of glucose. *Stroke* **24,** 855–863.
38. Hochachka, P. W. and Mommsen, T. P. (1983) Protons and anaerobiosis. *Science* **219,** 1391–1397.
39. Tang, C. M., Dichter, M. and Morad, M. (1990) Modulation of the N-methyl-D-aspartate channel by extracellular H+. *Proc. Natl. Acad. Sci. U.S.A.* **87,** 6445–6449.
40. Traynelis, S. F. and Cull-Candy, S. G. (1990) Proton inhibition of N-methyl-D-aspartate receptors in cerebellar neurons. *Nature* **345,** 347–350.
41. Giffard, R. G., Monyer, H., Christine, C. W., and Choi, D. W. (1990) Acidosis reduces NMDA receptor activation, glutamate neurotoxicity, and oxygen-glucose deprivation neuronal injury in cortical cultures. *Brain Res.* **506,** 339–342.
42. Tombaugh, G. C. and Sapolsky, R. M. (1990) Mild acidosis protects hippocampal neurons from injury induced by oxygen and glucose deprivation. *Brain Res.* **506,** 343–5.
43. Schurr, A., West, C. A. and Rigor, B. M. (1988) Lactate-supported synaptic function in the rat hippocampal slice preparation. *Science* **240,** 1326–1328.
44. Fern, R., Davis, P., Waxman, S. G., and Ransom, B. R. (1998) Axon conduction and survival in CNS white matter during energy deprivation: a developmental study. *J. Neurophysiol.* **79,** 95–105.
45. Kraig, R. P., Petito, C. K., Plum, F., and Pulsinelli, W. A. (1987) Hydrogen ions kill brain at concentrations reached in ischemia. *J. Cereb. Blood Flow Metab.* **7,** 379–386.
46. Nedergaard, M., Goldman, S. A., Desai, S., and Pulsinelli, W. A. (1991) Acid-induced death in neurons and glia. *J. Neurosci.* **11,** 2489–2497.
47. O'Donnell, B. R. and Bickler, P. E. (1994) Influence of pH on calcium influx during hypoxia in rat cortical brain slices. *Stroke* **25,** 171–177.
48. Katsura, K., Kristian, T., Smith, M. L., and Siesjo, B. K. (1994) Acidosis induced by hypercapnia exaggerates ischemic brain damage. *J. Cereb. Blood Flow Metab.* **14,** 243–250.
49. Pulsinelli, W. A., Levy, D. E., Sigsbee, B., Scherer, P., and Plum, F. (1983) Increased damage after ischemic stroke in patients with hyperglycemia with or without established diabetes mellitus. *Am. J. Med.* **74,** 540–544.
50. Plum, F. (1983) What causes infarction in ischemic brain?: The Robert Wartenberg Lecture. *Neurology* **33,** 222–233.
51. Ginsberg, M. D., Prado, R., Dietrich, W. D., Busto, R., and Watson, B. D. (1987) Hyperglycemia reduces the extent of cerebral infarction in rats. *Stroke,* **18,** 570–574.
52. Zasslow, M. A., Pearl, R. G., Shuer, L. M., Steinberg, G. K., Lieberson, R. E. and Larson, C. P., Jr. (1989) Hyperglycemia decreases acute neuronal ischemic changes after middle cerebral artery occlusion in cats. *Stroke* **20,** 519–523.
53. Benveniste, H., Drejer, J., Schousboe, A., and Diemer, N. H. (1984) Elevation of the extracellular concentrations of glutamate and aspartate in rat hippocampus during transient cerebral ischemia monitored by intracerebral microdialysis. *J. Neurochem.* **43,** 1369–1374.
54. Schousboe, A. and Hertz, L. (1981) Role of astroglial cells in glutamate homeostasis. *Adv. Biochem. Psychopharmacol.* **27,** 103–113.
55. Chaudhry, F. A., Lehre, K. P., van Lookeren Campagne, M., Ottersen, O. P., Danbolt, N. C., and Storm-Mathisen, J. (1995) Glutamate transporters in glial plasma membranes: highly differentiated localizations revealed by quantitative ultrastructural immunocytochemistry. *Neuron* **15,** 711–720.

56. Bouvier, M., Szatkowski, M., Amato, A., and Attwell, D. (1992) The glial cell glutamate uptake carrier countertransports pH-changing anions. *Nature* **360,** 471–474.
57. Sonnewald, U., Westergaard, N. and Schousboe, A. (1997) Glutamate transport and metabolism in astrocytes. *Glia* **21,** 56–63.
58. Nicholls, D. and Attwell, D. (1990) The release and uptake of excitatory amino acids. *Trends Pharmacol. Sci.* **11,** 462–468.
59. Longuemare, M. C. and Swanson, R. A. (1995) Excitatory amino acid release from astrocytes during energy failure by reversal of sodium-dependent uptake. *J. Neurosci. Res.* **40,** 379–386.
60. Takahashi, M., Billups, B., Rossi, D., Sarantis, M., Hamann, M. and Attwell, D. (1997) The role of glutamate transporters in glutamate homeostasis in the brain. *J. Exp. Biol.* **200,** 401–409.
61. Ogata, T., Nakamura, Y., Shibata, T., and Kataoka, K. (1992) Release of excitatory amino acids from cultured hippocampal astrocytes induced by a hypoxic-hypoglycemic stimulation. *J. Neurochem.* **58,** 1957–1959.
62. Levi, G. and Raiteri, M. (1993) Carrier-mediated release of neurotransmitters. *Trends Neurosci.* **16,** 415–419.
63. Fern, R., Waxman, S. G., and Ransom, B. R. (1994) Modulation of anoxic injury in CNS white matter by adenosine and interaction between adenosine and GABA. *J. Neurophysiol.* **72,** 2609–2616.
64. Fern, R., Waxman, S. G., and Ransom, B. R. (1995) Endogenous GABA attenuates CNS white matter dysfunction following anoxia. *J. Neurosci.* **15,** 699–708.
65. Johnson, J. W. and Ascher, P. (1987) Glycine potentiates the NMDA-response in cultured mouse brain neurons. *Nature* **325,** 529–531.
66. Petito, C. K., Morgello, S., Felix, J. C., and Lesser, M. L. (1990) The two patterns of reactive astrocytosis in postischemic rat brain. *J. Cereb. Blood Flow Metab.* **10,** 850–859.
67. Ridet, J. L., Malhotra, S. K., Privat, A., and Gage, F. H. (1997) Reactive astrocytes: cellular and molecular cues to biological function [published erratum appears in Trends Neurosci 1998 Feb;21(2):80]. *Trends Neurosci.* **20,** 570–577.
68. Kraig, R. P. and Jaeger, C. B. (1990) Ionic concomitants of astroglial transformation to reactive species. *Stroke* **21,** III184–187.
69. Reier, P. J. (1986) Gliosis following neuronal injury: The anatomy of astrocyte scars and their influence on axonal elongation. In: (Federoff, S. and Vernadakis, A. eds.) Astrocytes, Vol 3. New York: Academic Press, 263–324.
70. Hatten, M. E., Liem, R. K., Shelanski, M. L., and Mason, C. A. (1991) Astroglia in CNS injury. *Glia* **4,** 233–243.
71. Gage, F. H., Olejniczak, P., and Armstrong, D. M. (1988) Astrocytes are important for sprouting in the septohippocampal circuit. *Exp. Neurol.* **102,** 2–13.
72. David, S. (1988) Neurite outgrowth from mammalian CNS neurons on astrocytes in vitro may not be mediated primarily by laminin [published erratum appears in J. Neurocytol. (1988) Aug; 17(4): 581]. *J. Neurocytol.* **17,** 131–144.
73. Kliot, M., Smith, G. M., Siegal, J. D., and Silver, J. (1990) Astrocyte-polymer implants promote regeneration of dorsal root fibers into the adult mammalian spinal cord. *Exp. Neurol.* **109,** 57–69.
74. Li, Y. and Raisman, G. (1995) Sprouts from cut corticospinal axons persist in the presence of astrocytic scarring in long-term lesions of the adult rat spinal cord. *Exp. Neurol.* **134,** 102–111.
75. Neal, J. W., Singhrao, S. K., Jasani, B., and Newman, G. R. (1996) Immunocytochemically detectable metallothionein is expressed by astrocytes in the ischaemic human brain. *Neuropathol. Appl. Neurobiol.* **22,** 243–247.
76. Takizawa, S., Matsushima, K., Shinohara, Y. et al. (1994) Immunohistochemical localization of glutathione peroxidase in infarcted human brain. *J. Neurol. Sci.* **122,** 66–73.
77. Braun, N., Lenz, C., Gillardon, F., Zimmermann, M., and Zimmermann, H. (1997) Focal cerebral ischemia enhances glial expression of ecto-5′-nucleotidase. *Brain Res.* **766,** 213–226.

78. Lipton, S. A., Singel, D. J. and Stamler, J. S. (1994) Nitric oxide in the central nervous system. *Prog. Brain Res.* **103,** 359–364.
79. Endoh, M., Maiese, K., and Wagner, J. (1994) Expression of the inducible form of nitric oxide synthase by reactive astrocytes after transient global ischemia. *Brain Res.* **651,** 92–100.
80. Clemens, J. A., Stephenson, D. T., Smalstig, E. B., et al. (1996) Reactive glia express cytosolic phospholipase A2 after transient global forebrain ischemia in the rat [see comments]. *Stroke,* **27,** 527–535.
81. Macciocchi, S. N., Diamond, P. T., Alves, W. M., and Mertz, T. (1998) Ischemic stroke: relation of age, lesion location, and initial neurologic deficit to functional outcome. *Arch. Phys. Med. Rehabil.* **79,** 1255–1257.
82. Dziewulska, D. (1997) Age-dependent changes in astroglial reactivity in human ischemic stroke. Immunohistochemical study. *Folia Neuropathol.* **35,** 99–106.
83. Manganaro, F., Chopra, V. S., Mydlarski, M. B., Bernatchez, G., and Schipper, H. M. (1995) Redox perturbations in cysteamine-stressed astroglia: implications for inclusion formation and gliosis in the aging brain. *Free. Radic. Biol. Med.* **19,** 823–835.
84. Schipper, H. M. (1991) Gomori-positive astrocytes: biological properties and implications for neurologic and neuroendocrine disorders. *Glia* **4,** 365–377.
85. Finn, S. F., Hyman, B. T., Storey, E., Miller, J. M., and Beal, M. F. (1991) Effects of aging on quinolinic acid lesions in rat striatum. *Brain Res.* **562,** 276–280.
86. Kesslak, J. P., Yuan, D., Neeper, S., and Cotman, C. W. (1995) Vulnerability of the hippocampus to kainate excitotoxicity in the aged, mature and young adult rat. *Neurosci. Lett.* **188,** 117–120.
87. Massieu, L. and Tapia, R. (1997) Glutamate uptake impairment and neuronal damage in young and aged rats in vivo. *J. Neurochem.* **69,** 1151–1160.
88. Curti, D. and Benzi, G. (1990) Role of synaptosomal enzymatic alterations and drug treatment in brain aging. *Clin. Neuropharmacol.* **13,** S59–72.
89. Ransom, B. R., Walz, W., Davis, P. K., and Carlini, W. G. (1992) Anoxia-induced changes in extracellular K+ and pH in mammalian central white matter. *J. Cereb. Blood Flow Metab.* **12,** 593–602.

23

Glial-Neuronal Interactions during Oxidative Stress

Implications for Parkinson's Disease

Catherine Mytilineou

1. PARKINSON'S DISEASE

Idiopathic Parkinson's disease (PD) is a common neurological disorder of adult onset (estimated prevalence 1:1000 of the population), with tremor, rigidity, bradykinesia, and postural instability as the primary symptoms. The etiology of PD is not known. The majority of the cases are sporadic although inherited parkinsonism is reported in several families. Recently, it was discovered that a mutation in the α-synuclein gene (A53T) is associated with autosomal dominant PD in several families *(1,2)*. There is also evidence suggesting a systemic defect in mitochondrial respiration in PD *(3–7)*, possibly transmitted through the mitochondrial genome *(8)*. This defect results in decreased activity of complex I of the mitochondrial respiratory chain *(3)* and links idiopathic and l-methyl-4-phenyl-1,2,3,6-tetrahydropyridine (MPTP)-induced parkinsonism *(9)*. MPTP causes degeneration of dopamine neurons through its metabolism by monoamine oxidase B (MAO B) to the neurotoxin l-methyl-4-phenylpyridine (MPP+), which inhibits complex I activity *(10)*. Other environmental factors may also influence the risk for PD *(11–13)*. Endogenous neurotoxins, such as products of dopamine catabolism and/or isoquinolines, are also thought to potentially contribute to the pathogenesis of PD *(14,15)*.

Neuropathologically PD is characterized by the loss of dopamine neurons in the substantia nigra pars compacta, with consequent severe reductions in the dopamine content of the striatum *(16)*. Other areas of the brain (norepinephrine neurons in the locus ceruleus, cholinergic neurons in the nucleus basalis of Mynert) are also affected in PD, but to a lesser degree. The cause of the selective degeneration of dopamine neurons in PD is not understood at present. Several theories have been proposed to explain this selective loss. Data obtained from postmortem studies, as well as in vivo and in vitro experiments support a role of oxidative stress in the selective degeneration of dopamine neurons in PD.

2. EVIDENCE OF OXIDATIVE STRESS IN PARKINSON'S DISEASE

Several observations point to the existence of conditions that either favor or are the result of oxidative stress in the substantia nigra of PD brains. Glutathione (GSH), one

From: *Neuroglia in the Aging Brain*
Edited by: Jean S. de Vellis © Humana Press Inc., Totowa, NJ

of the major cellular antioxidants, has been shown to be decreased in the substantia nigra of PD brains *(17,18)*. GSH is usually present in high concentrations in the brain where it acts as a reducing agent, removing hydrogen peroxide and maintaining the thiol groups of intracellular proteins *(19)*.

$$2\ GSH + H_2O_2 \xrightarrow{\text{GSH peroxidase}} GSSG + 2\ H_2O \quad (1)$$

$$GSSG + NADPH + H^+ \xrightarrow{\text{GSSG reductase}} 2\ GSH + NADP^+ \quad (2)$$

The enzyme GSH peroxidase participates in the reduction of H_2O_2 (equation 1) and GSSG reductase in the regeneration of GSH from GSSG (equation 2). Decreased levels of cellular GSH in PD would create vulnerability to oxidative stress. Dopamine neurons may be particularly sensitive to GSH loss, in view of the production of H_2O_2 during the metabolism of dopamine by MAO (equation 3).

$$RNH_2 + O_2 + H_2O \xrightarrow{\text{MAO}} ROH + H_2O_2 + NH_3 \quad (3)$$

$$H_2O_2 + Fe^{2+} \rightarrow Fe^{2+} + {}^{\bullet}OH + OH^- \quad (4)$$

In the presence of iron, H_2O_2 can generate the highly cytotoxic hydroxyl radical (OH^-) through the Fenton reaction (equation 4; for reviews on the subject *see* refs. *14,20*). There are several reports of increased iron concentration in the substantia nigra of PD brains associated with melanin-containing dopamine neurons *(21–25)*. Increased levels of iron, in combination with reduced GSH levels, would result in additional free radical generation, which could damage lipids, proteins, and nucleic acids. Increased lipid peroxidation, as well as DNA damage have been reported in PD brains *(26,27)*.

Depletion of GSH can cause mitochondrial damage *(28,29)*. A reduction in the activity of complex I of the mitochondrial respiratory chain, such as shown in PD substantia nigra *(4)*, can also generate conditions of oxidative stress *(30)*.

3. ROLE OF ASTROCYTES IN OXIDATIVE STRESS

Brain cells, which depend mainly on molecular oxygen for supply of energy, are equipped with safeguards for protection against oxidative stress. In addition to the antioxidant activity of the GSH-GSH peroxidase system described previously, there is also a high concentration of the antioxidant ascorbic acid in the brain *(31)* and ascorbic acid levels can be maintained in the brain under conditions of reduced supply *(32)*. Other antioxidant defenses of the brain include catalase, which is present at relatively low concentrations *(33)*, superoxide dismutase, α-tocopherol, and metal chelators such as metallothionins *(34–36)*.

Glial cells appear to play a very important role in the protection of neurons from oxidative stress. Several in vitro studies have shown that the levels of GSH are higher in astrocytes than in neurons *(37–40)*. In the brain, GSH concentration appears low in neuronal somata, whereas the neuropil, which includes glial processes, is enriched in GSH *(41,42)*. Sagara and colleagues *(39,43)* have shown that glial cells, in contrast to neurons, can effectively accumulate cystine from the extracellular space for use in GSH synthesis. Neurons, on the other hand, have a more efficient uptake for cysteine, which however is low in the plasma and culture media compared to cystine. Thus neurons depend on astrocytes for the export of cysteine into the extracellular fluid, which they can use for the synthesis of GSH *(43)*.

Many studies have shown that glial cells have potent neurotrophic and neuroprotective activities toward dopamine neurons (44–46). In the presence of glial cells dopamine neurons become resistant to the neurotoxins MPP+ *(47)* and 6-hydroxydopamine (6-OHDA; *48,49*). MPP+ uses the dopamine transporter to accumulate in the dopamine neurons *(50,51)* and then enters the mitochondria, where it inhibits NADH-ubiquinone oxidoreductase (complex I) activity of the electron transport chain *(10,52)*. The interaction of MPP+ with complex I involves the generation of free radicals, which contribute to the irreversible inhibition of the enzyme complex and the toxic effects of MPP+ *(53)*. 6-OHDA is a classic oxidative stress-inducing neurotoxin *(54,55)*. Our studies showed that glial cells confer very powerful protection to dopamine neurons exposed to 6-OHDA *(49)*. Glial growth was stimulated in mesencephalic cells in culture by treatment with basic fibroblast growth factor (bFGF). Treatment with bFGF in the presence of the mitotic inhibitor fluorodeoxyuridine (FUDR) completely abolished the protection against 6-OHDA by bFGF (Fig. 1A). When GSH levels were assessed in control and bFGF-treated cultures, it was found that in addition to the higher GSH levels, cultures containing glial cells were able to double their GSH content following 6-OHDA treatment (Fig. 1B). The upregulation of GSH levels following oxidative stress probably contributed to the protection from 6-OHDA toxicity. Inhibition of glial proliferation prevented the increase in the basal levels of GSH and also prevented the upregulation following 6-OHDA (Fig. 1C).

L-Dopa and dopamine are potent toxins to neurons when applied in culture *(38,56–60)*. Toxicity is due to the autoxidation of the catecholamine moiety of the compounds *(38)* and is prevented by antioxidants *(38,60)*. The presence of glial cells offers significant protection to the neurons against this oxidative stress paradigm *(38,61)*. The mechanisms by which glial cells protect neurons against catecholamine toxicity are not well defined, but several properties of glial cells are consistent with a neuroprotective role. A study by Mena et al. *(61)* showed that glial conditioned medium was protective against L-dopa toxicity to dopamine neurons in culture, suggesting that glial cells may secrete antioxidants and/or other protective molecules. Both high and low molecular weight fractions (higher or lower than 10kD) of the glial conditioned medium had neuroprotective effects, as did directly applied glial cell line derived neurotrophic factor (GDNF) and GSH *(61)*. Ascorbic acid, at concentrations present normally in the extracellular space (200 μM), completely prevents L-dopa toxicity *(58)*. Astrocytes can accumulate ascorbic acid, as well as its oxidized form, dehydroascorbic acid, from the medium *(62–64)* and could thus maintain the antioxidant status of brain cells.

Direct application of hydrogen peroxide to cultured neurons results in degeneration *(65–67)*. The toxicity of hydrogen peroxide is attenuated when neurons are cultured in the presence of astrocytes *(67,68)*. A study by Desagher et al. *(67)* showed that the degree of protection depended on the number of astrocytes and that significant protection against hydrogen peroxide could be achieved when one astrocyte was present per

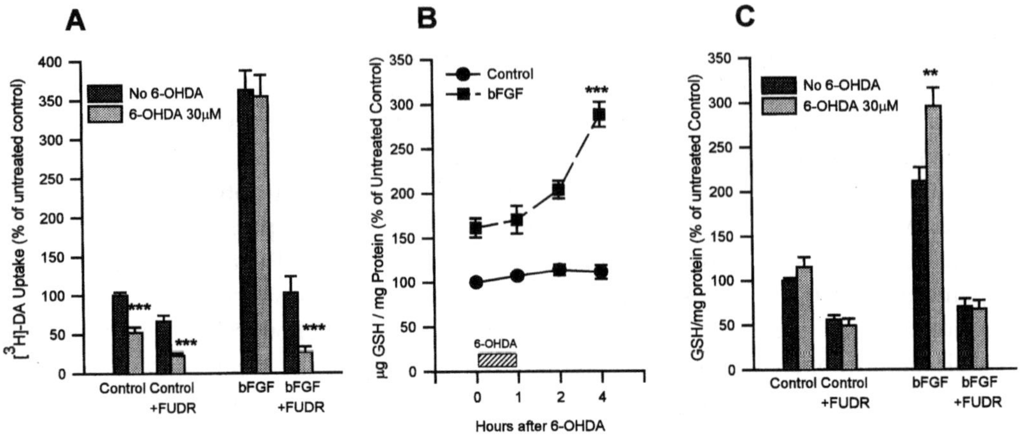

Fig. 1. Effect of glial cells on GSH levels and 6-OHDA toxicity in mesencephalic cultures. Glial growth was induced by treatment with 10 ng/mL bFGF. The mitotic inhibitor FUDR was used to prevent glial growth in the presence of bFGF. **(A)** [^{3}H]Dopamine uptake after treatment with 6-OHDA. Glial growth completely prevented the toxicity of 6-OHDA. **(B)** GSH levels in enriched neuronal cultures and in cultures treated with bFGF following exposure to 6-OHDA. bFGF-treated cultures had higher basal GSH levels and were able to upregulate GSH content after 6-OHDA. **(C)** Effect of inhibition of glial proliferation on GSH content. Treatment with FUDR completely prevented the increases in GSH levels caused by bFGF. ***Significantly different from corresponding control not treated with 6-OHDA $p<0.001$. **$p<0.01$. Modified from *(49)*.

20 neurons *(67)*. Catalase and glutathione peroxidase are the major enzymes protecting from hydrogen peroxide toxicity. Glutathione peroxidase is exclusively localized in glial cells (astrocytes) in the brain *(69)*, which probably explains the highly protective role of astrocytes against hydrogen peroxide toxicity. On the other hand Cu^{++}/Zn^{++} superoxide dismutase, which generates hydrogen peroxide from superoxide, is mainly localized within neurons *(70,71)*, indicating the need for glial-neuronal interactions for protection from oxygen free radicals.

4. INFLAMMATION AND GLIAL CELLS IN PARKINSON'S DISEASE

Reactive microglia is found in the substantia nigra of PD patients *(72,73)*, often associated with extraneuronal melanin granules and in some cases seen engulfing dopaminergic neurons *(72)*. Proliferation of astrocytes (astrogliosis) is also present in the substantia nigra in PD *(74)*. Whether these indices of an inflammatory reaction occur as a consequence of the continuous degeneration of dopamine neurons in the parkinsonian brain or they are playing a role in the process of neurodegeneration is not known. Recently, it has been proposed that activation of glial cells may contribute to the degeneration of dopamine neurons in PD *(75)*. The authors hypothesize that activation of glial cells, caused by an as yet unknown agent, induces the synthesis and secretion of mediators of inflammation (cytokines, nitric oxide), which then interact with specific receptors and induce death of dopamine neurons *(75)*. Several observations lend support to this hypothesis. Glial cells immunoreactive to interleukin 1β (IL-1β) and interferon-γ are present in the substantia nigra of PD patients *(76)*. Tumor necrosis factor-α (TNF-α) is also expressed in glial cells in the substantia nigra of PD patients, although it is essentially undetected in control brains *(77)*. In addition, increased expression of TNF-α is found in the striatum and cerebrospinal fluid of PD patients *(78)*. Furthermore, several other cytokines (IL-1β, IL-2, IL-4, IL-6, transforming growth factor-α) known to be secreted by activated astrocytes and microglia are found in higher concentrations in the cerebrospinal fluid of PD patients *(79)*.

5. MICROGLIA AND ASTROCYTES AS POTENTIAL CONTRIBUTORS TO NEURONAL TOXICITY

Microglia serve as the resident immune system of the brain and are essential for defense against inflammation and for wound repair *(80)*. Upon activation they secrete a large variety of compounds including growth factors, cytokines, complement factors, lipid mediators, free radicals, and neurotoxins (*see* review by Minghetti and Levy; *80*). Although microglia-derived compounds have been shown to enhance the survival of DA neurons in enriched neuronal cultures *(81)*, activated microglia also secrete potential neurotoxins such as proinflammatory cytokines, glutamate, quinolinic acid, and nitric oxide (NO; *82*).

As discussed earlier in this chapter, astrocytes play an important role in neuronal survival by releasing growth factors, providing important nutrients and GSH to neurons, sequestering and degrading excessive amounts of neurotransmitters and other potentially toxic molecules, and regulating ion homeostasis (see review by Tsacopoulos and Magistretti; *83*). However, astrocytes may also contribute to neuronal degeneration. Two mechanisms potentially linking astrocytes to neurotoxicity are increased NO release and PLA_2 activation. Two major forms of nitric oxide synthase (NOS) have been identified. cNOS, present in the neurons, is a constitutive form of the enzyme activated by elevated intracellular Ca^{2+} levels *(84)*. An inducible form (iNOS), is a Ca^{2+} independent enzyme expressed in cells related to the immune system and in astrocytes and microglia *(85,86)*. cNOS activated by physiological agents causes release of NO, which is terminated within seconds or minutes, probably reflecting the return of intracellular Ca^{2+} to baseline. However, once iNOS is induced, NO release can continue for days *(87)*. NO can cause oxidative damage and impair mitochondrial respiration *(88,89)* and it has been suggested that NO may play an important role in the pathogenesis of PD *(90)*.

Astrocytes also contain phospholipase A_2 (PLA_2). The products of PLA_2-catalyzed reactions include free fatty acids such as arachidonic acid (AA). AA is one of the most important fatty acids in the brain, where it can serve as a precursor of prostaglandins, leukotrienes, and thromoboxanes, biologically active substances important in signal transduction *(91,92)*. However, it is believed that increased PLA_2 activity contributes to neurodegeneration in many disease states, including ischemia, Alzheimer's disease, schizophrenia, and head and spinal cord injury (reviewed in ref. *93*). The cytokines TNF-α and IL-1 enhance the expression of PLA_2 in astrocytes *(94)*. Increased PLA_2 activity can result in the accumulation of free fatty acids, which may initiate an "arachidonic acid cascade" resulting in the formation of ROS and membrane lipid peroxidation. In addition, activation of PLA_2 can lead to increased Ca^{2+} influx, which can result in apoptosis *(95)*.

There is experimental evidence that, under certain conditions, glial cells may contribute positively toward neuronal toxicity. In a study using mesencephalic cultures,

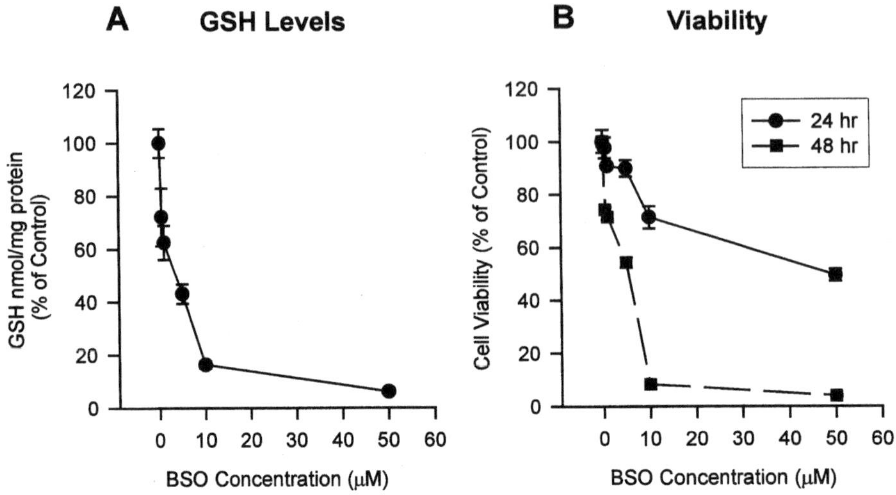

Fig. 2. Effect of treatment with increasing concentrations of BSO on GSH levels **(A)** and cell viability **(B)** in mesencephalic cultures containing glial cells. The growth of glial cells was stimulated by the use of serum in the medium. There was a dose dependent decrease in GSH levels 24 h after treatment with BSO. Viability also showed a dose dependent decrease following BSO, which was more pronounced 48 h after treatment.

Bronstein et al., *(48)* showed that a glial substrate protected dopamine neurons from 6-OHDA toxicity, but it promoted dopamine neuron degeneration in the presence of the endotoxin lipopolysaccharide (LPS). It is known that LPS causes activation of microglia and astrocytes. Inhibition of NOS activity by N^G-nitro-L-arginine methyl ester (NAME) completely prevented the effect of LPS, suggesting that NO, secreted by activated glial cells, was responsible for LPS toxicity *(48)*. Similarly, a selective toxicity to dopamine neurons was observed in co-cultures of mesencephalic neurons and microglial cells, when they were stimulated with LPS or zymosan granules *(96)*. Only the LPS-induced toxicity was blocked by inhibition of NOS, whereas zymosan toxicity resulted from protein kinase C activation *(96)*. A selective increase in the survival of dopamine neurons in cultures treated with low concentrations of the antimitotic agent cytocine arabinoside has also been attributed to the elimination of glial influences that promote neuronal apoptosis *(97)*.

We have recently shown that glial cells are the mediators of neuronal cell death in mesencephalic cultures in a model of oxidative stress *(40)*. Cultures were depleted of GSH using the GSH synthesis inhibitor L-buthionine sulfoximine (BSO). In cultures containing glial cells, treatment with BSO caused a dose dependent reduction in GSH content and cell death, which paralleled the loss of GSH but occurred with a time lag of about 24 h (Fig. 2). Similar reduction in GSH levels in enriched neuronal cultures caused no loss in cell viability (Fig. 3). The toxicity caused by GSH depletion was prevented by treatment with the antioxidant ascorbic acid, indicating that oxidative stress caused by the depletion of GSH was initiating events that led to cell degeneration. In addition to ascorbic acid, complete protection from toxicity was achieved by inhibition of lipoxygenase activity with nordihydroguaiaretic acid (NDGA), indicating an involvement of arachidonic acid metabolism in the toxic events *(40)*. The fact that severe GSH

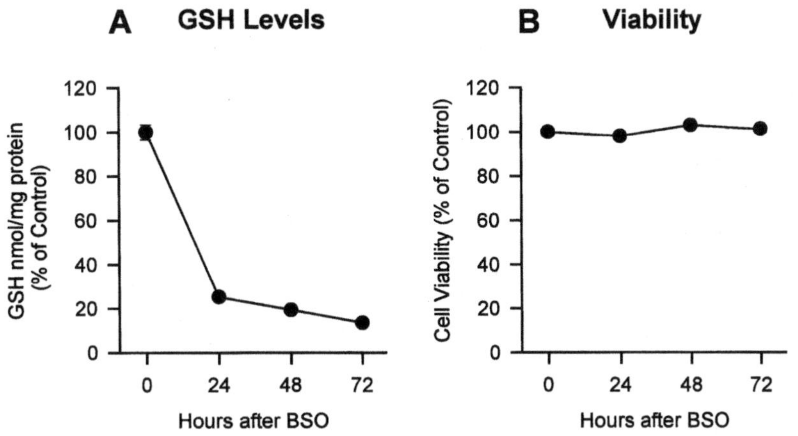

Fig. 3. Effect of BSO treatment on GSH levels and cell viability in enriched neuronal mesencephalic cultures. The growth of glial cells was prevented by the use of serum-free, chemically defined medium. Treatment with 50 μM BSO resulted in increasing depletions of GSH over 72 h **(A).** However, no loss in cell viability resulted from this treatment at any time point **(B).**

depletion causes cell death only in the presence of glial cells suggests that oxidative damage resulting from loss of GSH may activate glial cells and this activation may lead to neuronal degeneration.

6. SUMMARY: GLIAL CELLS AS MEDIATORS OF PROTECTION AND TOXICITY IN PARKINSON'S DISEASE

The major neuropathological feature of PD is the severe loss of dopamine neurons of the substantia nigra pars compacta (76% loss; *98*). Dopamine neurons in the pars lateralis, ventral tegmental area, and the peri- and retro-rubral regions are less affected (31–55% loss) and there is only minimal loss (3%) in the central gray *(98)*. Even within the substantia nigra pars compacta there is differential sensitivity, with dopamine neurons on the dorsal part being less affected than those in the ventral tier *(99)*. In a study of the distribution of astrocytes in the different parts of the substantia nigra of control human brains it was found that there was an inverse relationship between the density of astrocytes present and the extent of loss of dopamine neurons in PD *(75)*. The authors propose that dopamine neurons in astrocyte-rich areas may be protected by neurotrophic factors secreted by glial cells or by the ability of astrocytes to catabolize dopamine and protect from free radical damage *(75)*.

Microglia represent a substantial fraction of all glial cells in the brain (10–20%; *100*) and the substantia nigra is the area with the highest concentration of microglia in the brain *(101)*. Reactive microglia is observed in the substantia nigra of PD patients *(72)* suggesting the existence of an inflammatory response. Activated microglia and reactive astrocytes may play a role in neurodegenerative disorders through the secretion of proinflammatory cytokines, excitotoxins, and free radicals *(80,102,103)*. The presence of proinflammatory cytokines, as well as increased levels of iNOS in glial cells in the substantia nigra in PD *(75)*, suggest that an inflammatory response could be contributing to the pathogenesis of the disease. If GSH depletion in the substantia nigra is indeed an early event in PD as suggested *(104),* one could hypothesize that oxidative stress-induced protein alterations may initiate a glial inflammatory reaction that could contribute to neurodegeneration. Our finding that depletion of GSH in mesencephalic cultures can induce neuronal damage, only in the presence of glial cells *(40),* supports this hypothesis.

In conclusion, glial cells provide protection and support neuronal survival. However, under special conditions, as during oxidative stress or inflammation, they might also secrete toxic compounds that promote neurodegeneration. A balance between neuroprotective and toxic influences is necessary for the maintenance of neuronal survival and function.

REFERENCES

1. Polymeropoulos, M. H., Lavedan, C., Leroy, E. et al. (1997) Mutation in the alpha-synuclein gene identified in families with Parkinson's disease. *Science* **276,** 2045–2047.
2. Polymeropoulos, M. H. (1998) Autosomal dominant Parkinson's disease. *J. Neurol.* **245,** 1–3.
3. Schapira, A. H., Cooper, J. M., Dexter, D., Jenner, P., Clark, J. B., and Marsden, C. D. (1989) Mitochondrial complex I deficiency in Parkinson's disease. *Lancet* **1,** 1269.
4. Schapira, A. H., Cooper, J. M., Dexter, D., Clark, J. B., Jenner, P., and Marsden, C. D. (1990) Mitochondrial complex I deficiency in Parkinson's disease. *J. Neurochem.* **54,** 823–827.
5. Parker, W. D., Jr., Boyson, S. J., and Parks, J. K. (1989) Abnormalities of the electron transport chain in idiopathic Parkinson's disease. *Ann. Neurol.* **26,** 719–723.
6. Hattori, N., Ikebe, S., Tanaka, M., Ozawa, T., and Mizuno, Y. (1993) Immunohistochemical studies on complexes I, II, III, and IV of mitochondria in Parkinson's disease. *Adv. Neurol.* **60,** 292–296.
7. Mytilineou, C., Werner, P., Molinari, S., Di Rocco, A., Cohen, G., and Yahr, M. D. (1994) Impaired oxidative decarboxylation of pyruvate in fibroblasts from patients with Parkinson's disease. *J. Neural Transm. Park. Dis. Dement. Sect.* **8,** 223–228.
8. Swerdlow, R. H., Parks, J. K., Davis, J. N., 2nd, et al. (1998) Matrilineal inheritance of complex I dysfunction in a multigenerational Parkinson's disease family. *Ann. Neurol.* **44,** 873–881.
9. Langston, J. W., Ballard, P., Tetrud, J. W., and Irwin, I. (1983) Chronic Parkinsonism in humans due to a product of meperidine-analog synthesis. *Science* **219,** 979–980.
10. Nicklas, W. J., Vyas, I., and Heikkila, R. E. (1985) Inhibition of NADH-linked oxidation in brain mitochondria by 1-methyl-4- phenyl-pyridine, a metabolite of the neurotoxin, 1-methyl-4-phenyl- 1,2,5,6-tetrahydropyridine. *Life Sci.* **36,** 2503–2508.
11. Barbeau, A., Roy, M., Cloutier, T., Plasse, L., and Paris, S. (1987) Environmental and genetic factors in the etiology of Parkinson's disease. *Adv. Neurol.* **45,** 299–306.
12. Semchuk, K. M., Love, E. J., and Lee, R. G. (1992) Parkinson's disease and exposure to agricultural work and pesticide chemicals. *Neurology* **42,** 1328–1335.
13. Fall, P. A., Fredrikson, M., Axelson, O., and Granerus, A. K. (1999) Nutritional and occupational factors influencing the risk of Parkinson's disease: a case-control study in southeastern Sweden [In Process Citation]. *Mov. Disord.* **14,** 28–37.
14. Cohen, G. (1983) The pathobiology of Parkinson's disease: biochemical aspects of dopamine neuron senescence. *J. Neural Transm. Suppl.* **19,** 89–103.
15. McNaught, K. S., Carrupt, P. A., Altomare, C., et al. (1998) Isoquinoline derivatives as endogenous neurotoxins in the aetiology of Parkinson's disease. *Biochem. Pharmacol.* **56,** 921–933.
16. Hornykiewicz, O., (1966) Metabolism of brain dopamine in human parkinsonism: Neurochemical and clinical aspects. In *Biochemistry and Pharmacology of the Basal Ganglia,* (E. Costa, L. J. C., M. D. Yahr, ed.,) Raven Press, New York, pp. 171–181.

17. Sian, J., Dexter, D. T., Lees, A. J., et al. (1994) Alterations in glutathione levels in Parkinson's disease and other neurodegenerative disorders affecting basal ganglia. *Ann. Neurol.* **36,** 348–355.
18. Perry, T. L., Godin, D. V., and Hansen, S. (1982) Parkinson's disease: a disorder due to nigral glutathione deficiency? *Neurosci. Lett.* **33,** 305–310.
19. Meister, A. (1991) Glutathione deficiency produced by inhibition of its synthesis, and its reversal; applications in research and therapy. *Pharmacol. Ther.* **51,** 155–194.
20. Cohen, G. (1990) Monoamine oxidase and oxidative stress at dopaminergic synapses. *J. Neural Transm. Suppl.* **32,** 229–238.
21. Dexter, D. T., Wells, F. R., Lees, A. J., et al. (1989) Increased nigral iron content and alterations in other metal ions occurring in brain in Parkinson's disease. *J. Neurochem.* **52,** 1830–1836.
22. Riederer, P., Sofic, E., Rausch, W. D., et al. (1989) Transition metals, ferritin, glutathione, and ascorbic acid in parkinsonian brains. *J. Neurochem.* **52,** 515–520.
23. Jellinger, K., Kienzl, E., Rumpelmair, G., et al. (1992) Iron-melanin complex in substantia nigra of parkinsonian brains: an x-ray microanalysis. *J. Neurochem.* **59,** 1168–1171.
24. Hirsch, E. C., Brandel, J. P., Galle, P., Javoy-Agid, F., and Agid, Y. (1991) Iron and aluminum increase in the substantia nigra of patients with Parkinson's disease: an X-ray microanalysis. *J. Neurochem.* **56,** 446–451.
25. Good, P. F., Olanow, C. W., and Perl, D. P. (1992) Neuromelanin-containing neurons of the substantia nigra accumulate iron and aluminum in Parkinson's disease: a LAMMA study. *Brain Res.* **593,** 343–346.
26. Dexter, D. T., Carter, C. J., Wells, F. R., et al. (1989) Basal lipid peroxidation in substantia nigra is increased in Parkinson's disease. *J. Neurochem.* **52,** 381–389.
27. Sanchez-Ramos, J., Overvik, E., and Ames, B. (1994) A marker of oxyradical-mediated DNA damage (8-hydroxy-2′deoxyguanosine) is increased in nigro-striatum of Parkinson's disease brains. *Neurodegeneration* **3,** 197–204.
28. Hillered, L., and Ernster, L. (1983) Respiratory activity of isolated rat brain mitochondria following in vitro exposure to oxygen radicals. *J. Cereb. Blood Flow Metab.* **3,** 207–214.
29. Jain, A., Martensson, J., Stole, E., Auld, P. A., and Meister, A. (1991) Glutathione deficiency leads to mitochondrial damage in brain. *Proc. Natl. Acad. Sci. USA* **88,** 1913–1917.
30. Beal, M. F. (1998) Excitotoxicity and nitric oxide in Parkinson's disease pathogenesis. *Ann. Neurol.* **44,** S110–S114.
31. Milby, K., Oke, A., and Adams, R. N. (1982) Detailed mapping of ascorbate distribution in rat brain. *Neurosci. Lett.* **28,** 169–174.
32. Hughes, R. E., Hurley, R. J., and Jones, P. R. (1971) The retention of ascorbic acid by guinea-pig tissues. *Br. J. Nutr.* **26,** 433–438.
33. Skrzydlewska, E., Witek, A., and Farbiszewski, R. (1998) The comparison of the antioxidant defense potential of brain to liver of rats after methanol ingestion. *Comp. Biochem. Physiol. C. Pharmacol. Toxicol. Endocrinol.* **120,** 289–294.
34. Dalton, T., Pazdernik, T. L., Wagner, J., Samson, F., and Andrews, G. K. (1995) Temporalspatial patterns of expression of metallothionein-I and -III and other stress related genes in rat brain after kainic acid-induced seizures. *Neurochem. Int.* **27,** 59–71.
35. Ebadi, M., Leuschen, M. P., el Refaey, H., Hamada, F. M., and Rojas, P. (1996) The antioxidant properties of zinc and metallothionein. *Neurochem. Int.* **29,** 159–166.
36. Hussain, S., Slikker, W., Jr., and Ali, S. F. (1996) Role of metallothionein and other antioxidants in scavenging superoxide radicals and their possible role in neuroprotection. *Neurochem. Int.* **29,** 145–152.
37. Pileblad, E., Eriksson, P. S., and Hansson, E. (1991) The presence of glutathione in primary neuronal and astroglial cultures from rat cerebral cortex and brain stem. *J. Neural Transm. Gen. Sect.* **86,** 43–49.

38. Han, S. K., Mytilineou, C., and Cohen, G. (1996) L-DOPA up-regulates glutathione and protects mesencephalic cultures against oxidative stress. *J. Neurochem.* **66,** 501–510.
39. Sagara, J., Miura, K., and Bannai, S. (1993) Cystine uptake and glutathione level in fetal brain cells in primary culture and in suspension. *J. Neurochem.* **61,** 1667–1671.
40. Mytilineou, C., Kokotos Leonardi, E. T., Kramer, B. C., Jamindar, T., and Olanow, C. W. (1999) Glial cells mediate toxicity in glutathione-depleted mesencephalic cultures. *J. Neurochem. in press.*
41. Slivka, A., Mytilineou, C., and Cohen, G. (1987) Histochemical evaluation of glutathione in brain. *Brain Res.* **409,** 275–284.
42. Philbert, M. A., Beiswanger, C. M., Waters, D. K., Reuhl, K. R., and Lowndes, H. E. (1991) Cellular and regional distribution of reduced glutathione in the nervous system of the rat: histochemical localization by mercury orange and o-phthaldialdehyde-induced histofluorescence. *Toxicol. Appl. Pharmacol.* **107,** 215–227.
43. Sagara, J. I., Miura, K., and Bannai, S. (1993) Maintenance of neuronal glutathione by glial cells. *J. Neurochem.* **61,** 1672–1676.
44. Prochiantz, A. (1985) Neuronal growth and shape. *Dev. Neurosci.* **7,** 189–198.
45. Engele, J., Schubert, D., and Bohn, M. C. (1991) Conditioned media derived from glial cell lines promote survival and differentiation of dopaminergic neurons in vitro: role of mesencephalic glia. *J. Neurosci. Res.* **30,** 359–371.
46. O'Malley, E. K., Sieber, B. A., Black, I. B., and Dreyfus, C. F. (1992) Mesencephalic type I astrocytes mediate the survival of substantia nigra dopaminergic neurons in culture. *Brain Res.* **582,** 65–70.
47. Park, T. H. and Mytilineou, C. (1992) Protection from l-methyl-4-phenylpyridinium (MPP+) toxicity and stimulation of regrowth of MPP(+)-damaged dopaminergic fibers by treatment of mesencephalic cultures with EGF and basic FGF. *Brain Res.* **599,** 83–97.
48. Bronstein, D. M., Perez-Otano, I., Sun, V., et al. (1995) Glia-dependent neurotoxicity and neuroprotection in mesencephalic cultures. *Brain Res.* **704,** 112–116.
49. Hou, J. G., Cohen, G., and Mytilineou, C. (1997) Basic fibroblast growth factor stimulation of glial cells protects dopamine neurons from 6-hydroxydopamine toxicity: involvement of the glutathione system. *J. Neurochem.* **69,** 76–83.
50. Heikkila, R. E., Youngster, S. K., Manzino, L., Cabbat, F. S., and Duvoisin, R. C. (1985) Effects of l-methyl-4-phenyl-1,2,5,6-tetrahydropyridine and related compounds on the uptake of [3H]3,4-dihydroxyphenylethylamine and [3H]5-hydroxytryptamine in neostriatal synaptosomal preparations. *J. Neurochem.* **44,** 310–313.
51. Javitch, J. A., D'Amato, R. J., Strittmatter, S. M., and Snyder, S. H. (1985) Parkinsonism-inducing neurotoxin, N-methyl-4-phenyl-1,2,3,6 - tetrahydropyridine: uptake of the metabolite N-methyl-4-phenylpyridine by dopamine neurons explains selective toxicity. *Proc. Natl. Acad. Sci. USA* **82,** 2173–2177.
52. Ramsay, R. R., Dadgar, J., Trevor, A., and Singer, T. P. (1986) Energy-driven uptake of N-methyl-4-phenylpyridine by brain mitochondria mediates the neurotoxicity of MPTP. *Life Sci.* **39,** 581–588.
53. Cleeter, M. W., Cooper, J. M., and Schapira, A. H. (1992) Irreversible inhibition of mitochondrial complex I by l-methyl-4-phenylpyridinium: evidence for free radical involvement. *J. Neurochem.* **58,** 786–789.
54. Cohen, G., Allis, B., Winston, B., Mytilineou, C., and Heikkila, R. (1975) Prevention of 6-hydroxydopamine neurotoxicity. *Eur. J. Pharmacol.* **33,** 217–221.
55. Heikkila, R. E., Mytilineou, C., Cote, L. J., and Cohen, G. (1973) The biochemical and pharmacological properties of 6-aminodopamine: similarity with 6-hydroxydopamine. *J. Neurochem.* **21,** 111–116.
56. Rosenberg, P. A. (1988) Catecholamine toxicity in cerebral cortex in dissociated cell culture. *J. Neurosci.* **8,** 2887–2894.

57. Mena, M. A., Pardo, B., Casarejos, M. J., Fahn, S., and Garcia de Yebenes, J. (1992) Neurotoxicity of levodopa on catecholamine-rich neurons. *Mov. Disord.* **7,** 23–31.
58. Mytilineou, C., Han, S. K., and Cohen, G. (1993) Toxic and protective effects of L-dopa on mesencephalic cell cultures. *J. Neurochem.* **61,** 1470–1478.
59. Melamed, E., Offen, D., Shirvan, A., Djaldetti, R., Barzilai, A., and Ziv, I. (1998) Levodopa toxicity and apoptosis. *Ann. Neurol.* **44,** S149–154.
60. Offen, D., Ziv, I., Sternin, H., Melamed, E., and Hochman, A. (1996) Prevention of dopamine-induced cell death by thiol antioxidants: possible implications for treatment of Parkinson's disease. *Exp. Neurol.* **141,** 32–39.
61. Mena, M. A., Casarejos, M. J., Carazo, A., Paino, C. L., and Garcia de Yebenes, J. (1997) Glia protect fetal midbrain dopamine neurons in culture from L-DOPA toxicity through multiple mechanisms. *J. Neural. Transm.* **104,** 317–328.
62. Wilson, J. X. (1989) Ascorbic acid uptake by a high-affinity sodium-dependent mechanism in cultured rat astrocytes. *J. Neurochem.* **53,** 1064–1071.
63. Siushansian, R., Tao, L., Dixon, S. J., and Wilson, J. X. (1997) Cerebral astrocytes transport ascorbic acid and dehydroascorbic acid through distinct mechanisms regulated by cyclic AMP. *J. Neurochem.* **68,** 2378–2385.
64. Siushansian, R. and Wilson, J. X. (1995) Ascorbate transport and intracellular concentration in cerebral astrocytes. *J. Neurochem.* **65,** 41–49.
65. Abe, K. and Saito, H. (1998) Characterization of t-butyl hydroperoxide toxicity in cultured rat cortical neurones and astrocytes. *Pharmacol. Toxicol.* **83,** 40–46.
66. Mischel, R. E., Kim, Y. S., Sheldon, R. A., and Ferriero, D. M. (1997) Hydrogen peroxide is selectively toxic to immature murine neurons in vitro. *Neurosci. Lett.* **231,** 17–20.
67. Desagher, S., Glowinski, J., and Premont, J. (1996) Astrocytes protect neurons from hydrogen peroxide toxicity. *J. Neurosci.* **16,** 2553–2562.
68. Langeveld, C. H., Jongenelen, C. A., Schepens, E., Stoof, J. C., Bast, A., and Drukarch, B. (1995) Cultured rat striatal and cortical astrocytes protect mesencephalic dopaminergic neurons against hydrogen peroxide toxicity independent of their effect on neuronal development. *Neurosci. Lett.* **192,** 13–16.
69. Damier, P., Hirsch, E. C., Zhang, P., Agid, Y., and Javoy-Agid, F. (1993) Glutathione peroxidase, glial cells and Parkinson's disease. *Neuroscience* **52,** 1–6.
70. Delacourte, A., Defossez, A., Ceballos, I., Nicole, A., and Sinet, P. M. (1988) Preferential localization of copper zinc superoxide dismutase in the vulnerable cortical neurons in Alzheimer's disease. *Neurosci. Lett.* **92,** 247–253.
71. Moreno, S., Nardacci, R., and Ceru, M. P. (1997) Regional and ultrastructural immunolocalization of copper-zinc superoxide dismutase in rat central nervous system. *J. Histochem. Cytochem.* **45,** 1611–1622.
72. McGeer, P. L., Itagaki, S., Boyes, B. E., and McGeer, E. G. (1988) Reactive microglia are positive for HLA-DR in the substantia nigra of Parkinson's and Alzheimer's disease brains. *Neurology* **38,** 1285–1291.
73. Banati, R. B., Daniel, S. E., and Blunt, S. B. (1998) Glial pathology but absence of apoptotic nigral neurons in long-standing Parkinson's disease. *Mov. Disord.* **13,** 221–227.
74. Elizan, T. S. and Casals, J. (1991) Astrogliosis in von Economo's and postencephalitic Parkinson's diseases supports probable viral etiology. *J. Neurol. Sci.* **105,** 131–134.
75. Hirsch, E. C., Hunot, S., Damier, P., and Faucheux, B. (1998) Glial cells and inflammation in Parkinson's disease: a role in neurodegeneration? *Ann. Neurol.* **44,** S115–120.
76. Hunot, S., Betard, C., Faucheux, B., Agid, Y., and Hirsch, E. C. (1997) Immunohistochemical analysis of interferon-gamma and interleukin-1 beta in the substantia nigra of Parkinsonian patients. *Movement Disorders* **12 (Suppl. 1),** 20.

77. Boka, G., Anglade, P., Wallach, D., Javoy-Agid, F., Agid, Y., and Hirsch, E. C. (1994) Immunocytochemical analysis of tumor necrosis factor and its receptors in Parkinson's disease. *Neurosci. Lett.* **172,** 151–154.
78. Mogi, M., Harada, M., Riederer, P., Narabayashi, H., Fujita, K., and Nagatsu, T. (1994) Tumor necrosis factor-alpha (TNF-alpha) increases both in the brain and in the cerebrospinal fluid from parkinsonian patients. *Neurosci. Lett.* **165,** 208–210.
79. Mogi, M., Harada, M., Narabayashi, H., Inagaki, H., Minami, M., and Nagatsu, T. (1996) Interleukin (IL)-1 beta, IL-2, IL-4, IL-6 and transforming growth factor-alpha levels are elevated in ventricular cerebrospinal fluid in juvenile parkinsonism and Parkinson's disease. *Neurosci. Lett.* **211,** 13–16.
80. Minghetti, L. and Levi, G. (1998) Microglia as effector cells in brain damage and repair: focus on prostanoids and nitric oxide. *Prog. Neurobiol.* **54,** 99–125.
81. Nagata, K., Takei, N., Nakajima, K., Saito, H., and Kohsaka, S. (1993) Microglial conditioned medium promotes survival and development of cultured mesencephalic neurons from embryonic rat brain. *J. Neurosci. Res.* **34,** 357–363.
82. Chao, C. C., Hu, S., and Peterson, P. K. (1996) Glia: the not so innocent bystanders. *J. Neurovirol.* **2,** 234–239.
83. Tsacopoulos, M. and Magistretti, P. J. (1996) Metabolic coupling between glia and neurons. *J. Neurosci.* **16,** 877–885.
84. Bredt, D. S. and Snyder, S. H. (1992) Nitric oxide, a novel neuronal messenger. *Neuron* **8,** 3–11.
85. Simmons, M. L. and Murphy, S. (1992) Induction of nitric oxide synthase in glial cells. *J. Neurochem.* **59,** 897–905.
86. Stuehr, D. J., Cho, H. J., Kwon, N. S., Weise, M. F., and Nathan, C. F. (1991) Purification and characterization of the cytokine-induced macrophage nitric oxide synthase: an FAD- and FMN-containing flavoprotein. *Proc. Natl. Acad. Sci. USA* **88,** 7773–7777.
87. Nathan, C. (1992) Nitric oxide as a secretory product of mammalian cells. *Faseb J.* **6,** 3051–3064.
88. Bolanos, J. P., Peuchen, S., Heales, S. J., Land, J. M., and Clark, J. B. (1994) Nitric oxide-mediated inhibition of the mitochondrial respiratory chain in cultured astrocytes. *J. Neurochem.* **63,** 910–916.
89. Dawson, V. L., Dawson, T. M., Bartley, D. A., Uhl, G. R., and Snyder, S. H. (1993) Mechanisms of nitric oxide-mediated neurotoxicity in primary brain cultures. *J. Neurosci.* **13,** 2651–2661.
90. Bolanos, J. P., Almeida, A., Stewart, V., et al. (1997) Nitric oxide-mediated mitochondrial damage in the brain: mechanisms and implications for neurodegenerative diseases. *J. Neurochem.* **68,** 2227–2240.
91. Korystov, Y. N., Dobrovinskaya, O. R., Shaposhnikova, V. V., and Eidus, L. (1996) Role of arachidonic acid metabolism in thymocyte apoptosis after irradiation. *FEBS Lett.* **388,** 238–241.
92. Tang, D. G., Chen, Y. Q., and Honn, K. V. (1996) Arachidonate lipoxygenases as essential regulators of cell survival and apoptosis. *Proc. Natl. Acad. Sci. USA* **93,** 5241–5246.
93. Farooqui, A. A., Yang, H. C., Rosenberger, T. A., and Horrocks, L. A. (1997) Phospholipase A2 and its role in brain tissue. *J. Neurochem.* **69,** 889–901.
94. Oka, S. and Arita, H. (1991) Inflammatory factors stimulate expression of group II phospholipase A2 in rat cultured astrocytes. Two distinct pathways of the gene expression. *J. Biol. Chem.* **266,** 9956–9960.
95. Farooqui, A. A. and Horrocks, L. A. (1991) Excitatory amino acid receptors, neural membrane phospholipid metabolism and neurological disorders. *Brain Res. Rev.* **16,** 171–191.
96. McMillian, M. K., Vainio, P. J., and Tuominen, R. K. (1997) Role of protein kinase C in microglia-induced neurotoxicity in mesencephalic cultures. *J. Neuropathol. Exp. Neurol.* **56,** 301–307.
97. Michel, P. P., Ruberg, M., and Agid, Y. (1997) Rescue of mesencephalic dopamine neurons by anticancer drug cytosine arabinoside. *J. Neurochem.* **69,** 1499–1507.

98. Hirsch, E., Graybiel, A. M., and Agid, Y. A. (1988) Melanized dopaminergic neurons are differentially susceptible to degeneration in Parkinson's disease. *Nature* **334,** 345–348.
99. Fearnley, J. M. and Lees, A. J. (1991) Ageing and Parkinson's disease: substantia nigra regional selectivity. *Brain* **114,** 2283–2301.
100. Perry, V. H. (1994) Modulation of microglia phenotype. *Neuropathol. Appl. Neurobiol.* **20,** 177.
101. Lawson, L. J., Perry, V. H., Dri, P., and Gordon, S. (1990) Heterogeneity in the distribution and morphology of microglia in the normal adult mouse brain. *Neuroscience* **39,** 151–170.
102. Benveniste, E. N. (1998) Cytokine actions in the central nervous system. *Cytokine Growth Factor Rev.* **9,** 259–275.
103. Aschner, M. (1998) Astrocytes as mediators of immune and inflammatory responses in the CNS. *Neurotoxicology* **19,** 269–281.
104. Dexter, D. T., Sian, J., Rose, S., et al. (1994) Indices of oxidative stress and mitochondrial function in individuals with incidental Lewy body disease. *Ann. Neurol.* **35,** 38–44.

24
Astrocytic Changes Associated with Epileptic Seizures

Angélique Bordey and Harald Sontheimer

1. INTRODUCTION

Throughout the central nervous system (CNS), most neurons are closely surrounded by astrocytes. In electron microscopic sections, astrocytes appear to encapsulate neuronal cell bodies and astrocytic processes reach close into the vicinity of synapses. This anatomical proximity gives astrocytes privileged access to the neuronal microenvironment and it is believed that astrocytes actively regulate the extracellular space surrounding neurons. Such regulation appears essential to ensure normal neuronal excitability, since even small changes in extracellular potassium (K^+), pH, or the accumulation of neurotransmitters can alter or compromise neuronal function. For example, recordings from hippocampal brain slices showed that modest elevations of extracellular K^+ results in hyperexcitability of hippocampal neurons *(1)*, demonstrating properties that are reminiscent of epileptic seizures. Similarly, compromised astrocytic glutamate uptake, as observed for example in mice lacking astrocytic glutamate transporters, induces epileptic seizures in vivo *(2)*. Thus, fine control of the microenvironment is essential for the maintenance of normal neuronal signaling. To accomplish this role, astrocytes express a number of transport systems, ion channels, and neurotransmitter receptors that are believed to jointly participate in the fine tuning of the neuronal microenvironment (for review, 3–11).

Epilepsy is characterized by spontaneous, synchronized discharges of neuronal populations. Episodes of such activity, also termed epileptic seizures, can originate from a number of conditions including acute trauma, brain oxygen deprivation, tumors, infections, or genetic abnormalities. In at least 50% of patients, no such causal factors can be identified that may have initiated the disease. Although epilepsy is generally considered to be a neuronal disease, astrocytes have long been recognized to show conspicuous morphological changes in conjunction with epilepsy. Thus, hypertrophic astrocytes (astrogliosis) were observed by Chasling in 1889 *(12)*, and in 1927, Penfield *(13)* showed that reactive astrocytes are commonly present in brain regions exhibiting epileptiform activity. To the present date, however, there is no conclusive evidence that astroglial malfunction can induce or contribute to the generation of epileptic seizures. Nonetheless, accumulating evidence suggests that changes in the astrocytes associated

From: *Neuroglia in the Aging Brain*
Edited by: Jean S. de Vellis © Humana Press Inc., Totowa, NJ

with epileptic seizures, may affect the composition of the neuronal microenvironment, which in turn may contribute to the long-term progression of the disease.

In this chapter, we will briefly review some of the most conspicuous morphological and physiological changes of astrocytes that have been associated with epileptic seizures, and will discuss the proposed influence that these changes may have on neuronal function in epilepsy. This is by no means a comprehensive review of the current knowledge of astrocytic changes in epilepsy but instead highlights some of the most evident changes that have been described. This chapter will review how reactive astrocytes contribute to neuronal hyperexcitability by means of chemical imbalances at synapses. We will further describe how reactive astrocytes may contribute to synaptic reorganization and neuronal loss, and how posttraumatic scar formation may contribute to hyperexcitability.

2. ASTROCYTIC CONTROL OF THE NEURONAL MICROENVIRONMENT AND CONTRIBUTIONS TO NEURONAL HYPEREXCITABILITY

For several decades, epilepsy has been studied in animal models or tissue slice models of epilepsy in which seizures were chemically induced. Essentially all these acute models induce a chemical imbalance that alters synaptic transmission in ways that induce repetitive and synchronous neuronal discharges. The simplest of these models chronically depolarizes neurons by bathing slices in elevated K^+ *(1)*. Alternatively, hyperexcitability can be induced by interfering at the level of synaptic communication between neurons, as for example in the bicucculine model of epilepsy, in which gamma-aminobutyric acidergic (GABAergic) inhibition is reduced by blockade of $GABA_A$ receptors *(14,15)*.

In these models, ionic imbalances cannot only induce seizures, but more importantly, the resulting neuronal excitability can itself lead to increases in extracellular K^+ concentration ($[K^+]_o$) and neurotransmitter concentrations. For example, a single action potential can transiently change $[K^+]_o$ by 0.75 m*M*. Repetitive stimulation can result in increases in excess of 10 m*M* *(16)*. Similarly, it is estimated that glutamate concentrations rise to hundreds of micromoles following intense neuronal activity. These changes are transient and ion homeostasis is reestablished within seconds by the combined activity of both neuronal and glial transport systems and ion channels. Any perturbation of the ion homeostatic transport processes are expected to greatly alter neuronal signaling and have long been speculated to contribute to epileptic seizures. Maintenance of extracellular K^+ and glutamate has been the focus of numerous studies aimed at delineating the possible contributions of astrocytes to epileptic seizures, and these studies will be discussed henceforth.

2.1. *Control of Extracellular K^+ by Astrocytes*

The spatial buffer hypothesis for K^+, dating back to the early physiological studies by Kuffler and colleagues *(17)* in the nervous system of the leech, has been extensively reviewed *(18–25)*. This hypothesis proposes the diffusional uptake of K^+ by glial cells and its subsequent spatial redistribution to areas where K^+ concentrations are low. In theory, any elevation of $[K^+]_o$ in the extracellular space (ECS) following neuronal discharge can be dissipated via different mechanisms:

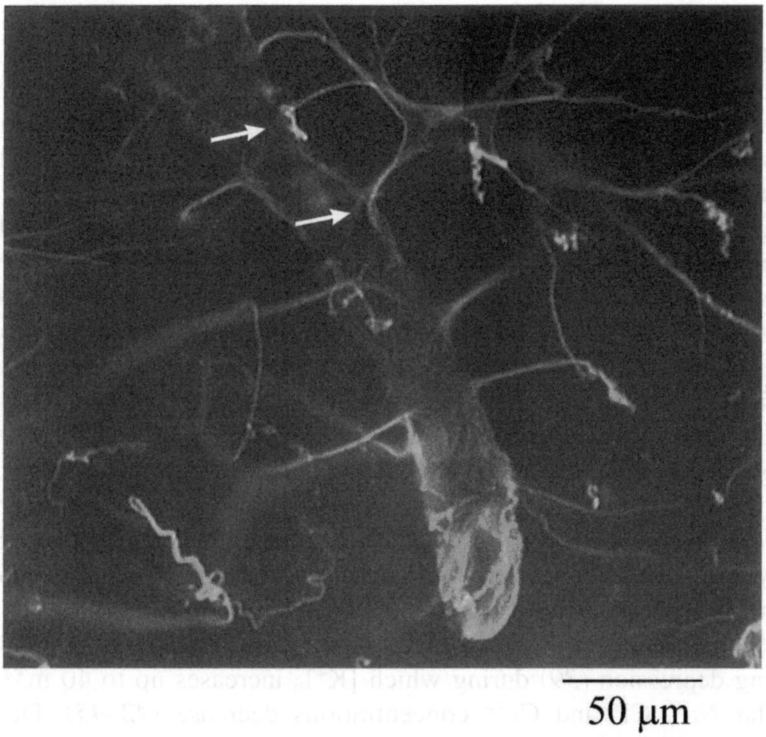

Fig. 1. GFAP antibodies stain astrocytes in CA1 rat hippocampus. Astrocytes display endfeet (white arrows) contacting blood vessels. The staining was performed 14 d after intraperitoneal kainic acid injection.

1. K^+ can simply diffusion through the ECS;
2. K^+ can enter astrocytes via channels or pumps, and
3. K^+ can be taken-up into neurons.

On theoretical grounds, the diffusional dissipation of K^+ through the extracellular space appeared insufficient and has since been rejected experimentally *(26,27)*. However, excellent support remains for the notion that the majority of K^+ enters glial cells rather than neurons. K^+ uptake may occur through ion channels, co-transporters, or the Na^+/K^+-pump and, indeed, evidence exists to implicate all of these mechanisms. For example, in retinal glial cells, K^+ uptake through inwardly rectifying K^+ (Kir) channels, which are abundantly expressed by astrocytes and retinal glial cells, has been demonstrated to be a key mechanism in controlling activity-dependent changes in K^+ *(28,29)*. In cultured astrocytes, a significant uptake also occurs via the $Na^+/K^+/2Cl^-$ co-transporter (for review, *30*.) Activity of the latter results in a net influx of water leading to astrocyte swelling *(27,31–33)*, which is a consistent feature in hyperexcitable tissue. Unquestionably, uptake into both astrocytes and neurons may occur via the Na^+/K^+-ATPase and at least one study has implicated this pathway in the K^+ uptake into axons in optic nerve *in situ (34)*.

Intuitively, neuronal uptake of K^+ would be the most logical mechanism, as K^+ ultimately needs to be replenished to the cell from which it is released. Glial K^+ uptake, however, may have some distinct advantages over neuronal uptake. Most notably, the gap-junction coupling of glial cells allows them to share a large common cytoplasm and promotes the rapid dilution of any ions taken up *(35)*. Moreover gap junctions also facilitate the redistribution of K^+ within the glial syncytium, specifically the transport of K^+ away from the site of uptake, and toward astrocytic endfeet onto blood vessels, which are believed to function as K^+ release sites into the blood stream (Fig.1). Consistent with this theory, clustering of K^+ channels has been observed on astrocytic endfeet *(28)*.

Although astrocytes were long considered to be exclusively permeable to potassium ions, accumulating evidences show that they express a whole series of voltage-activated (VA) ion channels for sodium (Na^{+}), calcium (Ca^{2+}), and chloride (Cl^-) (for review, *5,6,36,37)*. Any local increase in K^+ will thus depolarize the astrocytic membrane possibly activating these voltage-dependent channels. How and if VA ion channels contribute to K^+ buffering is, as of yet, not clear. It has been proposed that entry of Na^+ ions through VA Na^+ channels is required to fuel the astrocytic Na^+/K^+ pump *(38)*.

The notion that a failure to buffer K^+ may underlie epileptic seizures has been around since the late 50s though various studies employing different approaches have come to very divergent conclusions. The concept of high K^+ as a cause epilepsy came from early work with ion-selective microelectrodes demonstrating that physiological increases in $[K^+]_o$ reach a ceiling of abount 10 m*M* *(16)*. When this level is reached, the control mechanisms involving neuroglial interactions breakdown *(19,21–23)*, resulting in spreading depression *(39)* during which $[K^+]_o$ increases up to 40 m*M* *(40,41)* and extracellular Na^+, Cl^- and Ca^{2+} concentrations decrease *(42–45)*. During in vivo seizures in cats, $[K^+]_o$ also increased from 3 m*M* to 10–12 m*M* *(40,46–48)*. These data lead us to ask the two following questions.

2.2. *Are Increases in K^+ Alone Sufficient to Induce Seizures?*

Traynelis and Dingledine (1988) reported that bath application of 8 m*M* K^+ can trigger seizures in the CA1 region of the hippocampus. In 30 to 60 seconds prior to seizure onset, both astrocytes and neurons in CA1 depolarize and there is a concomitant shrinkage of the extracellular space. Seizures can be blocked by the addition of hyperosmotic medium or by application of NMDA receptor antagonists *(49,50)*. It is thought that elevated K^+ depolarizes neurons thereby moving them closer to the threshold for spiking *(51)*. However, elevations of K^+ trigger a whole cascade of events, outlined below, which may contribute to the hyperexcitability.

1. In astrocytes, K^+ influx through transport systems leads to the net influx of K^+, Cl^-, and water resulting in cell swelling and shrinkage of the extracellular space *(20,21,31,32,52–54)*. These effects are also observed during neuronal hyperactivity *(43,55–57)*. The reduction of extracellular space will alter the concentration of extracellular Na^+ and Cl^- ions *(57,58)* and may increase neuronal synchronization due to ephaptic communication *(59)*.
2. Activation of VA ion channels in astrocytes leads to influx of Ca^{2+} and Na^+. VA sodium channels are necessary to fuel the Na^+/K^+ ATPase *(38)* required to extrude K^+ from astrocytes during high neuronal activity *(60,61)*. Astrocytic calcium influx may contribute to the reported decrease in $[Ca^{2+}]_o$ in the interstitial spaces from 1.2 to 0.7 m*M* reported during seizures *(57,62,63)*. The importance of these changes in Ca^{2+} are not entirely known, but several studies have shown that perfusion of hippocampal slices with low Ca^{2+} is sufficient to induce seizures *(64,65)*. Interestingly, several clinically used anticonvulsants including phenytoin, valproate, and ethosuximide attenuate or

inhibit the high potassium-induced Ca^{2+} increases in astrocytes via blockade of astrocytic VA Ca^{2+} channels *(36,66,67).*

3. Depolarization of astrocytes induces the release of aspartate and glutamate by reversal of glutamate transporters, the contribution of which to epilepsy is further discussed below. Briefly, any increases in extracellular glutamate would further activate both neuronal and glial glutamatergic receptors. Moreover, glutamate can also induce cell swelling *(68–70),* which in astrocytes would further enhance glutamate release *(71,72).* Astrocytic swelling appears to be a key element in this cycle and can be mimicked by hypo-osmotic conditions that also induce seizures (for review, *73*). Further evidence that astrocyte swelling is of crucial importance to these events is the finding that furosemide, a chloride co-transporter inhibitor can inhibit seizure activity in hippocampal slices *(74).*

2.3. Are Astrocytes Responsible for the K^+ Imbalance?

The participation of posttraumatic reactive astrocytes in K^+ disturbance and epileptogenesis was first proposed by Pollen and Trachtenberg *(75),* and it has been recently reported that a reduction of K^+ uptake in glial cells can induce epileptiform activity *in situ (76).* In this study, the authors recorded hippocampal astrocytes while monitoring network electrical activity. When they blocked astrocytic potassium channels, in particular Kir channels by cesium, stimulation-induced $[K^+]_o$ increased and synchronous interictal bursting occurred. Interestingly, we reported that these channels were lost in some reactive astrocytes recorded in human epileptic tissue, *in situ (77).* In this study, astrocytes in the epileptic foci also had a depolarized membrane potential because of the loss of Kir channels. This result is in agreement with the early work of Picker et al. who showed that reactive glial cells from patients with intractable, focal epilepsy were less permeable to K^+ than normal glial cells *(78).* As a result, they proposed that gliotic astrocytes associated with epileptic seizure foci are intrinsically less capable of buffering K^+ than normal astrocytes.

These results, however, differ from the conclusions drawn by two earlier studies *(79,80).* Burnard et al. (1990) reported no changes in the resting membrane potentials of reactive astrocytes in kainate-lesioned hippocampal slices, and Glötzner (1973) found that in epileptic foci induced by aluminum hydroxide, feline glial cells could transport even more K^+ away from sites of K^+ release than they do under normal conditions. It is possible that different models of epilepsy could account for this discrepancy. In addition, Glötzner only reported cells displaying a resting potential less than –50 mV although lower potentials were encountered more frequently in gliotic astrocytes. Hence, these studies may have only reported on a subpopulation of astrocytes. In fact, we have reported two populations of reactive astrocytes in epileptic tissue, one with depolarized resting potentials and no Kir channels and another one with increased passive K^+ currents and more hyperpolarized potentials *(77).*

The information reported above, which are summarized in Figure 2 provides little evidence that astrocytes directly initiate seizures. However, they strongly suggest a more indirect participation of astrocytes in the generation of seizures, primarily due to a compromised ability to buffer K^+ within seizure foci.

2.4. Astrocytic Control of Glutamate and Potential Consequences for Epilepsy

Glutamate is the major excitatory neurotransmitter in the CNS. The threshold for activation of glutamate receptors at the postsynaptic membrane is only marginally

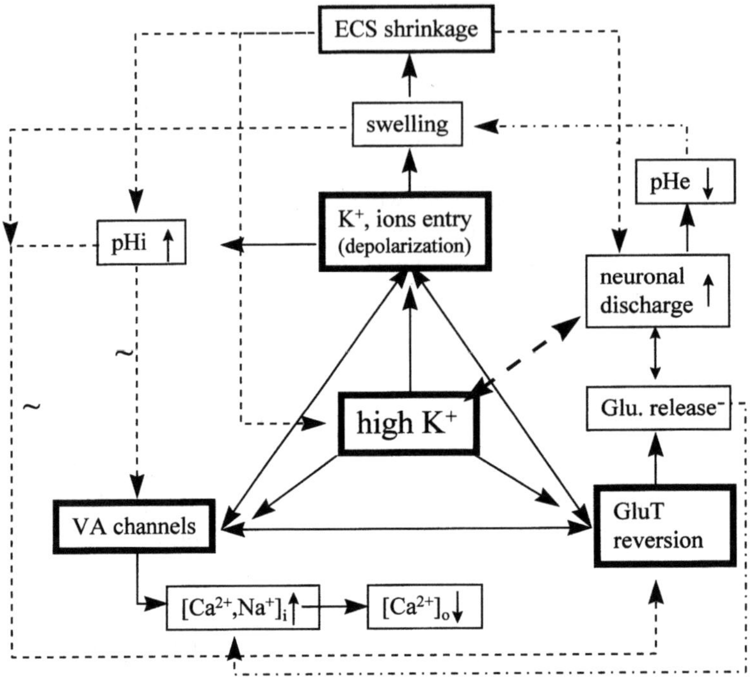

Fig. 2. This schematic displays a hypothesis of how high potassium might initiate hyperexcitability and amplify seizure activity.

above the normal levels of free glutamate in the interstitial space (1 μM) *(81)*. Moreover, even a transient exposure of neurons to elevated glutamate levels has been shown to lead to widespread neurotoxicity *(82–84)*. It is thus essential to keep glutamate concentrations in the interstitial space tightly controlled in order to assure normal synaptic transmission and to avoid excitotoxicity.

Glutamate has been implicated in the pathogenesis of epilepsy *(85–87)* and altered levels of glutamate and aspartate have been reported in patients with epilepsy by microdialysis *(88–91)*. Moreover, a number of experimental animal models for epilepsy present with increased interstitial glutamate levels *(92–98)* supporting the notion that increases in interstitial glutamate may contribute to epilepsy. Increased glutamate levels may be attributed to increased activity of glutamatergic neurons. Yet, under normal physiological conditions, glutamate levels in the brain are tightly controlled by both neuronal and glial uptakes, both expressing Na^+-dependent glutamate transporters. Astrocytes express primarily GLT1 (also called Excitatory Amino Acid Transporter 2, EAAT2) and GLAST (or EAAT1), whereas neurons appear to express primarily EAAC1 (or EAAT3) *(99–102)*. The implication of astrocytic glutamate transporters in epilepsy is supported by recent knockout studies in which the selective loss of GLT1 in vivo leads to markedly increased interstitial glutamate levels that were associated with excitotoxicity *(103)*. In addition, mice lacking the glutamate transporter, GLT-1, exhibited epileptic seizures *(2)*. Neither seizures nor excitotoxicity was observed in EAAC1 knockouts. Thus if compromised transporter function is involved in human epilepsy, it appears most likely that the astrocytic glutamate transporters function is impaired. Changes in the expression levels of astrocytic glutamate transporters are reported in some studies of experimental epilepsy. Thus, in kindling-induced epilepsy in rats, the astrocytic glutamate transporter, GLAST, was downregulated in the piriform cortex/amygdala but not in the hippocampus and GLT-1 was unchanged *(104,105)*. In genetically epilepsy-prone rats, GLT-1 and EAAC1 displayed differences in mRNA levels but there was no reported obvious changes in the expression of the corresponding proteins *(106)*.

Impairment of glutamate transport may not require any changes in the expression levels of the transporter but may be due to functional modulation of uptake. Interestingly, all the cloned glutamate transporters show a strong dependence on transmembrane voltage and gradients for Na^+, K^+, and protons (H^+) (for review *see* refs. *101,102,107–109*). Thus, the above mentioned accumulation of $[K^+]_o$ during seizure activity would be sufficient to greatly reduce glutamate uptake and indeed may even lead to nonvesicular glutamate release through depolarization-induced reversal of the astrocytic glutamate transporters *(110–112)*. A number of other factors have been shown to decrease the glutamate transporters expression and/or activity (for review, *101*). For example, glutamate transport is inhibited by arachidonic acid *(113)*, which is released by neurons during seizure activity *(114)*. Oxygen free radicals and in particular nitric oxide or its metabolites which are formed following NMDA receptor activation, can similarly inhibit astrocytic glutamate transport *(115–119)*. A number of cytokines that have been shown to be activated in animal models of epilepsy are also potent inhibitors of astrocytic glutamate transport *(119,120)*. Finally both cyclic adenosine-monophosphate (cAMP) *(121,122)* and protein kinase C (PKC) *(123,125)* can modulate glial glutamate transporters, and both have been shown to be activated following neurotransmitter receptor activation. Thus collectively, these data suggest that hyperexcitability would, in many ways, negatively feed back on astrocytic glutamate transport, thereby enhancing interstitial glutamate levels.

When one considers the role that astrocytes play in the metabolism/catabolism of glutamate, the contribution of glutamate uptake by astrocytes to the formation of seizures, becomes even more complex. The neuronal metabolism of glutamate depends on the astrocytic supply of glutamine *(126,127)* and tricarboxyl acid (TCA) intermediates (see *128*), which are synthesized by the astrocytic enzymes, glutamine synthetase (GS) *(129,130)* and pyruvate carboxylase *(131–133)*, respectively. Concerning the expression of GS in experimental epilepsy, the data are controversial and vary with species and epileptic model. In the Mongolian seizure prone-gerbils, there was a deficiency in cerebral GS *(134)*, and likewise, in the genetically epilepsy prone-rat, GS activity was significantly lower than in control brain *(135)*. In experimental models of $FeCl_2$-induced focal and kindled amygdaloid seizures, there was an increase and a decrease in GS activity, respectively *(136)*. The authors concluded that increased GS activity is associated with submaximal seizure development, whereas decreased enzyme activity is characteristic of mature, chronic seizures such as those of kindling or epilepsy-prone animals. Concerning the pyruvate carboxylase, no changes have been reported in epilepsy. However, the activity of another enzyme, the aspartic acid aminotransferase, involved in the aspartate metabolism in both astrocytes and neurons, is increased in human epileptic cortical tissue *(137)*.

Another aspect of neurotransmitter accumulation in the synaptic cleft, is the activation of their receptors on the astrocytic membranes. Astrocytes possess receptors for

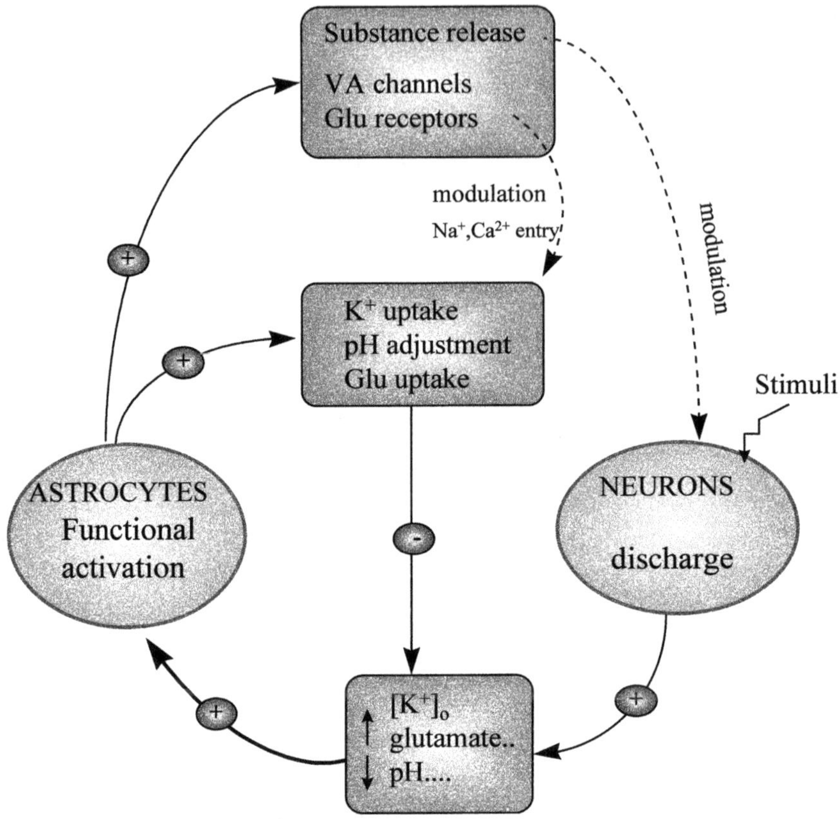

Fig. 3. Simplified and summarized schematic describing the regenerative cycle leading to epileptic seizures.

almost every neurotransmitter present at synapses *(7–10)*. Although the role of these receptors is currently unclear, it has been proposed that they primarily allow astrocytes to sense synaptic activity. It is now clear that activation of glutamate receptors in astrocytes can induce calcium increases and spreading Ca^{2+} waves (discussed in Subheading 2.5.).

The diagram in figure 3 summarizes the activation of glial cells following neuronal discharge and their possible role in restoring normal extracellular conditions suitable for proper neuronal function. This cycle is oversimplified since not only glutamate is released by neurons but many other neurotransmitters can be involved depending on the CNS region. Nevertheless, it sets the starting point from which to discuss the role of neuron-glial interactions in the context of epilepsy.

2.5. Astrocytic Ca^{2+} Waves Triggered by Glutamate

An increasing amount of evidence suggests that neuronally released transmitters and peptides can induce Ca^{2+} responses in astrocytes, and may even lead to spreading Ca^{2+} waves within the glial syncytium. This exciting area of glial physiology is still poorly understood but provides "food for thought" as to the design of future studies aimed at delineating the role of astrocytes in epilepsy. Many transmitters that are released by enhanced neuronal activity, can produce increases in astrocytic intracellular calcium levels *(7–9,138,139)*. For example, glutamate receptor induced depolarization triggers influx of Ca^{2+} through VA Ca^{2+} channels *(37,140–144)*. These channels are upregulated in kainic acid induced-epilepsy *(145)*. Astrocytic muscarinic acetylcholine receptors are likewise upregulated after amygdala kindling *(146)* and induce increases in Ca^{2+} via PKC activation. Any rise in astrocytic Ca^{2+} would have marked and immediate effects on K^+ buffering, astrocyte swelling, and glutamate release, each further enhancing neuronal excitability (for more details about Ca^{2+} roles, see *147*). In addition, these Ca^{2+} rises spread in the form of slowly traveling astrocytic calcium waves *(35,139,147–151)*. The spread of these waves may be facilitated in epileptic tissue as astrocytes from epileptic patients demonstrate increased gap junctional coupling *(152,153)*. The functional implications of Ca^{2+} rises are many-fold and include those typically associated with this important messenger molecule. In the context of astrocyte function in epilepsy, it is easy to envision how increases in intracellular Ca^{2+} ($[Ca^{2+}]_i$) may affect the reactive phenotype, stimulate glycogenolysis, and modulate gene expression. Indeed, an induction of transcription factors and immediate-early genes by increases in $[Ca^{2+}]_i$ has been demonstrated in astrocytes (for review *see* refs. *139,154,155*). These may subsequently induce the long-term changes in gene expression typical of reactive astrocytes that characterize the glial scar often associated with seizure foci.

Overall these data suggest that under seizure conditions astrocytes appear to lose their tight control over extracellular glutamate levels that they maintain in the normal brain. Although this may not be causative for the disease, it certainly may be an important contributing factor to the long-term pathology associated with epilepsy. Increases in Ca^{2+} influx into astrocytes and the spreading of Ca^{2+} waves, may induce expression of a reactive astrocytic phenotype.

3. MORPHOLOGICAL CHANGES INVOLVING ASTROCYTES

In the hippocampi of epileptic patients with medial temporal lobe epilepsy, neuronal reorganizations occur in conjunction with astrocytic hypertrophy. Neuronal cell death has been documented in the hippocampal CA3 (Ammon's horn) and dentate hilus *(12,156,157)*. In addition, mossy fiber sprouting e.g., the sprouting of granule cell axons into the inner molecular layer of the dentate gyrus, and the formation of recurrent excitatory circuits by innervation of granule cell dendrites is a conspicuous feature *(158–160)*. These morphological changes can be partially induced in two experimental models, namely kindling of limbic structures *(161,162)*, and the administration of kainic acid *(163–165)*.

3.1. *Morphological Changes in Astrocytes*

In medial temporal lobe epilepsy, a glial scar is frequently associated with the epileptic seizure focus, and these scars resemble those seen after acute brain injury. Surgical removal of this scar is essential to surgically cure patients from seizures. These scars contain astrocytes that are characterized by hypertrophy and cytoskeleton changes, of which accumulation of intermediate filaments is most conspicuous. The principal protein constituting astrocytic intermediate filaments is glial fibrillary acidic protein (GFAP) *(166)* and its intense immunoreactivity is commonly used to identify astrogliosis (Fig. 4, 5). Such

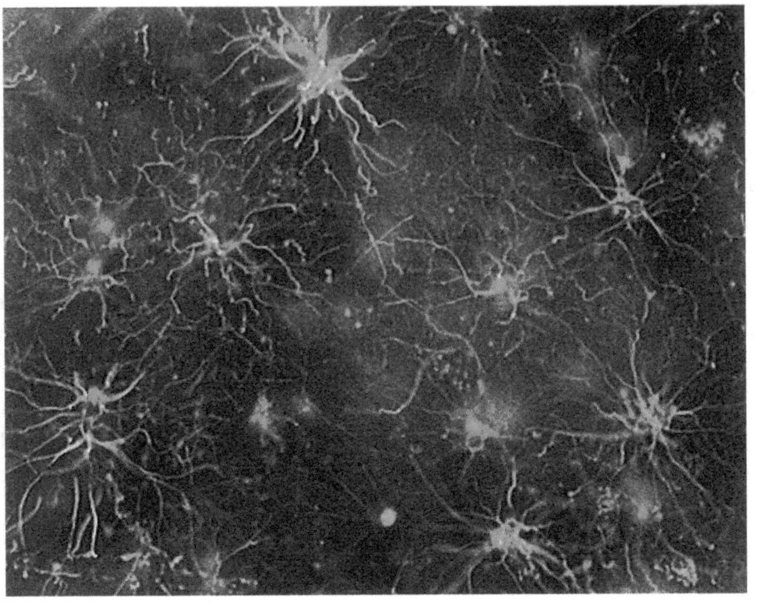

Fig. 4. Astrocytes in slices from a patient biopsy stained intensively for GFAP. The patient was operated for intractable mesio-temporal lobe epilepsy.

astrocytes are termed reactive astrocytes and include at least some proliferative cells *(167,168)*. Interestingly, two distinct populations of reactive astrocytes, one nonproliferative and another proliferative have been observed in the kainic acid-induced seizure model (for review, *163,169,170*). In the hippocampal region associated with neuronal degeneration (CA3 and hilus), astrocytes proliferate whereas in regions with only axonal degeneration (e.g., molecular layer) astrocytes do not proliferate *(163,167,168)*. It has been shown that hypertrophy of astrocytes can result from abnormal neuronal activity and in the dentate gyrus leads to increased GFAP mRNA *(171,172)* and GFAP protein levels *(169,173)*. Astrogliosis seems to be an early event that precedes neuronal degeneration. By contrast, astrocytosis (astroglial proliferation) occurs once the degeneration of neurons has been initiated *(167,168,174,175)*.

3.2. *Trophic Role*

The role that these glial scars play in the vicinity of epileptic seizure foci is not entirely clear. In addition to filling spaces left vacant by dead neurons, astrocytes might release neurotrophic factors (NTF) and cytokines, the levels of which are significantly altered in epileptic tissue (*176–182*, for review *see* ref. *183*). It has been postulated that the sprouting of mossy fibers in conjunction with seizures involves seizure-induced expression of neurotrophic genes *(184,185)*. Increased levels of glial cell-line derived neurotrophic factor (GDNF) released by astrocytes have been reported *(186,187)*, suggesting a potent role for astrocytes in synaptic reorganization *(188,189)*. Other cytokines like IL-6 and S100β display increased levels in cerebrospinal fluid from epileptic patients and in human epileptic cortical tissue *(190,191)*, respectively. Since

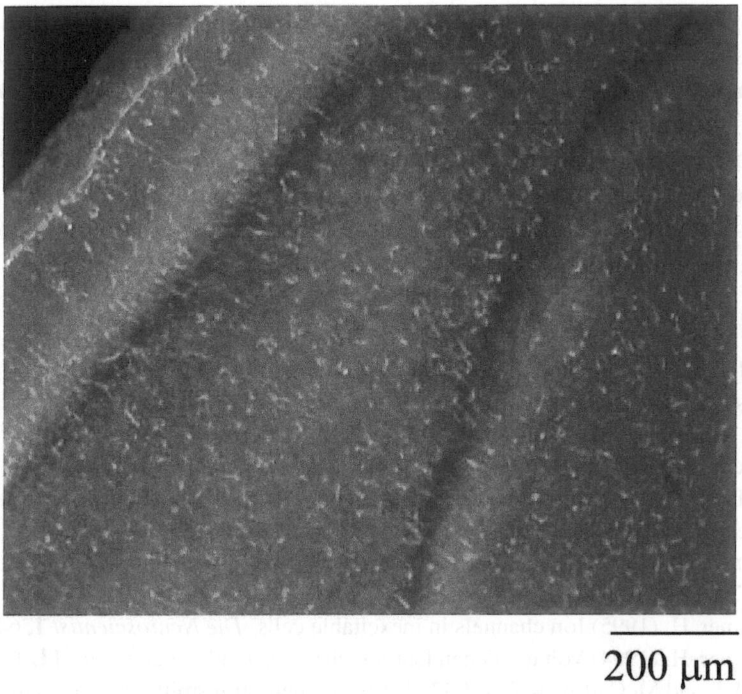

Fig. 5. Distribution of GFAP-immunoreactive astrocytes in the hippocampus (dentate gyrus) of a rat, 14 d after intraperitoneal injection of kainic acid.

cytokines are trophic factors for neurons, enhancing neuronal survival and neurite outgrowth *(192–195),* the over-expression of cytokines by astrocytes may have deleterious consequences, including excessive growth of neurites. In addition, a release of NTF and other cytokines can have an autocrine effect on astrocytes resulting in morphological changes and enhanced proliferation *(196).*

Thus, neuronal reorganizations that occur in the context of epileptic seizures, appear to occur in conjunction with gliosis, and some of the reorganization may be greatly influenced by trophic factors and cytokines released by reactive astrocytes.

An active role of the scar tissue which surrounds the area of cortical damage in epileptogenesis was proposed by Penfield in 1927. He considered the scar tissue as a structural basis of epilepsy leading to the "irritation" of neurons. This concept was also stated by Ward in 1961 *(see 197),* who proposed that astrogliosis caused a mechanical deformation of dendrites, resulting in neuronal membrane depolarization and hyperexcitability. Although this reasoning cannot be discounted entirely, more plausible explanations may exist for neuronal hyperexcitability that include changes in ion homeostasis in the immediate vicinity of the seizure focus. This topic has already been discussed in detail previously.

4. CONCLUSION

Surgical management of intractable epilepsy includes, if possible, the removal of the glial scar associated with the epileptic seizure focus. Much evidence suggests that astrocytes within epileptic tissue show markedly altered abilities in K^+ buffering and in their ability to handle extracellular glutamate. Taken together, these data suggest that astrocytes surrounding epileptic seizure foci may be ill equipped to deal with the enhanced neuronal activity and may thus contribute to the progression of epileptic pathology. The release of neurotrophic factors and cytokines could further contribute to some of the morphological rearrangements commonly observed in epilepsy, and thus, may expand the role astrocytes play beyond that of extracellular space homeostasis. Further studies will hopefully elucidate further evidence that astrocytes are indeed protagonists in this disease.

REFERENCES

1. Traynelis, S.F. and Dingledine, R. (1988) Potassium-induced spontaneous electrographic seizures in the rat hippocampal slice. *J. Neurophysiol.* **59,** 259–276.
2. Tanaka, K., Watase, K., Manabe, T., et al. (1997) Epilepsy and exacerbation of brain injury in mice lacking the glutamate transporter GLT-1. *Science* **276,** 1699–1702.
3. Hertz, L. (1992) Autonomic control of neuronal-astrocytic interactions, regulating metabolic activities, and ion fluxes in the CNS. *Brain Res. Bull.* **29,** 303–313.
4. Clausen, T. (1992) Potassium and sodium transport and pH regulation. *Can. J. Physiol. Pharmacol.* **70,** S219–S222.
5. Sontheimer, H. (1995) Ion channels in inexcitable cells. *The Neuroscientist* **1,** 64–67.
6. Sontheimer, H. (1994) Voltage-dependent ion channels in glial cells. *Glia* **11,** 156–172.
7. Porter, J.T. and McCarthy, K.D. (1997) Astrocytic neurotransmitter receptors *in situ* and *in vivo. Progr. Neurobiol.* **51,** 439–455.
8. Hansson, E. (1989) Co-existence between receptors, carriers, and second messengers on astrocytes grown in primary cultures. *Neurochem. Res.* **14,** 811–819.
9. Wilkin, G.P., Marriott, D.R., Cholewinski, A.J., et al. (1991) Receptor activation and its biochemical consequences in astrocytes. *Ann N.Y. Acad. Sci.* **633,** 475–488.
10. Kimelberg, H.K. (1995) Receptors on astrocytes-what possible functions? *Neurochem. Int.* **26,** 27–40.
11. Deschepper, C.F. (1998) Peptide receptors on astrocytes. *Front. Neuroendocrinol.* **19,** 20–46.
12. Meldrum, B.S. and Bruton, C.J. (1992) Epilepsy. In: *Greenfield's Neuropathology* (Adams, J.H. and Duchen, L.W., ed.), Oxford University Press, New York, pp. 1246–1283.
13. Penfield, W. (1927) The mechanism of cicatricial contraction. *Brain* **50,** 499–517.
14. Hablitz, J.J. (1987) Spontaneous ictal-like discharges and sustained potential shifts in the developing rat neocortex. *J. Neurophysiol.* **58,** 1052–1065.
15. Swann, J.W. and Brady, R.J. (1984) Penicillin-induced epileptogenesis in immature rat CA3 hippocampal pyramidal cells. *Dev. Brain Res.* **12,** 243–254.
16. Heinemann, U. and Lux, H.D. (1977) Ceiling of stimulus induced rises in extracellular potassium concentration in the cerebral cortex of cat. *Brain Res.* **120,** 231–249.
17. Orkand, R.K., Nicholls, J.G. and Kuffler, S.W. (1966) Effect of nerve impulses on the membrane potential of glial cells in the central nervous system of amphibia. *J. Neurophysiol.* **29,** 788–806.
18. Hertz, L. (1965) Possible role of neuroglia: a potassium-mediated neuronal–neuroglial–neuronal impulse transmission system. *Nature* **206,** 1091–1094.
19. Walz, W. and Hertz, L. (1983) Functional interactions between neurons and astrocytes. II. Potassium homeostasis at the cellular level. *Prog. Neurobiol.* **20,** 133–183.
20 Walz, W. and Hertz, L. (1983) Intracellular ion changes of astrocytes in response to extracellular potassium. *J. Neurosci. Res.* **10,** 411–423.
21. Sykova, E. and Chvatal, A. (1993) Extracellular ionic and volume changes: the role of glia-neuron interaction. *J. Chem. Neuroanat.* **6,** 247–260.

22. Sykova, E. (1991) Activity-related ionic and volume changes in neuronal microenvironment. In: *Volume transmission in the brain: novel mechanisms for neural transmission* (Fuxe, K. and Agnati, L.F., ed.), Raven Press, New York, pp. 317–335.
23. Walz, W. (1989) Role of glial cells in the regulation of the brain ion microenvironment. *Prog. Neurobiol.* **33,** 309–333.
24. Sontheimer, H. (1995) Glial neuronal interactions: a physiological perspective. *The Neuroscientist* **1,** 328–337.
25. Amedee, T., Robert, A., and Coles, J., A. (1997) Potassium homeostasis and glial energy metabolism. *Glia* **21,** 46–55.
26. Gardner-Medwin, A.R. (1983) Analysis of potassium dynamics in mammalian brain tissue. *J. Physiol.* **335,** 393–426.
27. Gardner-Medwin, A.R., Coles, J.A., and Tsacopoulos, M. (1981) Clearance of extracellular potassium: evidence for spatial buffering by glial cells in the retina of the drone. *Brain Res.* **209,** 452–457.
28. Newman, E.A. (1993) Inward-rectifying potassium channels in retinal (Muller) cells. *J. Neurosci.* **13,** 3333–3345.
29. Newman, E.A. and Reichenbach, A. (1996) The Muller cell: a functional element of the retina. *Trends Neurosci.* **19,** 307–312.
30. Walz, W. (1992) Role of Na/K/C1 cotransport in astrocytes. *Can. J. Physiol. Pharmacol.* **70,** S260–S262.
31. Walz, W. (1992) Mechanism of rapid $K^{(+)}$-induced swelling of mouse astrocytes. *Neurosci. Letters* **135,** 243–246.
32. Walz, W. (1987) Swelling and potassium uptake in cultured astrocytes. *Can. J. Physiol. Pharmacol.* **65,** 1051–1057.
33. Walz, W. and Hinks, E.C. (1985) Carrier-mediated KCl accumulation accompanied by water movements is involved in the control of physiological K^+ levels by astrocytes. *Brain Res.* **343,** 44–51.
34. Ransom, C.B., Ransom, B.R., and Sontheimer, H. (2000) Activity-dependent K^+ accumulation in rat optic nerve the role of glial and axonal Na+ pumps. *J. Physiol.* **522.3,** 427–442.
35. Giaume, C. and Venance, L. (1998) Intercellular calcium signalling and gap junctional communication in astrocytes. *Glia* **24,** 50–64.
36. White, H.S., Skeen, G.A., and Edwards, J.A. (1994) Pharmacological regulation of astrocytic calcium channels: implications for the treatment of seizure disorders. *Prog. Brain Res.* **94,** 77–87.
37. MacVicar, B.A. (1984) Voltage-dependent calcium channels in glial cells. *Science* **226,** 1345–1347.
38. Sontheimer, H., Fernandez-Marques, E., Ullrich, N., Pappas, C.A., and Waxman, S.G. (1994) Astrocyte Na^+ channels are required for maintenance of $Na^+/K(^+)$- ATPase activity. *J. Neurosci.* **14,** 2464–2475.
39. Leao, A.A.P. (1944) Spreading depression of activity in the cerebral cortex. *J. Neurophysiol.* **7,** 359–390.
40. Moody, W.J., Futamachi, K.J., and Prince, D.A. (1974) Extracellular potassium activity during epileptogenesis. *Exp. Neurol.* **42,** 248–263.
41. Yaari, Y., Konnerth, A., and Heinemann, U. (1986) Nonsynaptic epileptogenesis in the mammalian hippocampus in vitro. II. Role of extracellular potassium. *J. Neurophysiol.* **56,** 424–438.
42. Krnjevic, K., Morris, M.E., and Reiffenstein, R.J. (1980) Changes in extracellular Ca^{2+} and K^+ activity accompanying hippocampal discharges. *Can. J. Physiol. Pharmacol.* **58,** 579–583.
43. Dietzel, I., Heinemann, U., and Lux, H.D. (1989) Relations between slow extracellular potential changes, glial potassium buffering, and electrolyte and cellular volume changes during neuronal hyperactivity in cat brain. *Glia* **2,** 25–44.

44. Nicholson, C. and Kraig, R.P. (1981) The behavior of extracellular ions during spreading depression. In: *The application of ion-selective microelectrodes* (Zeuthen, T., Ed.), Elsevier, amsterdam, pp. 217–238.
45. Somjen, G.G., Aitken, P.G., Giacchino, J.L., and McNamara, J.O. (1986) Interstitial ion concentrations and paroxysmal discharges in hippocampal formation and spinal cord. *Adv. Neurol.* **44,** 663–680.
46. Fisher, R.S., Pedley, T.A., Moody, W.J., and Prince, D.A. (1976) The role of extracellular potassium in hippocampal epilepsy. *Arch. Neurol.* **33,** 76–83.
47. Lothman, E.W. and Somjen, G.G. (1976) Functions of primary afferents and responses of extracellular K^+ during spinal epileptiform seizures. *Electroencephalogr. Clin. Neurophysiol.* **41,** 253–267.
48. Somjen, G.G. (1979) Extracellular potassium in the mammalian central nervous system. *Annu. Rev. Physiol.* **41,** 159–177.
49. Traynelis, S.F. and Dingledine, R. (1989) Role of extracellular space in hyperosmotic suppression of potassium-induced electrographic seizures. *J. Neurophysiol.* **61,** 927–938.
50. Traynelis, S.F. and Dingledine, R. (1989) Modification of potassium-induced interictal bursts and electrographic seizures by divalent cations. *Neurosci. Lett.* **98,** 194–199.
51. Balestrino, M., Aitken, P.G., and Somjen, G.G. (1986) The effects of moderate changes of extracellular K^+ and Ca^{2+} on synaptic and neural function in the CA1 region of the hippocampal slice. *Brain Res.* **377,** 229–239.
52. Kimelberg, H.K., Bourke, R.S., Stieg, P.E., et al. (1982) Swelling of astroglia after injury to the central nervous system: mechanisms and consequences. In: *Head injury: basic and clinical aspects* (Grossman, R.G. and Gilbenberg, P.L., ed.), Raven, New York, pp. 31–44.
53. Kimelberg, H.K. and Frangakis, M.V. (1985) Furosemide- and bumetanide-sensitive ion transport and volume control in primary astrocyte cultures from rat brain. *Brain Res.* **361,** 125–136.
54. Dietzel, I., Heinemann, U., Hofmeier, G., and Lux, H.D. (1980) Transient changes in the size of the extracellular space in the sensorimotor cortex of cats in relation to stimulus-induced changes in potassium concentration. *Exp. Brain Res.* **40,** 432–439.
55. Dietzel, I. and Heinemann, U. (1986) Dynamic variations of the brain cell microenvironment in relation to neuronal hyperactivity. *Ann. N. Y. Acad. Sci.* **481,** 72–85.
56. Dudek, F.E., Obenaus, A., and Tasker, J.G. (1990) Osmolality-induced changes in extracellular volume alter epileptiform burts independent of chemical synapses in the rat: importance of non-synaptic mechanisms in hippocampal epileptogenesis. *Neurosci. Letters* **120,** 267–270.
57. Lux, H.D. and Heinemann, U. (1978) Ionic changes during experimentally induced seizure activity. *Electroencephalogr. Clin. Neurophysiol.* **45,** 289–297.
58. Dietzel, I., Heinemann, U., Hofmeier, G., and Lux, H.D. (1982) Stimulus-induced changes in extracellular Na+ and Cl-concentration in relation to changes in the size of the extracellular space. *Exp. Brain Res.* **46,** 73–84.
59. Jefferys, J.G.R. (1995) Nonsynaptic modulation of neuronal activity in the brain: electric currents and extracellular ions. *Physiol. Rev.* **75,** 689–723.
60. Walz, W. and Hertz, L. (1984) Intense furosemide-sensitive potassium accumulation in astrocytes in the presence of pathologically high extracellular potassium levels. *J. Cereb. Blood Flow Metab.* **4,** 301–304.
61. Walz, W. and Hertz, L. (1982) Ouabain-sensitive and ouabain-resistant net uptake of potassium into astrocytes and neurons in primary cultures. *J. Neurochem.* **39,** 70–77.
62. Heinemann, U., Lux, H.D., and Gutnick, M.J. (1977) Extracellular free calcium and potassium during paroxysmal activity in the cerebral cortex of the cat. *Exp. Brain Res.* **27,** 237–248.
63. Heinemann, U., Konnerth, A., and Lux, H.D. (1981) Stimulation-induced changes in extracellular free calcium in normal cortex and chronic alumina cream foci of cats. *Brain Res.* **213,** 246–250.

64. Jefferys, J.G. and Haas, H.L. (1982) Synchronized bursting of CAl hippocampal pyramidal cells in the absence of synaptic transmission. *Nature* **300,** 448–450.
65. Konnerth, A., Heinemann, U., and Yaari, Y. (1986) Nonsynaptic epileptogenesis in the mammalian hippocampus in vitro. I. Development of seizurelike activity in low extracellular calcium. *J. Neurophysiol.* **56,** 409–423.
66. Macdonald, R.L. (1991) Antiepileptic drug actions on neurotransmitter receptors and ion channels. In: *Neurotransmitters and Epilepsy* (Fisher, R.S. and Coyle, J.T., ed.), Wiley-Liss, New York, pp. 231–245.
67. Edwards, J.A., Woodbury, D.M., and White, H.S. (1991) Anticonvulsants block voltagegated Ca^{2+} channels in astrocytes. *Trans. Am. Soc. Neurochem.* **22,** 138.
68. Bender, A.S., Schousboe, A., Reichelt, W., and Norenberg, M.D. (1998) Ionic mechanisms in glutamate-induced astrocyte swelling: role of K^+ influx. *J. Neurosci. Res.* **52,** 307–321.
69. Hansson, E., Blomstrand, F., Khatibi, S., Olsson, T., and Ronnback, L. (1997) Glutamate induced astroglial swelling-methods and mechanisms. *Acta Neurochir.* **70,** 148–151.
70. Yuan, F. and Wang, T. (1996) Glutamate-induced swelling of cultured astrocytes is mediated by metabotropic glutamate receptor. *Series C. Life Sci.* **39,** 517–522.
71. Rutledge, E.M., Aschner, M., and Kimelberg, H.K. (1998) Pharmacological characterization of swelling-induced D-[^{3}H]aspartate release from primary astrocyte cultures. *Am. J. Physiol.* **273,** C1511–C1520.
72. Rutledge, E.M. and Kimelberg, H.K. (1996) Release of [3H]-D-aspartate from primary astrocyte cultures in response to raised external potassium. *J. Neurosci.* **16,** 7803–7811.
73. Schwartzkroin, P.A., Baraban, S.C., and Hochman, D.W. (1998) Osmolarity, ionic flux, and changes in brain excitability. *Epilepsy Res.* **32,** 275–285.
74. Hochman, D.W., D'Ambrosio, R., Janigro, D., and Schwartzkroin, P.A. (1999) Extracellular chloride and the maintenance of spontaneous epileptiform activity in rat hippocampal slices. *J. Neurophysiol.* **81,** 49–59.
75. Pollen, D.A. and Trachtenberg, M.C. (1970) Neuroglia: gliosis and focal epilepsy. *Science* **167,** 1252–1253.
76. Janigro, D., Gasparini, S., D'Ambrosio, R., McKhann II, G., and DiFrancesco, D. (1997) Reduction of K^+ uptake in glia prevents long-term depression maintenance and causes epileptiform activity. *J. Neurosci.* **17,** 2813–2824.
77. Bordey, A. and Sontheimer, H. (1998) Properties of human glial cells associated with epileptic seizure foci. *Epilepsy Res.* **32,** 286–303.
78. Picker, S., Pieper, C.F., and Goldring, S. (1981) Glial membrane potentials and their relationship to $[K^+]_o$ in man and guinea pig. *J. Neurosurg.* **55,** 347–363.
79. Burnard, D.M., Crichton, S.A., MacVicar, B.A. (1990) Electrophysiological properties of reactive glial cells in the kainate-lesioned hippocampal slice. *Brain Res.* **510,** 43–52.
80. Glotzner, F.L. (1973) Membrane properties of neuroglia in epileptogenic gliosis. *Brain Res.* **55,** 159–171.
81. Zorumski, C.F., Mennerick, S., and Que, J. (1996) Modulation of excitatory synaptic transmission by low concentrations of glutamate in cultured rat hippocampal neurons. *J. Physiol.* **494–2,** 465–477.
82. Cheung, N.S., Pascoe, C.J., Giardina, S.F., John, C.A., and Beart, P.M. (1998) Micromolar L-glutamate induces extensive apoptosis in an apoptotic- necrotic continuum of insult-dependent, excitotoxic injury in cultured cortical neurones. *Neuropharmacol.* **37,** 1419–1429.
83. Sohn, S., Kim, E.Y., and Gwag, B.J. (1998) Glutamate neurotoxicity in mouse cortical neurons: atypical necrosis with DNA ladders and chromatin condensation. *Neurosci. Lett.* **240,** 147–150.
84. Choi, D.W., Maulucci-Gedde, M., and Kriegstein, A.R. (1987) Glutamate neurotoxicity in cortical cell culture. *J. Neurosci.* **7,** 357–368.

85. Engelson, B. (1986) Neurotransmitter glutamate: its clinical importance. *Acta Neurol. Scand.* **74,** 337–355.
86. Meldrum, B.S. (1995) Excitatory amino acid receptors and their role in epilepsy and cerebral ischemia. *Ann. N. Y. Acad. Sci.* **757,** 492–505.
87. Meldrum, B.S. (1994) The role of glutamate in epilepsy and other CNS disorders. *Neurol.* **44,** S14–23.
88. Carlson, H., Ronne-Engstrom, E., Ungerstedt, U., and Hillered, L. (1992) Seizure related elevations of extracellular amino acids in human focal epilepsy. *Neurosci. Letters* **140,** 30–32.
89. Ronne-Engstrom, E., Hillered, L., Flink, R., Spannare, B., Ungerstedt, U., and Carlson, H. (1992) Intracerebral microdialysis of extracellular amino acids in the human epileptic focus. *J. Cereb. Blood Flow Metab.* **12,** 873–876.
90. Perry, T.L. and Hansen, S. (1981) Amino acid abnormalities in epileptogenic foci. *Neurol.* **31,** 872–876.
91. Chapman, A.G., Elwes, R.D., Millan, M.H., Polkey, C.E., and Meldrum, B.S. (1996) Role of glutamate and aspartate in epileptogenesis; contribution of microdialysis studies in animal and man. *Epilepsy Res.* **12,** 239–246.
92. Dodd, P.R. and Bradford, H.F. (1976) Release of amino acids from the maturing cobalt-induced epileptic focus. *Brain Res.* **111,** 377–388.
93. Dodd, P.R., Bradford, H.F., Abdul-Ghani, A.S., Cox, D.W., and Continho-Netto, J. (1980) Release of amino acids from chronic epileptic and subepileptic foci *in vivo. Brain Res.* **193,** 505–517.
94. Wade, J.V., Samson, F.E., Nelson, S.R., and Pazdernik, T.L. (1987) Changes in extracellular amino acids during soman- and kainic acid- induced seizures. *J. Neurochem.* **49,** 645–650.
95. Lehmann, A. (1989) Abnormalities in the levels of extracellular and tissue amino acids in the brain of the seizure-susceptible rat. *Epilepsy Res.* **3,** 130–137.
96. Nilsson, P., Hillered, L., Ponten, U., and Ungerstedt, U. (1990) Changes in cortical extracellular levels of energy-related metabolites and amino acids following concussive brain injury in rats. *J. Cereb. Blood Flow Metab.* **10,** 631–637.
97. Nilsson, P., Ronne-Engstrom, E., Flink, R., Ungerstedt, U., Carlson, H., and Hillered, L. (1994) Epileptic seizure activity in the acute phase following cortical impact trauma in rat. *Brain Res.* **637,** 227–232.
98. Nakase, H., Tada, T., Hashimoto, H., et al. (1994) Experimental study of the mechanism of seizure induction: changes in the concentrations of excitatory amino acids in the epileptic focus of the cat amygdaloid kindling model. *Neurol. Med. Chir. (Tokyo)* **34,** 418–422.
99. Rothstein, J.D., Martin, L., Levey, A.I., et al. (1994) Localization of neuronal and glial glutamate transporters. *Neuron* **13,** 713–725.
100. Lehre, K.P., Levy, L.M., Ottersen, O.P., Storm-Mathisen, J., and Danbolt, N.C. (1995) Differential expression of two glial glutamate transporters in the rat brain: quantitative and immunocytochemical observations. *J. Neurosci.* **15,** 1835–1853.
101. Gegelashvili, G. and Schousboe, A. (1997) High affinity glutamate transporters: regulation of expression and activity. *Mol. Pharmacol.* **52,** 6–15.
102. Gegelashvili, G. and Schousboe, A. (1998) Cellular distribution and kinetic properties of high-affinity glutamate transporters. *Brain Res. Bull.* **45,** 233–238.
103. Rothstein, J.D., Dykes-Hoberg, M., Pardo, C.A., et al. (1996) Knockout of glutamate transporters reveals a major role for astroglial transport in excitotoxicity and clearance of glutamate. *Neuron* **16,** 675–686.
104. Akbar, M.T., Torp, R., Danbolt, N.C., Levy, L.M., Meldrum, B.S., and Ottersen, O.P. (1997) Expression of glial glutamate transporters GLT-1 and GLAST is unchanged in the hippocampus in fully kindled rats. *Neurosci.* **78,** 351–359.
105. Miller, H.P., Levey, A.I., Rothstein, J.D., Tzingounis, A.V., and Conn, P.J. (1997) Alterations in glutamate transporter protein levels in kindling-induced epilepsy. *J. Neurochem.* **68,** 1564–1570.

106. Akbar, M.T., Rattray, M., Williams, R.J., Chong, N.W., and Meldrum, B.S. (1998) Reduction of GABA and glutamate transporter messenger RNAs in the severe-seizure genetically epilepsy-prone rat. *Neurosci.* **85,** 1235–1251.
107. Attwell, D. and Mobbs, P. (1994) Neurotransmitter transporters. *Curr. Opin. Neurobiol.* **4,** 353–359.
108. Zerangue, N. and Kavanaugh, M.P. (1996) Flux coupling in a neuronal glutamate transporter. *Nature* **383,** 634–637.
109. Bouvier, M., Szatkowski, M., Amato, A., and Attwell, D. (1992) The glial cell glutamate uptake carrier countertransports pH-changing anions. *Nature* **360,** 471–474.
110. Szatkowski, M., Barbour, B., and Attwell, D. (1990) Non-vesicular release of glutamate from glial cells by reversed electrogenic glutamate uptake. *Nature* **348,** 443–446.
111. Nichols, D. and Attwell, D. (1990) The release and uptake of excitatory amino acids. *Trends Pharmacol. Sci.* **11,** 462–468.
112. Attwell, D., Barbour, B., and Szatkowski, M. (1993) Nonvesicular release of neurotransmitter. *Neuron* **11,** 401–407.
113. Zerangue, N., Arriza, J.L., Amara, S.G., and Kavanaugh, M.P. (1995) Differential modulation of human glutamate transporter subtypes by arachidonic acid. *J. Biol. Chem.* **270,** 6433–6435.
114. Bazan, N.G., Birkle, D.L., Tang, W., and Reddy, T.S. (1986) The accumulation of free arachidonic acid, diacylglycerols, prostaglandins, and lipoxygenase reaction products in the brain during experimental epilepsy. In: *Advances in Neurology* (Degado-Escueta, A.V., Ward, A.A.J., Woodbury, D.M., and Porter, R.J., ed.), Raven, New York, pp. 879–902.
115. Volterra, A., Trotti, D., and Racagni, G. (1994) Glutamate uptake is inhibited by arachidonic acid and oxygen radicals via two distinct and additive mechanisms. *Mol. Pharmacol.* **46,** 986–992.
116. Volterra, A., Trotti, D., Tromba, C., Floridi, S., and Racagni, G. (1994) Glutamate uptake inhibition by oxygen free radicals in rat cortical astrocytes. *J. Neurosci.* **14,** 2924–2932.
117. Trotti, D., Rossi, D., Gjesdal, O., et al. (1996) Peroxynitrite inhibits glutamate transporter subtypes. *J. Biol. Chem.* **271,** 5976–5979.
118. Volterra, A., Trotti, D., Floridi, S., and Racagni, G. (1994) Reactive oxygen species inhibit high-affinity glutamate uptake: molecular mechanism and neuropathological implications. *Ann. N. Y. Acad. Sci.* **738,** 153–162.
119. Ye, Z.-C. and Sontheimer, H. (1996) Cytokine modulation of glial glutamate uptake: a possible involvement of nitric oxide. *Neuroreport* **7,** 2181–2185.
120. Fine, S.M., Angel, R.A., Perry, S.W., Epstein, L.G., Rothstein, J.D., Dewhurst, S., and Gelbard, H.A. (1996) Tumor necrosis factor alpha inhibits glutamate uptake by primary human astrocytes. Implications for pathogenesis of HIV-1 dementia. *J. Biol. Chem.* **271,** 15303–15306.
121. Eng, D.L., Lee, Y.L., and Lal, P.G. (1997) Expression of glutamate uptake transporters after dibutyryl cyclic AMP differentiation and traumatic injury in cultured astrocytes. *Brain Res.* **778,** 215–221.
122. Schlag, B.D., Vondrasek, J.R., Munir, M., et al. (1998) Regulation of the glial Na^+-dependent glutamate transporters by cyclic AMP analogs and neurons. *Mol. Pharmacol.* **53,** 355–369.
123. Casado, M., Bendahan, A., Zafra, F., et al. (1993) Phosphorylation and modulation of brain glutamate transporters by protein kinase C. *J. Biol. Chem.* **268,** 27313–27317.
124. Casado, M., Zafra, F., Aragon, C., and Gimenez, C. (1991) Activation of high-affinity uptake of glutamate by phorbol esters in primary glial cell cultures. *J. Neurochem.* **57,** 1185–1190.
125. Conradt, M. and Stoffel, W. (1997) Inhibition of the high-affinity brain glutamate transporter GLAST-1 via direct phosphorylation. *J. Neurochem.* **68,** 1244–1251.
126. Balazs, R., Patel, A.J., and Richter, D. (1972) Metabolic compartments in the brain: their properties and relation to morphological structures. In: *Metabolic Compartmentation in the brain.* (Balazs, R. and Cremer, J.E., eds.), MacMillan, New York, pp. 167–184.

127. Benjamin, A.M. and Quastel, J.H. (1974) Fate of L-glutamate in the brain. *J. Neurochem.* **23,** 457–464.
128. Shank, R.P. and Aprison, M.H. (1988) Glutamate as a neurotransmitter. In: *Glutamine and Glutamate in Mammals* (Kvamme, E., ed.). CRC Press, Boca Raton, FL, pp. 3–19.
129. Yamamoto, H., Konno, H., Yamamoto, T., Ito, K., Mizugaki, M., and Iwasaki, Y. (1987) Glutamine synthetase of the human brain: purification and characterization. *J. Neurochem.* **49,** 603–609.
130. Martinez-Hernandez, A., Bell, K.P., and Norenberg, M.D. (1977) Glutamine synthetase: glial localization in brain. *Science* **195,** 1356–1358.
131. Gamberino, W.C., Berkich, D.A., Lynch, C.J., Xu, B., and LaNoue, K.F. (1997) Role of pyruvate carboxylase in facilitation of synthesis of glutamate and glutamine in cultured astrocytes. *J. Neurochem.* **69,** 2312–2325.
132. Yu, A.C., Drejer, J., Hertz, L., and Schousboe, A. (1983) Pyruvate carboxylase activity in primary cultures of astrocytes and neurons. *J. Neurochem.* **41,** 1484–1487.
133. Shank, R.P., Bennett, G.S., Freytag, S.O., and Campbell, G.L. (1985) Pyruvate carboxylase: an astrocyte-specific enzyme implicated in the replenishment of amino acid neurotransmitter pools. *Brain Res.* **329,** 364–367.
134. Laming, P.R., Cosby, S.L., and O'Neill, J.K. (1989) Seizures in the Mongolian gerbil are related to a deficiency in cerebral glutamine synthetase. *Comp. Biochem. Physiol.* **94,** 399–404.
135. Carl, G.F., Thompson, L.A., Williams, J.T., Wallace, V.C., and Gallagher, B.B. (1992) Comparison of glutamine synthetases from brains of genetically epilepsy prone and genetically epilepsy resistant rats. *Neurochem. Res.* **17,** 1015–1019.
136. Tiffany-Castiglioni, E.C., Peterson, S.L., and Castiglioni, A.J. (1990) Alterations in glutamine synthetase activity by $FeCl_2$-induced focal and kindled amygdaloid seizures. *J. Neurosci. Res.* **25,** 223–228.
137. Kish, S.J., Dixon, L.M., and Sherwin, A.L. (1988) Aspartic acid aminotransferase activity is increased in actively spiking compared with non-spiking human epileptic cortex. *J. Neurol. Neurosurg. Psychiatry* **51,** 552–556.
138. Duffy, S. and MacVicar, B.A. (1995) Adrenergic calcium signaling in astrocyte networks within the hippocampal slice. *J. Neurosci.* **15,** 5535–5550.
139. Finkbeiner, S.M. (1993) Glial calcium. *Glia* **9,** 83–104.
140. Barres, B.A., Chun, L.L., and Corey, D.P. (1989) Calcium current in cortical astrocytes: induction by cAMP and neurotransmitters and permissive effect of serum factors. *J. Neurosci.* **9,** 3169–3175.
141. MacVicar, B.A., Hochman, D., Delay, M.J., and Weiss, S. (1991) Modulation of intracellular Ca^{2+} in cultured astrocytes by influx through voltage-activated Ca^{2+} channels *Glia* **4,** 448–455.
142. Duffy, S. and MacVicar, B.A. (1994) Potassium-dependent calcium influx in acutely isolated hippocampal astrocytes. *Neurosci.* **61,** 51–61.
143. Duffy, S. and MacVicar, B.A. (1996) *In vitro* ischemia promotes calcium influx and intracellular calcium release in hippocampal astrocytes. *J. Neurosci.* **16,** 71–81.
144. Akopian, G., Kressin, K., Derouiche, A., and Steinhauser, C. (1996) Identified glial cells in the early postnatal mouse hippocampus display different types of Ca^{2+} currents. *Glia* **17,** 181–194.
145. Westenbroek, R.E., Bausch, S.B., Lin, R.C., Franck, J.E., Noebels, J.L., and Catterall, W.A. (1998) Upregulation of L-type Ca^{2+} channels in reactive astrocytes after brain injury, hypomyelination, and ischemia. *J. Neurosci.* **18,** 2321–2334.
146. Beldhuis, H.J., A., Everts, H.G., J., Van der Zee, E.A., Luiten, P.G., M., and Bohus, B. (1992) Amydala kindling-induced seizures selectively impair spatial memory. 2. Effects on hippocampal neuronal and glial muscarinic acetylcholine receptor. *Hippocampus* **2,** 411–420.
147. Cooper, M.S. (1995) Intercellular signaling in neuronal-glial networks. *Biosystems* **34,** 65–85.

148. Jensen, A.M. and Chiu, S.Y. (1990) Fluorescence measurements of changes in intracellular calcium induced by excitatory amino acids in cultured cortical astrocytes. *J. Neurosci.* **10,** 1165–1175.
149. Cornell-Bell, A.H., Finkbeiner, S.M., Cooper, M.S., and Smith, S.J. (1990) Glutamate induces calcium waves in cultured astrocytes: long-range glial signaling. *Science* **247,** 470–473.
150. Cornell-Bell, A.H. and Finkbeiner, S.M. (1991) Calcium waves in astrocytes. *Cell Calcium* **12,** 185–204.
151. Froes, M.M. and de Carvalho, A.C. (1998) Gap junction-mediated loops of neuronal-glial interactions. *Glia* **24,** 97–107.
152. Lee, S.H., Magge, S., Spencer, D.D., Sontheimer, H., and Cornell-Bell, A.H. (1995) Human epileptic astrocytes exhibit increased gap junction coupling. *Glia* **15,** 195–202.
153. Manning, T.J.J. and Sontheimer, H. (1997) Spontaneous intracellular calcium oscillations in cortical astrocytes from a patient with intractable childhood epilepsy (Rasmussen's encephalitis). *Glia* **21,** 332–337.
154. Roche, E. and Prentki, M. (1994) Calcium regulation of immediate-early response genes. *Cell Calcium* **16,** 331–338.
155. Hisanaga, K., Sagar, S.M., and Sharp, F.R. (1992) C-fos induction occurs in cultured cortical neurons and astrocytes via multiple signaling pathways. *Prog. Brain Res.* **94,** 189–195.
156. McNamara, J.O. (1994) Cellular and molecular basis of epilepsy. *J. Neurosci.* **14,** 3413–3425.
157. Bouchet, C. and Cazauveilh, C. (1825) De l'epilepsie consideree dans ses rapports avec l'alienation mentale. *Arch. G. M.* **9,** 510–542.
158. Castiglioni, A.J., Peterson, S.L., Sanabria, E.L., and Tiffany-Castiglioni, E. (1990) Structural changes in astrocytes induced by seizures in a model of temporal lobe epilepsy. *J. Neurosci. Res.* **26,** 334–341.
159. Houser, C.R. (1992) Morphological changes in the dentate gyrus in human temporal lobe epilepsy. *Epilepsy Res.* **7,** 223–234.
160. Cavazos, J.E., Golarai, G., and Sutula, T.P. (1991) Mossy fiber synaptic reorganization induced by kindling: time course of development, progression, and permanence. *J. Neurosci.* **11,** 2795–2803.
161. Goddard, G.V., McIntyre, D.C., and Leech, C.K. (1969) A permanent change in brain function resulting from daily electrical stimulation. *Exp. Neurol.* **25,** 295–330.
162. Goddard, G.V. (1967) Development of epileptic seizures through brain stimulation at low intensity. *Nature* **214,** 1020–1021.
163. Represa, A., Niquet, J., Pollard, H., and Ben-Ari, Y. (1995) Cell death, gliosis, and synaptic remodelling in the hippocampus of epileptic rats. *J. Neurobiol.* **26,** 413–425.
164. Nadler, J.V. (1981) Minireview. Kainic acid as a tool for the study of temporal lobe epilepsy. *Life Sci.* **29,** 2031–2042.
165. Ben-Ari, Y. (1985) Limbic seizure and brain damage produced by kainic acid: mechanisms and relevance to human temporal lobe epilepsy. *Neurosci.* **14,** 375–403.
166. Eng, L.F. (1985) Glial fibrillary acidic protein (GFAP): the major protein of glial intermediate filaments in differentiated astrocytes. *J. Neuroimmunol.* **8,** 203–214.
167. Niquet, J., Ben-Ari, Y., and Represa, A. (1994) Glial reaction after seizure induced hippocampal lesion: immunohistochemical characterization of proliferating glial cells. *J. Neurocytol.* **23,** 641–656.
168. Niquet, J., Jorquera, I., Ben-Ari, Y., and Represa, A. (1994) Proliferative astrocytes may express fibronectin-like protein in the hippocampus of epileptic rats. *Neurosci. Letters* **180,** 13–16.
169. Adams, B., Von Ling, E., Vaccarella, L., Ivy, G.O., Fahnestock, M., and Racine, R.J. (1998) Time course for kindling-induced changes in the hilar area of the dentate gyrus: reactive gliosis as a potential mechanism. *Brain Res.* **804,** 331–336.

170. Khurgel, M. and Ivy, G.O. (1996) Astrocytes in kindling: relevance to epileptogenesis. *Epilepsy Res.* **26,** 163–175.
171. Kelley, M.S. and Steward, O. (1993) The role of neuronal activity in upregulating GFAP mRNA levels after electrolytic lesions of the entorhinal cortex. *Int. J. Dev. Neurosci.* **11,** 105–115.
172. Torre, E.R., Lothman, E.W., and Steward, O. (1993) Glial response to neuronal activity: GFAP-mRNA and protein levels are transiently increased in the hippocampus after seizures. *Brain Res.* **631,** 256–26.
173. Hansen, A., Jorgensen, O.S., Bolwig, T.G., and Barry, D.I. (1991) Hippocampal kindling in the rat is associated with time-dependent increases in the concentration of glial fibrillary acidic protein. *J. Neurochem.* **57,** 1716–1720.
174. Represa, A., Niquet, J., Charriaut-Marlangue, C., and Ben-Ari, Y. (1993) Reactive astrocytes in the kainic acid-damaged hippocampus have the phenotypic features of type-2 astrocytes. *J. Neurocytol.* **22,** 299–310.
175. Represa, A., Jorquera, I., Le Gal La Salle, G., and Ben-Ari, Y. (1993) Epilepsy induced collateral sprouting of hippocampal mossy fibers: does it induce the development of ectopic synapses with granule cell dendrites? *Hippocampus* **3,** 257–268.
176. Mathern, G.W., Babb, T.L., Micevych, P.E., Blanco, C.E., and Pretorius, J.K. (1997) Granule cell mRNA levels for BDNF, NGF, and NT-3 correlate with neuron losses or supragranular mossy fiber sprouting in the chronically damaged and epileptic human hippocampus. *Mol. Chem. Neuropathol.* **30,** 53–76.
177. Akoev, G.N., Chalisova, N.I., Ludino, M.I., Terent'ev, D.A., Yatsuk, S.L., and Romanjuk, A.V. (1996) Epileptiform activity increases the level of nerve growth factor in cerebrospinal fluid of epileptic patients and in hippocampal neurons in tissue culture. *Neurosci.* **75,** 601–605.
178. Van Der Wal, E.A., Gomez-Pinilla, F., and Cotman, C.W. (1994) Seizure-associated induction of basic fibroblast growth factor and its receptor in the rat brain. *Neurosci.* **60,** 311–323.
179. Bugra, K., Pollard, H., Charton, G., Moreau, J., Ben-Ari, Y., and Khrestchatisky, M. (1994) aFGF, bFGF and flg mRNAs show distinct patterns of induction in the hippocampus following kainate-induced seizures. *Eur. J. Neurosci.* **6,** 58–66.
180. Takahashi, M., Hayashi, S., Kakita, A., et al. (1999) Patients with temporal lobe epilepsy show an increase in brain-derived neurotrophic factor protein and its correlation with neuropeptide Y. *Brain Res.* **818,** 579–582.
181. Liang, F., Le, L.D., and Jones, E.G. (1998) Reciprocal up- and down-regulation of BDNF mRNA in tetanus toxin-induced epileptic focus and inhibitory surround in cerebral cortex. *Cereb. Cortex.* **8,** 481–491.
182. Kar, S., Seto, D., Dore, S., Chabot, J.G., and Quirion, R. (1997) Systemic administration of kainic acid induces selective time dependent decrease in [^{25}I]insulin-like growth factor I, [^{125}I]insulin-like growth factor II and [125I]insulin receptor binding sites in adult rat hippocampal formation. *Neurosci.* **80,** 1041–1055.
183. Gall, C.M., Lauterborn, J.C., Guthrie, K.M., and Stinis, C.T. (1997) Seizures and the regulation of neurotrophic factor expression: associations with structural plasticity in epilepsy. *Adv. Neurol.* **72:9–24,** 9–24.
184. Watanabe, Y., Johnson, R.S., Butler, L.S., et al. (1996) Null mutation of c-fos impairs structural and functional plasticities in the kindling model of epilepsy. *J. Neurosci.* **16,** 3827–3836.
185. Adams, B., Sazgar, M., Osehobo, P., et al. (1997) Nerve growth factor accelerates seizure development, enhances mossy fiber sprouting, and attenuates seizure-induced decreases in neuronal density in the kindling model of epilepsy. *J. Neurosci.* **17,** 5288–5296.
186. Schmidt-Kastner, R., Tomac, A., Hoffer, B., Bektesh, S., Rosenzweig, B., and Olson, L. (1994) Glial cell-line derived neurotrophic factor (GDNF) mRNA upregulation in striatum and cortical areas after pilocarpine-induced status epilepticus in rats. *Brain Res. Mol. Brain Res.* **26,** 325–330.

187. Reeben, M., Laurikainen, A., Hiltunen, J.O., Castren, E., and Saarma, M. (1998) The messenger RNAs for both glial cell line-derived neurotrophic factor receptors, c-ret and GDNFRalpha, are induced in the rat brain in response to kainate-induced excitation. *Neuroscience* **83,** 151–159.
188. Ebendal, T., Tomac, A., Hoffer, B.J., and Olson, L. (1995) Glial cell line-derived neurotrophic factor stimulates fiber formation and survival in cultured neurons from peripheral autonomic ganglia. *J. Neurosci. Res.* **40,** 276–284.
189. Widenfalk, J., Nosrat, C., Tomac, A., Westphal, H., Hoffer, B., and Olson, L. (1997) Neurturin and glial cell line-derived neurotrophic factor receptor-beta (GDNFR-beta), novel proteins related to GDNF and GDNFR-alpha with specific cellular patterns of expression suggesting roles in the developing and adult nervous system and in peripheral organs. *J. Neurosci.* **17,** 8506–8519.
190. Peltola, J., Hurme, M., Miettinen, A., and Keranen, T. (1998) Elevated levels of interleukin-6 may occur in cerebrospinal fluid from patients with recent epileptic seizures. *Epilepsy Res.* **31,** 129–133.
191. Griffin, W.S., Yeralan, O., Sheng, J.G., et al. (1995) Overexpression of the neurotrophic cytokine S100 beta in human temporal lobe epilepsy. *J. Neurochem.* **65,** 228–233.
192. Ebadi, M., Bashir, R.M., Heidrick, M.L., et al. (1997) Neurotrophins and their receptors in nerve injury and repair. *Neurochem. Int.* **30,** 347–374.
193. Barker, P.A. and Murphy, R.A. (1992) The nerve growth factor receptor: a multicomponent system that mediates the actions of the neurotrophin family of proteins. *Mol. Cell Biochem.* **110,** 1–15.
194. Casaccia-Bonnefil, P., Kong, H., and Chao, M.V. (1998) Neurotrophins: the biological paradox of survival factors eliciting apoptosis. *Cell Death Differ.* **5,** 357–364.
195. Maness, L.M., Kastin, A.J., Weber, J.T., Banks, W.A., Beckman, B.S., and Zadina, J.E. (1994) The neurotrophins and their receptors: structure, function, and neuropathology. *Neurosci. Biobehav. Rev.* **18,** 143–159.
196. Otero, G.C. and Merrill, J.E. (1994) Cytokine receptors on glial cells. *Glia* **11,** 117–129.
197. Tower, D.B. (1992) A century of neuronal and neuroglial interactions, and their pathological implications: an overview. *Prog. Brain Res.* **94,** 3–17.

25
Synaptic and Neuroglial Pathobiology in Acute and Chronic Neurological Disorders

Lee J. Martin

1. INTRODUCTION

Pathology at the molecular and cellular level is the fundamental basis of disorders that effect the brain or spinal cord of humans. A pathophysiological event can target specific molecules and groups of cells, causing neuronal dysfunction and degeneration and resulting in acute and chronic neurological and behavioral disabilities ranging from memory loss to paralysis. For example, Alzheimer's disease (AD) is associated with loss of forebrain neurons and the formation of brain lesions consisting of abnormal deposits of glial- or -neuronal generated amyloid β protein, possibly in response to gene mutations, synaptic perturbations, oxidative stress, or neuronal cytoskeletal defects *(1–5)*. The degeneration of motor neurons in amyotrophic lateral sclerosis (ALS) is associated with impairments in glutamate reuptake by astroglia *(6,7)*, gene mutations in superoxide dismutase-1 *(8)*, and abnormal apoptosis of motor neurons that appears to be mediated by programmed cell death mechanisms *(9,10)*. The nerve cell damage in adults and children that have experienced cardiac arrest, asphyxiation, strokes, and head or spinal cord trauma may be caused by failure of astroglial glutamate transport, excessive stimulation of glutamate receptors, abnormal activation or impaired function of intracellular signaling pathways, toxic generation of free oxygen radicals, and structural damage to target molecules within selectively vulnerable populations of neurons *(11)*. Epidemiological studies reveal the impact that neurodegenerative disorders have on our society. For example, AD affects ~4 million adults (most are >65-yr-of-age) and is the fourth-leading cause of death in the United States, accounting for >100,000 deaths annually *(12)*. ALS affects approx 30,000 Americans (4–6 people in 100,000) *(13)*. Stroke is the third leading cause of death in industrialized populations and is a major cause of long term neurological disability *(14)*.

No cures or therapies (e.g., synthetic drugs or biological factors) are available yet that can prevent the degeneration of neurons in the human brain and spinal cord and that can improve the quality of life for individuals with these neurological disorders. The current inability to effectively and rationally manage and treat individuals with these nervous system abnormalities is due to insufficient information on how neurons and glia become dysfunctional or degenerate and the synaptic and molecular mecha-

From: *Neuroglia in the Aging Brain*
Edited by: Jean S. de Vellis © Humana Press Inc., Totowa, NJ

nisms for this pathobiology. This chapter will focus on synaptic and neuroglial perturbations in two broad settings of neurodegeneration: Acute neuronal cell death after cerebral ischemia and chronic, progressive neuronal degeneration that occurs in age-related disorders of the nervous system, such as AD and ALS. Treatments for these disorders will be forthcoming eventually through a more complete understanding of the normal cellular and molecular organization of the nervous system and of cellular and molecular neuropathology.

2. ORGANIZATION OF THE CENTRAL SYNAPSE

2.1. Structural and Molecular Design of the Synapse

Synapses are the principal units of intercellular communication in neural circuits. The structural organization of a central nervous system (CNS) synapse consists of a presynaptic component (the axon terminal), a synaptic cleft (~20–40 nm in width), a postsynaptic component (a neuronal cell body or process), and an astroglial sheath (Fig. 1). Each of these components has its own elaborate structural organization and molecular specifications. The presynaptic and postsynaptic membranes have thickenings on their cytoplasmic surfaces (Fig. 1), and these specialized membranes together with the synaptic cleft form the synaptic junction. These membrane specializations are the presynaptic grid or active zone and the postsynaptic density.

The presynaptic nerve terminal (Fig. 1A), containing several hundred neurotransmitter vesicles, is the principal site of regulated release of excitatory or inhibitory neurotransmitters. A complex ensemble of proteins (Fig. 1B) within the presynaptic nerve terminal functions in exocytosis of neurotransmitter-containing vesicles *(15)*. Synaptic vesicles dock at the active zone. Specific proteins function in neurotransmitter vesicle docking, priming, fusion, and exocytosis and endocytosis. This process is called the synaptic vesicle cycle *(15)*.

Dendrites receive the vast majority of excitatory synapses in the mammalian CNS (Figs. 1 and 2). Like the presynaptic terminal, the postsynaptic element contains an equally staggering array of proteins that function in inter- and intracellular signaling. These proteins are neurotransmitter receptors (ionotropic and metabotropic receptors), ion channels (Ca^{2+}, Na^{+}, K^{+}, and Cl^{-} channels), and other signal transduction molecules, including protein kinases, protein phosphatases, heterotrimeric GTP-binding proteins (G-proteins), phospholipases, nitric oxide synthase (NOS), calmodulin, and cytoskeletal proteins (Fig. 1B). At the postsynaptic site, the postsynaptic density is an

Fig. 1. ***(Right)*** Organization of a central nervous system synapse **(A)** Electron micrograph illustrating the synaptic complex. Two presynaptic axon terminals (t), containing numerous neurotransmitter vesicles (arrows) which are ~50 nm in diameter, form synaptic junctions with dendrites (d). The dark thickening at the dendritic side of the junction is the postsynaptic density. The synaptic cleft of one junction is identified (arrowheads). Astroglial processes (asterisks) ensheathe the synaptic complex. **(B)** Simplified synaptic organization in forebrain as represented by an presynaptic axon terminal (with synaptic vesicles and mitochondria) forming an asymmetrical synapse with a dendrite (seen in cross sectional profile). An astrocyte partially envelops the synapse. Each component of the central synapse has a complex ensemble of proteins integral to their function. Although the lists are incomplete, some important proteins are identified.

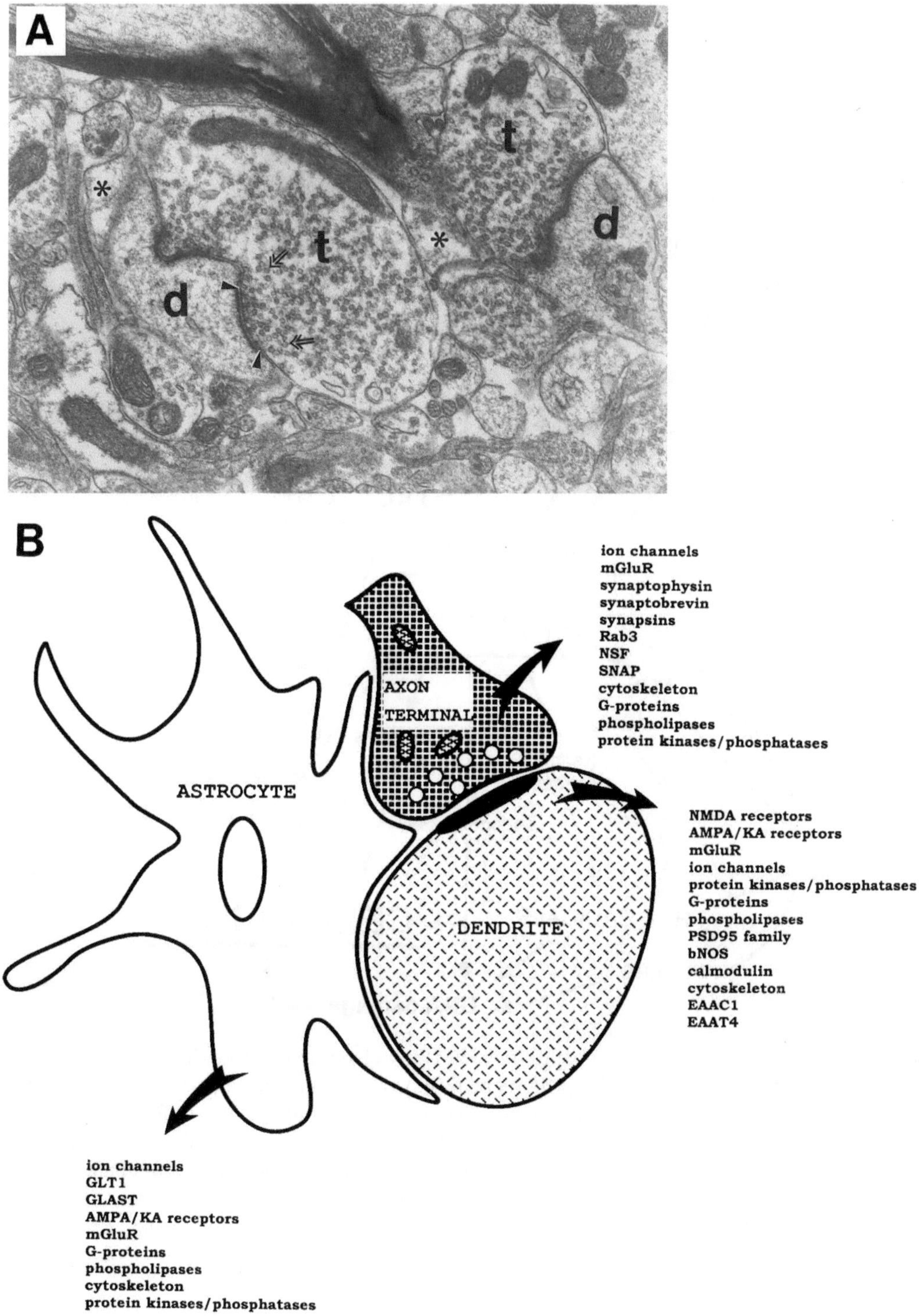
A
t
t
d
d
*
*
B
ASTROCYTE
AXON TERMINAL
DENDRITE
ion channels
mGluR
synaptophysin
synaptobrevin
synapsins
Rab3
NSF
SNAP
cytoskeleton
G-proteins
phospholipases
protein kinases/phosphatases
NMDA receptors
AMPA/KA receptors
mGluR
ion channels
protein kinases/phosphatases
G-proteins
phospholipases
PSD95 family
bNOS
calmodulin
cytoskeleton
EAAC1
EAAT4
ion channels
GLT1
GLAST
AMPA/KA receptors
mGluR
G-proteins
phospholipases
cytoskeleton
protein kinases/phosphatases

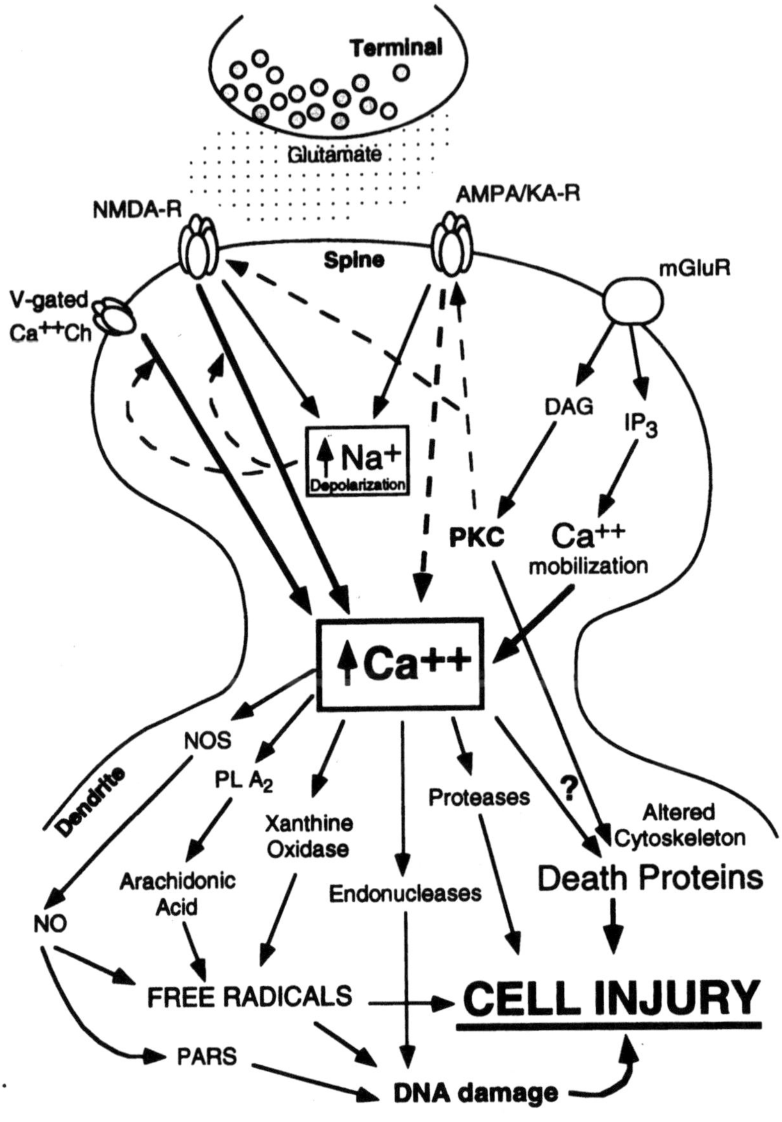

Fig. 2. Summary diagram of synaptic mechanisms related to excitotoxic neuronal degeneration. The diagram summarizes the prominent postsynaptic intracellular pathways that lead to neuronal injury and death resulting from excitotoxic activation of GluR. Abbreviations: AMPA/KA-R, α-amino-3-hydroxy-5-methyl-4-isoxazole propionate and kainate receptors; DAG, diacylglycerol; Ip_3, inositol trisphosphate; mGluR, metabotropic glutamate receptor; NMDA-R, *N*-methyl-*D*-aspartate receptor; NO, nitric oxide, NOS, NO synthase, PARS, poly (ADP-ribose) synthetase also known as PARP, poly(ADP-ribose) polymerase; PKC protein kinase C; PLA_2, phospholipase A_2; V-gated C^{++} ch, voltage-gated Ca^{2+} channel.

electron-dense meshwork of fine filaments that underlies the plasma membrane and is thought to anchor and cluster neurotransmitter receptors *(16–18)*.

Astroglia have a critical role in synaptic function and neuronal survival. Astroglia are the most numerous cellular element of the mammalian CNS. They outnumber neurons by ten to one and comprise up to 50% of the total tissue volume in some brain regions, with the volume of the astroglial compartment increasing with phylogenetic development *(19,20)*. Astroglial processes ensheathe the cell bodies and processes of neurons, envelop synapses (Fig. 1), invest the nodes of Ranvier of myelinated axons, and form glial limiting membranes over the meninges and parenchymal vasculature. The known functions of astrocytes are numerous, including regulation of synaptic neurotransmission, extracellular pH, ion concentration and osmolarity, production and secretion of extracellular matrix proteins, neurotrophic factors and cytokines, detoxification of toxic metabolites (e.g., ammonia, glutamate, and free radicals), and immune/inflammatory response mechanisms *(19,20)*. Astrocytes express functional ligand-gated and metabotropic receptors to neurotransmitters, including glutamate receptors, and they express voltage-gated and other ion channels, including Ca^{2+}, Na^+, K^+, and Cl^- channels. Astroglia also possess uptake systems for neurotransmitters for terminating synaptic neurotransmission. Astrocytes respond to a wide variety of potentially injurious stimuli in an effort to restore normal physiological set points. Perturbations in astroglial functions may participate directly in the pathobiology of neuronal degeneration in disease. Important topics that will be discussed in this chapter are astroglial uptake of glutamate and production of amyloid β protein.

2.2. Glutamate Receptor Subtypes

The amino acid glutamate is the major excitatory neurotransmitter in the CNS *(21,22)*. In normal circumstances, glutamatergic synaptic transmission occurs by regulated release of glutamate from presynaptic axon terminals (Figs. 1 and 2). Concentrations of neurotransmitter glutamate at the synaptic cleft have been estimated to be ~1 mM, whereas the concentration of interstitial glutamate is ~1 μM *(23)*. At synapses, glutamate binds and activates several molecular subtypes of glutamate receptors (GluRs) located on the plasma membranes of neurons and some glial cells *(24)*. These receptors are categorized as members of one of two families, the ionotropic receptors and metabotropic receptors, which differ structurally (distinct molecular compositions) and functionally (distinct signal transduction mechanisms) *(24,25)*.

Ionotropic GluRs form monovalent cation (Na^+, K^+)-conducting channels, but the different subtypes have differences in their permeabilities to divalent cations (Ca^{2+}) *(24,25)*. The activation of ionotropic GluRs directly changes conductance of specific ions through the receptor-ion channel complex, thereby inducing membrane depolarization. The ionotropic GluRs are the *N*-methyl-*D*-aspartate (NMDA) receptors and the non-NMDA receptors (α-amino-3-hydroxy-5-methyl-4-isoxazole propionate [AMPA] and kainate receptors). These receptors are oligomers, most likely pentameric heterooligomers, of homologous subunits encoded by distinct genes. The NMDA receptor subunits are NR1 and NR2A-NR2D, the AMPA receptor subunits are GluRl-GluR4 (or GluRA-GluRD), and the kainate receptor subunits are GluR5-GluR7 and KA1-2 *(24,25)*. Variants of AMPA receptor subunits (designated as "flip" and "flop") are generated by alternate splicing of GluR1-GluR4 mRNA *(26)*. These two versions of subunits are different minimally in a 38-amino acid transmembrane sequence, but the "flip" variant shows less desensitization and larger currents *(26)*. Additional diversity in GluR function is generated by RNA editing *(25)*. GluR2 is a negative regulator of Ca^{2+} permeability through non-NMDA receptors (i.e., its presence in the pentameric receptor assembly blocks Ca^{2+} flux) *(25,27–29)*. The molecular mechanism for the Ca^{2+} impermeability of GluR2 is RNA editing. The genomic sequence of GluR2 differs from the cDNA in the coding sequence between transmembrane domains 1 and 2, and this editing of mRNA transcripts results in an arginine instead of a glutamine, thereby changing the channel properties *(27–29)*.

In most neurons in the CNS, activation of glutamate-gated ion channels produces fast, short-lived, excitatory postsynaptic currents with latency periods ranging from 1–10 msec; excitatory postsynaptic currents of non-NMDA receptors have greater amplitudes and shorter durations than NMDA receptors. AMPA and kainate receptor channels are voltage-independent, whereas the NMDA receptor is voltage-dependent because of a Mg^{2+} block at resting membrane potentials *(24,25)*. Because the non-NMDA receptor channels are permeable to monovalent cations, Na^+ enters the cell following its electrochemical gradient, thereby depolarizing the postsynaptic plasma membrane and activating voltage-sensitive ion channels, including NMDA receptors, Ca^{2+} channels, and Na^+ channels (Fig. 2).

Other GluRs do not form ion channels but are instead linked to signal transduction molecules within the plasma membrane. These metabotropic GluRs (mGluRs) are G protein-coupled receptors that are single proteins encoded by single genes *(30,31)*. mGluRs have slower electrophysiological characteristics (latencies >100 msec) than ion channel GluRs. Group I mGluRs (mGluR1 and mGluR5) operate through activation of phospholipase C (PLC) by G_q proteins, phosphoinositide hydrolysis and generation of inositol-1,4,5 triphosphate and diacylglycerol, and subsequent mobilization of Ca^{2+} from nonmitochondrial intracellular stores *(31)*. Group II mGluRs (mGluR 2 and 3) and Group III mGluRs (mGluR 4 and 6–8) function by G_I or G_O protein-mediated inhibition of adenylyl cyclase and modulation of ion channel activity *(30,31)*.

2.3. Glutamate Transporter Subtypes

Excitatory synaptic transmission is largely terminated by high-affinity, Na^+-dependent transport of glutamate into cells (neurons and astroglia), thereby preventing extracellular concentrations of synaptic glutamate from reaching neurotoxic levels (Fig. 1B). Five distinct high affinity, Na^+-dependent glutamate transporters have been cloned from animal and human tissue (GLAST, GLT1, EAAC1, EAAT4, and EAAT5), and these proteins differ in structure, pharmacological properties, and tissue distribution *(32–34)*. Under physiological conditions in normal adult CNS, immunohistochemical studies have shown that GLAST and GLT1 are expressed primarily in astrocytes, whereas EAAC1 is widely distributed in neurons *(35–38)*. EAAT4 is expressed mainly in cerebellar Purkinje cells *(37,39)*, and EAAT5 is primarily a retinal glutamate transporter *(34)*. Thus, the dominant astroglial and neuronal glutamate transporters in cerebellum are GLAST and EAAT4, respectively, whereas GLT1 and EAAC1 are the primary astroglial and neuronal glutamate transporters, respectively, in forebrain, brainstem, and spinal cord.

The different molecular subtypes of glutamate transporters have very distinct, but coordinated, regional patterns of expression during CNS development. The cell-type specificity in the expression of distinct glutamate transporters subtypes is generally similar in the developing and adult CNS. However, one major exception to this conclusion is that GLT1 is also expressed transiently in a variety of neurons in the developing CNS *(37,38,40)*.

2.4. Synaptic Pathobiology: GluR-Mediated Excitotoxicity and Glutamate Toxicity

Abnormalities in synaptic function (Fig. 2) can kill neurons in animal models and in humans. Although glutamate and GluR activation are critical for normal brain function, glutamate is toxic to neurons at abnormally high concentrations *(41,42)*. When GluRs are activated excessively, a process called excitotoxicity ensues *(41,42)*. Acute excitotoxicity causes degeneration in cultures of neurons from animal brain and spinal cord and after in vivo delivery of GluR agonists into the CNS of experimental animals *(42–45)*. In addition, excitotoxicity participates in the mechanisms for neuronal degeneration in animal models of cerebral ischemia and traumatic brain injury *(11,42,43)* and in the neurotoxicity of humans resulting from ingestion of mussels contaminated with the GluR agonist domoic acid *(46)*. In humans, GluR-mediated excitotoxic mechanisms have also been suspected to be responsible for epilepsy, for the neuronal death in the brain resulting from cardiac arrest and stroke, and for neuronal degeneration occurring in individuals with AD, ALS, Huntington's disease, and Parkinson's disease *(11,14,42,43,47)*.

The excessive interaction of glutamate or chemical analogs of glutamate with subtypes of GluRs produces abnormalities in intracellular ionic concentrations, pH, protein phosphorylation, molecular stability, and energy metabolism *(42,43)*. Excitotoxically challenged cells (Fig. 2) undergo rapid osmotic perturbations and swell, reflecting the influx of Na^+, Cl^-, and H_2O *(43)*. An increase in cytosolic free Ca^{2+}, resulting from activation of Ca^{2+}-permeable AMPA receptors and voltage-sensitive ion channels including NMDA receptors and Ca^{2+} channels, causes increased enzymatic activity of Ca^{2+}-sensitive proteases, protein kinases, and phosphatases, endonucleases, and phospholipases (Fig. 2). Through the activation of proteins with DNase activity, excitotoxicity results in internucleosomal DNA fragmentation in cultures of cortical neurons *(48,49)* and cerebellar granule cells *(50,51)*, although others have not found internucleosomal DNA fragmentation in cerebellar granule cell cultures *(52)*. In vivo, internucleosomal and random fragmentation of genomic DNA occur 12–48 h after intracerebral injections of excitotoxins *(44,45,53,54)*, and this pattern can persist to 5 d postlesion *(54)*. The activation of other Ca^{2+}-dependent enzyme systems, such as phospholipase A_2, cyclo-oxygenase, and xanthine oxidase, can generate highly toxic reactive oxygen species (ROS) *(55,56)*. Another pathway for the production of ROS, namely nitric oxide (NO), is activation of Ca^{2+}/calmodulin-dependent NOS *(57)*.

Oxidative stress is a potent stimulus for neuronal death in cultured neurons *(58,59)*. ROS (e.g., superoxide anion radical, hydrogen peroxide, hydroxyl radical, NO, and peroxynitrite) can cause direct oxidative damage to macromolecules, including DNA, protein, and lipid membranes *(55–57)*. The balance between the formation of endogenous free radicals and antioxidant defense mechanisms is important for cellular survival. ROS are also products of oxidative metabolism. The mitochondrial electron-transfer chain is a primary generator of superoxide and peroxide, and damaged mitochondria are believed to produce greater amounts of superoxide ion *(60)*. Damage to mitochondrial DNA caused by ROS may lead to protein conformational changes usually associated with an inefficient electron transfer to cytochrome c oxidase *(61)* and, hence, enhanced superoxide and peroxide formation. Other critical molecular targets for oxidative damage include glutamate transporters *(62),* Na^+/K^+ ATPase *(63),* glucose transporters *(64),* cytoskeletal proteins *(65,66),* and superoxide dismutase-1 *(67).* ROS can also mediate trophic factor deprivation-induced apoptosis of sympathetic neurons in vitro *(68),* and we have implicated ROS in target deprivation-induced apoptosis of brain and spinal cord neurons in vivo *(69,70).* Because neuronal survival depends on trophic factors *(71,72),* some of which are glial-derived, abnormalities in neurotrophic support may result in apoptotic death of neurons by inducing a PCD mechanism involving the generation of ROS *(68–70).*

Another mechanism through which glutamate can cause death of neuronal cells is oxidative glutamate toxicity. In this process, glutamate alters cellular metabolism by interacting with the cystine-glutamate antiporter, resulting in depletion of intracellular cystine/cysteine and reduced levels of the cysteine-containing tripeptide glutathione *(73).* Oxidative stress results from depletion of intracellular pools of the antioxidant glutathione *(73).* Activation of phosphoinositide-liked mGluRs generates a cellular response that protects against oxidative glutamate toxicity; in contrast, group I mGluR antagonists potentiate glutamate toxicity *(74).*

3. NEUROGLIAL AND SYNAPTIC MECHANISMS FOR NEURONAL DEGENERATION AFTER CEREBRAL ISCHEMIA

3.1. Abnormalities in Glutamate Transport and Astroglia Occur after Global Cerebral Ischemia

Numerous overlapping mechanisms for neuronal injury are likely to be operative in ischemic brain damage. Here, the focus will be on impaired glutamate transport, excitotoxicity, and the associated perturbations in intracellular signal transduction. In models of transient global cerebral ischemia, extracellular glutamate levels are increased transiently *(75–78),* possibly resulting in excitotoxic activation of neuronal GluR *(79).* Abnormally high concentrations of extracellular glutamate may be the result of vesicular exocytosis of glutamate *(80),* reversed glutamate transporter function *(81),* defective uptake of glutamate *(82–84),* or astrocyte swelling *(85).* Of these possible mechanisms, impaired glutamate transporter function has received considerable attention.

Impaired glutamate transport is thought to cause neurodegeneration. For example, intrastriatal delivery of the glutamate transport inhibitor DL-threo-3-hydroxyaspartate causes neuronal degeneration in rat *(86).* Lethal spontaneous seizures and increased susceptibility to acute cortical injury occur in some mice deficient in GLT1 *(87),* and GLAST gene-ablation exacerbates retinal damage after ischemia in mice *(88).* However, we have shown that transient reductions in glutamate transporter function and astroglial glutamate transporter (GLAST and GLT1) protein expression can occur in the absence of neuronal degeneration in models of axotomy and deafferentation *(89,90).*

Alterations occur in glutamate transporters after cerebral ischemia. In adult rat, *D*-[^{3}H]aspartate binding sites are increased in hippocampus within 5 min after forebrain ischemia *(91),* but the levels of GLT1 mRNA and protein are reduced in hippocampus at 3–6 h postischemia *(92).* In rat pups (7-d-of-age), high-affinity glutamate transport is transiently reduced in striatum during hypoxia-ischemia and after one hour of recovery *(82)*. In hypoxic-ischemic piglets (1-wk-old), high-affinity glutamate transport in striatal synaptosomes is defective functionally by 6 to 12 h recovery *(84)*, and GLT1 and EAAC1 protein levels are reduced in striatum at 24 h recovery and thereafter *(83)*. In contrast, GLAST protein expression is maintained in striatum after hypoxic-ischemic injury *(84)*. Therefore, the evolution of neuronal degeneration in some brain regions after hypoxia-ischemia is paralleled by sustained abnormalities in glutamate transport during the first 24 h of recovery. These defects in glutamate transporters are molecular subtype specific *(83,84)*.

A functional defect in glutamate uptake could be mediated by inactivation of glutamate transporters. For example, covalent modification of glutamate transporters by glycosylation influences their function. *N*-linked glycosylation of EAAC1 *(93)*, but not *N*-glycosylation of GLAST *(94)*, appears to be required for transporter activity. We have shown that the Golgi apparatus within striatal neurons undergoes fragmentation and vesiculation by 3–6 h after hypoxia-ischemia. Furthermore, during this time course, tubulin undergoes extensive oxidative damage (i.e., protein nitration) through a NO-associated toxic pathway. Therefore, posttranslational processing as well as targeting and transport of proteins may be abnormal early after hypoxia-ischemia *(11,95)*.

Covalent modification of glutamate transporters by phosphorylation has subtype-specific modulatory effects on activity. Protein kinase C (PKC) activation results in direct phosphorylation of GLAST and inhibits GLAST function *(96)*; in contrast, phosphorylation of GLT1 by PKC stimulates GLT1 function *(97)*. We have found increases in PKC activity as well as increases in membrane protein phosphorylation at serine sites in vulnerable brain regions at 24 h after global ischemia *(98)*. Thus, phosphorylation of astroglial GLAST, but not GLT1, by PKC may be a mechanism for reduced glutamate uptake in vulnerable brain regions after hypoxia-ischemia *(84)*.

Another possibility is that glutamate transporter function may be impaired by direct structural damage to the proteins. In cell culture systems, oxygen radicals decrease high-affinity glutamate transport *(99)*. Peroxynitrite (formed by the combination of superoxide and NO) is a potent inhibitor of GLAST, GLT1, and EAAC1 function *(62,100)*. We have found evidence for peroxynitrite-mediated oxidative damage to membrane proteins within the piglet striatum early (3 to 6 h) after hypoxia-ischemia *(95)*. Some of these proteins have not yet been identified. It is interesting that the molecular mass of some of these proteins is in the range of glutamate transporters (i.e., ~65–73 kDa). These abnormalities in protein nitration and in the expression or function of glutamate transporters may participate in striatal neurotoxicity after hypoxia-ischemia.

Abnormal control of extracellular glutamate concentrations by astroglia could be an early mechanism for neuronal degeneration after ischemia. In our piglet model of hypoxia-ischemia, striatal astrocytes are damaged early, as evidenced by their swelling, fragmentation, and death as well as by the loss of GLT1 protein at 24 h recovery *(83)*. Astroglia in culture survive exposures to excitotoxins at concentrations sufficient to kill most neurons, suggesting that astrocytes are more resistant than neurons to excitotoxicity; however, overactivation of AMPA receptors is lethal to astrocytes, when receptor desensitization is blocked *(101)*. In addition, astrocytes in vivo are damaged by acidosis *(102)*, and prolonged, severe acidosis is lethal to astrocytes in vitro *(103)*.

Astroglial swelling occurs within minutes of hypoxia-ischemia, suggesting that initial astrocytic abnormalities may contribute to the evolution of neuronal damage *(104)*. In piglets, principal striatal neurons and astrocytes have a similar time course of injury and death at one to two days following hypoxia-ischemia *(83,84,95)*. Cell culture studies show that neurons in an astrocyte-poor environment are more vulnerable to excitotoxicity than neurons cultured with astrocytes *(105)* and that uptake of glutamate by astrocytes improves neuronal survival in an excitotoxic environment *(106)*. The finding that astroglial and neuronal injury occur concurrently within brain further strengthens the concept that abnormalities in astroglia could participate in the evolution of neuronal degeneration after hypoxia-ischemia *(83,104)*. Astroglial damage and defective glutamate transport would favor excitotoxic neuronal cell death after hypoxia-ischemia, and oxidative damage to astroglial proteins could be a potential molecular mechanism for this defect *(83,84,95)*.

3.2. Glutamate Receptor-Mediated Excitotoxicity as a Mechanism for Neuronal Cell Death After Cerebral Ischemia

Neuronal cell death after ischemia involves perturbations in intracellular Ca^{2+} homeostasis and impairments in protein synthesis. The GluRs that modulate intracellular Ca^{2+} levels within neurons are the ion channel receptors (NMDA and AMPA) and the group I mGluRs *(24–31)*. In models of transient global cerebral ischemia a possible role for excessive activation of GluRs in the mechanisms for delayed neuronal death is supported by studies showing that blockade of AMPA receptors is neuroprotective following forebrain ischemia in adult rodents *(107)* and that abnormalities in phosphoinositide signaling pathways occur in hippocampus after ischemia *(108)*. In models of focal ischemia (stroke), antisense oligodeoxynucleotide knockdown of NR1 production *(109)*, targeted disruption of the NR2A gene *(110)*, and selective pharmacological antagonism of NR2B *(111)* attenuate forebrain damage, thus a decrease in the number of functional NMDA receptors appears to mediate this neuroprotection.

We have demonstrated that excitotoxic activation of GluRs causes an apoptosis-necrosis continuum for cell death in which the structure of neuronal degeneration is influenced by the subtype of GluR that is activated and the age (maturity) of the CNS at the time of the insult *(44,45)*. In the developing rat CNS, excitotoxic activation of NMDA and non-NMDA GluRs causes neuronal death with phenotypes ranging from apoptosis to necrosis. Excitotoxic neuronal death occurs as three structurally different forms: Classic apoptosis, classic necrosis, and a hybrid of apoptosis and necrosis *(11,44,15,47)*. In contrast, in the adult rat CNS, the degeneration of neurons caused by NMDA receptor activation is necrosis morphologically; however, the neuronal death produced by non-NMDA receptor activation appears to be a structural hybrid of apoptosis and necrosis and is distinct from the death caused by NMDA receptor stimulation.

Interestingly, neuronal death after cerebral ischemia may not follow the apoptosis-necrosis continuum for excitotoxic neuronal death. We have shown with three different animal models of transient global ischemia that neuronal cell death within selectively vulnerable brain regions (e.g., CA1 pyramidal neurons, cerebellar Purkinje cells, and striatal neurons) is identical structurally to excitotoxic neuronal cell necrosis mediated by NMDA receptor activation, and, in the vulnerable neuronal populations, classic apoptosis and hybrids of apoptosis-necrosis rarely occur *(11,47,112)*. However, in several different paradigms of global ischemia, we and others have been unable to rescue these neurons from degeneration using noncompetitive antagonists to NMDA receptors *(113,114)* or competitive antagonists to NMDA receptors *(115,116),* AMPA receptors *(117–119),* and group I mGluRs *(120).* Some of our data even suggest that antagonists to NMDA receptors *(116),* AMPA receptors *(118,119)* and mGluRs *(120)* worsen neurologic outcome after ischemia. Yet, other studies have shown that AMPA receptor antagonism is neuroprotective in adult rodent models of global ischemia *(107,114).* The lack of neuroprotection and the deleterious effects of GluR antagonists in in vivo animal models are difficult to interpret and may be related to several variables associated with experimental design, including the dosage, solubility, delivery, administration route, and systemic pharmacokinetics of the drug. Although GluR-mediated excitotoxicity and glutamate toxicity continue to be potential mechanisms for neurodegeneration after hypoxia-ischemia, the pathobiology of this neuronal cell death remains poorly understood. The future clarification of the structural and molecular mechanisms of neuronal death after ischemia and the neuroglial participation in these mechanisms will lead to a rational and effective design of experiments for the protection of neurons and glia.

3.3. *Possible Roles for Protein Kinases in Neurodegeneration After Ischemia*

Protein kinases function in cell growth and differentiation, and, in nervous tissue, additional functions are regulation of neurotransmitter receptors and ion channels as well as modulation of neuronal excitability, neurotransmitter release, and synaptic plasticity (Fig. 2; for review *see* refs. *121,122*). Protein kinase activation may thus play a role in GluR-mediated excitotoxic neuronal death *(123,124)* and in neuronal degeneration after ischemia *(98,122,125).* GluR stimulation causes increased intracellular Ca^{2+} and activation of Ca^{2+}/calmodulin protein kinase II (CaM kinase II) and causes increased activity of phospholipase A_2 and PLC (Fig. 2), leading to the accumulation of diacylglycerol and activation of PKC *(24,31,43).* Mild ischemia induces a persistent translocation of CaM kinase II from the cytosol to synaptic plasma membranes *(126,127).* We have identified a sustained increase in PKC activity, an isoform-specific translocation to the plasma membrane, as well as increased protein phosphorylation at serine residues in plasma membranes of selectively vulnerable brain regions after ischemia *(98).* Astroglia may contribute to some of these changes *(98).* Other studies have suggested that protein kinases that are membrane-translocated after ischemia function improperly *(126,128).* However, changes in brain protein kinase activity may depend on the severity of ischemia and energy failure *(98).*

Several studies have shown that phosphorylation of AMPA, kainate, and NMDA receptors by CAM kinase II, PKC, or cAMP-dependent protein kinase A causes positive modulation of receptor function by potentiating activation *(129–134).* This enhancement of receptor function is transient because of the activity of postsynaptic protein phosphatases *(134).* In contrast, PKC-mediated phosphorylation of PLC-coupled mGluRs causes desensitization to glutamate *(135,136).* The integrative actions of different molecular subtypes of GluRs are also being revealed. Activation of PLC-coupled mGluRs increases NMDA receptor function through PKC activation *(137,138).* PKC potentiates NMDA receptor function by increasing the probability of channel openings and by reducing the voltage-dependent Mg^{2+} block *(139).*

The possible relevance of these observations to neuronal and astroglial degeneration after ischemia is provocative but still only speculative. We have found increases in PLC and PKC after ischemia *(98,140)*, and other groups have found abnormalities in phosphoinositide signaling pathways after ischemia *(108)*. Electrophysiological studies of neuronal activity in vulnerable regions after transient ischemia are discrepant. By extracellular recording in vivo, studies have shown that spontaneous firing rate in CA1 is increased *(141)*, whereas other studies show that spontaneous firing rate and evoked responses are suppressed *(142,143)*. By intracellular recording, CA1 neuronal activities and evoked synaptic potentials are suppressed after in vitro hypoxia-ischemia *(144)*; similarly, in spiny striatal neurons in vivo, spontaneous activities and evoked postsynaptic potentials are suppressed and excitability is decreased after transient ischemia *(145)*. However, whole-cell recordings reveal that CA1 pyramidal neurons have abnormal excitatory postsynaptic currents produced by activation of non-NMDA receptors that mediate Ca^{2+} influx *(146)*. It is still uncertain whether changes in protein kinases after ischemia mediate these functional abnormailties and contribute directly to the mechanisms of neuronal death or whether they reflect secondary degenerative or regenerative processes in injured or surviving neurons and/or changes due to activation of glial cells. Based on the knowledge that protein kinases modulate GluR receptor and glutamate transporter function, ion channel function, Na^+/K^+ ATPase activity, and neurotransmitter release (to name only a few critical functions for protein kinases), much more work is needed to identify the possible contributions of protein kinases to the mechanisms for neuronal degeneration after ischemia (Fig. 2).

3.4. Participation of NO-Mediated Mechanisms in the Pathobiology of Neurodegeneration After Ischemia

NO is a short-lived, membrane-permeant ROS that functions in synaptic neurotransmission, synaptic plasticity, cell-mediated immune response, and cerebrovascular regulation *(57,147,148)*. NO is generated by the catalytic activity of NOS, of which three molecular isoforms exist. Brain or neuronal NOS (NOS1) is constitutively expressed in subsets of neurons, inducible NOS (NOS2) is an inflammatory/glial cell isoform, and endothelial NOS (NOS3) is constitutively expressed in vascular endothelial cells *(57,148–150)*. Neuronal NOS activity is regulated by intracellular Ca^{2+} *(149)*. NMDA receptor activation results in NOS activation (Fig. 2; *151*), and mGluR activation enhances NO production, possibly involving a Ca^{2+}-induced Ca^{2+} release amplification of mGluR-associated intracellular signals *(152)*. NO has presynaptic actions as well by stimulation of synaptic vesicle exocytosis in a process involving cGMP *(153)*. NO functions by forming covalent linkages and redox interactions with intracellular proteins *(154)*. Excessive generation of NO is thought to be toxic to cells in vivo and may be an important mediator of neuronal degeneration resulting from cerebral ischemia *(57,155–157)*, CNS inflammation *(158)*, and axotomy/target deprivation *(70)*. A predominant mechanism by which NO toxicity occurs is by the diffusion-limited reaction of NO with superoxide to generate peroxynitrite anion, which is a strong and relatively long-lived oxidant that directly damages the structure of a variety of proteins *(57)*. Peroxynitrite decomposes to form secondary oxidants, including the particularly toxic hydroxyl radical *(57)*, which we have shown to damage neuronal DNA and RNA after cerebral ischemia *(95,159)* and axonal injury *(69,70)*.

In studies of cerebral ischemia, the results supporting a role for neurotoxic actions of NO in vivo are discrepant. In settings of focal ischemia, mice deficient in neuronal NOS *(155)* or inducible NOS *(156)* have slightly smaller infarct volumes than controls, and pharmacological inhibition of NOS has been shown to decrease brain damage volume in some studies *(160)* but not in other experiments *(161,162)*. In settings of global ischemia, mice with targeted disruption of the neuronal NOS gene have less CA1 damage than wild-type mice *(163)*. However, in rat and gerbil, pharmacological inhibition of NOS is not protective *(162,164)*. We have been unable to show that an inhibitor of NOS results in neuroprotection in hippocampus and cerebellum after global ischemia *(165)*. However, we have found that peroxynitrite-mediated damage to striatal membrane proteins occurs strongly by 6 h after hypoxic-asphyxic cardiac arrest *(95,159)*. In addition, we discovered that the critical cytoskeletal protein tubulin is a target of peroxynitrite *(159)*. It is therefore possible that the participation of NO-mediated toxicity is different in acute vs delayed neuronal degeneration after cerebral ischemia, and, thus, onset and severity of oxidative stress may influence rate of neuronal death on a regional basis, because striatal damage evolves faster than CA1 neuron and Purkinje cell damage *(11)*.

4. SYNAPTIC PATHOBIOLOGY IN PROGRESSIVE, AGE-RELATED NEURODEGENERATIVE DISEASES

4.1. Synaptic Dysfunction in AD

AD is the most common type of dementia occurring in middle and late life *(166)*, and it affects 7–10% of individuals >65 years of age and possibly 40% of people >80 years of age *(12,167)*. The prevalence of AD is increasing proportionally to increased life expectancy (estimates predict that ~25% of the population will be >65 years of age in the year 2050). AD now affects >4 million people in the United States *(168)*. Although most cases of AD have unknown etiologies and are called sporadic, some cases of AD, particularly those with early onset, are familial and are inherited as autosomal dominant disorders linked to mutations in the gene that encodes amyloid precursor protein *(169–171)* or genes that encode for proteins called presenilins *(172,173)*. For late onset sporadic cases, a variety of risk factors have been identified in addition to age. The apolipoprotein E (apoE) allele is a susceptibility locus with the apoE4 type showing dose-dependent contributions *(174)*. Cardiovascular disease and head trauma are additional risk factors for AD *(166)*.

The mechanisms that cause the profound neuronal degeneration and the progressive impairments in memory and intellect that occur with sporadic and familial AD are not understood. Neuronal survival and normal memory and cognition depend on synaptic function. Regulated exocytosis of neurotransmitter-containing vesicles (Fig. 1A) is obligate for normal synaptic function *(15)*. Synaptophysin (p38), an integral membrane glycoprotein of small synaptic vesicles in the presynaptic terminal *(15)*, is one protein of many that functions in regulated exocytosis (Fig. 1B). We have found that synaptophysin is reduced in the hippocampus of individuals with AD who have moderate to severe deficits in memory *(5)*. This finding is not surprising in light of the vulnerability of the hippocampus in advanced AD *(1,3,4,166)*. A more exciting finding is the loss of synaptic marker in individuals with early AD (or possible AD) who had no detectable cognitive impairment *(5)*. The severity of this abnormality in synaptophysin in hippocampus correlates strongly with the severity of memory impairment in individuals with AD. Defects in other presynaptic components for regulated exocytosis of neurotransmitters (Fig. 1B) may also occur in the hippocampus of individuals with early AD. We also have interesting new data suggesting that synaptobrevin, synaptotagmin, syntaxin, and Rab3a are abnormal in the hippocampus of AD, but synapsin I, SNAP25, and SV2 levels are normal *(175)*. Thus, presynaptic vesicles appear to be more vulnerable in AD than the presynaptic membrane that contains the active zone where vesicle docking occurs. In addition, we have discovered in individuals with AD selective abnormalities in the levels of the NR2B subunit of the NMDA receptor in the hippocampus as well as abnormalities in NMDA receptor phosphorylation *(176)*. These synaptic molecule defects in subjects with early AD may foreshadow the clinical appearance of memory/cognitive impairment. Ultimately, structural loss of the entire synaptic complex could contribute to the atrophy of the cerebral cortex (i.e., cortical volume loss) which correlates strongly with cognitive decline *(177)*.

4.2. Cellular Dynamics of Senile Plaque Formation and Amyloid Deposition within the Brain During Aging

Senile plaques (SP) are brain lesions that occur in individuals with AD, patients with Down's syndrome (DS), and, less frequently, in people aging normally *(1,4,178,179, see ref. 2* for additional references). These lesions have a complex composition, consisting of dystrophic neurites (damaged and swollen dendrites or axon terminals), activated astrocytes and microglia, and extracellular deposits of insoluble amyloid fibrils (*179, see ref. 2* for additional references). These fibrils are composed of a 4-kDa peptide (Aβ) *(180–182)* consisting of 40–42 amino acid residues *(183)*. This protein fragment is derived proteolytically from the amyloid precursor protein (APP), a cell surface protein with a large N-terminal extracellular domain that contains 22 residues of Aβ, a hydrophobic membrane-spanning region including the transmembrane portion of the Aβ region, and a short C-terminal cytoplasmic segment *(184–186)*. The APP gene is located on human chromosome 21 *(187–189)*, and mRNA gene products are spliced alternatively to generate ~5 different forms of APP transcripts and protein isoforms *(184,189)*. A role for APP in the pathogenesis of AD is supported further by the identification of mutations in the APP gene linked to early-onset AD in some families *(169–171)*.

The functions of APP are still not well defined, although it appears that APP functions at synapses *(2,190)*. APP is an abundant and ubiquitous protein within CNS and other tissues. However, mice deficient in APP show no major neurological or neuropathological abnormalities *(191)*, possibly due to homologous proteins. APP has structural features similar to some cell surface receptors *(184)* and may be a G protein-coupled receptor *(192)*. Secreted and nonsecreted forms of APP exist *(193,194)*, with different APP derivatives showing neurotrophic or neurotoxic actions. APP is incorporated into the extracellular matrix *(195)* and, thus, may have roles in cell-cell and cell-substrate adhesion *(193,196–199)*. Furthermore, APP may function in the regulation of neurite outgrowth *(200–201)*, perhaps by mediating the effects of nerve growth factor *(202)* and in neuronal and glial responses to brain injury *(203–205)*.

In cell culture, APP normally undergoes constitutive proteolytic cleavage *(186,190,193)* by an α-secretase, an enzyme that cleaves APP within the Aβ region at or near the plasma membrane *(206–208),* thereby generating secreted forms of APP and precluding the formation of full-length Aβ peptide fragments *(206,208).* APP is also metabolized by an endosomal-lysosomal pathway that, unlike the α-secretase pathway, yields amyloidogenic fragments of Aβ *(209,210).* Aβ can be formed normally in vivo and in vitro *(211,212),* and studies of cultured human cells show that it is generated intracellularly *(213,214).* Although Aβ has been shown to be neurotoxic in cell culture, a causal role for Aβ in neuronal degeneration in vivo remains speculative. A β-secretase cleaves APP at the N-terminus of Aβ, and a γ-secretase cleaves APP at the C-terminus of Aβ, causing the formation of Aβ that is either 40 amino acids or 42 amino acids long *(215).* This pathway for APP metabolism is found within the endoplasmic reticulum and Golgi apparatus of neurons *(216,217).* Interestingly, presenilins, which are present at relatively low levels in brain *(218),* localize to the endoplasmic reticulum and Golgi apparatus *(219),* and mutant presenilins promote Aβ42 generation *(220).* Mutant presenilin is processed differently than normal presenilin, and fragments that are normally subject to endoproteolytic cleavage tend to accumulate *(221).* Thus, metabolism of APP through the β- and γ-secretase pathways may be promoted by presenilin-1 and presenilin-2 gene mutations linked to early onset familial AD.

We and others have shown that APP is expressed by neurons and by subsets of astroglia, microglia, and vascular endothelial cells *(2,196,222–224).* The most prominent neuronal localization of APP is within cell bodies and dendrites and is particularly enriched postsynaptically at subsets of synapses *(2,205).* The expression of APP in nonneuronal cells in brain is low in comparison to the dominant expression of APP within neurons and their processes. It appears that astroglia and microglia constitutively express APP at low levels in the resting state *(2,222).* However, an important finding is that the relative enrichment of APP within these neuroglial cells changes in response to brain injury and synaptic abnormalities *(2,204).* This idea is supported by our finding that APP is expressed prominently by activated astroglia and microglia within SP of aged nonhuman primates *(2,224)* and by other reports *(225)* showing that APP is localized to astrocytes in SP in cases of AD. Other studies have shown that APP isoforms containing the Kunitz protease inhibitor domain are expressed in reactive astrocytes in early stages of brain damage *(204).* Because levels of APP in some neurons and nonneuronal cells are increased by the cytokine interleukin-1 *(226),* it is likely that the expression of APP is inducible in glia when these cells are activated in response to neuronal injury.

Several hypotheses for Aβ deposition and SP formation have been presented. The genesis of SP may begin with the formation of extracellular Aβ before the degeneration of cellular elements within these lesions *(227,228).* Alternatively, Aβ may be derived from degenerating axonal nerve terminals or dendrites containing APP that evolve into neurite-rich foci that form Aβ at the cell plasma membrane by aberrant processing of APP within neurons *(181,223,229).* In addition, invading reactive microglia *(2,179,230,231)* and astroglia *(2)* as well as capillaries *(232)* may actively produce Aβ from APP. In one scheme for SP formation, Aβ (in a nonfibrillar form and then a fibrillar form) is deposited before neuronal neuritic damage occurs, and thus is a cause of neuronal degeneration *(233,234).* We have not identified extracellular Aβ (even as sparse, scattered bundles of fibrils) before the appearance of synaptic abnormalities and dendritic/axonal neurites *(2,224)*. We believe that astroglia and microglia are primary generators of Aβ deposits in the aging brain *(2,224)*. In early diffuse plaques, we have identified Aβ-containing transformed astrocytes and microglia after neuritic defects have appeared. Our experiments with aging nonhuman primates are concordant with studies of SP in the cerebral cortex from cases of AD *(179,230)*.

These studies reveal that SP are dynamic brain lesions that evolve from early synaptic defects within the neuropil to mature plaques and extracellular deposits of Aβ. The staging of these lesions is thought to be the degeneration of neuritic structures, followed by the attraction of reactive glia, and the subsequent deposition of extracellular Aβ derived from microglia *(2,179,230)* or astrocytes *(2)*. Our studies demonstrate that structural and biochemical perturbations within neuronal and nonneuronal cells occur before the deposition of extracellular Aβ fibrils *(2,224)*. Furthermore, our results suggest that focal abnormalities in synaptic contacts within the neuropil (synaptic disjunction) may initiate this complex series of events resulting in the formation of diffuse SP and deposits of Aβ. In response to synaptic disjunction in the aged brain, astroglia and microglia produce Aβ *(2)*. The molecular pathology that we and others have identified at the synaptic level in humans with AD *(5,175,176,235)* may be related to SP lesions and Aβ deposits. For example, we found a strong inverse correlation between synaptophysin loss and the density of neuritic and diffuse plaques in hippocampus. It is not surprising to find an inverse correlation between SP density and synaptophysin immunoreactivity, because synaptic disjunction in the neuropil and abnormal APP processing within neuroglia may be early events in the formation of SP *(2,224)*.

4.3. *Oxidative Stress in the AD Brain*

Neurons have a particularly high risk for structural damage caused by ROS, because of high rates of oxygen consumption and high contents of polyunsaturated fatty acids that are susceptible to lipid peroxidation; therefore, a free radical hypothesis has been suggested as a mechanism for compromised CNS functioning with aging *(3,11,47,66)*. Some experimental evidence is available to support this argument. For example, superoxide dismutase-1 (SOD1) activity is elevated in fibroblasts from individuals with familial AD *(236)*, and the SOD1 gene is located on human chromosome 21 *(237)*, which individuals with DS possess an extra copy, and a gene dosage effect on lipid peroxidation has been observed in brain tissue of patients with DS *(238)*. ROS have been implicated in the induction of neuronal death by apoptosis in DS *(239)* and in in vitro *(58,59)* and in vivo models *(11,69,70)*. Mitochondria may participate in the effector stage of apoptosis by directly providing a rich source of ROS or by changes in mitochondrial membrane cell death proteins that prevent apoptosis by an antioxidant mechanism, or that supress apoptosis by blocking release of cytochrome c (a required molecule for apoptosis in cell culture systems) or by regulating membrane potential and volume homeostasis of mitochondria (for references see *10,11,69,71*).

Cellular oxidative stress and antioxidant enzyme regulation are relevant to age-related neurodegenerative disorders because abnormalities in antioxidant enzyme systems (e.g., superoxide dismutase) and oxidative injury may provide mechanisms for neuronal degeneration. Cytosolic copper/zinc-superoxide dismutase (SOD1) and mitochondrial manganese-superoxide dismutase (SOD2) are localized predominantly to neocortical and hippocampal pyramidal neurons, but are scarcely seen in glial cells in normal human brain *(3)*. In AD and DS, the cellular expressions of these two forms of SOD are very different from normal individuals *(3)*. SOD1 is highly enriched in pyramidal neurons undergoing degeneration, whereas SOD2 is more enriched in reactive astrocytes than in neurons. In SP, astrocytes highly enriched in SOD2 surround SOD1-enriched neurites. Some pyramidal neurons coexpress SOD and the cytoskeletal protein tau, and some SOD1-enriched structures in SP are tau-positive *(3)*. Microglia infrequently show expression of high levels of superoxide dismutase. These findings in AD and DS support a role for oxidative stress in neuronal degeneration and SP formation. The differential localizations of SOD1 and SOD2 in sites of degeneration in diseased brain suggest that the responses of cells to oxidative stress is antioxidant enzyme-specific and cell type-specific and that these two forms of superoxide dismutase may have different functions in brain antioxidant mechanisms in response to injury.

A glial expression of SOD has been reported in AD brain *(240)*. Our studies have shown that SOD protein is rarely seen in glial cells in control brain *(3)*. However, in AD and DS brain, SOD2 is induced strongly in reactive astrocytes in cerebral cortex and white matter. This change in the astroglial expression of SOD2 in the brains of AD and DS cases is a major difference between control and diseased tissue, and is possibly due to metabolic activation of reactive astrocytes, because SOD2 is localized within the mitochondrial matrix *(241)*. SOD1-expressing astrocytes are also seen, but this expression is weaker than that of SOD2. Because SOD2 mRNA is induced by tumor necrosis factor-α and interleukin-1, SOD2 is thought to have a cytoprotective function against various microglia/macrophage-derived inflammatory mediators *(242,243)* that may participate in the pathogenesis of AD and DS *(244)*. Our findings that SOD2 may be downregulated in pyramidal neurons and upregulated in reactive astrocytes in AD and DS brain *(3)* may relate to the selective cytotoxicity of microglia-derived cytokines (e.g., tumor necrosis factor-α and interleukin-1) *(245)*. Activated microglia secrete superoxide radicals *(246)*, and stimulated neutrophils disrupt SOD structure in cell culture *(247)*, but reactive microglia-macrophage cells that are widely distributed in AD and DS cases, show scarce evidence for expression of SOD *(3)*. Thus, modulation of glial expression of SOD appears to be relatively selective for astrocytes, because in AD and DS, astroglia show the greatest increase in SOD2 expression.

4.4. Synaptic and Astroglial Pathobiology in ALS

ALS is a disease that causes progressive weakness, muscle atrophy, and eventual paralysis. Death occurs within 3 to 5 years of onset and is characterized neuropathologically by progressive degeneration of upper and lower motor neurons in the brain and spinal cord *(13,248)*. The mechanisms leading to the selective degeneration of motor neurons in ALS are not known. Some forms of ALS are inherited *(8,9)*. Mutant forms of SOD1 have been identified in a small subset (10–20%) of individuals with familial ALS which occurs in 5–10% of all patients with ALS *(8,9)*. Forced expression of mutant forms of the gene encoding SOD1 results in degeneration of motor neurons in mice *(249,250)*. A toxic gain in function of mutant SOD1 might initiate this neurodegeneration *(249,251)*. In cultured cells, expression of mutant SOD1 can induce abnormalities in the production of ROS and neuronal apoptosis *(251,252)*. Oxidative stress resulting from downregulation of normal SOD1 can also cause apoptosis in cell culture *(253)*. Other possible causes for neuronal death in ALS are neurofilament abnormalities *(254–256)*, defective uptake of glutamate by astroglia *(6,7)* leading to GluR-mediated excitotoxic neuronal death, or deficiencies in neurotrophic factors *(257)*.

We have found *(10)* that the degeneration of motor neurons in sporadic and familial ALS is different structurally from the motor neuron degeneration found in transgenic mice overexpressing the familial ALS mutant forms of SOD1 *(249,250)* and in transgenic mice overxpressing normal and mutant neurofilament proteins *(254–256)*. SOD1 and neurofilament transgenic mouse models of ALS have not revealed structural or biochemical changes for apoptotic death of motor neurons similar to those we have discovered in humans with ALS *(10)*. In fact, it is still uncertain whether motor neurons in these mouse models die or whether they remain in a severely atrophic state. The survival of mice with familial ALS mutations is prolonged when crossed with mice overexpressing Bcl-2 *(258)* or a dominant negative inhibitor of caspase-1 *(259)*, although the degeneration of motor neurons is not prevented *(258)*, suggesting that neuronal degeneration in these mice is not apoptosis controlled by programmed cell death mechanisms. The prominent vacuolar and edematous degeneration of motor neurons in mice overexpressing mutant SOD1 *(249,250)* or neurofilament protein *(254–257)* more closely resembles excitotoxic neurodegeneration *(45)* or transsynaptic neuronal atrophy (but not death) in response to deafferentation *(260)*.

We and others have proposed that excitotoxicity may explain the selective neuronal cell death in ALS *(6,7,11,42,47)*. This hypothesis was formulated based on background information showing that abnormal activation of GluR can kill neurons *(41–43)* and that exogenous glutamate analogs may be responsible for the damage to upper motor neurons in lathyrism via actions at specific GluRs *(42,261)*. Subsequently, studies revealed that serum and cerebrospinal fluid concentrations of glutamate are increased in some patients with sporadic ALS *(262,263)*, but this change was not found in other studies *(264,265)*. In more recent studies, it has been shown that spinal cord and affected brain regions of patients with sporadic ALS have reduced high-affinity glutamate uptake *(6)* and a selective reduction in GLT1 *(7)*. It has also been reported that aberrant splicing of astroglial GLT1 mRNA is ALS-specific and is the cause for reduced expression of GLT1 protein and even possibly the cause of motor neuron degeneration *(266)*. However, defective astroglial glutamate uptake and loss of GLT1 occur also in brain regions selectively vulnerable in AD *(267)*, in brain regions vulnerable to hypoxia-ischemia *(83,84)*, and even in brain regions that do not develop neuronal degeneration after axotomy and deafferentation in experimental animals *(89,90)*. In addition, splicing of GLT1 mRNA transcripts is highly variable in human CNS tissue, and variant transcipts considered to be "aberrant" are found in normal individuals and in other species *(268–270)*. Furthermore, the degeneration of motor neurons in individuals with ALS is structurally very different from excitotoxic neuronal degeneration in the adult CNS *(10)*. Thus, it is unlikely that the abnormalities in astroglial GLT1 and GluR-mediated excitotoxicity are the specific causes for motor neuron degeneration in ALS.

Over the years it has become increasingly clear that selective neuronal vulnerability in ALS is not simply related to excitotoxic processes. Neither differential subunit compositions of ion channel GluRs (AMPA and NMDA receptors) nor differential expressions of glutamate transporters appear to explain sufficiently the mechanisms that dictate the selective vulnerability of neurons in motor cortex and motor neurons in brainstem and spinal cord in ALS. This conclusion is based partly on glutamate receptor/transporter expression and cellular localization studies in animal and human CNS performed by us *(35,37,38,40,271,273)* and by others *(36,274,275)*. The foundation for the excitotoxicity theory in the pathogenesis of ALS is built on experiments showing a selective loss of the astroglial glutamate transporter GLT1 in vulnerable regions in ALS *(7)*. Nevertheless, this observation is still indirect evidence for GluR-mediated excitotoxicity, and the explanations for this abnormality and the disease specificity for motor neurons are controversial. It is also still uncertain whether this abnormality is related to the causes or the consequences of the disease. Most patients with ALS are maintained on mechanical ventilation, as a means of relieving symptoms of chronic hypoventilation and for prolonging life *(276)*, but patients with ALS die eventually from respiratory insufficiency *(10,276)*. In fact, it has been argued that riluzole, a drug that may improve the survival of ALS patients, acts as a Na^+ channel blocker and increases resistance to hypoxia (by reducing energy demand), not by anti-excitotoxic actions *(277)*. In support of this recent interpretation *(277)*, we have observed in a physiologically well-characterized animal model of cerebral hypoxia-asphyxia that regionally selective abnormalities occur in astroglial GLT1 expression *(83,84)*. Moreover, our new data reveal that non-NMDA and NMDA GluR toxicity in adult CNS neurons is structurally different from the pattern seen in ALS *(10,11,45)*. Thus, it is possible that changes in astroglial glutamate transporters in individuals with ALS are consequences of neurodegeneration, or, alternatively, these changes are related to premortem agonal state.

We have found recently in adult animal models of peripheral nerve avulsion that motor neuron degeneration is induced by oxidative stress and is apoptosis *(70)*. We have also found that motor neuron death in ALS is apoptosis that may be mediated by programmed cell death mechanisms *(9,10,282)*. The neuronal degeneration in ALS closely resembles neuronal apoptosis induced by target deprivation in the mature CNS *(10,11,69,70)* and could be related to the deficiency in neurotrophic factors that is found in individuals with ALS *(257)*. Some of these neurotrophic factors are produced by neuroglia. Trophic factor deprivation-induced motor neuron apoptosis in cell culture and in developing animals *(278–280)* is caspase- Bax-, and p53-dependent, as might be the case for motor neuron apoptosis in ALS *(10,281,282)*. Our model of motor neuron apoptosis in the spinal cord will be useful for identifying the synaptic and molecular pathobiology of neuronal apoptosis in the adult CNS and thereby will provide insight into the mechanisms, including the roles for neuroglial-derived trophic factors, for the neuronal degeneration in ALS *(281,282)*.

ACKNOWLEDGMENTS

This research was supported by grants from the U.S. Public Health Service (NS 34100, and A61682) and the U.S. Army Medical Research and Materiel Command (DAMD17-99-1-9553). The author thanks Ann Price and Adeel Kaiser for technical assistance and Drs. Carlos Portera-Cailliau, Stephen Ginsberg, Nael Al-Abdulla, Chun-I Sze, Akiko Furuta, Frances Northington, JoAnne Natale, Ansgar Brambrink, Frederick Sieber, Jeffrey Kirsch, and Stephen Hays for their contributions in his laboratory.

REFERENCES

1. Price, D.L., Martin, L.J., Clatterbuck, R.E., et al. (1992) Neuronal degeneration in human diseases and animal models. *J. Neurobiol.* **23,** 1277–1294.
2. Martin, L.J., Pardo, C.A., Cork, L.C., and Price, D.L. (1994) Synaptic pathology and glial responses to neuronal injury precede the formation of senile plaques and amyloid deposits in the aging cerebral cortex. *Am. J. Pathol.* **145,** 1358–1381.
3. Furuta, A., Price, D.L., Pardo, C.A., et al. (1995) Localization of superoxide dismutases in Alzheimer's disease and Down's syndrome neocortex and hippocampus. *Am. J. Pathol.* **146,** 357–367.
4. Terry, R.D. (1996) The pathogenesis of Alzheimer disease: an alternative to the amyloid hypothesis. *J. Neuropathol. Exp. Neurol.* **55,** 1023–125.
5. Sze, C.-I., Troncoso, J.C., Kawas, C., Mouton, P., Price, D.L., and Martin, L.J. (1997) Loss of the presynaptic vesicle protein synaptophysin in hippocampus correlates with cognitive decline in Alzheimer's disease. *J. Neuropathol. Exp. Neurol.* **56,** 933–994.
6. Rothstein, J.D., Martin, L.J., and Kuncl, R.W. (1992) Decreased glutamate transport by the brain and spinal cord in amyotrophic lateral sclerosis. *N. Engl. J. Med.* **326,** 1464–1468.
7. Rothstein, J.D., Van Kammen, M., Levey, A.I., Martin, L.J., and Kuncl, R.W. (1995) Selective loss of glial glutamate transporter GLT-1 in amyotrophic lateral sclerosis. *Ann. Neurol.* **38,** 73–84.
8. Rosen, D.R., Siddique, T., Patterson, D., et al. (1993) Mutations in Cu/Zn superoxide dismutase gene are associated with familial amyotrophic lateral sclerosis. *Nature* **362,** 59–62.
9. Martin, L.J. (2000) p53 is abnormally elevated and active in the CNS of patients with amyotrophic lateral sclerosis. *Neurobiol. Disease* **7,** 613–622.
10. Martin, L.J. (1999) Neuronal death in amyotrophic lateral sclerosis is apoptosis: possible contribution of a programmed cell death mechanism. *J. Neuropathol. Exp. Neurol.* **58,** 459–471.
11. Martin, L.J., Al-Abdulla, N.A., Brambrink, A.M., Kirsch, J.R., Sieber, F.E., and Portera-Cailliau, C. (1998) Neurodegeneration in excitotoxicity, global cerebral ischemia, and target deprivation: a perspective on the contributions of apoptosis and necrosis. *Brain Res. Bull.* **46,** 281–309.
12. Evans, D.A., Funkenstein, H.H., Albert, M.S. et al. (1989) Prevalence of Alzheimer's disease in a community population of older persons. Higher than previously reported. *JAMA* **262,** 2551–2556.
13. Williams, D.B. and Windebank, A.J. (1993) Motor neuron disease, in *Peripheral Neuropathy* (Dyck, P.J., Thomas, P.K., Griffin, J.W., Low, P.A., and Poduslo, J.F. eds.), Saunders, Philadelphia, pp. 1028–1050.
14. Kalimo, H., Kaste, M., and Haltia, M. (1997) Vascular diseases, in *Greenfields Neuropathology,* (Graham, D.I. and Lantos P.L., eds.), Arnold, London, pp. 315–396.
15. Sudhof, T.C. (1995) The synaptic vesicle cycle: a cascade of protein-protein interactions. *Nature* **375,** 645–653.
16. Kim, E., Naisbitt, S., Hsueh Y.-P., et al. (1997) GKAP, a novel synaptic protein that interacts with the guanylate kinase-like domain of the PSD95/SAP90 family of channel clustering molecules. *J. Cell Biol.* **136,** 669–678.
17. Ehlers, M.D., Fung, E.T., O'Brien, R.J., and Huganir, R.L. (1998) Splice variant-specific interaction of the NMDA receptor subunit NR1 with neuronal intermediate filaments. *J. Neurosci.* **18,** 720–730.
18. Wechsler, A. and Teichberg V.I. (1998) Brain spectrin binding to the NMDA receptor is regulated by phosphorylation, calcium and calmodulin. *EMBO J.* **17,** 3931–3939.
19. Montgomery, D.L. (1994) Astrocytes: form, functions, and roles in disease. *Vet. Pathol.* **31,** 145–167.

20. Norenberg, M.D. (1994) Astrocyte responses to CNS injury. *J. Neuropathol. Exp. Neurol.* **53,** 213–220.
21. Curtis, D.R., Phillis, J.W., and Watkins, J.C. (1959) Chemical excitation of spinal neurons. *Nature* **183,** 611–612.
22. Watkins, J.C. and Evans, R.H. (1981) Excitatory amino acid transmitters. *Ann. Rev. Pharmcol. Toxicol.* **21,** 165–204.
23. Clements, J.D., Lester, R.A.J., Tong, G., Jahr, C.E., and Westbrook, G.L. (1992) The time course of glutamate in the synaptic cleft. *Science* **258,** 1498–1501.
24. Nakanishi, S. (1992) Molecular diversity of glutamate receptors and implications for brain function. *Science* **258,** 597–603.
25. Seeburg, P.H. (1993) The molecular biology of mammalian glutamate receptor channels. *TINS* **16,** 359–365.
26. Sommer, B., Keinänen, K., Verdoorn, T.A., et al. (1990) Flip and flop: a cell-specific functional switch in glutamate-operated channels of the CNS. *Science* **249,** 1580–1585.
27. Hollman, M., Hartley, M. and Heinemann, S. (1991) Ca^{2+} permeability of KA-AMPA-gated glutamate receptor channels depends on subunit composition. *Science* **252,** 851–853.
28. Sommer, B., Köhler, M., Sprengel, R., and Seeburg, P.H. (1991) RNA editing in brain controls a determinant of ion flow in glutamate-gated channels. *Cell* **67,** 11–19.
29. Burnashev, N., Monyer, H., Seeburg, P.H., and Sakmann, B. (1992) Divalent ion permeability of AMPA receptor channels is dominated by the edited form of a single subunit. *Neuron* **8,** 189–198.
30. Tanabe, Y., Masu, M., Ishii, T., Shigemoto, R., and Nakanishi, S. (1992) A family of metabotropic glutamate receptors. *Neuron* **8,** 169–179.
31. Pin, J.-P., and Duvoisin, R. (1995) The metabotropic glutamate receptors: structure and functions. *Neuropharmacol.* **34,** 1–26.
32. Kanai, Y., Smith, C.P. and Hediger, M.A. (1993) A new family of neurotransmitter transporters: the high-affinity glutamate transporters. *FASEB J.* **7,** 1450–1459.
33. Danbolt, N.C. (1994) The high affinity uptake system for excitatory amino acids in the brain. *Prog. Neurobiol.* **44,** 377–396.
34. Arriza, J.L., Eliasof, S., Kavanaugh, M.P., and Amara, S.P. (1997) Excitatory amino acid transporter 5, a retinal glutamate transporter coupled to a chloride conductance. *Proc. Natl. Acad. Sci. USA* **94,** 4155–4160.
35. Rothstein, J.D., Martin, L., Levey, A.I., et al. (1994) Localization of neuronal and glial glutamate transporters. *Neuron* **13,** 713–725.
36. Chaudhry, F.A., Lehre, K.P., van Lookeren Campagne, M., Ottersen, O.P., Danbolt, N.C., and Storm-Mathisen, J. (1995) Glutamate transporters in glial plasma membranes: highly differentiated localizations revealed by quantitative ultrastructural immunocytochemistry. *Neuron* **15,** 711–720.
37. Furuta, A., Rothstein, J.D., and Martin, L.J. (1997) Glutamate transporter protein subtypes are expressed differentially during rat CNS development. *J. Neurosci.* **17,** 8363–8375.
38. Northington, F.J., Traystman, R.J., Koehler, R.C., Rothstein, J.R., and Martin, L.J. (1998) Regional and cellular expression of glial (GLT1) and neuronal EAAC1 glutamate transporter proteins in ovine fetal brain. *Neuroscience* **85,** 1183–1194.
39. Furuta, A., Martin, L.J., C.-L. G., Lin, Dykes-Hoberg, M., and Rothstein, J.D. (1997) Cellular and synaptic localization of the neuronal glutamate transporters, excitatory amino acid transporters 3 and 4. *Neuroscience* **81,** 1031–1042.
40. Northington, F.J., Traystman, R.J., Koehler, R.C., and Martin, L.J. (1999) GLT1, glial glutamate transporter, is transiently expressed in neurons and develops astrocyte specificity only after midgestation in the ovine fetal brain. *J. Neurobiol.* **39,** 515–526.

41. Lucas, D.R. and Newhouse, J.P. (1957) The toxic effect of sodium *L*-glutamate on the inner layers of the retina. *AMA Arch. Opthalmol.* **58,** 193–201.
42. Olney, J.W. (1994) Excitatory transmitter neurotoxicity. *Neurobiol. Aging* **15,** 259–260.
43. Choi, D.W. (1992) Excitotoxic cell death. *J. Neurobiol.* **23,** 1261–1276.
44. Portera-Cailliau, C., Price, D.L., and Martin, L.J. (1997) Excitotoxic neuronal death in the immature brain is an apoptosis-necrosis morphological continuum. *J. Comp. Neurol.* **378,** 70–87.
45. Portera-Cailliau, C., Price, D.L., and Martin, L.J. (1997) Non-NMDA and NMDA receptor-mediated excitotoxic neuronal deaths in adult brain are morphologically distinct: further evidence for an apoptosis-necrosis continuum. *J. Comp. Neurol.* **378,** 88–104.
46. Stewart, G.R., Zorumski, C.F., Price, M.T., and Olney, J.W. (1990) Domoic acid: a dementia-inducing excitotoxic food poison with kainic acid receptor specificity. *Exp. Neurol.* **110,** 127–138.
47. Martin, L.J., Portera-Cailliau, C., Ginsberg, S.D., and Al-Abdulla, N.A. (1998) Animal models and degenerative disorders of the human brain. *Lab Animal* **27,** 18–25.
48. Gwag, B.J., Koh, J.Y., DeMaro, J.A., Ying, H.S., Jacquin, M., and Choi, D.W. (1997) Slowly triggered excitotoxicity occurs by necrosis in cortical cultures. *Neuroscience* **77,** 393–401.
49. Kure, S., Tominaga, T., Yoshimoto, T., Tada, K., and Narisawa, K. (1991) Glutamate triggers internucleosomal DNA cleavage in neuronal cells. *Biochem. Biophys. Res. Commun.* **179,** 39–45.
50. Ankarcrona, M., Dypbukt, J.M., Bonfoco, E., et al. (1995) Glutamate-induced neuronal death: a succession of necrosis or apoptosis depending on mitochondrial function. *Neuron* **15,** 961–973.
51. Simonian, N.A., Getz, R.L., Leveque, J.C., Konradi, C., and Coyle, J.T. (1996) Kainate induces apoptosis in neurons. *Neuroscience* **74,** 675–683.
52. Dessi, F., Charriaut-Marlangue, C., Khrestchatisky, M., and Ben-Ari, Y. (1993) Glutamate-induced neuronal death is not a programmed cell death in cerebellar culture. *J. Neurochem.* **60,** 1953–1955.
53. Ferrer, I., Martin, F., Serrano, T., et al. (1995) Both apoptosis and necrosis occur following intrastriatal administration of excitotoxins. *Acta Neuropathol.* **90,** 504–510.
54. van Lookeren Campagne, M., Lucassen, P.J., Vermeulen, J.P., and Balázs, R. (1995) NMDA and kainate induced internucleosomal DNA cleavage associated with both apoptotic and necrotic cell death in the neonatal rat brain. *Eur. J. Neurosci.* **7,** 1627–1640.
55. Halliwell, B. and Gutteridge, J.M.C. (1986) Oxygen free radicals and iron in relation to biology and medicine: some problems and concepts. *Arch. Biochem. Biophys.* **246,** 501–514.
56. McCord, J.M. (1985) Oxygen-derived free radicals in postischemic tissue injury. *N. Engl. J. Med.* **312,** 159–163.
57. Beckman, J.S., Chen, J., Ischiropoulos, H., and Conger, K.A. (1992) Inhibition of nitric oxide synthesis and cerebral neuroprotection, in *Pharmacology of Cerebral Ischemia* (Krieglstein, J. and Oberpichler-Schwenk, H., eds.), Wissenschaftliche Verlagsgesellschaft, Stuttgart, pp. 383–394.
58. Ratan, R.R., Murphy, T.H., and Baraban, J.M. (1994) Macromolecular synthesis inhibitors prevent oxidative stress-induced apoptosis in embryonic cortical neurons by shunting cysteine from protein synthesis to glutathione. *J. Neurosci.* **14,** 4385–4392.
59. Bonfoco, E., Krainc, D., Ankarcrona, M., Nicotera, P., and Lipton, S.A. (1995) Apoptosis and necrosis: two distinct events induced respectively by mild and intense insults with NMDA or nitric oxide/superoxide in cortical cell cultures. *Proc. Natl. Acad. Sci. USA* **92,** 72162–72166.
60. Boveris, A. and Cadenas, E. (1997) Cellular sources and steady-state levels of reactive oxygen species, in *Oxygen, Gene Expression, and Cellular Function* (Biadasz Clerch, L. and Massaro, D.J., eds.), Marcel Dekker, New York, pp. 1–25.

61. Bandy, B. and Davison, A.J. (1990) Mitochondrial mutations may increase oxidative stress: implications for carcinogenesis and aging? *Free Rad. Biol. Med.* **8,** 523–539.
62. Trotti, D., Danbolt, N.C., and Volterra, A. (1998) Glutamate transporters are oxidant-vulnerable: a molecular link between oxidative and excitotoxic neurodegeneration. *TIPS* **19,** 328–334.
63. Sato, T., Kamata, Y., Irifune, M., and Nishikawa, T. (1997) Inhibitory effects of several nitric oxide-generating compounds on purified Na^+, K^+-ATPase activity from porcine cerebral cortex. *J. Neurochem.* **68,** 1312–1318.
64. Mark, R.J., Pang, Z., Gedded, J.W., Uchida, K., and Mattson, M.P. (1997) Amyloid β-peptide impairs glucose transport in hippocampal and cortical neurons: involvement of membrane lipid peroxidation. *J. Neurosci.* **17,** 1046–1054.
65. Mirabelli, F., Salis, A., Perotti, M., Taddei, F., Bellomo, G., and Orrenius, S. (1998) Alterations of surface morphology caused by the metabolism of menadione in mammalian cells are associated with the oxidation of critical sulfhydryl groups in cytoskeletal proteins. *Biochem. Pharmacol.* **37,** 3423–3427.
66. Troncoso, J.C., Costello, A.C., Kim, J.H., and Johnson, G.V.W. (1995) Metal-catalyzed oxidation of bovine neurofilaments in vitro. *Free Rad. Biol. Med.* **18,** 891–899.
67. Salo, D.C., Pacifici, R.E., Lin, S.W., Giulivi, C., and Davies, K.J.A. (1990) Superoxide dismutase undergoes proteolysis and fragmentation following oxidative modification and inactivation. *J. Biol. Chem.* **265,** 11919–11927.
68. Greenlund, L.J.S., Deckwerth, T.L., and Johnson, E.M. (1995) Superoxide dismutase delays neuronal apoptosis: a role for reactive oxygen species in programmed neuronal death. *Neuron* **14,** 303–315.
69. Al-Abdulla, N.A. and Martin, L.J. (1998) Apoptosis of retrogradely degenerating neurons occurs in association with the accumulation of perikaryal mitochondria and oxidative damage to the nucleus. *Am. J. Pathol.* **153,** 447–456.
70. Martin, L.J., Kaiser, A., and Price, A.C. (1999) Motor neuron degeneration after sciatic nerve avulsion in adult rat evolves with oxidative stress and is apoptosis. *J. Neurobiol.* **40,** 185–201.
71. Hamburger, V. (1975) Cell death in the development of the lateral motor column of the chick embryo. *J. Comp. Neurol.* **160,** 535–546.
72. Yan, Q., Elliot, J.L., Matheson C, et al. (1993) Influences of neurotrophins on mammalian motoneurons in vivo. *J. Neurobiol.* **24,** 1555–1577.
73. Murphy, T.H., Miyamoto, M., Sastre, A., Schnaar, R.L., and Coyle, J.T. (1989) Glutamate toxicity in a neuronal cell line involves inhibition of cystine transport leading to oxidative stress. *Neuron* **2,** 1547–1558.
74. Sagara, Y. And Schubert, D. (1998) The activation of metabotropic glutamate receptors protects nerve cells from oxidative stress. *J. Neurosci.* **18,** 6662–6671.
75. Benveniste, H., Drejer, J., Schousboe, A., and Diemer, N.H. (1984) Elevation of the extracellular concentrations of glutamate and aspartate in rat hippocampus during transient cerebral ischemia monitored by intracerebral microdialysis. *J. Neurochem.* **43,** 1369–1374.
76. Drejer, J., Benveniste, H., Diemer, N.H., and Schousboe, A. (1985) Cellular origin of ischemia-induced glutamate release from brain tissue in vivo and in vitro. *J. Neurochem.* **45,** 145–151.
77. Globus, M.Y.-T., Busto, R., Dietrich, W.D., Martinez, E., Valdes, I., and Ginsberg, M.D. (1988) Intra-ischemic extracellular release of dopamine and glutamate is associated with striatal vulnerability to ischemia. *Neurosci. Lett.* **91,** 36–40.
78. Mitani, A., Andou, Y., and Kataoka, K. (1992) Selective vulnerability of hippocampal CA1 neurons cannot be explained in terms of an increase in glutamate concentration during ischemia in the gerbil: brain microdialysis study. *Neuroscience* **48,** 307–313.
79. Diemer, N.H., Valente, E., Bruhn, T., Berg, M., Jørgensen, M.B., and Johansen, F.F. (1993) Glutamate receptor transmission and ischemic nerve cell damage: evidence for involvement of excitotoxic mechanisms. *Prog. Brain Res.* **96,** 105–123.

80. Rothman, S.M. (1984) Synaptic release of excitatory amino acid neurotransmitter mediates anoxic neuronal death. *J. Neurosci.* **4,** 1884–1891.
81. Szatkowski, M. and Attwell, D. (1994) Triggering and execution of neuronal death in brain ischaemia: two phases of glutamate release by different mechanisms. *TINS* **17,** 359–365.
82. Silverstein, F.S., Buchanan, K., and Johnston, M.V. (1986) Perinatal hypoxia-ischemia disrupts striatal high-affinity ^{3}H-glutamate uptake into synaptosomes. *J. Neurochem.* **47,** 1614–1619.
83. Martin, L.J., Brambrink, A.M., Lehmann, C., Portera-Cailliau, C., Koehler, R., Rothstein, J., and Traystman, R.J. (1997) Hypoxia-ischemia causes abnormalities in glutamate transporters and death of astroglia and neurons in newborn striatum. *Ann. Neurol.* **42,** 335–348.
84. Martin, L.J. (2001) Mechanisms of brain damage in animal models of hypoxia-ischemia in newborns. In: Fetal and neonatal brain injury: Mechanisms, management, and the risks of practice (Stevenson, D.K. and Sunshine, P., eds.), 2nd edition. Oxford University Press, *in press.*
85. Kimelberg, H.K., Goderie, S.K., Higman, S., Pang, S., and Waniewski, R.A. (1990) Swelling-induced release of glutamate, aspartate, and taurine from astrocyte cultures. *J. Neurosci.* **10,** 1583–1591.
86. McBean, G.J. and Roberts, P.J. (1985) Neurotoxicity of L-glutamate and DL-threo-3-hydroxy-aspartate in the rat striatum. *J. Neurochem.* **44,** 247–254.
87. Tanaka, K., Watase, K., Manabe, T., et al. (1997) Epilepsy and exacerbation of brain injury in mice lacking the glutamate transporter GLT-1. *Science* **276,** 1699–1702.
88. Harada, T., Harada, C., Watanabe, M., et al. (1998) Functions of the two glutamate transporters GLAST and GLT-1 in the retina. *Proc. Natl. Acad. Sci. U.S.A.* **95,** 4663–4666.
89. Ginsberg, S.D., Martin, L.J., and Rothstein, J.D. (1995) Regional deafferentation down-regulates subtypes of glutamate transporter proteins. *J. Neurochem.* **65,** 2800–2803.
90. Ginsberg, S.D., Rothstein, J.D., Price, D.L., and Martin, L.J. (1996) Fimbria-fornix transections selectively down-regulate subtypes of glutamate transporter and glutamate receptor proteins in septum and hippocampus. *J. Neurochem.* **67,** 1208–1216.
91. Anderson, K.J., Nellgård, B., and Wieloch, T. (1993) Ischemia-induced upregulation of excitatory amino acid transport sites. *Brain Res.* **662,** 93–98.
92. Torp, R., Lekieffre, D., Levy, L.M., et al. (1995) Reduced postischemic expression of a glial glutamate transporter, GLT1, in the rat hippocampus. *Exp. Brain Res.* **103,** 51–58.
93. Ferrer-Martinez, A., Felipe, A., Nicholson, B., Casado, J., Pastor-Anglada, M., and McGivan, J. (1995) Induction of the high-affinity Na^+-dependent glutamate transport system XAG by hypertonic stress in the renal epithelial cell line NBL-1. *Biochem. J.* **310,** 689–692.
94. Conradt, M., Storck, T., and Stoffel W. (1995) Localization of N-glycosylation sites and functional role of the carbohydrate units of GLAST-1, a cloned rat brain L-glutamate/L-aspartate transporter. *Eur. J. Biochem.* **229,** 682–687.
95. Martin, L.J., Brambrink, A.M., Price A.C., et al. (2000) Neuronal death in newborn striatum after hypoxia-ischemia is necrosis and evolues with oxidative stress. *Neurobiol. Disease* **7,** 169–191.
96. Conradt, M. and Stoffel, W. (1997) Inhibition of the high-affinity brain glutamate transporter GLAST-1 via direct phosphorylation. *J. Neurochem.* **68,** 1244–1251.
97. Casado, M., Bendahan, A., Zafra, F., et al. (1993) Phosphorylation and modulation of brain glutamate transporters by protein kinase *C. J. Biol. Chem.* **268,** 27313–27317.
98. Sieber, F.E., Traystman, R.J., Brown, P.R., and Martin, L.J. (1998) Protein kinase C expression and activity after global incomplete cerebral ischemia in dogs. *Stroke* **29,** 1445–1453.
99. Volterra, A., Trotti, D., Tromba, C., Floridi, S., and Racagni, G. (1994) Glutamate uptake inhibition by oxygen free radicals in rat cortical astrocytes. *J. Neurosci.* **14,** 2924–2932.
100. Trotti, D., Rossi, D., Gjesdal, O., et al. (1996) Peroxynitrite inhibits glutamate transporter subtypes. *J. Biol. Chem.* **271,** 5976–5979.

101. David, J.C., Yamada, K.A., Bagwe, M.R., and Goldberg, M.P. (1996) AMPA receptor activation is rapidly toxic to cortical astrocytes when desensitization is blocked. *J. Neurosci.* **16,** 200–209.
102. Kalimo, H., Rehncrona, S., Soderfeldt, B., Olsson, Y., and Siesjo, B.K. (1981) Brain lactic acidosis and ischemic cell damage: 2. Histopathology. *J. Cereb. Blood Flow Metabol.* **1,** 313–327.
103. Norenberg, M.D., Mozes, L.W., Gregorios, J.B., and Norenberg, L.B. (1987) Effects of lactic acid on astrocytes in primary culture. *J. Neuropathol. Exp. Neurol.* **46,** 154–166.
104. Brown, A.W. and Brierley, J.B. (1973) The earliest alterations in rat neurones and astrocytes after anoxia-ischaemia. *Acta Neuropathol.* **23,** 9–22.
105. Rosenberg, P.A., Amin, S., and Leitner, M. (1992) Glutamate uptake disguises neurotoxic potency of glutamate agonists in cerebral cortex in dissociated cell culture. *J. Neurosci.* **12,** 56–61.
106. Sugiyama, K., Brunori, A., and Mayer, M.L. (1989) Glial uptake of excitatory amino acids influences neuronal survival in cultures of mouse hippocampus. *Neuroscience* **32,** 779–791.
107. Sheardown, M.J., Nielsen, E.Ø., Hansen, A.J., Jacobsen, P., and Honoré, T. (1990) 2,3-Dihydroxy-6-nitro-7-sulfamoyl-benzo(F)quinoxaline: a neuroprotectant for cerebral ischemia. *Science* **247,** 571–574.
108. Kirino, T., Robinson, H.P.C., Miwa, A., Tamura, A., and Kawai N. (1992) Disturbance of membrane function preceding ischemic delayed neuronal death in the gerbil hippocampus. *J. Cereb Blood Flow Metabol* **12,** 408–417.
109. Wahlestedt, C., Golanov, E., Yamamoto, S., et al. (1993) Antisense oligodeoxynucleotides to NMDA-R1 receptor channel protect cortical neurons from excitotoxicity and reduce focal ischaemic infarctions. *Nature* **363,** 260–263.
110. Morikawa, E., Mori, H., Kiyama, Y., Mishina, M., Asano, T., and Kirino, T. (1998) Attenuation of focal ischemic brain injury in mice deficient in the ε1 (NR2A) subunit of NMDA receptor. *J. Neurosci.* **18,** 9727–9732.
111. Di, X., Bullock, R., Watson, J., et al. (1997) Effect of CP101,606, a novel NR2B subunit antagonist of the *N*-methyl-*D*-aspartate receptor, on the volume of ischemic brain damage and cytotoxic brain edema after middle cerebral artery occlusion in the feline brain. *Stroke* **28,** 2244–2251.
112. Martin, L.J., Sieber, F.E., and Traystman, R.J. (2000) Apoptosis and necrosis occur in separate neuronal populations in hippocampus and cerebellum after ischemia and are associated with alterations in metabotropic glutamate receptor signaling pathways. *J. Cereb. Blood Flow Metab.* **20,** 153–167.
113. Buchan, A. and Pulsinelli, W.A. (1990) Hypothermia but not the *N*-methyl-*D*-aspartate antagonist, MK-801, attenuates neuronal damage in gerbils subjected to transient global ischemia. *J. Neurosci.* **10,** 311–316.
114. Sheardown, M.J., Suzdak, P.D., and Nordholm, L. (1993) AMPA, but not NMDA, receptor antagonism is neuroprotective in gerbil global ischaemia, even when delayed 24 h. *Eur. J. Pharmacol.* **236,** 347–353.
115. LeBlanc, M.H., Vig, V., Smith, B., Parker, C.C., Evans, O.B., and Smith, E.E. (1991) MK-801 does not protect against hypoxic-ischemic brain injury in piglets. *Stroke* **22,** 1270–1275.
116. Helfaer, M.A., Ichord, R.N., Martin, L.J., Hurn, P.D., Castro, A., and Traystman, R.J. (1998) Treatment with the competitive NMDA receptor antagonist GPI 3000 does not improve outcome after cardiac arrest in dog. *Stroke* **29,** 824–829.
117. LeBlanc, M.H., Li, X.Q., Huang, M., Patel, D.M., and Smith, E.E. (1995) AMPA antagonist LY293558 does not affect the severity of hypoxic-ischemic injury in newborn pigs. *Stroke* **26,** 1908–1915.
118. Martin, L.J., Brambrink, A., Koehler, R.C., and Traystman, R.J. (1997) Neonatal asphyxic brain injury is neural system preferential and targets sensory-motor networks, in *Fetal and Neonatal*

Brain Injury: Mechanisms, Management, and the Risks of Practice (Stevenson, D.K. and Sunshine, P. eds.), Oxford University Press, pp 374–399.

119. Brambrink, A.M., Martin, L.J., Hanley, D.F., Becker, K.J., Koehler, R.C., and Traystman, R.J. (1999) Effects of the AMPA receptor antagonist NBQX on outcome of newborn pigs after asphyxic cardiac arrest. *J. Cereb. Blood Flow Metab.* **19,** 927–938.
120. Sieber, F.E., Brown, P.R., Ichord, R.N., Traystman, R.J., and Martin, L.J. (1998) Antagonists to metabotropic glutamate receptors worsen neurologic outcome following global ischemia. *Soc. Neurosci.* **24,** 897.
121. Kaczmarek, L.K. (1987) The role of protein kinase C in the regulation of ion channels and neurotransmitter release. *TINS* **10,** 30–34.
122. Saitoh, T., Masliah, E., Jin, L.-W., Cole, G.M., Wieloch, T., and Shapiro, I.P. (1991) Protein kinases and phoshorylation in neurologic disorders and cell death. *Lab. Invest.* **64,** 596–616.
123. Candeo, P., Favaron, M., Lengyel, I., Manev, R.M., Rimland, J.M., and Manev, H. (1992) Pathological phosphorylation causes neuronal death: effect of okadaic acid in primary culture of cerebellar granule cells. *J. Neurochem.* **59,** 1558–1561.
124. Hajimohammadreza, I, Probert, A.W., Coughenour, L.L., et al. (1995) A specific inhibitor of calcium/calmodulin-dependent protein kinase-II provides neuroprotection against NMDA- and hypoxia/hypoglycemia-induced cell death. *J. Neurosci.* **15,** 4093–4101.
125. Hara, H., Onodera, H, Yoshidomi, M., Matsuda, Y., and Kogure, K. (1990) Staurosporine, a novel protein kinase C inhibitor, prevents postischemic neuronal damage in the gerbil and rat. *J. Cereb. Blood Flow Metab.* **10,** 646–653.
126. Hu, B.-R. and Wieloch, T. (1995) Persistent translocation of Ca^{2+}/calmodulin-dependent protein kinase II to synaptic junctions in the vulnerable hippocampal CA1 region following transient ischemia. *J. Neurochem.* **64,** 277–284.
127. Hu, B.-R., Kamme, F., and Wieloch, T. (1995) Alterations of Ca^{2+}/calmodulin-dependent protein kinase II and its messenger RNA in the rat hippocampus following normo- and hypothermic ischemia. *Neuroscience* **68,** 1003–1016.
128. Wieloch, T., Cardell, M., Bingren, H., Zivin, J., and Saitoh, T. (1991) Changes in the activity of protein kinase C and the differential subcellular redistribution of its isozymes in the rat striatum during and following transient forebrain ischemia. *J. Neurochem.* **56,** 1227–1235.
129. Wang, L.-Y., Taverna, F.A., Huang, X.-P., MacDonald, J.F., and Hampson, D.R. (1993) Phosphorylation and modulation of a kainate receptor (GluR6) by cAMP-dependent protein kinase. *Science* **259,** 1173–1175.
130. Raymond, L.A., Blackstone, C.D., and Huganir, R.L. (1993) Phosphorylation and modulation of recombinant GluR6 glutamate receptors by cAMP-dependent protein kinase. *Nature* **361,** 637–641.
131. Tingley, W.G., Roche, K.W., Thompson, A.K., and Huganir, R.L. (1993) Regulation of NMDA receptor phosphorylation by alternative splicing of the C-terminal domain. *Nature* **364,** 70–73.
132. Blackstone, C., Murphy, T.H., Moss, S.J., Baraban, J.M., and Huganir, R.L. (1994) Cyclic AMP and synaptic activity-dependent phosphorylation of AMPA-preferring glutamate receptors. *J. Neurosci.* **14,** 7585–7593.
133. Tan, S.-E., Wenthold, R.J., and Soderling, T.R. (1994) Phosphorylation of AMPA-type glutamate receptors by calcium/calmodulin-dependent protein kinase II and protein kinase C in cultured hippocampal neurons. *J. Neurosci.* **14,** 1123–1129.
134. Wyllie, D.J.A. and Nicoll, R.A. (1994) A role for protein kinases and phosphatases in the Ca^{2+}-induced enhancement of hippocampal AMPA receptor-mediated synaptic responses. *Neuron* **13,** 635–643.
135. Desai, M.A., Burnett, J.P., Mayne, N.G., and Schoepp, D.D. (1996) Pharmacological characterization of desensitization in a human mGluR1 α-expressing non-neuronal cell line co-transfected with a glutamate transporter. *Br. J. Pharmacol.* **118,** 1558–1564.

136. Gereau, R.W., IV and Heinemann, S.F. (1998) Role of protein kinase C phosphorylation in rapid desensitization of a metabotropic glutamate receptor 5. *Neuron* **20,** 143–151.
137. Kelso, S.R., Nelson, T.E., and Leonard, J.P. (1992) Protein kinase C-mediated enhancement of NMDA currents by metabotropic glutamate receptors in *Xenopus oocytes. J. Physiol.* **449,** 705–718.
138. Aniksztejn, L., Otani, S., Ben-Ari, Y. (1992) Quisqualate metabotropic receptors modulate NMDA currents and facilitate induction of long-term potentiation through protein kinase C. *Eur. J. Neurosci.* **4,** 500–505.
139. Chen, L., and Huang, L.-Y. M. (1992) Protein kinase C reduces Mg^{2+} block of NMDA-receptor channels as a mechanism of modulation. *Nature* **356,** 521–523.
140. Sieber, F.E., Traystman, R.J., and Martin, L.J. (1997) Delayed neuronal death following global incomplete ischemia in dogs is accompanied by changes in phospholipase C protein expression. *J. Cereb. Blood Flow Metab.* **17,** 527–533.
141. Chang, H.S., Sasaki, T., and Kassel, N.F. (1989) Hippocampal unit activity after transient ischemia in rats. *Stroke* **20,** 1951–1058.
142. Omon, H., Mitani, A., Andou, Y., Arai, T., and Kataoka, K. (1991) Delayed neuronal death is induced without postischemia hyperexcitability: continuous multiple-unit recording from ischemic CA1 neurons. *J. Cereb. Blood Flow Metab.* **11,** 819–823.
143. Mitani, A., Imon, H., Iga, K., Kubo, H., and Kataoka, K. (1990) Gerbil hippocampal extracellular glutamate and neuronal activity after transient ischemia. *Brain Res. Bull.* **25,** 319–324.
144. Urban, L., Neill, K.H., Crain, B.J., Nadler, J.V., and Somjen, G.G. (1989) Postischemic synaptic physiology in area CA1 of the gerbil hippocampus studied *in vitro. J. Neurosci.* **9,** 3966–3975.
145. Xu, Z.C. (1995) Neurophysiological changes of spiny neurons in rat neostriatum after transient forebrain ischemia: an in vivo intracellular recording and staining study. *Neuroscience* **67,** 823–836.
146. Tsubokawa, H., Ogura, K., Masuzawa, T., and Kawai, N. (1994) Ca^{+}-dependent non-NMDA receptor-mediated synaptic currents in ischemic CA1 hippocampal neurons. *J. Neurophysiol.* **71,** 1190–1196.
147. Schmidt, H.H.H.W. and Walter, U. (1994) NO at work. *Cell* **78,** 919–925.
148. Vincent, S.R. (1994) Nitric oxide: a radical neurotransmitter in the central nervous system. *Prog. Neurobiol.* **42,** 129–160.
149. Bredt, D.S., and Snyder, S.H. (1992) Nitric oxide, a novel neuronal messenger. *Neuron* **8,** 3–11.
150. Northington, F.J., Koehler, R.C., Traystman, R.J., and Martin, L.J. (1996) Nitric oxide synthase 1 and nitric oxide synthase 3 protein expression is regionally and temporally regulated in fetal brain. *Dev. Brain Res.* **95,** 1–14.
151. Garthwaite, J., Garthwaite, G., Palmer, R.M.J., and Moncada, S. (1989) NMDA receptor activation induces nitric oxide synthesis from arginine in rat brain slices. *Eur. J. Pharmacol.* **172,** 413–416.
152. Bhardwaj, A., Northington, F.J., Martin, L.J., Hanley, D.F., Traystman, R.J., and Koehler, R.C. (1997) Characterization of metabotropic glutamate receptor mediated nitric oxide production *in vivo. J. Cereb. Blood Flow Metab.* **17,** 153–160.
153. Meffert, M.K., Calakos, N.C., Scheller, R.H., and Schulman, H. (1996) Nitric oxide modulates synaptic vesicle docking/fusion reactions. *Neuron* **16,** 1229–1236.
154. Stamler, J.S. (1994) Redox signaling: nitrosylation and related target interactions of nitric oxide. *Cell* **78,** 931–936.
155. Huang, A., Huang, P.L., Panahian, N., Dalkara, T., Fishman, M.C., and Moskowitz, M.A. (1994) Effects of cerebral ischemic in mice deficient in neuronal nitric oxide synthase. *Science* **265,** 1883–1885.
156. Iadecola, C., Zhang, F., Casey, R., Nagayama, M., and Ross, M.E. (1997) Delayed reduction of ischemic brain injury and neurological deficits in mice lacking the inducible nitric oxide synthase gene. *J. Neurosci.* **17,** 9157–9164.

158. Hooper, D.C., Bagasra, O., Marini, J.C., et al. (1997) Prevention of experimental allergic encephalomyelitis by targeting nitric oxide and peroxynitrite: implications for the treatment of multiple sclerosis. *Proc. Natl. Acad. Sci. USA* **94,** 2528–2533.
159. Martin, L.J. (2001) The apoptosis-necrosis continuum in CNS development, injury, and diseases: contributions and mechanisms. In: Neuroprotection (Lo, E. and Marwah, J., eds.) Prominent Press, in press.
160. Yoshida, T., Limmroth, V., Irkura, K., and Moskowitz, M.A. (1994) The NOS inhibitor, 7-nitroindazole, decreases focal infarct volume but not the response to topical acetylcholine in pial vessels. *J. Cereb. Blood Flow Metab.* **14,** 924–929.
161. Kuluz, J.W., Prado, R.J., Dietrich, W.D., Schleien, C.L., and Watson, B.D. (1993) The effect of nitric oxide synthase inhibition on infarct volume after reversible focal cerebral ischemia in conscious rats. *Stroke* **24,** 2023–2029.
162. Buchan, A.M., Gertler, S.Z., Huang, Z.-G., Li, H., Chaundy, K.E., and Xue, D. (1994) Failure to prevent selective CA1 neuronal death and reduce cortical infarction following cerebral ischemia with inhibition of nitric oxide synthase. *Neuroscience* **61,** 1–11.
163. Panahian, N., Yoshida, T., Huang, P.L., et al. (1996) Attenuated hippocampal damage after global cerebral ischemia in mice mutant in neuronal nitric oxide synthase. *Neuroscience* **72,** 343–354.
164. Sancesario, G., Iannone, M., Morello, M., Nisticò, G., and Bernardi, G. (1994) Nitric oxide inhibition aggravates ischemic damage of hippocampal but not of NADPH neurons in gerbils. *Stroke* **25,** 436–444.
165. Kirsch, J.R., Bhardwaj, A., Martin, L.J., Hanley, D.F., and Traystman, R.J. (1997) Neither L-arginine nor L-NAME affect neurological outcome after global ischemia in cats. *Stroke* **28,** 2259–2264.
166. Katzman, R. (1993) Education and the prevalence of dementia and Alzheimer's disease. *Neurology* **43,** 13–20.
167. McKhann, G., Drachman, D., Folstein, M., Katzman, R., Price, D., and Stadlan, E.M. (1984) Clinical diagnosis of Alzheimer's disease: report of the NINCDS-ADRDA work group under the auspices of the Department of Health and Human Services task force on Alzheimer's disease. *Neurology* **34,** 939–944.
168. Olshansky, S.J., Carnes, B.A., and Cassel, C.K. (1993) The aging of the human species. *Sci. Am.* **268,** 46–52.
169. Chartier-Harlin, M.-C., Crawford, F., Houlden, H., et al. (1991) Early-onset Alzheimer's disease caused by mutations at codon 717 of the β-amyloid precursor protein gene. *Nature* **353,** 844–846.
170. Goate, A., Chartier-Harlin, M.-C., Mullan, M., et al. (1991) Segregation of a missense mutation in the amyloid precursor protein gene with familial Alzheimer's disease. *Nature* **349,** 704–706.
171. Naruse, S., Igarashi, S., Kobayashi, H., et al. (1991) Mis-sense mutation Val->Ile in exon 17 of amyloid precursor protein gene in Japanese familial Alzheimer's disease. *Lancet* **337,** 978–979.
172. Campion, D., Flaman, J.M., Brice, A., et al. (1995) Mutations of the presenilin 1 gene in families with early-onset Alzheimer's disease. *Hum. Mol. Genet.* **4,** 2373–2377.
173. Sherrington, R., Rogaev, E.I., Liang, Y., et al. (1995) Cloning of a gene bearing missense mutations in early-onset familial Alzheimer's disease. *Nature* **375,** 754–760.
174. Roses, A.D. (1996) Apolipoprotein E alleles as risk factors in Alzheimer's disease. *Annu. Rev. Med.* **47,** 387–400.
175. Sze, C.-I., Bi, H., Kleinschmidt-DeMasters, B.K., Filley, C.M., and Martin, L.J. (2000) Seletive regional loss of exocytotic presynaptic vesicle proteins in Alzheimer's disease. *J. Neurol. Sci.* **175,** 81–90.

176. Sze, C.-I., Bi, H., Filley, C.M., Kleinschmidt-DeMasters, B.K., and Martin, L.J. (2001) *N*-Methyl-*D*-aspartate receptor subunits and their phosphorylation status are altered selectively in Alzheimer's disease brain. *J. Neurol. Sci.* **182,** 151–159.
177. Mouton, P.R., Martin, L.J., Calhoun, M.E., Dal Forno, G., and Price, D.L. (1998) Cognitive decline strongly correlates with cortical atrophy in Alzheimer's disease. *Neurobiol. Aging* **19,** 371–377.
178. Alzheimer, A. (1907) Über eine eigenartige Erkrankung der Hirnrinde *Allg. Z. Psychiatrie Psychisch-Gerichlich Med.* **64,** 146–148.
179. Wisniewski, H.M. and Terry, R.D. (1973) Reexamination of the pathogenesis of the senile plaque, in *Progress in Neuropathology* (Zimmerman, H.M., ed.), Grune & Stratton, New York, pp. 1–26.
180. Allsop, D., Landon, M., and Kidd, M. (1983) The isolation and amino acid composition of senile plaque core protein. *Brain Res.* **259,** 348–352.
181. Masters, C.L., Simms, G., Weinman, N.A., Multhaup, G., McDonald, B.L., and Beyreuther, K. (1985) Amyloid plaque core protein in Alzheimer disease and Down syndrome. *Proc. Natl. Acad. Sci. USA* **82,** 4245–4249.
182. Wong, C.W., Quaranta, V., and Glenner, G.G. (1985) Neuritic plaques and cerebrovascular amyloid in Alzheimer disease are antigenically related. *Proc. Natl. Acad. Sci. USA* **82,** 8729–8732.
183. Müller-Hill, B. and Beyreuther, K. (1989) Molecular biology of Alzheimer's disease. *Annu. Rev. Biochem.* **58,** 287–307.
184. Kang, J., Lemaire, H.-G., Unterbeck, A., et al. (1987) The precursor of Alzheimer's disease amyloid A4 protein resembles a cell-surface receptor. *Nature* **325,** 733–736.
185. Dyrks, T., Weidemann, A., Multhaup, G., et al. (1988) Identification, transmembrane orientation and biogenesis of the amyloid A4 precursor of Alzheimer's disease. *EMBO J.* **7,** 949–957.
186. Weidemann, A., König, G., Bunke, D., et al. (1989) Identification, biogenesis, and localization of precursors of Alzheimer's disease A4 amyloid protein *Cell* **57,** 115–126.
187. Goldgaber, D., Lerman, M.I., McBride, O.W., Saffiotti, U., and Gajdusek, D.C. (1987) Characterization and chromosomal localization of a cDNA encoding brain amyloid of Alzheimer's disease. *Science* **235,** 877–880.
188. Tanzi, R.E., Gusella, J.F., Watkins, P.C., et al. (1987) Amyloid β protein gene: cDNA, mRNA distribution, and genetic linkage near the Alzheimer locus. *Science* **235,** 880–884.
189. Kitaguchi, N., Takahashi, Y., Tokushima, Y., Shiojiri, S., and Ito, H. (1998) Novel precursor of Alzheimer's disease amyloid protein shows protease inhibitory activity. *Nature* **331,** 530–532.
190. Schubert, W., Prior, R., Weidemann, A., et al. (1991) Localization of Alzheimer βA4 amyloid precursor protein at central and peripheral synaptic sites. *Brain Res.* **563,** 184–194.
191. Zheng, H., Jiang, M.-H., Trumbauer, M.E., et al. (1995) β-amyloid precursor protein-deficient mice show reactive gliosis and decreased locomotor activity. *Cell* **81,** 525–531.
192. Nishimoto, I., Okamoto, T., Matsuura, Y., et al. (1993) Alzheimer amyloid protein precursor complexes with brain GTP-binding protein G_O. *Nature* **362,** 75–79.
193. Schubert, D., Jin, L.-W., Saitoh, T., and Cole, G. (1989) The regulation of amyloid β protein precursor secretion and its modulatory role in cell adhesion. *Neuron* **3,** 689–694.
194. Kametani, F., Haga, S., Tanaka, K., and Ishii, T. (1990) Amyloid β-protein precursor (APP) of cultured cells: secretory and non-secretory forms of APP *J. Neurol. Sci.* **97,** 43–52.
195. Klier, F.G., Cole, G., Stalleup, W., and Schubert, D. (1990) Amyloid β-protein precursor is associated with extracellular matrix. *Brain Res.* **515,** 336–342.
196. Shivers, B.D., Hilbich, C., Multhaup, G., Salbaum, M., Beyreuther, K., and Seeburg, P.H. (1988) Alzheimer's disease amyloidogenic glycoprotein: expression pattern in rat brain suggests a role in cell contact. *EMBO J.* **7,** 1365–1370.

197. Octave, J.-N., de Sauvage, F., and Maloteaux, J.-M. (1989) Modification of neuronal cell adhesion affects the genetic expression of the A4 amyloid peptide precursor. *Brain Res.* **486,** 369–371.
198. Breen, K.C., Bruce, M., and Anderton, B.H. (1991) Beta amyloid precursor protein mediates neuronal cell-cell and cell-surface adhesion. *J. Neurosci. Res.* **28,** 90–100.
199. Chen, M. and Yankner, B.A. (1991) An antibody to β amyloid and the amyloid precursor protein inhibits cell-substratum adhesion in many mammalian cell types. *Neurosci. Lett.* **125,** 223–226.
200. Saitoh, T., Sundsmo, M., Roch, J.-M., et al. (1989) Secreted form of amyloid β protein precursor is involved in the growth regulation of fibroblasts. *Cell* **58,** 615–622.
201. Masliah, E., Mallory, M., Ge, N., and Saitoh, T. (1992) Amyloid precursor protein is localized in growing neurites of neonatal rat brain. *Brain Res.* **593,** 323–328.
202. Milward, E.A., Papadopoulos, R., Fuller, S.J., et al. (1992) The amyloid protein precursor of Alzheimer's disease is a mediator of the effects of nerve growth factor on neurite outgrowth. *Neuron* **9,** 129–137.
203. Siman, R., Card, J.P., Nelson, R.B., and Davis, L.G. (1989) Expression of β-amyloid precursor protein in reactive astrocytes following neuronal damage. *Neuron* **3,** 275–285.
204. Kawarabayashi, T., Shoji, M., Harigaya, Y., Yamaguchi, H., and Hirai, S. (1991) Expression of APP in the early stage of brain damage. *Brain Res.* **563,** 334–338.
205. Shigematsu, K., McGeer, P.L., and McGeer, E.G. (1992) Localizaiton of amyloid precursor protein in selective postsynaptic densities of rat cortical neurons. *Brain Res.* **593,** 353–357.
206. Esch, F.S., Keim, P.S., Beattie, E.C., et al. (1990) Cleavage of amyloid β peptide during constitutive processing of its precursor. *Science* **248,** 1122–1124.
207. Wang, R., Meschia, J.F., Cotter, R.J., and Sisodia, S.S. (1991) Secretion of the β/A4 amyloid precursor protein. Identification of a cleavage site in cultured mammalian cells. *J. Biol. Chem.* **266,** 16960–16964.
208. S.S. Sisodia (1992) β-amyloid precursor protein cleavage by a membrane-bound protease. *Proc. Natl. Acad. Sci. USA* **89,** 6075–6079.
209. Cole, G.M., Huynh, T.V., and Saitoh, T. (1989) Evidence for lysosomal processing of amyloid β-protein precursor in cultured cells. *Neurochem. Res.* **14,** 933–939.
210. Golde, T.E., Estus, S., Younkin, L.H., Selkoe, D.J., and Younkin, S.G. (1992) Processing of the amyloid protein precursor to potentially amyloidogenic derivatives. *Science* **255,** 728–730.
211. Haass, C., Schlossmacher, M.G., Hung, A.Y., et al. (1992) Amyloid β-peptide is produced by cultured cells during normal metabolism. *Nature* **359,** 322–325.
212. Shoji, M., Golde, T.E., Ghiso, J., et al. (1992) Production of the Alzheimer amyloid β protein by normal proteolytic processing. *Science* **258,** 126–129.
213. Knauer, M.F., Soreghan, B., Burdick, D., Kosmoski, J., and Glabe, C.G. (1992) Intracellular accumulation and resistance to degradation of the Alzheimer amyloid A4/β protein. *Proc. Natl. Acad. Sci. USA* **89,** 7437–7441.
214. Turner, R.S., Suzuki, N., Chyung, A.S.C., Younkin, S.G., and Lee, V.M.-Y. (1996) Amyloids β40 and β42 are generated intracellularly in cultured human neurons and their secretion increases with maturation. *J. Biol. Chem.* **271,** 8966–8970.
215. Seubert, P., Oltersdorf, T., Lee, M.G., et al. (1993) Secretion of β-amyloid precursor protein cleaved at the amino terminus of the β-amyloid peptide. *Nature* **361,** 260–263.
216. Yang, Y., Turner, R.S., and Gaut, J.R. (1998) The chaperone BiP/GRP78 binds to amyloid precursor protein and decreases Aβ40 and Aβ42 secretion. *J. Biol. Chem.* **273,** 25552–25555.
217. Doan, A., Thinakaran, G., Borchelt, D.R., et al. (1996) Protein topology of presenilin 1. *Neuron* **17,** 1023–1030.
218. Lee, M.K., Slunt, H.H., Martin, L.J., et al. (1996) Expression of presenilin 1 and 2 (PS1 and PS2) in human and murine tissues. *J. Neurosci.* **16,** 7513–7525.

219. Thinakaran, G., Borchelt, D.R., Lee, M.K., et al. (1996) Endoproteolysis of presenilin 1 and accumulation of processed derivatives in vivo. *Neuron* **17,** 181–190.
220. Borchelt, D.R., Thinakaran, G., Eckman, C.B., et al. (1996) Familial Alzheimer's disease-linked presenilin 1 variants elevate Aβ1-42/1–40 ratio in vitro and in vivo. *Neuron* **17,** 1005–1013.
221. Lee, M.K., Borchelt, D.R., Kim, G., et al. (1997) Hyperaccumulation of FAD-linked Presenilin 1 variants in vivo. *Nature Med.* **3,** 756–760.
222. Card, J.P., Meade, R.P. and Davis, L.G. (1998) Immunocytochemical localization of the precursor protein for β-amyloid in the rat central nervous system. *Neuron* **1,** 835–846.
223. Martin, L.J., Sisodia, S.S., Koo, E.H., et al. (1991) Amyloid precursor protein in aged nonhuman primates. *Proc. Natl. Acad. Sci. USA* **88,** 1461–1465.
224. Martin, L.J. (1993) Cellular dynamics of senile plaque formation and amyloid deposition in cerebral cortex: an ultrastructural study of aged nonhuman primates. *Neurobiol. Aging* **14,** 681–683.
225. Yamaguchi, H., Yamazaki, T., Ishiguro, K., Shoji, M., Nakazato, Y., and Hirai, S. (1992) Ultrastructural localization of Alzheimer amyloidβ/A4 protein precursor in the cytoplasm of neurons and senile plaque-associated astrocytes. *Acta Neuropathol.* **85,** 15–22.
226. Goldgaber, D., Harris, H.W., Hla, T., et al. (1898) Interleukin 1 regulates synthesis of amyloid-protein precursor mRNA in human endothelial cells. *Proc. Natl. Acad. Sci. USA* **86,** 7606–7610.
227. Allsop, D., Haga, S.-I., Haga, C., Ikeda, S.-I., Mann, D.M.A., and Ishii, T. (1989) Early senile plaques in Down's syndrome brains show a close relationship with cell bodies of neurons. *Neuropathol. Appl. Neurobiol.* **15,** 531–542.
228. Yamaguchi, H., Nakazato, Y., Hirai, S., Shoji, M., and Harigaya, Y. (1989) Electron micrograph of diffuse plaques. Initial stage of senile plaque formation in the Alzheimer brain. *Am. J. Pathol.* **135,** 593–597.
229. Probst, A., Langui, D., Ipsen, S., Robakis, N., and Ulrich, J. (1991) Deposition of β/A4 protein along neuronal plasma membranes in diffuse senile plaques. *Acta Neuropathol.* **83,** 21–29.
230. Wisniewski, H.M., Wegiel, J., Wang, K.C., Kujawa, M., and Lach, B. (1989) Ultrastructural studies of the cells forming amyloid fibers in classical plaques. *Can. J. Neurol. Sci.* **16,** 535–542.
231. Frackowiak, J., Wisniewski, H.M., Wegiel, J., Merz, G.S., Iqbal, K., and Wang, K.C. (1992) Ultrastructure of the microglia that phagocytose amyloid and the microglia that produce β-amyloid fibrils. *Acta Neuropathol.* **84,** 225–233.
232. Miyakawa, T., Shimoji, A., Kuramoto, R., and Higuchi, Y. (1982) The relationship between senile plaques and cerebral blood vessels in Alzheimer's disease and senile dementia. Morphological mechanism of senile plaque production. *Virchows Arch. (Cell Pathol.)* **40,** 121–129.
233. Yamaguchi, H., Nakazato, Y., Hirai, S., and Shoji, M. (1990) Immunoelectron microscopic localization of amyloid β protein in the diffuse plaques of Alzheimer-type dementia. *Brain Res.* **508,** 320–324.
234. Yamaguchi, H., Nakazato, Y., Shoji, M., Takatama, M., and Hirai, S. (1991) Ultrastructure of diffuse plaques in senile dementia of the Alzheimer type: comparison with primitive plaques. *Acta Neuropathol.* **82,** 13–20.
235. Troncoso, J.C., Martin, L.J., Dal Forno, G., and Kawas, C.H. (1996) Neuropathology in controls and demented subjects from the Baltimore Longitudinal Study of Aging. *Neurobiol. Aging* **17,** 365–371.
236. Zemlan, F.P., Thienhaus, O.J., and Bosmann, H.B. (1989) Superoxide dismutase activity in Alzheimer's disease: possible mechanism for paired helical filament formation. *Brain Res.* **476,** 160–162.
237. Tan, Y.H., Tischfield, J., and Ruddle, F.H. (1973) The linkage of genes for the human interferon induced antiviral protein and indophenol oxidase-B traits to chromosome G-21. *J. Exp. Med.* **137,** 317–330.

238. Brooksbank, B.W.L. and Balazs, R. (1984) Superoxide dismutase, glutathione peroxidase and lipoperoxidation in Down's syndrome fetal brain. *Dev. Brain Res.* **16,** 37–44.
239. Busciglio, J. and Yankner, B.A. (1995) Apoptosis and increased generation of reactive oxygen species in Down's syndrome neurons in vitro. *Nature* **378,** 776–779.
240. Pappolla, M.A., Omar, R.A., Kim, K.S., and Robakis, N.K. (1992) Immunohistochemical evidence of antioxidant stress in Alzheimer's disease. *Am. J. Pathol.* **140,** 621–628.
241. Weisiger, R.A. and Fridovich, I. (1973) Superoxide dismutase, organelle specificity. *J. Biol. Chem.* **248,** 3582–3592.
242. Wong, G.H.W. and Goeddel, D.V. (1988) Induction of manganous superoxide dismutase by tumor necrosis factor: possible protective mechanism. *Science* **242,** 941–944.
243. Masuda, A., Longo, D.L., Kobayashi, Y., Appella, E., Oppenheim, J.J., and Matsushima, K. (1988) Induction of mitochondrial manganese superoxide dismutase by interleukin 1. *FASEB J.* **2,** 3087–3090.
244. McGeer, E.D. and McGeer, P.L. (1998) Inflammation in the brain in Alzheimer's disease: implications for therapy. *NeuroScience News* **1,** 29–35.
245. Sawada, M., Kondo, N., Suzumura, A., and Marunouchi, T. (1989) Production of tumor necrosis factor-α by microglia and astrocytes in culture. *Brain Res.* **491,** 394–397.
246. Colton, C.A. and Gilbert, D.L. (1987) Production of superoxide anions by a CNS macrophage, the microglia. *FEBS Lett.* **223,** 284–288.
247. Sharonov, B.P. and Churilova, I.V. (1992) Inactivation and oxidative modification of Cu,Zn superoxide dismutase by stimulated neutrophils: the appearance of new catalytically active structures. *Biochem. Biophys. Res. Commun.* **189,** 1129–1135.
248. Kuncl, R.W., Crawford, T.O., Rothstein, J.D., and Drachman, D.B. (1992) Motor neuron diseases, in *Diseases of the Nervous System. Clinical Neurobiology* (Asbury, A.K., McKhann, G.M., and McDonald, W.I., eds.), WB Saunders, Philadelphia, pp. 1179–1208.
249. Gurney, M.E., Pu, H., Chiu, A.Y., et al. (1994) Motor neuron degeneration in mice that express a human Cu,Zn superoxide dismutase mutation. *Science* **264,** 1772–1775.
250. Wong, P.C., Pardo, C.A., Borchelt, D.R., et al. (1995) An adverse property of a familial ALS-linked SOD1 mutation causes motor neuron disease characterized by vacuolar degeneration of mitochondria. *Neuron* **14,** 1105–1116.
251. Rabizadeh, S., Butler Gralla, E., Borchelt, D.R., et al. (1995) Mutations associated with amyotrophic lateral sclerosis convert superoxide dismutase from an antiapoptotic gene to a proapoptotic gene: studies in yeast and neural cells. *Proc. Natl. Acad. Sci. USA* **92,** 3024–3028.
252. Ghadge, G.D., Lee, J.P., Bindokas, V.P., et al. (1997) Mutant superoxide dismutase-1-linked familial amyotrophic lateral sclerosis: molecular mechanisms of neuronal death and protection. *J. Neurosci.* **17,** 8756–8766.
253. Troy, C.M. and Shelanski, M.L. (1994) Down-regulation of copper/zinc superoxide dismutase causes apoptotic death in PC12 neuronal cells. *Proc. Natl. Acad. Sci. USA* **91,** 6384–6387.
254. Xu, Z., Cork, L.C., Griffin, J.W., and Cleveland, D.W. (1993) Increased expression of neurofilament subunit NF-L produces morphological alterations that resemble the pathology of human motor neuron disease. *Cell* **73,** 23–33.
255. Côté, F., Collard, J.-F., and Julien, J.-P. (1993) Progressive neuronopathy in transgenic mice expressing the human neurofilament heavy gene: a mouse model of amyotrophic lateral sclerosis. *Cell* **73,** 35–46.
256. Lee, M.K., Marszalek, J.R., and Cleveland, D.W. (1994) A mutant neurofilament subunit causes massive, selective motor neuron death: implications for the pathogenesis of human motor neuron disease. *Neuron* **13,** 975–988.
257. Anand, P., Parrett, A., Martin, J., et al. (1995) Regional changes of ciliary neurotrophic factor and nerve growth factor levels in post mortem spinal cord and cerebral cortex from patients with motor disease. *Nature Med.* **1,** 168–172.

258. Kostic, V., Jackson-Lewis, V., de Bilbao, F., Dubois-Dauphin, M., and Przedborski, S. (1997) Bcl-2: prolonging life in a transgenic mouse model of familial amyotrophic lateral sclerosis. *Science* **277,** 559–562.
259. Friedlander, R.M., Brown, R.H., Gagliardini, V., Wang, J., and Juan, J. (1997) Inhibition of ICE slows ALS in mice. *Nature* **388,** 31.
260. Ginsberg, S.D., Portera-Cailliau, C., and Martin, L.J. (1999) Fimbria-fornix transection and excitotoxicity produce similar neurodegeneration in the septum. *Neuroscience* **88,** 1059–1071.
261. Chase, R.A., Pearson, S., Nunn, P.B., and Lantos, P.L. (1985) Comparative toxicities of α- and β-N-oxalyl-L-a,β-diaminopropionic acids to rat spinal cord. *Neurosci. Lett.* **55,** 89–94.
262. Plaitakis, A. (1990) Glutamate dysfunction and selective motor neuron degeneration in amyotrophic lateral sclerosis: a hypothesis. *Ann. Neurol.* **28,** 3–8.
263. Rothstein, J.D., Tsai, G., Kuncl, R.W., et al. (1990) Abnormal excitatory amino acid metabolism in amyotrophic lateral sclerosis. *Ann. Neurol.* **28,** 18–25.
264. Camu, W., Billiard, M., and Baldy-Moulinier, M. (1993) Fasting plasma and CSF amino acid levels in amyotrophic lateral sclerosis: a subtype analysis. *Acta Neurol. Scand.* **88,** 51–55.
265. Perry, T.L., Krieger, C., Hansen, S., and Eisen, A. (1990) Amyotrophic lateral sclerosis: amino acid levels in plasma and cerebrospinal fluid. *Ann. Neurol.* **28,** 12–17.
266. Lin, C.-L., Bristol, L.A., Jin, L., et al. (1998) Aberrant RNA processing in neurodegenerative disease: the cause for absent EAAT2, a glutamate transporter, in amyotrophic lateral sclerosis. *Neuron* **20,** 589–602.
267. Shi, L., Mallory, M., Alford, M., Tanaka, S., and Masliah, E. (1997) Glutamate transporter alterations in Alzheimer's disease are possibly associated with abnormal APP expression. *J. Neuropathol. Exp. Neurol.* **56,** 901–911.
268. Nagai, M., Abe, K., Okamoto, K., and Itoyama, Y. (1998) Identification of alternative splicing forms of GLT-1 mRNA in the spinal cord of amyotrophic lateral sclerosis patients. *Neurosci. Lett.* **244,** 165–168.
269. Meyer, T., Münch, C., Knappenberger, B., Liebau, S., Völkel, H., and Ludolph, A.C. (1998) Alternative splicing of the glutamate transporter EAAT2. *Neurosci. Lett.* **241,** 68–70.
270. Meyer, T., Münch, C., Liebau, S., et al. (1998) Splicing of the glutamate transporter EAAT2: a candidate gene of amyotrophic lateral sclerosis. *J. Neurol. Neurosurg. Psychiatry* **65,** 920.
271. Blackstone, C.D., Levey, A.I., Martin, L.J., Price, D.L., and Huganir, R.L. (1992) Immunological detection of glutamate receptor subtypes in human central nervous system. *Ann. Neurol.* **31,** 680–683.
272. Martin, L.J., Blackstone, C.D., Levey, A.I., Huganir, R.L., and Price, D.L. (1993) AMPA glutamate receptor subunits are differentially distributed in rat brain. *Neuroscience* **53,** 327–358.
273. Portera-Cailliau, C., Price, D.L., and Martin, L.J. (1996) *N*-methyl-*D*-aspartate receptor proteins NR2A and NR2B are differentially distributed in the developing rat central nervous system as revealed by subunit-specific antibodies. *J. Neurochem.* **66,** 692–700.
274. Williams, T.L., Ince, P.G., Oakley, A.E., and Shaw, P.J. (1996) An immunocytochemical study of the distribution of AMPA selective glutamate receptor subunits in the normal human motor system. *Neuroscience* **74,** 185–198.
275. Petralia, R.S., Wang, Y.-X., and Wenthold, R.J. (1994) The NMDA receptor subunits NR2A and NR2B show histological and ultrastructural localization patterns similar to those of NR1. *J. Neurosci.* **14,** 6102–6120.
276. Escarrabill, J., Estopa, R., Farrero, E., Monasterio, C., and Manresa, F. (1998) Long-term mechanical venilation in amyotrophic lateral sclerosis. *Respir. Med.* **92,** 438–441.
277. Obrenovitch, T.P. (1998) Amytrophic lateral sclerosis, excitotoxicity and riluzole. *TIPS* **19,** 9.
278. Milligan, C.E., Prevette, D., Yaginuma, H., et al. (1995) Peptide inhibitors of the ICE protease family arrest programmed cell death of motoneurons in vivo and in vitro. *Neuron* **15,** 385–393.

279. Deckwerth, T.L., Elliott, J.L., Knudson, C.M., Johnson, E.M., Snider, W.D., and Korsmeyer, S.J. (1996) Bax is required for neuronal death after trophic factor deprivation and during development. *Neuron* **17,** 401–411.
280. Estévez, A.G., Spear, N., Manuel, S.M., et al. (1998) Nitric oxide and superoxide contribute to motor neuron apoptosis induced by trophic factor deprivation. *J. Neurosci.* **18,** 923–931.
281. Martin, L.J., Price, A.C., Kaiser, A., Shaikh, A.Y., and Kiu, Z. (2000) Mechanisms for neuronal degeneration in amyotrophic lateral sclerosis and in models of motor neuron death. *Int. J. Mol. Med.* **5,** 3–13.
282. Martin, L.J., (2001) Neuronal cell death in nervous system development, disease, and injury. *Int. J. Mol. Med.* **7,** 455–478.

26
Astrocytes and Ammonia in Hepatic Encephalopathy

Michael D. Norenberg

1. INTRODUCTION

Hepatic encephalopathy (HE; hepatic coma) refers to a complex neuropsychiatric syndrome resulting from severe liver failure. It is probably the best example of a neurological condition in which astrocytes appear to play a dominant role in its pathogenesis. This disorder gave genesis to the concept of a primary gliopathy, whereby an initial disturbance in astroglial function leads to abnormal neuronal activity *(1,2)*. Studies in HE have additionally provided important insights into the role of astrocytes in ammonia and amino acid neurotransmitter metabolism, and greatly contributed to the evolution of the concept of glial-neuronal interactions/trafficking.

This chapter will review the role of astrocytes in the pathogenesis of HE. Because ammonia is a prime candidate as the neurotoxin in HE, the effects of ammonia on astrocytes will be highlighted with emphasis on its role in glutamatergic and GABAergic neurotransmission and its involvement in astrocyte swelling, a major complication of acute HE. This article also has relevance to clinical conditions associated with hyperammonemia, such as urea cycle disorders and Reyes syndrome.

2. CLINICAL CONSIDERATIONS

HE is often referred to as portal-systemic encephalopathy, which usually occurs in the setting of alcoholic liver cirrhosis. Early manifestations may be quite subtle and only established with formal neuropsychologic tests. As the process progresses patients manifest disturbances in behavior, personality, intellectual capacity, and sleep patterns. They also may display lethargy, hyperventilation, and abnormal motor function (incoordination, ataxia, asterixis). In severe cases there is a depression in the level of consciousness, including coma and death. HE is associated with characteristic electroencephalographic changes *(3)* and hyperintensities of the globus pallidus on T1-weighted magnetic resonance images (MRI) *(4,5)*.

HE is usually precipitated by gastrointestinal hemorrhage, high protein diet, constipation, infection, excessive diuresis, electrolyte imbalance, or the use of sedatives and hypnotics. With appropriate treatment, chronic HE is usually reversible. However, following repeated bouts of encephalopathy, irreversible anatomical changes occur, especially in the basal ganglia and cerebral cortex (acquired chronic hepatocerebral

From: *Neuroglia in the Aging Brain*
Edited by: Jean S. de Vellis © Humana Press Inc., Totowa, NJ

degeneration), or in the spinal cord (shunt myelopathy or portal systemic myelopathy). These patients present with dementia, ataxia, dysarthria, intention tremor, choreoathetosis, and spastic paraparesis.

HE may also present in an acute form (fulminant hepatic failure; FHF) that generally occurs following viral hepatitis, drug hepatotoxicity (acetaminophen, halothane, nonsteroidal antiinflammatory agents), or after exposure to various hepatotoxins (mushroom/Amanita phalloides, volatile hydrocarbons). Acute HE manifests itself with the abrupt onset of delirium, seizures, and coma. The principal cause of death in acute HE is brain edema associated with increased intracranial pressure.

3. PATHOLOGY

The histopathology of HE in humans and experimental animals is dominated by astroglial changes. Acute HE is characterized by massive astrocyte swelling. This may be accompanied by neuronal necrosis secondary to cerebral ischemia which is mediated by the elevated intracranial pressure.

Alzheimer type II astrocytosis (protoplasmic astrocytosis, metabolic gliosis) is the histological hallmark of chronic HE *(3,6)*. This change is characterized by an apparent increase in the number of astrocytes with enlarged, pale nuclei, peripheral margination of chromatin and often prominent nucleoli. Excessive amounts of lipofuscin granules are occasionally identified. Ultrastructural studies in experimental models of HE have shown an increase in the number of mitochondria, smooth and rough endoplasmic reticulum, and cytoplasmic glycogen, suggesting that astrocytes are metabolically activated. Although unproven, it is generally believed that these glial changes are reversible. Eventually, degenerative changes ensue characterized by cell swelling and the presence of cytoplasmic vacuoles and degenerated mitochondria *(7)*. No significant or consistent neuronal changes have been identified.

Acquired chronic hepatocerebral degeneration is associated with irreversible damage and is characterized by cortical laminar necrosis, microcavitation in the cortex and basal ganglia, and cerebral and cerebellar atrophy, along with degeneration of the white matter columns in the spinal cord. Characteristic features include the presence of Alzheimer type II astrocytes, Alzheimer type I astrocytes (large atypical reactive astrocytes), and nuclear glycogen inclusions *(8)*. These changes are very similar to those seen in Wilson's disease. These abnormalities may be due to an accumulation of manganese *(9)* (*see* **Subheading 4.**).

4. PATHOPHYSIOLOGY

The basis for the neurological disorder in liver failure remains elusive. The dominant view over many decades has been that gut-derived nitrogenous products are not detoxified by the diseased liver or are not extracted by the liver as a result of vascular (portal-systemic) shunts that commonly occur in chronic liver disease. These toxins then enter the central nervous system (CNS) and exert deleterious effects. These vascular shunts may occur spontaneously as a consequence of liver disease, or may be iatrogenically induced by the surgical construction of a portacaval anastomosis or a transjugular intrahepatic portosystemic shunt (TIPS) performed to reduce portal hypertension. The presence of a normal liver in these cases highlights the importance of systemic shunting in the pathogenesis of HE.

Various toxins have been invoked over the years, including ammonia, short-chain fatty acids, mercaptans, phenols, biogenic amines, endotoxin, and "middle molecules," which may act synergistically *(10)*. More recently a role for manganese has been proposed *(11)*. Additionally, abnormalities in GABAergic and glutamatergic neurotransmission have been invoked. These prevailing hypotheses are not necessarily mutually exclusive, as will be developed below. Abnormalities in monoamines and monoaminergic neurotransmission *(12)*, as well as opiate-mediated neurotransmission *(13)* have also been proposed.

4.1. Ammonia

Of all of the toxins implicated in the pathogenesis of HE, ammonia has received the greatest emphasis for the following reasons:

1. Factors that lead to increased levels of blood or brain ammonia worsen HE *(10)*;
2. Almost all of the effective therapies in HE bring about a reduction in blood ammonia levels *(14)*;
3. Procedures that increase blood or brain ammonia in experimental animals reproduce the clinical and pathological changes of HE *(6)*;
4. Patients with hereditary hyperammonemia have similar clinical and pathological findings as in HE *(6)*;
5. Administration of ammonium chloride to cultured astrocytes reproduces the pathological changes observed in HE *(15,16)*;
6. Blood *(14)*, CSF *(17)*, and particularly brain *(17)* ammonia levels correlate well with the clinical state. At present, no other factor can better explain the clinical, pathological and neurochemical features of HE.

Ammonia exists either as the nonionized ammonia gas (NH_3), or as the protonated NH_4^+ ion. The nonionized form is a diffusible molecule *(18)* that readily crosses the blood-brain-barrier *(19)* whereas the ionized form is far less permeable. The pKa of ammonia is 9.0, so that at physiologic pH (7.4) over 98% of ammonia exists as the NH_4^+ ion. At alkaline pH, a greater amount of ammonia is present in the more diffusible form (NH_3) and is thus potentially more toxic.

Ammonia is largely generated from protein degradation and is principally detoxified in the liver by urea synthesis. In brain, however, ammonia is metabolized/detoxified into the electrophysiologically inert glutamine through the action of glutamine synthetase (GS), as the urea cycle is absent in brain *(20)*. Increased brain and CSF levels of glutamine are among the neurochemical hallmarks of HE *(19)*. The synthesis of glutamine is carried out in astrocytes *(21)* where GS is primarily localized *(22)*. The glutamate required for this reaction is mostly derived by uptake from the extracellular space, whereas a smaller amount of ammonia may be utilized in the reductive amination of α-ketoglutarate, catalyzed by glutamate dehydrogenase *(19)*. Glutamine is then released from astrocytes and taken up by nerve endings where it is converted to glutamate through the action of glutaminase *(23)*. Glutamate released from nerve endings is later taken up by astrocytes thereby completing the so-called glutamate-glutamine cycle *(24)* (Fig. 1). The key components of the cycle, namely, glutamate uptake, glutamine synthesis, and glutamine release have all been convincingly shown in astrocyte cultures *(25,26)*. As expected, treatment of cultured astrocytes with ammonia cause a reduction of glutamate and an increase in glutamine *(21,27)*. As will be developed later, HE can be considered as a prototype disorder of the glutamate-glutamine cycle (*see* Fig. 1).

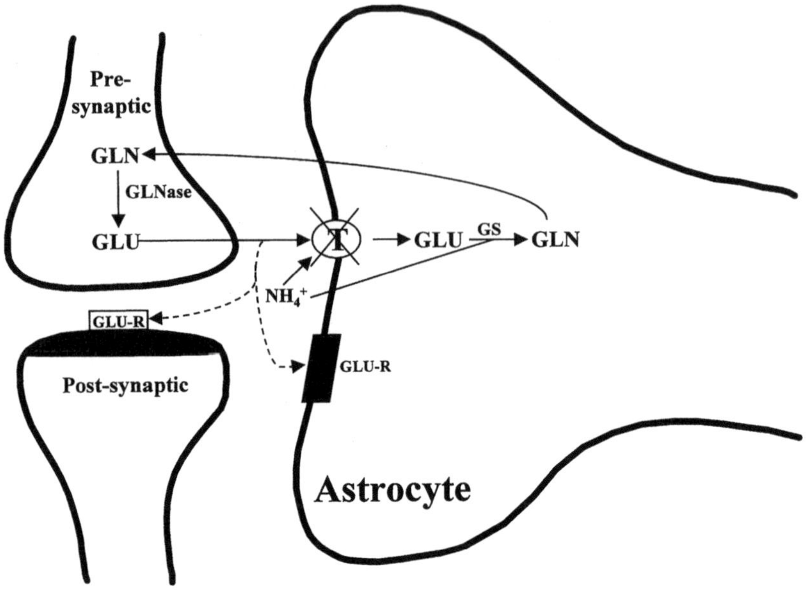

Fig. 1. Diagram depicting derangement in the glutamate-glutamine cycle and the role of ammonia-induced interference in glutamate uptake on glutamatergic neurotransmission. Failure of astrocytes to take up glutamate will lead to elevated levels in the synaptic space resulting in the activation of both neuronal and glial glutamate receptors (GLU-R). GLN, glutamine; GLNase, glutaminase; GLU, glutamate; GS, glutamine synthetase; T, glutamate transporter.

Ammonia is clearly a neurotoxin *(19)* which, in high concentrations (>1 m*M*), can result in seizures and death, while stupor and coma occur at lower concentrations. The fundamental mechanisms of ammonia toxicity are poorly understood. Views have ranged from altered bioenergetics, electrophysiological effects, altered intracellular pH, and effects on the membrane potential.

4.1.1. Energy Metabolism

The Bessmans *(28)* introduced the concept of energy failure as a factor in ammonia neurotoxicity. This view was based on the fact that the ammonia detoxification pathway (glutamine synthesis) consumes ATP. This view, however, has been controversial (*see* refs. *19,29* for reviews). In general, it appears that there are no acute losses of high energy metabolites; however, chronically, there is evidence of energy failure. One possible mechanism for such failure is the inability of the malate-aspartate shuttle to generate reducing equivalents due to an ammonia-induced fall in glutamate. This concept was originally introduced by Hindfelt et al. *(30)* and more recently supported by Ratnakumari and Murthy *(31)* and Faff-Michalak and Albrecht *(32)*.

Another ammonia-induced energy demanding process is glutamate uptake which is coupled with Na^+ influx. The latter is pumped out of the cell through the action of Na^+, K^+-ATPase. The activity of Na^+, K^+-ATPase is the major energy consuming process in the CNS *(33)*, and the activity of this enzyme has been shown to be elevated in hyperammonemia *(34–36)*. This view also goes along with the fact that energy demand in brain is tightly coupled to the recycling of glutamate to glutamine *(37)*.

In view of the prominent role of astrocytes in HE, it is possible that energy failure may be compartmentalized in these cells. Yet, knowledge regarding bioenergetics in HE/ammonia toxicity in astrocytes is rudimentary. Fitzpatrick et al. *(38)* showed that 3 m*M* NH_4Cl decreased pyruvate oxidation in cultured astrocytes but had no effects on cultured neurons. Using relatively high concentrations of NH_4Cl (7.5–15 m*M*), Haghighat and McCandless *(39)* observed increase in lactate and decreases in ATP and phosphocreatine. Ammonia was also observed to impair the operation of the malate-aspartate shuttle in cultured astrocytes *(40)*.

4.1.2. Electrophysiologic Effects

Effects on excitatory and inhibitory neurotransmission have been ascribed to ammonia, possibly due to inactivation of Cl^- extrusion pumps *(41)*. Since NH_4^+ can substitute for K^+ *(42)*, ammonia is also believed to interrupt nerve conduction *(43)*. Ammonia has also been recently shown to depolarize astrocytes in culture *(44)*, which can lead to inactivation of ionic conductance. This can contribute to cell swelling and disturbances in neurotransmitter uptake (see Subheading 4.4.).

4.1.3. pH

The importance of ammonia-induced changes in pH as a factor in ammonia neurotoxicity has received scant attention. This is surprising as ammonia has major effects on pH, and changes in pH have been shown to exert a profound effect on cell function by influencing the activity enzymes, the state of various ion channels, receptors and transporters *(45,46)*, as well as by influencing lysosomal activity *(47)*.

In most cells, ammonia (NH_3+ NH_4^+) produces a transient rise in pH *(18)*. The situation in brain is not clear. Portacaval-shunted rats show increased pH in astrocytes *(48)*. However, nuclear magnetic resonance (NMR) studies have either shown no change in pH *(49)*, a rise *(50)*, or a fall *(51)*, presumably due to the generation of lactic acid. Although in high doses (20 m*M*) ammonia causes intracellular alkalosis in cultured astrocytes *(52)*, in lower pathophysiologically relevant doses (0.1–5.0 m*M*), Nagaraja and Brookes *(53)* and ourselves *(54)* have found that ammonia causes intracellular acidification, suggesting the presence of NH_4^+ transporting channels in astrocytes. Similar NH_4^+ channels have been identified in kidney tubule cells *(55)* and Xenopus oocytes *(56)*. The subsequent dissociation of NH_4^+ into NH_3 and H^+ would result in intracellular acidification.

4.1.4. Free Radicals

Oxidative stress is an evolving concept in the pathogenesis of HE/ammonia neurotoxicity *(57)*. There is evidence of lipid peroxidation in cultured astrocytes after ammonia treatment *(58)*. Ammonia was also recently shown to be capable of generating free radicals in cultured astrocytes *(59)*. A dose-dependent response was observed, reaching its peak 2.5 min after exposure and declining to baseline levels by 15 min. The generation was blocked by the simultaneous addition of catalase and superoxide dismutase. Methionine sulfoximine (MSO) also blocked the production of free radicals, suggesting that glutamine may be involved in this process. Studies showing that antioxidants have a beneficial effect in experimental HE/hyperammonemia *(60,61)* and in patients with fulminant hepatic failure *(62)* further support the role of free radicals in HE

The free radical nitric oxide (NO) has also been implicated in hyperammonemia/HE. Nitric oxide synthase (NOS) activity has been shown to be elevated in experimental models of HE *(63,64),* and increased brain NO production was shown in portacaval shunted rats given ammonia infusions *(65).* Consistent with these observations is the ammonia-induced astrocyte uptake of arginine, the precursor for NO *(66).* The NOS inhibitor nitroarginine has been shown to attenuate ammonia toxicity *(67)*

4.1.5. General Effects of Ammonia on Astrocytes

In addition to effects on energy metabolism, membrane potential, pH and free radicals, ammonia has been shown to result in the morphological changes in astrocytes similar to those seen in HE; decreased GFAP content and mRNA levels; decreased glycogen content; altered protein phosphorylation; decreased cyclic AMP levels after stimulation with a β-adrenergic-receptor agonist; and decreased uptake of potassium, myo-inositol, and calcium. These and other effects on cultures astrocytes have been summarized *(68).* Similar, although not identical, findings have been reported by Reichenbach et al. *(69)* in cultured Müller (glial) cells of the neonatal rabbit retina. For review on the effect of ammonia on bulk-derived astrocytes from animals with acute HE, *see* ref. *70.*

4.2. Glutamate Transport

The detoxification of ammonia consumes glutamate, so that abnormalities in glutamate metabolism and glutamatergic neurotransmission have long been suspected to contribute to the pathogenesis of HE *(12,29).* Indeed, total brain glutamate levels are decreased in various models of HE, in hyperammonemia, postmortem tissue from patients with HE, ammonia-treated brain slices, and ammonia-treated cultured astrocytes (reviewed in ref. *71*). Interestingly, however, extracellular levels of glutamate are elevated in HE. Although increased release of glutamate has been reported in various models of HE and hyperammonemia, such "release" may be due to impaired glutamate uptake, either by nerve endings or by adjacent astroglial cells (*see* refs. *12,29* for further references).

Astrocytes are critically involved in glutamatergic neurotransmission *(72,73)* which is greatly dependent on glutamate uptake *(74,75),* and recent studies have shown a predominant role of glia in the clearing of glutamate after synaptic release *(76).* The uptake of glutamate is accomplished by various high-affinity, Na^+-dependent transporters. Five subtypes of transporters have been cloned: EAAT1 (GLAST), EAAT2 (GLT-1), EAAT3 (EAAC1), EAAT4 and EAAT5. GLT-1 and GLAST are principally found in astrocytes whereas EAAC1 is present mainly in neurons. GLT-1 appears to be the main transporter in vivo. The status in astrocyte culture is less clear as some have found mainly GLAST with little to no GLT-1, whereas others have found significant amounts of GLT-1. EAAC1 is chiefly neuronal, most likely postsynaptic. EAAT4 and EAAT5 are additionally Cl^--dependent, and have been identified in neurons of the cerebellum and retina, respectively (for reviews, *see* refs. *77,78*).

Treatment of cultured astrocytes with ammonia decreases glutamate uptake by 30–45% *(71),* which was reversible on discontinuation of ammonia treatment. Decreased uptake was also observed by Albrecht and colleagues *(79)* in bulk-isolated astrocytes derived from rats with acute liver failure. Depressed glutamate transport has likewise been described in synaptosomes derived from thioacetamide-treated rats *(80),* synaptosomes treated with ammonia *(81),* and hippocampal slices treated with sera and CSF from patients with HE *(82).*

The effect of ammonia on GLAST (EAAT1) mRNA has been examined in cultured astrocytes derived from cerebral cortex, striatum, and cerebellum *(83).* GLAST was

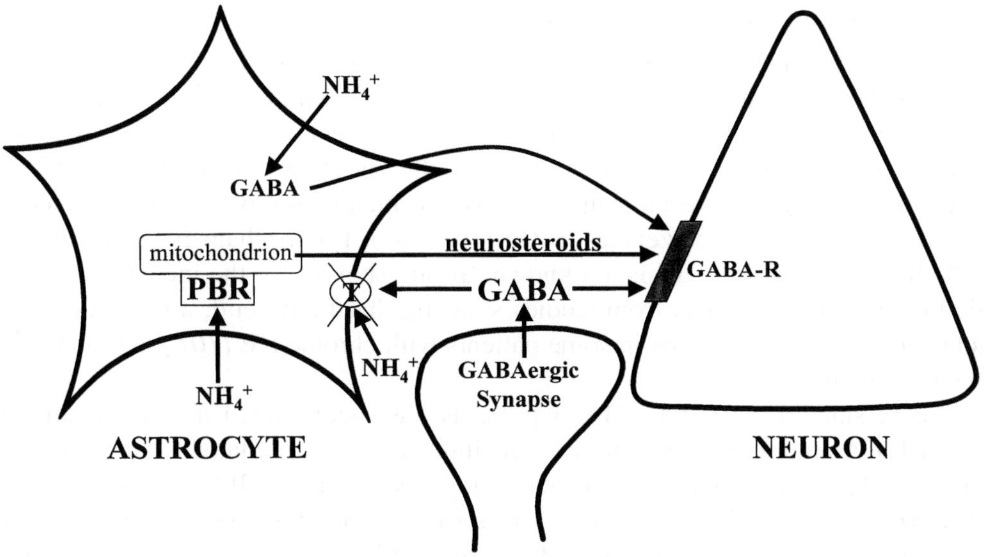

Fig. 2. Involvement of ammonia and astrocytes in enhanced GABAergic tone. Ammonia inhibits astroglial uptake of GABA and stimulates its release. This will result in elevated synaptic levels of GABA which can activate $GABA_A$ receptors (GABA-R). Ammonia also upregulates the peripheral-type benzodiazepine receptor (PBR) leading to enhanced production of neurosteroids which positively modulate the $GABA_A$ receptor.

selected for the study as it represents the principal transporter in cultured astrocytes *(84,85)*. Decreased content of GLAST mRNA was observed in all regions, with the greatest effect noted in the striatum. The greater involvement of astrocytes derived from striatum is of interest because there is growing evidence of major involvement of the basal ganglia in HE *(86)*, as evidenced by hyperintensities in the globus pallidus on MRI *(4,5)*, which are likely because of the accumulation of manganese *(11)*. In vivo studies have also shown a downregulation of the GLT-1 mRNA both in rats with hyperammonemia and with acute liver failure *(29,87)*. As in the culture studies, the striatum showed the largest loss.

The mechanisms by which ammonia impairs glutamate uptake in astrocytes are not well understood. As noted, ammonia affects the transcription of transporter genes. Ammonia is known to cause cell swelling (*see* **Subheading 4.4**), membrane depolarization *(44)* nitric oxide production *(65)*, and oxidative stress *(57)*, which have all been shown to acutely impair astroglial glutamate uptake *(88–91)*. Whether any or all of these factors are involved in the failure of astrocytes to take up glutamate is not known. Manganese, which has been implicated in HE (see above), was shown to decrease glutamate uptake, and had additive effects when combined with ammonia *(92)*.

Failure of astrocytes to clear glutamate from the extracellular space may result in aberrant glutamatergic neurotransmission and excitotoxicity, as well as impairment in energy metabolism. In this regard, it is noteworthy that NMDA receptor antagonists are capable of ameliorating ammonia neurotoxicity *(93,94)*. As glutamate receptors are located on astrocytes, excessive activation of these receptors may also lead to glial functional changes *(95,96)*. Figure 2 depicts consequences of glutamate uptake failure.

4.3. GABAergic Neurotransmission

In the 1980s the increased GABAergic tone hypothesis dominated pathogenetic views of HE *(97),* and the fact that HE is associated with excessive neuroinhibition (stupor, coma) was certainly consistent with this concept. As the GABA hypothesis evolved, several factors were incriminated, including hypersensitivity of the $GABA_A$ receptor, and/or excessive amounts of allosteric modulators of the $GABA_A$ receptor (e.g., endogenous benzodiazepines; BZDs) *(98–100).* The mechanism for the increased BZD-like compounds in HE is not known. Strong support for the increased GABAergic tone hypothesis comes from studies showing that flumazenil, a $GABA_A$ receptor antagonist, can reverse coma in some patients with chronic HE *(101–103)* and experimental animals *(102).*

Recent support for the GABA hypothesis has incorporated a role for ammonia *(104,105).* Ammonia has been shown to facilitate GABA-stimulated Cl^- currents *(106),* and to enhance the binding of benzodiazepines to the $GABA_A$ receptor complex *(107,108).* Additionally, recent studies have shown that ammonia decreases the ability of astrocytes to take up GABA as well as enhance its release *(109,110).* The temporal response is in sharp contrast to that observed with the effect of ammonia on glutamate uptake; i.e., the effects on GABA uptake occurred within minutes, whereas the effects on glutamate uptake took at least one day to evolve. These effects of ammonia on astrocyte GABA transport may elevate synaptic levels of GABA sufficient to cause excessive GABAergic neurotransmission.

The peripheral-type benzodiazepine receptor (PBR) further links the GABA hypothesis to astrocytes. In contrast to the central benzodiazepine receptor, the PBR is primarily located on the outer mitochondrial membrane *(111).* The PBR also has a different rank order of ligand binding affinities than the "central" receptor. In the CNS, PBRs are not found on neurons but rather on astrocytes *(112,113)* and microglia *(114,115).* Although the function of the PBR is not completely understood, its best known function is steroidogenesis as it serves to transfer cholesterol from the outer to the inner mitochondrial membrane *(116).* By the action of cytochrome P450scc cholesterol is subsequently converted to pregnenolone, the parent compound for all neurosteroids. The steroid products are predominantly derivatives of progesterone, some of which (e.g., tetrahydroprogesterone, THP; tetrahydrodeoxycorticosterone, THDOC) are among the most potent positive modulators of the $GABA_A$ receptor known (for review, *see* ref. *117*).

Ammonia has been shown to upregulate the PBR in cultured astrocytes *(118).* Butterworth and colleagues have shown increased numbers of PBR receptors in human postmortem tissue from encephalopathic patients *(119),* and increased binding sites and mRNA levels were found in portacaval-shunted rats *(120,121).* Increased numbers of PBR binding sites were similarly identified in hyperammonemic and thioacetamide-treated mice *(122).* Interestingly, manganese, another purported HE-related toxin has been shown to upregulate the PBR *(123).* We have also shown that animals with HE/hyperammonemia had elevated levels of THDOC, THP, and pregnenolone *(124).*

Neurosteroids and benzodiazepines may also exert direct effects on astrocytes that are potentially relevant to HE. Some neurosteroids decrease the ability of astrocytes to take up glutamate, GABA and K^+. They also exert behavioral and neuropathologic changes similar to HE (Alzheimer type II change). These studies have been summarized in a recent publication *(124)*.

The involvement of the PBR and neurosteroids in the pathogenesis of HE may have therapeutic implications. Treatment with PK 11195, a putative antagonist of the PBR, significantly attenuated ammonia toxicity *(125)*. Pregnenolone sulfate whose electrophysiological effects are opposite to that of allopregnanolone *(126)* also had a protective effect, although of lesser magnitude *(125)*.

4.4. Astrocyte Swelling

Brain edema/swelling leading to increased intracranial pressure is a major cause of death (20–70%) in patients with FHF *(127)*. There is currently no satisfactory treatment for this condition other than liver transplantation. Although the basis for the intracranial hypertension is not clear, astroglial swelling plays an important role in this process *(7,128,129)*.

There are several consequences of astrocyte swelling. Most important is the mass effect resulting in increased intracranial pressure and brain herniation. Swelling also results in cell depolarization *(130)*, thereby interfering with the cell's ability to maintain ionic gradients and uptake of neurotransmitters. The reduction in the size of the extracellular space following astrocyte swelling may elevate extracellular ionic concentrations which can affect neuronal excitability *(131)*. Astrocyte swelling may also compress capillaries, contributing to a reduction in cerebral blood flow *(132)*. As suggested by Häussinger *(133)*, even small volume changes in astrocytes may affect cell function similar to those previously shown in cultured hepatocytes *(134)*.

The mechanism of astroglial swelling in FHF is not completely understood. Clearly ammonia plays a major role as it has been shown to cause swelling of astrocytes in culture *(135)*, brain slices *(136,137)*, and in vivo *(138–140)*. Furthermore, the extent of intracranial pressure correlates well with arterial ammonia levels in patients with acute liver failure *(141)*. Brusilow and Traystman *(142)* have proposed that the generation of glutamine from glutamate and ammonia by the action of glutamine synthetase (GS) may create a sufficient osmotic load to create astroglial swelling. Support for this view was derived from studies showing that treatment of hyperammonemic rats with MSO, an inhibitor of GS, reduced the amount of brain edema *(143,144)* and diminished the extent of astrocyte swelling *(140)*. Similarly, ammonia-induced swelling of cultured astrocytes could be abolished by treatment with MSO *(145)*. Additionally, MSO induces a massive release of glutamine from glial cells *(146)* which may contribute to its beneficial effects in HE. Thus, although the synthesis of glutamine is generally regarded as the principal means of detoxifying ammonia, this "protective" process also causes glial swelling and brain edema. It appears that the detoxification of ammonia occurs at the expense of developing astroglial swelling.

It is possible that other osmolytes may contribute to glial swelling. Ammonia-treated astrocytes showed an elevation in glutathione (GSH) levels *(147,148)*. The magnitude of GSH increase observed in this study was comparable to elevations in glutamine occurring after ammonia treatment *(27)*. As in the case of glutamine, GSH concentration in astrocytes is also in millimolar range *(149)*, and there is evidence that it may also act as an osmolyte *(150,151)*. It is thus possible that part of the swelling after ammonia treatment may be due to GSH. This premise is supported by a recent report

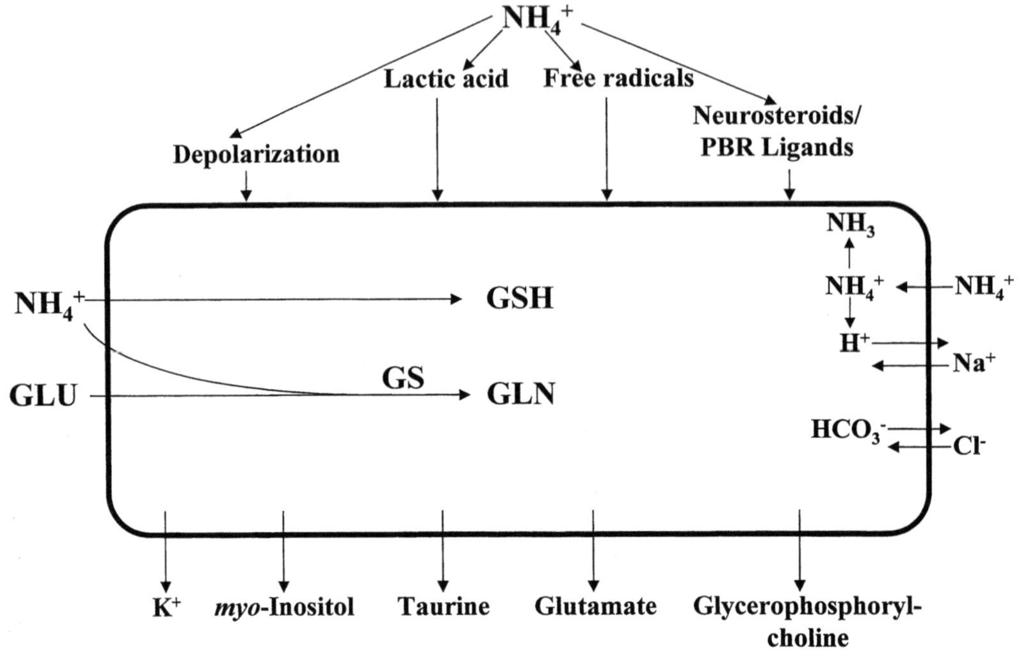

Fig. 3. Schematic illustration of mechanisms and consequences of ammonia-induced astroglial swelling. Factors known to be associated with ammonia, and known to cause swelling, are shown on top. Additionally, ammonia-mediated increase in glutamine and glutathione (GSH), which can act as osmolytes, are shown. Also, the influx of NH_4^+ through NH_4^+/K^+ channels will result in the influx of NaCl along with obliged water. As part of the volume regulatory events, inorganic (K^+) and organic (myo-inositol, taurine, glutamate, and glycerophosphorylcholine) osmolytes are released. These released osmolytes may have potent effects on CNS function. GLN, glutamine; GLU, glutamate; GS, glutamine synthetase; PBR, peripheral-type benzodiazepine receptor.

noting that BSO, which inhibits γ-glutamylcysteine synthetase activity *(152),* the GSH catabolizing enzyme, decreased ammonia-induced astrocyte swelling *(147).*

As noted previously, there is emerging data on the potential role of free radicals in ammonia toxicity. Since free radicals may contribute to astrocyte swelling *(153,154),* it is possible that ammonia-induced oxidative stress may also be involved in astrocyte swelling. Similarly, ammonia has been shown to cause membrane depolarization in cultured astrocytes *(44).* Such depolarization can lead to passive, Donnan-mediated KCl and water fluxes into cells *(155).* Also, the unidirectional entry of NH_4^+ into astrocytes leads to an intracellular acidosis *(53);* this would stimulate the Na^+/H^+ exchanger which, along with activation of the HCO_3^-/Cl^- exchanger, will result in NaCl and obliged water entering the cell *(156,157).* The accumulation of lactic acid in HE/hyperammonemia *(158–160)* may also be a factor as lactic acid has been shown to cause glial swelling *(161,162).* Additionally, PBR ligands and neurosteroids may contribute to astrocyte swelling *(163).* Potential mechanisms responsible for ammonia-mediated astroglial swelling are summarized in figure 3.

In response to swelling, a number of volume regulatory systems become activated *(164),* particularly the release of inorganic (K^+) and organic (glutamate, taurine, *myo-*

inositol, glycerophosphorylcholine) osmolytes *(165)*. The release of these factors by swollen astrocytes, however, may have profound effects on CNS excitability and function.

myo-Inositol is an important regulatory organic osmolyte *(166)* and is one of the main organic osmolytes in glial cells *(167)*. Brain *myo*-inositol levels has been shown to be decreased in chronic HE *(133,168–170)*. Treatment of astrocytes with ammonium chloride (5 m*M*) decreased *myo*-inositol uptake, increased its release and decreased its intracellular content *(171,172)*. As *myo*-inositol is important in the metabolism of various second messenger and lipids, loss of this metabolite may have major consequences on cell function.

Brain taurine levels are decreased in rats six weeks after a portacaval anastomosis *(170)*. This decrease is likely due to increased release as ammonia has been shown to stimulate taurine release in cultured astrocytes *(173)*, and in vivo *(174)*. While this release would have volume regulatory effects, taurine has inhibitory neuromodulatory effects *(175)* which could exacerbate the neuroinhibition in HE. On the other hand, taurine may exert a neuroprotective effect through its known antioxidant properties *(176)*. This may be particularly pertinent as there is recent evidence of excessive free radical formation in hyperammonemia (*see* **Subheading 4.**).

5. CONCLUDING REMARKS

The pathogenesis of HE and the mechanism of ammonia neurotoxicity remain enigmatic. Nevertheless, it is clear that astrocytes are a target of ammonia toxicity. Such toxicity results in a "primary gliopathy" leading to a disturbance in glial-neuronal interactions/trafficking, disruption of the glutamate-glutamine cycle, bioenergetic failure and potential abnormalities in glutamatergic and GABAergic neurotransmission, as well as in cell swelling. The initial event triggering these phenomena is not clear. A number of likely candidates are emerging, including oxidative stress and changes in membrane potential and pH.

This review has presented evidence that abnormalities in glutamatergic, and GABAergic neurotransmission are key considerations in HE. More importantly, it has attempted to integrate available data implicating ammonia neurotoxicity as a major factor in these neurotransmitter defects. Inadequate astroglial glutamate transport may lead to abnormalities in glutamatergic neurotransmission, and possible excitotoxic injury. Although the role of GABA in HE has been controversial, ammonia-induced decrease in astrocytic GABA uptake and increased release, as well as enhanced production of neurosteroids known to have a positive modulatory effect on the $GABA_A$ receptor are supportive of this hypothesis.

Astrocyte swelling in fulminant hepatic failure is a major clinical problem. The mechanisms for such swelling are unclear but likely include the generation of excessive osmolytes, chiefly glutamine and possibly glutathione. Additionally, oxidative stress and abnormalities in intracellular pH and membrane potential can also disrupt in ionic homeostasis and lead to cell swelling.

A fuller understanding of the role of astrocytes in HE and ammonia toxicity will undoubtedly yield critical information on glial function and glial-neuronal interactions, and hopefully generate novel therapeutic strategies.

ACKNOWLEDGMENTS

The author is indebted to Drs. Alex Bender, Yossef Itzhak, and Ch.R.K. Murthy who contributed greatly to the ideas and work presented in this article. This work was supported by the Department of Veterans Affairs, GRECC, and USPH grants NS-30291 and DK-38153.

REFERENCES

1. Norenberg, M.D. (1986) Hepatic encephalopathy: a disorder of astrocytes, in *Astrocytes* **Vol. 3,** (Fedoroff, S. and Vernadakis, A. eds.), Academic Press, Orlando, 425–460.
2. Norenberg, M.D., Neary, J.T., Bender, A.S., and Dombro, R.S. (1992) Hepatic encephalopathy: a disorder in glial-neuronal communication, in *Neuronal-Astrocytic Interactions. Implications for Normal and Pathological CNS Function* (Yu, A.C.H., Hertz, L., Norenberg, M.D., Sykova, E., and Waxman, S.G. eds.), Elsevier, Amsterdam, 261–269.
3. Adams, R.D. and Foley, J.M. (1953) The neurological disorder associated with liver disease. *Assoc. Res. Nerv. Ment. Dis. Proc.* **32,** 198–237.
4. Inoue, E., Hori, S., Narumi, Y., et al. (1991) Portal-systemic encephalopathy: presence of basal ganglia lesions with high signal intensity on MRI images. *Radiology* **179,** 551–555.
5. Pujol, A., Graus, F., Peri, J., Mercader, J.M., and Rimola, A. (1991) Hyperintensity in the globus pallidus on T_1-weighted and inversion-recovery MRI: A possible marker of advanced liver disease. *Neurology* **41,** 1526–1527.
6. Norenberg, M.D. (1981) The astrocyte in liver disease, in *Advances in Cellular Neurobiology* **Vol. 2** (Fedoroff, S. and Hertz, L. eds.), Academic Press, New York, 303–352.
7. Norenberg, M.D. (1977) A light and electron microscopic study of experimental portal-systemic (ammonia) encephalopathy. *Lab. Invest.* **36,** 618–627.
8. Victor, M., Adams, R.D., and Cole, M. (1965) The acquired (non-Wilsonian) type of chronic hepatocerebral degeneration. *Medicine* **44,** 345–396.
9. Hauser, R.A., Zesiewicz, T.A., Rosemurgy, A.S., Martinez, C., and Olanow, C.W. (1994) Manganese intoxication and chronic liver failure. *Ann. Neurol.* **36,** 871–875.
10. Zieve, L. (1987) Pathogenesis of hepatic encephalopathy. *Metab. Brain Dis.* **2,** 147–165.
11. Layrargues, G.P., Rose, C., Spahr, L., Zayed, J., Normandin, L., and Butterworth, R.F. (1998) Role of manganese in the pathogenesis of portal-systemic encephalopathy. *Metab. Brain Dis.* **13,** 311–317.
12. Hazell, A.S. and Butterworth, R.F. (1999) Hepatic encephalopathy: An update of pathophysiologic mechanisms. *Proc. Soc. Exp. Biol. Med.* **222,** 99–112.
13. Yurdaydin, C., Li, Y., Ha, J.-H., Jones, E.A., Rothman, R., and Basile, A.S. (1995) Brain and plasma levels of opioid peptides are altered in rats with thioacetamide-induced fulminant hepatic failure: Implications for the treatment of hepatic encephalopathy with opioid antagonists. *J. Pharmacol. Exp. Ther.* **273,** 185–192.
14. Conn, H.O. (1993) Hepatic encephalopathy, in *Diseases of the Liver* (Schiff, L. and Schiff, E.R. eds.), Lippincott, Philadelphia, 1036–1061.
15. Gregorios, J.B., Mozes, L.W., Norenberg, L.O.B., and Norenberg, M.D. (1985) Morphologic effects of ammonia on primary astrocyte cultures. I. Light microscopic studies. *J. Neuropathol. Exp. Neurol.* **44,** 397–403.
16. Gregorios, J.B., Mozes, L.W., and Norenberg, M.D. (1985) Morphologic effects of ammonia on primary astrocyte cultures. II. Electron microscopic studies. *J. Neuropathol. Exp. Neurol.* **44,** 404–414.
17. Plum, F. and Hindfelt, B. (1976) The neurological complications of liver disease, in *Handbook of Clinical Neurology* (Vol. 27) (Vinken, P.J. and Bruyn, G.W. eds.), North-Holland Publ., Amsterdam, 349–377.

18. Roos, A. and Boron, W.F. (1981) Intracellular pH *Physiol. Rev.* **61,** 296–434.
19. Cooper, A.J.L. and Plum, F. (1987) Biochemistry and physiology of brain ammonia. *Physiol. Rev.* **67,** 440–519.
20. Sadasivudu, B. and Hanumantharao, T.I. (1974) Studies on the distribution of urea cycle enzymes in different regions of rat brain. *J. Neurochem.* **23,** 267–269.
21. Waniewski, R.A. (1992) Physiological levels of ammonia regulate glutamine synthesis from extracellular glutamate in astrocyte cultures. *J. Neurochem.* **58,** 167–174.
22. Norenberg, M.D. (1983) Immunohistochemistry of glutamine synthetase, in *Glutamine, Glutamate, and GABA in the Central Nervous System* (Hertz, L., Kvamme, E., McGeer, E.G., and Schousboe, A. eds.), Alan R. Liss, New York, 95–111.
23. Bradford, H.F. and Ward, H.K. (1976) On glutaminase activity in mammalian synaptosomes. *Brain Res.* **110,** 115–125.
24. Shank, R.P. and Aprison, M.H. (1981) Present status and significance of the glutamine cycle in neural tissue. *Life Sci.* **28,** 837–842.
25. Waniewski, R.A. and Martin, D.L. (1986) Exogenous glutamate is metabolized to glutamine and exported by rat primary astrocyte cultures. *J. Neurochem.* **47,** 304–313.
26. Farinelli, S.E. and Nicklas, W.J. (1992) Glutamate metabolism in rat cortical astrocyte cultures. *J. Neurochem.* **58,** 1905–1915.
27. Lai, J.C.K., Murthy, C.R.K., Cooper, A.J.L., Hertz, E., and Hertz, L. (1989) Differential effects of ammonia and beta-methylene-DL-aspartate on the metabolism of glutamate and related amino acids by astrocytes and neurons in primary cultures. *Neurochem. Res.* **14,** 377–389.
28. Bessman, S.P. and Bessman, A.N. (1955) The cerebral and peripheral uptake of ammonia in liver disease with an hypothesis for the mechanism of hepatic coma. *J. Clin. Invest.* **34,** 622–628.
29. Norenberg, M.D., Huo, Z., Neary, J.T., and Roig-Cantisano, A. (1997) The glial glutamate transporter in hyperammonemia and hepatic encephalopathy: relation to energy metabolism and glutamatergic neurotransmission. *Glia* **21,** 124–133.
30. Hindfelt, B., Plum, F., and Duffy, T.E. (1977) Effect of acute ammonia intoxication on cerebral metabolism in rats with portacaval shunts. *J. Clin. Invest.* **59,** 386–396.
31. Ratnakumari, L. and Murthy, C.R.K. (1989) Activities of pyruvate dehydrogenase, enzymes of citric acid cycle and aminotransferases in subcellular fractions of cerebral cortex in normal and hyperammonemic rats. *Neurochem. Res.* **14,** 221–228.
32. Faff-Michalak, L. and Albrecht, J. (1991) Aspartate aminotransferase, malate dehydrogenase and pyruvate dehydrogenase activities in rat cerebral synaptic and nonsynaptic mitochondria: effects of in vitro treatment with ammonia, hyperammonemia and hepatic encephalopathy. *Metab. Brain Dis.* **6,** 187–197.
33. Siesjö, B.K. (1978) *Brain Energy Metabolism.* John Wiley & Sons, Chichester, p 42.
34. Albrecht, J., Wysmyk-Cybula, U., and Rafalowska, U. (1985) Na^+/K^+-ATPase activity and GABA uptake in astroglial cell-enriched fractions and synaptosomes derived from rats in the early stage of experimental hepatogenic encephalopathy. *Acta Neurol. Scand.* **72,** 317–320.
35. Kosenko, E., Kaminsky, Y., Grau, E., et al. (1994) Brain ATP depletion induced by acute ammonia intoxication in rats is mediated by activation of the NMDA receptor and Na^+, K^+-ATPase. *J. Neurochem.* **63,** 2172–2178.
36. Ratnakumari, L., Audet, R., Qureshi, I.A., and Butterworth, R.F. (1995) Na^+, K^+-ATPase activities are increased in brain in both congenital and acquired hyperammonemic syndromes. *Neurosci. Lett.* **197,** 89–92.
37. Sibson, N.R., Dhankhar, A., Mason, G.F., Rothman, D.L., Behar, K.L., and Shulman, R.G. (1998) Stoichiometric coupling of brain glucose metabolism and glutamatergic neuronal activity. *Proc. Nat. Acad. Sci. USA* **95,** 316–321.
38. Fitzpatrick, S.M., Cooper, A.J., and Hertz, L. (1988) Effects of ammonia and beta-methylene-DL-aspartate on the oxidation of glucose and pyruvate by neurons and astrocytes in primary culture. *J. Neurochem.* **51,** 1197–1203.

39. Haghighat, N. and McCandless, D.W. (1997) Effect of ammonium chloride on energy metabolism of astrocytes and C6-glioma cells in vitro. *Metab. Brain Dis.* **12,** 287–298.
40. Murthy, C.R. and Hertz, L. (1988) Pyruvate decarboxylation in astrocytes and in neurons in primary cultures in the presence and the absence of ammonia. *Neurochem. Res.* **13,** 57–61.
41. Raabe, W.A. (1989) Neurophysiology of ammonia intoxication, in *Hepatic Encephalopathy: Physiology and Treatment* (Butterworth, R.F. and Pomier Layrargues, G. eds.), Humana Press, Totowa, New Jersey, 49–77.
42. Latorre, R. and Miller, C. (1983) Conduction and selectivity in potassium channels. *J. Memb. Biol.* **71,** 11–30.
43. Binstock, L. and Lecar, H. (1969) Ammonium ion currents in the squid giant axon. *J. Gen. Physiol.* **53,** 342–361.
44. Allert, N., Köller, H., and Siebler, M. (1998) Ammonia-induced depolarization of cultured rat cortical astrocytes. *Brain Res.* **782,** 261–270.
45. Busa, W.B. and Nuccitelli, R. (1984) Metabolic regulation via intracellular pH. *Amer. J. Physiol.* **246,** R409.
46. Chesler, M. and Kaila, K. (1992) Modulation of pH by neuronal activity. *Trends Neurosci.* **15,** 396–402.
47. Segelen, P.O. (1983) Inhibitors of lysosomal function. *Meth. Enzymol.* **96,** 737–765.
48. Swain, M.S., Blei, A.T., Butterworth, R.F., and Kraig, R.P. (1991) Intracellular pH rises and astrocytes swell after portacaval anastomosis in rats. *Am. J. Physiol. Regul. Integr. Comp. Physiol.* **261,** R1491–R1496.
49. Fitzpatrick, S.M., Hetherington, H.P., Behar, K.L., and Shulman, R.G. (1989) Effects of acute hyperammonemia on cerebral amino acid metabolism and pH_i in vivo, measured by ^1H and ^{31}P nuclear magnetic resonance. *J. Neurochem.* **52,** 741–749.
50. Kanamori, K. and Ross, B.D. (1997) Glial alkalinization detected in vivo by ^1H-^{15}N heteronuclear multiple-quantum coherence-transfer NMR in severely hyperammonemic rat. *J. Neurochem.* **68,** 1209–1220.
51. Brooks, K.J., Kauppinen, R.A., Williams, S.R., Bachelard, H.S., Bates, T.E., and Gadian, D.G. (1989) Ammonia causes a drop in intracellular pH in metabolizing cortical brain slices. A [^{31}P]- and [^1H]nuclear magnetic resonance study. *Neuroscience* **33,** 185–192.
52. Boyarsky, G., Ransom, B.R., Schlue, W.R., Davis, M.B., and Boron, W.F. (1993) Intracellular pH regulation in single cultured astrocytes from rat forebrain. *Glia* **8,** 241–248.
53. Nagaraja, T.N. and Brookes, N. (1998) Intracellular acidification induced by passive and active transport of ammonium ions in astrocytes. *Am. J. Physiol. Cell Physiol.* **274,** C883–C891.
54. Norenberg, M.D., Bender, A.S., and Vastag, M. (1998) Effect of ammonia on intracellular pH in cultured astrocytes and endothelial cells. *Soc. Neurosci. Abstr.* **24,** 2013.
55. Kikeri, D., Sun, A., Zeidel, M.L., and Hebert, S.C. (1989) Cell membranes impermeable to NH_3. *Nature* **339,** 478–480.
56. Humphreys, B.D., Chernova, M.N., Jiang, L.W., Zhang, Y., and Alper, S.L. (1997) NH_4Cl activates AE2 anion exchanger in *Xenopus* oocytes at acidic pH_i. *Am. J. Physiol. Cell Physiol.* **272,** C1232–C1240.
57. Kosenko, E., Kaminski, Y., Lopata, O., Muravyov, N., and Felipo, V. (1999) Blocking NMDA receptors prevents the oxidative stress induced by acute ammonia intoxication. *Free Radic. Biol. Med.* **26,** 1369–1374.
58. Murphy, M.G., Jollimore, C., Crocker, J.F.S., and Her, H. (1992) Beta-Oxidation of [1-^{14}C]palmitic acid by mouse astrocytes in primary culture: Effects of agents implicated in the encephalopathy of Reye's syndrome. *J. Neurosci. Res.* **33,** 445–454.
59. Murthy, C.R.K., Liu, H., and Norenberg, M.D. (2000) Ammonia-induced free radical formation in primary cultures of astrocytes. *J. Neurochem.* **74,** Suppl 580.

60. Guerrini, V.H. (1994) Effect of antioxidants on ammonia induced CNS-renal pathobiology in sheep. *Free Radic. Res.* **21,** 35–43.
61. Bruck, R., Aeed, H., Shirin, H., et al. (1999) The hydroxyl radical scavengers dimethylsulfoxide and dimethylthiourea protect rats against thioacetamide-induced fulminant hepatic failure. *J. Hepatol.* **31,** 27–38.
62. Wendon, J.A., Harrison, P.M., Keays, R., and Williams, R. (1994) Cerebral blood flow and metabolism in fulminant hepatic failure. *Hepatology* **19,** 1407–1413.
63. Rao, V.L.R., Audet, R.M., and Butterworth, R.F. (1995) Increased nitric oxide synthase activities and L-[^{3}H]arginine uptake in brain following portacaval anastomosis. *J. Neurochem.* **65,** 677–681.
64. Norenberg, M.D. and Itzhak, Y. (1995) Acute liver failure and hyperammonemia increase nitric oxide synthase in mouse brain. *Soc. Neurosci. Abstr.* **21,** 869.
65. Master, S., Gottstein, J., and Blei, A.T. (1999) Cerebral blood flow and the development of ammonia-induced brain edema in rats after portacaval anastomosis. *Hepatology* **30,** 876–880.
66. Hazell, A.S. and Norenberg, M.D. (1998) Ammonia and manganese increase arginine uptake in cultured astrocytes. *Neurochem. Res.* **23,** 869–873.
67. Kosenko, E., Kaminsky, Y., Grau, E., Miñana, M.-D., Grisolía, S., and Felipo, V. (1995) Nitroarginine, an inhibitor of nitric oxide synthetase, attenuates ammonia toxicity and ammonia-induced alterations in brain metabolism. *Neurochem. Res.* **20,** 451–456.
68. Norenberg, M.D. (1995) Hepatic encephalopathy, in *Neuroglia* (Kettenmann, H. and Ransom, B.R. eds.), Oxford, New York, 950–963.
69. Reichenbach, A., Stolzenburg, J.-U., Wolburg, H., Härtig, W., El-Hifnawi, E., and Martin, H. (1995) Effects of enhanced extracellular ammonia concentration on cultured mammalian retinal glial (Müller) cells. *Glia* **13,** 195–208.
70. Albrecht, J. and Faff, L. (1994) Astrocyte-neuron interactions in hyperammonemia and hepatic encephalopathy. *Adv. Exp. Med. Biol.* **368,** 45–54.
71. Bender, A.S. and Norenberg, M.D. (1996) Effects of ammonia on L-glutamate uptake in cultured astrocytes. *Neurochem. Res.* **21,** 567–573.
72. Hansson, E. and Rönnbäck, L. (1995) Astrocytes in glutamate neurotransmission. *FASEB J.* **9,** 343–350.
73. Schousboe, A., Sonnewald, U., Civenni, G., and Gegelashvili, G. (1997) Role of astrocytes in glutamate homeostasis – Implications for excitotoxicity. *Adv. Exp. Med. Biol.* **429,** 195–206.
74. Schousboe, A. and Westergaard, N. (1995) Transport of neuroactive amino acids in astrocytes, in *Neuroglia* (Kettenmann, H. and Ransom, B.R. eds.), Oxford University Press, New York, 246–258.
75. Rothstein, J.D., Dykes-Hoberg, M., Pardo, C.A., et al. (1996) Knockout of glutamate transporters reveals a major role for astroglial transport in excitotoxicity and clearance of glutamate. *Neuron* **16,** 675–686.
76. Bergles, D.E. and Jahr, C.E. (1998) Glial contribution to glutamate uptake at Schaffer collateral-commissural synapses in the hippocampus. *J. Neurosci.* **18,** 7709–7716.
77. Gegelashvili, G. and Schousboe, A. (1998) Cellular distribution and kinetic properties of high-affinity glutamate transporters. *Brain Res. Bull.* **45,** 233–238.
78. Danbolt, N.C. (1994) The high affinity uptake system for excitatory amino acids in the brain. *Prog. Neurobiol.* **44,** 377–396.
79. Albrecht, J., Hilgier, W., Lazarewicz, J.W., Rafalowska, U., and Wysmyk-Cybula, U. (1988) Astrocytes in acute hepatic encephalopathy: metabolic properties and transport functions, in *The Biochemical Pathology of Astrocytes* (Norenberg, M.D., Hertz, L., and Schousboe, A. eds.), Alan R. Liss, New York, 465–476.
80. Oppong, K.N.W., Bartlett, K., Record, C.O., and Al Mardini, H. (1995) Synaptosomal glutamate transport in thioacetamide-induced hepatic encephalopathy in the rat. *Hepatology* **22,** 553–558.

81. Mena, E.E. and Cotman, C.W. (1985) Pathologic concentrations of ammonium ions block L-glutamate uptake. *Exp. Neurol.* **89,** 259–263.
82. Schmidt, W., Wolf, G., Grungreiff, K., Meier, M., and Reum, T. (1990) Hepatic encephalopathy influences high-affinity uptake of transmitter glutamate and aspartate into the hippocampal formation. *Metab. Brain Dis.* **5,** 19–31.
83. Zhou, B.G. and Norenberg, M.D. (1999) Ammonia downregulates GLAST mRNA glutamate transporter in rat astrocyte cultures. *Neurosci. Lett.* **276,** 145–148.
84. Kondo, K., Sashimoto, H., Kitanaka, J., et al. (1995) Expression of glutamate transporters in cultured glial cells. *Neurosci. Lett.* **188,** 140–142.
85. Gegelashvili, G., Danbolt, N.C., and Schousboe, A. (1997) Neuronal soluble factors differentially regulate the expression of the GLT1 and GLAST glutamate transporters in cultured astroglia. *J. Neurochem.* **69,** 2612–2615.
86. Weissenborn, K. and Kolbe, H. (1998) The basal ganglia and portal-systemic encephalopathy. *Metab. Brain Dis.* **13,** 261–272.
87. Knecht, K., Michalak, A., Rose, C., Rothstein, J.D., and Butterworth, R.F. (1997) Decreased glutamate transporter (GLT-1) expression in frontal cortex of rats with acute liver failure. *Neurosci. Lett.* **229,** 201–203.
88. Volterra, A., Trotti, D., Tromba, C., Floridi, S., and Racagni, G. (1994) Glutamate uptake inhibition by oxygen free radicals in rat cortical astrocytes. *J. Neurosci.* **14,** 2924–2932.
89. Pogun, S., Dawson, V., and Kuhar, M.J. (1994) Nitric oxide inhibits ^{3}H-glutamate transport in synaptosomes. *Synapse* **18,** 21–26.
90. Kimelberg, H.K., Rutledge, E., Goderie, S., and Charniga, C. (1995) Astrocytic swelling due to hypotonic or high K^+ medium causes inhibition of glutamate and aspartate uptake and increases their release. *J. Cereb. Blood Flow Metab.* **15,** 409–416.
91. Billups, B., Rossi, D., and Attwell, D. (1996) Anion conductance behavior of the glutamate uptake carrier in salamander retinal glial cells. *J. Neurosci.* **16,** 6722–6731.
92. Hazell, A.S. and Norenberg, M.D. (1997) Manganese decreases glutamate uptake in cultured astrocytes. *Neurochem. Res.* **22,** 1443–1447.
93. Vogels, B.A.P.M., Maas, M.A.W., Daalhuisen, J., Quack, G., and Chamuleau, R.A.F.M. (1997) Memantine, a noncompetitive NMDA receptor antagonist improves hyperammonemia-induced encephalopathy and acute hepatic encephalopathy in rats. *Hepatology* **25,** 820–827.
94. Felipo, V., Hermenegildo, C., Montoliu, C., Llansola, M., and Miñana, M.D. (1998) Neurotoxicity of ammonia and glutamate: Molecular mechanisms and prevention. *Neurotoxicology* **19,** 675–681.
95. Chan, P.H., Chu, L. and Chen, S. (1990) Effects of MK-801 on glutamate-induced swelling of astrocytes in primary cell culture. *J. Neurosci. Res.* **25,** 87–93.
96. Bridges, R.J., Hatalski, C.G., Shim, S.N., et al. (1992) Gliotoxic actions of excitatory amino acids. *Neuropharmacology.* **31,** 899–907.
97. Jones, E.A., Yurdaydin, C. and Basile, A.S. (1994) The GABA hypothesis—state of the art. *Adv. Exp. Med. Biol.* **368,** 89–101.
98. Mullen, K.D., Martin, J.V., Mendelson, W.B., Bassett, M.L., and Jones, E.A. (1988) Could an endogenous benzodiazepine ligand contribute to hepatic encephalopathy. *Lancet* **i,** 457–459.
99. Rothstein, J.D., McKhann, G., Guarneri, P., Barbaccia, M.L., Guidotti, A., and Costa, E. (1989) Cerebrospinal fluid content of diazepam binding inhibitor in chronic hepatic encephalopathy. *Ann. Neurol.* **26,** 57–62.
100. Basile, A.S., Hughes, R.D., Harrison, P.M., et al. (1991) Elevated brain concentrations of 1,4-benzodiazepines in fulminant hepatic failure. *N. Engl. J. Med.* **325,** 473–478.
101. Ferenci, P., Grimm, G., Meryn, S., and Gangl, A. (1989) Successful long-term treatment of portal-systemic encephalopathy by the benzodiazepine antagonist flumazenil. *Gastroenterology* **96,** 240–243.

102. Gammal, S.H., Basile, A.S., Geller, D., Skolnick, P., and Jones, E.A. (1990) Reversal of the behavioral and electrophysiological abnormalities of an animal model of hepatic encephalopathy by benzodiazepine receptor ligands. *Hepatology* **11,** 371–378.
103. Pomier-Layrargues, G., Giguère, J.F., Lavoie, J., et al. (1994) Flumazenil in cirrhotic patients in hepatic coma: A randomized double-blind placebo-controlled crossover trial. *Hepatology* **19,** 32–37.
104. Basile, A.S. and Jones, E.A. (1997) Ammonia and GABA-ergic neurotransmission: Interrelated factors in the pathogenesis of hepatic encephalopathy. *Hepatology* **25,** 1303–1305.
105. Jones, E.A. and Basile, A.S. (1998) Does ammonia contribute to increased GABA-ergic neurotransmission in liver failure? *Metab. Brain Dis.* **13,** 351–360.
106. Takahashi, K., Kameda, H., Kataoka, M., Sanjou, K., Harata, N., and Akaike, N. (1993) Ammonia potentiates $GABA_A$ response in dissociated rat cortical neurons. *Neurosci. Lett.* **151,** 51–54.
107. Branchey, L., Branchey, M., Worner, T.M., Zucker, D., Shaaw, S., and Lieber, C.S. (1985) Association between amino acid alterations and hallucinations in alcoholic patients. *Biol. Psychiat.* **20,** 1167–1173.
108. Sato, M. and et al. (1984) Antiepileptic effects of thyrotropin-releasing hormone and its new derivative, DN-1417, examined in feline amygdaloid kindling preparation. *Epilepsia* **25,** 537–544.
109. Aguila-Mansilla, N., Bender, A.S., and Norenberg, M.D. (1998) Effect of ammonia, peripheral benzodiazepine ligands and neurosteroids on GABA uptake in cultured astrocytes. *J. Neurochem.* **70, (Suppl 1)** S28.
110. Bender, A.S. and Norenberg, M.D. (2000) Effect of ammonia on GABA uptake and release in cultured astrocytes. *Neurochem. Internat.* **36,** 385–395.
111. Anholt, R.R.H., Pedersen, P.L., DeSouza, E.B., and Snyder, S.H. (1986) The peripheral-type benzodiazepine receptor: localization to the mitochondrial outer membrane. *J. Biol. Chem.* **261,** 576–583.
112. Bender, A.S. and Hertz, L. (1985) Binding of (3H) RO5-4864 in primary cultures of astrocytes. *Brain Res.* **341,**41–9.
113. Itzhak, Y., Baker, L., and Norenberg, M.D. (1993) Characterization of the peripheral-type benzodiazepine receptor in cultured astrocytes: evidence for multiplicity. *Glia* **9,** 211–218.
114. Myers, R., Manjil, L.G., Cullen, B.M., Price, G.W., Franckowiak, R.S., and Cremer, J.E. (1991) Macrophage and astrocyte populations in relation to (3H)PK 11195 binding following a local ischemic lesion. *J. Cereb. Blood Flow Metab.* **11,** 314–322.
115. Park, C.H., Carboni, E., Wood, P.L., and Gee, K.W. (1996) Characterization of peripheral benzodiazepine type sites in a cultured murine BV-2 microglial cell line. *Glia* **16,** 65–70.
116. Papadopoulos, V. (1993) Peripheral-type benzodiazepine/diazepam binding inhibitor receptor: Biological role in steroidogenic cell function. *Endocr. Rev.* **14,** 222–240.
117. Lambert, J.J., Belelli, D., Hill-Venning, C., and Peters, J.A. (1995) Neurosteroids and $GABA_A$ receptor function. *Trends Pharmacol. Sci.* **16,** 295–303.
118. Itzhak, Y. and Norenberg, M.D. (1994) Ammonia-induced upregulation of peripheral-type benzodiazepine receptors in cultured astrocytes labeled with [^{3}H]PK 11195. *Neurosci. Lett.* **177,** 35–38.
119. Lavoie, J., Layrargues, G.P., and Butterworth, R.F. (1990) Increased densities of peripheral-type benzodiazepine receptors in brain autopsy samples from cirrhotic patients with hepatic encephalopathy. *Hepatology* **11,** 874–878.
120. Giguere, J.F., Hamel, E., and Butterworth, R.F. (1992) Increased densities of binding sites for the 'peripheral-type' benzodiazepine receptor ligand [^{3}H]PK 11195 in rat brain following portacaval anastomosis. *Brain Res.* **585,** 295–298.
121. Desjardins, P., Bandeira, P., Rao, V.L.R., Ledoux, S., and Butterworth, R.F. (1997) Increased expression of the peripheral-type benzodiazepine receptor isoquinoline carboxamide binding protein mRNA in brain following portacaval anastomosis. *Brain Res.* **758,** 255–258.

122. Itzhak, Y., Roig-Cantisano, A., Dombro, R.S., and Norenberg, M.D. (1995) Acute liver failure and hyperammonemia increase peripheral-type benzodiazepine receptor binding and pregnenolone synthesis in mouse brain. *Brain Res.* **705,** 345–348.
123. Hazell, A.S., Desjardins, P., and Butterworth, R.F. (1999) Chronic exposure of rat primary astrocyte cultures to manganese results in increased binding sites for the 'peripheral-type' benzodiazepine receptor ligand ^{3}H-PK 11195. *Neurosci. Lett.* **271,** 5–8.
124. Norenberg, M.D., Itzhak, Y., and Bender, A.S. (1997) The peripheral benzodiazepine receptor and neurosteroids in hepatic encephalopathy, in *Cirrhosis, Hyperammonemia, and Hepatic Encephalopathy* (Felipo, V. ed.), Plenum Press, 95–111.
125. Itzhak, Y. and Norenberg, M.D. (1994) Attenuation of ammonia toxicity in mice by PK 11195 and pregnenolone sulfate. *Neurosci. Lett.* **182,** 251–254.
126. Majewska, M.D. (1992) Neurosteroids: endogenous bimodal modulators of the $GABA_A$ receptor: mechanism of action and physiological significance. *Prog. Neurobiol.* **38,** 379–395.
127. Córdoba, J. and Blei, A.T. (1996) Brain edema and hepatic encephalopathy. *Semin. Liver Dis.* **16,** 271–280.
128. Traber, P.G., Dal Canto, M.C., Ganger, D., and Blei, A.T. (1987) Electron microscopic evaluation of brain edema in rabbits with galactosamine-induced fulminant hepatic failure: ultrastructure and integrity of the blood-brain barrier. *Hepatology* **7,** 1272–1277.
129. Kato, M., Hughes, R.D., Keays, R.T., and Williams, R. (1992) Electron microscopic study of brain capillaries in cerebral edema from fulminant hepatic failure. *Hepatology* **15,** 1060–1066.
130. Kimelberg, H.K. and O'Connor, E.R. (1988) Swelling-induced depolarization of astrocyte potentials. *Glia* **1,** 219–224.
131. Smith, S.J. (1992) Do astrocytes process neural information? in *Neuronal-Astrocytic Interactions: Implications for Normal and Pathological CNS Function* (Yu, A.C.H., Hertz, L., Norenberg, M.D., Sykova, E., and Waxman, S. eds.), Elsevier, Amsterdam, 119–136.
132. Garcia, J.H., Liu, K.F., Yoshida, Y., Chen, S., and Lian, J. (1994) The brain microvessels: factors altering their patency after the occlusion of a middle cerebral artery (Wistar rat). *Am. J. Pathol.* **145,** 1–13.
133. Häussinger, D., Laubenberger, J., Vom Dahl, S., et al. (1994) Proton magnetic resonance spectroscopy studies on human brain *myo*-inositol in hypo-osmolarity and hepatic encephalopathy. *Gastroenterology* **107,** 1475–1480.
134. Häussinger, D. (1996) The role of cellular hydration in the regulation of cell function. *Biochem. J.* **313,** 697–710.
135. Norenberg, M.D., Baker, L., Norenberg, L.-O.B., Blicharska, J., Bruce-Gregorios, J.H., and Neary, J.T. (1991) Ammonia-induced astrocyte swelling in primary culture. *Neurochem. Res.* **16,** 833–836.
136. Benjamin, A.M., Okamoto, K., and Quastel, J.H. (1978) Effects of ammonium ions on spontaneous action potentials and on contents of sodium, potassium, ammonium and chloride ions in brain in vitro. *J. Neurochem.* **30,** 131–143.
137. Ganz, R., Swain, M., Traber, P., Dal Canto, M.C., Butterworth, R.F., and Blei, A.T. (1989) Ammonia-induced swelling of rat cerebral cortical slices: implications for the pathogenesis of brain edema in acute hepatic failure. *Metab. Brain Dis.* **4,** 213–223.
138. Cole, M., Rutherford, R.B., and Smith, F.O. (1972) Experimental ammonia encephalopathy in the primate. *Arch. Neurol.* **26,** 130–136.
139. Voorhies, T.M., Ehrlich, M.E., Duffy, T.E., Petito, C.K., and Plum, F. (1983) Acute hyperammonemia in the young primate. Physiologic and neuropathological correlates. *Pediatr. Res.* **17,** 970–975.
140. Willard-Mack, C.L., Koehler, R.C., Hirata, T., et al. (1996) Inhibition of glutamine synthetase reduces ammonia-induced astrocyte swelling in rat. *Neuroscience* **71,** 589–599.

141. Clemmesen, J.O., Larsen, F.S., Kondrup, J., Hansen, B.A., and Ott, P. (1999) Cerebral herniation in acute liver failure is correlated with arterial ammonia concentration. *Hepatology* **29,** 648–653.
142. Brusilow, S.W. and Traystman, R.J. (1986) Letter to editor. *New Engl. J. Med.* **314,** 786
143. Takahashi, H., Koehler, R.C., Brusilow, S.W., and Traystman, R.J. (1991) Inhibition of brain glutamine accumulation prevents cerebral edema in hyperammonemic rats. *Am. J. Physiol.* **261,** H825–H829.
144. Blei, A.T., Olafsson, S., Therrien, G., and Butterworth, R.F. (1994) Ammonia-induced brain edema and intracranial hypertension in rats after portacaval anastomosis. *Hepatology* **19,** 1437–1444.
145. Norenberg, M.D. and Bender, A.S. (1994) Astrocyte swelling in liver failure: role of glutamine and benzodiazepines. *Acta Neurochir.* **60 (Suppl),** 24–27.
146. Albrecht, J. and Norenberg, M.D. (1990) L-Methionine-DL-sulfoximine induces massive efflux of glutamine from cortical astrocytes in primary culture. *Eur. J. Pharmacol.* **182,** 587–590.
147. Bender, A.S., Dombro, R.S., and Norenberg, M.D. (1998) Glutathione as a factor in ammonia-induced swelling. *Soc. Neurosci. Abstr.* **24,** 2013.
148. Murthy, C.R.K., Bender, A.S., Dombro, R.S., Bai, G., and Norenberg, M.D. (2000) Elevation of glutathione levels by ammonium ions in primary cultures of rat astrocytes. *Neurochem. Internat.* **37,** 255–268.
149. Yudkoff, M., Pleasure, D., Cregar, L., Lin, Z.P., Nissim, I., and Stern, J. (1990) Glutathione turnover in cultured astrocytes: studies with [15N]glutamate. *J. Neurochem.* **55,** 137–145.
150. Häussinger, D., Lang, F., Bauers, K., and Gerok, W. (1990) Control of hepatic nitrogen metabolism and glutathione release by cell volume regulatory mechanisms. *Eur. J. Biochem.* **193,** 891–898.
151. Clark, E.C., Thomas, D., Baer, J., and Sterns, R.H. (1996) Depletion of glutathione from brain cells in hyponatremia. *Kidney Internat.* **49,** 470–476.
152. Griffith, O.W. (1982) Mechanism of action, metabolism and toxicity of buthionine sulphoximine and its higher homologues, potent inhibitors of glutathione synthesis. *J. Biol. Chem.* **257,** 13704–13712.
153. Chan, P.H., Longar, S., Chen, S., et al. (1989) The role of arachidonic acid and oxygen radical metabolites in the pathogenesis of vasogenic brain edema and astrocytic swelling. *Ann. N. Y. Acad. Sci.* **559,** 237–247.
154. Staub, F., Winkler, A., Peters, J., Kempski, O., Kachel, V., and Baethmann, A. (1994) Swelling, acidosis, and irreversible damage of glial cells from exposure to arachidonic acid in vitro. *J. Cereb. Blood Flow Metab.* **14,** 1030–1039.
155. Walz, W. (1997) Role of astrocytes in the spreading depression signal between ischemic core and penumbra. *Neurosci. Biobehav. Rev.* **21,** 135–142.
156. Jakubovicz, D.E., Grinstein, S., and Klip, A. (1987) Cell swelling following recovery from acidification in C6 glioma cells: an in vitro model of postischemic brain edema. *Brain Res.* **435,** 138–146.
157. Kempski, O., Staub, F., Jansen, M., and Baethmann, A. (1990) Molecular mechanisms of glial cell swelling in acidosis. *Adv. Neurol.* **52,** 39–45.
158. de Graaf, A.A., Deutz, N.E., Bosman, D.K., Chamuleau, R.A., De Haan, J.G., and Bovee, W.M. (1991) The use of in vivo proton NMR to study the effects of hyperammonemia in the rat cerebral cortex. *NMR. Biomed.* **4,** 31–37.
159. Yao, H., Sadoshima, S., Fujii, K., et al. (1987) Cerebrospinal fluid lactate in patients with hepatic encephalopathy. *Eur. Neurol.* **27,** 182–187.
160. Ross, B., Kreis, R., and Ernst, T. (1992) Clinical tools for the 90s: magnetic resonance spectroscopy and metabolite imaging. *Eur. J. Radiol.* **14,** 128–140.

161. Jakubovicz, D. and Klip, A. (1989) Lactic acid-induced swelling in C6 glial cells via Na^+/H^+ exchange. *Brain Res.* **485,** 215–224.
162. Staub, F., Peters, J., Kempski, O., Schneider, G.-H., Schürer, L., and Baethmann, A. (1993) Swelling of glial cells in lactacidosis and by glutamate: Significance of Cl^--transport. *Brain Res.* **610,** 69–74.
163. Bender, A.S. and Norenberg, M.D. (1998) Effect of benzodiazepines and neurosteroids on ammonia-induced swelling in cultured astrocytes. *J. Neurosci. Res.* **54,** 673–680.
164. Hoffmann, E.K. and Dunham, P.B. (1996) Membrane mechanisms and intracellular signalling in cell volume regulation. *Internat. Rev. Cytol* **161,** 173–262.
165. Bluml, S., Zuckerman, E., Tan, J., and Ross, B.D. (1998) Proton-decoupled ^{31}P magnetic resonance spectroscopy reveals osmotic and metabolic disturbances in human hepatic encephalopathy. *J. Neurochem.* **71,** 1564–1576.
166. Burg, M.B. (1995) Molecular basis of osmotic regulation. *Am. J. Physiol* **268,** F9983–F9996.
167. Isaacks, R.E., Bender, A.S., Kim, C.Y., Prieto, N.M., and Norenberg, M.D. (1994) Osmotic regulation of *myo*-inositol uptake in primary astrocyte cultures. *Neurochem. Res.* **19,** 331–338.
168. Kreis, R., Ross, B.D., Farrow, N.A., and Ackerman, Z. (1992) Metabolic disorders of the brain in chronic hepatic encephalopathy detected with H-1 MR spectroscopy. *Radiology.* **182,** 19–27.
169. Taylor-Robinson, S.D., Sargentoni, J., Marcus, C.D., Morgan, M.Y., and Bryant, D.J. (1994) Regional variations in cerebral proton spectroscopy in patients with chronic hepatic encephalopathy. *Metab. Brain Dis.* **9,** 347–359.
170. Córdoba, J., Gottstein, J., and Blei, A.T. (1996) Glutamine, myo-inositol, and organic brain osmolytes after portacaval anastomosis in the rat: Implications for ammonia-induced brain edema. *Hepatology* **24,** 919–923.
171. Isaacks, R.E., Bender, A.S., and Norenberg, M.D. (1995) Rate of myo-inositol uptake and its content in primary astrocyte cultures after exposure to ammonia. *J. Neurochem.* **64 (Suppl),** 66.
172. Isaacks, R.E., Bender, A.S., Kim, C.Y., Shi, Y.F., and Norenberg, M.D. (1999) Effect of ammonia and methionine sulfoximine on *myo*-inositol uptake in cultured astrocytes. *Neurochem. Res.* **24,** 51–59.
173. Albrecht, J., Bender, A.S., and Norenberg, M.D. (1994) Ammonia stimulates the release of taurine from cultured astrocytes. *Brain Res.* **660,** 288–292.
174. Hilgier, W., Olson, J.E., and Albrecht, J. (1996) Relation of taurine transport and brain edema in rats with simple hyperammonemia or liver failure. *J. Neurosci. Res.* **45,** 69–74.
175. Kuriyama, K. (1980) Taurine as an neuromodulator. *Fed. Proc.* **39,** 2680–2684.
176. Aruoma, O.I., Halliwell, B., Hoey, B.Y., and Butler, J. (1988) The antioxidant action of taurine, hypotaurine and their metabolic precursors. *Biochem. J.* **256,** 251–255.

Index

G

R

MIX
Papier aus verantwortungsvollen Quellen
Paper from responsible sources
FSC® C105338

If you have any concerns about our products,
you can contact us on
ProductSafety@springernature.com

In case Publisher is established outside the EU,
the EU authorized representative is:
Springer Nature Customer Service Center GmbH
Europaplatz 3, 69115 Heidelberg, Germany

Printed by Libri Plureos GmbH
in Hamburg, Germany